Edited by
Kuiling Ding and Li-Xin Dai

Organic Chemistry –
Breakthroughs
and Perspectives

Related Titles

Christmann, M., Bräse, S. (eds.)

Asymmetric Synthesis

More Methods and Applications

2012
ISBN: 978-3-527-32900-7

Hanessian, S., Giroux, S., Merner, B.L.

Design and Strategy in Organic Synthesis

From the Chiron Approach to Catalysis

2012
ISBN: 978-3-527-31964-0

Nicolaou, K. C., Chen, J. S.

Classics in Total Synthesis III

Further Targets, Strategies, Methods

2011
ISBN: 978-3-527-32958-8

Carreira, E. M., Kvaerno, L.

Classics in Stereoselective Synthesis

2009
ISBN: 978-3-527-29966-9

Steinborn, D.

Fundamentals of Organometallic Catalysis

2011
ISBN: 978-3-527-32716-4

Edited by Kuiling Ding and Li-Xin Dai

Organic Chemistry – Breakthroughs and Perspectives

WILEY-VCH

WILEY-VCH Verlag GmbH & Co. KGaA

The Editors

Prof. Dr. Kuiling Ding
Chinese Academy of Sciences
Shanghai Institute of Organic Chemistry
345 Ling Ling Road
Shanghai 200032
China

Prof. Dr. Li-Xin Dai
Chinese Academy of Sciences
Shanghai Institute of Organic Chemistry
345 Ling Ling Road
Shanghai 200032
China

Library of Congress Card No.: applied for

British Library Cataloguing-in-Publication Data
A catalogue record for this book is available from the British Library.

Bibliographic information published by the Deutsche Nationalbibliothek
The Deutsche Nationalbibliothek lists this publication in the Deutsche Nationalbibliografie; detailed bibliographic data are available on the Internet at <http://dnb.d-nb.de>.

© 2012 Wiley-VCH Verlag & Co. KGaA, Boschstr. 12, 69469 Weinheim, Germany

ISBN Hardcover: 978-3-527-33377-6
ISBN Softcover: 978-3-527-32963-2
ISBN Online: 978-3-527-66480-1

Cover Design   Formgeber, Eppelheim, Germany
Typesetting   Laserwords Private Limited, Chennai, India
Printing and Binding   Markono Print Media Pte Ltd, Singapore

Printed on acid-free paper

Contents

List of Contributors

Guillermo C. Bazan
University of California,
Santa Barbara
Department of Chemistry and
Biochemistry
Santa Barbara, CA 93106
USA

Matthias Beller
Universität Rostock
Leibniz-Institut für Katalyse
Albert Einstein Strasse 29a
18059 Rostock
Germany

Albert Boddien
Universität Rostock
Leibniz-Institut für Katalyse
Albert Einstein Strasse 29a
18059 Rostock
Germany

Ronald Breslow
Columbia University
Department of Chemistry
3000 Broadway
New York, NY 10027
USA

Yong Cao
South China University of
Technology
Institute of Polymer Optoelectronic
Materials and Devices
State Key Laboratory of Luminescent
Materials and Devices
Wushan Road
Guangzhou 510640
China

Tak Hang Chan
McGill University
Department of Chemistry
801 Sherbrooke Street West
Montreal, QC H3A 2K6
Canada

Eugene Y.-X. Chen
Colorado State University
Department of Chemistry
1301 Centre Av.
Fort Collins, CO 80523-1872
USA

David Crich
Centre de Recherche de Gif CNRS
Institut de Chimie des Substrates
Naturelles
Avenue de la Terrasse
91198 Gif-sur-Yvette
France

Li-Xin Dai
Chinese Academy of Sciences
Shanghai Institute of Organic
Chemistry
345 Ling Ling Road
Shanghai 20032
China

Sam Danishefsky
Columbia University
Department of Chemistry
3000 Broadway
New York, NY 10027
USA

Huw M. L. Davies
Emory University
Department of Chemistry
201 Dowman Drive
Atlanta, GA 30322
USA

Guo-Jun Deng
Xiangtan University
College of Chemistry
XiangDa New Road
Xiangtan
Hunan 411105
China

Zi-Xin Deng
Shanghai Jiaotong University
School of Life Sciences and
Biotechnology
1954 Huashan Road
Shanghai 200030
China

Ke Ding
Guangzhou Institute of Life Science
190 Kaiyuan Road
Guangzhou Science Park
Guangzhou 510530
China

Kuiling Ding
Chinese Academy of Sciences
Shanghai Institute of Organic
Chemistry
345 Ling Ling Road
Shanghai 20032
China

Chun-Hui Duan
South China University of
Technology
Institute of Polymer Optoelectronic
Materials and Devices
State Key Laboratory of Luminescent
Materials and Devices
Wushan Road
Guangzhou 510640
China

Keary M. Engle
The Scripps Research Institute
Department of Chemistry
10550 North Torrey Pines Road
La Jolla, CA 92037
USA

Christopher Federsel
Universität Rostock
Leibniz-Institut für Katalyse
Albert Einstein Strasse 29a
18059 Rostock
Germany

Valery V. Fokin
The Scripps Research Institute
Department of Chemistry
10550 North Torrey Pines Road
La Jolla, CA 92037
USA

Michael Foley
Broad Institute of MIT and Harvard
301 Binney Street
Cambridge, MA 02142
USA

Terunori Fujita
Mitsui Chemicals Singapore R&D
Center Pte, Ltd
50 Science Park Road
#06-08 The Kendall Singapore
Science Park II
Singapore 117406
Singapore

Felix Gärtner
Universität Rostock
Leibniz-Institut für Katalyse
Albert Einstein Strasse 29a
18059 Rostock
Germany

Liu-Zhu Gong
University of Science and
Technology of China
Hefei National Laboratory for
Physical Sciences at the Microscale
and Department of Chemistry
96 Jinzhai Road
Hefei, Anhui 230026
China

Xiong Gong
University of Akron
College of Polymer Science and
Polymer Engineering
Goodyear Polymer Center
185 East Mill Street
Akron, OH 44325
USA

Robert Grubbs
California Institute of Technology
Division of Chemistry and Chemical
Engineering
1200 E. California Boulevard
Pasadena, CA 91125
USA

Xue-Long Hou
Chinese Academy of Sciences
Shanghai Institute of Organic
Chemistry
345 Ling Ling Road
Shanghai 200032
China

K. N. Houk
University of California Los Angeles
Department of Chemistry and
Biochemistry
Los Angeles, CA 90095-1569
USA

Jinbo Hu
Chinese Academy of Sciences
Shanghai Institute of Organic
Chemistry
345 Ling Ling Road
Shanghai 20032
China

Fei Huang
South China University of
Technology
Institute of Polymer Optoelectronic
Materials and Devices
State Key Laboratory of Luminescent
Materials and Devices
Wushan Road
Guangzhou 510640
China

Takao Ikariya
Tokyo Institute of Technology
Graduate School of Science and
Engineering
Department of Applied Chemistry
2-12-1 Ookayama
Meguro-ku
Tokyo 152-8550
Japan

Ralf Jackstell
Universität Rostock
Leibniz-Institut für Katalyse
Albert Einstein Strasse 29a
18059 Rostock
Germany

Li-Qun Jin
Wuhan University
College of Chemistry and
Molecular Sciences
Luo-jia-shan, Wuchang
Wuhan
Hubei Province 430072
China

Henrik Junge
Universität Rostock
Leibniz-Institut für Katalyse
Albert Einstein Strasse 29a
18059 Rostock
Germany

Motomu Kanai
University of Tokyo
Graduate School of Pharmaceutical
Sciences
Laboratory of Synthetic Organic
Chemistry
7-3-1 Hongo
Bunkyo-ku
Tokyo 113-003
Japan

Hiromu Kaneyoshi
Mitsui Chemical, Inc.
Research Center
Molecular Catalysis Unit
Catalysis Science Laboratory
580-32 Nagaura
Sodegaura
Chiba 299-0265
Japan

Gábor Laurenczy
École Polytechnique Fédérale de
Lausane (EPFL)
SB ISIC LCOM, BCH 2405
Bâtiment de Chimie UNIL
10015 Lausanne
Switzerland

Ai-Wen Lei
Wuhan University
College of Chemistry and
Molecular Sciences
Luo-jia-shan, Wuchang
Wuhan
Hubei Province 430072
China

Chao-Jun Li
McGill University
Department of Chemistry
801 Sherbrooke Street West
Montreal, QC H3A2K6
Canada

Wei-Dong Li
Nankai University
State Key Laboratory and Institute of
Elemento-Organic Chemistry
94 Weijin Road
Tianjin 300071
China

Yongfang Li
Chinese Academy of Sciences
Institute of Chemistry
CAS Key Laboratory of
Organic Solids
Zhongguancun
Beijing 100190
China

Zhan-Ting Li
Fudan University
Department of Chemistry
Shanghai 200433
China

Yong Liang
Peking University
College of Chemistry and Molecular
Engineering
202 Chengfu Road
Beijing 100871
China

Yu-Fan Liang
Peking University
Shenzhen Graduate School
Department of Chemistry
Li Shui Road
Shenzhen 518055
China

Sai-Hu Liao
Max-Planck-Institut für
Kohlenforschung
Kaiser-Wilhelm-Platz 1
45470 Mülheim an der Ruhr
Germany

Benjamin List
Max-Planck-Institut für
Kohlenforschung
Kaiser-Wilhelm-Platz 1
45470 Mülheim an der Ruhr
Germany

Guo-Sheng Liu
Chinese Academy of Sciences
Shanghai Institute of Organic
Chemistry
345 Ling Ling Road
Shanghai 20032
China

Lei Liu
Tsinghua University
Department of Chemistry
Beijing 100084
China

Wen Liu
Chinese Academy of Sciences
Shanghai Institute of Organic
Chemistry
State Key Laboratory of Bioorganic
and Natural Products Chemistry
345 Ling Ling Road
Shanghai 200032
China

Tien Yau Luh
National Taiwan University
Department of Chemistry
Roosevelt Road
Taipei 10617
Taiwan (Republic of China)

Da-Wei Ma
Chinese Academy of Sciences
Shanghai Institute of Organic
Chemistry
345 Ling Ling Road
Shanghai 200032
China

Haruyuki Makio
Mitsui Chemicals Singapore R&D
Centre Pte, Ltd
50 Science Park Road
#06-08 The Kendall Singapore
Science Park II
Singapore 117406
Singapore

Seth R. Marder
Georgia Institute of Technology
901 Atlantic Drive
Atlanta, GA 30332
USA

Keiji Maruoka
Kyoto University
Graduate School of Science
Department of Chemistry
Kyoto 606-8502
Japan

Krzysztof Matyjaszewski
Carnegie Mellon University
Department of Chemistry
4400 Fifth Avenue
Pittsburgh, PA 15213
USA

David Milstein
Weizmann Institute of Science
Department of Organic Chemistry
Rehovot 76100
Israel

Xin Mu
Chinese Academy of Sciences
Shanghai Institute of Organic
Chemistry
345 Ling Ling Road
Shanghai 20032
China

Kyriacos C. Nicolaou
The Scripps Research Institute
Department of Chemistry
10550 N. Torrey Pines Road
La Jolla, CA 92037
USA

and

University of California
Department of Chemistry and
Biochemistry
9500 Gilman Drive
San Diego
La Jolla, CA 92093
USA

Ryoji Noyori
Nagoya University
Department of Chemistry and
Research Center for Materials
Science
Furo-cho
Chikusa-ku
Nagoya 464-8601
Japan

David O'Hagan
University of St Andrews
Centre for Biomolecular Sciences
North Haugh
St. Andrews
Fife KY16 9ST
UK

Jun Okuda
RWTH Aachen University
Institut für Anorganische Chemie
Landoltweg 1
52074 Aachen
Germany

Takashi Ooi
Nagoya University
Graduate School of Engineering
Department of Applied Chemistry
Chikusa
Nagoya 464-8603
Japan

Jian Pei
Peking University
College of Chemistry and Molecular
Engineering
202 Chengfu Road
Beijing 100871
China

Andreas Pfaltz
University of Basel
Department of Chemistry, Organic
Chemistry
St. Johanns-Ring 19
4056 Basel
Switzerland

Irene Piras
Universität Rostock
Leibniz-Institut für Katalyse
Albert Einstein Strasse 29a
18059 Rostock
Germany

G. K. Surya Prakash
University of Southern California
Loker Hydrocarbon Research
Institute
Department of Chemistry
University Park
Los Angeles, CA 90089
USA

Xu-Dong Qu
Chinese Academy of Sciences
Shanghai Institute of Organic
Chemistry
State Key Laboratory of Bioorganic
and Natural Products Chemistry
345 Ling Ling Road
Shanghai 200032
China

Christian A. Sandoval
Chinese Academy of Sciences
Shanghai Institute of Organic
Chemistry
345 Ling Ling Road
Shanghai 200032
China

Niyazi Serdar Sariciftci
Johannes Kepler University of Linz
Linz Institute for Organic Solar
Cells (LIOS)
Altenberger Strasse 69
4040 Linz
Austria

Roger A. Sheldon
Delft University of Technology
Faculty of Applied Sciences
Lorentzweg 1
2628 CJ Delft
The Netherlands

Ben Shen
University of Wisconsin-Madison
Microbiology Doctoral Training
Program
777 Highland Avenue
Madison, WI 53705
USA

and

University of Wisconsin-Madison
School of Pharmacy
Division of Pharmaceutical Sciences
777 Highland Avenue
Madison, WI 53705
USA

and

Scripps Research Institute
Scripps Florida
Departments of Chemistry and
Molecular Therapeutics and Natural
Products Library Initiative
130 Scripps Way, #3A1
Jupiter, FL 33458
USA

Qi-Long Shen
Chinese Academy of Sciences
Shanghai Institute of Organic
Chemistry
345 Ling Ling Road
Shanghai 20032
China

Zhang-Jie Shi
Peking University
College of Chemistry and Molecular
Engineering
202 Chengfu Road
Beijing 100871
China

Min Shi
Chinese Academy of Sciences
Shanghai Institute of Organic
Chemistry
345 Ling Ling Road
Shanghai 20032
China

Seiji Shirakawa
Kyoto University
Graduate School of Science
Department of Chemistry
Kyoto 606-8502
Japan

Michael J. Smanski
University of Wisconsin-Madison
Microbiology Doctoral Training
Program
777 Highland Avenue
Madison, WI 53705
USA

Scott A. Snyder
Columbia University
Department of Chemistry
3000 Broadway
New York, NY 10027
USA

Peter J. Stang
University of Utah
Department of Chemistry
315 South 1400 East
Salt Lake City, UT 84112
USA

Yi Tang
University of California, Los Angeles
420 Westwood Plaza
Los Angeles, CA 90096
USA

Chen-Ho Tung
Chinese Academy of Sciences
Technical Institute of Physics and
Chemistry
2 Beiyitiao Street
Zhongguancun
Haidian District
Beijing 100190
China

Fang Wang
University of Southern California
Loker Hydrocarbon Research
Institute
Department of Chemistry
University Park
Los Angeles, CA 90089
USA

Lai-Xi Wang
University of Maryland
Institute of Human Virology
Department of Biochemistry and
Molecular Biology
725 West Lombard Street
Baltimore, MD 21201
USA

Mei-Xiang Wang
Tsinghua University
Department of Chemistry
MOE Key Laboratory of Bioorganic
Phosphorus Chemistry and
Chemical Biology
Zhongguancun North First Street 2
Beijing 100084
China

Qian Wang
Swiss Federal Institute of
Technology
EPFL SB ISIC LSPN
BCH 5304 (Bâtiment de Chimie
UNIL)
1015 Lausanne
Switzerland

Ren-Xiao Wang
Chinese Academy of Sciences
Shanghai Institute of Organic
Chemistry
State Key Laboratory of Bioorganic
and Natural Products Chemistry
345 Ling Ling Road
Shanghai 200032
China

Zhao-Hui Wang
Chinese Academy of Sciences
Institute of Chemistry
CAS Key Laboratory of
Organic Solids
Zhongguancun North First Street 2
Beijing 100190
China

Henry N.C. Wong
The Chinese University of
Hong Kong
University Administration Building
2/F, Room 215
Shatin, New Territories
Hong Kong SAR
China

Yun-Dong Wu
Peking University
Shenzhen Graduate School
Laboratory of Chemical Genomics
Li Shui Road
Shenzhen 518055
China

Zhenfeng Xi
Peking University
College of Chemistry and Molecular
Engineering
202 Chengfu Road
Beijing 100871
China

Wen-Jing Xiao
Central China Normal University
College of Chemistry
152 Luoyu Road
Wuhan
Hubei 430079
China

Ling-Min Xu
Peking University
Shenzhen Graduate School
Department of Chemistry
Li Shui Road
Shenzhen 518055
China

Hisashi Yamamoto
University of Chicago
Department of Chemistry
5735 South Ellis Avenue
Chicago, IL 60635
USA

Yoshinori Yamamoto
Tohoku University
WPI-AIMR (Advanced Institute for
Materials Research)
Sendai 980-8577
Japan

and

Dalian University of Technology
The State Key Laboratory of
Fine Chemicals
Dalian 116023
China

Zhen Yang
Peking University
Shenzhen Graduate School
Laboratory of Chemical Genomics
Li Shui Road
Shenzhen 518055
China

Qin-Da Ye
Peking University
Shenzhen Graduate School
Department of Chemistry
Li Shui Road
Shenzhen 518055
China

Shu-Li You
Chinese Academy of Sciences
Shanghai Institute of Organic
Chemistry
345 Ling Ling Road
Shanghai 200032
China

Biao Yu
Chinese Academy of Sciences
Shanghai Institute of Organic
Chemistry
State Key Laboratory of Bioorganic
and Natural Products Chemistry
345 Ling Ling Road
Shanghai 200032
China

Jin-Quan Yu
The Scripps Research Institute
Department of Chemistry
10550 North Torrey Pines Road
La Jolla, CA 92037
USA

Yi Yu
Wuhan University
School of Pharmaceutical Sciences
185 East Lake Road
Wuhan 430071
China

Zhi-Xiang Yu
Peking University
College of Chemistry and Molecular
Engineering
202 Chengfu Road
Beijing 100871
China

Jun-Ying Yuan
Harvard University
Department of Cell Biology
240 Longwood Avenue
Boston, MA 02115
USA

Xiao-Wei Zhan
Chinese Academy of Sciences
Institute of Chemistry
CAS Key Laboratory of Organic
Solids
Zhongguancun North First Street 2
Beijing 100190
China

De-Qing Zhang
Chinese Academy of Sciences
Institute of Chemistry
CAS Key Laboratory of
Organic Solids
Zhongguancun North First Street 2
Beijing 100190
China

Guan-Xin Zhang
Chinese Academy of Sciences
Institute of Chemistry
CAS Key Laboratory of
Organic Solids
Zhongguancun North First Street 2
Beijing 100190
China

Li-He Zhang
Peking University
College of Chemistry and Molecular
Engineering
202 Chengfu Road
Beijing 100871
China

Xin-Hao Zhang
Peking University
Shenzhen Graduate School
Laboratory of Chemical Genomics
Li Shui Road
Shenzhen 518055
China

Liang Zhao
Tsinghua University
Department of Chemistry
MOE Key Laboratory of Bioorganic
Phosphorous and Chemical Biology
Beijing 100084
China

Qi-Lin Zhou
Nankai University
State Key Laboratory of
Elemento-Organic Chemistry
94 Weijin Road
Tianjin 300071
China

Jie-Ping Zhu
Swiss Federal Institute of
Technology
EPFL SB ISIC LSPN
BCH 5304 (Bâtiment de Chimie
UNIL)
1015 Lausanne
Switzerland

Dao-Ben Zhu
Chinese Academy of Sciences
Institute of Chemistry
CAS Key Laboratory of
Organic Solids
Zhongguancun North First Street 2
Beijing 100190
China

Introduction

Kuiling Ding and Li-Xin Dai

From the 1950s to the Second Decade of the Twenty-First Century

About half a century ago, two books were published. The first, entitled *Perspectives in Organic Chemistry*, 1956 [1], was dedicated to Sir Robert Robinson on the occasion of his 70th birthday. Twenty-two years later, another book, with the same theme as the first, with the title *Further Perspectives in Organic Chemistry*, 1978 [2], appeared. This book contains all the lectures given at a Symposium in memory of the late Sir Robert Robinson. These two books share a similar relationship with the great organic chemist, Sir Robert, and recorded the progress of organic chemistry at that time. Now we are in a new century and living in a rapidly changing world. The stimulus for us to edit this book, *Organic Chemistry – Breakthroughs and Perspectives*, is the tremendous achievements of organic chemistry in the last two decades. After the Second World War, we had experienced basically a sufficiently long peaceful and stable period that engendered the steady growth of chemistry. A relatively good financial situation to support the efforts of talented chemists, and also mutual interactions with neighboring sciences, all contributed to the tremendous achievements of organic chemistry. Organic chemistry is thus endowed with vital forces. The emergence of many new disciplines marks the vigorous new faces of organic chemistry. These disciplines include chemical biology, organocatalysis, supramolecular chemistry, green or sustainable chemistry, combinatorial chemistry, and flow chemistry, to name just a few.

The renaming of the Department of Chemistry as the Department of Chemistry and Chemical Biology by Harvard University at the end of the last century marked the emergence of chemical biology, an interface of chemistry and life sciences. To manifest the maturity of this new discipline, several new journals, namely *Nature Chemical Biology*, *ChemBioChem*, *Chemical Biology*, *BMC Chemical Biology*, *Chemical Biology and Drug Design*, and *Chemistry and Biology*, with the sole aim of reporting developments in chemical biology, were launched in the last couple of years. Chemical biology is now a distinct discipline in understanding science at the intersections of chemistry and biology. Chemical biologists believe that they are standing at the doorstep of an exciting era [3].

Similarly, green chemistry or sustainable chemistry is another new discipline, and as such also has its specialized new journals. Sixteen years ago, the establishment of the

Green Chemistry Award at the Presidential level in the United States demonstrated to us the importance of green chemistry. In 1998, Paul T. Anastas coined the term "green chemistry" in his book *Green Chemistry, Theory and Practice* [4], and therefore defined this new discipline. Some new research areas are involved in green chemistry, such as ionic liquid chemistry, fluorous chemistry, and synthetic reactions in aqueous medium.

Supramolecular chemistry and organocatalysis are further examples. Organocatalysis can really be regarded as the chemistry of this century. In the last century, we usually said that there were two ways to carry out an asymmetric catalytic reaction, that is, metal-catalyzed asymmetric reaction and enzymatic asymmetric reaction. Now, in the twenty-first century, we should add a third: organocatalysis.

As can be seen, the emergence in a not too long period of so many new disciplines and new research areas has evidenced the rapid and exciting growth of organic chemistry. Thus, organic chemistry has become a matured as well as a dynamic science. From this status of organic chemistry, probably it is a good time to survey the breakthroughs in this subject and to make a perspective view of this dynamic science. To speak frankly, neither of the Editors is in a perfect position to do this job. However, fortunately, many of our friends and colleagues promised to support this project, and with their great support this project is now realized. Furthermore, the publisher has also shown great interest and lasting support, which were indeed a strong motivation behind this project.

It has been a great privilege and an honor for us to assemble a magnificent international team of outstanding organic chemists – over 100 organic chemists from around 10 nations. They have made great efforts to write excellent reviews and to present insightful comments during their very busy hours between two important holidays, Christmas and the Chinese New Year. We owe them a great deal and thank them wholeheartedly.

In this book, there are 21 chapters in total and an introduction. The last chapter gives us overall remarks on the perspectives if organic chemistry in this new century. Not only is the last not the least, we would also say that the last chapter is of the utmost importance. We are extremely grateful to Professor Ronald Breslow: without his agreement to write that chapter, we would not have been confident to undertake such an ambitious task as to edit this book. The other chapters may be divided into four groups. First, Chapters 1–7 deal with total synthesis of natural products and chemical biology. The second group covers synthetic methodology, encompassing Chapters 8–12. Chapters 13, 15, and 16 involve physical organic chemistry. The last group, preceding Prof. Breslow's essay in Chapter 21, comprises Chapters 14 and 17–20 and is dedicated to chemistry in meeting the urgent needs of human beings.

Total Synthesis of Natural Products and Chemical Biology

Total syntheses of natural products have long been regarded as a mainstream of organic chemistry. The milestones in this area often represent the advancement of organic chemistry to a certain extent. Friedrich Wöhler's synthesis of urea from inorganic compounds is frequently cited because it is regarded as the first example of total synthesis. The total synthesis of vitamin B_{12}, palytoxin, brevetoxin, and many others are milestones in total syntheses. The total synthesis of vitamin B_{12} by Robert Woodward and Albert Eschenmoser not only has manifested a chemical synthesis of an important naturally occurring and structurally complex cobalt-containing chemical compound, but also bestowed on our scientific community the beautiful Woodward–Hoffmann rules on conservation of molecular orbital symmetry [5].

Vitamin B$_{12}$

Brevetoxin

Palytoxin

The total synthesis of palytoxin had once been regarded as the Mount Everest of total synthesis owing to its molecular weight of 2860 Da and 64 stereogenic centers [6]. This was true in the past, but on terms of the molecular complexity, this molecule is really not complex enough! In this regard, brevetoxin is more complex, with many fused alicyclic rings, ranging from six- to nine-membered, and the associated stereochemistry stems from this polycyclic structure [7].

These molecules are just a very few examples among the vast array of important natural products that have been synthesized in recent decades. In a way, the total syntheses of these molecules may help us to answer the question "Can we make it?" This question, however, is now not worth posing any longer. And then comes the second question, as addressed in Chapter 2, written by Qian Wang and Jieping Zhu, which has now shifted to "Why should we make it?" The answer to this question may be taken from the first sentence of Chapter 1, written by Zhen Yang and his co-authors: "Natural products have proven to be valuable sources for the identification of new drug candidates." Christopher T. Walsh and Michael A. Fischbach recently answered the question "Why do natural products still matter?" with four rationales: they continue to inspire synthetic and analytical chemists; they remain a major source of human medicines; they have led to important biological insights; and there are many more natural products to discover [8a]. In addition, Alexander Todd once urged organic chemists "to devote much thought to the function of natural products" [2]. When we turn our attention to the function of natural products, "diversity-oriented synthesis" (DOS) would emerge. Chapter 1 talks about diversity-oriented synthesis and also diverted total synthesis (DTS) and function-oriented synthesis (FOS). Although "Can we make it?" is no longer a valid question, "Can we make it efficiently?" still remains a real problem. Chapter 2 also addresses the synergy of synthetic methodology with the total synthesis of natural products. Samuel Danishefsky once said, "the opportunities for total synthesis are realizable only in the context of continuing advances in the allied field of synthetic methodology" [8b].

Scott Snyder, in concluding his commentary on Chapter 1, inspiringly raises three thought-provoking queries: (i) the reluctance to invest in natural products isolation efforts by major pharmaceutical companies, (ii) the reduction in the ease of federal funding, and (iii) the increasing complexity of natural products isolated. The reduction in investment by pharmaceutical companies probably is a reflection of the global economic crisis. With the increasing fruitful results from natural products, it is believed that pharmaceutical companies will regain their interest in natural products. The recent launch of two highly potent (at the sub-nanomolar level) anti-cancer drugs, ET-743 or Yondelis® by Elias Corey and E7389 or Halaven® by Yoshito Kishi (see Chapters 1 and 2) are good examples to reconstruct their confidence. Scott Snyder also pointed out the rich history of traditional Chinese medicine. The recent award to You-You Tu, the inventor of anti-malaria compound artemisinin, of the Lasker–Debakey Clinical Medical Research Award, is a response to Scott Snyder's observation. Traditional Chinese medicine is the wealth of the whole world, and it is gratifying that the language barrier associated with the use of this wealth may be somewhat alleviated by the publishing of several English versions of relevant literature [9].

When chemists care more about the physiological functions in our bodies, they are coming very close to chemical biology. Ren-Xiao Wang wrote Chapter 3, delineating the interplay between the chemical space and the biological space with special emphasis on drug discovery. Chapter 4, entitled "Biosynthesis of Pharmaceutical Natural Products and Their Pathway Engineering," was written by Ben Shen, Wen Liu and their co-authors. Despite the fact that research in biosynthesis is a field with a long and rich history, modern biosyntheses and especially those in the post-genome era have shown a wholly different new facet. Nowadays, the genome sequence of a microorganism can be determined within a fairly short time and at acceptable cost. The whole gene sequence of 1891 bio-species has already been determined and 11 446 targets are currently being determined. Hence this rich genomic information may facilitate the understanding of the metabolic pathway, the modification of the pathway or the engineering of a brand new pathway. In this sense, synthetic biology will probably initially blossom in the area of microorganisms. It is now clear that the gene sequence will boost this field to a new height. This chapter wholly reflects this new facet in gene cluster analysis, genome sequencing, scanning, and mining so as to allow us to understand and to reconstruct the biosynthetic pathway at the molecular level.

In addition to proteins and nucleic acids, carbohydrates are recognized as one of the three principal biomolecular materials. Carbohydrates are not simply molecular structural materials or means for energy storage, but also play important functions in cellular communication. Chapter 5, written by Biao Yu and Lai-Xi Wang, presents an overview of the synthesis of this important class of compounds. Although the empirical formula of carbohydrates is extremely simple, $C_n(H_2O)_n$, that is, hydrates of carbon, the synthesis of these molecules is of tremendous complexity. We can synthesize peptide chains and deoxynucleic acids routinely with an automated machine, but the automated synthesis of oligosaccharides has only recently been realized and still with some limitations. This may be explained by a paper by Hans Paulsen, who observed in 1982 [10] that "... it should be emphasized that each oligosaccharide synthesis remains an independent problem, whose resolution requires considerable systematic research and a good deal of know-how. There are no universal reaction conditions for oligosaccharide syntheses." Indeed, through systematic research, two major breakthroughs in this field are elegantly addressed in this chapter. They both involve chemistry belonging to this century. Now, a branched dodecasaccharide of the phytoalexin elicitor β-glucan can be synthesized by an automated synthesizer. The other breakthrough is one-pot sequential glycosylation. A computer database recording the relative reactivities of a large number of diversely protected sugars has been established by Chi-Huey Wong that allows the choice of appropriate reactants for a sequential glycosylation in one pot. By this methodology, several hexa- and dodecaoligosaccharides were synthesized. It has been claimed that over 90% of human proteins are predicted to be glycosylated [11]. Hence the chemical synthesis of glycoprotein is highly needed in SAR (structure–activity relationship) studies. Furthermore, carbohydrate synthesis is highly relevant to vaccine research. In this connection, we have more reasons to expect further breakthroughs in this field in the coming decades.

In Chapter 6, Lei Liu elegantly addresses the chemical synthesis of proteins and recalled the total synthesis of crystalline bovine insulin in 1965, a great event of science at that time in China. Insulin was then the only protein whose structure had been established (by Frederick Sanger), which contains 51 amino acid residues with two disulfide linkages. The realization of bovine insulin was the first synthesis of a protein, and inevitably it took an exceedingly long time and much effort to achieve the goal. Later, thanks to the invention and elaboration of solid-phase peptide synthesis by Robert Merrifield, and to the invention of native chemical ligation by Stephen Kent for the efficient coupling of unprotected peptide segments, the chemical synthesis of proteins has now reached a new height. The list of authors on the paper that announced the total synthesis of bovine insulin is a very long one, whereas only two authors were included in the phenomenal 2011 paper reporting the synthesis of the covalent dimer of HIV-1 protease enzyme that features 203 amino acid residues. The advancement of science in these 50 years is truly amazing.

Lei Liu also describes several applications of chemical synthesis in biological science and pharmaceutical discovery. With the advancement of proteomics and structural biology, the number of crystalline structures of proteins revealed has boomed to 70 209. This huge number of protein structures is apparently a great challenge to organic chemists. Although native chemical ligation (NCL), invented by Stephen Kent, is a great breakthrough in the last 20 years, the development of more efficient methods is still needed. In this chapter, Lei Liu reports the efforts in overcoming the cys limitation of NCL. A recent paper by Lei Liu in tackling this problem with a peptide hydrazide is yet another efficient complement to the current NCL. In meeting this grand challenge, further improvement of NCL and other protocols will certainly appear.

There is an obvious direction at present and will be in the future that synthetic chemistry should combine continuously with other subjects in order to create more interdisciplinary sciences. From this point of view, synthetic chemistry requires an extremely high level of scientific creativity and insight to explore its limitless possibilities. In 2001, Barry Sharpless introduced a chemical philosophy termed "click chemistry," emphasizing that synthetic chemical reactions must strive for a high reaction rate, high selectivity, and excellent tolerance to various functional groups and reaction conditions in order to generate substances quickly and reliably by joining two units together. As one of the frontier concepts in synthetic chemistry, click chemistry has provided a brand new idea, methodology, and substance basis for the life and materials sciences, and has been widely used in the development of new medicines and new materials and in the research areas of molecular biology, chemical biology, and related disciplines [12], including modification of peptides, DNA and nucleotides, natural products, and pharmaceuticals, and also carbohydrate clusters and bioconjugates. Chapter 7, written by Valery Fokin, highlights the concept of click chemistry and its applications in numerous areas. Clearly, be it in biology, polymers, or materials science, click chemistry is starting to click [13a]. As a further elaboration of click chemistry, two very recent papers are worth mentioning here. Carolyn Bertozzi developed a copper-free AAC (azide–alkyne cycloaddition) reaction [13b] in order to alleviate the copper ion effect in living systems, and another paper reports the use of mechano-chemistry to unclick this reaction back [13c].

Chemistry Is a Central Science and Synthesis Plays an Essential Role in Chemistry

Chemistry has a central role in science, and organic synthesis plays an essential role in chemistry [14]. As a "central, useful, and creative" science, chemistry has played and will continue to play an immense and irreplaceable role to improve the quality of life and health condition of human, and also to promote the development of other disciplines and solve the problems of our society [15]. Synthetic substances and materials have demonstrated a significant impact in determining the quality of our life in the last century. Although chemical synthesis has now reached an extraordinary level of sophistication, there is still vast room for improvement. In this century, to be more tightly interlinked with other fields such as materials and life sciences, and to generate more interdisciplinary areas, are two inevitable trends that synthetic chemistry must follow. Nowadays, synthetic chemistry should pursue "green" processes. In this context, the atom economy, the E-factor, and 3Rs (reduction, recycling, and reuse) of resources must be taken into account in industrial synthetic processes. In Chapter 19, entitled "Synthetic Chemistry with an Eye on Future Sustainability," Guo-Jun Deng and Chao-Jun Li present a general introduction to green chemistry and in particular instruct us how to make a synthetically useful reaction, namely the cross-dehydrogenative coupling reaction, greener and greener. They also address the catalytic nucleophilic additions of terminal alkynes in water and metal-catalyzed A^3 coupling. Different from other disciplines, the most remarkable feature of synthetic chemistry is its powerful creativity. The synthetic chemist is able not only to realize the synthesis of substances that already exist in Nature, but is also ready to create new materials with predicted properties and functions. By serving satisfactorily the primary mission for creating novel substances with new structures and new functions, synthetic chemistry lies in the central position and is situated at the frontier of chemical science. By integrating with other disciplines, more and more cutting-edge interdisciplinary areas will emerge, thereby providing more and more new opportunities for the further development of synthetic chemistry. Enriched by extensive research targets, complicated chemical processes, and diverse demands on structures and properties, synthetic chemistry is provided with more space and higher requirements in respect of methods and theory developments. In addition, the development of synthetic chemistry provides possibilities for elucidating the structure–function relationship and for the synthesis of new materials with structural diversity and excellent properties. Synthetic chemistry therefore has provided us with an avenue to explore its infinite possibilities by tapping into its extremely high level of scientific creativity.

The 2010 Nobel Prize in Chemistry was awarded to Richard Heck, Ei-ichi Negishi, and Akira Suzuki for their contribution to palladium-catalyzed cross-coupling reactions used in organic synthesis. These reactions represent one of the most important methods for the construction of C–C bonds in modern synthetic organic chemistry. These methods have had a significant impact on the development of new drugs and materials, and have been widely used in the industrial production of agrochemicals, pharmaceuticals, and organic materials. The new generation of chemical transformations based on transition metal-catalyzed C–H bond activation and functionalization is obviously one of the most fundamental and challenging issues to impart molecular complexity, and is currently undergoing rapid development in the area of chemical synthesis. This thematic research area was recognized as one of the "Holy Grails" in chemistry [16]. In Chapter 8, Keary Engle and Jin-Quan Yu highlight the history, the state-of-art research, and future challenges of this exciting area in synthetic organic chemistry. Direct conversion of hydrocarbons to value-added chemicals (first functionalization) and further transformations of molecules with chemical functionality (including precursors of drugs, agrochemicals, and fine chemicals) to realize higher level molecular complexity by the rapid construction of new stereocenters and functional group patterns (further functionalization) through transition metal-catalyzed C–H activation and functionalization are the main topics of this chapter. For both the first and further functionalizations, metal complexes play a critically important role. The authors emphasize that future efforts will focus on the discovery of (i) new reactivities of transition metal complexes, (ii) new reaction patterns to achieve high levels of efficiency (including chemo-, regio-, and enantioselectivity) for the catalysis of C–H functionalization, and (iii) cost-effective processes.

The 2001 Nobel Prize in Chemistry was awarded to William S. Knowles, Ryoji Noyori, and Barry Sharpless for their great contributions to the area of asymmetric catalysis. After that, asymmetric catalysis experienced rapid development in the last decade. Numerous new chiral catalysts and reactions have been developed by chemists. As highlighted by Christian Sandoval and Ryoji Noyori in Chapter 9, metal-catalyzed asymmetric transformations are still the mainstream of this research area. In comparison with the relative maturity of asymmetric hydrogenation and oxidation, the past decade witnessed many breakthroughs in catalytic asymmetric carbon–carbon and carbon–heteroatom bond formation processes, although many challenges, including selectivity, activity, and applicability, still remain to be overcome. In comparison with metal-catalyzed asymmetric transformations, organocatalysis, on the other hand, has emerged as one of the most active and fascinating areas of asymmetric catalysis since the beginning of this century [17]. Chapter 10, by Benjamin List and Sai-Hu Liao, in combination with the commentary by Keiji Maruoka, Liu-Zhu Gong, and Wen-Jing Xiao, highlights the significant advances in this fantastic research area. The concepts of enamine catalysis, iminium catalysis, amine catalysis, hydrogen-bonding catalysis, singly occupied molecular orbital (SOMO) activation, ketone catalysis, phase-transfer catalysis, nucleophilic catalysis, base catalysis, cooperative multifunctional catalysis, and multicomponent domino or cascade catalysis have been the lodestar for new catalyst and new reaction design. In particular, the methodology of organocatalysis has provided an efficient and convenient approach to access molecular complexity with diverse chemical functionality through one-pot multicomponent reaction patterns, which is expected to have significant impacts on the generation of molecular diversity for drug discovery [17] (see also Chapters 1 and 2) and process development for drug production. Two recent examples have been reported concerning the development of an efficient process for the synthesis of Tamiflu (oseltamivir phosphate), a neuraminidase inhibitor for the treatment of human influenza. Yujiro Hayashi used β-nitro acrylate as a Michael receptor through three "one-pot" operations (nine steps) and realized the synthesis of the target molecule [18a]. A similar strategy was employed by Da-Wei Ma [18b], but he started from (Z)-2-nitroethenamine and completed the process in only four steps (three steps in one pot). Organocatalysis was used in these two processes. The development of new activation modes, understanding of the activation mechanism, and the discovery of new organocatalysts to address the problem of catalytic efficacy hold the future trends of organocatalysis. The combination of metal complex catalysis and organocatalysis seems to be a promising complementary approach to tackle problems that could not be settled by using either metal catalysis or organocatalysis alone.

The concept of cooperative catalysis has become a well-developed and generally accepted strategy in catalysis, which originated from the inspiration of Nature's enzymatic catalysis. As a consequence, how to create a catalyst system by learning from Nature's principles and to use this system to mimic or exceed enzymes represents one of the significant challenges for chemists. Chapter 11 by Motomu Kanai highlights the topic of "cooperative catalysis" by focusing on several of the most important and fundamental chemical transformations, including asymmetric catalysis, polymerization, and hydrogen activation/generation, with a significant contribution of "cooperative catalysis" concept to the breakthroughs in this area. The complementary comments by Takao Ikariya, Takashi Ooi, Kuiling Ding and David Milstein further emphasize the importance of this fundamental concept in catalysis by demonstrating more examples of catalytic systems involving cooperative effects. It is reasonable to expect that the concept of "cooperative catalysis" will continue to stimulate innovations to address the challenging issues of catalysis, in particular the efficiency, on the basis of deep insights into the underlying mechanism of the processes, and accordingly to revolutionize many chemical processes. The field of "systems catalysis" (analogous to "systems biology") with the core ideology for the design of several coexisting, integrated catalytic cycles which can perform several parallel and consecutive tasks, is emphasized by David Milstein in his commentary. Obviously, the compatibility and cooperativity of these distinctive processes and cycles are critically important in systems catalysis. Some recent examples clearly indicate that catalysis could be of high-level sophistication but at the same time is also cooperative, synergistic, and systematic. Alan Goldman and Maurice Brookhart used a catalyst combination, a pincer-ligated iridium complex (for dehydrogenation and hydrogenation) together with a molybdenum-based Schrock alkene metathesis catalyst, to transform hexane into higher hydrocarbons [19a]. Robert Grubbs reported a triple relay catalysis system to convert a terminal alkene into a primary alcohol. The catalysis system couples together palladium-catalyzed oxidation, acid-catalyzed hydrolysis, and ruthenium-catalyzed reduction. This is a formal anti-Markovnikov hydration of a terminal alkene similar to Herbert Brown's stoichiometric hydroboration pathway [19b]. David Milstein realized a formal reduction of CO_2 to methanol under mild reaction conditions with a ruthenium catalyst by metal–ligand cooperation [19c].

Over the past decades, organofluorine chemistry has been receiving significant attention because of its extreme importance in biology and advanced materials sciences. The increasing demand for structurally diverse fluoroorganics triggers the impetus to develop reliable synthetic methodologies for synthesizing fluorine-containing molecules with higher degrees of complexity and functionality. In Chapter 12, Surya Prakash and Fang Wang give a comprehensive review of the developments in synthetic organofluorine chemistry with particular attention to the advances in the area in the past two decades. These areas include C–F bond formation, trifluoromethylation, difluoromethylation, and monofluoromethylation reactions. The breakthroughs, including new reagents, new reactions, and a mechanistic understanding of various transformations, are highlighted. Special attention is particularly drawn to the asymmetric synthesis of fluoroorganics and transition metal-catalyzed fluorinations and fluoroalkylations. As stated by authors and commentators, organofluorine chemistry will remain fascinating and dynamic, as evidenced by the emergence of various new breakthroughs very recently after the submission of this chapter. It is really astonishing that the number of publications disclosing new catalysts and new reactions exceeded 20 in this specific field in the top journals in 2011. Many of these articles are related to organometallic methodology [20]. The authors mentioned the "extraordinary stability of palladium-CF_3 bonds," hence the aryl-CF_3 bond-forming reductive elimination step is very difficult. Vladimir Grushin [20x] and Stephen Buchwald [20y] independently overcame this difficulty by using large sterically demanding ligands. Jin-Quan Yu [20v] and Melanie Sanford [20w] separately used an oxidatively induced aryl-CF_3 bond formation with a Pd(II)/Pd(IV)-catalyzed reaction to realize this step. Melanie Sanford recently reported an elegant mechanistic study on arene trifluoromethylation at room temperature [20e]. One of the future challenges of this area will include the development of innovative and robust protocols for the precise assembly of fluorinated motifs to generate molecular diversity efficiently with extraordinary functions. Obviously, organometallic chemistry will play a significant role in tackling these challenges.

Physical Organic Chemistry in its Modern Context

Since it was first conceptualized by Jean-Marie Lehn in 1973 [21], supramolecular chemistry, chemistry beyond the molecule, has become one of the most important and dynamic research areas in chemistry. Supramolecular chemistry has primarily found its inspiration from biological molecules, such as proteins and lipids, and their interactions. However, it is highly interdisciplinary in nature and, as a result, attracts not only chemists but also biochemists, environmental scientists, engineers, physicists, theoreticians, mathematicians, and a whole host of other researchers. In fact, the last 20 years have witnessed a rapid growth of this fantastic area. The Nobel Prize for Chemistry in 1987 was awarded to Donald Cram, Jean-Marie Lehn, and Charles Pedersen for their development and use of molecules with structure-specific interactions of high selectivity, which significantly stimulated the development of supramolecular chemistry. Its concepts, principles, methodologies, and materials have not only been extensively established in chemistry, but have also been widely used in biology, medicine [22], materials science [23], and catalysis [24]. Chapter 13, by Zhan-Ting Li, highlights the advances in supramolecular organic chemistry with the focus on foldamer chemistry, including their molecular design, synthesis, supramolecular recognition, assembly, and structure–function relationships. In a supplementary commentary, Mei-Xiang Wang attempts to sketch the breakthroughs in supramolecular chemistry in the last decade from an alternative angle by exemplifying three noteworthy macrocyclic molecules and three significant noncovalent interactions. In their commentaries, Peter Stang and Chen-Ho Tung, on the other hand, further emphasize separately the potential impact of supramolecular chemistry on other scientific fields, and also point out the future challenges and ongoing directions of this area.

From the viewpoint of synthetic chemistry, catalysis is a central theme in order to make processes greener. The discovery of a new generation of chemical transformations with extremely high efficiency and selectivity, and also high atom economy, with the aid of catalysis is essential to make the process more economical, energy saving, and environmentally friendly. To approach the ideal realm is not only the issue of technological improvement of the chemical process, but also more importantly, the fundamental scientific advances and innovations. The deep understanding of the underlying mechanistic issues of the process will be critically important to gain information about why and how a reaction proceeds, and thus would be beneficial to the design of new catalysts and reactions with ideal performance. Chapter 15 by Zhi-Xiang Yu and Yong Liang exemplifies the importance of computational chemistry in mechanistic study and the impacts of computational organic chemistry on the development of synthetic organic chemistry by highlighting a variety of excellent examples of computational and theoretical studies. The authors try to emphasize the instructive role of computational chemistry and the "coupling effect" of computational and experimental study on the development of new catalysts, new reactions, and ideal synthetic strategies for target molecule preparation. Matthew Sigman recently developed a three-dimensional free-energy relationship correlating steric and electronic effects to design and optimize a ligand class for the asymmetric propargylation of ketones [25]. It can be expected that computational chemistry will become one of the indispensable means in the twenty-first century in most synthetic organic chemistry laboratories. Therefore, chemistry can become "chem is try" that can be done computationally and experimentally, so is organic chemistry! In line with this philosophy, Chapter 16 by Ai-Wen Lei and Li-Qun Jin, and the commentaries by Xin Mu, Guo-Sheng Liu, and Qi-Long Shen and by Yoshinori Yamamoto demonstrate some successful experimental studies on organic reaction mechanisms in the last decade with the focus on a variety of palladium-catalyzed C–C and C–N bond-forming reactions and aerobic oxidations. The experimental approaches employed for the mechanistic understanding include kinetic study, reaction intermediate isolation, and structure determination by X-ray diffraction or *in situ* spectroscopic methods, and also the stereo and electronic effects of the possible intermediates and solvent or salt effects on the outcome of the reactions. The examples given in this chapter are convincing in illustrating how to develop more efficient and selective catalytic processes for realizing better chemistry through step-by-step mechanistic understanding.

Chemistry – to Meet the Urgent Needs of Human Beings

We usually maintain that "chemistry is a central, useful, and creative science." As chemists ourselves, we have the responsibility to solve societal needs. Richard Smalley once mentioned 10 important items of urgent human needs, and emphasized that energy is our number one problem. Among all the renewable energies, solar energy is of the utmost importance. In Chapter 17, Chun-Hui Duan, Fei Huang and Yong Cao review the recent progress in materials design and synthesis and applications for solar energy conversion and their application in organic photovoltaic devices and, in particular, the bulk heterojunction (BHJ) polymer solar cell. The BHJ polymer solar cell is a new innovation developed during the last two decades. This field was initiated by Ching Tang, and a *Science* paper by Alan Heeger and co-workers (Ref. [7] in the chapter) may be regarded as an important milestone in this new area. The authors also showcase how organic chemistry impacted the growth of the BHJ solar cell in many respects: metal-catalyzed coupling reactions in the construction of donor polymers; fullerene-derivatized chemistry in exploring fullerene-based acceptors; how to calculate or predict the exciton dissociation at the donor–acceptor interface; the broadening of the absorption spectrum to match that of solar light; living polymerization to prepare block copolymers; and many others. Of course, organic chemistry is just a part of it. The realization of practical, large-area, high power conversion efficiency (PCE), stable, and low-cost BHJ devices will be accomplished only by the joint efforts of organic chemists, polymer scientists, physicists, electronic engineers, and others. This is really an interdisciplinary project, as pointed out by Niyazi Sariciftci in his commentary. The authors hope in their Outlook to improve the performance of BHJ solar cells "with the aim of 10% PCE for commercial applications." In their Introduction, they mention the more than 8% PCE of Solarmer Energy (Ref. [9] in the chapter). Anyhow, Lu-Ping Yu reported a 7.4% PCE of a BHJ in 2010 [26], and this is a step closer to the above aim.

De-Qing Zhang and his co-authors have contributed Chapter 20, with the title "Organic π-Conjugated Molecules for Organic Semiconductors and Photovoltaic Materials." These π-conjugated compounds are not only useful in energy problems but are also important materials for light-emitting diodes and transistors. Many other uses of conjugated molecules are not included. Their discussion is arranged as conjugated molecules for p-type and n-type semiconductors and for photovoltaic materials. In their Conclusion and Outlook they urge that further attention should be directed to (i) new air-stable, high-mobility, solution-processable n-type organic semiconductors, (ii) new 3D small-molecular architectures, and (iii) heteroatom-containing conjugated molecules (heteroatoms other than S, Se, N, and B, such as Si or even metals).

We appreciate these three points, which are interesting challenges for organic chemists and materials scientists. These three challenges have provided a vast area for us to imagine, to speculate, to design, and to synthesize.

Carbon dioxide is considered to be the major cause of climate change, because of its greenhouse properties and continuous accumulation in the atmosphere. It has been predicted that carbon-based fossil fuels will continue to provide 80–85% of the world energy consumption at least until 2030. Today, the need to control CO_2 production and emissions is at the center of attention of both the scientific and industrial worlds and, as a consequence, several methods have been developed to achieve this goal. From the viewpoint of chemistry, the utilization of CO_2 by chemical methods may lead to one of the options for the reduction of CO_2 emission, while the use of CO_2 in this way will not solve the problem of atmospheric CO_2 accumulation that is on the scale of hundreds of billions of tons. It is also worth noting that the fraction of CO_2 produced via the use of chemicals is $\sim$10% of the total, the remainder being derived from energy products. Therefore, if efficient technologies capable of converting CO_2 into energy-rich products (fuels) were to be developed, a much greater amount of CO_2 could be converted into usable products. This would result in a much more significant step in the direction of chemicals and energy production, with a close-to-zero carbon-emission level [27]. Chapter 18 by Matthias Beller and co-authors presents a comprehensive review of the state-of-the-art in the catalytic utilization of CO_2 and highlights the future vision of further research from the perspective of organic chemists. It is obvious that future efforts will be made to develop suitable processes for the transformation of CO_2 into useful materials and energy-rich products, and organometallic catalysis will definitely play a key role [28]. Kyoko Nozaki and co-workers [28a] recently reported the hydrogenation of CO_2 to formic acid by using a pincer iridium complex, with a surprisingly high turnover number of 3.5×10^6. This catalyst efficiency is among the top three in homogeneous catalysis. Formic acid is a promising source and storage material for hydrogen. In this respect, Matthias Beller reported an efficient dehydrogenation of formic acid using an iron catalyst, $Fe(BF_4)_2$, and a phosphine ligand [28d]. The efficiency of this catalyst is also very high with turnover numbers up to 92 000.

Organic polymeric materials have made a great impact on the betterment of human life. The huge capacity of the petrochemical industry is now exceeding that of the steel industry in terms of volume production. Economic prosperity in the 1970s was greatly promoted by polyalkene production that was in turn motivated by the invention of Ziegler–Natta catalysis. The development of Kaminsky's metallocene catalyst surprised the community with its extremely high activity and tailor-made polyalkene properties. The metallocene catalyst is characterized by its single-sited catalysis mechanism. Metallocene catalysts, bearing organic ligands that can be tailored to influence the polymerization process, have promoted research on the discovery of new catalyst systems for alkene polymerization, from ligand design to metal screening. Subsequently, the highly efficient catalyst invented independently by Maurice Brookhardt [29a,b] and Vernon Gibson [29c] shifted the metal complex of the catalyst to late transition metal (Ni, Fe, Co) catalysts. This success benefited from the design of highly sterically demanding ligands. Since then, the surge in research activity on catalysts for the polymerization of alkenes was soon spread over academic and industrial institutions with the aim of seeking non-metallocene catalysts for alkene polymerization. The metals used cover almost the whole spectrum of transition metals, from the early to the later transition metals. For the later transition metals, one example is an Ni-complexed catalyst for the copolymerization of ethylene with a comonomer bearing an oxygen functionality, alcohol, or ester. This work, by Robert Grubbs [29d] and Maurice Brookhart [29a], fully revealed the lower oxophilicity compared with the early transition metals and tolerance to heteroatoms. Zhi-Bin Guan on the other hand, devised a new strategy to control polymer topology with a Pd–α-diimine complex catalyst [29g]. He tried to control the propagation process against the chain-transfer (chain-walking) process, and found that the ethylene pressure may control the competition between these two processes, and then the topology of the polyethylene (PE) may vary from linear PE with moderate branches to hyperbranched PE. For the early transition metals, Terunori Fujita and co-workers at Mitsui Chemicals survey that in Chapter 14, with special emphasis on the FI catalyst system. We appreciate the subtitle of this chapter – "A Gateway to New Polyalkenes." With the unique properties of the PE synthesized using FI catalysts and the advances made by many other groups in this field, we believe that the gateway is open and many new polyalkenes will soon appear. Polyalkenes are the world's most common synthetic polymer, and their monomers are the simplest. It seems that there is little or no room for further breakthroughs with polyalkenes. Anyhow, several recent publications have shown that some new polyalkenes will be realized. One of them is an alkene block copolymer claimed by Dow Chemicals [29e]. The synthesis of block copolymers from simple alkenes is a longstanding and not easy goal for polymer chemists. Dow chemists approach this goal by using a sophisticated chain shuttling process [29f]. These block copolymers have high melting temperatures and low glass transition temperatures, and therefore they maintain excellent elastomeric properties at high temperatures. The second example is SURPASS resin by Nova Chemicals, a copolymer with octene, used for blown films, with very good mechanical strength and processability. The third example is a report by Sanjay Rastogi in 2010 [29h] showing an ultra-high molecular weight polyethylene (UHMWPE), prepared using a perfluorinated FI catalyst at low temperature, characterized by disentanglement and nonexistence of branches up to 10 000 C atoms. A UHMWPE with high linearity (<1 branch/100 000) was also developed by Yong Tang and co-workers at the Shanghai Institute of Organic Chemistry in 2010 [30a,b] by single-site Ziegler–Natta catalysts [30b,c] In particular, the UHMWPE materials synthesized in the above two laboratories are both easily processable, and this characteristic will certainly expand their application to a much broader area. Additionally [O⁻, N, X] tridentate catalysts developed by Yong Tang's group are versatile catalysts, excellent for the homopolymerization of ethylene [31] and for copolymerization of ethylene with α-alkenes [31, 32] and with functionalized alkenes [33]. The combination of these characteristics and the easy preparation and thermostability makes them suitable for industrial use.

Epilog: Better Chemistry, Better Life

With the above summary, we are now coming to the end of this Introduction. When we re-read the contents, we realized that we had missed some important breakthroughs in organic chemistry. Concerning micro-RNA and the production of bulk or fine chemicals from a renewable feedstock, just to mention a couple, we should have given them independent chapters. The alkene metathesis reaction should have had more space in an appropriate chapter. "To meet the urgent needs of human beings" seems to need more to be written. When chemists pay more attention to the function of matter, such a function should be the function to serve the society. For materials science, the physical properties of matter, that is, the conductivity, semiconductivity, superconductivity; the field effect, nonlinear optical properties, the thermochromic effect, and many others are important topics. For biological science, there are health-related problems, pharmaceuticals, agrochemicals, various clinical diagnostic reagents, crop production, and many others to be considered. Therefore, we are really proud of our central, useful, and creative chemistry.

We would like to use a epigram of Prof. Yu-Liang Yang of Fudan University as the epilog of this chapter. The epigram is: "Better Chemistry, Better Life," which imitates the slogan of Shanghai World Expo 2010, "Better City, Better Life." Of course, not every city can endow people with better life, only a better city could do that, so is the chemistry. Even a single explosion or a fire accident, or a single case of pollution of a river, will downgrade the credibility of chemistry in the mind of the general public. It is therefore the responsibility of chemists and chemical engineers in this century to transform all chemical manufacturing industries into greener ones from the origin to downstream, without endangering the people who are living in the neighborhood of a chemical industry, and without scaring students who probably will choose chemistry as their professional career. The solar cell is a green energy, but the production of solar photovoltaic cells would cause serious pollution. In this connection, we should consider this case as a whole system.

Better chemistry is chemistry that should fulfill all the criteria of green chemistry, that is, high yielding, high selectivity, high efficiency, no or little byproduct, from a green starting material, and with low energy consumption.

For the last chapter, Ronald Breslow wrote an essay about the future of organic chemistry. In his essay, he inspiringly addresses seven challenges to organic chemists. He is looking forward to seeing the solutions by organic chemists. Therefore, let us strive to do good chemistry, to do better chemistry for "meeting and solving" these challenges.

Last but not the least, we sincerely thank all the contributors for their great support in preparing this book. We are extremely indebted to Elias James Corey for promising us to put his beautiful molecule, phthalascidin on cover of this book. We would like to acknowledge Professors Henry Wong and Shu-Li You for their many suggestions and corrections and for valuable information. Thanks are also due to Ms. Di Chen, Dr. Fang Lin, and Dr Shao-Xu Huang for their help with many editing tasks.

References

1. Todd, A. (ed.) (1956) *Perspectives in Organic Chemistry*, Interscience Publishers, New York.
2. Ciba Foundation (1978) *Further Perspectives in Organic Chemistry*, Ciba Foundation Symposium, vol. 53, Excerpta Medica, Amsterdam.
3. Bucci, M., Goodman, C., and Sheppard, T.L. (2010) *Nat. Chem. Biol.*, **6**, 847–854.
4. Anastas, P.T. and Warner, J. (1998) *Green Chemistry, Theory and Practice*, Oxford University Press, Oxford.
5. (a) Woodward, R.B. (1973) *Pure Appl. Chem.*, **33**, 145–177; (b) Eschenmoser, A. and Wintner, C.E. (1977) *Science*, **190**, 1410.
6. (a) Armstrong, R.W., Beau, J.M., Cheon, S.H., Christ, W.J., Fujioka, H., Ham, W.H., Hawkins, L.D., Jim, H., Kang, S.H., Kishi, Y., Martinelli, M.J., McWhorter, W.W. Jr., Mizuno, M., Nakata, M., Stutz, A.E., Talamas, F.X., Taniguchi, M., Tino, J.A., Ueda, K., Uenishi, J., White, J.B., and Yonaga, M. (1989) *J. Am. Chem. Soc*, **111**, 7530–7533; (b) Edward, M.S. and Kishi, Y. (1994) *J. Am. Chem. Soc.*, **116**, 11205–11206.
7. Nicolaou, K.C., Yang, Z., Shi, G.Q., Gunzner, J.L., Agrios, K.A., and Gärtner, P. (1998) *Nature*, **392**, 264–269.
8. (a) Walsh, C.T. and Fischbach, M.A. (2010) *J. Am. Chem. Soc.*, **132**, 2469–2493; (b)Wilson, R.M. and Danishefsky, S.J. (2006) *J. Org. Chem*, **71**, 8329–8351.
9. Li, S.-Z. (2003) *Compendium of Materia Medica*, vols. 1–6, Foreign Languages Press, Beijing.
10. Paulsen, H. (1982) *Angew. Chem. F01. Ed. Engl.*, **21**, 155–173.
11. Wu, C.Y. and Wong, C.H. (2011) *Chem. Commun.*, **47**, 6201–6207.
12. Lahann, J. (ed.) (2009) *Click Chemistry for Biotechnology and Materials Sciences*, John Wiley & Sons, Inc., New York.
13. (a) Service, R.F. (2008) *Science*, **320**, 868–869. (b) Baskin, J. M., Prescher,

J. A., Laughlin, S. T., Agard, N. J., Chang, P. V., Miller, I. A., Lo, A., Codelli, J. A., and Bertozzi, C. R. (2007) *PNAS*, **104**, 16793–16797. (c) Leibfarth, F. A., and Hawker, C. J. (2011) *Science*, **333**, 1582–1583.

14. Noyori, R. (2009) *Nat. Chem.*, **1**, 5–6.
15. (a) Breslow, R. (1997) *Chemistry Today and Tomorrow: the Central, Useful and Creative Science*, Jones and Bartlett Publishers, Sudbury, MA; (b) Committee on Challenges for the Chemical Sciences in the 21st Century (2003) *Beyond the Molecular Frontier: Challenges for Chemistry and Chemical Engineering*, National Academies Press, Washington, DC.
16. Arndtsen, B.A., Bergman, R.G., Mobley, T.A., and Peterson, T.H. (1995) *Acc. Chem. Res.*, **28**, 154–162.
17. Dalko, P.I. and Moisan, L. (2004) *Angew. Chem. F01. Ed.*, **43**, 5138–5175.
18. (a) Ishikawa, H., Suzuki, T., and Hayashi, Y. (2009) *Angew. Chem. F01. Ed.*, **48**, 1304–1307; (b) Zhu, S., Yu, S., Wang, Y., and Ma, D. (2010) *Angew. Chem. F01. Ed.*, **49**, 4656–4660.
19. (a) Goldman, A.S., Roy, A.H., Huang, Z., Ahuja, R., Schinski, W., and Brookhart, M. (2006) *Science*, **312**, 257–261; (b) Dong, G., Teo, P., Wickens, Z.K., and Grubbs, R.H. (2011) *Science*, **333**, 1609–1612; (c) Balaraman, E., Gunanathan, C., Zhang, J., Shimon, L.J.W., and Milstein, D. (2011) *Nat. Chem.*, **3**, 609–614.
20. For recent reviews, see: (a) Furuya, T., Kamlet, A.S., and Ritter, T. (2011) *Nature*, **473**, 470–477; (b) Tomashenko, O.A. and Grushin, V.V. (2011) *Chem. Rev.*, **111**, 4475–4521; (c) Lundgren, R.J. and Stradiotto, M. (2011) *Angew. Chem. F01. Ed.*, **50**, 9322–9324. For examples of trifluoromethylation, see: (d) Ji, Y., Brueckl, T., Baxter, R.D., Fujiwara, Y., Seiple, I.B., Su, S., Blackmond, D.G., and Baran, P.S. (2011) *Proc. Natl. Acad. Sci. U. S. A.*, **108**, 14411–14415; (e) Ball, N.D., Gary, J.B., Ye, Y., and Sanford, M.S. (2011) *J. Am. Chem. Soc.*, **133**, 7577–7584; (f) Huang, C.H., Liang, T., Harada, S.J., Lee, E.S., and Ritter, T. (2011) *J. Am. Chem. Soc.*, **133**, 13308–13310; (g) Parsons, A.T. and Buchwald, S.L. (2011) *Angew. Chem. F01. Ed.*, **50**, 9120–9123; (h) Morimoto, H., Tsubogo, T., Litvinas, N.D., and Hartwig,

J.F. (2011) *Angew. Chem. F01. Ed.*, **50**, 3793–3798; (i) Pham, P.V., Nagib, D.A., and MacMillan, D.W.C. (2011) *Angew. Chem. F01. Ed.*, **50**, 6119–6122; (j) Zhang, C.P., Wang, Z.L., Chen, Q.Y., Zhang, C.T., Gu, Y.C., and Xiao, J.C. (2011) *Angew. Chem. F01. Ed.*, **50**, 1896–1900; (k) Niedermann, K., Früh, N., Vinogradova, E., Wiehn, M.S., Moreno, A., and Togni, A. (2011) *Angew. Chem. F01. Ed.*, **50**, 1059–1063.; (l) Morandi, B. and Carreira, E.M. (2011) *Angew. Chem. F01. Ed.*, **50**, 9085–9088; (m) Tomashenko, O.A., Escudero-Adán, E.C., Belmonte, M.M., and Grushin, V.V. (2011) *Angew. Chem. F01. Ed.*, **50**, 7655–7659; (n) Kawai, H., Tachi, K., Tokunaga, E., Shiro, M., and Shibata, N. (2011) *Angew. Chem. F01. Ed.*, **50**, 7803–7806; (o) Xu, J., Fu, Y., Luo, D.F., Jiang, Y.Y., Xiao, B., Liu, Z.J., Gong, T.J., and Liu, L. (2011) *J. Am. Chem. Soc.*, **133**, 15300–15303; (p) Wang, X., Ye, Y., Zhang, S., Feng, J., Xu, Y., Zhang, Y., and Wang, J. (2011) *J. Am. Chem. Soc.*, **133**, 16410–16413; (q) Chu, L.L. and Qing, F.L. (2010) *J. Am. Chem. Soc.*, **132**, 7262–7263; (r) Mu, X., Chen, S., Zhen, X., and Liu, G. (2011) *Chem. Eur. J.*, **17**, 6039–6042. For new examples of C–F bond formation, see: (s) Engle, K.M., Mei, T.S., Wang, X., and Yu, J.Q. (2011) *Angew. Chem. F01. Ed.*, **50**, 1478–1491; (t) Chan, K.S.L., Wasa, M., Wang, X., and Yu, J.Q. (2011) *Angew. Chem. F01. Ed.*, **50**, 9081–9084; (u) Choi, J., Wang, D.Y., Kundu, S., Choliy, Y., Emge, T.J., Krogh-Jespersen, K., and Goldman, A.S. (2011) *Science*, **332**, 1545–1548; (v) Wang, X., Truesdale, L., and Yu, J.Q. (2010) *J. Am. Chem. Soc.*, **132**, 3648–3649; (w) Ye, Y., Ball, N.D., Kanpf, J.W., and Sanford, M.S. (2010) *J. Am. Chem. Soc.*, **132**, 14682–14687; (x) Grushin, V.V. and Marshall, W.J. (2006) *J. Am. Chem. Soc.*, **128**, 12644; (y) Cho, E.J., Senecal, T.D., Kinzel, T., Zhang, Y., Watson, D.A., and Buchwald, S.L. (2010) *Science*, **328**, 1679–1681.
21. Lehn, J.-M. (1973) *Struct. Bond.*, **16**, 1–69.
22. (a) Uhlenheuer, D.A., Petkau, K., and Brunsveld, L. (2010) *Chem. Soc. Rev.*, **39**, 2817–2826; (b) Kurihara, K., Tamura, M., Shohda, K., Toyota, T., Suzuki, K., and Sugawara, T. (2011) *Nat. Chem.*, **3**, 775–781.

23. Steed, J.W. and Gale, P.A. (eds.) (2012) *Supramolecular Chemistry: from Molecules to Nanomaterials*, Wiley-Blackwell, Oxford.

24. (a) van Leeuwen, P.W.N.M. (ed.) (2008) *Supramolecular Catalysis*, Wiley-VCH Verlag GmbH, Weinheim; (b) Meeuwissen, J. and Reek, J.N.H. (2010) *Nat. Chem.*, **2**, 615–621.

25. Harper, K.C. and Sigman, M.S. (2011) *Science*, **333**, 1875–1878.

26. Liang, Y.Y., Xu, Z., Xia, J.B., Tsai, S.T., Wu, Y., Li, G., Ray, C., and Yu, L.P. (2010) *Adv. Mater.*, **22**, E135–E138.

27. Aresta, M. (ed.) (2010) *Carbon Dioxide as Chemical Feedstock*, Wiley-VCH Verlag GmbH, Weinheim.

28. (a) Tanaka, R., Yamashita, M., and Nozaki, K. (2009) *J. Am. Chem. Soc.*, **131**, 14168–14169; (b) Schmeier, T.J., Dobereiner, G.E., Crabtree, R.H., and Hazari, N. (2011) *J. Am. Chem. Soc.*, **133**, 9274–9277; (c) Sujith, S., Min, J.K., Seong, J.E., Na, S.J., and Lee, B.Y. (2008) *Angew. Chem. F01. Ed.*, **47**, 7306–7309; (d) Boddien, A., Mellmann, D., Gärtner, F., Jackstell, R., Junge, H., Dyson, P.J., Laurenczy, G., Ludwig, R., and Beller, M. (2011) *Science*, **333**, 1733–1736.

29. (a) Johnson, L.K., Killian, C.M., and Brookhart, M. (1995) *J. Am. Chem. Soc.*, **117**, 6414–6415; (b) Small, B.L., Brookhart, M., and Bennett, A.M.A. (1998) *J. Am. Chem. Soc.*, **120**, 4049–4050; (c) Britovsek, G.J.P., Gibson, V.C., Kimberley, B.S., Maddox, P.J., McTavish, S.J., Solan, G.A., White, A.J.P., and Williams, D.J. (1998) *Chem. Commun.*, 849–850; (d) Younkin, T.R., Connor, E.F., Henderson, J.I., Friedrich, S.K., Grubbs, R.H., and Bansleben, D.A. (2000) *Science*, **287**, 460–462; (e) Hustad, P.D. (2009) *Science*, **325**, 704–707; (f) Arriola, D.J., Carnahan, E.M., Hustad, P.D., Kuhlman, R.L., and Wenzel, T.T. (2006) *Science*, **312**, 714–719; (g) Guan, Z.B., Cotts, P.M., McCord, E.F., and Mclain, S.J. (1999) *Science*, **283**, 2059–2062; (h) Talebi, S., Duchateau, R., Rastogi, S., Kaschta, J., Peters, G.W.M., and Lemstra, P.J. (2010) *Macromolecules*, **43**, 2780–2788, and references therein.

30. (a) Tang, Y., Wei, B., Li, J.F., and Sun, X.I. (2010) CN 201010554473.8; (b) Tang, Y., Ma, Z., Yang, X.H., Liu, B., Sun, X.L., Gao, Y., and Wang, C. (2006) CN200610026766; (c) Tang, Y., Yang, X.H., Liu, B., Sun, X.L., Ma, Z., Gao, Y., and Wang, C. (2007) WO2007134537.

31. (a) Hu, W.Q., Sun, X.L., Wang, C., Gao, Y., Tang, Y., Shi, L.P., Xia, W., Sun, J., Dai, H.L., Li, X.Q., Yao, X.L., and Wang, X.R. (2004) *Organometallics*, **23**, 1684–1688; (b) Wang, C., Ma, Z., Sun, X.L., Gao, Y., Guo, Y.H., Tang, Y., and Shi, L.P. (2006) *Organometallics*, **25**, 3259–3266; (c) Wang, C., Sun, X.L., Gao, Y.H., Gao, Y., Liu, B., Ma, Z., Shi, L.P., and Tang, Y. (2005) *Macromol. Rapid Commun.*, **26**, 1609–1614; (d) Gao, M.L., Gu, Y.F., Wang, C., Yao, X.L., Sun, X.L., Li, C.F., Qian, C.T., Liu, B., Ma, Z., Tang, Y., Xie, Z., Bu, S.Z., and Gao, Y. (2008) *J. Mol. Catal. A: Chem.*, **292**, 62–66; (e) Yang, X.H., Wang, Z., Sun, X.L., and Tang, Y. (2009) *Dalton Trans.*, 8945.

32. (a) Gao, M.L., Wang, C., Sun, X.L., Qian, C.T., Ma, Z., Bu, S.Z., Tang, Y., and Xie, Z.W. (2007) *Macromol. Rapid Commun.*, **28**, 1511–1516; (b) Gao, M.L., Sun, X.L., Gu, Y.F., Yao, X.L., Li, C.F., and Qian, J.T. (2008) *J. Polym. Sci., Part A: Polym. Chem.*, **46**, 2807–2819.

33. Yang, X.H., Liu, C.R., Wang, C., Sun, X.L., Guo, Y.H., Wang, X.K., Wang, Z., Xie, Z.W., and Tang, Y. (2009) *Angew. Chem. F01. Ed.*, **48**, 8099–8012.

1
Diversity-Oriented Syntheses of Natural Products and Natural Product-Like Compounds

Ling-Min Xu, Yu-Fan Liang, Qin-Da Ye, and Zhen Yang

1.1
Introduction

Natural products have proven to be valuable sources for the identification of new drug candidates, and also as tools for chemical biology and medicinal chemistry research [1]. In fact, if we trace back human history, many natural products have also long been used to treat various human disorders and distinguished by their drug-like properties [2]. To date, tens of thousands of bioactive compounds have been isolated from plants, microbes, marine invertebrates, and other sources [3]. Consequently, these chemical structures have been employed by chemists as references to scan the diversity space for drug discovery efforts [4]. It is estimated that 50–70% of launched drugs in the marketplace are either natural products themselves or natural product-derived molecules [5].

So, what features of natural products make them effective drug candidates? Natural products play important roles in biomedical research and drug discovery largely attributable to their structural complexity and diversity, which Nature has engineered to facilitate optimal functions of living systems. Structural complexity and diversity enable natural products to modulate biological targets of human diseases. The drug-likeness of natural products often possesses common factors including molecular complexity, ability to bind to proteins, structural rigidity, and three-dimensionality [6]. Compared with synthetic molecules, the chemical structures of natural products are more constrained, which allows the accumulation of reliable structure–activity relationship (SAR) data for studying protein target–chemical ligand interactions [7]. Recently, much attention has been devoted to the natural products which are an obvious violation of "Rule-of-Five" [8], but still possess reasonable biopharmaceutical and pharmacokinetic properties, for example, ciclosporin A. Thus, exploitation of such types of natural products could not only significantly broaden the chemical space beyond the "Rule-of-Five" domain for the purpose of drug discovery, but also give us the opportunity to appreciate and investigate how Nature selects and optimizes natural products which are able to bind to and disrupt the function of biological targets through evolution.

We should also recognize that natural products were optimized in living systems presumably not for the same purpose as is desired to serve in a biomedical research

Organic Chemistry – Breakthroughs and Perspectives, First Edition. Edited by Kuiling Ding and Li-Xin Dai.
© 2012 Wiley-VCH Verlag GmbH & Co. KGaA. Published 2012 by Wiley-VCH Verlag GmbH & Co. KGaA.

setting. Their residence in host systems determines that natural products can be optimized to the largest extent that a balance remains within the host systems, and that feasible biosynthetic pathways are in existence. These constraints might in theory explain why many natural products exhibit intolerable toxicity and certain other pharmacokinetics which are undesirable for human therapeutic use. In addition to the above deficiencies, many natural products are scarce or difficult to obtain from natural sources. Also, many possess highly complex structures and can hardly be synthesized in a practical fashion to impact supply.

For those in the chemistry field who are enthusiastic toward natural products, the complexity and diversity of natural products should mean opportunities in the search for new strategies and methodologies to sharpen insights into natural products and also to advance synthetic innovation.

Historically, natural product synthesis has been a very challenging area due to the structural complexity inherent in these molecules. Although synthetic chemists have long been fascinated by natural products, for the most part they have focused on developing the chemistry in order to make precise replicates of the compounds purified from natural sources. Recently, synthetic targets concerning natural products have not been limited to precise replication of the naturally occurring compounds. The accumulation of insights and learning in total synthesis over the last few decades should enable organic chemists to "aim higher," to integrate natural products more closely with advances in biomedical research. Today, chemists can develop synthetic strategies to make both natural products and natural product-like compounds that are comparable to true natural products in size and complexity. The ability to synthesize *in vitro* complex natural products, combined with the strategy of diversity-oriented synthesis (DOS) of natural product-like molecules, which allows very large numbers of natural product-based compound libraries to be made quickly, has made it possible for chemists to accelerate evolution *in vitro* in this process.

For this chapter, we selected several typical examples, such as DOS, diverted total synthesis (DTS), and function-oriented synthesis (FOS), to illustrate the impact of these synthetic strategies on the efficient synthesis of natural product-like compounds.

1.2
Diversity-Oriented Synthesis (DOS)

DOS is aimed at the efficient synthesis of a collection of structurally complex and diverse small molecules, which are screened for their ability to modulate a biological pathway in cells or organisms, without regard to any particular protein target. In other words, DOS is a means to identify simultaneously therapeutic protein targets and small molecules that can modulate the functions of these therapeutic targets.

As a strategy, DOS is based on forward-synthetic analysis [9] to guide its library synthesis. DOS allows many structurally complex and diverse compounds to be prepared efficiently in a flexible and modular way for biological assays. Because compounds generated from DOS by altering stereochemistry and skeletal arrays would display diverse chemical information in three-dimensional space, screening of such compounds would likely generate more hits poised for optimization. Screening of these would generate more hits [10].

It should be noted that DOS as a synthetic strategy was not originated specifically for natural product synthesis. Yet it goes without saying that the DOS approach can be used to generate analogs of natural products and thus enrich life science discovery. Therefore, DOS can obviously maximize the value of natural products in biomedical research by addressing unfavorable features of natural products. For instance, natural products tend to modulate biological targets that have general functions, but do not seem to modulate more specialized targets and processes. This challenge can perhaps be tackled by means of DOS.

DOS is an important synthetic strategy in the light of advances in genomics studies which have identified many biological pathways and processes as ideal points for therapeutic intervention. Today, chemical biology still requires thousands of compounds to be available for screening in biological assays, and DOS shows tremendous promise in this respect.

1.2.1
Diversity-Oriented Synthesis of Skeletally and Stereochemically Diverse Small Molecules [10a]

A typical example in the build–couple–pair strategy [9c] developed by Schreiber and co-workers is illustrated in Scheme 1.1. In the build phase, both R and S stereoisomers of building blocks are readily available. The couple phase employs the diastereoselective Petasis three-component reaction followed by propargylation to yield densely function- alized template **1**. This enables four stereoisomers to be synthesized and is regarded as stereochemical diversification. In the pair phase from **1**, a series of reagent-controlled reactions which selectively pair the nonpolar or the polar functional groups allow skele- tal diversification to be achieved and afford 15 structurally complex and diverse small molecules as shown in Scheme 1.1. Appendage diversification may also be achieved by variation of the lactol and amine building blocks.

1.2.2
Biomimetic Diversity-Oriented Synthesis of Galanthamine-Like Molecules

By application of the strategy of biomimetic solid-phase synthesis, Shair and co-workers developed a DOS of galanthamine-like molecules inspired by biosynthesis of galan- thamine (Figure 1.1) [10]. The chemistry developed allowed four diversity-generating reactions to be performed.

The general transformations are shown in Figure 1.1. Accordingly, Mitsunobu reaction was employed to introduce R^1 diversity. Conjugated addition of thiols was utilized for diversification of R^2. The last two diversity steps involved acylation/alkylation of the amine and imine formation. Finally, the compounds were detached from the solid support by desilylation. Figure 1.2 shows the building blocks used in the library synthesis of galanthamine-based molecules, which reached a total of 2527.

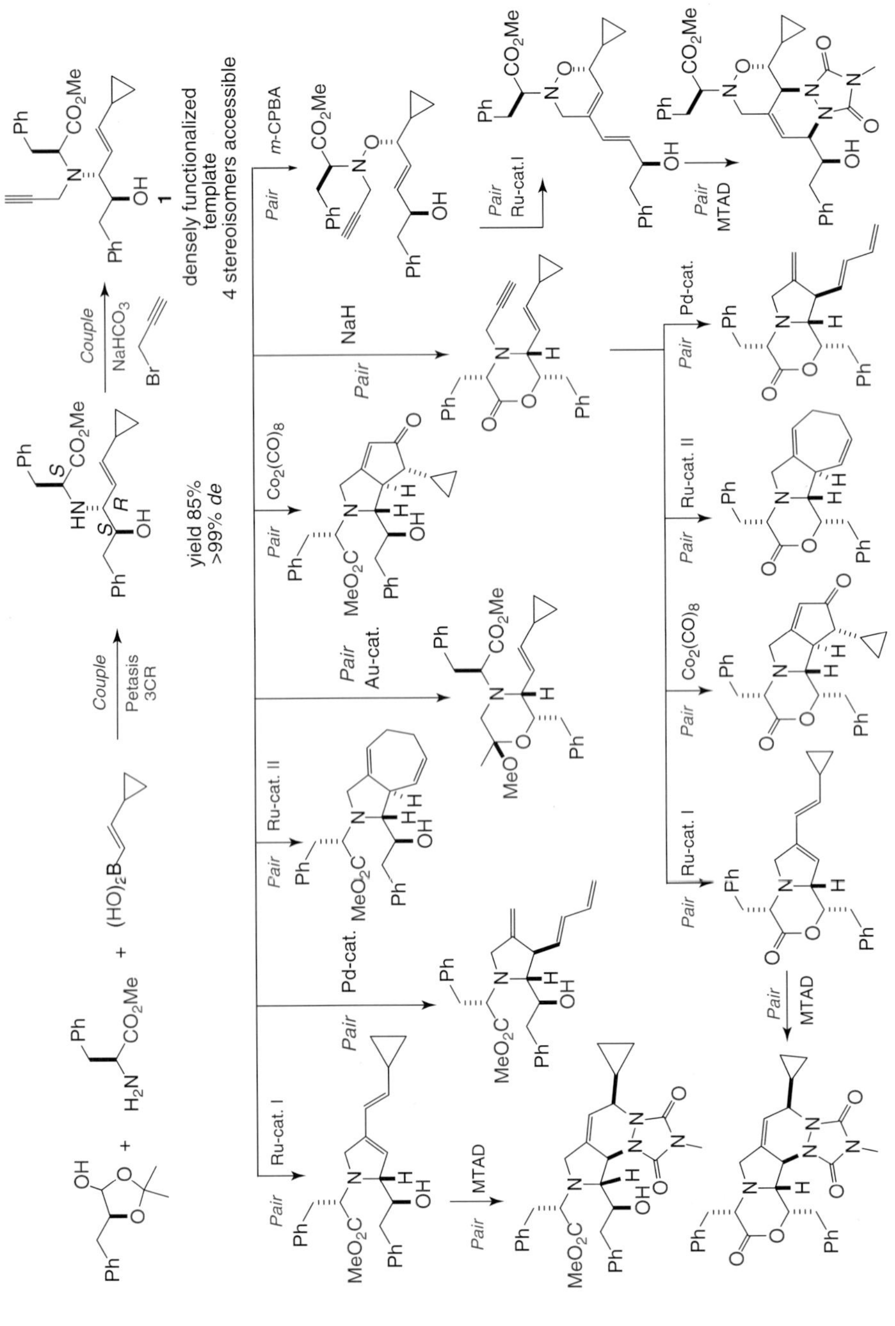

Scheme 1.1 Build–couple–pair strategy to access skeletally and stereochemically diverse small molecules.

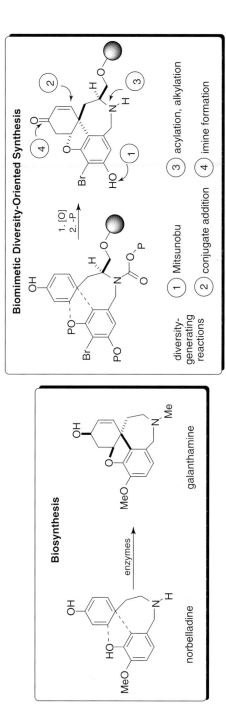

Figure 1.1 Biomimetic diversity-oriented synthesis parallels the biosynthesis of galanthamine.

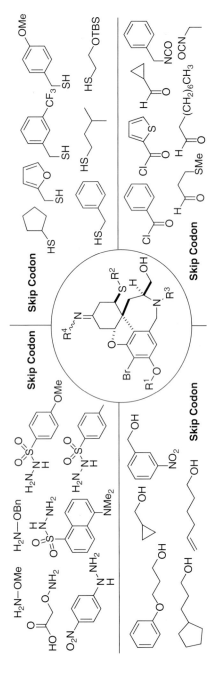

Figure 1.2 Building blocks used in the library synthesis.

1.3
Diverted Total Synthesis (DTS)

Natural products, although highly valuable, are not optimized in a pharmaceutical sense. First, natural products are resident in host systems, therefore they presumably can only be optimized to the extent that a balance can still be maintained within the host systems; second, optimization within host systems requires the existence of feasible biosynthetic pathways. Thus, the strategy of DTS is aimed at tackling these constraints existing in the biosynthetic setting by way of chemical synthetic methods. Relying on the "advanced intermediate" yielded in classic total synthesis, DTS may be employed to reach chemical space having higher or lower complexity than that of the natural product itself. In other words, the employment of the DTS method can achieve "molecular editing" of unnecessary or even undesirable structural features, the kind of optimization which cannot be achieved biosynthetically. On the other hand, these diverted structures often cannot be arrived at from the natural product, owing to chemical limitations. Hence DTS could be used to permit the development of natural product-like biological probes and pharmaceutical agents without recourse to the natural product itself [11].

1.3.1
Diverted Total Synthesis of the Migrastatins

Migrastatin (**2**), isolated from a cultured broth of *Streptomyces*, is a 14-membered macrolactone that reportedly inhibits tumor cell migration [11a,b]. The first asymmetric total synthesis of (+)-migrastatin was accomplished by Danishefsky and co-workers.

As shown in Scheme 1.2, the key step for the synthesis of intermediate **3** is a Lewis acid-catalyzed diene aldehyde condensation (LAC–DAC) with the formation of three contiguous stereocenters. The construction of a glutarimide-containing side chain in intermediate **4** was achieved by an *anti*-selective aldol addition of Evans propionyloxa-zolidinone to aldehyde **5**, and a Horner–Wadsworth–Emmons (HWE) reaction as the key steps. Finally, Yamaguchi esterification followed by a highly *E*-selective ring-closing metathesis led to macrocyclization and afforded migrastatin (**2**) after desilylation.

Despite migrastatin's rather modest inhibitory activity, Danishefsky and co-workers attempted to apply as a lead compound to yield more potent analogs. The total synthesis of migrastatin, which was developed efficiently in a convergent way, allows variations of different regions of migrastatin. These variable regions are highlighted in different colors in Figure 1.3.

The C2–C3 double bond is a potential site for deactivation by 1,4-addition of nucleophiles. The lactone functionality renders the molecule susceptible to hydrolysis and thus diminishes its stability. Furthermore, the highly functionalized C6–C12 portion of migrastatin is regarded as biologically relevant, and the glutarimide moiety might be indispensable for activity.

With these considerations in mind, a set of structurally simplified analogs were designed and synthesized with variations at these regions. Such a DTS strategy takes advantage of the advanced key intermediates **4** and **5**, utilizing them as branching points to yield a diverse set of analogs which could not easily be accessed from the natural

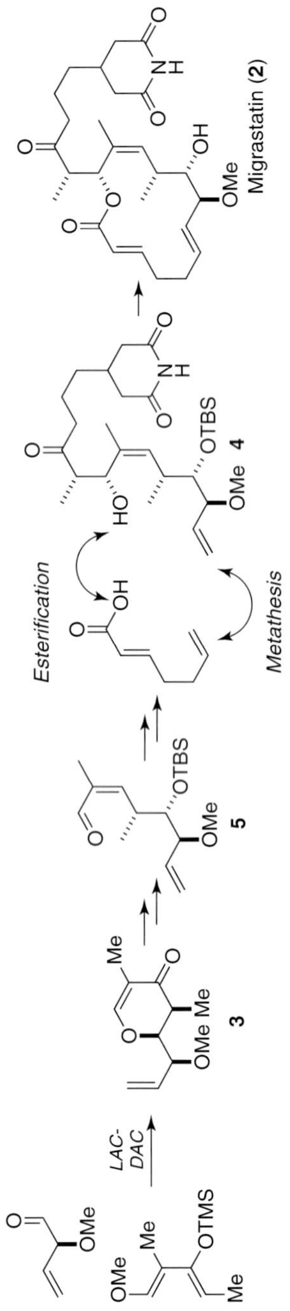

Scheme 1.2 Key steps in the total synthesis of migrastatin (2).

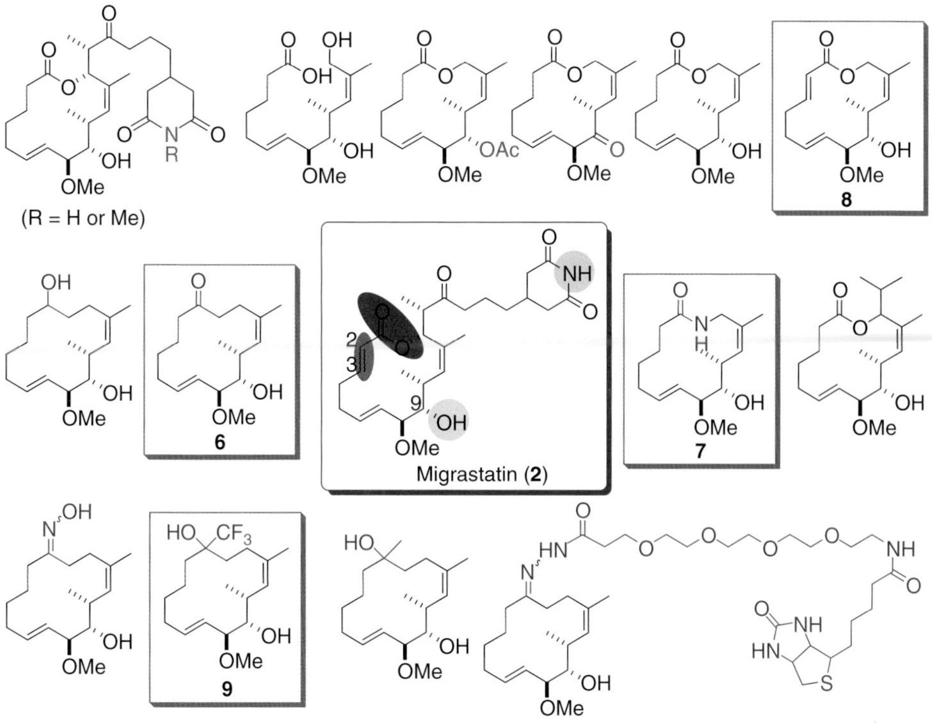

Figure 1.3 Syntheses of migrastatin analogs through diverted total synthesis.

product itself. In total 15 analogs have reportedly been generated, among which **6–9** are the most promising candidates which have demonstrated increased activities compared with the natural product, but without the appending glutarimide moiety.

1.4
Function-Oriented Synthesis (FOS)

The aim of FOS is to address several common problems associated with natural product leads: natural products are often too complex to be made in a practical fashion, and many carry undesired side effects given they are not "designed" for human therapeutic use. Use of the FOS strategy focuses on the function of a natural product, rather than its total structure, to design structurally simplified analogs by incorporating the activity-determining structural features (or their equivalent) of the natural product. Employment of FOS yields natural product analogs with comparable or superior function that could be prepared in a step-economic (i.e., cost-effective) manner. Also, by focusing on a specific function, the FOS can be used to enhance beneficial activities and also to minimize off-target activities better suited for biomedical research or medicinal use [12–14].

1.4.1
Syntheses of Novel and Highly Potent Analogs of Bryostatin

Bryostatins 1–14 are marine-sourced natural products that it is hoped to guide the discovery of medicines to cure cancer and Alzheimer's disease [12]. Unfortunately, their availability from marine sources is extremely limited. Before Wender applied FOS to the synthesis of bryostatin analogs, total syntheses of bryostatins had required more than 70 steps. Recently, Trost's group completed a similar synthesis in 39 steps.

Wender and co-workers believed bryostatins to be ideally suited for FOS considering that its therapeutic activity was connected to only a subset of its structure, and that a pharmacophore could be designed into a simplified target more readily accessed through synthesis. It was inferred from their previous study that the C1 carbonyl, C19 and C26 alcohols, and a corresponding lipophilic region of the bryostatin were likely to be key binding elements. Analog 10 was thus designed with simplified A- and B-rings and an intact C-ring (Figure 1.4). Moreover, the C26 methyl group was deleted to emulate other C1 binders and allow for closer association in the receptor binding pocket.

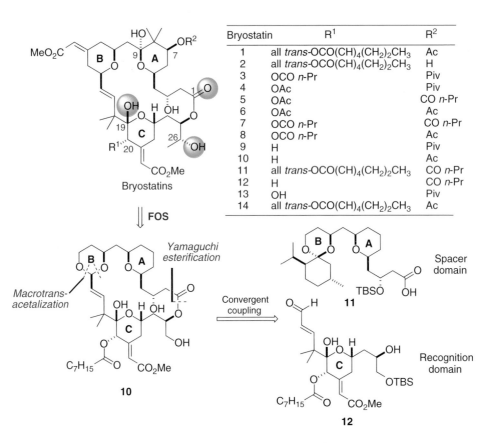

Bryostatin	R^1	R^2
1	all *trans*-OCO(CH)$_4$(CH$_2$)$_2$CH$_3$	Ac
2	all *trans*-OCO(CH)$_4$(CH$_2$)$_2$CH$_3$	H
3	OCO *n*-Pr	Piv
4	OAc	Piv
5	OAc	CO *n*-Pr
6	OAc	Ac
7	OCO *n*-Pr	CO *n*-Pr
8	OCO *n*-Pr	Ac
9	H	Piv
10	H	Ac
11	all *trans*-OCO(CH)$_4$(CH$_2$)$_2$CH$_3$	CO *n*-Pr
12	H	CO *n*-Pr
13	OH	Piv
14	all *trans*-OCO(CH)$_4$(CH$_2$)$_2$CH$_3$	Ac

Figure 1.4 Bryostatins and their more active analog **10** derived from FOS.

These have met the requirements for the development of a convergent synthesis. Retrosynthetic analysis of **10** led to the C1–C13 spacer domain **11** and C15–C26 recognition domain **12**, which featured Yamaguchi esterification and macrotransacetalization. The FOS approach led to the realization of bryostatin analogs that bind to protein kinase C (PKC) with affinities comparable, and in some cases superior, to bryostatins (K_i value of 0.3 versus 1.4 nM). Moreover, these analogs can be made in 29 steps, significantly shorter than the previous pathway for the total synthesis of bryostatins.

Further modification of bryostatins on the fragments of the A-, B-, and C-rings led to the efficient synthesis of a library of 31 analogs. Importantly, the analogs show greater potency than bryostatin when tested for growth inhibitory activity against the US National Cancer Institute panel of human cancer cell lines.

In summary, FOS led to more potent analogs that can be supplied in quantity and tuned for performance and at the same time allowed for the development of effective methods for convergent macrolide formation. An advantage of FOS was demonstrated in this study, namely that superior function can be achieved in fewer steps with simplified structures while enabling synthetic innovation.

1.4.2
Discovery of Potent and Practical Antiangiogenic Agents Inspired by Cortistatin A

Among 10 related natural products of the cortistatin family, cortistatin A (**13**) and cortistatin J (**14**) show the most promising antiangiogenic activity [14]. Rather than embarking on the total synthesis of these natural products, Corey's group started the journey by designing analogs with less complexity for access while still possessing equal or even enhanced biological activity.

After analyzing the structures of **13** and **14**, they drew the conclusion that the dimethylamino group at the C3 position and the isoquinoline appendage at C17 were required for biological activity, whereas the hydroxyl groups at C1 and C2 were not essential. The above information led to the design of analog **15** possessing a steroidal core and a double bond at C16–C17 (Figure 1.5).

By varying the C17 appendage and C3 amino group, a series of analogs were synthesized, as shown in Figure 1.5. The antiangiogenic activity of these analogs was evaluated *in vitro*, based on which the data indicate that the most active compounds are those underlined.

1.5
Target-Oriented Synthesis (TOS)

Target-oriented synthesis (TOS) is an effective synthetic strategy to generate natural products and natural product-like molecules in a concise, convergent, and systematic manner. One major deficit in natural product-based research is the inaccessibility of identified natural products, because they exist in only trace amounts in Nature, which prevents biological evaluation of natural products in a systematic manner. Therefore, accessing by way of chemical synthesis these valuable gifts of Nature derived from

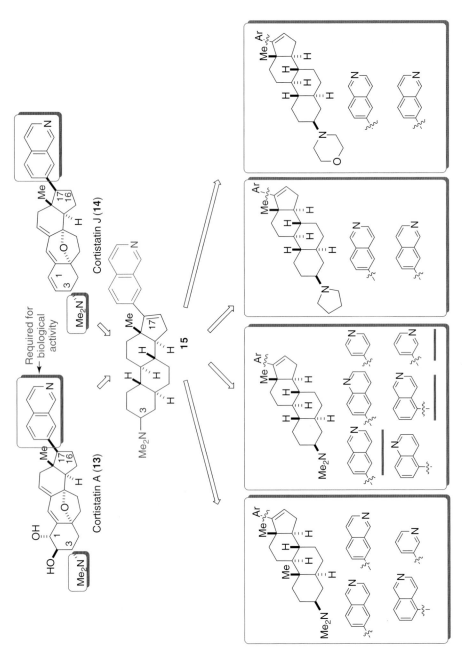

Figure 1.5 Design and syntheses of structurally simple analogs of cortastatin A.

millions of years of evolution represents a sizable challenge to organic chemistry, yet equally holds tremendous promise in biomedical research and drug discovery. TOS of natural products, which has been fueled by the methodology of retrosynthetic analysis since the mid-1960s, is an effective strategy to produce scarce and inaccessible natural products in laboratories. Today, many organic chemists are passionate about the total synthesis of newly discovered natural products and optimizing known synthetic steps. These synthetic missions continuously generate new experimental and intellectual challenges to organic chemists. As natural products are often biologically relevant, a principle of biomimetic synthesis is usually applied for guidance, which allows access to natural products with high efficiency.

1.5.1
Synthetic Studies and Biological Evaluation of Ecteinascidin 743

Ecteinascidin 743 (**16**) is an exceedingly potent and rare marine-derived antitumor agent that acts by inducing DNA–protein cross-linking [15b], and is now in advanced clinical trials (Figure 1.6) [15].

In the past 20 years, Corey's group has made tremendous efforts to optimize the synthetic route to ecteinascidin 743 (**16**). The concise and convergent route for the construction of its core structure was eventually developed successfully and is summarized in Scheme 1.3.

Key steps in this synthesis involve the following: (i) amide bond formation between **18** and **19**, followed by allylation to give **20**; (ii) selective reduction of lactone to its lactal was achieved by the reduction with $LiAlH_2(OMe)_2$, and **21** was obtained after desilylation with KF–MeOH; and (iii) the pentacyclic core was formed by treatment of **21** with TfOH in $H_2O-CF_3CH_2OH$ mixed solvents in the presence of butylated hydroxytoluene (BHT), and the formed pentacyclic lactam was first reduced with $LiAlH_2(OMe)_2$ to afford the corresponding cyclic aminal, which upon exposure to HCN provided pentacyclic aminonitrile **22** in high yield. The significance of this synthesis is its overall high yield (57% in six steps) and its reliability in a large-scale synthesis. A critical element to the

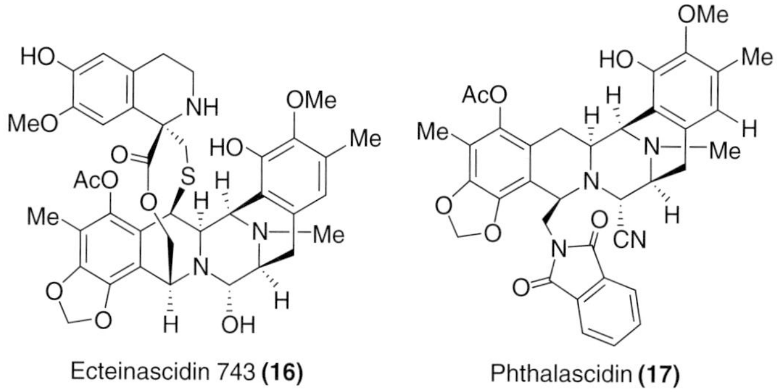

Figure 1.6 Ecteinascidin 743 (**16**) and phthalascidin (**17**).

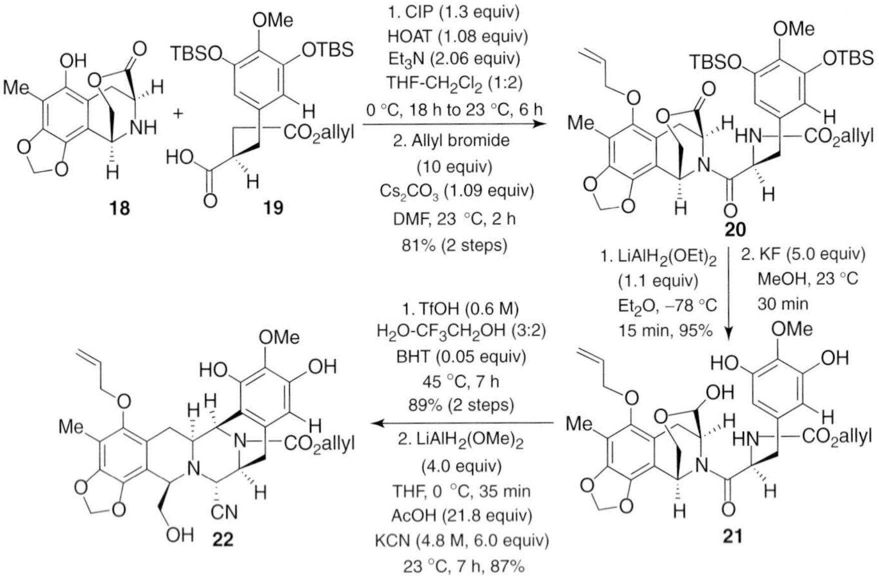

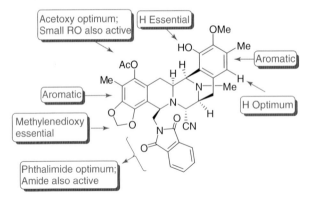

Scheme 1.3 The concise and convergent route to the core structure (**22**) for ecteinascidin 743 (**16**).

Figure 1.7 SAR of phthalascidin (**17**).

success of this synthesis was the utilization of LiAlH$_2$(OMe)$_2$ as an efficient reducing agent, which had not been used frequently previously.

From the common intermediate **22**, a series of totally synthetic molecules that are structurally related to phthalascidin (**17**) were prepared and evaluated as antitumor agents [15b] (Figure 1.7).

From this study, phthalascidin (**17**) was identified as the most active antitumor agent, and is now in advanced clinical trials. Its antiproliferative activity ($IC_{50} = 0.1–1$ nM) is greater than that of the agents of paclitaxel (Taxol), camptothecin, adriamycin, mitomycin C, cisplatin, bleomycin, and etoposide by 1–3 orders of magnitude. Because phthalascidin

(17) is more stable than ecteinascidin 743 (16) and considerably easier to make, it could potentially be a more practical therapeutic agent.

1.5.2
Total Synthesis and Biological Evaluation of Biyouyanagin A and Analogs

Biyouyanagin A was isolated from the *Hypericum* species *H. chinese* L. var. *salicifolium* and was originally assigned as structure 24a based on NMR spectroscopic analysis (Figure 1.8) [16].

As ambiguity about the stereocenters at C24 and around the cyclobutane ring exists, retrosynthetic analysis featuring a photoinduced [2 + 2]-cycloaddition (a potential

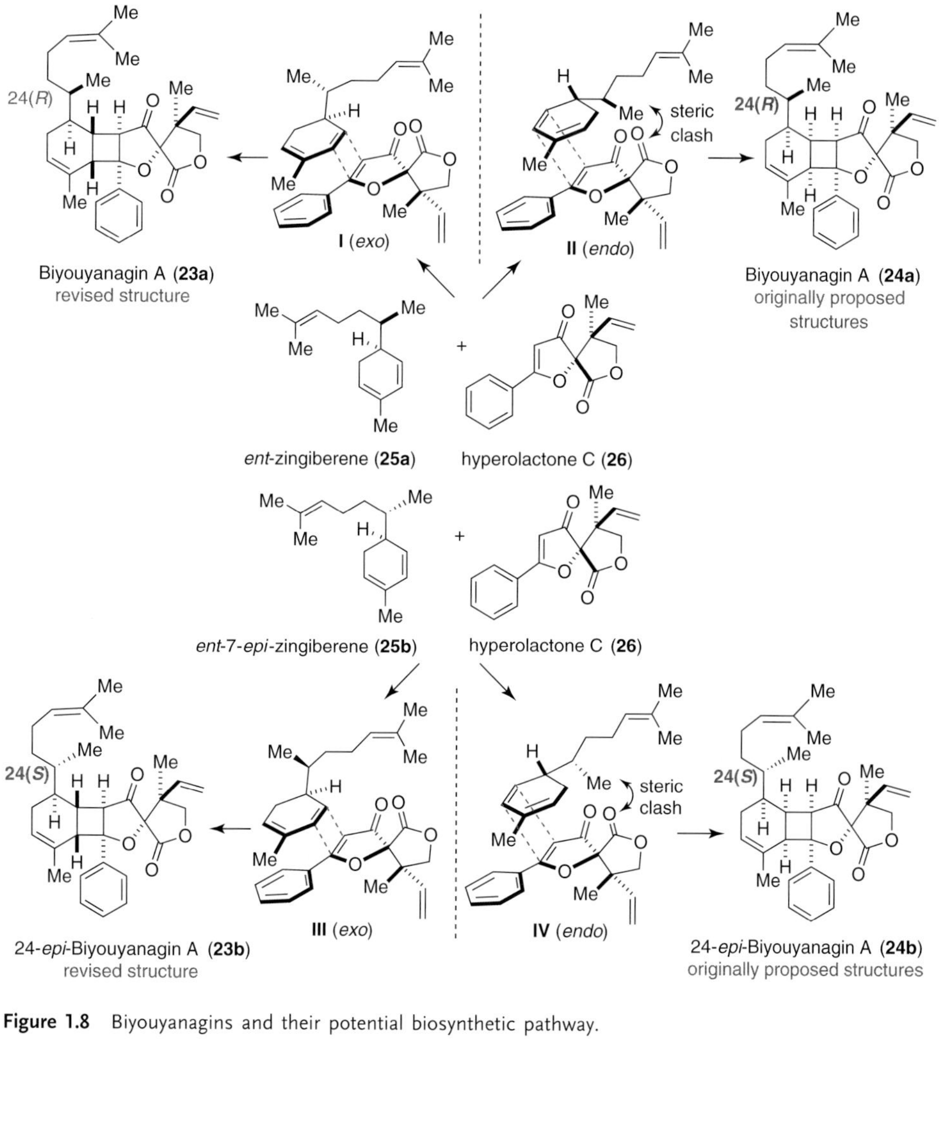

Figure 1.8 Biyouyanagins and their potential biosynthetic pathway.

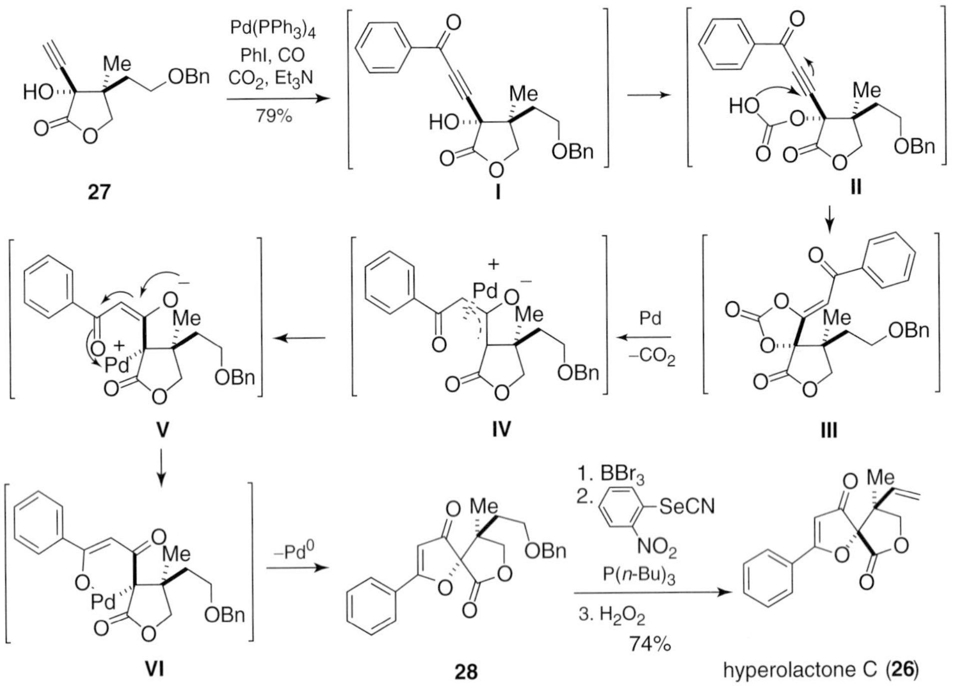

Scheme 1.4 Synthesis of hyperolactone C (**26**) through Pd-catalyzed reactions.

biosynthetic pathway) constitutes a perfect solution to this problem. Two transition states could be envisioned in this key reaction, with the more favorable *exo* transition state **I**, leading to **23a** as the natural product.

The journey began with the construction of *ent*-zingiberene (**25a**) and hyperolactone C (**26**), considering they are naturally occurring substances. Scheme 1.4 shows the details of the synthesis of hyperolactone C (**26**), which could be derived from **27** through palladium-catalyzed cascade sequences to deliver **28**, the stereochemistry of which was confirmed by X-ray crystallographic analysis (Scheme 1.4). The completion of the synthesis of hyperolactone C (**26**) required debenzylation (BBr$_3$), selenenylation, and oxidation/*syn*-elimination.

The key photoinduced [2 + 2]-cycloaddition of **26** with **25a** proceeded with the desired regio- and diastereoselectivity, leading to **23a** (Figure 1.8). NMR spectroscopic data for **23a** were consistent with those reported for the natural product biyouyanagin A, the structure of which was further verified by X-ray crystallographic analysis.

Having established [2 + 2]-cycloaddition as an efficient and concise way to build up the scaffold of biyouyanagin A, a further nine biyouyanagin A analogs (Figure 1.9) were synthesized via the same synthetic strategy as illustrated in Figure 1.8, and their biological activities were profiled.

Accordingly, all the newly synthesized compounds **23b** (Figure 1.8) and analogs in Figure 1.9 displayed similar activities against HIV-1 replication and cytotoxicities against

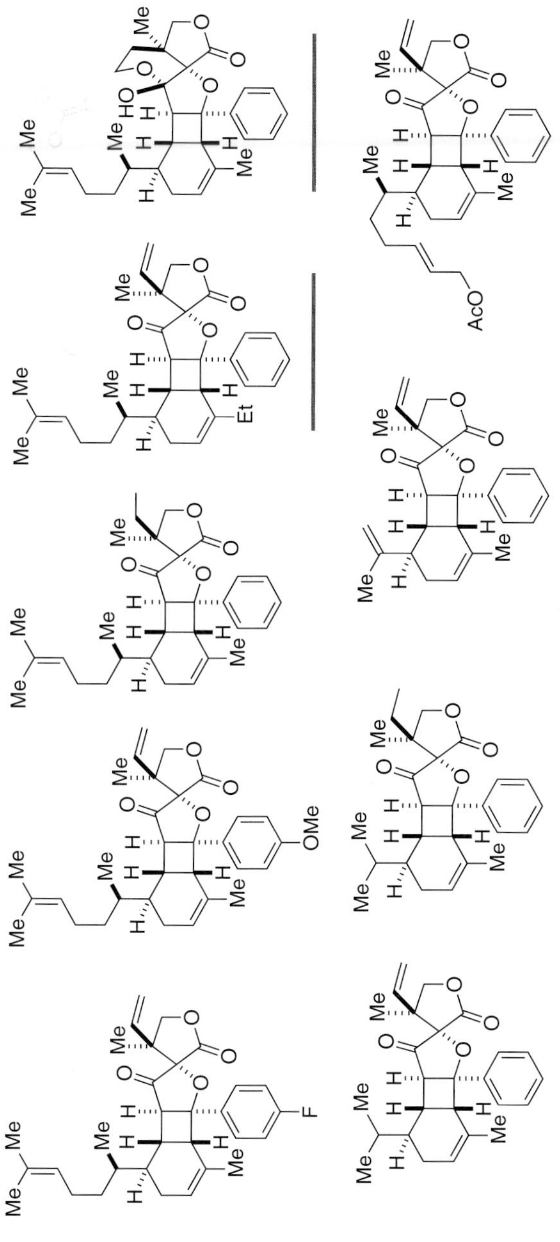

Figure 1.9 Synthesis of additional analogs of biyouyanagin A.

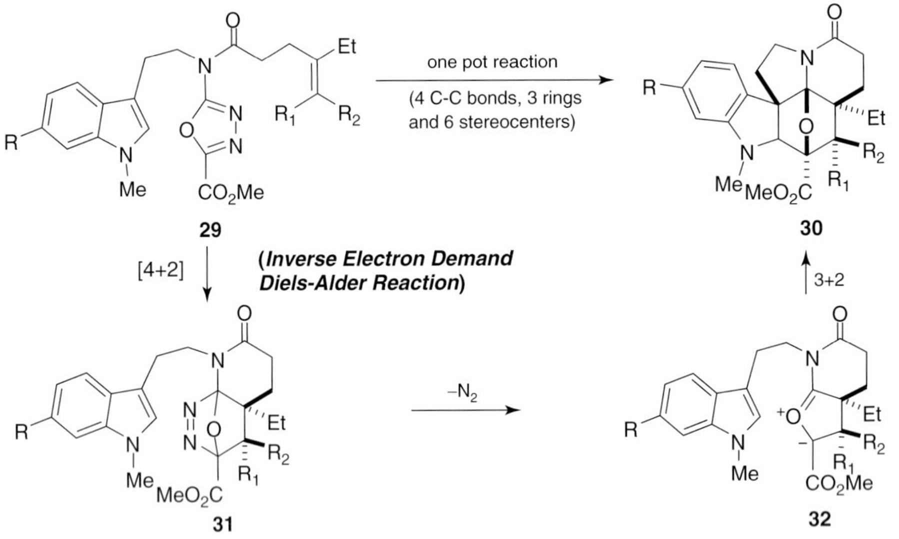

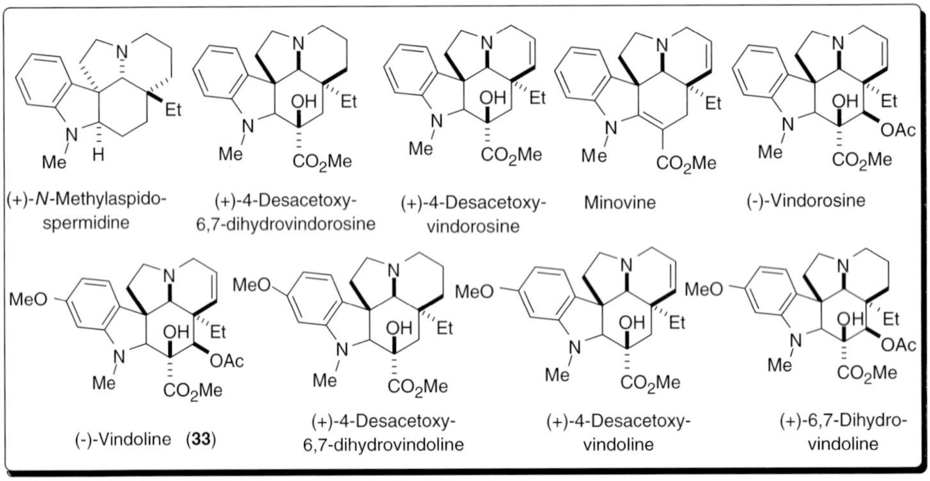

Figure 1.10 Synthesis of vindoline-related natural products via a [4 + 2]/[3 + 2]-cycloaddition cascade.

the MT-2 lymphocytes, with the two underlined analogs being the most potent against HIV-1.

1.5.3
Total Synthesis of Vindoline, Related Natural Products, and Structural Analogs

Boger's group developed a tandem intramolecular [4 + 2]/[3 + 2]-cycloaddition cascade of a 1,3,4-oxadiazole (**29**) to construct the vindoline skeleton (**30**), in which three

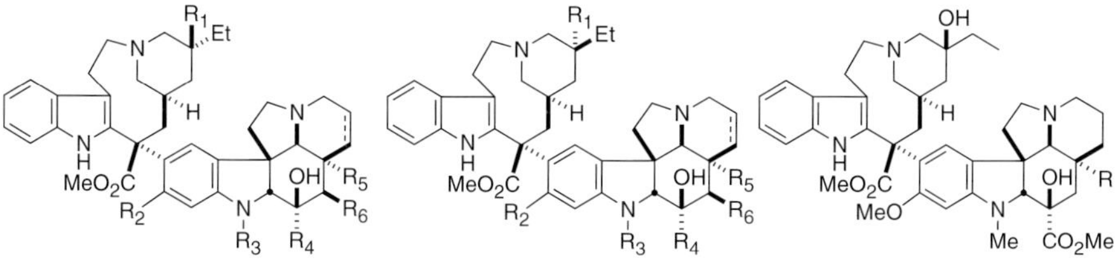

Type A, 19 compounds
R_1 = H, OH, TEMPO
R_2 = H, OMe
R_3 = H, Me, CHO
R_4 = CO$_2$Me, CONH$_2$, CONHNH$_2$
R_5 = H, Et
R_6 = H, OH, OAc
C6-C7 : C=C, C-C

Type B, 17 compounds
R_1 = H, OH, TEMPO, NO, NH$_2$, N$_3$
R_2 = H, OMe
R_3 = H, Me
R_4 = CO$_2$Me
R_5 = H, Et
R_6 = H, OH, OAc
C6-C7 : C=C, C-C

Type C
R = H, Me, Pr, CCH, CHO,
    CH$_2$OH, CH(OH)CH$_2$OH

Figure 1.11 Design of structural analogs.

rings, four C–C bonds, and six stereocenters are formed in a single step (Figure 1.10) [17].

The success of this cascade was based on the precise design of reacting partners, in which the electron-rich enol ether is matched with the electron-deficient oxadiazole in the inverse electron demand Diels-Alder reaction. Loss of N$_2$ in intermediate **31** provides the 1,3-dipole product **32**, which is stabilized by the electron-deficient substituent. A diastereoselective *endo* [3 + 2]-cycloaddition then takes place at the less hindered face to deliver **30**.

[4 + 2]- and [3 + 2]-cycloadditions belong to the type of pericyclic reactions, which usually proceed selectively. As a result, the asymmetric syntheses of nine naturally occurring products were achieved as a single isomer, which allowed their stereochemistry to be established (Figure 1.10).

The chemistry developed eventually led to the synthesis of a compound library of more than 40 members based on the scaffolds of vinblastine and leurosidine (Figure 1.11).

Vinblastine (**35**) and leurosidine (**36**) are bisindole alkaloids, and are believed to be made through direct coupling of catharanthine (**34**) with vindoline (**33**) under oxy-reduction conditions (Scheme 1.5). In this event, when catharanthine amine was oxidatively treated with FeCl$_3$, radical cation **A** could be generated, which then underwent further oxidation, followed by biomimetic coupling with vindoline (**33**) to give intermediate **C**, a single diastereoisomer. Intermediate **C** was subjected to reduction with NaBH$_4$ to give anhydrovinblastine, which was then oxidatively converted to vinblastine (**35**) and leurosidine (**36**) in 12 steps.

1.5.4
Total Synthesis of Eudesmane Terpenes by Site-Selective C–H Oxidations

The total synthesis of eudesmane terpenes by Baran's group was inspired by the biosynthesis of terpenes, which often comprise a "two-phase" process (cyclase phase and oxidase phase) [18]. Among the eudesmane family, dihydrojuenol (**37**) constitutes one of the lowest oxidized members.

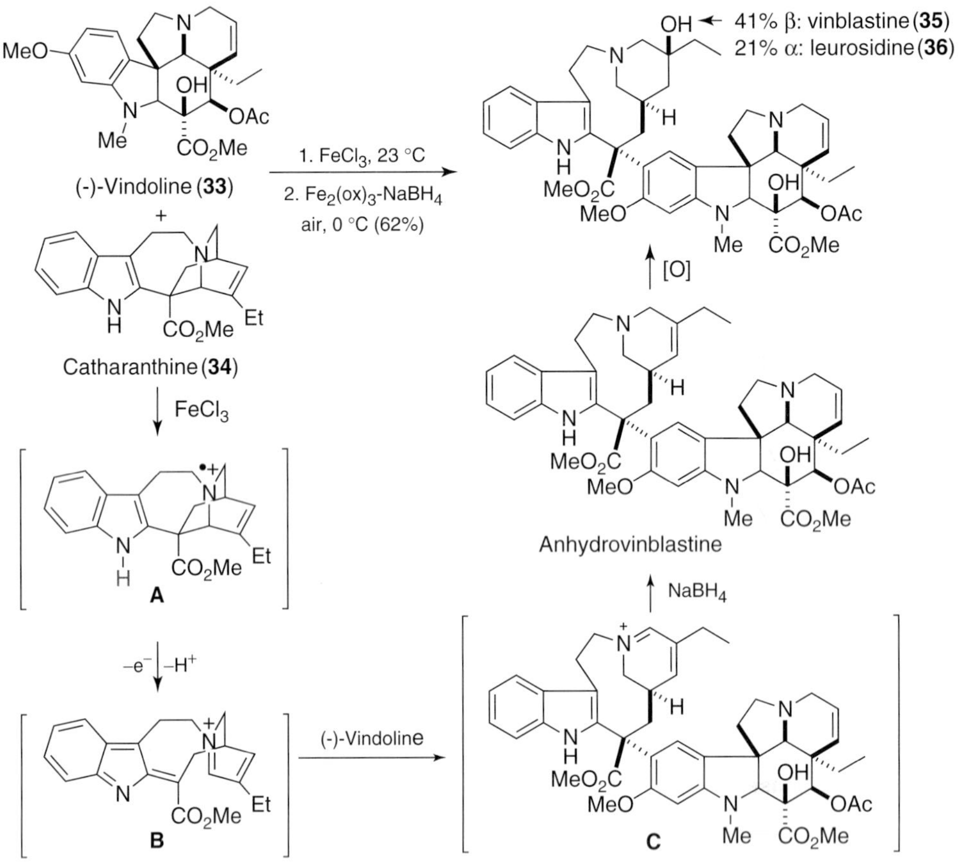

Scheme 1.5 Biomimetic synthesis of vinblastine (**35**) and leurosidine (**36**).

Starting from **37**, terpenes possessing different levels of oxidation state can be accessed by sequential, site-selective C–H oxidations (Figure 1.12). A simple and superior route to **37** (corresponding to the cyclase phase) has been developed which involves a nine-step sequence and allows enantioselective and gram-scale preparation.

To plan a biomimetic synthesis, the trifluoroethyl carbamate directing group was introduced to deliver **43**. Its electron-withdrawing property renders H1 and H5 most prone to be oxidized. Thus, site-specific oxidation and halogenation were realized to afford intermediates **43** and **44**, respectively, which finally gave 4-epiajanol (**38**) and dihydroxyeudesmane (**39**), the structures of which were verified by X-ray crystallography.

To access the higher oxidized targets in the retrosynthesis pyramid diagram, **43** was converted into bromide **46** via the similar site-specific halogenation (Figure 1.13). Subsequent silver-assisted cyclization to carbonate **48** and hydrolysis afforded pygmol (**40**). Alternatively, **46** was subjected to basic elimination conditions to deliver the

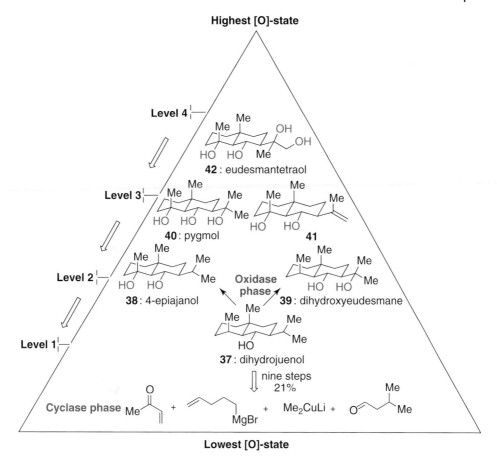

Figure 1.12 Pyramid diagram for the retrosynthetic planning of terpene synthesis using a "two-phase" approach.

alkene **47**. As the direct dihydroxylation of **47** proved ineffective, the neighboring group participate oxybromination followed by hydrolysis generated **48**, which was exposed to dilute acid to afford eudesmantetraol (**42**) by net inversion.

1.5.5
Total Synthesis of Bipinnatin J, Rubifolide, and Coralloidolides A, B, C, and E

The furanocembranoids are a family of diterpenoid natural products with a diverse level of oxidation states [19]. As bipinnatin J (**49**) was identified as a key precursor in the biomimetic synthesis, its first total synthesis was targeted as shown in Scheme 1.6. It features a ruthenium-catalyzed Alder−ene reaction, a Stille cross-coupling, and an intramolecular Nozaki−Hiyama−Kishi allylation as key steps.

With bipinnatin J (**49**) in hand, biomimetic transformations of it into other natural products were examined (Scheme 1.7). Luckily, reductive deoxygenation of the hydroxyl

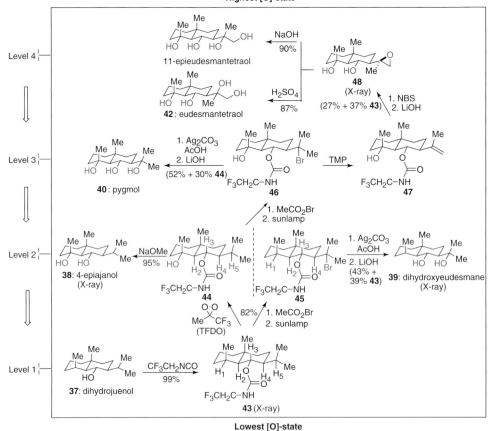

Figure 1.13 Biomimetic synthesis of eudesmantetraol.

group in bipinnatin J (**49**) delivered rubifolide (**50**), which was subjected to *m*-CPBA oxidative cleavage of the furan ring to yield isoepilophodione B (**51**) smoothly. Alternatively, Achmatowicz oxidation of bipinnatin J (**49**) proceeded cleanly to afford the sensitive hydroxypyranone **56** as a single stereoisomer which was then protected as its acetate followed by elimination of acetic acid to undergo 1,3-dipolar cycloaddition to afford intricarene (**57**). Rubifolide (**50**) could also be selectively epoxidized to afford coralloidolide A (**52**). Subsequent oxidative cleavage of the furan ring generated coralloidolide E (**53**) with dienedione functionality. Upon treatment with scandium triflate hydrate, it was ultimately converted into coralloidolide B (**54**). A transannular aldol reaction employing *N*,*N*′-dibutylurea (DBU) as the base generated coralloidolide C (**55**) in modest yield. Hence the total syntheses of natural products of furanocembranoids were realized in no more than 13 steps through this biomimetic synthesis without protecting groups.

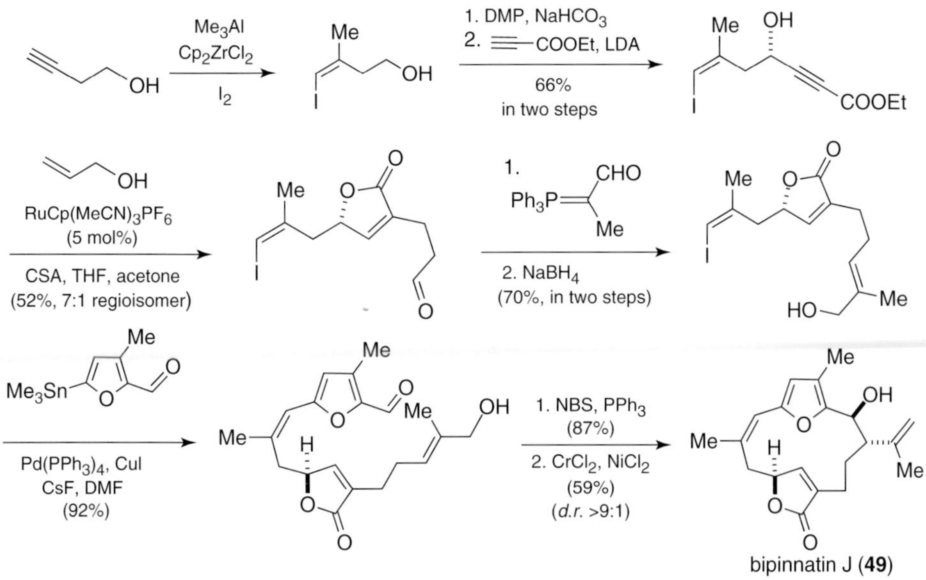

Scheme 1.6 Total synthesis of (±)-bipinnatin J (**49**).

1.5.6
Total Synthesis of Diverse Carbogenic Complexity Within the Resveratrol Class from a Common Building Block

The resveratrol family represents an important group of natural products with potential clinical applications [20]. However, owing to the structural complexity and liability, most members of the family could not be accessed in large quantities, either through synthesis or from natural resources.

To probe a biosynthetic pathway, Snyder's group identified a common building block, **58**, distinct from Nature's monomers. With alteration of the substituents in the aromatic rings, and careful control of the reaction conditions, cascade sequences initiated by relatively simple reagents could deliver one of the natural products selectively, as delineated in Scheme 1.8.

Exposure of **58** to acid initiated a proton-induced cyclization followed by capture with a thiol nucleophile to give intermediate **59**, which could be converted smoothly to **60**. Utilizing the skeleton of **60** as a branching point, a halogen electrophile-based cascade involving bromination and Friedel–Crafts alkylation delivered two new types of skeletons after dehalogenation [including natural products ampelopsin F (**61**) and pallidol (**62**)]. If **58** is first oxidized to ketone before cyclization, subsequent Friedel–Crafts alkylation induced by bromonium or dioxirane would lead to the seven-membered ring as represented by **63–65**. Alternatively, treatment with acid triggered a Friedel–Crafts/retro Friedel–Crafts sequence to deliver **66**. Subsequent

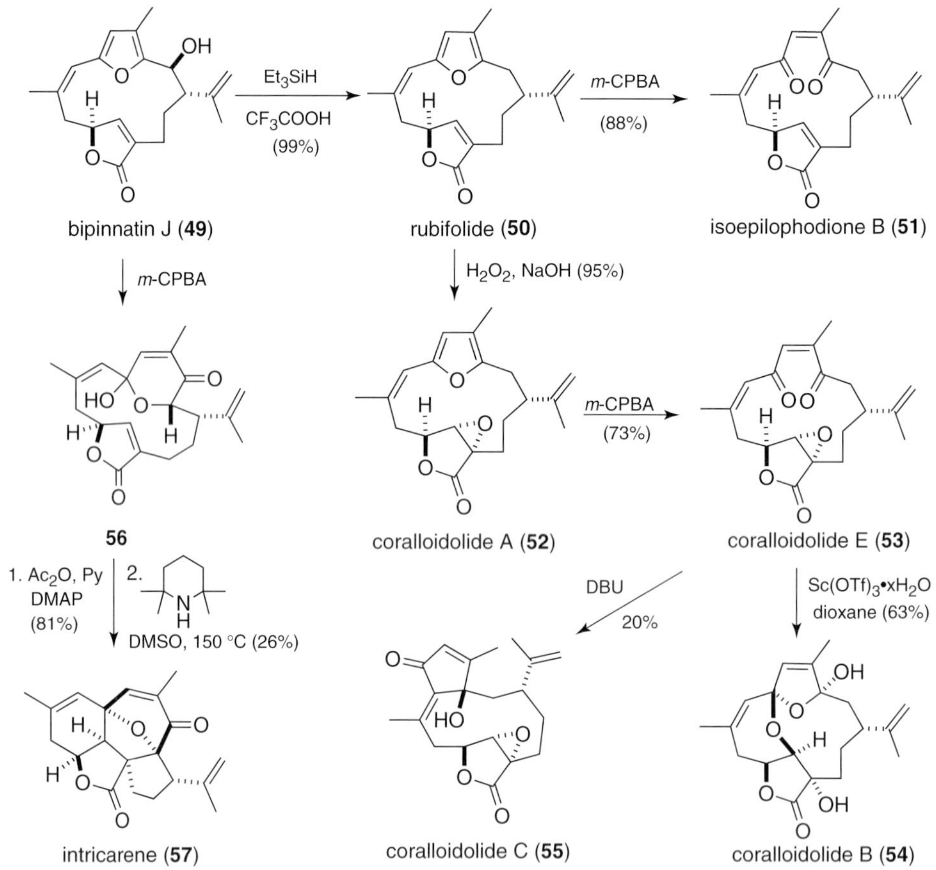

Scheme 1.7 Biomimetic synthesis of furanocembranoids.

reduction followed by acidic quenching allows a second Friedel–Crafts reaction to provide the [3.2.2]-bicycle (**67**). Overall, a series of natural products and analogs have been synthesized on a relatively large scale and in fewer than 10 steps, including 11 natural products (nine of which were synthesized for the first time) and 14 natural product-like analogs.

1.6
Conclusion and Perspectives

The need to identify fast and efficient synthetic methodologies and strategies for the generation of both natural products and natural product-like compounds has revolutionized the state-of-the-art of synthetic chemistry, demonstrating organic chemists' tireless

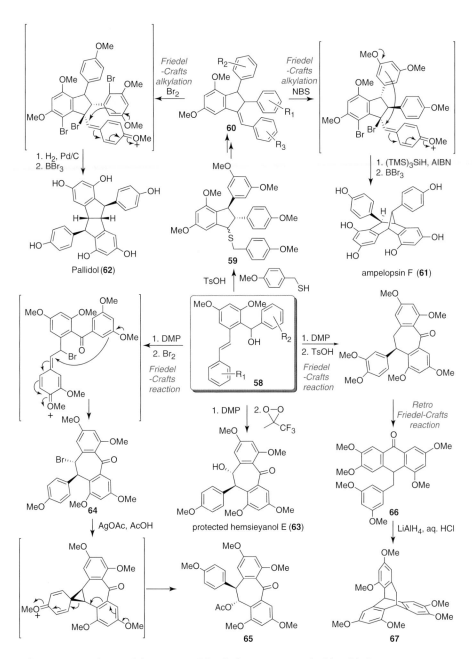

Scheme 1.8 Syntheses of the resveratrol family from a common building block.

efforts to design and synthesize compound collections more efficiently, practically, and with a greater chance of further optimization at a later research stage, compared with previous approaches. Among those efforts, the synthesis of natural products remains of great value in providing teaching and inspiration traceable to their biological activities, structural complexity and diversity for organic chemists to advance synthetic strategies and methodologies.

Synthetic targets concerning natural products are not limited to the precise replication of the naturally occurring compounds. The accumulation of insights and learning in total synthesis over the last few decades should enable organic chemists to "aim higher," to integrate natural products more closely with advances in life science discovery. During that process, synthetic chemists will aim to find a new paradise to refine continuously the beauty of synthetic art.

Acknowledgments

We thank for the Professor Ye-Feng Tang for his helpful comments. Financial support was provided by the National Science Foundation of China.

References

1. (a) Clardy, J. and Walsh, C. (2004) *Nature*, **432**, 829–837; (b) Nicolaou, K.C. and Montagnon, T. (2008) *Molecules that Changed the World*, Wiley-VCH Verlag GmbH, Weinheim.

2. (a) Breinbauer, R., Vetter, I.R., and Waldmann, H. (2002) *Angew. Chem. Int. Ed.*, **41**, 2878–2890; (b) Kumar, K. and Waldmann, H. (2009) *Angew. Chem. Int. Ed.*, **48**, 3224–3242.

3. Weber, L. (2002) *Drug Discov. Today*, **7**, 143.

4. Koch, M.A., Schuffenhauer, A., Scheck, M., Wetzel, S., Casaulta, M., Odermatt, A., Ertl, P., and Waldmann, H. (2005) *Proc. Natl. Acad. Sci. U. S. A.*, **102**, 17272–17277.

5. (a) Grifo, F., Newman, D., Fairfield, A.S., Bhattacharya, B., and Grupenhoff, J.T. (1997) in *The Origins of Prescription Drugs* (eds. F. Grifo and J. Rosenthal), Island Press, Washington, DC, p. 131; (b) Arvigo, R. and Balick, M. (1993) *Rainforest Remedies*, Lotus Press, Twin Lakes, WI.

6. Arya, P., Joseph, R., and Chou, D.T.H. (2002) *Chem. Biol.*, **9**, 145.

7. (a) Langer, T. and Hoffman, R.D. (2001) *Curr. Pharm. Des.*, **7**, 509; (b) Woolfrey, J.R. and Weston, G.S. (2002) *Curr. Pharm. Res.*, **8**, 1527; (c) Klavunde, T. and Hessler, G. (2002) *ChemBioChem*, **3**, 928.

8. Lipinski, C. and Hopkins, A. (2004) *Nature*, **432**, 855–861.

9. (a) Schreiber, S.L. (2000) *Science*, **287**, 1964–1969; (b) Burke, M.D. and Schreiber, S.L. (2004) *Angew. Chem. Int. Ed.*, **43**, 46–58; (c) Nielsen, T.E. and Schreiber, S.L. (2008) *Angew. Chem. Int. Ed.*, **47**, 48–56.

10. (a) Kumagai, N., Muncipinto, G., and Schreiber, S.L. (2006) *Angew. Chem. Int. Ed.*, **45**, 3635–3638; (b) Goess, B.C., Hannoush, R.N., Chan, L.K., Kirchhausen, T., and Shair, M.D. (2006) *J. Am. Chem. Soc.*, **128**, 5391–5403; (c) Pelish, H.E., Westwood, N.J., Feng, Y., Kirchhausen, T., and Shair, M.D. (2001) *J. Am. Chem. Soc.*, **123**, 6740–6741; (d) Lindsley, C.W., Chan, L.K., Goess, B.C., Joseph, R., and Shair, M.D. (2000) *J. Am. Chem. Soc.*, **122**, 422–423.

11. (a) Wilson, R.M. and Danishefsky, S.J. (2006) *J. Org. Chem.*, **71**, 8329–8351;

(b) Gaul, C., Njardarson, J.T., Shan, D., Dorn, D.C., Wu, K.D., Tong, W.P., Huang, X.Y., Moore, M.A.S., and Danishefsky, S.J. (2004) *J. Am. Chem. Soc.*, **126**, 11326–11337; (c) Paterson, I., Gardner, N.M., Poullennec, K.G., and Wright, A.E. (2008) *J. Nat. Prod.*, **71**, 364–369; (d) Fürstner, A., Kirk, D., Fenster, M.D.B., Aïssa, C., Souza, D.D., Nevado, C., Tuttle, T., Thiel, W., and Müller, O. (2007) *Chem. Eur. J.*, **13**, 135–149.

12. (a) Wender, P., Verma, V.A., Paxton, T.J., and Pillow, T.H. (2008) *Acc. Chem. Res.*, **41**, 40–49; (b) Wender, P.A. and Hinkle, K.W. (2000) *Tetrahedron Lett.*, **41**, 6725–6729; (c) Wender, P.A., Baryza, J.L., Bennett, C.E., Bi, F.C., Brenner, S.E., Clarke, M.O., Horan, J.C., Kan, C., Lacôte, E., Lippa, B., Nell, P.G., and Turner, T.M. (2002) *J. Am. Chem. Soc.*, **124**, 13648–13649; (d) Wender, P.A., Hilinski, M.K., and Mayweg, A.V.W. (2005) *Org. Lett.*, **7**, 79–82; (e) Wender, P.A., Clarke, M.O., and Horan, J.C. (2005) *Org. Lett.*, **7**, 1995–1998; (f) Wender, P.A. and Horan, J.C. (2006) *Org. Lett.*, **8**, 4581–4584; (g) Wender, P.A. and Verma, V.A. (2008) *Org. Lett.*, **10**, 3331–3334; (h) Wender, P.A., Horan, J.C., and Verma, V.A. (2006) *Org. Lett.*, **8**, 5299–5302; (i) Wender, P.A., DeChristopher, B.A., and Schrier, A.J. (2008) *J. Am. Chem. Soc.*, **130**, 6658–6659; (j) Wender, P.A. and Baryza, J.L. (2005) *Org. Lett.*, **7**, 1177–1180; (k) Wender, P.A., Hegde, S.G., Hubbard, R.D., and Zhang, L. (2002) *J. Am. Chem. Soc.*, **124**, 4956–4957; (l) Wender, P.A., Hilinski, M.K., Skaanderup, P.R., Soldermann, N.G., and Mooberry, S.L. (2006) *Org. Lett.*, **8**, 4105–4108; (m) Wender, P.A., Hegde, S.G., Hubbard, R.D., Zhang, L., and Mooberry, S.L. (2003) *Org. Lett.*, **5**, 3507–3509.

13. (a) Smith, A.B. III, Razler, T.M., Meis, R.M., and Pettit, G.R. (2008) *J. Org. Chem.*, **73**, 1201–1208; (b) Smith, A.B. III, Razler, T.M., Meis, R.M., and Pettit, G.R. (2006) *Org. Lett.*, **8**, 797–799.

14. Czakó, B., Krti, L., Mammoto, A., Ingber, D.E., and Corey, E.J. (2009) *J. Am. Chem. Soc.*, **131**, 9014–9019.

15. (a) Corey, E.J., Gin, D.Y., and Kania, R.S. (1996) *J. Am. Chem. Soc.*, **118**, 9202–9203; (b) Martinez, E.J., Owa, T., Schreiber, S.L., and Corey, E.J. (1999) *Proc. Natl. Acad. Sci. U. S. A.*, **96**, 3496–3501; (c) Martinez, E.J. and Corey, E.J. (2000) *Org. Lett.*, **2**, 993.

16. (a) Nicolaou, K.C., Wu, T.R., Sarlah, D., Shaw, D.M., Rowcliffe, E., and Burton, D.R. (2008) *J. Am. Chem. Soc.*, **130**, 11114–11121; (b) Nicolaou, K.C., Lister, T., Denton, R.M., and Gelin, C.F. (2007) *Angew. Chem. Int. Ed.*, **46**, 7501–7505; (c) Nicolaou, K.C., Li, A., Edmonds, D.J., Tria, G.S., and Ellery, S.P. (2009) *J. Am. Chem. Soc.*, **131**, 16905–16918.

17. (a) Ishikawa, H., Elliott, G.I., Velcicky, J., Choi, Y., and Boger, D.L. (2006) *J. Am. Chem. Soc.*, **128**, 10596–10612; (b) Ishikawa, H., Colby, D.A., Seto, S., Va, P., Tam, A., Kakei, H., Rayl, T.J., Hwang, I., and Boger, D.L. (2009) *J. Am. Chem. Soc.*, **131**, 4904–4916; (c) Va, P., Campbell, E.L., Robertson, W.M., and Boger, D.L. (2010) *J. Am. Chem. Soc.*, **132**, 8489–8495; (d) Sasaki, Y., Kato, D., and Boger, D.L. (2010) *J. Am. Chem. Soc.*, **132**, 13533.

18. (a) Chen, K. and Baran, P.S. (2009) *Nature*, **459**, 824–828; (b) Richter, J.M., Ishihara, Y., Masuda, T., Whitefield, B.W., Llamas, T., Pohjakallio, A., and Baran, P.S. (2008) *J. Am. Chem. Soc.*, **130**, 17938–17954; (c) Su, S., Seiple, I.B., Young, I.S., and Baran, P.S. (2008) *J. Am. Chem. Soc.*, **130**, 16490–16491; (d) O'Malley, D.P., Yamaguchi, J., Young, I.S., Seiple, I.B., and Baran, P.S. (2008) *Angew. Chem. Int. Ed.*, **47**, 3581–3583; (e) Seiple, I.B., Su, S., Young, I.S., Lewis, C.A., Yamaguchi, J., and Baran, P.S. (2010) *Angew. Chem. Int. Ed.*, **49**, 1095–1098.

19. (a) Kimbrough, T.J., Roethle, P.A., Mayer, P., and Trauner, D. (2010) *Angew. Chem. Int. Ed.*, **49**, 2619–2621; (b) Roethle, P.A. and Trauner, D. (2006) *Org. Lett.*, **8**, 345–347; (c) Roethle, P.A., Hernandez, P.T., and Trauner, D. (2006) *Org. Lett.*, **8**, 5901–5904; (d) Malerich, J.P., Maimone, T.J., Elliott, G.I., and Trauner, D. (2005) *J. Am. Chem. Soc.*, **127**, 6276–6283; (e) Miller, A.K. and Trauner, D. (2005) *Angew. Chem. Int. Ed.*, **44**, 4602–4606; (f) Sofiyev, V., Navarro, G., and Trauner, D. (2008) *Org. Lett.*, **10**, 149–152.

20. Snyder, S.A., Breazzano, S.P., Ross, A.G., Lin, Y.Q., and Zografos, A.L. (2009) *J. Am. Chem. Soc.*, **131**, 1753–1765.

Commentary Part

Comment 1

Michael Foley

Ling-min Xu, Yu-Fan Liang, Qin-da Ye, and Zhen Yang present an excellent overview of strategies developed by modern synthetic organic chemists to create natural products and collections of compounds that are natural product-like. The strategies include DOS, DTS, FOS, and TOS. Natural products have been an important source of new drugs and continue to be an inspiration to chemists charged with creating collections of compounds that will be used for drug discovery. Despite their proven value in drug discovery, large collections of discrete purified natural products are not widely available for high-throughput screening (HTS) because the structural complexity of natural products makes it very difficult to replicate their structure precisely on a scale that is useful for drug discovery even for a single compound, much less for thousands of compounds. However, as the authors point out, it is structural complexity that makes natural products so useful in drug discovery. Until recently, it has been difficult to describe structural complexity in a straightforward way that can be coupled to the strategies described in this review. There is a great need for new chemical matter to modulate the function of the new biological targets emerging from modern biology and perhaps structural complexity is the key to success.

New insights into how genomic changes contribute to disease have rapidly expanded the list of important targets and pathways that are relevant to drug discovery. Although the strength of the evidence is increasing that transcription factors, regulatory RNAs, protein–protein, and protein–DNA interactions are involved in disease processes, such targets are often described as "undruggable" because when they are screened against current HTS compound collections few, if any, hits are found. These new targets represent a great opportunity for the treatment of disease, but can the chemistry community provide high-quality hits, leads, and drugs against them to improve treatment outcomes for patients in the future?

Unfortunately, the initial results do not look very promising. A particularly striking example of the failure of a screening collection to provide high-quality hits in genomics-derived, target-based, HTS was published by researchers at GSK in 2007 [C1]. GSK screened its entire compound collection in an effort to find hits against 70 separate antibacterial targets derived from their genomic information. Disappointingly, this effort delivered few leads and, remarkably, most screens failed to produce any hits at all. The authors cited the need for new chemical matter and new approaches to attack undruggable targets in order to overcome these lackluster results.

In order to modulate the function of new biological targets with small molecules, HTS compound collections must be updated regularly with new chemical matter. Screening compounds can come from a variety of sources, including medicinal chemistry programs, commercial vendor libraries commercial vendor library (CVL), and natural products. Although most synthetic compounds created for use in HTS obey Lipinski's "Rule-of-Five" [C2], many of the most important natural product drugs do not obey this rule [C3]. Since the "Rule-of-Five" was published, medicinal chemists have looked for correlations between the physical properties of their compounds and success in transitioning through the various steps of the drug discovery process. In parallel with the development of the "Rule-of-Five," new technologies that allowed the creation of large collections of compounds for HTS emerged. These technologies, coupled with a strict focus on the "Rule-of-Five," produced achiral aromatic compounds that were high in sp^2-hybridized carbon content and lacked the complexity of natural products. The majority of compounds used in HTS are purchased from CVLs, and produced with the new technologies and with strict adherence to the "Rule-of-Five," so it should not be surprising that they have failed to deliver high-quality hits against the important targets and pathways illuminated by modern biology because they are so dissimilar to natural products in terms of their complexity.

Complexity can be quantified by the fraction of sp^3-hybridized carbons (Fsp^3), where $Fsp^3 =$ number of sp^3-hybridized carbons/total carbon count. Interestingly, compounds with a greater sp^3 carbon content have higher success rates in progressing from screening to drug approval [C4]. We applied this measure of complexity to five different compound collections used in HTS and compared the average Fsp^3 for the Analyti-Con Discovery Natural Product collection, Broad

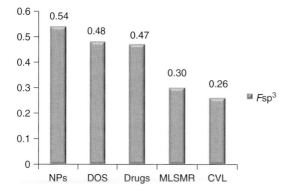

NPs = Natural products from AnalytiCon Discovery
DOS = DOS compounds in Broad collection
Drugs = Drugs from Drug Banks
MLSMR = NIH Molecular Libraries Small-Molecule Repository
CVL = Commercial Ven dor Libary compounds in Broad collection

Figure C1 Complexity analysis of different compound collections as measured by Fsp^3, where Fsp^3 = number of sp^3-hybridized carbons/total carbon count.

Institute DOS collection, Approved Drug collection from Drug Bank, NIH Molecular Libraries Small Molecule Repository (MLSMR) collection, and the Broad Institute CVL collection (Figure C1). The results revealed that natural products, DOS compounds, and drugs share a similar degree of complexity, with natural products being the most complex. The MLSMR collection and the CVL, which are most similar to typical HTS collections used in pharmaceutical companies and major academic screening centers, were significantly less complex and predicted by this analysis to be less likely to succeed in drug discovery. We speculate that if a similar analysis were performed on compounds emerging from the DTS, FOS, and TOS strategies described in this review, those compounds would look more like natural products, drugs, and DOS compounds in their Fsp^3.

The clear correlation between complexity and success in the drug discovery process demands that modern synthetic chemists strive to increase the efficiency with which they can prepare both natural products and natural product-like compounds. The strategies described in this review provide tools for chemists to begin to create collections of compounds with both the complexity and desirable physicochemical properties not only to modulate the function of the new targets emerging from modern biology but also to become drugs.

Comment 2

Scott A. Snyder

If organic chemists are to play a role in addressing the largest biomedical challenges facing society, our ability both to identify and to create functional small molecules rapidly with unique properties must improve dramatically. The question, of course, is how best to achieve this lofty objective, a problem with which the field has been grappling for many years, particularly over the course of the past two decades as new and more ominous health threats such as drug-resistant strains of bacteria and HIV have been identified. One avenue that has attracted much attention is the structural modification of natural products. Given that the architectures of these materials have been honed over millennia to achieve specific biochemical activity, they afford highly attractive starting points for investigations to improve upon their potency, selectivity, and pharmacology within humans.

The problem is the overall time and effort required for these investigations to reach fruition. Although simple adjustment of the natural product itself through derivatization of outlying functional handles sometimes succeeds, such as was the case for the creation of aspirin from salicylic acid in 1897 [C5], recent clinical leads by altering the side-chain of Taxol [C6], and converting an epoxide linkage within the epothilones to an aziridine [C7], these outcomes are unfortunately all too rare. Instead, fully fledged medicinal chemistry efforts involving the synthesis of hundreds, if not thousands, of analogs with major structural modifications over the course of many years are typically needed to identify suitable properties. Medicines such as Lipitor (atorvastatin) is a recent example of highly successful programs along these lines [C8], one that has saved thousands of lives. The goal for the future, however, has to be far more of these success stories achieved in a much shorter period of time.

This chapter by Zhen Yang and co-workers explores efforts to achieve that goal through alternate paradigms for natural product structure alteration. Through seven examples encompassing concepts from DOS, DTS, and FOS, and efforts to divert or co-opt biogenetic pathways to achieve much structural diversity from a single starting material, the authors provide a compelling array of potential approaches that could convert natural products more quickly into clinical agents. Rather than offer any recap of the stories they provide, in which I am honored to have some of my group's work included, I just wish to note a couple other examples along these lines that have proven inspirational to me in my thinking on natural product synthesis within a drug discovery paradigm. The first is the Corey group's identification of phthalascidin [C9], a derivative of the large and highly complex natural product ecteinascidin 743 [C10], which is now in advanced clinical trials. The second is the recent rapid assembly of vinblastine and related molecules by Boger's group through a cascade of pericyclic reaction processes [C11], efforts which not only afford an expedient synthesis of the drug itself, but also provide a number of analogs that may well lead to unique activity and new thoughts on the biogenesis of this important material. In my opinion, the main message of these, and also all the endeavors mentioned in this chapter, is that the process of identifying clinical agents can likely be accelerated, given that each of these programs was achieved in a relatively short period of time by only a few researchers. It is hoped that the fruits of these, and other, approaches will be harvested in the near future. I am, however, left with some questions for the future, and I will close my commentary with these queries:

1) Despite the promise shown for these approaches, they are all dependent at the most basic level on the existence of a unique natural product architecture to serve as a forum for exploration. Given the reluctance of major pharmaceutical companies to continue investment in natural product isolation efforts, together with a decrease in the ease of federal funding for such programs and the increasing challenge of structural elucidation for materials of increasing complexity found in ever more minute scales, how can the future of natural products in the drug discovery process be secured?

2) To what extent can China potentially lead in this endeavor, not only because of enhanced governmental and private sector funding, but also because of the rich history of traditional medicines which are only now beginning to be thoroughly explored?

3) How should synthetic chemists, natural product isolation scientists, and those at the frontiers of biosynthetic engineering best work together to achieve shared goals in identifying clinical candidates?

Comment 3

Da-Wei Ma

For a long time, synthetic chemists in academia have focused their attention on how to construct the complex structures of natural products and how to complete the total synthesis of these molecules. The achievements within this challenging area during the last century have been astonishing. These efforts are well recognized as the driving force for organic chemistry development because during the total synthesis numerous new synthetic reactions and methods have been discovered. However, in addition to its inherent beauty, total synthesis also opens the gates widely to a better understanding of biological processes and the development of pharmaceutically interesting substances. Indeed, natural products have been proven to be a rich source for the development of therapeutic agents and pharmaceutical tools, as is evident from the fact that over 50% of clinically used drugs came from natural products themselves, natural product derivatives, and natural product mimics. In recent years, more and more synthetic chemists have realized that they could combine their synthetic efforts with biological studies to make total synthesis more valuable.

This trend has stimulated the birth of concepts such as diversity-oriented organic synthesis, DTS, FOS, and, TOS, although the differences between some of them are not very obvious.

In this chapter, Zhen Yang and co-workers give a comprehensive summary of recent achievements in this emerging field. Through these results one can find three remarkable characters. The first is that via synthetic efforts, accessible derivatives can be used to modify the biological activity of the natural products and to shed light on their role in biological systems, as exampled by Danishefsky's DTS of the tumor cell migration inhibitor migrastatin that delivered several migrastatin mimics with increased activities, and Wender's FOS of the PKC activator bryostatin that provided potent analogs with simplified structures. The second is that cycloaddition methods are powerful tools for quickly generating natural product mimics, as indicated by Nicolaou's TOS of the HIV-1 replication inhibitor biyouyanagin A via [4 + 2]- and [2 + 2] cycloadditions, and Boger's total synthesis of the antitumor agent vindoline and its analogs via a [4 + 2]/[3 + 2]-cycloaddition cascade. The third is that biomimetic transformations are frequently considered in the DOS of natural products, as demonstrated by Shair's synthesis of galanthamine-like molecules, Baran's total synthesis of eudesmane terpenes by site-selective C-H oxidation, Trauner's biomimetic syntheses of furanocembranoids from the natural product bipinnatin J, and Snyder's syntheses of the resveratrol family from a common building block.

Authors' Response to the Commentaries

We greatly appreciate the reviewers' comments and are very encouraged by their support of the work presented. We have revised the text based on the reviewers' comments. Following Prof. Snyder's comments, we have added Corey's group's identification of phthalascidin in the text. Since the synthesis of vinblastine by Boger was already included, we have provided additional references recommended by Prof. Snyder. We believe that these revisions, as suggested by the reviewers, improve the quality of our chapter and thanks are due for their efforts.

References

C1. Payne, D.J., Gwynn, M.N., Holmes, D.J., and Pomplano, D.L. (2007) *Nat. Rev. Drug Discov.*, **6**, 29–40.

C2. Lipinski, C.A., Lombardo, F., Dominy, B.W., and Feeney, P. (1997) *J. Adv. Drug Deliv. Rev.*, **23**, 3–25.

C3. Lipinski, A.H. (2004) *Nature*, **432**, 855–861.

C4. Lovering, F., Bikker, J., and Humblet, C. (2009) *J. Med. Chem.*, **52**, 6752–6756.

C5. For a review on this molecule, see: Nicolaou, K.C. and Montagnon, T. (2008) *Molecules that Changed the World*, Wiley-VCH Verlag GmbH, Weinheim, pp. 21–28.

C6. For a recent review on approaches to the taxanes, see: Ishihara, Y. and Baran, P.S. (2010) *Synlett*, 1733.

C7. Regueiro-Ren, A., Borzilleri, R.M., Zheng, X., Kim, S.H., Johnson, J.A., Fairchild, C.R., Lee, F.Y.F., Long, B.H., and Vite, G.D. (2001) *Org. Lett.*, **3**, 2693.

C8. Roth, B.D. (2002) *Prog. Med. Chem.*, **40**, 1.

C9. Martinez, E.J., Owa, T., Schreiber, S.L., and Corey, E.J. (1999) *Proc. Natl. Acad. Sci. U. S. A.*, **96**, 3496.

C10. (a) Corey, E.J., Gin, D.Y., and Kania, R.S. (1996) *J. Am. Chem. Soc.*, **118**, 9202; (b) Martinez, E.J. and Corey, E.J. (2000) *Org. Lett.*, **2**, 993.

C11. (a) Sasaki, Y., Kato, D., and Boger, D.L. (2010) *J. Am. Chem. Soc.*, **132**, 13533; (b) Va, P., Campbell, E.L., Robinson, W.M., and Boger, D.L. (2010) *J. Am. Chem. Soc.*, **132**, 8489; (c) Ishikawa, H., Colby, D.A., Seto, S., Va, P., Tam, A., Kakei, H., Rayl, T.J., Hwang, I., and Boger, D.L. (2009) *J. Am. Chem. Soc.*, **131**, 4904.

2
Total Synthesis of Natural Products and the Synergy with Synthetic Methodology

Qian Wang and Jie-Ping Zhu

2.1
Introduction

"Because it's there," the reply of British alpinist George Mallory to the question, "why do you want to climb Mount Everest?," became famous in mountaineering. In the synthetic community, the phrase of Woodward, "because nobody else could do it," has also been a good reason to attack a target natural product. However, there are fundamental differences between climbing a mountain and a total synthesis exercise. Mount Everest will remain the highest mountain summit forever, whereas molecular summits keep on coming, higher and trickier year after year. Indeed, the size and complexity of natural products synthesized today bear no resemblance to those targeted in earlier days. This can be easily seen by a quick glance at the structures of vitamin B_{12}, palytoxin, vancomycin, brevetoxin B, palau'amine, and so on. The most challenging pathway to climb Mount Everest remains, and the new generation of alpinist can follow the route traced by their predecessors. After all, to get there is already a heroic human achievement. On the other hand, synthetic routes to reach a given molecular summit evolve constantly and most of the best syntheses do not last long as a consequence of the development of new reactions and new synthetic strategies. The different synthesis routes for reaching strychnine [1] developed over the last half a century testify to how the evolution of synthetic strategies can shorten the length of a synthesis, a nice showcase of synergy between synthesis and synthetic methodologies.

In its simplest definition, the goal of total synthesis is to get there. Therefore, the race for a molecular summit was prevalent even in the very earlier stages of development and become phenomenal in the last decade of the twentieth century [2]. Natural products such as paclitaxel (Taxol), calichimycin, vancomycin, and bryostatin, due to the beauty and the complexity of their molecular architecture, and their medical importance, have attracted intensive synthetic efforts. Like any race, to be the first to plant a flag on the molecular mountain was of critical importance. The assault has led to the fall of these summits one by one at an impressive pace in spite of their molecular complexity. While the philosophy of being the first was not necessary to the fundamental discoveries along the way, the results obtained in this period were fantastic not only in terms of the purely intellectual challenge and the sheer excitement of the endeavors, but also

Organic Chemistry – Breakthroughs and Perspectives, First Edition. Edited by Kuiling Ding and Li-Xin Dai.
© 2012 Wiley-VCH Verlag GmbH & Co. KGaA. Published 2012 by Wiley-VCH Verlag GmbH & Co. KGaA.

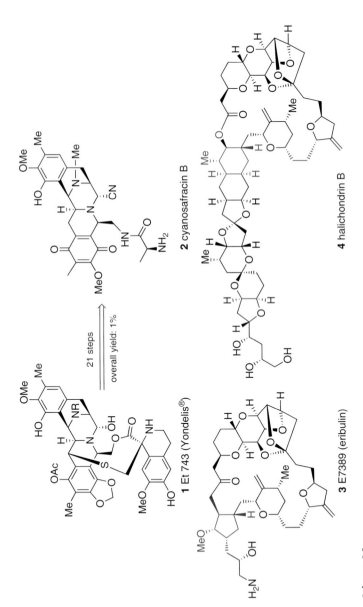

Scheme 2.1

for their impact on the supply of natural products that are either rare or difficult to isolate in abundance. Through these extraordinary achievements in conjunction with the advances in synthetic tools, methods, and theories, the question "can we make it?" has gradually given way to "why should we make it and, if so, can we make it with elegance and utility that impact science and supply?" As a consequence, synthesis emerges as the central resource in bridging the gap between natural products and drug discovery. This point is readily illustrated by following two recently approved anticancer drugs. Ecteinascidin 743 (Et 743, **1**), isolated from the Caribbean tunicate *Ecteinascidia turbinate* in 1990 by the groups of Rinehart [3] and Wright [4], displays potent antitumor and antimicrobial activities. However, its restricted natural availability (1 g from 1 ton of tunicate) prevented further evaluation of its potential clinical application. Based on Corey's elegant total synthesis route [5], PharmaMar developed a 21-step synthesis of this natural product from cyanosafracin B (**2**), a more easily accessible natural product via fermentation technology (Scheme 2.1), and therefore solved the supply issue [6]. Et 743 or trabectedin has been approved for use in Europe, Russia, and South Korea since 2007 for the treatment of advanced soft tissue sarcoma under the brand name Yondelis. The recent US Food and Drug Administration (FDA)-approved anticancer drug eribulin (**3**) is yet another illustrative example of the power of total synthesis in drug discovery endeavors. Thanks to Kishi and co-workers' highly modular and efficient total synthesis of halichondrin B (**4**) [7] at Eisai Pharmaceuticals, they were able to synthesize a collection of simplified natural product analogs. One of them, originally called E7389, now eribulin (trade name Halaven) [8], displayed potent antitumor activities at least equal to that of the natural substance and received marketing authorization in November 2010. Yondelis and eribulin are probably two most complicated molecules ever made on a commercial scale. It is safe to say that without laboratory synthesis, these two drug candidates would not be of value to humanity.

If one accepted the notion that virtually all structures can be synthesized in the laboratory given enough time and human resources, the imperative of being the first to cross the line would be downregulated. Indeed, the major determinants in contemporary total synthesis that have evolved are becoming more subtle and sophisticated. One might pay particular attention to the selection of the target, the planning of the synthetic strategy, and the science that was discovered along the way. Fortunately, there is a creative synergism between target pursuits and methodology/strategy development. Thus, total synthesis endeavors helped to gain insights into chemical reactivity and selectivity principles and served as the testing ground for the development of novel synthetic methodologies and strategies. Conversely, new strategies (methodologies) developed during the pursuit of one specific target can find application in completely unrelated problems, allowing the synthesis of other previously inaccessible natural products.

The criteria for appreciating the beauty of a given synthesis are also evolving. In addition to being chemo-, regio-, and stereoselective in each individual step, a synthesis has to be atom [9], step [10], and redox economic [11] to meet current standards. In addition, the concept of protecting group-free synthesis has also been advanced in searching for an ideal synthesis, an ultimate goal in organic synthesis [12].

In this chapter, we summarize some recent total synthesis examples to demonstrate the synergism between target pursuits and the development of new synthetic strategies.

2.2
Domino Process

2.2.1
Introduction

A traditional approach to the synthesis of organic compounds has been a stepwise construction of individual chemical bonds found in the target molecules, with a work-up procedure after each transformation. The increase in molecular complexity with such a strategy is thus incremental, leading to a lengthy synthesis. A domino process, as defined by Tietze *et al.* [13], is a combination of two or more bond-forming reactions under identical conditions wherein the subsequent reactions result as a consequence of the functionality formed in the previous step. The quality of a domino reaction can be correlated with the number of chemical bonds formed, and also to the increase in complexity and its general applicability. The greater the number of bonds formed, which usually goes hand-in-hand with an increase in complexity of the product, the more useful the process might be. The ingenious implementation of a domino process in a synthesis design could therefore not only shorten the synthetic sequence, but also enhance the esthetic appeal of the synthesis endeavor.

The benefit of applying a domino process to organic synthesis has been recognized for many years. Johnson *et al.*'s synthesis of (±)-progesterone (**5**) [14] and Nicolaou and co-workers' synthesis of (±)-endiandric acid C [15] (**6**) (Figure 2.1) are classical examples that were inspired by a biosynthetic hypothesis. The application of truly designed or serendipitously discovered domino processes to target pursuits has appeared more and more frequently in recent years [16]. Below we present a few recent examples to illustrate the power of these processes in the construction of complex natural products.

2.2.2
Total Synthesis of Hirsutellone B

Hirsutellone B (**7**), isolated from the insect pathogenic fungus *Hirsutella nivea* BCC 2594, belongs to a growing family of fungal secondary metabolites. It displays potent activities against *Mycobacterium tuberculosis* (MIC $= 0.78\,\mu g\,ml^{-1}$), the causative pathogen of tuberculosis. Structurally, it contains a 6,5,6-*trans* fused tricyclic core, a γ-lactam embedded in a 13-membered *p*-cyclophane motif with an *endo* aryl–alkyl ether bond. A retro-synthetic presentation of Nicolaou *et al.*'s synthesis [17] is shown in Scheme 2.2. The key strategic

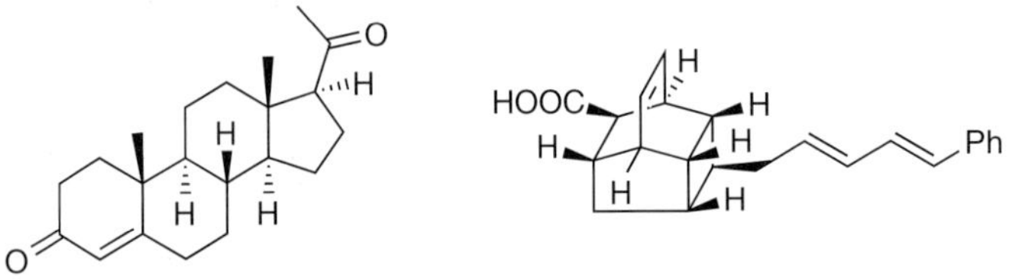

5 (±)-progesterone   **6** (±)-endiandric acid C

Figure 2.1   Structures of compounds **5** and **6**.

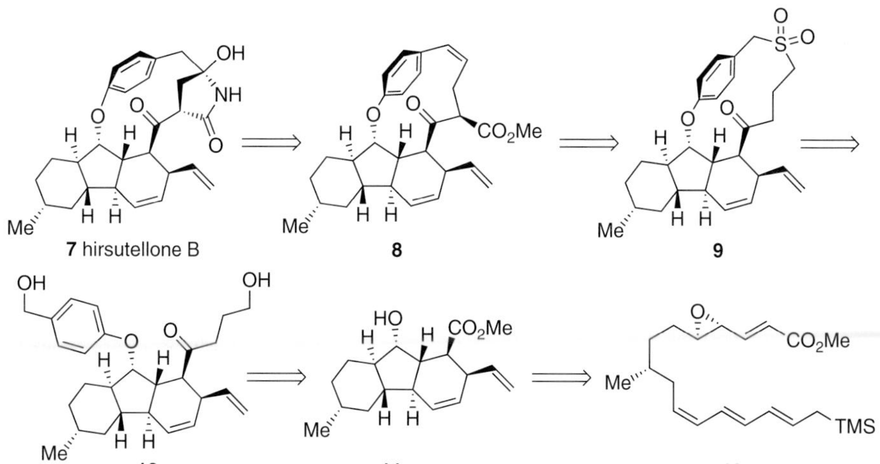

Scheme 2.2

steps included: (i) formation of a strained 13-membered styrene-containing *p*-cyclophane (**8**) by ring contraction from the less strained 14-membered sulfone through the Ramberg–Bäcklund reaction and (ii) formation of tricyclic core **11** from linear polyene epoxide **12** via a sequence of intramolecular epoxide opening–Diels–Alder cycloaddition.

The forward synthesis is detailed in Scheme 2.3. The (*R*)-citronellal (**13**) was converted to *cis*-vinyl iodide (**14**), which was then epoxidized to **15** following Jørgensen's organocatalytic conditions. Stille coupling between **15** and vinyltin (**16**) in the presence of CuTC provided cyclization precursor **17**. Treatment of **17** with Et$_2$AlCl afforded the expected tricyclic core **11** in 50% yield as a single diastereomer. The reaction went through an intramolecular epoxide ring opening via a preferred transition-state conformation **18** to afford, in a highly stereoselective manner, the intermediate **19**, which then underwent Diels–Alder reaction via an ester-*endo* transition state to provide **11**.

A series of functional group manipulations converted **11** to **10**, which was then exposed to AcSH and ZnI$_2$ to afford **24**. In this latter transformation, the benzylic alcohol and the primary aliphatic alcohol were converted to the thioester and alkyl iodide, respectively. Again, a domino process involving intermediates **20–23** was proposed (Scheme 2.4). Hydrolysis of thioacetate followed by macrocyclization via an intramolecular S$_N$2 reaction proceeded in a one-pot manner to provide, after oxidation of sulfide, the sulfone **9** in 79% overall yield. Ramberg–Bäcklund reaction under Chan's conditions followed by acylation furnished ring-contracted product **8** [(*Z*)-alkene only].

Conversion of **8** to a natural product is depicted in Scheme 2.5. Sharpless dihydroxylation of **8** afforded diol **25**. Selective deoxygenation of the benzylic alcohol via thionocarbonate under Barton's conditions followed by oxidation of the remaining secondary alcohol afforded the β-keto ester **26**. Heating of **26** with NH$_3$ in MeOH–H$_2$O (1:1 v/v) at 120 °C for 1 h led to hirsutellone B (**7**) in 50% yield through a domino sequence of amidation, epimerization (desired!), and cyclization (Scheme 2.5).

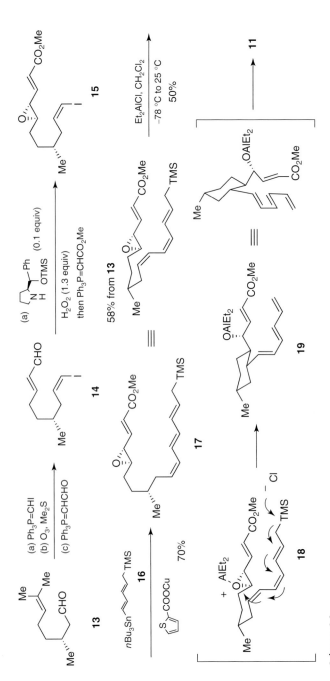

Scheme 2.3

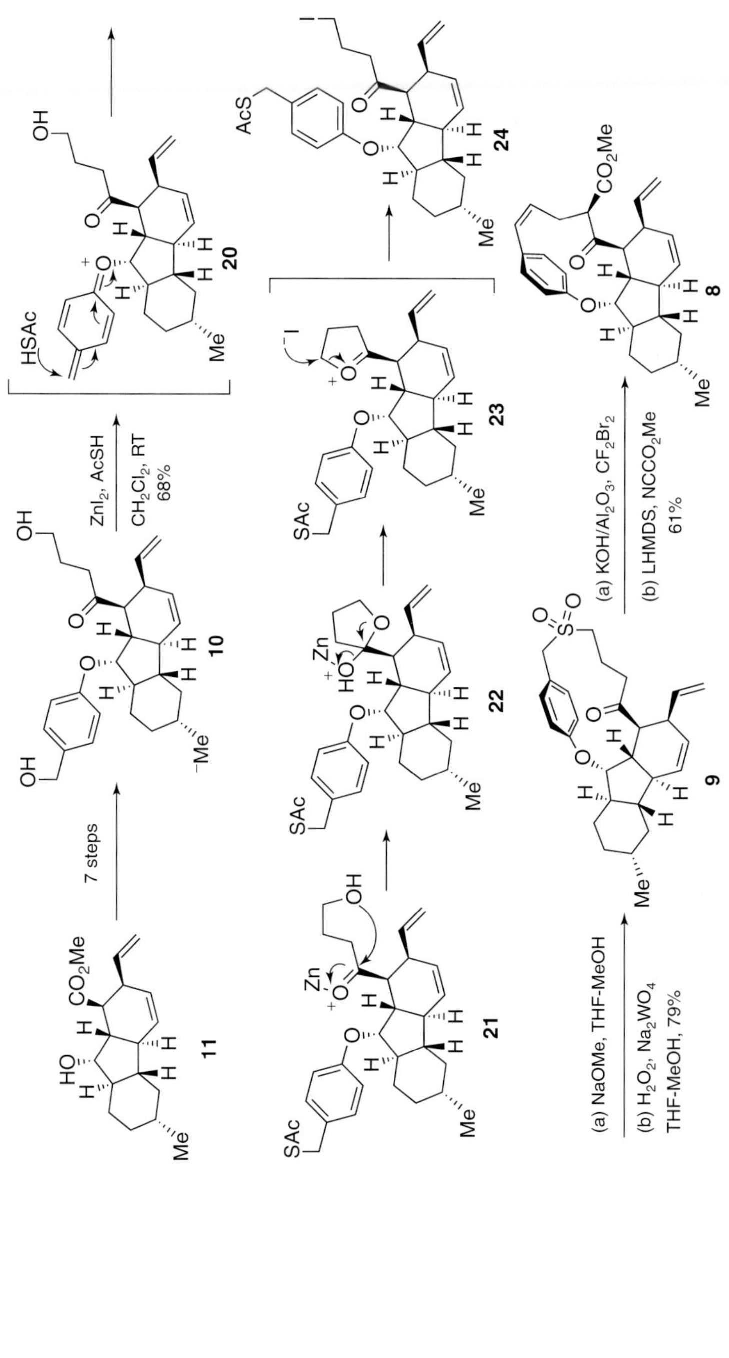

Scheme 2.4

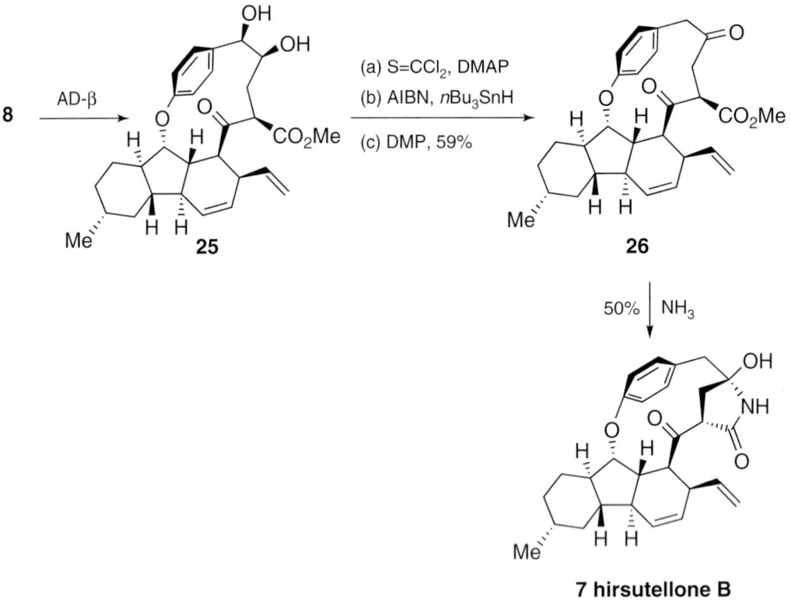

Scheme 2.5

The remarkable domino intramolecular epoxide opening–internal Diels–Alder cyclization (**17**–**11**) that established three carbocyclic rings of **7** with the proper relative and absolute configuration laid the foundation for this elegant total synthesis.

2.2.3
Total Synthesis of (±)-Minfiensine, (−)-Phalarine, and Aspidophytine

Minfiensine (**27**), isolated from *Strychnos minfiensis* [18], is a member of the akuammiline family of alkaloids. It possesses a highly congested pentacyclic ring system. A highly efficient synthesis involving a one-step triple domino process for the construction of the tetracyclic ring system was developed by Qin and co-workers as shown in Scheme 2.6 [19].

Treatment of diazo ketone **30a** under optimized conditions [CuOTf, CH$_2$Cl$_2$, RT] afforded the tetracycle **33a**, isolated exclusively in an enol form, in 50% yield (Scheme 2.7). The reaction was assumed to proceed via an initial cyclopropanation to give intermediate **31**. Ring opening of cyclopropane assisted by the lone-pair of the indole nitrogen would provide the indolenium salt **32**, which was in turn trapped by the tethered sulfonamide via a 5-*exo*-trig cyclization to afford the observed tetracycle **33a**. The yield of the process depended on the nature of the N-protecting group. Using **30b** as starting material, **33b** was isolated in 81% yield. Elaboration of **33a** to the natural product was straightforward. Decarboxylation using Krapucho's conditions, followed by one-step N-tosyl deprotection–ketone reduction, afforded **34**. Selective N-allylation and oxidation of alcohol to ketone provided **28**. Palladium-catalyzed intramolecular α-vinylation of

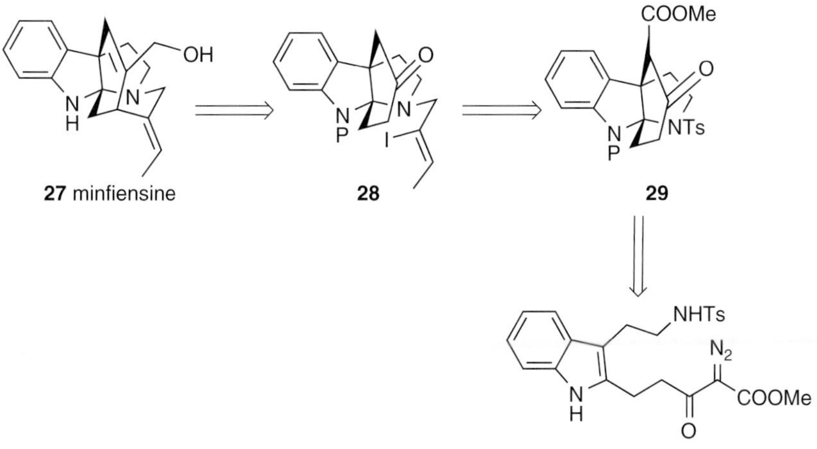

Scheme 2.6

ketone furnished pentacycle **29**, which was converted to allylic alcohol through a vinyl triflate intermediate. Finally, removal of the *N*-Boc function provided (±)-minfiensine (**27**).

In Qin and co-workers' synthesis, the electrophilic C1 carbon was trapped by a nucleophile attached at C2 (Scheme 2.8a). Alternatively, it is also possible to generate a C2-cationic species and to capture it by a nucleophile tethered at C1 (Scheme 2.8b). This latter case is illustrated by Danishefsky and co-workers' recent enantioselective synthesis of phalarine (**36**) [20].

The 2,3-disubstituted indole **37** was prepared using Suzuki–Miyaura coupling between an L-2-iodotryptophan derivative and 4-methoxy-2-methoxymethoxyphenyl boronate as a key step (Scheme 2.9). Reaction of **37** with formalin in the presence of camphorsulfonic acid (CSA) afforded the tetracyclic core of phalarine in 91% yield as a single diastereomer. This domino sequence went through an iminium intermediate **39**, which may undergo the Mannich reaction to provide the indolenium **40**. The Wagner–Meerwein rearrangement would then provide C3 cationic intermediate **41**, which would be attacked by the internal phenol group to afford **38**. Alternatively, the Pictet–Spengler reaction from **39** would furnish directly the intermediate **41**, which could then cyclize to **38**. Owing to the mechanistic ambiguity regarding the nature of the cyclization (C$_2$ versus C$_3$ cyclization) and also the proved reversibility of the first cyclization step (**39** and **40**), the origin of the high diastereoselectivity remains to be clarified.

The completion of synthesis is shown in Scheme 2.10. Reaction of **38** with azodicarboxylate followed by reduction of the resulting hydrazine afforded aniline **42**. Gassman oxindole synthesis converted **42** to **45** via intermediates **43** and **44** [21]. Finally, conversion of oxindole to indole under reductive conditions followed by its reaction with *N,N*-dimethylmethyleneammonium chloride and cleavage of the sulfonamide afforded phalarine (**36**) [22]. This synthesis also established the absolute configuration of natural product as indicated in Scheme 2.10.

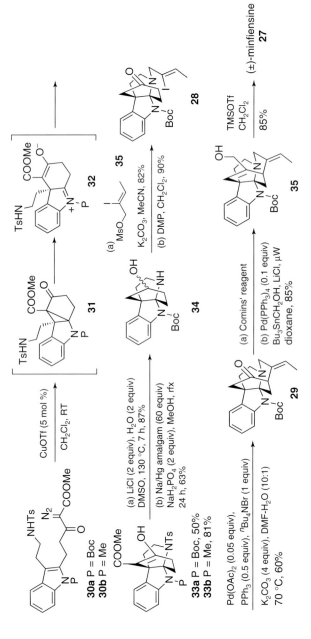

Scheme 2.7

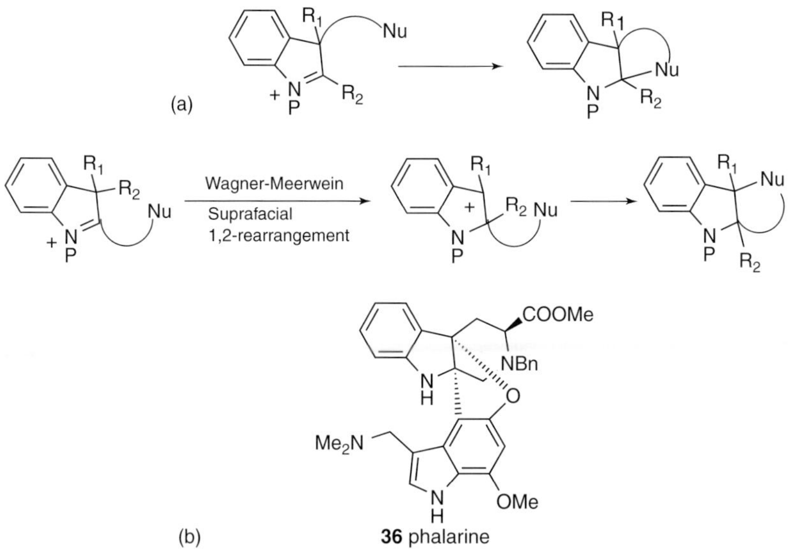
(a)
R1
Nu
+N
P
R2
Nu
N
P
R1
R2
R1
R2
Nu
+N
P
Wagner-Meerwein
Suprafacial
1,2-rearrangement
R1
+
R2 Nu
N
P
R1
Nu
N
P
R2
COOMe
NBn
N
H
O
Me2N
N
H
OMe
(b)
36 phalarine

Scheme 2.8

Corey and co-workers exploited the reactivity of indole C_2 and C_3 carbons in their memorable synthesis of aspidophytine (**47**) [23], an acid-degradation product of haplophytine [24] (**48**) (Figure 2.2). Reaction of indole (**49**) and enantiomerically enriched dialdehyde (**50**, 97% *ee*) under reductive conditions afforded the pentacycle **55** in 66% yield (Scheme 2.11). This domino process started with the bimolecular condensation between **49** and **50** to afford the dihydropyridinium salt **51**, which then underwent Friedel–Crafts reaction to give iminium intermediate **52**. Intramolecular nucleophilic addition of tethered allylsilane to iminium then provided pentacycle **53**. Enamine–imine tautomerization under acidic conditions afforded the iminium **54**, that was subsequently reduced by sodium cyanoborohydride to **55**. Hydrolysis of isopropyl ester to acid followed by regio- and chemoselective oxidation of tertiary amine to iminium and trapping of the latter by the tethered carboxylic acid function afforded the lactone **56**. Functional group manipulation then provided aspidophytine (**47**).

The one-step construction of pentacycle **55** from **49** and **50** is truly impressive and illustrated fully the power of domino processes in building complex molecular structures.

2.3
Multicomponent Reactions

2.3.1
Introduction

Multicomponent reactions (MCRs) are processes in which three or more reactants are combined in a single chemical step to produce a compound that incorporates substantial portions of all the components [25]. They are by definition sustainable chemistry and

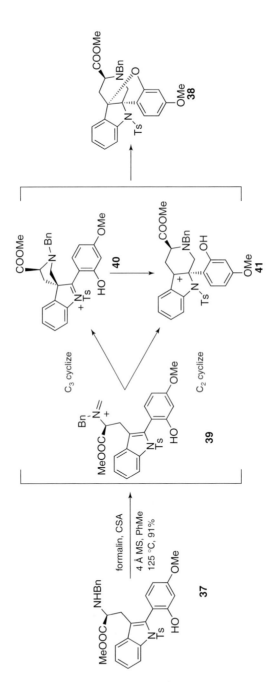

Scheme 2.9

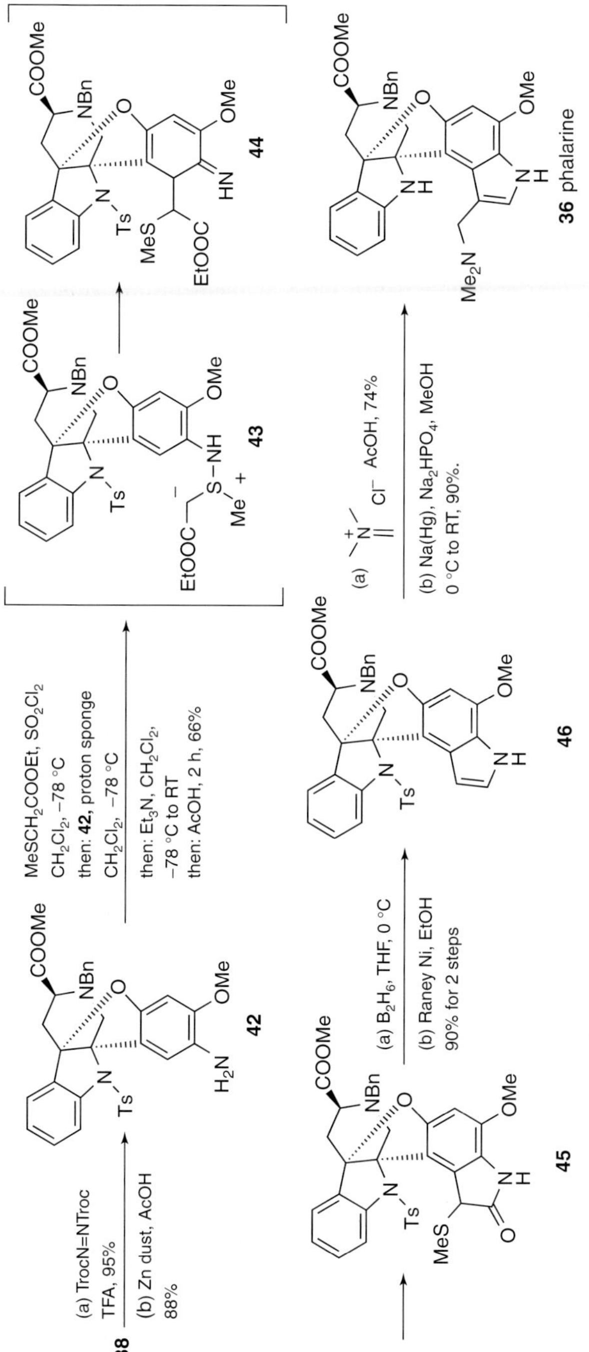
38
(a) TrocN=NTroc
TFA, 95%
(b) Zn dust, AcOH
88%
42
MeSCH2COOEt, SO2Cl2
CH2Cl2, −78 °C
then: 42, proton sponge
CH2Cl2, −78 °C
then: Et3N, CH2Cl2,
−78 °C to RT
then: AcOH, 2 h, 66%
43
44
45
(a) B2H6, THF, 0 °C
(b) Raney Ni, EtOH
90% for 2 steps
46
(a) Cl− AcOH, 74%
(b) Na(Hg), Na2HPO4, MeOH
0 °C to RT, 90%.
36 phalarine

Scheme 2.10

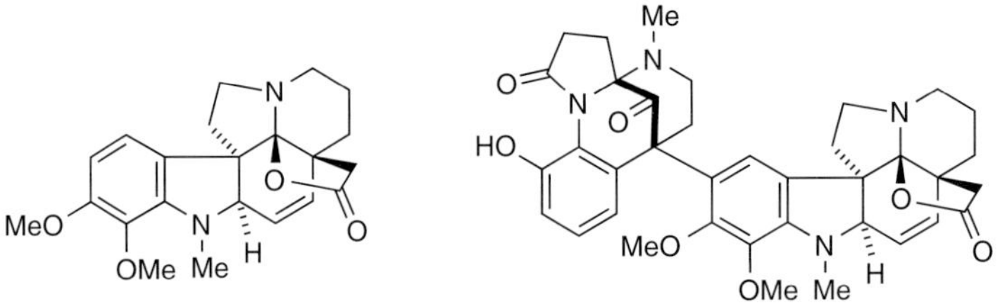

Figure 2.2 Structures of compounds **47** and **48**.

are inherently (i) atom economic, since most of them involve addition rather than substitution reactions; and (ii) step efficient, since they create at least two chemical bonds in one operation, consequently decreasing the consumption of energy, time, and resources. They reduce waste production significantly by minimizing the number of costly end-of-pipe treatments. In short, MCRs constitute a gateway to the ideal organic synthesis in which the target molecule is made from readily available starting materials in one simple, safe, environmentally acceptable, and resource-effective operation. As an enabling technology, the development and application of MCRs have attracted much attention in recent years in both diversity-oriented and target-oriented synthesis [26].

Robinson's synthesis of tropinone (**58**) is probably one of the first applications of MCRs in natural product synthesis and it remains as a milestone in the art of organic synthesis (Scheme 2.12) [27]. Reaction of succinaldehyde (**59**), methylamine, and acetonedicarboxylic acid (**60**) in water afforded in one operation tropinone (**58**) in 92% yield. The reaction was initiated by condensation of **59** with methylamine to afford iminium **61**, which was then attacked by enolate derived from **60** (Mannich reaction) to furnish intermediate **62**. A subsequent intramolecular Mannich reaction afforded the bicyclic compound **63**, which upon double decarboxylation afforded the target product. Note that the synthesis is protecting group free and exceptional in being finished in only one step.

Robinson's synthesis of tropinone was guided by his insight into alkaloid biosynthesis. In designing a multicomponent synthesis of complex natural products, the capability of chemists to trace back a complex structure to a seemingly irrelevant MCR is determinant. This is best illustrated by Fukuyama and co-workers' total synthesis of Et 743 (**1**, Scheme 2.13) [28]. Thus, Fukuyama and co-workers were able to disconnect the structure of Et 743 to the dipeptide **64**, which could in turn be constructed in one step by the Ugi four-component reaction of amine **65**, acid **66**, isocyanide **67**, and acetaldehyde (**68**). Although this is not the short synthesis known for Et 743 (**1**) [29], it is certainly one of the most ingenious applications of the Ugi four-component reaction in natural product synthesis.

2.3.2
Total Synthesis of (−)-Spirotryprostatin B

Spirotryprostatin B (**69**), isolated from the fermentation broth of *Aspergillus fumigatus*, has been shown to inhibit completely the G2/M progression of mammalian tsFT210 cells

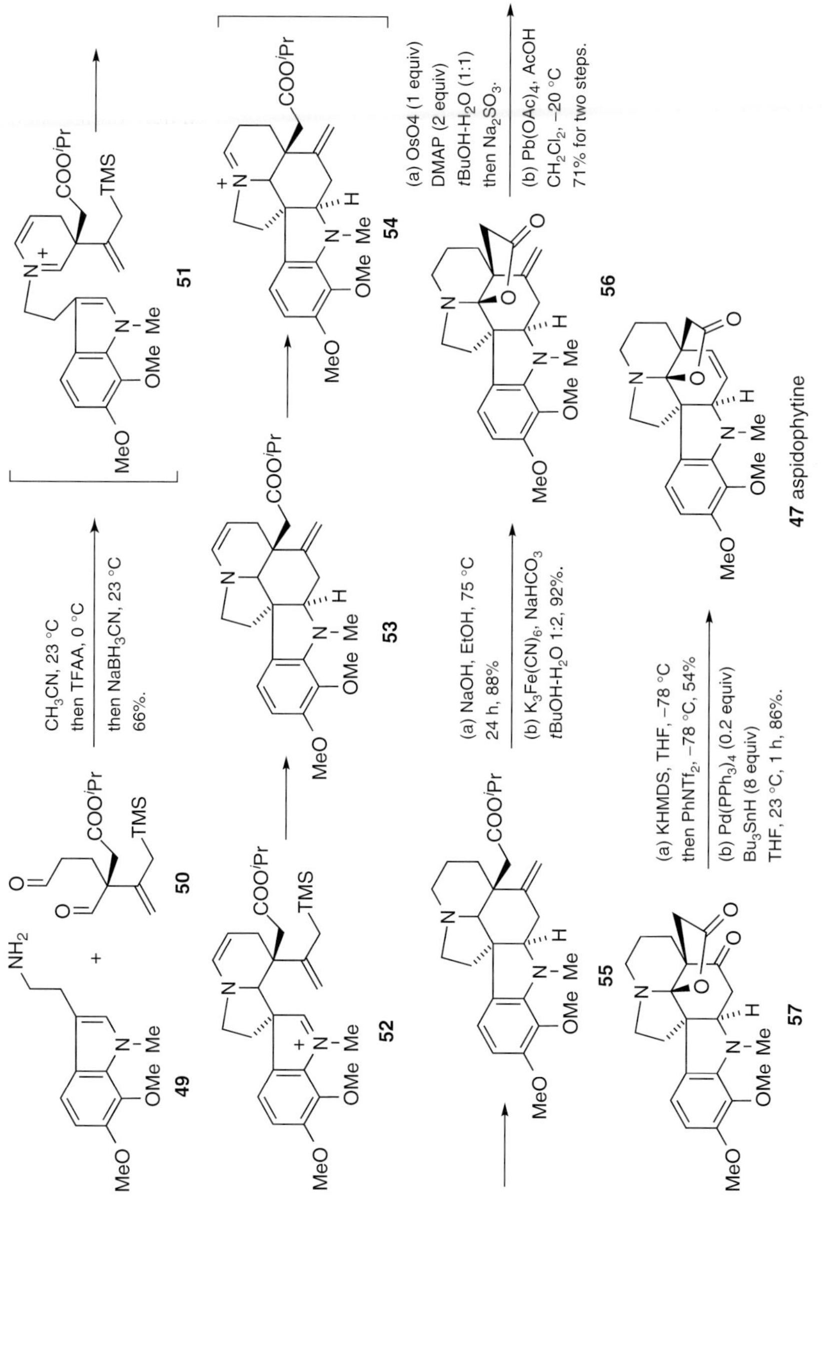

49
50
51
52
53
54
55
56
57
47 aspidophytine
MeO
OMe
Me
NH2
COO^{i}Pr
TMS
CH$_3$CN, 23 °C
then TFAA, 0 °C
then NaBH$_3$CN, 23 °C
66%.
(a) OsO4 (1 equiv)
DMAP (2 equiv)
tBuOH-H$_2$O (1:1)
then Na$_2$SO$_3$.
(b) Pb(OAc)$_4$, AcOH
CH$_2$Cl$_2$, −20 °C
71% for two steps.
(a) NaOH, EtOH, 75 °C
24 h, 88%
(b) K$_3$Fe(CN)$_6$, NaHCO$_3$
tBuOH-H$_2$O 1:2, 92%.
(a) KHMDS, THF, −78 °C
then PhNTf$_2$, −78 °C, 54%
(b) Pd(PPh$_3$)$_4$ (0.2 equiv)
Bu$_3$SnH (8 equiv)
THF, 23 °C, 1 h, 86%.

Scheme 2.11

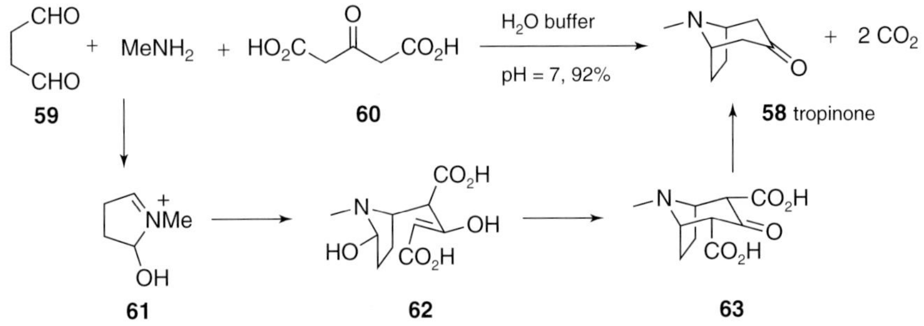

Scheme 2.12

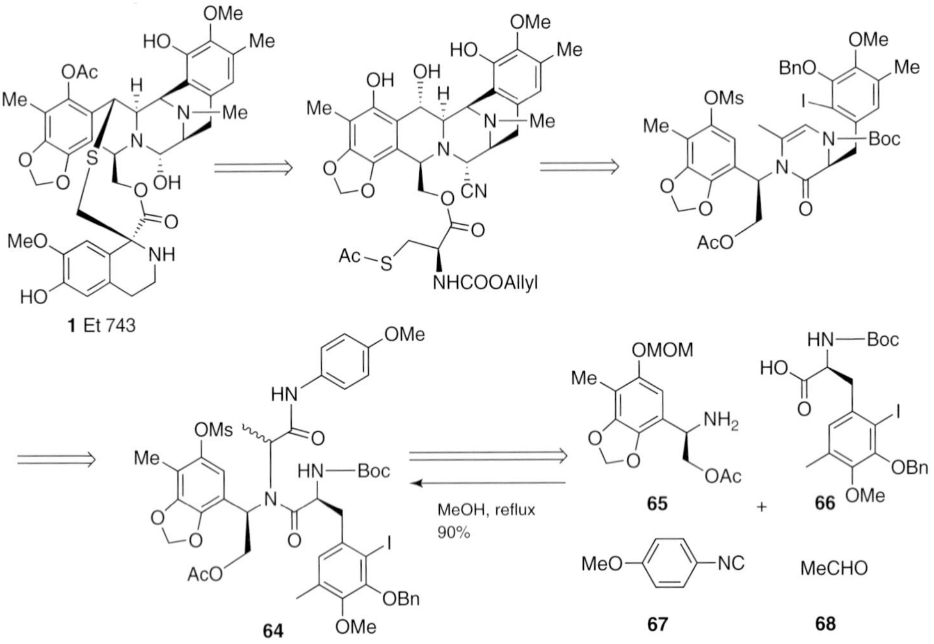

Scheme 2.13

with an IC_{50} of 14 µg ml^{-1}. Sebahar and Williams developed a concise synthesis featuring a key three-component azomethine ylide formation–[3 + 2]-cycloaddition (Scheme 2.14) [30]. From a retro-synthetic viewpoint, disconnecting the diketopiperazine unit and introduction at C8 of an ester group as a handle for the late introduction of the double bond is the key. From **70**, installing the 5,6-diphenylmorpholinone unit as a chiral inducer gave intermediate **71**, which can be disassembled to three simple building blocks **72**–**74**.

   In practice, the reaction of aldehyde **72** with oxazinone **73** and oxindole **74** in toluene at room temperature in the presence of 3 Å molecular sieves afforded cycloadduct **71** (82% yield), the relative and absolute stereochemistry of which were established

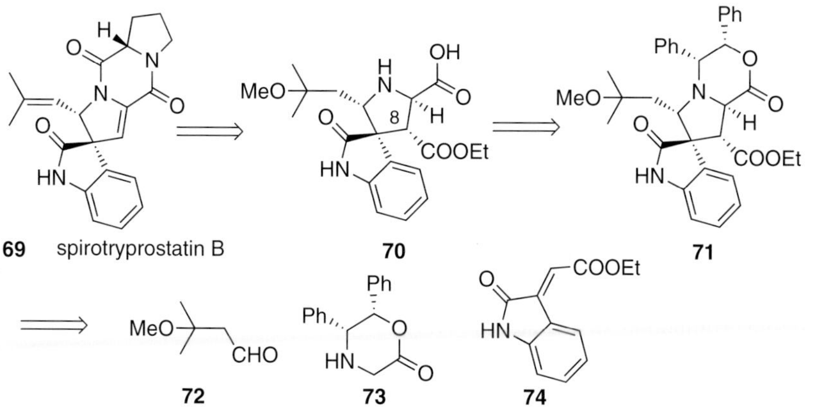

Scheme 2.14

through single-crystal X-ray analysis. This three-component reaction constructed the entire prenylated tryptophyl moiety of spirotryprostatin B with complete control of four contiguous stereogenic centers (Scheme 2.15). Reductive cleavage of bibenzyl from oxazinone **71** proceeded in essentially quantitative yield, affording the amino acid **70**. Coupling of **70** with D-proline benzyl ester (BOP reagent, MeCN, triethylamine, 74%) furnished the requisite dipeptide. It is interesting that the steric bulkiness around the amino group of **70** obviated the need for a protecting group during the peptide coupling and the free amino acid **70** was directly and effectively used in the reaction. Deprotection of the benzyl ester under standard conditions followed by BOP-mediated cyclization generated the diketopiperazine **75** in 94% yield in two steps. Note that the natural product contains an L-Pro unit. The reason for selecting a D-Pro here as a coupling partner was that using L-Pro gave low yield of the diastereomer of **75**. This, of course,

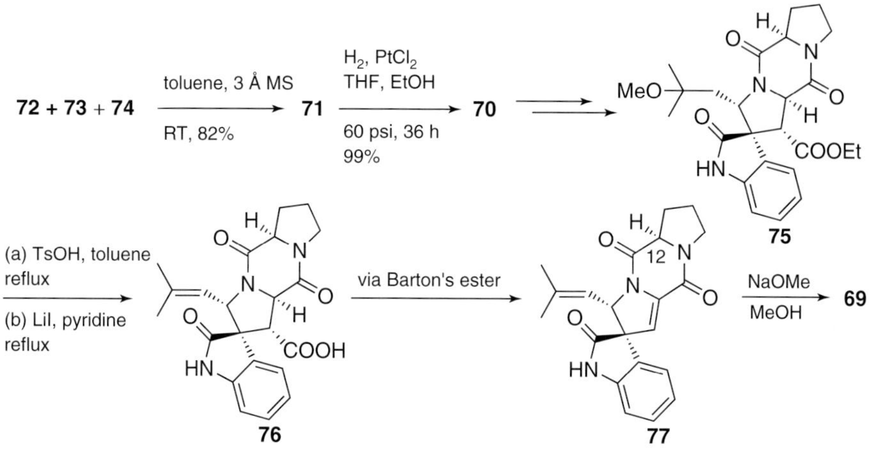

Scheme 2.15

mandated an epimerization step at the end of the synthesis. Installation of the isoprenyl unsaturation (TsOH, toluene, reflux) and hydrolysis of ethyl ester (LiI, pyridine, reflux) afforded the carboxylic acid **76**. Oxidative decarboxylation under Barton's conditions afforded 12-*epi*-spirotryprostatin B (**77**), which was then epimerized with NaOMe in MeOH to give a 2:1 ratio of natural product **69** and its epimer **77**.

2.3.3
Total Synthesis of Hirsutine

Tietze *et al.* developed an efficient three-component reaction of an aldehyde, a 1,3-dicarbonyl compound **78** and an enol ether **79** leading to functionalized dihydropyran. The reaction went through Knoevenagel condensation followed by the Diels–Alder reaction of the resulting reactive 1-oxa-1,3-butadiene with enol ether (Scheme 2.16) [31].

Using this three-component reaction as a key reaction step, Tietze *et al.* were able to design a short synthesis of hirsutine (**82**), a strong inhibitor against the influenza A virus (subtype H3N2) with an EC_{50} value of 0.40–0.57 µg ml^{-1} (Scheme 2.17) [32]. At first glance, this natural product, without a dihydropyran unit, seems irrelevant to Tietze *et al.*'s three-component process. However, two seemingly simple and logical bond disconnections traced hirsutine down to the lactone **85**, which in turn could be prepared by applying authors' own reaction.

Condensation of **86** with Meldrum's acid (**87**) and 4-methoxybenzyl butenyl ether (**88**) (R = *p*-methoxybenzyl, $E:Z = 1:1$) led to the cycloadduct **85** in 84% yield (*dr* > 20:1, Scheme 2.17). Reaction proceeded via Knoevenagel condensation product **89**

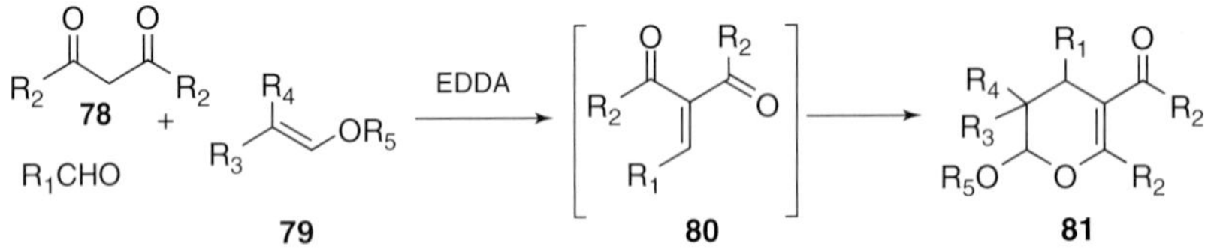

Scheme 2.16

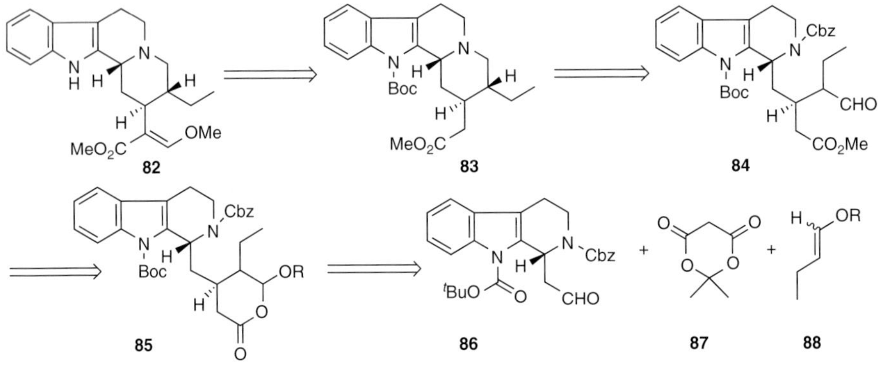

Scheme 2.17

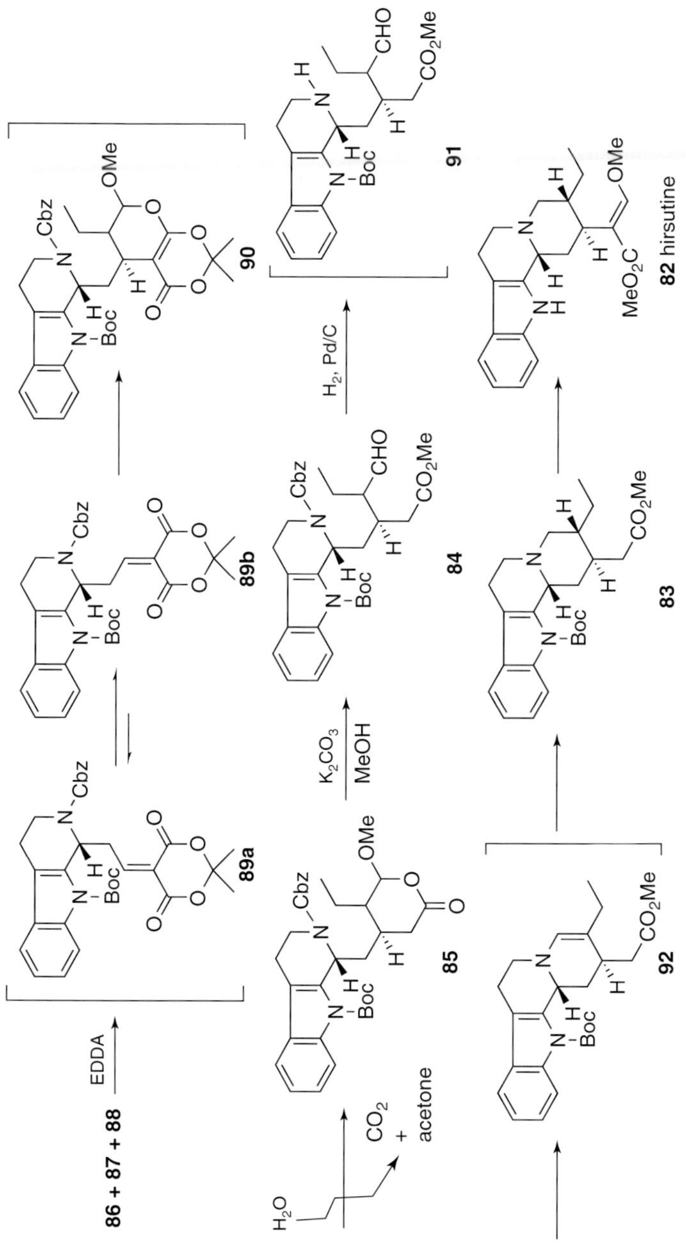

Scheme 2.18

and the observed high stereoselectivity during the subsequent hetero-Diels–Alder reaction was explained by the preferred conformer **89b** (Scheme 2.18). Direct solvolysis (methanol–K$_2$CO$_3$) of **85** without further purification followed by hydrogenation (10% Pd/C, 1 bar H$_2$) afforded the tetracycle **83** as a single diastereoisomer in 67% yield. All stereogenic centers in **83** have the desired absolute configurations of hirsutine (**82**). Cleavage of the *tert*-butoxycarbonyl group in **83** followed by condensation with methyl formate and treatment with diazomethane gave the desired enantiomerically pure hirsutine (**82**).

2.4
Oxidative Anion Coupling

2.4.1
Direct Coupling of Indole with Enolate, Total Synthesis of Hapalindoles, Fischerindoles, and Welwitindolinones

Hapalindole-type natural products have been isolated from soil samples in a myriad of habitats around the world. Since the first example isolated from the Stigonemataceae family of cyanobacteria in 1984 by Moore and co-workers [33], over 60 members are now known [34]. The structures of three representative examples are shown in Figure 2.3.

To access this family of natural products, Baran and co-workers reported in 2004 the first examples of the oxidative coupling of enolates with indoles [35] and pyrroles [36] (Scheme 2.19). These reactions gave the heterocoupling products in good to excellent yields under their carefully optimized conditions. Although oxidative dimerization of enolates has been known for many years and has been developed into a useful method for the preparation of symmetric 1,4-dicarbonyl compounds [37], application of this reaction in a complex molecular setting and heterocoupling of enolates were previously unknown.

Based on this hetero-coupling reaction, Baran and Richter developed a highly efficient synthesis of hapalindoline as shown in Scheme 2.20 [35]. Addition of LHMDS (3 equiv.) to a tetrahydrofuran (THF) solution of indole (**96**, 2 equiv.) and carvone (**97**, 1 equiv.) at −78 °C followed by addition of 1.5 equiv. of copper(II) 2-ethylhexanoate provided the desired diastereomer **98** in 53% isolated yield. A sequence involving

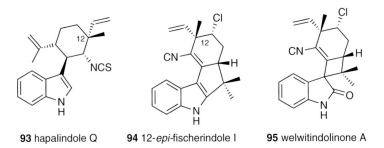

93 hapalindole Q   **94** 12-*epi*-fischerindole I   **95** welwitindolinone A

Figure 2.3 Structures of compounds **93–95**.

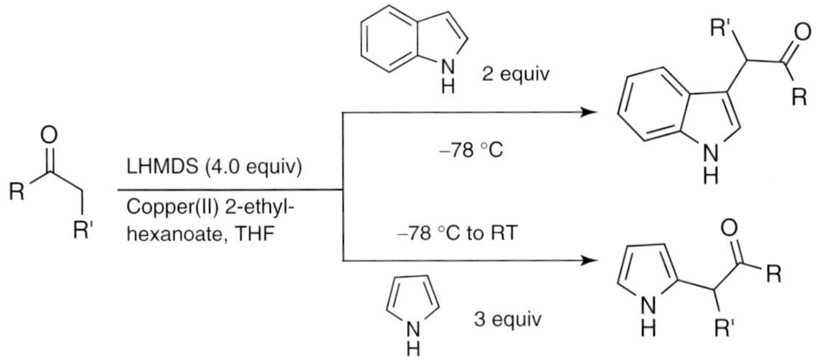

Scheme 2.19

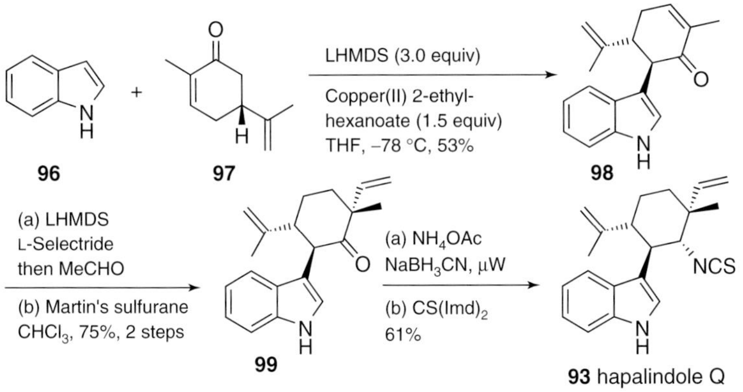

Scheme 2.20

deprotonation of the indole N–H of **98** (LHMDS, THF, −78 °C), 1,4-reduction of enone (L-Selectride), trapping of the resulting enolate with acetaldehyde and dehydration (Martin's sulfurane) converted **98** to vinylated compound **99**. Microwave-accelerated reductive amination of ketone afforded the corresponding primary amine ($dr = 6:1$ in favor of the desired α-stereoisomer), which was then converted to hapalindole Q (**93**) in 61% overall yield.

Since a substoichiometric amount of Cu(II) salt (1.5 equiv.) relative to anion (3 equiv.) is sufficient to promote the heterocoupling between **96** and **97**, the direct radical coupling mechanism was eliminated. Evidence was accumulated to support one of the plausible mechanisms (Scheme 2.21). Formation of copper chelate **100** followed by single-electron transfer would produce the chelated α-keto radical **101**. Nucleophilic attack of indole nucleophile on the radical would afford the radical anion **102**. This high-energy intermediate could then be further oxidized by the proximal copper(I) center, leading, after tautomerization, to the coupled product and copper(0) [38].

Using this coupling as a key step, Baran and co-workers also accomplished the total syntheses of a number of hapalindole-type alkaloids such as 12-*epi*-fischerindole (**94**) and welwitindolinone (**95**) [39].

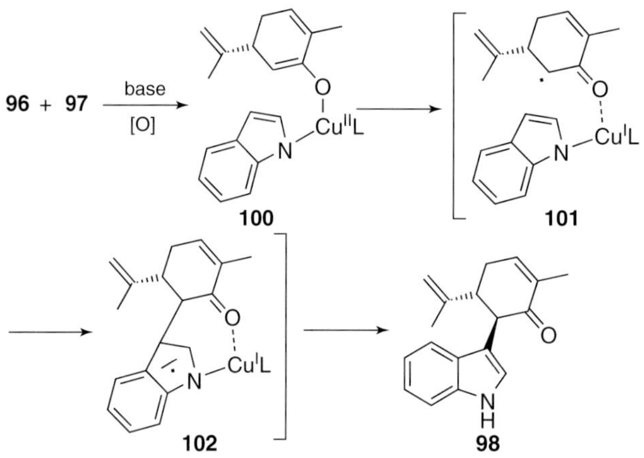

Scheme 2.21

Using intramolecular oxidative coupling of amide and ester enolates as a key step, concise syntheses of (+)-stephacidin A, (−)-stephacidin B, and (+)-avrainvillamide were developed by Baran's group [40].

2.4.2
Total Synthesis of (±)- and (−)-Actinophyllic Acid

Actinophyllic acid was isolated from the leaves of the tree *Alstonia actinophylla* in 2005 by Quinn and co-workers [41]. The presence of the 1-azabicyclo[4.4.2]dodecane and 1-azabicyclo[4.2.1]nonane fragments made this natural product structurally unique. More recently, Overman and co-workers [42] reported a concise synthesis of this indole alkaloid featuring two key steps: (i) an intramolecular oxidative coupling of dienolate and (ii) a sequence of aza-Cope–Mannich reaction. The retro-synthesis is highlighted in Scheme 2.22. Initial disconnection of hemiketal revealed the pentacyclic ketone **104**. After oxidation state adjustment, **104** could be prepared from formamidinium ion derivative **106** via an aza-Cope–Mannich sequence [43]. The hexahydro-1,5-methanoazocino[4,3-*b*]indole ring system of the ketone **107**, the precursor of allylic alcohol **106**, was thought to be formed by intramolecular oxidative coupling of dienoates derived from indole malonate **108**.

In practice, the indole **109** was prepared in two steps by reaction of the magnesium enolate of di-*tert*-butyl malonate with 2-nitrophenylacetyl chloride followed by reductive cyclization under hydrogenolysis conditions. Simply stirring a DMF solution of **109** and the crude bromopiperidone (**110**), generated by bromination of *N*-Boc-3-piperidone afforded **108** in 85% yield on a multigram scale. After screening of different reaction conditions, varying the bases and the oxidants, the best results for the oxidative coupling were obtained with a combination of LDA and [Fe(DMF)$_3$Cl$_2$][FeCl$_4$], a complex formed by simply combining FeCl$_3$ with DMF. Thus, deprotonation of indolepiperidone *rac*-**108** with 3.2 equiv. of LDA in THF at −78 °C followed by adding a THF solution of 3.5 equiv. of [Fe(DMF)$_3$Cl$_2$][FeCl$_4$] and allowing the reaction to warm to room temperature over 60–90 min provided crystalline tetracyclic ketone *rac*-**107** in 60 – 63% yield on scales

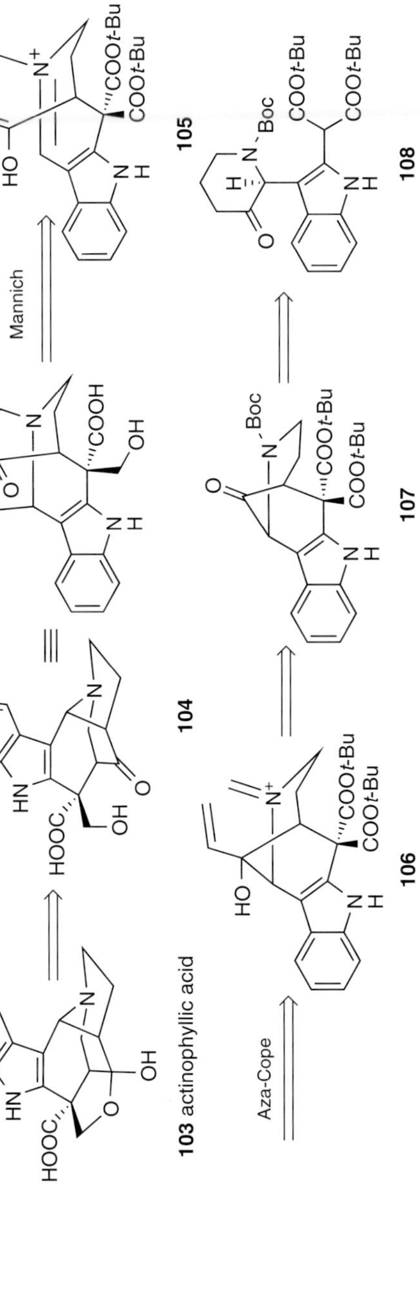

Scheme 2.22

up to 10 g (Scheme 2.23). This represented the first example of intramolecular oxidative coupling of ketone and malonate Enolate. Reaction of *rac*-**107** with 2.5 equiv. each of cerium(III) chloride and vinylmagnesium bromide at −78 °C in THF afforded the allylic alcohol **111**, which is perfectly stable and can be purified. However, taking advantage of the vicinity of the tertiary alcohol and one of the two *tert*-butyl ester functions, a lactonization reaction allowing the differentiation of the two ester groups was devised. Therefore, after complete consumption of the *rac*-**107** (established by thin-layer chromatography), 1.5 equiv. of acetic acid was added to quench the excess organometallic reagent. Warming the reaction mixture to −20 °C promoted the lactonization, leading to pentacyclic lactone *rac*-**112** in 83% yield. Chemoselective reduction of the lactone to diol **113** followed by its exposure to aqueous HCl (5 M) at 60 °C removed both the *N*-Boc and *tert*-butyl ester groups. Concentration of this reaction mixture, dissolution of the crude product in acetonitrile–water (5:1 v/v), addition of paraformaldehyde (1.1 equiv.), and heating to 70 °C promoted the aza-Cope − Mannich reaction to furnish (±)-actinophyllic acid hydrochloride (*rac*-**103**·HCl) in 93% yield. The reaction course of this ingenious transformation can be seen in the retro-synthetic analysis shown in Scheme 2.22 (conversion of **106** to **103** in one step). In summary, Overman and co-workers accomplished the first total synthesis of (±)-actinophyllic acid (**103**) in 22% overall yield from commercially available di-*tert*-butyl malonate and o−nitrophenylacetic acid.

An enantioselective synthesis of (-)-actinophyllic acid was reported by the same group featuring a scandium triflate-catalyzed diastereoselective nucleophilic addition of indole **109** to hemiaminal **114** (Scheme 2.24) [44]. The 2,3-*trans* adduct **115** was obtained in 88% yield with a *dr* of 17:1. Chemoselective reduction of acetate with DIBAL followed by Swern oxidation of the resulting secondary alcohol afforded (*S*)-**108**, which was then converted to (−)-actinophyllic acid (91% *ee*) following the sequence detailed for (±)-actinophyllic acid. This synthesis confirmed the absolute configuration of the natural product, previously hypothesized based on the possible biosynthesis pathway.

2.4.3
Total Synthesis of (−)-Communesin F

(−)-Communesin F (**116**), isolated from a marine fungal strain of *Penicillium* species, possesses significant cytotoxicity and insecticidal activities. Eight structurally related natural products are now known (communesins A−H) and two total syntheses have been achieved by the groups of Qin [45] and Weinreb [46]. Ma and co-workers' total synthesis of (−)-communesin F [47] is shown retro-synthetically in Scheme 2.25. The G and A rings were planned to be constructed via a nucleophilic substitution and Staudinger-type reaction, respectively, from pentacycle **117**, which in turn could be obtained by alkylation of **118**. The latter was prepared from **119** by an intramolecular oxidative coupling of indole and amide enolate.

Reduce of this strategy to practice is shown in Scheme 2.26 [47]. Oxidation of 4-bromotryptophol (**120**) followed by reductive amination with TBS-protected (*S*)-phenylglycinol afforded secondary amine **121**, which was acylated with 2-(2-nitrophenyl)acetic acid (**122**) to provide the desired amide **119**. Intramolecular oxidative coupling of **119** under optimized conditions (LHMDS, THF, HMPA, −78 °C,

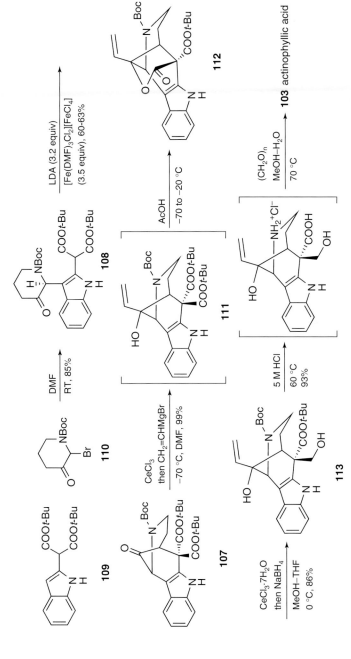

Scheme 2.23

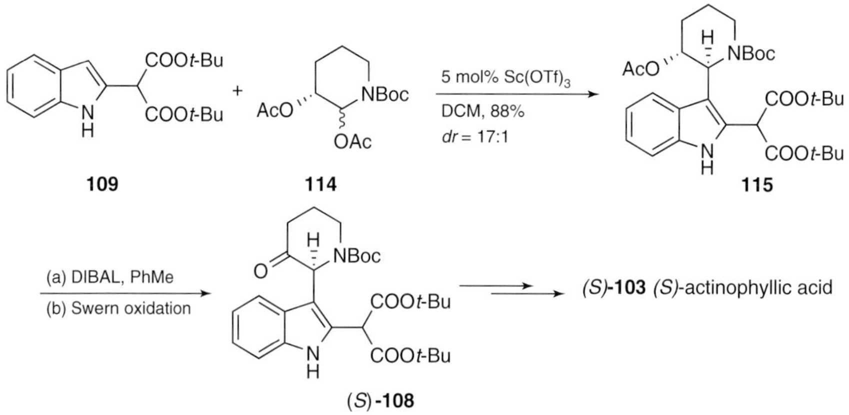

Scheme 2.24

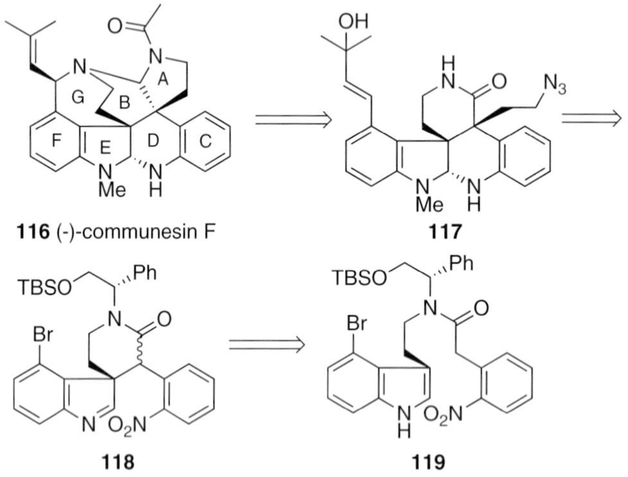

Scheme 2.25

then iodine) furnished the desired spiroindoline **118** as a diastereomeric mixture. Reduction of the nitro group was followed by spontaneous cyclization to afford a pentacyclic intermediate. Selective N-methylation under standard conditions afforded aminal **123** in 50% yield over three steps. A diastereomer of **123** was also isolated in 16% yield (structure not shown). After insertion of an N-Boc function and C-allylation (**124**), the chiral auxiliary was removed using Ennis's conditions (LiOH, DMSO, 100 °C, then HCl) to afford **125** [48]. The reaction was thought to proceed via the enamide intermediate, which was then hydrolyzed under acidic conditions.

Oxidative cleavage of the double bond followed by the Heck reaction converted **125** to **126**. Treatment of the latter with 2 equiv. of mesyl chloride and Et₃N followed by NaN₃ afforded directly the hexacyclic azide **127**. The ring G was formed during the first mesylation step. It was assumed that the projected Staudinger reaction for the formation

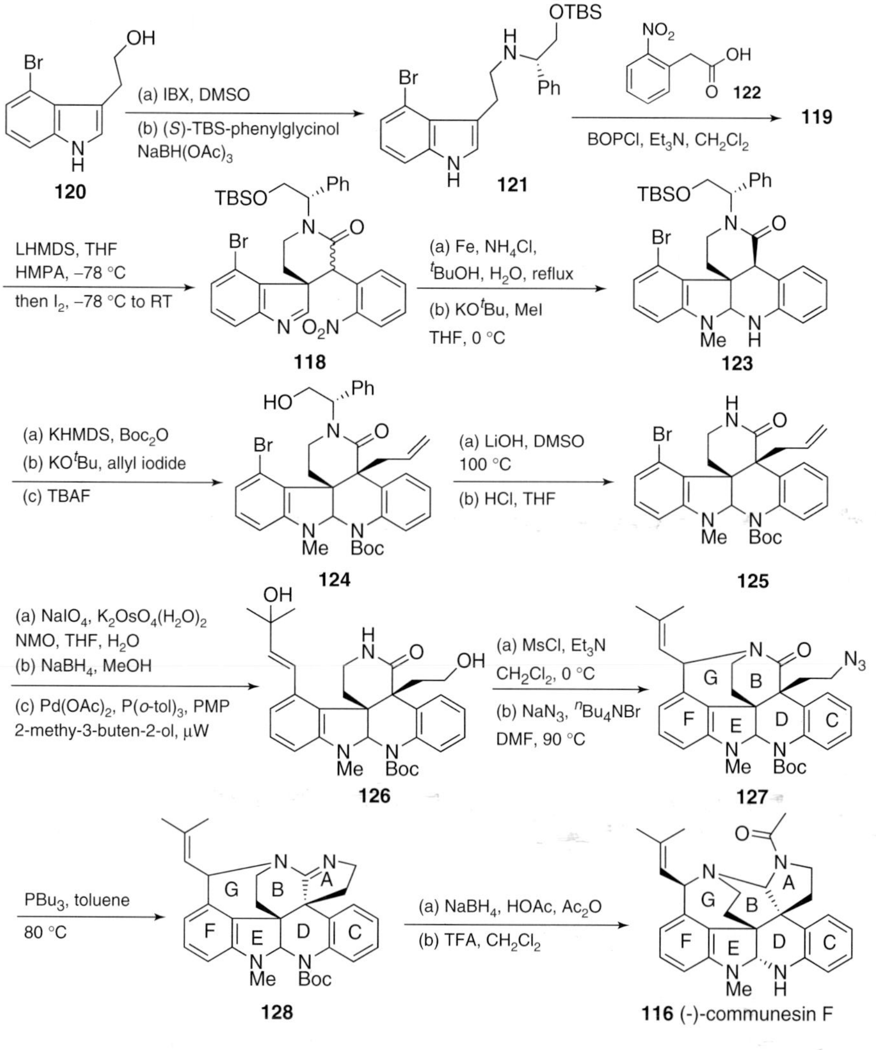

Scheme 2.26

of amidine should be facilitated by the fact that the amide adopted a twisted conformation wherein the conjugation between the C=O π-bond and amide nitrogen was substantially diminished. In the event, treatment of **127** with PBu$_3$ in toluene afforded the heptacyclic amidine **128** in 82% yield. Finally, a sequence involving reduction of the amidine group in **128**, acetylation, and removal of the N-Boc function afforded (−)-communesin F (**116**). In summary, (−)-communesin F was synthesized in 19 steps from 4-bromotryptophol in about 6% overall yield. In addition, comparison of the sign of optical rotation allowed authors to assign the absolute configuration of the natural product as shown in the scheme.

2.5
Pattern Recognition

2.5.1
Introduction

In retro-synthetic analysis, one proceeds via strategic bond identification and simplifies stepwise the target molecule through bond disconnections. Another modality of retro-synthetic analysis, termed "*pattern recognition*" by Wilson and Danishefsky, seeks to discover an exploitable substructural motif around which to organize the thought process and then the synthesis [49]. There are many levels of opportunity for creativity in deeply disguised patterns. The most fascinating cases arise when the target itself must actually be modified, leading to a more complex compound before a staple pattern is revealed.

[2 + 2]-Photoaddition of an enol ether with an alkene followed by retro-aldol reaction of the resulting cyclobutanol for the synthesis of 1,5-dicarbonyl compounds is named the De Mayo reaction (Scheme 2.27a) [50]. With a deep understanding of this reaction and insight into the molecular structure of longifolene, Oppolzer and Godel were able to implement this reaction as a key step in their landmark total synthesis of longifolene [51]. Retro-synthetically, the key thinking process is connecting (rather than disconnecting) C_a and C_b in **130** to produce a seemingly more complex compound **131**. However, **131** with added molecular complexity can subsequently be unfolded to give an extremely simple starting material, the cyclopentanone **132** by the De Mayo reaction (Scheme 2.27b). With this strategy, the authors were able to realize the total synthesis of longifolene in only eight steps with 25% overall yield.

2.5.2
Total Synthesis of (±)-Aplykurodinone-1

Aplykurodinone-1 (**133**) was isolated from the sea hare *Synphonota geographica* in 2005 [52]. Zhang and Danishefsky's synthesis [53] is presented retro-synthetically in Scheme 2.28. Disconnection of the side chain and the γ-lactone converted the natural

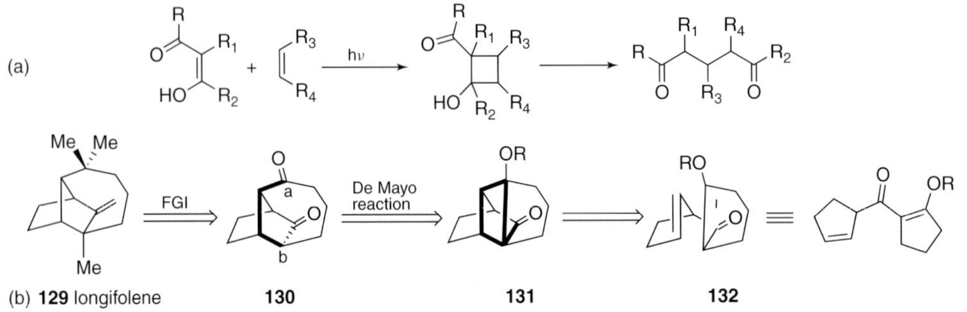

(a)

(b) **129** longifolene   **130**   **131**   **132**

Scheme 2.27

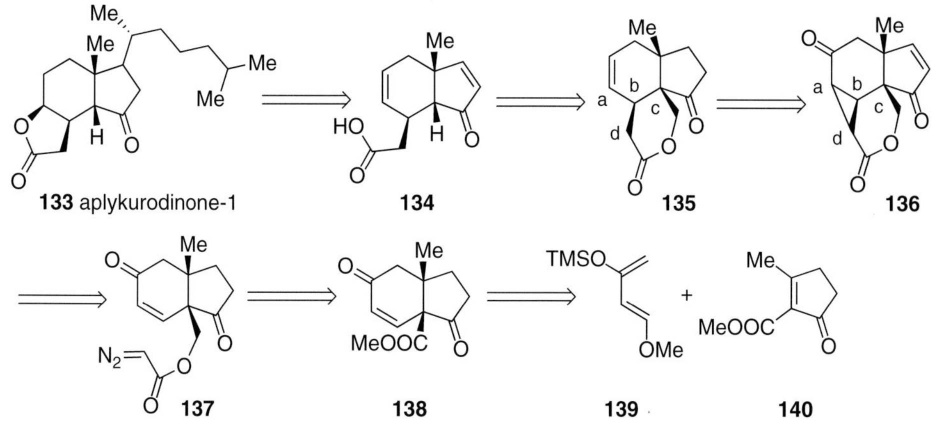

Scheme 2.28

product to the simplified *cis*-fused hydrindane **134**. They subsequently transformed this simplified bicycle into a very complex bridged tetracyclic compound **136** through a "pattern recognition" thinking process (pattern mapping). In this *"moving backwards"* retro-synthetic analysis, a new C–C bond (C_a–C_d) and a quaternary center at C_c were created with concurrent creation of a cyclopropane ring. However, the increased complexity in **136** actually allowed the authors simultaneously to disconnect the C_a–C_d and C_b–C_d bonds to give diazo ester **137**, which could in turn be prepared by a standard Diels–Alder reaction.

The forward synthesis of aplykurodinone-1 (**133**) is shown in Scheme 2.29 [53]. Cycloaddition of lithium enolate, generated *in situ* from **139** (MeLi), with cyclopentenone **140** afforded **138** in 73% yield after acidic treatment. Following a five-step sequence, bicycle **141** was obtained, which was converted to diazo ester **142** in 88% yield. Intramolecular cyclopropanation of **142** took place in the presence of bis(*N-tert*-butylsalicylaldiminato)copper(II) to furnish tetracycle **143** in moderate yield. Conversion of ketone to iodide followed by samarium iodide-initiated radical fragmentation afforded tricycle **144** in good overall yield. Exposure of ketone **135**, obtained from **144** in two conventional steps, to aqueous potassium carbonate solution afforded the keto acid **145** via a hydrolysis–retro-aldol sequence. Without isolation, **145** underwent *in situ* iodolactonization to provide **146**, which was subsequently converted to **147** under standard conditions. The methyl ester function, strategically incorporated in the Diels–Alder adduct **138**, not only served as a handle to facilitate stereoselectively the creation of the C_b–C_d bond, but was also traceless as it was removed at this stage via a retro-aldol process.

Michael addition of organocuprate, generated *in situ* from vinyl bromide **148**, afforded **149** with excellent diastereoselectivity ($dr = 10:1$). Catalytic hydrogenation of the trisubstituted alkene in the presence of Crabtree's catalyst afforded **150** with the desired stereochemistry in 50% yield ($dr = 10:1$). Compound **150** was converted to the natural product via a sequence of deprotection–oxidation–Julia olefination.

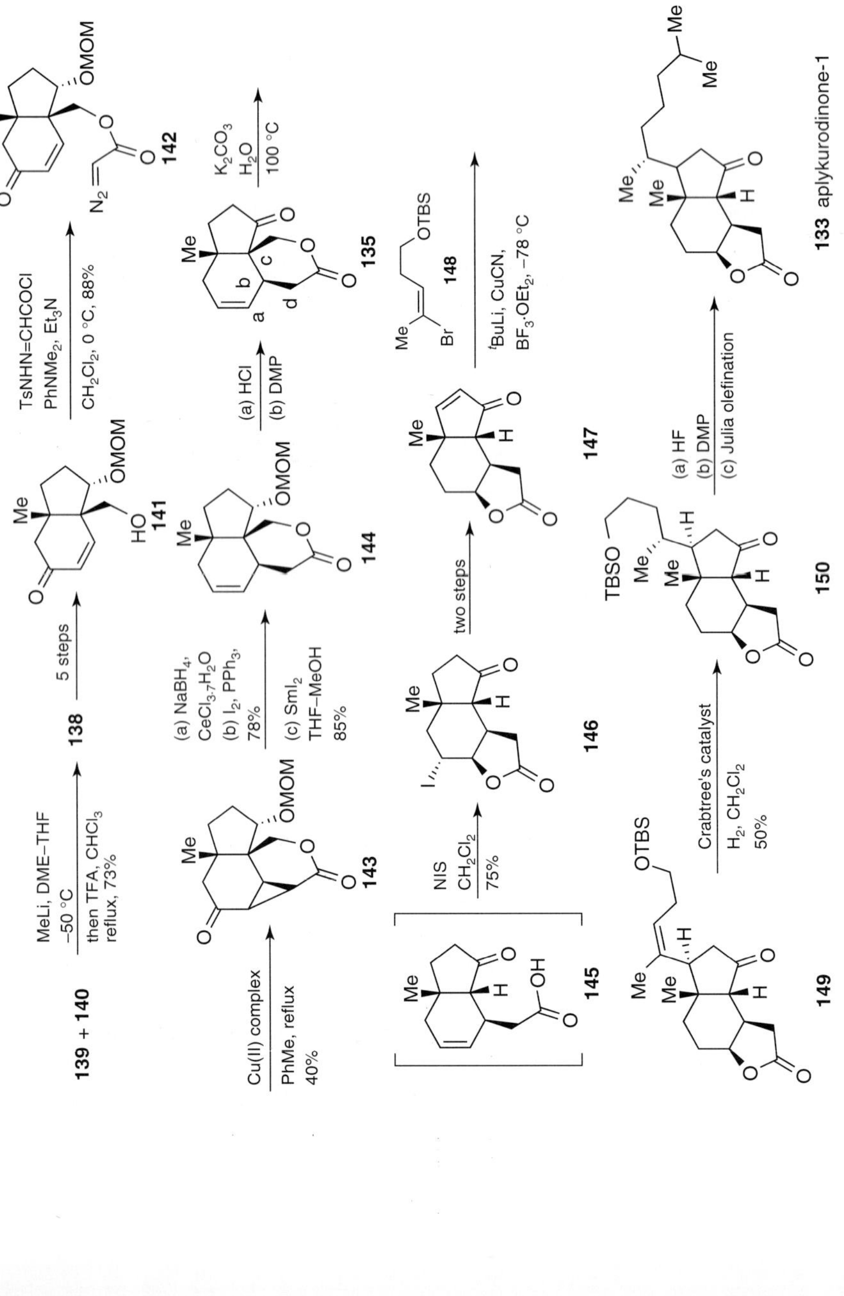

139 + 140
MeLi, DME–THF
–50 °C
then TFA, $CHCl_3$
reflux, 73%
138
5 steps
141
TsNHN=CHCOCl
$PhNMe_2$, Et_3N
CH_2Cl_2, 0 °C, 88%
142
Cu(II) complex
PhMe, reflux
40%
143
(a) $NaBH_4$, $CeCl_3 \cdot 7H_2O$
(b) I_2, PPh_3, 78%
(c) SmI_2
THF–MeOH
85%
144
(a) HCl
(b) DMP
135
K_2CO_3
H_2O
100 °C
145
NIS
CH_2Cl_2
75%
146
two steps
147
148
${}^tBuLi$, CuCN,
$BF_3 \cdot OEt_2$, –78 °C
149
Crabtree's catalyst
H_2, CH_2Cl_2
50%
150
(a) HF
(b) DMP
(c) Julia olefination
133 aplykurodinone-1

Scheme 2.29

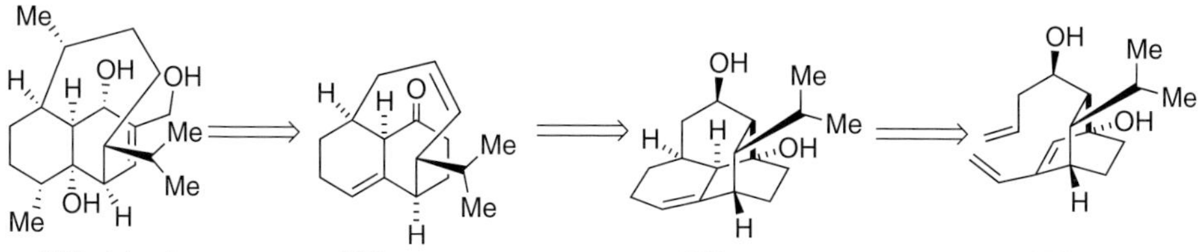

151 vinigrol   **152**   **153**   **154**

Scheme 2.30

2.5.3
Total Synthesis of (±)-Vinigrol

Vinigrol (**151**) was isolated in 1987 by Hashimoto and co-workers from the fungal strain *Virgaria nigra* F-5408 [54]. This novel diterpenoid is the only natural product containing the decahydro-1,5-butanonenaphthalene carbon skeleton. The promising bioactivities combined with its unique molecular architecture have attracted significant attention from the synthetic community. After 20 years of intensive research by different groups, the first total synthesis was recently accomplished by Baran's group. As shown in Scheme 2.30, the key retro-synthesis step in their elegant synthesis is the conversion of the bridged tricycle **152** to the more complex tetracycle **153**, with a structural pattern conducive to Grob fragmentation for the formation of the difficult to access bridged eight-membered ring. The access to the seemingly more complex tetracycle **153** is in fact much easier than access to tricycle **152**, since **153** can now be prepared by a sequence of intermolecular followed by intramolecular Diels–Alder reaction.

Baran and co-workers' successful synthesis is summarized in Scheme 2.31 [55]. The *endo*-selective Diels–Alder reaction between diene **155** and (*E*)-methyl 4-methyl-2-pentenoate (**156**) afforded bicyclic ketone **157** in 65% yield (*dr* = 2:1). Formation of the enol triflate followed by Stille coupling produced the diene **158** (78%). Conversion of ester to aldehyde (LAH then DMP) and subsequent addition of allylmagnesium chloride furnished the intermediate alkoxide (*dr* = 6:1), which, upon heating to 105 °C, underwent proximity-induced intramolecular Diels–Alder reaction to afford the alcohol **161** in 75% yield. Oxidation of the secondary alcohol to ketone followed by methylation, *O*-TBS deprotection and Evans's $Me_4NBH(OAc)_3$-mediated, hydroxy-directed reduction of ketone [56] furnished diol **163** as a single diastereomer in 72% yield. The relative stereochemistry of this diol is of the utmost importance as only the *anti*-diol shown in **163** is properly oriented for the Grob fragmentation.

Conversion of the secondary alcohol to mesylate followed by treatment with a base trigged the key Grob fragmentation to afford the desired tricycle **164** in an impressive 85% yield. Selective dipolar cycloaddition between **164** and bromonitrile oxide, generated *in situ* from **165**, produced **166**, which was subsequently converted to **167** via a two-step sequence. Chugaev elimination of xanthate derived from alcohol **167** afforded the alkene **168** (85% yield). Reduction of bromoisoxazole with LAH furnished the 1,3-amino alcohol. The primary amine was converted to a methyl group according to the Saegusa deamination sequence (*N*-formylation, dehydration, and reduction of the resulting isocyanide) [57]. Dihydroxylation of **169** and chemoselective oxidation of the resulting

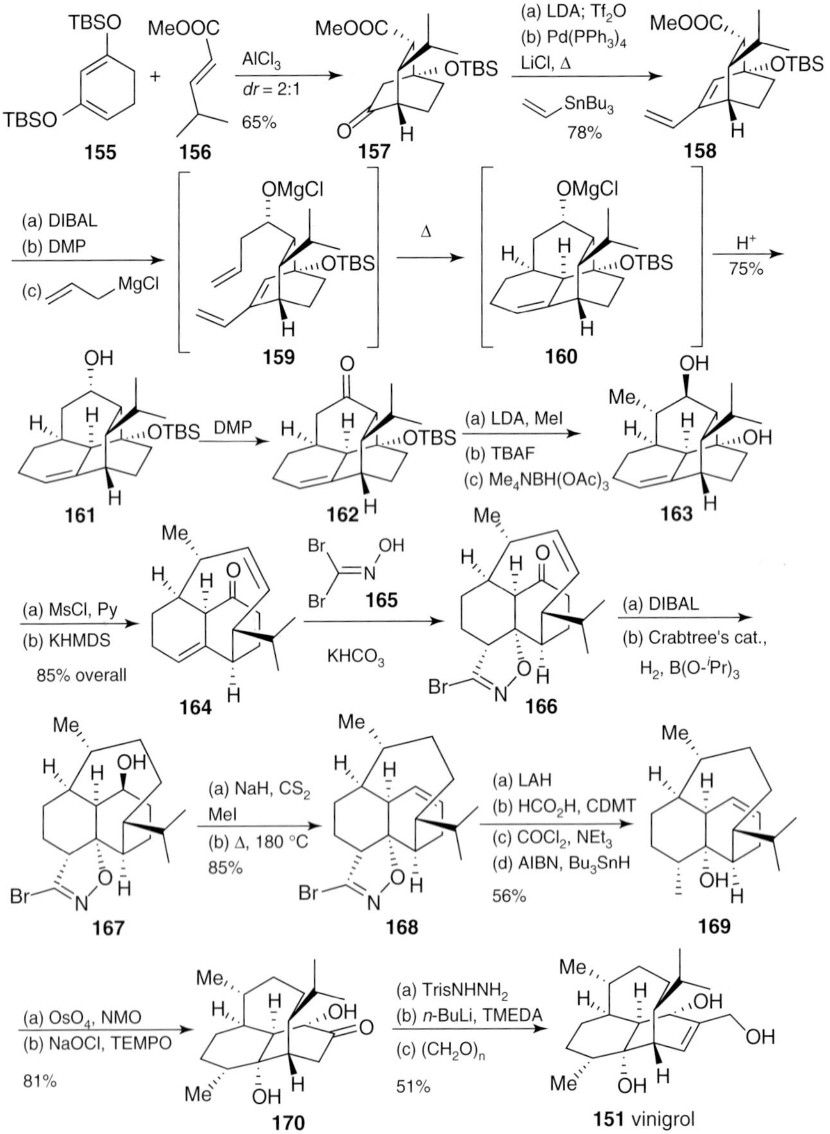

Scheme 2.31

diol (NaOCl, TEMPO) led to the hydroxy ketone **170** in 81% overall yield. Finally, a Shapiro reaction via trisylhydrazone and trapping of the trianionic intermediate by formaldehyde accomplished the long-awaited total synthesis of vinigrol (**151**).

The recognition of the Grob fragmentation is the central piece of Baran and coworkers' exciting total synthesis of vinigrol that ended the 20 years' struggle by synthetic chemists. With this elegant strategy in hand, Baran and co-workers were able to make this daunting molecule in only 23 steps with 3% overall yield from commercially available starting materials.

2.6
Conformation-Directed Cyclization

2.6.1
Introduction

Macrocycles, by virtue of their widespread occurrence in Nature and their intrinsic three-dimensional structures, play an important role in chemistry and biology and are medicinally relevant [58]. Indeed various important drugs, such as cyclosporin, macrolides (erythromycin), and the vancomycin family of glycopeptides, contain macrocyclic elements. Furthermore, turning linear molecules into macrocyclic structures is an important tool for manipulating the properties of compounds. Indeed, bioactive linear peptides can exist in a myriad of different conformations, very few of which are able to bind to their receptor [59]. Cyclization is a common approach to force peptides to adopt bioactive conformations and to access the important structural and dynamic properties of peptides [60]. In addition, cyclopeptides are much more resistant to *in vivo* enzymatic degradation than their linear counterparts. To address the synthesis of macrocyclic natural products or any designed macrocycles with a specific purpose, ring closure is naturally the key step that will determine the efficacy of the overall synthetic strategy. Although the success of a given macrocyclization is still largely a matter of trial and error, the importance of conformational preorganization in a given cyclization has been recognized ever since Eschenmoser's synthesis of vitamin B_{12} and Woodward's synthesis of erythromycin.

The activation energy for a ring closure can be lowered by the preorganization of the two reacting termini into close proximity before the actual cyclization step. Naturally, the energy for bringing the reaction centers together and restraining their motion, for example, the rotational freedom of the molecular framework, in order to facilitate the formation of the ring-bond, has to be paid for in the preorganization steps. The forces responsible for favoring one conformation over another are covalent bonds, hydrogen bonding, and steric and electronic interactions of different nature, such as electrostatic interactions, repulsive forces, polarization, and charge transfer. The interplay of these factors of different strengths and to different extents contributes to the conformation of molecules and hence also to the preorganization of reactive centers.

2.6.2
Cyclosporin and Ramoplanin A2

Hydrogen bonding interactions are one of the key elements responsible for the vast diversity of spatial arrangements of peptides and proteins. The secondary structures are induced by a multitude of directional changes and folding patterns stabilized by hydrogen bonds across the chains, the so-called turn-units. Obviously, H-bonding is also of special importance in the context of designing effective conformation-directed macrocyclization

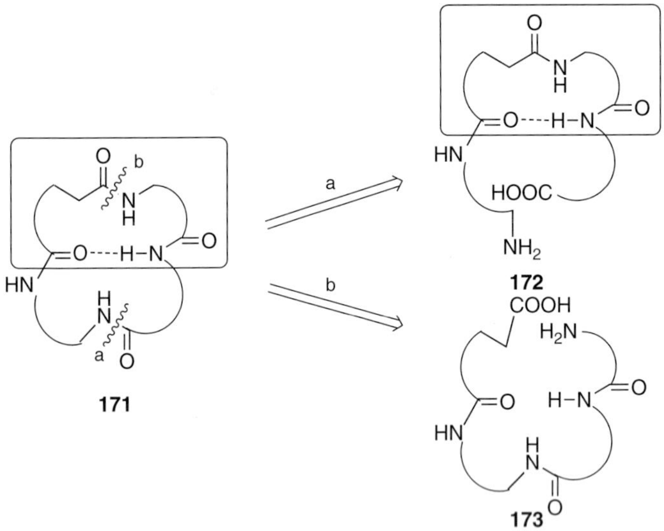

Scheme 2.32

reactions. Intramolecular H-bonding can favor one conformation over another, thereby enabling or preventing the two cyclization sites to come into productive proximity.

In the total synthesis of cyclopeptides, the ring disconnection carries significant strategic importance, having a large impact on the overall efficiency of the synthetic approach. Generally, for target molecules displaying H-bonding interactions, the incorporation of a turn-fragment into the linear precursor is the obvious choice in order to obtain a conformation-biased substrate. Considering the general structure **171** with the turn motif in the top part of the molecule, the cyclization of **172** (disconnection a) is more promising than the same process for substrate **173**, evoking the potential of the conformational principle (Scheme 2.32).

The practical importance of considering H-bonding interactions was highlighted by Wenger's total synthesis of cyclosporin, an immunosuppressant (Scheme 2.33) [61]. The amino acids 1–6 of this cyclic undecapeptide are engaged in an antiparallel β-pleated sheet conformation which contains three transannular H-bonds. The remaining amino acids form an open loop that contains a D-alanine in position 8 and the only *cis*-amide linkage between the two adjacent *N*-methylleucine residues 9 and 10. Wenger selected the peptide bond between the L-alanine (residue 7) and the D-alanine (residue 8) as the strategic bond leading to linear substrate **175**. One of the major reasons for such a strategic choice is that the linear undecapeptide **175** may adopt a folded conformation by the formation of intramolecular H-bonds, as found in the cyclic structure of cyclosporin. Indeed, cyclization of **175** under appropriate conditions provided the 33-membered macrocycle **174** in 62% yield. The presumed conformation of the linear precursor as a "cyclosporin"-like, H-bond-stabilized folded structure was supported by NMR analysis. Additionally, this cyclization may also be facilitated by the fact that it involves a D-amine terminus (D-Ala) as a nucleophilic partner and the small side chains of two alanine residues [62].

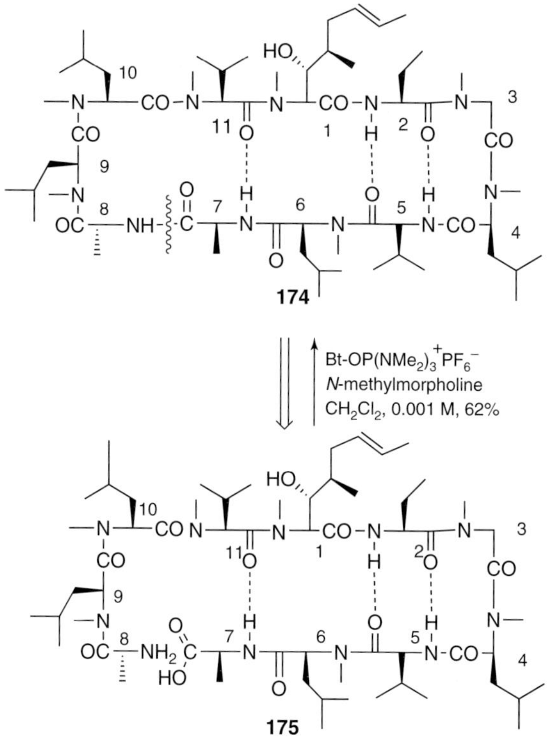

174

Bt-OP(NMe$_2$)$_3$$^+PF_6$$^-$
N-methylmorpholine
CH$_2$Cl$_2$, 0.001 M, 62%

175

Scheme 2.33

The total synthesis of the antibiotic ramoplanin A2 by Boger's group is another example for the outstanding importance of preorganization by H-bonding interactions [63]. The convergent approach towards the 49-membered cyclodepsipeptide envisaged two different sites for the final cyclization, both benefiting potentially from preorganization, although to different extents. Whereas the cyclization site between the Gly (14th amino acid) and Leu (15th amino acid) units **(176)** is less sterically hindered, it does not include ring closure at a D-amino acid site and is situated at the more flexible loop of the macrocycle. In contrast, the macrolactamization between Phe (9th amino acid) and D-Orn (10th amino acid) **(177)** is situated next to the β-turn [Thr(8)–Phe(9)], including a D-amino acid, and the linear precursor is stabilized by the antiparallel β-strands (Scheme 2.34). Both cyclizations took place smoothly to afford the desired macrocycle **178** in 50–89% yields. Both sites are obviously preorganized by the H-bonding pattern of the precursors and vary in this aspect only in the degree of organization.

By far most of the macrocyclizations reported in the literature are effected under high-dilution conditions in order to avoid the competitive oligomerization/polymerization processes. However, if a linear precursor contains structural elements that strongly favor the folded conformer, then macrocyclization could be performed even in a high concentration with minimum interference of the intermolecular process. As shown in Scheme 2.35, stirring a DMF solution of **179** in the presence of CsF (0.01 M) provided

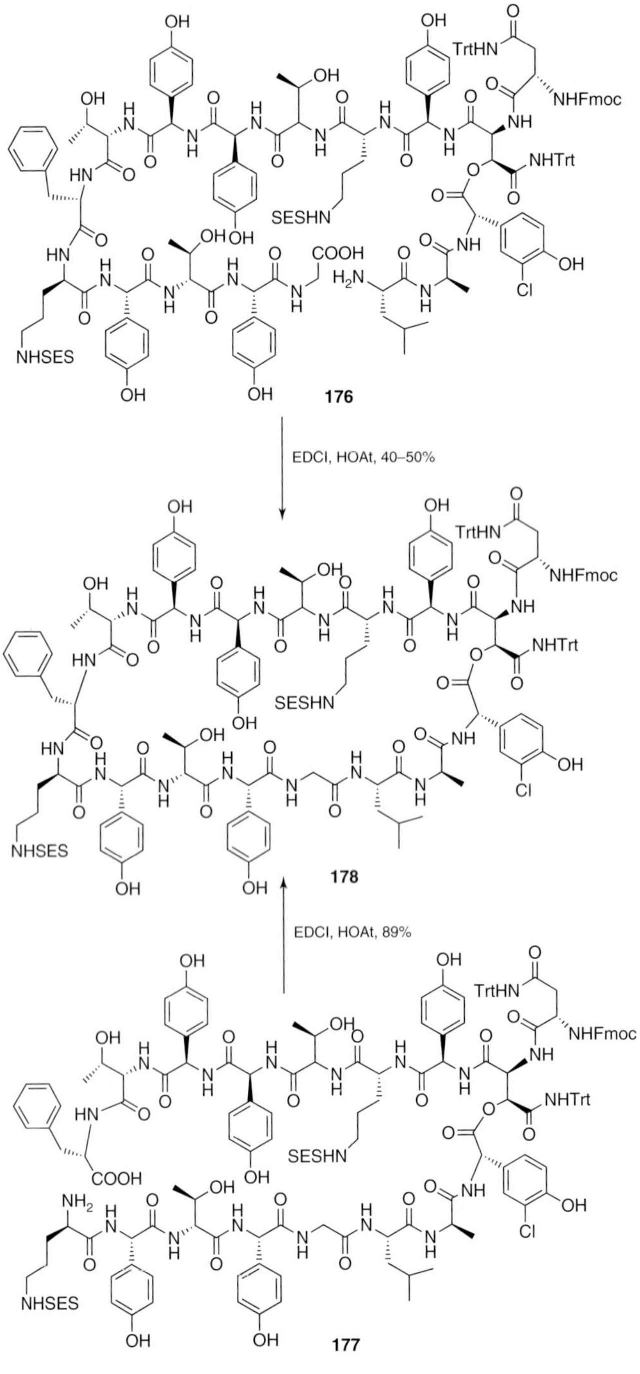

176

EDCI, HOAt, 40–50%

178

EDCI, HOAt, 89%

177

Scheme 2.34

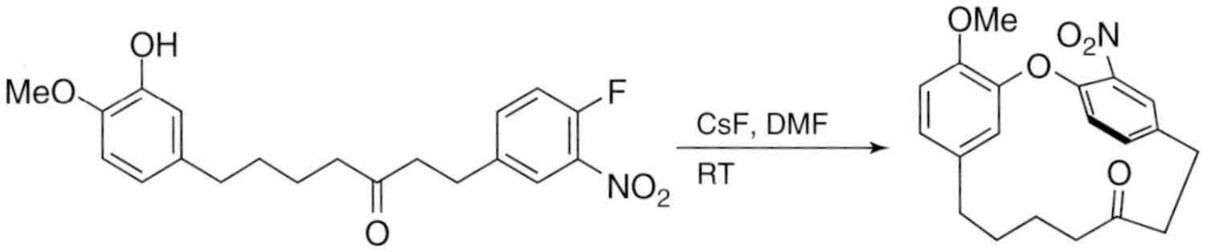

Scheme 2.35

concentration of **179** in DMF	isolated yield of **180**
0.01 M	90–95%
0.05 M	90–95%
0.1 M	75–80%
1 M	45–50%

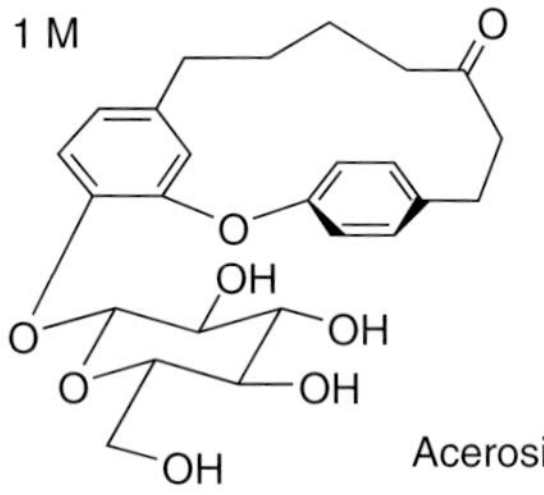

the 15-membered m,p-cyclophane **180** in over 90% yield. Interestingly, even at 1.0 M, the desired macrocyclization of **179** still took place as a major reaction pathway to afford **180** in about 50% yield. The propensity of **179** for macrocyclization is therefore indeed extraordinary and it was assumed that **179** adopted a bent conformation that is particularly conducive to the macrocyclization [64]. The macrocycle **180** has subsequently been converted to aceroside IV by a sequence of standard transformations [65].

2.7
Conclusion and Perspectives

The total synthesis of quinine by Woodward and Doering in the 1940s marked the debut of modern multi-step total syntheses of natural products. The field subsequently evolved at such a rapid pace that it is now assumed that almost all structures could be made in the laboratory given sufficient time and resources. However, total synthesis remains a fundamentally healthy wellspring of chemical discovery and the ideal training ground for chemists in the pharmaceutical industry. Synthetic chemists are now more cautious about the selection of their targets, and pay more attention to how to make them and what it brings after the completion of the synthesis. From the viewpoint of a hardcore synthetic chemist, the development of highly efficient synthetic strategies would certainly be a key point to be associated with any total synthesis endeavors. The synergism between total synthesis and synthetic strategy has played and will continue to play key roles in advancing chemical science and in educating the new generation of chemists.

Acknowledgments

Financial supports from CNRS, France, and the Ecole Polytechnique Fédérale de Lausanne (EPFL), Switzerland, are gratefully acknowledged.

We are most grateful to Professor K.C. Nicoulaou (Scripps Research Institute), Professor H.N.C Wong (Chinese University of Hong Kong), and Professor W.-D. Li (Nankai University) for their thoughtful comments and their encouragement.

References

1. Bonjoch, J. and Solé, D. (2000) *Chem. Rev.*, **100**, 3455–3482.
2. Service, R.F. (1999) *Science*, **285**, 184–187.
3. Rinehart, K.L., Holt, T.G., Fregeau, N.L., Stroh, J.G., Keifer, P.A., Sun, F., Li, L.H., and Martin, D.G. (1990) *J. Org. Chem.*, **55**, 4512–4515.
4. Wright, A.E., Forleo, D.A., Gunawardana, G.P., Gunasekera, S.P., Koehn, F.E., and McConnell, O.J. (1990) *J. Org. Chem.*, **55**, 4508–4512.
5. Corey, E.J., Gin, D.Y., and Kania, R.S. (1996) *J. Am. Chem. Soc.*, **118**, 9202–9203.
6. Menchaca, R., Martínez, V., Rodríguez, A., Rodríguez, N., Flores, M., Gallego, P., Manzanares, I., and Cuevas, C. (2003) *J. Org. Chem.*, **68**, 8859–8866.
7. Aicher, T.D., Buszek, K. R., Fang, F.G., Forsyth, C.J., Jung, S.H., Kishi, Y., Matelich, M.C., Scola, P.M., Spero, D.M., and Yoon, S.K. (1992) *J. Am. Chem. Soc.*, **114**, 3162–3164.
8. Kim, D.S., Dong, C.G., Kim, J.T., Guo, H., Huang, J., Tiseni, P.S., and Kishi, Y. (2009) *J. Am. Chem. Soc.*, **131**, 15636–15641.
9. Trost, B.M. (1995) *Angew. Chem.*, **107**, 285–307; *Angew. Chem. Int. Ed. Engl.*, **34**, 259–281.
10. (a) Wender, P.A., Handy, S., and Wright, D.L. (1997) *Chem. Ind. (London)*, 765–769; (b) Wender, P.A. and Miller, B.L. (2009) *Nature*, **460**, 197–201.
11. Burns, N.Z., Baran, P.S., and Hoffmann, R.W. (2009) *Angew. Chem.*, **128**, 2896–2910; *Angew. Chem. Int. Ed.*, **48**, 2854–2867.
12. (a) Young, I.S. and Baran, P.S. (2009) *Nat. Chem.*, **1**, 193–205; (b) Hoffmann, R.W. (2006) *Synthesis*, 3531–3541.
13. (a) Tietze, L.F. (1996) *Chem. Rev.*, **96**, 115–136; (b) Tietze, L.F., Brasche, G., and Gericke, K. (eds.) (2006) *Domino Reactions in Organic Synthesis*, Wiley-VCH Verlag GmbH, Weinheim.
14. Johnson, W.S., Gravestock, M.B., and McCarry, B.E. (1971) *J. Am. Chem. Soc.*, **93**, 4332–4334.
15. (a) Nicolaou, K.C., Petasis, N.A., Zipkin, R.E., and Uenishi, J. (1982) *J. Am. Chem. Soc.*, **104**, 5555–5557; (b) Nicolaou, K.C., Petasis, N.A., Uenishi, J., and Zipkin, R.E. (1982) *J. Am. Chem. Soc.*, **104**, 5557–5558; (c) Nicolaou, K.C., Zipkin, R.E., and Petasis, N.A. (1982) *J. Am. Chem. Soc.*, **104**, 5558–5560; (d) Nicolaou, K.C., Petasis, N.A., and Zipkin, R.E. (1982) *J. Am. Chem. Soc.*, **104**, 5560–5562.
16. Nicolaou, K.C., Edmonds, D.J., and Bulger, P.G. (2006) *Angew. Chem.*, **118**, 7292–7344; *Angew. Chem. Int. Ed.*, **45**, 7134–7186.
17. Nicolaou, K.C., Sarlah, D., Wu, T.R., and Zhan, W. (2009) *Angew. Chem.*, **121**, 7002–7006; *Angew. Chem. Int. Ed.*, **48**, 6870–6874.
18. Massiot, G., Thépenier, P., Jacquier, M.J., Le Men-Oliver, L., and Delaude, C. (1989) *Heterocycles*, **29**, 1435–1438.
19. Shen, L., Zhang, M., Wu, Y., and Qin, Y. (2008) *Angew. Chem.*, **120**, 3674–3677; *Angew. Chem. Int. Ed.*, **47**, 3618–3621.
20. Trzupek, J.D., Lee, D., Crowley, B.M., Marathias, V.M., and Danishefsky, S.J. (2010) *J. Am. Chem. Soc.*, **132**, 8506–8512.
21. Gassman, P.G. and van Bergen, T.J. (1973) *J. Am. Chem. Soc.*, **95**, 2718–2719.
22. Kinast, G. and Tietze, L.F. (1976) *Angew. Chem.*, **88**, 261–262; *Angew. Chem. Int. Ed.*, **15**, 239–240.

23. He, F., Bo, Y., Altom, J.D., and Corey, E.J. (1999) *J. Am. Chem. Soc.*, **121**, 6771–6772.

24. Total synthesis of haplophytine has been realized recently, see: (a) Ueda, H., Satoh, H., Matsumoto, K., Sugimoto, K., Fukuyama, T., and Tokuyama, H. (2009) *Angew. Chem.*, **121**, 7736–7739; *Angew. Chem. Int. Ed.*, **48**, 7600–7603; (b) Nicolaou, K.C., Dalby, S.M., Li, S., Suzuki, T., and Chen, D.Y.K. (2009) *Angew. Chem.*, **121**, 7752–7756; *Angew. Chem. Int. Ed.*, **48**, 7616–7620.

25. Zhu, J., Bienaymé, H. (eds.) (2005) *Multicomponent Reaction*, Wiley-VCH Verlag GmbH, Weinheim.

26. Touré, B.B. and Hall, D.G. (2009) *Chem. Rev.*, **109**, 4439–4486.

27. Robinson, R. (1917) *J. Chem. Soc.*, **111**, 762–768.

28. Endo, A., Yanagisawa, A., Abe, M., Tohma, S., Kan, T., and Fukuyama, T. (2002) *J. Am. Chem. Soc.*, **124**, 6552–6554.

29. For another total synthesis, see: Chen, J., Chen, X., Bois-Choussy, M., and Zhu, J. (2006) *J. Am. Chem. Soc.*, **128**, 87–89.

30. Sebahar, P.R. and Williams, R.M. (2000) *J. Am. Chem. Soc.*, **122**, 5666–5667.

31. For a review, see: Tietze, L.F., Kettschau, G., Gewert, J.A., and Schuffenhauer, A. (1998) *Curr. Org. Chem.*, **2**, 19–62.

32. Tietze, L.F. and Zhou, Y. (1999) *Angew. Chem.*, **111**, 2076–2078; *Angew. Chem. Int. Ed.*, **38**, 2045–2047.

33. Moore, R.E., Cheuk, C., and Patterson, G.M. (1984) *J. Am. Chem. Soc.*, **106**, 6456–6457.

34. (a) Jimenez, J.I., Huber, U., Moore, R.E., and Patterson, G.M.L. (1999) *J. Nat. Prod.*, **62**, 569–572; (b) Becher, P.G., Keller, S., Jung, G., Süssmuth, R.D., and Jüttner, F. (2007) *Phytochemistry*, **68**, 2493–2497.

35. Baran, P.S. and Richter, J.M. (2004) *J. Am. Chem. Soc.*, **126**, 7450–7451.

36. Baran, P.S., Richter, J.M., and Lin, D.W. (2005) *Angew. Chem.*, **117**, 615–618; *Angew. Chem. Int. Ed.*, **44**, 609–612.

37. (a) Rathke, M.W. and Lindert, A. (1971) *J. Am. Chem. Soc.*, **93**, 4605–4606; (b) Dessau, R.M. and Heiba, E.I. (1974) *J. Org. Chem.*, **39**, 3457–3459; (c) Ito, Y., Konoike, T., Harada, T., and Saegusa, T. (1977) *J. Am. Chem. Soc.*, **99**, 1487–1493; (d) Csákÿ, A.G. and Plumet, J. (2001) *Chem. Soc. Rev.*, **30**, 313–320.

38. Richter, J.M., Whitefield, B.W., Maimone, T.J., Lin, D.W., Castroviejo, M.P., and Baran, P.S. (2007) *J. Am. Chem. Soc.*, **129**, 12857–12869.

39. (a) Baran, P.S. and Richter, J.M. (2005) *J. Am. Chem. Soc.*, **127**, 15394–15396; (b) Richter, J.M., Ishihara, Y., Masuda, T., Whitefield, B.W., Llamas, T., Pohjakallio, A., and Baran, P.S. (2008) *J. Am. Chem. Soc.*, **130**, 17938–17954.

40. Baran, P.S., Hafensteiner, B.D., Ambhaikar, N.B., Guerrero, C.A., and Gallagher, J.D. (2006) *J. Am. Chem. Soc.*, **128**, 8678–8693.

41. Carroll, A.R., Hyde, E., Smith, J., Quinn, R.J., Guymer, G., and Foster, P.I. (2005) *J. Org. Chem.*, **70**, 1096–1099.

42. Martin, C.L., Overman, L.E., and Rohde, J.M. (2008) *J. Am. Chem. Soc.*, **130**, 7568–7569.

43. Overman, L.E. (2009) *Tetrahedron*, **65**, 6432–6446.

44. Martin, C.L., Overman, L.E., and Rohde, J.M. (2010) *J. Am. Chem. Soc.*, **132**, 4894–4906.

45. Yang, J., Wu, H., Shen, L., and Qin, Y. (2007) *J. Am. Chem. Soc.*, **129**, 13794–13795.

46. Liu, P., Seo, J.H., and Weinreb, S.M. (2010) *Angew. Chem.*, **122**, 2044–2047; *Angew. Chem. Int. Ed.*, **49**, 2000–2003.

47. Zuo, Z., Xie, W., and Ma, D. (2010) *J. Am. Chem. Soc.*, **132**, 13226–13228.

48. Ennis, M.D., Hoffman, R.L., Ghazal, N.B., Old, D.W., and Mooney, P.A. (1996) *J. Org. Chem.*, **61**, 5813–5817.

49. Wilson, R.M. and Danishefsky, S.J. (2007) *J. Org. Chem.*, **72**, 4293–4305.

50. (a) De Mayo, P. (1971) *Acc. Chem. Res.*, **4**, 41–47; (b) Oppolzer, W. (1982) *Acc. Chem. Res.*, **15**, 135–141.

51. Oppolzer, W. and Godel, T. (1978) *J. Am. Chem. Soc.*, **100**, 2583–2584.

52. Gavagnin, M., Carbone, M., Nappo, N., Mollo, E., Roussis, V., and Cimino, G. (2005) *Tetrahedron*, **61**, 617–621.

53. Zhang, Y. and Danishefsky, S.J. (2010) *J. Am. Chem. Soc.*, **132**, 9567–9569.

54. Uchida, I., Ando, T., Fukami, N., Yoshida, K., Hashimoto, M., Tada, T., Koda, S., and Morimoto, Y. (1987) *J. Org. Chem.*, **52**, 5292–5293.

55. (a) Maimone, T.J., Voica, A.F., and Baran, P.S. (2008) *Angew. Chem.*, **122**, 3097–3099;

Angew. Chem. Int. Ed., **47**, 3054–3056;
(b) Maimone, T.J., Shi, J., Ashida, S., and
Baran, P.S. (2009) *J. Am. Chem. Soc.*, **131**,
17066–17067.

56. Evans, D.A. and Chapman, K.T. (1986)
Tetrahedron Lett., **27**, 5939–5942.

57. Saegusa, T., Kobayashi, S., Ito, Y., and
Yasuda, N. (1968) *J. Am. Chem. Soc.*, **90**,
4182.

58. Wessjohann, L.A., Ruijter, E.,
Garcia-Rivera, D., and Brandt, W. (2005)
Mol. Diversity, **9**, 171–186.

59. (a) Kessler, H. (1982) *Angew. Chem.*, **94**,
509–520; *Angew. Chem. Int. Ed. Engl.*, **21**,
512–523; (b) Lambert, J.N., Mitchell, J.P.,
and Roberts, K.D. (2001) *J. Chem. Soc.,
Perkin Trans. 1*, 471–484.

60. Fairlie, D.P., Abbenante, G., and
March, D.R. (1995) *Curr. Med. Chem.*,
2, 654–686.

61. Wenger, R.M. (1984) *Helv. Chim. Acta*, **67**,
502–525.

62. (a) Rich, D.H., Bhatnagar, P.,
Mathiaparanam, P., Grant, J.A., and Tam,
J.P. (1978) *J. Org. Chem.*, **43**, 296–302;
(b) Brady, S.F., Varga, S.L., Freidinger,
R.M., Schwenk, D.A., Mendlowski, M.,
Holly, F.W., and Veber, D.F. (1979) *J. Org.
Chem.*, **44**, 3101–3105.

63. (a) Jiang, W., Wanner, J., Lee, R.J.,
Bounaud, P.Y., and Boger, D.L. (2003)
J. Am. Chem. Soc., **125**, 1877–1887; (b)
Jiang, W., Wanner, J., Lee, R.J., Bounaud,
P.Y., and Boger, D.L. (2002) *J. Am. Chem.
Soc.*, **124**, 5288–5290.

64. For a review on conformation-directed
macrocyclization, see: Blankenstein, J.
and Zhu, J. (2005) *Eur. J. Org. Chem.*,
1949–1964.

65. González, G.I. and Zhu, J. (1999) *J. Org.
Chem.*, **64**, 914–924.

Commentary Part

Comment 1

Kyriacos C. Nicolaou

Organic molecules are key participants in biology, medicine, agriculture, energy, and materials (among other fields). These diverse and wide-ranging industries have differing requirements for the properties of organic molecules. Ultimately, these characteristics are the result of molecular and/or supramolecular structure. Organic synthesis offers access to materials with virtually limitless structural and functional variation, and a means to deliver molecules that either do not exist naturally or cannot be accessed in sufficient quantities. It also serves to validate the assigned structures of complex secondary metabolites, frequently leading to the determination of the absolute configuration and sometimes to structural revisions. In view of the growing demand for useful compounds that can only be accessed through synthesis, such as scarce medicines from the oceans (e.g., ecteinascidin 743 and E7389, two recently approved anticancer drugs) and high-performance designed materials (e.g., organic light-emitting diodes and organic photovoltaics), organic synthesis promises to play an increasing role in the supply of valuable products.

It is little wonder, then, that the art and science of synthesis has been a central discipline of organic chemistry throughout the history of the field, and continues to attract many new practitioners.

Total synthesis is a key inspiration for the discovery and development of new synthetic methods and strategies, and an important yardstick for progress in the field. Although the field has made tremendous strides in the past few decades, there is still significant room for improvements to the state-of-the-art. In order for organic synthesis to address current and future needs, synthetic chemists not only need to be able to produce the desired compounds, but also must do so in an economical and environmentally responsible fashion. The continued development of cascade reactions, catalytic transformations, and novel reactivity, and their application to complex molecule synthesis, will ensure that ever more capable organic molecules can be delivered in a practical and efficient manner.

However, despite the many utilitarian reasons to pursue organic synthesis, total synthesis should not become merely a powerful tool for the delivery of useful products and the verification of structural assignments. Synthesis also provides its own challenges and excitement, and is an art form that deserves to be advanced for its own sake in directions that may or may not have immediate applications. Through advancing the art and science of synthesis, we sharpen our understanding of structure and reactivity and, ultimately, of function and compound properties. Such knowledge is worth pursuing not only because it permits the design of compounds with desirable properties, but also because it provides a clearer understanding of the world around us.

In this chapter, Qian Wang and Jieping Zhu highlight a number of modern aspects of total synthesis through examples that demonstrate its power, elegance and applications, and also current trends in its pursuit. They should be congratulated for their splendid account.

Comment 2

Henry N.C. Wong

It gives me great pleasure to write a Commentary for this chapter by Wang and Zhu, because Qian Wang obtained her PhD degree from the Chinese University of Hong Kong under my supervision. Consequently, her husband Jieping Zhu may therefore be regarded as my academic son-in-law! As indicated by the title, this chapter outlines the total syntheses of several naturally occurring molecules inside the synergistic scope of synthetic methodologies. The five principal themes are the domino process [hirsutellone B, (±)-minfiensine, (−)-phalarine and aspidophytine], multicomponent reaction [(−)-spirotryprostatin B and hirsutine], oxidative anion coupling [hapalindoles, fischerindoles, welwitindolinones, (±)-actinophyllic, (−)-actinophyllic acid and (−)-communesin F], pattern recognition [(±)-aplykurodinone-1 and (±)-vinigrol] and conformation-directed cyclization [cyclosporin and ramoplanin A2]. They are utilized as fundamental routes towards the total syntheses of the natural products listed. Wang and Zhu begin by referring to the famous question, "Why do you want to climb Mount Everest?" In other words, they aimed to give a rational foundation to justify why we want to spend time, manpower, and funding to tackle the usually lengthy total syntheses of structurally very complex molecules. In this connection, I could not agree more with them when they state in the Introduction section, "One might pay particular attention on the selection of target, the planning of the synthetic strategy, and the science that was discovered along the way."

It goes without saying that the total synthesis of natural products is of tremendous value in the training of competent researchers in the pharmaceutical, agricultural, electronic, and food industries. In the process of a total synthesis, we will learn a great deal about carbon–carbon bond formation reactions, execute many functional group interchanges, use a plethora of chemical reagents, and learn the ways to manage and administer logically not only the research undertakings, but also when to start the preparation of starting materials all over again, and when to order more reagents.

After reading Wang and Zhu's chapter, two notable synthetic landmarks immediately strike a chord, namely Nicolaou and co-workers' total synthesis of endiandric acids A–D by a domino process [C1], and Still's total synthesis of periplanone B through conformational control considerations [C2]. Endiantric acids and periplanone B, with relatively little exploration of their chemistry in the 1970s and 1980s, were the very first molecules chosen by Nicolaou and Still in their early careers. History later told us that Nicolaou and Still both turned out to become world renowned after they successfully complete these total syntheses. Lessons to be learnt here are that (i) young synthetic chemists should make a very prudent choice of a novel molecule to work on in their early career and (ii) this synthetic work must also be combined with a sound synthetic methodology. Here I shall try to provide a brief description to explain my viewpoints on how the total syntheses of endiandric acid A and periplanone B were accomplished all the way through amazing strategies (Figure C1).

Endiandric acids are very intriguing naturally occurring molecules. Despite their structural complexity, these molecules are optically inactive. In order to explain this phenomenon, Black and co-workers, who had isolated these molecules [C3], proposed a nonenzymatic biosynthetic pathway that features sequential electrocyclization of achiral polyenes. Nicolaou and co-workers successfully completed the total synthesis of endiandric acid A and its related molecules making good use of Black and co-workers' hypothesis. The key steps are shown in Scheme C1.

Periplanone B, a constituent of excretions from female American cockroaches, was structurally elucidated by Persoons and co-workers in 1976 [C4]. The total synthesis of periplanone B by Still demonstrates aptly his insightful conformational analysis of carbocyclic 10-membered rings [C2].

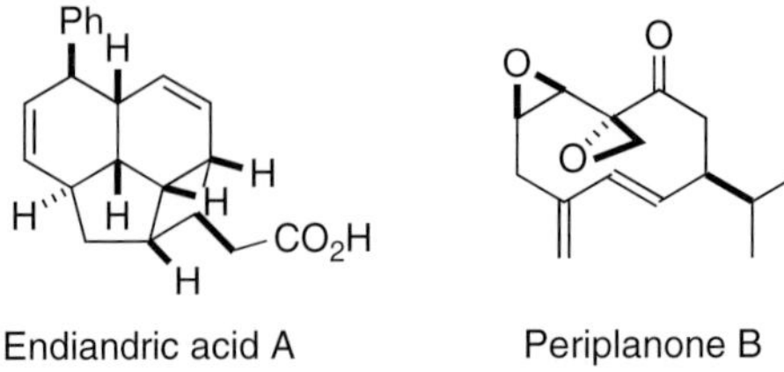

Endiandric acid A   Periplanone B

Figure C1   Structures of endiandric acid A and periplanone B.

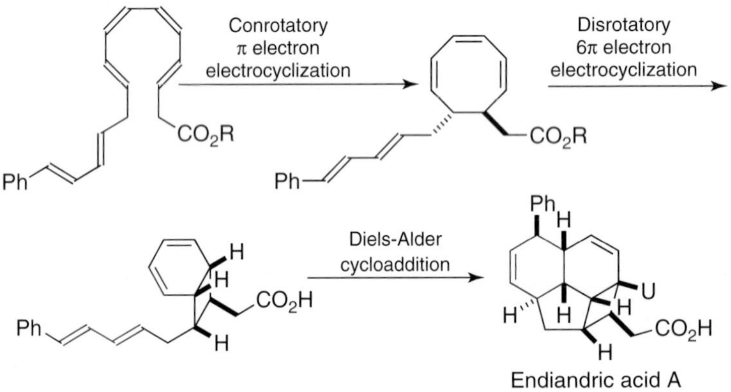

Conrotatory
π electron
electrocyclization

Disrotatory
6π electron
electrocyclization

Diels-Alder
cycloaddition

Endiandric acid A

Scheme C1

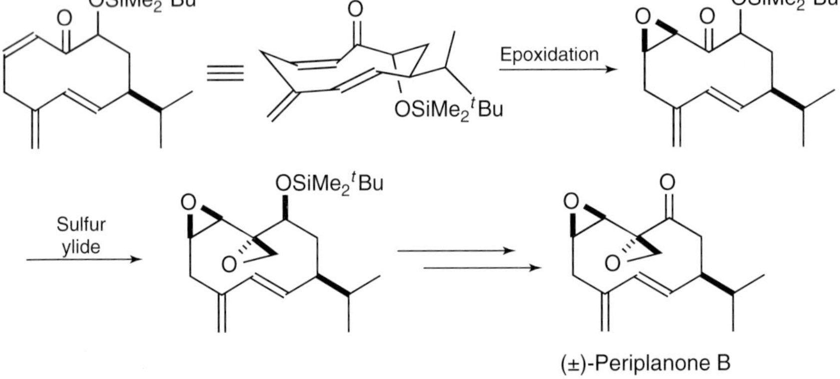

Epoxidation

Sulfur
ylide

(±)-Periplanone B

Scheme C2

As can be seen in Scheme C2, the stereospecific introduction of epoxy groups was based on the conformational preferences of the 10-membered ring substrate.

The aforementioned examples are unquestionably classics in organic synthesis. These syntheses were meticulously steered by very careful planning of the essential steps employing pertinent information on electrocyclization and conformational analysis in the utmost artistic and imaginative way. As shown in this chapter, Wang and Zhu have nonetheless iterated the notion of synergy between total syntheses of natural products and synthetic methodology, and even touched upon physical principles. Also from a historical perspective, we must not forget that Woodward–Hoffmann rules were devised during the course of the total synthesis of vitamin B_{12}. In the light of this fact, it is only natural that the intense research on total syntheses of naturally occurring molecules that has characterized the past century will still play a leading role in chemical science in the years to come. Still in its prime, organic synthesis can be looked upon as the central science within the domain of chemistry, which itself is the central science amongst other basic science subjects such as physics and biology.

Comment 3

Wei-Dong Li

It is my great pleasure to give a Commentary for this chapter by Dr. Wang and Dr. Zhu. I should first congratulate them for this timely and excellent account on the total synthesis of natural products. The account focuses on the synergistic interplay of the target total synthesis and the corresponding synthetic methodology development in five distinct categories: (i) the domino process; (ii) multicomponent analysis; (iii) oxidative anion coupling; (iv) pattern recognition; and (v) conformation-directed cyclization. Each category is discussed in depth with some selected target total synthesis examples accomplished in recent decades. Together with their insightful introduction, this account highlights some valuable perspectives on the future trend of natural product synthesis.

Despite the striking advances in recent decades, significant challenges remain in achieving ideal total synthesis. There are great opportunities for further enabling the science of chemical synthesis in the future. Total synthesis is still the most effective and constitutional tool to (i) gain a greater understanding of the chemical structure and reactivity correlations and (ii) probe the biosynthetic origin of natural product(s), which modern structural instrumental tools cannot reveal equally well [C5]. It is appropriate to emphasize that innovative strategic design is the essential starting point for a rewarding experimental exploration in total synthesis. Ingenious ideas could originate from intuitive pattern recognition of target molecule(s) or orthogonally, biogenetic inspirations (biogenetically patterned), which set a platform for (i) discoveries of new reactivity, mechanistic insights, and even new reactions, and (ii) developments of new synthetic methods, more efficient syntheses, and even new theories of organic chemical reactions and biogenesis of natural product(s) (Figure C2).

I would like to recall here a few historical examples which might illustrate the above viewpoints. Depicted in Scheme C3 is the Woodward total synthesis of erythromycin [C6]. The essence of this synthesis is the recognition of a symmetrically patterned common synthon, which was synthesized enantioselectively via an organocatalytic dynamic asymmetric aldol annulation of a mixture of diastereomeric racemic precursors.

The second example is the Woodward synthesis of prostaglandins via the Corey synthon as shown in Scheme C4 [C7]. Although the initially attempted cyclooctatetraene strategy did not deliver the equivalent Corey synthon, a facile intramolecular Rauhut–Currier annulation (or vinylogous intramolecular Morita–Baylis–Hillman condensation) was discovered (Scheme C4a) [C8]. A remarkably stereocontrolled and efficient *cis*-1,3,5-cyclohexanetriol strategy was evolved subsequently (Scheme C4b) in view of a structurally patterned equivalent tricyclic synthon.

Serial synthetic studies on prostaglandins by Corey's group have resulted in biogenetically patterned strategies as illustrated in Scheme C5 [C9]. Although the exact mechanistic pathway for the formal cyclopentadienyl cation cyclization has still not been ascertained (Scheme C5a, cation–ene versus Nazarov) [C9a], the implication for further investigation is clear. Interestingly, the biomimetic study of the synthesis of the marine prostanoid preclavulone [C9b] via an oxy-Nazarov

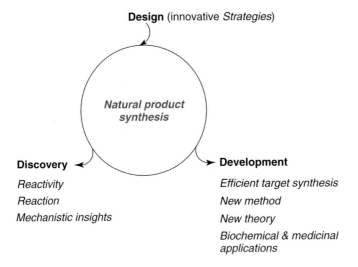

Design (innovative Strategies)
Natural product synthesis
Discovery
Reactivity
Reaction
Mechanistic insights
Development
Efficient target synthesis
New method
New theory
Biochemical & medicinal applications

Figure C2 Strategy-driven Discovery and Development.

Diastereomeric mixture
D-Proline
70%
36% ee
36% ee
– H2O
recryst.
99% ee
OBn
RO
Erythromycin

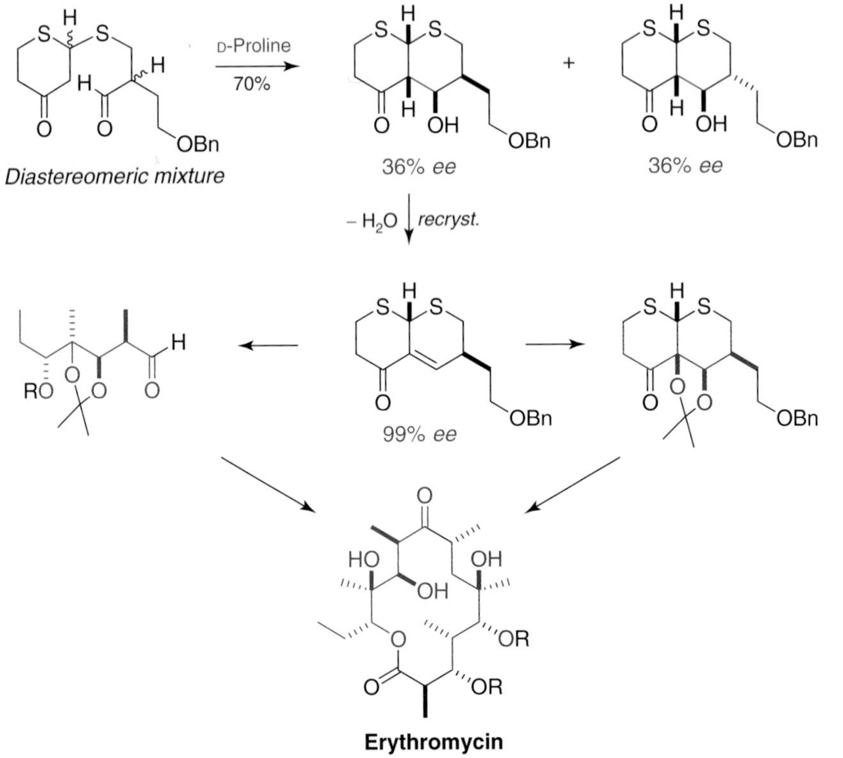

Scheme C3 Woodward synthesis of erythromycin (1981).

cyclization [C10] of a vinylallene oxide model substrate provided firm evidence on the biogenetic origins of such marine prostanoids (Scheme C5b).

References

C1. (a) Nicolaou, K.C., Petasis, N.A., Zipkin, R.E., and Ueishi, J. (1982) *J. Am. Chem. Soc.*, **104**, 5555; (b) Nicoloau, K.C., Petasis, N.A., Uenishi, J., and Zipkin, R.E. (1982) *J. Am. Chem. Soc.*, **104**, 5557; (c) Nicolaou, K.C., Zipkin, R.E., and Petasis, N.A. (1982) *J. Am. Chem. Soc.*, **104**, 5558; (d) Nicolaou, K.C., Petasis, N.A., and Zipkin, R.E. (1982) *J. Am. Chem. Soc.*, **104**, 5560.

C2. Still, W.C. (1979) *J. Am. Chem. Soc.*, **101**, 2493.

C3. (a) Bandaranayake, W.M., Banfield, J.E., Black, D.St.C., Fallon, G.D., and Gatehouse, B.M. (1980) *J. Chem. Soc., Chem. Commun.*, 162; (b) Bandaranayake, W.M., Banfield, J.E., and Black, D.St.C. (1980) *J. Chem. Soc., Chem. Commun.*, 902; (c) Bandaranayake, W.M., Banfield, J.E., Black, D.St.C., Fallon, G.D., and Gatehouse, B.M. (1981) *Aust. J. Chem.*, **34**, 1655; (d) Bandaranayake, W.M., Banfield, J.E., and Black, D.St.C. (1982) *Aust. J. Chem.*, **35**, 557; (e) Bandaranayake, W.M., Banfield, J.E., Black, D.St.C., Fallon, G.D., and Gatehouse, B.M. (1982) *Aust. J. Chem.*, **35**, 567; (f) Banfield, J.E.,

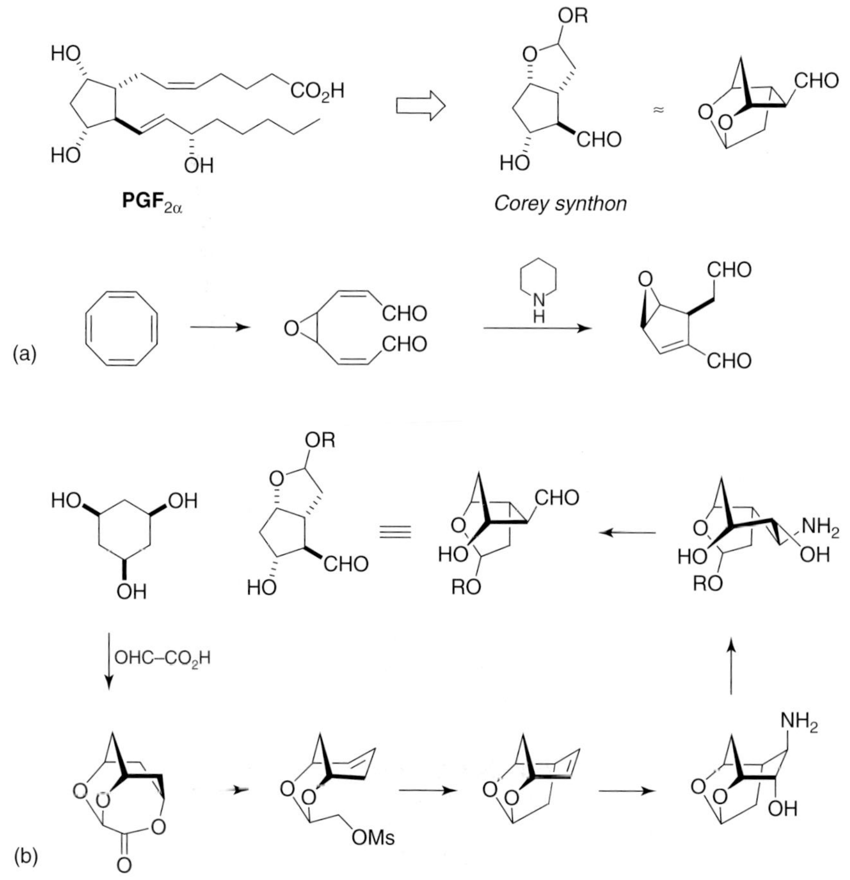

Scheme C4 Woodward synthesis of prostaglandins (1973).

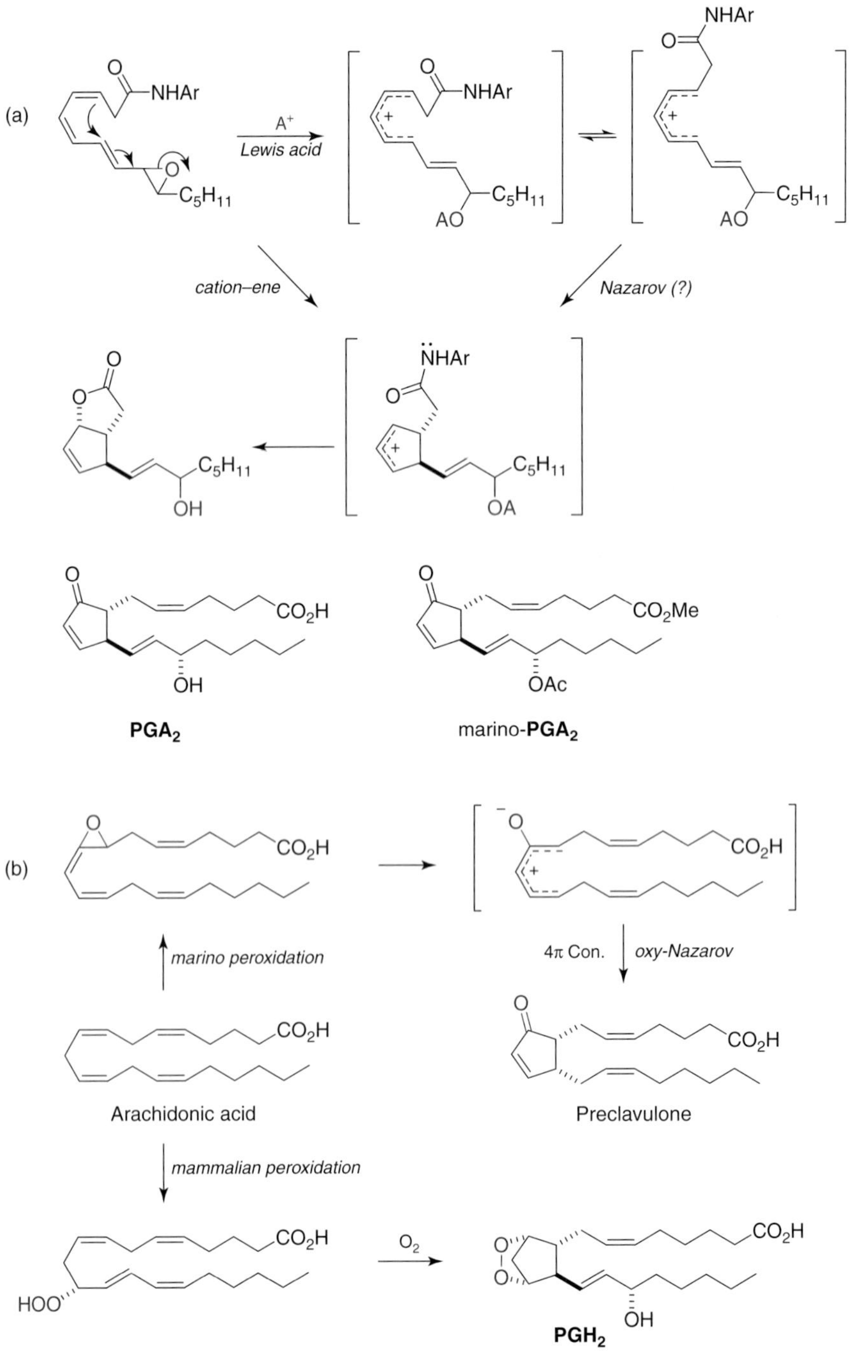

Scheme C5 Corey biomimetic syntheses of prostanoids (1973, 1987).

Black, D.St.C., Johns, S.R., and Willing, R.I. (1982) *Aust. J. Chem.*, **35**, 2247.

C4. (a) Persoons, C.J., Verwiel, P.J.F., Ritter, F.J., Talman, F.E., Nooijen, P.F.J., and Nooijen, W.J. (1976) *Tetrahedron Lett.*, 2055; (b) Talman, E., Verwiel, P.E.J., Ritter, F.J., and Persoons, C.J. (1978) *Isr. J. Chem.*, **17**, 227; (c) Persoons, C.J., Verwiel, P.E.J., Talman, E., Ritter, F.J. (1979) *J. Chem. Ecol.*, **5**, 219.

C5. Woodward, R.B. (1956) in *Perspectives in Organic Chemistry* (ed. A. Todd), Interscience, New York, pp. 155–184.

C6. Woodward, R.B., Logusch, E., Nambiar, K.P., Sakan, K., Ward, D.E., Au-Yeung, B.W., Balaram, P., Browne, L.J., Card, P.J., Chen, C.H., Chênevert, R.B., Fliri, A., Frobel, K., Gais, H.J., Garratt, D.G., Hayakawa, K., Heggie, W., Hesson, D.P., Hoppe, D., Hoppe, I., Hyatt, J.A., Ikeda, D., Jacobi, P.A., Kim, K.S., Kobuke, Y., Kojima, K., Krowicki, K., Lee, V.J., Leutert, T., Malchenko, S., Martens, J., Matthews, R.S., Ong, B.S., Press, J.B., Rajan Babu, T.V., Rousseau, G., Sauter, H.M., Suzuki, M., Tatsuta, K., Tolbert, L.M., Truesdale, E.A., Uchida, I., Ueda, Y., Uyehara, T., Vasella, A.T., Vladuchick, W.C., Wade, P.A., Williams, R.M., and Wong, H.N.C. (1981) *J. Am. Chem. Soc.*, **103**, 3210–3213.

C7. Ernest, I. (1976) *Angew. Chem. Int. Ed. Engl.*, **15**, 207–214.

C8. For relevant discoveries, see: (a) Wang, L.C., Luis, A.L., Agapiou, K., Jang, H.Y., and Krische, M.J. (2002) *J. Am. Chem. Soc.*, **124**, 2402–2403; (b) Frank, S.A., Mergott, D.J., and Roush, W.R. (2002) *J. Am. Chem. Soc.*, **124**, 2404–2405.

C9. (a) Corey, E.J., Fleet, G.W.J., and Kato, M. (1973) *Tetrahedron Lett.*, **14**, 3963–3966; (b) Corey, E.J., Ritter, K., Yus, M., and Nájera, C. (1987) *Tetrahedron Lett.*, **28**, 3547–3550.

C10. Li, W-D.Z., Duo, W-G., and Zhuang C-H. (2011) *Org. Lett.*, **13**, 3538–3541.

3
Interplay Between the Chemical Space and the Biological Space

Ren-Xiao Wang

3.1
Chemical Biology: Historical and Philosophical Aspects

3.1.1
The Chemical Space and the Biological Space

By definition, the chemical space encompasses all possible organic and inorganic substances and compounds. The total number of possible small organic molecules that populate the chemical space has been estimated to exceed 10^{60} [1]. In comparison with the number of such molecules that scientists have identified or made, this amount is so vast that it may be considered as virtually infinite. In fact, there are just around 64 million ($<10^8$) organic and inorganic substances registered with the Chemical Abstracts Service (*http://www.cas.org/*) as of January 2012. Hence our exploration of the possible chemical space has so far been extremely limited.

The middle of the twentieth century witnessed the birth of modern molecular biology, which was marked by the historical discovery of the double-helix structure of DNA molecules by James D. Watson and Francis Crick. This paved the way for scientists who started to study biological systems and biological events at the molecular level. It also effectively broke the conceptual boundary between biological systems and chemical systems. A major concern in molecular biology is to understand the interactions between various biological molecules, such as nucleic acids and proteins, and how these interactions are *regulated*. The most outstanding finding is the so-called "central dogma" (Figure 3.1). The central dogma of molecular biology describes the transfer of sequence information between sequential information-carrying biopolymers in living organisms. Generally, DNA can be copied to DNA (DNA replication), DNA information can be copied into mRNA (transcription), and proteins can be synthesized using the information in mRNA as a template (translation). This information flow generally exists in living organisms except in some viruses or under specific conditions.

The missing chapter in the central dogma, from the viewpoint of a chemist, is small molecules. In fact, small molecules play critical roles at all levels of biological complexity. They are key elements of a range of topics in the life sciences, including the origins of life, memory and cognition, sensing and signaling, understanding cell circuitry, and

Organic Chemistry – Breakthroughs and Perspectives, First Edition. Edited by Kuiling Ding and Li-Xin Dai.
© 2012 Wiley-VCH Verlag GmbH & Co. KGaA. Published 2012 by Wiley-VCH Verlag GmbH & Co. KGaA.

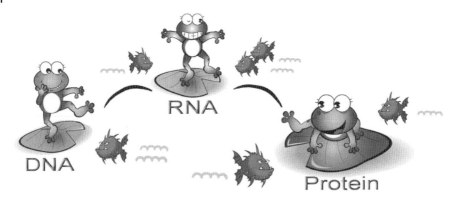

Figure 3.1 From the viewpoint of a chemical biologist, small molecules fill up the space left by the central dogma.

treating disease [2]. Life would not exist with biological macromolecules alone. There are also a range of connections between small molecules and the three major families of biological macromolecules. Small molecules have been synthesized that bind and modulate DNA [3], and DNA has been used as a template to synthesize small molecules [4], small-molecule riboswitches regulate RNA function [5], and certain small-molecule antibiotics bind and inhibit RNA [6]. Small molecules are synthesized in cells by protein-based catalysis machines [7], and both natural and non-natural small molecules have served as powerful probes in biology and as drugs in medicine [8]. Chemical biology, as a multidisciplinary field acting on the interplay between the chemical space and the biological space, has substantially expanded our understanding of the elements of life.

Just as much of the astronomical space is void, so much of the chemical space contains nothing of biological interest [9]. The small-molecule compounds used by biological systems represent a staggeringly small fraction of the total possible number of chemical compounds with molecular weights in the same range as those in biological systems. It seems that the total number of such small-molecule compounds within our own bodies could be just a few thousand [10]. It is remarkable that so many complex processes can be carried out with such a limited number of molecules. It is also clear that the "biologically relevant" chemical space is only a tiny fraction of the complete chemical space. hence it is rather challenging to identify chemical compounds that can effectively modulate biological processes. It is challenging because the enormous size of the chemical space makes a thorough exploration impossible. It is challenging also because we still lack an essential understanding of many aspects of the biological space. After all, molecular biology was born only about half of a century ago, and sequencing the entire genome became possible less than 20 years ago. Today, mankind has reached the deep space by a distance of many light years, but we are still at a very early stage towards the ultimate goal of revealing the secret of life.

Given all these difficulties, a key question for chemical biology is how to direct our efforts towards certain regions of chemical space that are most likely to contain molecules with useful biological implications. An answer to this question will help us best use the powerful new methods that are emerging to probe biological systems, both to understand the fundamental processes of life and to develop new strategies to treat disease. Obviously, joint efforts by biologists and chemists across traditional disciplines are necessary in order to seek the answer.

3.1.2
Historical Aspects of Chemical Biology

The term "chemical biology" officially became popular in 1990s, mainly due to the contributions of Stuart L. Schreiber, Peter Schultz, and a number of other distinguished scientists. Since then, chemical biology has quickly grown as a research frontier on an international scale. Interestingly, chemical biology actually has historical roots back to the birth of chemistry and biology as distinct sciences at least two centuries ago. Early organic chemistry focused on synthesizing compounds isolated from living systems, and that was exactly why the term "organic" was used to name this discipline. An accidental discovery was made by Friedrich Wöhler (1800–1882) in 1828. He heated a solution of silver cyanate and ammonium chloride in an attempt to obtain ammonium cyanate. Separately, he also heated lead cyanate and aqueous ammonia. In both cases, he obtained not the expected product but urea (Scheme 3.1). Wöhler's experiment showed that inorganic starting materials could be used to synthesize substances previously associated only with living organisms. This remarkable discovery has been respected by most people as the birth of modern organic chemistry. More importantly, it revealed the chemical basis of life, which is still the essential basis of the chemical biology studies today.

Another notable example illustrating the early history of chemical biology is the synthesis of aniline dyes in the nineteenth century. With the technical advances in microscopy at that time, these chemical dyes helped to uncover fine structures within cells, which contributed directly to the development of early cell biology. As a consequence, Rudolf Virchow (1821–1902) published *Die Cellularpathologie* in 1858. In this book, he established two key theories: first, all cells descend from other living cells; and second, cellular changes can result in disease. Virchow's theory of cellular pathology transformed the scientific understanding of biology and revolutionized the field of medicine.

By the early twentieth century, it was substantiated that pure chemical substances may have biological activity. Some researchers began to suggest that plant extracts that affect animals contain a single active ingredient that acts on a discrete part of the animal.

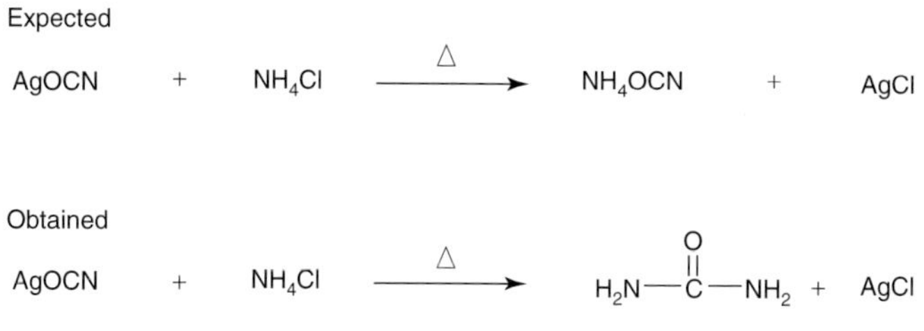

Expected

$$AgOCN \quad + \quad NH_4Cl \quad \xrightarrow{\Delta} \quad NH_4OCN \quad + \quad AgCl$$

Obtained

$$AgOCN \quad + \quad NH_4Cl \quad \xrightarrow{\Delta} \quad H_2N-\overset{\overset{\displaystyle O}{\|}}{C}-NH_2 \quad + \quad AgCl$$

Scheme 3.1 Wöhler's synthesis of urea: yielding a chemical previously isolated only from living organisms, this work represented a landmark achievement in both organic synthesis and chemical biology.

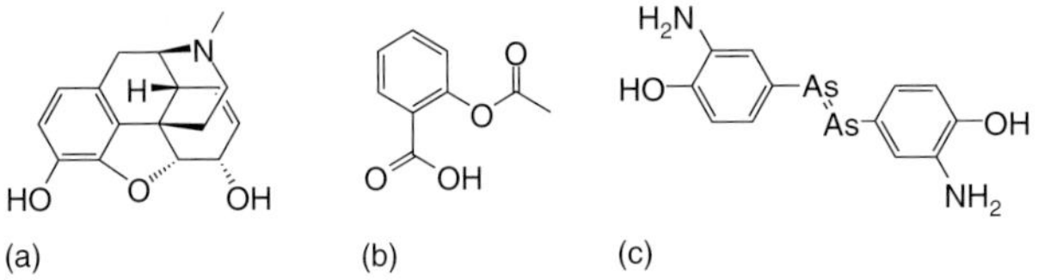

(a)   (b)   (c)

Figure 3.2 Examples of small-molecule active compounds extracted from natural substances: (a) morphine and (b) a derivatized natural product, aspirin. As a synthesized compound, Salvarsan (c) demonstrated the efficacy of the magic bullet approach to treating diseases. It is also known as "Ehrlich's 606," named after 605 preceding failed compounds studied by Paul Ehrlich.

Subsequently, Frederick Serturner succeeded in isolating the first active compound, morphine (Figure 3.2a), from opium. Several useful small molecules, such as aspirin (Figure 3.2b), were also discovered, showing that even very simple organic molecules could have a range of interesting and useful biological effects. Before long, it was accepted that simple organic molecules could profoundly affect cellular and organismal systems, paving the way for a more comprehensive chemical genetic approach to understanding biological systems.

However, the crucial breakthrough came at the beginning of the twentieth century, when Paul Ehrlich (1854–1915) developed both the term for, and the concept of, a "receptor" – the single, specific protein target of a small molecule [11]. Ehrlich also devoted his early work to applying chemical approaches to the visualization of living cells. He experimented with the newly discovered aniline dyes derived from coal tar and found that some dyes could differentially stain specific cells and tissues. He correctly interpreted that this difference resulted from a chemical reaction of the dyes with specific substances within the cells. Applying the aniline methylene blue dyes for diagnostics, Ehrlich identified a tiny rod-shaped bacterium responsible for tuberculosis. However, what made Ehrlich outstanding compared with other researchers at that time was that his insight into the intracellular reduction of various synthetic dyes led to amazing leaps in medicine, pioneering the earliest form of chemotherapy. For example, Ehrlich proposed the development of "magic bullets" or toxins capable of targeting specific pathogens. In the first example of a magic bullet, Ehrlich discovered Salvarsan (arsphenamine) to treat syphilis (Figure 3.2c). Salvarsan soon became a blockbuster and replaced mercury, which was previously used as the therapy. The success of Salvarsan opened up a new era of drug discovery through synthesized chemicals.

The selected examples given above illustrate the historical aspects of chemical biology. This type of research was redefined in the 1990s and still continues to expand. This field, of course, has been completely innovated by new technologies, but it seems that the basic philosophy established in those early studied has not altered much. Many current chemical biologists emphasize the use of small molecules to regulate cellular processes. This approach to chemical biology is familiar to its pioneers, such as the magic bullet approach put forward by Ehrlich. Similarly, using chemistry to enhance imaging has been a longstanding interest of chemical biologists since Anna Atkins (1799–1871). Reviewing the history of chemical biology thus helps us to understand the essential trends in this field and to acknowledge the effectiveness of chemical biology approaches which have advanced both chemistry and biology.

As indicated above, chemistry has always been deeply interested in understanding the nature of life. However, this problem proved to be too complex for several centuries, leaving most chemists dwelling in the chemical space, that is, exploration of individual reactions and simple systems that could be studied in molecular detail. This situation has changed. The great technical advances made in the past several decades, such as reliable protein expression and purification techniques, characterization of proteins by mass spectrometry, high-resolution structural determination by X-ray and NMR techniques, biomolecular syntheses, and many others, have enabled chemists to manipulate and characterize biological molecules with greater facility than small molecules could be manipulated and characterized 50 years ago. Genome sequence, transcriptome and proteome analysis, and powerful new visualization techniques make it possible to obtain information in a cellular context. These advances allow chemists to study the cell in unprecedented detail. Not only organic chemists and medicinal chemists but also physical, inorganic, and analytical chemists find themselves in the mainstream of chemical biology research. Collectively, their efforts have changed the way in which we perceive biology and have opened up almost infinite opportunities into the future.

As a result, over the past 20 years there has been a steady increase in biologically related research carried out in chemistry departments. Numerous universities have developed chemical biology programs that cross traditional departmental lines. Funding agencies worldwide have identified the chemistry–biology interface as a target for investment. For example, the US National Institutes of Health (NIH) Roadmap (*http://nihroadmap.nih.gov*) underscores the importance of collaborations between chemists and biologists to the future of biomedical research. Several high-quality professional journals, such as *Nature Chemical Biology*, *ACS Chemical Biology*, and *Current Opinion in Chemical Biology*, have been published since early 2000s. The increasing number of chemical biology conferences also highlights the importance of this field not only for the researchers in academia but also those in industry.

3.1.3
Scope of Today's Chemical Biology

Before we can proceed with the remainder of this chapter, it is necessary to discuss the scope of chemical biology. *Nature Chemical Biology*, one of the leading journals in chemical biology, defines chemical biology as *"both the use of chemistry to advance a molecular understanding of biology and the harnessing of biology to advance chemistry"* [12]. Indeed, chemical biology spans the fields of chemistry and biology, which involves the application of chemical techniques and tools, often compounds produced through synthetic chemistry, to the study and manipulation of biological systems. Many of today's chemical biology studies could be given the conventional labels of bioorganic chemistry, medicinal chemistry, pharmacology, cellular biology, or even genetics. However, if one remembers that chemical biology aims at the interplay between the chemical space and the biological space, one will understand the philosophic difference between chemical biology and those overlapping disciplines.

In particular, the difference between chemical biology and overlapping disciplines needs to be explained very clearly. It is not just a linguistic challenge but serves as a good showcase for illustrating the focus of chemical biology. Bioorganic chemistry, which applies synthetic and physical organic chemistry to biological questions, is the primary disciplinary precursor of chemical biology. Bioorganic chemists used their ability to synthesize almost any desired molecule as a means to understand chemical mechanisms in biological systems with great success. For example, they provided robust automated methods for the synthesis of nucleotides, peptides, and oligosaccharides and new fundamental insights into enzymatic catalysis through the use of "biomimetic" enzyme models. Along with bioorganic chemistry, biological chemistry (or biochemistry) stands as one of the most influential intellectual threads in modern chemical biology. In brief, chemical biology deals with how chemistry can be applied to study and solve biological problems; whereas biological chemistry studies the chemistry of biological systems. In general, chemical biology focuses on small molecules; whereas biological chemistry focuses on proteins, nucleic acids, and other typical organic molecules found in living systems. In the pharmaceutical and biotech industries, a chemical biologist would be the person who designs and synthesizes the drug candidates, whereas a biological chemist would be the person who thinks more about biological sense and designs assays to test the drug candidates.

Some other terms are also widely used by researchers to describe chemical biology studies. The most common one is "chemical genetics." This term emphasizes the study of gene-product function in a cellular or organismal context using exogenous chemical ligands [13], which is named after classic genetics. Classical genetics explores the function of unknown genes by comparing the normal phenotype with that of knockouts, or through the overexpression of target genes. In chemical genetics, a chemical compound may specifically inhibit or activate single or multiple target proteins. Therefore, the compound is equivalent in conventional genetics to a gene knockout or to the overexpression of a gene. Similarly, if chemical genetics is conducted on the genome level, it then becomes "chemical genomics." Exploring gene functions is, of course, the central goal of chemical biology research. However, one should keep in mind that chemical biology has a broader scope by definition. For the sake of convenience, we will not always try to differentiate "chemical biology" and "chemical genetics" in this chapter. The more appropriate term will be used according to the context.

In the literature, chemical genetics is classified into two basic approaches, that is, forward chemical genetics and reverse chemical genetics, which are named after classical genetics (Figure 3.3). Forward genetics operates from phenotype (physically apparent characteristic) to genotype (genetic sequence) and requires no prior knowledge of specific gene(s). It studies changes in phenotype such as morphology, growth, or behavior resulting from random genomic mutations or deletions induced from radioactive or chemical mutagenesis, and then it identifies the gene responsible through mutation mapping. Forward chemical genetics mimics this approach by substituting random mutagenesis with a library of chemical compounds as protein function regulators. The first step in both forward genetics and forward chemical genetics is to screen for changes induced by either inhibition or stimulation of the function of a protein and both go on to explore the genetic origin. Reverse genetics is a relatively later development in genetics

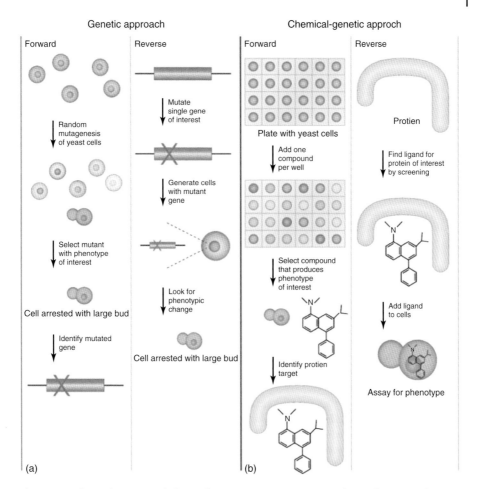

Figure 3.3 Classical genetic and chemical genetics approaches identify genes and proteins that regulate biological processes. (a) Forward genetics entails introducing random mutations into cells, screening mutant cells for a phenotype of interest, and identifying mutated genes in affected cells. Reverse genetics entails introducing a mutation into a specific gene of interest and studying the phenotypic consequences of the mutation in a cellular or organismal context. (b) Forward chemical genetics entails screening exogenous chemical compounds in cells, selecting compounds that induce a phenotype of interest, and identifying the protein target of such compounds. Reverse chemical genetics entails overexpressing a protein of interest, screening for compounds for the protein, and using the ligand to determine the phenotypic consequences of altering the function of this protein in a cellular context. Reproduced from [13].

with the advances in molecular biology techniques. In contrast to forward genetics, reverse genetics operates from genotype to phenotype (Figure 3.3). Reverse genetics begins with a selected gene, manipulates it to produce an organism harboring the mutated gene, and characterizes the phenotypic differences between the mutant and the wild-type organisms. Similarly, reverse chemical genetics begins with a known protein.

An appropriate assay related to this known protein is then employed to screen a vast pool of chemical compounds to identify those that can either inhibit or stimulate the target protein. Once some active compounds have been identified, they can be introduced into a cell or organism, as an analog to genetic mutation, and the resulting changes in phenotype are studied.

3.1.4
Forward Chemical Genetics

In reverse chemical genetics, selected chemical libraries are mainly tested on a single target protein (or a group of target proteins) which are selected based on known information. In contrast, forward chemical genetics studies are conducted on whole cells or organisms, and therefore compounds are screened against multiple potential targets simultaneously. Ideally, a successful forward chemical genetics study will identify a novel functional gene (the target protein) and its on/off switch (the small-molecule regulator), representing an efficient "two-birds-with-one-stone" approach.

Compared with classical genetics, forward chemical genetics has several obvious advantages. First, the use of chemical tools offers a greater level of ease and flexibility than classical genetic manipulations. Classical genetic techniques are relatively difficult to employ, especially in mammals, because of their diploid genome, physical size, and slow reproduction rate. In contrast, forward chemical genetics approaches may be applied to any complex cellular or animal models without time-consuming genetic modifications. Chemical genetics also allows one to practice "multiple knockouts" by adding multiple specific ligands, which is a situation often described as a "nightmare" by geneticists [14]. Very importantly, one can even operate in the relevant context of human cells/tissues under physiological condition with chemical tools. This task is often impractical for conventional genetics techniques for technical and sometimes ethical reasons.

Second, classical genetic knockouts in principle delete the protein entirely from the organism under experiment [15]. The organism may mask the consequential phenotype through related gene functional compensation. Therefore, it can be difficult to separate the effects that develop from the deletion from those that develop from merely a particular function of the gene [16]. It is possible that one protein is associated with multiple biological functions. Chemical genetics, ideally, can isolate and dissect particular functions of that protein while leaving others intact [17]. In addition, if a gene is essential for survival or development, a total knockout may abolish the chance to study the later stage function of that gene because the deletion may be lethal [18]. On the other hand, chemical genetics in principle allows the use of chemical probes in non-lethal doses to avoid full lethality [19].

Another advantage of chemical genetics is the possibility of real-time control. Chemical genetics allows for this control by the ability to introduce rapidly a cell-permeable chemical regulator at any stage that can yield the desired phenotype as quickly as diffusion-limited kinetics will allow. The chemical regulator acts as a "switch" that turns on/off the event under study almost in real time and allows for kinetic *in vivo* analysis, which is something usually not possible in classical genetics. Although temporal control is available in classical genetic studies through conditional alleles, such as temperature-sensitive mutations, they often have unwanted broad side effects that may interfere with the desired result [20]. The antisense oligonucleotide and RNA interference (RNAi) are other popular alternatives for conditional knockouts [16] that work by inhibiting the synthesis of the target protein from mRNA. However, as their effects are delayed until all existing proteins are degraded, they are particularly unsuitable for time-sensitive studies, such as signal transduction, that occur on the milliseconds to hours time scale.

Despite all of these advantages, chemical genetics cannot replace classical genetics completely. One of the greatest advantages of classical genetics is the incredible precision of a gene knockout. Although some chemical ligands can be specific switches with specificity close to that of a gene knockout, the low specificity of many ligands often gives "off-target" effects in which the probe may interact with proteins other than the protein targeted. This low specificity makes defining specific protein functions very difficult because these off-target effects may lead to toxicity or false or unwanted positive/negative biological results. In addition to this lack of specificity, chemical genetics cannot yet match the generality of genetics. Geneticists can, in theory, knock out any gene as long as the genome information is complete for the given species. At present, this goal is beyond the ability of chemical geneticist. Hence chemical genetics and classical genetics may complement each other well [21]. An effective integration of the two techniques will eventually lead to a thorough understanding of life.

A forward chemical genetics study not only will elucidate and characterize a target functional gene but also offers a potential lead compound that may eventually be developed into a candidate drug after necessary follow-up modifications by medicinal chemists. Nevertheless, there is a notable difference between a chemical probe applicable to chemical genetics and a successful candidate drug. Unlike drug development where specificity is critical, a chemical probe applied in a chemical genetics study does not need to be highly specific provided that they give an identifiable phenotype that allows for the deciphering of the function of target protein(s). Although one may desire compounds with affinity in the sub-nanomolar range capable of producing the desired effect, compounds of low micromolar affinities are actually often adopted as acceptable candidates in chemical genetics [22]. A lead compound developed in drug discovery that may not possess the desired pharmacokinetic properties may still be used as a qualified probe in chemical genetics studies [17]. In fact, the lower pharmacokinetic property requirements for chemical genetics probes compared with drug candidates allow for the use of a greater variety of functional groups and for maximization of the chemical space in library constituents [23]. Because of the notable difference between a chemical genetics probe and a successful candidate drug, pharmaceutical companies are not making a full commitment to forward chemical genetics studies, leaving the field to be developed by the academic community so far.

Forward chemical genetics studies require three basic components: a collection of organic compounds, a biological assay with a quantifiable phenotypic output, and a strategy for identifying the target(s) of the active compounds. Accordingly, the following sections describe these essential aspects, including the preparation of chemical toolboxes, phenotypic screening, and target identification and validation.

3.2
Preparation of Chemical Libraries

Collections of compounds (libraries) are the very starting point for chemical genetics studies. In many cases, such chemical libraries are derived from the natural products of different organisms, largely plants and bacteria, and their derivatives. Synthesized compounds are, of course, another major source for chemical libraries. With the immense developments in combinatorial methods over the past two decades, huge arrays of new molecules can be produced in relatively short periods of time [24].

3.2.1
Natural Product-Inspired Synthesis

Chemical substances derived from animals, plants, and microbes have been used to treat human disease since the dawn of medicine. For example, colchicum extract, which contains colchicine, has been used as a treatment for gout for over 2000 years. Natural products are still major sources of innovative therapeutic agents for infectious diseases, cancer, lipid disorders, and immunomodulation. It is generally accepted that collections of natural products have a higher probability of delivering hits than a typical synthetic combinatorial library. The rich structural diversity and complexity of natural products have prompted synthetic chemists to produce them in the laboratory, often with therapeutic applications in mind, and many drugs used today are natural products or their derivatives. For example, among the 983 new chemical entities approved as drugs by the US Food and Drug Administration (FDA) between 1981 and 2006, about 5.7% were natural products, and a further 27.6% were natural product derivatives [25, 26].

Many academic groups consider natural products to be ideal targets for testing their synthetic methodology, and many remarkable achievements have been documented. This field is reviewed in-depth in Chapters 1 and 2, and therefore will not be repeated here. In the course of these heroic efforts, it is often possible to define the crucial structural elements of natural products required for maintaining the specific biological activity. For example, peptides are modular structures joined together by amide (peptide) bonds. Small peptides, which often contain non-protein amino acids, are typically assembled by multi-enzyme complexes referred to as non-ribosomal peptide synthases, and in many instances these compounds are cyclized. These peptides, which represent a major class of biologically active natural products, are particularly amenable to parallel synthetic methodologies owing to the repetitive nature of the bond-forming process. Systematic variation of the individual amino acid residues on their side chains therefore allows the pinpointing of structural features essential for biological activity. An elegant application of a combination of such processes was used to probe the crucial features of HUN-7293, a naturally occurring cyclic depsipeptide that is a potent and selective inhibitor of cell-adhesion molecule expression. In an approach referred to as systematic chemical mutagenesis (Figure 3.4), Boger and co-workers explored the effect on biological activity (both potency and selectivity) of simplifying each of the seven residues of HUN-7293, including removal of N-methyl groups [27].

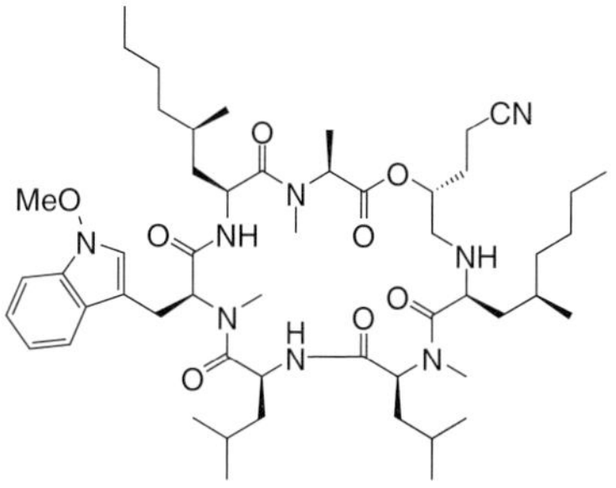

Figure 3.4 HUN-7293, a naturally occurring cyclic depsipeptide.

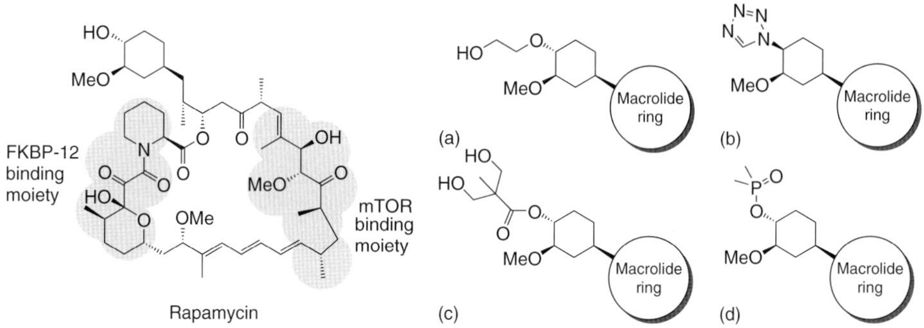

HUN-7293

Figure 3.5 Chemical modifications on rapamycin.

Another well-known case is the chemical modification on rapamycin (Figure 3.5), which has led to three clinical drugs [(a) sirolimus; (b) everolimus; (c) temsirolimus] and one in phase II clinical trials [(d) deforolimus]. In all cases, modifications were made at one site, the C43 alcoholic hydroxyl group that avoids both the FKBP-12 and the target of rapamycin (mTOR) binding sites, since modifications in other areas would disable the basic biological activity of this molecule [28].

Understanding the essential interactions of the natural products and their target can also lead to novel mimetics. An example of this approach was reported by Hirschmann and co-workers, who developed mimetics of cryptophycin, a natural product with antitumor potency, utilizing an azepine scaffold to resemble the overall geometry of cryptophycin [29]. A synthetic strategy was developed that allowed a class of peptidomimetic (Figure 3.6)

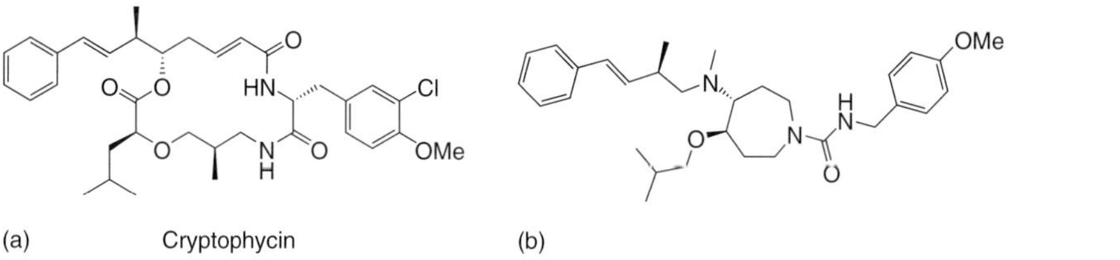

(a)   Cryptophycin   (b)

Figure 3.6 (a) Cryptophycin and (b) its mimetics developed by Hirschmann and co-workers [29].

to be prepared in reasonable overall yield. These compounds were designed to replace the 16-membered macrolide ring with a seven-membered azepine ring for attachment of the cryptophycin side chains with the required spatial orientation to mimic the conformation of the relevant region of cryptophycin. The stereochemical arrangement of the side chains and side chain composition were explored to optimize the desired biological responses. Some compounds obtained were tested for *in vitro* cytotoxicity against six human cancer cell lines, and the results suggested that bioactive cryptophycin analogs can be designed by altering the central cyclic structure of cryptophycin.

3.2.2
Diversity-Oriented Synthesis

Chemists have probably accepted it as a basic rule that natural products are "different" from synthetic compounds. As a notable new concept, diversity-oriented synthesis (DOS) has now become popular with a number of synthetic chemists, involving compounds resembling natural products in terms of their complexity. The precise origin of the term "diversity-oriented synthesis" is a matter of dispute. Certainly it was used in Schreiber's group in the late 1990s [24], and the concept was used by Nicolaou and co-workers in the same period as exemplified by their reports on benzopyran libraries [30–32]. However, what is certain is that DOS has been very popular since then. To give an idea of the great popularity of this approach, over 700 articles published from 2000 to 2010, among which 160 are reviews, have the term "diversity-oriented synthesis" in the title or key words according to the search results given by the ISI Web of Science (*http://apps.isiknowledge.com/*). DOS is also reviewed in Chapter 1. Here, we only mention the biological implications of some representative DOS studies.

Currently, DOS-sourced molecules have been tested in a large number and variety of biological assays in order to determine their potential as chemical probes or drug candidates. This is probably best demonstrated by the work of Nicolaou's group, reported in a series of papers in *Journal of the American Chemical Society* in 2000. Based on the core structure of benzopyran or partially reduced benzopyran, a series of iterative molecules were obtained through combinatorial syntheses and were tested in a wide range of biological assays (Figure 3.7). To date, four distinct, previously unrecognized biological activities have been reported from this relatively small series of compounds. These currently include an inhibitor **3** of NADH/ubiquinone oxidoreductase with cytostatic activity against specific cell lines [33], a compound **4** with antibacterial activity against methicillin-resistant *Staphylococcus aureus* [34], nonsteroidal farnesoid X receptor (FXR) agonists **5a–c**, which helped define the interactions within this receptor for the first time [35, 36], and an inhibitor **6** of hypoxia-inducible factor-1R (HIF-1R) [37]. These are probably the most diverse activities yet shown from a single-base natural product structure, and it will be very interesting to see how many more biological results will be reported from these series in due course. The potential for such synthetic strategies is further exemplified by another review from the same group that expands the natural products under consideration to polysaccharides, the eleutherobin/sarcodictyin derivatives, and glycopeptide antibiotics such as vancomycin and epothilones [38]. All of these can be considered to be part of the collection of privileged structures.

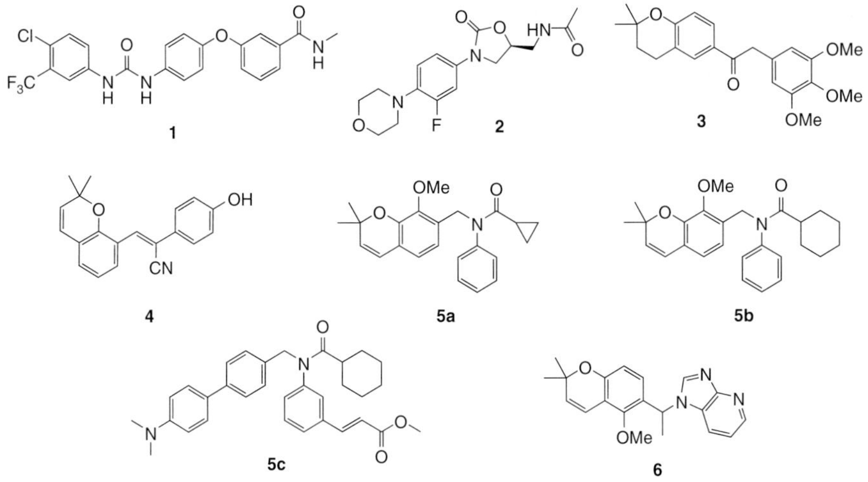

Figure 3.7 Benzopyran derivatives and related leads prepared through diversity-oriented synthesis by Nicolaou and co-workers.

Note that in addition to diversity-oriented libraries, the "focused library" approach is also widely employed in the preparation of chemical libraries [39]. Focused libraries are designed around a specific structural scaffold and are used to target a specific class of proteins. Often, such scaffolds may be chemically related to endogenous ligands for particular protein classes. In contrast, diversity-oriented libraries are not targeted to any specific protein class and are often used in broad screens in which the target proteins are not known. Each approach to chemical-library design has its advantages and disadvantages. Compounds in focused libraries are more likely than random compounds to be active, but they only target proteins in a known class. Diversity-oriented libraries, in contrast, offer the possibility of targeting entirely new classes of proteins, but any individual compound has a lower probability of activity. The pharmaceutical industry favors the use of focused libraries since fewer compounds of greater quality and with a greater probability of becoming drugs are more valuable than larger libraries with compounds that are not likely to become drugs. Academic groups, however, without the same pressure as industry, are pursuing high-risk strategies centered on diversity-oriented approaches. The two approaches are ultimately complementary: a ligand to a new protein class discovered from a diversity-oriented library can serve as the lead for a future focused library that further explores the structure–function relationships for the compounds under study.

3.2.3
Available Chemical Libraries

For researchers who are not interested in preparing compounds themselves, large and diverse compound collections can be obtained from academic and governmental agencies or purchased from commercial vendors. These compounds are generally of high purity and formatted into multi-well plates for high-throughput screening (HTS). The Drug Testing Program (DTP) through the National Cancer Institute (NCI) has a large repository of compounds that have been gathered for evaluation as possible anti-cancer or anti-HIV agents (*http://dtp.nci.nih.gov/*). Numerous investigators have submitted these diverse small molecules over the past several decades. The DTP will send specially plated sets of these compounds to users, free of charge. The only requirement is the completion of appropriate paperwork, consisting of a short proposal, with relevance to cancer or HIV, and other information about the user. Approved researchers can obtain the "diversity set" of 1990 compounds, the "mechanistic set" of 879 compounds, the "challenge set" of 57 cytotoxins with unknown mechanisms of action, and the "natural product set" of 235 compounds. All are sent as 1 or 10 mM stock solutions in dimethyl sulfoxide (DMSO), formatted in 96-well plates. A cottage industry has also arisen based on the sale of compound collections to industrial and academic researchers. Multiple companies provide chemical libraries to the market, such as ChemBridge, Maybridge, Chemical Diversity Laboratories, and Asinex. Sigma-Aldrich now also has a drug-like screening collection for sale, the MyriaScreen Diversity Collection. In general, compounds can be shipped as 1–10 mM solutions in DMSO in 96-well plates, and the price is normally acceptable.

Many successes have been reported from chemical genetic screens involving commercial compound collections. For example, a library containing ~16 600 compounds from ChemBridge was used as input into a phenotypic assay for mitotic spindle disruptors, leading to the identification of the kinesin inhibitor monastrol [40]. Torrance *et al.* used different green fluorescent protein (GFP) variants to distinguish between cells transformed with mutant Ras and wild-type cells and screened a collection from ChemBridge for differential toxicity in the two cell lines [41]. They identified a new class of compounds specifically toxic to the Ras-transformed cells and were able to trace the inhibitors to known targets in the Ras pathway. Phenotypic screens using commercial collections have also led to compounds that inhibit Sir2-mediated transcriptional silencing in yeast [42], alter the developmental program of zebrafish [43], and modulate transforming growth factor-β (TGF-β) signaling in mammalian cells [44].

A major benefit of screening commercial collections is that, in principle, the compounds are of known structure, avoiding the troublesome process of structural decoding. The screening usually provides a wealth of both negative and positive data for each screen, so that structure–activity relationships (SARs) can be determined without the need to synthesize and test many derivatives. However, it should be mentioned that the samples provided by commercial vendors are often associated with unforeseen impurities or decompositions, which may have an uncertain impact on the outcomes of subsequent biological screening. In addition, the limited quantity of desired compounds available from commercial vendors may also be a bottleneck for follow-up studies.

A special note should be made regarding the discovery of new therapeutic potentials of the drugs already on the market. The major bottleneck in the drug discovery process is gaining FDA approval for a new chemical entity. Depending on the therapeutic area, this process can extend over years and at considerable expense: current estimates put the cost of developing a novel therapeutic at $800 million [45]. As a consequence, the fastest way to move a compound to market is not to develop a novel compound at all, but to obtain a new indication for an already-approved drug. Recognizing the inherent value in FDA-approved compounds, the majority of which are off-patent, a number of different high-throughput screens have been performed that exclusively use FDA-approved drugs [46]. The obvious advantage of this strategy is that a wealth of information is already available on the bioavailability, toxicity, and dosing of any "hit" molecules that arise from such screens. To facilitate this process, collections of FDA-approved drugs are available for purchase in formats suitable for screening, and several interesting stories are emerging from this line of research. Companies such as Prestwick Chemical (*http://www.prestwickchemical.com/*), MicroSource Discovery Systems (*http://www.msdiscovery.com/*), and Sequoia Research Products (*http://www.seqchem.com/*) sell collections of various on- and off-patent medications for screening purposes. The potential of β-lactam antibiotics to treat amyotrophic lateral sclerosis (ALS), nonsteroidal anti-inflammatory drugs (NSAIDs) to treat Alzheimer's disease, and the dopamine receptor ligand fluphenzine against multiple myeloma are just some examples to demonstrate that FDA-approved drugs have been identified as having additional therapeutic indications [46].

3.3
Screening Strategies

Once compound libraries have been assembled, the next step in chemical genetics is screening, that is, testing their biological activities in certain assays. Generally, forward chemical genetics screens for an observable response from the assay system and goes on to identify the specific protein that is targeted. For this purpose, there are two major types of assays: phenotypic assays and *in vitro* binding assays. Ideally, these assays should be sensitive, selective, and well reproducible. Cost-effectiveness is another factor that must be taken into account.

Considering the size of the available chemical libraries nowadays, HTS techniques are much desired. Assay volumes were reduced to $100-300$ µl on 96-well plates in the late 1970s, and to $20-80$ µl on 384-well plates in the 1990s. At present, ultra-HTS has already been performed on "nano-well" plates, such as 3456-well plates, and their limits have not yet been reached. This fulfills two purposes: first, decreasing the length of time that is required per assay, and second, decreasing the cost of each assay. The combined effect has allowed the testing of hundreds of thousands of compounds each day in a fully automated HTS laboratory [47]. Therefore, the rate-limiting step in applying this approach to a biological question is the development of a robust HTS for the process of interest.

3.3.1
Phenotypic Assays

Such assays test the activity of the given compounds to induce a specific phenotypic outcome in a cell or organism. Phenotype-based chemical genetic screens have been performed on single-cell organisms, such as bacteria [48, 49] and fungi [50], and in cells from multicellular organisms, such as mice and humans [40, 51]. Chemical genetic screens performed on whole-animal models used to be rare owing to technical difficulties and poor reproducibility of biological responses. Nevertheless, a few well-studied animal models, including zebrafish, *Drosophila*, *Caenorhabditis elegans*, and some special types of plants, have been applied to chemical genetic screenings. TA comprehensive review by Walsh and Chang [52], which summarizes a number of active compounds identified in various assays, is recommended.

Detection methods for phenotype-based screens include the use of markers, functional assays, and microscopic imaging of cells. Assays using marker genes or proteins measure the abundance of a specific gene or protein epitope as a marker of a broader cellular phenotype, such as reporter-gene assays and antibody-based cellular immunoassays. For example, TGF-β induces numerous transcriptional and post-translational changes in target cells, but the expression level of a single gene, that is, plasminogen activator inhibitor type I, is often used as a marker for active TGF-β signaling [44]. Functional assays directly measure cellular phenotypes affected by small molecules, including cell proliferation, apoptosis, metabolism, chemotaxis, or adhesion. In recent years, a number of high-throughput cellular phenotypic assays have been reported, including assays that measure cell viability or proliferation [13, 53, 54]. Such assays measure the presence of intact cell membranes, the abundance of cellular energy (ATP concentration), or the presence of cellular reductases or esterases, which are found in nearly all cells. Such viability assays have been extended to the analysis of synthetic lethal effects: a compound is tested for its ability to kill cells in the presence, but not in the absence, of a defined element, such as another compound or a gene of interest [55].

Gene-expression signatures have been developed into high-throughput phenotypic assays [56]. In this approach, a gene expression profile is measured using DNA microarrays for two cell states of interest, such as undifferentiated neutrophil (a type of granular white blood cell) precursors and differentiated neutrophils. Then, the profiles are compared and a gene signature is created which determines whether the cell is in one state or the other. By measuring the effects of small molecules on the appearance of this gene signature, it is possible to determine whether each compound changes the cell state, for example, inducing differentiation of neutrophil precursors into neutrophils.

Another emerging trend in high-throughput phenotypic assays involves imaging cells using an automated microscope. Such an approach allows for the detection of phenotypes that can be measured using microscopy. For example, Yarrow *et al.* [57] used an imaging-based screen to identify compounds that affect cell migration during wound healing, and Kau *et al.* [58] used this technique to screen for compounds that prevent nuclear export of FOXO transcription factors. Image-analysis algorithms then allow for the automated processing of these images so that conclusions regarding the effects of compounds on these phenotypes can be derived. Imaging-based phenotypes could allow for digitization and clustering of otherwise unrelated phenotypes. Because any image consists of a series of pixels with distinct values, the relationship between any two images can be quantified mathematically.

3.3.2
Binding Assays

Since Ehrlich's invention of the protein receptor concept, it has been understood that small molecules exert their effects on biological systems by interacting with specific target(s), often protein molecules. *In vitro* binding assays are therefore used to identify ligands for a specific protein of interest. Subsequently, active ligands identified in such assays may be used to inhibit or activate the function of this protein *in vivo*. Obviously, this procedure is also widely applied to reverse chemical genetics studies.

A variety of experimental methods have been created to measure the affinity of small molecules to bind to specific proteins. A brief description of the main assay formats of each type can be found in Table 3.1. These protein-binding assays can be divided into two types: those that use labeled compounds and those that are label free. Here, a label refers to a fluorescent or radioactive group that is added to the test compound. Although labels make protein–ligand interactions easier to observe, they can be difficult to introduce into a compound, which increases the time and expense associated with measuring protein binding. Label-free detection methods are preferred because they do not require the extra synthetic chemistry for introducing a label, and because introducing a label may change the properties of a molecule. However, such measurements can be more difficult to perform: without a label, a larger amount of both protein and compound must often be produced, and the instruments used for label-free measurements are often slow.

Recent attempts to create high-throughput assays for measuring protein–ligand interactions require the use of labels. One class of high-throughput assay involves immobilizing each test compound (e.g., peptide, carbohydrate, drug-like molecule, and natural product) on a surface and then incubating these immobilized compounds with a soluble labeled protein [59, 60]. This method, so-called small-molecule microarrays (Figure 3.8a), can measure thousands of protein–small-molecule interactions. Instruments are now commercially available for spotting high-density arrays on glass microscope slides. Incubating the microarray probe sites with the protein of interest, washing and then detecting sites with retained proteins, identifies protein–small molecule interactions. The protein–small molecule interaction can be confirmed then by other techniques to

Table 3.1 Methods for measuring the affinity of small molecules for proteins[a].

Method	Phase	Label	Throughput	Protein amount
Fluorescence polarization	Solution	Yes	High	High
Fluorescence perturbation	Solution	No	Low	High
Fluorescence correlation spectroscopy	Solution	Yes	High	Medium
Radio-ligand binding assay	Solid	Yes	High	High
NMR spectroscopy	Solution	No	Low	High
Mass spectrometry	Solution	No	Medium	Medium
Surface plasmon resonance	Solid	No	Low/medium	High
Isothermal titration calorimetry	Solution	No	Low	High
Differential scanning calorimetry	Solution	No	Low	High
Small-molecule microarray	Solid	Yes	High	Medium
Protein microarray	Solid	Yes	High	Low/none
Three-hybrid system	In cell	Yes	High	None

[a]This table is taken from [8]. Rreaders are encouraged to check the references cited therein for descriptions of the assays.

verify that the interaction is reproducible and that the small molecule does not bind to the label on protein. Once the small molecule has been identified and verified, phenotypic assays are employed to characterize the effect of modulating the function of the protein.

Small-molecule microarrays have been applied successfully to a number of chemical genetics studies. For example, Kuruvilla *et al.* [61] employed DOS to prepare 3780 structurally complex and diverse small molecules. These compounds were printed on to a microscope slide to make a small-molecule microarray to screen for those that bind to the transcriptional repressor Ure2p. One compound, uretupamine, was found to bind to Ure2p. Experiments with uretupamine showed that it affects only a subset of genes controlled by Ure2p, that is, it only affects a subset of Ure2p function. This level of detail cannot be replicated by traditional genetics, where Ure2 would simply be deleted, and this example highlights the flexibility of chemical genetics in deciphering the individual functions of multifunctional proteins.

It is possible to invert these surface-based methods and to immobilize thousands of proteins side-by-side on a surface [62, 63]. A small molecule with a label, such as a fluorescent or radioactive group, can be applied to the surface, washed away, and detected by measuring the remaining label (Figure 3.8b). Some applications of protein microarrays include the use of an array of most yeast proteins to assess the global pattern of protein activities found in yeast cells [64], the discovery of novel protein–protein interactions in human cells [65], and an analysis of interactions between leucine zipper transcription factors [66].

A variation of this technology involves creating arrays of expression plasmids, which encode the information required to produce each protein of interest. Creating DNA arrays has become routine in the past decade and is preferable to creating arrays of proteins directly, primarily because DNA can be amplified and because thousands of different DNA-expression constructs will have similar chemical properties, such as solubility and stability. In contrast, thousands of different proteins will show idiosyncratic properties that are unique to each protein. It is possible either to place cells on this DNA array and cause proteins to be produced inside the cells [67], or to use a cell lysate to produce an array of proteins *in vitro* (Figure 3.8c) [68]. In either case, the net result is a protein array without the added complication of purifying and immobilizing each protein. However, post-translational modifications and protein complexes that are physiologically relevant will not be captured in these formats. So far, only proof-of-principle experiments have been performed with these more recent technologies.

Another high-throughput method for measuring the binding of many proteins to one or more small molecules also has the advantage of not requiring protein purification. This is the three-hybrid system, which is typically carried out in yeast or bacterial cells [69]. In such systems, a test protein is fused to the activation domain of a transcription activator, and the test small molecule is synthetically linked to an "anchor" compound that will interact with a protein containing a DNA-binding domain (Figure 3.8d). Therefore, if the test small molecule is able to interact with the test protein, the transcription activation domain will be brought into close proximity with the DNA-binding domain, and expression of the reporter gene that is controlled by the system will be activated. This method was used successfully by Liberles *et al.* [70] to create a mutant version of the

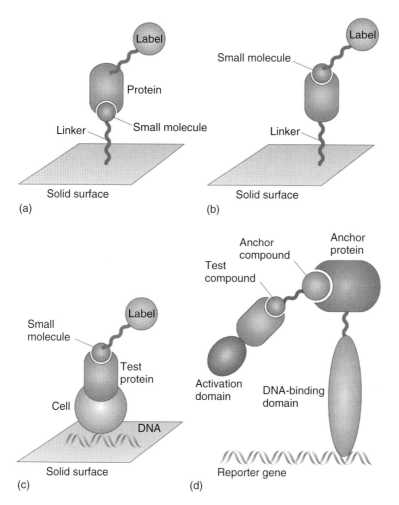

Figure 3.8 High-throughput assays for detecting the binding between small-molecule compounds and proteins. (a) Small molecules can be covalently linked to a surface, and a test protein with a tag is brought into contact with the surface. (b) Similarly, proteins can be immobilized on a surface and brought into contact with a labeled small molecule. (c) DNA expression plasmids can be arrayed on a surface and cells subsequently plated on top of these expression plasmids. The cells take up the DNA and produce the proteins encoded by each plasmid. A labeled compound localizes to where cells overexpress their high-affinity binding proteins. (d) Yeast three-hybrid system. Transcription factors that regulate gene expression can be divided into DNA-binding domains and transcription-activation domains. It is possible to fuse the complementary DNA sequence of a DNA-binding domain to the cDNA of an anchor protein that interacts with a known small molecule (anchor compound). The anchor compound is then chemically fused to a new test compound. If the cDNA of an activation domain is fused to the cDNA of a test protein, it is possible to determine whether the test protein interacts with the test compound with high affinity by determining whether transcription of a reporter gene has been activated. Reproduced from [8].

FKBP–rapamycin binding domain (FRB), which binds to a modified, non-toxic version of rapamycin.

Although several high-throughput methods have been developed for measuring protein–ligand interactions, many desirable features are not found in these systems. First, a technique measuring the binding of small molecules to target proteins in solution is preferable to that relying on a surface-based method since protein–ligand binding may be interfered with in the latter scenario. Unfortunately, most high-throughput methods involve immobilizing either the ligand or the protein on a solid surface to allow parallel processing of all samples with a single solution. Second, it is better to avoid the use of labels on both small molecules and proteins because of the added time and expense needed to introduce such labels into thousands of compounds or proteins, and because the labels may change the activity of the compound or protein. Third, it is easier to use only minute quantities of protein to manipulate only the corresponding DNA sequences and allow the system to produce the desired proteins *in situ*. This obviates the need to purify many proteins, each with its own solubility requirements. Fourth, it would be useful to have a system that is "scalable," in terms of the binding both of small molecules and of the proteins; ideally, it should be possible to automate the detection of the binding of hundreds of proteins to thousands of ligands without the need for idiosyncratic modifications to the system for each ligand or each protein. Finally, all of these technologies require significant investment in equipment and knowledge bases, which limit their application by many researchers. Therefore, although each of these problems may ultimately be solved, significant barriers will prevent the widespread adoption of these technologies in the near future.

3.3.3
Challenges in Screening

One limitation of small molecules is their frequent lack of specificity for a single target protein. This can be problematic when using small molecules both as therapeutic agents and as chemical probes: a lack of specificity can lead to unexpected toxicity, preventing the development of an otherwise promising compound into a drug, and can also confound the interpretation of the effects of a compound. This problem of nonspecificity is often dose dependent. At higher concentrations, compounds interact with additional proteins. In addition, specific functional groups and scaffolds have been found to be promiscuous, in the sense that they allow binding to a wide range of proteins or nonspecific killing of a wide-range of cell types. Such chemical functions need to be identified and removed from future library designs.

There are several strategies for overcoming the problem of specificity. First, it is preferable to identify and use potent compounds, that is, compounds that are likely to modulate a target protein at low nanomolar concentrations, because at such low concentrations they are less likely to affect other proteins. Second, measuring the binding specificity of compounds in the type of large-scale protein-binding assays described above should identify some of the alternative protein targets of compounds. Third, it is always critical to confirm the putative mechanism of action of a compound using either additional compounds or other reagents, such as small interfering RNAs (siRNAs) [71, 72]. Although the phenotypic consequences of an RNAi reagent and a small molecule targeting the corresponding mRNA are not always the same, their effects are often sufficiently similar to make this comparison useful. RNAi itself can lack specificity, and sometimes it is necessary to test numerous RNAi reagents designed against a target mRNA sequence. A large collection of RNAi reagents can be a useful tool for HTS [73]. By using such collections, it should eventually be possible to measure the phenotypic consequences of turning off expression of each gene in an organism.

Building redundancy into a set of probe molecules is an effective way of dealing with the problem of specificity. That is, it is desirable to have not just one compound that inhibits each protein, but rather dozens of compounds that inhibit each protein. If inhibition of protein X causes phenotype Y, we would expect, in an ideal world, all of the small molecules in our collection that inhibit protein X to cause phenotype Y. In the real world, not every protein X inhibitor will be effective, because some will bind protein X in slightly different ways or be metabolized differently in different cell types. Nonetheless, our confidence that the modulation of protein X causes phenotype Y should be proportional to the percentage of our protein X inhibitors that cause phenotype Y. Hence the problem of specificity can be overcome by assembling a sufficiently redundant set of probe compounds. Even if no single compound is specific for one target protein, the collection as a whole contains the requisite information on the effects of modulating each target protein.

Finally, given that compounds have different specificities at different concentrations, it would be preferable to collect information on the effects of each compound at multiple concentrations. A full dose–response curve for each compound would be ideal. Unfortunately, the added time and expense associated with collecting this additional information usually make it impractical. Therefore, new technologies that allow an increase in the number of tests performed per unit time would be valuable. Alternatively, a smaller number of compounds may be tested with more replicates and a full dose–response curve. This trade-off between the number of compounds tested and the quality and completeness of the data set collected for each compound needs to be optimized in each project.

Another major challenge in screening is quality control. When collecting large-scale data sets, attention to quality control is crucial. However, there is an inherent trade-off between the level of throughput and data quality in large-scale data collection. A minimum level of quality is necessary to ensure that reliable conclusions are extracted from such data sets. However, attention to data quality has not been a priority for many researchers engaged in high-throughput chemical screens, simply because the data quality required for a screen is much lower than that required for a global analysis [74]. In addition, it is important to eliminate artifacts through the use of counter screens for properties that could interfere with the assay readout, such as intrinsic compound fluorescence or compound aggregation. In general, a counter screen is performed on the compounds that emerge from an initial screen, and compounds that are active in the counter screen are not taken further. For example, in a screen that uses the fluorescent dye calcein for detection, any compound that shows the same color of fluorescence as calcein will appear to be a positive compound from the screen; a counter screen would involve testing each compound for its intrinsic fluorescence to eliminate those compounds that were falsely active because of this property. Finally, it is important to assess the solubility and stability of each tested compound or protein, and to confirm that the chemical being tested is the desired one. Solubility can be measured using nephelometry, which detects insoluble particles in solution, and compound identities can be confirmed using liquid chromatography and mass spectrometry. All these methods of improving data quality increase the time and expense associated with large-scale data collection but are crucial if meaningful conclusions are to be drawn.

3.3.4
Data Management and Informatics Analysis

All screening efforts generate a large amount of data. Therefore, a significant challenge in screening, especially HTS, is the development of tools that allow for data management and analysis in chemical genetics [75]. With increasingly diverse, reliable, and accessible databases of information about the effects of new chemical compounds on specific biochemical processes, we shall be able to understand much more about the nature of biologically relevant chemical space. In this regard, among the most exciting recent developments are efforts to generate public databases of chemical information [76]. Several databases are listed below as examples. Many more databases of this type exist on the Internet. Other common databases, such as SciFinder and MDL Crossfire, also allow direct structure searches and refinement of results by biological activity.

PubChem (*http://pubchem.ncbi.nlm.nih.gov/*) is a database that contains small-molecule information linked to biochemical data. The database is maintained by the National Center for Biotechnology Information (NCBI) of the NIH, as a component of the NIH Molecular Libraries Roadmap Initiative. Modeled after the popular PubMed, PubChem allows users to search chemical structures and to view full data sets from HTS. Millions of structures and associated descriptive datasets and also HTS results can be accessed free through a web interface.

ChemBank (*http://chembank.broad.harvard.edu/*) is a public, web-based informatics environment created by the Broad Institute's Chemical Biology Program, part of the National Cancer Institute Initiative for Chemical Genetics. ChemBank data are also available through PubChem. ChemBank intends to guide chemists who are synthesizing novel compounds or libraries, to assist the development of small molecule probes that perturb specific biological pathways, and to catalyze drug development.

DrugBank (*http://www.drugbank.ca/*) is a database available at the University of Alberta that combines detailed chemical, pharmacological, and pharmaceutical data combined with drug target information (sequence, structure, pathway). The database contains roughly 4800 entries that include FDA-approved small-molecule drugs, FDA-approved biotech (protein/peptide) drugs, nutraceuticals, and experimental drugs. In addition, more than 2500 protein sequences are linked to these drug entries.

Although a rich array of data are accumulating from large-scale screening, analysis of such data, that is, data mining, represents both opportunities and challenges. The flourishing bioinformatics and cheminformatics communities have been providing computational tools for this purpose, which deepen our understanding of chemical biology and may eventually lead to the discovery of promising therapeutic targets and compounds [77, 78]. For example, Stockwell and co-workers have developed a freely available software tool termed the small laboratory information management system (SLIMS) [79]. SLIMS aids users in data collection and analysis, together with relevant literature retrieval for aid in determining an active compound's mechanism of action. Leonetti and co-workers have also made freely available a data processing tool called the ELISA data manager that aids in the treatment and archiving of small- to medium-sized libraries [80]. Another popular approach is to employ informatics tools in library and screening design [77, 81].

Although progress is now being made in developing tools for data mining, such progress is often limited by the difficulty in accessing much of the data of interest. Some estimates suggest that only about 1% of the information useful for chemical biology studies is in public domains [82]. In contrast, the majority of many forms of "pure" biological data, for example, from gene sequences to protein structures, is freely accessible to scientists in both academia and industry. One of the reasons for the inaccessibility of so much chemical information is concerned with issues of intellectual property. However, one can be optimistic that solutions will eventually be found to overcome the various hurdles preventing these resources from being used in the most effective ways.

3.3.5
Chemical Approaches to Stem Cell Biology

Small molecules have provided the key to many biological discoveries. Recent work has translated these successes to the field of stem cell biology [83–85]. Owing to the significance of this field, a separate section in this chapter needs to be devoted to the progresses made therein.

Breakthroughs in stem cell biology have unraveled the possibility of tissue and organ repair or regeneration. These advances include insights into the intrinsic mechanisms [86–89] and the niche interactions [90] that regulate the fate of embryonic stem (ES) cells and adult stem cells. Other recent advances are the ability to manipulate and control stem cell fate and function in a more precise and defined manner [91–94] and the ability to reprogram lineage-committed cells to revert to more primitive multipotent states [95] or even pluripotent states [96]. Not only are various types of stem cell excellent model systems for studying the fundamental biology of human development and diseases, but they also provide superior vehicles compared with conventional cell lines for drug discovery. They may serve for disease modeling, target validation, and HTS to the development of clinical candidates. Furthermore, stem cells can be used to replace cells that have been damaged as a result of disease and injury. They can also be endogenous targets for conventional small-molecule or biological therapeutics that stimulate a body's own cells to regenerate *in vivo*.

Application of small molecules to the field of stem cell biology can also be executed through target-based or phenotype-based approaches. A target-based approach relies on the use of known mechanistic chemical compounds to probe their effects in stem cell systems. Consequently, specific pathways and proteins are implicated as important in stem cell regulation. For example, the complex regulatory pathways that control self-renewal in ES cells have been an area of study where known small-molecule

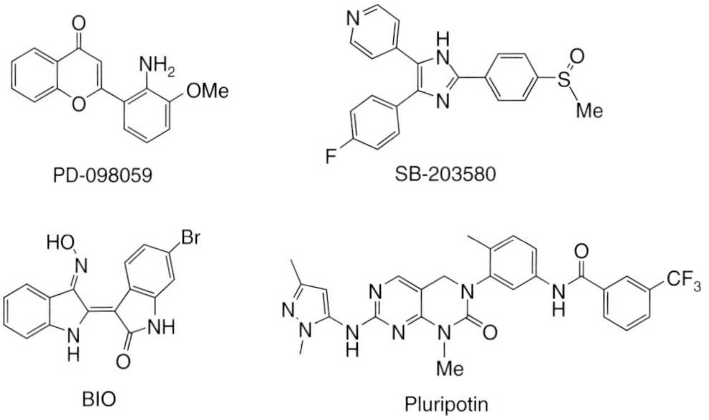

Figure 3.9 Examples of small molecules shown to control the fate of stem cells.

modulators have proved useful [97, 98]. In murine embryonic stem (mES) cells, leukemia inhibitory factor (LIF) has been used to maintain an undifferentiated pluripotent state. LIF acts through the heterodimerization of gp130 and LIF receptor, which leads to activation of Janus-associated tyrosine kinases (JAKs) and subsequent activation of STAT3 (signal transducer and activator of transcription 3). Recent studies have also elucidated that BMP4 in combination with LIF can support long-term self-renewal of mES cells in the absence of feeder cells and serum in a chemically defined media condition. A major role of BMP4 in the promotion of mES cell self-renewal is its inhibition of extracellular-signal-regulated kinase (ERK) and p38 mitogen-activated protein kinase (MAPK) pathways, which have differentiation-inducing activities. This mechanism was validated using PD098059 [99] and a p38 MAPK inhibitor SB203580 [100] (Figure 3.9).

Another small molecule that is widely used in stem cell biology is 6-bromoin-dirubin-30-oxime (BIO) [101, 102], which is a glycogen synthase kinase-3 (GSK3) inhibitor (Figure 3.9). This molecule was used to show that activation of the Wnt–β-catenin pathway also promotes self-renewal of mES and hES cells. In addition to the Wnt pathway, BIO also plays a role in suppressing the GSK3b-mediated Myc T58 phosphorylation to sustain stability of Myc, which is also regulated by STAT3 at the transcriptional level.

The use of small molecules against known targets highlights the significant role of small molecules in facilitating and providing new understandings of the complex regulatory circuits that are involved in stem cell biology. However, limitations of this target-based method include the need for a reasonable knowledge of the molecular pathways under study. This is particularly problematic in stem cell biology, where biological mechanisms are still poorly understood. The lack of previously characterized small-molecule modulators of these pathways presents an additional difficulty. There is also the concern of the specificity of a small molecule to its target, and the problem that a specific phenotype, or lack of thereof, could be a result of an off-target effect.

A phenotype-based approach relies on the screening of chemical libraries in high-throughput functional assays. This is done on either cells or whole organisms to identify compounds that produce the desired phenotype. These compounds are subsequently characterized to elucidate their mechanism(s) of action. Given the current poor understanding and control of stem cell fate, unbiased phenotypic screens of chemical libraries hold great promise for the generation of desired cell types in a controlled manner and also for providing useful tools to study the underlying mechanisms involved in stem cell regulation. Chen *et al.*'s study [91] provides a good example of this strategy. A high-throughput screen of 50 000 synthetic small molecules was carried out using a transgenic reporter mouse ES cell line expressing GFP under control of the *Oct4* gene, which is specifically expressed in pluripotent stem cells. The screen was conducted in the absence of feeder cells, serum, and LIF, to search for small molecules that can maintain self-renewal of ES cells in chemically defined conditions. Oct4–GFP expression and the characteristic compact domed colony morphology of mouse ES cells were used as criteria to select primary hits in the screen. Secondary confirmation assays and SAR studies led to the identification of a novel compound named pluripotin (Figure 3.9). Pluripotin sustains homogeneous self-renewal of mouse ES cells long term in chemically defined medium conditions in the absence of feeder cells, serum, LIF and bone morphogenetic proteins (BMPs). The cells remain pluripotent in serially passaged culture *in vitro* without losing their germline transmission ability *in vivo*. Furthermore, pluripotin functions independently of exogenous activation of LIF–STAT3, BMP–SMAD-ID, and Wnt–β-catenin pathways. Through affinity pull-down experiments using a pluripotin-immobilized matrix, the molecular targets of pluripotin were identified as RasGAP and ERK1, two endogenously expressed proteins with differentiation-inducing activity. Additional biochemical, genetic, and pharmacological studies have shown that simultaneous inhibition of ERK1 and RasGAP by pluripotin or other independent methods is sufficient for long-term self-renewal of mouse ES cells.

Phenotype-based screens have thus been very successful in the identification of small molecules that are able to control stem cell fate efficiently and also uncover new molecular mechanisms in stem cell systems. In order to ensure that the screen is a success, emphasis must also be placed on the method of screening. The development of permissive assay conditions with positive and negative controls is critical for obtaining meaningful results from a screen. In terms of reading for phenotypic output, the utilization of more functional criteria in addition to reporter systems would be desirable. Developments in imaging techniques [103] allow for the monitoring of increasingly complex phenotypic changes in cell morphology and multiple biomarker expression and localization. Concerning engineering cell assays (e.g., reporter lines), additional experimental advances with respect to more efficient and precise marking of cells, especially with respect to hES cells, are needed. As opposed to gene perturbation, which is often difficult in ES cells, especially hES cells (for a review, see [39]), small molecules offer the additional advantage of being able to elicit a particular phenotype without the need for complicated cell manipulations.

In conclusion, stem cells present great opportunities for basic research in chemical biology and drug discovery. Conventional small-molecule or biological therapeutics will probably become a more convenient form of regenerative medicine, working to unleash the human body's own regenerative capacities by promoting survival, migration, proliferation, differentiation, or reprogramming of endogenous cells. Continued development and application of chemical approaches to stem cells will undoubtedly lead to the identification of additional small molecules and more precisely defined and individualized conditions for controlling cell fate both *in vitro* and *in vivo*.

3.4
Target Elucidation and Validation

Once a compound that induces a certain phenotypic response in a cell culture or *in vivo* experiment has been found, several techniques may be employed to identify its protein target. Target elucidation is by far the most challenging part of a chemical genetics study. To date, there is no general, systematic process to discover the molecular target or action mechanism of an active small-molecule compound. However, two major categories of approaches can be applied to tackle this problem. The classical approach to target identification uses the small molecule as a bait in order to trap the protein target, which is frequently referred to as the "pull-down" approach. A critical assumption here is that the phenotype induced by an active compound arises from the cellular consequence of a covalent or noncovalent interaction between the small-molecule compound and its target protein(s). This approach requires the small molecule to be chemically modified using a fluorescent, photoaffinity, or biotin label. Alternatively, the small molecule can be attached to a solid-phase matrix via a linker. The protein target may be retained on the affinity column in a chromatographic experiment. Then, the retained protein can be characterized by mass spectrometric microsequencing and translated back to its gene sequence. The second category of approaches, including hypothesis-driven and genetics-based approaches, do not require the small molecule of interest to be chemically modified. These two categories of approaches are reviewed in the following section. Note that this tutorial review provides a conceptual framework, supplied with brief descriptions of a few selected examples. Extensive reviews on the studies in this field are available [104–106].

3.4.1
Strategies Employing Affinity Reagents

3.4.1.1 Methods Employing Affinity Chromatography
The simplest and oldest of these techniques involves the tethering of modified versions of lead compounds to a solid basis, followed by incubation with a lysate of a desired cell line, organ, or organism. Affinity-based separation of the binding protein from all other proteins in the cell can be conducted by filtration, washing, and elution. This technique is often referred to as the "pull-down" method. The solid support, or resin, for these techniques is typically an agarose- or Sepharose-based polymer that is functionalized with reactive functional groups such as amines, thiols, or carboxylic acids. Activated resins functionalized with *N*-hydroxysuccinimidyl (NHS) esters are by far the most commonly used, which are readily coupled with amino and hydroxyl functional groups. NHS-coated resins are available from many commercial vendors, such as Bio-Rad and Aldrich. In a typical affinity chromatographic purification, a cellular lysate or tissue homogenate is incubated with the affinity matrix under conditions designed

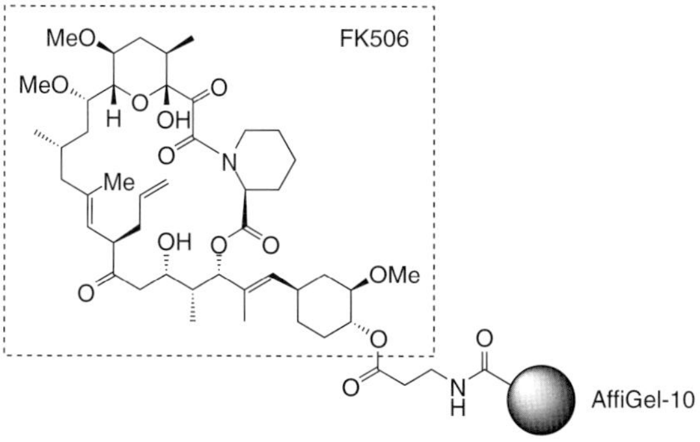

Figure 3.10 An agarose-supported derivative of FK506.

to minimize protein degradation. After incubation, the affinity resin is washed extensively with an aqueous buffer to elute any nonbinding proteins from the resin, either by serially centrifuging the resin and aspirating the supernatant, or by washing the resin in fritted glass or plastic chromatographic columns. At this point, the proteins remaining on the column should be those binding to the ligand or to some part of the affinity matrix. These proteins are then eluted from the affinity matrix under denaturing conditions, or by incubation with the free ligand, and resolved by electrophoresis on a sodium dodecyl sulfate polyacrylamide gel electrophoresis (SDS-PAGE) gel. This purified target protein can then be identified by peptide microsequencing, immunohistochemical analysis, or, most commonly, modern degradative mass spectrometric techniques.

The protein targets of many biologically important natural products and small molecules have been discovered in this way. A classical study was performed by Schreiber and co-workers [107]. They demonstrated that a derivative of the immunosuppressant FK506 (Figure 3.10), when incubated with the cellular lysate of bovine thymus or human spleen, was able to purify selectively FKBP12, a *cis–trans* prolyl isomerase. FKBP12 was competitively eluted from the matrix using soluble FK506, and was not observed to bind to a control resin which had been functionalized with ethanolamine instead of FK506, indicating that FKBP12 was isolated on the affinity resin because it bound specifically to FK506. This study was certainly successful owing to the extremely high affinity between the target and the ligand, but it serves as the archetypical example of affinity pull-down experiments, especially in the light of the broad use of the FK506–FKBP12 ligand pair in chemical genetics.

In order to identify the target of bioactive compounds rapidly, the concept of tagged library screening was introduced to reduce the time spent on SAR analysis [108–110]. As a demonstration of this concept, a combinatorial library of over 1500 multifunctionalized triazines (Figure 3.11) was synthesized containing aminotriethylene glycol linkers and was screened for phenotypic effects in zebrafish. Active compounds which influence skin pigmentation and brain/eye development were subjected to Boc deprotection, and the resulting amine was then coupled to AffiGel-10, an NHS ester-functionalized agarose resin, to generate affinity matrices with known activity. This approach allowed the rapid isolation of the target proteins by facile affinity matrix preparation and elucidated the

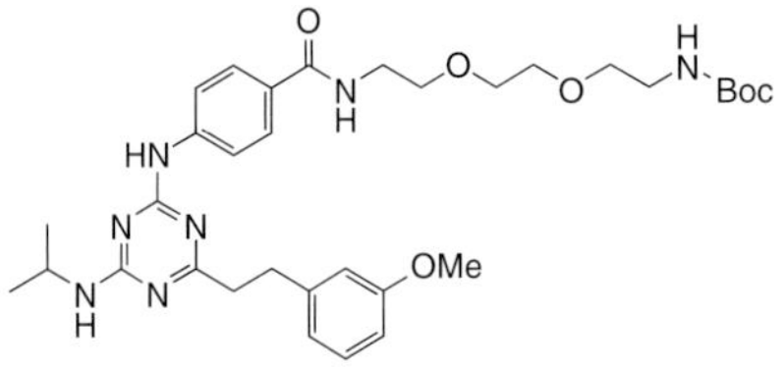

Figure 3.11 A compound from a tagged triazine library used directly in affinity purification experiments.

first small-molecule inhibitors for several ribosomal accessory proteins or their complex as the target.

Two critical concerns in affinity chromatographic experiments are the affinity of a ligand for its target and the ability to detect specific over nonspecific interactions. The affinity of a ligand for its target is a central concern, as the strength of this interaction has been the largest factor for success in target identification experiments. Protein binders must remain bound to affinity resins during the extensive washing required to remove nonspecific binding partners in order to be detected. Natural products and drug-like molecules often bind tightly to their targets. However, chemical genetics studies more often deal with compounds with binding affinity at the micromolar level. Two promising strategies have been proposed for target identification using low-affinity ligands. The first involves combinatorial assembly of two or more modest-affinity lead molecules in an effort to discover high-affinity bivalent noncompetitive ligands [111, 112]. The second involves incorporation of a photoreactive group into a lead scaffold to turn a weak-affinity ligand into an irreversible binder.

In addition, detection of the binding of the compound and its plausible target proteins must be treated with extreme care. Identification of protein targets is dependent on the ability to enrich the concentration of ligand binding partners over background nonspecific binding by attaining a balance between maximal target binding and minimal nonspecific interactions. As organic small molecules are typically uncharged and hydrophobic, a high ligand loading on the resin can contribute to nonspecific interactions with the ligand. However, the use of excessive amounts of affinity matrix may lead to increased nonspecific interactions with the resin. In practice, it is advisable to construct a matrix with low ligand loading, and then using the minimal amount of resin to isolate a detectable amount of protein. Since detection of a target is dependent on both the affinity and abundance of target proteins, it is imperative to maximize the ratio of total protein to ligand to ensure that high-affinity but low-abundance binding partners can be detected over low-affinity but high-abundance proteins. This means that it is imperative to use as much protein as possible, as it should be assumed that the target is of low abundance until evidence suggests otherwise. Selection of an analysis technique with a low limit of detection enables smaller amounts of protein to be used.

3.4.1.2 Methods Employing Biotinylated Probes

Another common technique for target identification is the use of biotinylated probes. Biotin binds to the proteins avidin, streptavidin, and neutravidin with dissociation constants of 10^{-15} M (Figure 3.12), making it one of the strongest noncovalent protein–ligand interactions known. Thus, avidin-functionalized resins, when incubated with a biotinylated probe, will create a noncovalent affinity matrix that is stable under all but

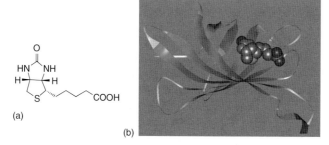

Figure 3.12 (a) Chemical structure of biotin and (b) crystal structure of the biotin–avidin complex (Protein Data Bank entry 1AVD).

the harshest conditions. Avidin's extensive glycosylation is responsible for its high charge ($pI = 10$) and propensity for nonspecific binding. The more expensive deglycosylated variants streptavidin ($pI = 5$) and neutravidin ($pI = 6.3$) are commonly used in order to minimize nonspecific binding. Biotin and several derivatives are commercially available, allowing them to be easily appended to most common functional groups.

Biotinylated probes can be utilized in various ways in target identification strategies: (i) they can be precomplexed with a streptavidin resin, creating a noncovalent affinity matrix; (ii) they can be incubated with a cell lysate, and then the streptavidin resin can be added to pull out the biotinylated probe together with any bound protein; (iii) they can be incubated with live cells and, after cell lysis, can be pulled out of solution with streptavidin resin, which will then pull out any bound protein. Either way, biotinylated probes and their attached proteins may be eluted from avidin resins by 8 M guanidine·HCl (pH 1.5), or by boiling in a typical denaturing SDS-PAGE loading dye. A major advantage over the use of affinity matrices is that biotinylated probes are discrete small molecules that can be fully characterized. Since the binding of biotin to avidin is quantitative and essentially irreversible, the loading of a known amount of probe on to a solid support is possible. In contrast, when coupling affinity probes directly to commercially available solid-phase resins, analysis of coupling efficiency and chemical characterization of the matrix can be problematic. However, biotinylation methods also have certain limitations since biotinylation of a target molecule often significantly decreases its aqueous solubility. In addition, the solubility of biotin conjugates in standard organic solvents can also be poor. The introduction of biotin and further manipulation of biotin conjugates must be carried out in high-boiling solvents such as dimethylformamide (DMF) or DMSO. Consequently, biotinylation reactions are usually carried out on a small scale in the final step of a synthetic route.

Whitesell and co-workers determined that the natural product withaferin A and its biotinylated probe (Figure 3.13) covalently bind to annexin II, which induces actin microfilament aggregation [113]. In order to decrease the nonspecific binding when using avidin-based resins, they "precleared" the protein sample of nonspecific binders by passing the cellular lysate though one of the control resins before incubation with the affinity matrix. Protein detection shows that many proteins isolated by probe (b)

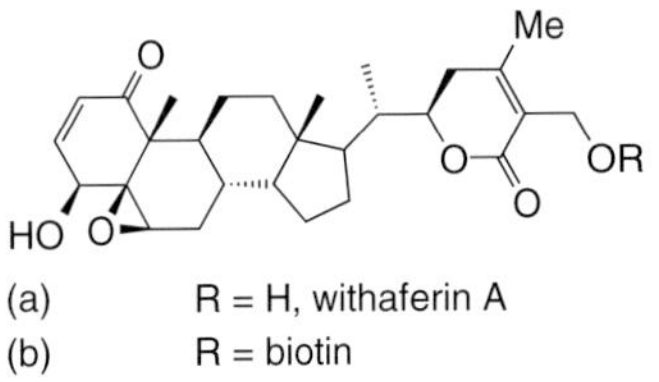

Figure 3.13 (a) Withaferin A and (b) biotinylwithaferin A.

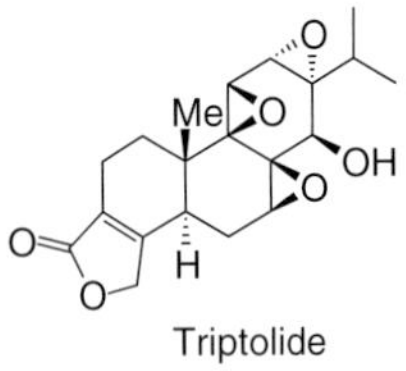

(a)   R = H, withaferin A
(b)   R = biotin

Triptolide

Figure 3.14 Chemical structure of triptolide, a natural product extracted from a Chinese medicinal herb, *lei gong teng*.

complexed with neutravidin beads were still present when the experiment was performed in the presence of excess of (a). These proteins bind to the resin, and not to the probe, as confirmed by their presence in the "preclear" lane on the SDS-PAGE gel. The band that is absent in the presence of (a) but present in the control lane (DMSO) is the target, which was determined to be annexin II. This set of experiments indicates that even nonglycosylated variants of avidin can exhibit significant nonspecific binding.

3.4.1.3 Methods Employing Radiolabeled/Fluorescent and Photoaffinity Probes

If the compound under study binds to its target protein through a covalent linkage and thus is still linked to the target under the denaturing conditions used for SDS-PAGE analysis, a group of methods employing radiolabeled or fluorescent probes are applicable to target elucidation [114]. Crews and co-workers identified the target of triptolide, a diterpene component of the Chinese medicinal herb *lei gong teng* (Figure 3.14), with this method [115, 116]. A sample of the natural product was tritiated and incubated with HeLa cells. After cellular lysis, the proteome was fractionated via fast protein liquid chromatography (FPLC) over an anion-exchange column and radioactive fractions were analyzed by SDS-PAGE and stained with Coomassie Blue. The intensity of Coomassie Blue staining of a band at 110 kDa was proportional to the radioactive intensity of the fractions, indicating that [^{3}H]triptolide bound specifically to this protein, which mass spectrometric analysis determined to be the calcium channel protein PC2.

However, the compound of interest binds to its molecular target more often in a noncovalent fashion. In addition, in many cases the compound has only a moderate affinity to its target. In order to prevent the dissociation of weak binders during the course of an affinity purification experiment, crosslinking reagents can be used to attach a small molecule irreversibly to its target. An effective strategy is to functionalize the compound of interest with a photoreactive group. In these experiments, the compound with a photoreactive group is incubated with a lysate or whole cells to allow the probe to bind to its target protein. In the absence of UV light, the photoreactive group is stable, allowing the unactivated probe to come into binding contact with its target.

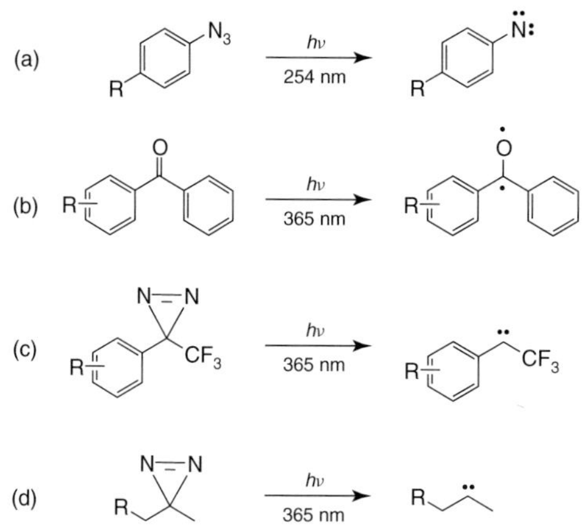

Figure 3.15 Photoreactive groups commonly employed in photocrosslinking experiments. Arylazides (a) are excited initially to singlet nitrenes using potentially damaging short-wavelength UV light. Irradiation of benzophenones (b) or diazirines (c, d) with long-wavelength UV light forms radicals or carbenes, respectively.

Irradiation of the photoprobe at a specific wavelength will generate a reactive species that rapidly reacts with a variety of amino acid residues, forming a covalent linkage between the probe and its target protein. The three most common photophores used in target identification studies are arylazides, benzophenones, and diazirines (Figure 3.15) [117].

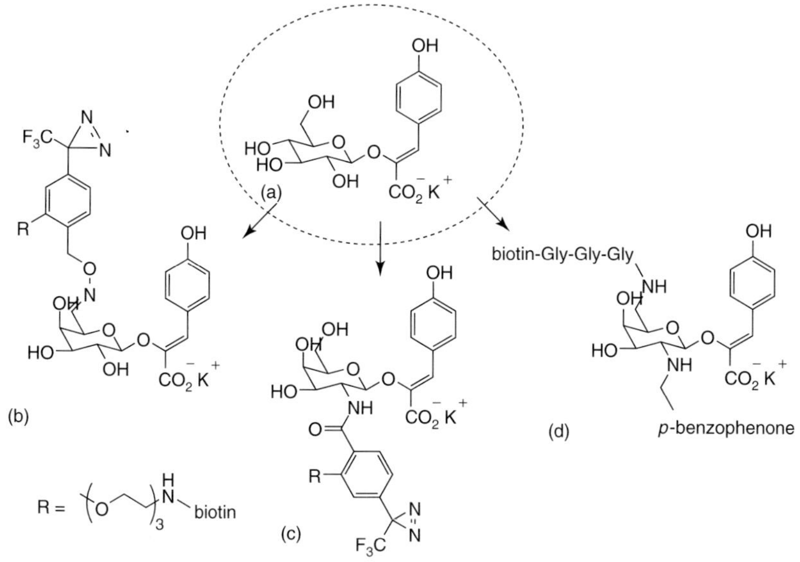

Figure 3.16 Detection of receptor proteins of the leaf-movement factor potassium lespedezate. (a) Potassium lespedezate and (b–d) photoprobes.

Fujii *et al.* used biotinylated photocrosslinking probes to discover receptors for the leaf-movement factor potassium lespedezate, a ligand whose concentration governs circadian rhythmic leaf-movement in leguminous plants (Figure 3.16) [118]. Three different probes were synthesized in order to examine the relationship between bioactivity and photocrosslinking efficiency of the probes, as systematic studies of such relationships are lacking. Probes were incubated with plasma membrane proteins from motor cells in plant leaves and irradiated with long-wavelength UV light for 10 min to crosslink receptor proteins. After gel electrophoretic separation of the lysates, immunoblotting for biotinylated–coupled proteins identified a 210 kDa protein as a potential target. Probe (b), which positions the photoreactive group furthest away from the pharmacophore for increased bioactivity, showed no specific crosslinking. Diazirine (c) and benzophenone (d) both label an unknown 210 kDa protein, which is a potential receptor for (a). The authors reported that (d), the least bioactive of the three photoprobes synthesized, produced the most efficient photolabeling, which they attributed to the proximity of the benzophenone photophore to the presumed binding pocket of (a). However, in the absence of a binding model, it is unclear whether the difference in crosslinking efficiency between probes (c) and (d) is due to the proximity of these two groups to the binding pocket of the target, or the difference in the photoreactivity of the photophores themselves. This highlights an important point, that is, there may be an elegant balance between bioactivity and crosslinking efficiency, and photophore selection may have a significant effect on crosslinking efficiency.

3.4.2
Hypothesis-Driven Approaches

The methods mentioned in the above section all require chemical modification of the small molecule of interest, which can be technically difficult. In addition, a critical concern is that it is not clear where the tag or linker should be attached to the small molecule so that it will not disturb the binding to the target protein. Hence target identification approaches that do not require the small molecule to be chemically modified are inherently more attractive. Two major methods have this advantage: the hypothesis-driven approaches and the genetics-based approaches.

A hypothesis-driven approach requires the greatest degree of chemical and biological insight as it involves studying closely the chemical structure and the phenotypic effect of the small molecule, then inventing several hypotheses regarding its mechanism of action and testing each hypothesis. If the small molecule or its analogs are already known to have biological effects in other systems, then this information should help to choose likely modes of action. Databases of screening data such as PubChem and ChemBank can help in this approach. There are also databases available publicly that can be used to search the literature for chemical structures and their pharmacological activity, such as SciFinder Scholar and Beilstein CrossFire. In this process, the "profiling" technique is often very helpful, which involves monitoring simultaneously the expression level of genes (transcriptomics) or proteins (proteomics) in an organism. For example, treating cells with a small-molecule compound may result in changes in the pattern of gene expression. This pattern may reveal clues to the mechanism of action of the compound.

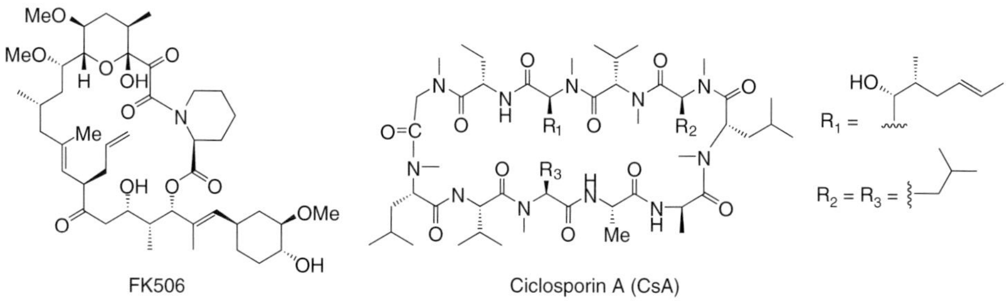

Figure 3.17 Chemical structures of FK506 and ciclosporin A.

Moreover, the pattern can be used as a fingerprint and matched with transcriptional profiles for specific gene deletion, if such data are available.

Small molecules or peptides discovered in phenotype-based screens can be used to identify proteins regulating a specific phenotype. The discovery of calcineurin as the target of the immunosuppressant FK506 (Figure 3.17) in 1987 represents the first successful application of the three steps of the chemical genetic process. Before the discovery of this small molecule, the structurally unrelated cyclic peptide ciclosporin A (CsA) (Figure 3.17) was used to treat solid-organ transplant rejection. CsA was known to inhibit the production of T-cell-derived cytokines such as interleukin-2 (IL-2). Using a phenotype-based screen, Kino and co-workers tested fermentation broth extracts from *Streptomyces* for their ability to phenocopy CsA by blocking IL-2 production, which led to the discovery of FK506 [119, 120]. Although the immediate receptors (immunophilins) of the immunosuppressants CsA and FK506 are distinct, their similar mechanisms of inhibition of cell signaling suggest that their associated immunophilin complexes interact with a common target. Thus, a later study by Schreiber and coworkers [121] found that the complexes cyclophilin–CsA and FKBP–FK506, but not cyclophilin, FKBP, FKBP–rapamycin, or FKBP–506BD, competitively bind to and inhibit Ca^{2+} and calmodulin-dependent phosphatase calcineurin, although the binding and inhibition of calcineurin do not require calmodulin. These results suggest that (i) calcineurin is involved in a common step associated with T-cell receptor and IgE receptor signaling pathways and that (ii) cyclophilin and FKBP mediate the actions of CsA and FK506, respectively, by forming drug-dependent complexes with and altering the activity of calcineurin–calmodulin. This study placed the molecular target of FK506, calcineurin, in the T-cell receptor-initiated membrane-to-nucleus signaling pathway.

Another notable example of the hypothesis-based approach is the identification of Eg5 as the molecular target of monastrol by Schreiber and co-workers [40]. In order to identify cell-permeable small molecules that target mitotic proteins other than tubulin, they first used the versatile whole-cell immunodetection (cytoblot) assay to identify compounds that increase the phosphorylation of nucleolin. Nucleolin is a nucleolar protein that is specifically phosphorylated in cells entering mitosis, and compounds that cause mitotic arrest would be expected to show the phenotype of increased amounts of phosphonucleolin. Using this assay, they screened a library of 16 320 small molecules and selected 139 from it. Because many known antimitotics target tubulin, they tested

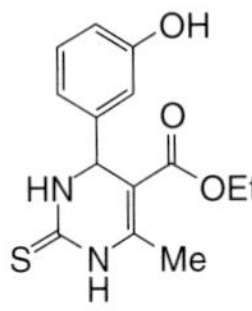

Figure 3.18 Monastrol, a small molecule affecting mitosis by targeting the mitotic kinesin Eg5.

each of these 139 compounds for the ability to affect tubulin polymerization *in vitro*. Fifty-two compounds inhibited tubulin polymerization, and one stimulated tubulin polymerization. The remaining 86 small molecules that increased the mitotic index presumably targeted other proteins involved in mitosis. These molecules were then examined through a series of visualizations of microtubules, actin, and chromatin in fixed cells by fluorescence microscopy to identify those that affect mitosis and to eliminate several other possibilities. Finally, one compound, monastrol (Figure 3.18), was observed to arrest mammalian cells in mitosis with monopolar spindles. *In vitro*, monastrol specifically inhibited the motility of the mitotic kinesin Eg5, a motor protein required for spindle bipolarity. All previously known small molecules that specifically affect the mitotic machinery target tubulin. Monastrol will therefore be a particularly useful tool for studying mitotic mechanisms.

3.4.3
Genetics-Based Approach

Since the ultimate goals of forward chemical genetics and classical genetics are identical, that is, elucidation of the function of certain genes, genetic techniques are naturally helpful for target identification in forward chemical genetic screens. For example, mutant cell lines can be generated, if the system is amenable, that are resistant to the effects of the small molecule. Then the problem becomes the identification of the mutant gene product that was responsible for resistance. The genes to be mutated can be selected based on an understanding of the phenotype induced by the small molecule of interest. Alternatively, random mutant genes can be transfected into wild-type cells. These can be identified by transfecting random mutant genes into wild-type cells and selecting for resistance. The transfected gene product responsible for resistance can be sequenced and therefore identified.

cDNA microarrays, a common and prolific genomic tool nowadays, have been used in target identification. However, the results provide clues and indirect evidence of the target, rather than the direct validation seen in many other approaches A representative use of this approach was presented by Marton *et al.* in the identification of the transcription factor Gcn4 as a secondary target of FK506 [122]. Tanaka *et al.* introduced an interesting new method for isolating the genes encoding cellular drug-binding proteins, called "drug western" [123]. This method is based on the use of the drug conjugated with a marker molecule as a probe for the screening of a cDNA library. Unlike the other methods, this method allows one to identify the genes for trace amounts of cellular drug-binding proteins without purification. They used this method to isolate human cDNA clones encoding binding proteins of HMN-154, a novel benzenesulfonamide anticancer compound. The proteins encoded by two of the isolated clones are identical with NF-YB, the B subunit of nuclear transcription factor NF-Y, and thymosin β-10, respectively. Recombinants of both proteins bind specifically to HMN-154 *in vitro*. Comparison of amino acid sequences between these proteins shows the sequence similarity in a short amino acid stretch [K(X)AKXXK]. Deletion or mutation of this region causes a significant loss of binding of both proteins to HMN-154. Furthermore, HMN-154 inhibits DNA binding of NF-Y to the human major histocompatibility complex class II human leukocyte antigen DRA Y-box sequence in a dose-dependent manner. Interestingly, other binding proteins identified by this method also possess the same or a similar motif. These results clearly demonstrate that NF-YB and thymosin β-10 are specific cellular binding proteins of HMN-154 and that this shared region is necessary for the binding to HMN-154.

Phage display was originally developed for the generation of antibody libraries and the selection of antibodies that bind strongly to a specific antigen. Now it can also be configured to identify drug-binding proteins. A significant advantage of phage display over other technologies is that the iterative amplification steps allow the identification of low-abundance targets and of targets of low-affinity ligands. For this purpose, a library of DNA sequences is cloned into phage genomes as fusions with a phage coat protein gene, in order to display potential target proteins on the phage surface. This phage library is then exposed to the immobilized small molecule, resulting in the capture of phages that have the appropriate affinity for the drug molecule. Non-binding phages are washed off, and the binding phages are eluted and transfected into bacterial host cells for amplification. This amplified phage population is greatly enriched in phages that display proteins that bind to the small molecule. After iterative rounds of affinity enrichment, a monoclonal phage population can be selected, and then DNA isolation and sequencing can be used to identify the binding protein based on sequence-similarity searches.

Recently, phage display was successfully used to identify the molecular target of the curcumin derivative HBC, a molecule which inhibits the proliferation of several human cancer cells [124]. In this study, Shim *et al.* employed phage display biopanning analysis from genome-wide human cDNA libraries. Consequently, they isolated and identified Ca^{2+}/CaM as a putative target protein of HBC. Direct interaction between HBC and Ca^{2+}/CaM was confirmed using both phage display binding assay and surface plasmon resonance. In biological systems, HBC induces sustained phosphorylation of ERK1/2 and activates p21 expression, resulting in the suppression of the cell cycle progression of HCT15 colon cancer cells. These biological activities of HBC are similar to those of other known Ca^{2+}/CaM antagonists, suggesting that Ca^{2+}/CaM is a biologically relevant target of HBC.

Kahne and co-workers used a mixed chemical genetic and genetic approach to identify the gene target responsible for the activity difference between vancomycin and gly-colipid derivatives of vancomycin [125]. Vancomycin is a drug used in the treatment of resistant Gram-positive bacterial infections. Vancomycin binds to D-Ala-D-Ala in peptidoglycan precursors and inhibits the transglycosylation step in peptidoglycan synthesis. Vancomycin-resistant strains express a D-Ala-D-Lac motif to which vancomycin cannot bind. However, hydrophobic substitution to the carbohydrate moiety on vancomycin overcomes this resistance through a different pathway and does so more rapidly. Based on this, the group sought to determine the genetic basis. This was accomplished by employing a standard genetic screening method in which they screened for genetic mutations that confer resistance. Of the resistant strains identified, they all contained the mutation *yfgL* with no known gene function. Through follow-up experiments, they implicated *yfgL* as conferring resistance to vancomycin derivatives, but not to vancomycin itself. Interestingly, a genetic approach would have failed, since the *yfgL* deletion is lethal. Since it was shown that *ygfL* gives rise to a discernible phenotype only in the presence of small molecules that perturb transglycosylation in peptidoglycan synthesis, they proposed that *yfgL* is involved in regulating that process.

3.5
Conclusion and Perspectives

As mentioned at the beginning of this chapter, the chemical space, which encompasses all possible small organic molecules including those present in biological systems, is enormous, and so far only a tiny fraction has been explored. Nevertheless, the discovery of new bioactive molecules, facilitated by a deeper understanding of the nature of the regions of chemical space that are relevant to biology, will advance our knowledge of biological processes and eventually lead to new therapies to treat diseases. Much of our current knowledge of biology has been derived from genetics. However, there are limitations to this approach alone. The chemical genetic approach uses cell-permeable and selective small molecules to perturb protein function rapidly, reversibly, and conditionally with temporal and quantitative control in any biological system. Clearly, this approach is powerful especially when combined with genetic and other biochemical information. The use of natural products such as Taxol (paclitaxel), colchicine, and cyclopamine in biological investigations put this point beyond doubt, and highlights the strength of the chemical genetics approach in studying dynamic processes that are often intractable otherwise.

It is evident that the field of chemical genetics is growing rapidly. Chemical genetics has now become a standard tool of many researchers, including chemists and biologists. The proof of principle stage of chemical genetics has long passed. Many are seeking to optimize and expand upon the existing tools. In order to generalize the chemical genetics approach, the three essential aspects discussed in this chapter, that is, generation of structurally diverse and complex small molecules, reliable large-scale screening assays, and systematic methods for target elucidation, still remain significant challenges. The first aspect stresses the importance of synthetic chemistry (such as DOS) in materializing the full potential of the approach toward chemical genomics. More comprehensive compound libraries will allow us to perturb an increasing percentage of the macromolecules that make up living systems. Both the second and third aspects actually require more complex and sophisticated phenotypic assays and protein-binding measurements that can be performed on vast arrays of molecules. While technologies such as small-molecule microarrays and protein microarrays will surely find a wide range of applications, it will be interesting to see what technologies will be further integrated into chemical genetic studies. It is expected that more and more in the future, simple reports of an inhibitor being identified from one type of screening may well be replaced by "full-story" studies that incorporate inhibitor identification with in-depth biochemical and genetic analysis.

Given the significance of chemical biology and the multitude of challenges ahead, there is no doubt that chemical biology will continue to deliver ground-breaking new findings in the next decade. So, what is the future of chemical biology, what should its objectives be, and what should one try to accomplish? Although it is difficult to formulate a general answer to these questions, it seems that identification of highly specific small-molecule modulators of each individual gene function in the cell and creation of an effective bridge between basic and clinical research by systematically linking genetic variation in cells to the ability of small molecules to effect phenotypic changes in the cell should be among the ultimate goals of future chemical biology. These goals were actually narrated by Stuart Schreiber in a commentary published in 2005 [2]. Thus, chemical biology studies in the future should continue to investigate the biological impact of structurally modified versions of all types of cellular components, it should continue to develop and refine chemistry for the direct covalent modification of cells, and it should expand the scope of small-molecule-based functional modulation from proteins to nucleic acids. In addition, chemical biology should further strengthen its link with therapy-oriented biomedical research, not only at the level of target identification, but also in areas such as medicinal chemistry and pharmacology.

References

1. Bohacek, R.S., McMartin, C., and Guida, W.C. (1996) *Med. Res. Rev.*, **16**, 3–50.
2. Schreiber, S.L. (2005) *Nat. Chem. Biol.*, **1**, 64–66.
3. Dervan, P.B. (2005) *Bioorg. Med. Chem.*, **9**, 2215–2235.
4. Li, X. and Liu, D.R. (2004) *Angew. Chem. Int. Ed.*, **43**, 4848–4870.
5. Winkler, W.C. and Breaker, R.R. (2003) *ChemBioChem*, **4**, 1024–1032.
6. Clardy, J. and Walsh, C. (2004) *Nature*, **432**, 829–837.
7. Pfeifer, B.A. and Khosla, C. (2001) *Microbiol. Mol. Biol. Rev.*, **65**, 106–118.
8. Stockwell, B.R. (2004) *Nature*, **432**, 846–854.
9. Dobson, C.M. (2004) *Nature*, **432**, 824–828.
10. Goto, S., Okuno, Y., Hattori, M., Nishioka, T., and Kanehisa, M. (2002) *Nucleic Acids Res.*, **30**, 402–404.
11. Ehrlich, P. (1913) *Lancet*, **ii**, 445–451.
12. Morrison, K.L. and Weiss, G.A. (2006) *Nat. Chem. Biol.*, **2**, 3–6.
13. Stockwell, B.R. (2000) *Nat. Rev. Genet.*, **1**, 116–125.
14. MacBeath, G. (2001) *Genome Biol.*, **2**, 2005.1–2005.6.
15. Darvas, F., Dorman, G., Krajcsi, P., Puskas, L.G., Kovari, Z., Lorincz, Z., and Urge, L. (2004) *Curr. Med. Chem.*, **11**, 3119–3145.
16. Shokat, K. and Velleca, M. (2002) *Drug Discov. Today*, **7**, 872–879.
17. Alaimo, P.J., Shogren-Knaak, M.A., and Shokat, K.M. (2001) *Curr. Opin. Chem. Biol.*, **5**, 360–367.
18. Ward, G.E., Carey, K.L., and Westwood, N.J. (2002) *Cell. Microbiol.*, **4**, 471–482.
19. Zheng, X.S., Chan, T.F., and Zhou, H.H. (2004) *Chem. Biol.*, **11**, 609–618.
20. Zheng, X.F. and Chan, T.F. (2002) *Drug Discov. Today*, **7**, 197–205.
21. Strausberg, R.L. and Schreiber, S.L. (2003) *Science*, **300**, 294–295.
22. Burdine, L. and Kodadek, T. (2004) *Chem. Biol.*, **11**, 593–597.

23. Zanders, E.D., Bailey, D.S., and Dean, P.M. (2002) *Drug Discov. Today*, **7**, 711–718.

24. Schreiber, S.L. (2000) *Science*, **287**, 1964–1969.

25. Newman, D.J. and Cragg, G.M. (2007) *J. Nat. Prod.*, **70**, 461–477.

26. David, J.N. (2008) *J. Med. Chem.*, **51**, 2589–2599.

27. Chen, Y., Bilban, M., Foster, C.A., and Boger, D.L. (2002) *J. Am. Chem. Soc.*, **124**, 5431–5440.

28. Tsang, C.K., Qi, H., Liu, L.F., and Zheng, X.F. (2007) *Drug Discov. Today*, **12**, 112–124.

29. Smith, A.B. III, Cho, Y.S., Pettit, G.R., and Hirschmann, R. (2003) *Tetrahedron*, **59**, 6991–7009.

30. Nicolaou, K.C., Pfefferkorn, J.A., Roecker, A.J., Cao, G.Q., Barluenga, S., and Mitchell, H.J. (2000) *J. Am. Chem. Soc.*, **122**, 9939–9953.

31. Nicolaou, K.C., Pfefferkorn, J.A., Mitchell, H.J., Roecker, A.J., Barluenga, S., Cao, G.Q., Affleck, R.L., and Lillig, J.E. (2000) *J. Am. Chem. Soc.*, **122**, 9954–9967.

32. Nicolaou, K.C., Pfefferkorn, J.A., Barluenga, S., Mitchell, H.J., Roecker, A.J., and Cao, G.Q. (2000) *J. Am. Chem. Soc.*, **122**, 9968–9976.

33. Nicolaou, K.C., Pfefferkorn, J.A., Schuler, F., Roecker, A.J., Cao, G.Q., and Casida, J.E. (2000) *Chem. Biol.*, **7**, 979–992.

34. Nicolaou, K.C., Roecker, A.J., Barluenga, S., Pfefferkorn, J.A., and Cao, G.Q. (2001) *ChemBioChem*, **2**, 460–465.

35. Nicolaou, K.C., Evans, R.M., Roecker, A.J., Hughes, R., Downes, M., and Pfefferkorn, J.A. (2003) *Org. Biomol. Chem.*, **1**, 908–920.

36. Downes, M., Verdecia, M.A., Roecker, A.J., Hughes, R., Hogenesch, J., Kast-Woelbern, H.R., Bowman, M.E., Ferrer, J.L., Anisfeld, A.M., Edwards, P.A., Rosenfeld, J.M., Alvarez, J.G., Noel, J.P., Nicolaou, K.C., and Evans, R.M. (2003) *Mol. Cell*, **11**, 1079–1092.

37. Tan, C., de Noronha, R.G., Roecker, A.J., Pyrzynska, B., Khwaja, F., Zhang, Z., Zhang, H., Teng, Q., Nicholson, A.C., Giannakakou, P., Zhou, W., Olson, J.J., Pereira, M.M., Nicolaou, K.C., and Van Meir, E.G. (2005) *Cancer Res.*, **65**, 605–612.

38. Nicolaou, K.C. and Pfefferkorn, J.A. (2001) *Biopolymers*, **60**, 171–193.

39. Young, S.S. and Ge, N. (2004) *Curr. Opin. Drug Discov. Dev.*, **7**, 318–324.

40. Mayer, T.U., Kapoor, T.M., Haggarty, S.J., King, R.W., Schreiber, S.L., and Mitchison, T.J. (1999) *Science*, **286**, 971–974.

41. Torrance, C.J., Agrawal, V., Bert, V., and Kinzler, K.W. (2001) *Nat. Biotechnol.*, **19**, 940–945.

42. Grozinger, C.M., Chao, E.D., Blackwell, H.E., Moazed, D., and Schreiber, S.L. (2001) *J. Biol. Chem.*, **276**, 38837–38843.

43. Peterson, R.T., Link, B.A., Dowling, J.E., and Schreiber, S.L. (2000) *Proc. Natl. Acad. Sci. U. S. A.*, **97**, 12965–12969.

44. Stockwell, B.R., Hardwick, J.S., Tong, J.K., and Schreiber, S.L. (1999) *J. Am. Chem. Soc.*, **121**, 10662–10663.

45. Rawlins, M.D. (2004) *Nat. Rev. Drug Discov.*, **3**, 360–364.

46. O'Connor, K.A. and Roth, B.L. (2005) *Nat. Rev. Drug Discov.*, **4**, 1005–1014.

47. Sundberg, S.A. (2000) *Curr. Opin. Biotechnol.*, **11**, 47–53.

48. Dolle, R.E. (1998) *Mol. Diversity*, **3**, 199–233.

49. Dolle, R.E. and Nelson, K.H. Jr. (1999) *J. Comb. Chem.*, **1**, 235–282.

50. Norman, T.C., Smith, D.L., Sorger, P.K., Drees, B.L., O'Rourke, S.M., Hughes, T.R., Roberts, C.J., Friend, S.H., Fields, S., and Murray, A.W. (1999) *Science*, **285**, 591–595.

51. Rosania, G.R., Chang, Y.T., Perez, O., Sutherlin, D., Dong, H., Lockhart, D.J., and Schultz, P.G. (2000) *Nat. Biotechnol.*, **18**, 304–308.

52. Walsh, D.P. and Chang, Y.T. (2006) *Chem. Rev.*, **106**, 2476–2530.

53. Stockwell, B.R. (2000) *Trends Biotechnol.*, **18**, 449–455.

54. Stockwell, B.R. (2002) *Neuron*, **36**, 559–562.

55. Dolma, S., Lessnick, S.L., Hahn, W.C., and Stockwell, B.R. (2003) *Cancer Cell*, **3**, 285–296.

56. Stegmaier, K., Ross, K.N., Colavito, S.A., O'Malley, S., Stockwell, B.R., and Golub, T.R. (2004) *Nat. Genet.*, **36**, 257–263.

57. Yarrow, J.C., Perlman, Z.E., Westwood, N.J., and Mitchison, T.J. (2004) *BMC Biotechnol.*, **4**, 21–29.

58. Kau, T.R., Schroeder, F., Ramaswamy, S., Wojciechowski, C.L., Zhao, J.J., Roberts, T.M., Clardy, J., Sellers, W.R., and Silver, P.A. (2003) *Cancer Cell*, **4**, 463–476.

59. MacBeath, G., Koehler, A.N., and Schreiber, S.L. (1999) *J. Am. Chem. Soc.*, **121**, 7967–7968.

60. Winssinger, N., Ficarro, S., Schultz, P.G., and Harris, J.L. (2002) *Proc. Natl. Acad. Sci. U. S. A.*, **99**, 11139–11144.

61. Kuruvilla, F.G., Shamji, A.F., Sternson, S.M., Hergenrother, P.J., and Schreiber, S.L. (2002) *Nature*, **416**, 653–657.

62. Jona, G. and Snyder, M. (2003) *Curr. Opin. Mol. Ther.*, **5**, 271–277.

63. MacBeath, G. (2002) *Nat. Genet.*, **32**, 526–532.

64. Zhu, H., Bilgin, M., Bangham, R., Hall, D., Casamayor, A., Bertone, P., Lan, N., Jansen, R., Bidlingmaier, S., Houfek, T., Mitchell, T., Miller, P., Dean, R.A., Gerstein, M., and Snyder, M. (2001) *Science*, **293**, 2101–2105.

65. Espejo, A., Cote, J., Bednarek, A., Richard, S., and Bedford, M.T. (2002) *Biochem. J.*, **367**, 697–702.

66. Newman, J.R. and Keating, A.E. (2003) *Science*, **300**, 2097–2101.

67. Ziauddin, J. and Sabatini, D.M. (2001) *Nature*, **411**, 107–110.

68. Ramachandran, N., Hainsworth, E., Bhullar, B., Eisentein, S., Rosen, B., Lau, A.Y., Walter, J.C., and LaBear, J. (2004) *Science*, **305**, 86–90.

69. Lefurgy, S. and Cornish, V. (2004) *Chem. Biol.*, **11**, 151–153.

70. Liberles, S.D., Diver, S.T., Austin, D.J., and Schreiber, S.L. (1997) *Proc. Natl. Acad. Sci. U. S. A.*, **94**, 7825–7830.

71. Tuschl, T. (2002) *Nat. Biotechnol.*, **20**, 446–448.

72. Elbashir, S.M., Harborth, J., Lendeckel, W., Yalcin, A., Weber, K., and Tuschl, T. (2001) *Nature*, **411**, 494–498.

73. Hannon, G.J. (2002) *Nature*, **418**, 244–251.

74. Root, D.E., Kelley, B.P., and Stockwell, B.R. (2002) *Curr. Opin. Drug Discov. Dev.*, **5**, 355–360.

75. King, R. and Elands, J. (2002) *Mod. Drug Discov.*, **5**, 21.

76. Schreiber, S.L. (2003) *Chem. Eng. News*, **81**, 51–61.

77. Agrafiotis, D.K., Lobanov, V.S., and Salemme, F.R. (2002) *Nat. Rev. Drug Discov.*, **1**, 337–346.

78. Buckingham, S. (2003) *Nature*, **425**, 209–215.

79. Kelley, B.P., Lunn, M.R., Root, D.E., Flaherty, S.P., Martino, A.M., and Stockwell, B.R. (2004) *Chem. Biol.*, **11**, 1495–1503.

80. Leblanc, B., Coulet, A., Andre, E., Molina, F., and Leonetti, J.P. (2004) *Biotechniques*, **37**, 223–225.

81. Bleicher, K.H., Bohm, H.J., Muller, K., and Alanine, A.I. (2003) *Nat. Rev. Drug Discov.*, **2**, 369–378.

82. Townsend, J.A., Adams, S.E., Waudby, C.A., de Souza, V.K., Goodman, J.M., and Murray-Rust, P. (2004) *Org. Biomol. Chem.*, **22**, 294–300.

83. Xu, Y., Shi, Y., and Ding, S. (2008) *Nature*, **453**, 338–344.

84. Emre, N., Coleman, R., and Ding, S. (2007) *Curr. Opin. Chem. Biol.*, **11**, 252–258.

85. Ao, A., Hao, J., and Hong, C.C. (2011) *Chem. Biol.*, **18**, 413–424.

86. Boyer, L.A., Lee, T.I., Cole, M.F., and Young, R.A. (2005) *Cell*, **122**, 947–956.

87. Lee, T.I., Jenner, R.G., Boyer, L.A., and Young, R.A. (2006) *Cell*, **125**, 301–313.

88. Bernstein, B.E., Mikkelsen, T.S., Xie, X., and Lander, E.S. (2006) *Cell*, **125**, 315–326.

89. Mikkelsen, T.S., Ku, M., Jaffe, D.B., and Bernstein, B.E. (2007) *Nature*, **448**, 553–560.

90. Scadden, D.T. (2006) *Nature*, **441**, 1075–1079.

91. Chen, S., Do, J.T., Zhang, Q., Yao, S., Yan, F., Peters, E.C., Schöler, H.R., Schultz, P.G., and Ding, S. (2006) *Proc. Natl. Acad. Sci. U. S. A.*, **103**, 17266–17271.

92. Ludwig, T.E., Levenstein, M.E., Jones, J.M., and Thomson, J.A. (2006) *Nat. Biotechnol.*, **24**, 185–187.

93. Yao, S., Chen, S., Clark, J., Hao, E., Beattie, G.M., Hayek, A., and Ding, S. (2006) *Proc. Natl. Acad. Sci. U. S. A.*, **103**, 6907–6912.

94. D'Amour, K.A., Bang, A.G., Eliazer, S., Kelly, O.G., Agulnick, A.D., Smart, N.G., Moorman, M.A., Kroon, E., Carpenter,

M.K., and Baetge, E.E. (2006) *Nat. Biotechnol.*, **24**, 1392–1401.

95. Chen, S.B., Zhang, Q.S., Wu, X., Schultz, P.G., and Ding, S. (2004) *J. Am. Chem. Soc.*, **126**, 410–411.

96. Takahashi, K. and Yamanaka, S. (2006) *Cell*, **126**, 663–676.

97. Burdon, T., Smith, A., and Savatier, P. (2002) *Trends Cell Biol.*, **12**, 432–438.

98. Avery, S., Inniss, K., and Moore, H. (2006) *Stem Cells Dev.*, **15**, 729–740.

99. Dudley, D.T., Pang, L., Decker, S.J., Bridges, A.J., and Saltiel, A.R. (1995) *Proc. Natl. Acad. Sci. U. S. A.*, **92**, 7686–7689.

100. Wang, X.Z. and Ron, D. (1996) *Science*, **272**, 1347–1349.

101. Leclerc, S., Garnier, M., Hoessel, R., Marko, D., Bibb, J.A., Snyder, G.L., Greengard, P., Biernat, J., Wu, Y.Z., and Mandelkow, E.M. (2001) *J. Biol. Chem.*, **276**, 251–260.

102. Meijer, L., Skaltsounis, A.L., Magiatis, P., Polychronopoulos, P., and Dajani, R. (2003) *Chem. Biol.*, **10**, 1255–1266.

103. Eggert, U.S. and Mitchison, T.J. (2006) *Curr. Opin. Chem. Biol.*, **10**, 232–223.

104. Wang, S., Sim, T.B., Kim, Y.S., and Chang, Y.T. (2004) *Curr. Opin. Chem. Biol.*, **8**, 371–377.

105. Terstappen, G.C., Schlupen, C., Raggiaschi, R., and Gaviraghi, G. (2007) *Nat. Rev. Drug Discov.*, **6**, 891–903.

106. Leslie, B.J. and Hergenrother, P.J. (2008) *Chem. Soc. Rev*, **37**, 1347–1360.

107. Harding, M.W., Galat, A., Uehling, D.E., and Schreiber, S.L. (1989) *Nature*, **341**, 758–760.

108. Jung, D.W., Williams, D., Khersonsky, S.M., Kang, T.W., Heidary, N., Chang, Y.T., and Orlow, S.J. (2005) *Mol. BioSyst.*, **1**, 85–92.

109. Snyder, J.R., Hall, A., Ni-Komatsu, L., Khersonsky, S.M., Chang, Y.T., and Orlow, S.J. (2005) *Chem. Biol.*, **12**, 477–484.

110. Khersonsky, S.M., Jung, D.W., Kang, T.W., Walsh, D.P., Moon, H.S., Jo, H., Jacobson, E.M., Shetty, V., Neubert, T.A., and Chang, Y.T. (2003) *J. Am. Chem. Soc.*, **125**, 11804–11805.

111. Maly, D.J., Choong, I.C., and Ellman, J.A. (2000) *Proc. Natl. Acad. Sci. U. S. A.*, **97**, 2419–2424.

112. Reddy, M.M., Bachhawat-Sikder, K., and Kodadek, T. (2004) *Chem. Biol.*, **11**, 1127–1137.

113. Falsey, R.R., Marron, M.T., Gunaherath, G.M.K.B., Shirahatti, N., Mahadevan, D., Gunatilaka, A.A.L., and Whitesell, L. (2006) *Nat. Chem. Biol.*, **2**, 33–38.

114. Zhu, S., McHenry, K.T., Lane, W.S., and Fenteany, G. (2005) *Chem. Biol.*, **12**, 981–991.

115. Leuenroth, S.J. and Crews, C.M. (2005) *Chem. Biol.*, **12**, 1259–1268.

116. Leuenroth, S.J., Okuhara, D., Shotwell, J.D., Markowitz, G.S., Yu, Z., Somlo, S., and Crews, C.M. (2007) *Proc. Natl. Acad. Sci. U. S. A.*, **104**, 4389–4394.

117. Dorman, G. and Prestwich, G.D. (2000) *Trends Biotechnol.*, **18**, 64–77.

118. Fujii, T., Manabe, Y., Sugimoto, T., and Ueda, M. (2005) *Tetrahedron*, **61**, 7874–7893.

119. Kino, T., Hatanaka, H., Hashimoto, M., Nishiyama, M., Goto, T., Okuhara, M., Kohsaka, M., Aoki, H., and Imanaka, H. (1987) *J. Antibiot.*, **40**, 1249–1255.

120. Kino, T., Hatanaka, H., Miyata, S., Inamura, N., Nishiyama, M., Yajima, T., Goto, T., Okuhara, M., Kohsaka, M., and Aoki, H. (1987) *J. Antibiot.*, **40**, 1256–1265.

121. Liu, J., Farmer, J.D. Jr., Lane, W.S., Friedman, J., Weissman, I., and Schreiber, S.L. (1991) *Cell*, **66**, 807–815.

122. Marton, M.J., DeRisi, J.L., Bennett, H.A., Iyer, V.R., Meyer, M.R., Roberts, C.J., Stoughton, R., Burchard, J., Slade, D., Dai, H.Y., Bassett, D.E., Hartwell, L.H., Brown, P.O., and Friend, S.H. (1998) *Nat. Med.*, **4**, 1293–1301.

123. Tanaka, H., Ohshima, N., and Hidaka, H. (1999) *Mol. Pharmacol.*, **55**, 356–363.

124. Shim, J.S., Lee, J., Park, H.J., Park, S.J., and Kwon, H.J. (2004) *Chem. Biol.*, **11**, 1455–1463.

125. Eggert, U.S., Ruiz, N., Falcone, B.V., Branstrom, A.A., Goldman, R.C., Silhavy, T.J., and Kahne, D. (2001) *Science*, **294**, 361–364.

Commentary Part

Comment 1

Ke Ding

Chemical biology has a long history but has developed rapidly as an interdisciplinary field that has been widely applied in the biological and biomedical sciences. Given the significant impact of chemical biology and the multitude of challenges in understanding living life, there is no doubt that chemical biology will continue to deliver ground-breaking new findings in the next decade.

In this important chapter, Renxiao Wang summarized the history of chemical biology and clearly defined chemical biology as *"both the use of chemistry to advance a molecular understanding of biology and the harnessing of biology to advance chemistry."* The fundamental difference between chemical biology and biological chemistry was also clearly delineated. Furthermore, this chapter focused on the use of chemical approaches to solve problems in living systems by emphasizing the three main components of chemical biology: chemical library preparation; screening strategies and target elucidation and validation.

In the chemical libraries preparation part, the author outlined the major methods for preparing chemical libraries, which included natural product-inspired synthesis and DOS. Several available chemical libraries were also listed. With the development of both chemistry and biology, many screening strategies have been developed for testing compounds' biological activities. Among them, phenotypic assay and *in vitro* binding assays are two major approaches. With the significant technical advances, gene expression signatures and living cell image have become emerging trends in high-throughput phenotypic screening. Both labeled and label-free technologies have been widely applied to direct ligand binding screening methods. The three-hybrid system, which does not require protein purification, has also been successfully developed and applied to the identification of new chemical probes. However, these screening approaches have their own limitations. All these technologies require significant investment in equipment and knowledge bases for HTS. False-positive results and specificity of the compounds were other limitations. Therefore,

a secondary method is essential for activity validation. Quality control and later data management are also significant challenges in screening.

Target elucidation is the most challenging aspect of chemical biology. This chapter provides a conceptual framework and two major categories of approaches for target identification. Employing chemically modified small molecule as bait to trap the protein target is one classical approach, and hypothesis-driven and genetics-based approaches are another. However, the latter approaches do not need chemically modified small molecules. A few simple examples were also given to demonstrate the strategies of target identification.

I firmly believe that this chapter covers most of the interesting aspects in chemical biology area and provides valuable information for those scientists working in the field of chemical biology. In addition, the reader will find classical examples, current technical limitations and significant challenges in chemical biology for library preparation, screening, and target identification.

Since stem cell chemical biology is an emerging field, is it possible to report some advances in this area?

Comment 2

Li-He Zhang

In this chapter, the general methodology for study of chemical biology was well described. I should like to emphasize that despite a diversity of opinions on how to define chemical biology, modern chemical biology is a new interdisciplinary field that is focused on the insights of scientific questions for a complex biological system. Chemical biologists are interested not only in developing chemical tools and methods but also in applying them to make profound discoveries.

First, synthetic and natural bioactive compounds continue to provide a starting point for chemical biologists to investigate basic biological processes, many of which have important implications for exploring new signaling pathways and drug development. However, there is a recent tendency to develop new chemical tools for the study of complex biological systems. For example, the burgeoning area of stem cell biology encourages chemical biologists to provide more specific probes to induce pluripotent stem cells and regulate the process instead of the regulation by Oct3/4, Sox2, c-Myc, and Klf4 [C1]. The same challenge also arises in the non-coding RNA field [C2]. To obtain new insights into the complex networks in a biological system, chemical biologists will have increasing collaboration with the related research laboratories.

Second, chemical tools with the mechanistic thinking of chemical biologists are more important. For example, Cu(I)-catalyzed Huisgen [3 + 2]-cycloaddition is a well-known organic reaction; however, with intellectual and technological threads from molecular biology, click reactions were applied to resolve many interesting biological problems [C3]. Recently, Bertozzi and co-workers investigated the glycosidation in live zebrafish embryos by using a click reaction to attach a fluorescent probe to GalNAc-containing glycans [C4]. Further, he and co-workers applied click reactions to reveal the genome-wide distribution of 5-hydroxymethylcytosine [C5].

Finally, modern chemical biology makes multidisciplinary areas focus on the main scientific questions in human biology and changes the traditional definitions of both chemistry and biology.

Comment 3

Jun-Ying Yuan

The completion of the entire human genome sequence revealed the entire booklet of our genetic code. Our challenge now is to understand this booklet written in four-letter language, ATGC. There are mega-proteomics projects under way to elucidate the structures of the entire proteome and studies about the global protein post-translational modifications, and so on. Concerning functional studies, projects to create genetic mutations for all of the mouse genes are also near completion. There has been tremendous progress in each of these areas and our knowledge about how our genetics work to control protein functions, development, and adult homeostasis has been increasing in an exponential manner. However, what is unclear or uncertain is the extent to which we will be able to utilize such information for the benefit of improving the quality of our lives and treating diseases. As pointed out by Renxiao Wang in this chapter, from the viewpoint of a chemist, we are lagging behind in developing small molecules that can target specific protein functions. As a biologist, I strongly concur with this opinion. Developing drugs is notoriously difficult and costly, as demonstrated by the decades of time and billions of dollars required to develop a drug. However, this is not because of the lack of chemical diversity: the chemistry universe is vast and highly diverse, with currently around 64 million ($<10^8$) organic and inorganic substances. This is in contrast to 20 000+ genes in the human genome, and each gene might encode several types of proteins due to splicing and post-translational modification. The numbers of different types of proteins in our cells are certainly much less than 64 million. Thus, with the help of modern chemistry, including DOS and focused chemical syntheses, it should theoretically be possible, given sufficient numbers of graduate students and funds, to discover chemical probes for all of the proteins. This is the ultimate goal of chemical biology.

The chapter by Wang provides a good overview of the past accomplishments and the current state in the field of chemical biology. However, we are clearly very far from reaching the ultimate goal of chemical biology. Why? A major obstacle to accomplishing this goal, in my opinion, is the lack of true collaboration between chemists and biologists. Chemistry and biology are two very different disciplines of science. Although it is possible for someone to be able to understand and utilize the concepts of both chemistry and biology, it is impossible, or almost impossible, for someone to be a visionary chemist and a visionary biologist at the same time. Most, if not all, of the current "chemical biology" laboratories consist of chemists who are trying to do biology with the help of some postdocs with biology training, or vice versa, and biologists who are trying to do some basic chemistry with the help of some postdocs with training in chemistry, which is not sufficient to create true ground-breaking discoveries in chemical biology. The success of chemical biology needs the close collaboration of chemists and biologists. I urge like-minded chemists and biologists to work together more effectively.

Author's Response to the Commentaries

I greatly appreciate the comments made by Profs. Ding, Zhang, and Yuan. Following Prof. Ding's suggestion, a new section, Chemical Approaches to Stem Cell Biology, has been added to the final version of this chapter. Prof. Zhang cited several very important research studies in this field, which I believe will benefit readers. I fully agree with Prof. Yuan that the success of chemical biology requires in-depth collaboration between chemists and biologists.

References

C1. Takahashi, K. and Yamanaka, S. (2006) *Cell*, **126** (4), 663–676.

C2. (a) Young, D.D., Connelly, C.M., Grohmann, C., and Deiters, A. (2010) *J. Am. Chem. Soc.*, **132**, 7976–7981; (b) Melo, S., Villanueva, A., Moutinho, C., Davalos, V., Spizzo, R., Ivan, C., Rossi, S., Setien, F., Casanovas, O., Simo-Riudalbas, L., Carmona, J., Carrere, J., Vidal, A., Aytes, A., Puertas, S., Ropero, S., Kalluri, R., Croce, C.M., Calin, G.A., and Esteller, M. (2011) *Proc. Natl. Acad. Sci. U. S. A.*, **108**, 4394–4399.

C3. Lewis, W.G., Green, L.G., Grynszpan, F., Radic, Z., Carlier, P.R., Taylor, P., Finn, M.G., and Sharpless, K.B. (2002) *Angew. Chem. Int. Ed.*, **41**, 1053–1057.

C4. (a) Baskin, J.M., Prescher, J.A., Laughlin, S.T., Agard, N.J., Chang, P.V., Miller, I.A., Lo, A., Codelli, J.A., and Bertozzi, C.R. (2007) *Proc. Natl. Acad. Sci. U.S.A.*, **104**, 16793–16797; (b) Laughlin, S.T., Baskin, J.M., Amacher, S.L., and Bertozzi, C.R. (2008) *Science*, **320**, 664–667.

C5. Song, C.-X., Szulwach, K.E., Fu, Y., Dai, Q., Yi, C., Li, X., Chen, C.-H., Zhang, W., Jian, X., Wang, J., Zhang, L., Looney, T.J., Zhang, B., Godley, L.A., Hicks, L.M., Lahn, B.T., Jin, P., and He, C. (2011) *Nat. Biotechnol.*, **29**, 68–72.

4
Biosynthesis of Pharmaceutical Natural Products and Their Pathway Engineering

Michael J. Smanski, Xu-Dong Qu, Wen Liu, and Ben Shen

4.1
Introduction

Natural products (NPs) are small molecules of stunning chemical complexity and molecular diversity that are produced by a wide range of living organisms. Although the benefits that NPs provide to human society are enormous and multifaceted, they are especially recognized for their critical role in drug discovery and development [1]. The study of NP biosynthetic pathways is situated at the interface of chemistry and biology and benefits from technological and conceptual advances in both fields. In the past 10 years, perhaps nothing has had a greater impact in the field of biology than the exponential growth of DNA sequencing technologies. It has truly revolutionized the way in which scientists view and interact with the natural world. From the technology used to sequence the first human genomes at the turn of the century [2, 3] until today, the cost of DNA sequencing has dropped by an astounding three orders of magnitude [4]. The relative ease of amassing millions of base pairs of sequence information from an organism or community of interest has left its mark on each of the life sciences, and NP chemistry is no exception. During the first decade of the twenty-first century, the explosion of DNA sequencing technologies has greatly benefited NP biosynthetic studies, where it has led to expanded paradigms of biosynthetic logic, new approaches to solving problems in NP biosynthesis, and a broadened understanding of NP diversity. Our growing knowledge of these biosynthetic pathways coupled with technological advancements in the field has facilitated the engineering of new pathways to produce molecules with improved pharmaceutical characteristics.

This chapter examines these areas of growth in more depth by providing specific examples to highlight the conceptual and technological leaps in NP biosynthetic studies during the past 10 years. We start by examining how the plethora of NP gene clusters that have been identified have revealed countless nuances and variations to the previously described paradigms of biosynthetic logic. It now seems that NP biosynthetic pathways lie on a continuous spectrum, with every imaginable permutation of the originally characterized prototypes manifested somewhere in Nature. This is perhaps not surprising given that Nature has been evolving production pathways to NPs that serve a multitude of functions and display an unimaginable diversity of molecular architectures for millennia.

Organic Chemistry – Breakthroughs and Perspectives, First Edition. Edited by Kuiling Ding and Li-Xin Dai.
© 2012 Wiley-VCH Verlag GmbH & Co. KGaA. Published 2012 by Wiley-VCH Verlag GmbH & Co. KGaA.

Next, we highlight some of the ways in which NP biosynthesis research has changed to accommodate newly available technologies and subsequent data sets. This is a very exciting topic, as it is apparent that our current ability to generate DNA sequence data exceeds our ability to analyze and explore fully the chemical potential hidden within them. The coming years will surely see continued growth in this respect. Finally, we touch upon some of the ways in which DNA sequence information has helped to convey the extent of NP diversity and evolution, along with the implications that this may have for future NP research. It is important to note that this chapter is not intended to provide a comprehensive review of the past decade of NP biosynthesis; the incredible volume of research that has been completed during this timeframe precludes such an undertaking. Instead, we will draw predominantly from the area with which we are most familiar, that of bacterial NP biosynthetic pathways, to highlight some of the major advances that have taken place over the past 10 years and that clearly demonstrate that the times have never been better for NPs drug discovery.

4.2
Expanded Paradigms in Biosynthetic Logic

4.2.1
Thio-Template Biosynthesis

Polyketides and nonribosomal peptides (NRPs), originating from short carboxylic acids and amino acids, respectively, are structurally complex and biologically important families of NPs. Many of these metabolites have been successfully exploited as clinically used pharmaceuticals, veterinary agents, and agrochemicals. Whereas polyketide synthases (PKSs), in general, catalyze C–C bond formation and use simple carboxylic acids to assemble polyketides, nonribosomal peptide synthetases (NRPSs) often incorporate amino acids into peptides by forming C–N bond linkages [5–8]. Both PKSs and NRPSs share a common thio-template mechanism to perform the biosynthesis via a 20 Å long phosphopantetheinyl group derived from coenzyme A (CoA). This group is flexible and terminated with a cysteamine thiol, which serves as the site of covalent attachment for the activated monomers and for the growing acyl chain intermediates (Scheme 4.1). Over the past decade, remarkable discoveries at the genetic and enzymatic levels have revealed striking variants of the PKS and NRPS catalysts distinct in catalytic logic, setting the stage for understanding the biosynthesis of polyketides, NRPs, and their hybrids showing various structural diversity. These advances have consequently paved the way for new drug development via the application of combinatorial biosynthesis methods.

4.2.1.1 **Archetypical PKS Paradigms**
PKSs biosynthesize polyketides via the decarboxylative condensation of short carboxylic acid precursors, in a manner chemically similar to that of fatty acid synthases (FASs). Three types of PKSs were established by the end of the twentieth century (Figure 4.1) [9]. Starting with purification of the first microbial PKS 6-methylsalicylic acid synthase (6-MSAS) in 1970 [10], and also uncovering of the PKS genes encoding

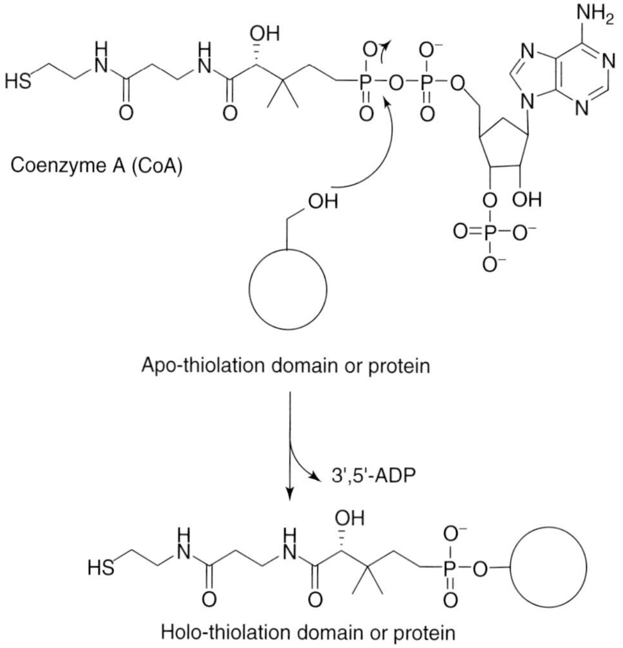

Scheme 4.1 Conversion of the thiolation domain or protein from an apo-form to a holo-form. Transfer of the phosphopantetheinyl group depends on nucleophilic attack of the hydroxyl group of the conserved serine residue on to the β-phosphate of CoA.

6-deoxyerythronolide B (6-dEB) (an intermediate of the antibiotic erythromycin) biosynthesis in 1990 [11, 12], a number of multifunctional PKSs, denoted type I, had been found. They resemble vertebrate FASs in organization, and harbor a set of distinct functional domains responsible for catalyzing the cycle of polyketide chain elongation and β-oxo functionality processing (Figure 4.1a). In 1984, cloning of the biosynthetic gene cluster of the aromatic polyketide actinorhodin provoked extensive investigations into the type II PKSs [13], which closely resemble bacterial FASs in architecture as they are multienzyme complexes that carry a single set of iteratively acting activities (Figure 4.1b). Following these advances, the type III PKSs [14], originally known only from plants, were found to be widely distributed in bacteria [15, 16], and type III PKSs were subsequently characterized as homodimeric proteins that act iteratively as condensing enzymes (Figure 4.1c) [17]. Type I and II PKSs use acyl carrier proteins (ACPs) to tether the CoA-activated acyl substrates and to channel the growing polyketide intermediates, whereas type III PKSs, independent of ACPs, act directly on the acyl-CoAs. Despite structural and mechanistic differences, all types of PKSs employ the ketoacylsynthase (KS) domain (for type I PKSs) or subunit (for types II and III PKSs) to catalyze C–C bond formation during the carbon chain assembly. Given the significant advances in microbial genome sequencing, the numbers of members belonging to each type of PKS have increased rapidly, providing the molecular basis to account for the vast structural diversities embodied by polyketide NPs [18]. Consistently, the growing number of structural analyses of PKSs has provided

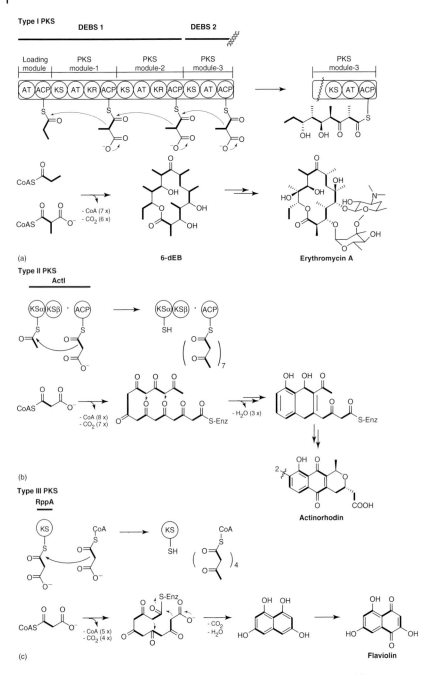

Figure 4.1 Archetypical PKSs in bacteria. (a) Modular type I PKS represented by DEBSs in erythromycin biosynthesis. (b) Type II PKS as exemplified by ActI in actinorhodin formation. (c) Type III PKS shown by RppA for affording the skeleton of flaviolin.

structural imperatives for the observed biochemistry and enriched our understanding of how these enzymes function as efficient catalysts [19]. However, recent progress in polyketide biosynthesis clearly indicates much greater diversity for PKS mechanism and structure. The emergence of new PKSs apparently has challenged the oversimplification of polyketide biosynthesis according to the types I, II, and III PKS paradigms [9].

4.2.1.2 Modular Type I PKSs and Their Broken Colinearity Rule

Members of the modular type I PKS class had long been considered to follow a colinearity rule, dictating that all enzyme activities required for chain extension and β-oxo functionality processing reside on a single module and that each active site is used only once, that is, noniteratively in a "one domain, one function" organization [20]. This PKS system is typically comprised of a loading module and multiple extension modules (Figure 4.1a), which are organized colinearly with their activities in the biosynthetic assembly line. Each module consists of the KS, acyltransferase (AT), and ACP domains for carbon chain extensio and, optionally, the dehydratase (DH), enoylreductase (ER), and ketoreductase (KR) domains for reductive modifications during chain extension cycle. This colinearity feature significantly expedites efforts to reprogram individual enzymatic activities on modular type I PKSs, as exemplified by engineering of 6-deoxyerythronolide B synthases (6-DEBSs) to produce the analogs utilizing most, if not all, of combinatorial biosynthesis strategies [21].

The growing number of exceptions to the colinearity rule have caused us to question the above well-established paradigm, as modular type I PKS systems do not always behave in a colinear fashion [22, 23]. On the one hand, certain PKS module activities can be bypassed from the biosynthetic machinery, termed "skipping," leading to the generation of the eventual product shorter in carbon length than that predicted on the basis of a modular constitution (Figure 4.2a). For instance, the pikromycin type I PKS system has the unique ability to produce two macrolactone products of differing ring size [24]. Although full extension through all elongation modules affords a heptaketide intermediate for cyclization to generate the 14-membered ring macrolactone narbonolide, premature termination of polyketide biosynthesis at the penultimate elongation module results in the production of the 12-membered macrolide 10-deoxymethynolide, the biosynthesis of which apparently does not involve the last module for carbon chain extension. On the other hand, some PKS modules can be used in more than one elongation cycle, termed "stuttering," to yield a polyketide intermediate longer than would normally be expected; such examples include the biosynthesis of the polyketides stigmatellin [25], borrelidin (Figure 4.2b) [26], and aureothin [27].

4.2.1.3 New Enzymology Complementing the Established Type I PKS Paradigms

Modular-type I PKS-programmed polyketide biosynthesis is usually initiated by an AT domain needed to load either an acyl-CoA or an α-carboxylated acyl-CoA (often followed by decarboxylation after loading) on to the ACP [28]. The variation of the starter unit is dependent on the AT selectivity (Figure 4.3a), such as propionyl-CoA is for erythromycin biosynthesis [29], isovaleryl-CoA is for avermectin biosynthesis [30], and cyclohexanoyl-CoA is for phoslactomycin biosynthesis [31]. The lack of this cognate AT may require a *trans*-loading system (Figure 4.3b), as has recently been suggested for

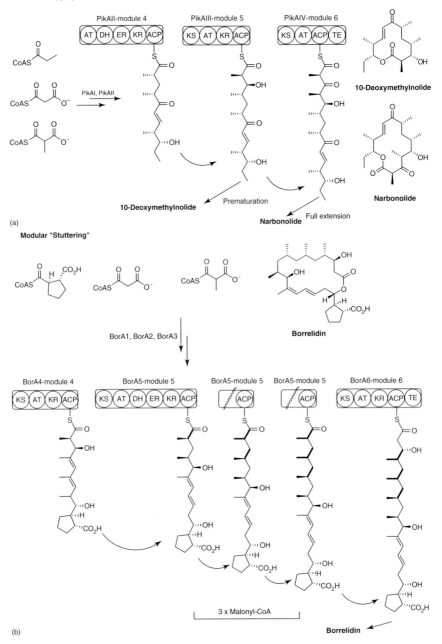

Figure 4.2 Modular type I PKSs acting against the colinearity rule. (a) "Skipping" as exemplified by "bypassing" the activity of module 6 in PikAIV to generate 10-deoxymethynolide. (b) "Stuttering" as exemplified by "repeating" the activity of module 5 in BorA5 to afford borrelidin.

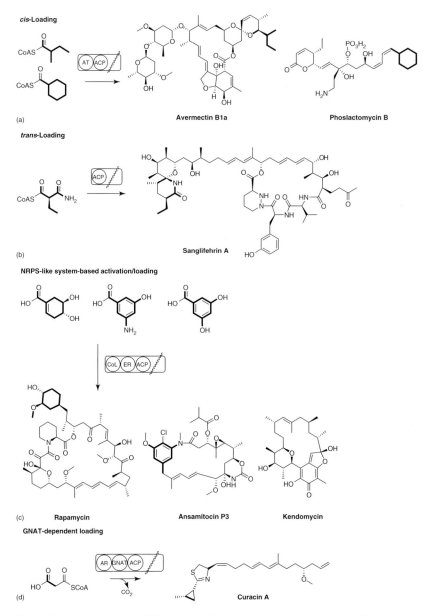

Figure 4.3 Starter units for PKS priming with variation of loading mechanism. (a) *cis*-Loading, shown as that of isovaleryl for avermectin B1a biosynthesis and that of cyclohexanoyl for phoslactomycin B biosynthesis. (b) *trans*-Loading as exemplified by that of (2*R*)-2-ethylmalonamyl in sanglifehrin A biosynthesis. (c) NRPS-like system-based activation/loading, as exemplified by those for incorporating dihydroxycyclohexenecarboxylic acid, 3-amino-5-hydroxybenzoic acid, and 3,5-dihydroxybenzoic acid into rapamycin, ansamitocin P3, and kendomycin, respectively. (d) GNAT-dependent loading shown by that for acetyl initiation in curacin A biosynthesis.

priming of sanglifehrin A biosynthesis by transferring (2R)-2-ethylmalonamyl to the single ACP-containing loading module [32]. If the loading starts with a free acid, the initiation of the polyketide assembly can involve an NRPS-like adenylation and thiolation system (Figure 4.3c), such as those seen in rapamycin/FK506 biosynthesis for incorporating dihydroxycyclohexenecarboxylic acid [33], in ansamitocin biosynthesis for 3-amino-5-hydroxybenzoic acid [34], and in kendomycin biosynthesis for 3,5-dihydroxybenzoic acid [35]. Understanding such priming mechanisms has helped to lay the foundation for pathway engineering [36] by showing that structural diversity can be achieved by enlarging the range of starter units. In addition, new functions can be integrated into the loading module to initiate polyketide biosynthesis. In curacin A/pederin biosynthesis, a domain related to general control of nonderepressible 5 *N*-acyltransferase (GNAT) sequentially catalyzes the decarboxylation of malonyl-CoA, yielding acetyl-CoA (Figure 4.3d), followed by direct transfer of the acetyl group on to the ACP loading domain [37, 38]. Thus, the GNAT domain is a bifunctional decarboxylase/*S*-acyltransferase that functions to initiate polyketide assembly.

Modular type I PKSs utilize α-carboxylated acyl-CoAs (or -ACPs) as the monomers, selection of which is controlled by the AT specificity, for a two-carbon unit extension in each elongation cycle [39]. Over the past 10 years, the pool of extensive units, deriving either from acetate or from non-acetate, has been greatly expanded, showing variable α-branching to the carbonyl group on the resulting polyketide products (Figure 4.4a). The reported examples include ethylmalonyl for concanamycin [40] and kirromycin [41], allylmalonyl for FK506 [42], chloroethylmalonyl for salinosporamide A [43], methoxymalonyl for FK520 [44], ansamitocin P-3 [45] and oxazolomycin [46], and hydroxymalonyl and aminomalonyl for zwittermicin A [47]. The formation of each of the unusual building blocks often requires pathway-specific enzyme(s) outside the PKS systems. For instance, ethylmalonyl-CoA [48], allylmalonyl-CoA [49], and chloroethylmalonyl-CoA [50] typically arise in crotonyl-CoA reductase/carboxylase-catalyzed α-carboxylation of acryloyl-CoA (or ACP) precursors, whereas hydroxymalonyl and methoxymalonyl units derive from D-1,3-bisphosphoglycerate [45, 51, 52], which is selected by a transferase/phosphatase bifunctional protein to generate glyceryl-ACP that then undergoes subsequent oxidative transformations (additional *O*-methylation for methoxymalonyl) to furnish the malonate group prior to incorporation into the polyketide chain. By contrast, an isoprenoid logic has recently been implicated in polyketide assembly showing that the alkylation for β-branching can take place at positions corresponding to former acetyl carboxyl groups (Figure 4.4b) [53]. Similarly to mevalonate biosynthesis, the β-substitutions, encoded by a set of pathway-specific isoprenoid branching genes, often require a key aldol addition of free-standing malonyl-ACP or methylmalonyl-ACP (via decarboxylation by a discrete KS) on to the ongoing modular ACP-tethered β-ketoacyl intermediate. The diversity in structure relies on further processing of the two- or three-carbon unit, including the methyl and cyclopropyl groups for curacin A [54] and bacillaene [55], ethyl or methoxymethyl groups for myxovirescins [56, 57], methylene group for pederin [37], acrylic ester group for bryostatin [58], and other moieties that have not yet been elucidated. Alternatively, the β-branch can be introduced presumably by an integrative branching domain in the modular type I PKS system via a Michael addition of an acetate unit on to the ACP-bound enoyl polyketide intermediate, such as the δ-lactone β-branch formation in rhizoxin [59].

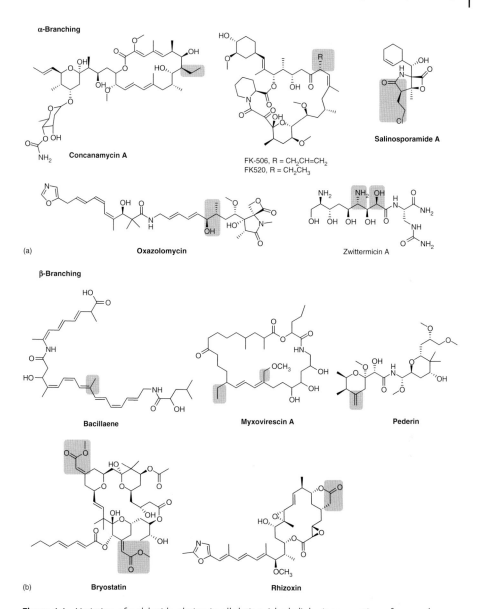

Figure 4.4 Variation of polyketide chains in alkylation (shaded) by incorporation of unusual extensive units. (a) α-Branching in the structures of concanamycin A, FK506/FK520, salinosporamide A, oxazolomycin, and zwittermicin A. (b) β-Branching in the structures of bacillaene, myxovirescin, pederin, bryostatin, and rhizoxin.

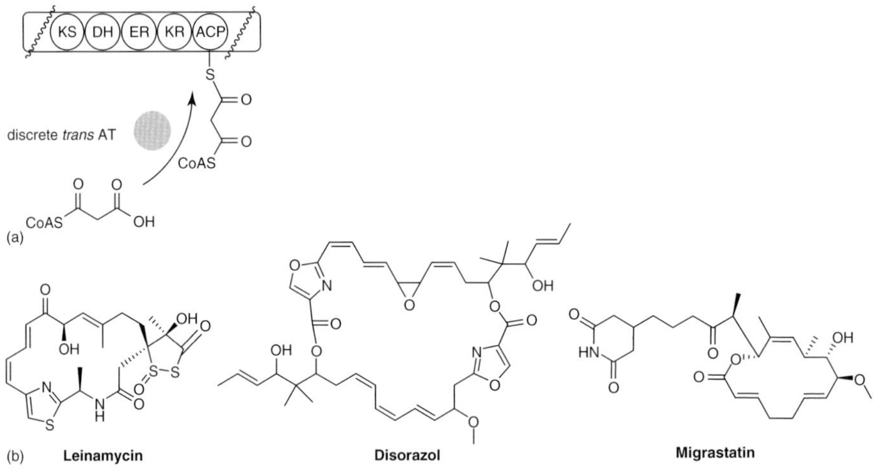

Figure 4.5 AT-less type I PKS system. Lack of the cognate *cis* AT domain can be functionally complemented with a discrete *trans* AT (a), as exemplified in the biosynthesis of leinamycin, disorazol, and migrastatin (b).

Modular type I PKSs are believed to function by a series of *cis*-acting catalytic domains to establish the polyketide backbone. However, the AT-less type I PKS systems have been identified in the biosynthesis of a number of polyketide NPs [60, 61], such as leinamycin [62], pederin [63], disorazol [64], and migrastatin [65], showing that the lack of the cognate AT domain can be functionally complemented with a discrete *trans* AT (Figure 4.5). This AT protein, acting iteratively, efficiently, and exclusively, loads malonyl-CoA *in trans* on to ACPs to initiate polyketide biosynthesis. In addition to the key AT domain for chain elongation in the PKS module, the reductive tailoring domain can also operate in a *trans* fashion, as exemplified by a recently characterized, general enoyl reduction pathway featuring a discrete *trans*-acting ER protein in polyketide biosynthesis [66, 67].

4.2.1.4 Iterative Type I PKSs in Bacteria
In bacteria, the biosynthesis of reduced polyketides (i.e., macrolides, polyethers, and polyenes) is normally catalyzed by modular type I PKSs, whereas the paradigm for aromatic polyketide biosynthesis had long been thought of as involving the iterative types II or III PKSs [9]. In contrast, PKSs that are categorized structurally as belonging to the type I class, but that act in an iterative manner for aromatic polyketide biosynthesis, are widely distributed in fungi [68]. For example, fungal 6-MSAS, along with other iterative PKSs such as the nonreducing member in aflatoxin biosynthesis and highly reducing member in lovastatin biosynthesis, have been extensively investigated as model systems by which to understand the mechanisms dictating chain-length control, regiospecific reduction, substrate tolerance, and subunit–domain interaction. The 6-MSAS-like, fungal-type PKSs have now been widely found in bacteria since the first example was observed in 1997 [69], suggesting their evolutionary relationship. These bacterial PKSs are typically organized in a KS–AT–DH–(KR)–ACP order (Figure 4.6a), catalyzing the biosynthesis of monocyclic moieties such as orsellinic acid in avilamycin [69] and calicheamicin [70]

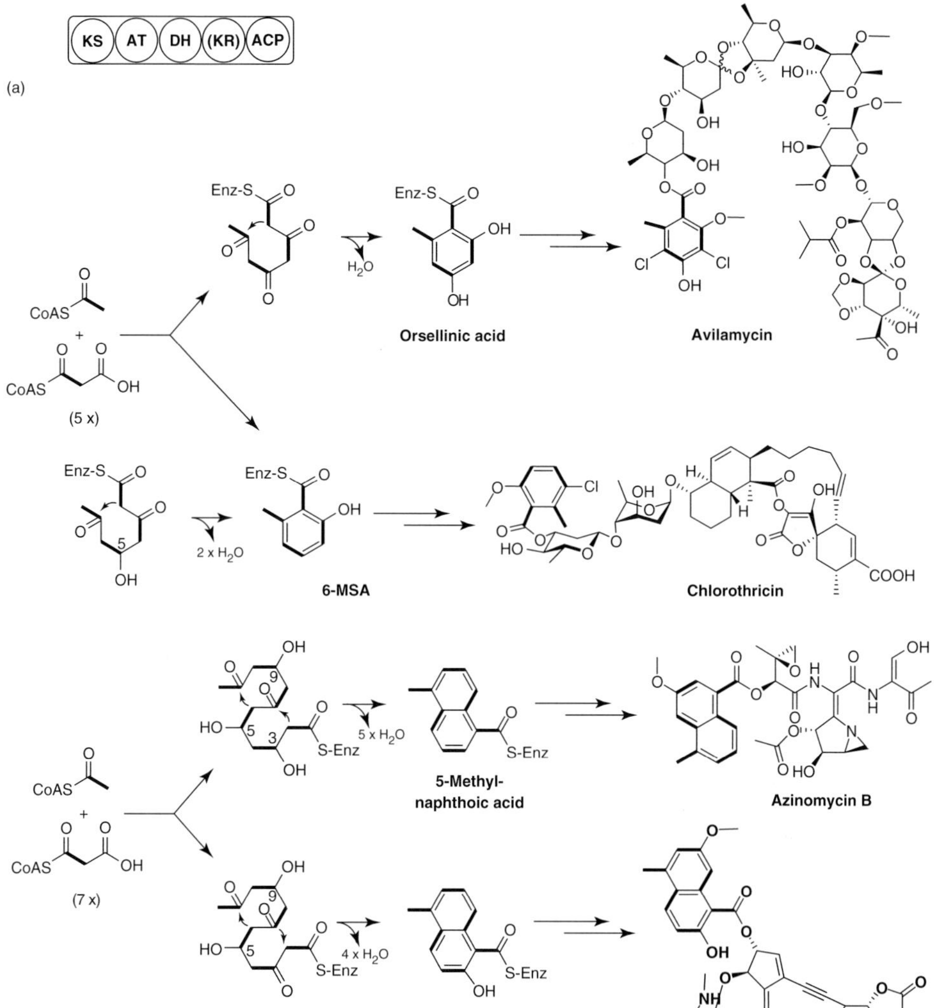

Figure 4.6 Type I PKSs for aromatic polyketide formation in bacteria. (a) Domain organization. (b) Types of condensation cycles and modes of selective reduction(s), as exemplified by forming orsellinic acid of avilamycin, 6-MSA of chlorothricin, 5-methylnaphthoic acid of azinomycin B, and 2-hydroxyl-5-methylnaphthoic acid of neocarzinostatin.

and 6-MSA in chlorothricin [71], maduropeptin [72], polyketomycin [73], and pactamycin [74], and the bicyclic moieties including 5-methylnaphthoic acid in azinomycin B [75] and its 2-hydroxylated analog in neocarzinostatin [76]. The formation of these aromatic polyketides differs not only in the two types of cyclization patterns and chain lengths affording mono- or bicyclic rings, but also in the four modes of selective reduction(s) necessary for establishing relevant substitution patterns (Figure 4.6b).

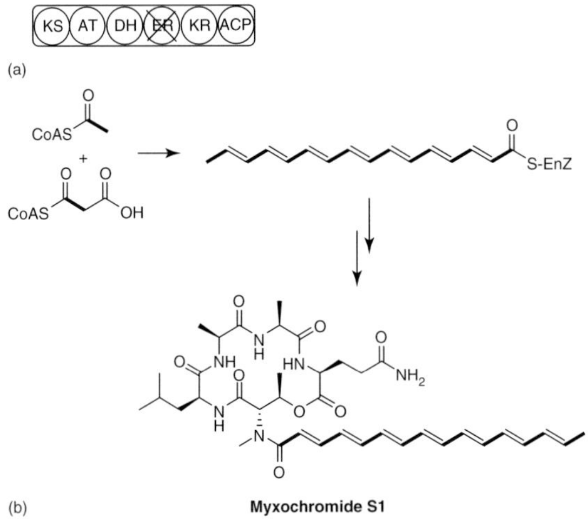

(a)

(b)　　　　　　　　　　Myxochromide S1

Figure 4.7 Bacterial type I PKS for assembling an aliphatic side chain. The example shown here is the domain organization of MchA (a) and its catalyzed reaction in myxochromide S biosynthesis (b).

Distinct from the above system for aromatic polyketide biosynthesis, an iterative type I PKS in myxochromide S biosynthesis was found to be responsible for the assembly of the aliphatic side chain (Figure 4.7) [77]. This enzyme, composed of the KS, AT, DH, ER (likely inactive), and ACP domain, can be regarded as an iterative modular PKS lacking perfect chain length control, given the fact that it accepts different types of starter units (acetyl- or propionyl-CoA) and that it occasionally performs additional chain extension cycles.

Enediynes are potent antitumor antibiotics that are classified as either nine- or 10-membered carbacycles according to the size of the enediyne core, which, bearing no structural resemblance to any characterized polyketides, is composed of two acetylenic groups conjugated by a double bond (Figure 4.8a). Consistently, biosynthetic characterization of a group of enediyne antibiotics, including the nine-membered C-1027 [78], neocarzinostatin [76], and maduropeptin [72] and the 10-membered calicheamicin [70] and dynemicin [79], revealed a new iterative type I PKS, termed PKSE (Figure 4.8b), for enediyne core formation with a mechanism that is shared among the entire enediyne family but distinct from all known PKSs [80]. PKSE contains four typical domains based on sequence homology and putative active site mapping: a KS, AT, KR, and DH, sequentially located on a single protein from the N- to the C-terminus. Recently, the remaining regions have been biochemically validated [81], as an atypical ACP domain central to the AT and KR domains and an unusual phosphopantetheinyl transferase (PPTase) domain presented at the C-terminus. The PPTase domain is responsible for *in situ* post-translational phosphopantetheinylation of a conserved Ser residue within the ACP domain. Various PKSEs, along with the associated thioesterase (TE) domains for the polyketide intermediate release, have been cross-complemented and are highly interchangeable [82], suggesting that the biosynthesis of the enediyne core, either for

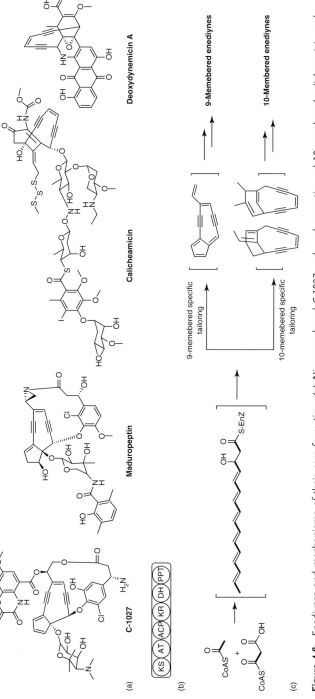

Figure 4.8 Enediynes and mechanisms of their core formation. (a) Nine-membered C-1027 and maduropeptin and 10-membered calicheamicin and dynemicin. (b) Domain organization of PKSE. (c) Generality of PKSE chemistry and divergence of the common intermediate in tailoring to afford nine- and 10-membered enediynes, respectively.

nine- or for 10-membered, occurs through a common polyene intermediate (Figure 4.8c). This intermediate, following subsequent modification with other enzyme activities, could be subsequently desaturated to furnish the two yne groups and cyclized to afford the enediyne core, consistent with the findings that a group of highly conserved genes is clustered with the PKSE gene [83, 84].

4.2.1.5 ACP-Independent, Noniterative Type II PKSs

Known type I and II PKSs employ ACP to activate the acyl-CoA substrates and channel the polyketide intermediates. This chemistry is distinct from that of the type III PKSs, which utilize acyl-CoAs as the substrates directly. All types of PKSs share the highly conserved Cys residue that is essential for PKS activity, but differ in the amino acid sequence of the Cys-containing motif. However, characterization of the biosynthetic machinery associated with nonactin, a macrotetrolide NP, revealed a total of five PKS proteins that share the sequence homology to KSs of types I or II PKSs but act in an ACP-independent manner [85, 86]. Three of them could be responsible for one of the three steps required for forming the nonactic acid units from four acyl-CoA precursors, representing a novel type II PKS that acts noniteratively, lacks ACP and utilizes acyl-CoA substrates directly for polyketide biosynthesis (Figure 4.9a). On the other hand, the remaining two proteins catalyze C–O bond-forming steps via the stereospecific dimerization between (−)- and (+)-nonactyl-CoA to form (−)-nonactyl-(+)-nonactyl-CoA and its subsequent stereospecific cyclodimerization to afford nonactin (Figure 4.9b). Given that all known PKSs catalyze C–C bond formation, the fact that these PKS proteins are responsible for cyclotetramerization via four C–O linkages is unprecedented.

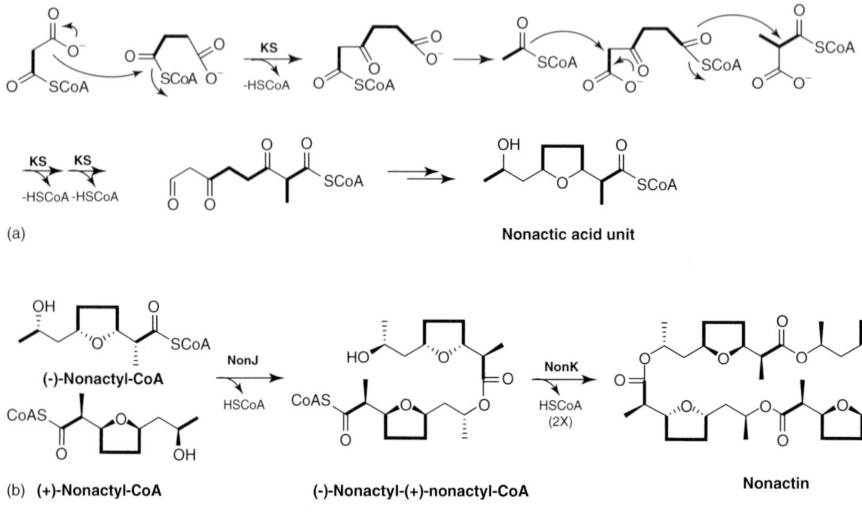

Figure 4.9 Mechanism of nonactin biosynthesis. (a) Nonactic acid formation from the acyl-CoA precursors catalyzed by three KSs in an ACP-independent, noniterative process. (b) The KSs NonJ- and NonK-catalyzed cyclotetramerization to furnish the macrotetrolide via C–O bond formation.

4.2.1.6 Archetypical NRPS Paradigms

In comparison with the PKS systems for polyketide biosynthesis, NRPSs are a large family of biocatalysts charged with assembling NRPs that have important medicinal applications [7, 87–89]. NRPSs usually possess a modular architecture similar to that for modular type I PKSs, and each module typically consists of three minimal domains for substrate selection and peptidyl-chain elongation (Figure 4.10a): (i) an adenylation (A) domain responsible for amino acid activation to generate an aminoacyl-AMP intermediate, (ii) a thiolation domain known as peptidyl carrier protein (PCP) for thioesterification of the activated amino acid, and (iii) a condensation (C) domain for transpeptidation between the aligned peptidyl and aminoacyl thioesters to elongate the growing peptide chain. Additional domains have also been identified for the modification of the aminoacyl and/or peptidyl substrates (i.e., *N*-, *C*-, and *O*-methylation, acylation, heterocyclic ring formation, and conversion of the substrates into their stereoisomers) during this process (Figure 4.10b), furnishing a dimension of structural diversity associated with peptide backbone formation. Remarkably, and in contrast to the relative paucity of building blocks for PKSs, the NRPS biosynthetic machineries utilize hundreds of different monomers, including not only the 20 canonical amino acids but also D-configured- and β-amino acids, methylated, glycosylated, and phosphorylated residues, heterocyclic elements, and even fatty acid units. This manner of building block diversity leads to highly diverse functionalizations, which are often essential for the biological activities of NRPs. Like PKSs, most NRPSs employ a TE domain, found at the C-terminus of the last module, to catalyze product release by hydrolysis, cyclization, or oligomerization (Figure 4.10c). Since solid-phase peptide chemistry has been well established to produce linear peptides in good yields, macrocyclization by the excised TE domains to synthesize cyclic peptides, which pose certain challenges to chemical synthesis, is becoming an effective chemoenzymatic approach to create libraries of new compounds [90, 91]. In the last 10 years, biochemical and structural studies on several NRPS assembly lines have contributed substantially to the understanding of molecular mechanisms and dynamics for recognition of the substrate and its shuffling among the different active sites, and also peptide bond formation and the regio- and stereoselective product release [92]. Of particular significance was the elucidation of the crystal structure of a four-domain-containing, entire NRPS termination module (with a C–A–PCP–TE organization) from the surfactin biosynthetic machinery which provided detailed insights into the overall NRPS structural conformation and domain–domain communication [93].

4.2.1.7 Atypical NRPS Paradigms

Similarly to typical modular type I PKSs, NRPSs that follow a colinearity rule are termed "linear NRPSs," in which the number of the catalytic modules and their order are completely consistent with the amino acid residues constituting the peptide backbone [87]. This linear biosynthetic logic has significantly facilitated the development of rationally redesigning NRPS templates to synthesize new peptide products [94]. However, given the number of type I PKSs that violate an analogous colinearity rule, it is not surprising that many atypical NRPS systems are now becoming apparent in which the amino acid sequence of peptide products does not match the arrangement of modules and/or domains within the NRPS assembly line [95]. These systems represent complicated

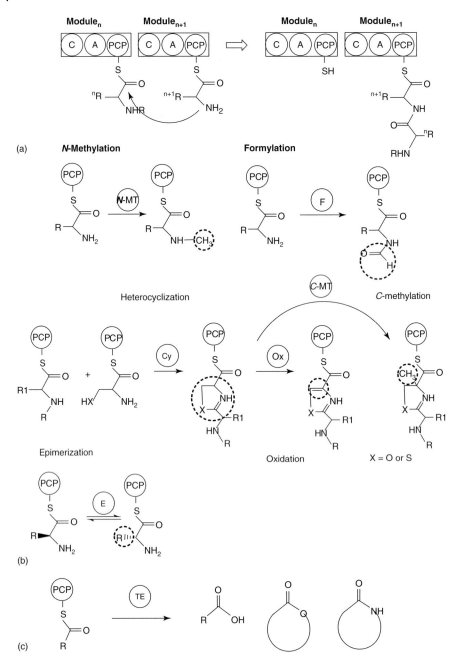

Figure 4.10 Modular organization of NRPS. (a) NRPS-catalyzed C–N bond formation in NRP biosynthesis. (b) Reactions catalyzed by selected modification domains that act in a *cis* fashion during the peptide backbone assembly, including methylation, formylation, heterocyclization, oxidation, and epimerization. (c) Terminal TE domain-catalyzed chain release.

NRPS variants and yet also increase the biosynthetic potential of these systems, since unusual products are often observed.

Some NRPS modules or domains can be used iteratively to produce a single product that usually contains repeated units, as exemplified by the biosynthesis of the antitumor antibiotic saframycin A [96]. The saframycin NRPS system has been suggested to assemble a tetrapeptidyl intermediate via a process involving iterative reactions catalyzed by SfmC as the termination module (Figure 4.11a). SfmC contains an additional reductase (R) domain, instead of the more commonly found TE domain, appended to C–A–PCP at the C-terminus responsible for reductively releasing the resultant peptidyl intermediates. Recently, *in vitro* reconstitution of the saframycin pentacyclic tetrahydroisoquinoline scaffold revealed that SfmC catalyzes a seven-step biotransformation requiring the communication of individual domains and the cryptic roles of the fatty acyl chain [97]. The process includes the activation and thioesterification of a tyrosine derivative, 3-hydoxy-5-methyl-*O*-methyltyrosine, by the A and PCP domains twice, reduction of various peptidyl thioesters by the R domain, and dual Pictet–Spengler reactions iteratively mediated by the C domain.

The NRPSs that deviate in their domain organization from the standard C–A–PCP architecture of linear NRPSs, often referred as to "nonlinear," constitute a considerable fraction of the NRPS repertoire. They feature at least one unusual arrangement of the minimal domains C, A, and PCP. Recently, characterization of the biosynthesis of ε-polylysine (ε-PL), one of only two amino acid homopolymers found in Nature, revealed that ε-PL synthetase acts as a nonlinear NRPS iteratively catalyzing peptide bond formation (Figure 4.11b) [98]. ε-PL synthetase was found to be a membrane protein with a N-terminal A domain and its associated PCP domain, characteristic of known NRPSs; however, it does not contain any typical C domain for extending the amino acid residues and TE domain for releasing the peptidyl chain. Followed by the A–PCP di-domain, it has three tandem soluble domains (denoted C1, C2, and C3) surrounded by six transmembrane (TM) domains, which may function as the requisite C domains to catalyze iteratively L-lysine polymerization with isopeptide linkages by using free L-lysine polymer as the acceptor and PCP-bound L-lysine as the donor, yielding the products varying in chain length (25–35 L-lysine residues).

4.2.1.8 Hybrid NRPS–PKS Paradigms

NRPSs and type I PKSs use a very similar strategy for the biosynthesis of two distinct classes of NPs [5, 8, 99, 100]. In addition to sharing a modular organization, both systems use carrier proteins (PCP for NRPS and ACP for PKS) to tether the growing chain. Both PCP and ACP are post-translationally modified by a 4′-PPTase (Scheme 4.1). During the entire elongation process, the growing intermediates remain covalently attached to the carrier proteins via a thioester linkage to the sulfhydryl moiety of the 4′-phosphopantetheinyl group. These striking structural and catalytic similarities between NRPS and PKS have inspired extensive attempts to search for hybrid PKS–NRPS systems integrating both PKS and NRPS chemistries. Indeed, during the past decade, cloning and sequence analysis of gene clusters for biosynthesizing a number of hybrid peptide–polyketide NPs have supported Nature's wisdom in this regard, showing that the PKS and NRPS systems are highly compatible (Figure 4.12). Hybrid NRPS–PKS

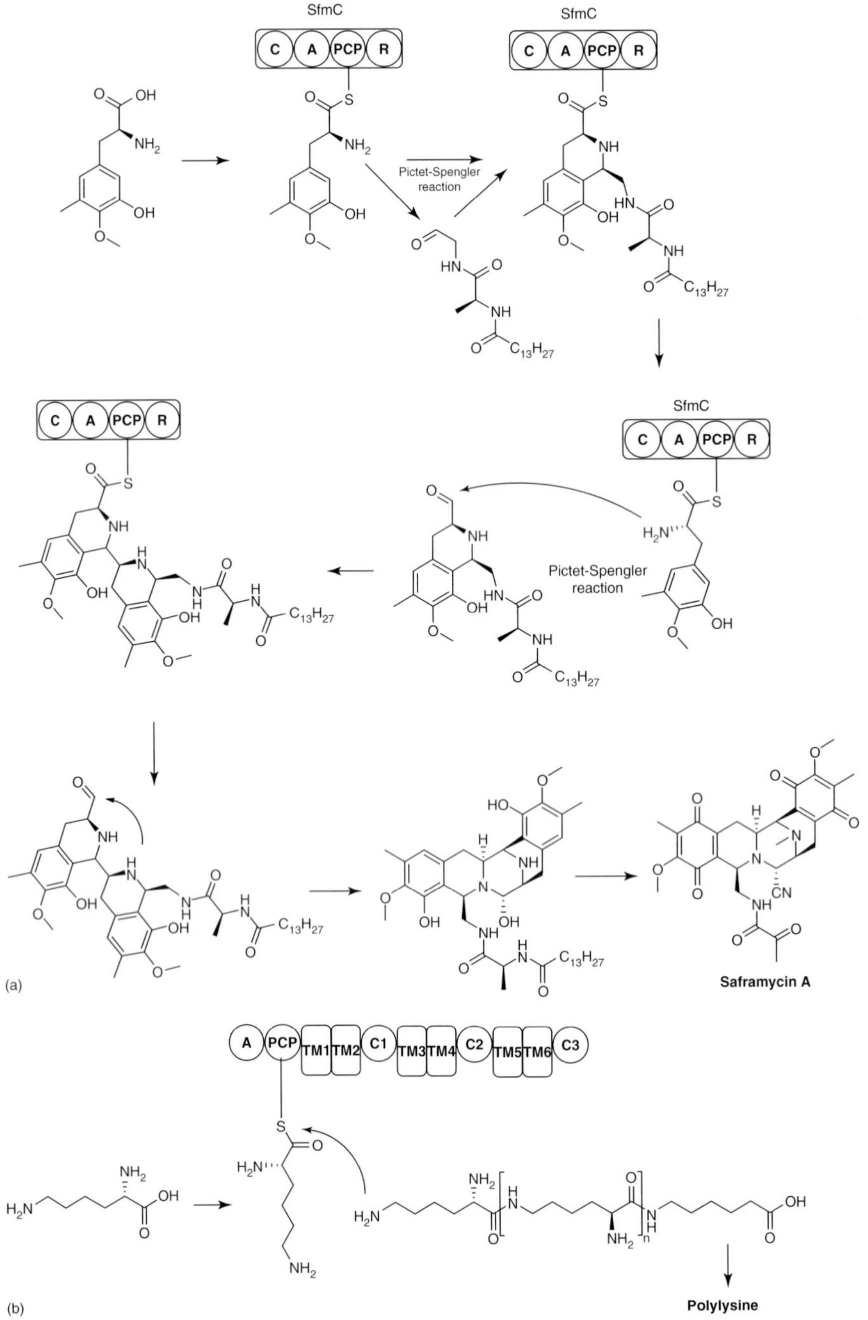

(a)

(b)

Figure 4.11 Atypical NRPSs in bacteria. (a) Iterative NRPS as exemplified by SfmC to assemble a tetrapeptidyl intermediate in saframycin A biosynthesis. (b) NRPS with unusual domain organization as exemplified by ε-polylysine synthetase.

NRPS–PKS

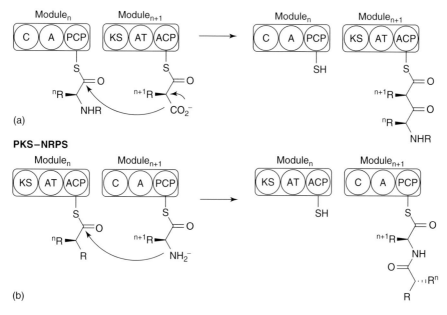

(a)

PKS–NRPS

(b)

Figure 4.12 Domain organization in hybrid paradigms. (a) NRPS/PKS-catalyzed C–C bond forma-tion in hybrid peptide–polyketide biosynthesis. (b) PKS/NRPS-catalyzed C–N bond formation in hybrid polyketide–peptide biosynthesis.

systems, combining the biosynthetic capacity of both NRPS and PKS machineries, allow the production of novel hybrid peptide–polyketide metabolites by incorporating both amino acids and short carboxylic acids, greatly expanding the size and diversity of the resulting combinatorial biosynthetic libraries. The genetic tools developed for engineering NRPS and PKS are applicable to the hybrid NRPS–PKS systems, significantly facilitating combinatorial biosynthesis of hybrid peptide–polyketide NP libraries.

4.2.2
Ribosomal Paradigms of Peptide NPs

In the post-genomic era, peptide NPs have served as the chemical inspiration for the discovery of new biosynthetic mechanisms in addition to NRPS paradigms. This has facilitated a profound understanding of several biosynthetic ribosomal paradigms, featuring post-translational modifications and proteolytic processing of a genetically encoded precursor peptide [101–103], with the equally remarkable ability to synthesize a wide variety of structurally diverse peptide NPs (Figure 4.13). During the past 10 years, a few key modifications have been well established, including the formation of lantionine bridges in lantibiotics [104, 105], aromatic heterocycles in certain bacteriocins [106] and cyanobactins [107], and macrolactam and/or macrolactone linkages in lasso peptides [108] and microviridins [109, 110]. Although the ribosome-based systems only employ

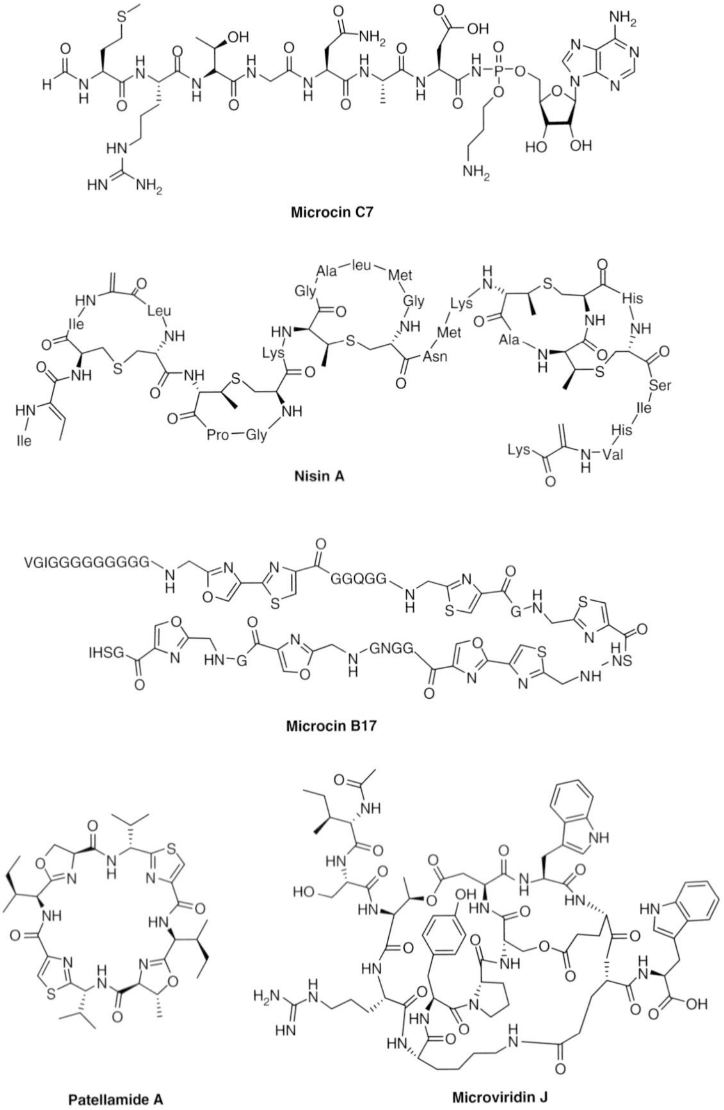

Figure 4.13 Selected peptide NPs with ribosomal origins.

substrates within the framework of the 20 proteinogenic amino acid building blocks compared with a much wider array of those found in NRPS systems, comparable structural diversity among final products can be achieved by the highly variable enzymatic machineries that process the ribosomally synthesized precursor peptides.

The investigation of thiopeptide biosynthesis revealed the complexity of post-translational modifications for peptide NPs with a ribosomal origin [111]. Thiopeptides are a class of clinically interesting and highly modified polythiazolyl antibiotics. The potent activity of many members against various drug-resistant bacterial pathogens has promoted extensive investigations into the development of new analogs; a main objective has been to overcome physicochemical drawbacks such as poor water solubility and rapid clearance. Thiopeptides possess a characteristic macrocyclic core, consisting of a nitrogen-containing, six-membered ring central to multiple thiazoles and dehydroamino acids, but vary in side chain (and/or ring) and pendant functionalities [112]. Although previous isotope labeling experiments established that all moieties derive exclusively from proteinogenic amino acids, the biosynthetic pathways affording thiopeptides remained elusive until recently. The application of genome mining and direct polymerase chain reaction (PCR) amplification approaches to clone and characterize biosynthetic gene clusters of several representatives in the thiopeptide family, including thiocillins [113], thiostrepton [114, 115], thiomuracins [116], nosiheptide [117], nocathiacins [118], and cyclothiazomycin [119], uncovered a common paradigm for thiopeptide framework

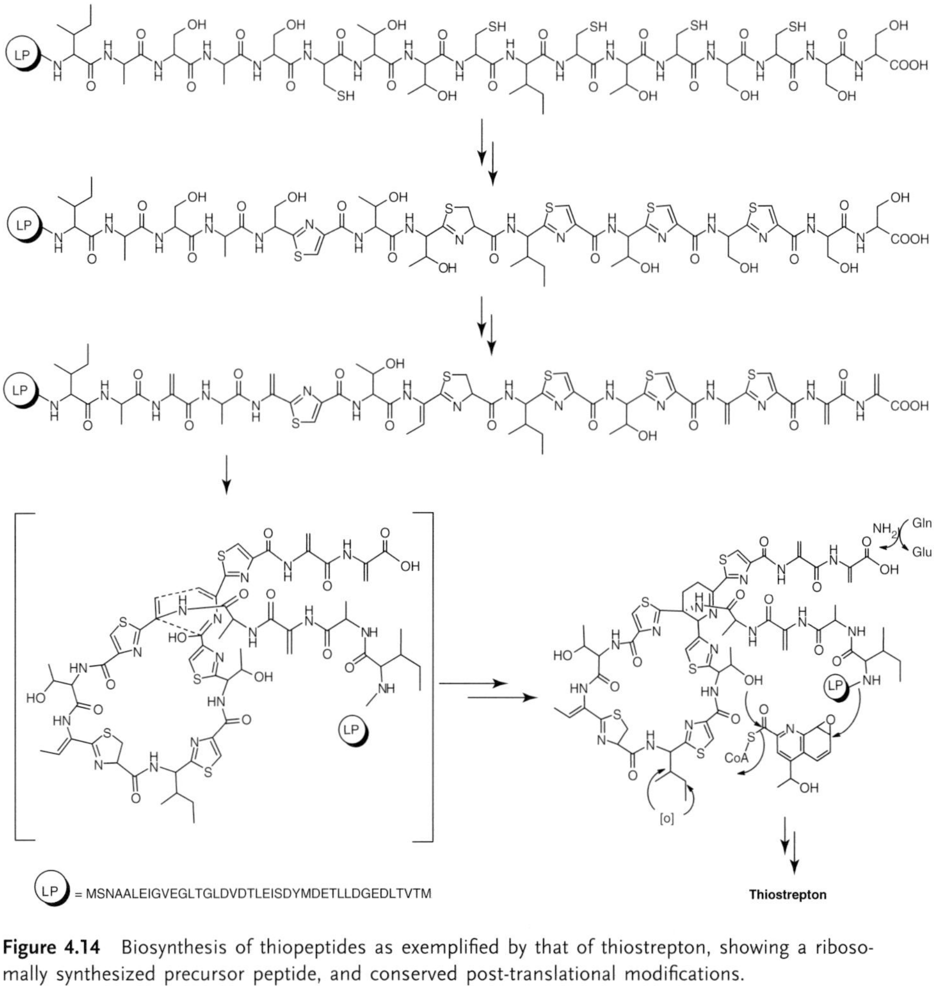

Figure 4.14 Biosynthesis of thiopeptides as exemplified by that of thiostrepton, showing a ribosomally synthesized precursor peptide, and conserved post-translational modifications.

construction that features conserved post-translational modifications on a ribosomally synthesized precursor peptide (Figure 4.14). In contrast to the heroic efforts of chemical synthesis, biosynthetic pathways for thiopeptide framework formation are remarkably concise, and are now well understood and exploitable. Some of the reactions executed by these biosynthetic machineries include, but are not limited to: (i) cyclodehydrations and subsequent dehydrogenations to form aromatic thiazoles characteristic to bacteriocins, (ii) dehydrations to generate dehydroamino acids such as those found in lantibiotics, and (iii) intramolecular cyclizations to afford the nitrogen heterocycles that characterize the thiopeptides. However, thiopeptide biosynthesis involves a number of pathway-specific enzymes for converting a shared framework into individual bioactive members, which differ in their decoration of the core system. Reactions executed in this system in order to diversify the mutual framework include (i) substitution of the central heterocycle domain, (ii) installation of the side ring system, and (iii) C-terminal functionalization of the extended side chain. For instance, two distinct ways have been found to process the peptide backbone to furnish the C-terminal amide moiety [120, 121], and an unprecedented, radical-mediated enzymatic carbon chain fragmentation–recombination was found to be involved in the indole side ring formation [122]. These findings can be applied to combinatorial biosynthesis, complementing recent advances in sequence permutation of the precursor peptide for structural diversity in developing new bioactive thiopeptide agents.

4.2.3
New Strategies for Peptide–Amide Bond Formation

Independent of NRPS or ribosomal templates, there are a few distinct ways to form peptide–amide bonds that have been newly characterized in NP biosynthesis. One representative reaction relies on a group of more than 80 acyladenylate-forming enzymes,

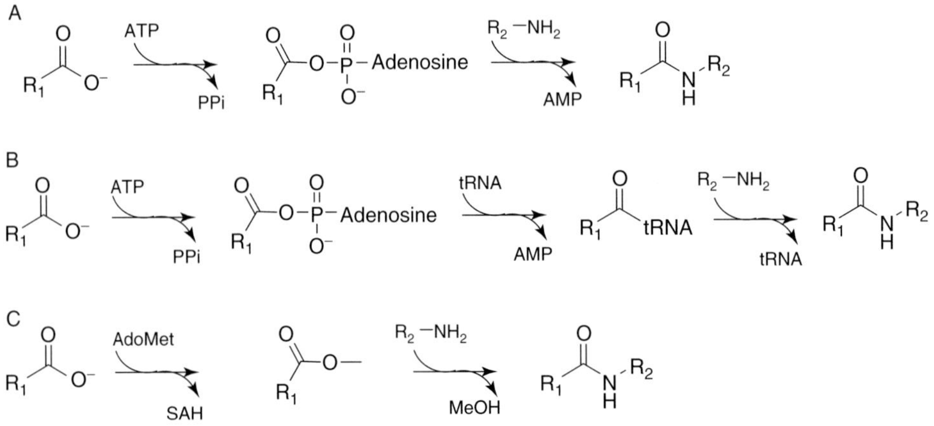

Scheme 4.2 Selected new strategies for peptide/amide bond formation distinct from NRPS or ribosomal paradigm. (a) Adenosine monophosphate as the leaving group. (b) Activation of amino acid in the form of aminoacyl-tRNA. (c) ATP-independent ester–amide exchange.

which are believed to catalyze the specific condensation of various carboxylic acids with a wide variety of amines (or alcohols in some cases) [123]. To form the amide bond, the carboxylic acid substrate is first activated by ATP to form an enzyme-bound acyl adenylate intermediate (Scheme 4.2). Displacement of adenosine monophosphate by an amine via an addition–elimination mechanism affords the amide linkage. Recently, structural investigations of AcsD, a member of this class of enzymes which is involved in the biosynthesis of the siderophore achromobactin, provided a mechanistic rationale for the stereospecific formation of AcsD-bound (3R)-citryladenylate, which reacts with L-serine to afford the amide product [124].

Cyclodipeptides and their derivatives are a broad array of NPs that belong to the diketopiperazine (DKP) family with highly diverse biological activities. In most known cases, NRPS systems act as the templates to furnish the cyclodipeptide scaffold. However, studies on the biosynthesis of albonoursin, a DKP product with antibacterial activity, showed that the formation of its cyclo(Phe–Leu) intermediate is catalyzed by AlbC, a small protein unrelated to NRPSs [125]. AlbC contains no predicted ATP binding motif. In fact, AlbC uses activated amino acids in the form of aminoacyl-tRNAs (aa-tRNAs) as the substrates en route to the cyclic peptide. AlbC homologs that have subsequently been found from various microorganisms display similar activities and synthesize different cyclodipeptides, thus constituting a new family of tRNA-dependent peptide bond-forming enzymes, termed cyclodipeptide synthases (CDPSs). Given that aa-tRNAs are well known as the substrates in normal peptide assembly, the escape of these products from the ribosome to the NP "shunt" biosynthetic machinery is intriguing [126]. Recently, the three-dimensional structure of Rv2275, an AlbC-like CDPS catalyzing cyclo(Tyr–Tyr) formation, has been determined, revealing that it is structurally similar to the class Ic aa-tRNA synthetase family of enzymes [127]. This finding suggests their relationship in evolution.

Amide bond formation in NP biosynthesis typically necessitates the activation of the carboxyl acid group in the form of an acyladenylate or acylphosphate. The process occurs at the expense of high-energy bonds within ATP, as described above in complex ribosomally or nonribosomally encoded peptides and other metabolites containing a single amide moiety. Recently, a class C β-lactamase-like protein, CapW, was identified in the biosynthetic gene cluster of A-503083 B, a capuramycin-type antibiotic containing an L-aminocaprolactam and an unsaturated hexuronic acid that are linked via an amide bond [128]. In contrast to the expected β-lactamase activity, CapW catalyzes an unprecedented ester–amide exchange reaction to afford A-503083 B by acting on a methyl ester intermediate, which is the product of an S-adenosylmethionine-dependent carboxymethyltransferase CapS. This ATP-independent amide bond formation using methyl esterification followed by an ester–amide exchange reaction represents an alternative strategy to furnish the amide moiety and may be a much more common phenomenon in NP biosynthesis than has been previously appreciated.

4.3
New Approaches to NP Biosynthesis Research

Efforts to characterize NP biosynthetic pathways during the past decade still rely heavily on time-tested *in vivo* and *in vitro* strategies. However the relative ease of DNA sequencing and bioinformatics has altered the way in which many NP biosynthesis researchers approach problems [129]. Research in the 1990s was dominated by in-depth analyses of a limited number of gene clusters that served as paradigms for the major classes of PKS and NRPS biosynthetic machineries. Researchers today more commonly have multiple gene clusters of biosynthetically related NPs that can be used in comparative bioinformatic analyses to help guide bench-top experiments and to reveal intricacies of the molecular architectures and routes of evolution that were not apparent at first glance. *In vitro* analyses of biosynthetic enzymes or proteins that are difficult to work with can often be facilitated by turning to closely related homologs from other organisms. Investigations of the routes and strategies that Nature has come up with to evolve new NP scaffolds have become possible through phylogenetic analyses of large numbers of related gene clusters. This information has been applied in combinatorial biosynthesis efforts to mimic natural evolution by generating structural diversity on a scale amenable to high-throughput screening campaigns. Lastly, the technological advancements that expedite genetic manipulations or obviate the need for tedious and time-consuming cloning steps have allowed experimentation on a scale not previously possible. In the following section, we highlight selected examples that showcase the power of these strategies.

4.3.1
Comparative Gene Cluster Analyses Facilitate Biochemical Characterization

The power of comparative cluster analysis for guiding *in vitro* and *in vivo* experimentation is exemplified in our recent efforts to decipher the mechanisms allowing for the dual production of antibiotics platensimycin and platencin in the native producing organism, *Streptomyces platensis* (Figure 4.15) [130]. Originally working with *S. platensis* MA7327, a strain capable of producing both antibiotics under certain conditions [131], we cloned and sequenced a ∼42 kb gene cluster (denoted the *ptm* gene cluster) containing 40 open reading frames predicted to be involved in the biosynthesis of both platensimycin and platencin (Figure 4.15a). With a genetic system in place and bioinformatic predictions for many of the gene functions, a complete *in vivo* and *in vitro* characterization of the gene cluster was experimentally feasible. However, the reported isolation of numerous producing strains from all over the world by researchers at Merck [132], including strains that produced only platencin, offered an excellent opportunity for comparative cluster analysis. We reasoned that given two closely related strains with distinct chemical profiles, we should be able to find differences in the producing gene clusters to account for the unique chemotypes, and that this will expedite the characterization of the molecular branch point in the biosynthesis of platensimycin and platencin. The ∼37 kb platencin-only producing gene cluster (the *ptn* gene cluster) from *S. platensis* MA7339 was cloned and sequenced in a matter of weeks (Figure 4.15a). It was found to bear a remarkably high level of sequence identity at a DNA level to the *ptm* gene cluster but was missing a five-gene cassette from the middle of the gene cluster. These genes were then targeted in a combination of *in vitro* and *in vivo* experiments to show they are both necessary and sufficient for platensimycin production [130]. Furthermore, the presence of this five-gene cassette led to the characterization of two unique diterpene cyclases that provide the carbon scaffolds of platensimycin and platencin and represent

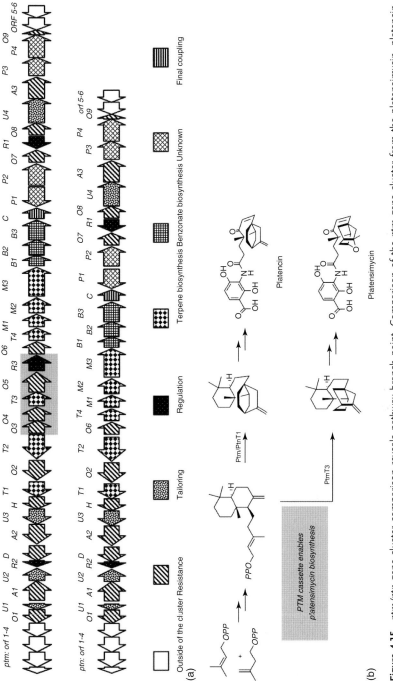

Figure 4.15 *ptm/ptn* gene cluster comparison reveals pathway branch point. Comparison of the *ptm* gene cluster from the platensimycin–platencin dual producing strain *S. platensis* MA7327 to the *ptn* gene cluster from the platencin-only producing strain *S. platensis* MA7339 (a) revealed the presence of a five-gene cassette that was found to be necessary and sufficient to confer platensimycin production to a platencin-producing strain. The key branch point in pathway (b) involves dedicated terpene synthases that convert a common substrate, *ent*-copalyl pyrophosphate, into either *ent*-atiserene or *ent*-kaurene for platencin and platensimycin biosynthesis, respectively.

the branch point in the biosynthesis (Figure 4.15b). The *ent*-atiserene cyclase was not recognized based on the bioinformatic analysis of the *ptm* gene cluster alone and it is unlikely that it would have been targeted in initial knockout experiments without the insight that the second gene cluster provided. Additional analysis of the two clusters provided clues to explain how Nature evolved these unique scaffolds that can be exploited for the production of new analogs with unique terpene scaffolds. In this example, our ability to rapidly sequence and compare the two gene clusters from organisms with different chemotypes allowed for the characterization of the pathway branch point with

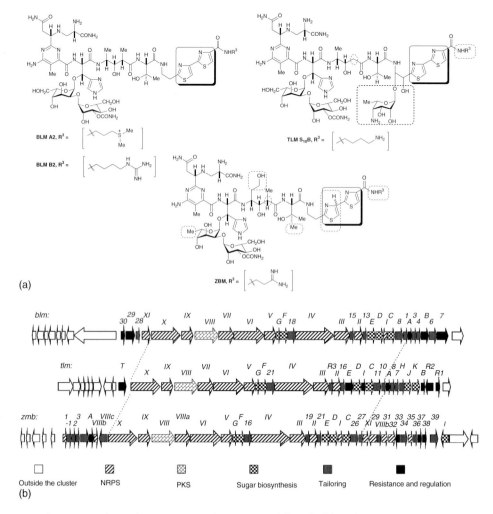

(a)

(b)

Figure 4.16 Chemical structures (a) of bleomycin (BLM), tallysomycin (TLM), and zorbamycin (ZBM), with differences highlighted within dashed lines. Of particular interest is the biosynthesis of the bithiazole (BLM and TLM) and thiazolinylthiazole (ZBM) moieties highlighted within solid lines. (b) Comparison of *blm*, *tlm*, and *zmb* gene clusters, showing consistent overall organization.

an economy of experimentation that would not have been possible working in just the original gene cluster.

A second example from our laboratory where comparative gene cluster analysis has yielded priceless insights involves the characterization of the biosynthetic machinery of the bleomycin-like group of anticancer agents, which includes bleomycin, tallysomycin, and zorbamycin (Figure 4.16). As opposed to the platensimycin example, comparative analysis of these clusters was required owing to difficulties in working within each system individually: the bleomycin producer, *Streptomyces verticillus* ATCC15003, has a low frequency of genetic recombination, the tallysomycin producer, *Streptoalloteichus hindustanus* E465-94 ATCC31158, does not produce abundant spores, again complicating genetic manipulation, and the zorbamycin producer, *Streptomyces flavoviridis* ATCC21892, is genetically amenable but fails to produce zorbamycin or related products in titers conducive to *in vivo* characterization. Despite these shortcomings, much progress has been made towards understanding this fascinating biosynthetic pathway through the creative design of experiments that intertwine the three gene clusters [133]. One intriguing question that arises from these NPs involves the biosynthesis of the bithiazole moiety in bleomycin and tallysomycin and the thiazolinylthiazole moiety in zorbamycin, despite an identical domain organization in the three clusters (Figure 4.16a). These differences could possibly be accounted for by (i) an oxidation domain that functions twice for bleomycin/tallysomycin biosynthesis but only once for zorbamycin biosynthesis (Figure 4.17a), (ii) an oxidation domain that always functions once but is coupled with a discrete oxidase located elsewhere in the bleomycin/tallysomycin gene clusters (Figure 4.17b), or (iii) an oxidation domain that functions twice in each pathway coupled with a discrete reductase in the zorbamycin gene cluster to reduce one of the thiazole rings back to a thiazoline (Figure 4.17c). An interesting caveat is that the *R*-configuration of the thiazoline moiety in zorbamycin requires that, for explanation (i) or (ii) to be true, either the A domain would need to activate one L-cysteine and one D-cysteine or a discrete epimerase would need to invert a loaded L-cysteine during the biosynthesis. Biochemical characterization of bleomycin NRPS-1 and -0 suggests that L-cysteine is, in fact, loaded to each PCP, and their epimerases are absent from the zorbamycin gene cluster, pointing towards option (iii) as the most likely scenario. Evaluating the zorbamycin gene cluster in the light of this knowledge allowed the identification of Zbm−Orf2 as a candidate reductase, which proved to be necessary for zorbamycin biosynthesis [134]. Both this example and the platensimycin example above illustrate how comparing multiple gene clusters can lead to insights that would not have otherwise been possible by studying each one in isolation. Many NP biosynthesis groups have taken advantage of this concept in the past decade to clone and sequence numerous gene clusters for many classes of NPs.

4.3.2
Unique Combinatorial Strategies for Different Pathways

Although the full potential of combinatorial biosynthesis has yet to be realized [135, 136], the last decade has seen great advances in our ability to diversify NP scaffolds using genetic engineering. It is now clear that different classes of biosynthetic pathways

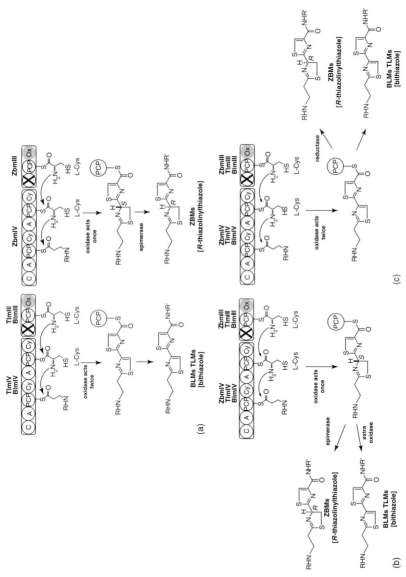

Figure 4.17 Possible biosynthetic routes to bithiazole or (*R*)-thiazolinylthiazole moieties of BLM, TLM, and ZBM. (a) Differential activity of the homologous oxidase domains produces a bithiazole moiety in BLM and TLM and an (*S*)-thiazolinylthiazole moiety on ZBM, which must be epimerized. (b) In each case, the oxidase domain produces an (*S*)-thiazolinylthiazole moiety, that is subject to extra tailoring by an oxidase in BLM/TLM biosynthesis or an epimerase in ZBM biosynthesis. (c) In each case, the oxidase domain produces a bithiazole system, that is reduced to (*R*)-thiazolinylthiazole in ZBM biosynthesis.

require distinct combinatorial strategies. For instance, engineering protein–protein interactions is paramount for many thio-templated biosyntheses but is often unnecessary in non-templated pathways that utilize discrete soluble enzymes. The following paragraphs describe several examples of combinatorial biosynthesis efforts during the past decade that employ unique strategies for success.

The indolocarbazole alkaloids are a family of NPs that have proven to be particularly amenable to combinatorial biosynthesis [137]. Indolocarbazoles contain either a bisindolylmaleimide or indolo[2,3-a]carbazole core scaffold that is decorated by various tailoring reactions such as oxidations, halogenations, and glycosylations [138]. Much of the interest surrounding this class of compounds is due to their potent anticancer activity stemming from their ability to inhibit protein kinases and eukaryotic DNA topoisomerase I, or to intercalate into DNA. However, many natural indolocarbazoles also display anti-infective activities [138]. Characterization of the biosynthetic gene clusters for the indolocarbazoles rebeccamycin [139–141], staurosporine [142], and AT2433 [143] has revealed (i) a conserved set of enzymes responsible for forming the core scaffold from two molecules of tryptophan and (ii) a diverse set of enzymes responsible for differential tailoring. Many congeners have been produced by co-expressing genes from distinct pathways in a single operon driven by a strong promoter [144], including those with either the rebeccamycin or staurosporine core scaffold and new halogenated analogs. Furthermore, new sugar moieties have been appended to the staurosporine scaffold by the addition of "sugar vectors" that encode the biosynthetic enzymes for a diverse set of sugars [145]. In total, over 50 new indolocarbazoles have been generated using combinatorial biosynthetic methods, all via the straightforward approach of co-expressing genes from disparate origins in a heterologous host where the gene products are required to work together to form a functional biochemical pathway [137]. The success of this strategy suggests that enzymes in various indolocarbazole pathways tend to display relaxed substrate specificities and do not rely on specific protein–protein interactions between enzymes in the pathway for proper function.

Alternatively, the enzymes and domains involved in polyketide biosynthesis require both intrapolypeptide and interpolypeptide interactions to function together as a cohesive macromolecular machine (Figure 4.18). In many cases, these interactions have evolved to be quite specific, possibly to minimize spurious interactions between different PKSs

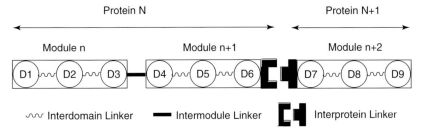

Figure 4.18 Intrapolypeptide (including intradomain and intramodule) and interpolypeptide interactions in type I PKSs or linear NRPSs required for constituting a macromolecule biosynthetic machinery. D, domain.

present within a single cell. Intermodule linker domains present on a single polypeptide are important for mediating the transfer of a growing acyl chain from the ACP of one module to the KS of the proceeding module [146]. When adjacent domains are present on separate polypeptides, complementary 40 amino acid α-helical capping domains are needed at the respective N- and C-termini to facilitate the interaction [147, 148]. The observation that both of the above types of interactions can be exploited to form new cooperative pairs gave birth to the strategy of engineering new polyketide pathways by considering entire modules, as opposed to individual domains, as the most basic functional units. Modules meant to cooperate chemically must be fitted with appropriate linker regions to ensure that appropriate protein–protein interactions can occur. Using such a strategy, researchers at Kosan Biosciences constructed 14 modules from several distinct type I PKSs and used them to build over 150 bimodular complexes, half of which proved to be functional [149]. Compared with the proposed discrete enzymatic reactions of the indolocarbazole system, efficient combinatorial biosynthesis in assembly-line type pathways exemplified by type I PKSs requires an extra level of complexity. Similar complementary linkers are known to exist in NRPS [150] and hybrid NRPS–PKS systems [151] to direct intermodule interactions and have been exploited for combinatorial biosynthesis [152, 153], albeit on a smaller scale than described above.

A third NP class ideally suited for combinatorial biosynthesis is the small ribosomal peptides (RPs), including cyanobactins, microcins, lantibiotics, and thiopeptides [154, 155]. Although ~3 kb of DNA sequence is required to incorporate one amino acid into an NRP, RPs can accomplish the same goal with only three nucleotides [154]. This means that the amino acid sequence can be manipulated by simply mutating the gene that encodes the prepeptide. Indeed, this strategy seems to have been used by Nature to allow the production of dozens of unique small RPs using a minimal amount of genomic information [156]. RP biosynthetic pathways have also proven amenable to the design of unnatural NP-like molecules [157, 158], although some systems appear to be more tolerant than others in accepting a wide range of precursor peptides.

Although great strides have been made in the combinatorial biosynthesis of many classes of NPs during the past decade, great opportunities for advancement still exist at every level. At the transcription level, a more detailed understanding of the principles that govern the complex regulation of genes in secondary metabolism will lead to the design of intricate genetic circuits to optimize the timing and level of gene expression in combinatorial pathways [159]. For instance, knowing more about important protein–protein interactions between discrete ATs and their cognate ACPs would enable researchers to apply the engineering principles described in the PKS example above to new types of pathways. Lastly, efficient platforms that allow for the synthesis and evaluation of combinatorial gene clusters in a high-throughput manner [160] need to be utilized in drug discovery efforts to sift through inactive products and to realize fully the power of combinatorial biosynthesis. Continuing to look to Nature for guidance is a good start, as unique routes of evolution that have been followed by different classes of NP gene clusters will likely require similarly unique combinatorial strategies for their directed evolution [161]. For example, domains from canonical type I PKSs display phylogenetic clustering according to the NP pathway from which they derive [162]. In other words, the ATs from each module in the erythromycin PKS are more closely related to each other than to ATs from other pathways. This suggests that the primary mechanism of evolution that produced these gene clusters relied on gene duplication within the cluster followed by functional diversification. This contrasts with the phylogenetic clustering of ketosynthase domains in *trans*-AT PKS gene clusters according to the chemistry of their substrates rather than their pathways [163, 164]. The latter implies that a "mix-and-match" route of evolution has occurred to build these gene clusters, which is more in line with current strategies for combinatorial biosynthesis.

4.3.3
Synthetic Metagenomics for Improved Methyl Halide Production

The tremendous amount of DNA sequence information available and the low cost of DNA synthesis now allow for experimentation on a scale not previously possible. This is made evident by the recent strategy used by Voigt and co-workers to find more efficient methyl halide transferases (MTHs) for the biological production of methyl halides [165]. Methyl halides have a variety of industrial uses as substrates for biofuels, agricultural fumigants, solvents, and more. Although putative MTHs can be found in all three domains of life, previously studied producing organisms emit methyl halides at low titers that are not suitable for industrial purposes. Voigt's group identified 89 genes in the NCBI databases that display homology to known MTHs, including genes from metagenomic sequencing projects and uncharacterized organisms, and proceeded to synthesize codon-optimized versions of each homolog for expression in *Escherichia coli* and yeast [165]. *In vivo* methyl halide production was measured for each recombinant gene to discover a highly active MTH derived from the halophylic plant *Batis maritima*, which could readily methylate bromide, chloride, and iodide ions. This was then used to engineer a co-culture of the cellulosic *Actinotalea fermentans* with MTH-expressing *Saccharomyces cerevisiae* capable of converting unprocessed biomass to methyl halides with promising efficiency.

The power of functional metagenomics to discover enzymes or small molecules from unculturable organisms has been recognized for some time (Figure 4.19) [166]. However, previous efforts employ non-specific libraries and often require high throughput screens in order to sort through hundreds of thousands of clones to find the function of interest. Without a doubt, a large jump is needed to reliably use metagenomic experiments to discover and functionally express the large (up to 100 kb) gene clusters required for the biosynthesis of many classes of pharmaceutically relevant NPs, but much progress has been made toward this goal in the past decade [167]. The most attractive application described above for methyl halide production wherein the metagenomic library consists of a collection of synthetic genes originating from a sequence database allows for a more thorough and efficient investigation of candidate enzymes (Figure 4.19). As the number of sequenced genomes continues to rise and the cost of DNA synthesis continues to drop, synthetic NP gene cluster libraries will be generated in studies that mirror the methyl halide example.

Functional metagenomics

Synthetic metagenomics

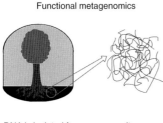

1. DNA is isolated from a community of organisms

1. Homologous genes are selected from an electronic database

ATTCGCTGACGAGCG...

2. DNA is fragmented and randomly cloned into plasmid vectors.

2. Using DNA synthesis, selected genes are 'printed' into plasmid vectors

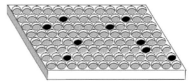

3. Library members are screened or selected for the desired property

Figure 4.19 Comparison of functional metagenomics and synthetic metagenomics. Improvements in DNA synthesis technology provide easy access to a large number of genes from electronic databases and permit the creation of targeted libraries that represent a diverse collection of organisms.

4.4
Better Understanding of the Scope and Diversity of NP Production

The wealth of genome sequences now available has broadened our understanding of NP diversity and scope. Early actinomycete genome sequences revealed a surprising number of secondary metabolite gene clusters, up to 37 in *Streptomyces avermitilis* MA-4680, suggesting that only ~10% of the NP potential has been accessed even from culturable organisms [168, 169]. Although the mere sequencing and annotation of NP gene clusters help to inform the biology of NP production, the ultimate goal for many is to transform this sequence information into new chemical structures that can be used for drug discovery. The extensive exploitation of genome sequencing in twenty-first century NP research distinguishes it from past decades, as many groups have turned to genome mining for the discovery of new NPs. In the following section, selected examples that illustrate the potential of genome mining in NP research are followed by a more in-depth look at the major avenues by which NP researchers have been able to translate DNA sequence information successfully into new NPs. Also covered are examples of how DNA sequence information allows for the rapid screening of new microorganisms and a broadened understanding of the distribution of certain biosynthetic pathways.

4.4.1
Genome Sequencing, Scanning, and Screening for Chemical Potential

Developing sequence-based techniques for determining the chemical potential of an organism early on in the drug discovery process can save time and resources. Additionally, such techniques provide an insight into the relative abundances of specific biosynthetic pathways in an organism or ecosystem. Although major human pathogens were the first microorganisms to be selected for genome sequencing projects, it was not long before drug-producing microbes were targeted for sequencing campaigns [170–172]. A 2009 review found 61 complete whole genome sequences from the order *Actinomycetales* [169], but many more remain in draft form. Analyses of these genome sequences suggest that the ability to produce a wide range of NPs is not shared equally by different genera of actinomycetes. Members of *Streptomyces*, *Salinispora*, and *Saccharopolyspora* can contain upwards of 30 NP gene clusters, yet members of other genera contain fewer than five. This is not surprising given the incredible biological diversity encompassed by the order *Actinomycetales*, with a number of the genome sequences in this group coming from human pathogens. Perhaps the most salient information taken from these early genomic data is that the number of metabolites isolated from an organism generally falls short of that organism's genetic potential. This realization helped to spur a renewed interest in NP drug discovery, which had previously been tempered by the high rediscovery rate of known compounds and the belief by some that the majority of NP scaffolds had already been discovered. Another important result from whole genome sequencing is the perception that it gives of the linear streptomycete chromosome, with a large essential core region flanked on both sides with variable arms contained mostly nonessential genes [173]. This knowledge gave rise to genome minimization efforts described below. Lastly, the first whole genome sequences illuminated the stunning complexity of gene regulation in the streptomycetes. The 65 sigma factors in *Streptomyces coelicolor* dwarf the previous record of 23 in *Mesorhizobium loti* [170]. A global understanding of transcriptional activation promises to improve our ability to awaken cryptic NP gene clusters in drug discovery efforts.

Although whole genome sequences have been valuable in shaping our understanding of the overall biology of the producing organism, researchers interested in genome mining need to access only the NP gene clusters present in the genome. Because of this, a technique known as "genome scanning" was developed to provide a minimally biased method for identifying NP gene clusters (Figure 4.20) [83, 173]. In this method, genomic DNA is fragmented and shotgun sequenced to afford several thousand ~700 bp sequence tags that can be searched against sequence databases to identify genes involved in NP biosynthesis. Knowing the size of a typical gene cluster and the total size

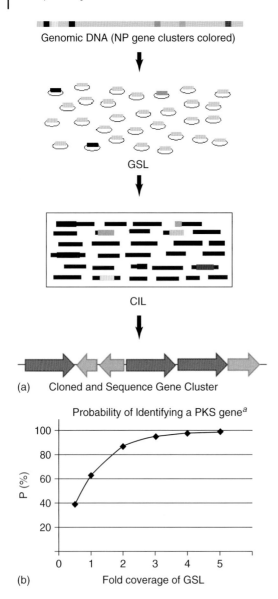

(a)   Cloned and Sequence Gene Cluster

(b)

Figure 4.20 Genome scanning for high-throughput identification of NP gene clusters. (a) Schematic representation of genome scanning method [79]. Genomic DNA is randomly fragmented and cloned to create a genome sampling library (GSL). GSL plasmid inserts are sequenced and "hits" are used to probe a cluster identification library (CIL), composed of large genomic fragments that have been cloned into a cosmid or BAC vector. CIL members that hybridize to the hit genes are completely sequenced to afford the complete gene cluster. The probability of identifying a random PKS gene for a given fold coverage of the genome by the GSL (b) shows that a very high probability of identifying a PKS gene can be achieved with a relatively small GSL [160]. [a]Determined by Poisson distribution and assumes that each PKS gene is readily identifiable in sequence data.

of the genome, the Poisson distribution can be used to estimate the probability of identifying a given PKS, NRPS, or other gene of interest depending on the number of sequences generated. This strategy allows the researcher to (i) "fish" for gene clusters of interest, (ii) obtain perfect probes that can be used to locate colonies of interest in a genomic library, and (iii) estimate the total number of unique gene clusters in an organism of interest. In an effort to find novel NPs bearing the enediyne warhead, a group of researchers at Ecopia BioSciences performed high-throughput genome scanning of 50 actinomycete strains reported to produce a variety of NP scaffolds, but not enediynes. Surprisingly, 16% of the strains were found to contain the enediyne warhead gene cassette. Fermenting these hit strains in specialized enediyne production medium activated the production of putative enediynes in many of these strains [83]. Environmental metagenomic sequences have similarly been used to identify genes involved in secondary metabolism from entire communities [156], but it can be difficult to estimate the frequency of such genes in the community, as the total size of the metagenome is often unknown.

Lastly, PCR has been recognized for the past few decades as an efficient tool to probe for genes of interest in an organism or environment. The utility of PCR screening is dependent on the availability of previously sequenced and characterized genes from which primers are designed, and so it follows that the increase in sequenced gene clusters has led to an increase in probe sequences for PCR screens in recent years. PCR-dependent screens can be used to assess the chemical-producing potential of a strain or to estimate the frequency of a gene cluster in the environment [174, 175]. In a recent study, degenerate primers capable of amplifying NRPSs, ketosynthases, and enediyne PKSs were used to determine the frequencies of these genes in 60 marine actinobacteria [175]. Whereas genes involved in enediyne biosynthesis were detected in only five strains (8%), much higher hit rates where seen for ketosynthases (43%) and NRPSs (63%). This technique is likely to return an underestimate of the actual frequency due to poor amplification of only distantly related genes due to poor primer annealing, as shown by its application to an organism for which the complete genome was known. The major advantages to using PCR screening strategies in NP research are the relative ease and scalability of the method. The major drawback is that the observed bias towards amplifying genes similar to known sequences can mask new and unique gene clusters.

Whether through genome sequencing/scanning or genetic screens, it is clear that our ability to find NP biosynthetic gene clusters has improved dramatically over the past decade. However, finding them is only half of the struggle. Transforming this sequence information into new chemical entities that can be assayed for drug discovery often represents an imposing technological barrier. This goal of genome mining will perhaps grow to define NP biosynthesis research in the post-genomic era. Although detailed reviews have been published on genome mining [176, 177], here we will describe a particular success story concerning the complete characterization of the terpene production potential in the first two streptomycete genomes to be completely sequenced. Following this, we will highlight a few of the biological and chemical strategies for genome mining that have proven successful during the past decade.

4.4.2
Genome Mining for Terpene Biosynthesis

The feasibility of genome mining for NP discovery has been realized by many groups [176]. Perhaps the most promising illustration of the utility of genome mining for NP discovery involves investigations into terpene production in the model actinomycetes *S. coelicolor* and *S. avermitilis*. *S. coelicolor* was the first streptomycete to have its genome completely sequenced [170], and it was found to harbor at least 30 identifiable secondary metabolite gene clusters, five of which are involved in terpenoid production [169]. The *S. avermitilis* genome sequence [171] suggests that this strain has slightly broader NP production potential, with 37 secondary metabolite gene clusters including six for terpenoid production [169]. Terpenoids constitute the largest family of compounds found in Nature, and have been extensively studied in plants and fungi where their production is better understood [178]. Although several species of actinomycetes were known to produce triterpene hopanoids [179], carotenoid pigments [180], and volatile, odorous mono- and sesquiterpenes [181], the extent of microbial terpene production was not fully appreciated. Prior to the sequencing of these streptomycete genomes, only the hopanoid-producing gene cluster in *S. coelicolor* had been identified [182].

In the past decade, the biosynthetic pathways of all terpene NPs encoded by these two genomes have been characterized (Figure 4.21). The carotenoid gene clusters were predicted during the annotation of the genomes and validated with knockout mutations [183], and were found to produce not only isorenieratene but also β-carotene. The remaining three- and four-terpene-producing gene clusters in *S. coelicolor* and *S. avermitilis*, respectively, contain terpene synthases, and therefore were predicted to produce mono-, sesqui-, or diterpenes. One pair of orthologs, *sco6073* in *S. coelicolor* and *sva2163* in *S. avermitilis*, were found to be bifunctional enzymes whose N-terminal domains catalyze the conversion of farnesyl pyrophosphate to germacradienol and whose C-terminal domains convert the latter to geosmin [184–187]. This is the first example of a bifunctional monoterpene synthase, and it is notable that the order of types I and II carbocation-generating lyase reactions is opposite to that seen previously in bifunctional diterpene synthases [187]. A second volatile monoterpene produced by *S. coelicolor*, methyl-2-isoborneol, is produced by the terpene synthase *sco7700* following methylation of geranyl pyrophosphate by *sco7701* [188].

The above examples describe the elucidation of the biosynthetic pathways of known metabolites of *S. coelicolor* and *S. avermitilis*. However, the true promise of genome mining is to find metabolites that were not previously known (Figure 4.21). Orthologous gene clusters in *S. coelicolor* and *S. avermitilis* encoded by *sco5222–5223* and *sav3031–3032*, respectively, contain closely related terpene synthases and P450 monooxygenases. In both cases, the terpene synthase was found to produce a novel tricyclic sesquiterpene, *epi*-isozizaene, which is then converted to the known antibiotic albaflavenone [189–191]. The *S. avermitilis* gene cluster also produces a new, double-oxygenated *epi*-isozizaene derivative [191]. Although none of these compounds were previously isolated from either strain, a more direct search by gas chromatography–mass spectrometry identified albaflavenone in wild-type *S. coelicolor* cultures. Similar analyses along with real-time PCR of the gene cluster in *S. avermitilis* indicate that this cluster is silent in all

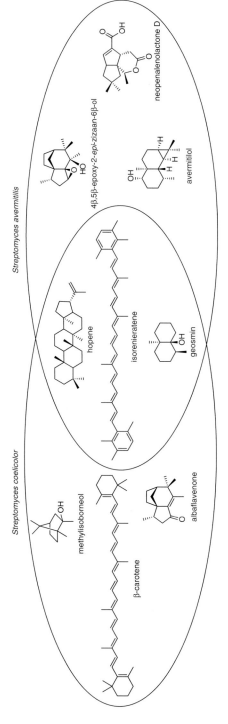

Figure 4.21 Genome mining for terpenoids in *S. coelicolor* and *S. avermitilis*. The complete genome sequences of *S. coelicolor* A3(2) and *S. avermitilis* ATCC31267 were found to contain five and six putative gene clusters, respectively. In the past decade, the metabolites produced by all of these gene clusters have been characterized.

culture media, although it was shown to be functional following overexpression in an engineered host [191]. One of the two remaining terpene synthesis gene clusters present in the *S. avermitilis* genome, corresponding to *sav2990–3002*, was originally assigned to pentalenolactone production based on the functional assignment of *sav2998* as a pentalenene synthase and the identification of known pentalenolactone resistance genes in close proximity [192]. However, further characterization of the gene cluster revealed a new Baeyer–Villiger monooxygenase that converts the pathway intermediate pentalenolactone D to neopentalenolactone D, representing a new branch in this family of compounds [193]. The final terpene synthase to be characterized was determined to produce a new sesquiterpene, avermitilol, *in vitro* using recombinant enzyme [194].

The elucidation of the complete terpene production potential of *S. coelicolor* and *S. avermitilis* represents a major milestone in genome mining. A survey of other successful examples of genome mining suggests that this approach has been most successful in producing NPs from relatively small gene clusters, with a number of RPs [155], small polyketides synthesized by type III PKSs [176], and various dipeptides [195–197] also having been isolated from different organisms. As technological advances continue, we predict that the mining and functional characterization of larger gene clusters responsible for the production of complicated macromolecular machines will become more routine [198]. A variety of chemical and biological strategies have emerged in the past 10 years to allow the discovery of new molecules from cryptic or orphan gene clusters. As these methods continue to develop, the number of isolable NPs from a given organism can be expected to increase by up to an order of magnitude. The following sections describe several successful approaches that have gained popularity recently.

4.4.3
Genomisotopic Approach for Orphan Gene Clusters

To mine microbial genomes readily for the discovery of new NPs will require contributions from both chemists and biologists towards developing novel approaches to NP discovery [199]. One such chemical approach was proposed and validated by Gerwick and co-workers [200] that relied on structural information that could be discerned from the genome sequence to tailor creatively their isolation strategy towards the molecule of interest (Figure 4.22). The observed colinearity of many PKS and NRPS complexes to the molecules they produce allows the prediction of the order and nature of extender units and also the tailoring chemistry performed on each unit [5]. In an effort to characterize an orphan NRPS-containing gene cluster discovered in the genome of *Pseudomonas fluorescens* Pf-5, the authors used this colinearity principle in conjunction with the predicted amino acid specificity of each module's adenylation domain to propose a chemical structure for the molecule produced. The structural prediction called for the incorporation of four leucine residues. Consequently, ^{15}N-labeled leucine was fed to production cultures. Because of the low natural abundance of ^{15}N, metabolites bearing the labeled leucines were strongly enriched in the ^{15}N NMR spectrum of the crude extract and isotopically enriched materials could be followed through the fractionation process. This novel strategy, dubbed the "genomisotopic approach," led to the isolation of orfamide A and several congeners, the first members of a new class of cyclic lipopeptides

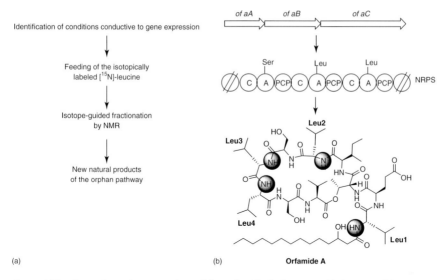

Figure 4.22 Genomisotopic approach used for orfamide A discovery. (a) Strategy. (b) NRPS-encoding gene cluster, selected domain organization (substrate specificity is shown for A domains) of NRPSs, and structure (labeled ^{15}N of the leucine residues is highlighted) for orfamide A.

[200]. Although others have tailored isolation protocols to target a particular class of molecule thought to be produced, this strategy goes one step further by prelabeling the NP using sequence-dependent knowledge of the biosynthetic pathway. One key limitation, however, is the requirement for a gene cluster that is already transcriptionally activated in known fermentation conditions. In this case, the authors verified expression of the NRPS gene by reverse transcription PCR prior to undertaking the isolation. A greater challenge comes from isolating NPs produced by gene clusters that fail to express under any conditions. These require new strategies to activate production before analytical chemistry can be of use.

4.4.4
Awakening Cryptic Gene Clusters through Global Regulators

Biological strategies for expanding the numbers of NPs that can be identified from a given strain generally fall into one of two categories: (i) awakening cryptic clusters in the native producing strain [201] or (ii) cloning the biosynthetic gene cluster for expression in a heterologous host [202]. Working in the native producing organism has the advantage of saving the potentially time/resource-consuming step of cloning the gene cluster into a heterologous expression construct, but also has the advantage of ensuring that a proper cytoplasmic environment for functional production of the biosynthetic enzymes exists. Manipulating the specific regulatory enzymes or promoter sequences located in a particular gene cluster can turn it on in genetically amenable microorganisms [203]; even better would be to manipulate global regulatory components that control numerous cryptic clusters. The benefit to this strategy is that nothing needs to be known about the

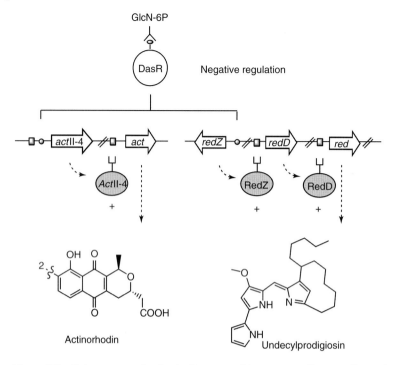

Figure 4.23 Enhancement of actinorhodin and undecylprodigiosin production by inactivating *dasR*. Glucosamine-6-phosphate (GlcN-6P), a metabolite originating from *N*-acetylglucosamine, inhibits the DNA binding of DasR as its allosteric effector, leading to the loss of transcriptional repression of *actII–ORF4* for actinorhodin and *redZ* for undecylprodigiosin, respectively.

regulation of specific gene clusters to turn them on, and ideally the same method could be used to activate cryptic gene clusters in numerous organisms. Recently, the pleiotropic regulator DasR was found to repress antibiotic production in *S. coelicolor* [204]. Knockout mutants overproduced known antibiotics actinorhodin and undecylprodigiosin and expressed the cryptic *kas* gene cluster (Figure 4.23). Preliminary work suggests that DasR plays a similar role in other actinomycetes and, importantly, its repression of secondary metabolism can be relieved chemically through the addition of *N*-acetylglucosamine to production media [204]. These results, along with our growing understanding of the importance of small-molecule signaling factors in activating secondary metabolic pathways [205], point towards a promising future combining chemical biology with NP biosynthesis to turn on pathways in genetically recalcitrant species.

4.4.5
Activation of NP Pathways Through Mixed Culturing

That many antibiotic-producing bacteria fail to activate the majority of their secondary metabolic gene clusters in typical laboratory growth media is not surprising when one considers just how different these conditions are from their natural environments. In

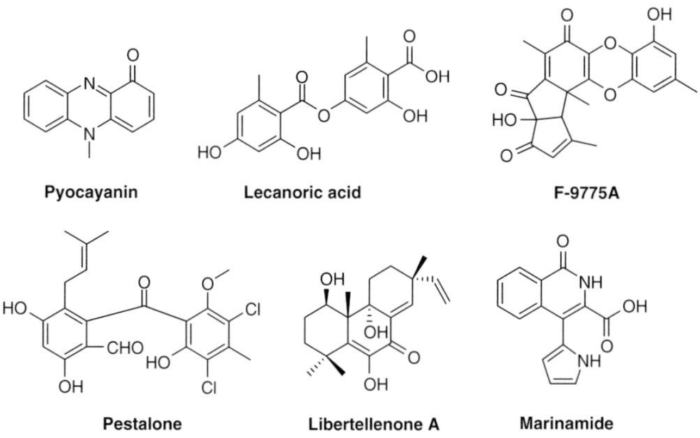

Figure 4.24 Selected compounds obtained from mixed culturing.

Nature, bacteria are more likely to face a variety of stresses, including but not limited to nutrient depletion, competition for resources and space, temperature, and osmotic fluctuations [206]. The last decade has witnessed renewed interest in creative culture techniques, including the mixed culturing of different species of bacteria (or fungi) with the hope of stimulating cryptic NP pathways (Figure 4.24) [207, 208]. Watanabe's group found that co-culturing an *Enterobacter* strain with *Pseudomonas aeruginosa* Pup14B elicited the production of pyocyanin, which was not produced in pure culture [209]. Follow-up experiments suggested that a two-way metabolite exchange between the co-cultured strains is necessary for pathway activation although direct cell–cell contact is not necessary. Conversely, other co-culture production systems do require an intimate physical interaction [210], with one study going a step further to implicate mycolic acid as the molecule responsible for induction in the membrane of the inducer strain of an obligate co-culture production system [211]. Over the past decade, mixed cultures have been used in the discovery of several new NPs [208].

4.4.6
Heterologous Production

An alternative to awakening the cryptic gene cluster in its native host is to express it heterologously in a different organism. In heterologous production, the NP gene cluster is cloned and moved into a suitable host with the hope that this host will be able to convert the information present in the DNA sequence to a functional biochemical pathway that produces the molecule of interest. Heterologous production holds the potential to access the chemical diversity produced by the 99% of bacteria that are currently unculturable. Heterologous production is also envisioned as a means to facilitate the production and engineering of molecules naturally synthesized by strains that are recalcitrant to genetic manipulation, slow growing, or generally difficult to study. For these reasons, the last decade has seen many groups focusing on how to advance the science of heterologous production of pharmaceutical NPs (Figure 4.25) [202]. Because technological advances in molecular biology and DNA sequencing over the past decades have made cloning and

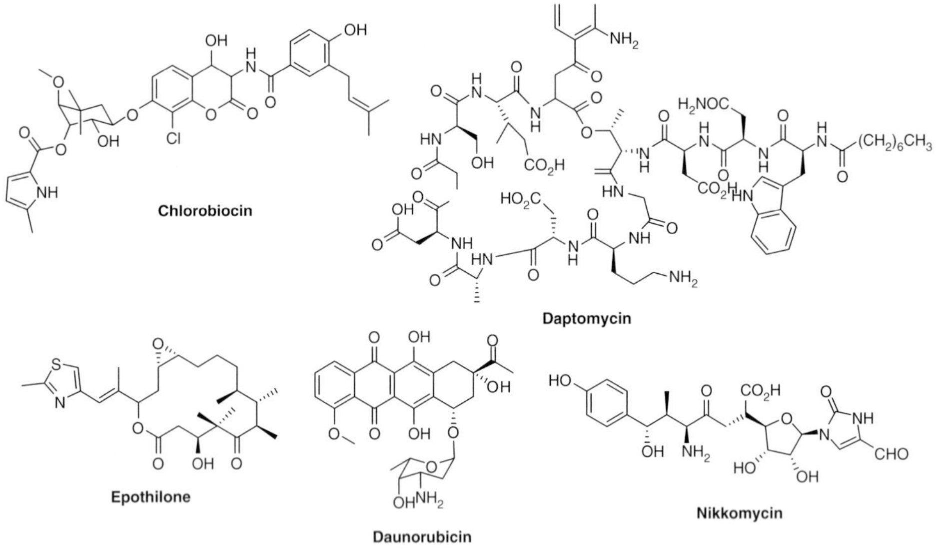

Figure 4.25 Selected structures of clinically relevant NPs that have been produced by heterologous expression.

sequencing of even large (>100 kb) gene clusters relatively straightforward, the largest hurdle to overcome now in any given project lies in finding an appropriate heterologous host. Engineering a single "universal host" that will be able to produce efficiently molecules encoded by any gene cluster is a goal in the field that has been approached from many different angles.

Actinomycete bacteria, including the genus *Streptomyces*, have proven to be a treasure trove of pharmaceutically important NPs and, as such, significant effort has gone into engineering universal hosts from various model streptomycete strains. Heterologous hosts that have diverged more recently from the producing organisms on an evolutionary time scale are more likely to have retained the cytoplasmic environments required for proper transcription, protein folding, and enzymatic function. Another justification for engineering a streptomycete universal host is that streptomycetes are known to possess complex metabolic pathways, including those that provide precursors to every major class of secondary metabolite [170]. *Streptomycetes* harbor large, linear chromosomes comprised of a central conserved core region that houses the essential genes and two variable arms that contain many of the secondary metabolite gene clusters and non-essential genes [172]. By removing large sections of the variable arms from the genome of the industrial strain of *S. avermitilis*, Ikeda and co-workers created genome-minimized variants that no longer produce any of the known endogenous secondary metabolites from the parent strain [212]. The strains produce a cleaner crude extract, facilitating the identification of new compounds produced, and have been shown to synthesize efficiently a number of non-native NPs, including streptomycin, cephamycin C, pladienolide, and amorphadiene, at higher titers than the native producers. A separate effort to create an actinomycete universal host by researchers at the John Innes Centre employed a more targeted approach. Specific gene clusters were deleted sequentially from a plasmid-less derivative of the model organism *S. coelicolor* A3(2). Additional mutations were made in the genes encoding RNA polymerase and the ribosomal protein S12 that were previously shown to increase antibiotic production in other strains [213]. Like the *S. avermitilis* host, these engineered *S. coelicolor* strains were shown to produce several non-native NPs efficiently [214]. Although both engineered hosts have proven effective in the heterologous expression of limited NP gene clusters, only time will tell the extent of their utility. One difficulty that is already apparent is how to ensure transcriptional activation of non-native gene clusters in engineered hosts [212]. One proven method to accomplish this is to manipulate the specific regulatory elements in the non-native gene clusters [203], but a more attractive alternative will be to engineer host strains that are able to turn on the expression of any non-native gene cluster. To achieve this will, without a doubt, require additional basic research into the regulatory mechanisms of NP biosynthesis.

Removing nonessential and superfluous genetic material from a large-genome strep-tomycete can be viewed as a top-down method towards engineering a universal host, but a second, bottom-up strategy has also seen success in the past decade. Specifically, the laboratory workhorse *E. coli* has been outfitted with the prerequisite tools to allow the biosynthesis of complex polyketides [215], terpenoids [216], peptides [217], and other NPs [218]. Some of the special challenges that must be addressed for functional expression of NP pathways in *E. coli* are (i) efficient transcription, translation, and post-translational modification of each biosynthetic protein, (ii) *in vivo* production of the required precursor metabolites, and (iii) proper coordination of precursor production and pathway induction [215]. Because of the falling cost of DNA synthesis, it is now economically feasible to synthesize even the largest of genes; designing codon-optimized genes outfitted with *E. coli* promoters facilitates transcription and translation in this bottom-up approach. Post-translational modification of biosynthetic proteins, for example, the phosphopan-tetheinylation of the carrier proteins of PKSs and NRPSs, is accomplished by engineered hosts that express recombinant enzymes such as *Sfp* from *Bacillus subtilis* [215, 217]. *In vivo* production of precursor metabolites in titers required for efficient production of the desire molecule can be accomplished by rewiring endogenous biosynthetic pathways [215], or by heterologously expressing complete precursor biosynthesis pathways [216]. The past decade has seen creative advances in improving the flux through such pathways that were borrowed from the field of engineering. For instance, the linear scaffolding of biosynthetic enzymes to ensure metabolic channeling from one active site to the next [219] and multiplexed automated genome engineering (MAGE) for the directed evolution of whole pathways [160] both lead to enhanced resource flux *en route* to target NP production. MAGE directs accelerated evolution by using automation to control iterative rounds of mutagenesis and cell growth. Multiple loci in the genome can be targeted simultaneously to manipulate the desired phenotype, and this technique can generate an astonishing 4.3 billion bp of sequence variation per day. Lastly, ensuring proper coor-dination of the expression of each component can be controlled rudimentarily by using inducible promoters, but more sophisticated methods for controlling gene expression in *E. coli* are on the near horizon with inputs from the field of synthetic biology [220].

An excellent example that illustrates the multi-pronged approach to heterologous expression that has become possible during the last decade involves the anticancer agent epothilone. The epothilones are 16-membered macrolide NPs isolated from the cellulose-degrading myxobacterium *Sorangium cellulosum* SMP 44 that exhibit cytotoxicity through a mechanism similar to that of the blockbuster anticancer drug Taxol (paclitaxel) [221, 222]. A semisynthetic lactam analog of epothilone B, ixabepilone, was recently approved by the FDA for clinical use [223]. Because of the considerable potential for drug development from this class of NP, in the past decade much effort has been devoted to taking advantage of its biosynthetic gene cluster. The objectives have been very clear: (i) to engineer a more reliable source of the parent compounds and (ii) to engineer unnatural analogs with potentially superior activity or physicochemical properties. To expedite the localization of the epothilone gene cluster (*epo*) in *S. cellulosum*, genome scanning, similar to that described for the enediynes, was used to identify each of the PKS-containing gene clusters [173]. Sequencing and characterization of the *epo* gene cluster revealed that the core epothilone scaffold is formed by an NRPS–PKS hybrid system composed of one NRPS and nine PKS modules, and is subject to epoxidation by EpoK to form epothilones A and B [224, 225]. Because the low NP titer (20 mg l^{-1}) and long doubling time (16 h) of the native producer made production at industrial scales impractical, several efforts to produce epothilones in a recombinant organism soon followed. First, the genes were placed under the control of the *Streptomycete* promoter *act*I, and expressed in a two-plasmid system in *S. coelicolor* CH999, a very distant relative of the Gram-negative producing organism. That production of epothilones A–D could be observed at all indicates the versatility of actinomycetes as heterologous hosts. The 2 h doubling time of *S. coelicolor* greatly reduced the time required for fermentation; however, only low titers of 100 μg l^{-1} could be attained [226], prompting the search for alternative hosts. The model myxobacterium *Myxococcus xanthus* was chosen as a possible host for its genetic amenability, shorter doubling time, and closer relation to the producing strain [227]. Titers in *M. xanthus* were low at first, but were increased to 23 mg l^{-1} by fermentation optimization [228] to result in a genetically amenable epothilone production strain that grows faster and has slightly improved titers relative to the native producer. More recently, efforts have been made to produce epothilones in *E. coli* [229, 230]. Despite initial challenges concerning the functional expression of the larger enzymes in the pathway, epothilones can now be produced in this convenient host, although this system also suffers from low NP titers. Attempts to engineer novel epothilones through combinatorial biosynthesis are promising, as several new analogs have been isolated from mutated gene clusters expressed in the *M. xanthus* host [231].

4.5
Future Perspectives

Although traditional NP discovery programs provided diminished returns on their investment in the last two decades of the twentieth century, the "tank" of NPs certainly has not run dry. Recent progress in several aspects of NP research and microbial genomics clearly suggests that the potential of NP diversity and discovery is vastly underestimated, offering several promising alternatives to existing methods for the discovery of new NPs. First, whole-genome sequencing has undoubtedly revealed that there are far more biosynthetic gene clusters than there are currently known metabolites for any given organism, suggesting that the biosynthetic potential for NPs in microorganisms has been greatly underexplored by traditional methods of NP discovery. These so-called cryptic or silent gene clusters, indeed, encode new NPs, whose "sleeping" biosynthesis can now be "awoken" using innovative *in vivo* and *in vitro* strategies and methods. Second, only 1% of the microbial community is estimated to have been cultivated in the laboratory, implying that the overwhelming majority of NP biodiversity from microorganisms remains to be discovered and exploited. Newly emerging cultivation techniques, culture-independent methods, which involve expressing gene clusters in model heterologous hosts, and chemoenzymatic bioconversion strategies are starting to permit access to these previously inaccessible NP resources. Finally, the exponential growth in cloning and sequencing of NP biosynthetic gene clusters, in characterization of NP biosynthetic machineries, and in chemical and mechanistic studies of NP biosynthetic enzymes have fundamentally changed the landscape of NP research and discovery. Although these advances have clearly rejuvenated NP research, injected optimism and excitement into discovering new NPs, and inspired innovation in the past decade, their continued refinement and expansion surely will further permit the discovery and design of new NPs.

NPs remain the best sources of drugs and drug leads despite the fact that major pharmaceutical companies have de-emphasized NP research in favor of high-throughput screenings of combinatorial libraries during the last two decades. Several realizations support this view. First, NPs have an outstanding track record, unmatched by any other class of small molecules in the current inventory of clinically significant drugs, particularly for antibacterial, antifungal, and anticancer drugs. Second, NPs display an enormous array of structural diversity unsurpassed by synthetic libraries, and ~40% of the chemical scaffolds found in NPs are absent in today's medicinal chemistry. The rich functionality of NPs is, without a doubt, one of their greatest strengths, often providing remarkable potency and selectivity of biological activities. Third, NPs are evolutionarily optimized to serve as drug-like molecules. This is evident when realizing that NPs and drugs occupy approximately the same molecular space. It is no coincidence that both their construction and exertion of biological activity derive from an NP's ability to interact with various protein-based machineries. Consequently, the ability of a given NP to serve as a drug might be viewed as biologically validated *a priori*. In a very real sense, the genesis of an NP is the result of selective pressures applied at the molecular level whereas the survival and propagation of an NP producer may, in turn, be the result of selective pressures either enabled or resisted by the NP in question. NPs are, in truth, designed almost invariably to illicit some biological activity. The real question has more to do with whether or not the current biological activity is relevant to human health. Finally, NPs represent the richest source of novel molecular scaffolds and chemistry known. No-one can predict, in advance, the details of how a small molecule will interact with the myriad of targets that we now know drive fundamental biological processes. Indeed, the history of NP discovery is full of remarkable stories in which serendipitous biological functions were found to be attenuated by either crude or purified NP-containing research samples; refinement of these discoveries revealed discrete NPs with profound uses in biology and/or medicine. However, despite these time-tested attributes of NPs, they are gravely under-represented in current libraries of small molecules targeted for drug screenings. Among the many considerations contributing to the massive under-representation of NPs in these libraries are (i) extremely low yields of NPs, which translate to challenging isolation, (ii) limited resupply possibilities due to slow growth or adverse impacts on their native habitats and ecology upon overharvesting, and (iii) complicated molecular architectures that preclude practical organic synthesis and pose large hurdles to medicinal chemistry studies. Combinatorial biosynthesis, as illustrated in this chapter, now enables one to circumvent these challenges. Specific, rather than random, effects can often be achieved by precise manipulation of NP biosynthetic machinery. Targeted NP analogs can be produced by recombinant organisms amenable to large-scale fermentation, thus lowering production costs and reducing environmental concerns. Additionally, judicious application of genetic principles in metabolic pathway engineering to address titer improvement, production bottlenecks, and generation of focused libraries of NP scaffolds have been advanced from a limited set of proof-of-principle experiments into routine approaches. It is clear that combinatorial biosynthesis surely will play an increasingly important role in NP drug discovery and development.

NPs will continue to inspire the discovery of novel chemistry and enzymology. Most of the biosynthetic gene clusters discussed in this chapter were identified only within the last two decades. Despite this short time frame, a significant amount of unprecedented biosynthetic logic has been established, including the assembly-line enzymology exemplified by modular PKS and NRPS systems and also the myriad of tailoring enzymes that feature a plethora of novel enzymologies and catalytic mechanisms. Owing to these dramatic advances, NP biosynthesis is rapidly reclaiming its title as one of the most fertile grounds by which to discover new chemistry and enzymology. Additionally, the last two decades have witnessed the relevance of distinct paradigms for the biosynthesis of almost all major NP families. We can now attribute the vast structural diversity of NPs to variations of a few common biosynthetic machineries. Although the differences between these paradigms have been invaluable in delineating the biosynthesis of various classes of NPs, it also has become very clear that the plasticity of various biosynthetic machineries is much more profound than previously appreciated. These features of NP biosynthetic machineries, realized only in the last decade, will continue to inspire the search for and discovery of novel chemistries and enzymologies for NP biosynthesis and to illuminate novel opportunities and strategies for enhancing NP structural diversity and drug discovery via combinatorial biosynthetic strategies.

Acknowledgments

M.J.S. was supported in part by US NIH predoctoral training grant **GM08505**. X.Q. was supported by grant **30900021** from the Natural Science Foundation for Youth, China. Studies on natural product biosynthesis, engineering, and drug discovery in W.L.'s laboratories were supported in part by Chinese grants from NNSF (**20832009**), "973 program" (**2010CB833200**), and STCSM (**09QH1402700**), respectively, and in B.S's laboratories in part by U.S. NIH grants **AI51689, AI79070, CA78747, CA94426, CA106150, CA113297**, and **GM86184** and US NSF grant **MCB9733938**.

Abbreviations

The following abbreviations are used on the reaction schemes:

CoL   CoA ligase
A   adenylation
C   condensation
E   epimerization
F   formylation
Cy   condensation/cyclization
C-MT   *C*-methylation
N-MT   *N*-methylation
Ox   oxidation
R   reductase

References

1. Newsman, D.J. and Cragg, G.M. (2007) *J. Nat. Prod.*, **70**, 461–477.
2. International Human Genome Sequencing Consortium (2001) *Nature*, **409**, 860–921.
3. Venter, J.C. *et al.* (275 authors) (2001) *Science*, **291**, 1304–1351.
4. Metzker, M.L. (2005) *Genome Res.*, **15**, 1767–1776.
5. Fischbach, M.A. and Walsh, C.T. (2006) *Chem. Rev.*, **106**, 3468–3496.
6. Hertweck, C. (2009) *Angew. Chem. Int. Ed.*, **48**, 4688–4716.
7. Marahiel, M.A. (2009) *J. Pept. Sci.*, **15**, 799–807.
8. Walsh, C.T. and Fischbach, M.A. (2010) *J. Am. Chem. Soc.*, **132**, 2469–2493.
9. Shen, B. (2003) *Curr. Opin. Chem. Biol.*, **7**, 285–295.
10. Dimroth, P., Ringelmann, E., and Lynen, F. (1970) *Eur. J. Biochem.*, **13**, 98–110.
11. Cortes, J., Haydock, S.F., Roberts, G.A., Bebitt, D.J., and Leadlay, P.F. (1990) *Nature*, **348**, 176–178.
12. Donadio, S., Staver, M.J., McAlpine, J.B., Swanson, S.J., and Katz, L. (1991) *Science*, **252**, 675–679.
13. Malpartida, F. and Hopwood, D.A. (1984) *Nature*, **309**, 462–464.
14. Austin, M.B. and Noel, J.P. (2003) *Nat. Prod. Rep.*, **20**, 79–110.
15. Funa, M., Ohnishi, Y., Fujii, I., Shibuya, M., Ebizuka, Y., and Horinouchi, S. (1999) *Nature*, **400**, 897–899.
16. Bangera, M.G. and Thomashow, L.S. (1999) *J. Bacteriol.*, **181**, 3155–3163.
17. Moore, B.S. and Hopke, J.N. (2001) *Chem-BioChem*, **2**, 35–38.
18. van Lanen, S.G. and Shen, B. (2008) *Curr. Opin. Drug Discov. Dev.*, **11**, 186–195.
19. Khosla, C. (2009) *J. Org. Chem.*, **74**, 6416–6420.
20. Staunton, J. and Weissman, K.J. (2001) *Nat. Prod. Rep.*, **18**, 380–416.
21. McDaniel, R., Welch, M., and Hutchinson, C.R. (2005) *Chem. Rev.*, **105**, 543–558.
22. Moss, S.J., Martin, C.J., and Wilkinson, B. (2004) *Nat. Prod. Rep.*, **21**, 575–593.
23. Wenzel, S.C. and Muller, R. (2005) *Curr. Opin. Chem. Biol.*, **9**, 447–458.
24. Buchholz, T.J., Geders, T.W., Bartley, F.E. III, Reynolds, K.A., Smith, J.L., and Sherman, D.H. (2009) *ACS Chem. Biol.*, **4**, 41–52.
25. Gaitatzis, N., Silakowski, B., Kunze, B., Nordsiek, G., Blocker, H., Hofle, G., and Muller, R. (2002) *J. Biol. Chem.*, **277**, 13082–13090.
26. Olano, C., Wilkinson, B., Sanchez, C., Moss, S.J., Sheridan, R., Math, V., Weston, A.J., Brana, A.F., Martin, C.J., Oliynyk, M.,

Mendez, C., Leadlay, P.F., and Salas, J.A. (2004) *Chem. Biol.*, **11**, 87–97.

27. He, J. and Hertweck, C. (2003) *Chem. Biol.*, **10**, 12225–11232.

28. Bisang, C., Long, P.F., Corte, J., Westcott, J., Crosby, J., Matharu, A., Cox, R.J., Simpson, T.J., Staunton, J., and Leadlay, P.F. (1999) *Nature*, **401**, 502–505.

29. Weissman, K.J., Bycrofy, M., Staunton, J., and Leadlay, P.F. (1998) *Biochemistry*, **37**, 11012–11017.

30. Ikeda, H., Nonomiya, T., Usami, M., Ohta, T., and Omura, S. (1999) *Proc. Natl. Acad. Sci. U. S. A.*, **96**, 9509–9514.

31. Palaniappan, N., Kim, B.S., Sekiyama, Y., Osada, H., and Reynolds, K.A. (2003) *J. Biol. Chem.*, **278**, 35552–35557.

32. Qu, X., Jiang, N., Xu, F., Shao, L., Tang, G., Wilkinson, B., and Liu, W. (2011) *Mol. BioSyst.*, **7**, 852–861.

33. Gregory, M.A., Petkovic, H., Lill, R.E., Moss, S.J., Wilkinson, B., Gaisser, S., Leadlay, P.F., and Sheridan, R.M. (2005) *Angew. Chem. Int. Ed.*, **44**, 4757–4760.

34. Yu, T., Bai, L., Clade, D., Hoffmann, D., Toelzer, S., Trinh, K., Xu, J., Moss, S.J., Leistner, E., and Floss, H.G. (2002) *Proc. Natl. Acad. Sci. U. S. A.*, **99**, 7968–7973.

35. Wenzel, B.H. (2008) *ChemBioChem*, **9**, 2711–2721.

36. Moore, B.S. and Hertweck, C. (2002) *Nat. Prod. Rep.*, **19**, 70–99.

37. Piel, J., Wen, G., Platzer, M., and Hui, D. (2004) *ChemBioChem*, **5**, 93–98.

38. Gu, L., Geders, T.W., Wang, B., Gerwick, W.H., Hakansson, K., Smith, J.L., and Sherman, D.H. (2007) *Science*, **318**, 970–974.

39. Liou, G.F. and Khosla, C. (2003) *Curr. Opin. Chem. Biol.*, **7**, 279–284.

40. Haydock, S.F., Appleyard, A.N., Mironenko, T., Lester, J., Scott, N., and Leadlay, P.F. (2005) *Microbiology*, **151**, 3161–3169.

41. Weber, T., Laiple, K.J., Pross, E.K., Textor, A., Grond, S., Welzel, K., Pelzer, S., Vente, A., and Wohlleben, W. (2008) *Chem. Biol.*, **15**, 175–188.

42. Goranovic, D., Kosec, G., Mrak, P., Fujs, S., Horvat, J., Kuscer, E., Kopitar, G., and Petkovic, H. (2010) *J. Biol. Chem.*, **285**, 14292–14300.

43. Eustaquio, A.S. and Moore, B.S. (2008) *Angew. Chem. Int. Ed.*, **47**, 3936–3938.

44. Reeves, C.D., Chung, L.M., Liu, Y., Xue, Q., Carney, J.R., Revill, W.P., and Katz, L. (2002) *J. Biol. Chem.*, **277**, 9155–9159.

45. Wenzel, S.C., Williamson, R.M., Grunanger, C., Xu, J., Gerth, K. II, Martinez, R.A., Moss, S.J., Carroll, B.J., Grond, S., Unkefer, C.J., Muller, R., and Floss, H.G. (2006) *J. Am. Chem. Soc.*, **128**, 14325–14336.

46. Zhao, C., Coughlin, J.M., Ju, J., Zhu, D., Wendt-Pienkowski, E., Zhou, X., Wang, Z., Shen, B., and Deng, Z. (2010) *J. Biol. Chem.*, **285**, 20097–20108.

47. Emmert, E.A.B., Klimowicz, A.K., Thomas, M.G., and Handelsman, J. (2004) *Appl. Environ. Microbiol.*, **70**, 104–113.

48. Reyraud, R., Kiefer, P., Christen, P., Massou, S., Portais, J., and Vorholt, J.A. (2009) *Proc. Natl. Acad. Sci. U. S. A.*, **106**, 4846–4851.

49. Mo, S., Kim, D.H., Lee, J.H., Park, J.W., Basnett, D.B., Ban, Y.H., Yoo, Y.J., Chen, S., Park, S.R., Choi, E.A., Kim, E., Jin, Y., Lee, S., Park, J.Y., Liu, Y., Lee, M.O., Lee, K.S., Kim, S.J., Kim, D., Park, B.C., Lee, S., Kwon, H.J., Sun, J., Moore, B.S., Lim, S., and Yoon, Y.J. (2010) *J. Am. Chem. Soc.*, **133**, 976–985.

50. Eustaquio, A.S., McGlinchey, R.P., Liu, Y., Hazzard, C., Beer, L.L., Florova, G., Alhamadsheh, M.M., Lechner, A., Kale, A.J., Kobayashi, Y., Reynolds, K.A., and Moore, B.S. (2009) *Proc. Natl. Acad. Sci. U. S. A.*, **106**, 12295–12300.

51. Dorrestein, P.C., van Lanen, S.G., Li, W., Zhao, C., Deng, Z., Shen, B., and Kelleher, N.L. (2006) *J. Am. Chem. Soc.*, **128**, 10386–10387.

52. Chan, Y.A., Boyne, M.T. II, Podevels, A.M., Klimowicz, A.K., Handelsman, J., Kelleher, N.L., and Thomas, M.G. (2006) *Proc. Natl. Acad. Sci. U. S. A.*, **103**, 14349–14354.

53. Calderone, C.T. (2008) *Nat. Prod. Rep.*, **25**, 845–853.

54. Gu, L., Jia, J., Liu, H., Hakansson, K., Gerwick, W.H., and Sherman, D.H. (2006) *J. Am. Chem. Soc.*, **128**, 9014–9015.

55. Calderone, C.T., Kowtoniuk, W.E., Kelleher, N.L., Walsh, C.T., and Dorrestein, P.C. (2006) *Proc. Natl. Acad. Sci. U. S. A.*, **103**, 8977–8982.

56. Calderone, C.T., Iwig, D.F., Dorrestein, P.C., Kelleher, N.L., and Walsh, C.T. (2007) *Chem. Biol.*, **14**, 835–846.

57. Simunovic, V. and Muller, R. (2007) *ChemBioChem*, **8**, 1273–1280.

58. Sudek, S., Lopanik, N.B., Waggoner, L.E., Hildebrand, M., Anderson, C., Liu, H., Patel, A., Sherman, D.H., and Haygood, M.G. (2007) *J. Nat. Prod.*, **70**, 67–74.

59. Kusebauch, B., Busch, B., Scherlach, K., Roth, M., and Hertweck, C. (2009) *Angew. Chem. Int. Ed.*, **48**, 5001–5004.

60. Cheng, Y.Q., Coughlin, J.M., Lim, S.K., and Shen, B. (2009) *Methods Enzymol.*, **459**, 165–186.

61. Peil, J. (2010) *Nat. Prod. Rep.*, **27**, 996–1047.

62. Cheng, Y.Q., Tang, G.L., and Shen, B. (2003) *Proc. Natl. Acad. Sci. U. S. A.*, **100**, 3149–3154.

63. Peil, J. (2002) *Proc. Natl. Acad. Sci. U. S. A.*, **99**, 14002–14007.

64. Kopp, M., Irschik, H., Pradella, S., and Muller, R. (2005) *ChemBioChem*, **6**, 1277–1286.

65. Lim, S.K., Ju, J., Zazopoulos, E., Jiang, H., Seo, J.W., Chen, Y., Feng, Z., Rajski, S.R., Farnet, C.M., and Shen, B. (2009) *J. Biol. Chem.*, **284**, 29746–29756.

66. Bumpus, S.B., Magarvey, N.A., Kelleher, N.L., Walsh, C.T., and Calderone, C.T. (2008) *J. Am. Chem. Soc.*, **130**, 11614–11616.

67. Heneghan, M.N., Yakasai, A.A., Williams, K., Kadir, K.A., Wasil, Z., Bakeer, W., Fisch, K.M., Bailey, A.M., Simpson, T.J., Cox, R.J., and Lazarus, C.M. (2011) *Chem. Sci.*, **2**, 972–979.

68. Fujii, I. (2010) *J. Antibiot.*, **63**, 207–218.

69. Gaisser, S., Trefzer, A., Stockert, S., Kirschning, A., and Bechthold, A. (1997) *J. Bacteriol.*, **179**, 6271–6278.

70. Ahlert, J., Shepard, E.M., Lomovskaya, N., Zazopoulos, E., Staffa, A., Bachmann, B.O., Huang, K., Yang, X., Fonstein, L., Czisny, A., Whitwam, R.E., Farnet, C.M., and Thorson, J.S. (2002) *Science*, **197**, 1173–1176.

71. Jia, X., Tian, Z., Shao, L., Qu, X., Zhao, Q., Tang, J., and Liu, W. (2006) *Chem. Biol.*, **13**, 575–585.

72. van Lanen, S.G., Oh, T.J., Liu, W., Wendt-Pienkowski, E., and Shen, B. (2007) *J. Am. Chem. Soc.*, **129**, 13082–13094.

73. Daum, M., Peintner, I., Linnenbrink, A., Frerich, A., Weber, M., Paululat, T., and Bechthold, A. (2009) *ChemBioChem*, **10**, 1073–1083.

74. Ito, T., Roongsawang, N., Shirasaka, N., Lu, W., Flatt, P.M., Kasanah, N., Miranda, C., and Mahmud, T. (2009) *ChemBioChem*, **10**, 2253–2265.

75. Zhao, Q., He, Q., Ding, W., Tang, M., Kang, Q., Yu, Y., Deng, W., Zhang, Q., Fang, J., Tang, G., and Liu, W. (2008) *Chem. Biol.*, **15**, 693–705.

76. Liu, W., Nonaka, K., Nie, L., Zhang, J., Christenson, S.D., Bae, J., van Lanen, S.G., Zazopoulos, E., Farnet, C.M., Yang, C.F., and Shen, B. (2005) *Chem. Biol.*, **12**, 293–302.

77. Wenzel, S.C., Kunze, B., Hofle, G., Silakowski, B., Blocker, H., and Muller, R. (2005) *ChemBioChem*, **6**, 375–385.

78. Liu, W., Christenson, S.D., Standage, S., and Shen, B. (2002) *Science*, **197**, 1170–1173.

79. Gao, Q. and Thorson, J.S. (2008) *FEMS Microbiol. Lett.*, **282**, 105–114.

80. Horsman, G.P., van Lanen, S.G., and Shen, B. (2009) *Methods Enzymol.*, **459**, 97–112.

81. Zhang, J., van Lanen, S.G., Ju, J., Liu, W., Dorrestein, P.C., Li, W., Kelleher, N.L., and Shen, B. (2008) *Proc. Natl. Acad. Sci. U. S. A.*, **105**, 1460–1465.

82. Horsman, G.P., Chen, Y., Thorson, J.S., and Shen, B. (2010) *Proc. Natl. Acad. Sci. U. S. A.*, **107**, 11331–11335.

83. Zazopoulos, E., Huang, K., Staffa, A., Liu, W., Bachmann, B.O., Nonaka, K., Ahlert, J., Thorson, J.S., Shen, B., and Farnet, C.M. (2003) *Nat. Biotechnol.*, **21**, 187–190.

84. Liu, W., Ahlert, J., Gao, Q., Wendt–Pienkowski, E., Shen, B., and Thorson, J.S. (2010) *Proc. Natl. Acad. Sci. U. S. A.*, **100**, 11959–11963.

85. Kwon, H.J., Smith, W.C., Xiang, L., and Shen, B. (2001) *J. Am. Chem. Soc.*, **123**, 3385–3386.

86. Kwon, H.J., Smith, W.C., Scharon, A.J., Hwang, S.H., Kurth, M.J., and Shen, B. (2002) *Science*, **297**, 1327–1330.

87. Walsh, C.T., Chen, H., Keating, T.A., Hubbard, B.K., Losey, H.C., Luo, L., Marshall, C.G., Miller, D.A., and Patel, H.M. (2001) *Curr. Opin. Chem. Biol.*, **5**, 525–534.

88. Samel, S.A., Marahiel, M.A., and Essen, L.O. (2008) *Mol. BioSyst.*, **4**, 387–393.

89. Marahiel, M.A. and Essen, L.O. (2009) *Methods Enzymol.*, **458**, 337–351.

90. Copp, D. and Marahiel, M.A. (2007) *Curr. Opin. Biotechnol.*, **18**, 513–520.

91. Copp, F. and Marahiel, M.A. (2007) *Nat. Prod. Rep.*, **24**, 735–749.

92. Strieker, M., Tanovic, A., and Marahiel, M.A. (2010) *Curr. Opin. Struct. Biol.*, **20**, 234–240.

93. Tanovic, A., Samel, S.A., Essen, L.O., and Marahiel, M.A. (2008) *Science*, **321**, 659–663.

94. Sieber, S.A. and Maraheil, M.A. (2005) *Chem. Rev.*, **105**, 715–738.

95. Mootz, H.D., Schwarzer, D., and Marahiel, M.A. (2002) *ChemBioChem*, **3**, 490–504.

96. Lei, L., Deng, W., Song, J., Ding, W., Zhao, Q.F., Peng, C., Song, W.W., Tang, G.L., and Liu, W. (2008) *J. Bacteriol.*, **190**, 251–263.

97. Koketsu, K., Watanabe, K., Suda, H., Oguri, H., and Oikawa, H. (2010) *Nat. Chem. Biol.*, **6**, 408–410.

98. Yamanaka, K., Maruyama, C., Takagi, H., and Hamano, Y. (2008) *Nat. Chem. Biol.*, **4**, 766–772.

99. Walsh, C.T. (2008) *Acc. Chem. Res.*, **41**, 4–10.

100. Sattely, E., Fischbach, M.A., and Walsh, C.T. (2008) *Nat. Prod. Rep.*, **25**, 757–793.

101. Moore, B.S. (2008) *Angew. Chem. Int. Ed.*, **47**, 9386–9387.

102. Nolan, E.M. and Walsh, C.T. (2009) *ChemBioChem*, **10**, 34–53.

103. Oman, T.J. and van der Donk, W.A. (2010) *Nat. Chem. Biol.*, **1**, 9–18.

104. Willey, J.M. and van der Donk, W.A. (2007) *Annu. Rev. Microbiol.*, **61**, 477–501.

105. Li, B., Cooper, L.E., and van der Donk, W.A. (2009) *Methods Enzymol.*, **458**, 533–588.

106. Lee, S.W., Mitchell, D.A., Markley, A.L., Hensler, M.E., Gonzalez, D., Wohlrab, A., Dorrestein, P.C., Nizet, V., and Dixon, J.E. (2008) *Proc. Natl. Acad. Sci. U. S. A.*, **105**, 5879–5884.

107. McIntosh, J.A. and Schmidt, E.W. (2010) *ChemBioChem*, **11**, 1413–1421.

108. Duquesne, S., Destoumieux–Garzon, D., Zirah, S., Goulard, C., Peduzzi, J., and Rebuffat, S. (2007) *Chem. Biol.*, **14**, 793–803.

109. Ziemert, N., Ishida, K., Liaimer, A., Hertweck, C., and Dittmann, E. (2008) *Angew. Chem. Int. Ed.*, **47**, 7756–7759.

110. Philmus, B., Christiansen, G., Yoshida, W.Y., and Hemscheidt, T.K. (2008) *ChemBioChem*, **15**, 3066–3073.

111. Walsh, C.T., Acker, M.G., and Bowers, A.A. (2010) *J. Biol. Chem.*, **285**, 27525–27531.

112. Baglay, M.C., Dale, J.W., Merritt, E.A., and Xiong, X. (2005) *Chem. Rev.*, **105**, 685–714.

113. Wieland Brown, L.C., Acker, M.G., Clardy, J., Walsh, C.T., and Fischbach, M.A. (2009) *Proc. Natl. Acad. Sci. U. S. A.*, **106**, 2549–2553.

114. Kelly, W.L., Pan, L., and Li, C. (2009) *J. Am. Chem. Soc.*, **131**, 4327–4334.

115. Liao, R., Duan, L., Lei, C., Pan, H., Ding, Y., Zhang, Q., Chen, D., Shen, B., Yu, Y., and Liu, W. (2009) *Chem. Biol.*, **16**, 141–147.

116. Morris, R.P., Leeds, J.A., Naegeli, H.U., Oberer, L., Memmert, K., Weber, E., LaMarche, M.J., Parker, C.N., Burrer, N., Esterow, S., Hein, A.E., Schmitt, E.K., and Krastel, P. (2009) *J. Am. Chem. Soc.*, **131**, 5946–5955.

117. Yu, Y., Duan, L., Zhang, Q., Liao, R., Ding, Y., Pan, H., Wendt–Pienkowski, E., Tang, G., Shen, B., and Liu, W. (2009) *ACS Chem. Biol.*, **4**, 855–864.

118. Ding, Y., Yu, Y., Pan, H., Guo, H., Li, Y., and Liu, W. (2010) *Mol. BioSyst.*, **6**, 1180–1185.

119. Wang, J., Yu, Y., Tang, K., Liu, W., He, X., Huang, X., and Deng, Z. (2010) *Appl. Environ. Microbiol.*, **76**, 2335–2344.

120. Yu, Y., Guo, H., Zhang, Q., Duan, L., Ding, Y., Liao, R., Lei, C., Shen, B., and Liu, W. (2010) *J. Am. Chem. Soc.*, **132**, 16324–16326.

121. Liao, R. and Liu, W. (2011) *J. Am. Chem. Soc.*, **133**, 2852–2855.

122. Zhang, Q., Li, Y., Chen, D., Yu, Y., Duan, L., Shen, B., and Liu, W. (2011) *Nat. Chem. Biol.*, **7**, 154–160.

123. Challis, G.L. (2005) *ChemBioChem*, **6**, 601–611.

124. Schmelz, S., Kadi, N., McMahon, S.A., Song, L., Oves–Costales, D., Oke, M., Liu, H., Johnson, K.A., Carter, L.G., Botting, C.H., White, M.F., Challis, G.L., and

Naismith, J.H. (2009) *Nat. Chem. Biol.*, **5**, 174–182.

125. Gondry, M., Sauguet, L., Belin, P., Thai, R., Amouroux, R., Tellier, C., Tuphile, K., Jacquet, M., Braud, S., Courcon, M., Masson, C., Dubois, S., Lautru, S., Lecoq, A., Hashimoto, S.I., Genet, R., and Pernodet, J.L. (2009) *Nat. Chem. Biol.*, **5**, 414–420.

126. Lahoud, G. and Hou, T.M. (2010) *Nat. Chem. Biol.*, **6**, 795–796.

127. Vetting, M.W., Hegde, S.S., and Blanchard, J.S. (2010) *Nat. Chem. Biol.*, **6**, 797–799.

128. Funabashi, M., Yang, Z., Monaka, K., Hosobuchi, M., Fujita, Y., Shibata, T., Chi, X., and van Lanen, S.G. (2010) *Nat. Chem. Biol.*, **6**, 581–586.

129. van Lanen, S.G. and Shen, B. (2006) *Curr. Opin. Microbiol.*, **9**, 252–260.

130. Smanski, M.J., Yu, Z., Casper, J., Lin, S., Peterson, R.M., Chen, Y., Wendt-Pienkowski, E., Rajski, S.R., and Shen, B. (2011) *Proc. Natl. Acad. Sci. U. S. A.*, **108**, 13498–13503.

131. Smanski, M.J., Peterson, R.M., Rajski, S.R., and Shen, B. (2009) *Antimicrob. Agents Chemother.*, **53**, 1299–1304.

132. Jayasuriya, H., Herath, K.B., Zhang, C., Zink, D.L., Basilio, A., Genilloud, O., Teresa Diez, M., Vicente, F., Gonzalez, I., Salazar, O., Pelaez, F., Cummings, R., Ha, S., Wang, J., and Singh, S.B. (2007) *Angew. Chem. Int. Ed.*, **119**, 4768–4772.

133. Galm, U., Wendt-Pienkowski, E., Wang, L., Huang, S.X., Unsin, C., Tao, M., Coughlin, J.M., and Shen, B. (2011) *J. Nat. Prod.*, **74**, 526–536.

134. Galm, U., Wendt-Pienkowski, E., Wang, L., George, N.P., Oh, T.J., Yi, F., Tao, M., Coughlin, J.M., and Shen, B. (2009) *Mol. BioSyst.*, **5**, 77–90.

135. Hutchinson, C.R. (1998) *Curr. Opin. Microbiol.*, **1**, 319–329.

136. Walsh, C.T. (2002) *ChemBioChem*, **3**, 125–134.

137. Salas, J. and Mendez, C. (2009) *Curr. Opin. Chem. Biol.*, **13**, 152–160.

138. Sanchez, C., Mendez, C., and Salas, J. (2006) *Nat. Prod. Rep.*, **23**, 1007–1045.

139. Sanchez, C., Butovich, I.A., Brana, A.F., Rohr, J., Mendez, C., and Salas, J.A. (2002) *Chem. Biol.*, **9**, 519–531.

140. Onaka, H., Taniguchi, S., Igarashi, Y., and Furumai, T. (2003) *Biosci. Biotechnol., Biochem.*, **67**, 127–138.

141. Hyun, C.G., Bililign, T., Liao, J., and Thorson, J.S. (2003) *ChemBioChem*, **4**, 114–117.

142. Onaka, H., Taniguchi, S., Igarashi, Y., and Furumai, T. (2002) *J. Antibiot.*, **55**, 1063–1071.

143. Gao, Q., Zhang, C., Blanchard, S., and Thorson, J.S. (2006) *Chem. Biol.*, **13**, 733–743.

144. Sanchez, C., Zhu, L., Brana, A.F., Salas, A.P., Rohr, J., Mendez, C., and Salas, J.A. (2005) *Proc. Natl. Acad. Sci. U. S. A.*, **102**, 461–466.

145. Salas, A.P., Sanchez, C., Brana, A.F., Rohr, J., Mendez, C., and Salas, J.A. (2005) *Mol. Microbiol.*, **58**, 17–27.

146. Gokhale, R.S. (1999) *Science*, **284**, 482–485.

147. Tsuji, S.Y., Cane, D.E., and Khosla, C. (2001) *Biochemistry*, **40**, 2326–2331.

148. Broadhurst, R.W., Nietlispach, D., Wheatcroft, M.P., Leadlay, P.F., and Weissman, K.J. (2003) *Chem. Biol.*, **10**, 723–731.

149. Menzella, H.G., Reid, R., Carney, J.R., Chandran, S.S., Reisinger, S.J., Patel, K.G., Hopwood, D.A., and Santi, D.V. (2005) *Nat. Biotechnol.*, **23**, 1171–1176.

150. Hahn, M. and Stachelhaus, T. (2004) *Proc. Natl. Acad. Sci. U. S. A.*, **101**, 15585–15590.

151. Richter, C.D., Nietlispach, D., Broadhurst, R.W., and Weissman, K.J. (2007) *Nat. Chem. Biol.*, **4**, 75–81.

152. Hahn, M. and Stachelhaus, T. (2006) *Proc. Natl. Acad. Sci. U. S. A.*, **103**, 275–280.

153. Nguyen, K.T., Ritz, D., Gu, J.Q., Alexander, D., Chu, M., Miao, V., Brain, P., and Baltz, R.H. (2006) *Proc. Natl. Acad. Sci. U. S. A.*, **103**, 17462–17467.

154. McIntosh, J.A., Donia, M.S., and Schmidt, E.W. (2009) *Nat. Prod. Rep.*, **26**, 537–559.

155. Velasquez, J.E. and van der Donk, W.A. (2010) *Curr. Opin. Chem. Biol.*, **15**, 11–21.

156. Li, B., Sher, D., Kelly, L., Shi, Y., Huang, K., Knerr, P.J., Joewono, I., Rusch, D., Chisholm, S.W., and van der Donk, W.A. (2010) *Proc. Natl. Acad. Sci. U. S. A.*, **107**, 10430–10435.

157. Widdick, D.A., Dodd, H.M., Barraille, P., White, J., Stein, T.H., Chater, K.F.,

Gasson, M.J., and Bibb, M.J. (2003) *Proc. Natl. Acad. Sci. U. S. A.*, **100**, 4316–4321.

158. Donia, M.S., Hathaway, B.J., Sudek, S., Haygood, M.G., Rosovitz, M.J., Ravel, J., and Schmidt, E.W. (2006) *Nat. Chem. Biol.*, **2**, 729–735.

159. Medema, M.H., Breitling, R., Bovenberg, R., and Takano, E. (2010) *Nat. Rev. Microbiol.*, **9**, 131–137.

160. Wang, H.H., Isaacs, F.J., Carr, P.A., Sun, Z.Z., Xu, G., Forest, C.R., and Church, G.M. (2009) *Nature*, **460**, 894–898.

161. Fischbach, M., Walsh, C.T., and Clardy, J. (2008) *Proc. Natl. Acad. Sci. U. S. A.*, **105**, 4601–4608.

162. Jenke-Kodama, H., Sandmann, A., Muller, R., and Dittmann, E. (2005) *Mol. Biol. Evol.*, **22**, 2027–2039.

163. Nguyen, T., Ishida, K., Jenke-Kodama, H., Dittmann, E., Gurgui, C., Hochmuth, T., Taudien, S., Platzer, M., Hertweck, C., and Piel, J. (2008) *Nat. Biotechnol.*, **26**, 225–233.

164. Irschik, H., Kopp, M., Weissman, K.J., Buntin, K., Piel, J., and Muller, R. (2010) *ChemBioChem*, **11**, 1840–1849.

165. Bayer, T.S., Widmaier, D.M., Temme, K., Mirsky, E.A., Santi, D.V., and Voigt, C.A. (2009) *J. Am. Chem. Soc.*, **131**, 6508–6515.

166. Handelsman, J. (2004) *Microbiol. Mol. Biol. Rev.*, **68**, 669–685.

167. Banik, J.J. and Brady, S.F. (2010) *Curr. Opin. Microbiol.*, **13**, 603–609.

168. Omura, S., Ikeda, H., Ishikawa, J., Hanamoto, A., Takahashi, C., Shinose, M., Takahashi, Y., Horikawa, H., Nakazawa, H., Osonoe, T., Kikuchi, H., Shiba, T., Sakaki, Y., and Hattori, M. (2001) *Proc. Natl. Acad. Sci. U. S. A.*, **98**, 12215–12220.

169. Nett, M., Ikeda, H., and Moore, B.S. (2009) *Nat. Prod. Rep.*, **26**, 1362–1384.

170. Bentley, S.D. *et al.* (42 authors) (2002) *Nature*, **417**, 141–147.

171. Ikeda, H., Ishikawa, J., Hanamoto, A., Shinose, M., Kikuchi, H., Shiba, T., Sakaki, Y., Hattori, M., and Omura, S. (2003) *Nat. Biotechnol.*, **21**, 526–531.

172. Ohnishi, Y., Ishikawa, J., Hara, H., Suzuki, H., Ikenoya, M., Ikeda, H., Yamashita, A., Hattori, M., and Horinouchi, S. (2008) *J. Bacteriol.*, **190**, 4050–4060.

173. Santi, D.V., Siani, M.A., Julien, B., Kupfer, D., and Roe, B. (2000) *Gene*, **247**, 97–102.

174. Banik, J.J. and Brady, S.F. (2008) *Proc. Natl. Acad. Sci. U. S. A.*, **105**, 17273–17277.

175. Gontang, E.A., Gaudencio, S.P., Fenical, W., and Jensen, P.R. (2010) *Appl. Environ. Microbiol.*, **76**, 2487–2499.

176. Challis, G.L. (2008) *Microbiology*, **154**, 1555–1569.

177. Zerikly, M. and Challis, G.L. (2009) *ChemBioChem*, **10**, 625–633.

178. Dairi, T. (2005) *J. Antibiot.*, **58**, 227–243.

179. Kannenberg, E.L. and Poralla, K. (1999) *Naturwissenschaften*, **86**, 168–176.

180. Johnson, E.A. and Schroeder, W.A. (1996) *Adv. Biochem. Eng. Biotechnol.*, **53**, 119–178.

181. Gerber, N.N. and Lechevalier, H.A. (1965) *Appl. Microbiol.*, **13**, 935–938.

182. Poralla, K., Muth, G., and Hartner, T. (2000) *FEMS Microbiol. Lett.*, **189**, 93–95.

183. Takano, H., Obitsu, S., Beppu, T., and Ueda, K. (2005) *J. Bacteriol.*, **187**, 1825–1832.

184. Gust, B., Challis, G.L., Fowler, K., Kieser, T., and Chater, K.F. (2003) *Proc. Natl. Acad. Sci. U. S. A.*, **100**, 1541–1546.

185. Jiang, J., He, X., and Cane, D.E. (2006) *J. Am. Chem. Soc.*, **128**, 8128–8129.

186. Cane, D.E., He, X., Kobayashi, S., Omura, S., and Ikeda, H. (2006) *J. Antibiot.*, **59**, 471–479.

187. Jiang, J., He, X., and Cane, D.E. (2007) *Nat. Chem. Biol.*, **3**, 711–715.

188. Wang, C.M. and Cane, D.E. (2008) *J. Am. Chem. Soc.*, **130**, 8908–8909.

189. Lin, X. and Cane, D.E. (2009) *J. Am. Chem. Soc.*, **131**, 6332–6333.

190. Zhao, B., Lin, X., Lei, L., Lamb, D.C., Kelly, S.L., Waterman, M.R., and Cane, D.E. (2008) *J. Biol. Chem.*, **283**, 8183–8189.

191. Takamatsu, S., Lin, X., Nara, A., Komatsu, M., Cane, D.E., and Ikeda, H. (2011) *Microb. Biotechnol.*, **2**, 184–191.

192. Tetzlaff, C.N., You, Z., Cane, D.E., Takamatsu, S., Omura, S., and Ikeda, H. (2006) *Biochemistry*, **45**, 6179–6186.

193. Jiang, J., Tetzlaff, C.N., Takamatsu, S., Iwatsuki, M., Komatsu, M., Ikeda, H., and Cane, D.E. (2009) *Biochemistry*, **48**, 6431–6440.

194. Chou, W.K.W., Fanizza, I., Uchiyama, T., Komatsu, M., Ikeda, H., and Cane, D.E. (2010) *J. Am. Chem. Soc.*, **132**, 8850–8851.

195. Blasiak, L.C. and Clardy, J. (2010) *J. Am. Chem. Soc.*, **132**, 926–927.

196. Balskus, E.P. and Walsh, C.T. (2010) *Science*, **329**, 1653–1656.

197. Wyatt, M.A., Wang, W., Roux, C.M., Beasley, F.C., Heinrichs, D.E., Dunman, P.M., and Magarvey, N.A. (2010) *Science*, **329**, 294–296.

198. Lautru, S., Deeth, R.J., Bailey, L.M., and Challis, G.L. (2005) *Nat. Chem. Biol.*, **1**, 265–269.

199. Gross, H. (2007) *Appl. Microbiol. Biotechnol.*, **75**, 267–277.

200. Gross, H., Stockwell, V.O., Henkels, M.D., Nowak-Thompson, B., Loper, J.E., and Gerwick, W.H. (2007) *Chem. Biol.*, **14**, 53–63.

201. Scherlach, K. and Hertweck, C. (2009) *Org. Biolmol. Chem.*, **7**, 1753–1760.

202. Galm, U. and Shen, B. (2006) *Expert Opin. Drug Discov.*, **1**, 409–437.

203. Chen, Y., Smanski, M.J., and Shen, B. (2010) *Appl. Microbiol. Biotechnol.*, **86**, 19–25.

204. Rigali, S., Titgemeyer, F., Barends, S., Mulder, S., Thomae, A.W., Hopwood, D.A., and van Wezel, G.P. (2008) *EMBO Rep.*, **9**, 670–675.

205. Takano, E. (2006) *Curr. Opin. Microbiol.*, **9**, 287–294.

206. Roszak, D.B. and Cowell, R.R. (1987) *Appl. Environ. Microbiol.*, **53**, 2889–2893.

207. Cueto, M., Jensen, P.R., Kauffman, C., Fenical, W., Lobkovsky, E., and Clardy, J. (2001) *J. Nat. Prod.*, **64**, 1444–1446.

208. Pettit, R.K. (2009) *Appl. Microbiol. Biotechnol.*, **83**, 19–25.

209. Angell, S., Bench, B.J., Williams, H., and Watanabe, C.M.H. (2006) *Chem. Biol.*, **13**, 1349–1359.

210. Schroeckh, V., Scherlach, K., Nutzmann, H.W., Shelest, E., Schmidt–Heck, W., Schuemann, J., Martin, K., Herweck, C., and Brakhage, A.A. (2009) *Proc. Natl. Acad. Sci. U. S. A.*, **106**, 14558–14563.

211. Onaka, H., Mori, Y., Igarashi, Y., and Furumai, T. (2010) *Appl. Environ. Microbiol.*, **77**, 400–406.

212. Komatsu, M., Uchiyama, T., Omura, S., Cane, D.E., and Ikeda, H. (2010) *Proc. Natl. Acad. Sci. U. S. A.*, **107**, 2646–2651.

213. Hosaka, T., Ohnishi-Kameyama, M., Muramatsu, H., Murakami, K., Tsurumi, Y., Kodani, S., Yoshida, M., Fujie, A., and Ochi, K. (2009) *Nat. Biotechnol.*, **27**, 462–464.

214. Flinspach, K., Westrich, L., Kaysser, L., Siebenberg, S., Gomez–Escribano, J.P., Bibb, M., Gust, B., and Heide, L. (2010) *Biopolymers*, **93**, 823–832.

215. Pfeifer, B.A., Admiraal, S.J., Gramajo, H., Cane, D.E., and Khosla, C. (2001) *Science*, **291**, 1790–1792.

216. Martin, V.J., Pitera, D.J., Withers, S.T., Newman, J.D., and Keasling, J.D. (2003) *Nat. Biotechnol.*, **21**, 796–802.

217. Watanabe, K., Hotta, K., Praseuth, A.P., Koketsu, K., Migita, A., Boddy, C.N., Wang, C.C., Oguri, H., and Oikawa, H. (2006) *Nat. Chem. Biol.*, **2**, 423–428.

218. Minami, H., Kim, J.S., Ikezawa, N., Takemura, T., Katayama, T., Kumagai, H., and Sato, F. (2008) *Proc. Natl. Acad. Sci. U. S. A.*, **105**, 7393–7398.

219. Dueber, J.E., Wu, G.C., Malmirchegini, G.R., Moon, T.S., Petzold, C.J., Ullal, A.V., Prather, K.L.J., and Keasling, J.D. (2009) *Nat. Biotechnol.*, **27**, 753–759.

220. Clancy, K. and Voigt, C.A. (2010) *Curr. Opin. Biotechnol.*, **21**, 572–581.

221. Gerth, K., Bedorf, N., Hofle, G., Irschik, H., and Reichenbach, H. (1996) *J. Antibiot.*, **49**, 560–563.

222. Bollag, D.M., McQueney, P.A., Zhu, J., Hensens, O., Koupal, L., Liesch, J., Goetz, M., Lazarides, E., and Woods, C.M. (1995) *Cancer Res.*, **55**, 2325–2333.

223. Conlin, A., Fornier, M., Hudis, C., Kar, S., and Kirkpatrick, P. (2007) *Nat. Rev. Drug Discov.*, **6**, 953–954.

224. Julien, B., Shah, S., Ziermann, R., Goldman, R., Katz, L., and Khosla, C. (2000) *Gene*, **249**, 153–160.

225. Molnar, I., Schupp, T., Ono, M., Zirkle, R., Milnamow, M., Nowak-Thompson, B., Engel, N., Toupet, C., Stratmann, A., Cyr, D.D., Gorlach, J., Mayom, J.M., Hu, A., Goff, S., Schmid, J., and Ligon, J.M. (2000) *Chem. Biol.*, **7**, 97–109.

226. Tang, L., Shah, S., Chung, L., Carney, J., Katz, L., Khosla, C., and Julien, B. (2000) *Science*, **287**, 640–642.

227. Julien, B. and Shah, S. (2002) *Antimiocrob. Agents Chemother.*, **46**, 2772–2778.

228. Lau, J., Frykman, S., Regentin, R., Ou, S., Tsuruta, H., and Licari, P. (2002) *Biotechnol. Bioeng.*, **78**, 280–288.

229. Boddy, C.N., Hotta, K., Tse, M.L., Watts, R.E., and Khosla, C. (2004) *J. Am. Chem. Soc.*, **126**, 7436–7437.

230. Mutka, S.C., Carney, J.R., Liu, Y., and Kennedy, J. (2006) *Biochemistry*, **45**, 1321–1330.

231. Tang, L., Chung, L., Carney, J., and Katz, L. (2005) *J. Antibiot.*, **58**, 178–184.

Commentary Part

Comment 1

Yi Tang

This chapter provides an excellent overview of the recent research advances in bacterial NP biosynthesis, which focuses on the new, expanded paradigms in NP biosynthesis, emerging approaches in NP research, and a better appreciation of the NP diversity from a genomic perspective. Section 4.2 highlights the recent discoveries of many unusual NP biosynthetic systems, which have challenged our existing paradigms, especially on the linear biosynthetic logic established for the modular assembly line-like PKS and NRPS systems. Together with the wealth of unknown NP biosynthetic gene clusters uncovered by genome sequencing, it suggests that we may understand only a small fraction of the biosynthetic machineries employed by Nature and underlines the importance of continuous efforts in discovery of and research on NP biosynthetic pathways. The following sections discuss new approaches for understanding and generating NP structural diversity via bioinformatics and pathway engineering, and strategies to unveil the full chemical potential of microorganisms as found encoded in their genome.

In addition to bacteria, fungi and plants are the other two major sources of NPs, and play a similarly important role in drug discovery and development. Owing to their simpler prokaryotic nature, the bacterial NPs have been leading NP research, trailed by studies on the biosynthesis of fungal and plant NPs. The knowledge uncovered in bacterial NP research has its parallels in fungi and plants, and has served as an important foundation for the investigation of NP biosynthetic systems of non-bacterial origins. In recent years, we have seen a similar boom in non-bacterial NP research, especially in fungal NPs. A significant understanding of the fungal thiol template-based PKS–NRPS systems was achieved in the past decade, and at the same time recent research indicates that there is much more to be discovered. The falling cost of genome sequencing made available by the next-generation sequencing technologies will mean that the larger genome size of fungi is no longer a limitation. With the increasing number of fungal genomes being sequenced, we have also observed a surge of new pathways and biosynthetic mechanisms being discovered. With the continuous advances in -omics technologies, a significant growth in plant NP research can also be predicted in the near future. Research into the biosynthesis of NPs will continuously inspire innovative ways to harness the enormous chemical potential provided by Nature for pharmaceutical and other human applications.

Comment 2

Yi Yu and Zi-Xin Deng

The discovery of penicillin by Alexander Fleming in 1928, as a pharmaceutical legend, inspired massive mining of NPs as useful pharmaceutical drugs, many of which continue to be extensively used clinically. The interest in drug discovery in this respect had been flourishing far more than half of a century, and perhaps few subjects had been developing so rapidly and increasingly as a multidisciplinary area. The constant facing of the diminishing returns for the discovery of novel NPs as new drug candidates has been conflicting with the increasing need for new drug supplies as a result of the appearance of new diseases and increasing pathogenic resistances. Many scientific measures had been taken to combat the bottleneck for new or derivative structures of the NPs to appear, and here we have seen tremendous and progressive technological advances over recent years. Natural screening programs are now extended to search from the extremophiles in the ecosystems not searched before, and combined with uses of specific probing based on extensive bioinformatics alignments, and/or genomic searches. Biosynthetic mechanisms for increasing classes of variable NPs, especially those originating from actinomycetes, were extensively demonstrated as studies on NP pathways became routine practice in widespread laboratories. Such studies provided detailed information on how different classes of NPs were synthesized *in vivo*, and opened up routes for generating derivative or new structures by pathway engineering and/or combinatorial biosynthesis, leading further to the rational design of novel NPs by conceptual synthetic biology by making increasing numbers of catalytic enzymes or enzyme complexes as standardized parts or modules as compatible building blocks. Central to all of these was a deep, mechanistic understanding of NPs biosynthesis. Focusing on this, this chapter by Smanski *et al.* has done a great job by providing us with updated and comprehensive knowledge on the mechanisms and achievements of NPs biosynthesis over past decade or so, which exhibited both classical and novel paradigms for pharmaceutical NP biosynthetic logic and the application potential for those natural beauty.

The chapter begins with a brief overview of the great conceptual and technological leaps in NP biosynthesis studies during the last decade, and defines the objective of the chapter to argue that NP biosynthesis research has entered another golden period. Section 4.2, dealing with the expanded paradigms in NP biosynthetic logic, covers three major, most studied, and most popular types of NPs, including PKS, NRPS, and ribosomal peptide NPs. Readers will be truly impressed by the latest research achievements illustrating the novel chemistry and enzymology underlying the irregular or unusual NP biosynthetic logic. Instead of presenting extensive examples and schemes to approaches involving the use of combinatorial strategies and synthetic metagenomics, Section 4.3 comes with a detailed description of some of the modern, powerful, and readily accessible approaches to NP biosynthesis research, including comparative analysis of gene cluster encoding biosynthetic enzymes and biochemical characterization of these proteins, which were showcased by deciphering the mechanisms of platensimycin and platencin biosynthesis, allowing readers to take advantage of the technological advances in trail exploitation of their extreme potential in variable systems. Methodological advances are highlighted in Section 4.4 with emphasis on the use of the exponential growth of DNA

sequencing technologies and thriving microbial genome information, foreseeing arrival of a new era of NP biosynthesis treasure hunting.

The conclusion of the chapter is certainly apparent! NP biosynthesis has been one of the richest sources for the discovery of new chemistry and enzymology, and will continue to serve as the driving force for the development of novel drugs and/or drug leads. Furthermore, with the rapid growth of genome sequence databases coupled with analytical techniques and genetic tools such as combinatorial biosynthesis and metabolic engineering, our understanding of the diversity and application potential of NPs will be further broadened. This is undoubtedly a highly fascinating research area featuring in-depth cross-talk between chemistry and biology, which will continue to inspire the research enthusiasm of microbiologists, biochemists, and chemists to enter the age of synthetic biology for designed biosynthesis of NPs.

Authors' Response to the Commentaries

Response to Yi Tang

We have limited this chapter to NPs of bacterial origin. Plants and fungi are two other major sources of NPs, and many of them have been developed into clinically important drugs. Both plant and fungal NPs are equally rich in structural diversity, chemistry, and biology. Because of the remarkable advances in DNA sequencing, DNA synthesis, and bioinformatics, the large genome sizes and perceived genetic intractability of plants and fungi apparently no longer constitute insurmountable barriers in applying contemporary strategies and methods to NP biosynthesis in these species. Indeed, if the spectacular progress in the last decade is any indication, we are about to witness exponential growth in plant and fungal NP biosynthesis for the decade to come.

Response to Yi Yu and Zixin Deng

NP biosynthesis, traditionally a study of the chemical events for transforming readily available precursors into a myriad of fascinating molecular scaffolds most known to NPs, can now be rapidly correlated with genes encoding their production. In the contemporary sense, it is truly an interdisciplinary endeavor. We are now in anticipation of moving from the efforts at understanding how Nature makes these molecules to the efforts at improving Nature's biosynthetic machineries. These advances continuously incorporate NP-associated chemistry into the rational application of biotechnology, and will greatly benefit the pharmaceutical industry in new NP discovery, in the structural diversity of NPs, and in the design of new biosynthetic pathways by using the principles of synthetic biology.

5
Carbohydrate Synthesis Towards Glycobiology

Biao Yu and Lai-Xi Wang

5.1
Introduction

Recent advances in glycobiology have implicated essential roles of oligosaccharides and glycoconjugates in many important biological recognition processes, such as intracellular signaling, cell adhesion, cell differentiation, cancer progression, host–pathogen interactions, and immune responses [1–7]. Following nucleic acids and proteins, glycans are viewed as the third language of life, yet we are only just beginning to unravel the mysteries of this language. In contrast to nucleic acids and proteins that are biosynthetically assembled on templates and under direct transcriptional control, oligosaccharides are generally more complex in their biosynthesis. Oligosaccharides are assembled in the endoplasmic reticulum and/or Golgi apparatus through multistep sugar chain elongation and trimming via the action of a number of glycosyltransferases and glycosidases. As a result, the oligosaccharide components in glycoconjugates such as glycoproteins and glycolipids are usually heterogeneous in structures. In addition, the complexity of oligosaccharides also comes from the structural branching and varied stereochemistry of the glycosidic linkages. The functions of complex carbohydrates in biological processes also seem to be more diverse, ranging from simple structural components and energy storage to molecular recognition and signaling.

The structural complexity and functional diversity of complex carbohydrates have posed enormous challenges to chemists who are interested in structural characterization, synthesis, and correlation of fine structures with biological functions of oligosaccharides and glycoconjugates. Synthetic carbohydrates are highly demanded for basic studies in glycobiology and for drug discovery, as homogeneous oligosaccharides and glycoconjugates are very difficult to isolate from natural sources because of their structural heterogeneity and, in many cases, their natural scarcity. The past decade has witnessed tremendous progress in the development of efficient chemical and chemoenzymatic methods for synthesizing structurally well-defined complex carbohydrates, which were used for addressing critical biological problems that were otherwise difficult to tackle by genetic approaches [8–25]. For example, synthetic oligosaccharides are essential for identifying glycan-binding proteins and for diagnosis in glycan microarrays [26–30]; novel multivalent carbohydrate ligands have been designed and used for characterizing cell-surface

Organic Chemistry – Breakthroughs and Perspectives, First Edition. Edited by Kuiling Ding and Li-Xin Dai.
© 2012 Wiley-VCH Verlag GmbH & Co. KGaA. Published 2012 by Wiley-VCH Verlag GmbH & Co. KGaA.

carbohydrate–protein interactions involving in signaling and B-cell activation [31, 32]; chemical synthesis of multivalent, starfish-shaped oligosaccharide ligands based on the structure of the B-subunits of Shiga-like toxins has led to the discovery of a potent inhibitor that could neutralize Shiga-like toxins at subnanomolar concentrations [33]; synthetic unnatural substrates have frequently been used as probes via biosynthetic incorporation for cell surface imaging and engineering [34]; and sophisticatedly designed glyconanoparticles and glycodendrimers have been used for specific targeting of lectins for drug and vaccine delivery [35, 36]. In particular, a very active area of research in carbohydrate chemistry and glycobiology is the design and synthesis of carbohydrate-based vaccines against cancer, bacteria, and viruses [8, 11, 37–44]. For example, a carbohydrate-based synthetic vaccine was approved for human use for preventing infection by *Haemophilus influenza* type b that causes pneumonia and meningitis in infants and young children [45]; synthetic vaccines based on the tumor-associated carbohydrate antigens (TACAs) such as the glycolipid-related Globo-H, Lewis^Y, GM2, GD2, and GD3 antigens, and also the mucin glycoprotein-related Tn, STn, and Tf antigens, were designed, synthesized, and tested in animals and/or in clinical trials that showed promise for cancer treatment [8, 38, 39, 46, 47]; and a novel three-component vaccine construct that includes a TACA B cell epitope, a T helper epitope, and a Toll-like receptor (TLR) ligand was designed and proved to be effective in raising relatively high titers of IgG-type antibodies against TACA [48]. This novel vaccine construct holds promise for overcoming the tolerance to the weakly immunogenic carbohydrate antigens.

A number of excellent reviews have been published on chemical approaches to glycobiology [9, 19, 22, 49–52]. Indeed, synthetic oligosaccharides and glycoconjugates have proven indispensable for studies in glycobiology. On the other hand, the structural and functional complexity of oligosaccharides, glycoproteins, glycolipids, proteoglycans, and other complex carbohydrates has provided an excellent platform for synthetic chemists to pursue their natural curiosity. This chapter highlights some major advances in the last decade in the synthesis of complex carbohydrates. We provide selected examples on emerging technologies and new strategies that showcase our ability to assemble complex, biologically interesting oligosaccharides and glycoconjugates. It should be pointed out that it is impossible for a single chapter to cover all the advances in synthetic carbohydrate chemistry and glycobiology in recent years and we regret having to leave out many important examples. Additional examples on synthesis in glycobiology and some detailed discussions on classical glycosylation method developments can be found in a number of earlier reviews [9, 19, 22, 49–52].

5.2
Advances in Chemical Glycosylation

Numerous new glycosylation methods have been developed since the very early invention of glycosylation reactions by Michael in 1879, Fisher in 1893, and Koenigs and Knorr in 1901. Because of the branching structures and anomeric stereochemistry of oligosaccharides, chemical glycosylation is centered on the control of regio- and stereoselectivity in glycosidic bond formation. In general, chemical glycosylation relies on selective protection/deprotection strategies and stereoelectronic effects including neighboring group participations to achieve the regioselectivity and to control the anomeric stereochemistry in glycosylation, respectively. Common glycosyl donors, including glycosyl halides, glycosyl trichloroacetimidates, and thioglycosides, are still widely used today for oligosaccharide synthesis. It should be pointed out that many factors work together to determine the outcome of a glycosylation reaction and, thus, the synthesis of each new, complex oligosaccharide or glycoconjugate may pose a new synthetic challenge. Gratifyingly, the last decade has witnessed continuous advances in this field. These include the invention of new glycosyl donors with novel leaving groups for enhancing glycosylation efficiency, armed–disarmed strategies for controlling glycosylation sequence, and novel neighboring group auxiliary for controlling the anomeric stereochemistry. The general aspects of chemical glycosylation and major glycosylation methods have been reviewed and discussed in several excellent reviews [9, 10, 22, 53, 54]. In this chapter, we highlight only a few recent methodology developments.

5.2.1
New Glycosyl Donors with Novel Leaving Groups

Classical glycosylation methods, such as those with glycosyl trichloroacetimidates and thioglycosides as donors, have been widely used. Nevertheless, new glycosyl donors are still highly valuable to meet the challenge of complex oligosaccharide synthesis in terms of efficiency and selectivity. In particular, three new glycosylation donors, the glycosyl N-phenyltrifluoroacetimidate (PTFAI) donors, the 2′-carboxybenzyl (CB) glycoside donors, and the o-alkynylbenzoate donors have shown promising features in glycosylation.

5.2.1.1 Glycosylation with PTFAI Donors
Glycosyl trichloroacetimidates, owning to their easy preparation, high coupling efficiency, and mild activation conditions, continue to be the most popular donors for glycosidation. Evolved from trichloroacetimidates, the PTFAI donors, which can be easily prepared, are found to be superior to the glycosyl trichloroacetimidates in a few respects in glycosylation [55] (Scheme 5.1): (i) the PTFAI donors are generally more stable than the corresponding trichloroacetimidates and are therefore easier to handle and more efficient in slower glycosylation reactions, such as for solid-phase oligosaccharide synthesis where glycosylation proceeds more slowly than in solution [56]; (ii) ketosyl trichloroacetimidates are inaccessible owing to the reversible nature of the preparation procedure,

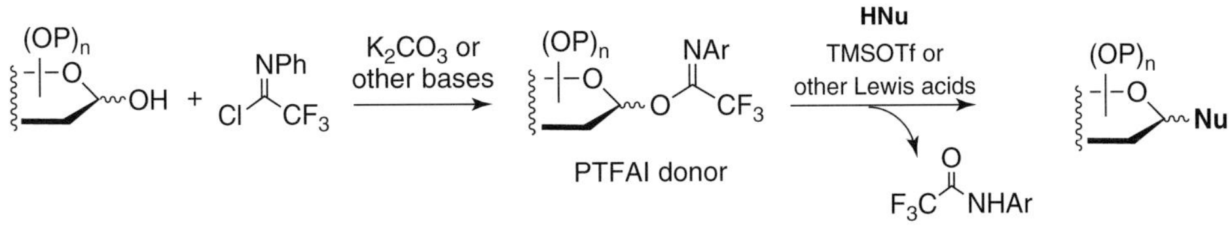

Scheme 5.1 Synthesis of glycosyl PTFAIs as glycosyl donors.

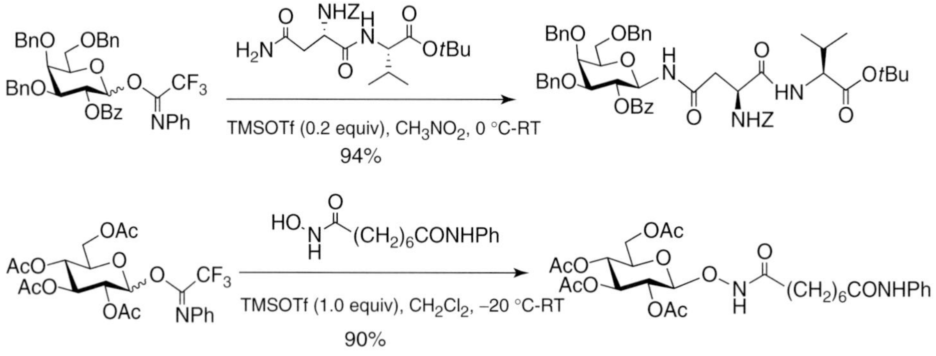

Scheme 5.2 Direct glycosylation of amides and hydroxyamino acids with glycosyl PTFAIs.

whereas ketosyl PTFAIs can be readily prepared and used in glycosylation, such as in sialylation [57]; and (iii) when encountered with poorly nucleophilic acceptors, glycosyl trichloroacetimidates could lead to trichloroacetamide side products via back-attack of the leaving trichloroacetamide, whereas the *N*-phenyltrifluoroacetamide leaving from PTFAIs is much less competitive. Thus, glycosylation gives better yield of the desired coupling product. For example, the less active amide can be *N*-glycosylated and the hydroxyamino acid derivative can be *O*-glycosylated in excellent yields by the PTFAI donors (Scheme 5.2) [58, 59]. As a test case, a highly efficient synthesis of a glycopeptide carrying a tetra-*N*-acetyllactosamine-containing core 2-decasaccharide was accomplished by iterative use of the PTFAI donors [60].

5.2.1.2 Glycosylation with 2'-Carboxybenzyl Glycosides (CB Donors)

CB glycosides are readily derived from 2-(benzyloxycarbonyl)benzyl (BCB) glycosides via selective hydrogenolysis of the benzyl ester. This transformation does not affect the benzyl and benzylidene protecting groups. Upon activation with Tf$_2$O, CB donors can be converted to active species such as the glycosyl triflate which could undergo glycosidation efficiently [61] (Scheme 5.3). Highly stereoselective synthesis of β-mannopyranosides and 2-deoxyglycosides employing this method has been demonstrated [61, 62]. In addition, BCB glycosides are fairly stable. Therefore, protecting group manipulations could be performed at this stage, allowing subsequent conversion into the active CB glycosides in an iterative manner in oligosaccharide synthesis. This method was recently applied to the total synthesis of agelagalastatin, an antineoplastic glycosphingolipid [63]. Notably, the CB glycoside method permitted completely α-stereoselectivity in the formation of α-galactofuranoside.

5.2.1.3 Glycosylation with Glycosyl *o*-Alkynylbenzoates

Glycosyl *o*-alkynylbenzoates are easily prepared via condensation of sugar hemiacetals with *o*-hexynylbenzoic acid under a variety of ester-forming conditions. These donors are stable but can be selectively activated with a catalytic amount of an Au(I) complex, such as Ph$_3$PAuOTf and Ph$_3$PAuNTf$_2$, for glycosidation under mild conditions (Scheme 5.4) [64]. The Au(I) catalysts possess little oxophilic character, and therefore display good

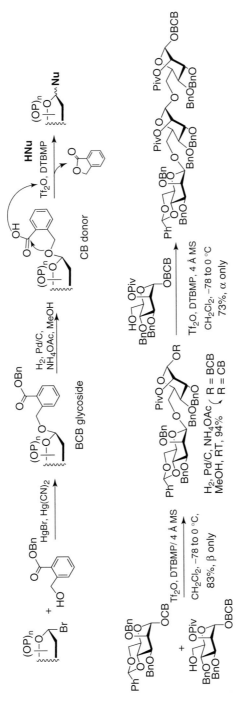

Scheme 5.3 Use of 2'-carboxybenzyl glycosides for glycosylation.

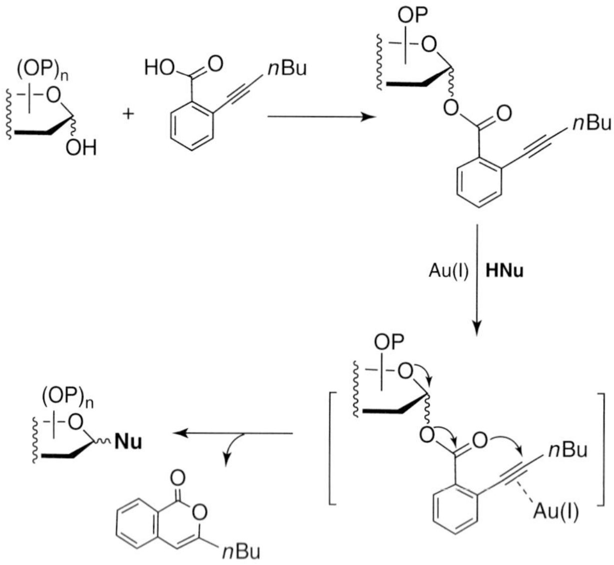

Scheme 5.4 Use of glycosyl *o*-alkynylbenzoates for glycosylation.

functional group compatibility in carbohydrate synthesis [64]. This glycosylation reaction is catalytic, neutral, and orthogonal to other glycosylation conditions. The glycosylation yield is generally high, as demonstrated for the glycosylation of flavonol and triterpene derivatives [64, 65]. This glycosylation protocol was used as key steps for the total synthesis of TMG-chitotriomycin, a tetrasaccharide derivative that is a potent and selective inhibitor of insect and fungal β-*N*-acetylglucosaminidases [66].

5.2.2
New Methods for Controlling the Stereochemistry in Glycosylation

In addition to the issue of coupling efficiency in a glycosylation reaction, stereoselectivity in the formation of α- or β-glycosidic bonds presents another challenging task for synthetic chemists [9]. In this regard, the glycosidic linkages could be categorized into three types: (i) the 1,2-*trans* glycosidic linkages (e.g., β-D-glucoside and α-D-mannoside); (ii) the 1,2-*cis* glycosidic linkages (e.g., α-D-glucoside and β-D-mannoside); and (iii) the glycosidic linkages devoid of a neighboring substituent (e.g., 2-deoxyglucoside and sialoside). The first (1,2-*trans*) type of glycosidic linkages can be synthesized confidently with the assistance of a neighboring participating group at the C2 position of the glycosyl donors (Scheme 5.5a). However, the synthesis of the other two types of glycosidic linkages requires special recipes that should consider all possible factors affecting the glycosylation outcome, including the protective patterns of the coupling partners, the leaving group of the donor, the promoter, the solvent, the reaction temperature, and the sequence of mixing the substrates and reagents. In the absence of a participating neighboring group, formation of an α-isomer of *O*-glycoside (e.g., of *O*-glucoside and *O*-galactoside)

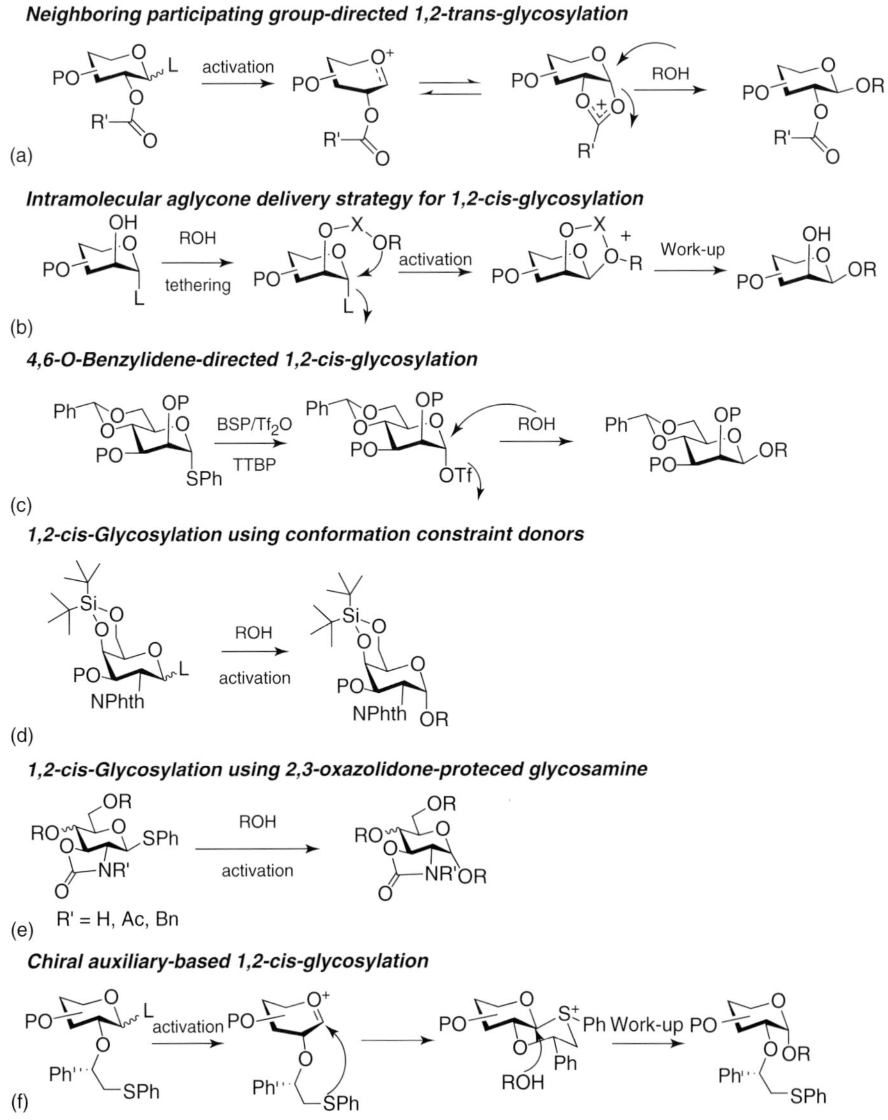

Neighboring participating group-directed 1,2-trans-glycosylation

(a)

Intramolecular aglycone delivery strategy for 1,2-cis-glycosylation

(b)

4,6-O-Benzylidene-directed 1,2-cis-glycosylation

(c)

1,2-cis-Glycosylation using conformation constraint donors

(d)

1,2-cis-Glycosylation using 2,3-oxazolidone-proteced glycosamine

R' = H, Ac, Bn

(e)

Chiral auxiliary-based 1,2-cis-glycosylation

(f)

Scheme 5.5 Methods for controlling the stereoselectivity in glycosylation.

is usually favored because of the anomeric effect. However, complete control of the
α-isomer formation is still difficult to achieve relying solely on the anomeric effect. An
innovative approach to making a 1,2-*cis*-glycosidic bond is via intramolecular aglycone
delivery (IAD) [67, 68]. In this approach, a sugar alcohol is temporarily attached to the
neighboring position of the glycosyl donor via an acetal or silicon tether. Subsequent
activation of the glycosyl donor with this construct forces the aglycone to be delivered
from the same face of the neighboring group, leading to the formation of 1,2-*cis*-glycosidic

linkage (Scheme 5.5b). In addition to the synthesis of the difficult β-mannoside, IAD has been also applied to the synthesis of α-D-glucopyranoside, β-L-rhamnopyranoside, β-D-arabinofuranoside, β-D-fructofuranoside, and α-D-fucopyranoside [68].

Another innovative approach to 1,2-*cis*-glycosylation was reported by Crich and Sun, who applied the cyclic 4,6-*O*-benzylidene acetal to direct β-mannosylation (Scheme 5.5c) [69, 70]. Subsequent in-depth studies of this phenomenon have resulted in a better understanding of the glycosylation mechanism [71–73]. The glycosylation goes through a glycosyl triflate intermediate, and the constraint cyclic 4,6-*O*-benzylidene provides an essential stereoelectronic effect that destabilizes the formation of the anomeric oxacarbenium, thus allowing direct S_N2-type substitution by the acceptor to form β-mannoside. Based on the mechanistic observation that the electron-withdrawing property of the rigid cyclic 4,6-*O*-benzylidene plays a major role in directing the β-glycosidic bond formation, van der Marel and co-workers examined the glycosylation of the corresponding 6-carboxyl derivative of the glycosyl donor and found that the presence of the remotely attached carboxylic ester in the 1-thiomannuronic acid esters could sufficiently influence the electronic environment to permit excellent β-selectivity [74].

The strategies employing steric and conformational constraints in the glycosyl donors have been explored for directing 1,2-*cis*-glycosidic bond formation in other cases. As a notable example, glycosylation of 4,6-di-*O-tert*-butylsilyl-protected galactosyl donors led to the formation of the corresponding galactoside with very high α-anomeric selectivity, even when a neighboring participating group such as the *N*-phthaloyl group was present [75] (Scheme 5.5d). In addition to a fixed conformation, the bulky *tert*-butylsilyl groups might block the β-face of the sugar, thus allowing only the attack by the acceptor from the α-face. Similar strategies were used for the stereoselective synthesis of β-arabinofuranosides [76, 77].

2-Amino-2-deoxyglycosides are major components of glycosaminoglycans such as heparin and heparan sulfate. They are also commonly present in glycoproteins, glycolipids, and bacterial polysaccharides. Chemical synthesis of β-linked 2-amino-2-deoxyglycosides is relatively straightforward by using a suitable *N*-protecting group that can participate in directing (1,2-*trans*) β-glycosidic bond formation during glycosylation. However, the synthesis of 1,2-*cis*-glycosides of 2-amino-2-deoxy sugars is a difficult task. To address this problem, several groups investigated the use of 2,3-oxazolidone-protected glucosamine derivatives as glycosyl donors and found that the 2,3-oxazolidone-protected glycosyl donors gave excellent α-anomeric selectivity when a sufficient amount of the acid promoter was used [78–80] (Scheme 5.5e). Mechanistic studies suggest that the predominant formation of α-isomer involves acid-promoted anomerization of the initially formed β-isomer. The cyclic 2,3-oxazolidone structure promotes the endocyclic C–O bond cleavage and favors the formation of the α-isomer [81]. In another related study, Mensah and Nguyen carried out nickel-catalyzed stereoselective glycosylation of C2-*N*-substituted benzylidene D-glucosamine and galactosamine trichloroacetimidates, leading to highly selective formation of the corresponding 1,2-*cis*-2-amino-2-deoxyglycosides [82, 83]. This method was successfully applied to the synthesis of heparin disaccharides, glycophosphatidylinositol (GPI) anchor pseudodisaccharides, and α-GalNAc-linked amino acids.

Boons and co-workers introduced another innovative approach towards 1,2-*cis*-glycosides, based on neighboring group participation by an (*S*)-(phenylthiomethyl)benzyl group at the C2 position of a glycosyl donor [84, 85] (Scheme 5.5f). Activation of the glycosyl imidate led to the formation of a quasi-stable *trans*-decalinsulfonium ion. The β-sulfonium ion species was preferentially formed because of its much more favorable steric and electronic arrangement than the *cis*-decalin structure. Displacement of the sulfonium ion by the acceptor resulted in the formation of 1,2-*cis*-glycosides. The (S)-(phenylthiomethyl)benzyl group can be readily introduced at C2. After glycosylation, it can be removed by conversion into acetate by treatment with $BF_3 \cdot Et_2O$ in acetic anhydride. Recently, this chiral auxiliary-based glycosylation method was successfully applied to the solid-phase synthesis of a biologically important branched α-glucan [86].

Stereoselective synthesis of α-sialosides poses a special challenge because of the lack of neighboring participating groups and the presence of an electron-withdrawing carboxyl group at the anomeric position [53]. Nevertheless, important progress has been made in the synthesis of this type of glycoside. For example, acylation of the 5-acetamido moiety in a thioglycoside donor of a sialic acid could significantly enhance the glycosylation yield and the α-anomeric selectivity [87]. In addition, modifications on the *N*-acetyl group in the thioglycoside donor, such as its change to *N*-trifluoroacetyl, *N*-trichloroacetyl, and *N*-phthaloyl moieties, or its conversion to an *N*-acyloxazolidinone derivative, all led to dramatic improvements in the glycosylation efficiency and stereoselectivity [88–90]. It is amazing to see how a small change at a site remote to the anomeric center can have a significant impact on the efficiency and stereochemistry of glycosylation.

5.3
New Strategies in Oligosaccharide Assembly

5.3.1
Automated Oligosaccharide Synthesis

The last decade has witnessed major advances in the strategies in oligosaccharide assembly. The success in the automated solid-phase synthesis of polypeptides and nucleic acids has motivated organic chemists to take the challenge of developing practical protocols for automated oligosaccharide synthesis. However, because of the complex branching structures and the stereochemistry involved in the glycosidic linkages, oligosaccharides are much more difficult to make than the linear polypeptides and oligonucleotides. The lack of a general, high-yielding (>99%) glycosylation method, and the difficulties in 100% control of the anomeric stereochemistry in glycosylation, have been major roadblocks in automatic synthesis. Solid-phase-based oligosaccharide synthesis has been attempted since the 1970s. However, the first practical automated synthesis of oligosaccharides on a solid support was reported by Seeberger and co-workers in 2001 [91]. In their approach, glycosyl phosphates were used as efficient glycosyl donors, an octenediol-functionalized resin was used as the solid support that could be selectively removed at the end of the synthesis by catalytic cleavage, and an automated oligosaccharide synthesizer was invented based a solid-phase peptide synthesizer. The automatic oligosaccharide synthesis followed the same principle as automatic solid-phase peptide synthesis (SPSS). Thus, the first sugar acceptor was covalently linked to a polystyrene resin through an octenediol linker. The functionalized resin was then mixed with reagents and the glycosyl donor for the glycosylation. Excess reagents, including the glycosyl donors, were used to ensure the completion of the glycosylation reaction. After the reaction, excess reagents were removed by simple filtration and washing. Selective deprotection would be performed to expose the new nucleophilic hydroxyl group (the acceptor), which would enter the next glycosylation cycle (Scheme 5.6). After the repeats of the glycosylation cycles, the final assembled oligosaccharide could be cleaved from the resin by alkene cross-metathesis using a Grubbs catalyst under an ethylene atmosphere. This mild retrieval procedure would provide the synthetic oligosaccharide in the form of pentenyl glycoside, which could be further functionalized by radical reaction, used for further glycosylation, or simply hydrolyzed to provide the free reducing oligosaccharide. The phytoalexin elicitor (PE) hexasaccharide was obtained in 89% yield as judged by high-performance liquid chromatographic (HPLC) analysis. In comparison with conventional oligosaccharide synthesis, a major advantage of the solid-phase synthesis is that only one simple purification step is required after the final product has been retrieved from the solid support. Very impressively, a more complex, branched dodecasaccharide of the PE β-glucan was synthesized in 17 h and in 50% total yield using the same automated synthetic cycle. This remarkable initial success proved that an automated synthesis of complex oligosaccharides was conceptually feasible. Nevertheless, a number of problems still remain when various complex oligosaccharides with diverse structures are involved. It should be pointed out that many initial drawbacks faced in the original automated oligosaccharide synthesis have been addressed in the past decade [92]. For example, a new synthesizer was designed that allows glycosylations at very low temperatures [93]; some difficult glycosidic linkages such as α-galactoside and β-mannoside could be installed in the automatic protocol [56, 94].

Recently, Nishimura and co-workers developed a method for automated glycan synthesis that applies enzymatic glycosylations and a globular protein-like dendrimer as an ideal support [95]. The dendritic polymer support possesses high solubility and low viscosity in water, and allows excellent recovery during the repetitive reaction–separation steps. In contrast, the use of random polyacrylamide-based supports resulted in a significant loss of the polymer during the filtration. The automatic synthesizer was designed based on an HPLC-style system that permits automatic reagent mixing and filtration. A fully automated enzymatic synthesis of sialyl Lewis X tetrasaccharide was achieved in 4 days using the respective glycosyltransferases. In a related study, Linhardt and co-workers designed a so-called "artificial Golgi" for enzymatic oligosaccharide modifications, which has potential for automation [96]. They immobilized a sugar primer on magnetic nanoparticles and elegantly used a digital microfluidic platform to control the enzymatic reactions. In addition, the fluorous tag-assisted synthesis of oligosaccharides that allows solution-phase glycosylation and protecting group manipulations with solid-phase work-up to remove the reagents and by-products quickly holds promise for automated protocol development [97].

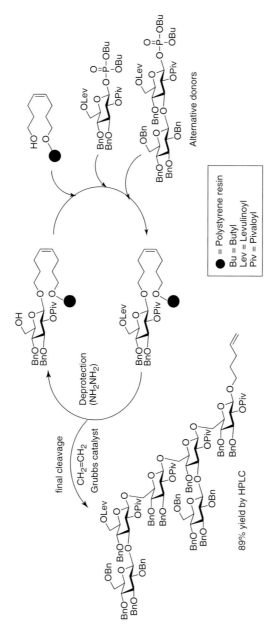

Scheme 5.6 Automated solid-phase synthesis of a phytoalexin elicitor oligosaccharide.

5.3.2
One-Pot Sequential Glycosylations

To avoid the tedious procedures in intermediate purifications in oligosaccharide synthesis, several strategies have been developed that permit multiple sequential glycosylations in one pot, without the need for isolation of the reaction intermediates. Wong and co-workers devised the so-called programmed oligosaccharide synthesis strategy that relies on the chemoselectivity and fine-tuning of the reactivity of latent glycosyl donors [10, 98]. For example, the anomeric reactivity of a thioglycoside can be fine-tuned with a variety of protecting groups and protecting patterns. Thus different thioglycosides can be distinctly activated by using different activators and reaction conditions. A computer database was established that allows a choice of appropriate reactants for sequential glycosylations in one pot without cross-reactivity [99]. As a notable example, a total synthesis of the sialic acid-containing antigenic epitope fucosylated GM1 oligosaccharide was achieved by this reactivity-based one-pot strategy. Based on a thioglycoside reactivity database, three thioglycoside building blocks were designed and prepared to incorporate a descending order of reactivity towards thiophilic activation [99]. The three-component coupling in one pot led to the assembly of the oligosaccharide, with only a single final-step purification.

Another one-pot multiple glycosylation strategy uses preactivation of glycosyl donors in the absence of an acceptor. In this approach, stable glycosyl donors such as thioglycosides

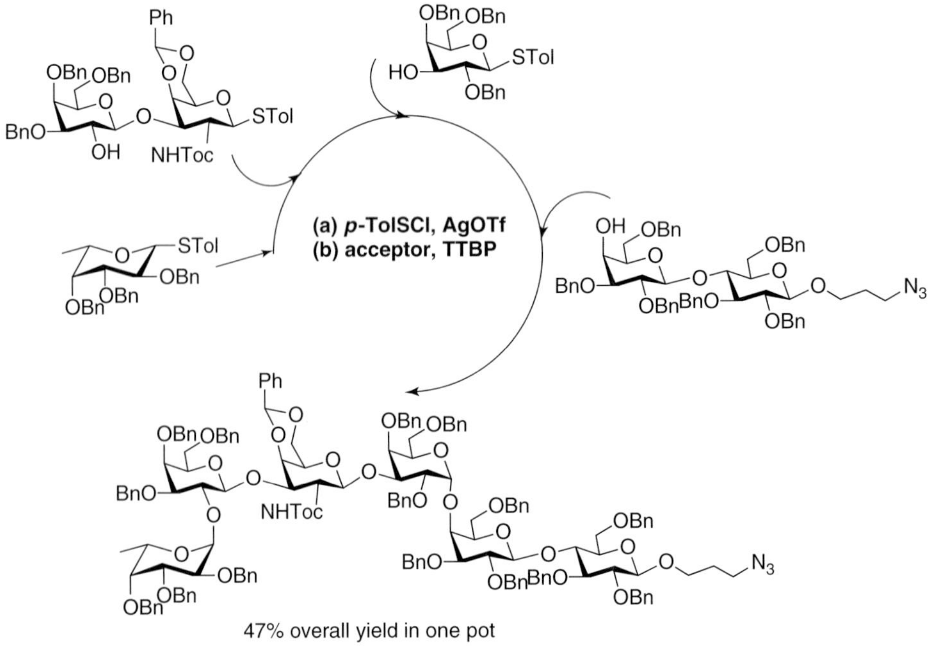

47% overall yield in one pot

Scheme 5.7 One-pot synthesis of the tumor-associated antigen Globo-H by sequential preactivation glycosylation.

with similar activities can be used for sequential glycosylations without work-up and intermediate purification [100–102]. Thus, a glycosyl donor, for example a thioglycoside, can be preactivated with an activator such as *p*-toluenesulfenyl triflate to generate a reactive intermediate (glycosyl triflate). Then an acceptor that could be another thioglycoside with a free hydroxyl group is added and reacts with the activated glycosyl donor to form a disaccharide thioglycoside. The new thioglycoside, without separation, can be preactivated and then react with another thioglycoside acceptor to form a higher oligosaccharide. The steps can be repeated in one flask to lead to the final product. The one-pot synthesis of an α-2,3-sialylated and core-fucosylated complex type biantennary *N*-glycan dodecasaccharide [101], and also the one-pot combinatorial synthesis of heparin hexasaccharides [102], highlighted the great potential of this strategy for the expeditious construction of complex oligosaccharides. A one-pot synthesis of the TACA Globo-H by this preactivation strategy [103] is demonstrated in Scheme 5.7. Sequential coupling of four glycosyl building blocks in one pot led to the synthesis of the Globo-H hexasaccharide skeleton in 47% overall yield. Despite impressive progress in this area, however, the successful implementation of this strategy requires an exact stoichiometric addition of the activator so that no residual activator will be left that may activate the subsequently mixed glycosyl acceptor. In addition, high-yield conversion of the reactants in each glycosylation step is crucial for success, as any reactants left from the previous cycle, and also any byproducts with free hydroxyl groups, would certainly complicate the next step of glycosylation.

5.4
Enzymatic and Chemoenzymatic Methods

Enzymatic and chemoenzymatic syntheses are emerging as an exciting area of research in carbohydrate chemistry. In contrast to conventional pure chemical synthesis that usually requires tedious protection/deprotection manipulations in order to achieve regioselectivity and special recipes to control the anomeric configuration, enzymatic glycosylation usually takes place in aqueous solution and provides perfect control of the anomeric stereochemistry and high regioselectivity, without the need for protecting groups. In particular, incorporation of enzymatic transformations as key steps in a chemical synthesis, the so-called chemoenzymatic approach, can significantly simplify the synthetic scheme and enhance the overall efficiency of the synthesis. Both glycosyltransferases and glycosidases have been vigorously studied as synthetic tools [15, 23–25]. Glycosyltransferases are the natural enzymes for making glycosidic bonds, whereas glycosidases are involved in breaking glycosidic linkages in carbohydrate metabolism.

Using natural glycosyltransferases for oligosaccharide synthesis has hitherto met with restrictions such as limited availability, the requirement for expensive sugar nucleotides as the donor substrates, and the usually stringent substrate specificity. However, substantial advances have been made in addressing the limitations. For example, numerous glycosyltransferases from bacterial and mammalian sources have been cloned and expressed, novel sugar nucleotide regeneration systems have been invented, making it possible to use only catalytic amounts of the expensive donor substrates [104–106], and mechanistic and mutational studies have led to mutant enzymes that demonstrate much broader substrate tolerance and are capable of accepting modified and altered substrates [25, 107]. Assembly of complex N- and O-glycans of glycoproteins has been achieved by *in vitro* sequential enzymatic glycosylations using a series of cloned enzymes [108, 109]. These structurally well-defined glycans are extremely valuable for functional glycomics studies. Thorson and co-workers developed a novel "glycorandomization" method that takes advantage of the promiscuity of bacterial glycosyltransferases to diversify the glycosylation patterns of complex natural products [110–113].

On the other hand, chemoenzymatic synthesis using glycosidase-based transglycosylation has attracted considerable attention in recent years, because of the apparent advantages of glycosidase-based synthesis, namely the use of readily available donor substrates and the relaxed substrate specificity for acceptors [18]. In addition, a large number of glycosidases are readily available. To address the issue of product hydrolysis associated with glycosidase-catalyzed reactions, a conceptual breakthrough was made in the field by the invention of glycosynthases, a class of novel glycosidase mutants that lack product hydrolysis activity, but can promote glycosidic bond formation when a suitable activated glycosyl donor is provided [15, 114]. For example, Withers and co-workers reported the first glycosynthase by engineering the β-glycosidase from *Agrobacterium* sp. via mutating the nucleophile residue E358 to a non-nucleophilic residue alanine [115]. The E358A mutant could take an activated glycosyl donor with an opposite anomeric configuration such as the corresponding α-glycosyl fluoride as donor for transglycosylation to form a new glycosidic linkage. However, the mutant was unable to hydrolyze the product because of the lack of the nucleophilic residue, thus permitting the accumulation of the product. Independently, Malet and Planas reported the first *endo*-acting glycosynthase by site-directed mutation of the nucleophile E134 of the retaining 1,3−1,4-β-glucanase from *Bacillus licheniformis* [116], and Moracci *et al.* described another interesting glycosynthase approach that used an activated glycosyl species as substrate and rescued the transglycosylation activity of the non-hydrolyzing glycosidase mutant with sodium formate that acts as an external nucleophile [117].

Following these pioneering studies, various new glycosynthases have been discovered from glycosidases belonging to more than a dozen glycoside hydrolase families. Moreover, protein engineering, including directed evolution coupled with elegant screening methodology, has led to the discovery of an expanding number of novel glycosynthases with enhanced transglycosylation activity and/or altered substrate specificity [118–122]. The glycosynthase approach has been extended to the synthesis of diverse oligosaccharides and glycoconjugates, including thiooligosaccharides [123–125], cell-surface oligosaccharide antigens [126], glycolipids [127], and glycoproteins [128–130]. Of particular note is the highly efficient chemoenzymatic synthesis of glycosphingolipids using an endoglycoceramidase (EGC) mutant (Scheme 5.8). Through engineering the EGC II from *Rhodococcus* sp, a glycosphingolipid-hydrolyzing enzyme, Withers and co-workers generated the first glycosynthase for glycosphingolipid synthesis by deleting the catalytic nucleophilic residue E351 [127]. It was found that an E351S mutant could efficiently take synthetic 3′-O-sialyl-α-D-lactosyl fluoride as a donor substrate to glycosylate D-*erythro*-sphingosine and related lipids, but it lacked activity to hydrolyze the product, permitting a high-yield (>90%) synthesis of the corresponding glycosphingolipids. To

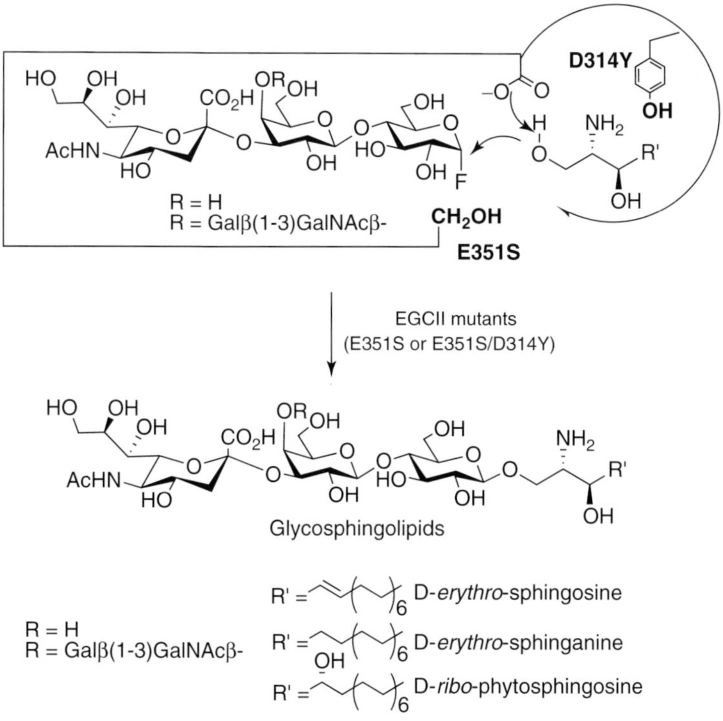

Scheme 5.8 Glycosynthase-catalyzed synthesis of glycosphingolipids.

extend the substrate adaptability, Withers and co-workers performed directed evolution of EGC II [131]. Using an enzyme-linked immunosorbent assay (ELISA)-based assay to screen the EGC mutant libraries, they successfully identified a double mutant, D314Y/E351S, which showed several thousand-fold enhancement of the catalytic activity towards another type of acceptor, phytosphingosine, while maintaining the same ability as the E351S mutant to recognize sphingosine as the lipid acceptor. This study showcases the power of directed evolution in improving on glycosynthase for glycoconjugate synthesis.

5.5
Synthesis of Heparin and Heparan Sulfate Oligosaccharides

Heparin and heparan sulfate (HS) are naturally occurring, polyanionic, linear polysaccharides. Heparin is found mainly in the granules of mast cells, whereas HS is a major component of the extracellular matrix in most mammalian cells. Heparin and HS are composed of glucosamine, D-glucuronic acid, and L-iduronic acid connected with specific glycosidic linkages. However, they are highly heterogeneous polysaccharides because of their highly varied O-sulfation, N-sulfation, and/or N-acetylation patterns. Heparin isolated from animal organs has been widely used as an anticoagulant since the 1940s, making it one of the oldest carbohydrate-based drugs. It exerts its activity by binding to the serine protease inhibitor antithrombin III (ATIII) in the plasma, leading to its activation. The activated ATIII then inactivates thrombin and other proteases such as factor Xa that are involved in the coagulation cascade essential for blood clotting.

5.5.1
Chemical Synthesis of Heparin Oligosaccharides

Chemical synthesis has played a key role in determining the structure–activity relationships in heparin–ATIII interactions. After the identification in the early 1980s of a specific pentasaccharide as the minimum sequence responsible for heparin's binding for ATIII and its anticoagulant property [132], a large and bold effort was launched in the 1980s and 1990s to achieve the chemical synthesis of the heparin pentasaccharide and its derivatives, aiming at understanding the molecular mechanism of its anticoagulant activity [133, 134]. The first total chemical synthesis of the highly complex heparin pentasaccharide structure about three decades ago represents a milestone in oligosaccharide synthesis [135]. An important feature of the total chemical synthesis, with more than 60 steps, is the implementation of an elegant orthogonal protection strategy. The hydroxyl groups to be sulfated were protected by ester groups, while those corresponding to the free hydroxyl groups in the final product were protected with the "permanent" benzyl ether groups. Moreover, after global deprotection, selective N-sulfation of the free amino groups over free hydroxyl groups was achieved to afford the target pentasaccharide (Scheme 5.9). It should be pointed out that installing a methyl group at the reducing end anomeric center led to a pentasaccharide derivative that was much more stable and much simpler to handle than the original reducing oligosaccharide, and yet the biological activity remained the same. The orthogonal protection synthetic strategy laid a solid foundation for subsequent systematic modifications of the structures, leading to a series of structurally well-defined heparin oligosaccharides. The chemical synthesis not only verified the fine structure of the identified pentasaccharide sequence, but also led to a detailed understanding of the structure–anticoagulant activity relationships of the heparin pentasaccharide and the mode of its recognition by ATIII [136]. This tremendous effort eventually led to the development of a fully synthetic heparin oligosaccharide analog, Arixtra (fondaparinux), as an anticoagulant agent for the treatment of deep vein thrombosis [134, 137]. It should be emphasized that optimization of the synthetic procedures allowed the production of the heparin pentasaccharide on a multi-kilogram scale for clinical applications. In addition to Arixtra, several other synthetic heparin oligosaccharide analogs are also being developed for anticoagulation therapeutics. A very promising analog is idraparinux, a simplified version of Arixtra, in which all the free hydroxyl groups are protected with methyl ethers and the N-sulfates are replaced with O-sulfates. This modified oligosaccharide is much easier to synthesize than Arixtra. Interestingly, idraparinux has shown an increased half-life, and has demonstrated improved anticoagulation activity [134].

Over 100 metric tons of heparin isolated from pig intestine are used worldwide for anticoagulation therapy annually, indicating a huge market. However, side effects

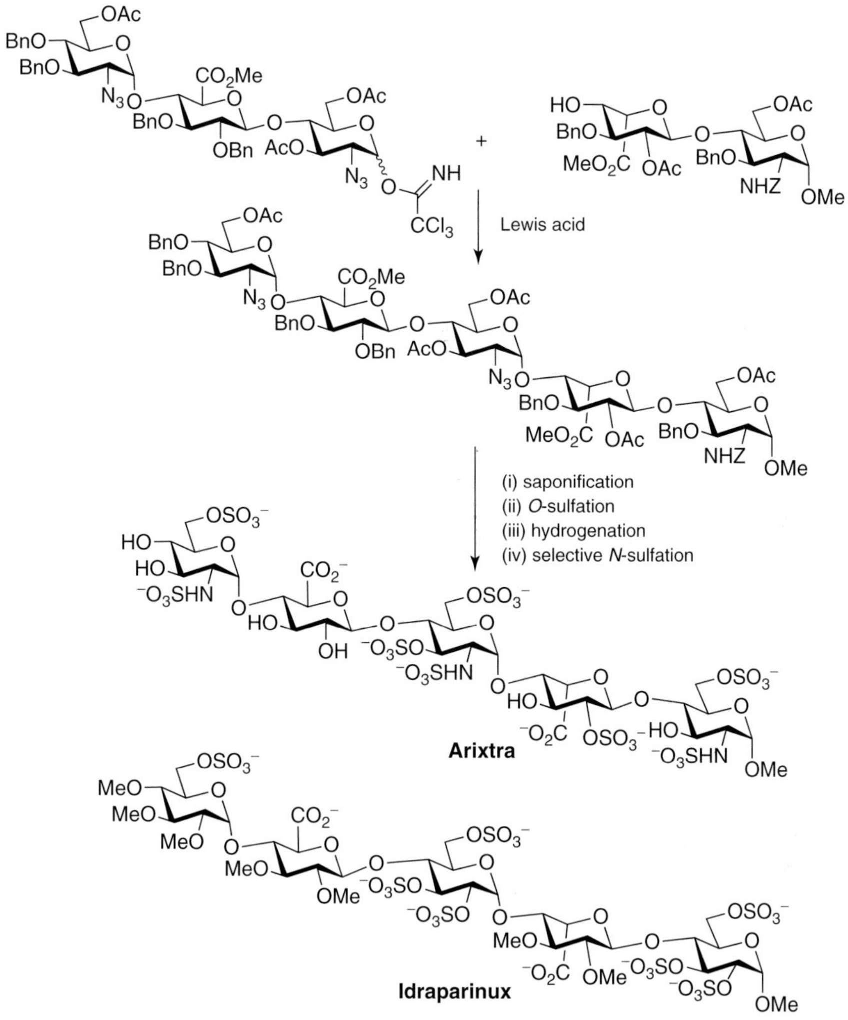

Scheme 5.9 Chemical synthesis of heparin pentasaccharide.

associated with natural heparin, together with the recent clinical crisis associated with the use of heparin contaminated with over-sulfated glucosaminoglycans [138, 139], highlight the importance of developing synthetic versions of heparin and its derivatives for clinical use. Moreover, it has been implicated that alterations to the expression and/or fine structures of heparan sulfate are associated with various diseases such as cancer, neurological disorders, and infections [140]. More than 100 heparan sulfate binding proteins such as growth factors have been identified. An expanded library of structurally well-defined heparin and heparan sulfate oligosaccharides is still highly desirable in order to decipher the structure–activity relationships in binding with known proteins and to search for new heparin/heparan sulfate-binding proteins for new functions [141]. Recently, a modular chemical synthesis of heparan sulfate oligosacchrides was described, which allows a parallel combinatorial synthesis of oligosaccharides using a relatively small number of selectively protected disaccharide building blocks [142]. The synthetic heparan sulfate tetra- and hexasaccharides were used for microarray analysis with a selected proteinase and the results indicated that the fine structures of heparan sulfate oligosaccharides are important for high-affinity interactions. Moreover, heparin and heparan sulfate oligosaccharide microarrays have been fabricated and used for screening various glycan-binding proteins, including fibroblast growth factors (FGFs), natural cytotoxicity receptors (NCRs), and chemokines [143].

5.5.2
Enzymatic Synthesis of Heparin Oligosaccharides

In addition to pure chemical synthesis, enzymatic and chemoenzymatic methods have been explored to make selectively decorated heparin and heparan sulfate. This approach takes advantage of the high selectivity of enzymatic transformations involved in the biosynthesis of the polysaccharides. A beautiful example is an efficient *in vitro* enzymatic conversion of a nonsulfated *N*-acetylheparonan polysaccharide produced by *E. coli* strain K5 to an ATIII-binding heparan sulfate pentasaccharide, using a series of enzymes cloned from the heparin biosynthetic pathway [144]. Thus, the polysaccharide precursor was converted into an *N*-sulfated hexasaccharide in two steps using *N*-deacetylase-*N*-sulfotransferase (NDST2), a bifunctional enzyme, and heparitinase I, a bacterial lyase enzyme. Selective *O*-sulfation of the hexasaccharide with three specific *O*-sulfotransferases and C5-epimerization at the second glucuronic acid residue by C5-epimerase, followed by removal of the nonreducing terminal $\Delta^{4,5}$-glucuronic acid residue with a bacterial $\Delta^{4,5}$-glycuronidase, gave the key pentasaccharide intermediate. Finally, introduction of an *O*-sulfate group at the third glucosamine residue by a 3-*O*-sulfotransferase afforded the target ATIII-binding pentasaccharide (Scheme 5.10). This enzymatic approach was also used to convert the *E. coli*-expressed *N*-acetylheparonan precursor into stable isotope-enriched heparin polysaccharide with anticoagulant activity [145]. This stable isotope (^{13}C and ^{15}N) labeled heparin was extremely useful for NMR solution conformational studies of this class of important biomolecules. More recently, a chemoenzymatic approach was established for making various heparan sulfate oligosaccharides through sugar chain elongations on a disaccharide primer by specific glycosyltransferases with concurrent site-specific enzymatic modifications (*O*- and *S*-sulfations) [146].

It is expected that these chemical and chemoenzymatic methods will be further developed to provide various heparin and heparan sulfate oligosaccharides with predicted structures in the future, which would be extremely valuable for functional glycomics studies in determining the fine structures in heparin/heparan sulfate–protein interactions and in identifying new glycan-binding proteins.

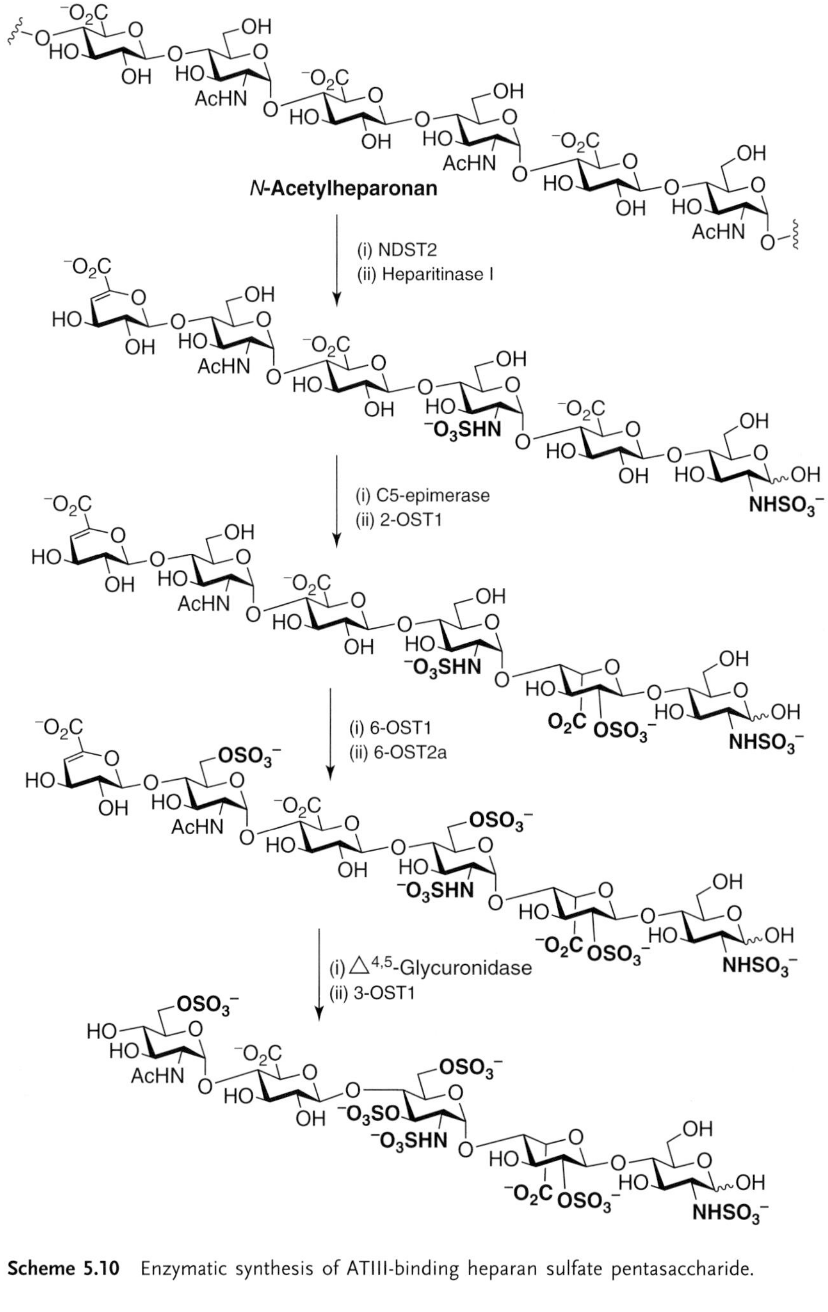

Scheme 5.10 Enzymatic synthesis of ATIII-binding heparan sulfate pentasaccharide.

5.6
Synthesis of Homogeneous Glycoproteins

Glycosylation can profoundly affect a protein's structure and function, such as its folding and half-life in serum. Recent advances in glycobiology have demonstrated that the carbohydrate components of glycoproteins play crucial roles in a number of important biological processes, such as intracellular sorting, cell adhesion, development, host–pathogen interactions, and immune responses [1–4]. Abnormal glycosylation is associated with various disease states such as cancer progression, inflammation, and immune disorders [5]. In the drug discovery arena, glycosylated proteins represent a major class of biotechnology products, which account for more than two-thirds of the total therapeutic proteins [147–149].

Because of their structural microheterogeneity, pure glycoforms of glycoproteins are usually difficult to obtain from natural sources using current chromatographic techniques. Hence the development of methods to produce homogeneous glycoproteins carrying defined glycans has become a major focus of research both in academia and in the biopharmaceutical industry. Engineering of the glycan biosynthetic pathways in a host expression system (e.g., yeast and CHO cell lines) has been an area of active research, which has led to the production of some selected human glycoproteins carrying defined sugar chains [150–152]. Nevertheless, the ultimate goal of synthetic chemists is to make homogeneous glycoproteins via selective chemical and enzymatic reactions *in a flask*. From the chemistry point of view, glycoproteins represent one of the most challenging synthetic targets that organic chemists could face: extremely large size, complex branching structures of the oligosaccharide component, and multifunctionality from both the oligosaccharide and polypeptide portions. Many synthetic methods have been explored, as summarized in a number of excellent reviews [10, 19, 24, 153–158]. There are three major synthetic strategies that hold promise for constructing full-size homogeneous glycoproteins valuable for structural and functional studies (Scheme 5.11). These strategies elegantly combine various techniques, such as convergent coupling, native chemical ligation (NCL), expressed protein ligation (EPL), chemoselective ligation, and enzymatic ligation, to achieve the goal.

5.6.1
Convergent Glycopeptide Synthesis Coupled with Native Chemical Ligation

In general, oligosaccharide synthesis requires a number of selective transformations including glycosylation reactions under strictly anhydrous conditions that are incompatible with unprotected polypeptides and proteins; on the other hand, the standard deprotection and retrieval of polypeptide from resin in an SPPS requires strong acid treatment (trifluoroacetic or hydrofluoric acid), which may destroy the acid-labile glycosidic bonds when oligosaccharides are attached. The conventional solid-phase glycopeptide synthesis using glycosylamino acids as building blocks have been very successful for preparing glycopeptides carrying relatively small oligosaccharides and/or relatively short polypeptides [159, 160], but a linear assembly of amino acid or glycoamino acid building blocks on a solid support is less attractive for making a natural full-size glycoprotein. Towards

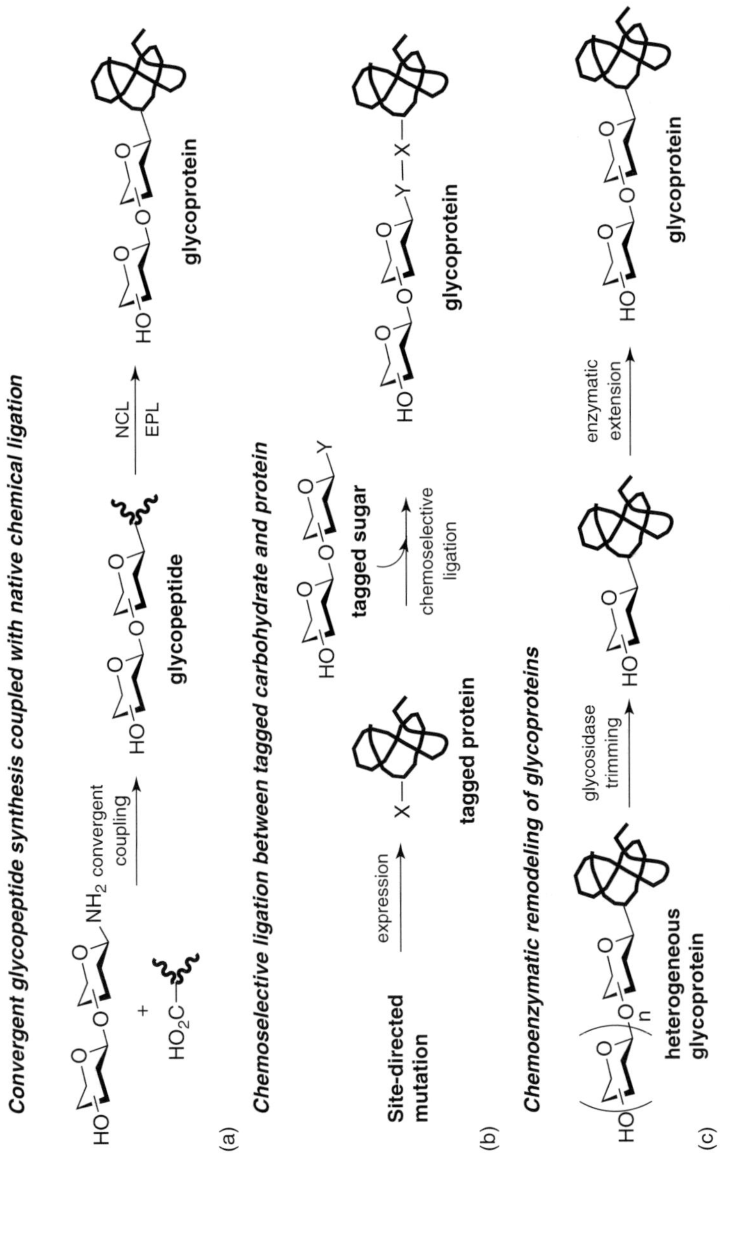

Convergent glycopeptide synthesis coupled with native chemical ligation

(a)

Chemoselective ligation between tagged carbohydrate and protein

(b)

Chemoenzymatic remodeling of glycoproteins

(c)

Scheme 5.11 Strategies for glycoprotein synthesis.

N-glycoprotein synthesis, an interesting strategy is to perform a convergent coupling between a synthetic oligosaccharide (in the form of glycosylamine) and a protected polypeptide containing a free or selectively activated aspartyl side chain to make relatively large N-glycopeptides [161–167]. The glycopeptide thus synthesized could be designed for extension to larger glycopeptides or glycoproteins via NCL [168, 169]. The original version of NCL relies on the highly efficient chemoselective reaction between a peptide thioester and an N-terminal cysteine in aqueous solutions without the need for any protecting groups. A number of complex glycopeptides and glycoproteins containing N- or O-linked glycans have been made by this synthetic strategy [158, 170–177]. To avoid the intrinsic limitation of NCL that requires a cysteine residue at the ligation junction, several auxiliary-based cysteine-free NCL methods were developed and applied to glycoprotein synthesis. For example, a 2-mercaptobenzyl-based auxiliary was used by Danishefsky and co-workers for NCL of glycopeptides and peptides, which can be removed by mild reaction conditions after the ligation [178–180]. An elegant sugar-assisted ligation with a thiol auxiliary being installed in the sugar residue was developed by Wong and co-workers for glycopeptide synthesis [181–183]. Furthermore, EPL [184] has been elegantly applied to glycoprotein synthesis, which takes advantage of the powerful protein expression technology [185–188]. In this approach, either a protein thioester is prepared by expression of a corresponding protein–intein fusion protein followed by thiolysis, or an N-terminal cysteine-containing protein fragment is obtained by site-specific cleavage of an expressed protein to generate a free cysteine at the N-terminus. As a notable example, Macmillan and Bertozzi used the EPL method to prepare GlyCAM-1 glycoforms with multiple N-acetylgalactosamine residues at the predetermined glycosylation sites [188]. Thus, a cysteine-containing protein fragment was obtained by factor Xa protease digestion of an expressed protein containing the IEGR recognition site. The glycopeptide thioester was prepared using the safety-catch linker method. Finally, the ligation between the two fragments led to the synthesis of the complex glycoproteins. It is expected that different native ligation methods will continue to play a major role in constructing large glycopeptides and glycoproteins useful for elucidating the biological functions of glycoproteins.

5.6.2
Site-Selective Glycosylation via a Protein "Tag and Modify" Strategy

In addition to sequential NCL of glycopeptides and peptide fragments, another efficient approach is to perform selective protein glycosylation by a combination of site-directed mutagenesis and subsequent chemoselective reaction to attach oligosaccharides. In this approach, a tag such as a cysteine residue is usually introduced at the predetermined glycosylation site by site-directed mutagenesis. The free cysteine residue in the expressed protein is then selectively modified with a glyco derivative that contains a thiol-reactive functional group, such as glycosyl iodoacetamide, glycosyl methanethiosulfonate (GlycoMTS), and the second-generation glycosyl phenylthiosulfonate (GlycoPTS) to provide a synthetic glycoprotein. The oligosaccharide is usually linked to the cysteine residue through a disulfide or a thioether linkage. For example, Flitsch and co-workers prepared glycosylated erythropoietin (EPO) by introducing cysteines at the conserved Asn residues of the N-glycosylation sites and then attaching synthetic oligosaccharides at the sites by reacting them with glycosyliodoacetamide [189]; Boons and co-workers selectively introduced sugar chain at the Asn297 of IgG-Fc through the N297C mutation and subsequent disulfide formation with glycothiols [190]. The synthetic glycoforms of IgG-Fc were used to study the effects of the glycosylation on antibodies' effector functions; and Davis and co-workers used this "tag and modify" strategy to make novel antibacterial glycodendriproteins that contain branched sugar chains at predetermined sites of conjugation in the protein [191].

To expand the scope of the "tag and modify" approach, additional site-specific tagging methods were introduced. Schultz and co-workers employed amber codon suppression technology to introduce unnatural amino acids containing ketone "handles" into a protein [192]. Sugars could be specifically attached to the ketone handle through chemoselective reaction with a glycoacylhydrazide. Additional unnatural amino acids such as azidohomoalanine (Aha) and homopropargylglycine (Hpg) can be incorporated into proteins by employing a Met(−)auxotropic strain, E. coli B834(DE3), to express the target protein in the presence of the corresponding unnatural amino acids instead of methionine [193–195]. Thus, in addition to cysteine, the introduction of ketone, azide, and alkyne groups can permit orthogonal modifications of a given protein to introduce multiple functional groups at predetermined sites. Recently, Davis and co-workers elegantly applied the "tag and modify" strategy to construct a synthetic glycoprotein that functionally mimics the P-selectin glycoprotein ligand-1 (PSGL-1) [196]. Two post-translational modifications, including a sulfate group at Tyr48 and a sialylated glycan such as sialyl Lewis X attached at the Ser57, are essential for the binding of PSGL-1 to P-selectin in the primary rolling/adhesion phases of the inflammatory response. Based on the structural parameters of PSGL-1's recognition with P-selectin, Davis and co-workers used the LacZ-type reporter enzyme SSβG as a bacterial scaffold protein to introduce a sulfotyrosine mimic group at position 439 and a sialylated glycan (siaLacNAc or sialyl Lewis X) at position 43 of the enzyme SSβG, so that the two groups could spatially and functionally mimic the two post-translational modifications found in PSGL-1. This was achieved by expression of a 10-point (Met)10(Cys)1 to (Met43)1(Ile)9(Ser)1 SSβG mutant, which also contains an additional mutation at position 439 to introduce a Cys residue, in the Met-auxotrophic E. coli strain B834(DE3) in the presence of a Met analog. The expression provided a tagged TIM-barrel protein SSβG-Aha43-Cys439, which contains a thiol tag at site 439 and an azide tag at site 43. The tagged protein then was selectively reacted with a novel Cys-modifying reagent, Tys-MTS, to introduce the sulfotyrosine mimic at position 439, followed by a second orthogonal click chemistry with sialyl Lewis X-alkyne to attach a sialylated glycan at position 43 (Scheme 5.12). The highly efficient orthogonal click reactions allowed selective incorporation and systematic analysis of altered glycans in the context of the sulfotyrosine mimic. Binding studies demonstrated functional mimicking of the modified SSβG to PSGL-1 and also revealed a clear synergistic effect between the sulfotyrosine mimic and the sialylated glycan in the scaffold for P-selectin recognition. The resulting differentially modified SSβG protein, which also maintains a LacZ-type enzyme activity, was successfully used to detect *in vivo* inflammatory brain lesions by its specific recognition of P-selectin and subsequent enzymatic reactions for X-Gal tissue staining [196].

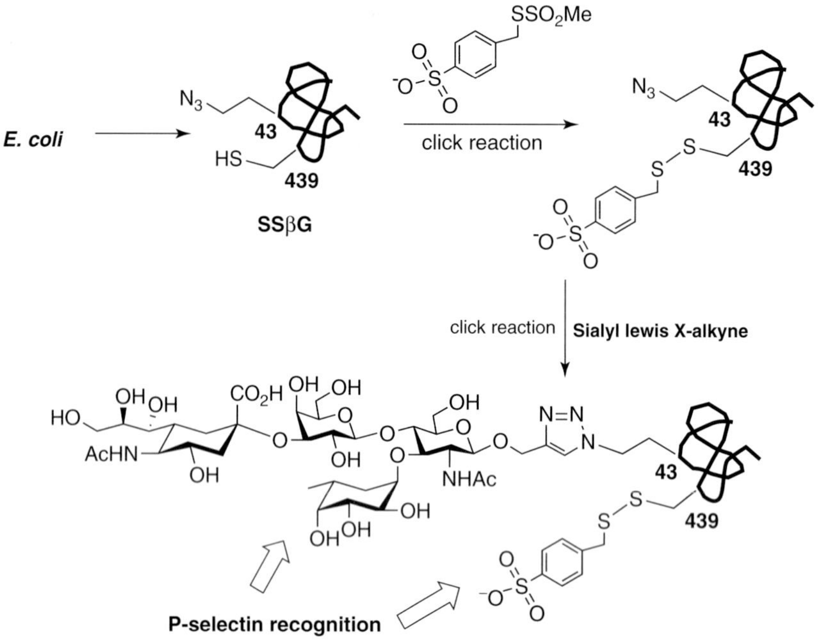

Scheme 5.12 "Tag and modify" strategy for functionalizing proteins.

5.6.3
Chemoenzymatic Glycosylation Remodeling of Glycoproteins

A drawback for direct chemoselective glycosylation of proteins is the introduction of unnatural linkages, which might not precisely mimic the functions of natural glycoproteins and could be immunogenic for human use. To address this problem, an alternative approach is to perform *in vitro* enzymatic glycan remodeling of natural or recombinant glycoproteins to make homogeneous glycoproteins having all natural glycosidic linkages. This chemoenzymatic method has been particularly useful for preparing N-linked glycoproteins. This approach consists of two key steps. First, the heterogeneous *N*-glycans of natural and recombinant glycoproteins are trimmed down to the innermost *N*-acetylglucosamine (GlcNAc) using endoglycosidases, thus converting a heterogeneous population to a homogeneous population in which each glycosylation site has only a single GlcNAc attached. Then the sugar chains are extended enzymatically to increase the size and complexity of the glycan by sequential glycosylation with glycosyltranferases or by transglycosylation with endoglycosidases (Scheme 5.13). For example, Wong and co-workers successfully remodeled the heterogeneous high-mannose-type *N*-glycans in bovine ribonuclease B (RNase B) into a homogeneous sialyl Lewis X moiety [197]. In this approach, the high-mannose *N*-glycan in RNase B was removed by Endo-H to give a homogeneous GlcNAc–RNase B. Following this step, β-1,4-galactosyltransferase, α-2,3-sialyltransferase, and α-1,3-fucosyltransferase were used sequentially to introduce a sialyl Lewis X oligosaccharide to make a novel homogeneous glycoprotein.

Extension of sugar chain by sequential glycosyltransferase-catalyzed reactions

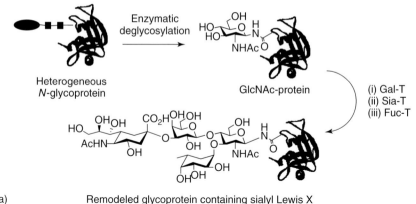

Heterogeneous
N-glycoprotein

GlcNAc-protein

(i) Gal-T
(ii) Sia-T
(iii) Fuc-T

(a)   Remodeled glycoprotein containing sialyl Lewis X

Extension of sugar chain by endoglycosidase-catalyzed transglycosylation

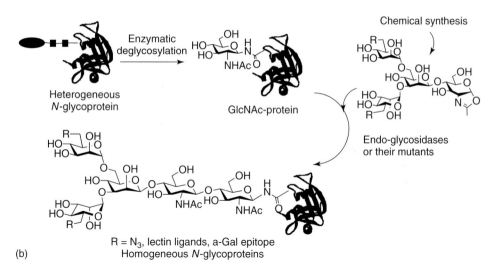

Chemical synthesis

Heterogeneous
N-glycoprotein

GlcNAc-protein

Endo-glycosidases
or their mutants

R = N₃, lectin ligands, a-Gal epitope
Homogeneous N-glycoproteins

(b)

Scheme 5.13 Chemoenzymatic approach to glycosylation remodeling of glycoproteins.

An alternative approach for sugar chain elongation is endoglycosidase-catalyzed trans-glycosylation, which attaches a large oligosaccharide *en bloc* to the GlcNAc primer in a single step, thus avoiding the heterogeneity that could arise from the sequential glycosyla-tions by glycosyltransferases [198, 199]. Two *endo-β-N*-acetylglucosaminidases (ENGases), the Endo-A from *Arthrobacter protophormiae* and the Endo-M from *Mucor hiemalis*, have been particularly useful for transglycosylation [155]. To address the low efficiency of the original endoglycosidase-catalyzed transglycosylation, Wang and co-workers explored synthetic sugar oxazolines, the presumed enzymatic reaction intermediate generated from a substrate-assisted mechanism, as donor substrates for transglycosylation [200, 201]. The use of sugar oxazolines as substrates not only extended the substrate availability, but also significantly enhanced the efficiency for enzymatic synthesis, as the transglycosylation with the highly active sugar oxazolines was more favored than the product hydrolysis by the endoglycosidases. Using this approach, natural ribonuclease B was efficiently remodeled to various homogeneous glycoforms carrying core N-glycans, azido-tagged N-glycans, and other large oligosaccharide ligands [201, 202]. As an application of the method, glycosylation engineering of human IgG1-Fc was achieved [203]. First, human IgG1-Fc was expressed in yeast *Pichia pastoris*. After enzymatic deglycosylation of the heterogeneous yeast N-glycans, defined synthetic N-glycans were then attached to the GlcNAc primer in the IgG-Fc domain by Endo-A-catalyzed transglycosylation to afford homogeneous glycoforms of IgG1-Fc. A remarkable observation in this study was that the Endo-A could efficiently glycosylate the native GlcNAc-containing IgG1-Fc homodimer without the need to denature the protein.

Another significant advance in the endoglycosidase-catalyzed transglycosylation approach is the successful generation of novel glycosynthase mutants that can promote transglycosylation with the highly activated sugar oxazolines, but lack product hydrolysis activity [128–130, 204]. A special mutant of Endo-M, N175A, was found to be able to transfer both high-mannose-type and complex-type N-glycans, while the corresponding Endo-A mutant, N171A, was particularly useful for transferring high-mannose-type N-glycans [130]. With these glycosynthases in hand, homogeneous glycoproteins carrying natural N-glycans such as sialylated N-glycans can now be efficiently prepared [205]. Recently, a novel combined method for producing homogeneous glycoproteins with eukaryotic N-glycosylation was reported by Wang and co-workers [206]. The method involves the engineering and functional transfer of the *Campylobacter jejuni* glycosylation machinery in *E. coli* to express glycosylated proteins in which a key GlcNAc–Asn linkage was engineered. The bacterial glycans were then trimmed and remodeled *in vitro* by the endoglycosidase-catalyzed transglycosylation to achieve a eukaryotic N-glycosylation. This method combines the power of protein expression in *E. coli*, biotechnology's workhorse, and the flexibility of the *in vitro* endoglycosidase-catalyzed glycosylation remodeling system, thus providing a potentially general platform for producing eukaryotic N-glycoproteins.

5.7
Synthesis of Carbohydrate-Containing Complex Natural Compounds

Carbohydrate-containing compounds are ubiquitous in Nature as secondary metabolites, which possess various biological functions. These compounds, displaying extremely diverse structures in both the carbohydrate and the aglycone parts, continuously challenge the synthetic limits of chemists. Achievements in the synthesis of the most complex ones reflect the frontier of synthetic carbohydrate chemistry. Sophisticated protecting group arrangement and efficient glycosylation methods are required to accommodate the diverse functionality presented in the aglycone. The total synthesis of a number of sugar-containing complex natural products has been achieved, some of which are among

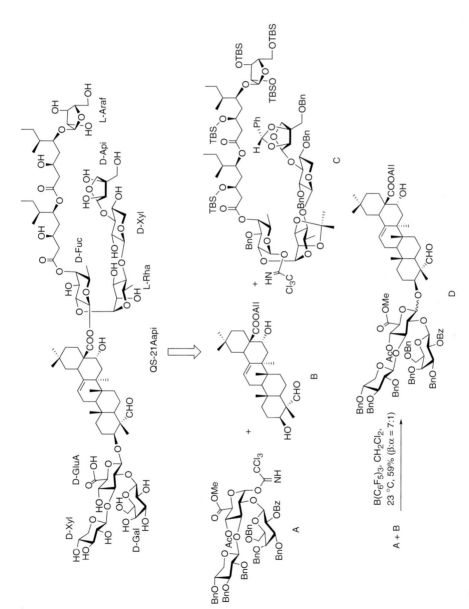

Scheme 5.14 Chemical synthesis of QS-21A.

the most challenging synthetic targets [9, 16, 207]. Here we highlight just the chemical synthesis of four sugar-containing complex natural products that showcase the power of modern synthetic chemistry.

5.7.1
Total Synthesis of Carbohydrate Immune-Adjuvant QS-21A$_{api}$

QS-21A$_{api}$, a minor component of quillajasaponins, is a potent compound for immune response potentiation and dose sparing in vaccine therapy. This complex plant glycoside contains a quillaic acid triterpene core which is flanked by a branched trisaccharide and a linear tetrasaccharide on each side; the tetrasaccharide is further modified by a dimeric fatty acyl moiety which is capped by an arabinose residue. Surprisingly, all eight monosaccharide residues in QS-21A$_{api}$ are different. The first total chemical synthesis of QS-21A$_{api}$ was achieved by Gin and co-workers in 2005 [208]. The retrosynthetic disconnection of the target led to three major advanced building blocks, A, B, and C (Scheme 5.14). Among a number of difficult connections, the glycosidic coupling of a branched trisaccharide to a triterpene 3-OH derivative in a β-selective manner (in the absence of neighboring group participation) was found to be extremely difficult. After extensive studies, this was finally accomplished with α-imidate **A** as a glycosyl donor under the promotion of an unusual promoter, $B(C_6F_5)_3$, leading to the formation of the trisaccharide glycoside D in 59% yield with a β/α ratio of 7:1.

5.7.2
Total Synthesis of Lobatoside E

Lobatoside E, a member of the triterpene saponins named cyclic bisdesmosides with potent antitumor activities, has a disaccharide and a trisaccharide flanked on a pentacyclic triterpene which is bridged with 3-hydroxy-3-methylglutarate. Total synthesis of this complex plant glycoside was achieved by Yu and co-workers in 2008 with 73 overall steps [209]. A retrosynthetic disconnection of lobatoside E led to the aglycone and the corresponding mono- and disaccharide building blocks (Scheme 5.15). Stepwise assembly of the sugar residues allowed the highly stereoselective formation of the respective glycosidic bonds with the auxiliary of neighboring group participation. However, coupling of the galactosyl imidate D to the 2-OH of the glucose residue in building block G was found to be difficult owing to steric hindrance. A 65% yield of the desired product H was obtained at the expense of 5 equiv. of the donor D. Similar glycosylation of a substrate without the proximal 2,23-dibenzyloxy group with 1.2 equiv. of donor D led to the coupling product in 95% yield.

5.7.3
Total Synthesis of Moenomycin A

Moenomycin A is the only natural inhibitor known to bind directly to the transg-lycosylases that catalyze elongation of the carbohydrate chains of peptidoglycan in bacterial cell walls. This complex glycoconjugate consists of a highly functionalized

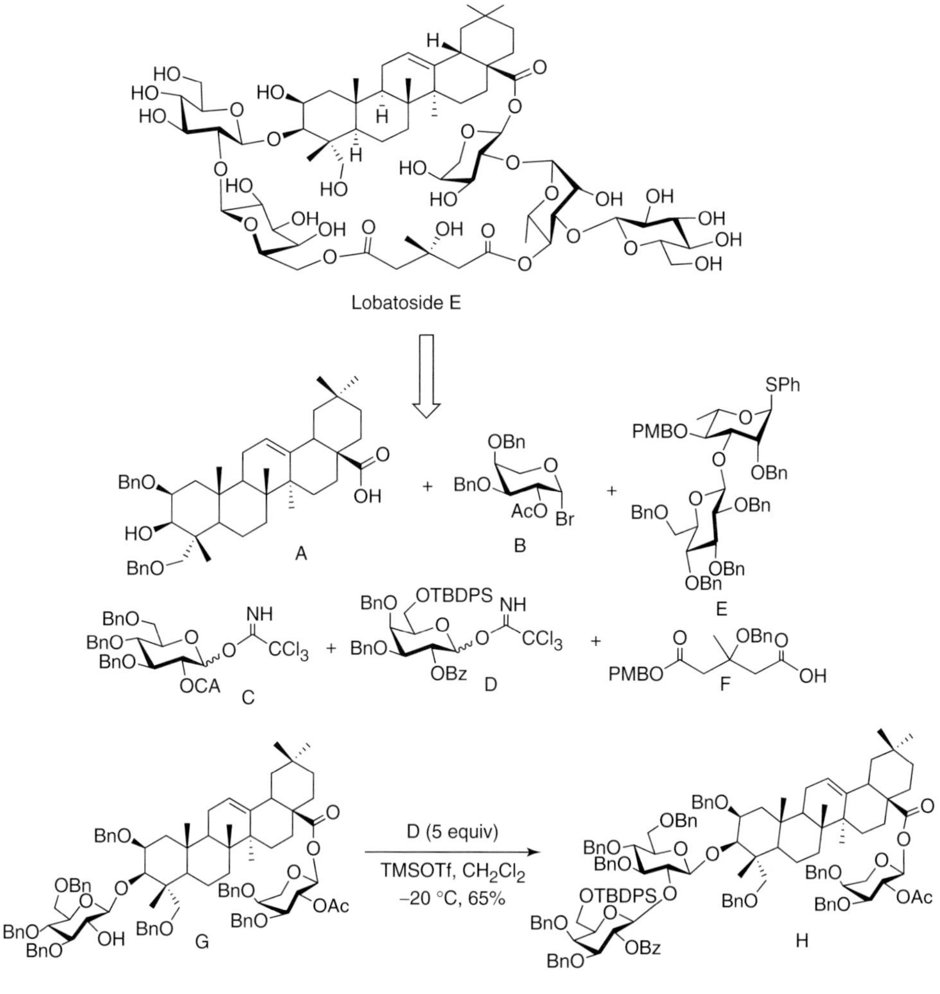

Scheme 5.15 Chemical synthesis of lobatoside E

pentasaccharide attached via a unique phosphoglycerate linkage to a polyprenyl chain. Synthetic studies on moenomycin A have lasted over 30 years. The first total synthesis of moenomycin A was accomplished by Kahne and co-workers in 2006 [210]. The retrosynthetic disconnection is shown in Scheme 5.16. Notably, all four glycosidic linkages between sugars were constructed by the powerful glycosylation method with glycosyl sulfoxides as the donors. To achieve the coupling of disaccharide sulfoxide A with disaccharide acceptor B, "inverse addition" was required. That is, sulfoxide A was added to a solution containing acceptor B and the promoter triflic anhydride to ensure the capture of the oxacarbenium intermediate by the acceptor before it underwent decomposition. Additional scavengers (2,6-di-*tert*-butyl-4-methylpyridine and

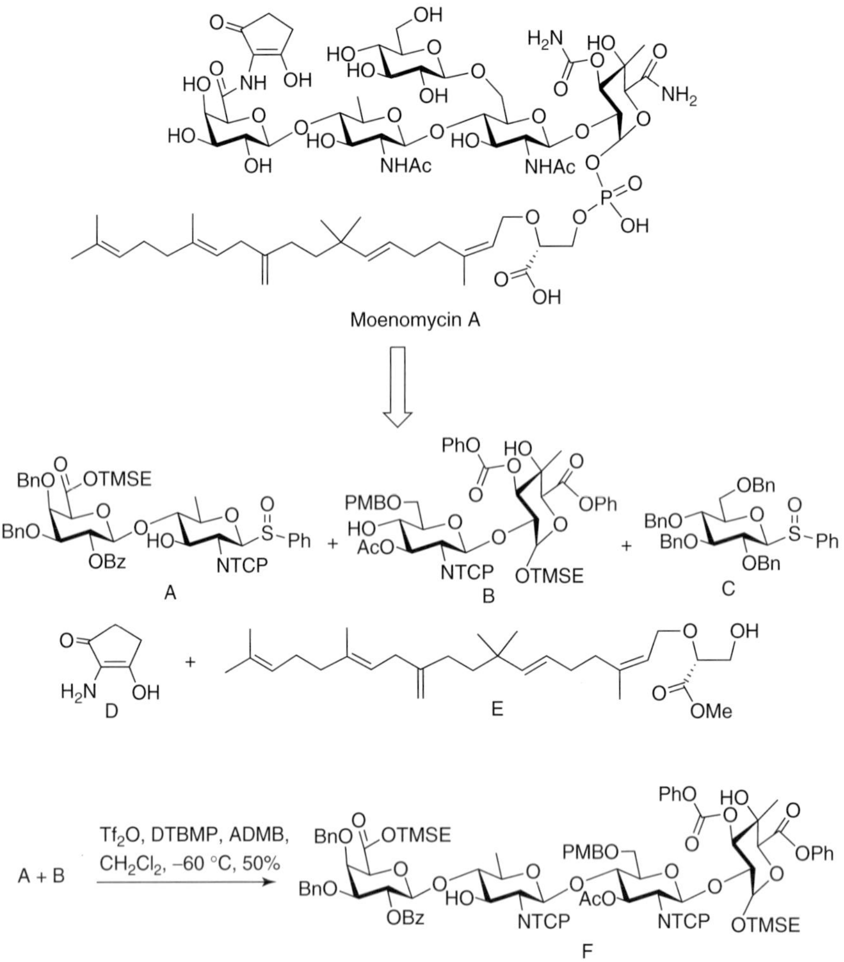

Scheme 5.16 Chemical synthesis of moenomycin A.

4-allyl-1,2-dimethoxybenzene) for deteriorative species were also required to enable this [2 + 2]-glycosylation to furnish the desired tetrasaccharide F in 50% yield.

5.7.4
Total Synthesis of Lipoteichoic Acid

Lipoteichoic acids (LTAs), anchoring in the cell membrane of Gram-positive bacteria, are recognized by the innate immune system, thereby stimulating the release of proinflammatory cytokines. The LTA shown in Scheme 5.17 from *Streptococcus pneumoniae* consists of a diacylglycerol trisaccharide connected with a phosphocholine-modified tetrasaccharide via a ribitolphosphate linker. This complex glycolipid was synthesized recently by Schmidt and co-workers in overall 88 steps from widely available monosaccharides [211].

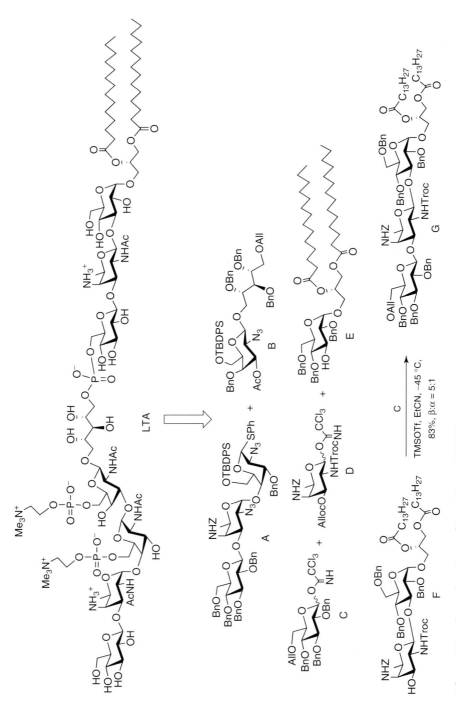

Scheme 5.17 Chemical synthesis of lipoteichoic acid (LTA).

The synthetic design is shown in Scheme 5.17. Initial attempt to couple the glucosyl lipid derivative E with a CD disaccharide donor failed to afford the glycosylation product. An alternative approach to couple disaccharide lipid derivative F with glucosyl imidate C was successful, providing the trisaccharide G. In the absence of a neighboring participation group, the solvent effect (EtCN) was exploited to give the β-glycoside as the major isomer (β : α = 5 : 1).

5.8
Conclusion and Perspectives

Recent advances in synthetic carbohydrate chemistry have provided new strategies for tackling many interesting problems in glycobiology. Now even very complex oligosaccharides and glycoconjugates can be synthesized using various sophisticated chemical and chemoenzymatic methods. These synthetic carbohydrates have proven indispensable for elucidating the diverse functions of carbohydrates in biological systems and for developing carbohydrate-based drugs. On the other hand, the structural and functional complexity of carbohydrates also provides an excellent platform for chemists to develop new synthetic and analytical chemistry. Despite tremendous progress in the past decade, there is still a rich chemistry to be explored associated with carbohydrate synthesis. These will include the development of new methods for rapid and, in the long term, automated assembly of oligosaccharides and new strategies for constructing glyco-proteins and glycolipids. The interplay between carbohydrate synthesis and glycobiology will continue to play essential roles in our understanding of the diverse functions of car-bohydrates in biological systems and in the development of efficient carbohydrate-based therapeutics to combat human diseases.

Acknowledgments

This work was supported in part by the National Institutes of Health (NIH grants **R01 GM080374** to L.-X.W.) and the National Natural Science Foundation of China (grant **20932009** to B.Y.).

References

1. Varki, A. (1993) *Glycobiology*, 3, 97–130.
2. Dwek, R.A. (1996) *Chem. Rev.*, **96**, 683–720.
3. Helenius, A. and Aebi, M. (2001) *Science*, 291, 2364–2369.
4. Haltiwanger, R.S. and Lowe, J.B. (2004) *Annu. Rev. Biochem.*, **73**, 491–537.
5. Dube, D.H. and Bertozzi, C.R. (2005) *Nat. Rev. Drug Discov.*, 4, 477–488.
6. Nimmerjahn, F. and Ravetch, J.V. (2008) *Nat. Rev. Immunol.*, **8**, 34–47.
7. Jefferis, R. (2009) *Nat. Rev. Drug Discov.*, 8, 226–234.
8. Danishefsky, S.J. and Allen, J.R. (2000) *Angew. Chem. Int. Ed.*, **39**, 836–863.
9. Nicolaou, K.C. and Mitchell, H.J. (2001) *Angew. Chem. Int. Ed.*, **40**, 1576–1624.
10. Sears, P. and Wong, C.H. (2001) *Science*, 291, 2344–2350.
11. Dziadek, S. and Kunz, H. (2004) *Chem. Rec.*, 3, 308–321.

12. Seeberger, P.H. and Werz, D.B. (2005) *Nat. Rev. Drug Discov.*, **4**, 751–763.

13. Smoot, J.T. and Demchenko, A.V. (2009) *Adv. Carbohydr. Chem. Biochem.*, **62**, 161–250.

14. Kajihara, Y., Yamamoto, N., Miyazaki, T., and Sato, H. (2005) *Curr. Med. Chem.*, **12**, 527–550.

15. Hancock, S.M., Vaughan, M.D., and Withers, S.G. (2006) *Curr. Opin. Chem. Biol.*, **10**, 509–519.

16. Galonic, D.P. and Gin, D.Y. (2007) *Nature*, **446**, 1000–1007.

17. Takeda, Y., Totani, K., Matsuo, I., and Ito, Y. (2009) *Curr. Opin. Chem. Biol.*, **13**, 582–591.

18. Wang, L.X. and Huang, W. (2009) *Curr. Opin. Chem. Biol.*, **13**, 592–600.

19. Gamblin, D.P., Scanlan, E.M., and Davis, B.G. (2009) *Chem. Rev.*, **109**, 131–163.

20. Hinou, H. and Nishimura, S. (2009) *Curr. Top. Med. Chem.*, **9**, 106–116.

21. Andre, S., Kozar, T., Kojima, S., Unverzagt, C., and Gabius, H.J. (2009) *Biol. Chem.*, **390**, 557–565.

22. Boltje, T.J., Buskas, T., and Boons, G.J. (2009) *Nat. Chem.*, **1**, 611–622.

23. Crout, D.H. and Vic, G. (1998) *Curr. Opin. Chem. Biol.*, **2**, 98–111.

24. Bennett, C.S. and Wong, C.H. (2007) *Chem. Soc. Rev.*, **36**, 1227–1238.

25. Shaikh, F.A. and Withers, S.G. (2008) *Biochem. Cell Biol.*, **86**, 169–177.

26. Blixt, O., Head, S., Mondala, T., Scanlan, C., Huflejt, M.E., Alvarez, R., Bryan, M.C., Fazio, F., Calarese, D., Stevens, J., Razi, N., Stevens, D.J., Skehel, J.J., van Die, I., Burton, D.R., Wilson, I.A., Cummings, R., Bovin, N., Wong, C.H., and Paulson, J.C. (2004) *Proc. Natl. Acad. Sci. U. S. A.*, **101**, 17033–17038.

27. Stevens, J., Blixt, O., Paulson, J.C., and Wilson, I.A. (2006) *Nat. Rev. Microbiol.*, **4**, 857–864.

28. Paulson, J.C., Blixt, O., and Collins, B.E. (2006) *Nat. Chem. Biol.*, **2**, 238–248.

29. Horlacher, T. and Seeberger, P.H. (2008) *Chem. Soc. Rev.*, **37**, 1414–1422.

30. Liang, P.H., Wu, C.Y., Greenberg, W.A., and Wong, C.H. (2008) *Curr. Opin. Chem. Biol.*, **12**, 86–92.

31. Courtney, A.H., Puffer, E.B., Pontrello, J.K., Yang, Z.Q., and Kiessling, L.L.

32. O'Reilly, M.K., Collins, B.E., Han, S., Liao, L., Rillahan, C., Kitov, P.I., Bundle, D.R., and Paulson, J.C. (2008) *J. Am. Chem. Soc.*, **130**, 7736–7745.

33. Kitov, P.I., Sadowska, J.M., Mulvey, G., Armstrong, G.D., Ling, H., Pannu, N.S., Read, R.J., and Bundle, D.R. (2000) *Nature*, **403**, 669–672.

34. Laughlin, S.T. and Bertozzi, C.R. (2009) *Proc. Natl. Acad. Sci. U. S. A.*, **106**, 12–17.

35. van Kasteren, S.I., Campbell, S.J., Serres, S., Anthony, D.C., Sibson, N.R., and Davis, B.G. (2009) *Proc. Natl. Acad. Sci. U. S. A.*, **106**, 18–23.

36. Wang, S.K., Liang, P.H., Astronomo, R.D., Hsu, T.L., Hsieh, S.L., Burton, D.R., and Wong, C.H. (2008) *Proc. Natl. Acad. Sci. U. S. A.*, **105**, 3690–3695.

37. Brocke, C. and Kunz, H. (2002) *Bioorg. Med. Chem.*, **10**, 3085–3112.

38. Ouerfelli, O., Warren, J.D., Wilson, R.M., and Danishefsky, S.J. (2005) *Expert Rev. Vaccines*, **4**, 677–685.

39. Liakatos, A. and Kunz, H. (2007) *Curr. Opin. Mol. Ther.*, **9**, 35–44.

40. Buskas, T., Thompson, P., and Boons, G.J. (2009) *Chem. Commun.*, 5335–5349.

41. Pozsgay, V. (2000) *Adv. Carbohydr. Chem. Biochem.*, **56**, 153–199.

42. Wang, L.X. (2006) *Curr. Opin. Drug Discov. Dev.*, **9**, 194–206.

43. Scanlan, C.N., Offer, J., Zitzmann, N., and Dwek, R.A. (2007) *Nature*, **446**, 1038–1045.

44. Astronomo, R.D. and Burton, D.R. (2010) *Nat. Rev. Drug Discov.*, **9**, 308–324.

45. Verez-Bencomo, V., Fernandez-Santana, V., Hardy, E., Toledo, M.E., Rodriguez, M.C., Heynngnezz, L., Rodriguez, A., Baly, A., Herrera, L., Izquierdo, M., Villar, A., Valdes, Y., Cosme, K., Deler, M.L., Montane, M., Garcia, E., Ramos, A., Aguilar, A., Medina, E., Torano, G., Sosa, I., Hernandez, I., Martinez, R., Muzachio, A., Carmenates, A., Costa, L., Cardoso, F., Campa, C., Diaz, M., and Roy, R. (2004) *Science*, **305**, 522–525.

46. Zhu, J., Wan, Q., Ragupathi, G., George, C.M., Livingston, P.O., and Danishefsky, S.J. (2009) *J. Am. Chem. Soc.*, **131**, 4151–4158.

47. Kaiser, A., Gaidzik, N., Westerlind, U., Kowalczyk, D., Hobel, A., Schmitt, E., and Kunz, H. (2009) *Angew. Chem. Int. Ed.*, **48**, 7551–7555.

48. Ingale, S., Wolfert, M.A., Gaekwad, J., Buskas, T., and Boons, G.J. (2007) *Nat. Chem. Biol.*, **3**, 663–667.

49. Bertozzi, C.R. and Kiessling, L.L. (2001) *Science*, **291**, 2357–2364.

50. Seeberger, P.H. (2009) *Nat. Chem. Biol.*, **5**, 368–372.

51. Lepenies, B., Yin, J., and Seeberger, P.H. (2010) *Curr. Opin. Chem. Biol.*, **14**, 404–411.

52. Kiessling, L.L. and Splain, R.A. (2010) *Annu. Rev. Biochem.*, **79**, 619–653.

53. Boons, G.J. and Demchenko, A.V. (2000) *Chem. Rev.*, **100**, 4539–4566.

54. Zhu, X. and Schmidt, R.R. (2009) *Angew. Chem. Int. Ed.*, **48**, 1900–1934.

55. Yu, B. and Sun, J. (2010) *Chem. Commun.*, **46**, 4668–4679.

56. Werz, D.B., Castagner, B., and Seeberger, P.H. (2007) *J. Am. Chem. Soc.*, **129**, 2770–2771.

57. Cai, S. and Yu, B. (2003) *Org. Lett.*, **5**, 3827–3830.

58. Tanaka, H., Iwata, Y., Takahashi, D., Adachi, M., and Takahashi, T. (2005) *J. Am. Chem. Soc.*, **127**, 1630–1631.

59. Thomas, M., Gesson, J.P., and Papot, S. (2007) *J. Org. Chem.*, **72**, 4262–4264.

60. Ueki, A., Takano, Y., Kobayashi, A., Nakahara, Y., Hojo, H., and Nakahara, Y. (2010) *Tetrahedron*, **66**, 1742–1759.

61. Kim, K.S., Kim, J.H., Lee, Y.J., and Park, J. (2001) *J. Am. Chem. Soc.*, **123**, 8477–8481.

62. Kim, K.S., Park, J., Lee, Y.J., and Seo, Y.S. (2003) *Angew. Chem. Int. Ed.*, **42**, 459–462.

63. Lee, Y.J., Lee, B.Y., Jeon, H.B., and Kim, K.S. (2006) *Org. Lett.*, **8**, 3971–3974.

64. Li, Y., Yang, X., Liu, Y., Zhu, C., Yang, Y., and Yu, B. (2010) *Chem. Eur. J.*, **16**, 1871–1882.

65. Yang, W., Sun, J., Lu, W., Li, Y., Shan, L., Han, W., Zhang, W.D., and Yu, B. (2010) *J. Org. Chem.*, **75**, 6879–6888.

66. Yang, Y., Li, Y., and Yu, B. (2009) *J. Am. Chem. Soc.*, **131**, 12076–12077.

67. Cumpstey, I. (2008) *Carbohydr. Res.*, **343**, 1553–1573.

68. Ishiwata, A., Lee, Y.J., and Ito, Y. (2010) *Org. Biomol. Chem.*, **8**, 3596–3608.

69. Crich, D. and Sun, S. (1996) *J. Org. Chem.*, **61**, 7200–7201.

70. Crich, D. and Sun, S. (1998) *Tetrahedron*, **54**, 8321–8348.

71. Jensen, H.H., Nordstrom, L.U., and Bols, M. (2004) *J. Am. Chem. Soc.*, **126**, 9205–9213.

72. Crich, D. and Chandrasekera, N.S. (2004) *Angew. Chem. Int. Ed.*, **43**, 5386–5389.

73. Crich, D. (2010) *Acc. Chem. Res.*, **43**, 1144–1153.

74. van den Bos, L.J., Dinkelaar, J., Overkleeft, H.S., and van der Marel, G.A. (2006) *J. Am. Chem. Soc.*, **128**, 13066–13067.

75. Imamura, A., Ando, H., Korogi, S., Tanabe, G., Muraoka, O., Ishida, H., and Kiso, M. (2003) *Tetrahedron Lett.*, **44**, 6725–6728.

76. Zhu, X., Kawatkar, S., Rao, Y., and Boons, G.J. (2006) *J. Am. Chem. Soc.*, **128**, 11948–11957.

77. Imamura, A. and Lowary, T.L. (2010) *Org. Lett.*, **12**, 3686–3689.

78. Benakli, K., Zha, C., and Kerns, R.J. (2001) *J. Am. Chem. Soc.*, **123**, 9461–9462.

79. Boysen, M., Gemma, E., Lahmann, M., and Oscarson, S. (2005) *Chem. Commun.*, 3044–3046.

80. Manabe, S., Ishii, K., and Ito, Y. (2006) *J. Am. Chem. Soc.*, **128**, 10666–10667.

81. Manabe, S., Ishii, K., Hashizume, D., Koshino, H., and Ito, Y. (2009) *Chem. Eur. J.*, **15**, 6894–6901.

82. Mensah, E.A. and Nguyen, H.M. (2009) *J. Am. Chem. Soc.*, **131**, 8778–8780.

83. Mensah, E.A., Yu, F., and Nguyen, H.M. (2010) *J. Am. Chem. Soc.*, **132**, 14288–14302.

84. Kim, J.H., Yang, H., Park, J., and Boons, G.J. (2005) *J. Am. Chem. Soc.*, **127**, 12090–12097.

85. Kim, J.H., Yang, H., and Boons, G.J. (2005) *Angew. Chem. Int. Ed.*, **44**, 947–949.

86. Boltje, T.J., Kim, J.H., Park, J., and Boons, G.J. (2010) *Nat. Chem.*, **2**, 552–557.

87. Demchenko, A.V. and Boons, G.J. (1999) *Chem. Eur. J.*, **5**, 1278–1283.

88. De Meo, C. and Priyadarshani, U. (2008) *Carbohydr. Res.*, **343**, 1540–1552.

89. Crich, D. and Wu, B. (2008) *Org. Lett.*, **10**, 4033–4035.

90. Tanaka, H., Nishiura, Y., and Takahashi, T. (2006) *J. Am. Chem. Soc.*, **128**, 7124–7125.

91. Plante, O.J., Palmacci, E.R., and Seeberger, P.H. (2001) *Science*, **291**, 1523–1527.

92. Seeberger, P.H. (2008) *Chem. Soc. Rev.*, **37**, 19–28.

93. Routenberg Love, K. and Seeberger, P.H. (2004) *Angew. Chem. Int. Ed.*, **43**, 602–605.

94. Codee, J.D., Krock, L., Castagner, B., and Seeberger, P.H. (2008) *Chem. Eur. J.*, **14**, 3987–3994.

95. Matsushita, T., Nagashima, I., Fumoto, M., Ohta, T., Yamada, K., Shimizu, H., Hinou, H., Naruchi, K., Ito, T., Kondo, H., and Nishimura, S. (2010) *J. Am. Chem. Soc.*, **132**, 16651–16656.

96. Martin, J.G., Gupta, M., Xu, Y., Akella, S., Liu, J., Dordick, J.S., and Linhardt, R.J. (2009) *J. Am. Chem. Soc.*, **131**, 11041–11048.

97. Jaipuri, F.A. and Pohl, N.L. (2008) *Org. Biomol. Chem.*, **6**, 2686–2691.

98. Zhang, Z., Ollmann, I.R., Ye, X.S., Wischnat, R., Baasov, T., and Wong, C.H. (1999) *J. Am. Chem. Soc.*, **121**, 734–753.

99. Mong, T.K., Lee, H.K., Duron, S.G., and Wong, C.H. (2003) *Proc. Natl. Acad. Sci. U. S. A.*, **100**, 797–802.

100. Huang, X., Huang, L., Wang, H., and Ye, X.S. (2004) *Angew. Chem. Int. Ed.*, **43**, 5221–5224.

101. Sun, B., Srinivasan, B., and Huang, X. (2008) *Chem. Eur. J.*, **14**, 7072–7081.

102. Wang, Z., Xu, Y., Yang, B., Tiruchinapally, G., Sun, B., Liu, R., Dulaney, S., Liu, J., and Huang, X. (2010) *Chem. Eur. J.*, **16**, 8365–8375.

103. Wang, Z., Zhou, L., El-Boubbou, K., Ye, X.S., and Huang, X.F. (2007) *J. Org. Chem.*, **72**, 6409–6420.

104. Ichikawa, Y., Look, G.C., and Wong, C.H. (1992) *Anal. Biochem.*, **202**, 215–238.

105. Chen, X., Zhang, J., Kowal, P., Liu, Z., Andreana, P.R., Lu, Y., and Wang, P.G. (2001) *J. Am. Chem. Soc.*, **123**, 8866–8867.

106. Chen, X., Fang, J., Zhang, J., Liu, Z., Shao, J., Kowal, P., Andreana, P., and Wang, P.G. (2001) *J. Am. Chem. Soc.*, **123**, 2081–2082.

107. Lairson, L.L., Henrissat, B., Davies, G.J., and Withers, S.G. (2008) *Annu. Rev. Biochem.*, **77**, 521–555.

108. Serna, S., Etxebarria, J., Ruiz, N., Martin-Lomas, M., and Reichardt, N.C. (2010) *Chem. Eur. J.*, **16**, 13163–13175.

109. Ito, H., Chiba, Y., Kameyama, A., Sato, T., and Narimatsu, H. (2010) *Methods Enzymol.*, **478**, 127–149.

110. Williams, G.J., Goff, R.D., Zhang, C., and Thorson, J.S. (2008) *Chem. Biol.*, **15**, 393–401.

111. Williams, G.J., Zhang, C., and Thorson, J.S. (2007) *Nat. Chem. Biol.*, **3**, 657–662.

112. Zhang, C., Griffith, B.R., Fu, Q., Albermann, C., Fu, X., Lee, I.K., Li, L., and Thorson, J.S. (2006) *Science*, **313**, 1291–1294.

113. Fu, X., Albermann, C., Jiang, J., Liao, J., Zhang, C., and Thorson, J.S. (2003) *Nat. Biotechnol.*, **21**, 1467–1469.

114. Perugino, G., Trincone, A., Rossi, M., and Moracci, M. (2004) *Trends Biotechnol.*, **22**, 31–37.

115. MacKenzie, L.F., Wang, Q., Warren, R.A.J., and Withers, S.G. (1998) *J. Am. Chem. Soc.*, **120**, 5583–5584.

116. Malet, C. and Planas, A. (1998) *FEBS Lett.*, **440**, 208–212.

117. Moracci, M., Trincone, A., Perugino, G., Ciaramella, M., and Rossi, M. (1998) *Biochemistry*, **37**, 17262–17270.

118. Kim, Y.W., Lee, S.S., Warren, R.A., and Withers, S.G. (2004) *J. Biol. Chem.*, **279**, 42787–42793.

119. Lin, H., Tao, H., and Cornish, V.W. (2004) *J. Am. Chem. Soc.*, **126**, 15051–15059.

120. Ben-David, A., Shoham, G., and Shoham, Y. (2008) *Chem. Biol.*, **15**, 546–551.

121. Kone, F.M., Le Bechec, M., Sine, J.P., Dion, M., and Tellier, C. (2009) *Protein Eng. Des. Sel.*, **22**, 37–44.

122. Kittl, R. and Withers, S.G. (2010) *Carbohydr. Res.*, **345**, 1272–1279.

123. Kim, Y.W., Lovering, A.L., Chen, H., Kantner, T., McIntosh, L.P., Strynadka, N.C., and Withers, S.G. (2006) *J. Am. Chem. Soc.*, **128**, 2202–2203.

124. Mullegger, J., Chen, H.M., Chan, W.Y., Reid, S.P., Jahn, M., Warren, R.A., Salleh, H.M., and Withers, S.G. (2006) *ChemBioChem*, **7**, 1028–1030.

125. Mullegger, J., Chen, H.M., Warren, R.A., and Withers, S.G. (2006) *Angew. Chem. Int. Ed.*, **45**, 2585–2588.

126. Champion, E., Andre, I., Moulis, C., Boutet, J., Descroix, K., Morel, S.,

Monsan, P., Mulard, L.A., and Remaud-Simeon, M. (2009) *J. Am. Chem. Soc.*, **131**, 7379–7389.

127. Vaughan, M.D., Johnson, K., DeFrees, S., Tang, X., Warren, R.A., and Withers, S.G. (2006) *J. Am. Chem. Soc.*, **128**, 6300–6301.

128. Umekawa, M., Huang, W., Li, B., Fujita, K., Ashida, H., Wang, L.X., and Yamamoto, K. (2008) *J. Biol. Chem.*, **283**, 4469–4479.

129. Heidecke, C.D., Ling, Z., Bruce, N.C., Moir, J.W., Parsons, T.B., and Fairbanks, A.J. (2008) *ChemBioChem*, **9**, 2045–2051.

130. Huang, W., Li, C., Li, B., Umekawa, M., Yamamoto, K., Zhang, X., and Wang, L.X. (2009) *J. Am. Chem. Soc.*, **131**, 2214–2223.

131. Hancock, S.M., Rich, J.R., Caines, M.E., Strynadka, N.C., and Withers, S.G. (2009) *Nat. Chem. Biol.*, **5**, 508–514.

132. Petitou, M., Casu, B., and Lindahl, U. (2003) *Biochimie*, **85**, 83–89.

133. Capila, I. and Linhardt, R.J. (2002) *Angew. Chem. Int. Ed.*, **41**, 391–412.

134. Petitou, M. and van Boeckel, C.A. (2004) *Angew. Chem. Int. Ed.*, **43**, 3118–3133.

135. Sinay, P. Jacquinet, J.C. Petitou, M. Duchaussoy, P. Lederman, I. Choay, J. Torri, G. (1984) *Carbohydr. Res.*, **132**, C5–C9.

136. Jin, L., Abrahams, J.P., Skinner, R., Petitou, M., Pike, R.N., and Carrell, R.W. (1997) *Proc. Natl. Acad. Sci. U. S. A.*, **94**, 14683–14688.

137. de Kort, M., Buijsman, R.C., and van Boeckel, C.A. (2005) *Drug Discov. Today*, **10**, 769–779.

138. Guerrini, M., Beccati, D., Shriver, Z., Naggi, A., Viswanathan, K., Bisio, A., Capila, I., Lansing, J.C., Guglieri, S., Fraser, B., Al-Hakim, A., Gunay, N.S., Zhang, Z., Robinson, L., Buhse, L., Nasr, M., Woodcock, J., Langer, R., Venkataraman, G., Linhardt, R.J., Casu, B., Torri, G., and Sasisekharan, R. (2008) *Nat. Biotechnol.*, **26**, 669–675.

139. Liu, H., Zhang, Z., and Linhardt, R.J. (2009) *Nat. Prod. Rep.*, **26**, 313–321.

140. Bishop, J.R., Schuksz, M., and Esko, J.D. (2007) *Nature*, **446**, 1030–1037.

141. de Paz, J.L., Noti, C., and Seeberger, P.H. (2006) *J. Am. Chem. Soc.*, **128**, 2766–2767.

142. Arungundram, S., Al-Mafraji, K., Asong, J., Leach, F.E. III, Amster, I.J., Venot,

A., Turnbull, J.E., and Boons, G.J. (2009) *J. Am. Chem. Soc.*, **131**, 17394–17405.

143. Yin, J. and Seeberger, P.H. (2010) *Methods Enzymol.*, **478**, 197–218.

144. Kuberan, B., Lech, M.Z., Beeler, D.L., Wu, Z.L., and Rosenberg, R.D. (2003) *Nat. Biotechnol.*, **21**, 1343–1346.

145. Zhang, Z., McCallum, S.A., Xie, J., Nieto, L., Corzana, F., Jimenez-Barbero, J., Chen, M., Liu, J., and Linhardt, R.J. (2008) *J. Am. Chem. Soc.*, **130**, 12998–13007.

146. Liu, R., Xu, Y., Chen, M., Weiwer, M., Zhou, X., Bridges, A.S., DeAngelis, P.L., Zhang, Q., Linhardt, R.J., and Liu, J. (2010) *J. Biol. Chem.*, **285**, 34240–34249.

147. Sethuraman, N. and Stadheim, T.A. (2006) *Curr. Opin. Biotechnol.*, **17**, 341–346.

148. Aggarwal, S. (2009) *Nat. Biotechnol.*, **27**, 987–993.

149. Aggarwal, S. (2010) *Nat. Biotechnol.*, **28**, 1165–1171.

150. Stanley, P. (1992) *Glycobiology*, **2**, 99–107.

151. Wildt, S. and Gerngross, T.U. (2005) *Nat. Rev. Microbiol.*, **3**, 119–128.

152. Hamilton, S.R. and Gerngross, T.U. (2007) *Curr. Opin. Biotechnol.*, **18**, 387–392.

153. Davis, B.G. (2002) *Chem. Rev.*, **102**, 579–601.

154. Buskas, T., Ingale, S., and Boons, G.J. (2006) *Glycobiology*, **16**, 113R–136R.

155. Wang, L.X. (2008) *Carbohydr. Res.*, **343**, 1509–1522.

156. Rich, J.R. and Withers, S.G. (2009) *Nat. Chem. Biol.*, **5**, 206–215.

157. Bernardes, G.J., Castagner, B., and Seeberger, P.H. (2009) *ACS Chem. Biol.*, **4**, 703–713.

158. Yuan, Y., Chen, J., Wan, Q., Wilson, R.M., and Danishefsky, S.J. (2010) *Biopolymers*, **94**, 373–384.

159. Grogan, M.J., Pratt, M.R., Marcaurelle, L.A., and Bertozzi, C.R. (2002) *Annu. Rev. Biochem.*, **71**, 593–634.

160. Seitz, O. (2000) *ChemBioChem*, **1**, 214–246.

161. Cohen-Anisfeld, S.T. and Lansbury, P.T. Jr. (1993) *J. Am. Chem. Soc.*, **115**, 10531–10537.

162. Roberge, J.Y., Beebe, X., and Danishefsky, S.J. (1995) *Science*, **269**, 202–204.

163. Miller, J.S., Dudkin, V.Y., Lyon, G.J., Muir, T.W., and Danishefsky, S.J. (2003) *Angew. Chem. Int. Ed.*, **42**, 431–434.

164. Kaneshiro, C.M. and Michael, K. (2006) *Angew. Chem. Int. Ed.*, **45**, 1077–1081.

165. Mandal, M., Dudkin, V.Y., Geng, X., and Danishefsky, S.J. (2004) *Angew. Chem. Int. Ed.*, **43**, 2557–2561.

166. Geng, X., Dudkin, V.Y., Mandal, M., and Danishefsky, S.J. (2004) *Angew. Chem. Int. Ed.*, **43**, 2562–2565.

167. Dudkin, V.Y., Miller, J.S., and Danishefsky, S.J. (2004) *J. Am. Chem. Soc.*, **126**, 736–738.

168. Dawson, P.E. and Kent, S.B. (2000) *Annu. Rev. Biochem.*, **69**, 923–960.

169. Dawson, P.E., Muir, T.W., Clark-Lewis, I., and Kent, S.B. (1994) *Science*, **266**, 776–779.

170. Marcaurelle, L.A., Mizoue, L.S., Wilken, J., Oldham, L., Kent, S.B., Handel, T.M., and Bertozzi, C.R. (2001) *Chem. Eur. J.*, **7**, 1129–1132.

171. Warren, J.D., Miller, J.S., Keding, S.J., and Danishefsky, S.J. (2004) *J. Am. Chem. Soc.*, **126**, 6576–6578.

172. Hojo, H., Matsumoto, Y., Nakahara, Y., Ito, E., Suzuki, Y., Suzuki, M., and Suzuki, A. (2005) *J. Am. Chem. Soc.*, **127**, 13720–13725.

173. Yamamoto, N., Tanabe, Y., Okamoto, R., Dawson, P.E., and Kajihara, Y. (2008) *J. Am. Chem. Soc.*, **130**, 501–510.

174. Yuan, Y., Chen, J., Wan, Q., Tan, Z., Chen, G., Kan, C., and Danishefsky, S.J. (2009) *J. Am. Chem. Soc.*, **131**, 5432–5437.

175. Nagorny, P., Fasching, B., Li, X., Chen, G., Aussedat, B., and Danishefsky, S.J. (2009) *J. Am. Chem. Soc.*, **131**, 5792–5799.

176. Piontek, C., Ring, P., Harjes, O., Heinlein, C., Mezzato, S., Lombana, N., Pohner, C., Puttner, M., Varon Silva, D., Martin, A., Schmid, F.X., and Unverzagt, C. (2009) *Angew. Chem. Int. Ed.*, **48**, 1936–1940.

177. Piontek, C., Varon Silva, D., Heinlein, C., Pohner, C., Mezzato, S., Ring, P., Martin, A., Schmid, F.X., and Unverzagt, C. (2009) *Angew. Chem. Int. Ed.*, **48**, 1941–1945.

178. Wu, B., Chen, J., Warren, J.D., Chen, G., Hua, Z., and Danishefsky, S.J. (2006) *Angew. Chem. Int. Ed.*, **45**, 4116–4125.

179. Offer, J., Boddy, C.N., and Dawson, P.E. (2002) *J. Am. Chem. Soc.*, **124**, 4642–4646.

180. Macmillan, D. and Anderson, D.W. (2004) *Org. Lett.*, **6**, 4659–4662.

181. Yang, Y.Y., Ficht, S., Brik, A., and Wong, C.H. (2007) *J. Am. Chem. Soc.*, **129**, 7690–7701.

182. Payne, R.J., Ficht, S., Tang, S., Brik, A., Yang, Y.Y., Case, D.A., and Wong, C.H. (2007) *J. Am. Chem. Soc.*, **129**, 13527–13536.

183. Bennett, C.S., Dean, S.M., Payne, R.J., Ficht, S., Brik, A., and Wong, C.H. (2008) *J. Am. Chem. Soc.*, **130**, 11945–11952.

184. Muir, T.W. (2003) *Annu. Rev. Biochem.*, **72**, 249–289.

185. Tolbert, T.J., Franke, D., and Wong, C.H. (2005) *Bioorg. Med. Chem.*, **13**, 909–915.

186. Tolbert, T.J. and Wong, C.H. (2000) *J. Am. Chem. Soc.*, **122**, 5421–5428.

187. Hackenberger, C.P., Friel, C.T., Radford, S.E., and Imperiali, B. (2005) *J. Am. Chem. Soc.*, **127**, 12882–12889.

188. Macmillan, D. and Bertozzi, C.R. (2004) *Angew. Chem. Int. Ed.*, **43**, 1355–1359.

189. Macmillan, D., Bill, R.M., Sage, K.A., Fern, D., and Flitsch, S.L. (2001) *Chem. Biol.*, **8**, 133–145.

190. Watt, G.M., Lund, J., Levens, M., Kolli, V.S., Jefferis, R., and Boons, G.J. (2003) *Chem. Biol.*, **10**, 807–814.

191. Rendle, P.M., Seger, A., Rodrigues, J., Oldham, N.J., Bott, R.R., Jones, J.B., Cowan, M.M., and Davis, B.G. (2004) *J. Am. Chem. Soc.*, **126**, 4750–4751.

192. Liu, H., Wang, L., Brock, A., Wong, C.H., and Schultz, P.G. (2003) *J. Am. Chem. Soc.*, **125**, 1702–1703.

193. Kiick, K.L., Saxon, E., Tirrell, D.A., and Bertozzi, C.R. (2002) *Proc. Natl. Acad. Sci. U. S. A.*, **99**, 19–24.

194. Wang, A., Winblade Nairn, N., Johnson, R.S., Tirrell, D.A., and Grabstein, K. (2008) *ChemBioChem*, **9**, 324–330.

195. Wiltschi, B., Wenger, W., Nehring, S., and Budisa, N. (2008) *Yeast*, **25**, 775–786.

196. van Kasteren, S.I., Kramer, H.B., Jensen, H.H., Campbell, S.J., Kirkpatrick, J., Oldham, N.J., Anthony, D.C., and Davis, B.G. (2007) *Nature*, **446**, 1105–1109.

197. Witte, K., Sears, P., Martin, R., and Wong, C.H. (1997) *J. Am. Chem. Soc.*, **119**, 2114–2118.

198. Takegawa, K., Tabuchi, M., Yamaguchi, S., Kondo, A., Kato, I., and Iwahara, S. (1995) *J. Biol. Chem.*, **270**, 3094–3099.

199. Fujita, K., Tanaka, N., Sano, M., Kato, I., Asada, Y., and Takegawa, K. (2000)

Biochem. Biophys. Res. Commun., **267**, 134–138.

200. Li, B., Zeng, Y., Hauser, S., Song, H., and Wang, L.X. (2005) *J. Am. Chem. Soc.*, **127**, 9692–9693.

201. Li, B., Song, H., Hauser, S., and Wang, L.X. (2006) *Org. Lett.*, **8**, 3081–3084.

202. Ochiai, H., Huang, W., and Wang, L.X. (2008) *J. Am. Chem. Soc.*, **130**, 13790–13803.

203. Wei, Y., Li, C., Huang, W., Li, B., Strome, S., and Wang, L.X. (2008) *Biochemistry*, **47**, 10294–10304.

204. Umekawa, M., Li, C., Higashiyama, T., Huang, W., Ashida, H., Yamamoto, K., and Wang, L.X. (2010) *J. Biol. Chem.*, **285**, 511–521.

205. Huang, W., Yang, Q., Umekawa, M., Yamamoto, K., and Wang, L.X. (2010) *ChemBioChem*, **11**, 1350–1355.

206. Schwarz, F., Huang, W., Li, C., Schulz, B.L., Lizak, C., Palumbo, A., Numao, S., Neri, D., Aebi, M., and Wang, L.X. (2010) *Nat. Chem. Biol.*, **6**, 264–266.

207. Nicolaou, K.C. (2005) *J. Org. Chem.*, **70**, 7007–7027.

208. Wang, P., Kim, Y.J., Navarro-Villalobos, M., Rohde, B.D., and Gin, D.Y. (2005) *J. Am. Chem. Soc.*, **127**, 3256–3257.

209. Zhu, C., Tang, P., and Yu, B. (2008) *J. Am. Chem. Soc.*, **130**, 5872–5873.

210. Taylor, J.G., Li, X., Oberthur, M., Zhu, W., and Kahne, D.E. (2006) *J. Am. Chem. Soc.*, **128**, 15084–15085.

211. Pedersen, C.M., Figueroa-Perez, I., Lindner, B., Ulmer, A.J., Zahringer, U., and Schmidt, R.R. (2010) *Angew. Chem. Int. Ed.*, **49**, 2585–2590.

Commentary Part

Comment 1

Sam Danishefsky

The chapter entitled Carbohydrate Synthesis Towards Glycobiology is well done, and shows excellent judgment in identifying the main areas of interest. Also, I want to compliment the authors on their excellent use of language, though I am sure there will be further improvements in the editorial process. Let me turn to possible improvements to the text. One would involve providing better insights into a Kotchetkov amination and Lansbury aspartylation. These are important components in the synthesis of N-linked glycopolypeptides. There should also be an account of the cassette method for synthesizing α-O-linked glycoproteins. There could be better mechanistic descriptions of the factors governing α- and β-glycosylation ratios. There could also be additional coverage of some of the key classical methods, which still persist. It is true that they have been treated in earlier review articles but, nonetheless, the key ones should be presented again here. However, overall, the chapter is commendable.

Comment 2

David Crich

As is clear from this chapter by Yu and Wang, carbohydrate chemistry has made extraordinary progress in the last two decades, to the extent that the synthesis of complex oligosaccharides employing more than 50 steps can be both envisaged and achieved. The chapter also underlines the fact that this progress has been driven by the ever-expanding list of critical roles that carbohydrates play in biology and whose study demands the preparation of structurally homogeneous oligosaccharides and their conjugates such as are rarely available by isolation from natural sources. However, perusal of the superlative synthetic work presented also reveals the largely empirical nature of the glycosidic bond-forming reactions. Indeed, oligosaccharide and glycoconjugate synthesis, for all the apparent simplicity of its central reaction of glycosidic bond formation, is perhaps the most empirical branch of the discipline of organic synthesis [C1]. Perhaps because of this empirical nature and the fact that the central reaction involves C–O and not C–C bond formation, most synthetic groups, with certain obvious exceptions, curtail their endeavors to obtain the aglycone of any particular natural product target. On the other hand, most traditional carbohydrate research groups shy away from the aglycone part. This is a great shame because the fully stereocontrolled synthesis of a glycosidic bond is every bit as challenging and important as that of most C–C bonds; increased cross-fertilization between the disparate branches of organic chemistry would be of benefit to all, and especially to glycoscience.

This chapter essentially characterizes glycosidic bond-forming reactions as belonging to one of two classes: those which are accessible by neighboring group participation and traditionally considered easy to make, and those not accessible by such means and considered harder to prepare. However, are these widespread generalizations truly valid? The answer has to be "no," as numerous examples are known, particularly in the synthesis of the β-(1 → 3)-glucans, in which neighboring group participation fails to give the desired stereochemistry, and because for more than a dozen years now it has been possible to make a β-mannopyranoside as easily and predictably as a β-glucopyranoside. Part of the problem arises from the lack of attention paid by many in the carbohydrate chemistry community to the glycosyl acceptor when considering a glycosylation reaction – stereochemical mismatching is widely appreciated and feared in mainstream organic synthesis, but its role in glycosylation reactions is only slowly coming to be recognized [C2]. The other main facet of the problem is the rather poorly understood mechanism of glycosidic bond-forming reactions – the mechanisms of glycosidase enzymes have been far more deeply studied – which arises in part from the ease with which one writes a glycosyl oxocarbenium ion as the key intermediate. In reality, the existence of the glycosyl oxocarbenium ion in solution has yet to be demonstrated, and well-respected workers in the field have suggested that it will have no existence in organic solution [C3]. An appraisal of the recent literature, however, suggests that glycosyl oxocarbenium ions probably do exist in organic solution in some cases and that their visualization only awaits the discovery of a suitable rapid and clean method of generation [C4].

In order that the full potential of glycoconjugate synthesis can be realized and routinely applied to the solution of biomedical problems, the present level of empiricism needs to be considerably reduced. This can best be achieved by removing the imaginary barriers that separate mainstream organic chemistry and physical organic chemistry from the art of carbohydrate chemistry. There are many difficult and worthwhile challenges to be met, which are not likely to be solved by specialists working in isolation, and which merit the much broader attention of the organic chemistry community.

Authors' Response to the Commentaries

We thank Prof. Danishefsky and Prof. Crich for their insightful comments on this chapter. The last decade has witnessed tremendous progress both in glycosylation method development and in the synthesis of complex carbohydrates for glycobiology and medicine. Instead of trying to give a comprehensive review of synthetic carbohydrate chemistry, which would be difficult for a single chapter, we have attempted to focus the discussions on emerging synthetic technologies and new strategies that showcase our ability to assemble complex oligosaccharides and glycoconjugates related to glycobiology. Hence the examples described in this short chapter are only selective, and many important examples had to be left out. Nevertheless, we agree that there could have been better coverage of mechanistic aspects of glycosylation and on those classical glycosylation methods that still prevail today. Readers are recommended to read those excellent previous reviews covering specific topics that are cited in this chapter. It is expected that carbohydrate synthesis will continue to play essential roles in deciphering biological functions and in developing efficient carbohydrate-based therapeutics.

References

C1. Barresi, F. and Hindsgaul, O. (1995) *J. Carbohydr. Chem.*, **14**, 1043–1087.

C2. (a) Spijker, N.M. and van Boeckel, C.A.A. (1991) *Angew. Chem. Int. Ed. Engl.*, **30**, 180–183 ; (b) Bohé, L. and Crich, D. (2010) *Trends Glycosci. Glycotechnol.*, **22**, 1–15.

C3. Sinnott, M.L. (2007) *Carbohydrate Chemistry and Biochemistry*, RSC Publishing, Cambridge.

C4. Bohé, L. and Crich, D. (2011) *C. R. Chim.*, **14**, 3–16.

6
Chemical Synthesis of Proteins

Lei Liu

6.1
Introduction

Proteins are the basic molecular units of all the life forms known to humans. Almost all biological functions and processes involve proteins, such as the enzymes catalyzing metabolic transformations and the antibodies necessary in the body's defense systems. Many regulatory hormones and neurotransmitters are also proteins, and proteins are responsible for muscle movement and material transportation in the body. Finally, proteins are widely employed in structural roles and energy storage. The understanding of the life phenomenon demands a precise knowledge of the structure–activity relationships of protein molecules. By studying and engineering proteins, we can also develop tools and technologies to solve the problems of human health and wellbeing.

It is important to emphasize that proteins are just organic molecules, without any mysterious "vital force." A typical protein molecule contains about 300 amino acid residues with a molecular weight of about 30 000 Da. Because each amino acid residue can have 20 different choices, the potential numbers of protein isomers are almost beyond calculation. To increase further the diversity of the proteins found in Nature, a variety of post-translational modifications can take place at specific amino acids after they have been incorporated into a protein's polypeptide chain. It has been estimated that a single cell can express more than 100 000 different proteins. The study of all these protein molecules still presents an enormous challenge in today's research.

The generation of a sufficient quantity of a target protein with a certain purity is the first step in most studies of proteins. This task can be achieved with three approaches. The first is extraction and isolation of target proteins from natural resources such as animals, plants, and microbes. This method is limited by the low abundance of the target protein in the natural resources. Additionally, only the proteins with natural sequences can be obtained through this approach.

The second approach involves recombinant DNA techniques or genetic engineering. This method incorporates exogenous genes that are generated through *in vitro* recombination into easily cultivatable cells (e.g., bacteria, yeast, or even mammalian cells). The replication, transcription, translation, and expression of the exogenous gene will produce a large quantity of the target protein. An important advantage of the recombinant DNA

Organic Chemistry – Breakthroughs and Perspectives, First Edition. Edited by Kuiling Ding and Li-Xin Dai.
© 2012 Wiley-VCH Verlag GmbH & Co. KGaA. Published 2012 by Wiley-VCH Verlag GmbH & Co. KGaA.

technique is that the target proteins can be generated at relatively low cost. Further, the recombinant DNA technique enables one to mutate the amino acids in the target protein by changing the corresponding precursor nucleic acid sequence. The changes in the properties (e.g., folding, binding, and catalytic activity) of the mutant proteins are analyzed on the basis of the amino acid sequence. By using this method, we can even develop methods (i.e., molecular evolution in principle) to "optimize" the properties and function of a target protein. The recombinant DNA technique is the core method in current molecular biology for the generation of proteins. However, this method has an important limitation, namely that only the 20 genetically encoded amino acids can be routinely incorporated into the target protein.

The third approach is that of total chemical synthesis of proteins by using the principles of synthetic organic chemistry. In comparison with the previous two biological methods, a major advantage of the chemical approach is that total synthesis promises unlimited variation of the covalent structure of a protein. In other words, chemistry holds promise for tuning the properties of a protein molecule at atomic resolution in a general fashion. It permits a level of control of protein composition beyond that attainable by protein expression and it can provide otherwise difficult to access insights into a protein's structure and function. Up to now, protein chemical synthesis has been intensively explored in the study of chemical biology and its successful applications have demonstrated the importance of modern synthetic organic chemistry to cutting-edge research in biomedicine. In this chapter, we briefly summarize the history of the development, the current status, and applications of the chemical synthesis of proteins.

6.2
Brief History

6.2.1
The Beginning

The synthesis of peptides can be dated back to as early as 1882, when Theodor Curtius [1] made a dipeptide, Bz-Gly-Gly, by mixing benzoyl chloride and the silver salt of glycine. Nonetheless, Emil Fischer [2] was generally deemed to be the first scientist who proposed that the total chemical synthesis of proteins will represent one of the "grand challenges" in organic chemistry. In 1905, Fischer wrote, "My entire yearning is directed toward the first synthetic enzyme. If its preparation falls into my lap with the synthesis of a natural protein material, I will consider my mission fulfilled." In 1907 [3], Fischer synthesized an octadecapeptide consisting of 15 glycine and three leucine residues, which constituted a milestone achievement in the history of synthetic chemistry.

6.2.2
Synthesis of Small Peptides

An important advance in peptide synthesis after Fischer's era was made by Bergmann and Zervas [4]. In early 1930s, they introduced a reversible protecting group for the α-amino

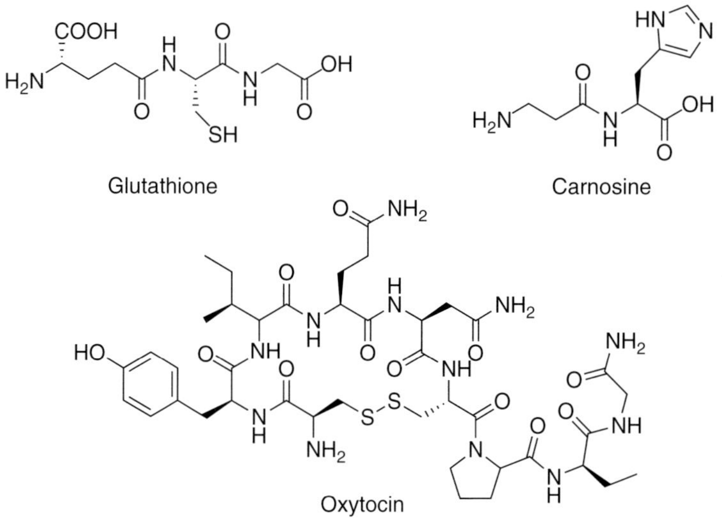

Figure 6.1 Synthesis of small peptides.

groups [i.e., carbobenzoxy (Cbz)]. This breakthrough not only solved the problem of sequence control in peptide synthesis, but also prevented racemization of the amino acid in the amide coupling. With the emergence of protecting group strategies and also the discovery of carbodiimide coupling methods [5], the solution-phase synthesis of small peptides became feasible by using the strategies that had been used by natural product chemists. A number of small peptide hormones such as glutathione [6] and carnosine [7] were successfully synthesized before and after the Second World War. A culminating achievement was made in 1954 by du Vigneaud *et al.*, who developed the synthesis of an octapeptide hormone, oxytocin (Figure 6.1) [8].

6.2.3
Total Synthesis of Insulin

The success of small peptide synthesis encouraged scientists to study the total synthesis of functional proteins with full biology activities. Unlike the total synthesis of small peptide hormones, an important yet unsolved scientific question at that time was whether or not the synthetic proteins would form correctly folded structures and, therefore, exhibit full biological activities. As a landmark study in the development of protein chemical synthesis, Chinese scientists accomplished the first total synthesis of crystalline bovine insulin (with 51 amino acid residues and three disulfide bonds) on 17 September 1965 [9]. This synthesis was started in 1958 and completed by scientists at the Biochemistry and Organic Chemistry Institutes of the Chinese Academy of Sciences and Peking University. Through 6 years of arduous work, they successfully prepared the A and B chains through the fragment condensation approach. They also discovered a method for combining the A and B chains to produce the crystalline insulin. The synthetic insulin was determined to have the same biological activity as the natural protein, showing that proteins are synthetically accessible molecules without any inexplicable "vital force." Hence the total synthesis of insulin not only was a milestone achievement in the history of organic chemistry, but also represented a pioneering exploration in the field of synthetic biology.

6.2.4
Solid-Phase Peptide Synthesis

The above solution-phase approaches to peptide and protein chemical synthesis all belong to classical organic chemistry. They suffer from a number of shortcomings, including the racemization of amino acids in the amide coupling steps, the low efficiency of the isolation and purification of synthetic intermediates, and the poor solubility and handling properties of the peptide fragments. To solve these problems, in 1963 Merrifield devised an ingenious technique that profoundly changed the field, that is, solid-phase peptide synthesis (SPPS) [10]. In SPPS, the C-terminal amino acid of the target peptide is covalently attached to an insoluble resin (Figure 6.2). Subsequent amino acid residues are introduced by removal of the N(α)-protecting group (usually Boc or Fmoc) [11] of the attached residue, washing the resin-bound amino acid, and addition of the next amino acid in N(α)- and side chain-protected, carboxyl-activated form. After the formation of the desired peptide bond, the excess reagents and soluble byproducts are removed by filtration and washing. These steps are repeated until the full peptide chain is assembled on the resin. In the last step, all the protecting groups are removed by using HF or trifluoroacetic acid (TFA). In this process, the covalent link to the resin is also cleaved to release the crude peptide that is subjected to various purification methods.

In comparison with the solution-phase methods, SPPS has two major advantages. First, the simplicity of removing soluble reagents and byproducts allows the use of large excesses of activated amino acid for each amide coupling step. As a result, the couplings are rapid and nearly quantitative. Second, the synthetic intermediate is always bound to the resin. Therefore, the handling losses are minimized and the synthesis is much less labor intensive. For these reasons, SPPS has allowed the routine chemical synthesis of peptides containing up to 30 amino acids or more. This accomplishment fundamentally changed the history of peptide and protein chemistry. Nonetheless, due to the statistical accumulation of resin-bound byproducts, SPPS is still limited to the synthesis of about 40–50 amino acids even by using the most carefully optimized methods. Therefore, most proteins cannot be directly synthesized by the stepwise assembly of amino acids through SPPS.

6.2.5
Chemical Ligation

To synthesize larger proteins with more than 50 residues, it is necessary to carry out convergent condensation of peptide segments because this method can avoid the accumulation of resin-bound impurities. Earlier condensation methods [12] through amide formation require the use of protected peptides, because these amidation methods are not chemoselective and, therefore, may cause reactions at peptide side chains (e.g., Lys side chain NH_2 group). Unfortunately, the use of protected peptides brings many

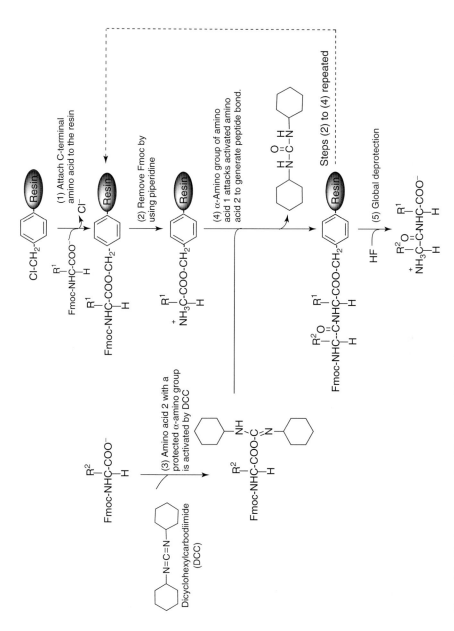

Figure 6.2 The key process of solid-phase peptide synthesis.

operational problems because protected peptides (particularly when the peptides have more than 20 amino acid residues) are not readily soluble in any solvent. Hence it is difficult to accomplish the condensation of protected peptides in a homogeneous solution. In addition, it is difficult to purify and characterize the intermediates and products (which are also protected peptides) of the condensation reactions. On the other hand, most unprotected peptides are readily soluble in both water–acetonitrile–0.1% TFA and neutral aqueous 6 M guanidine·HCl solutions [13]. Hence unprotected peptides are easier to handle in the solution and they can be readily purified by several modern techniques such as preparative reversed-phase high-performance liquid chromatography (HPLC).

For an unambiguous covalent condensation of unprotected peptides, it is necessary to use a chemoselective condensation reaction. In such a reaction, two mutually reactive functional groups are employed that react with each other, but not with any other functional group in the reaction system, thus giving a single product. Some earlier chemoselective condensation reactions included the nucleophilic reaction of a thiocarboxylate with a bromoacetyl group [14] and oxime-forming ligation between an aminooxyacetyl group and a glyoxylyl functionality [15]. Unfortunately, these methods do not produce an amide bond and, therefore, introduce non-native structures. To overcome this problem, Kent and co-workers [16] developed the native chemical ligation (NCL) method. This ligation involves a chemoselective reaction between a C-terminal peptide thioester and an N-terminal Cys to generate a peptide bond at the ligation site. It is significant that NCL permits the efficient covalent coupling of even very large peptide fragments. Also, this method surmounts the solubility and racemization problems encountered in earlier condensation methods with protected peptides.

Since then, the NCL method has become the most robust and practical ligation method for covalently joining two unprotected peptides. So far, NCL has been used to synthesize hundreds of proteins ranging in size up to more than 300 amino acids [17], and to synthesize protein analogs with masses of up to 50 825 Da [18]. This method has greatly facilitated general synthetic access to various proteins. The success of protein chemical synthesis also confirms that there is no inherent barrier to the coupling of two large polypeptides. It is important to point out that in addition to NCL, there are several other recent methods for the condensation of peptides. The earlier Ag^+-mediated thioester ligation method developed by Hojo and Aimoto [19] still requires the protection of side chain amino groups. On the other hand, the traceless Staudinger ligation developed by Raines [20] and Saxon and Bertozzi [21] and the decarboxylative amide ligation method developed by Bode *et al.* [22] are fully chemoselective in nature. These methods show some utilities that are complementary to NCL, but they also have various limitations.

6.2.6
Expressed Protein Ligation

Chemical ligation of synthetic peptides provides a practical method for preparing relatively small proteins. For larger protein targets it is often desirable to produce one of the two coupling partners through biological expression. The use of recombinant peptides in protein chemical synthesis is generally termed expressed protein ligation

[23]. The importance of expressed protein ligation is that very large ligation partners can be generated through recombinant methods. This approach greatly increases the size and complexity of the synthetic target. It has revolutionized our ability to generate proteins that contain unnatural amino acid moieties.

Production of the C-terminal coupling partner is relatively straightforward. It often involves introduction of an N-terminal protease [such as factor Xa [24] and tobacco etch virus (TEV) protease [25] recognition sequence followed by a Cys residue into a peptide fragment which is recombinantly expressed. After cleavage of the protease recognition sequence, an N-terminal Cys residue is unveiled which can undergo NCL with a peptide thioester.

On the other hand, production of an N-terminal coupling partner (i.e., peptide thioester) for use in NCL is more difficult. At present, this task is usually achieved by using the intein-based thioester technology developed by Muir *et al.* [23]. An intein is an internal protein domain that excises itself out of a precursor protein in an autocatalytic reaction called *protein splicing*. This autocatalytic reaction proceeds through a sequence of rearrangement reactions that involve the formation of a thioester intermediate. The mutation of a key amino acid in the intein can stop the rearrangement at the thioester stage. Incubation of this modified intein with thiols (e.g., MESNa) releases the corresponding C-terminal peptide thioester through transthioesterification. Subsequent ligation of the peptide thioester leads to the formation of a semi-synthetic protein.

6.3
Current Technology

6.3.1
General Protocol

The standard protocol and basic mechanism of NCL are illustrated in Figure 6.3. The first step in the transformation is thiol–thioester exchange, which is reversible. A highly

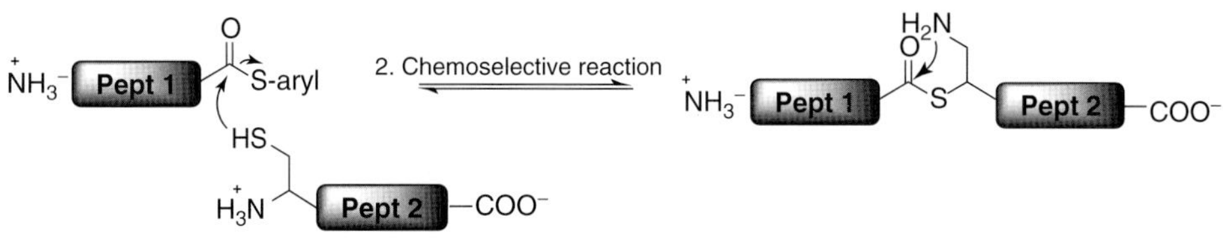

Figure 6.3 Standard native chemical ligation.

reactive peptide thiophenyl ester is produced by transthioesterfication of an alkyl thioester with an arylthiol. No accumulation of the exchange product has been observed, indicating that the thiol–thioester exchange is the rate-limiting step in the overall reaction. The second step generates a thioester-linked intermediate through exchange of the peptide thiophenyl ester with the thiol moiety of the Cys-peptide. Finally, S-to-N acyl transfer occurs irreversibly, leading to the formation of the desired product with a native peptide bond at the ligation site.

In order to attain a high reaction efficiency, it is critical to conduct the ligation under optimized conditions. First, sterically hindered amino acids (such as Thr, Val, and Ile) are usually not used as the ligation sites. When the ligation site is Asp-Cys or Glu-Cys, protection of the side-chain carboxyl group is necessary because otherwise intramolecular cyclization may occur between the thioester and C-terminal side chain. Second, the ligation rate is dependent on the nature and concentration of the exogenous thiol additive. 4-Mercaptophenylacetic acid (MPAA) has been shown to be a highly efficient promoter at concentrations of 50–250 mM. Third, the optimal pH value for NCL is 6.2–7.0. Lower pH values dramatically slow the ligation, and higher pH values cause the loss of chemoselectivity and the hydrolysis of the thioester (in the presence of divalent metal ions). Finally, the use of a chaotrope such as 6 M Gn·HCl is favorable in NCL to improve peptide solubility (typically >1 mM). However, for the synthesis of membrane proteins, standard ligation conditions must be adjusted by adding detergents such as DDM (*n*-dodecyl β-D-maltoside) or SDS (sodium dodecyl sulfate).

6.3.2
Thioester Synthesis

Peptide thioesters are important substrates in protein chemical synthesis. The preparation of peptide thioesters is straightforward by Boc SPPS using *in situ* neutralization protocols. Essentially all peptide thioesters for use in protein chemical synthesis are still made in this way [26]. However, the use of toxic HF in Boc SPPS often causes problems due to safety and/or regulation concerns. In addition, repeated treatment with strongly acidic TFA in Boc SPPS limits the functional group tolerance. Glycosylated and phosphorylated peptide thioesters are difficult to synthesize by using Boc SPPS [27]. Therefore, it is important to develop methods for the synthesis of peptide thioesters through Fmoc SPPS. Note that peptide thioesters are sensitive to piperidine, which is used for repeated Fmoc deprotection [28]. The development of methods to solve this problem has been intensively studied in the last few years.

Generally, there are four types of methods available for the Fmoc SPPS of peptide thioesters. The first is to use non-nucleophilic bases (e.g., 1-methylpyrrolidine) to replace piperidine in the Fmoc deprotection step [29]. Even under these conditions, aminolysis of the thioester linkage is still observed, particularly in the first two cycles of peptide couplings. The use of a tertiary alkanethiol may partly solve the problem. Nonetheless, the approach of using non-nucleophilic bases has not been widely used by researchers in the field.

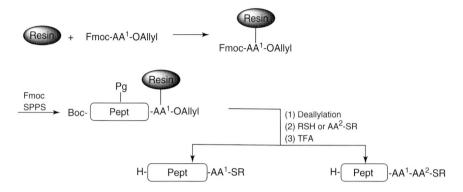

Figure 6.4 Conversion of peptide acid to peptide thioester.

The second method is to convert the C-terminal acid of the protected peptide to thioester at a late stage of SPPS. This method has two different versions. In the first, a fully protected peptide acid is cleaved from the resin by using highly acid-sensitive resins such as chlorotrityl resin [30]. Then through coupling with an external thiol in an organic solvent, the peptide acid is converted to peptide thioester with all the side-chain groups protected. Finally, the side-chain groups are deprotected to yield the desired peptide thioester. In the second version, the peptide is anchored to the resin through a side-chain group (e.g., the OH group in Ser) or its amide backbone, whereas its C-terminus is protected by the allyl group (Figure 6.4) [31]. After the peptide chain assembly is completed on the resin, the C-terminal allyl ester group is deprotected and then coupled with a thiol (or an amino acid thioester) to form the C-terminal thioester on the resin. Finally, the peptide is cleaved from the resin to give the desired peptide thioester. A limitation of the above approaches is that racemization may occur during the thioester formation step.

The third method is to activate the peptide C-terminus after the chain assembly, which is followed by thioesterification of the activated C-terminus with an external thiol. To date, the most popular approach in this category is the "safety catch" method with a sulfonamide linker (Figure 6.5) [32]. In this method, the peptidyl–sulfonamide bond

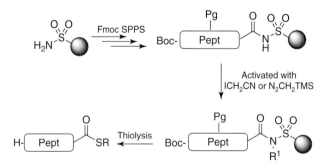

Figure 6.5 Peptide C-terminus activation followed by thiolysis.

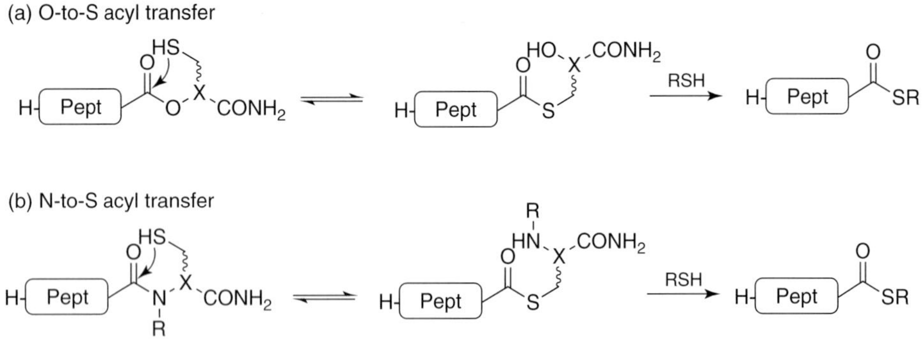

(a) O-to-S acyl transfer

(b) N-to-S acyl transfer

Figure 6.6 Peptide thioester synthesis through intramolecular acyl transfer.

is alkylated with either iodoacetonitrile or trimethylsilyldiazomethane after the peptide chain assembly. This alkylation step activates the *N*-acylsulfonamide, so that it will react with an external thiol through nucleophilic thiolysis. A fully protected peptide thioester is thus produced, which is then treated with TFA for global deprotection to yield the desired peptide thioester. A major problem with this method is that the activation step may cause side reactions with other functional groups in the peptide. To improve the method, Blanco-Canosa and Dawson [33] developed a 3,4-diaminobenzoyl linker and used it to generate peptide α-arylbenzimidazolones as precursors for peptide thioesters.

The fourth method is to convert a peptide ester or amide to the thioester by taking advantage of the O-to-S [34] or N-to-S [35] acyl transfer reaction (Figure 6.6). In this method, the peptide is anchored to the resin through an ester or amide bond. A protected thiol group is incorporated into the C-terminus of the peptide, which after deprotection can undergo the O-to-S or N-to-S acyl transfer through a five- or six-membered ring intermediate. This acyl transfer reaction can occur in the presence of an external thiol, which will lead to the formation of a thioester through an intermolecular thioester exchange reaction. Alternatively, the acyl transfer reaction can occur *in situ* in the NCL, which will lead directly to the formation of the ligation product. The acyl transfer method appears to be the most promising for solving the thioester synthesis problem. The recent development of bis(β-mercaptoethyl)amides [36, 37] is a significant advance as it nicely circumvents the barrier to rearrangement present in the more usual secondary amides with their *trans* configuration. More recently, our group developed an operationally simple method for the synthesis of peptide thioesters using standard Fmoc SPPS procedures [38]. The method also relies on N-to-S acyl transfer. However, it uses a pre-made enamide-containing amino acid, which, in the final TFA cleavage step, renders the desired peptide thioester through an irreversible intramolecular N-to-S acyl transfer.

Finally, we recently developed ligation of peptide hydrazides that is complementary to the NCL (Figure 6.7) [39]. An important advantage of this new ligation method is that peptide hydrazides can be readily synthesized through both Boc and Fmoc SPPS. In addition, peptide hydrazides can be easily generated through recombinant expression.

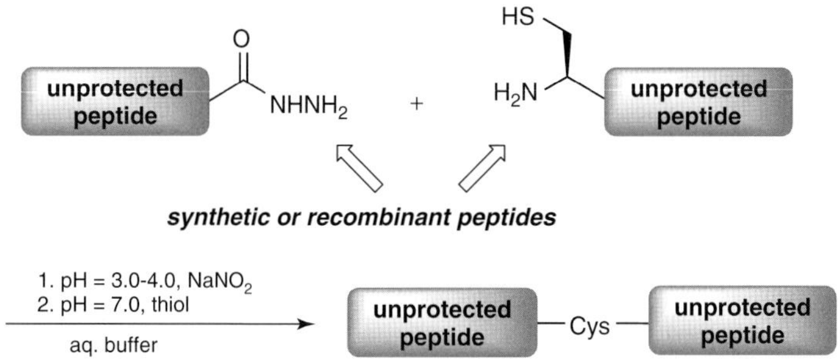

Figure 6.7 Ligation of peptide hydrazides.

Both total and semi-synthesis of proteins can be achieved with the ligation of peptide hydrazides. The ligation of peptide hydrazides also permits sequential ligation in the N-to-C direction. In essence, the ligation of peptide hydrazides is a modified version of NCL with *in situ* generation of a peptide thioester from a peptide hydrazide. It is expected that the combination of NCL and the ligation of peptide hydrazides may allow more convergent synthesis of proteins in the future.

6.3.3
Overcome the Cys Limitation

NCL is chemoselective to Cys. However, the requirement of Cys in NCL may bring about problems because Cys is the least common amino acid found in proteins [40]. An often encountered difficulty is that in the design of protein chemical synthesis, the Cys residues are not located close to the desirable ligation sites. To overcome the Cys limitation, two categories of methods have been developed. The first is to attach an auxiliary group to the terminal α-amino group of the C-terminal fragment (Figure 6.8) [41]. This auxiliary group possesses a cleavable thiol unit to imitate the Cys structure and it can be selectively removed from the newly formed amide bond after the ligation reaction. Unfortunately, it has been found that the use of thiol-containing auxiliaries is highly sensitive to the nature of the amino acids at the ligation site. Hence with almost all the auxiliaries, it is necessary to select the coupling site with participation of at least one Gly residue. Therefore, the thiol auxiliary method changes the Cys requirement to the Gly requirement.

An alternative approach to overcome the Cys limitation is to use a β-mercaptoamino acid at the N-terminus of the C-terminal fragment. The β-mercapto group can be selectively desulfurized after the ligation reaction by using reductants such as Raney nickel [42]. More recently, Danishefsky and co-workers developed a more efficient desulfurization method by using trialkyl phosphites. In comparison with the thiol auxiliary method, the "ligation–desulfurization" approach shows much less sensitivity to the steric hindrance at the ligation site (Figure 6.9). So far, the β-mercaptoamino acids corresponding to many amino acids, including Ala [43], Phe [44], Val [45, 46], Thr [47], and Leu [48], have been successfully used in the ligation. This method greatly extends the utility of NCL

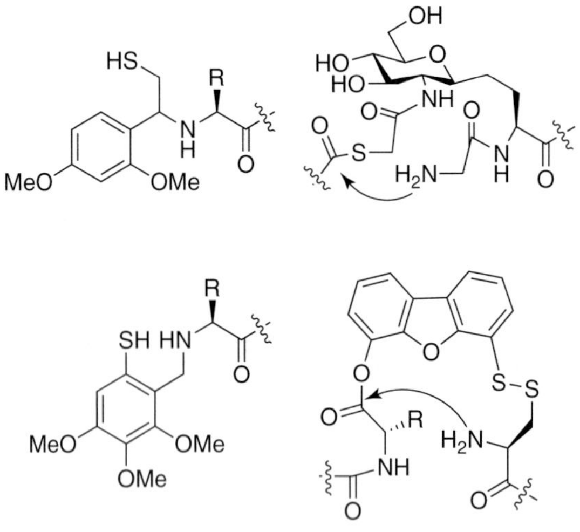

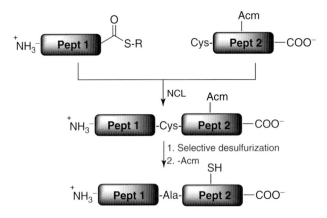

Figure 6.8 Thiol auxiliaries used for native chemical ligation.

Figure 6.9 The ligation–desulfurization method.

by allowing ligation at almost any amino acid–amino acid junction. Nonetheless, to protect nonparticipating Cys residues from desulfurization, they must be protected with an acetamidomethyl (Acm) group [49].

6.3.4
Multiple Fragment Condensation

Considering that typical proteins found in Nature consist of about 300 amino acids, efficient methods remained to be developed to condense multiple synthetic peptide fragments [50]. Two issues are important for the design of a synthetic strategy for relatively large proteins. First, an efficient synthesis should avoid purification of the synthetic intermediate as far as possible, because the purification steps are tedious and can lead to significant handling losses. Second, it is preferable to use a convergent approach to assemble the target proteins. Sequential ligation of multiple peptide fragments can cause the accumulation of byproducts. Also, as the peptide chain length increases, the desired product and reactant become difficult to separate.

At present there are several basic strategies available for the condensation of multiple peptide fragments [46]. By combining these strategies, one can design various versions of convergent peptide condensation methods. These basic strategies are briefly described below.

1) **One-pot sequential ligation [51].** Sequential ligation of multiple peptide fragments in one reaction mixture is expected to reduce the handling losses. To achieve this goal, an easily removable protecting group for the N-terminal Cys is needed. Currently, the one-pot sequential ligation usually employs the Thz (1,3-thiazolidine-4-carboxo) protecting group, which can be readily removed by using methoxyamine. After the first ligation reaction between the Thz-peptide thioester and Cys-peptide, the Thz group is removed by adding methoxyamine at pH 4 (Figure 6.10). This process enables further ligation steps to be carried out by changing the pH value back to 7. This strategy has been used in the total synthesis of several proteins, including crambin and ubiquitin. A major shortcoming of the method is the accumulation of unreacted peptide segments and byproducts during the process.

2) **His tag-assisted sequential ligation [52].** To remove the unreacted peptide fragments and byproducts rapidly, a His tag-assisted method has been developed in which a His_6 tag is attached to the C-terminal peptide. The use of excess peptide thioester drives the ligation reaction to near completion and improves the ligation yield. After the ligation is finished, the unreacted peptide thioester can be easily removed from

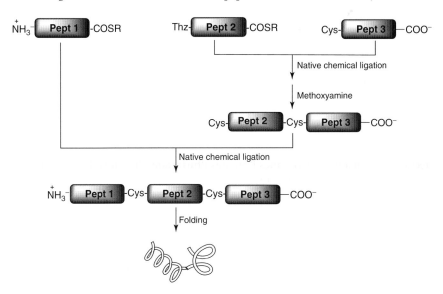

Figure 6.10 One-pot condensation of multiple fragments.

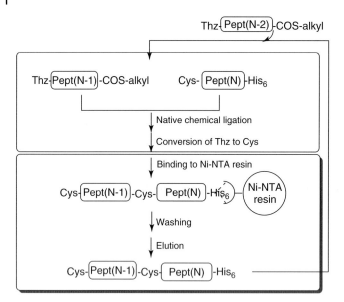

Figure 6.11 His$_6$ tag-assisted ligation.

the ligation mixture by washing through an Ni-NTA (nitrilotriacetic acid) agarose column (Figure 6.11). The key advantage of the His tag-assisted sequential ligation is that all the synthetic intermediates can be readily purified. This method has been used to synthesize several proteins, including crambin and a modular tetratricopeptide repeat protein.

3) **Solid-phase sequential ligation [53].** To reduce the handling losses further, the principles of the SPPS can be adopted for peptide ligation. Thus, a resin is used to attach the peptide fragment that is subject to sequential ligation reactions. Excess ligation partners are used to drive each of the ligations to completion, while the unreacted reactants and byproducts are readily removed by filtration and washing. When all the ligation steps are finished, the final product is cleaved from the resin (Figure 6.12). This method has been used to synthesize several proteins, including human group V phospholipase A2 and EETI-II [54]. A problem associated with the method is that little is known about what type of resin is generally applicable.

4) **Kinetically controlled ligation [55].** The above methods mostly involve the sequential ligation in the C-to-N direction. For the N-to-C direction, kinetically controlled ligation can be applied. This method was developed on the basis of the observation that a peptide thioarylester reacts with a Cys-peptide much more rapidly than a thioalkylester. Hence it is possible to ligate selectively a peptide thioarylester and a Cys-peptide thioalkylester without affecting the thioalkylester group (Figure 6.13). The kinetically controlled ligation method has been successfully used in the total synthesis of several proteins, including human lysozyme [56] and a covalent dimer of HIV-1 protease [57]. A limitation of the method is that only three fragments can be condensed by this approach.

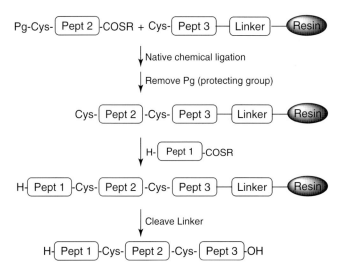

Figure 6.12 Solid-phase peptide ligation.

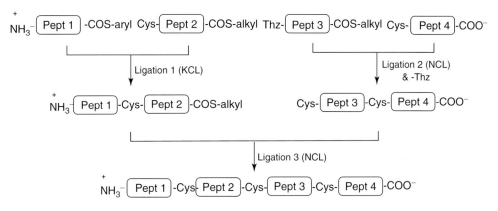

Figure 6.13 Kinetically controlled ligation.

6.3.5
An Illustrative Example

As an important illustrative example, the total synthesis of a 203 amino acid covalent dimer of HIV-1 protease enzyme by Torbeev and Kent [57] shows the power of protein chemical synthesis. This target molecule was divided into four segments, (A1–A40)-(thioarylester), Cys-(A42–A99)-(thioalkylester), Thz-Gly4-(B1–B40)-(thioalkylester), and Cys-(B42–B99). These four segments were first convergently condensed into two fragments, (A1–CysA41–A99)-(thioalkylester) and Cys-Gly4-(B1–B99). Then, the ligation between (A1–CysA41–A99)-(thioalkylester) and Cys-Gly4-(B1–B99) formed a final polypeptide chain consisting of 203 amino acids. Note that kinetically controlled ligation

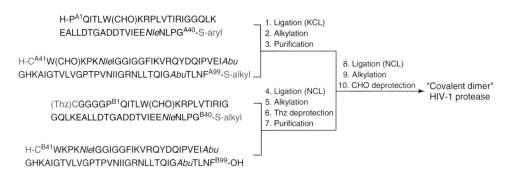

Figure 6.14 Total synthesis of HIV protease dimer.

was employed to ligate (A1–A40)-(thioarylester) with Cys-(A42–A99)-(thioalkylester), producing (A1–CysA41–A99)-(thioalkylester). On the other hand, the ligation between Thz-Gly4-(B1–B40)-(thioalkylester) and Cys-(B42–B99) produces Thz-Gly4-(B1–B99), which is converted to Cys-Gly4-(B1–B99) upon treatment with MeONH$_2$ (Figure 6.14). Moreover, the Cys residue at the ligation site was converted into ψ-Gln by treatment with 2-bromoacetamide.

The synthetic peptide containing 203 amino acid residues was characterized by liquid chromatography–mass spectrometry and Fourier transform ion cyclotron resonance mass spectrometry. The overall yield of the isolated product was 6.7% based on the limiting peptide segment. The synthetic peptide was then folded by dialysis against acetate buffer at pH 5.6 to give the HIV-1 protease enzyme with full biological activity with the expected k_{cat} and K_M values. Furthermore, through X-ray structural analysis, it was shown that the synthetic protein had the correct three-dimensional fold of the HIV-1 protease covalent dimer enzyme.

6.4
Applications

6.4.1
Biophysics and Structural Biology

Protein biophysics and structural biology are important fields, where recombinant proteins are subjected to various physical characterization methods [58]. The use of synthetic proteins may introduce predesigned biophysical probes into the protein targets with atom-to-atom precision, facilitating investigations on protein folding, protein structure, protein dynamics, and the interaction of proteins with other molecules. Here we use two recent examples to highlight the application of protein chemical synthesis to solving problems in biophysics and structural biology.

The first example is the study of the dynamics of "flap" structures in HIV-1 protease–inhibitor complexes by using pulse electron paramagnetic resonance spectroscopy [59]. For this purpose, both active HIV-1 protease and its inactive [D25N] analog were prepared in nitroxide spin-labeled form by total chemical synthesis. Solutions of the spin-labeled enzymes complexed with inhibitors for double electron electron resonance (DEER) measurements were then prepared, and inhibitors were added at 30–100-fold molar excess to ensure saturation of the enzymes. Through the analysis of the DEER signals, it was found that at the different stages of the catalytic reaction the flaps may adopt different dynamic properties. A dominant "closed/semi-open" conformer was proposed on the basis of the experimental data, where a mechanistic role of flap–substrate interactions was to stabilize a transition state via the formation of hydrogen bonds.

The second example is racemic protein crystallography, a technique that is made possible only by using chemical protein synthesis [60]. This technique is based on an earlier theoretical prediction that for globular proteins, a racemic protein mixture will crystallize more readily. It can be used to solve the difficulty that some proteins may be difficult to crystallize or may form crystals unsuitable for the collection of single-crystal X-ray diffraction data. The application of racemic protein crystallography has been shown for the snow flea antifreeze protein (sfAFP), which is an 81 amino acid protein that can be synthesized in both L- and D-forms. It was found that crystal formation occurred much more readily from a racemic mixture of D- and L-sfAFP than with the L-protein alone. This approach was used to solve the structure of L-sfAFP to a resolution of 0.98 Å. Hence the chemical synthesis of mirror image proteins for racemic crystallization is going to be a powerful tool in structural biology.

6.4.2
Post-Translational Modifications

Post-translational modification is one of the later steps in protein biosynthesis. It extends the range of functions of the protein by attaching to it other functional groups such as acetate, phosphate, and various lipids and carbohydrates. Post-translational modification influences the structure, stability, function, interacting partners, and localization of the protein within the cell. The study of protein post-translational modification is one of the most active areas in protein research. Unlike the proteins composed of 20 native amino acids, post-translationally modified proteins are usually difficult to obtain through purely biological methods. Therefore, total and semi-protein synthesis has played an important role in the study of protein post-translational modification. Here we use protein glycosylation and lipidation as examples to describe the applications in this direction.

First, protein glycosylation is one of the most abundant post-translational modifications in mammalian cells [61]. It plays an important role in various cellular processes such as cell–cell recognition, adhesion, and signaling. Aberrant glycosylation of proteins is also involved in a number of diseases, including autoimmune diseases, infectious diseases, and cancer. For these reasons, glycoproteins have potential utility for the development of therapeutics, diagnostics, and vaccines. At present, the study of glycoproteins is often limited by the difficulty of obtaining homogeneous

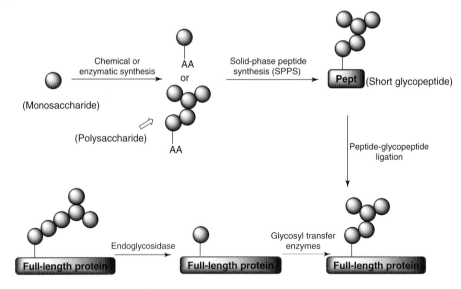

Figure 6.15 Glycoprotein synthesis.

glycoproteins, because glycoproteins are naturally expressed as complexed mixtures of glycoforms that differ in both the nature and site of glycosylation. Protein chemical synthesis provides a practical way to solve the accessibility problem in the study of protein glycosylation. Chemical synthesis may also allow the introduction of unnatural amino acids or carbohydrates for structure–activity relationship studies and therapeutic development.

A special feature of the chemical synthesis of glycoproteins is the preparation of glycosylated peptide fragments (Figure 6.15). In principle, there are three strategies for the synthesis of a glycopeptide with oligosaccharide side chains: (i) the use of glycosylamino acids with monosaccharide side chains in the SPPS to assemble the glycopeptide, followed by enzymatic or chemical glycosylation reactions to expand the glycan further; (ii) the use of glycosylamino acids with full oligosaccharide side chains to assemble the glycopeptide; and (iii) the conjugation of an oligosaccharide side chain with a peptide to generate the glycopeptide. It is important to note that O-glycosidic and inter-saccharidic linkages are relatively labile towards acid. Therefore, the Fmoc SPPS is usually favored in the synthesis of glycopeptides.

Once the glycopeptides have been obtained, the target glycoprotein can be synthesized by using chemoselective ligation reactions. As an illustrative example, Bertozzi and co-workers [62] accomplished the total synthesis of diptericin, which is an 82-residue antimicrobial glycoprotein from insects. The whole glycoprotein was split into two fragments, that is, a 58-mer Cys-glycopeptide and a 24-mer glycopeptide thioester. The former was synthesized by using a standard SPPS protocol. For the glycopeptide thioester, Fmoc-based SPPS was performed on a sulfonamide safety-catch resin through post-chain assembly activation and nucleophilic thiolysis. Finally, the ligation of the

24-mer thioester and the 58-mer Cys-glycopeptide was accomplished to afford the 82-mer glycoprotein.

In another example, Wong and co-workers [63] accomplished the semi-synthesis of a chemically defined version of a glycoprotein, human interleukin-2. This glycoprotein is a T-cell growth factor that has been used therapeutically to treat renal cell carcinoma and metastatic melanoma. Bacterial expression was carried out to produce a fusion protein His-tagged interleukin-2 (amino acids 6–133), which was purified and cleaved by the TEV protease to produce a truncated version of interleukin-2 containing an N-terminal Cys residue. This Cys-containing truncated interleukin-2 was then ligated with a synthetic glycopeptide thioester under denaturing conditions to afford the desired glycoprotein.

Second, lipid-based protein post-translation modification plays a vital role in the regulation of cellular processes such as trafficking and signaling. Common protein lipidations include myristoylation, palmitoylation, and prenylation. The strategy for the preparation of semi-synthetic proteins via expressed protein ligation has proven to be a powerful method to obtain access to post-translationally lipid-modified proteins in sufficient quantities for biological studies. A key difficulty in the synthesis of lipidated protein is the preparation of lipidated peptides because the lipid chains are sensitive to strong acids. Furthermore, lipidated peptides are poorly soluble in aqueous media, which causes some handling problems in the ligation step. Despite these difficulties, Waldmann's group carried out elegant studies on protein lipidation. They studied Rab/Ypt guanosine triphosphatases (GTPases) that represent a family of key membrane traffic regulators in eukaryotic cells [64]. These lipidated proteins require double modification with two covalently linked geranylgeranyl lipid moieties at their C-terminus. Because prenylated proteins are difficult to obtain by recombinant or enzymatic methods, Waldmann and co-workers used chemical methods to prepare prenylated Rab GTPases. Solution- and solid-phase strategies were developed for the preparation of peptides corresponding to the prenylated C-terminus of Rab7 GTPase. The functionalized Rab7 C-terminal peptides were then ligated with the truncated Rab7/Ypt1 thioester. The semi-synthetic prenylated Rab proteins were used to crystallize the monoprenylated Ypt1:RabGDI complex. The structure of this complex provides a structural basis for the ability of RabGDI to inhibit the release of nucleotide by Rab proteins.

6.4.3
Protein Probes

Protein chemical synthesis, in particular through the use of expressed protein ligation, is an exciting technology that allows the design and generation of protein probes with fluorophores, isotopes, and photo-cages. The protein-based bioprobes provide a useful strategy to investigate, perturb, or sample biological phenomena with the requisite control to address many of the organizational complexities with spatial and temporal resolution. So far just a few caged variants of proteins have been synthesized.

As an exampled, Hahn and Muir [65] studied the photocontrol of Smad2, which is a multiphosphorylated cell-signaling protein. A caged version of Smad2 was constructed by attaching a doubly caged phosphopeptide to the C-terminal end of the protein. Because a free C-terminus is required for Smad2 activity, the presence of a photocleavable moiety at this position renders the synthetic protein light activatable. In addition, a fluorophore and a fluorescence quencher were positioned on the opposite sides of the photocleavable moiety. As a result, activation of the protein is intimately linked with a dramatic intensification in fluorescence, thereby providing an immediate visual confirmation of photolysis. It was found that caged Smad2 was excluded from the nucleus, whereas uncaging of the protein with the UV light led to dramatic nuclear accumulation. Thus the use of photocaged Smad2 can yield quantitative insight into the kinetics of Smad2 nuclear import and export. These results indicate that protein chemical synthesis can be used for molecular photocontrol over protein function with live cell-imaging techniques.

6.4.4
Protein Biopharmaceuticals

Therapeutic proteins are proteins for pharmaceutical use and include monoclonal antibodies, interferons, insulin, erythropoietins, and blood clotting factors [66]. They may replace deficiencies in critical blood-borne growth factors or strengthen the immune system to fight cancer and infectious disease. They may also relieve patients suffering from many conditions, including heart attacks, strokes, cystic fibrosis and Gauchers disease, diabetes, anemia, and hemophilia. The majority of therapeutic proteins are recombinant human proteins manufactured using engineered non-human mammalian cell lines. However, recent studies have also shown the possible application of protein chemical synthesis to therapeutic proteins. The central idea of this application is that chemical synthesis or semi-synthesis can provides methods for the modification and optimization of proteins in a manner analogous to those widely used to explore and improve the medicinal properties of small molecule drugs.

Here we present one example to describe the studies in this direction, namely the study of the pharmaceutical optimization of an anti-HIV protein, CCL-5 (RANTES), carried out by Miranda *et al.* (Figure 6.16) [66]. Its central strategy involved a combination of coded and noncoded amino acid mutagenesis, peptide backbone engineering, and site-specific polymer attachment. The target protein was divided into three synthetic fragments, CCL-5(1–33)-thioester, CCL-5(33–70), and an aminooxy-functionalized polymer. CCL-5(33–70) was first ligated with the aminooxy-functionalized polymer through the formation of an oxime, followed by NCL with CCL-5(1–33)-thioester to produce the full-length protein. This synthetic protein was subjected to folding conditions and then purified by reversed-phase HPLC. Through several rounds of chemical synthesis and activity examination, two highly potent anti-HIV polymer-conjugated CCL-5 analogs were identified. These analogs also showed significantly improved CCR-5 selectivity over CCR-1 and CCR-3, better *in vivo* circulation, and a well-defined homodimeric structure. As demonstrated in this study, the ability to change specific protein residues and the control of the chemical modifications allows the exploration of specific molecular designs of therapeutic proteins. This technology may result in novel protein derivatives with significant variations in their respective biochemical and pharmaceutical properties.

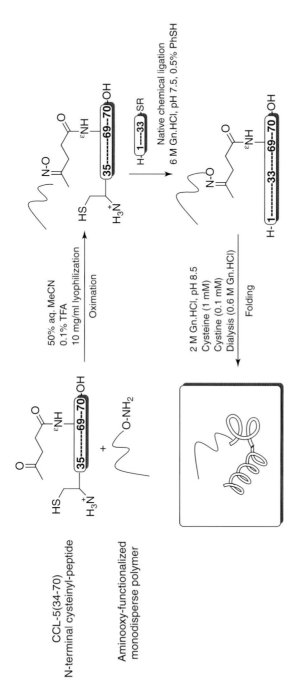

Figure 6.16 Chemical synthesis of CCL-5 derivatives.

6.5
Conclusion and Perspectives

Protein chemical synthesis has been a "grand challenge" in organic chemistry and it is also an important subject in synthetic biology. The generation of proteins with predesigned modifications and changes without the limitation of Earth-based biology is an enabling technology for elucidation of the molecular basis of protein function. At present, the efficiency of protein chemical synthesis is still relatively low. Therefore, more efficient methods need to be developed to improve the efficiency and flexibility of protein chemical synthesis. Furthermore, the application of protein chemical synthesis is expected to become more popular in the field of chemical biology and it will make important contributions to future studies in biology and medicine.

References

1. Curtius, T. (1882) *J. Prakt. Chem.*, **26**, 145.
2. Fischer, E. (1905) *Ber. Dtsch. Chem. Ges.*, **38**, 605.
3. Fischer, E. (1907) *J. Chem. Soc., Trans.*, **91**, 1749.
4. Bergmann, M. and Zervas, L. (1932) *Ber. Dtsch. Chem. Ges.*, **65**, 1192.
5. Sheehan, J.C. and Hess, G.P. (1955) *J. Am. Chem. Soc.*, **77**, 1067.
6. Harington, C.R. and Mead, T.H. (1935) *Biochem. J.*, **29**, 1602.
7. Sifferd, R.H. and du Vigneaud, V. (1935) *J. Biol. Chem.*, **108**, 753.
8. du Vigneaud, V., Ressler, B., Swan, J.M., Roberts, J.M., and Katsoyannis, C.W. (1954) *J. Am. Chem. Soc.*, **76**, 3115.
9. Kong, Y.T., Du, Y.C., Huang, W.T., Chen, C.C., and Ke, L.T. (1966) *Sci. Sin.*, **15**, 544.
10. Merrifield, R.B. (1963) *J. Am. Chem. Soc.*, **85**, 2149.
11. Wellings, D.A. and Atherton, E. (1997) *Methods Enzymol.*, **289**, 44.
12. Hirschmann, R., Nutt, R.F., Veber, D.F., Vitali, R.A., Varga, S.L., Jacob, T.A., Holly, F.W., and Denkewalter, R.G. (1969) *J. Am. Chem. Soc.*, **91**, 507.
13. Schnolzer, M. and Kent, S.B.H. (1992) *Science*, **256**, 221.
14. Tam, J.P., Lu, Y.A., Liu, C.F., and Shao, A.J. (1995) *Proc. Natl. Acad. Sci. U. S. A.*, **92**, 12485.
15. Rose, K. (1994) *J. Am. Chem. Soc.*, **116**, 30.
16. Dawson, P.E., Muir, T.W., Clark-Lewis, I., and Kent, S.B.H. (1994) *Science*, **266**, 776.
17. Kumar, K.S.A., Bavikar, S.N., Spasser, L., Moyal, T., Ohayon, S., and Brik, A. (2011) *Angew. Chem. Int. Ed.*, **50**, 6137.
18. Kochendoerfer, G.G., Chen, S.Y., Mao, F., Cressman, S., Traviglia, S., Shao, H., Hunter, C.L., Low, D.W., Cagle, E.N., Carnevali, M., Gueriguian, V., Keogh, P.J., Porter, H., Stratton, S.M., Wiedeke, M.C., Wilken, J., Tang, J., Levy, J.J., Miranda, L.P., Crnogorac, M.M., Kalbag, S., Botti, P., Schindler-Horvat, J., Savatski, L., Adamson, J.W., Kung, A., Kent, S.B., and Bradburne, J.A. (2003) *Science*, **299**, 884.
19. Hojo, H. and Aimoto, S. (1991) *Bull. Chem. Soc. Jpn.*, **64**, 111.
20. Raines, R.T. (2000) *Org. Lett.*, **2**, 1939.
21. Saxon, E. and Bertozzi, C.R. (2000) *Science*, **287**, 2007.
22. Bode, J.W., Fox, R.M., and Baucom, K.D. (2006) *Angew. Chem. Int. Ed.*, **45**, 1248.
23. Muir, T.W., Sondhi, D., and Cole, P.A. (1998) *Proc. Natl. Acad. Sci. U. S. A.*, **95**, 6705.
24. Erlanson, D.A., Chytil, M., and Verdine, G.L. (1996) *Chem. Biol.*, **3**, 981.
25. Tolbert, T. and Wong, C.-H. (2002) *Angew. Chem. Int. Ed.*, **41**, 2171.
26. Schnölzer, M., Alewood, P., Jones, A., Alewood, D., and Kent, S.B.H. (2007) *Int. J. Pept. Res. Ther.*, **13**, 31.
27. Gamblin, D.P., Scanlan, E.M., and Davis, B.G. (2009) *Chem. Rev.*, **109**, 131.
28. Bennett, C.S. and Wong, C.H. (2007) *Chem. Soc. Rev.*, **36**, 1227.

29. Li, X.Q., Kawakami, T., and Aimoto, S. (1998) *Tetrahedron Lett.*, **39**, 8669.

30. Futaki, S., Sogawa, K., Maruyama, J., Asahara, T., Niwa, M., and Hojo, H. (1997) *Tetrahedron Lett.*, **38**, 6237.

31. Nakamura, K., Hanai, N., Kanno, M., Kobayashi, A., Ohnishi, Y., Ito, Y., and Nakahara, Y. (1999) *Tetrahedron Lett.*, **40**, 515.

32. Backes, B.J., Virgilio, A.A., and Ellman, J.A. (1996) *J. Am. Chem. Soc.*, **118**, 3055.

33. Blanco-Canosa, J.B. and Dawson, P.E. (2008) *Angew. Chem. Int. Ed.*, **47**, 6851.

34. Zheng, J.-S., Cui, H.-K., Fang, G.-M., Xi, W.-X., and Liu, L. (2010) *ChemBioChem*, **11**, 511.

35. Kawakami, T., Sumida, M., Nakamura, K., Vorherr, T., and Aimoto, S. (2005) *Tetrahedron Lett.*, **46**, 8805.

36. Hou, W., Zhang, X., Li, F., and Liu, C.-F. (2011) *Org. Lett.*, **13**, 386.

37. Ollivier, N., Dheur, J., Mhidia, R., Blanpain, A., and Melnyk, O. (2010) *Org. Lett.*, **12**, 5238.

38. Zheng, J.-S., Chang, H.-N., Wang, F.-L., and Liu, L. (2011) *J. Am. Chem. Soc.*, **133**, 11080.

39. Fang, G.-M., Li, Y.-M., Shen, F., Huang, Y.-C., Li, J.-B., Lin, Y., Cui, H.-K., and Liu, L. (2011) *Angew. Chem. Int. Ed.*, **50**, 7645.

40. Nilsson, B.L., Soellner, M.B., and Raines, R.T. (2005) *Annu. Rev. Biophys. Biomol. Struct.*, **34**, 91.

41. Offer, J., Boddy, C.N.C., and Dawson, P.E. (2002) *J. Am. Chem. Soc.*, **124**, 4642.

42. Yan, L.Z. and Dawson, P.E. (2001) *J. Am. Chem. Soc.*, **123**, 526.

43. Wan, Q. and Danishefsky, S.J. (2007) *Angew. Chem. Int. Ed.*, **46**, 9248.

44. Crich, D. and Banerjee, A. (2007) *J. Am. Chem. Soc.*, **129**, 10064.

45. Chen, J., Wan, Q., Yuan, Y., Zhu, J.L., and Danishefsky, S.J. (2008) *Angew. Chem. Int. Ed.*, **47**, 8521.

46. Haase, C., Rohde, H., and Seitz, O. (2008) *Angew. Chem. Int. Ed.*, **47**, 6807.

47. Chen, J., Wang, P., Zhu, J.L., Wan, Q., and Danishefsky, S.J. (2010) *Tetrahedron Lett.*, **66**, 2277.

48. Harpaz, Z., Siman, P., Ajish, K.S., and Brik, A. (2010) *ChemBioChem*, **11**, 1232.

49. Pentelute, B.L. and Kent, S.B.H. (2007) *Org. Lett.*, **9**, 687.

50. Lee, J.Y. and Bang, D. (2010) *Biopolymers*, **94**, 441.

51. Bang, D. and Kent, S.B.H. (2004) *Angew. Chem. Int. Ed.*, **43**, 2534.

52. Bang, D., Pentelete, B.L., and Kent, S.B.H. (2005) *Proc. Natl. Acad. Sci. U. S. A.*, **102**, 5014.

53. Canne, L.E., Botti, P., Simon, R.J., Chen, Y., Dennis, E.A., and Kent, S.B.H. (1999) *J. Am. Chem. Soc.*, **121**, 8720.

54. Johnson, E.C., Durek, T., and Kent, S.B.H. (2006) *Angew. Chem. Int. Ed.*, **45**, 3283.

55. Bang, D., Pentelute, B.L., and Kent, S.B.H. (2006) *Angew. Chem. Int. Ed.*, **45**, 3985.

56. Durek, T., Torbeev, V.Y., and Kent, S.B.H. (2007) *Proc. Natl. Acad. Sci. U. S. A.*, **104**, 4846.

57. Torbeev, V.Y. and Kent, S.B.H. (2007) *Angew. Chem. Int. Ed.*, **46**, 1667.

58. Matthews, B.W. (2009) *Protein Sci.*, **18**, 1135.

59. Torbeev, V.Y., Raghuraman, H., Mandal, K., Senapati, S., Perozo, E., and Kent, S.B.H. (2009) *J. Am. Chem. Soc.*, **131**, 884.

60. Pentelute, B.L., Gates, Z.P., Dashnau, J.L., Vanderkooi, J.M., and Kent, S.B.H. (2008) *J. Am. Chem. Soc.*, **130**, 9702.

61. Liu, L., Bennett, C.S., and Wong, C.-H. (2006) *Chem. Commun.*, 21.

62. Shin, Y., Winans, K.A., Backes, B.J., Kent, S.B.H., Ellman, J.A., and Bertozzi, C.R. (1999) *J. Am. Chem. Soc.*, **121**, 11684.

63. Tolbert, T.J., Franke, D., and Wong, C.-H. (2005) *Bioorg. Med. Chem.*, **13**, 909.

64. Chen, Y.X., Koch, S., Uhlenbrock, K., Weise, K., Das, D., Gremer, L., Brunsveld, L., Wittinghofer, A., Winter, R., Triola, G., and Waldmann, H. (2010) *Angew. Chem. Int. Ed.*, **49**, 6090.

65. Hahn, M.E. and Muir, T.W. (2004) *Angew. Chem. Int. Ed.*, **43**, 5800.

66. Miranda, L.P., Shao, H., Williams, J., Chen, S.Y., Kong, T., Garcia, R., Chinn, Y., Fraud, N., O'Dwyer, B., Ye, J., Wilken, J., Low, D.E., Cagle, E.N., Carnevali, M., Lee, A., Song, D., Kung, A., Bradburne, J.A., Paliard, X., and Kochendoerfer, G.G. (2007) *J. Am. Chem. Soc.*, **129**, 13153.

Commentary Part

Comment 1

Sam Danishefsky

This chapter entitled "Chemical Synthesis of Proteins" is extremely well done, and I enjoyed reading it immensely. I think that the topic selection is excellent, and the contents will prove to be very valuable. I have put in some representative changes in language to bring it more into line with clear English. The author should review the chapter and make some additional changes of this type. I congratulate Professor Liu on doing an excellent job, and my only suggestion, in terms of change, would be to include a graphic representation of expressed ligation. I think that the average chemist would have difficulty figuring it out from the description presented in this text.

Comment 2

David Crich

The enormous difficulties encountered in the synthesis of peptides by the pioneers in the field provoked the development of the SPPS method by Merrifield, thereby heralding the modern era of polymer-supported synthesis and that of the parallel technology of supported reagents. It is ironic, therefore, that it is the limitations of the SPPS method that have forced a return to the use of solution-phase methods for what is termed chemical protein synthesis. Kent's technique of NCL permits the combination by purely chemical means of two peptide segments, produced by SPPS, into one much larger unit of a size such as would not itself be accessible by SPPS methods.

As described in this chapter, the chemistry of NCL is deceptively simple and involves an intermolecular displacement of a thiol from a thioester by the sulfhydryl group of an N-terminal cysteine group. This reaction joins the two segments together in the form of a thioester, and is followed by an intramolecular S → N shift that affords the native amide bond. In reality, the chemistry involved is finely balanced and relies on the exquisite chemoselectivity of the thioester unit to achieve success. Thus, the key intermolecular step that brings the

two segments together is a transthioesterification reaction, which, because of the correct reactivity match between the thioester electrophile and the thiol nucleophile, enables this reaction to win out in the face of competing nucleophilic attack by the aqueous reaction medium and a variety of side-chain functionality present in the mostly unprotected peptidyl side chains.

The very success of the NCL method has, as is the case for all breakthrough technologies, focused attention on its limitations and deficiencies, thereby opening a large window of opportunity for the creativity of the organic chemist.

As discussed in the text, the requirement for an N-terminal cysteine residue has been addressed first by the use of auxiliaries bound to other N-terminal amines and more recently by the synthesis of a variety of β-mercaptoamino acids from which the extraneous sulfur atom can be removed post-ligation. Such methods are clever, but if they are to become widely applicable the synthesis of the β-mercaptoamino acid building blocks will need to be simplified to the extent that they can be made available commercially.

The second challenge, again as discussed in the chapter, is that of the efficient synthesis of C-terminal thioesters by SPPS methods and preferably by SPPS methods employing Fmoc chemistry. Among the various creative solutions deployed by organic chemists for this problem, some of the more direct and simple involve the N → S shifts of mercaptoalkyl and -arylamides leading to transient thiols that are then captured by an excess of a second thiol present in the reaction medium. The very recent development of the bis(β-mercaptoethyl)amides by two independent groups (Figure C1) is a simple but significant advance here as it very nicely circumvents the barrier to rearrangement present in the more usual secondary amides with their *trans* configuration [C1].

The exquisite chemoselectivity, coupled with the use of aqueous media and the very minimal recourse to protecting groups, make the NCL process an outstanding example of the levels to which modern synthetic chemistry is capable of rising. The question remains, however, as to the sufficiency of the process: can we do better and even need we do better? Existing solutions to the problem of thioester synthesis by Fmoc-chemistry

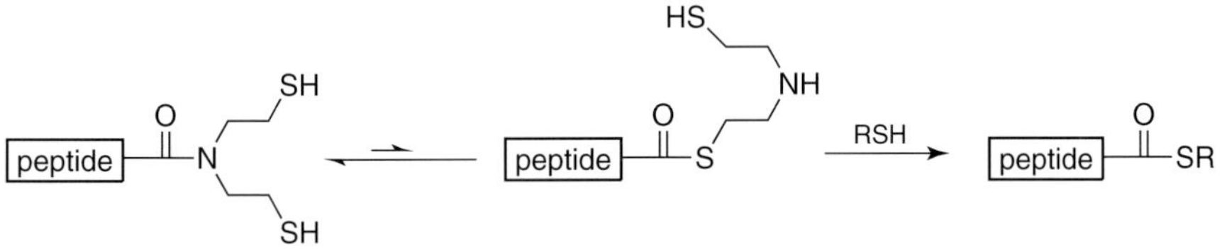

Figure C1 Synthesis of thioesters from bis(β-mercaptoethyl)amides.

SPPS can likely be refined and rendered more practical, but the existing methods to generalize away from cysteine remain cumbersome, particularly with respect to the desulfurization step. With NCL, the bar has been set very high, but clearly there is still scope for alternative and general methods for peptide segment coupling that do not rely on the presence of any one particular residue.

Reference

C1. (a) Hou, W., Zhang, X., Li, F., and Liu, C.-F. (2011) *Org. Lett.*, **13**, 386–389; (b) Ollivier, N., Dheur, J., Mhidia, R., Blanpain, A., and Melnyk, O. (2010) *Org. Lett.*, **12**, 5238–5241.

7
CuAAC: the Quintessential Click Reaction

Valery V. Fokin

7.1
Introduction

The birth of click chemistry is often associated with the seminal 2001 paper by Kolb, Finn, and Sharpless [1]. Although that was the first written account on the subject, the concept of click chemistry as a way of finding new and useful function through creating new chemical matter using only a handful of the best reactions began to take shape in the 1990s, and the term *"click chemistry"* was coined by Sharpless as early as 1995. It was the culmination of the group's work on the selective oxidation of alkenes. In retrospect, the rationale seems clear: alkenes are the most commonly available starting materials, they are easy to oxidize selectively, producing a variety of even more reactive intermediates, such as epoxides, aziridines, and episulfonium ions, and these energetic intermediates react stereospecifically and irreversibly with nucleophiles to give nonhydrolyzable products (Figure 7.1). These insights formed the foundation of click chemistry as the most efficient way to approach the ultimate goal of all synthetic methods endeavors: the discovery of powerful, highly reliable, and selective transformations which allow the rapid preparation of useful compounds, when necessary in large numbers and quantities, from readily available starting materials. Click transformations accomplish this goal: they are easy to perform, give rise to their intended products in very high yields with little or no byproducts, work well under many conditions, and are unaffected by the nature of the groups being connected to each other.

Among processes that qualified for the click status were several families of condensation reactions involving the carbonyl group and pericyclic reactions, including hetero-Diels–Alder and 1,3-dipolar cycloadditions. The potential of organic azides as highly energetic yet very selective functional groups in organic synthesis was highlighted in the 2001 review, and their dipolar cycloadditions with alkenes and alkynes were placed among the reactions fulfilling the click criteria [1]. However, the inherently low reaction rate of the azide–alkyne cycloaddition did not make it very useful in the click context, and its potential was not revealed until the discovery of the reliable and broadly useful catalysis by copper(I).

The copper-catalyzed reaction was reported simultaneously and independently by the groups of Meldal in Denmark [2] and Fokin in the USA [3]. It transforms organic azides

Organic Chemistry – Breakthroughs and Perspectives, First Edition. Edited by Kuiling Ding and Li-Xin Dai.
© 2012 Wiley-VCH Verlag GmbH & Co. KGaA. Published 2012 by Wiley-VCH Verlag GmbH & Co. KGaA.

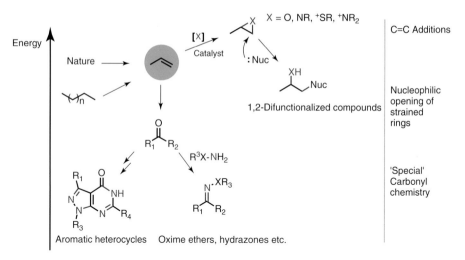

Figure 7.1 Click chemistry: energetically highly favorable linking reactions. Unsaturated compounds provide the carbon framework. New groups are introduced via carbon–heteroatom bonds.

and terminal alkynes exclusively into the corresponding 1,4-disubstituted 1,2,3-triazoles, in contrast to the uncatalyzed reaction, which requires much higher temperatures and provides mixtures of 1,4- and 1,5-triazole regioisomers. Meldal and co-workers used the transformation for the synthesis of peptidotriazoles in organic solvents starting from alkynylated amino acids attached to solid supports, emphasizing the requirement for immobilization of the alkyne on a resin for the success of the reaction, whereas Fokin and co-workers immediately turned to aqueous systems and devised a straightforward and practical procedure for the covalent "stitching" of virtually any fragments containing an azide and an alkyne functionality, noting the broad utility and versatility of the novel process "for those organic synthesis endeavors which depend on the creation of covalent links between diverse building blocks" [3].

Numerous applications of the copper catalyzed azide–alkyne cycloaddition (CuAAC) reaction reported during the last decade have been regularly reviewed [4–11], and are continually enriched by investigators in many fields. Indeed, as of the time of the writing, a SciFinder search returned 2829 papers out of 3050 that directly cited the 2001 click chemistry article [1] which were in fact using CuAAC. Unsurprisingly, it is often called *click reaction* and is frequently equated with click chemistry. However, it should be stressed that the CuAAC reaction is only a representative of this family of transformations. Its success in enabling the creation of new and useful functions serves for us, and we hope for others, as an inspiration for the development of new click reactions.

This chapter discusses fundamental aspects of CuAAC, the leading click reaction, its mechanistic features, and several representative applications with emphasis on the qualities of *in situ*-generated copper(I) acetylides that permit this unique mode of their reactivity. It is an updated and expanded version of recently published reviews on the topic [12, 13].

7.2
Azide–Alkyne Cycloaddition: the Basics

Among the most energetic hydrocarbons, alkynes have built-in reactivity that can be utilized for their transformation into useful intermediates and products. Transition metals allow selective and controlled manipulation of the triple bond, opening the door to a wealth of reliable reactivity. For example, η^2-coordination to π-acidic transition metals activates the acetylenic triple bond towards nucleophilic additions. Such cyclizations have been widely utilized in the synthesis of hetero- and carbocycles [14]. The chemistry of η^1-complexes of terminal alkynes (σ-acetylides) is dominated by the well-known reactions with electrophiles at the α-carbon of the metal-bound alkyne. Activation of alkynes by π-basic metals, which results in the formation of synthetically useful vinylidene complexes that are β-nucleophilic, is also known and is best represented by the complexes of group 6 elements (Cr, Mo, W) and ruthenium. In contrast, vinylidene complexes of group 11 metals, in particular copper(I), have neither been isolated nor implicated in catalysis. However, this first-period electron-rich d^{10} metal should be an ideal candidate for revealing β-nucleophilic properties of its acetylide ligands. Appreciation of the dynamics of exceedingly rapid ligand exchange of copper(I) acetylides in aqueous solutions and developing methods for the generation of reactive forms of these well-known organometallic species were key to the success of the CuAAC reaction.

Organic azides [15, 16] are unique reactive partners for copper(I) acetylides. They can be easily introduced into organic molecules and can usually be handled and stored as stable reagents. The N_3 is a small and relatively nonpolar functional group that is devoid of acid–base reactivity (with the exception of strong acids and organo-lithium/organomagnesium reagents). Organic azides remain nearly "invisible" to and unreactive with most other functional groups present in complex biological systems and encountered under laboratory synthetic conditions. The product of the union of azides and alkynes, 1,2,3-triazoles, are remarkably stable aromatic heterocycles that can serve, among other roles, as structural analogs of a peptide linkage. Azides can thereby be installed on structures that one wishes to link together and kept in place through many operations of synthesis and elaboration of the molecule. Their presence can be revealed when required *if* effective methods to transform them selectively to stable covalent connections are available. The complementary reactivity of *in situ*-generated copper(I) acetylides has allowed this general scheme to be put in operation for a wide array of applications and settings.

The thermal reaction of terminal or internal alkynes with organic azides (Figure 7.2a) has been known for more than a century, the first 1,2,3-triazole being synthesized by Michael from phenyl azide and diethyl acetylenedicarboxylate in 1893. The reaction was most thoroughly investigated by Huisgen's group in the 1950–1970s in the course of their studies of the larger family of 1,3-dipolar cycloaddition reactions [17]. Although the reaction is highly exothermic (ΔG° between -50 and -65 kcal mol^{-1}), its high activation barrier (~25 kcal mol^{-1} for methyl azide and propyne [18]) results in exceedingly low reaction rates for unactivated reactants even at elevated temperature. Furthermore, since the differences in HOMO–LUMO energy levels for both azides and alkynes are of

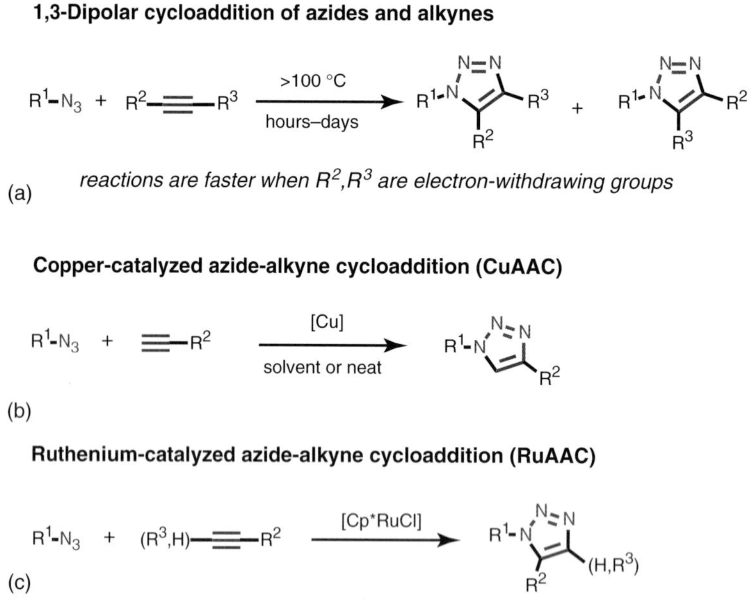

1,3-Dipolar cycloaddition of azides and alkynes

(a)   *reactions are faster when R²,R³ are electron-withdrawing groups*

Copper-catalyzed azide-alkyne cycloaddition (CuAAC)

(b)

Ruthenium-catalyzed azide-alkyne cycloaddition (RuAAC)

(c)

Figure 7.2 Thermal cycloaddition of azides and alkynes usually requires prolonged heating and results in mixtures of both 1,4- and 1,5-regioisomers (a), whereas CuAAC produces only 1,4-disubstituted-1,2,3-triazoles at room temperature in excellent yields (b). The RuAAC reaction (c) proceeds with both terminal and internal alkynes and gives 1,5-disubstituted and fully 1,4,5-trisubstituted-1,2,3-triazoles.

similar magnitude, both dipole–HOMO- and dipole–LUMO-controlled pathways operate in these cycloadditions. As a result, a mixture of regioisomeric 1,2,3-triazole products is usually formed when an alkyne is unsymmetrically substituted.

Copper catalysts (Figure 7.2b) drastically change the mechanism and the outcome of the reaction, converting it to a sequence of discrete steps culminating in the formation of a 5-triazolyl copper intermediate (Scheme 7.1). The key C–N bond-forming event takes place between the nucleophilic, vinylidene-like β-carbon of copper(I) acetylide and the electrophilic terminal nitrogen of the coordinated organic azide. As discussed below, the reaction is far more complex than this simplified representation, which is just intended to illustrate the observed reactivity of azides with copper(I) acetylides.

The rate of CuAAC is increased by a factor of 10^7 relative to the thermal process [18], making it conveniently fast at and below room temperature. The reaction is not

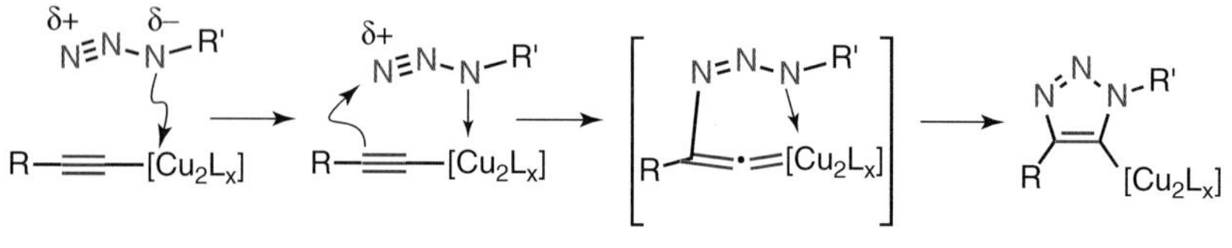

Scheme 7.1 Simplified representation of the proposed C–N bond-making steps in the reaction of copper(I) acetylides with organic azides.

significantly affected by the steric and electronic properties of the groups attached to the azide and alkyne centers, and primary, secondary, and even tertiary electron-deficient and electron-rich aliphatic, aromatic, and heteroaromatic azides usually react well with variously substituted terminal alkynes. The reaction proceeds in many protic and aprotic solvents, including water, and is unaffected by most organic and inorganic functional groups, therefore all but eliminating the need for protecting group manipulations. The 1,2,3-triazole heterocycle has the advantageous properties of high chemical stability (generally inert to severe hydrolytic, oxidizing, and reducing conditions, even at high temperature), a strong dipole moment (4.8–5.6 D), aromatic character, and hydrogen bond-accepting ability [19, 20]. Thus it can interact productively in several ways with biological molecules and organic and inorganic surfaces and materials, serving, for example, as a hydrolytically stable replacement for the amide bond.

To date, copper stands out as the only metal for the reliable, facile, and regiospecific catalysis of the azide–alkyne cycloaddition. Indeed, with the exception of ruthenium (see below), other metals known to catalyze various transformations of alkynes have not so far yielded effective catalysts for the conversion of azides and terminal alkynes to 1,4-triazoles. Surveys in our laboratories of complexes of all of the first-row transition elements and also complexes of Ag(I), Pd(0/II), Pt(II), Au(I/III), and Hg(II), among others, have all failed to produce triazoles in synthetically useful yields; their effect on the rate and selectivity of the cycloaddition was at best marginally noticeable. The unique catalytic function of Cu(I) may be explained by the fortuitous combination of its ability to engage terminal alkynes in both σ- and π-interactions *and* the rapid exchange of these and other ligands in its coordination sphere, especially in aqueous conditions. When an organic azide is a ligand, the synergistic nucleophilic activation of the alkyne and electrophilic activation of the azide drive the formation of the first carbon–nitrogen bond, C2-N3.

In 2005, ruthenium cyclopentadienyl complexes were found to catalyze the formation of the complementary 1,5-disubstituted triazole from azides and terminal alkynes, and also to engage internal alkynes in the cycloaddition (Figure 7.2c) [21]. As one would imagine from these differences, this sister process, designated RuAAC (ruthenium-catalyzed azide–alkyne cycloaddition), is mechanistically distinct from its copper cousin, although the underlying activation of the alkyne component appears to be fundamentally similar: the nucleophilicity of its π-system is increased by the back-donation from the ruthenium center. Although the scope and functional group compatibility of RuAAC are excellent [22–24], the reaction is more sensitive to solvents and to the steric demands of the azide substituents than CuAAC.

7.3
CuAAC: Catalysts and Ligands

A wide range of experimental conditions for CuAAC have been employed since its discovery, underscoring the robustness of the process and its compatibility with most functional groups, solvents, and additives regardless of the source of the catalyst. The choice of the catalyst is dictated by the particular requirements of the experiment, and usually many combinations will produce desired results. The most commonly used protocols and their advantages and limitations are discussed below.

Different copper(I) sources can be utilized in the reaction, as recently summarized in necessarily partial fashion by Meldal and Tornøe [6]. Copper(I) salts (iodide, bromide, chloride, acetate) and coordination complexes such as $[Cu(MeCN)_4]PF_6$ and $[Cu(MeCN)_4]OTf$ have commonly been employed. In general, however, we recommend against the use of cuprous iodide because of the ability of the iodide anion to act as a bridging ligand for the metal, resulting in the formation of polynuclear acetylide complexes which interfere with the productive catalytic cycle by tying up the catalyst. Furthermore, under certain conditions copper(I) iodide may result in the formation of 1-iodoalkynes and, consequently, 5-iodotriazoles [25]. High concentrations of chloride ion in water (0.5 M or above) can also be deleterious, although the inhibitory effect of chloride is far less pronounced compared with iodide. Therefore, for reactions performed in aqueous solvents, cuprous bromide and acetate are favored, as is the sulfate from *in situ* reduction of $CuSO_4$; for organic reactions, the acetate salt is generally a good choice.

Copper(II) salts and coordination complexes are not competent catalysts, and reports describing Cu(II)-catalyzed cycloadditions [26–28] are not accurate. Cupric salts and coordination complexes are well-known oxidizing agents for organic compounds [29]. Alcohols, amines, aldehydes, thiols, phenols, and even carboxylic acids may readily be oxidized by the cupric ion, reducing it to the catalytically active copper(I) species in the process. Especially relevant is the family of oxidative acetylenic couplings catalyzed by the cupric species [30], with the venerable Glaser coupling being the most studied example. Since terminal alkynes are necessarily present in CuAAC reactions, their oxidation is an inevitable side process which could, in turn, produce the needed catalytically active copper(I) species.

Among the three most common oxidation states of copper (0, +1, and +2), +1 is least thermodynamically stable. Cuprous ion can be oxidized to catalytically inactive Cu(II) species, or can disproportionate to a mixture of Cu(II) and Cu(0). The standard potential of the Cu^{2+}/Cu^+ couple is 159 mV, but can vary widely depending on the solvent and the ligands coordinated to the metal, and is especially complex in water [31]. When present in significant amounts, the ability of Cu(II) to mediate the aforementioned Glaser-type alkyne coupling processes can result in the formation of undesired byproducts while impairing triazole formation. When the cycloaddition is performed in organic solvents using copper(I) halides as catalysts, the reaction is plagued by the formation of oxidative coupling byproducts **4a–d** unless the alkyne is bound to a solid support (**5**) (Scheme 7.2a,b) [2]. When the alkyne is dissolved and the azide is immobilized on the resin, only traces of the desired triazole product are formed. Therefore, when copper(I) catalyst is used directly, whether by itself or in conjunction with amine ligands, exclusion of oxygen may be required. In contrast, performing the reaction in the presence of mild oxidants allows the capture of the triazolyl copper intermediates and isolation of the products derivatized at the 5-position in synthetically useful yields [32, 33].

Ascorbate, a mild reductant, was introduced by Fokin and co-workers [3] as a convenient and practical alternative to oxygen-free conditions. Its combination with a copper(II) salt, such as the readily available and stable copper(II) sulfate pentahydrate or copper(II) acetate, has been quickly accepted as the method of choice for preparative synthesis of

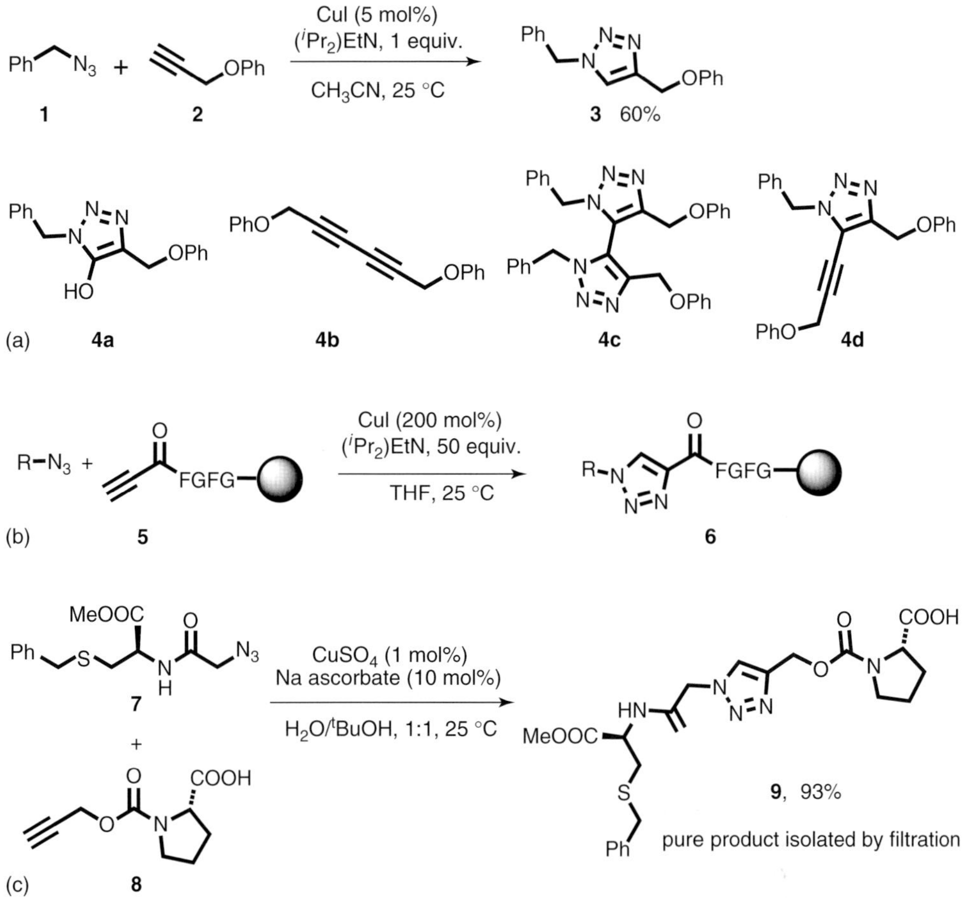

Scheme 7.2 (a) Oxidative coupling byproducts in the CuAAC reactions catalyzed by copper(I) salts in the presence of oxygen; (b) CuAAC with immobilized alkyne avoids the formation of the oxidative byproducts but requires a large excess of the catalyst; reactions with immobilized azide fail; and (c) solution-phase CuAAC in the presence of sodium ascorbate.

1,2,3-triazoles. Water is an ideal solvent capable of supporting copper(I) acetylides in their reactive state, especially when they are formed *in situ* (in fact, examination of the reactivity of *in situ*-generated, and hence less aggregated copper acetylides was the main impetus for the discovery and development of the ascorbate procedure). The formation of byproducts resulting from copper-mediated oxidative side reactions is suppressed as any dissolved dioxygen is rapidly reduced. The "aqueous ascorbate" procedure often furnishes triazole products in nearly quantitative yield and greater than 90% purity, without the need for ligands or additives or protection of the reaction mixture from oxygen (Scheme 7.2c). Of course, copper(I) salts can also be used in combination with ascorbate, wherein it converts any oxidized copper(II) species back to the catalytically active +1 oxidation state.

The reaction can also be catalyzed by Cu(I) species supplied by elemental copper, thus further simplifying the experimental procedure – a small piece of copper metal (wire or turning) is all that is added to the reaction mixture, followed by shaking or stirring for 12–48 h [3, 18, 34]. Aqueous alcohols (methanol, ethanol, *tert*-butanol), tetrahydrofuran (THF), and dimethyl sulfoxide (DMSO) can be used as solvents in this procedure. Cu(II) sulfate may be added to accelerate the reaction; however, this is not necessary in most cases, as copper oxides and carbonates, the patina on the metal surface, are sufficient to initiate the catalytic cycle. Although the procedure based on copper metal requires longer reaction times when performed at ambient temperature, it usually provides access to very pure triazole products with low levels of copper contamination. Alternatively, the reaction can be performed under microwave irradiation at elevated temperature, reducing the reaction time to 10–30 min [34, 35].

The copper metal procedure is experimentally very simple and is particularly convenient for the high-throughput synthesis of compound libraries for biological screening. Triazole products are generally isolated in >85–90% yields, and can often be submitted for biological assays without purification owing to the very low contamination with copper. When required, trace quantities of copper remaining in the reaction mixture can be removed with an ion-exchange resin or using solid-phase extraction techniques. Other heterogeneous copper(0) and copper(I) catalysts, such as copper nanoclusters [36], copper/cuprous oxide nanoparticles [37], and copper nanoparticles adsorbed on charcoal [38] have also shown good catalytic activity.

Although organic azides are generally stable and safe compounds, those of low molecular weight can spontaneously decompose and, therefore, could be difficult or dangerous to handle. This is especially true for small molecules with several azide functionalities that would be of much interest for the generation of polyfunctionalized structures. Indeed, small-molecule azides should never be isolated away from solvent, for example, by distillation, precipitation, or recrystallization. Fortunately, the CuAAC reaction is highly tolerant of all manner of additives and spectator compounds, including inorganic azides even in large excess. The process can therefore be performed in a one-pot, two-step sequence, whereby an *in situ*-generated organic azide is immediately consumed in a reaction with a copper acetylide (Scheme 7.2). We have implemented this process many times in our laboratories starting from alkyl halides or aryl sulfonates by an S_N2 reaction with sodium azide (Scheme 7.3a). In a recent example, Pfizer chemists developed a continuous-flow process wherein a library of 1,4-disubstituted 1,2,3-triazoles was synthesized from alkyl halides, sodium azide, and terminal alkynes, with the copper catalyst required for cycloaddition being supplied from the walls of the heated copper tubing through which the reaction solution was passed (Scheme 7.3b) [39].

Aryl and vinyl azides can also be accessed in one step from the corresponding halides or triflates via a copper-catalyzed reaction with sodium azide in the presence of catalytic amount of L-proline (Scheme 7.3c) [40]. In this fashion, a range of 1,4-disubstituted 1,2,3-triazoles can be prepared in excellent yields [41]. This reaction sequence can be performed at elevated temperature under microwave irradiation, reducing the reaction time to 10–30 min [34]. Anilines can be also converted to aryl azides by the reaction with *tert*-butyl nitrite and azidotrimethylsilane [42]. The resulting azides can be submitted to the CuAAC conditions without isolation, furnishing triazole products in excellent yields.

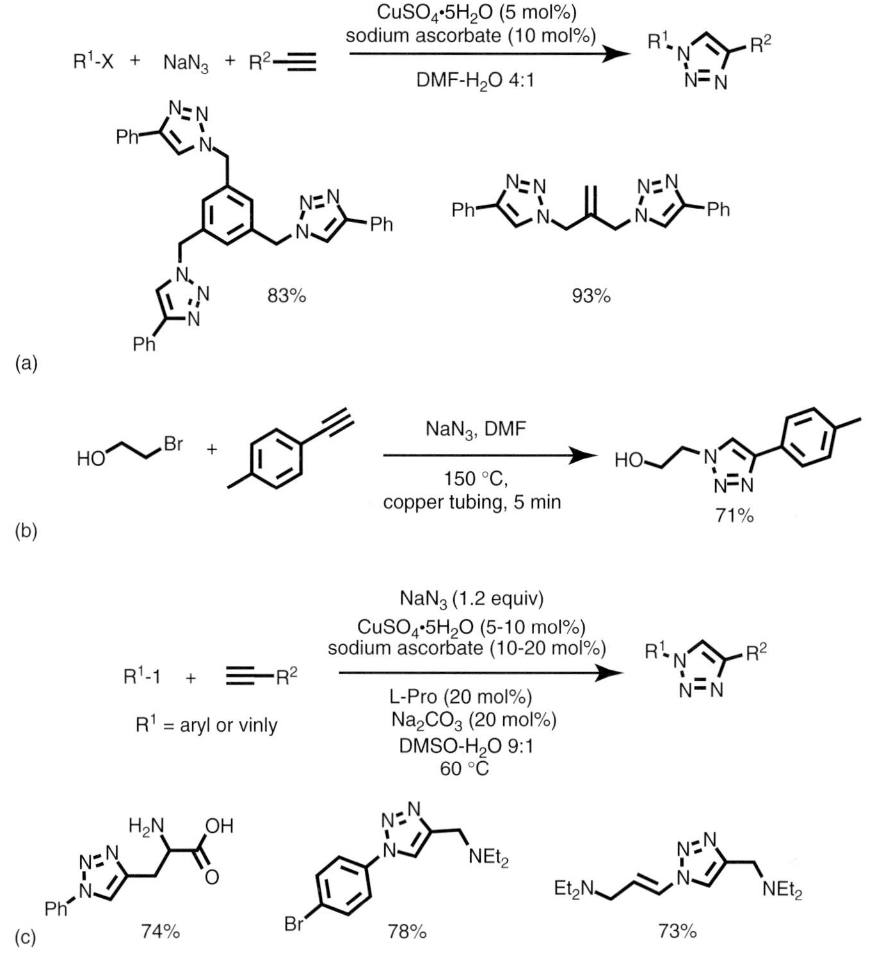

(a)

(b)

(c)

Scheme 7.3 One-pot syntheses of triazoles from halides at (a, b) sp³ and (c) sp² carbon centers. Reaction (b) was performed in a flow reactor in 0.75 mm diameter Cu tubing with no added copper catalyst.

Microwave heating further improves the reaction, significantly reducing the reaction time [35].

Many other copper complexes involving ligands have been reported as catalysts or mediators of the CuAAC reaction. It would be accurate to say that finding a copper(I) catalyst that is not active is more difficult than improving those that catalyze the reaction. Many reported reactions are performed under widely differing conditions, making quantitative comparisons of ligand performance difficult. Nevertheless, it may be useful to organize them into "soft" and "hard" classes by virtue of the properties of their donor centers [43]. Cu(I) is a "borderline soft" Lewis acid [44], and therefore can partner a wide variety of potentially effective ligands. A representative sample of reported systems, which is far from comprehensive, follows below.

The "soft" ligand class is exemplified by phosphine-containing catalytically active species such as the simple coordination complexes Cu[P(OMe)$_3$]$_3$Br [45] and Cu(PPh$_3$)$_3$Br [46, 47]. These species are often used in reactions in organic solvents, in which cuprous salts have limited solubility. A recent report described the bis(phosphine) complex Cu(PPh$_3$)$_2$OAc as an excellent catalyst for the CuAAC reaction in toluene and dichloromethane [48]. Monodentate phosphoramidite and related donors have also been evaluated [49]. Chelating complexes involving phosphines have not found favor, except for bidentate combinations of phosphine with the relatively weakly binding triazole unit [50]. Thiols are a potent poison of the CuAAC reaction in water, but thioethers show promise, although they remain an underexplored member of the "soft" ligand class [51].

Several Cu(I) complexes with *N*-heterocyclic carbene ligands have been described as CuAAC catalysts at elevated temperature in organic solvents, under heterogeneous aqueous conditions (when both reactants are not soluble in water), and under neat conditions [52]. These catalysts appear to be active under the solvent-free conditions, achieving turnover numbers as high as 20 000. However, their activity in solution-phase reactions is significantly lower than those of other catalytic systems [for example, a *stoichiometric* reaction of the isolated copper(I) acetylide–NHC complex with benzhydryl azide required 12 h to obtain a 65% yield of the product [53], whereas under standard solution conditions even a catalytic reaction would be complete within 1 h]. As an important footnote, *we strongly advise against running highly exothermic reactions without a solvent*, as runaway reactions and explosions can easily occur. Furthermore, performance of a catalyst under the solvent-free conditions is, at best, loosely correlated with its true activity, simply because most copper(I) species effortlessly catalyze such reactions.

The category of "hard" donor ligands of CuAAC-active systems is dominated by amines (see [6] and the Supporting Information in [54]). In many cases, amines are labeled as "additives" rather than "ligands," since it is often the intention to aid in the deprotonation of the terminal alkyne rather than to coordinate to the metal center. However, this assumption is not accurate, as the formation of copper(I) acetylides is so facile that it occurs even in strongly acidic media (up to 20–25% H$_2$SO$_4$) [55]. Instead, the primary role of amine ligands can be (i) to prevent the formation of unreactive polynuclear copper(I) acetylides; (ii) to facilitate the coordination of the azide to copper center at the ligand exchange step (see below); and (iii) to increase the solubility of the copper complex to deliver high solution concentrations of the necessary Cu(I) species. In several cases of amine-based chelates, it is probable that metal binding is at least part of the productive role of the polydentate ligand. An example is the use of a hydrophobic Tren ligand for CuAAC catalysis in an organic solvent at elevated temperature [56].

As befits the borderline nature of the Cu(I) ion, by far the largest and most successful class of ligands are those of intermediate character between "hard" and "soft," particularly those containing heterocyclic donors. With rare exceptions, these also contain a central tertiary amine center, which can serve as both a coordinating donor and a base. The need for these ligands was particularly evident for reactions involving biological molecules that are handled in water in low concentrations and are not stable to heating. Chemical transformations used in bioconjugations impose additional demands on the efficiency and selectivity. They must be exquisitely chemoselective, biocompatible, and fast. Despite the experimental simplicity and efficiency of the "ascorbate" procedure,

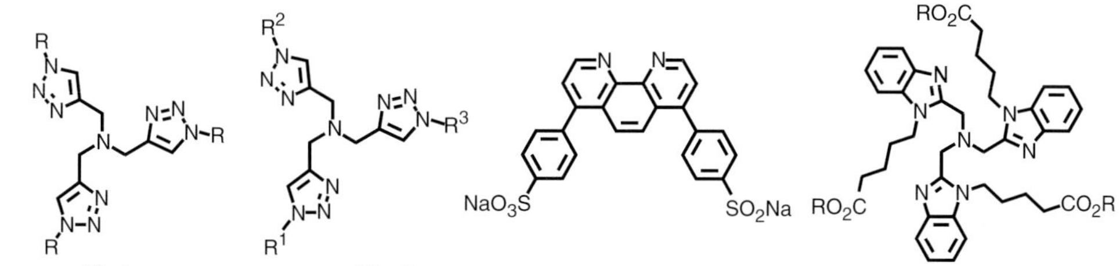

R = benzyl (**TBTA**)

10a,b

(a) R = benzyl (**TBTA**)
(b) R = tert-butyl (**TTTA**)

11a-d

(a) R = CH$_2$CH$_2$CH$_2$OH (**THPTA**)
(b) R = CH$_2$CH$_2$CO$_2$H
(c) R^1 = CH$_2$CH$_2$OH
    R^2, R^3 = tert-butyl
(d) R^1 = CH$_2$CH$_2$OSO$_3^-$
    R^2, R^3 = tert-butyl (**BTTES**)

12

13a,b

(a) R = H
(b) R = Et

Figure 7.3 CuAAC-accelerating ligands of choice: tris(1,2,3-triazolyl)methylamine (TBTA, **10a**) and its tert-butyl analog (TTTA, **10b**), water-soluble analogs (**11**), sulfonated bathophenanthroline (**12**), and tris(benzimidazole)methyl amine (TBIA) (**13**).

the CuAAC reaction in the forms described above is simply not fast enough when the concentrations of the reactants are low, particularly in aqueous media.

The first general solution to the bioconjugation problem was provided by the tris(benzyltriazolyl)methylamine (TBTA) ligand **10a** (Figure 7.3), prepared using the CuAAC reaction and introduced shortly after its discovery [57]. This ligand was shown to accelerate the reaction significantly and stabilize the Cu(I) oxidation state in water-containing mixtures. After its utility in bioconjugation had been demonstrated by the efficient attachment of 60 alkyne-containing fluorescent dye molecules to the azide-labeled cowpea mosaic virus [58], it was widely adopted for use with such biological entities as nucleic acids, proteins, *Escherichia coli* bacteria, and mammalian cells. A resin-immobilized version of TBTA [59] has also been shown to be very useful in library synthesis and conjugation experiments when contamination of products by copper needs to be minimized. The tris(tert-butyl) analog, TTTA (**10b**), shows superior activity to TBTA in both aqueous and organic solvents. A reliable procedure for the synthesis of TBTA, TTTA, and other analogs has been developed [60].

The poor solubility of TBTA in water prompted the development of more polar analogs such as **11a–d** (Figure 7.3) [61]. These ligands are particularly useful in bioconjugations involving sensitive biological molecules. In some cases, they allow CuAAC-based bioconjugations even with live organisms [62, 63]. At the same time, a combinatorial search for alternatives led to the identification of the commercially available sulfonated bathophenanthroline **12** as the ligand component of the fastest water-soluble CuAAC catalyst under dilute aqueous conditions [64]. However, Cu·**12** complexes are strongly electron-rich and are therefore highly susceptible to oxidation in air. Ascorbate can be used to keep the metal in the +1 oxidation state but, when exposed to air, reduction of O$_2$ is very fast and can easily use up all of the available reducing agent. Therefore, **12** is usually used under an inert atmosphere, which can be inconvenient, particularly when small amounts of biomolecule samples are used. A procedural solution was found in the use of an electrochemical cell to provide the reducing equivalents to scrub O$_2$ out of such reactions and maintain Cu·12 in the cuprous oxidation state, but this was again less than optimal owing to the need for extra equipment and electrolyte salts [65]. The polydentate trimethylamine theme has been extended to benzimidazole, benzothiazole, oxazoline, and pyridine substituents [54, 66]. Several have provided significantly faster catalysis when quantitative rates are measured, particularly the pendant ester and water-soluble acid derivatives of the tris(benzimidazole) motif (**13a,b**).

7.4
Mechanistic Aspects of the CuAAC

Before considering possible mechanistic pathways of the CuAAC process, it is important to highlight the fundamental reactivity of the players: organic azides and copper(I) acetylides. The reactivity of organic azides, with the exception of thermal and photochemical decomposition, is dominated by reactions with nucleophiles at the terminal N3 atom. Examples of reactions with electrophiles reacting at the proximal N1 (Figure 7.4) are also known, although they are less common. These reactivity patterns are in agreement with theoretical studies of the electronic structure of the azido group. The coordination chemistry of organic azides follows the same trend, and the azide usually behaves as an L-type σ-donor via its N1 nitrogen atom, with a few exceptions when electron-rich π-basic metals engage the terminal N3 atom in back-donation or when coordination via the terminal nitrogen is forced due to steric or geometric constraints [67].

The history of copper(I) acetylides dates back as far as Glaser's discovery in 1869 of oxidative dimerization of copper(I) phenylacetylide. However, the precise nature of the reactive alkynyl copper species in CuAAC [and also in many other reactions involving copper(I) acetylide complexes] is not well understood. The chief complications are the tendency of copper species to form polynuclear complexes [55, 68] and the great facility of the ligand exchange at the copper center. As a result, mixtures of Cu(I), terminal alkynes, and other ligands usually contain multiple organocopper species in rapid equilibrium with each other. Although this may make elucidation of the exact mechanism difficult, the dynamic nature of copper acetylides is undoubtedly a major contributor to the remarkable adaptability of the reaction to widely different conditions. Whatever the details of the interactions of Cu with an alkyne during the CuAAC reaction, it is clear that copper acetylide species are easily formed and are productive components of the reaction mechanism.

The initial computational treatment of CuAAC focused on the possible reaction pathways available to mononuclear copper(I) acetylides and organic azides; propyne and methyl azide were chosen for simplicity (Scheme 7.4). Formation of copper(I) acetylide (**15**) (step A) was calculated to be exothermic by 11.7 kcal mol^{-1}, consistent with the well-known facility of this step which probably occurs through a π-alkyne copper

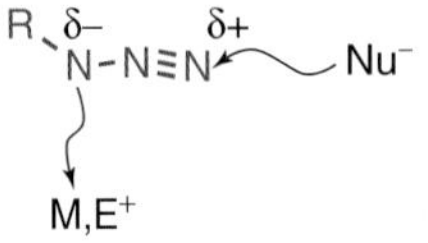

Figure 7.4 Common reactivity patterns of organic azides.

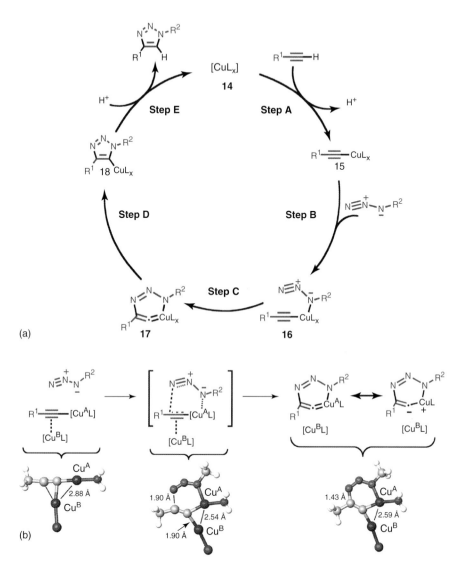

Scheme 7.4 (a) Early proposed catalytic cycle for the CuAAC reaction based on DFT calculations. (b) Introduction of a second copper(I) atom favorably influences the energetic profile of the reaction (L = H$_2$O in DFT calculations). At the bottom is shown the optimized structures for dinuclear Cu forms of the starting acetylide (left, corresponding to **15**), transition state for the key C-N bond-forming step (middle), and the metallacycle **17**. The calculated structures are essentially identical when acetylide instead of chloride is used as the ancillary ligand on the second copper center (Cu^B).

complex intermediate. π-Coordination of alkyne to copper significantly acidifies the terminal hydrogen of the alkyne, bringing it into the proper range to be deprotonated in an aqueous medium and resulting in the formation of a σ-acetylide. The azide is then activated by coordination to copper (step B), forming intermediate **16**. This ligand-exchange step is nearly thermoneutral computationally (2.0 kcal mol^{-1} uphill when L is water). This coordination is synergistic for both reactive partners: coordination of the azide reveals the β-nucleophilic, vinylidene-like properties of the acetylide, whereas the azide's terminus becomes even more electrophilic. In the next step (C), the first C-N bond-forming event takes place, and a strained copper metallacycle (**17**) forms. This step is endothermic by 12.6 kcal mol^{-1} with a calculated barrier of 18.7 kcal mol^{-1}, which corresponds roughly to the observed rate increase and is considerably lower than the barrier for the uncatalyzed reaction (∼26.0 kcal mol^{-1}), thus accounting for the observed rate acceleration accomplished by Cu(I). The formation of copper triazolide is very facile and energetically favorable. When protected by steric bulk, Cu–triazolyl complexes can be isolated from CuAAC reactions [53], and in rare cases of low catalyst loading and high catalytic rate, step E can be turnover-limiting [66]. Alternative pathways, including the concerted cycloaddition of azide and copper acetylide, were ruled out. The coordination of the azide to copper via the terminal nitrogen, proposed in a recent review by Meldal and Tornøe [6], had been also examined and found to be energetically unfavorable (no feasible intermediates could be identified). There is no experimental evidence to support this proposal either.

The density functional theory (DFT) investigation described above was soon followed by a study of the kinetics of the copper-mediated reaction between benzyl azide and phenylacetylene in DMSO under pseudo-first-order conditions. With low catalyst loadings and saturating alkyne concentrations, and in the presence of the triazole product, the initial rate law was found to be *second order* in metal [69]. A second-order dependence on catalyst concentration was also observed for CuAAC reactions accelerated by tris(benzimidazolyl)methylamine ligands such as **13** in Tris buffer solutions [53, 66]. The second-order rate law in the catalyst during the initial stages of the reaction did not, of course, unequivocally indicate the involvement of two copper atoms in the bond-forming events in the catalytic cycle. Indeed, we have learned since then that the mechanism of the reaction is more complex, and the reaction often exhibits discontinuous kinetic behavior (specifically, rate law orders in the catalyst and reagents change as the reaction progresses). However, the implication that multinuclear copper species were involved in catalysis prompted us to examine such a possibility both computationally and experimentally.

When the CuAAC pathway was investigated by DFT taking into account the possibility of the involvement of dinuclear copper(I) acetylides, a further fall in the activation barrier (by ∼3–6 kcal mol^{-}) was revealed (Scheme 7.4b) [70, 71]. In the transition state, a second copper(I) atom, Cu^B, strongly interacts with the proximal acetylide carbon (C^1), as indicated by the short Cu-C distances of 1.93 and 1.90 Å. A computational study reported by Straub compared dinuclear complexes with higher order aggregates and concluded that dinuclear intermediates were favored over the tetranuclear complexes [71].

Additional experiments further extended our mechanistic understanding of this process and helped us formulate experimental guidelines that we share here. Since isolated copper(I) acetylides normally exist in highly aggregated form, preparation of the reactive copper(I) acetylides *in situ*, that is, in the presence of an organic azide, is critical for the success of the reaction. Although organic azides are weak ligands for copper, their interaction with the metal center appears to be sufficient to prevent the formation of the unproductive polymeric acetylides and channel the acetylides into the productive catalytic cycle. This is demonstrated by a simple experiment which can be followed visually: the addition of 5 mol% of copper(II) sulfate to a 0.1 M solution of phenylacetylene in *tert*-butanol–water (2:1) results, within minutes, in the formation of dark-yellow phenyl copper acetylide. Removal of the precipitate after 15 min leaves the supernatant virtually copper free, and the isolated phenyl copper acetylide does not react with azides. In fact, once the dark-yellow acetylide precipitate has been formed, the addition of the azide does not revive the catalysis. When the same experiment is performed in the presence of 1 equiv. of benzyl azide, the light-yellow color that develops upon the addition of $CuSO_4$ disappears within about 20 min, but the reaction is still far from completion (about 40% conversion in the case at hand). It is clear that the catalyst is undergoing reorganization while the catalytic cycle is turning over during the initial 20 min after the start of the reaction. Before the catalytic cycle enters the steady-state regime, a number of turnovers occur, and both the identity and concentration of the catalyst undergo significant changes. Such behavior is not uncommon for catalytic reactions, but clearly complicates the interpretation of the initial rates results [72].

The multimodal kinetic profile of the reaction is confirmed by following its rate using heat flow calorimetry. A representative rate profile of a CuAAC reaction is shown in Figure 7.5a.

As already mentioned, the nature of the copper counter ion also has a dramatic effect on the rate and efficiency of the reaction. As Figure 7.5b demonstrates, the cuprous iodide-catalyzed reaction takes nearly 40 min to reach the maximum rate and over 100 min to reach full conversion, whereas the replacement of the iodide with a much weaker coordinating tetrafluoroborate propels the reaction to completion within minutes.

These mechanistic insights underscore the importance of taking into account every event affecting the fate of the catalytic cycle intermediates when analyzing complex catalytic processes involving multiple equilibria, competing pathways, and off-cycle dead ends (even if they are potentially reversible). Although the elementary steps for the key bond-making and bond-breaking events that were previously proposed and supported by DFT calculations [18, 70, 71] still stand, we now have a better quantitative measure of catalyst activation and deactivation pathways and their effect on the overall process. The details of our studies are beyond the scope of this review and have been published elsewhere. Nevertheless, in Scheme 7.5, we offer our current mechanistic proposal that takes into account both earlier studies and recent results.

The main observations and hypotheses are as follows. First, the formation of higher order polynuclear copper acetylides has a detrimental effect on the rate and outcome of the reaction. Therefore, solvents that promote ligand exchange (e.g., water and alcohols) are preferred over apolar, organic solvents which promote aggregation of copper species. Ill-defined catalysts perform well in the CuAAC precisely for this reason.

Second, dinuclear complexes may exhibit enhanced reactivity. Even though to date we have not seen an example of a reaction that exhibits uniform second order in the

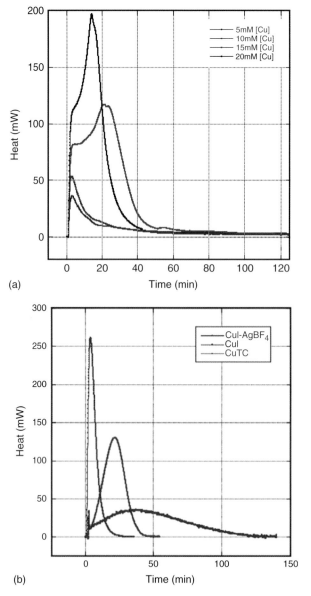

(a)

(b)

Figure 7.5 (a) Calorimetric trace of the reaction of benzyl azide and phenylacetylene [0.1 M in *tert*-butanol–water (1:1)], the indicated amount of copper sulfate, and 2 equiv. of ascorbate with respect to copper. The multimodal kinetic profile is especially easy to see in the black and green traces: the reactions take 12 and 25 min, respectively, to reach maximum rate. All four reactions proceeded to completion.

(b) Calorimetric trace of the reaction of benzyl azide with phenylacetylene [0.1 M in THF, 5 mM CuX as a 1:1 complex with TTTA ligand (**10b**)]. The reaction is significantly faster when the iodide is removed with silver(I) tetrafluoroborate (red trace) than in the presence of the iodide (blue trace). Copper thiophenecarboxylate (CuTC)-catalyzed trace, which shows intermediate behavior, is in green.

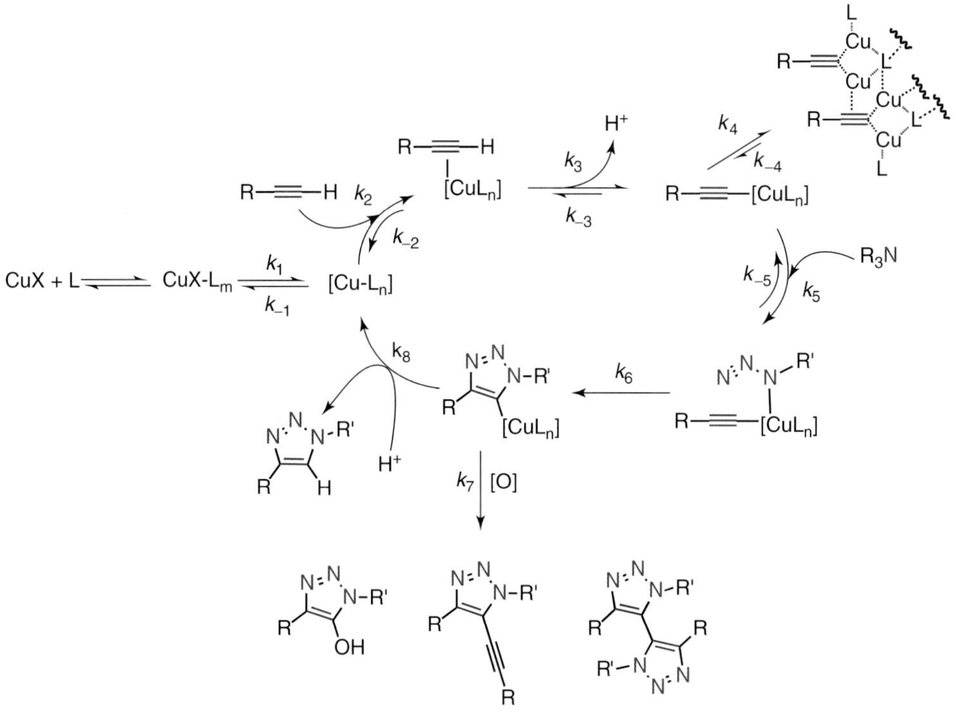

Scheme 7.5 Several key equilibria and irreversible off-cycle pathways affect the productive CuAAC catalytic cycle. For example, formation of oligomeric copper acetylides may become a key event under certain conditions: if most of the catalyst is occupied in the form of unproductive polymeric copper acetylides, the rate of the re-entry of reactive copper acetylide into the catalytic cycle (k_{-4}) would determine the overall rate of the process.

catalyst as it progresses, it is possible that nuclearity of the catalytic species is maintained throughout the catalytic cycle and, as a consequence, all elementary steps are effectively bimolecular, exhibiting the observed first order in the catalyst. As already mentioned, order of the reaction in catalyst is not an indicator of its molecularity or of the composition of the catalyst.

Third, successful CuAAC ligands need to balance the competing requirements of binding Cu(I) strongly enough to prevent the formation of unreactive polymeric complexes, yet allowing azide to access the coordination sphere of the σ-acetylide Cu center. Too potent a binder would tie up the necessary Cu coordination sites and too weak a ligand would not prevent the formation of higher order aggregates. Tris(triazolyl)amine ligands, such as TBTA (**10a**) and TTTA (**10b**), and related tripodal ligands successfully meet these challenges by combining weak donor ligands on the "arms" of the structure with the stronger central nitrogen donor. A possible composition of a reactive complex is exemplified by structure **19**. These ligands also accomplish the delicate task of providing enough electron density to the metal to promote the catalysis while not destabilizing the Cu(I) oxidation state.

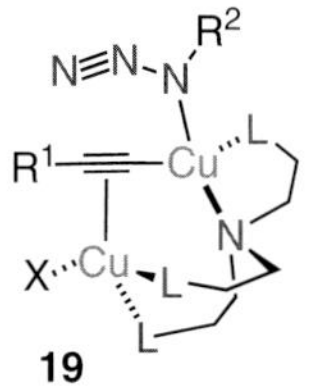

19

7.5
Reactions of 1-Iodoalkynes

1-Iodoalkynes are stable and readily accessible internal alkynes. As was disclosed recently, they are exceptionally reactive partners with organic azides under copper(I) catalysis conditions (Scheme 7.6) [73]. In fact, in some cases, their reactivity appears to surpass that of terminal alkynes. As an added benefit, the products of the reaction, 5-iodo-1,2,3-triazoles, are versatile synthetic intermediates amenable to further functionalization. The reaction is general and chemo- and regio-selective, and the experimental procedure is operationally simple. The 5-iodo-1,4,5-trisubstituted-1,2,3-triazoles are obtained in high yields from variously substituted organic azides and iodoalkynes. The catalysis is affected by copper(I) iodide in the presence of an amine ligand. In contrast to the CuAAC reaction, the observed rate and chemoselectivity of the reaction are strongly dependent on the nature of the ligand, and catalysis does not proceed in its absence. Both TBTA (**10a**) and its *tert*-butyl analog, TTTA (**10b**), give 5-iodotriazoles (**22**) as exclusive products in excellent yield.

The competing pathways, which arise from dehalogenation of the iodoalkyne and the triazole product and lead to 5-proto- (**23**) and 5-alkynyltriazoles (**24**), are far too slow in the presence of these ligands and do not interfere with the productive catalytic cycle.

The Cu(I)–TTTA catalyst system exhibits excellent functional group compatibility, and 5-iodotriazoles are obtained as exclusive products in high yield from structurally and functionally diverse azides and 1-iodoalkynes. Owing to the mild reaction conditions, high chemoselectivity, and low copper catalyst loading, reaction workup is usually as simple as trituration followed by filtration. The scale-up is also easy, and a number of 5-iodotriazoles have been prepared in multigram quantities.

1-Iodoalkynes are readily obtained from terminal alkynes by treating them with *N*-iodomorpholine (**25**) in the presence of CuI, giving the corresponding 1-iodoalkynes within 30–60 min. The products can be isolated by passing the reaction mixture through a pad of silica gel or alumina, yielding the desired 1-iodoalkyne in high yield. It can be

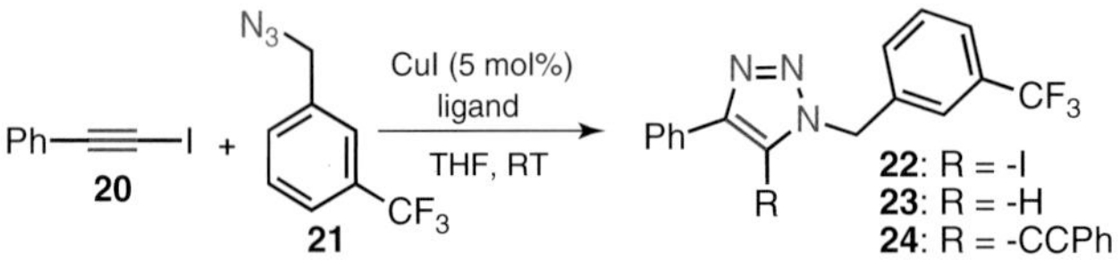

Scheme 7.6 The iodoalkyne version of the CuAAC reaction (ʰCuAAC).

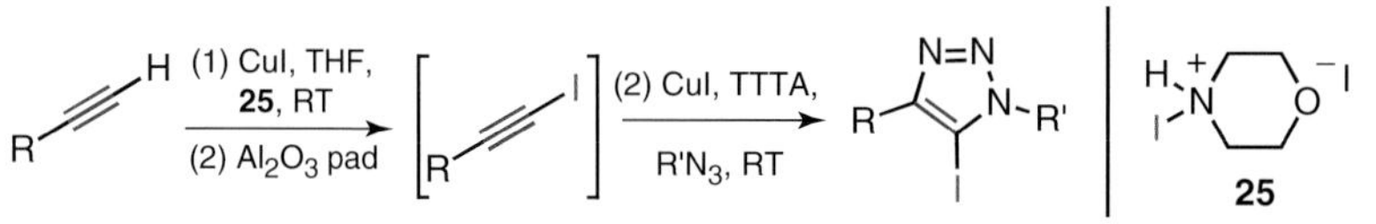

Scheme 7.7 One-pot, two-step synthesis of 5-iodo-1,2,3-triazoles.

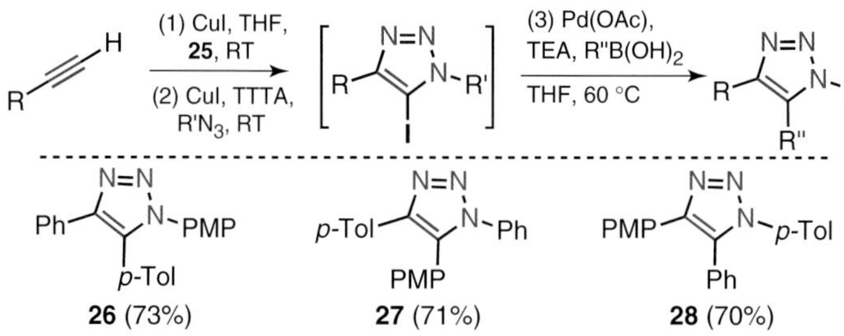

Scheme 7.8 One-pot, three-step synthesis of 1,4,5-triaryltriazoles. PMP = *p*-methoxyphenyl; *p*-Tol = *p*-tolyl (*p*-methylphenyl).

submitted to the reaction with an azide in a one-pot, two-step procedure (Scheme 7.7). The 1-iodoalkyne is partially purified via filtration through neutral alumina prior to the introduction of the azide component. This method gives 5-iodotriazoles with an efficiency comparable to that observed with the isolated 1-iodoalkynes.

This sequence could be further extended to the synthesis of 1,4,5-triaryl-1,2,3-triazoles (**26–28**) (Scheme 7.8) by assembling the 5-iodotriazole and immediately employing Pd(0)-catalyzed cross-coupling with an appropriate arylboronic acid. This simple stepwise construction obviates purification of any intermediates and simultaneously provides complete control over the placement of substituents around the 1,2,3-triazole nucleus, allowing facile access to all regioisomeric permutations of trisubstituted triazoles (**26–28**). Similar regiocontrolled synthesis would not be possible via thermal or ruthenium-catalyzed reaction due to the steric and electronic similarity of the aryl substituents (phenyl, tolyl, and *p*-methoxyphenyl).

In addition to the immediate synthetic utility of the iodo–CuAAC reactions, examination of the mechanism will provide a better understanding of both the iodo and the parent CuAAC processes. Although both reactions clearly share some common features, modes of activation of iodo- and terminal alkynes by copper are likely distinctly different. Our current mechanistic proposals are outlined in Scheme 7.8. One possible pathway is similar to that proposed for the CuAAC and involves the formation of the σ-acetylide complex **30** as the first key intermediate (Scheme 7.9, path *a*). Cu–I exchange via σ-bond metathesis with iodoalkyne **29** completes the cycle, liberating iodotriazole **34** and regenerating acetylide **30**.

Alternatively, copper may activate the iodoalkyne via the formation of a π-complex intermediate (Scheme 7.9, path *b*), which then engages the azide, producing complex **36**. Cyclization then proceeds via a now familiar vinylidene-like intermediate **37**, to give iodotriazole **34**. A similar transition state has been proposed to explain the involvement

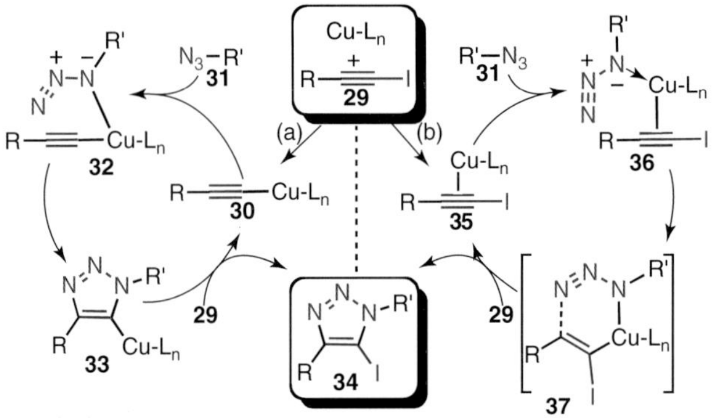

Scheme 7.9 Proposed mechanistic pathways for the Cu(I)-catalyzed azide–iodoalkyne cycloaddition.

of dicopper intermediates in the CuAAC reaction [70, 71]. The distinctive feature of this pathway is that the C-I bond is never severed during the catalysis.

As a result of extensive mechanistic and kinetic investigations, we have established that path *b* is the only operating pathway. The main argument in support of this hypothesis is the exclusive formation of the 5-iodotriazole even when the reaction is performed in protic solvents or with the substrates containing acidic protons. If path *a* were operational, the cuprated triazole intermediate **33** could be trapped with other electrophiles, including a proton, thereby producing a mixture of the 5-iodo- and 5-prototriazoles. Whatever the interactions of iodoalkynes with copper(I) are, they appear to dominate: when the reaction is performed on a mixture of a terminal and 1-iodoalkyne, the formation of the 5-prototriazole does begin until all the iodoalkyne has been consumed. This behavior indicates the complete catalyst monopoly by the iodoalkyne cycle and different catalyst resting states for the two competing processes.

7.6
Examples of Application of the CuAAC Reaction

The CuAAC reaction has been applied to a remarkable array of problems in synthetic chemistry, chemical biology, materials science, and other fields. A comprehensive or even representative list is beyond the scope of this chapter. Instead, two examples involving metallic copper as the source of CuAAC catalyst are highlighted. Although decidedly not typical of the body of CuAAC applications in the literature, it could well be the most convenient form of the catalyst and should receive greater attention in CuAAC. We also hope that these examples will give some indication of the facility with which the CuAAC process can be applied.

7.6.1
Synthesis of Compound Libraries for Biological Screening

The CuAAC process performs well in most common laboratory solvents and usually does not require protection from oxygen and water. Indeed, as noted above, aqueous

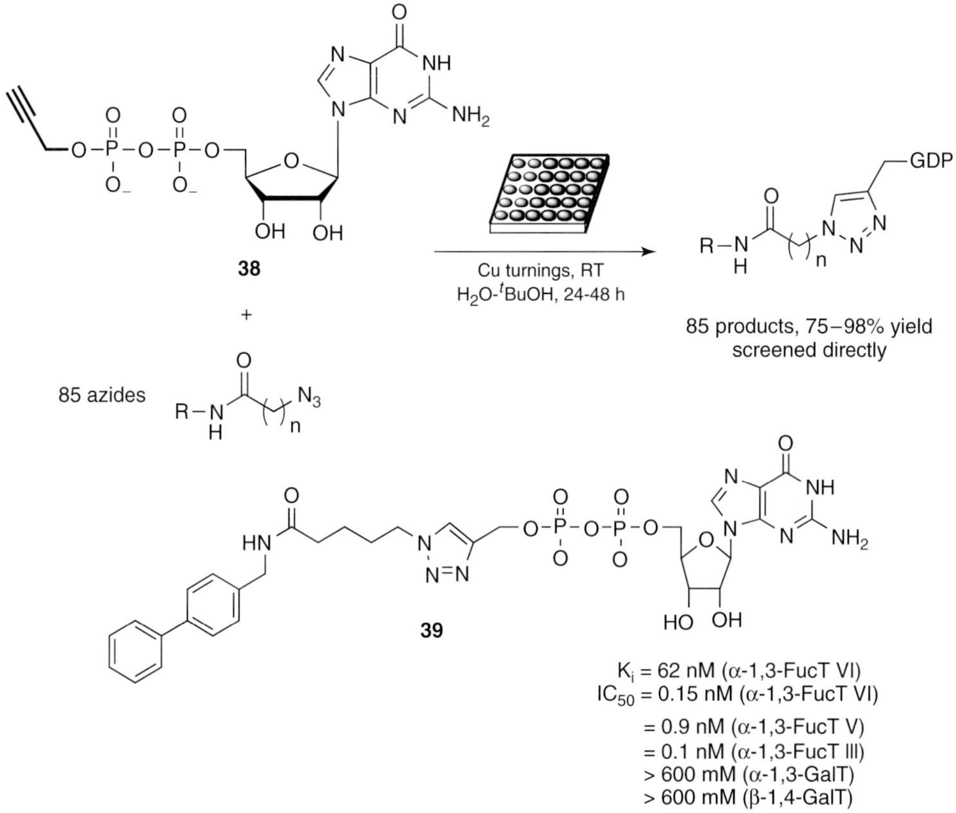

Scheme 7.10 Cu-mediated synthesis and direct screening of fucosyl transferase inhibitors.

solvents are commonly used and, in many cases, result in cleaner isolated products. The reaction is therefore an ideal tool for the synthesis of libraries for initial screening and also focused sets of compounds for structure–activity profiling. The lack of byproducts and high conversions often permit testing of CuAAC reaction mixtures without further purification. When necessary, traces of copper can be removed by solid-phase extraction utilizing a metal-scavenging resin or by simple filtration through a plug of silica gel.

In a study aimed at the discovery of inhibitors of human fucosyltransferase, CuAAC was used to link 85 azides to the GDP-derived alkyne **38** with excellent yields (Scheme 7.10) [74]. The library was screened directly, and a nanomolar inhibitor (**39**) was identified. Testing the purified hit compound **39** against several of glycosyltransferases and nucleotide-binding enzymes revealed that it was the most potent and selective inhibitor of human α-1,3-fucosyltransferase VI at that time.

Scheme 7.11 shows another example, in which a novel family of potent HIV-1 protease inhibitors was discovered using the CuAAC reaction [75]. A focused library of azide-containing fragments was united with a diverse array of functionalized alkyne-containing building blocks. After the direct screening of the crude reaction products, a lead structure (**40**) with $K_i = 98$ nM was identified. Optimization of both

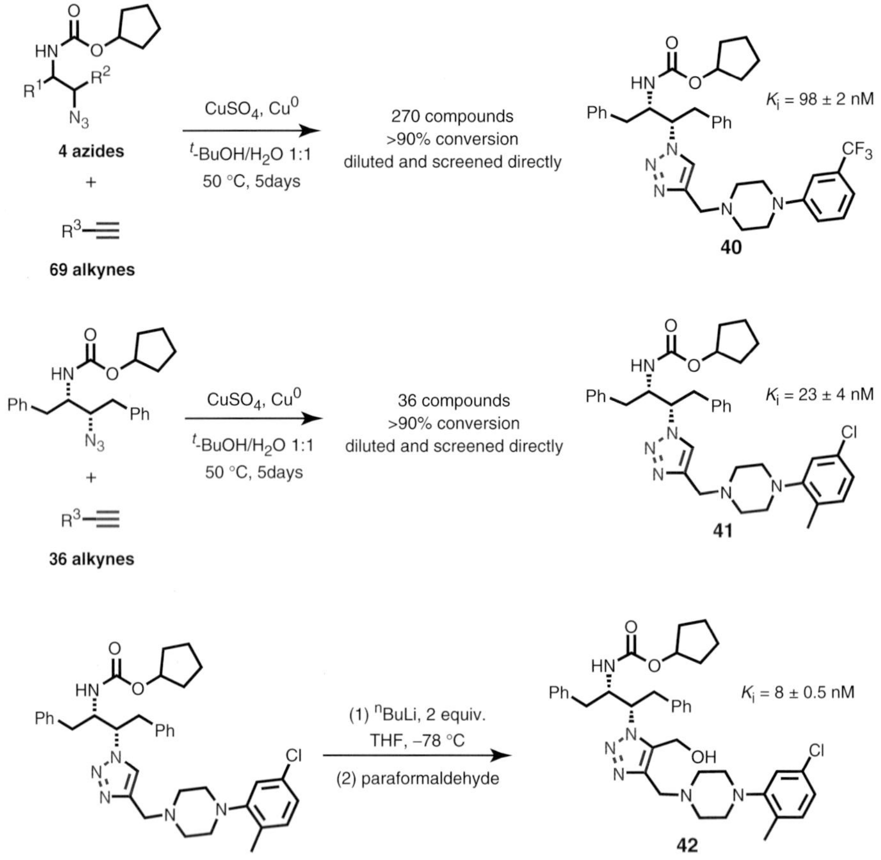

Scheme 7.11 CuAAC synthesis and direct screening of HIV protease inhibitors.

azide and alkyne fragments was equally facile (compound **41**, $K_i = 23$ nM). Further functionalization of the triazole at C5 gave a series of compounds with increased activity, exhibiting K_i values as low as 8 nM (compound **42**).

7.6.2
Copper-Binding Adhesives

Since Cu metal provides useful amounts of Cu(I) catalyst, and triazoles are major components of metal adhesives and coatings, we investigated the formation of crosslinked polymeric materials by the deposition of multivalent azides and alkynes on Cu-containing metallic surfaces [76–78]. The general design is shown in Scheme 7.12. Strong adhesives are formed over curing times ranging from many hours in the absence of added accelerating ligand to minutes in the presence of added Cu, ligands, and at elevated temperatures. Copper ions are extracted from the metal surface and distributed throughout the polymer matrix, presumably forming active catalytic sites as triazoles are created, which function

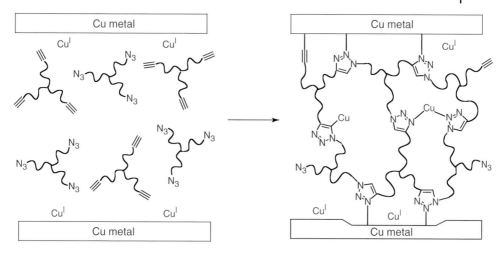

Scheme 7.12 Spontaneous formation of Cu-binding adhesives from polyvalent azides and alkynes placed between two metal surfaces.

as accelerating ligands in addition to stable linkages and adhesive units to the metal surface. This results in very high levels of crosslinking, giving rise to glass transition temperatures that are significantly higher than the curing temperature at which the adhesives are formed [78]. In the context of macroscopic materials, therefore, and also among competing molecules in solution, a kinetically labile metal catalyst can have significant advantages.

7.7
Reactions of Sulfonyl Azides

Sulfonyl azides participate in unique reactions with terminal alkynes under copper catalysis conditions. Depending on the conditions and reagents, products other than the expected triazole (43) [79] can be obtained, as shown in Scheme 7.13. These products are thought to derive from the cuprated triazole intermediate 47, which is destabilized by the strong electron-withdrawing character of the N-sulfonyl substituent. Ring–chain isomerization can occur to form the cuprated diazoimine 48, which, upon the loss of a molecule of dinitrogen, furnishes the N-sulfonylketenimine 50 [80]. Alternatively, copper(I) alkynamide 49 can be generated with a concomitant elimination of N_2 and, after protonation, would again generate the reactive ketenimine species 50. In the absence of a nucleophilic reagent, ketenimine dimers are isolated [80]. When the reaction is conducted in the presence of amines, N-sulfonylamidines (44) are formed [81]. In aqueous conditions, N-acylsulfonamides (45) are the major products [82, 83]. In addition to amines and water, ketenimine intermediates can be trapped with imines, furnishing N-sulfonylazetidinimines (46) (Scheme 7.13) [80].

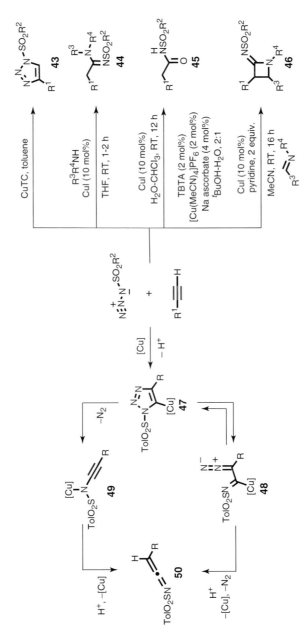

Scheme 7.13 Right, products of CuAAC reactions with sulfonyl azides. Left, possible pathways leading to ketenimine intermediates.

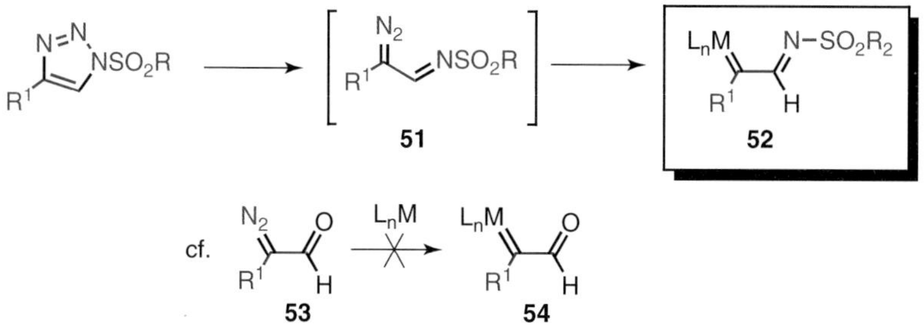

Scheme 7.14 Azavinyl carbenes from diazoimines.

7.7.1
1-Sulfonyl Triazoles: Convenient Precursors of Azavinyl Carbenes

As already mentioned, 1,2,3-triazoles are very stable heterocycles. However, those that bear a strong electron-withdrawing group at N1 are known to undergo ring–chain tautomerization. The ring–chain isomerism of 1-sulfonyl 1,2,3-triazoles can be exploited in synthesis. Thus, 1-sulfonyltriazoles could serve as precursors to the diazoimine species **51** that, in turn, could be converted to metal carbene complexes **52** (Scheme 7.14). Rhodium(II)-stabilized carbene complexes exhibit a wealth of reactivity, and this method of generating their diazo progenitors is particularly attractive considering that sulfonyltriazoles effectively become synthetic equivalents of α-diazoaldehydes (**53**), which are very unstable and cannot be converted to the corresponding metal carbenes (**54**).

In the presence of a dirhodium(II) tetraoctanoate catalyst, 1-sulfonyltriazoles react with nitriles, forming imidazoles [84]. The reaction proceeds at 60–80 °C with conventional heating or can be performed in a microwave reactor and generally provides imidazoles in good to excellent yields (Scheme 7.15a). The sulfonyl group can be readily removed, revealing the parent NH-imidazole. Alternatively, sulfonylimidazoles can be converted to 1,2,5-trisubstituted imidazoles by simple alkylation (Scheme 7.15b).

Another example of exquisite reactivity of rhodium(II) azavinyl carbenes is illustrated by their addition to alkenes, which is a very facile process that proceeds under mild conditions

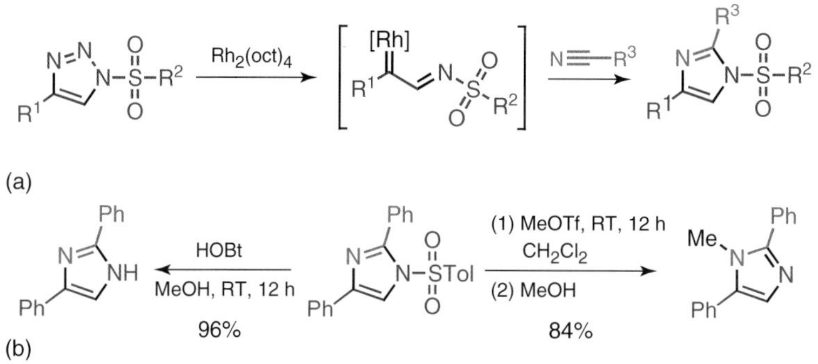

Scheme 7.15 Synthesis of imidazoles via rhodium-catalyzed transannulation of 1-sulfonyltriazoles.

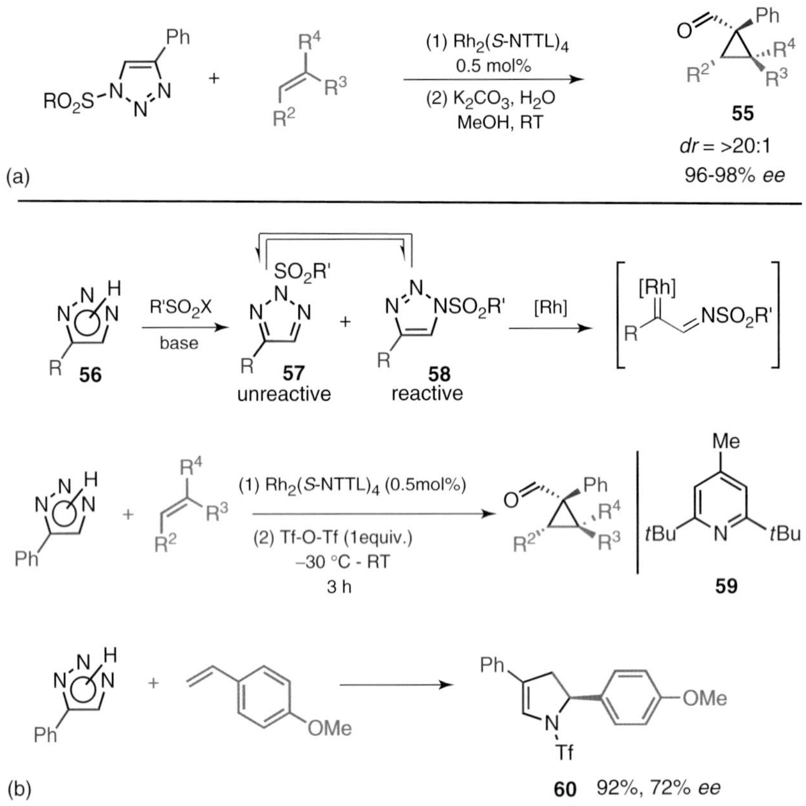

(a)

(b)

Scheme 7.16 Cyclopropanation of alkenes using 1-sulfonyl-1,2,3-triazoles.

with high diastereo- and enantioselectivity. Chiral rhodium(II) carboxylate complexes, such as $Rh_2(S\text{-}NTTL)_4$ and $Rh_2(S\text{-}NTV)_4$, are used in the reactions, which are usually performed at elevated temperature (Scheme 7.16a) when isolated N-sulfonyltriazoles are used as starting materials [85]. The N-sulfonylimine group can be easily removed, revealing virtually enantiopure cyclopropane carboxaldehydes (**55**). These useful synthetic intermediates are difficult to obtain using other methods.

In an alternative approach, very electron-deficient and highly reactive N-triflylazavinyl carbenes can be prepared by *in situ* sulfonylation of NH-triazoles (**56**) with triflic anhydride in the presence of a hindered pyridine base (**59**) (Scheme 7.16b) [86]. Although the sulfonylation step is not selective and results in the formation of both N1- and N2-sulfonylated triazoles (**57** and **58**, respectively), only the N1-isomer can undergo ring–chain isomerization and, therefore, form a reactive carbene intermediate. The resulting N-triflylazavinyl carbenes exhibit exceptional reactivity towards alkenes, resulting in the formation of cyclopropanes and 1,2-dihydropyrroles (when electron-rich alkenes are used; Scheme 7.16b) with excellent enantio- and diastereoselectivity. The ability to introduce extremely electron-withdrawing groups into the azavinyl carbene, thereby controlling its electrophilicity, is a valuable feature of this approach.

7.8
Outlook/Perspective

The quick acceptance which CuAAC gained by practitioners in many chemical disciplines illustrates the immense need for transformations that are experimentally simple, robust, and reliable. The click chemistry palette is, of course, not monochromatic, and recent examples of thiol–ene, thiol–yne, and strain-promoted cycloadditions have enriched the repertoire of click reactions. The efficiency and user friendliness of click reactions will continue to allow the preparation of new chemical functions, whether in organic synthesis, materials science, or biology. Just as importantly, CuAAC is a rare example of a reaction that thrives because of the highly adaptable, dynamic nature of its catalysts and the key intermediates, alkynyl copper complexes. Detailed examination of such complex catalytic processes requires unconventional approaches and unique tools for following reaction progress in real time under synthetically relevant conditions. Other superb catalysts are waiting to be found if we are adventurous enough to accept the uncertainty of not knowing the precise structure of the active catalytic species.

Acknowledgments

The author is grateful to many colleagues and co-workers who have contributed to the work cited here. This chapter is dedicated to Prof. K. Barry Sharpless, whose insatiable curiosity has been the inspiration for the development of click chemistry in general and this reaction in particular.

References

1. Kolb, H.C., Finn, M.G., and Sharpless, K.B. (2001) *Angew. Chem. Int. Ed.*, **40**, 2004–2021.
2. Tornøe, C.W., Christensen, C., and Meldal, M. (2002) *J. Org. Chem.*, **67**, 3057–3062.
3. Rostovtsev, V.V., Green, L.G., Fokin, V.V., and Sharpless, K.B. (2002) *Angew. Chem. Int. Ed.*, **41**, 2596–2599.
4. Tron, G.C., Pirali, T., Billington, R.A., Canonico, P.L., Sorba, G., and Genazzani, A.A. (2008) *Med. Res. Rev.*, **28**, 278–308.
5. Lutz, J.F. and Zarafshani, Z. (2008) *Adv. Drug Deliv. Rev.*, **60**, 958–970.
6. Meldal, M. and Tornøe, C.W. (2008) *Chem. Rev.*, **108**, 2952–3015.
7. Moses, J.E. and Moorhouse, A.D. (2007) *Chem. Soc. Rev.*, **36**, 1249–1262.
8. Fokin, V.V. (2007) *ACS Chem. Biol.*, **2**, 775–778.
9. Johnson, J.A., Koberstein, J.T., Finn, M.G., and Turro, N.J. (2008) *Macromol. Rapid Commun.*, **29**, 1052–1072.
10. Bock, V.D., Hiemstra, H., and van Maarseveen, J.H. (2006) *Eur. J. Org. Chem.*, 51–68.
11. Wu, P. and Fokin, V.V. (2007) *Aldrichim. Acta*, **40**, 7–17.
12. Fokin, V.V. (2011) New reactions of copper acetylides: catalytic dipolar cycloadditions and beyond, in *Catalyzed Carbon–Heteroatom Bond Formation* (ed. A.K. Yudin), Wiley-VCH Verlag GmbH, Weinheim, pp. 199–225.
13. Hein, J.E. and Fokin, V.V. (2010) *Chem. Soc. Rev.*, **39**, 1302–1315.
14. Larock, R.C. (2005) Synthesis of heterocycles and carbocycles by electrophilic cyclization of alkynes, in *Acetylene Chemistry: Chemistry, Biology, and Material Science*

(eds F. Diederich, P.J. Stang, and R.R. Tykwinski), Wiley-VCH Verlag GmbH, Weinheim, pp. 51–100.

15. Patai, S. (ed.) (1971) *Chemistry of the Azido Group*, John Wiley & Sons, Ltd., New York.

16. Braese, S., Gil, C., Knepper, K., and Zimmermann, V. (2005) *Angew. Chem. Int. Ed.*, **44**, 5188–5240.

17. Huisgen, R. (1963) *Angew. Chem. Int. Ed. Engl.*, **2**, 565–598.

18. Himo, F., Lovell, T., Hilgraf, R., Rostovtsev, V.V., Noodleman, L., Sharpless, K.B., and Fokin, V.V. (2005) *J. Am. Chem. Soc.*, **127**, 210–216.

19. Tomé, A.C. (2004) Product Class 13: 1,2,3-Triazolws, in *Five-Membered Het-arenes with Three or More Heteroatoms*, Houben–Weyl Methods of Molecular Transformations, *Science of Synthesis*, Vol. 13 (eds. R.C. Storr and T.L. Gilchrist), Georg Thieme, Stuttgart, pp. 415–601.

20. Krivopalov, V.P. and Shkurko, O.P. (2005) *Russ. Chem. Rev.*, **74**, 339–379.

21. Zhang, L., Chen, X., Xue, P., Sun, H.H.Y., Williams, I.D., Sharpless, K.B., Fokin, V.V., and Jia, G. (2005) *J. Am. Chem. Soc.*, **127**, 15998–15999.

22. Boren, B.C., Narayan, S., Rasmussen, L.K., Zhang, L., Zhao, H., Lin, Z., Jia, G., and Fokin, V.V. (2008) *J. Am. Chem. Soc.*, **130**, 8923–8930.

23. Rasmussen, L.K., Boren, B.C., and Fokin, V.V. (2007) *Org. Lett.*, **9**, 5337–5339.

24. Majireck, M.M. and Weinreb, S.M. (2006) *J. Org. Chem.*, **71**, 8680–8683.

25. Bernardes, G.J.L., Castagner, B., and Seeberger, P.H. (2009) *ACS Chem. Biol.*, **4**, 703–713.

26. Reddy, K.R., Rajgopal, K., and Kantam, M.L. (2006) *Synlett*, 957–959.

27. Reddy, K.R., Rajgopal, K., and Kantam, M.L. (2007) *Catal. Lett.*, **114**, 36–40.

28. Fukuzawa, S., Shimizu, E., and Kikuchi, S. (2007) *Synlett*, 2436–2438.

29. Nigh, W.G. (1973) Oxidations by CUpric Ion, in *Oxidation in Organic Chemistry, Part B* (ed. W.S. Trahanovsky), Academic Press, New York, pp. 1–95.

30. Siemsen, P., Livingston, R.C., and Diederich, F. (2000) *Angew. Chem. Int. Ed.*, **39**, 2632–2657.

31. Fahrni, C.J. (2007) *Curr. Opin. Chem. Biol.*, **11**, 121–127.

32. Angell, Y. and Burgess, K. (2007) *Angew. Chem. Int. Ed.*, **46**, 3649–3651.

33. Gerard, B., Ryan, J., Beeler, A.B., and Porco, J.A. Jr. (2006) *Tetrahedron*, **62**, 6405–6411.

34. Appukkuttan, P., Dehaen, W., Fokin, V.V., and van der Eycken, E. (2004) *Org. Lett.*, **6**, 4223–4225.

35. Moorhouse, A.D. and Moses, J.E. (2008) *Synlett*, 2089–2092.

36. Pachon, L.D., van Maarseveen, J.H., and Rothenberg, G. (2005) *Adv. Synth. Catal.*, **347**, 811–815.

37. Molteni, G., Bianchi, C.L., Marinoni, G., Santo, N., and Ponti, A. (2006) *New J. Chem.*, **30**, 1137–1139.

38. Lipshutz, B.H. and Taft, B.R. (2006) *Angew. Chem. Int. Ed.*, **45**, 8235–8238.

39. Bogdan, A.R. and Sach, N.W. (2009) *Adv. Synth. Catal.*, **351**, 849–854.

40. Zhu, W. and Ma, D. (2004) *Chem. Commun.*, 888–889.

41. Feldman, A.K., Colasson, B., and Fokin, V.V. (2004) *Org. Lett.*, **6**, 3897–3899.

42. Barral, K., Moorhouse, A.D., and Moses, J.E. (2007) *Org. Lett.*, **9**, 1809–1811.

43. Pearson, R.G. (1988) *J. Am. Chem. Soc.*, **110**, 7684–7690.

44. Broggi, J., Kumamoto, H., Berteina-Raboin, S., Nolan, S.P., and Agrofoglio, L.A. (2009) *Eur. J. Org. Chem.*, 1880–1888.

45. Perez-Balderas, F., Ortega-Munoz, M., Morales-Sanfrutos, J., Hernandez-Mateo, F., Calvo-Flores, F.G., Calvo-Asin, J.A., Isac-Garcia, J., and Santoyo-Gonzalez, F. (2003) *Org. Lett.*, **5**, 1951–1954.

46. Malkoch, M., Schleicher, K., Drockenmuller, E., Hawker, C.J., Russell, T.P., Wu, P., and Fokin, V.V. (2005) *Macromolecules*, **38**, 3663–3678.

47. Wu, P., Feldman, A.K., Nugent, A.K., Hawker, C.J., Scheel, A., Voit, B., Pyun, J., Frechet, J.M.J., Sharpless, K.B., and Fokin, V.V. (2004) *Angew. Chem. Int. Ed.*, **43**, 3928–3932.

48. Gonda, Z. and Novák, Z. (2010) *Dalton Trans.*, **39**, 726–729.

49. Campbell-Verduyn, L.S., Mirfeizi, L., Dierckx, R.A., Elsinga, P.H., and Feringa, B.L. (2009) *Chem. Commun.*, 2139–2141.

50. Detz, R.J., Arevalo Heras, S., de Gelder, R., van Leeuwen, P.W.N.M., Hiemstra, H., Reek, J.N.H., and van Maarseveen, J.H. (2006) *Org. Lett.*, **8**, 3227–3230.

51. Bai, S.Q., Koh, L.L., and Hor, T.S.A. (2009) *Inorg. Chem.*, **48**, 1207–1213.

52. Díez-González, S. and Nolan, S.P. (2008) *Angew. Chem. Int. Ed.*, **46**, 9013–9016.

53. Nolte, C., Mayer, P., and Straub, B.F. (2007) *Angew. Chem. Int. Ed.*, **46**, 2101–2103.

54. Rodionov, V.O., Presolski, S.I., Diaz, D.D., Fokin, V.V., and Finn, M.G. (2007) *J. Am. Chem. Soc.*, **129**, 12705–12712.

55. Mykhalichko, B.M., Temkin, O.N., and Mys'kiv, M.G. (2001) *Russ. Chem. Rev.*, **69**, 957–984.

56. Candelon, N., Lastecoueres, D., Diallo, A.K., Aranzaes, J.R., Astruc, D., and Vincent, J.M. (2008) *Chem. Commun.*, 741–743.

57. Chan, T.R., Hilgraf, R., Sharpless, K.B., and Fokin, V.V. (2004) *Org. Lett.*, **6**, 2853–2855.

58. Wang, Q., Chan, T.R., Hilgraf, R., Fokin, V.V., Sharpless, K.B., and Finn, M.G. (2003) *J. Am. Chem. Soc.*, **125**, 3192–3193.

59. Chan, T.R. and Fokin, V.V. (2007) *QSAR Comb. Sci.*, **26**, 1274–1279.

60. Hein, J.E., Krasnova, L.B., Iwasaki, M., and Fokin, V.V. (2011) *Org. Synth.*, **88**, 238–246.

61. Chan, T.R., Fokin, V.V., and Sharpless, K.B. (2004) Abstracts of Papers, 227th ACS National Meeting, Anaheim, CA, March 28–April 1, 2004, ORGN-041.

62. Hong, V., Presolski, S.I., Ma, C., and Finn, M.G. (2009) *Angew. Chem. Int. Ed.*, **48**, 9879–9883.

63. Soriano del Amo, D., Wang, W., Jiang, H., Besanceney, C., Yan, A.C., Levy, M., Liu, Y., Marlow, F.L., and Wu, P. (2010) *J. Am. Chem. Soc.*, **132**, 16893–16899.

64. Lewis, W.G., Magallon, F.G., Fokin, V.V., and Finn, M.G. (2004) *J. Am. Chem. Soc.*, **126**, 9152–9153.

65. Hong, V., Udit, A.K., Evans, R.A., and Finn, M.G. (2008) *ChemBioChem*, **9**, 1481–1486.

66. Rodionov, V.O., Presolski, S.I., Gardinier, S., Lim, Y.H., and Finn, M.G. (2007) *J. Am. Chem. Soc.*, **129**, 12696–12704.

67. Cenini, S., Gallo, E., Caselli, A., Ragaini, F., Fantauzzi, S., and Piangiolino, C. (2006) *Coord. Chem. Rev.*, **250**, 1234–1253.

68. Vrieze, K. and van Koten, G. (1987) *Comprehensive Coordination Chemistry*, vol. **2**, Pergamon Press, Oxford, pp. 189–245.

69. Rodionov, V.O., Fokin, V.V., and Finn, M.G. (2005) *Angew. Chem. Int. Ed.*, **44**, 2210–2215.

70. Ahlquist, M. and Fokin, V.V. (2007) *Organometallics*, **26**, 4389–4391.

71. Straub, B.F. (2007) *Chem. Commun.*, 3868–3870.

72. Rosner, T., Pfaltz, A., and Blackmond, D.G. (2001) *J. Am. Chem. Soc.*, **123**, 4621–4622.

73. Hein, J.E., Tripp, J.C., Krasnova, L.B., Sharpless, K.B., and Fokin, V.V. (2009) *Angew. Chem. Int. Ed.*, **48**, 8018–8021.

74. Lee, L.V., Mitchell, M.L., Huang, S.J., Fokin, V.V., Sharpless, K.B., and Wong, C.H. (2003) *J. Am. Chem. Soc.*, **125**, 9588–9589.

75. Whiting, M., Tripp, J.C., Lin, Y.C., Lindstrom, W., Olson, A.J., Elder, J.H., Sharpless, K.B., and Fokin, V.V. (2006) *J. Med. Chem.*, **49**, 7697–7710.

76. Diaz, D.D., Punna, S., Holzer, P., McPherson, A.K., Sharpless, K.B., Fokin, V.V., and Finn, M.G. (2004) *J. Polym. Sci., Part A: Polym. Chem.*, **42**, 4392–4403.

77. Liu, Y., Diaz, D.D., Accurso, A.A., Sharpless, K.B., Fokin, V.V., and Finn, M.G. (2007) *J. Polym. Sci., Part A: Polym. Chem.*, **45**, 5182–5189.

78. Le Baut, N., Diaz, D.D., Punna, S., Finn, M.G., and Brown, H.R. (2007) *Polymer*, **48**, 239–244.

79. Yoo, E.J., Ahlquist, M., Kim, S.H., Bae, I., Fokin, V.V., Sharpless, K.B., and Chang, S. (2007) *Angew. Chem. Int. Ed.*, **46**, 1730–1733.

80. Whiting, M. and Fokin, V.V. (2006) *Angew. Chem. Int. Ed.*, **45**, 3157–3161.

81. Bae, I., Han, H., and Chang, S. (2005) *J. Am. Chem. Soc.*, **127**, 2038–2039.

82. Cassidy, M.P., Raushel, J., and Fokin, V.V. (2006) *Angew. Chem. Int. Ed.*, **45**, 3154–3157.

83. Cho, S.H., Yoo, E.J., Bae, I., and Chang, S. (2005) *J. Am. Chem. Soc.*, **127**, 16046–16047.

84. Horneff, T., Chuprakov, S., Chernyak, N., Gevorgyan, V., and Fokin, V.V. (2008) *J. Am. Chem. Soc.*, **130**, 14972–14974.

85. Chuprakov, S., Kwok, S.W., Zhang, L., Lercher, L., and Fokin, V.V. (2009) *J. Am. Chem. Soc.*, **131**, 18034–18035.

86. Grimster, N.P., Zhang, L., and Fokin, V.V. (2010) *J. Am. Chem. Soc.*, **132**, 2510–2511.

Commentary Part

Comment 1

Krzysztof Matyjaszewski

CuAAC, reviewed by Valery Fokin, a top expert and practitioner of click chemistry, provides an excellent summary of the mechanistic features and phenomenology of these reactions and presents some selected applications of CuAAC in organic chemistry, bio-related areas, and also materials science.

A quick search for "click" and "polymer or polymerization" and "copper or Cu" resulted in 868 references in SciFinder, which is approximately one-third of all CuAAC papers. This indicates how important this chemistry is for modern polymer and materials science. Many excellent reviews have been published on applications of click chemistry to polymer science, and some of the most important are cited here (out of over 100!) [C1–C13].

In fact, very high yield, selectivity, orthogonality, and high rate are all critical for polymerization, chain end functionalization, and any post-polymerization transformation reactions. One can use these reactions not only to make linear polymers by a step-growth process [C14], but also prepare more sophisticated structures including cyclic polymers [C15], multisegmented block copolymers [C16], stars [C17], and even bottlebrushes [C18].

It has been observed previously that CuAAC is not fully compatible with DNA and RNA, but recently a successful click process has been reported in water and buffer in the presence of acetonitrile (as little as 2 vol.%), that stabilizes Cu(I) species sufficiently. Several bioconjugates with organic polymers have been prepared in this way [C19–C21].

Usually, CuAAC is considered an irreversible reaction, but recently it was demonstrated that one can convert triazines back to alkynes and azides in the presence of ultrasound. This shows that mechanochemistry may play an import role in azide–alkyne cycloaddition [C22]. It has been observed previously that CuAAC can give limited yields in the synthesis of densely grafted molecular brushes with sterically hindered side chains, suggesting a potential reversibility [C18].

In summary, CuAAC, the first and the most versatile click reaction, in spite of its relatively complex mechanism, is one of the simplest high-yield organic reactions and has been widely accepted as a versatile tool in bio-related systems and in polymer/materials science. I foresee the continued use of this process in various areas of chemistry and related disciplines.

References

C1. Harvison, M.A. and Lowe, A.B. (2011) *Macromol. Rapid Commun.*, **32**, 779–800.

C2. Akeroyd, N. and Klumperman, B. (2011) *Eur. Polym. J.*, **47**, 1207–1231.

C3. Fu, R. and Fu, G.D. (2011) *Polym. Chem.*, **2**, 465–475.

C4. Tasdelen, M.A., Kahveci, M.U., and Yagci, Y. (2011) *Prog. Polym. Sci.*, **36**, 455–567.

C5. Mansfeld, U., Pietsch, C., Hoogenboom, R., Becer, C.R., and Schubert, U.S. (2010) *Polym. Chem.*, **1**, 1560–1598.

C6. Qin, A., Lam, J.W.Y., and Tang, B.Z. (2010) *Chem. Soc. Rev.*, **39**, 2522–2544.

C7. Binder, W.H. and Sachsenhofer, R. "Click" chemistry on supramolecular materials, in *Click Chemistry for Biotechnology and Materials Science* (ed. J. Lahann), (2009) John Wiley & Sons, Ltd., Chichester, pp 119–175.

C8. Golas, P.L. and Matyjaszewski, K. (2010) *Chem. Soc. Rev.*, **39**, 1338–1354.

C9. Sumerlin, B.S. and Vogt, A.P. (2010) *Macromolecules*, **43**, 1–13.

C10. Binder, W.H. and Sachsenhofer, R. (2008) *Macromol. Rapid Commun.*, **29**, 952–981.

C11. Fournier, D., Hoogenboom, R., and Schubert, U.S. (2007) *Chem. Soc. Rev.*, **36**, 1369–1380.

C12. Binder, W.H. and Sachsenhofer, R. (2007) *Macromol. Rapid Commun.*, **28**, 15–54.

C13. Golas, P.L. and Matyjaszewski, K. (2007) *QSAR Comb. Sci.*, **26**, 1116–1134.

C14. Tsarevsky, N.V., Sumerlin, B.S., and Matyjaszewski, K. (2005) *Macromolecules*, **38**, 3558–3561.

C15. Laurent, B.A. and Grayson, S.M. (2006) *J. Am. Chem. Soc.*, **128**, 4238–4239.

C16. Golas, P.L., Tsarevsky, N.V., Sumerlin, B.S., Walker, L.M., and Matyjaszewski, K. (2007) *Aust. J. Chem.*, **60**, 400–404.

C17. Gao, H. and Matyjaszewski, K. (2006) *Macromolecules*, **39**, 4960–4965.

C18. Gao, H. and Matyjaszewski, K. (2007) *J. Am. Chem. Soc.*, **129**, 6633–6639.

C19. Averick, S., Paredes, E., Li, W., Matyjaszewski, K., and Das, S.R. (2011) *Bioconj. Chem.*, **22**, 2020–2037.

C20. Pan, P., Fujita, M., Ooi, W.Y., Sudesh, K., Takarada, T., Goto, A., and Maeda, M. (2011) *Polymer*, **52**, 895–900.

C21. Paredes, E. and Das, S.R. (2011) *ChemBioChem*, **12**, 125–131.

C22. Brantley, J.N., Wiggins, K.M., and Bielawski, C.W. (2011) *Science*, **333**, 1606–1609.

8

Transition Metal-Catalyzed C–H Functionalization: Synthetically Enabling Reactions for Building Molecular Complexity

Keary M. Engle and Jin-Quan Yu

8.1
Introduction

Ubiquitous in organic molecules, unactivated carbon–hydrogen (C–H) bonds are among the simplest chemical moieties found in Nature. Although C–H bonds are typically inert to chemical transformations, directly utilizing them as reaction partners is highly desirable from the perspective of organic chemistry. In target-oriented synthesis, for example, methods to convert C–H bonds to new carbon–carbon (C–C) or carbon–heteroatom bonds can expedite the synthesis of architecturally complex intermediates *en route* to drugs, agrochemicals, or other biologically active molecules [1–3]. Similarly, in diversity-oriented synthesis, an entire library of analogs can be synthesized directly from a lead compound via late-stage C–H functionalization [4–6]. In contrast, traditional methods would require one to return to the early stages to generate diversity, which is costly in terms of chemical and energy inputs.

C–H bonds are generally unreactive; however, there are several different approaches for activating and functionalizing C–H bonds. Free radicals are capable of abstracting the hydrogen atom from a C–H bond to form a new, highly reactive C• species, which can readily be functionalized. Additionally, it has long been established that enzymatic systems such as cytochrome P450 are capable of oxidizing unactivated C–H bonds [7]. For over a century, it has also been known that transition metals can mediate or catalyze C–H functionalization reactions [8–14]. For organic chemists, this last class of reactions is of particular interest because of the diverse reactivity inherent in the different transition metals and because the coordination environment around the metal, and thus the selectivity and reactivity, can be finely tailored through ligand design.

In this chapter, we discuss the subject of transition metal-catalyzed C–H functionalization as it pertains to generating molecular complexity in organic synthesis. We begin by giving a brief introduction to early studies with an inorganic and organometallic focus. This pioneering work paved the way for the dramatic upsurge of interest during the past few decades. Research in transition metal-catalyzed C–H

Organic Chemistry – Breakthroughs and Perspectives, First Edition. Edited by Kuiling Ding and Li-Xin Dai.
© 2012 Wiley-VCH Verlag GmbH & Co. KGaA. Published 2012 by Wiley-VCH Verlag GmbH & Co. KGaA.

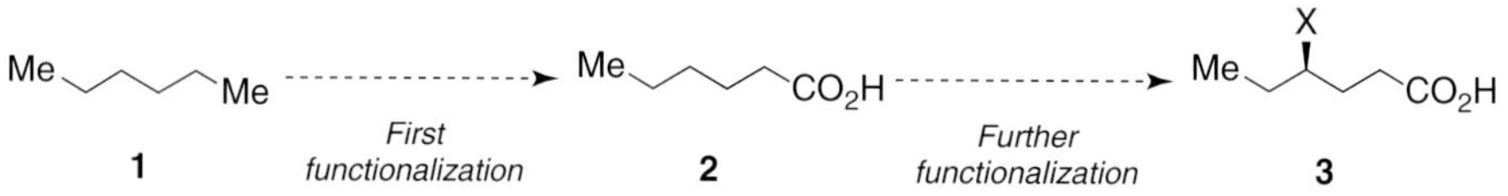

Scheme 8.1 Comparison of "first functionalization" and "further functionalization," two approaches that utilize C–H functionalization for different purposes.

functionalization can roughly be classified into two thematic subfields (Scheme 8.1). The first concerns the utilization of completely unfunctionalized aromatic and aliphatic hydrocarbons to install a single new functional group; we refer to this approach as "first functionalization." The second focuses on the modification of prefunctionalized substrates (compounds containing one or more reactive functional groups), which we call "further functionalization." The challenges and opportunities in each of these areas are very different and, realistically, a detailed discussion of either could easily fill an entire textbook. Our goal in this chapter is to give the reader a flavor for how further functionalization has emerged as a tool in organic synthesis.

This chapter is not intended to be comprehensive. Of the innumerable findings from the rich field of transition metal-catalyzed C–H functionalization, we have selected thematically linked examples that are pertinent to the discussion herein. In the interest of being concise, several topics are given only a cursory treatment; however, in these cases, we have attempted to provide references to review articles or primary literature for readers who are interested in further details. Inasmuch as possible, we have tried to limit our focus to the functionalization of *unactivated* C–H bonds. Thus, we do not cover reactions that effect the functionalization of other single bonds of C–H bonds, such as those that are acidic or those adjacent to heteroatoms. Following this logic, we also largely ignore C(allylic)–H oxidation, despite its prominence in organic synthesis. Enzymatic systems are mentioned only in the context of biomimetic catalysts, and heterogeneous transition metal-catalyzed processes are given minimal attention.

After presenting historical background information pertaining to transition metal-mediated C–H cleavage by stoichiometric complexes (Section 8.2), we shift our focus to catalytic transformations. We briefly highlight some of the recent progress on the first functionalization front (Section 8.3) and then move into a more thorough discussion of further functionalization, beginning with a review of the seminal work to establish and develop logic and strategy of applying C–H functionalization in complex settings (Section 8.4). Among the transition metal-catalyzed C–H functionalization reactions, we view those that proceed via metal insertion (Section 8.5) as particularly promising, and we therefore devote special attention to that topic. Other classes of transition metal-catalyzed C–H functionalization reactions are also covered (Section 8.6), albeit in less detail, including biomimetic oxidation, carbenoid insertion, and nitrenoid insertion. Throughout the text, we try to highlight future directions in the field to give practitioners of organic synthesis a sense of the untapped possibilities and potential opportunities that lie ahead.

8.2
Background and Early Work

8.2.1
The Challenges of Functionalizing C–H Bonds

Unactivated C–H bonds are challenging to functionalize owing to a confluence of factors that render them inert under most reaction conditions. First, the two atoms are held together by a strong covalent bond with a dissociation energy in the range 90–100 kcal mol^{-1}. Second, the bond acidity is low; pK_a values generally lie between 45 and 60, meaning that heterolytically cleaving an unactivated C–H bonds by treatment with strong base is generally not a viable approach. Third, C–H bonds possess unreactive molecular orbital profiles, true to their "paraffin" nature (in Latin, "paraffin" literally means "lacking affinity" or "lacking reactivity"). In other words, they have a high-lying lowest unoccupied molecular orbital (LUMO) σ^* and a low-lying highest occupied molecular orbital (HOMO) σ. This stands in contrast to other carbogenic fragments, such as alkenes, which have reactive π- and π^*-orbitals that can readily engage in chemical reactions, and to other bonds with similarly high dissociation energies, like C–O bonds, which can be easily broken, for instance, via activation of the oxygen atom with a Lewis acid and displacement with a nucleophile. Based on the molecular orbital considerations, devising systems for cleaving and functionalizing C(sp)–H, C(sp^2)–H, or C(allylic)–H bonds is generally more straightforward than in the case of unactivated C(sp^3)–H bonds because the occupied π-orbital(s) can serve to recruit transition metals (or other Lewis acidic species) to activate the C–H bond and lower the energetic barrier for bond breaking.

In systems where C–H functionalization can be performed, another major challenge is to control the selectivity. For instance, under the strongly oxidative conditions in which an unactivated C–H bond can be converted into a C–O bond, low levels of chemoselectivity often lead to over-oxidation. The most familiar example of this problem is in hydrocarbon (C$_n$H2$_{n+2}$) combustion, where the C–H bonds repeatedly undergo oxidative functionalization in the presence of O$_2$ until only CO$_2$ and H$_2$O remain (Scheme 8.2). Moreover, other problems of regio- and stereochemistry can also be manifest. Because C–H bonds are ubiquitous in carbogenic molecules, devising strategies that functionalize some but not others is difficult, particularly when they are disposed in similar spatial and electronic environments. Under the energetic conditions that are necessary for C–H functionalization, little meaningful selectivity can be achieved, which leads to intractable product mixtures, compromising utility.

Transition metals, which are capable of cleaving unactivated C–H bonds through a variety of different mechanisms, offer a potential avenue for overcoming the numerous aforementioned challenges. Indeed, during the past several decades, chemists have began to harness the power of transition metals to control these process with high levels of

$$C_nH_{2n+2} + \left(\frac{3n+1}{2}\right)O_2 \longrightarrow (n+1)\ H_2O + n\,CO_2 + \text{Energy}$$

Scheme 8.2 Hydrocarbon combustion, a familiar unselective C–H functionalization reaction.

selectivity and to take advantage of the synthetic potential of transition metal-catalyzed C–H functionalization methods.

Owing to the presence of their reactive d-orbitals, transition metals possess reactivity that is not typically observed in main group elements, including the ability to break unactivated C–H bonds in a controlled and predictable manner. Over the past several decades, several different mechanistic pathways for transition metal-mediated C–H cleavage have been elucidated, and this new knowledge has greatly expanded our understanding of these processes and the scientific principles that underlie them (Scheme 8.3). Because, the mechanistic aspects of C–H cleavage have been discussed thoroughly elsewhere [10, 13, 15], a detailed treatment of this topic is beyond the scope of this chapter. In this section, we merely highlight several of the common mechanistic pathways of transition metal-mediated C–H functionalization to offer the reader a suitable context for interpreting the reactivity patterns in the catalytic transformations below (Sections 8.5–8.7).

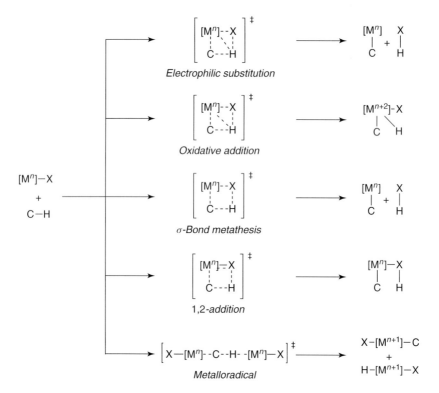

Scheme 8.3 Five mechanisms for inner-sphere transition metal-mediated C–H cleavage involving [M–C] intermediates.

At this stage, it is also important to clarify some of the terminology that we use throughout the text. We use the term "C–H cleavage" to denote to the general process of metal-mediated carbon–hydrogen bond breaking along any of a variety of mechanistic pathways and "C–H functionalization" to refer to the process of converting a C–H bond to a C–Y bond, irrespective of the method. The phrase "C–H activation" has been used in the literature to classify reactions that proceed via inner-sphere C(alkyl)–H bond breaking by a low-valent metal to generate a new intermediate [M–C] species [16, 17]. This traditional definition generally excludes metal-mediated reactions involving free radicals (e.g., Fenton [18–20] and Gif [21, 22] chemistry) and also C(sp^2)–H functionalization processes. Because this definition is not stringently adhered to in much of the modern literature, in this chapter we avoid the term "C–H activation" in an effort to obviate confusion.

Broadly speaking, transition metal-mediated methods for breaking C–H bonds can be classified into two categories: inner-sphere mechanisms that involve a new [M–C] species and outer-sphere mechanisms that do not. Within the first category, which we collectively refer to as "C–H functionalization via metal insertion," several possible operative mechanisms for transition metal-mediated C–H cleavage include (i) electrophilic activation, (ii) oxidative addition, (iii) σ-bond metathesis, (iv) 1,2-addition, and (v) metalloradical activation [13, 15]. Scheme 8.3 depicts general transition states for each of these processes and shows the corresponding organometallic species that are formed. Each of these mechanisms is potentially applicable for C(sp^2)–H and C(sp^3)–H cleavage. Among the five shown, the first two are the most common in synthetically useful catalytic transformations (Section 8.5), and we thus focus special attention on this class of reactions in this section and Section 8.3.

1) Electrophilic activation of arenes and alkenes leads to metallated carbon fragments in a process that is formally redox-neutral with concomitant loss of HX. Metals such as Hg(II), Pd(II), Pt(II), and Rh(III), have been shown to exhibit such reactivity. In the context of C(aryl)–H cleavage, these process can be seen as being roughly analogous to traditional electrophilic aromatic substitution; however, in C(alkyl)–H cleavage, this comparison obviously does not hold. Catalytic transformations using this mode of C–H cleavage require an external oxidant either during or after the functionalization step, in order to regenerate the catalytically active species.

2) Oxidative addition reactions between C–H bonds and transition metal complexes are typically observed with low-valent late transition metal species, including Ru(0), Rh(I), Ir(I), and Pt(II). A prerequisite is that the reactive metal species needs to possess an open coordination site for initial interaction with the C–H bond. In many of the early stoichiometric examples, these reactive metal species were generated by thermolysis or photolysis to induce the loss of a small molecule (e.g., CO, H$_2$, or CH$_4$) from a precursor complex. In more recent catalytic examples, alternative thermo-induced methods, such as alkene hydroarylation, are employed for the generation of the catalytically active low-valent metal species (see Section 8.5 for additional discussion).

3) σ-Bond metathesis is observed in the reaction between [M–R^1] or [M–H] species and alkanes (R^2–H), generally leading to alkyl group exchange, rather than the

more desirable formation of a new C–C bond (R^1–R^2) [23]. (A formal means of accomplishing this type of C–C bond formation in a catalytic sense, so-called "alkane metathesis," is discussed in Section 8.3.3.) This mechanism is typically observed with early transition metals in the lanthanide and actinide series.

4) 1,2-Addition of a C–H bond across an unsaturated [M=C] or [M=N] species can take place to generate a new M–C bond [24]. This reactivity has been observed with both alkanes and arenes using a variety of different metals, including Zr(IV), Ti(IV), and W(VI), with the active species generally generated *in situ*. The viability of this approach for catalytic transformations remains to be seen.

5) Metalloradical C–H cleavage has been established using [Rh(II)–porphyrin] complexes [25, 26]. The reactions are proposed to proceed via a four-center trimolecular transition state in which a metalloradical interacts at both the carbon and hydrogen sites (as depicted in Scheme 8.3). It is important to point out that these reactions are distinct from metal-mediated reactions involving alkyl-centered radicals, which are discussed below. Examples and applications of this mechanism remain limited in number.

In addition to inner-sphere C–H functionalization reactions that proceed via metal insertion, there are also transition metal-mediated C–H functionalization reactions that involve outer-sphere mechanisms and do not form new intermediate [M–C] species (Scheme 8.4). Two that are relevant for their utility as further functionalization methods in organic synthesis are biomimetic C–H oxidation reactions and carbenoid/nitrenoid C–H insertion reactions. In this chapter, we do not discuss C–H functionalization transformations where the active metal species is thought to generate diffusely free carbon- or heteroatom-centered radicals, such as those classified as Fenton or Gif chemistry.

1) Metal–oxo species can abstract a hydrogen atom from $C(sp^2)$–H or $C(sp^3)$–H bonds to generate a carbon-centered radical ($C^{\bullet}$) through homolytic cleavage. Recombination of the resulting $C^{\bullet}$ species with the oxygen-centered radical that is bound to the metal forges the new C–O bond. While the mechanistic fine features of these reactions vary from system to system, transformations of this type are common in enzymatic catalysis. The active sites of these biocatalysts contain transition metals,

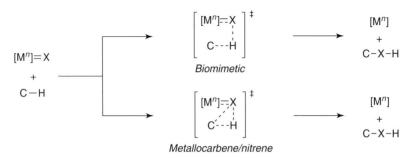

Scheme 8.4 Two mechanisms for outer-sphere transition metal-mediated C–H cleavage that do not involve [M–C] intermediates.

and their reactivity can be emulated by simplified small-molecule catalysts (so-called "biomimetic oxidation catalysts"). Many metals, including Fe, Cu, and Mn, in a variety of oxidation states, are known to participate in this C–H oxidation chemistry when situated in the appropriate coordination environment. Biomimetic C–H oxidation and its role in modern organic synthesis are covered in Section 8.6.1.

2) Similarly to the 1,2-addition mechanism described above, C–H functionalization reactions involving metallocarbenoid/nitrenoid employ a reactive [M=X] species (where X = CR^{1}R^2 or NR). However, unlike 1,2-addition, these reactions are thought to proceed through a three-center transition state, in which positive charge builds up on the X atom and negative charge on the M atom, and do not proceed via the formation of a new [M–C] intermediate. Suitable metals include Rh(II), Ru(II), Mn(III), and Fe(III). C–H functionalization by metallocarbenoids and -nitrenoids is the topic of Sections 8.6.2 and 8.6.3, respectively.

8.2.3
Early Work in Metal-Mediated C(Aryl)–H Cleavage

Among the earliest reports of transition metal-mediated C–H cleavage of a nonacidic, unactivated C–H bond are examples in which carbophilic soft Lewis acidic transition metals were found to react with electron-rich arenes (**4**), effectively exchanging one H-atom with the relevant transition metal (M^n) to form a new C–M bond (Scheme 8.5). The mechanism of this class of early reactions is thought to proceed via coordination of the electrophilic metal to the π-system of the aromatic ring to form an arenium species (**5** or **6**) (also known as a "Wheland intermediate") [27]. Subsequently, charged species of this type can undergo inter- or intramolecular deprotonation (**5** or **6**, respectively), to generate the resulting organometallic product **7**.

For example, in an early series of papers, electron-rich arenes were found to undergo direct mecuration upon treatment with HgX$_2$ (Scheme 8.6). In 1892, Pesci discovered that anilines (e.g., **8**) and Hg(OAc)$_2$ would react to form a new [Hg(II)–Ar] species **9** [28]. In the same year, Volhard showed that thiophene (**10**) could undergo mercuration selectively at the 2-position with HgCl$_2$ in the presence of NaOAc to form organomercury species **11**, which could react as a nucleophile with benzoyl chloride (**12**) to give phenyl

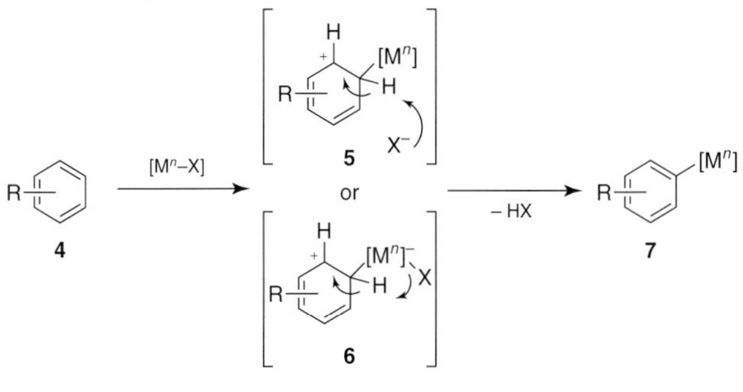

Scheme 8.5 Metallation of electron-rich arenes by carbophilic soft acidic transition metals.

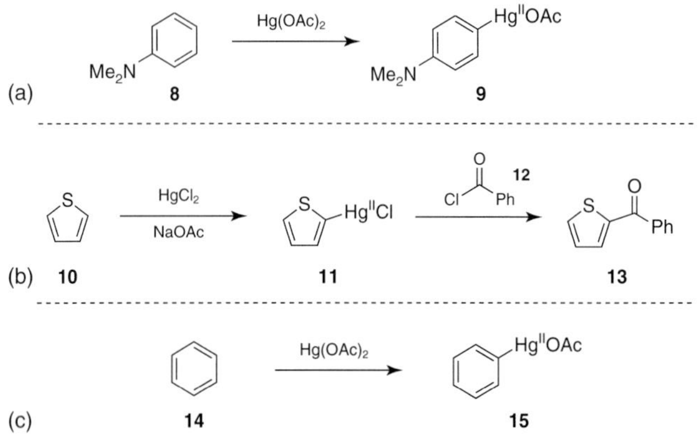

Scheme 8.6 (a) *para*-Mercuration of dimethylaniline (**8**) [28]. (b) C2-selective C−H mercuration of thiophene (**10**) and subsequent reaction with benzoyl chloride (**12**) [29]. (c) Mercuration of benzene (**14**) [30].

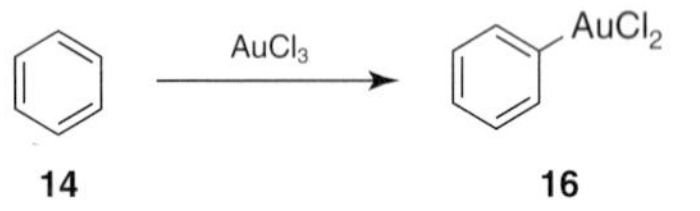

Scheme 8.7 Auration of benzene by AuCl₃ [31].

2-thienyl ketone (**13**) [29]. Dimroth subsequently found that treatment of benzene (**14**) with Hg(OAc)₂ led to phenylmercuric acetate (**15**) [30].

In 1931, Kharasch and Isbell described the first example of C−H auration (Scheme 8.7) [31]. In particular, they discovered that benzene (**14**) could be metallated by auric chloride under mild conditions to form phenylauric chloride (**16**), analogous to the reactivity observed in other electrophilic aromatic substitution reactions such as nitration and halogenation. The reactions were found to proceed rapidly and it was observed that the resulting organometallic complexes were unstable.

Several decades later, Fujiwara's group developed a series of transformations based on Pd(II)-mediated C−H insertion into aromatic solvents (Scheme 8.8). In the first report in 1967, dimeric [Pd^{II}Cl₂−styrene] complex **17** reacted with benzene solvent (**14**) in the presence of HOAc as the cosolvent, to generate *trans*-stilbene (**20**) [32]. It was later found that simple Pd(OAc)₂ could mediate the coupling of styrene (**19**) and arene solvents [33]. For monosubstituted arenes (e.g., **21**), regioisomeric product mixtures (**22**−**24**) were obtained [34]. Defined [Pd(II)−Ar] species **18** could be isolated and characterized by employing dialkyl sulfide stabilizing ligands [35].

In addition to these early electrophilic arene metallation studies, it was also found that transition metal complexes could activate C(aryl)−H bonds by oxidative addition (Scheme 8.9). This mode of reactivity relies on the generation of high-energy low-valent metal species with open coordination sites.

In 1962, Chatt and Watson undertook an investigation to stabilize a series of zerovalent transition metals with a bidentate diphosphine ligand, 1,2-bis(dimethylphosphino)ethane

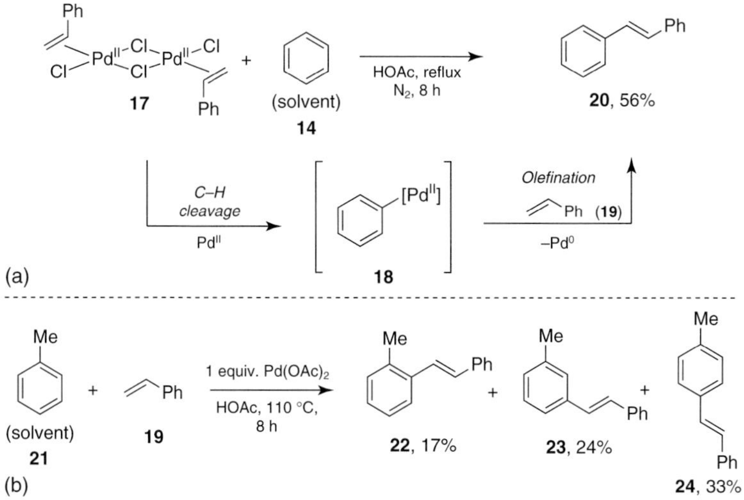

(a)

(b)

Scheme 8.8 (a) Stoichiometric reaction of [Pd(II)Cl$_2$–styrene] dimer (**17**) and benzene (**14**) as solvent [32]. (b) Pd(II)-mediated C–H olefination of monosubstituted arenes (**21**) and the resulting positional isomeric mixture of products [34].

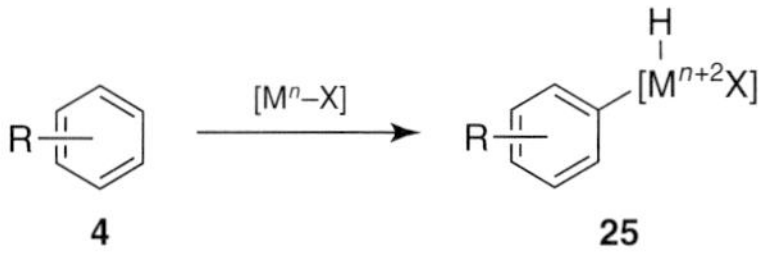

Scheme 8.9 Oxidative addition of arenes to low-valent transition metals.

(dmpe) [36]. M(0) was generated by *in situ* reduction of the corresponding metal halides with Na[C$_{10}$H$_8$] (**27**). Anomalous reactivity was observed in the case of Ru(0), and the authors commented in passing on the possible formation of a hydrido complex (Scheme 8.10). Later, in 1965, Chatt and Davidson revisited this observation with the aid of nuclear magnetic resonance (NMR) spectroscopy [37], and it was found that [Ru(0)(dmpe)$_2$–naphthalene] complex **28** underwent oxidative addition to generate [Ru(II)(dmpe)$_2$(H)(Ar)] complex **29**, which could be isolated and characterized. In 1970, Green and Knowles observed that under photolysis conditions, benzene was found to add oxidatively to Cp$_2$WH$_2$ (Cp = cyclopentadienyl) [38], reactivity that was later also observed with benzylic C(sp^3)–H bonds [39].

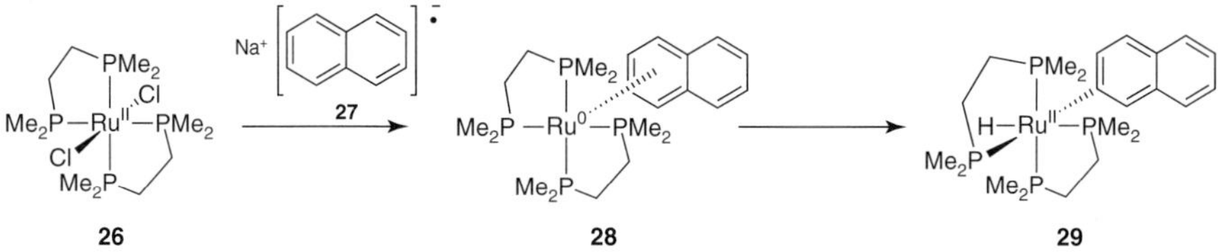

Scheme 8.10 Oxidative addition of a C(aryl)–H bond to an Ru(0) species [37].

These early arene metallation reactions set the stage for subsequent investigations where C–H cleavage could be controlled and facilitated by chelation of the metal to a proximate functional group, henceforth referred to as a "directing group" (DG) (Section 8.2.4). Moreover, as briefly alluded to at the end of the preceding paragraph, this work also initiated a wave of interest in unactivated $C(sp^3)$–H cleavage processes mediated by transition metals (Section 8.2.5).

8.2.4
C(aryl)–H Functionalization via Cyclometallation

From the perspective of developing synthetically useful catalytic transformations based on C–H cleavage via metal insertion (Section 8.5), two major limitations of the chemistry discussed in the previous section are (i) the fact that the arene is generally used in excess (often as the solvent) and (ii) the lack of positional selectivity in reactions. One strategy for overcoming these challenges is to use a proximate Lewis basic functional group as a DG to coordinate the metal species, which facilitates metallation both kinetically (by bringing the metal in closer proximity to the C–H bond of interest) and thermodynamically (by increasing the stability of the resulting organometallic intermediates) (Scheme 8.11).

In 1963, Kleiman and Dubeck reported that refluxing $NiCp_2$ in neat azobenzene led to the formation of a new compound, which they characterized at nickelacycle **33** (Figure 8.1) [40]. Cope and Siekman found that analogous reactivity with azobenzenes could be observed using Pt(II) or Pd(II) salts, allowing for the synthesis of dimeric metallacycles **34** and **35** [41]. Notably, C–H cleavage using Pd(II) was found to be more facile that with Pt(II).

Since these early examples, cyclometallation with Pd(II) has attracted a great deal of attention [42, 43], in large part because of its relevance to catalytic $C(sp^2)$–H and $C(sp^3)$–H functionalization reactions. A series of reaction mechanisms have been proposed in which AcO^- plays a crucial role as an internal base (Figure 8.2). Mechanisms involving

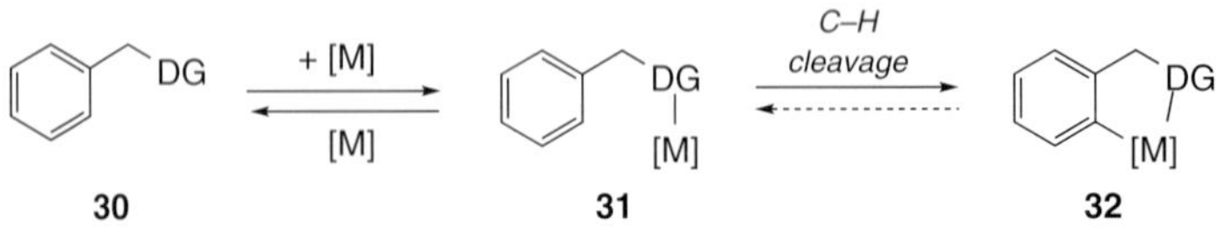

Scheme 8.11 General depiction of cyclometallation facilitated by the presence of a directing group (DG).

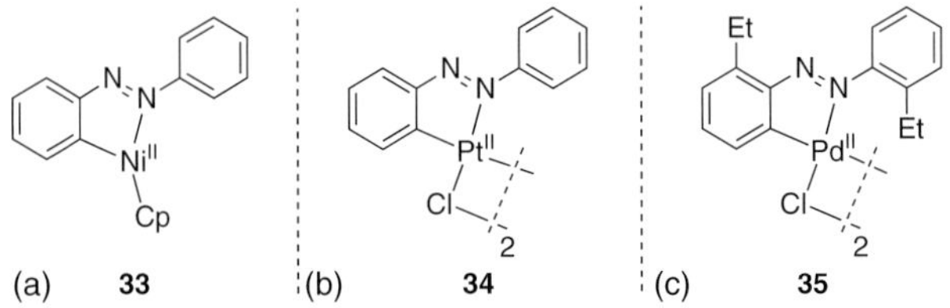

Figure 8.1 Metallacycles prepared via electrophilic metal-mediated C(aryl)–H cleavage of azobenzene. (a) Nickelacycle (**33**) [40]; (b) platinacycle (**34**) [41]; (c) palladacycle (**35**) [41].

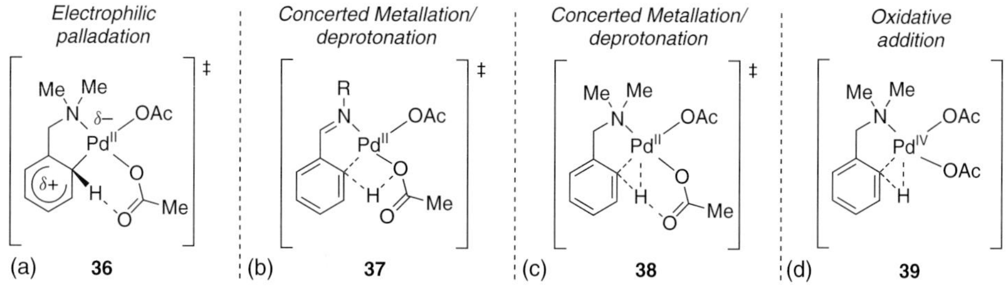

Figure 8.2 Transition states for different proposed mechanisms for C–H cleavage with Pd(II). (a) Electrophilic palladation (**36**) [44]; (b) concerted metallation/deprotonation via a four-membered transition state (**37**) [45]; (c) concerted metallation/deprotonation via a six-membered transition state (**38**) [46]; (d) oxidative addition (**39**) [47].

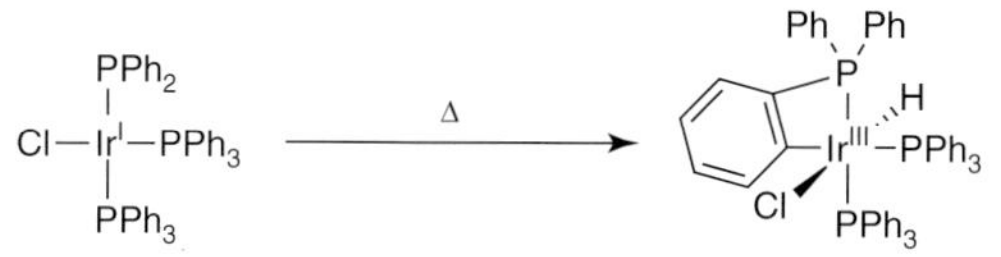

Scheme 8.12 Early example of metallacycles prepared via oxidative addition of a C(sp²)–H bond [48].

transition states **36** [44], **37** [45], and **38** [46] all fall within the broad category of "electrophilic activation" described above, while that involving **39** proceeds via oxidative addition [47]. The operative mechanism with Pd(II) is system-dependent, but is normally thought to proceed via either **36** or **38** in the case of C(sp²)–H cleavage and **38** in the case of C(sp³)–H cleavage.

Early on, examples of directed oxidative addition of C–H bonds to low-valent metal species were also reported. During their studies of Ir(I) complex **40**, Bennett and Milner found that upon heating, one of the o-C(sp²)–H bonds of a triphenylphosphine ligand could oxidatively add to the metal center to generate cyclometallated Ir(III) species **41** (Scheme 8.12) [48]. Importantly, this reaction differs mechanistically from the cases described in Figure 8.1 because there is a redox change at the metal center.

Taken together, the nondirected C(aryl)–H metallation studies in Section 8.2.3 and the cyclometallation reactions presented in this section represented compelling early evidence of the unique reactivity of transition metals in performing C–H cleavage. Beginning in the late 1960s and early 1970s, discoveries of inner-sphere metal-mediated C(sp³)–H cleavage were reported, ushering in a new era in the field.

8.2.5
Early Investigations of C(sp³)–H Cleavage

The reactions described in the previous two sections demonstrate the evolution of mechanistic understanding and strategies for regiochemical control in arene metallation. This section describes pioneering investigations of inner-sphere C(sp³)–H cleavage by

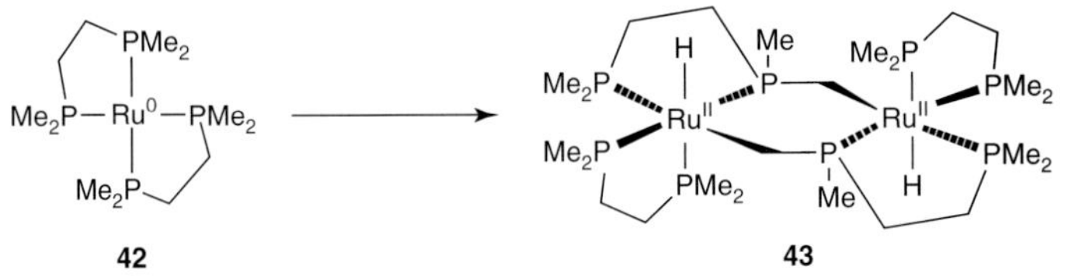

Scheme 8.13 C(sp^3)–H cleavage by a low-valent Ru(0) species [37, 49].

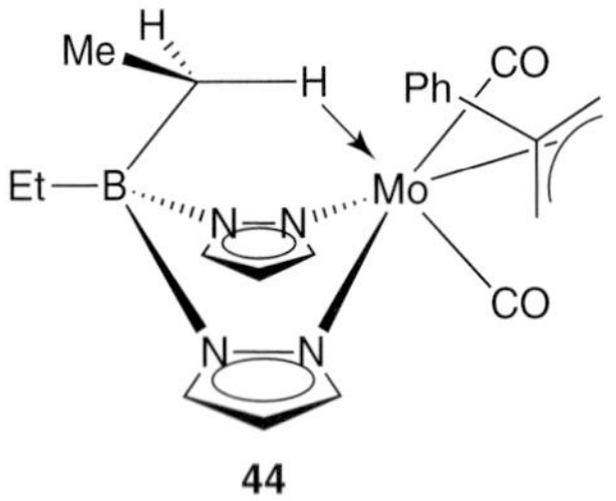

44

Figure 8.3 An early example of a C(alkyl)–H agostic interaction with Mo [58].

stoichiometric transition metal complexes and the mechanistic features of these reactions, with an eye towards potential applications in catalysis.

One early observation came in 1965 from Chatt and Davidson in their studies of Ru(0)(dmpe)$_2$ (**42**) (Scheme 8.13) [37]. Based on analytical data, they proposed a mononuclear species in which one of the C(sp^3)–H methyl groups on the phosphine ligand had oxidatively added to Ru(0) to form a new Ru(II)–H bond along with an [Ru–P–CH$_2$] three-membered ring. Ten years later, Cotton *et al.* were able to obtain X-ray crystallographic data to establish that the complex was actually dimeric species **43** [49].

In addition to this early example of chelation-assisted C(sp^3)–H cleavage, in the late 1960s Pt(II)-mediated H–D exchange in aromatic hydrocarbons and heterocycles was reported, including cases of H–D exchange on the alkyl chains of alkylaromatic compounds [50–53]. In the 1970s, Shilov and co-workers found that Pt(II) could mediate C–H cleavage and functionalization of simple alkanes and observed regioselective cleavage at the terminal positions [11, 54, 55]. These reactions, which can be rendered catalytic by employing an external reoxidant, are discussed in Section 8.3. Contemporaneously, other unusual interactions between C–H bonds and transition metal complexes were also made. For instance in a "five-coordinate" Ru(II)(PPh$_3$)$_3$Cl$_2$ complex, one of hydrogen atoms of a phenyl ring was found to occupy the sixth coordination site [56], a phenomenon that was later also observed with C(alkyl)–H bonds. In 1974, in structural studies of the Mo complex **44** [57], Cotton *et al.* [58] observed and characterized an usual [C–H···Mo] interaction, describing it as a three-center, two-electron bond (Figure 8.3).

Following the identification of additional structures of this type during the next decade (Figure 8.4) [59–62], in 1983 the term "agostic interaction" was coined by Brookhart and Green [63, 64]. The word "agostic" is Greek for "to hold close to oneself," and in the context of coordination chemistry this term describes the interaction of a coordinately unsaturated transition metal and an intramolecularly tethered C–H bond. The C–H bond donates its two electrons into an empty d-orbital of a transition metal,

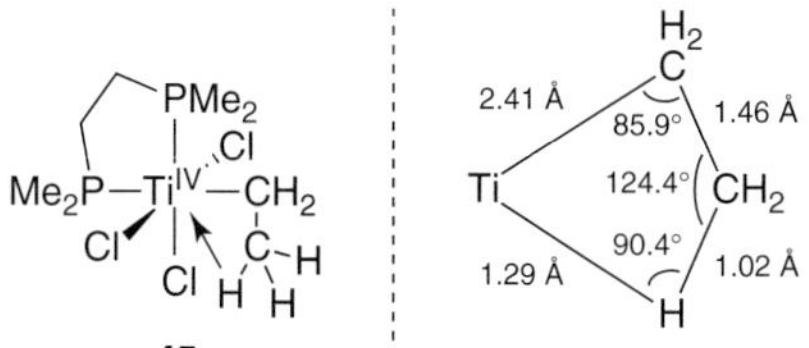

Figure 8.4 (a) An early example of a structurally characterized β-agostic [Ti–alkyl] species **45** [60]. (b) Pertinent bond distances and bond angles.

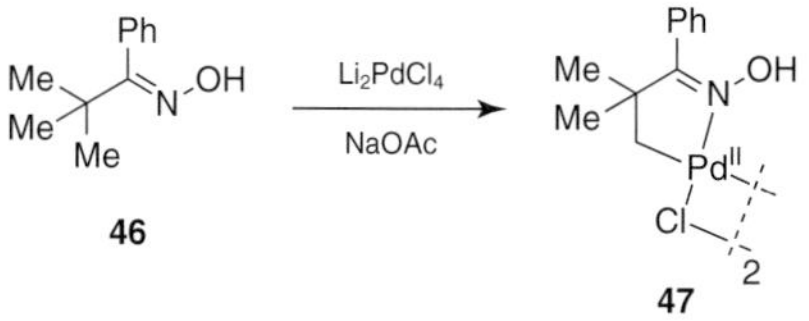

45

(a)   (b)

Scheme 8.14 C(sp^3)–H cleavage facilitated by an oxime directing group [67].

46   **47**

leading to a three-center, two-electron bond. These interactions have bond energies of 10–$15 \, \text{kcal mol}^{-1}$, which is weak in bonding terms, but strong when compared with intermolecular interactions, such as hydrogen bonds, which typically have energies between 1 and $10 \, \text{kcal mol}^{-1}$.

In contrast to transition metal-mediated C(sp^2)–H activation, which can proceed via an initial interaction between the π-system and the metal, in C(sp^3)–H activation, the same opportunity does not present itself [65]. In the latter case, these reactions are generally thought to proceed via initial formation of a C–H agostic interaction or a related alkane σ-complex [66]. Bond breaking can then proceed along any of the mechanism pathways depicted in Scheme 8.3. Indeed, this reactivity was harnessed throughout the late 1970s and 1980s in several reports in which unactivated C(sp^3)–H bonds were cleaved in the process of forming new [M–C] complexes.

McDonald, Shaw, and co-workers **46** could react in the presence of Pd(II) and NaOAc to form dimeric palladacycle **47** (Scheme 8.14) [67]. Importantly, this transformation served as the conceptual prerequisite for a series of recently developed Pd(II)-catalyzed C(sp^3)–H functionalization reactions. Falling within the broad category of electrophilic activation, the mechanism for this transformation is thought to proceed through a concerted metallation–deprotonation reaction, analogous to **38** in Figure 8.2.

Foley and Whitesides reported an example of intramolecular aliphatic C(sp^3)–H insertion by a coordinately unsaturated Pt(II) species **48** (Scheme 8.15) [68]. In the proposed mechanism, oxidative addition of the C(sp^3)–H bond leads to putative Pt(IV) species **49**, and subsequent reductive elimination of neopentane (**50**) gives **51**. The resulting isolable Pt(II) complex **51** could be treated with an oxidant such as I$_2$ to induce C–C reductive elimination of 1,1-dimethylcyclopropane.

At this stage, transition metal-mediated C(sp^3)–H cleavage in the absence of chelation assistance or covalent attachment still had little precedent. Then in 1979, Crabtree *et al.* described a seminal example of intramolecular C(sp^3)–H activation of completely saturated hydrocarbons using an Ir(III)-mediated transfer dehydrogenation reaction (Scheme 8.16) [69]. By employing *tert*-butylethylene (TBE) (**54**) as the hydrogen scavenger, cyclopentane (**52**) was found to undergo sequential dehydrogenation in refluxing

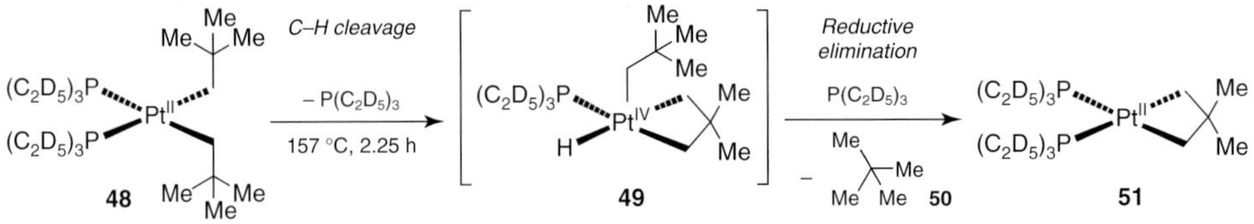

Scheme 8.15 Intramolecular C(sp³)–H cleavage by a coordinatively unsaturated Pt(II) species **48** [68].

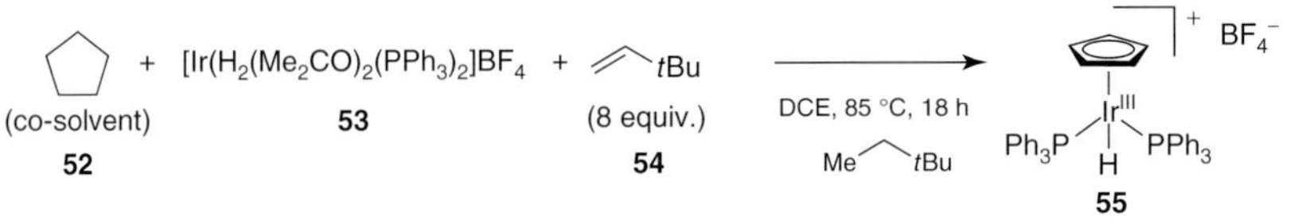

Scheme 8.16 Alkane dehydrogenation with Ir(I) complex **52** [69].

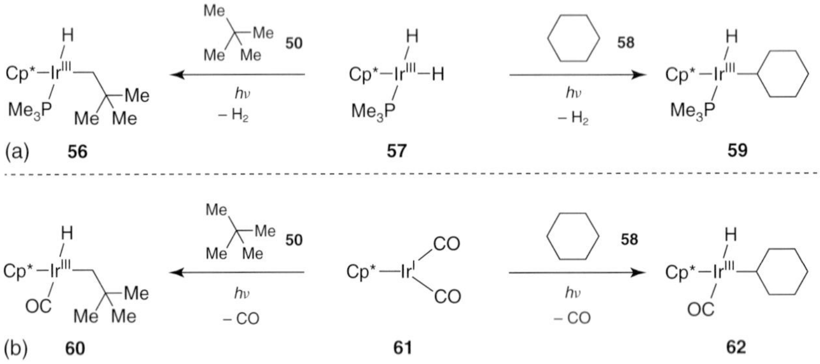

Scheme 8.17 (a) Photochemical C(sp³)–H cleavage of completely saturated alkanes by an *in situ*-generated low-valent Ir(I) species [70]. (b) A similar system that proceeds via initial extrusion of CO rather than H₂ [71].

1,2-dichloroethane (DCE), ultimately being converted into an η⁵-bound Cp ligand on newly generated Ir(III) complex **55**.

Another breakthrough came in 1982, when the research groups of Bergman [70] and Graham [71] independently reported the first examples of isolable hydridoalkylmetal complexes formed by oxidative addition of completely saturated hydrocarbons to low-valent metals, providing direct evidence of the crucial oxidative addition step (Scheme 8.17). In particular, these two groups observed that upon photolysis, Ir complexes **57** and **61** would undergo extrusion of CO or H₂ to generate high-energy Ir(I) intermediates bearing open coordination sites. In the presence of cyclohexane and neopentane, new hydridoalkyl–Ir(III) complexes could be isolated and characterized.

This work exploring the fundamental aspects of C(alkyl)–H cleavage by transition metals helped lay the foundation for many of the modern application of transition metal-mediated C–H functionalizations discussed throughout the rest of the text. The examples presented in this section involving completely saturated alkanes will be particularly relevant to Section 8.3. The reactions involving directed metallation of $C(sp^2)$–H and $C(sp^3)$–H are more applicable to modern efforts in organic synthesis; hence the lessons learned in those sections will be especially relevant to Sections 8.4 and 8.5. Overall, the insights gleaned from these fundamental studies have guided and will continue to guide catalyst development, which is a major driving force in the design and discovery of new C–H functionalization reactions via metal insertion.

8.3
First Functionalization: Challenges in Hydrocarbon Chemistry

Unfunctionalized hydrocarbons are abundant, inexpensive feedstock chemicals that are readily accessible from Nature. As such, they represent a potentially valuable source of raw carbogenic material for the synthesis of drugs, agrochemicals, and fine chemicals. However, the lack of existing reactive functional groups makes selective installation of new C–C, C–O, C–N, or C–H bonds a significant challenge. Therefore, generally, hydrocarbons are used primarily for combustion, an application which exploits their energy but not their synthetic potential (Scheme 8.2). However, even in fuel applications, low-molecular weight hydrocarbons, such as methane, are difficult to contain and transport because they are gaseous at ambient temperature and pressure. Hence the availability of methods for facile derivatization or chain growth would greatly enhance the utility of these compounds as fuels and chemical feedstock.

In the light of the discussion in Section 8.2.5, an exciting strategy for using hydrocarbons in this manner would be to carry out "first functionalization" using a transition metal catalyst (Scheme 8.1) [72]. In particular, one could envision taking advantage of transition metal-mediated $C(sp^3)$–H cleavage to generate a new [M–C] species, which could then reacted with a suitable nucleophile or electrophile to effect functionalization. Ideally, regeneration of the catalytically active species would accompany extrusion of the newly modified alkyl unit from the metal, completing a catalytic cycle. Indeed, transition metal-catalyzed first functionalization of hydrocarbons has attracted a great deal of interest during the past several decades. Nevertheless, developing truly practical catalytic systems for large-scale applications remains a challenge.

Although a detailed discussion of hydrocarbon functionalization is beyond the scope of this chapter, we offer here an overview of three popular research topics that have received tremendous attention during the past few decades, and we provide representative examples of recent progress in these areas. It is our hope that by presenting this chemistry, it will prompt the reader to think about related issues in catalysis involving molecules already bearing functional groups, what we call "further functionalization." Although first functionalization is appealing for the reasons stated above, ultimately, catalytic reactions of this type take simple starting materials to relatively simply products. Further functionalization, on the other hand, can offer quantum leaps in molecular complexity, which presents unique opportunities in organic synthesis, as discussed in Sections 8.4–8.6.

8.3.1
Selective Functionlization of Methane and Higher n-Alkanes

Natural gas is an abundant hydrocarbon resource, consisting primarily of methane. In industrial applications, natural gas is often converted to "syngas" (H_2 and CO) by steam reforming over a heterogeneous Ni catalyst. Subsequently syngas is used to meet a variety of different needs, including the production of methanol for applications as a fuel or as a chemical feedstock. Because steam reforming is energy intensive, requiring temperatures of 700−1100 °C, a more efficient and environmentally friendly method would be to convert methane directly to methanol using a transition metal catalyst to perform $C(sp^3)$−H functionalization, in analogy with the transformation that is elegantly performed by methane monooxygenase in biological systems. This catalyst would need to be highly reactive (in line with those described in Section 8.2.5), but would need to offer high levels of chemoselectivity to avoid over-oxidation of CH_4 to CO_2.

Above, we introduced one effective system for this selective alkane hydroxylation and chlorination originally developed by Shilov's group in the 1970s (Scheme 8.18) [11, 54, 55, 73]. In this reaction, an linear alkane (e.g., methane) undergoes Pt(II)-mediated C−H cleavage, possibly via an electrophilic activation mechanism [74]. The resulting [Pt(II)−alkyl] species **64** reacts with a Pt(IV) salt, which serves as an electrophilic halonium source to oxidize **64** to Pt(IV) species **65**. Reductive elimination, which is believed to proceed via nucleophilic attack of a water molecule at the Pt−C bond [75], then forges the new C−O bond and regenerates the catalytically active Pt(II) species **63**. The Shilov

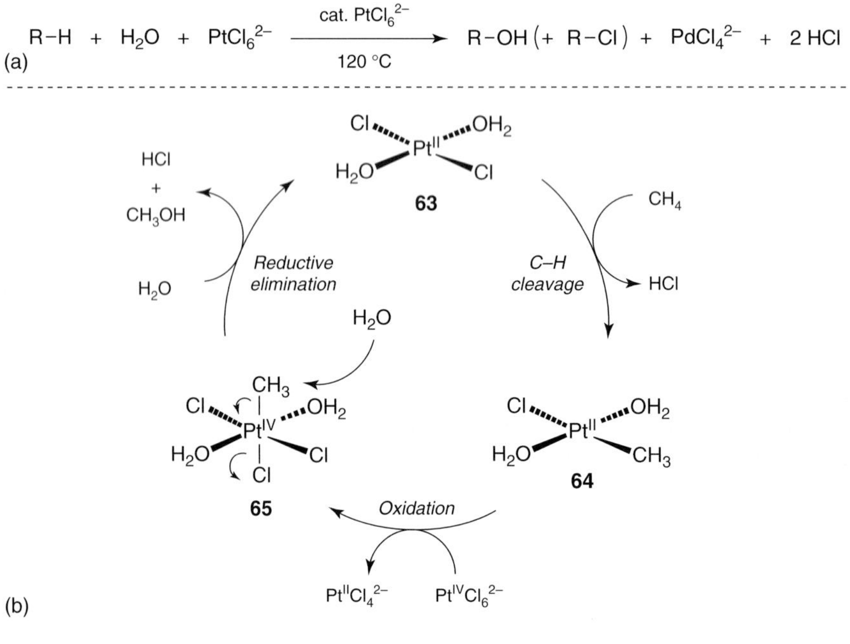

Scheme 8.18 (a) Shilov-type oxidation of alkanes. (b) A possible three-step catalytic cycle for the Pt(II)-catalyzed conversion of methane to methanol [75].

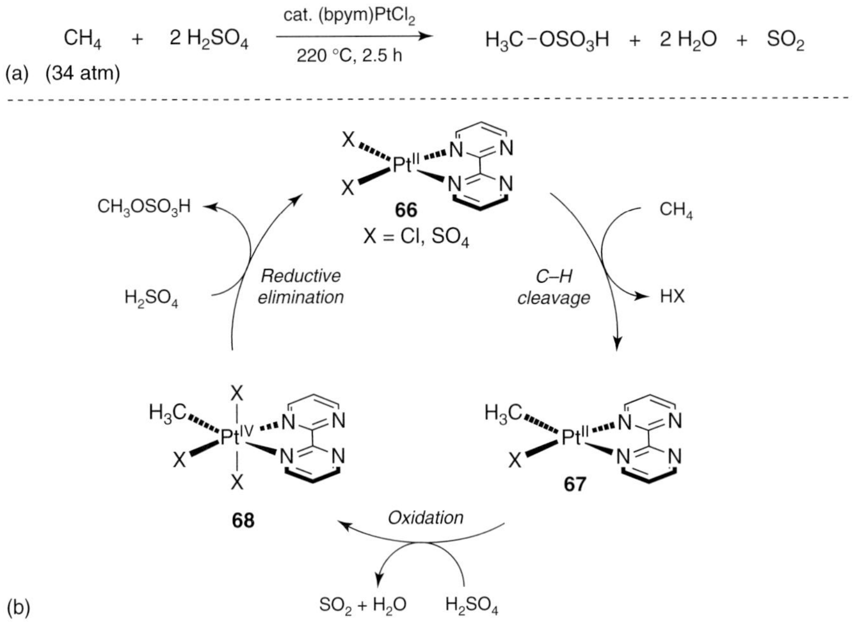

Scheme 8.19 (a) Catalytica system for conversion of methane to methyl bisulfate [76]. (b) Possible three-step catalytic cycle for the Pt(II)-catalyzed conversion of methane to a methanol derivative.

system is a rare example of a transition metal-catalyzed reaction capable of performing electrophilic C−H cleavage/functionalization on simple alkanes under relatively mild conditions. Disadvantages of this system include the need for stoichiometric quantities of Pt(IV) salts for catalyst oxidation during the catalyst cycle and the low stability of the Pt(II) catalyst, which is known to undergo decomposition to Pt metal or other polymeric Pt species. External ligands can be used to stabilize Pt(II) in order to address the latter of these two issues.

The Catalytica system reported by Periana et al. takes advantage of ligand stabilization to enhance the efficiency of Pt(II)-catalyzed methane oxidation and utilizes sulfuric acid as the oxidant to replace the Pt(IV) salts in the Shilov system (Scheme 8.19) [76]. Under the reaction conditions, the catalyst **66** is stable; the ligand does not undergo oxidative degradation, and precipitation of Pt metal is not observed. The catalytic system is capable of converting methane to methyl bisulfate with 72% one-pass yield and 81% selectivity. Overall, this reaction represents a promising alternative to the syngas route for natural gas processing described above; however, one limitation is that the product, methyl bisulfate, is not directly synthetically useful. Because it must be deprotected prior to use, it adds an additional step for potential applications.

Generally, alkane oxidation reactions catalyzed by late transition metals [e.g., Pt(II), Pd(II), and Hg(II)] are remarkably robust, tolerating aqueous conditions, strong acids, and oxidative environments. Although the above examples only highlight applications for methane oxidation, it is important to note that this class of catalysts is selective for

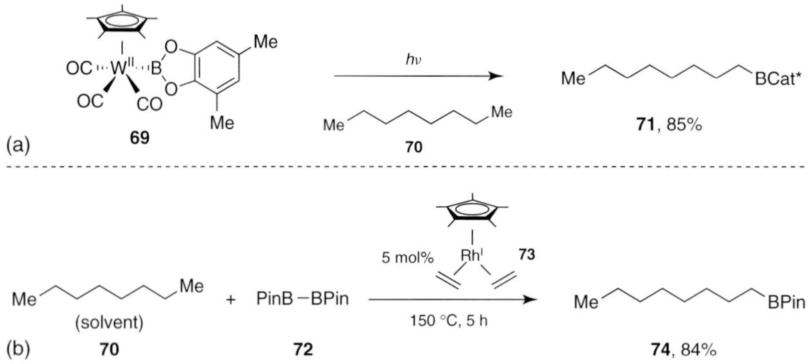

Scheme 8.20 (a) Photoinduced C(sp³)–H borylation of linear alkanes using stoichiometric W(II) complex **69** [77]. (b) Thermal C(sp³)–H borylation of linear alkanes using catalytic quantities of Rh(I) complex **73** [78].

oxidation of terminal methyl groups when used for oxidation of *n*-alkanes. This stands in sharp contrast to reactivity patterns observed in radical pathways: tertiary C–H > secondary C–H > primary C–H. Overall, Shilov-type chemistry offers good levels of regio- and chemoselectivity with linear alkanes. Even so, one drawback is that the method cannot generally be used to generate a single product for synthetic applications owing to the persistent formation of small quantities of overly oxidized byproducts or other regioisomers. Therefore, to realize the full potential of *n*-alkanes as versatile feedstock chemicals, complementary methods are needed.

Pioneering work by Hartwig's group established a protocol for C(sp³)–H borylation of linear alkanes with exclusive formation of the new C–B bond at the terminal position (Scheme 8.20). In the initial report, they reported a photoinduced reaction using stoichiometric quantities of the W(II) complex **69** [77]. In this transformation, photolysis opens a coordination site on the metal, and subsequently C–H cleavage (possibly through oxidative addition) takes place. C–B reductive elimination then extrudes the product **71**. Shortly thereafter, Hartwig's group developed a catalytic protocol using Rh complex **73** under thermal conditions [78]. (Bpin)₂ served as the boron source and the hydride acceptor. Interestingly, HBpin, which was generated during the course of the reaction, also proved to be an effective boron source and hydride acceptor, leading to the generation H₂. In this same study, they also found that benzene could undergo C(sp²)–H borylation using a similar catalyst. Regioselective arene borylation by metals such as Ir(I) and Rh(I) has been extensively explored by several groups, including most notably those of Smith [79–81], Miyaura, and Hartwig [82–85], leading ultimately to the development of a practical reaction with broad scope in which the regioselectivity is controlled entirely by steric considerations (i.e., 1,3-disubstituted arenes react selectively at the 5-position).

Overall, the practical utility of the methods presented in this section remains to be seen. They do, however, represent promising steps forward towards the goal of developing low-energy methods for selectively converting hydrocarbons to value-added chemicals. In the future, reducing the catalyst loading and improving the overall efficiency are important goals.

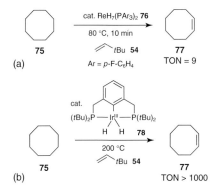

Scheme 8.21 Metal-catalyzed transfer hydrogenation reactions of cyclooctane (75) with TBE (54) as the hydrogen acceptor. (a) An early example using ReH$_7$(PR$_3$)$_2$ (76) [89]. (b) Highly efficient catalysis using an Ir(III) pincer complex (78) [91].

8.3.2
Alkane Dehydrogenation

Another attractive first functionalization approach is directly introducing a unit of unsaturation into a saturated hydrocarbon, and here the study of alkane dehydrogenation by homogeneous metal complexes has a rich history [86]. In Section 8.2.5, we presented Crabtree et al.'s early work using a stoichiometric Ir(III) complex to mediate transfer hydrogenation (Scheme 8.16). Later his group turned their attention to catalytic transformations under thermal and photochemical conditions [87, 88]. In these investigations, it was found that linear alkanes gave regioisomeric mixtures, with the thermodynamically favored internally alkenes as the major products, likely due to catalyst-facilitated product isomerization. In the early 1980s, Felkin and co-workers reported that ReH$_7$(PR$_3$)$_2$ (75) could catalyze the transfer dehydrogenation of cycloalkanes with TBE (54) as the hydrogen acceptor, giving turnover numbers (TONs) up to 9 (Scheme 8.21) [89]. When methylcyclohexane was used as a substrate, a mixture of monodehydrogenated products were obtained. Selective formation of 1-alkenes from n-alkanes was achieved by Felkin and co-workers via the corresponding stoichiometric Re–diene complex [90].

A major breakthrough came in 1996, when Jensen and co-workers discovered that Ir(III) "pincer" complex 78 offered unparalleled levels of reactivity in the catalytic transfer dehydrogenation of cycloalkanes (Scheme 8.21) [91]. Later, they utilized a modified Ir(III) pincer complex 79 to address the regioselectivity challenge in the catalytic dehydrogenation of n-alkanes (Scheme 8.22). For instance, with n-octane (70), 1-octene (80) could be obtained in ≥90% selectivity in the early stages of the reaction, evidencing high levels of regioselectivity resulting from kinetic favorability. Over time, the overall product distribution drifted to favor the more thermodynamically stable internal alkenes due to product isomerization. Ir(III) pincer complexes have proven to be remarkably effective in transfer dehydrogenation. For instance, Brookhart and co-workers reported a bisphosphonite pincer Ir(III) complex that gave TONs up to 2200 in the presence of TBE and NaOtBu [92].

Continued efforts centered on dehydrogenation catalyst development are likely to offer improved levels of reactivity and selectivity. Indeed, the decisive progress during the past two decades has stemmed largely from the emergence of Ir(III) pincer catalysts, which

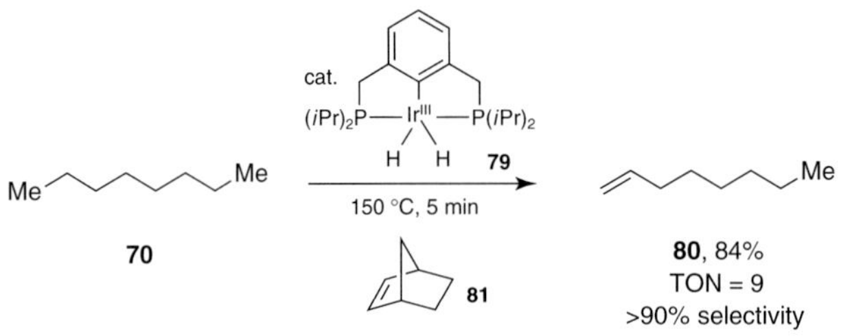

Scheme 8.22 Catalytic dehydrogenation of cyclic alkanes using an Ir(III) pincer complex [91].

speaks for the tremendous opportunities that still exist in ligand design and catalyst discovery.

8.3.3
Alkane Metathesis

As discussed in Section 8.3.1, a major issue in hydrocarbon chemistry is developing methods for exploiting the feedstock potential of low molecular weight hydrocarbons (below C9), which are characterized by their high volatility and lower ignition quality. These low-molecular weight hydrocarbons are commonly obtained during petrochemical refining or from Fischer–Tropsch coal refining. Hence developing methods to take advantage of them as a carbon source is important. Traditionally, chemists have relied upon hydrocarbon cracking to convert low molecular weight hydrocarbons into more tractable, higher molecular weight alkanes. However, this process is energy intense. An alternative route would be to grow the alkyl chains using alkane metathesis, with concomitant generation of shorter chain alkanes (or ideally H_2). Broadly, this process can be viewed as a C–C bond-forming first functionalization method.

As part of a wider effort to study surface organometallic chemistry, in 1997 Basset and co-workers reported a heterogeneous process using a silica-supported Ta catalyst which was found to be capable of transforming acyclic alkanes into lower and higher carbon number alkanes (Scheme 8.23) [94]. In particular, at temperatures of 25–250 °C, ethane could be converted to methane and propane, along with trace quantities of butane. More recent mechanistic studies of this single-site system found that the reaction proceeds via a mechanism involving metallacyclobutane intermediates, akin to alkene metathesis [95, 96]. As is the case in alkene metathesis, ΔG°_{273} is close to zero, meaning that in the absence of an additional driving force (e.g., liberation of a gas or release of ring strain) there is no thermodynamic pressure to drive equilibrium in favor of productive metathesis. In the case of transition metal-catalyzed alkane metathesis, this prevents quantitative conversions.

An alternative approach is to utilize a dual catalytic system, employing one metallic species capable of effecting alkane dehydrogenation and another that is reactive in alkene metathesis [97]. Goldman *et al.* presented an example of tandem catalysis using two homogeneous, molecularly defined catalysts (**83** and **84**) in a single pot to perform formal alkane metathesis (Scheme 8.24) [98]. Using this system, both self-metathesis and cross-metathesis were demonstrated. The system is selective for linear products and

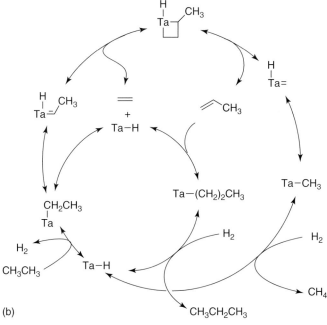

Scheme 8.23 (a) Alkane metathesis of ethane to give propane and methane [94]. (b) Proposed catalytic cycle [95].

is compatible with higher carbon hydrocarbons. For example, in the metathesis of 2 mol of *n*-hexane, decane was produced as the primary product.

Although the effectiveness of metal-catalyzed alkane metathesis reactions remains modest in comparison with the state-of-the-art in alkene and alkyne metathesis, this approach nonetheless represents an attractive strategy for accessing higher molecular weight hydrocarbons directly from their lower weight homologs. One pervasive limitation of the methods developed to date is that they lead to mixtures of products, which hampers their utility because further refinement or purification of the product streams is needed. In the future, the development of novel catalysts and fundamentally new strategies is necessary prior to large-scale applications.

Although transition metal-catalyzed first functionalization of hydrocarbons has received a great deal of attention, the true practical utility of this approach still remains to be seen. For large-scale industrial applications in the petroleum industry, exceptionally high levels of catalytic efficiency are needed in order to compete with other processes (e.g., heterogeneous catalysis, thermolytic cracking, and enzymatic oxidation), which current homogeneous C–H functionalization reactions are incapable of matching. Developing catalysts that offer high levels of reactivity coupled with precise selectivity in the C–H cleavage step is a major hurdle to be cleared towards widespread

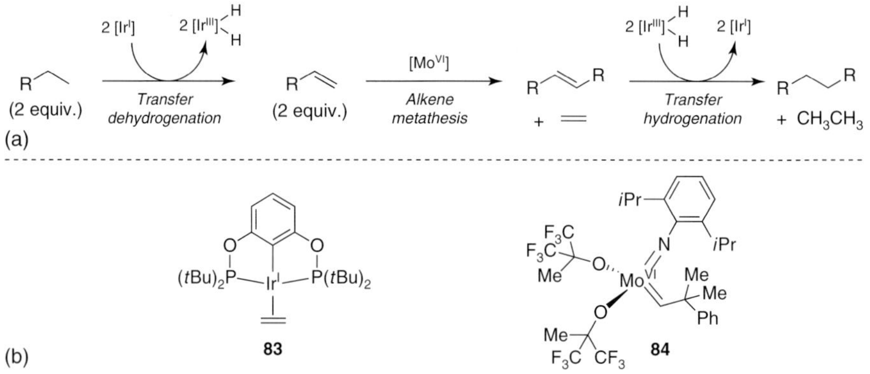

Scheme 8.24 (a) Alkane metathesis via dehydrogenation/alkene metathesis using a two-catalyst system [98]. (b) Catalyst structures.

application of this chemistry in petrochemical refining and fine chemical synthesis. Additionally, although the starting materials in first functionalization chemistry are inexpensive hydrocarbons, it is also important to keep in mind that the products also generally low-value, commercially available chemicals. Hence making these processes cost-effective is a major concern, particularly when precious metal catalysts are employed.

On the other hand, transition metal-catalyzed C−H functionalization has already proven to be a valuable synthetic process when applied to prefunctionalized substrates (i.e., further functionalization). In this context, C−H functionalization can rapidly advance the molecular complexity of simply feedstock chemicals for applications in the synthesis of natural products and other bioactive molecules. This topic is the focus of the remainder of the text.

8.4
Further Functionalization: C−H Bonds as Reaction Partners in Organic Synthesis

8.4.1
Philosophy

The target substrates encountered in organic synthesis are quite different from hydrocarbons. Typically they are prefunctionalized, meaning that in addition to C−H bonds the carbogenic skeleton is already adorned with one or more C−heteroatom bonds. Applying C−H functionalization methods to prefunctionalized substrates presents a unique set of challenges. Since natural products and natural product precursors contain reactive functional groups, it is of paramount importance that metal-mediated C−H functionalization methods are highly chemoselective. Additionally, there are added degrees of complexity in terms of controlling regioselectivity because each C−H bond will have a distinct steric and stereoelectronic environment. Moreover, in the interest of developing suitable methods for asymmetric synthesis, developing strategies to distinguish between diastereo- and enantiotopically disposed C−H bonds is also vital.

The power of further functionalization is that it installs a new functional group that has a specific relationship to existing functional groups in terms of reactivity and molecular geometry (Scheme 8.1, Section 8.1). Forging this new bond alters the functional group patterning around a carbon skeleton and thus drastically increases molecular complexity at a single stroke. Ultimately, this pattern can then be expressed in a target compound through further synthetic manipulations. Hence the key to further functionalization is developing catalysts that are reactive enough to cleave desired C–H bonds but have precisely controlled selectivity, such that defined molecular structures can be constructed with high levels of fidelity. Ideally, further functionalization reactions would take advantage of the inherent functionality of a given molecule to direct metal-mediated C–H cleavage, in line with the lessons learned in Section 8.2.5.

C–H functionalization logic changes synthetic plans because they offer new synthetic disconnection in retrosynthetic analysis by providing new modes of bond construction [3], including new classifications of functional group equivalencies. Methods of this type have the potential to streamline synthetic routes, enabling highly complex bioactive molecules to be synthesized from simple, low-cost, low-oxidation state starting materials in a minimum number of chemical steps.

8.4.2
Steroid Functionalization Using Free Radical Chemistry

The utility of C–H functionalization in the context of natural products synthesis has long been appreciated. The first wave of serious attention from the synthetic community came in reflecting on strategies to perform position-selective modification of defined steroid skeletons. Early work in this area exploited radical chemistry, setting the stage for the flurry of recent work concerning metal-mediated processes. For instance, in 1958, the research groups of Corey and Hertler [99] and Arigoni [100] independently developed a synthesis of dihydroconessine (86) by taking advantage of a Hofmann–Löffler–Freytag reaction [101, 102] to effect late-stage functionalization of the angular methyl group at C18 (Scheme 8.25). In this reaction, free-radical chain decomposition of the C20-N-chloramine (85), and subsequent H-atom abstraction, generates a primary carbon-based radical, which quickly reacts with $Cl^\bullet$. Nucleophilic displacement of Cl^- by the intramolecular amine nucleophile gives dihydroconessine (86).

Barton and Beaton also drew upon C–H functionalization logic in their landmark synthesis of aldosterol acetate (92) in 1961 (Scheme 8.26) [103]. Smooth conversion of corticosterone acetate to the corresponding 11β-nitrite (91) set the stage for a photoinduced $C(sp^3)$–H functionalization to generate the corresponding oxime. Hydrolysis of the oxime to the corresponding aldehyde with nitrous acid, followed by spontaneous ring closure, gave aldosterone acetate (92) in four steps and 15% overall yield from corticosterone acetate.

Breslow and co-workers explored methods for C–H oxidation chemistry using innovative molecular templates to effect regioselective remote oxidation of C–H bonds (Scheme 8.27) [105–108]. These reactions rely on a photoinduced free radical C–H abstraction between the template and the target methylene group (99 and 100). Subsequent dehydration and ozonolysis gave the ketone product 102. Reactions of this type

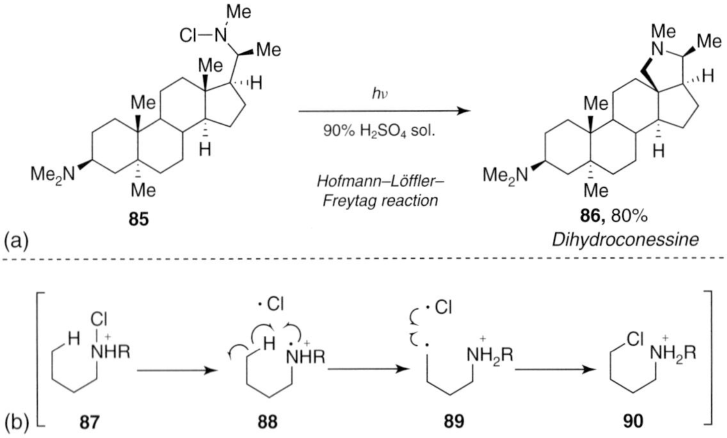

Scheme 8.25 (a) Radical-induced functionalization of an angular methyl group in the synthesis of dihydroconessine (**86**) [99]. (b) Mechanism of the Hofmann–Löffler–Freytag reaction [101, 102].

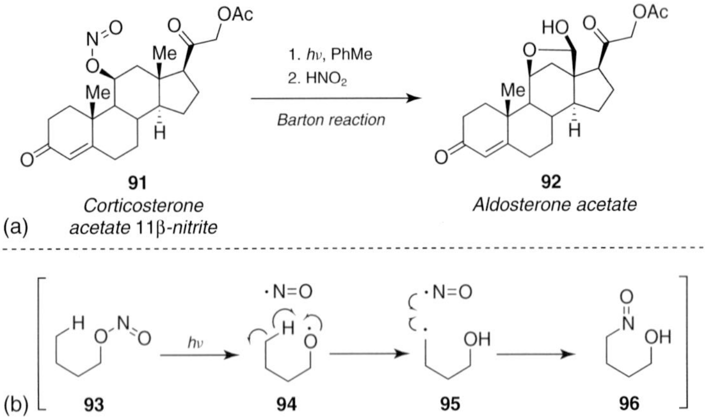

Scheme 8.26 Radical-induced functionalization of an angular methyl group in the synthesis of aldosterone acetate [103]. (b) Mechanism of the Barton reaction [104].

were generally found to give regioisomeric distributions of oxidized products, depending on the length of the tether and the nature of the target alkyl chain. However, when a p-benzoyl-β-phenylpropionic acid ester template was appended to a rigid steroid structure (**97**), it was found that high levels of selectivity could be achieved, albeit in modest overall yield [106]. The remote oxidation concept has found utility in other endeavors in steroid synthesis. For instance, drawing on Breslow and co-workers' findings [109], McMurry and co-workers completed a synthesis of digitoxigenin in which the key double bond was introduced using remote oxidation [110].

This early work on selective steroid functionalization using free radical-based processes for C–H oxidation set the stage for subsequent work utilizing the transition metal-based concepts introduced above.

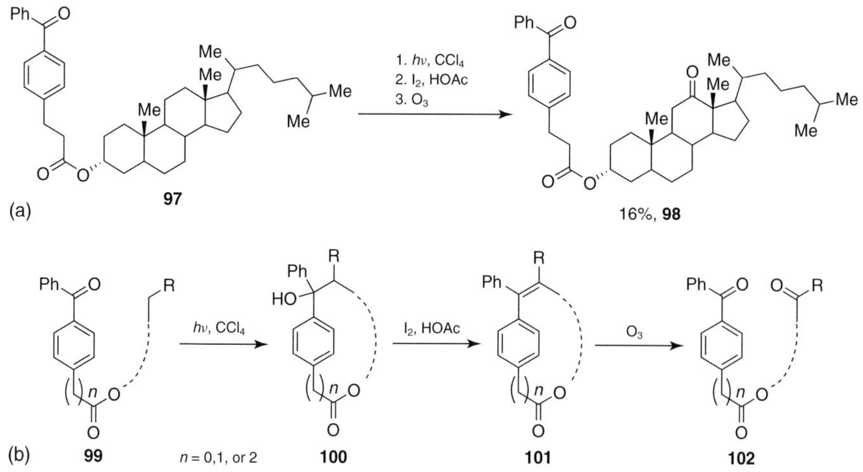

Scheme 8.27 (a) Template-controlled regioselective photochemical oxidation of a remote methylene group [106] and (b) stepwise overview of template-controlled remote oxidation [105].

8.4.3
Building Molecular Complexity Using Transition Metal-Mediated Reactions

Beyond utilizing C–H functionalization for late-stage modification of defined carbogenic cores, early on practitioners of organic synthesis recognized the opportunity to develop concise routes to the carbon skeletons of natural products by applying C–H functionalization logic. In 1978, Trost et al. completed a landmark synthesis of (+)-ibogamine (**110**) by utilizing a late-stage intramolecular Pd(II)-mediated C–H bond–alkene coupling of **107** to forge the intricate bridging tricyclic core of the natural product (Scheme 8.28) [111]. They also went on to apply this methodology in the total synthesis of (±)-catharanthine, showcasing its versatility in alkaloid synthesis [112, 113].

These early efforts set the stage for the modern wave of interest in developing further functionalization reactions for organic synthesis. On this front, one of the pervasive challenges that remain is to develop reactions that are compatible with commonly encountered substrates that install chemical functionality with defined selectivity rules to generate products with advanced molecular complexity. Modern efforts to bring metal-catalyzed C–H functionalization into the mainstream of organic synthesis are discussed below.

8.5
Catalytic C–H Functionalization via Metal Insertion

In Section 8.2, we described early work that showed that a variety of metals (e.g., Pd, Pt, Ru, Rh, and Ir) are capable of breaking C–H bonds along different mechanistic pathways, and in Section 8.3, we discussed *catalytic* C–H functionalization via metal insertion as it pertained to the idea of first functionalization of hydrocarbons. We

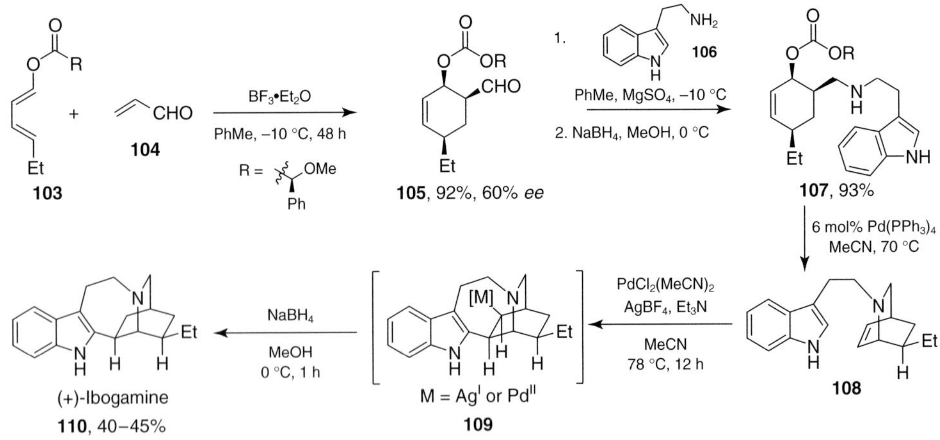

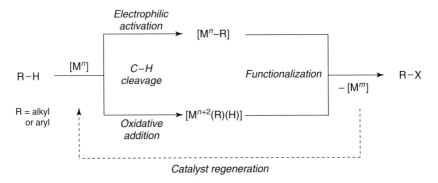

Scheme 8.28 Total synthesis of (+)-ibogamine (**110**) via intramolecular C−H olefination of an indole [111].

introduced the logic of further functionalization in the context of organic synthesis in Section 8.4. For the remainder of the chapter, we delve into the topic of metal-catalyzed further functionalization reactions and their evolving role in modern organic synthesis. In this section, we discuss C−H functionalization via metal insertion, and in Section 8.6, we highlight three other methods for further functionalization: biomimetic oxidation, carbenoid insertion, and nitrenoid insertion. Across these related areas, we focus on thematic commonalities, with the aim of painting a picture of how these catalytic reactions have evolved in terms of reactivity, selectivity, and substrate complexity.

In C−H functionalization via metal insertion, two distinct steps take place: (i) metal-mediated cleavage of a C−H bond to generate an [M^n−R] species and (ii) functionalization of the [M^n−R] intermediate with a suitable electrophile or nucleophile (Figure 8.5). To complete a catalytic cycle, the active transition metal species is then regenerated, which can require an external oxidant. Among the reactive metals for these transformations, Pd is one of the most versatile for synthetic applications. Hence throughout this section, we place a special emphasis on Pd(II)-catalyzed C−H functionalization [114–118].

Figure 8.5 Schematic depiction of C−H functionalization via metal insertion.

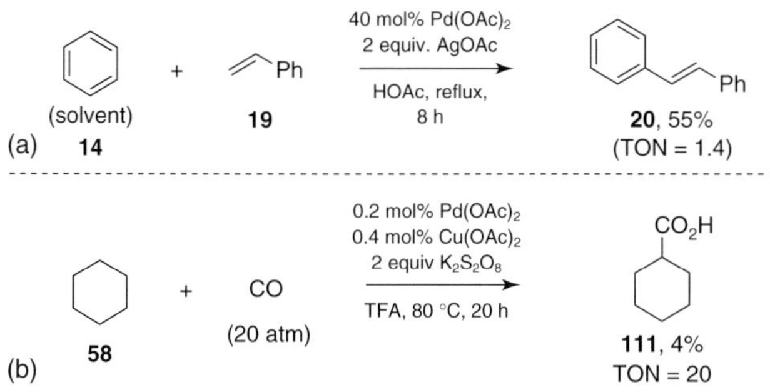

Scheme 8.29 (a) C–H olefination of benzene (**14**) using Pd(II)/Pd(0) catalysis [119]. (b) C–H carboxylation of cyclohexane (**58**) via Pd(II)/Pd(0) catalysis [121].

As discussed in Section 8.2.1, in 1967, Moritani and Fujiwara reported a Pd(II)-mediated C–H olefination reaction of aromatic hydrocarbons [32]. Shortly thereafter, Fujiwara *et al.* were able to develop a catalytic protocol based on Pd(II)/Pd(0) redox using stoichiometric quantities of transition metal reoxidants (Scheme 8.29), thereby establishing the viability of Pd(II)/Pd(0) catalysis for C–C bond formation [119]. Subsequent work led to the use of more practical oxidants, including Cu salts and peroxides [114]. Notably, shortly after Fujiwara and co-workers' initial report, Shue reported an aerobic protocol in which O_2 was used as the lone oxidant [120]. It was also established that other coupling partners could react with the corresponding [Pd(II)–R] species via Pd(II)/Pd(0) catalysis. For instance, in 1993, Fujiwara and co-workers reported a reaction in which cyclohexane could be carboxylated with CO (Scheme 8.29) [121].

This early work demonstrated the impressive reactivity of Pd(II) in activating C–H bonds, but the practical utility of this chemistry in organic synthesis has been limited for two reasons. First, a large excess of the substrate was required (often used as the solvent). Second, there was a lack of control of the regioselectivity when substituted arenes or alkanes were used as the substrate. One solution to both of these problems is to use DGs, which facilitate C–H cleavage by bringing the metal catalyst into close proximity to the target C–H bond and control the regioselectivity through assembly of a cyclometallated intermediate.

Building on the early concepts put forward in cyclometallation chemistry (Section 8.2.2), an attractive possibility would be to employ prefunctionalized substrates bearing DGs to coordinate the metal and position it for C–H cleavage. Not only would the DG control the regioselectivity, it would also improve the reactivity because the metal would be situated in close proximity to the target C–H bond. Moreover, the resulting products would have an added degree of molecular complexity, in line with the philosophy of further functionalization.

Pioneering work to establish the viability of this concept was reported by Fahey in 1971 in the Pd(II)-catalyzed *ortho*-C(sp²)–H chlorination of azobenzene (**112**) (Scheme 8.30) [122]. Reaction of the putative palladacycle intermediate [41] with Cl₂ leads to a high oxidation state Pd(III) or Pd(IV) intermediate, which then undergoes C–Cl reductive elimination to release the product, **113**. Another important finding came

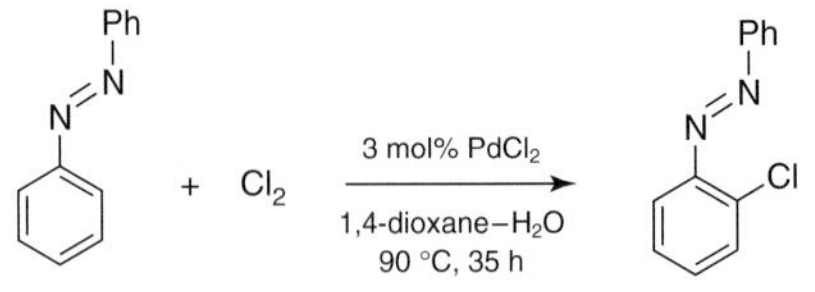

Scheme 8.30 Pd(II)-catalyzed *ortho*-C(sp^2)–H chlorination of azobenzene (**112**) [122].

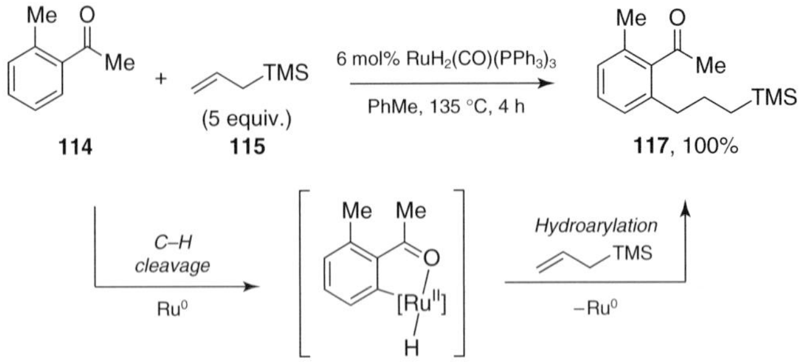

Scheme 8.31 Ru(0)-catalyzed *ortho*-C–H functionalization of aryl ketones (**114**) [124].

in 1984, when Tremont and Rahman found that acetanilide substrates could undergo *ortho*-C–H cleavage/C–C bond formation with MeI in good yields via Pd(II)/Pd(IV) catalysis, and they demonstrated that a TON of up to 10 could be achieved by using AgOAc as the reoxidant [123].

A seminal example of directed Ru(0)-catalyzed C–H cleavage/C–C bond formation came in 1993 from Murai *et al.* (Scheme 8.31) [124]. They reported that simple aryl ketones (**114**) could undergo selective C–H cleavage. Subsequently, the resulting [Ru(II)(H)(Ar)] can react with an alkene (**115**) by hydroarylation to generate the *ortho*-C–H functionalized product (**117**). In a related study, Trost *et al.* found that vinyl C–H bonds could also be functionalized using ketone or ester functional groups [125]. Interestingly, it was observed that esters gave improved reactivity, with shorter reaction times and higher yields. Employing the Shilov-type chemistry discussed in Section 8.3.1, Kao and Sen [126] and Sames and co-workers [127] were able to demonstrate directed C(sp^3)–H functionalization from simple carboxylic acid-containing substrates.

In the mid-2000s, a flurry of research concerning Pd(II)-catalyzed directed C–H functionalization appeared in the literature (Scheme 8.32), which attracted widespread interest from the chemical community. In 2004, Sanford and co-workers reported a C–H acetoxylation protocol using a series of different nitrogen-containing DGs (e.g., **118**) and PhI(OAc)$_2$ (**119**) as the oxidant [128]. Benzylic C(sp^3)–H bonds, and later unactivated C(sp^3)–H bonds, were reactive using this approach [129]. Early in 2005, Yu's group disclosed the first example of diastereoselective C(sp^3)–H iodination using a removable chiral oxazoline auxiliary [130], a method representing a potentially powerful approach for synthesizing all-carbon quaternary stereocenters. Later the same year, Daugulis and Zaitsev reported Pd(II)-catalyzed C(sp^2)–H arylation using acetanilide-type substrates

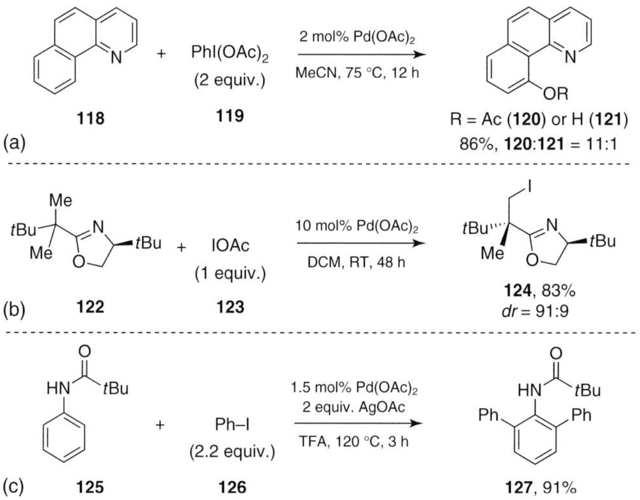

(a)

118   +   PhI(OAc)$_2$ →   R = Ac (120) or H (121)
                (2 equiv.)   86%, 120:121 = 11:1
       119

2 mol% Pd(OAc)$_2$
MeCN, 75 °C, 12 h

(b)

122   +   IOAc   →   124, 83%
              (1 equiv.)   dr = 91:9
       123

10 mol% Pd(OAc)$_2$
DCM, RT, 48 h

(c)

125   +   Ph–I   →   127, 91%
              (2.2 equiv.)
       126

1.5 mol% Pd(OAc)$_2$
2 equiv. AgOAc
TFA, 120 °C, 3 h

Scheme 8.32 C–H functionalization via Pd(II)/Pd(IV) or Pd(II)$_2$/Pd(III)$_2$ catalysis. (a) *ortho*-C(sp^2)–H acetoxylation [128]. (b) Diastereoselective C(sp^3)–H iodination using a removable chiral oxazoline auxiliary [130]. (c) *ortho*-C(sp^2)–H arylation [131].

(125) and aryl iodides (126) [131]. All of these reactions are proposed to proceed through Pd(II)/Pd(IV) or Pd(II)$_2$/Pd(III)$_2$ catalysis [132, 133].

Yu and co-workers went on to develop the first transformation to cross-couple C–H bonds with organometallic reagents via Pd(II)/Pd(0) catalysis (Scheme 8.33) [134]. 1,4-Benzoquinone (BQ) was found to be crucial for promoting reductive elimination from putative intermediate 131. Batchwise addition of the organotin reagent 129 lowered the rate of homocoupling such that Pd(II)-mediated C–H cleavage could take place. Generally, substrates containing synthetically restrictive DGs (e.g., pyridine and oxazoline) are disadvantageous because they limit the range of applications in organic synthesis [116]. For this reason, ubiquitous functional groups such as alcohols, amines, and carboxylic acids are more desirable directing elements. In line with this logic, Yu's group went on to develop a Pd(II)-catalyzed cross-coupling reaction of C(sp^2)–H and C(sp^3)–H bonds with organoboron reagents [135].

A fundamental challenge in the field of C–H functionalization via metal insertion is to control the stereoselectivity [136]. Broadly, there are two basic approaches in stereoselective C–H functionalization using metal insertion: (i) C–H cleavage/enantioselective functionalization and (ii) enantioselective C–H cleavage/functionalization. In comparing the two classes of transformations, one of the appealing features of enantioselective C–H cleavage/functionalization is that the chiral organometallic intermediate can potentially be coupled with any nucleophile or electrophile provided that stereocontrol during C–H cleavage is robust and that the chiral ligands are compatible with a variety of reaction conditions. Given the potential diversity of transformations and the possibility of performing enantioselective C(sp^3)–H cleavage on molecules containing *gem*-dimethyl (146) or methylene groups (147), the possibilities for constructing stereocenters are endless (Figure 8.6).

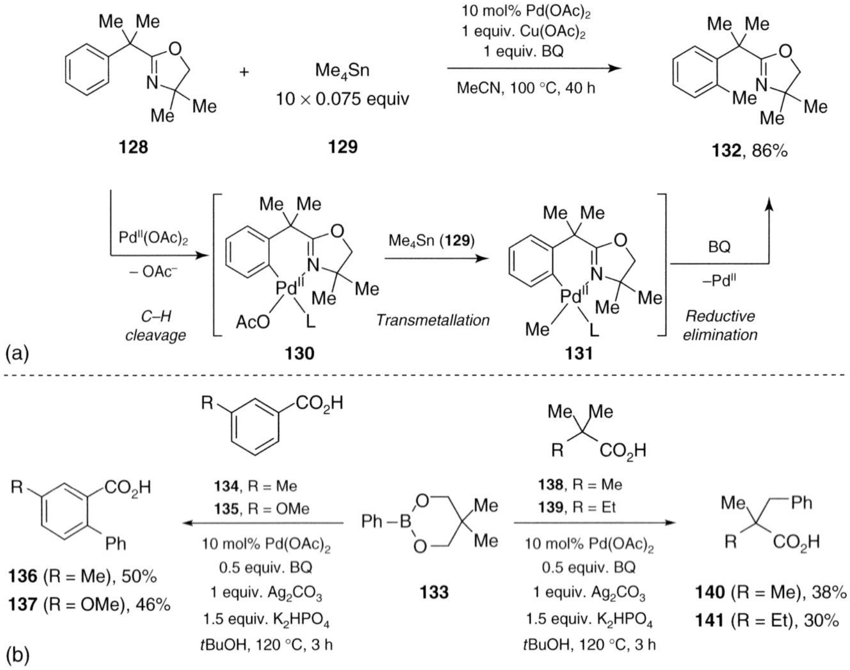

(a)

(b)

Scheme 8.33 (a) The first example of cross-coupling between C(sp²)–H bonds and organometallic reagents via Pd(II)/Pd(0) catalysis [134]. (b) C–C cross-coupling between C(sp²)–H bonds and organoboron reagents using simple carboxylic acid substrates [135].

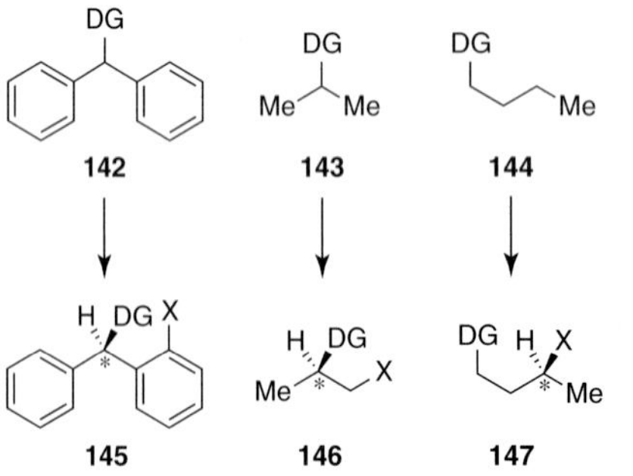

Figure 8.6 Enantioselective C–H functionalization of diphenyl compounds (**142**–**145**), *gem*-dimethyl moieties (**143**–**146**), and methylene groups (**144**–**147**) via metal insertion.

Many of the reported C–H cleavage/enantioselective functionalization protocols concern Rh(I)-catalyzed enantioselective alkene hydroarylation. In an early report, Murai and co-workers found that PPF-OMe (**149**), a chiral ferrocenylphosphino ligand, offered moderate levels of stereoinduction when used in an Rh(I)-catalyzed C–H cleavage/intermolecular hydroarylation (Scheme 8.34) [137]. In this reaction, directed C–H cleavage by Rh(I) generates the key putative intermediate **150**, which

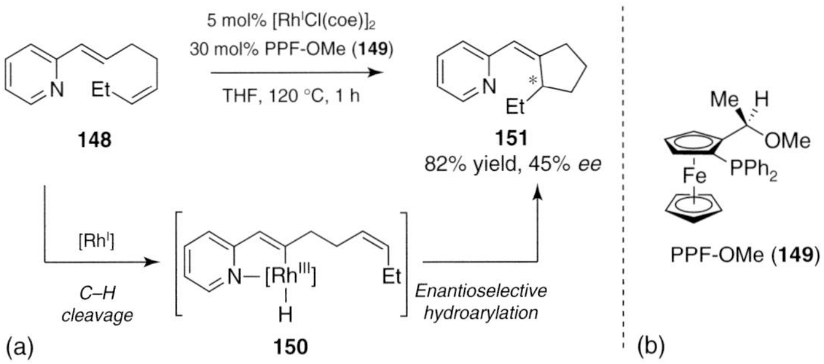

Scheme 8.34 (a) Rh(I)-catalyzed C–H cleavage (via oxidative addition)/enantioselective hydroary-lation [138]. (b) Ligand structure.

undergoes enantioselective intramolecular hydroarylation to give chiral five-membered carbocyclized products **151**. Murai and co-workers also reported a seminal example of atropselective C–H cleavage/hydroarylation to give axially chiral biaryl products in moderate enantiomeric excess (*ee*) [137].

In 2008, Yu's group published an example of Pd(II)-catalyzed enantioselective C–H cleavage/C–C cross-coupling with organoboron reagents by employing mono-*N*-protected amino acids as chiral ligands (Scheme 8.35) [139]. Mechanistically, the key feature of this transformation is the stereocontrolled C–H cleavage event, during which a new stereocenter is formed on the substrate to generate a chiral organometallic intermediate (**159**). In principal, treating this intermediate with any of a number of different nucleophiles or electrophiles would lead to new C–C or C–heteroatom bonds. In this initial report, success was achieved using nucleophilic organoboron reagents for C–C cross-coupling. The best combination of yield and enantioselectivity was achieved using the ligand (−)-Men-Leu-OH (**154**), which contains stereochemical information on both the amino acid backbone and the *N*-protecting group (PG). Importantly in substrates containing a *gem*-dimethyl moiety, enantioselective C(sp^3)–H cleavage could also be achieved using a more conformational restricted chiral cyclopropyl amino acid ligand, albeit in low *ee*. Taken together, these results represent an encouraging step forward in the development of methodology to construct the diverse collection of stereocenters depicted in Figure 8.6. Later, it was found that enantioselective diphenylacetic acid olefination [140] could also be carried out using mono-*N*-protected amino acid ligands (Scheme 8.35) [141]. In the working mechanistic model for these reactions, the amino acid side chain directs the spatial orientation of the protecting group on the nitrogen atom, rendering it chiral. The substrate is then oriented in an energetically favorable conformation by minimizing steric interactions with the bridgehead group (R$_s$, in **159** and **160**).

Further functionalization methods based on metal insertion have found application in natural products synthesis. Sames and co-workers showcased the power of sequential metal-mediated C–H functionalization in their elegant synthesis of the core of teleocidin B4 (**174**) (Scheme 8.36) [142]. By including an appended imine auxiliary in the starting material (**161**), C(sp^3)–H insertion by stoichiometric Pd(II) could be facilitated. The

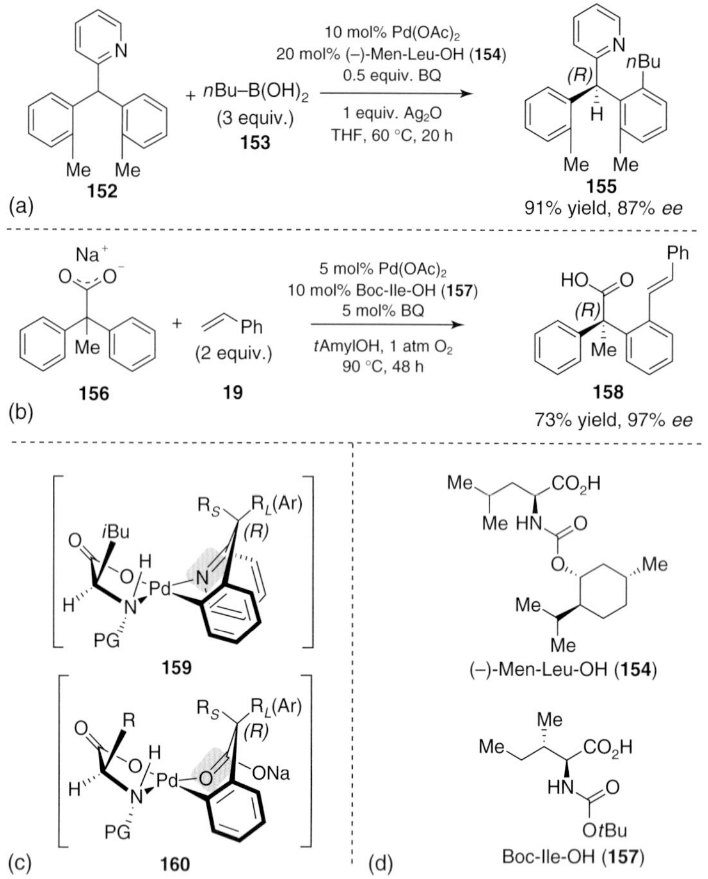

Scheme 8.35 Enantioselective C(sp²)–H activation/C–C cross coupling: (a) [139] and (b) [141]. (c) Proposed chiral cyclopalladated intermediates. (d) Ligand structures.

resulting [Pd(II)–alkyl] species **162** was coupled with boronic acid (**163**) to effect C–C bond formation. Following Friedel–Crafts cyclization to give **166**, an analogous Pd(II)-mediated C–H carbonylation gave **169**, which could be smoothly brought on to the desired core structure **174** following nucleophilic substitution on 2,3-dibromo-1-propene (**170**), deprotection, and Pd(0)-mediated intramolecular C–H vinylation.

Ellman and co-workers demonstrated the power of C–H cleavage/diastereoselective alkene hydroarylation in their total synthesis of (+)-lithospermic acid (**182**) (Scheme 8.37) [143]. The key step in their route utilized a chiral imine auxiliary to promote C(aryl)–H cleavage by Rh(I) and control the facial selectivity of addition of the resulting [Rh(III)–(Ar)(H)] to the tethered alkene. This reaction closed the dihydrobenzofuran ring and set the absolute stereochemical configuration of the *cis*-intermediate *cis*-**176**. Epimerization of the α-stereocenter gave the desired *trans*-stereochemical configuration. Three further steps, including a challenging global deprotection using optimized

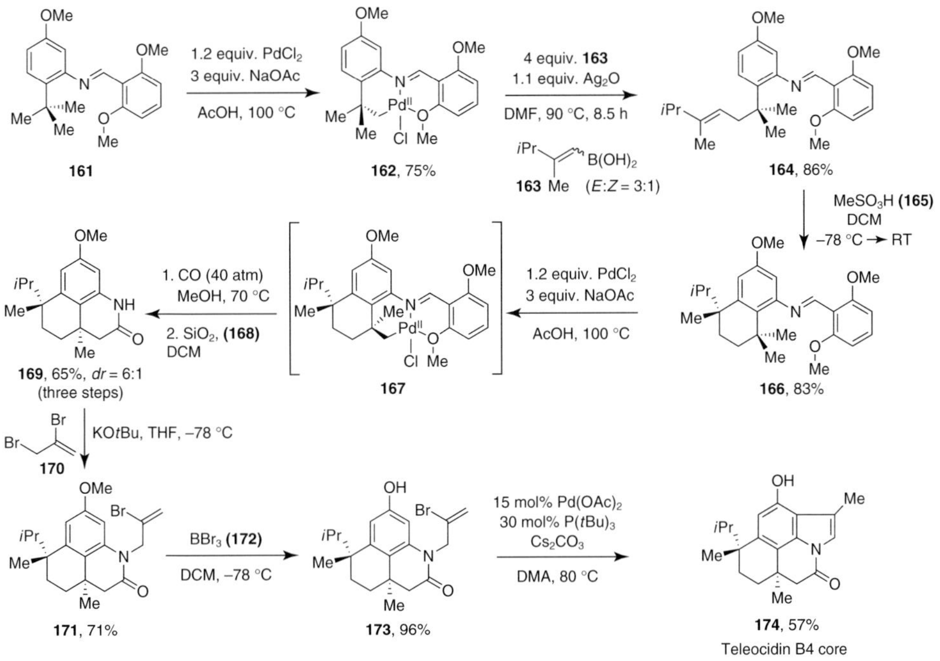

Scheme 8.36 Synthesis of the teleocidin B core via sequential C(sp³)–H functionalization [142].

conditions employing trimethylsilyl (TMS)-quinoline (181) gave (+)-lithospermic acid (182).

The work described in this chapter illustrates the exciting potential of C–H functionalization by metal insertion as a tool in organic synthesis. To realize the full power of these methods, continued research efforts devoted to expanding substrate scope and developing new ligand frameworks for enantioselective C–H cleavage/functionalization are crucial. Although elegant applications of C–H functionalization via metal insertion have already been reported, at the present time we are only scratching the surface of the true enabling qualities of C–H functionalization in expediting synthetic endeavors.

8.6
Other Emerging Metal-Catalyzed Further Functionalization Methods

In addition to the catalytic inner sphere C–H functionalization reactions via metal insertion discussed in the previous section, several outer-sphere metal-catalyzed C–H reactions offer exciting possibilities in organic synthesis. In this section, we take a brief look at three emerging classes of further functionalization reactions: biomimetic oxidation, metallocarbenoid insertion, and metallonitrenoid insertion. In highlighting advances in these different fields, we devote special attention to issues of reactivity, selectivity, and substrate scope.

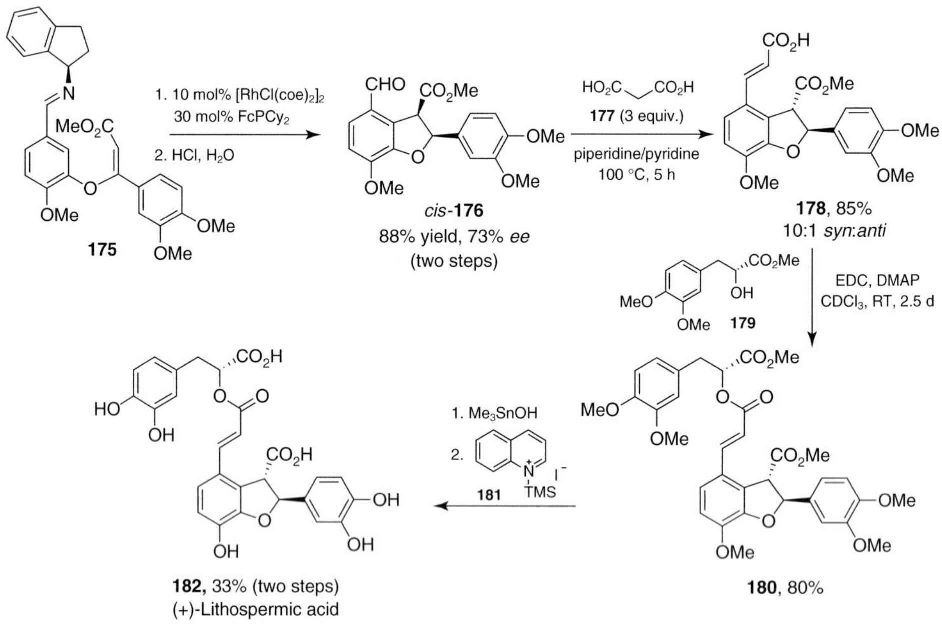

Scheme 8.37 Enantioselective total synthesis of (+)-lithospermic acid (**182**) using an Rh(I)-catalyzed C—H cleavage/diastereoselective alkene hydroarylation [143].

8.6.1
Biomimetic C—H Oxidation Methods

Nature has evolved enzymatic C—H oxidation catalysts that show precision in regio-, chemo-, and stereoselectivity that are difficult to emulate in the laboratory, including heme-containing cytochrome P450 and nonheme methane monooxygenase. As such, there is a great deal of interest among organic chemists in harnessing the catalytically active species in these transformations to develop reliable reactions for organic synthesis. The design and application of transition metal-based catalysts that exhibit analogous reactivity to natural enzymes (so-called "biomimetic oxidation") has a long history dating back to the late 1970s [144]. Following this early work, steady progress has been made in catalyst development, leading to an expanding pool of compatible substrates and several enantioselective C—H oxidation protocols.

As is the case with oxidase enzymes, the operative mechanisms for the transformations described in this section are complex and system dependent. High-valent [M(IV)=O] or [M(V)=O] could be at play, or they could proceed via initial formation of an [M(III)—OOH] which is capable of directly reacting with a C—H bond or can undergo O—O cleavage to generate a high-valent [M=O] species, which can react with C—H bonds. In biomimetic C(sp^3)—H oxidation reactions, the C—H cleavage step is based on a radical abstraction mechanism, hence the selectivity patterns fall in line with what is generally seen in traditional radical-based processes, with electron-rich C—H bonds being more reactive than electron-poor C—H bonds: tertiary C(sp^3)—H > secondary C(sp^3)—H >

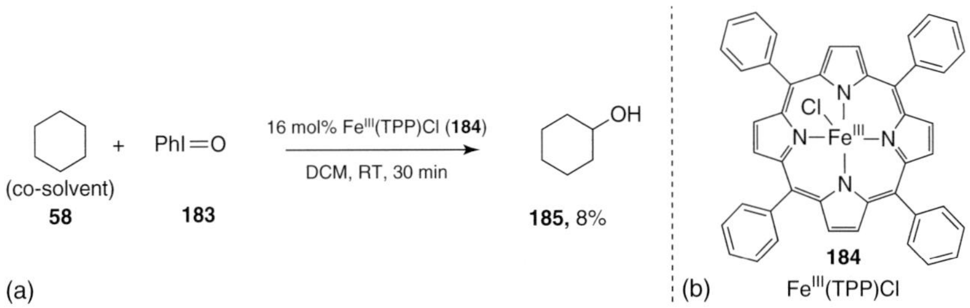

Scheme 8.38 (a) C(sp^3)−H oxidation catalyzed by [Fe(III)−porphyrin] complex **184** [145]. (b) Catalyst structure.

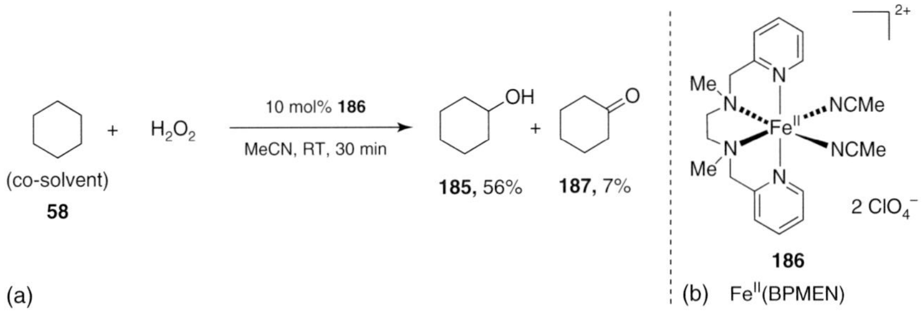

Scheme 8.39 (a) Oxidation of cyclohexane (**58**) using nonheme Fe(II) complex **186** with H_2O_2 as the oxidant [148]. (b) Catalyst structure.

primary C(sp^3)−H. Activated C(sp^3)−H, including benzylic and allylic C−H bonds, show particularly high reactivity.

In 1979, as part of an early effort to utilize simple heme-type iron complexes to mimic the oxidation of cytochrome P450, Groves *et al.* reported a seminal example of alkane hydroxylation using Fe(III)(TPP)Cl (**184**) (TPP = α,β,γ,δ-tetraphenylporphinato) as the catalyst and iodosylbenzene (**183**) as the oxidant (Scheme 8.38) [145]. For instance, with cyclohexane (**58**) as the substrate, cyclohexanol (**185**) could be obtained in 8% yield (based on PhIO).

Galvanized by this early success, several groups have studied other heme- and nonheme-type Fe catalysts, and also catalysts derived from other metals, in a search for better reactivity and selectivity and broader substrate scope [146, 147]. For instance, in 2001, Chen and Que reported an effective nonheme Fe(II) catalyst, **186**, that gave TONs up to 6.3 (Scheme 8.39) [148]. For instance, cyclohexane (**58**) could be oxidized using H_2O_2 as the oxidant, giving a mixture of the alcohol (**185**) and ketone (**187**) products.

Biomimetic oxidation catalysts have also found applications on more advanced prefunctionalized substrates. In these cases, the presence of multiple potential reaction sites can pose a problem due to the highly reactive nature of the active metal species. Success in controlling the regioselectivity of metalloporphyrin oxidation has been achieved using innovative strategies such as molecular scaffolding [149], covalent attachment [150, 151], and molecular recognition [152]. In more recent work, Chen and White made a thorough

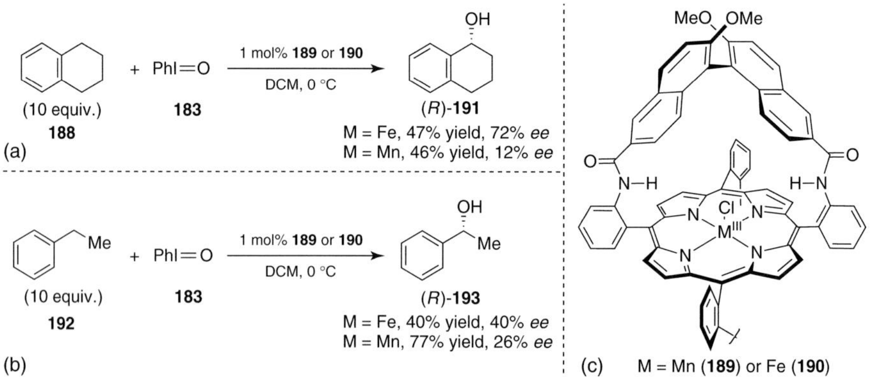

Scheme 8.40 Enantioselective C(sp³)–H oxidation by [M(III)–porphyrin] complexes **189** and **190** [155]. (a) Enantioselective oxidation of tetralin (**188**). (b) Enantioselective oxidation of ethylbenzene (**192**) (c) Catalyst structure.

study of the C(sp³)–H oxidation selectivity patterns using a nonheme Fe(II) catalyst with advanced substrates including natural products [153, 154].

One major challenge in this area is to control the stereoselectivity of biomimetic C–H oxidation reactions in unsymmetrical substrates. In pioneering work, Groves and Viski reported a porphyrin-derived ligand framework capable of effecting stereoinduction on cyclic and acyclic substrates bearing benzylic C(sp³)–H bonds [155]. Mn(III) complex **189** and Fe(III) complex **190** were found to give moderate yields and ee values (Scheme 8.40).

More recently, in 1994, Larrow and Jacobsen found that the salen-based Mn(III) complex **197** could catalyze the C(sp³)–H hydroxylation of chiral epoxides when NaOCl was employed as the oxidant (Scheme 8.41) [156]. The minor product (1S,2R)-**196**, formed

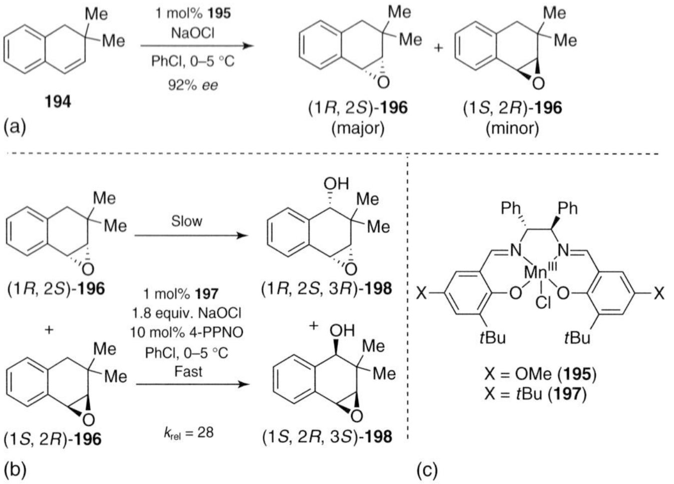

Scheme 8.41 (a) Enantioselective epoxidation of **194** [156]. (b) Application of stereoselective C(sp³)–H oxidation in the kinetic resolution of epoxides. (c) Catalyst structure.

following asymmetric epoxidation of **194**, was found to be more reactive in $C(sp^3)$–H hydroxylation. Using these two reactions in tandem, a kinetic resolution could be effected to give (1R,2S,3R)-**198** in $\geq$98% *ee* and 79% yield.

As discussed in Section 8.4.1, in the realm of total synthesis, strategic site-selective oxidation has long proven to be a powerful approach. Thus far, however, applications of biomimetic C–H oxidation catalysts have not found widespread utility. Nevertheless, reflections on biomimetic oxidation chemistry continue to inform and inspire endeavors in natural products synthesis. Recently, for example, Baran and co-workers showcased the power of a biomimetic two-phase (cyclase/oxidase) strategy for synthesizing terpene natural product families [157, 158].

To realize the full power of biomimetic C–H oxidation chemistry and further functionalization methods in organic synthesis, several lingering challenges must be addressed in terms of catalyst activity, stability, and substrate scope. Moreover, regioselectivity problems continue to plague biomimetic oxidation, hence new approaches (e.g., developing DG-based methods) are needed along with improved catalysts.

8.6.2
Metallocarbenoid Insertion

Since 1942, when Meerwein, Rathjen, and Werner first discovered that carbenes generated from photo-induced decomposition of diazo compounds could insert into unactivated $C(sp^3)$–H bonds [159], a great deal of attention has surrounded carbene C–H insertion as a means of dramatically advancing molecular complexity by appending new carbogenic motifs for further functionalization. In contemporary carbene chemistry, the standard method for generating a high-energy metallocarbenoid species is via metal-mediated diazo decomposition, with concomitant extrusion of gaseous N_2 (Scheme 8.42). Effective metals include silver, ruthenium, copper, and rhodium, the last of which is the most widely used [160–162].

The resulting metallocarbenoids are unstable and can react with an array of different functional groups, including alkenes (cyclopropanation), Lewis bases (ylide formation), and C–H bonds (insertion) [164]. In early work in this area, C–H functionalization using metallocarbenoids was found to be most effective when carried out in an intramolecular sense. Otherwise, the highly reactive metallocarbenoids displayed little to no regioselectivity. In the early 1980s, Taber's group explored methods for carbocyclic ring closure based on carbenoid C–H insertion with catalytic Rh (Scheme 8.43) [165]. In particular, they found that highly substituted cyclopentyl ring systems (e.g., **206**) could be constructed from the Rh-mediated decomposition of a tethered α-diazo carbonyl moiety and subsequent intramolecular $C(sp^3)$–H insertion. In these reactions, cyclization to form a five-membered ring is preferred and, because $C(sp^3)$–H insertion is stereoretentive, this methodology could be applied for constructing all-carbon stereocenters, a tactic that was drawn upon in the total synthesis of (+)-α-cuperenone [166]. This intramolecular cyclization based on metallocarbenoid C–H insertion has proven to be useful as a general strategy for constructing the carbogenic skeletons of natural products and other bioactive molecules [160].

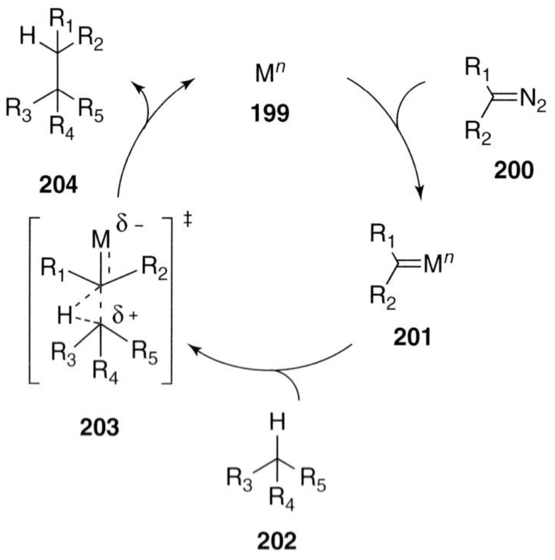

Scheme 8.42 Catalytic cycle for metal-mediated carbenoid insertion into a C−H bond with proposed transition state **203** [163].

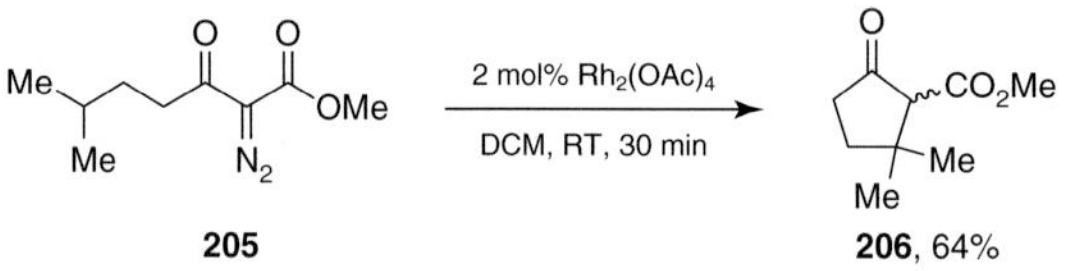

Scheme 8.43 Synthesis of highly substituted cyclopentane ring systems using Rh-catalyzed C−H insertion [165].

Chiral carboxamidate-containing catalysts have been developed for application in enantioselective C−H insertion [161]. Dating back to the early work by the groups of McKervey [167, 168] and Ikegami [169] in the early 1990s, enantioselective metallocarbenoid C−H insertion has been an active area of research. Among the most active catalysts developed to date for intramolecular metallocarbenoid C−H insertion are those developed by Doyle and co-workers, including $Rh_2(S\text{-MPPIM})_4$ (**208**). These intramolecular carbenoid insertion protocols have been applied to the enantioselective total syntheses of lignan natural products such as (−)-baclofen (**213**) [170] and (+)-imperanene (**214**) [171] (Scheme 8.44).

Although the impressively high reactivity of metallocarbenoids is useful in intramolecular transformations, these same energetic properties have proven to be deleterious to regio- and chemoselectivity in *intermolecular* reactions, a significant drawback. In this respect, one pervasive problem is that the majority of diazo compounds contain two electron-withdrawing groups at the α-position, so-called "acceptor only" carbenoids (i.e., both R^1 and R^2 are electron-withdrawing groups in Scheme 8.42). Carbenes are electrophilic by nature, hence the presence of two electron-withdrawing groups serves to destabilize the putative intermediate further, which ultimately decreases selectivity. One of the greatest breakthroughs in the development of robust intermolecular reactions has been the use of carbenoids bearing one electron-withdrawing group and one

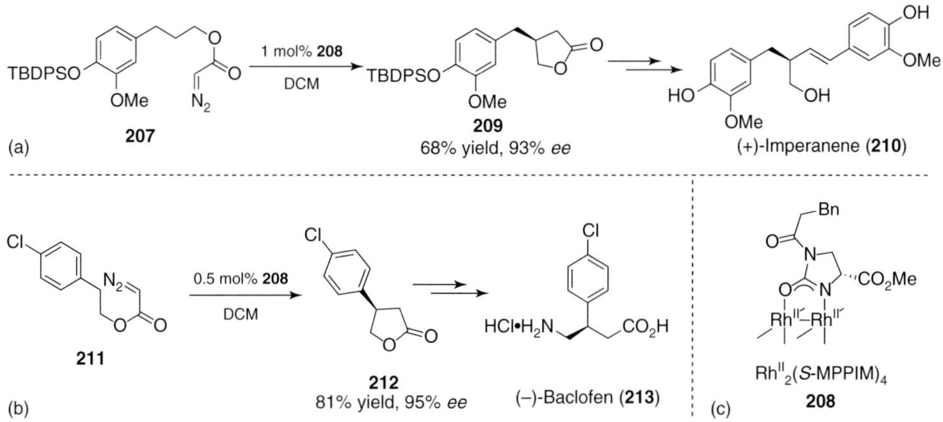

(a)

(b)

(c)

Scheme 8.44 Natural product synthesis employing enantioselective carbenoid C(sp³)–H insertion. (a) Total synthesis of (+)-imperanene (**210**) [171]. (b) Total synthesis of (–)-baclofen (**213**) [170]. (c) Catalyst structure.

electron-donating group at the α-position, dubbed "donor/acceptor" carbenoids, a tactic pioneered by Davies and Denton [172]. When donor/acceptor metallocarbenoids are used, the overall reactivity is attenuated, allowing for more precise regiochemical control. Mechanistically, donor/acceptor carbenoids stabilize positive charge buildup at the C–H insertion site by virtue of the electron-donating character of the appended group (i.e., R¹ is an electron-withdrawing group and R² is an electron-donating group such as N, O, aryl, or vinyl in Scheme 8.42) [163].

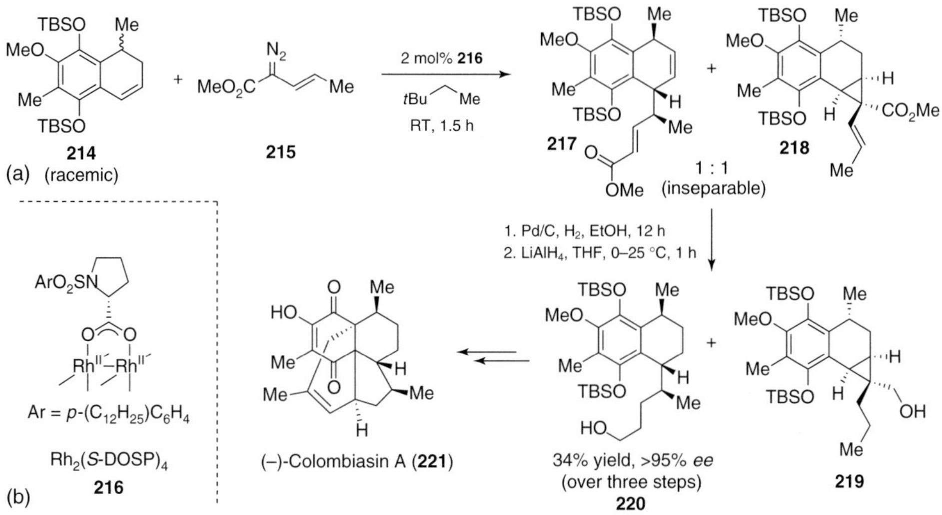

(a) (racemic)

(b)

Scheme 8.45 (a) Total synthesis of (–)-colombiasin (**221**) via an intermolecular asymmetric C–H insertion/Cope rearrangement strategy [173]. (b) Catalyst structure.

The power of this approach has been demonstrated by several applications in natural products total synthesis. For example, in Davies *et al.*'s expedient enantioselective synthesis of (−)-colombiasin A (**221**), the intricate carbogenic core structure could be prepared in a straightforward manner using an innovative asymmetric allylic C–H insertion/Cope rearrangement (Scheme 8.45) [173]. Donor/acceptor carbenoid **215** could be merged with racemic **214** in the presence of $Rh_2(R\text{-}DOSP)_4$ (**216**), leading to a kinetic resolution in which the desired enantiomer of **214** underwent a C–H insertion/Cope rearrangement to give **217** and the undesired enantiomer underwent cyclopropanation to give **218**. The resulting 1:1 mixture was hydrogenated and then reduced with $LiAlH_4$ to allow separation. Compound **220** could then be brought on to the (−)-colombiasin A (**221**).

Research to develop improved catalysts and to devise new synthetic strategies to exploit metallocarbenoid C–H insertion constitutes an exciting area of organometallic chemistry. Along with the recent surge of interest in applying metallocarbenoid C–H insertion in total synthesis, the discovery of effective chiral catalysts and the emergence of tactics for controllable intermolecular reactions using donor/acceptor carbenoids are among the most significant achievements during the past decade. In the future, devising methodology for controlling the regio- and chemoselectivity will render carbenoid C–H insertion increasingly useful for C–C bond formation in organic synthesis.

8.6.3
Metallonitrenoid Insertion

The development of metallonitrenoid C–H insertion (Scheme 8.46) as a tool for further functionalization in organic chemistry has similarly progressed at a brisk pace during the past decade, allowing for the expansion of substrate scope, application in total synthesis, and improvement in stereochemical control. In the 1980s, Breslow and Gellman established that Mn(III)– and Fe(III)–porphyrin complexes, which can be thought of as cytochrome P450 mimics, could effectively catalyze nitrenoid $C(sp^3)$–H insertion in both inter- [174] and intramolecular [175] settings (Scheme 8.47). This elegant work built upon the metal–oxo chemistry developed by Groves *et al.*, particularly their studies using iodosylbenzene and [Fe(III)–TPP] [145], as discussed above in Section 8.6.1. The reaction mechanism proceeds via initial reaction of the metalloporphyrin with (tosyliminoiodo)benzene (**228**) to generate a metallonitrenoid. Subsequent C–H insertion with cyclohexane (**58**) takes place to give the tosylamidated product **229**. A general catalytic cycle is depicted in Scheme 8.46.

Tremendous inroads have been made in the pursuit of practical and operationally simple conditions that are compatible with a variety of different substrates classes. In a broad sense, one of the major challenges in this area is that there are fewer options for modulating the reactivity of metallonitrenoids than there are for metallocarbenoid [173], given that the nitrogen atom bears only one substituent. Because the resulting electrophilic species are of high energy and highly reactive, controlling regioselectivity in intermolecular reactions with unsymmetrical substrates is a tremendous challenge. On the other hand, success has been achieved in developing high-yielding *intramolecular* reactions in pioneering work by Du Bois and co-workers. In 2001, Rh(II)-catalyzed $C(sp^3)$–H amination reactions of carbamates [176] and sulfamate esters [177] were

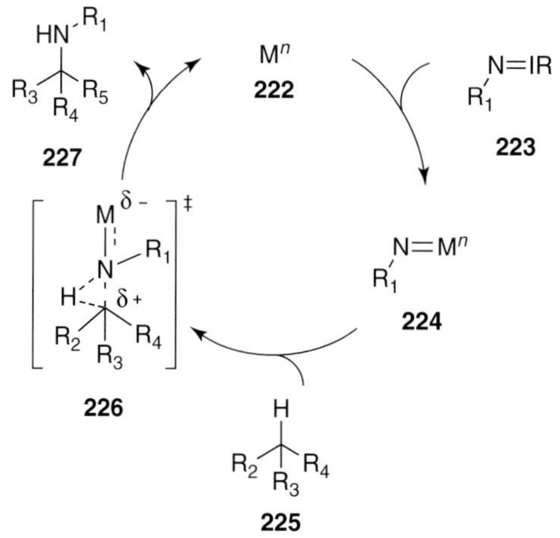

Scheme 8.46 Catalytic cycle for metal-mediated carbenoid insertion into a C−H bond with proposed transition state **226**.

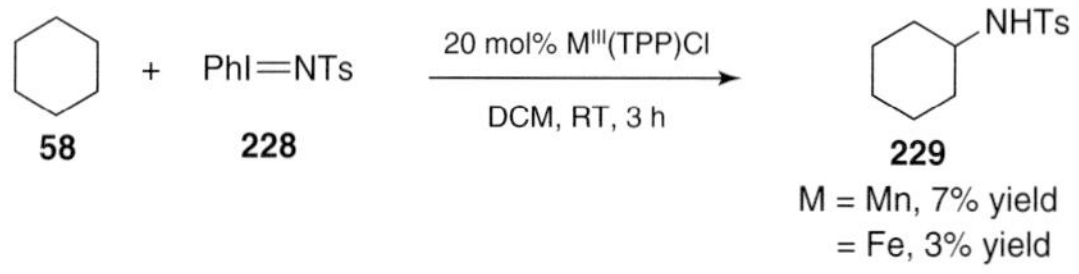

Scheme 8.47 Early work on metallonitrenoid C−H insertion [174].

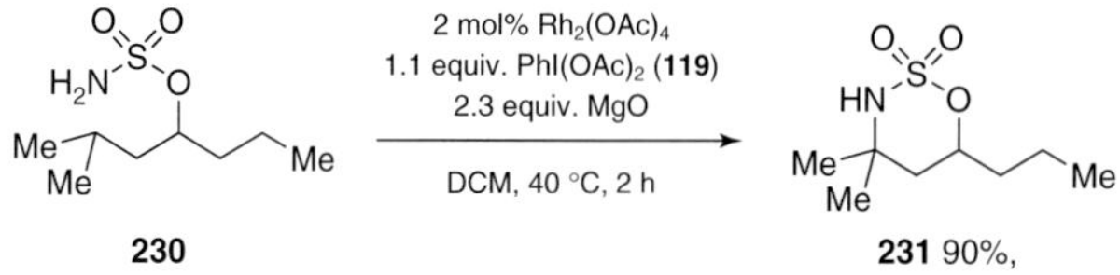

Scheme 8.48 Intramolecular $C(sp^3)$−H amination of sulfamate substrates [177].

reported (Scheme 8.48), chemistry that built upon early work on related systems [178]. The mechanism of these C−H amination reactions proceeds via initial generation of the iminophenyliodinanes *in situ*, followed by decomposition to form the metallonitrenoid, and subsequent C−H insertion. Beyond playing a pivotal role in the expansion of substrate scope, Du Bois and co-workers also reported a particularly effective catalyst for nitrenoid C−H insertion reactions, $Rh_2(esp)_2$, which contains a tethered dicarboxylate ligand [179]. With $Rh_2(esp)_2$, the scope of intra- and intermolecular $C(sp^3)$−H amination reactions has been improved, offering better yields and longer catalyst lifetimes.

As was the case in biomimetic C−H oxidation and C−H insertion by metallocarbenoids, serious efforts have been made to develop asymmetric C−H nitrenoid insertion methods (Scheme 8.49). Preliminary studies by Müller and co-workers [178] and later by Hashimoto and co-workers [180] demonstrated that low *ee* values could be obtained in these reactions.

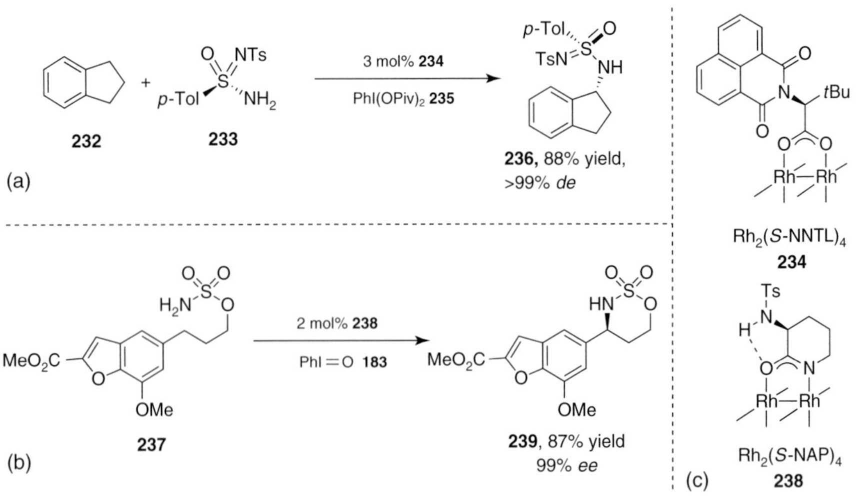

Scheme 8.49 (a) Diastereoselective C(sp³)–H amination of indenes via Rh(II)-catalyzed nitrenoid insertion [181]. (b) Enantioselective C(sp³)–H amination of a benzylic methylene group via Rh(II)-catalyzed nitrene insertion [182]. (c) Catalyst structures.

More recently, Müller and co-workers found that when a chiral tosylsulfonylimidamide (**233**) was used in conjunction with a chiral catalyst (**234**), high yields and high *de* values could be obtained for substrates containing benzylic or allylic C–H bonds [181]. In 2008, Zalatan and Du Bois reported that a novel ligand, (*S*)-NAP, gave *ee* values as high as 99% in enantioselective nitrenoid C–H insertions using sulfamate ester substrates (**237**) [182].

Other metals besides Rh(II) have also been shown to be effective for chiral induction. In 1999, Che and co-workers achieved modest enantioselectivity (54% *ee*) in benzylic C–H amidation reactions with TsN=IPh by employing a variety of chiral Ru(II)– and Mn(III)–porphyrin catalysts [183]. In 2001, Kohmura and Katsuki found that a chiral [Mn(III)–salen] complex bearing an electron-withdrawing group could be used in the intermolecular amination of allylic and benzylic C(sp³)–H bonds, offering high levels of enantioselectivity (89% *ee*) [184]. More recently, [Ru(II)–pybox] complexes have also emerged as a promising class of new catalysts [185].

The versatility of metallonitrenoid insertion chemistry was then demonstrated in the landmark asymmetric total synthesis of (−)-tetrodotoxin (**244**) by Hinman and Du Bois (Scheme 8.50) [186]. Two key steps of their strategy were an early-stage carbenoid C–H insertion to forge the central six-membered ring of **241** and a late-stage nitrenoid C–H insertion to set the C–N bond in **243**, which is ultimately expressed in the guanidinium moiety in the natural product **244**.

C(sp³)–H functionalization by metallonitrenoid insertion has advanced substantially in the past decade to the point where synthetically useful substrates can be functionalized in good yields with predictable selectivity. The power of these reactions has been demonstrated in natural product synthesis, and enantioselective processes (in both the intra- and intermolecular sense) are steadily improving. Taken together, these

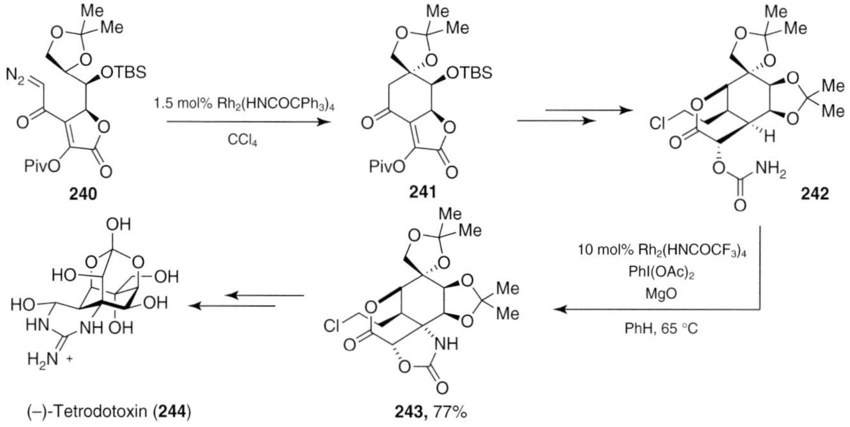

Scheme 8.50 Total synthesis of (−)-tetrodotoxin [186].

advancements represent progress towards a useful new class of synthetic transformations. Further catalyst development and expansion of substrate scope will continue to enhance the general applicability of this chemistry.

Through the outstanding contributions described above, biomimetic oxidation, metallocarbenoid insertion, and metallonitrenoid insertion have all begun to show great promise as further functionalization techniques in organic synthesis. Several key accomplishments since the initial discoveries in the 1970s and 1980s are important to highlight. First, the ability to use the substrate as the limiting reagent (rather than in excess, as the solvent or co-solvent) is tremendously important for expanding this chemistry to prefunctionalized starting materials. Of equal importance, catalytic systems have been developed such that the regio- and chemoselectivity can be controlled when advanced substrates are used, including natural product precursors. Enantioselective reactions have also emerged through the design of new chiral ligand scaffolds, an area of research that very likely will continue to be fruitful for long into the foreseeable future.

8.7
Outlook and Conclusion

In this chapter, we have attempted to give the reader an appreciation of modern efforts to develop "further functionalization" as a tool in organic synthesis, as compared with "first functionalization" for hydrocarbon chemistry. Through the selected examples, we hope that the reader will appreciate the power of C–H functionalization in permitting rapid assembly of molecular complexity, particular when enantioselectivity, regioselectivity, and chemoselectivity can be aptly controlled.

C–H functionalization by transition metals has a rich history dating back over a century. Early work in inorganic and organometallic chemistry helped establish the mechanistic underpinning of C–H cleavage processes, which helped inform the development of catalytic processes. One potential application of transition metal-catalyzed

C–H functionalization in the future is in converting hydrocarbons to value-added chemicals, so called "first functionalization." Although tremendous strides have been made on this front, the overall usefulness of this strategy remains to be seen, particularly when one considers the high levels of catalytic efficiency that are needed to make such processes cost-effective. On the other hand, for molecules that already contain some chemical functionality, including precursors for drugs, agrochemicals, and fine chemicals, C–H functionalization provides an opportunity for advancing molecular complexity through the construction of new stereocenters and functional group patterns, a process we call "further functionalization." This logic of C–H functionalization as a synthetic tool originated in the 1950s in investigations of steroid synthesis and, since that time, C–H functionalization has become increasingly prevalent in organic synthesis.

In this chapter, we have highlighted recent accomplishments and breakthroughs in inner-sphere metal-catalyzed C–H functionalization via metal insertion, with a special eye towards reactions catalyzed by Pd(II). Our focus then turned to three outer-sphere metal-catalyzed C–H functionalization methods, biomimetic oxidation, metallocarbenoid insertion, and metallonitrenoid insertion.

Throughout the text we have attempted to draw attention to unsolved problems in C–H functionalization, with the hope that this contribution will inspire the next generation of scientists to contemplate strategies and tactics for advancing the science of C–H functionalization for applications in organic chemistry.

Acknowledgments

We gratefully acknowledge TSRI, the US NSF (**NSF CHE-0615716**), the US NIH (**NIGMS, 1 R01 GM084019-03**), and Pfizer for financial assistance. Additional support was provided through the NSF Center for Stereoselective C–H Functionalization (**CHE-0943980**). Individual awards and fellowships were granted by TSRI, the US NSF, the US DOD, and the Skaggs Oxford Scholarship Program (K.M.E.), and the Dreyfus and Sloan Foundations (J.-Q.Y.). We are grateful to W. R. Gutekunst, P. S. Thuy-Boun, and K. S. L. Chan for their generous assistance in editing this chapter and for thoughtful advice during its preparation.

Abbreviations

The following abbreviations are used on the reaction schemes:

Ac	acetyl
Ar	aryl
Bn	benzyl
Boc	*tert*-butyloxycarbonyl
BQ	1,4-benzoquinone
cod	1,5-cyclooctadiene
coe	cyclooctene
Cp	cyclopentadienyl

Cp^*	1,2,3,4,5-pentamethylcyclopentadienyl
Cy	cyclohexyl
DCE	1,2-dichloroethane
DCM	dichloromethane
DMAP	4-dimethylaminopyridine
2,2-DMB	2,2-dimethylbutane
DMF	*N*,*N*-dimethylformamide
EDC	1-ethyl-3-(3-dimethylaminopropyl)carbodiimide
ESP	*a*, *a*, *a'*, *a'*-tetramethyl-1,3-benzenedipropionate
Fc	ferrocenyl
Fmoc	fluorenylmethyloxycarbonyl
For	formyl
Ph	phenyl
Pin	pinacolato
Piv	pivaloyl
4-PPNO	4-phenylpyridine *N*-oxide
Py	pyridine
RT	room temperature
TCBoc	β, β, β-trichloro-*tert*-butoxycarbonyl
Tf	trifluorosulfonyl
TFA	trifluoroacetic acid
THF	tetrahydrofuran
TMS	trimethylsilyl
TPP	α, β, γ, δ-tetraphenylporphinato
Tol	tolyl
Ts	toluenesulfonyl

References

1. Corey, E.J. and Cheng, X.-M. (1995) *The Logic of Chemical Synthesis*, John Wiley & Sons, Inc., New York.
2. Godula, K. and Sames, D. (2006) *Science*, **312**, 67–72.
3. Gutekunst, W.R. and Baran, P.S. (2011) *Chem. Soc. Rev.*, **40**, 1976–1991.
4. Boger, D.L. and Brotherton, C.E. (1984) *J. Org. Chem.*, **49**, 4050–4055.
5. Schreiber, S.L. (2000) *Science*, **287**, 1964–1969.
6. Wender, P.A., Hilinski, M.K., and Mayweg, A.V.W. (2005) *Org. Lett.*, **7**, 79–82.
7. Ortiz de Montellano, P.R. (1995) *Cytochrome P-450: Structure, Mechanism, and Biochemistry*, 2nd edn., Springer, New York.
8. Parshall, G.W. (1975) *Acc. Chem. Res.*, **8**, 113–117.
9. Crabtree, R.H. (1985) *Chem. Rev.*, **85**, 245–269.
10. Arndtsen, B.A., Bergman, R.G., Mobley, T.A., and Peterson, T.H. (1995) *Acc. Chem. Res.*, **28**, 154–162.
11. Shilov, A.E. and Shul'pin, G.B. (1997) *Chem. Rev.*, **97**, 2879–2932.
12. Dyker, G. (1999) *Angew. Chem. Int. Ed.*, **38**, 1698–1712.
13. Labinger, J.A. and Bercaw, J.E. (2002) *Nature*, **417**, 507–514.
14. Jazzar, R., Hitce, J., Renaudat, A., Sofack-Kreutzer, J., and Baudoin, O. (2010) *Chem. Eur. J.*, **16**, 2654–2672.
15. Periana, R.A., Bhalla, G., Tenn, W.J. III, Young, K.J.H., Liu, X.Y., Mironov, O.,

Jones, C.J., and Ziatdinov, V.R. (2004) *J. Mol. Catal. A: Chem.*, **220**, 7–25.

16. Goldman, A.S. and Goldberg, K.I. (eds.) (2004) *Activation and Functionalization of C–H Bonds*, ACS Symposium Series, vol. 885, American Chemical Society, Washington, DC.

17. Crabtree, R.H. (2004) *J. Organomet. Chem.*, **689**, 4083–4091.

18. Fenton, H.J.H. (1894) *J. Chem. Soc., Trans.*, **65**, 899–910.

19. Haber, F. and Weiss, J. (1932) *Naturwissenschaften*, **20**, 948–950.

20. Walling, C. (1975) *Acc. Chem. Res.*, **8**, 125–131.

21. Barton, D.H.R. and Doller, D. (1992) *Acc. Chem. Res.*, **25**, 504–512.

22. Stavropoulos, P., Çelenligil-Çetin, R., and Tapper, A.E. (2001) *Acc. Chem. Res.*, **34**, 745–752.

23. Thompson, M.E., Baxter, S.M., Bulls, A.R., Burger, B.J., Nolan, M.C., Santarsiero, B.D., Schaefer, W.P., and Bercaw, J.E. (1987) *J. Am. Chem. Soc.*, **109**, 203–219.

24. Cummins, C.C., Baxter, S.M., and Wolczanski, P.T. (1988) *J. Am. Chem. Soc.*, **110**, 8731–8733.

25. Sherry, A.E. and Wayland, B.B. (1990) *J. Am. Chem. Soc.*, **112**, 1259–1261.

26. Wayland, B.B., Ba, S., and Sherry, A.E. (1991) *J. Am. Chem. Soc.*, **113**, 5305–5311.

27. Wheland, G.W. (1942) *J. Am. Chem. Soc.*, **64**, 900–908.

28. (a) Pesci, L. (1892) *Gazz. Chim. Ital.*, **22**, 373–384; (b) Pesci, L. (1893) *Gazz. Chim. Ital.*, **23**, 521–529.

29. Volhard, J. (1892) *Justus Liebigs Ann. Chem.*, **267**, 172–185.

30. Dimroth, O. (1898) *Ber. Dtsch. Chem. Ges.*, **31**, 2154–2156.

31. Kharasch, M.S. and Isbell, H.S. (1931) *J. Am. Chem. Soc.*, **53**, 3053–3059.

32. Moritani, I. and Fujiwara, Y. (1967) *Tetrahedron Lett.*, **8**, 1119–1122.

33. Fujiwara, Y., Moritani, I., Matsuda, M., and Teranishi, S. (1968) *Tetrahedron Lett.*, **9**, 633–636.

34. Fujiwara, Y., Asano, R., Moritani, I., and Teranishi, S. (1976) *J. Org. Chem.*, **41**, 1681–1683.

35. Fuchita, Y., Hiraki, K., Kamogawa, Y., Suenaga, M., Tohgoh, K., and Fujiwara, Y. (1989) *Bull. Chem. Soc. Jpn.*, **62**, 1081–1085.

36. Chatt, J. and Watson, H.R. (1962) *J. Chem. Soc.*, 2545–2549.

37. Chatt, J. and Davidson, J.M. (1965) *J. Chem. Soc.*, 843–855.

38. Green, M.L. and Knowles, P.J. (1970) *J. Chem. Soc. D*, 1677.

39. Elmitt, K., Green, M.L.H., Forder, R.A., Jefferson, I., and Prout, K. (1974) *J. Chem. Soc., Chem. Commun.*, 747–748.

40. Kleiman, J.P. and Dubeck, M. (1963) *J. Am. Chem. Soc.*, **85**, 1544–1545.

41. Cope, A.C. and Siekman, R.W. (1965) *J. Am. Chem. Soc.*, **87**, 3272–3273.

42. Ryabov, A.D. (1985) *Synthesis*, 233–252.

43. Dupont, J., Consorti, C.S., and Spencer, J. (2005) *Chem. Rev.*, **105**, 2527–2572.

44. Ryabov, A.D., Sakodinskaya, I.K., and Yatsimirsky, A.K. (1985) *J. Chem. Soc., Dalton Trans.*, 2629–2638.

45. Gómez, M., Granell, J., and Martinez, M. (1997) *Organometallics*, **16**, 2539–2546.

46. Davies, D.L., Donald, S.M.A., and Macgregor, S.A. (2005) *J. Am. Chem. Soc.*, **127**, 13754–13755.

47. Canty, A.J. and van Koten, G. (1995) *Acc. Chem. Res.*, **28**, 406–413.

48. Bennett, M.A. and Milner, D.L. (1967) *Chem. Commun.*, 581–582.

49. Cotton, F.A., Hunter, D.L., and Frenz, B.A. (1975) *Inorg. Chim. Acta*, **15**, 155–160.

50. Hodges, R.J. and Garnett, J.L. (1968) *J. Phys. Chem.*, **72**, 1673–1682.

51. Hodges, R.J. and Garnett, J.L. (1969) *J. Phys. Chem.*, **73**, 1525–1539.

52. Hodges, R.J. and Garnett, J.L. (1969) *J. Catal.*, **13**, 83–98.

53. Garnett, J.L., Hodges, R.J., Kenyon, R.S., and Long, M.A. (1979) *J. Chem. Soc., Perkin Trans. 2*, 885–890.

54. Goldshleger, N.F., Shteinman, A.A., Shilov, A.E., and Eskova, V.V. (1972) *Zh. Fiz. Khim.*, **46**, 1353–1354.

55. Kushch, L.A., Lavrushko, V.V., Misharin, Y.S., Moravsky, A.P., and Shilov, A.E. (1983) *Nouv. J. Chim.*, **7**, 729–733.

56. La Placa, S.J. and Ibers, J.A. (1965) *Inorg. Chem.*, **4**, 778–783.

57. Trofimenko, S. (1968) *J. Am. Chem. Soc.*, **90**, 4754–4755.

58. Cotton, F.A., LaCour, T., and Stainislowski, A.G. (1974) *J. Am. Chem. Soc.*, **96**, 754–760.

59. Brown, R.K., Williams, J.M., Schultz, A.J., Stucky, G.D., Ittel, S.D., and Harlow, R.L. (1980) *J. Am. Chem. Soc.*, **102**, 981–987.

60. Dawoodi, Z., Green, M.L.H., Mtetwa, V.S.B., and Prout, K. (1982) *J. Chem. Soc., Chem. Commun.*, 802–803.

61. Dawoodi, Z., Green, M.L.H., Mtetwa, V.S.B., and Prout, K. (1982) *J. Chem. Soc., Chem. Commun.*, 1410–1411.

62. Dawoodi, Z., Green, M.L.H., Mtetwa, V.S.B., Prout, K., Schultz, A.J., Williams, J.M., and Koetzle, T.F. (1986) *J. Chem. Soc., Dalton Trans.*, 1629–1637.

63. Brookhart, M. and Green, M.L.H. (1983) *J. Organomet. Chem.*, **250**, 395–408.

64. Brookhart, M., Green, M.L.H., and Parkin, G. (2007) *Proc. Natl. Acad. Sci. U. S. A.*, **104**, 6908–6914.

65. Jones, W.D. and Feher, F.J. (1984) *J. Am. Chem. Soc.*, **106**, 1650–1663.

66. Hall, C. and Perutz, R.N. (1996) *Chem. Rev.*, **96**, 3125–3146.

67. Constable, A.G., McDonald, W.S., Sawkins, L.C., and Shaw, B.L. (1978) *J. Chem. Soc., Chem. Commun.*, 1061–1062.

68. Foley, P. and Whitesides, G.M. (1979) *J. Am. Chem. Soc.*, **101**, 2732–2733.

69. Crabtree, R.H., Mihelcic, J.M., and Quirk, J.M. (1979) *J. Am. Chem. Soc.*, **101**, 7738–7740.

70. Janowicz, A.H. and Bergman, R.G. (1982) *J. Am. Chem. Soc.*, **104**, 352–354.

71. Hoyano, J.K. and Graham, W.A.G. (1982) *J. Am. Chem. Soc.*, **104**, 3723–3725.

72. Sen, A. (1998) *Acc. Chem. Res.*, **31**, 550–557.

73. Stahl, S.S., Labinger, J.A., and Bercaw, J.E. (1998) *Angew. Chem. Int. Ed.*, **37**, 2180–2192.

74. Lersch, M. and Tilset, M. (2005) *Chem. Rev.*, **105**, 2471–2526.

75. Luinstra, G.A., Wang, L., Stahl, S.S., Labinger, J.A., and Bercaw, J.E. (1995) *J. Organomet. Chem.*, **504**, 75–91.

76. Periana, R.A., Taube, D.J., Gamble, S., Taube, H., Satoh, T., and Fujii, H. (1998) *Science*, **280**, 560–564.

77. Waltz, K.M. and Hartwig, J.F. (1997) *Science*, **277**, 211–213.

78. Chen, H., Schlecht, S., Semple, T.C., and Hartwig, J.F. (2000) *Science*, **287**, 1995–1997.

79. Iverson, C.N. and Smith, M.R. III (1999) *J. Am. Chem. Soc.*, **121**, 7696–7697.

80. Cho, J.-Y., Iverson, C.N., and Smith, M.R. III (2000) *J. Am. Chem. Soc.*, **122**, 12868–12869.

81. Cho, J.-Y., Tse, M.K., Holmes, D., Maleczka, R.E. Jr., and Smith, M.R. III (2002) *Science*, **295**, 305–308.

82. Ishiyama, T., Takagi, J., Ishida, K., Miyaura, N., Anastasi, N.R., and Hartwig, J.F. (2002) *J. Am. Chem. Soc.*, **124**, 390–391.

83. Takagi, J., Sato, K., Hartwig, J.F., Ishiyama, T., and Miyaura, N. (2002) *Tetrahedron Lett.*, **43**, 5649–5651.

84. Ishiyama, T., Nobuta, Y., Hartwig, J.F., and Miyaura, N. (2003) *Chem. Commun.*, 2924–2925.

85. Ishiyama, T., Takagi, J., Yonekawa, Y., Hartwig, J., and Miyaura, N. (2003) *Adv. Synth. Catal.*, **345**, 1103–1106.

86. Dobereiner, G.E. and Crabtree, R.H. (2010) *Chem. Rev.*, **110**, 681–703.

87. Burk, M.J., Crabtree, R.H., Parnell, C.P., and Uriarte, R.J. (1984) *Organometallics*, **3**, 816–817.

88. Burk, M.J. and Crabtree, R.H. (1987) *J. Am. Chem. Soc.*, **109**, 8025–8032.

89. Baudry, D., Ephritikhine, M., Felkin, H., and Holmes-Smith, R. (1983) *J. Chem. Soc., Chem. Commun.*, 788–789.

90. Baudry, D., Ephritikhine, M., Felkin, H., and Zakrzewski, J. (1982) *J. Chem. Soc., Chem. Commun.*, 1235–1236.

91. Gupta, M., Hagen, C., Flesher, R.J., Kaska, W.C., and Jensen, C.M. (1996) *Chem. Commun.*, 2083–2084.

92. Liu, F., Pak, E.B., Singh, B., Jensen, C.M., and Goldman, A.S. (1999) *J. Am. Chem. Soc.*, **121**, 4086–4087.

93. Göttker-Schnetmann, I., White, P., and Brookhart, M. (2004) *J. Am. Chem. Soc.*, **126**, 1804–1811.

94. Vidal, V., Théolier, A., Thivolle-Cazat, J., and Basset, J.-M. (1997) *Science*, **276**, 99–102.

95. Basset, J.M., Copéret, C., Lefort, L., Maunders, B.M., Maury, O., Le Roux, E., Saggio, G., Soignier, S., Soulivong, D., Sunley, G.J., Taoufik, M., and Thivolle-Cazat, J. (2005) *J. Am. Chem. Soc.*, **127**, 8604–8605.

96. Basset, J.-M., Copéret, C., Soulivong, D., Taoufik, M., and Cazat, J.T. (2009) *Acc. Chem. Res.*, **43**, 323–334.

97. Burnett, R.L. and Hughes, T.R. (1973) *J. Catal.*, **31**, 55–64.

98. Goldman, A.S., Roy, A.H., Huang, Z., Ahuja, R., Schinski, W., and Brookhart, M. (2006) *Science*, **312**, 257–261.

99. Corey, E.J. and Hertler, W.R. (1958) *J. Am. Chem. Soc.*, **80**, 2903–2904.

100. Buchschacher, P., Kalvoda, J., Arigoni, D., and Jeger, O. (1958) *J. Am. Chem. Soc.*, **80**, 2905–2906.

101. Hofmann, A.W. (1883) *Ber. Dtsch. Chem. Ges.*, **16**, 558–560.

102. Löffler, K. and Freytag, C. (1909) *Ber. Dtsch. Chem. Ges.*, **42**, 3427–3431.

103. Barton, D.H.R. and Beaton, J.M. (1961) *J. Am. Chem. Soc.*, **83**, 4083–4089.

104. Barton, D.H.R., Beaton, J.M., Geller, L.E., and Pechet, M.M. (1961) *J. Am. Chem. Soc.*, **83**, 4076–4083.

105. Breslow, R. and Winnik, M.A. (1969) *J. Am. Chem. Soc.*, **91**, 3083–3084.

106. Breslow, R. and Baldwin, S.W. (1970) *J. Am. Chem. Soc.*, **92**, 732–734.

107. Breslow, R., Dale, J.A., Kalicky, P., Liu, S.Y., and Washburn, W.N. (1972) *J. Am. Chem. Soc.*, **94**, 3276–3278.

108. Breslow, R. (1980) *Acc. Chem. Res.*, **13**, 170–177.

109. Breslow, R., Corcoran, R.J., Snider, B.B., Doll, R.J., Khanna, P.L., and Kaleya, R. (1977) *J. Am. Chem. Soc.*, **99**, 905–915.

110. Donovan, S.F., Avery, M.A., and McMurry, J.E. (1979) *Tetrahedron Lett.*, **20**, 3287–3290.

111. Trost, B.M., Godleski, S.A., and Genêt, J.P. (1978) *J. Am. Chem. Soc.*, **100**, 3930–3931.

112. Trost, B.M., Godleski, S.A., and Belletire, J.L. (1979) *J. Org. Chem.*, **44**, 2052–2054.

113. Trost, B.M. and Fortunak, J.M.D. (1982) *Organometallics*, **1**, 7–13.

114. Jia, C., Kitamura, T., and Fujiwara, Y. (2001) *Acc. Chem. Res.*, **34**, 633–639.

115. Alberico, D., Scott, M.E., and Lautens, M. (2007) *Chem. Rev.*, **107**, 174–238.

116. Chen, X., Engle, K.M., Wang, D.-H., and Yu, J.-Q. (2009) *Angew. Chem. Int. Ed.*, **48**, 5094–5115.

117. Daugulis, O., Do, H.-Q., and Shabashov, D. (2009) *Acc. Chem. Res.*, **42**, 1074–1086.

118. Lyons, T.W. and Sanford, M.S. (2010) *Chem. Rev.*, **110**, 1147–1169.

119. Fujiwara, Y., Moritani, I., Matsuda, M., and Teranishi, S. (1968) *Tetrahedron Lett.*, **9**, 3863–3865.

120. Shue, R.S. (1971) *J. Chem. Soc. D*, 1510–1511.

121. Nakata, K., Miyata, T., Jintoku, T., Kitani, A., Taniguchi, Y., Takaki, K., and Fujiwara, Y. (1993) *Bull. Chem. Soc. Jpn.*, **66**, 3755–3759.

122. Fahey, D.R. (1971) *J. Organomet. Chem.*, **27**, 283–292.

123. Tremont, S.J. and Rahman, H.U. (1984) *J. Am. Chem. Soc.*, **106**, 5759–5760.

124. Murai, S., Kakiuchi, F., Sekine, S., Tanaka, Y., Kamatani, A., Sonoda, M., and Chatani, N. (1993) *Nature*, **366**, 529–531.

125. Trost, B.M., Imi, K., and Davies, I.W. (1995) *J. Am. Chem. Soc.*, **117**, 5371–5372.

126. Kao, L.-C. and Sen, A. (1991) *J. Chem. Soc., Chem. Commun.*, 1242–1243.

127. Dangel, B.D., Johnson, J.A., and Sames, D. (2001) *J. Am. Chem. Soc.*, **123**, 8149–8150.

128. Dick, A.R., Hull, K.L., and Sanford, M.S. (2004) *J. Am. Chem. Soc.*, **126**, 2300–2301.

129. Desai, L.V., Hull, K.L., and Sanford, M.S. (2004) *J. Am. Chem. Soc.*, **126**, 9542–9543.

130. Giri, R., Chen, X., and Yu, J.-Q. (2005) *Angew. Chem. Int. Ed.*, **44**, 2112–2115.

131. Daugulis, O. and Zaitsev, V.G. (2005) *Angew. Chem. Int. Ed.*, **44**, 4046–4048.

132. Powers, D.C., Geibel, M.A.L., Klein, J.E.M.N., and Ritter, T. (2009) *J. Am. Chem. Soc.*, **131**, 17050–17051.

133. Deprez, N.R. and Sanford, M.S. (2009) *J. Am. Chem. Soc.*, **131**, 11234–11241.

134. Chen, X., Li, J.-J., Hao, X.-S., Goodhue, C.E., and Yu, J.-Q. (2006) *J. Am. Chem. Soc.*, **128**, 78–79.

135. Giri, R., Maugel, N., Li, J.-J., Wang, D.-H., Breazzano, S.P., Saunders, L.B., and Yu, J.-Q. (2007) *J. Am. Chem. Soc.*, **129**, 3510–3511.

136. Giri, R., Shi, B.-F., Engle, K.M., Maugel, N., and Yu, J.-Q. (2009) *Chem. Soc. Rev.*, **38**, 3242–3272.

137. Kakiuchi, F., Le Gendre, P., Yamada, A., Ohtaki, H., and Murai, S. (2000) *Tetrahedron: Asymmetry*, **11**, 2647–2651.

138. Fujii, N., Kakiuchi, F., Yamada, A., Chatani, N., and Murai, S. (1997) *Chem. Lett.*, **26**, 425–426.

139. Shi, B.-F., Maugel, N., Zhang, Y.-H., and Yu, J.-Q. (2008) *Angew. Chem. Int. Ed.*, **47**, 4882–4886.

140. Wang, D.-H., Engle, K.M., Shi, B.-F., and Yu, J.-Q. (2010) *Science*, **327**, 315–319.

141. Shi, B.-F., Zhang, Y.-H., Lam, J.K., Wang, D.-H., and Yu, J.-Q. (2010) *J. Am. Chem. Soc.*, **132**, 460–461.

142. Dangel, B.D., Godula, K., Youn, S.W., Sezen, B., and Sames, D. (2002) *J. Am. Chem. Soc.*, **124**, 11856–11857.

143. O'Malley, S.J., Tan, K.L., Watzke, A., Bergman, R.G., and Ellman, J.A. (2005) *J. Am. Chem. Soc.*, **127**, 13496–13497.

144. Que, L. Jr. and Tolman, W.B. (2008) *Nature*, **455**, 333–340.

145. Groves, J.T., Nemo, T.E., and Myers, R.S. (1979) *J. Am. Chem. Soc.*, **101**, 1032–1033.

146. Mansuy, D. (1993) *Coord. Chem. Rev.*, **125**, 129–141.

147. Costas, M., Chen, K., and Que, L. Jr. (2000) *Coord. Chem. Rev.*, **200–202**, 517–544.

148. Chen, K. and Que, L. Jr. (2001) *J. Am. Chem. Soc.*, **123**, 6327–6337.

149. Breslow, R., Zhang, X., and Huang, Y. (1997) *J. Am. Chem. Soc.*, **119**, 4535–4536.

150. Grieco, P.A. and Stuk, T.L. (1990) *J. Am. Chem. Soc.*, **112**, 7799–7801.

151. Moreira, R.F., Wehn, P.M., and Sames, D. (2000) *Angew. Chem. Int. Ed.*, **39**, 1618–1621.

152. Das, S., Incarvito, C.D., Crabtree, R.H., and Brudvig, G.W. (2006) *Science*, **312**, 1941–1943.

153. Chen, M.S. and White, M.C. (2007) *Science*, **318**, 783–787.

154. Chen, M.S. and White, M.C. (2010) *Science*, **327**, 566–571.

155. Groves, J.T. and Viski, P. (1990) *J. Org. Chem.*, **55**, 3628–3634.

156. Larrow, J.F. and Jacobsen, E.N. (1994) *J. Am. Chem. Soc.*, **116**, 12129–12130.

157. Chen, K. and Baran, P.S. (2009) *Nature*, **459**, 824–828.

158. Ishihara, Y. and Baran, P.S. (2010) *Synlett*, 1733–1745.

159. Meerwein, H., Rathjen, H., and Werner, H. (1942) *Ber. Dtsch. Chem. Ges.*, **75**, 1610–1622.

160. Taber, D.F. and Stiriba, S.-E. (1998) *Chem. Eur. J.*, **4**, 990–992.

161. Doyle, M.P. (2006) *J. Org. Chem.*, **71**, 9253–9260.

162. Davies, H.M.L. and Beckwith, R.E.J. (2003) *Chem. Rev.*, **103**, 2861–2904.

163. Davies, H.M.L. and Manning, J.R. (2008) *Nature*, **451**, 417–424.

164. Ye, T. and McKervey, M.A. (1994) *Chem. Rev.*, **94**, 1091–1160.

165. Taber, D.F. and Petty, E.H. (1982) *J. Org. Chem.*, **47**, 4808–4809.

166. Taber, D.F., Petty, E.H., and Raman, K. (1985) *J. Am. Chem. Soc.*, **107**, 196–199.

167. Kennedy, M., McKervey, M.A., Maguire, A.R., and Roos, G.H.P. (1990) *J. Chem. Soc., Chem. Commun.*, 361–362.

168. McKervey, M.A. and Ye, T. (1992) *J. Chem. Soc., Chem. Commun.*, 823–824.

169. Hashimoto, S., Watanabe, N., and Ikegami, S. (1990) *Tetrahedron Lett.*, **31**, 5173–5174.

170. Doyle, M.P. and Hu, W. (2002) *Chirality*, **14**, 169–172.

171. Doyle, M.P., Hu, W., and Valenzuela, M.V. (2002) *J. Org. Chem.*, **67**, 2954–2959.

172. Davies, H.M.L. and Denton, J.R. (2009) *Chem. Soc. Rev.*, **38**, 3061–3071.

173. Davies, H.M.L., Dai, X., and Long, M.S. (2006) *J. Am. Chem. Soc.*, **128**, 2485–2490.

174. Breslow, R. and Gellman, S.H. (1982) *J. Chem. Soc., Chem. Commun.*, 1400–1401.

175. Breslow, R. and Gellman, S.H. (1983) *J. Am. Chem. Soc.*, **105**, 6728–6729.

176. Espino, C.G. and Du Bois, J. (2001) *Angew. Chem. Int. Ed.*, **40**, 598–600.

177. Espino, C.G., Wehn, P.M., Chow, J., and Du Bois, J. (2001) *J. Am. Chem. Soc.*, **123**, 6935–6936.

178. Nägeli, I., Baud, C., Bernardinelli, G., Jacquier, Y., Moraon, M., and Müller, P. (1997) *Helv. Chim. Acta*, **80**, 1087–1105.

179. Espino, C.G., Fiori, K.W., Kim, M., and Du Bois, J. (2004) *J. Am. Chem. Soc.*, **126**, 15378–15379.

180. Yamawaki, M., Tsutsui, H., Kitagaki, S., Anada, M., and Hashimoto, S. (2002) *Tetrahedron Lett.*, **43**, 9561–9564.

181. Liang, C., Robert-Peillard, F., Fruit, C., Müller, P., Dodd, R.H., and Dauban, P. (2006) *Angew. Chem. Int. Ed.*, **45**, 4641–4644.

182. Zalatan, D.N. and Du Bois, J. (2008) *J. Am. Chem. Soc.*, **130**, 9220–9221.

183. Zhou, X.-G., Yu, X.-Q., Huang, J.-S., and Che, C.-M. (1999) *Chem. Commun.*, 2377–2378.

184. Kohmura, Y. and Katsuki, T. (2001) *Tetrahedron Lett.*, **42**, 3339–3342.
185. Milczek, E., Boudet, N., and Blakey, S. (2008) *Angew. Chem. Int. Ed.*, **47**, 6825–6828.
186. Hinman, A. and Du Bois, J. (2003) *J. Am. Chem. Soc.*, **125**, 11510–11511.

Commentary Part

Comment 1

Huw M.L. Davies

The field of C–H functionalization has undergone explosive growth in recent years. It is now widely recognized that the field represents one of the most exciting research frontiers in organic chemistry because it has the potential to offer a paradigm shift in how organic compounds could be made. New disconnection strategies could become available, leading to the direct synthesis of chemical targets, which otherwise would have been inaccessible using conventional functional group manipulations. The controlled derivatization of complex structures, such as natural products, drug candidates, and advanced materials, could be possible without requiring a redesign of the original synthetic route. The challenge is to achieve these C–H functionalizations in a controlled manner, such that site-selectivity and stereoselectivity can be manipulated at will. This review by Engle and Yu gives an excellent overview of the advances made in the development of methodologies for C–H functionalization and critically assesses the challenges that still need to be met before C–H functionalization becomes routinely employed in synthesis.

In a fast-moving research field, there is a tendency to forget the early contributors. A refreshing aspect of this chapter is a historical perspective of how C–H functionalization developed. Radical-induced C–H functionalization methodologies have been known for a long time, and seminal studies on the use of such chemistry in complex target synthesis were inspired in the 1950s–1960s in the realm of selective derivatization of steroids. The use of organometallic complexes to achieve selective C–H functionalization began in earnest in the 1960s. The early studies focused on stoichiometric metal complexes capable of inserting into unactivated C–H bonds. The development of catalytic variants of these processes has been challenging because catalyst turnover

was problematic, but a few exceptional catalytic processes have been reported. From a synthetic perspective, the recognition that certain types of coordinating groups could be used to direct the C–H functionalization to a specific site was a major advance. The details of these advances with a description of the underpinning mechanisms of these reactions are succinctly described in this review.

A complementary approach for C–H functionalization is to insert a metal-bound ligand, such as a carbene or a nitrene, into the C–H bond. This approach does not require a coordinating group on the substrate to achieve site selectivity, and a wide range of intramolecular and intermolecular transformations have been reported, many of which proceed with excellent control of diastereoselectivity and enantioselectivity. A few of the most interesting examples of the applications of this chemistry are described in this chapter.

Engle and Yu have chosen to discuss C–H functionalization according to the type of functionality that is present in the substrate. If the substrate is simply a hydrocarbon, the process is called "first functionalization," whereas if functional groups are already present is the substrate, the process is described as "further functionalization." The distinction is made because the challenges associated with each type of functionalization are different. In the case of "first functionalization," the challenge is to convert readily available hydrocarbons into more valuable products. Some of the classical challenges include the conversion of methane to methanol, the conversion of methane to higher molecular weight and less volatile hydrocarbons, and the selective functionalization of the higher molecular weight hydrocarbons. Some very important advances have been achieved in this field, such as Shilov's alkane hydroxylation, Periana's methane to methanol oxidation, and Hartwig, Miyara and Smith's primary borylation of *n*-alkanes, but it is uncertain how broadly "first functionalization" will impact synthetic strategies for synthesis. It should be noted, however, there are commercial processes that rely on selective C–H functionalization, such as Davey's process for the conversion of butane to maleic anhydride.

From a synthetic strategy viewpoint, "secondary C–H functionalization" will likely have a more immediate impact on complex target synthesis. Aromatic C–H functionalization by metal C–H insertion has already been demonstrated to be generally useful in cross-coupling chemistry and is likely to become widely employed in the pharmaceutical industry and materials science. Some recent notable examples by Bergman and Ellman and by Yu illustrate the potential of enantioselective variants of the metal C–H insertion process. C–H functionalization by metal carbenes and metal nitrenes has been widely used in total synthesis. Recent advances have shown that highly selective intermolecular reactions are feasible in this chemistry by appropriate choice of catalyst and reagents, which bodes well for the future development of this field.

Engle and Yu's chapter concludes with a summary of the accomplishments in C–H functionalization to date and the challenges for the future. The work over the last few decades demonstrates the promise of C–H functionalization, but also points towards the future challenges. Considerable work still needs to be done to develop even more efficient and selective catalysts, such that any C–H bond in a molecule can be selectively functionalized. The type of functionality that can be introduced needs to be expanded. A broader understanding of the selectivity issues needs to be developed so that it becomes clear to the uninitiated what types of transformations are feasible. With its exciting potential and considerable remaining challenges, it is expected that C–H functionalization with continue to be a very active research field for the foreseeable future.

Comment 2

Zhenfeng Xi

I have been very much interested in activation and synthetic applications of unreactive (or inert) chemical bonds, including C–H bonds, C–C bonds, N_2, and so on. Although I am so far not able to contribute much to such a topic, I like to use the term *"Holy Grails"* in chemistry when I talk about this research area, as the special issue of *Accounts of Chemical Research* (Issue 3, 1995)

did. I believe C–H activation (or functionalization) is one of the "Holy Grails" in chemistry.

I enjoyed very much reading this chapter on *Transition Metal-Catalyzed C–H Functionalization: Synthetically Enabling Reactions for Building Molecular Complexity* by Keary Engle and Jin-Quan Yu. This chapter, not intended to be comprehensive but rather concise, is mainly composed of seven sections: Background and Early Work, First Functionalization: Challenges in Hydrocarbon Functionalization, Further Functionalization: C–H Bonds as Reaction Partners in Organic Synthesis, Catalytic C–H Functionalization by Metal Insertion, and Other Emerging Metal-Catalyzed Further Functionalization Methods. I was especially impressed by the historical tracking of breakthroughs and developments, which should be useful for graduate students and researchers quickly to become informed about the history. The authors pointed out major challenges facing C–H functionalization and also described their possible solutions by highlighting, with insightful discussions and in-depth descriptions, several of the common mechanistic pathways of transition metal-mediated C–H functionalization.

Although I have no better idea, I cannot immediately accept the authors' classification of "first functionalization" and "further functionalization." Furthermore, "functionalized substrates," "unfunctionalized substrates," "prefunctionalized substrates," and "unactivated C–H bonds" are difficult to define. Similarly, terminologies such as "C–H activation," "C–H cleavage," and "C–H functionalization" are frequently seen in the literature and heard at oral presentations. All these different names often confuse me.

Two recent review articles [C1,C2] and a recent book [C3] could have been added for readers to appreciate other work in related areas.

A DG for arene metallation and other C–H functionalizations is often used to coordinate the metal species and facilitate the metallation. This has been a commonly used and very powerful strategy. Many (or most) groups in this area are trying to find different DGs. Is this a limitation?

The "agostic interaction" should attract more attention in this area. How can one make it predictable or synthetically more useful?

Many challenges remain, for instance, (i) development of truly practical catalytic systems for large-scale applications of unfunctionalized hydrocarbons; (ii) enantioselective C–H bond functionalization; (iii) applications of rare earth organometallic compounds, cheaper metals

(Fe, Mn, etc.), radical processes, and photochemistry; and (iv) the cooperation between photochemistry and transition metal complexes has been demonstrated to be very useful in some cases.

I believe this chapter will be as a key reference source in the area, where a number of "Holy Grails" are expected to be found in the future.

Comment 3

Shu-Li You

The direct functionalization of inert C–H bonds with controllable selectivity are very desirable and useful synthetic methods but highly challenging in organic synthesis. In the last decade, the development of C–H bond direct functionalization has undoubtedly been one of the hottest topics in the field of organic chemistry. Great progress has been witnessed, and numerous C–H bond activation methodologies have been developed with the aid by transition metal-based systems, significantly enabling the power of organic synthesis.

In this chapter, Engle and Yu, leading scientists in the field of C–H bond activation, have given an excellent overview of the development of transition metal-catalyzed C–H functionalization. The authors started with early studies on transition metal-mediated and catalyzed C–H cleavage and then categorized C–H functionalization by classifying it into "first functionalization" and "further functionalization." Discussions not only of the progress in both subfields and their applications by choosing suitable references, but also future directions in the field, have been well presented to give practitioners of organic synthesis a general sense of the remaining challenges and opportunities.

It was an enjoyable experience to read this fine review. I particularly appreciate the use of term "*C–H functionalization*" instead of "C–H activation." A flood of papers have appeared in the past 10 years bearing the C–H activation trademark, and unfortunately some of them do not really involve the "activation" process. C–H functionalization is a more accurate definition than C–H activation for all C–H bond-related transformations. Engle and Yu elegantly describe the full story, focusing on Pd-catalyzed reactions with inert C–H bonds, where the activation process generally takes place.

In addition to this well-balanced review chapter, several additional developments are mentioned or further discussed in this commentary. One aspect that should be pointed out is that this commentary represents more a personal preference of topics scattered in the field that have arisen in the past 10 years, rather than following the loop of the main chapter.

For most of the C–H functionalization processes involving Pd(II)/Pd(IV) or Pd(II)/Pd(III), the use of an external oxidant is necessary to regenerate the active catalyst. The most often used oxidants such as copper and silver salts will bring environmental and economic issues along with the synthesis. The use of air (or oxygen) is certainly a solution to this problem but still needs much more investigation. On the other hand, the Pd(0)/Pd(II) catalytic cycle is attractive in this regard since the related reactions are generally free of external oxidants. The halide (pseudo-halide) of the substrate serves as the internal oxidant. Another notable advantage is the availability of various ligands such as phosphines and N-heterocyclic carbenes that could allow the reaction to proceed under mild conditions and readily turn into an asymmetric version with chiral ligands. One representative example by Albicker and Cramer, depicted in Scheme C1, involves an enantioselective Pd-catalyzed direct arylation reaction, which allows the enantioselective synthesis of indanes with quaternary stereocenters [C4]. This enantioselective C–H functionalization reaction involves the Pd(0)/Pd(II) catalytic cycle and proceeds even at room temperature in the presence of a chiral phosphoramidite ligand.

Another trend in the past 10 years in C–H functionalization chemistry is the use of cheap metals such as Co, Fe, and Cu. The successful utilization of these cheap metals greatly expands the potential application of C–H functionalization in practical-scale syntheses. Elegant reviews on these topics are readily available so that further discussions here are unnecessary.

While chemists have paid enormous attention to developing C–H functionalization reactions, the mainstream progress has been made with the aid by transition metals. Remarkable recent discoveries simultaneously by several groups on a metal-free approach to functionalize C–H bonds are intriguing. The use of organic molecules such as DMEDA (*N,N'*-dimethylethylenediamine) and 1,10-phenanthroline together with an alkali metal base allowed the cross-coupling between aryl iodides and simple arenes (Scheme C2) [C5–C7]. Mechanistically, the reactions involve a radical process and fall in the category of homolytic

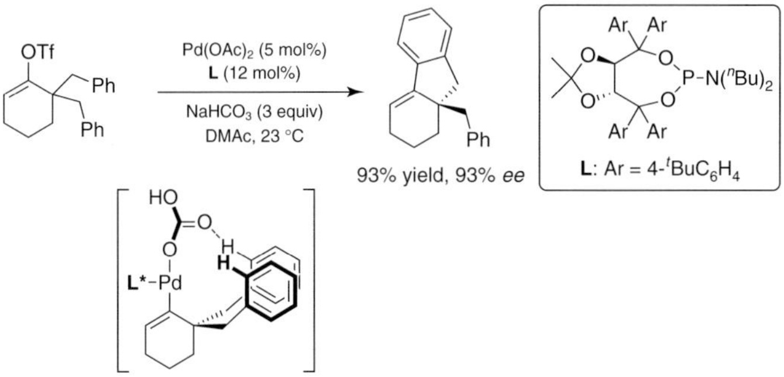

Scheme C1 Enantioselective C-H functionalization involving Pd(0)/Pd(II) catalytic cycle.

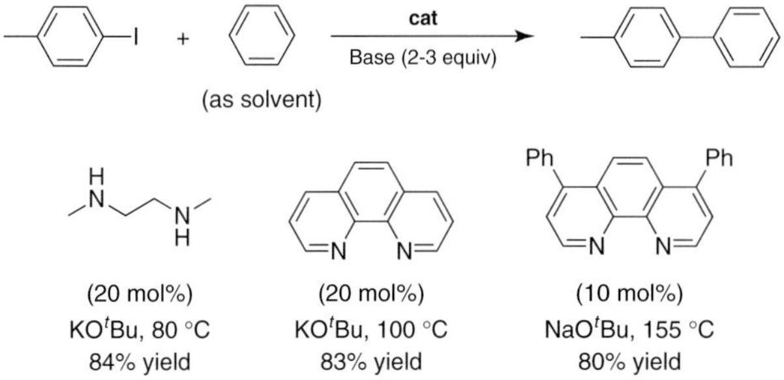

Scheme C2 Organo-catalyzed cross-coupling for the synthesis of biaryls.

aromatic substitution. There is always a doubt about whether the reaction indeed proceeds under metal-free conditions since a trace amount of a transition metal that is not even detectable might be able to catalyze the reaction. Nevertheless, regardless of whether these transformations are or are not transition metal free, their synthetic utility can be foreseen as long as the reproducibility is high in every chemist's hands. From the viewpoint of the pharmaceutical industry, the avoidance of purchasing expensive transition metal catalysts and subsequent removal of transition metal contamination is highly welcome.

As shown in the chapter, the transition metal-catalyzed C–H functionalization process has been utilized in the total synthesis of numerous natural products. In addition to many early examples involving either a radical process or relatively reactive C–H bonds such as α-functionalized C–H and electron-rich arenes, a recent notable example is presented here. As demonstrated by Feng and Chen in the total synthesis of celogentin C, diastereoselective Pd-catalyzed inert C(sp³)–H direct functionalization was employed as the key step to construct an unusual amino acid moiety

[C8] (Scheme C3). A clear conclusion that can be drawn is that the novel C–H functionalization reactions could completely change the logic in the total synthesis and lead to unprecedented retrosynthetic strategies.

The field of transition metal-catalyzed C–H bond functionalization reactions has expanded rapidly over the past decade. The discovery of versatile catalysts has led to synthetic methods for C–C and C–X bond formation that are highly regioselective and enantioselective with excellent functional group compatibility. The pioneering work by Yu's group has opened up exciting opportunities for catalytically asymmetric C–H bond functionalization reactions [C9]. Mechanistic studies are expected to continue to guide the design and discovery of novel catalysts. With continued mechanistic investigation and method development, the field of C–H bond functionalization promises to produce highly active catalyst systems for the mild and selective formation of C–C and C–X bonds. However, there are still many questions to be answered. Few current C–H functionalization methods can compete with existing synthetic methods, and many current C–H functionalization methods

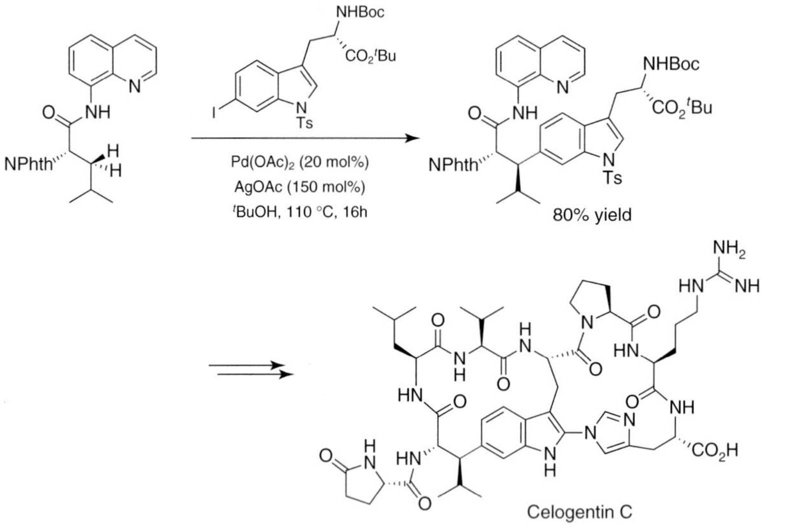

Scheme C3 Total synthesis of celogentin C via a diastereoselective Pd-catalyzed C(sp^3)-H functionalization.

require severe reaction conditions and are limited in substrate scope. Can the C–H functionalization methods compete with all the well-developed cross-coupling reactions in terms of efficiency, selectivity, and operational simplicity? Can the substrate scope for asymmetric C–H functionalization be further broadened to fulfill the main needs in organic synthesis? Selectivity control for the known C–H functionalization methods relies mainly on the DGs and specific substrates. Would the future development of C–H functionalization methods allow the controllable selectivity that is potentially suitable for target-oriented organic synthesis? Questions such as these and beyond will certainly motivate chemists in this field to develop more diverse catalytic systems and to understand the mechanism better. We are eager to see the exciting new developments in C–H functionalization in the years to come.

Comment 4

Zhang-Jie Shi

Much attention has recently been paid to direct C–H transformation since it meets the requirements of numerous chemicals while producing less environmentally harmful emissions. Since the 1870s, when the first direct C–H transformation was developed by Friedel and Crafts, chemists have never stopped pursuing efficient transformations to produce useful chemicals from readily available starting materials. Obviously, direct C–H functionalization could be the first choice to approach such a goal because of its simplicity, straightforwardness, atom and step economy, and environmental friendliness. However, chemists must meet several challenges: (i) most C–H bonds are thermodynamically stable owing to their high bond dissociation energy; (ii) the sigma orbital of the C–H bond is not sufficient to overlap with the approaching orbital to build up an interaction with suitable catalysts; and (iii) most importantly, the selectivity, including chemo-, regio-, and stereo-selectivity, is the greatest challenge for direct C–H transformations. In this chapter, Engle and Yu, among the leading chemists in the field of C–H functionalization, have summarized recent progress in C–H activation.

I enjoyed reading the overview of C–H transformations presented in this chapter. This review is presented with a clear approach to the different aspects. From the first example of transition metal-catalyzed C–H transformations to the recent successful development of the methodologies and their applications, the authors used the "first functionalization" and "further functionalization" classification to clarify the efficacy and value of C–H transformations at different stages in organic syntheses. Based on the understanding of the different mechanistic pathways of C–H transformations, transition metal-catalyzed C–H transformations are categorized into five pathways, which indicate the future direction to approach the final goals. Pd-catalyzed direct C–H transformation, and especially the successful developments in the authors' own laboratory, are explained with clarity, which will help chemists who are working in this field to find the most successful way to solve such challenging tasks.

Despite the abundance of successful examples in recent decades, there is still a long way to go in this field. Compared with the traditional organic transformations, in most cases the processes of C–H transformations require relatively complicated and severe reaction conditions. In general, a large amount of catalyst, especially noble transition metal catalysts, are essential to achieve high efficacy. Further, the class of C–H transformations is still limited. For example, C–H can hardly added to C=O/O as a ubiquitous nucleophile to construct C–C bonds, although recent advances indicate such a possibility [C10, C11]. Moreover, the challenge of controlling the stereoselectivity in C–H transformations still remains, in spite of some useful examples well demonstrated in this exciting chapter. The application of direct C–H transformation in synthesis and chemical production is very limited to date. There are difficult development challenges to be faced but there are unrivalled opportunities to find "gold" in this field. We are looking forward to further landmarks in this exciting area.

Authors' Response to the Commentaries

This chapter is intended to focus on transition metal-catalyzed C–H functionalization reactions that have already proven to be widely applicable in organic synthesis. The power of these transformations stems from their capacity to advance molecular complexity by providing unprecedented logic of retrosynthetic disconnections. Owing to limitations on space, we have necessarily given some key areas in the C–H functionalization literature only a cursory treatment and have left out others altogether. We are therefore grateful for the thoughtful comments from Professors Huw Davies, Zhangjie Shi, Shuli You, and Zhenfeng Xi. In their commentaries, they provide insights that nicely complement the main topics that we have chosen to present. We encourage readers to consult the references that they have provided for further details.

References

C1. Li, C.J. (2009) *Acc. Chem. Res.*, **42**, 335–344.

C2. Sun, C.-L., Li, B.-J., and Shi, Z.-J. (2011) *Chem. Rev.*, **111**, 1293–1314.

C3. Yu, J.-Q. and Shi, Z.-J. (eds.) (2010) *C–H Activation*. Topics in Current Chemistry, vol. 292, Springer, Berlin.

C4. Albicker, M.R. and Cramer, N. (2009) *Angew. Chem. Int. Ed.*, **48**, 9139–9142.

C5. Liu, W., Cao, H., Zhang, H., Zhang, H., Chung, K.H., He, C., Wang, H., Kwong, F.Y., and Lei, A. (2010) *J. Am. Chem. Soc.*, **132**, 16737–16740.

C6. Sun, C.-L., Li, H., Yu, D.-G., Yu, M., Zhou, X., Lu, X.-Y., Huang, K., Zheng, S.-F., Li, B.-J., and Shi, Z.-J. (2010) *Nat. Chem.*, **2**, 1044–1049.

C7. Shirakawa, E., Itoh, K.-i., Higashino, T., and Hayashi, T. (2010) *J. Am. Chem. Soc.*, **132**, 15537–15539.

C8. Feng, Y. and Chen, G. (2010) *Angew. Chem. Int. Ed.*, **49**, 958–961.

C9. Shi, B.-F., Maugel, N., Zhang, Y.-H., and Yu, J.-Q. (2008) *Angew. Chem. Int. Ed.*, **47**, 4882–4886.

C10. Tsai, A.S., Tauchert, M.E., Bergman, R.G., and Ellman, J.A. (2011) *J. Am. Chem. Soc.*, **133**, 1248–1250.

C11. Li, Y., Li, B.-J., Wang, W.-H., Huang, W.-P., Zhang, X.-S., Chen, K., and Shi, Z.-J. (2011) *Angew. Chem. Int. Ed.*, **50**, 2115–2119.

9

An Overview of Recent Developments in Metal-Catalyzed Asymmetric Transformations

Christian A. Sandoval and Ryoji Noyori

9.1
Introduction

Catalysis is the most efficient means to bring about a chemical transformation, since the catalyst used is in principle not consumed in the process and can continuously catalyze a given process to generate a product. When a metal catalyst is employed, the catalysis itself generally stems from the inherent properties of the given metal(s) – Lewis acidity, variable oxidation states, coordination sphere properties, and so on. For metal-based molecular catalysis, however, the ligand frameworks about the metal center can have a dramatic effect on the stability and catalytic performance, and/or actively participate through metal–ligand cooperative processes. The use of chiral ligands generates a chiral 3D environment that effectively controls the stereo-outcome of the chemical transformation, thus giving rise to asymmetric catalysis. Such strategies have been instrumental in the development of modern synthetic strategies at the academic and industrial levels, and constitute a cornerstone for modern organic syntheses.

This chapter aims to give a general overview of the types of metal-catalyzed processes currently possible, with emphasis on the catalytic transformation itself. Only representative examples from the last decade are given to highlight the catalytic process without aiming to be comprehensive [1]. More emphasis is given to C–C bond-forming metal-catalyzed asymmetric syntheses due to the rapid developments in this area in recent years. Mostly intermolecular processes are given as examples and typically the results presented are under optimum conditions and represent the highest efficiency and selectivity obtained. It should be mentioned that a similar chapter using different authors and examples may have been used to illustrate advances in a given area, which is a tribute to the excellence and scope of contemporary research advancing asymmetric catalysis.

Contributions where the metal is involved as a (chiral) reagent or systems where enantioselection stems from substrate control or use of chiral auxiliaries were not considered. Catalytically active supramolecular assemblies where metals act as linkers or transformations involving metal-containing enzymes were judged inappropriate. Similarly, selective metal-catalyzed polymerization processes and concurrent breakthroughs in organocatalysis are not included.

Organic Chemistry – Breakthroughs and Perspectives, First Edition. Edited by Kuiling Ding and Li-Xin Dai.
© 2012 Wiley-VCH Verlag GmbH & Co. KGaA. Published 2012 by Wiley-VCH Verlag GmbH & Co. KGaA.

9.2
Asymmetric Carbon–Carbon Bond Formation

The catalytic generation of C–C bonds offers a tremendous challenge since often regio-, stereo-, and chemoselective factors are involved. To date, the choice of chiral ligand(s) used largely determines the observed selectively for metal-catalyzed transformations. A variety of structural motifs have been developed to provide the necessary chiral environment about the metal center. Figure 9.1 shows some of the most successful and widely used chiral ligands for C–C bond formation which have also served as templates in the design of subsequent improved and/or more specialized generations (**L1–L9**).

9.2.1
Asymmetric Hydroformylations

Asymmetric hydroformylation (AHF) is a challenging transformation since it involves the efficient addition of carbon monoxide and dihydrogen to an alkene **1** in a chemo-, regio-, and enantioselective fashion to yield chiral branched aldehydes **2** (Scheme 9.1) [2, 3]. Thus, despite the tremendous synthetic potential, to date mostly simple functionalized terminal alkenes such as vinylarenes, vinyl acetates, and allyl cyanides have been employed as substrates in AHF. Good results have generally been achieved using rhodium catalysts containing a diphosphite such as **L10–L12**, although the Rh–phosphine-phosphite **L13** (BINAPHOS) combination has given optimum catalysis in terms of regio- and enantioselectivity and substrate scope (e.g., **1**, R = CF_3, C_6F_5, Et; and **1** is 2- or 3-vinylfuran, 2- or 3-vinylthiophene). Chiral bisphospholane ligands can also be used in Rh-catalyzed AHF [4]. For example, bis-3,4-diazaphospholane ligand **L14** yields branched products with 80–98% selectivity and up to 96% enantiomeric excess (*ee*) with very high catalyst activity for a number of terminal alkenes **1** [4c]. Further variation of BINAPHOS (**L13**) yields YANPHOS (**L15**), which has shown comparable selectivity for styrene derivatives (**1**, R = Ph) and vinyl acetates (**1**, R = AcO; 88% branched, *ee* = 98%) [5], and more recently N-allylamides (**1**, R = BocHNCH$_2$; 72% branched, *ee*

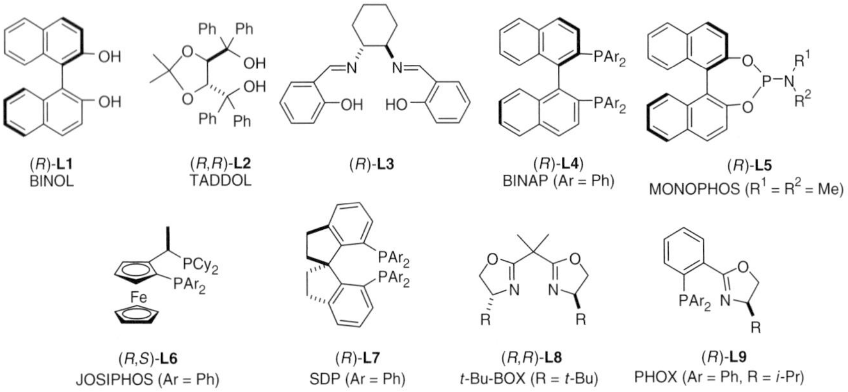

Figure 9.1 Representative privileged chiral ligands.

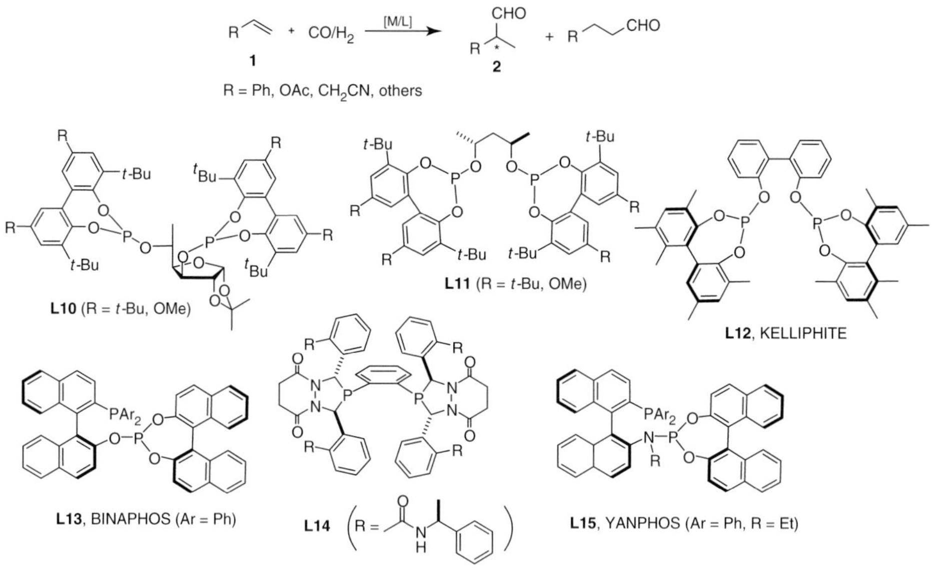

Scheme 9.1 AHF and effective ligands for the reaction.

= 96%) [5c]. Supramolecular interactions have also been utilized to generate useful chiral self-assembled pseudo-diphosphine ligands [6], and AHF tandem reactions have significantly expanded the synthetic scope and utility [7].

Although some degree of regio- and enantioselectivity has been obtained for selected cyclic alkenes [8], currently the regioselective hydroformylation of internal alkenes remains a challenge owing their low reactivity and tendency for isomerization under catalytic conditions.

9.2.2
Asymmetric Additions Involving Carbon Nucleophiles

The metal-catalyzed asymmetric addition of carbon nucleophiles or their equivalents to unsaturated polarized functionalities was developed in the late twentieth century [1, 9]. Among the major pathways for C–C bond formation, the utilization of preactivated nucleophiles or stabilized enol equivalents has allowed for a range of chiral compounds to be generated with high regio-, chemo-, and stereocontrol, as shown in Scheme 9.2. Here, metal catalysis stems predominantly from inherent Lewis acidity and transmetallation properties which allow for the generation of catalytically active intermediate(s). More recent advances involve catalysis under mild conditions which allow for the generation of functionalized, synthetically useful synthons capable of further transformation, manipulation of more sterically demanding components, and the generation of quaternary carbon centers in an enantioselective fashion [10]. Examples in Scheme 9.2 have been selected to showcase the diversity of ligand and catalyst design, while noting that the same catalysts are typically effective for different addition reaction types.

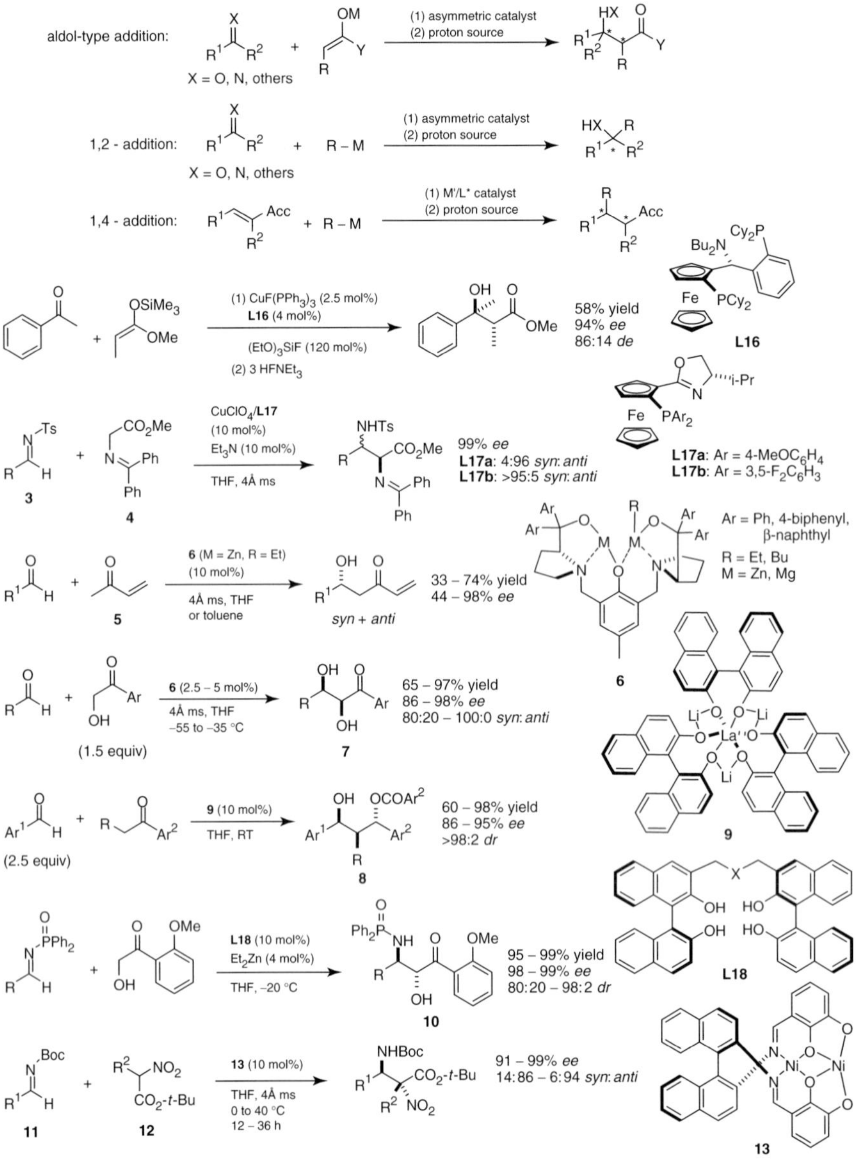

aldol-type addition:

X = O, N, others

(1) asymmetric catalyst
(2) proton source

1,2 - addition:

X = O, N, others

(1) asymmetric catalyst
(2) proton source

1,4 - addition:

(1) M'/L* catalyst
(2) proton source

(1) CuF(PPh$_3$)$_3$ (2.5 mol%)
L16 (4 mol%)

(EtO)$_3$SiF (120 mol%)

(2) 3 HFNEt$_3$

58% yield
94% ee
86:14 de

L16

CuClO$_4$/L17
(10 mol%)

Et$_3$N (10 mol%)

THF, 4Å ms

99% ee
L17a: 4:96 syn: anti
L17b: >95:5 syn: anti

L17a: Ar = 4-MeOC$_6$H$_4$
L17b: Ar = 3,5-F$_2$C$_6$H$_3$

6 (M = Zn, R = Et)
(10 mol%)

4Å ms, THF
or toluene

syn + anti

33 – 74% yield
44 – 98% ee

Ar = Ph, 4-biphenyl,
β-naphthyl
R = Et, Bu
M = Zn, Mg

6 (2.5 – 5 mol%)

4Å ms, THF
–55 to –35 °C

(1.5 equiv)

7

65 – 97% yield
86 – 98% ee
80:20 – 100:0 syn: anti

6

9 (10 mol%)

THF, RT

(2.5 equiv)

8

60 – 98% yield
86 – 95% ee
>98:2 dr

9

L18 (10 mol%)
Et$_2$Zn (4 mol%)

THF, –20 °C

95 – 99% yield
98 – 99% ee
80:20 – 98:2 dr

L18

10

13 (10 mol%)

THF, 4Å ms
0 to 40 °C
12 – 36 h

91 – 99% ee
14:86 – 6:94 syn: anti

11   12

13

Scheme 9.2 Asymmetric direct aldol and aldol-type reaction; ms, molecular sieve.

9.2.2.1 Direct Aldol and Aldol Type

Aldol and aldol-related transformations have been central to asymmetric C–C bond formation, and have thus enjoyed a long history [11]. In modern metal-catalyzed aldol-type reactions, the metal(s) function to facilitate enolization and/or activate the nucleophile. Good results have been obtained with a range of transition metals employing N-, O-, and P-based ligands, with earlier breakthroughs focusing on addition to aldehydes and aldimines [11, 12]. Bisphosphine and phosphorus or phosphorus/amine (P/N)-based ligands have been shown to provide efficient chiral scaffolds [12b]. For example, the copper/TANIAPHOS combination (Cu/**L16**) efficiently catalyzes the Mukaiyama aldol addition to ketones giving the adduct in up to 94% *ee* (86 : 14 *dr*) [13], while the addition of glycine imine to **3** yields *syn* or *anti* products with high selectivity depending on the electronic properties of the ligand **L17** employed (Scheme 9.2) [14]. Bimetallic proline-based catalyst **6** has been used for a range of addition C–C bond-forming reactions [15]. Notable examples are the addition of sensitive methyl vinyl ketones such as **5** catalyzed by **6** (M = Zn, R = Et) to give product in up to 98% *ee* [15c], and the generation of *syn*-1,2-diols **7** in 65–97% yield and 86–98% *ee* [15b]. The multifunctional lanthanum–lithium BINOL-based assembly **9** and a related complex with **L18** have proven effective catalysts for a number of aldol-type reactions [16]. For instance, generation of 1,3-diols via **8** from the aldol–Tischenko reaction catalyzed by **9** proceeds with high selectivity [17], while Zn/**L18** efficiently promotes Mannich addition to give *anti*-amino alcohols **10** in 99% *ee* and up to 98 : 2 *dr* [18]. More recently, the bishomometallic nickel complex **13** has been shown to catalyze efficiently addition to *N*-Boc-imines **11** by α-substituted nitroacetates **12** with high *anti* selectivity [19].

9.2.2.2 1,2-Additions

The 1,2-addition of carbanion equivalents to a C=O and C=N linkage is a straightforward approach to generate important functionalized chiral alcohols and amines [20]. Addition to carbonyls provides a simple route to synthetically versatile chiral propargylic alcohols [21] and cyanohydrins [22]. For the former, as shown in Scheme 9.3, catalyst **6** (M = Zn) efficiently forms **14** in high yield and *ee* [23], and cyanation of aldehydes to yield cyanohydrin products **16** was efficiently catalyzed by chiral complex **15** using TMSCN or NaCN [24]. Addition to ketones which gives tertiary alcohols deserves special mention [25]. For example, allylation of aromatic ketone **17** is effectively catalyzed by the chiral Ag–phosphine complex **18** to give chiral homoallylic alcohols in up to 96% *ee* [26]. In a different and highly atom-efficient strategy, partial hydrogenation has been employed to generate reactive carbon nucleophile equivalents [27]. For example, the Rh/**L19**-catalyzed hydrogenation of conjugated enynes **19** under mild conditions in the presence of aromatic or heteroaromatics aldehydes and ketones permits the direct formation of carbonyl addition products in up 96% yield and 98% *ee* [28].

Progress in additions to imines mirrors developments for C=O although with lower levels of success due to the inherent lower electrophilicity of the C=N linkage. Cu-based systems have proven to be efficient catalysts [20, 29]. For example, addition of dialkylzincs (or trimethylaluminum) to *N*-formylimines generated *in situ* from their corresponding sulfinate adducts **20** are efficiently catalyzed by a Cu/monodentate phosphoramidite system (Cu/**L20**) to give **21** in 81–99% yield and up to 99% *ee* [30]. The use of multifunctional Cu-based catalysts has been shown to promote efficiently the asymmetric addition of allyl cyanides **23** to activated imines **22** [31] and ketones [32] to yield **24** with a tetrasubstituted stereogenic center via an elegant proton-transfer/C–C bond-forming process.

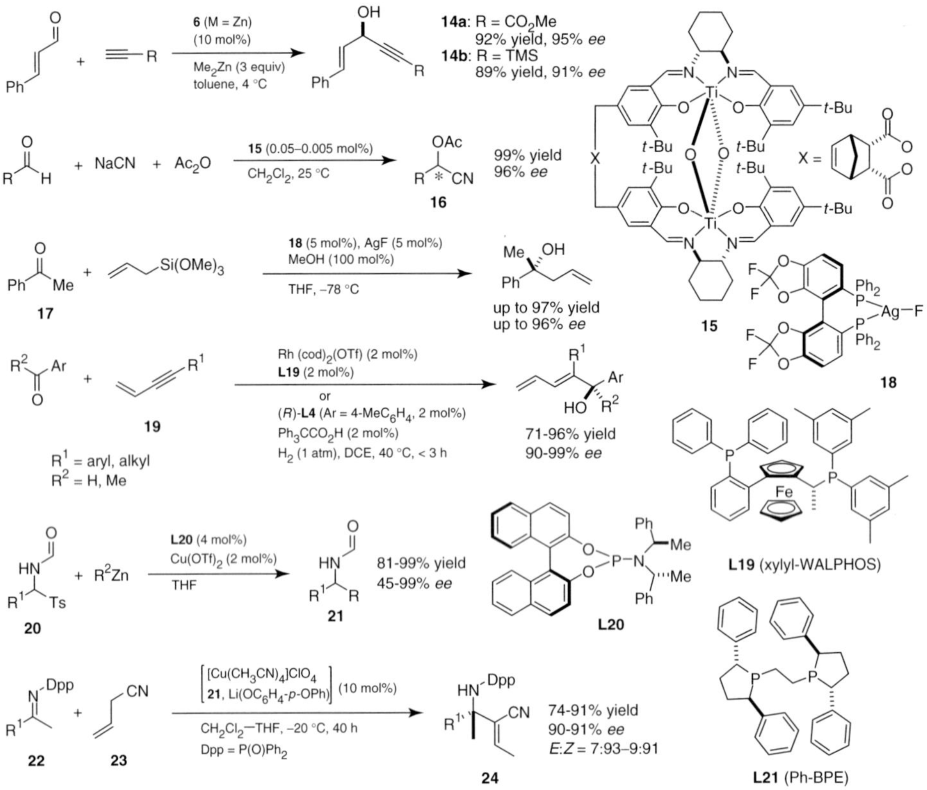

Scheme 9.3 1,2-Addition of C=O and C=N linkages.

9.2.2.3 1,4-Additions

Addition of readily available and functionalized metal-based reagents to sterically hindered and substituted Michael acceptors is noteworthy [33]. Dialkylzinc reagents are useful owing to the low reactivity of noncatalyzed reactions and high tolerance to functional groups, and excellent results have been obtained with a variety of Cu-based catalysts [34]. For instance, the conjugate addition of Et_2Zn to nitroalkene **25** proceeds in 87% yield giving product in 94% *ee* when catalyzed by the Cu/**L22** system (Scheme 9.4) [35]. Commercially available and readily prepared Grignard reagents are also usable [34b]. For example, the Cu-catalyzed conjugate addition of alkyl or aryl Grignards **27** to α,β-unsaturated methyl thioesters **26** proceeds with **L4** or **L6** as chiral ligand giving product in 85–99% *ee* [36]. The former in combination with Pd yields catalyst **28**, which efficiently catalyzes the Michael addition of both cyclic and acyclic β-dicarbonyl compounds **29** to enones **30** in good yields and up to 99% *ee* [37]. Carbenes [38] and alkenes [39] have also proven to be effective in a range of asymmetric metal-catalyzed processes. With respect to conjugate addition, good examples include the Cu/**L23**-catalyzed addition of a Grignard to cyclohexenone derivative **31**, giving **32** in 72% yield with 96% *ee* [40], and the Rh/**L24**-catalyzed addition of Ph_4BNa to give **33** in 83% yield with 98% *ee* [41].

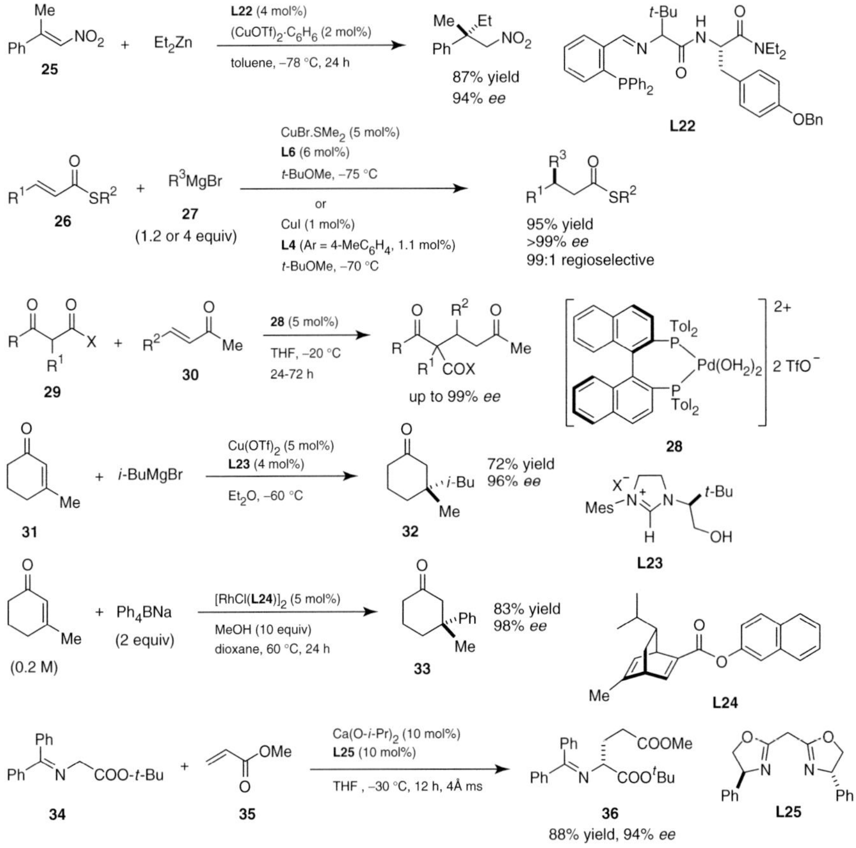

Scheme 9.4 1,4-Addition to Michael acceptors.

Alkaline earth metals have also been used effectively to catalyze Michael additions [42]. For example, bisoxazoline **L25** is used in the Ca-catalyzed addition of **34** to **35** giving functionalized amino acid **36** in 88% yield and 94% *ee* [43].

9.2.3
Cycloadditions

9.2.3.1 Cyclopropanation

Generation of chiral cyclopropane rings has been a long-standing challenge in organic synthesis. Despite tremendous advances in this area [44], catalytic asymmetric versions remain an important challenge. Cu-catalyzed cyclopropanations using chiral BOX **L8**, and a variety of related bisoxazoline ligands, have been the most commonly used [45]. In one development, boron-bridged **37** catalyzed the cyclopropanation of various alkenes with **38** to yield the corresponding cyclopropanes in up to 98% *ee* and *de* (Scheme 9.5) [46]. Similarly, aryliridium–salen ligand complex **39** catalyzes the cyclopropanation of a number of styrene derivatives **40** and cyclic alkenes **41** with high *cis*-selectivity (>98% *de*,

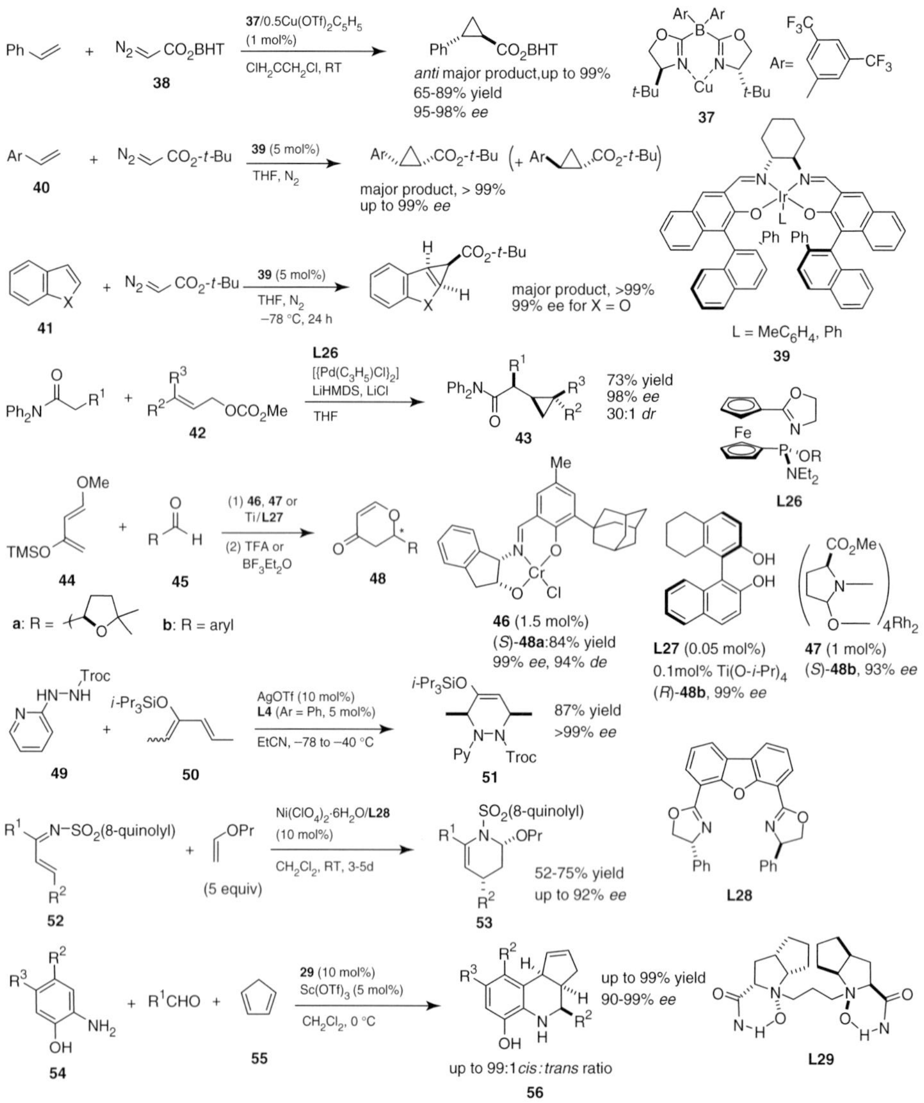

Scheme 9.5 Asymmetric cycloaddition reactions.

up to 99% *ee*) [47]. A recent attractive strategy utilizing **L26** (SiocPhox) involves attack of the central carbon atom of a π-allylpalladium intermediate, generated from **42**, to give cyclopropanes **43** in 89–98% *ee* [48].

9.2.3.2 Diels–Alder Reaction

The Diels–Alder (DA) reaction is a cornerstone for the construction of six-membered rings [49]. More recently, the hetero-Diels–Alder (HDA) variants have also become prominent [50]. For the latter, chiral ligand–Lewis acid-based catalysts using Cr, Cu, Ti, and Rh have been the most successful. HDA reaction between aldehydes **45** and Danishefsky's diene **44** is efficiently catalyzed by **46** [51], Ti(IV)/BINOL-based systems (Ti/**L27**) [52], or {Rh$_2$[(5S)-MEPY]$_4$}BF$_4$ (**47**), a cationic dirhodium complex (Scheme 9.5) [53]. In all these representative cases, chiral pyrone derivatives **48** were generated in 88–99% *ee* and up to 99% yield. Furthermore, catalyzed aza-HDA variants are known [54]. Thus, Ag/BINAP (Ag/**L4**)-catalyzed asymmetric cycloaddition of **49** and **50** gives cycloadduct **51** in 87% yield with 99% *ee* [55]. Building on earlier successes based on Cu/BOX (Cu/**L8**) catalysts [56], the Ni/**L28**-mediated inverse aza-HAD reaction of 1-azadienes provides functionalized piperidine derivatives **53** (*endo* product) in good yields and up to 92% *ee* [57]. In the presence of a Sc/**L29** catalyst, the asymmetric three-component inverse aza-HDA reaction of **54**, aldehyde, and cyclopentadiene **55** takes place to afford the ring-fused tetrahydroquinoline derivatives **56** in high yields and selectivity (99% *ee*, >99 : 1 *dr*) [58].

9.2.3.3 Other Addition Reaction

The asymmetric catalytic 1,3-dipolar cycloaddition (1,3-DC) of azomethine ylides and dipolarophiles such as alkenes provides one of the most powerful methods for the construction of highly substituted chiral five-membered heterocycles [59]. Notable examples include the use of Cu/PyBIDINE (Cu/**L30**) [60] and Ag/**L20** [61] systems, which catalyze the cycloaddition of imino esters to nitroalkenes and electron-deficient alkenes, respectively (Scheme 9.6). Here, *endo* prolines **57** are obtained in high yields and selectivity (99% *ee* and up to 99 : 1 *endo* : *exo*). In contrast, the 1,3-DC between α-aminophosphonate Schiff bases **58** and alkenes **59**, catalyzed by Ag/DTBM-SEGPHOS (Ag/**L31**) catalyst, yields the *exo* product **60** as the major diastereomer in 99% *ee* with >99 : 1 selectivity [62]. In a notable earlier example, the Mg/**L32**-catalyzed addition of functionalized nitrile imines **61** to alkenes **62** yielded dihydropyrazoles **63** in up to 99% yield and *ee* [63]. In an elegant novel route to substituted pyrrolidines, the Pd/**L33** combination catalyzes [3 + 2]-cycloaddition between trimethylenemethane (TMM), generated from **64**, and electron-deficient alkenes or imines [64]. Thus, asymmetric reaction of **64** with ketimines **65** provided highly substituted pyrrolidines **66** in 77–99% yields with 99% *ee* and >20 : 1 *dr*.

Thermal enantioselective [2 + 2]-additions lead to chiral cyclobutane derivatives [65]. Recent representative examples include cycloadditions of alkynes **67** and bicyclic strained alkenes **68** catalyzed by Ir/**L34** [66] or Rh/**L35** which proceed with good to excellent enantioselectivity [67]. Furthermore, a Cu complex with a planar-chiral bipyridine ligand **L36** catalyzes asymmetric [4 + 1]-cycloaddition [68]. Reaction of α,β-unsaturated ketones **69** and diazoacetates **70** yields highly substituted 2,3-dihydrofuran derivatives **71** with 93% *ee* in good yields and >20 : 1 diastereoselection. Transition metal-catalyzed [2 + 2 + 2]-cycloadditions are efficient and atom-economical methods for the construction of six-membered carbocycles and heterocycles [69]. The Rh/H$_8$-BINAP complex Rh/**L35** catalyzes the cycloaddition of 1,6-enynes with electron-deficient ketones to give a single fused diastereomer **72** with excellent regio- and enantioselectivity (up to 99% *ee*) [70].

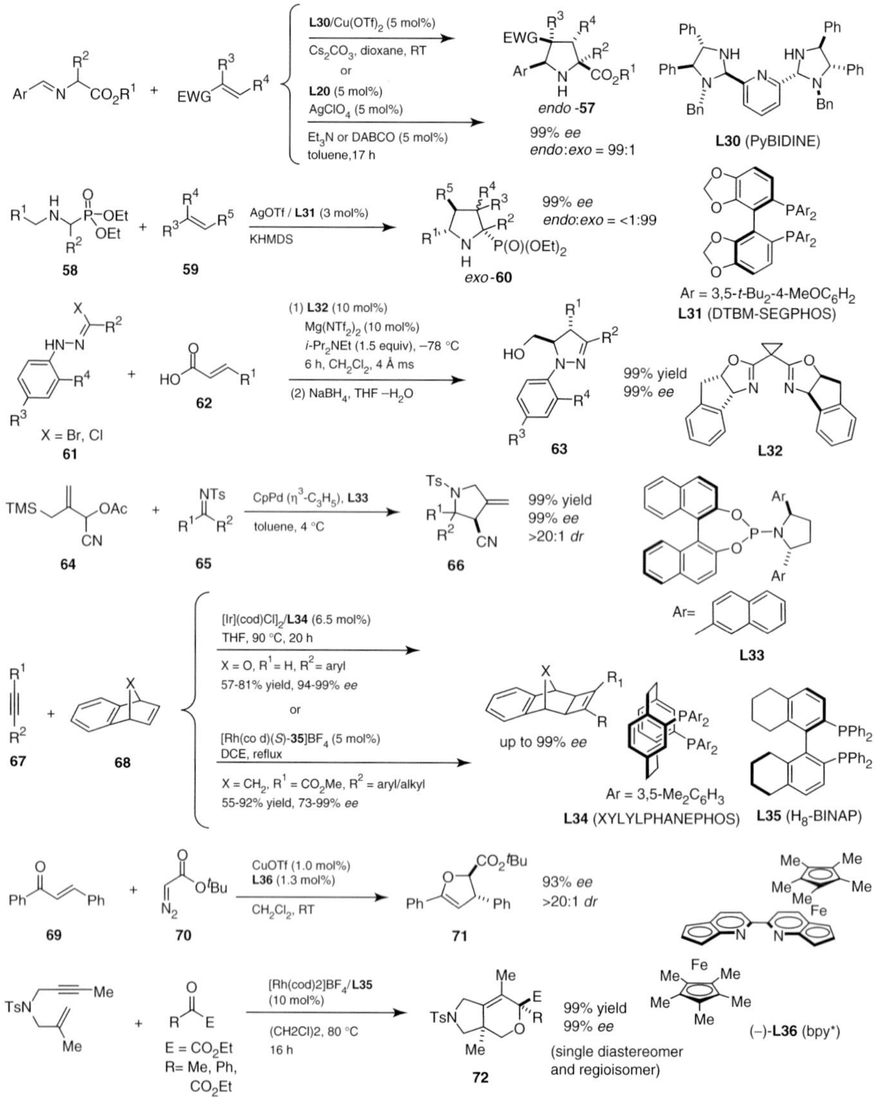

Scheme 9.6 Further asymmetric cycloaddition reactions.

9.2.4
Allylic Alkylations

Asymmetric allylic alkylation (AAA) is a powerful C–C bond-forming process due to the diversity and scope of the catalysis (note that C–H, C–N, C–O, and C–S bonds are also applicable). The most successful strategies have involved increasing either the electrophilic or nucleophilic character of the participating moieties. For the former, formation of a chiral metal–allyl intermediate has allowed the formation of products of high enantiomeric purity from achiral, prochiral, or racemic compounds [71]. Improvements in the regioselectivity have been achieved by careful ligand design of Pd-based catalysts showing a predominance for linear products. Soft nucleophiles may be employed under mild conditions, and the catalysis allows for kinetic resolution and desymmetrization strategies. Recent notable advancements include the Pd-catalyzed asymmetric allylation of ketone enolates generated via CO_2 extrusion under neutral conditions [72]. For example, using chiral ligand **L37**, compounds with quaternary stereogenic centers **73** [73] or synthetically versatile 2-imidazolo-substituted enol carbonates **74** [74] can be produced in excellent *ee* (Scheme 9.7). The ability to undergo dynamic kinetic asymmetric transformations has also been exploited [71]. For example, allylic alkylation of *meso-* and DL-1,2-divinylethylene carbonate using phthalimide, when catalyzed by Pd/**L38**, yields a single product **75** in 99% *ee* [75]. Enantioselective allylation catalyzed by Cu- or Ir-based catalysts, among others, generally favors branched products [71g,h]. Accordingly, the use of a preformed Ir(cod)(κ^2-**L20**)(**L20**) catalyst, comprised of a cyclometallated phosphoramidite **L20**, efficiently promoted the asymmetric allylation of acyclic aliphatic ketone enamines **76** to give branched allylated ketones **77** in excellent ratio and 97% *ee* [76].

AAA using hard organometallic nucleophiles, particularly organozinc, Grignard, and aluminum reagents, or lithium enolates has also seen progressive improvements [77]. Branched products are obtained with the best result using Cu-based systems. For example, the Cu/**L20**-catalyzed reaction between Grignard reagents and β-substituted allylic chloride **78** or endocyclic allylic chlorides **79** yielding products in high *ee* and ratio [78]. In a notable expansion to the substrate scope, Cu/**L39**-catalyzed regioselective allylations of **80** similarly occurred with γ-selectivity giving **81** in 98% *ee* with a > 99 : 1 γ : α ratio [79]. A further notable example is the addition of vinylaluminum reagents **82** to activated alkenes using a Cu/chiral *N*-heterocycle carbene catalyst generated from complex **84** [80]. The S_N2' reaction yields *E*-product **83** selectively in up to 98% *ee*. Interestingly, Cu-free AAA using Grignard reagents has also been shown to proceed in up to 91% *ee* [81].

9.2.5
Asymmetric Catalysis Involving Coupling Processes

The inherent ability of certain transition metals to undergo oxidative addition and reductive elimination has been elegantly exploited in the development of coupling catalyzed C–C bond formation [82]. In 2010, Richard F. Heck, Ei-Ichi Negishi, and Akira Suzuki were awarded the Nobel Prize in Chemistry for their development of "palladium-catalyzed cross-couplings in organic synthesis" [83]. Recent developments include asymmetric variations [84] and regioselective C–H activation for the generation of applicable active intermediates [85].

A significant recent breakthrough is the coupling of nonactivated secondary alkyl halides [84, 86]. Thus, Ni-catalyzed cross-coupling reactions [87] provide asymmetric alkyl–alkyl (or alkyl–aryl) Suzuki reactions of nonactivated secondary alkyls **85** [88] and acylated halohydrins **86** [89] with alkylborane reagents **87** in up to 90% yield and 98% *ee* with the aid of *in situ*-generated Ni/**L40** catalyst (Scheme 9.8). The use of an Ni/bisoxazoline catalyst (Ni/**L41**) achieves asymmetric cross-coupling of α-bromoketones **88** with aryl Grignard reagents (Kumada–Tamao coupling) in a high yield with up to

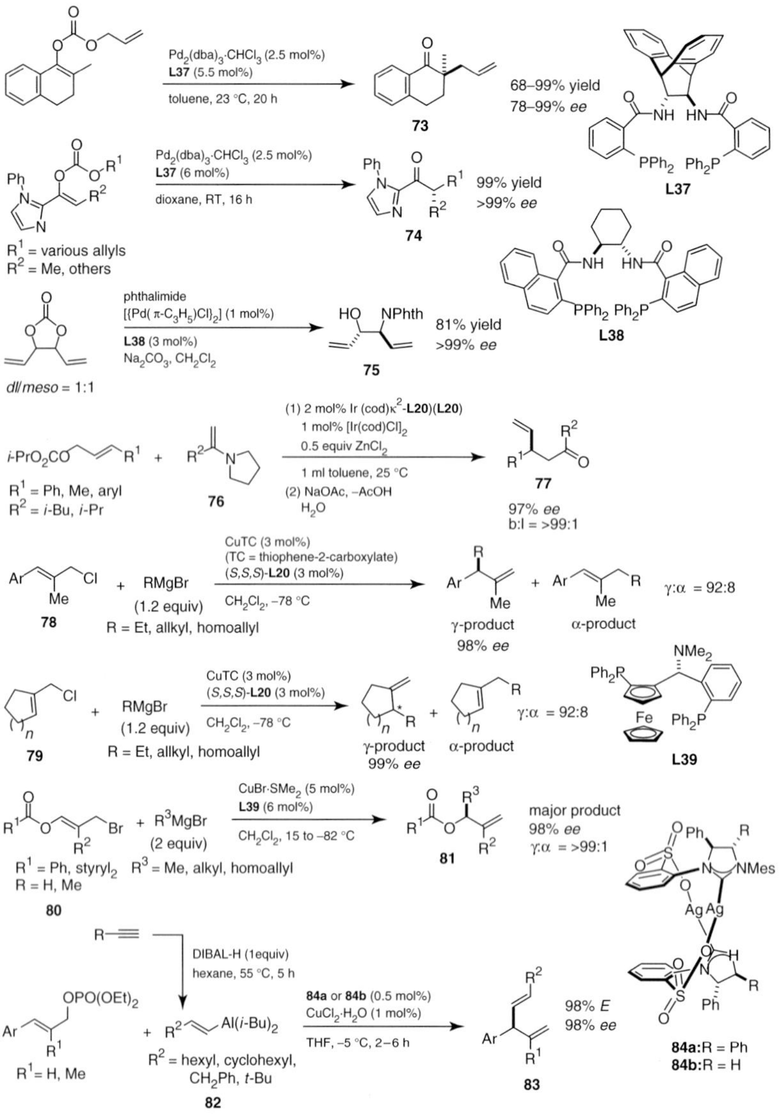

Scheme 9.7 Asymmetric allylic alkylations.

90% *ee* [90]. More atom-efficient is the generation of reactive intermediates by C–H activation [85], although the challenge is increased owing to the necessary concomitant control of regioselectivity. Notable examples include the intramolecular imine-directed C–H–alkene coupling of **89** using Rh/**L20** [91] and the more recent Pd-catalyzed carboxylate group-directed C–H olefination of diphenylacetic acids **90** using amino acid-derived ligand **L42** [92].

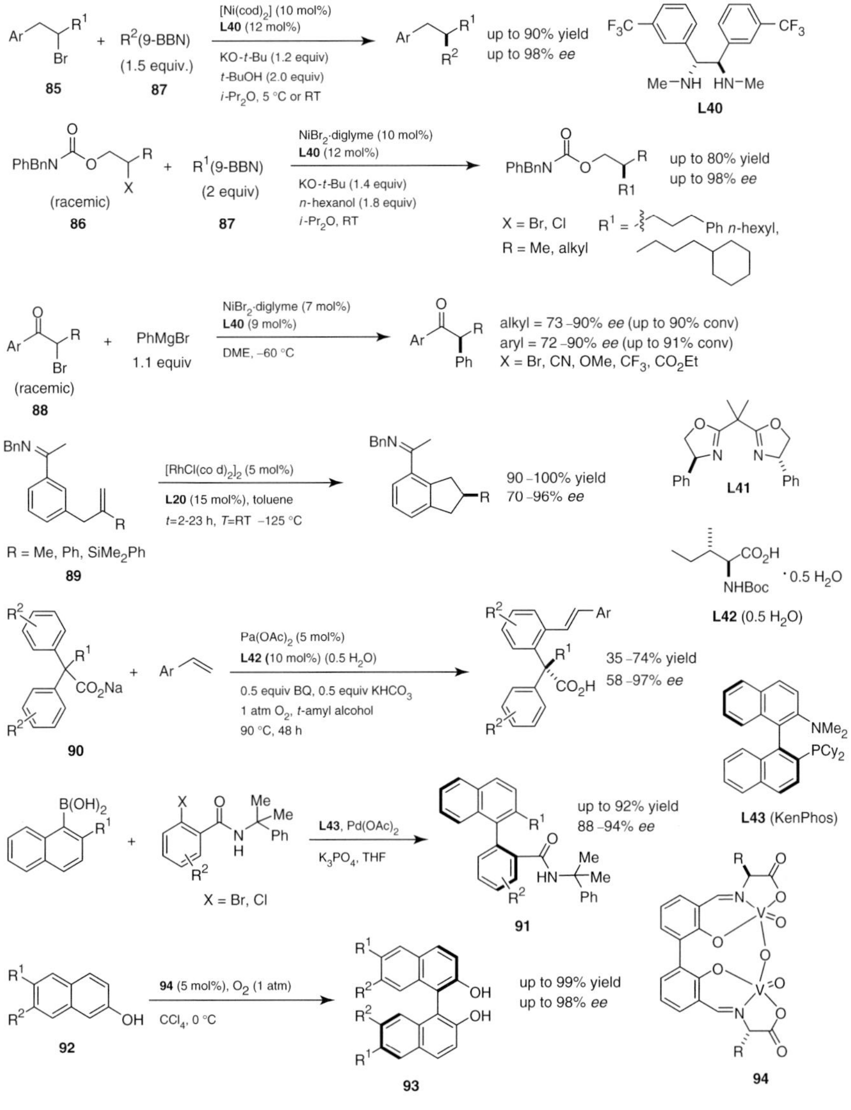

Scheme 9.8 Asymmetric coupling reactions.

The axially chiral biaryl structural motif features not only in a number of prominent ligands such as BINOL (**L1**) and BINAP (**L4**) [93], but also in numerous biologically active natural products [94]. Accordingly, there is growing interest in the catalytic generation of such chiral biaryl compounds [93, 94]. Notable recent advances include the further development of the Pd/KenPhos system (Pd/**L43**) for asymmetric Suzuki–Miyaura reaction [95], and oxovanadium–chiral Schiff base combination **94** for the air-oxidized coupling of 2-naphthols (**92**) [96]. The former yields products **91** in up to 92% yield and 88–94% ee, whereas the latter yields chiral binaphthols **93** in 99% yield and up to 98% ee.

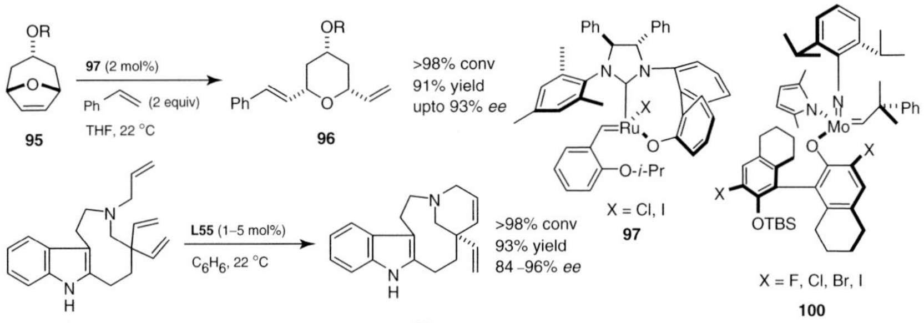

Scheme 9.9 Asymmetric reactions involving metathesis.

9.2.6
Asymmetric Catalysis Involving Metathesis

Alkene metathesis has become an important synthetic tool generally characterized by cross, ring-closing, and ring-opening metathesis [97]. The controlled redistribution of alkylidene fragments at C–C double bonds provides access to synthetically difficult alkenes which may then further undergo a wide range of transformations. In 2005, Yves Chauvin, Robert H. Grubbs, and Richard R. Schrock were awarded the Nobel Prize in Chemistry for their important contributions "for the development of the metathesis method in organic synthesis" [98]. Asymmetric ring-opening metathesis/cross metathesis (AROM/CM) and asymmetric ring-closing metathesis (ARCM) have also been developed [99]. Thus, as an example of AROM/CM, reaction of oxabicyclic alkenes **95** and styrene, yields a range of 2,6-disubstituted pyrans **96** in up to 98% *ee* when catalyzed by Ru-based complex **97** (Scheme 9.9) [100]. In addition, the efficiency of ARCM was exemplified in the total synthesis of quebrachamine where the Mo-based complex **100** catalyzed the generation of **99** from **98** in 84% yield and high selectivity, 96% *ee* and 98 : 2 *er* [101].

9.3
Asymmetric Reductions and Oxidations

Asymmetric reductions and oxidations are vital transformations in Nature and chemistry. Metal-catalyzed enantioselective catalytic processes have contributed greatly to the rapid development of modern synthetic methods. In 2001, the Nobel Prize in Chemistry was awarded jointly to William S. Knowles and Ryoji Noyori for "their work on chirally catalyzed hydrogenation reactions" and K. Barry Sharpless "for his work on chirally catalyzed oxidation reactions" [102].

9.3.1
Asymmetric Reductions

9.3.1.1 Asymmetric Hydrogenation (AH)
Metal-catalyzed asymmetric hydrogenation (AH) of unsaturated compounds to yield a variety of useful chiral products is the most abundantly employed asymmetric catalytic

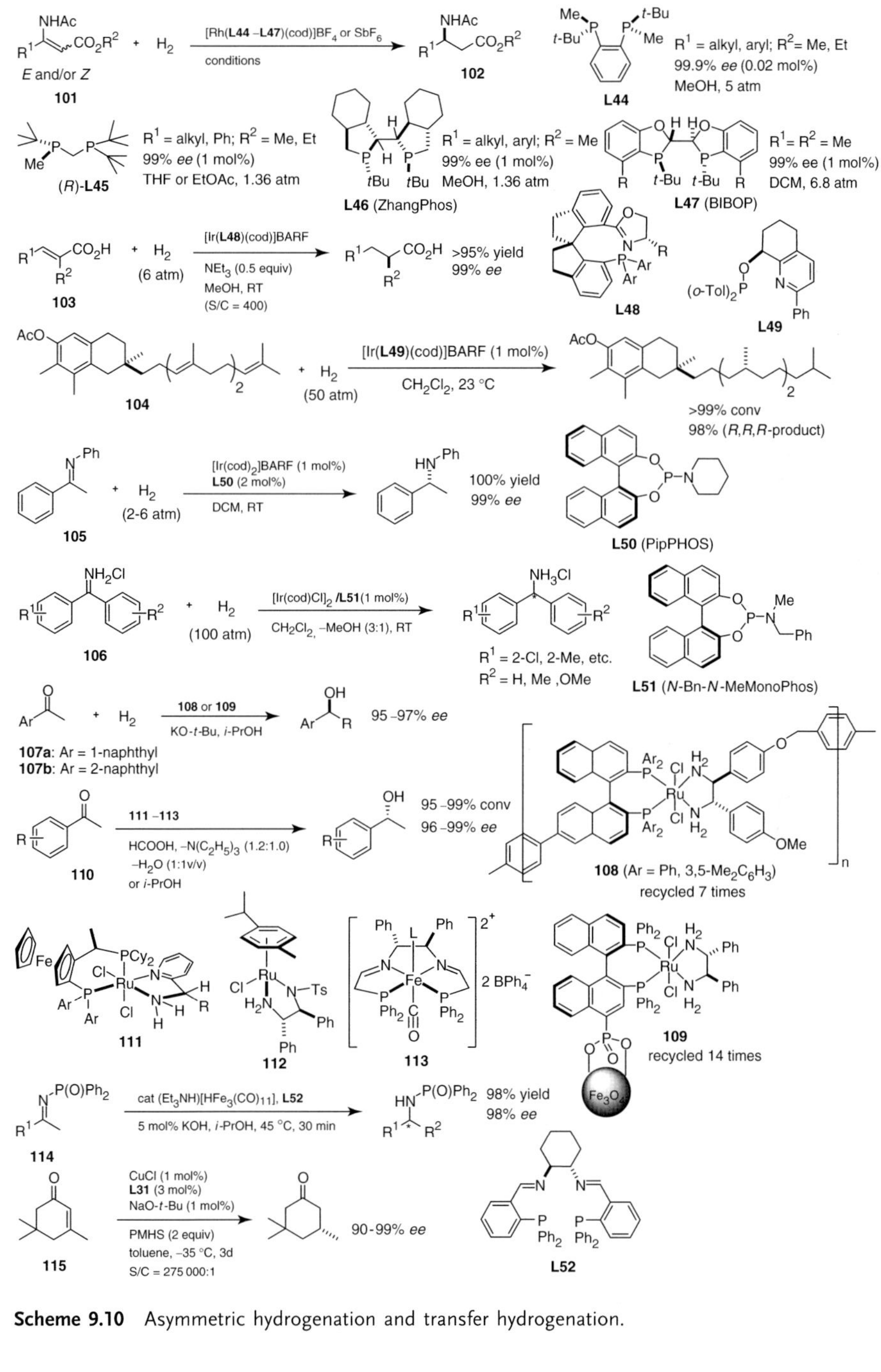

Scheme 9.10 Asymmetric hydrogenation and transfer hydrogenation.

process [103], and numerous industrial processes based on AH have been developed [104]. Recent advances in this core technology mostly involve further development of chiral ligands which have allowed for novel reactivity and/or fine tuning of catalytic efficiency, and expansion/improvement of the substrate scope. Representative examples are given in Scheme 9.10.

Some successful ligand designs based on P-stereogenic, spirocyclic, and oxazoline-based P,N-frameworks resulted in highly effective catalytic systems [12, 105]. For example, AH of both (*E*)- and (*Z*)-β-acetamidodehydroamino acids **101** catalyzed by Rh/**L44**–**L47** complexes yields chiral β-amino acid derivatives **102** in up to 99.9% *ee* [106], and AH of a variety α,β-unsaturated carboxylic acids **103** catalyzed by Ir/**48** gives the saturated products in up to 98% *ee* (Scheme 9.10) [107]. The breakthrough in AH of purely alkyl-substituted alkenes was achieved with Ir/**L49** catalysts [108]. Here, AH of the vitamin E constituent γ-tocotrienyl acetate (**104**) gave the product in 93–98% *ee* catalyzed by [Ir(**L49**)(cod)]BARF (BARF = [B[3,5-(CF$_3$)$_2$C$_6$H$_3$]$_4$]$^-$). Similarly, the development of monodentate phosphorus-based ligands have enjoyed a renaissance over the past decade [109]. The use of phosphoramidite ligands of type **L5** for Ir-catalyzed imine AH is particularly noteworthy [110, 111]. Thus, acyclic *N*-arylimines **105** give the amines in up to >99% *ee* under mild conditions using Ir/**L50** [111], and the AH of unprotected benzophenone imines **106** gave the product in up to 98% *ee* when catalyzed by an Ir/**L51** system [112]. Strategies for molecular catalyst heterogenization have also been actively developed [113]. For example, using established Ru catalysts for simple ketone AH as the template [114], a number of innovative approaches have yielded effective supported catalyst systems [115]. For instance, AH of **107** proceeded effectively when catalyzed by **108** [115c] or **109** [115d], which were readily recovered/recycled up to seven and 14 times, respectively, with minor decreases in enantioselectivity (95–97% *ee*).

9.3.1.2 Asymmetric Transfer Hydrogenation

The use alcohol or HCOOH–NEt$_3$ as the hydride source for the metal-catalyzed reduction of ketones and imines has enjoyed continued interest [116]. The advent of highly active catalytic systems [117], extended substrate and reaction parameter scope [118], and the use of Fe-based catalysts are noteworthy [119]. For example, efficient asymmetric transfer hydrogenation (ATH) of acetophenones **110** catalyzed by **111** proceeded with an exceptionally high turnover frequency (TOF) of 66 000 h^{-1} using *i*-PrOH as hydride source [117b]. Using a HCOOH–NEt$_3$ (1.2 : 1.0) mixture, reduction of **110** catalyzed by well-known **112** [116] proceeded efficiently in 50% (v/v) water giving chiral alcohols in up 97% *ee* [118a]. Here, careful adjustment of the reaction medium pH (5–8) and water to HCOOH–NEt$_3$ ratio allowed for exceptionally low catalyst loadings (S/C up to 10 000 : 1). Interestingly, **111** and **112** effectively catalyze AH under basic and slightly acidic conditions, respectively [120]. The promising use of Fe catalyst **113** for ATH of **110** was shown to proceed in 98% *ee* [119c], and a related Fe/**L52** system efficiently catalyzed the ATH of a number of *N*-(diphenylphosphinyl)imines **114** giving chiral products in up to 98% yield and *ee* [119d].

9.3.1.3 Other Asymmetric Reductions

Asymmetric hydroboration [1] and hydrosilylation [2] are important owing to their synthetic versatility, forming C–C, C–O, and C–N bonds, and broad applicability [121, 122]. For metal-catalyzed processes where reduction is required, the most notable is the recent advance in the use of Cu–phosphine combined systems [123]. For example, reduction of acyclic enones **115** using polymethylhydrosiloxane (PMHS) as hydride source gave reduced product in 90–99% *ee* when catalyzed by Cu/**L31** [124].

9.3.2
Asymmetric Oxidations

9.3.2.1 Asymmetric Oxidation

Enantioselective oxidation reactions are of fundamental importance [125]. Recent trends have been to use environmentally benign and abundant such as O_2 [126] and hydrogen peroxide (H_2O_2) [127]. For the latter, aqueous H_2O_2 is the oxidant of choice since it produces water as the sole byproduct. For example, the asymmetric oxidation (AO) of a variety of sulfides **116** was catalyzed by chiral Al complex **117** with good to excellent enantioselectivity (Scheme 9.11) [128]. The asymmetric dihydroxylation (AD) of alkenes **118** has enjoyed continued interest and development [129]. Enantioselective addition of two distinct nucleophiles across an alkene catalyzed by Pd/**L53** proceeds with high selectivity (98 : 2 *er* and 10 : 1 *dr*) for a number of nucleophiles [130]. A recently developed Fe-catalyzed selective oxidation via C–H activation is highly noteworthy, while the enantioselective version remained unexplored [131].

Enantio-enriched alcohols are important elements and building blocks for a range of natural products and pharmaceuticals. Catalytic AO of secondary alcohols is a way to obtain such highly enantiopure products efficiently [132]. The kinetic resolution of racemic cyclic and acyclic alcohols **119** proceeds in the presence of **120** or Pd/**L54** catalysts, yielding complementary enantiomers in up 65% yield and 99% *ee* [133].

9.3.2.2 Asymmetric Epoxidation

Catalytic asymmetric epoxidation (AE) of alkenes efficiently provides enantio-enriched products, which are versatile synthetic intermediates in a broad range of chemical transformations [134]. Metal–salen systems act as efficient catalysts for AE using H_2O_2 [135]. For instance, AE of a number of styrene derivatives **121** catalyzed by Ti/**L55** gives chiral epoxides in 84% yield and up to 98% *ee* [135c]. The original AE method for allylic alcohols [136] has been extended to include homoallylic and bishomoallylic alcohols [137]. Thus, Zr(IV)/**L56** and Hf(IV)/**L56** are able to catalyze the highly enantioselective epoxidation of **122** with high enantioselectivity [137g].

9.3.2.3 Asymmetric Amination and Halogenation

Asymmetric control in C–N bond formation is very important owing to their extensive occurrence in biologically active compounds. Accordingly, chiral aminohydroxylation [138] and various amination [139], and diamination [140] catalysts have gained increasing importance over the past decade. In a notable example, asymmetric amination of allylic alcohols catalyzed by an Ir/**L57** system provided branched allylic amine products **123** in up

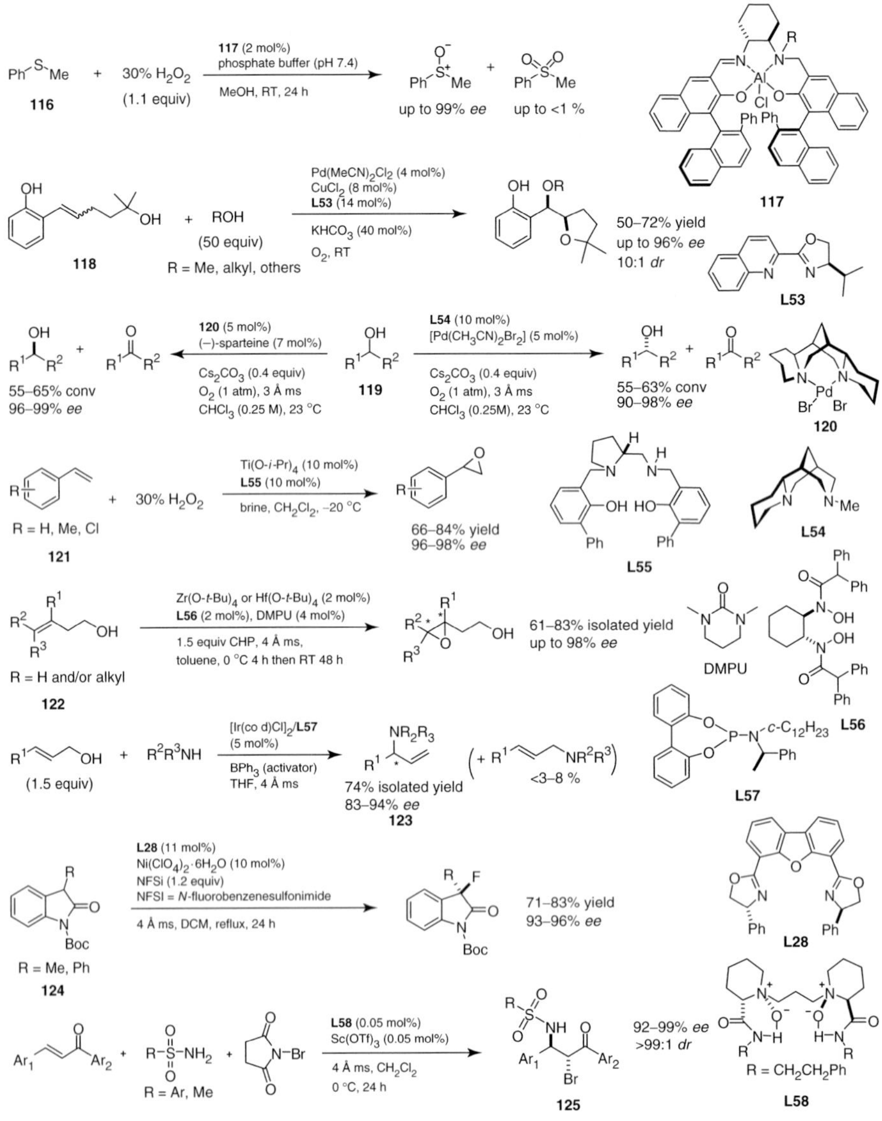

Scheme 9.11 Asymmetric oxidations.

to 94% *ee* [141]. Similarly, the development of catalytic asymmetric halogenation systems is an important recent development [142]. Catalytic enantioselective α-fluorination or -chlorination of amides **124** is achieved with Ni(II)/**L28**, proceeding in up to 99% *ee* [143]. Chiral scandium(III)/*N*,*N*′-dioxide complexes (Sc/**L58**) catalyze the asymmetric haloamination of alkenes with a sulfonamide and *N*-bromosuccinimide to afford α-bromo-β-amino ketone derivatives **125** in up to 99% *ee* (99 : 1 *dr* and 99% yield) [144].

9.4
Conclusion

Asymmetric catalysis is and will remain a changing and evolving science, an inter-disciplinary endeavor that continuously offers new possibilities. As the above sections illustrate, the inherent properties of metals make them an attractive platform from which to derive catalytic processes of high synthetic value. Notably, a number of more involved topics – tandem processes and cascade reactions, alternative reaction media such as CO_2, ionic liquids, and water, and heterogenized or phase-transfer systems – among others, were only touched upon or briefly mentioned, highlighting the current scope of metal-based asymmetric catalysis.

Although the future holds great promise, challenges remain. More environmentally benign catalysts are needed that provide efficient and sustainable synthetic and production methodologies – a fact particularly true for metal-catalyzed systems. Also, catalysts with higher productivity and more extensive substrate scope are required, particularly for asymmetric C–C bond formation. We are just beginning to unravel the tremendous potential that metals and ligand–metal combinations offer in catalysis. Thus, asymmetric catalysis will continue to be a fascinating and practical endeavor.

References

1. For selected comprehensive reviews, see: (a) Jacobsen, E.N., Pfaltz, A., and Yamamoto, H. (eds.) (1999) *Comprehensive Asymmetric Catalysis*, Springer, Berlin; (b) Jacobsen, E.N., Pfaltz, A., and Yamamoto, H. (eds.) (2004) *Comprehensive Asymmetric Catalysis, Supplement 1*, Springer, Berlin; (c) Jacobsen, E.N., Pfaltz, A., and Yamamoto, H. (eds.) (2004) *Comprehensive Asymmetric Catalysis, Supplement 2*, Springer, Berlin; (d) Ojima, I. (ed.) (2000) *Catalytic Asymmetric Synthesis*, 2nd edn., Wiley-VCH Inc., New York; (e) Ojima, I. (ed.) (2010) *Catalytic Asymmetric Synthesis*, 3rd edn., John Wiley & Sons, Inc., Hoboken, NJ; (f) Blaser, H.U. and Schmidt, E. (eds.) (2004) *Asymmetric Catalysis on Industrial Scale: Challenges, Approaches and Solutions*, Wiley-VCH Verlag GmbH, Weinheim; (g) Beller, M. and Bolm, C. (eds.) (2004) *Transition Metals for Organic Synthesis: Building Blocks and Fine Chemicals*, 2nd edn., Wiley-VCH Verlag GmbH, Weinheim; (h) Mikami, K. and Lautens, M. (eds.) (2007) *New Frontiers in Asymmetric Catalysis*, John Wiley & Sons, Inc., Hoboken, NJ; (i) Christmann, M. and Braese, S. (eds.) (2008) *Asymmetric Synthesis*, 2nd edn., Wiley-VCH Verlag GmbH, Weinheim.

2. For comprehensive reviews of hydro-formylation, see: (a) Claver, C. and van Leeuwen, P.W.N.M. (2000) *Rhodium Catalyzed Hydroformylation*, Kluwer, Dordrecht; (b) Nozaki, K. and Ojima, I. (1999) in *Comprehensive Asymmetric Catalysis* (eds. E.N. Jacobsen, A. Pfaltz, and H. Yamamoto), Springer, Berlin, p. 429; (c) Breit, B. (2005) *Angew. Chem. Int. Ed.*, **42**, 6816.

3. For reviews on AHF, see: (a) Breit, B. and Seiche, W. (2001) *Synthesis*, 1; (b) Breit, B. (2003) *Acc. Chem. Res.*, **36**, 264; (c) Dieguez, M., Pamies, O., and Claver, C. (2006) *Top. Organomet. Chem.*, **18**, 35; (d) Klosin, J. and Landis, C.R. (2007) *Acc. Chem. Res.*, **40**, 1251; (e) Gual, A., Godard, C., Castillón, S., and Claver, C. (2010) *Tetrahedron: Asymmetry*, **21**, 1135.

4. For selected references, see: (a) Clarkson, G.J., Ansell, J.R., Cole-Hamilton, D.J., Pogorzelec, P.J., Whittell, J., and Wills, M. (2004) *Tetrahedron: Asymmetry*, **15**, 1787; (b) Axtell, A.T., Cobley, C.J., Klosin, J., Whiteker, G.T., Zanotti-Gerosa, A., and

Abboud, K.A. (2005) *Angew. Chem. Int. Ed.*, **44**, 5834; (c) Clark, T.P., Landis, C.R., Freed, S.L., Klosin, J., and Abboud, K.A. (2005) *J. Am. Chem. Soc.*, **127**, 5040; (d) Axtell, A.T., Klosin, J., Whiteker, G.T., Cobley, C.J., Fox, M.E., Jackson, M., and Abboud, K.A. (2009) *Organometallics*, **28**, 2993.

5. (a) Yan, Y. and Zhang, X. (2006) *J. Am. Chem. Soc.*, **128**, 7198; (b) Zhang, X., Cao, B., Yan, Y., Yu, S., Ji, B., and Zhang, X. (2010) *Chem. Eur. J.*, **16**, 871; (c) Zhang, X., Cao, B., Yu, S., and Zhang, X. (2010) *Angew. Chem. Int. Ed.*, **49**, 4047.

6. (a) Breit, B. and Seiche, W. (2005) *Angew. Chem. Int. Ed.*, **44**, 1640; (b) Kuil, M., Goudriaan, P.E., van Leeuwen, P.W.N.M., and Reek, J.N.H. (2006) *Chem. Commun.*, 4679; (c) Kuil, M., Goudriaan, P.E., Kleij, A.W., Tooke, D.M., Spek, A.L., van Leeuwen, P.W.N.M., and Reek, J.N.H. (2007) *Dalton Trans.*, 2311.

7. Eilbracht, P. and Schmidt, A.M. (2004) in *Transition Metals for Organic Synthesis: Building Blocks and Fine Chemicals* (eds. M. Beller and C. Bolm), 2nd edn., Wiley-VCH Verlag GmbH, Weinheim, p. 57.

8. Selected examples: (a) Diéguez, M., Pàmies, O., and Claver, C. (2005) *Chem. Commun.*, 1221; (b) Mazuela, J., Coll, M., Pàmies, O., and Diéguez, M. (2009) *J. Org. Chem.*, **74**, 5440; (c) Chikkali, S.H., Bellini, R., Berthon-Gelloz, G., van der Vlugt, J.I., Bruin, B., and Reek, J.N.H. (2010) *Chem. Commun.*, **46**, 1244.

9. See also: (a) Yamamoto, H. (ed.) (2000) *Lewis Acids in Organic Synthesis*, Wiley-VCH Verlag GmbH, Weinheim; (b) Krause, N. and Hoffmann-Röder, A. (2001) *Synthesis*, 171.

10. For selected reviews on enantioselective quaternary carbon center generation, see: (a) Douglas, C.J. and Overman, L.E. (2004) *Proc. Natl. Acad. Sci. U. S. A.*, **101**, 5363; (b) Trost, B.M. and Jiang, C. (2006) *Synthesis*, 369; (c) Cozzi, P.G., Hilgraf, R., and Zimmermann, N. (2007) *Eur. J. Org. Chem.*, 5969; (d) Hawner, C. and Alexakis, A. (2010) *Chem. Commun.*, 7295.

11. Selected reviews. Direct aldol reaction: (a) Carreira, E.M. (2000) in *Catalytic Asymmetric Synthesis* (ed. I. Ojima), 2nd edn., Wiley-VCH Inc., New York, p. 513; (b) Mahrwald R. (ed.) (2004) *Modern Aldol Reactions*, Wiley-VCH Verlag GmbH, Berlin; (c) Trost, B.M. and Brindle, C.S. (2010) *Chem. Soc. Rev.*, **39**, 1600. Henry (nitroaldol) reaction: (d) Shibasaki, M., Gröger, H., and Kanai, M. (2004) in *Comprehensive Asymmetric Catalysis, Supplement 1* (eds. E.N. Jacobsen, A. Pfaltz, and H. Yamamoto), Springer, Berlin, p. 131; (e) Boruwa, J., Gogoi, N., Saikia, P.P., and Barua, N.C. (2006) *Tetrahedron: Asymmetry*, **17**, 3315; (f) Palomo, C., Oiarbide, M., and Laso, A. (2007) *Eur. J. Org. Chem.*, 2561. Mannich reaction: (g) Friestad, G.K. and Mathies, A.K. (2007) *Tetrahedron*, **63**, 2541; (h) Arrayas, R.G. and Carretero, J.C. (2009) *Chem. Soc. Rev.*, **38**, 1940; (i) Córdova, A. (2004) *Acc. Chem. Res.*, **37**, 102. Ene reactions: (j) Clarke, M.L. and France, M.B. (2008) *Tetrahedron*, **64**, 9003. See also: (k) Johnson, J.S. and Evans, D.A. (2000) *Acc. Chem. Res.*, **33**, 325.

12. (a) Börner A. (ed.) (2008) *Phosphorus Ligands in Asymmetric Catalysis: Synthesis and Application*, Wiley-VCH Verlag GmbH, Weinheim; (b) Pfaltz, A. (2008) in *Asymmetric Synthesis*, 2nd edn. (eds. M. Christmann and S. Braese), Wiley-VCH Verlag GmbH, Weinheim, p. 139.

13. (a) Oisaki, K., Suto, Y., Kanai, M., and Shibasaki, M. (2003) *J. Am. Chem. Soc.*, **125**, 5644; (b) Oisaki, K., Zhao, D., Suto, Y., Kanai, M., and Shibasaki, M. (2005) *Tetrahedron Lett.*, **46**, 4325. See also: (c) Oisaki, K., Zhao, D., Kanai, M., and Shibasaki, M. (2006) *J. Am. Chem. Soc.*, **128**, 7164.

14. Yan, X.X., Peng, Q., Li, Q., Zhang, K., Yao, J., Hou, X.L., and Wu, Y.D. (2008) *J. Am. Chem. Soc.*, **130**, 14362.

15. For selected examples, see: (a) Trost, M. and Ito, H. (2000) *J. Am. Chem. Soc.*, **122**, 12003; (b) Trost, B.M., Ito, H., and Silcoff, E.R. (2001) *J. Am. Chem. Soc.*, **123**, 3367; (c) Trost, B.M., Shin, S.H., and Sclafani, J.A. (2005) *J. Am. Chem. Soc.*, **127**, 8602; (d) Trost, B.M., Malhotra, S., and Fried, B.A. (2009) *J. Am. Chem. Soc.*, **131**, 1674.

16. Shibasaki, M. and Yoshikawa, N. (2002) *Chem. Rev.*, **102**, 2187.

17. Gnanadesikan, V., Horiuchi, Y., Ohshima, T., and Shibasaki, M. (2004) *J. Am. Chem. Soc.*, **126**, 7782.

18. (a) Matsunaga, S., Kumagai, N., Harada, S., and Shibasaki, M. (2003) *J. Am. Chem. Soc.*, **125**, 4712; (b) Matsunaga, S., Yoshida, T., Morimoto, H., Kumagai, N., and Shibasaki, M. (2004) *J. Am. Chem. Soc.*, **126**, 8777; (c) Yoshida, T., Morimoto, H., Kumagai, N., Matsunaga, S., and Shibasaki, M. (2005) *Angew. Chem. Int. Ed.*, **44**, 3470.

19. Xu, Y., Lu, G., Matsunaga, S., and Shibasaki, M. (2009) *Angew. Chem. Int. Ed.*, **48**, 3353.

20. For selected reviews, see: (a) Noyori, R. and Kitamura, M. (1991) *Angew. Chem. Int. Ed.*, **30**, 49; (b) Soai, K. and Niwa, S. (1992) *Chem. Rev.*, **92**, 833; (c) Pu, L. and Yu, H.B. (2001) *Chem. Rev.*, **101**, 757; (d) Ramon, D.J. and Yus, M. (2004) *Angew. Chem. Int. Ed.*, **43**, 284.

21. For selected reviews, see: (a) Trost, B.M. and Weiss, A.H. (2009) *Adv. Synth. Catal.*, **351**, 963; (b) Cozzi, P.G., Hilgraf, R., and Zimmermann, N. (2004) *Eur. J. Org. Chem.*, 4095.

22. For selected reviews, see: (a) North, M., Usanov, D.L., and Young, C. (2008) *Chem. Rev.*, **108**, 5146; (b) Brunel, J.M. and Holmes, I.P. (2004) *Angew. Chem. Int. Ed.*, **43**, 2752.

23. Trost, B.M., Weiss, A.H., and von Wangelin, A.J. (2006) *J. Am. Chem. Soc.*, **128**, 8.

24. Zhang, Z., Wang, Z., Zhang, R., and Ding, K. (2010) *Angew. Chem. Int. Ed.*, **49**, 6746.

25. For a comprehensive review on Cu-catalyzed processes, see: Shibasaki, M. and Kanai, M. (2008) *Chem. Rev.*, **108**, 2853.

26. Wadamoto, M. and Yamamoto, H. (2005) *J. Am. Chem. Soc.*, **127**, 14556.

27. (a) Jang, H.Y. and Krische, M.J. (2004) *Acc. Chem. Res.*, **37**, 653; (b) Iida, H. and Krische, M.J. (2007) *Top. Curr. Chem.*, **279**, 77; (c) Skucas, E., Ngai, M.Y., Komanduri, V., and Krische, M.J. (2007) *Acc. Chem. Res.*, **40**, 1394.

28. Komanduri, V. and Krische, M.J. (2006) *J. Am. Chem. Soc.*, **128**, 16448.

29. (a) Friestad, G.K. and Mathies, A.K. (2007) *Tetrahedron*, **63**, 2541; (b) Yamada, K. and Tomioka, K. (2008) *Chem. Rev.*, **108**, 2874.

30. Pizzuti, M.G., Minnaard, A.J., and Feringa, B.L. (2008) *J. Org. Chem.*, **73**, 940.

31. Yazaki, R., Nitabaru, T., Kumagai, N., and Shibasaki, M. (2008) *J. Am. Chem. Soc.*, **130**, 14477.

32. Yazaki, R., Kumagai, N., and Shibasaki, M. (2009) *J. Am. Chem. Soc.*, **131**, 3195.

33. For reviews, see: (a) Alexakis, A. and Benhaim, C. (2002) *Eur. J. Org. Chem.*, 3221; (b) Lopez, F., Minnaard, A.J., and Feringa, B.L. (2007) *Acc. Chem. Res.*, **40**, 179; (c) Lopez, F. and Feringa, B.L. (2007) in *Asymmetric Synthesis – The Essentials* (eds. M. Christmann and S. Bräse), Wiley-VCH Verlag GmbH, Weinheim, p. 78.

34. (a) Christoffers, J., Koripelly, G., Rosiak, A., and Rössle, M. (2007) *Synthesis*, 1279; (b) Harutyunyan, S.R., den Hartog, T., Geurts, K., Minnaard, A.J., and Feringa, B.L. (2008) *Chem. Rev.*, **108**, 2824; (c) Alexakis, A., Bäckvall, J.E., Krause, N., Pàmies, O., and Diéguez, M. (2008) *Chem. Rev.*, **108**, 2796.

35. (a) Mampreian, D.M. and Hoveyda, A.H. (2004) *Org. Lett.*, **6**, 2829; (b) Wu, J., Mampreian, D.M., and Hoveyda, A.H. (2005) *J. Am. Chem. Soc.*, **127**, 4584.

36. (a) Mazery, R.D., Pullez, M., López, F., Harutyunyan, S.R., Minnaard, A.J., and Feringa, B.L. (2005) *J. Am. Chem. Soc.*, **127**, 9966; (b) Ruiz, B.M., Geurts, K., Fernández-Ibáñez, M.A., ter Horst, B., Minnaard, A.J., and Feringa, B.L. (2007) *Org. Lett.*, **9**, 5123; (c) ter Horst, B., Feringa, B.L., and Minnaard, A.J. (2007) *Chem. Commun.*, 489.

37. (a) Hamashima, Y., Hotta, D., and Sodeoka, M. (2002) *J. Am. Chem. Soc.*, **124**, 11240; (b) Hamashima, Y., Hotta, D., Umebayashi, N., Tsuchiya, Y., Suzuki, T., and Sodeoka, M. (2005) *Adv. Synth. Catal.*, **347**, 1576.

38. (a) Herrmann, W.A. (2002) *Angew. Chem. Int. Ed.*, **41**, 1290; (b) Cesar, V., Bellemin-Laponnaz, S., and Gade, L.H. (2004) *Chem. Soc. Rev.*, **33**, 619.

39. (a) Glorius, F. (2004) *Angew. Chem. Int. Ed.*, **43**, 3364; (b) Defieber, C., Grutzmacher, H., and Carreira, E.M. (2008) *Angew. Chem. Int. Ed.*, **47**, 4482.

40. Martin, D., Kehrli, S., D'Augustin, M., Clavier, H., Mauduit, M., and Alexakis, A. (2006) *J. Am. Chem. Soc.*, **128**, 8416.

41. Shintani, R., Takeda, M., Nishimura, T., and Hayashi, T. (2010) *Angew. Chem. Int. Ed.*, **49**, 3969.

42. For representative examples, see: (a) Kumaraswamy, G., Jena, N., Sastry, M.N.V., Padmaja, M., and Markondaiah, B. (2005) *Adv. Synth. Catal.*, **347**, 867; (b) Yamatsugu, K., Yin, L., Kamijo, S., Kimura, Y., Kanai, M., and Shibasaki, M. (2009) *Angew. Chem. Int. Ed.*, **48**, 1070; (c) Tsubogo, T., Saito, S., Seki, K., Yamashita, Y., and Kobayashi, S. (2008) *J. Am. Chem. Soc.*, **130**, 13321; (d) Agostinho, M. and Kobayashi, S. (2008) *J. Am. Chem. Soc.*, **130**, 2430.

43. Saito, S., Tsubogo, T., and Kobayashi, S. (2007) *J. Am. Chem. Soc.*, **129**, 5364.

44. (a) Lebel, H., Marcoux, J.F., Molinaro, C., and Charette, A.B. (2003) *Chem. Rev.*, **103**, 977; (b) Pfaltz, A. (2004) in *Transition Metals for Organic Synthesis: Building Blocks and Fine Chemicals*, 2nd edn. (eds. M. Beller and C. Bolm), Wiley-VCH Verlag GmbH, Weinheim, p. 157; (c) Garcia, P., Martin, D.D., Anton, A.B., Garrido, N.M., Marcos, I.S., Basabe, P., and Urones, J.G. (2006) *Mini Rev. Org. Chem.*, **3**, 291; (d) Pellissier, H. (2008) *Tetrahedron*, **64**, 7041.

45. (a) Helmchen, G. and Pfaltz, A. (2000) *Acc. Chem. Res.*, **33**, 336; (b) McManus, H.A. and Guiry, P.J. (2004) *Chem. Rev.*, **104**, 4151.

46. Mazet, C., Köhler, V., and Pfaltz, A. (2005) *Angew. Chem. Int. Ed.*, **44**, 4888.

47. Kanchiku, S., Suematsu, H., Matsumoto, K., Uchida, T., and Katsuki, T. (2007) *Angew. Chem. Int. Ed.*, **46**, 3889.

48. Liu, W., Chen, D., Zhu, X.Z., Wan, X.L., and Hou, X.L. (2009) *J. Am. Chem. Soc.*, **131**, 8734.

49. For reviews, see: (a) Oppolzer, W. and Roush, W.R. (1991) in *Comprehensive Organic Synthesis*, vol. 5 (eds. B.M. Trost and I. Fleming), Pergamon Press, Oxford, Chapters 4.1 and 4.4; (b) Evans, D.A. and Johnson, J.S. (1999) in *Comprehensive Asymmetric Catalysis*, vol. 3 (eds. E.N. Jacobsen, A. Pfaltz, and H. Yamamoto), Springer, New York; (c) Corey, E.J. (2002) *Angew. Chem. Int. Ed.*, **41**, 1650; (d) Nicolaou, K.C., Snyder, S.A., Montagnon, T., and Vassilikogiannakis, G. (2002) *Angew. Chem. Int. Ed.*, **41**, 1668.

50. For reviews of enantioselective hetero-Diels–Alder reactions, see: (a) Wasserman, H.H. (ed.) (1987) *Hetero-Diels–Alder Methodology in Organic Synthesis*, Academic Press, San Diego, CA; (b) Buonora, P., Olsen, J.C., and Oh, T. (2001) *Tetrahedron*, **57**, 6099; (c) Rowland, G.B., Rowland, E.B., Zhang, Q., and Antilla, J.C. (2006) *Curr. Org. Chem.*, **10**, 981; (d) Pellissier, H. (2009) *Tetrahedron*, **65**, 2839.

51. (a) Joly, G.D. and Jacobsen, E.N. (2002) *Org. Lett.*, **4**, 1795; (b) for further examples, see ref. 51d.

52. (a) Long, J., Hu, J., Shen, X., Ji, B., and Ding, K. (2002) *J. Am. Chem. Soc.*, **124**, 10. See also: (b) Du, H., Zhang, X., Wang, Z., Bao, H., You, T., and Ding, K. (2008) *Eur. J. Org. Chem.*, 2248.

53. Wang, Y., Wolf, J., Zavalij, P., and Doyle, M.P. (2008) *Angew. Chem. Int. Ed.*, **47**, 1439.

54. For aza-HDA, see: (a) Jørgensen, K.A. (2000) *Angew. Chem. Int. Ed.*, **39**, 3559; (b) Gouverneur, V. and Reiter, M. (2005) *Chem. Eur. J.*, **11**, 5806.

55. Kawasaki, M. and Yamamoto, H. (2006) *J. Am. Chem. Soc.*, **128**, 16482.

56. (a) Audrain, H., Thorhauge, J., Hazell, R.G., and Jorgensen, K.A. (2000) *J. Org. Chem.*, **65**, 4487; (b) Zhuang, W., Thorhauge, J., and Jorgensen, K.A. (2000) *Chem. Commun.*, 459; (c) Johnson, J.S. and Evans, D.A. (2000) *Acc. Chem. Res.*, **33**, 325; (d) Evans, D.A., Johnson, J.S., and Olhava, E.J. (2000) *J. Am. Chem. Soc.*, **122**, 1635.

57. Esquivias, J., Arrayás, R.G., and Carretero, J.C. (2007) *J. Am. Chem. Soc.*, **129**, 1480.

58. Xie, M., Chen, X., Zhu, Y., Gao, B., Lin, L., Liu, X., and Feng, X. (2010) *Angew. Chem. Int. Ed.*, **49**, 3799.

59. (a) Padwa, A. and Pearson, W.H. (2002) *Synthetic Applications of 1,3-Dipolar Cycloaddition Chemistry Toward Heterocycles and Natural Products*, John Wiley & Sons, Inc., New York; (b) Pellissier, H. (2007) *Tetrahedron*, **63**, 3235; (c) Stanley, L.M. and Sibi, M.P. (2008) *Chem. Rev.*, **108**, 2887.

60. Nájera, C., de Gracia Retamosa, M., and Sansano, J.M. (2008) *Angew. Chem. Int. Ed.*, **47**, 6055.

61. Arai, T., Mishiro, A., Yokoyama, N., Suzuki, K., and Sato, H. (2010) *J. Am. Chem. Soc.*, **132**, 5338.

62. Yamashita, Y., Guo, X.X., Takashita, R., and Kobayashi, S. (2010) *J. Am. Chem. Soc.*, **132**, 3262.

63. Sibi, M.P., Stanley, L.M., and Jasperse, C.P. (2005) *J. Am. Chem. Soc.*, **127**, 8276.

64. Trost, B.M. and Silverman, S.M. (2010) *J. Am. Chem. Soc.*, **132**, 8238.

65. For a review on asymmetric [2 + 2] reactions, see: Lee-Ruff, E. and Mladenova, G. (2003) *Chem. Rev.*, **103**, 1449.

66. Fan, B.M., Li, X.J., Peng, F.Z., Zhang, H.B., Chan, A.S.C., and Shao, Z.H. (2010) *Org. Lett.*, **12**, 304.

67. Shibata, T., Takami, K., and Kawachi, A. (2006) *Org. Lett.*, **8**, 1343.

68. Son, S. and Fu, G.C. (2007) *J. Am. Chem. Soc.*, **129**, 1046.

69. (a) Heller, B. and Hapke, M. (2007) *Chem. Soc. Rev.*, **36**, 1085; (b) Chopade, P.R. and Louie, J. (2006) *Adv. Synth. Catal.*, **348**, 2307; (c) Gandon, V., Aubert, C., and Malacria, M. (2006) *Chem. Commun.*, 2209; (d) Yamamoto, Y. (2005) *Curr. Org. Chem.*, **9**, 503; (e) Varela, J.A. and Saá, C. (2003) *Chem. Rev.*, **103**, 3787.

70. Tanaka, K., Otake, Y., Sagae, H., Noguchi, K., and Hirano, M. (2008) *Angew. Chem. Int. Ed.*, **47**, 1312.

71. Selected reviews: (a) Trost, B.M. and van Vranken, D.L. (1996) *Chem. Rev.*, **96**, 395; (b) Agrofoglio, L.A., Gillaizeau, I., and Saito, Y. (2003) *Chem. Rev.*, **103**, 1875; (c) Trost, B.M. and Crawley, M.L. (2003) *Chem. Rev.*, **103**, 2921; (d) Trost, B.M. (2004) *J. Org. Chem.*, **69**, 5813; (e) Yorimitsu, H. and Oshima, K. (2005) *Angew. Chem. Int. Ed.*, **44**, 4435; (f) Miyabe, H. and Takemoto, Y. (2005) *Synlett*, 1641; (g) Helmchen, G., Dahnz, A., Dübon, P., Schelwies, M., and Weihofen, R. (2007) *Chem. Commun.*, 675; (h) Lu, Z. and Ma, S. (2008) *Angew. Chem. Int. Ed.*, **47**, 258.

72. (a) Behenna, D.C. and Stoltz, B.M. (2004) *J. Am. Chem. Soc.*, **126**, 15044; (b) You, S.L. and Dai, L.X. (2006) *Angew. Chem. Int. Ed.*, **45**, 5346; (c) Braun, M. and Meier, T. (2006) *Angew. Chem. Int. Ed.*, **45**, 6952; (d) Stoltz, B.M. and Mohr, J.T. (2007) *Chem. Asian J.*, **2**, 1476.

73. (a) Trost, B.M. and Xu, J. (2005) *J. Am. Chem. Soc.*, **127**, 2846; (b) Trost, B.M., Xu, J., and Schmidt, T. (2009) *J. Am. Chem. Soc.*, **131**, 18343.

74. Trost, B.M., Lehr, K., Michaelis, D.J., Xu, J., and Buckl, A.K. (2010) *J. Am. Chem. Soc.*, **132**, 8915.

75. Trost, B.M. and Aponick, A. (2006) *J. Am. Chem. Soc.*, **128**, 3931.

76. Weix, D.J. and Hartwig, J.F. (2007) *J. Am. Chem. Soc.*, **129**, 7720.

77. (a) Woodward, S. (2005) *Angew. Chem. Int. Ed.*, **44**, 5560; (b) Sofia, A., Karlström, E., and Bäckvall, J.E. (2002) in *Modern Organocopper Chemistry* (ed. N. Krause), Wiley-VCH Verlag GmbH, Weinheim, p. 259.; (c) Paquin, J.F. and Lautens, M. (2004) in *Comprehensive Asymmetric Catalysis, Supplement 2* (eds. E.N. Jacobsen, A. Pfaltz, and H. Yamamoto), Springer, Berlin, p. 73; (d) Tomioka, K. (2004) in *Comprehensive Asymmetric Catalysis, Supplement 2* (eds. E.N. Jacobsen, A. Pfaltz, and H. Yamamoto), Springer, Berlin, p. 109.

78. Falciola, C.A., Tissot-Croset, K., and Alexakis, A. (2006) *Angew. Chem. Int. Ed.*, **45**, 5995.

79. Geurts, K., Fletcher, S.P., and Feringa, B.L. (2006) *J. Am. Chem. Soc.*, **128**, 15572.

80. Lee, Y., Akiyama, K., Gillingham, D.G., Brown, M.K., and Hoveyda, A.H. (2008) *J. Am. Chem. Soc.*, **130**, 446.

81. For examples, see: (a) Lee, Y. and Hoveyda, A.H. (2006) *J. Am. Chem. Soc.*, **128**, 15604; (b) Lee, Y., Li, B., and Hoveyda, A.H. (2009) *J. Am. Chem. Soc.*, **131**, 11625; (c) Jackowski, O. and Alexakis, A. (2010) *Angew. Chem. Int. Ed.*, **49**, 3346.

82. For selected reviews, see: (a) Heck, R.F. (1979) *Acc. Chem. Res.*, **12**, 146; (b) Miyaura, N. and Suzuki, A. (1995) *Chem. Rev.*, **95**, 2457; (c) Negishi, E.I. (ed.) (2002) *Handbook of Organopalladium Chemistry for Organic Synthesis*, Wiley-Interscience, New York; (d) Hassan, J., Sevignon, M., Gozzi, C., Schulz, E., and Lemaire, M. (2002) *Chem. Rev.*, **102**, 1359; (e) Tsuji, J. (ed.) (2003) *Palladium Reagents and Catalysts: New Perspectives for the 21st Century*, John Wiley & Sons, Inc., New York; (f) de Meijere, A. and Diederich, F.

(eds.) (2004) *Metal-Catalyzed Cross-Coupling Reactions*, Wiley-VCH Verlag GmbH, New York; (g) Suzuki, A. (2005) *Chem. Commun.*, 4759; (h) Negishi, E.I., Huang, Z., Wang, G., Mohan, S., and Wang, C. (2008) *Acc. Chem. Res.*, **41**, 1474.

83. For published announcement, see: Heck, R.F., Negishi, E.I., and Suzuki, A. (2010) *Angew. Chem. Int. Ed.*, **49**, 8300.

84. (a) Rudolph, A. and Lautens, M. (2009) *Angew. Chem. Int. Ed.*, **48**, 2656; (b) Glorius, F. (2008) *Angew. Chem. Int. Ed.*, **47**, 8347.

85. (a) Thalji, R.K., Ellman, J.A., and Bergman, R.G. (2004) *J. Am. Chem. Soc.*, **126**, 7192; (b) Shi, B.F., Maugel, N., Zhang, Y.H., and Yu, J.Q. (2008) *Angew. Chem. Int. Ed.*, **47**, 4882; (c) Chen, X., Engle, K.M., Wang, D.H., and Yu, J.Q. (2009) *Angew. Chem. Int. Ed.*, **48**, 5094; (d) Albicker, M.R. and Cramer, N. (2009) *Angew. Chem. Int. Ed.*, **48**, 9139; (e) Colby, D.A., Bergman, R.G., and Ellman, J.A. (2010) *Chem. Rev.*, **110**, 624; (f) Lyons, T.W. and Sanford, M.S. (2010) *Chem. Rev.*, **110**, 1147; (g) Peng, H.M., Dai, L.X., and You, S.L. (2010) *Angew. Chem. Int. Ed.*, **49**, 5826. For discussion on atom-efficient catalysis, see: (h) Trost, B.M. (1991) *Science*, **254**, 1471; (i) Trost, B.M. (1995) *Angew. Chem. Int. Ed. Engl.*, **34**, 259. See also: (j) Whited, M.T. and Grubbs, R.H. (2009) *Acc. Chem. Res.*, **42**, 1607.

86. For selected Ni-catalyzed examples, see: (a) Zhou, J. and Fu, G.C. (2004) *J. Am. Chem. Soc.*, **126**, 1340; (b) Fischer, C. and Fu, G.C. (2005) *J. Am. Chem. Soc.*, **127**, 4594; (c) Gonzáles-Bobes, F. and Fu, G.C. (2006) *J. Am. Chem. Soc.*, **128**, 5360; (d) Saito, B. and Fu, G.C. (2007) *J. Am. Chem. Soc.*, **129**, 9602; (e) Lou, S. and Fu, G.C. (2010) *J. Am. Chem. Soc.*, **132**, 5010.

87. See also: (a) Altenhoff, G., Würtz, S., and Glorius, F. (2006) *Tetrahedron Lett.*, **47**, 2925; (b) He, A. and Falck, J.R. (2010) *J. Am. Chem. Soc.*, **132**, 2524; (c) Thaler, T., Haag, B., Gavryushin, A., Schober, K., Hartmann, E., Gschwind, R.M., Zipsel, H., Mayer, P., and Knochell, P. (2010) *Nat. Chem.*, **2**, 125.

88. Saito, B. and Fu, G.C. (2008) *J. Am. Chem. Soc.*, **130**, 6694.

89. Owston, N.A. and Fu, G.C. (2010) *J. Am. Chem. Soc.*, **132**, 11908.

90. Lou, S. and Fu, G.C. (2010) *J. Am. Chem. Soc.*, **132**, 1264.

91. Thalji, R.K., Ellman, J.A., and Bergman, R.G. (2004) *J. Am. Chem. Soc.*, **126**, 7192.

92. Shi, B.F., Zhang, Y.H., Lam, J.K., Wang, D.H., and Yu, J.Q. (2010) *J. Am. Chem. Soc.*, **132**, 460.

93. (a) Kocovsky, P., Vyskocil, S., and Smrcina, M. (2003) *Chem. Rev.*, **103**, 3213; (b) Chen, Y., Yekta, S., and Yudin, A.K. (2003) *Chem. Rev.*, **103**, 3155; (c) Brunel, J.M. (2005) *Chem. Rev.*, **105**, 857; (d) Berthod, M., Mignani, G., Woodward, G., and Lemaire, M. (2005) *Chem. Rev.*, **105**, 1801.

94. For reviews, see: (a) Kozlowski, M.C., Morgan, B.J., and Linton, E.C. (2009) *Chem. Soc. Rev.*, **38**, 3193; (b) Bringmann, G., Mortimer, A.J.P., Keller, P.A., Gresser, M.J., Garner, J., and Breuning, M. (2005) *Angew. Chem. Int. Ed.*, **44**, 5384; (c) Tanaka, K. (2009) *Chem. Asian J.*, **4**, 508; (d) Ogasawara, M. and Watanabe, S. (2009) *Synthesis*, 1761.

95. (a) Yin, J.J. and Buchwald, S.L. (2000) *J. Am. Chem. Soc.*, **122**, 12051; (b) Shen, X., Jones, G.O., Watson, D.A., Bhayana, B., and Buchwald, S.L. (2010) *J. Am. Chem. Soc.*, **132**, 11278.

96. (a) Guo, Q., Wu, Z., Luo, Z., Liu, Q., Ye, J., Luo, S., Cun, L., and Gong, L. (2007) *J. Am. Chem. Soc.*, **129**, 13927. See also: (b) Luo, Z., Liu, Q., Gong, L., Cui, X., Mi, A., and Jiang, Y.Z. (2002) *Chem. Commun.*, 914; (c) Luo, Z.B., Liu, Q.Z., Gong, L.Z., Cui, X., Mi, A.Q., and Jiang, Y.Z. (2002) *Angew. Chem. Int. Ed.*, **41**, 4532.

97. For general reviews, see: (a) Grubbs, R.H. (ed.) (2003) *Handbook of Metathesis*, Wiley-VCH Verlag GmbH, Weinheim; (b) Hoveyda, A.H. and Zhugralin, A.R. (2007) *Nature*, **450**, 243; (c) Nicolaou, K.C., Bulger, P.G., and Sarlah, D. (2005) *Angew. Chem. Int. Ed.*, **44**, 4490; (d) Grubbs, R.H. (2004) *Tetrahedron*, **60**, 7117; (e) Furstner, A. (2000) *Angew. Chem. Int. Ed.*, **39**, 3013; (f) Vougioukalakis, G.C. and Grubbs, R.H. (2010) *Chem. Rev.*, **110**, 1746.

98. (a) Chauvin, Y. (2006) *Angew. Chem. Int. Ed.*, **45**, 3740; (b) Schrock, R.R. (2006) *Angew. Chem. Int. Ed.*, **45**, 3748; (c) Grubbs, R.H. (2006) *Angew. Chem. Int. Ed.*, **45**, 3760.

99. For reviews, see: (a) Hoveyda, A.H. and Schrock, R.R. (2001) *Chem. Eur. J.*, **7**, 945; (b) Schrock, R.R. and Hoveyda, A.H. (2003) *Angew. Chem. Int. Ed.*, **42**, 4592; (c) Hoveyda, A.H. (2003) in *Handbook of Metathesis*, vol. **2**, (ed. R.H. Grubbs), Wiley-VCH Verlag GmbH, Weinheim, Chapter 2.3, p. 128; (d) Hoveyda, A.H. and Schrock, R.R. (2004) in *Comprehensive Asymmetric Catalysis, Supplement 1* (eds. E.N. Jacobsen, A. Pfaltz, and H. Yamamoto), Springer, Berlin, p. 207; (e) Schrock, R.R. (2009) *Chem. Rev.*, **109**, 3211.

100. Gillingham, D.G., Kataoka, O., Garber, S.B., and Hoveyda, A.H. (2004) *J. Am. Chem. Soc.*, **126**, 12288.

101. (a) Malcolmson, S.J., Meek, S.J., Sattely, E.S., Schrock, R.R., and Hoveyda, A.H. (2008) *Nature*, **456**, 933; (b) Sattely, E.S., Meek, S.J., Malcolmson, S.J., Schrock, R.R., and Hoveyda, A.H. (2009) *J. Am. Chem. Soc.*, **131**, 943.

102. (a) Knowles, W.S. (2002) *Angew. Chem. Int. Ed.*, **41**, 1998; (b) Noyori, R. (2002) *Angew. Chem. Int. Ed.*, **41**, 2008; (c) Sharpless, K.B. (2002) *Angew. Chem. Int. Ed.*, **41**, 2024.

103. For selected reviews, see: (a) Ohkuma, T. and Noyori, R. (1999) in *Comprehensive Asymmetric Catalysis* (eds. E.N. Jacobsen, A. Pfaltz, and H. Yamamoto), Springer, Berlin, p. 1; (b) Ohkuma, T., Kitamura, M., and Noyori, R. (2000) in *Catalytic Asymmetric Synthesis* (ed. I. Ojima), 2nd edn., Wiley-VCH Inc., New York, p. 1; (c) de Vries, J.G. and Elsevier, C.J. (eds.) (2007) *Handbook of Homogeneous Hydrogenation*, Wiley-VCH Verlag GmbH, Weinheim; (d) Shang, G., Li, W., and Zhang, X. (2010) in *Catalytic Asymmetric Synthesis* (ed. I. Ojima), 3rd edn., John Wiley & Sons, Inc., Hoboken, NJ, p. 343; (e) Ohkuma, T., Kitamura, M., and Noyori, R. (2007) in *New Frontiers in Asymmetric Catalysis* (eds. K. Mikami and M. Lautens), John Wiley & Sons, Inc., Hoboken, NJ, p. 1. See also: (f) Knowles, W.S. and Noyori, R. (2007) *Acc. Chem. Res.*, **40**, 1238.

104. (a) Blaser, H.U., Malan, C., Pugin, B., Spindler, F., Steiner, H., and Studer, M. (2003) *Adv. Synth. Catal.*, **345**, 103; (b) see also ref. 1f.

105. See also: (a) Tang, W. and Zhang, X. (2003) *Chem. Rev.*, **103**, 3029; (b) Pfaltz, A. and Bell, S. (2007) in *Handbook of Homogeneous Hydrogenation* (eds. J.G. de Vries and C.J. Elsevier), Wiley-VCH Verlag GmbH, Weinheim, p. 1029; (c) Freixa, Z. and van Leeuwen, P.W.N.M. (2008) *Coord. Chem. Rev.*, **252**, 1755.

106. (a) Tamura, K., Sugiya, M., Yoshida, K., Yanagisawa, A., and Imamoto, T. (2010) *Org. Lett.*, **12**, 4400; (b) Wu, H.P. and Hoge, G. (2004) *Org. Lett.*, **6**, 3645; (c) Zhang, X., Huang, K., Hou, G., Cao, B., and Zhang, X. (2010) *Angew. Chem. Int. Ed.*, **49**, 6421; (d) Tang, W., Qu, B., Capacci, A.G., Rodriguez, S., Wei, X., Haddad, N., Narayanan, B., Ma, S., Grinberg, N., Yee, N.K., Krishnamurthy, D., and Senanayake, C.H. (2010) *Org. Lett.*, **12**, 176.

107. (a) Li, S., Zhu, S.F., Zhang, C.M., Song, S., and Zhou, Q.L. (2008) *J. Am. Chem. Soc.*, **130**, 8584. See also: (b) Xie, J.H. and Zhou, Q.L. (2008) *Acc. Chem. Res.*, **41**, 581.

108. (a) Bell, S., Wüstenberg, B., Kaiser, S., Menges, F., Netscher, T., and Pfaltz, A. (2006) *Science*, **311**, 642. See also: (b) Pfaltz, A. and Bell, S. (2007) in *Handbook of Homogeneous Hydrogenation* (eds. J.G. de Vries and C.J. Elsevier), Wiley-VCH Verlag GmbH, Weinheim, p. 1049; (c) Roseblade, S.J. and Pfaltz, A. (2007) *Acc. Chem. Res.*, **40**, 1402.

109. For selected reviews, see: (a) Jerphagnon, T., Renaud, J.L., and Bruneau, C. (2004) *Tetrahedron: Asymmetry*, **15**, 2101; (b) Minnaard, A.J., Feringa, B.L., Lefort, L., and de Vries, J.G. (2007) *Acc. Chem. Res.*, **40**, 1267; (c) Reetz, M.T. (2008) *Angew. Chem. Int. Ed.*, **47**, 2556; (d) Bruneau, C. and Renaud, J.L. (2008) in *Phosphorus Ligands in Asymmetric Catalysis: Synthesis and Application* (ed. A. Börner), Wiley-VCH Verlag GmbH, Weinheim, p. 5.

110. For reviews on AH of imines, see: (a) Kobayashi, S. and Ishitani, H. (1999) *Chem. Rev.*, **99**, 1069; (b) Blaser, H.U., Malan, C., Pugin, B., Spinder, F., Steiner, H., and Studer, M. (2003) *Adv. Synth. Catal.*, **345**, 103; (c) Ohkuma, T. and Noyori, R. (2004) in *Comprehensive Asymmetric Catalysis, Supplement 2* (eds. E.N. Jacobsen, A. Pfaltz, and H. Yamamoto),

Springer, Berlin, p. 43; (d) Blaser, H.U. and Spindler, F. (2007) in *Handbook of Homogeneous Hydrogenation* (eds. J.G. de Vries and C.J. Elsevier), Wiley-VCH Verlag GmbH, Weinheim, p. 1193; (e) Fleury-Brégeot, N., de la Fuente, V., Castillón, S., and Claver, C. (2010) *ChemCatChem*, **2**, 1346.

111. Mrşiæ, N., Minnaard, A.J., Feringa, B.L., and de Vries, J.G. (2009) *J. Am. Chem. Soc.*, **131**, 8358.

112. (a) Hou, G., Tao, R., Sun, Y., Zhang, X., and Gosselin, F. (2010) *J. Am. Chem. Soc.*, **132**, 2124. See also: (b) Zhang, W., Chi, Y., and Zhang, X. (2007) *Acc. Chem. Res.*, **40**, 1278.

113. For selected reviews, see: (a) De Vos, D.E., Van-kelecom, I.F.J., and Jacobs, P.A. (eds.) (2000) *Chiral Catalyst Immobilization and Recycling*, Wiley-VCH Verlag GmbH, Weinheim; (b) Fan, Q.H., Li, Y.M., and Chan, A.S.C. (2002) *Chem. Rev.*, **102**, 3385; (c) McMorn, P. and Hutchings, G.J. (2004) *Chem. Soc. Rev.*, **33**, 108; (d) Dai, L.X. (2004) *Angew. Chem. Int. Ed.*, **43**, 5726; (e) Thomas, J.M., Raja, R., and Lewis, D.W. (2005) *Angew. Chem. Int. Ed.*, **44**, 6456; (f) Sandoval, C.A. and Ding, K. (2008) in *Phosphorous Ligands in Asymmetric Catalysis: Synthesis and Application* (ed. A. Börner), Wiley-VCH Verlag GmbH, Weinheim, p. 985; (g) Ding, K. and Uozumi, Y. (eds.) (2008) *Handbook of Asymmetric Heterogeneous Catalysis*, Wiley-VCH Verlag GmbH, Weinheim.

114. (a) Noyori, R. and Ohkuma, T. (2001) *Angew. Chem. Int. Ed.*, **40**, 40. See also: (b) Ohkuma, T. and Noyori, R. (2004) in *Transition Metals for Organic Synthesis: Building Blocks and Fine Chemicals* (eds. M. Beller and C. Bolm), 2nd edn., Wiley-VCH Verlag GmbH, Weinheim, p. 29.

115. For selected examples, see: (a) Hu, A., Ngo, H.L., and Lin, W. (2003) *J. Am. Chem. Soc.*, **125**, 11490; (b) Hu, A., Ngo, H.L., and Lin, W. (2003) *Angew. Chem. Int. Ed.*, **42**, 6000; (c) Liang, Y., Jing, Q., Li, X., Shi, L., and Ding, K. (2005) *J. Am. Chem. Soc.*, **127**, 7694; (d) Hu, A., Yee, G.T., and Lin, W. (2005) *J. Am. Chem. Soc.*, **127**, 12486.

116. For selected reviews, see: (a) Noyori, R. and Hashiguchi, S. (1997) *Acc. Chem. Res.*, **30**, 97; (b) Everaere, K., Mortreux, A.,

and Carpentier, J.F. (2003) *Adv. Synth. Catal.*, **345**, 67; (c) Serafino Gladiali, S. and Alberico, E. (2006) *Chem. Soc. Rev.*, **35**, 226; (d) Joseph, S.M., Samec, J.S., Bäckvall, J.E., Andersson, P.G., and Brandt, P. (2006) *Chem. Soc. Rev.*, **35**, 237; (e) Ikariya, T. and Blacker, A.J. (2007) *Acc. Chem. Res.*, **40**, 1300.

117. For selected recent examples, see: (a) Zweifel, T., Naubron, J.V., Büttner, T., Ott, T., and Grützmacher, H. (2008) *Angew. Chem. Int. Ed.*, **47**, 3245; (b) Baratta, W., Chelucci, G., Herdtweck, E., Magnolia, S., Siega, K., and Rigo, P. (2007) *Angew. Chem. Int. Ed.*, **46**, 7651.

118. For selected representative examples, see: (a) Wu, X.F., Li, X.G., King, F., and Xiao, J. (2005) *Angew. Chem. Int. Ed.*, **44**, 3407; (b) Reetz, M.T. and Li, X. (2006) *J. Am. Chem. Soc.*, **128**, 1044.

119. For selected examples, see: (a) Sui-Seng, C., Freutel, F., Lough, A.J., and Morris, R.H. (2008) *Angew. Chem. Int. Ed.*, **47**, 940; (b) Tondreau, A.M., Darmon, J.M., Wile, B.M., Floyd, S.K., Lobkovsky, E., and Chrik, P.J. (2009) *Organometallics*, **28**, 3928; (c) Mikhailine, A., Lough, A.J., and Morris, R.H. (2009) *J. Am. Chem. Soc.*, **131**, 1394; (d) Zhou, S., Fleischer, S., Junge, K., Das, S., Addis, D., and Beller, M. (2010) *Angew. Chem. Int. Ed.*, **49**, 8121.

120. For AH catalyzed by **111**, see: (a) Sandoval, C.A., Li, Y., Ding, K., and Noyori, R. (2008) *Chem. Asian J.*, **3**, 1801. See also: (b) Baratta, W., Ballico, M., Chelucci, G., Siega, K., and Rigo, P. (2008) *Angew. Chem. Int. Ed.*, **47**, 4362. For AH catalyzed by **112**, see: (c) Ohkuma, T., Utsumi, N., Tsutsumi, K., Murata, K., Sandoval, C.A., and Noyori, R. (2006) *J. Am. Chem. Soc.*, **128**, 8724; (d) Sandoval, C.A., Ohkuma, T., Utsumi, N., Tsutsumi, K., Murata, K., and Noyori, R. (2006) *Chem. Asian J.*, **1**, 102; (e) Ohkuma, T., Tsutsumi, K., Utsumi, N., Arai, N., Noyori, R., and Murata, K. (2007) *Org. Lett.*, **9**, 255.

121. For selected reviews using boron compounds, see: (a) Matteson, D.S. (ed.) (1995) *Stereodirected Synthesis with Organoboranes*, Springer, Berlin, p. 48; (b) Hall, D.G. (ed.) (2005) *Boronic Acids: Preparation and Applications in Organic Synthesis and Medicine*, Wiley-VCH Verlag GmbH

Verlag, Weinheim; (c) Burkhardt, E.R. and Matos, K. (2006) *Chem. Rev.*, **106**, 2617; (d) Schiffner, J.A., Müther, K., and Oestreich, M. (2010) *Angew. Chem. Int. Ed.*, **49**, 1194. For hydroboration of carbonyl groups, see: (e) Ohkuma, T. and Noyori, R. (2004) in *Comprehensive Asymmetric Catalysis, Supplement 2* (eds. E.N. Jacobsen, A. Pfaltz, and H. Yamamoto), Springer, Berlin, p. 7. See also: (f) Corey, E.J. (2009) *Angew. Chem. Int. Ed.*, **48**, 2100.

122. For selected reviews using silicon compounds, see: (a) Brook, M.A. (ed.) (2000) *Silicon in Organic, Organometallic, and Polymer Chemistry*, John Wiley & Sons, Inc., New York, Chapter 12; (b) Denmark, S.E. and Beutner, G.L. (2008) *Angew. Chem. Int. Ed.*, **47**, 1560. For hydrosilylation of carbonyl and imino groups, see: (c) Ohkuma, T. and Noyori, R. (2004) in *Comprehensive Asymmetric Catalysis, Supplement 2* (eds. E.N. Jacobsen, A. Pfaltz, and H. Yamamoto), Springer, Berlin, p. 7.

123. (a) Rendler, S. and Oestreich, M. (2007) *Angew. Chem. Int. Ed.*, **46**, 498; (b) Deutsch, C., Krause, N., and Lipshutz, B.H. (2008) *Chem. Rev.*, **108**, 2916.

124. (a) Lipshutz, B.H. and Servesko, J.M. (2003) *Angew. Chem. Int. Ed.*, **42**, 4789; (b) Lipshutz, B.H., Tanaka, N., Taft, B.R., and Lee, C.T. (2006) *Org. Lett.*, **8**, 1963; (c) Lipshutz, B.H., Lee, C.T., and Taft, B.R. (2007) *Synlett*, **20**, 3257.

125. (a) Singh, H.S. (1986) in *Organic Synthesis by Oxidation with Metal Compounds* (eds. W.J. Mijs and C.R.H.I. De Jonge), Plenum Press, New York; (b) Katsuki, T. (ed.) (2001) *Asymmetric Oxidation Reactions: a Practical Approach in Chemistry*, Oxford University Press, New York; (c) Bäckvall, J.E. (ed.) (2004) *Modern Oxidation Methods*, Wiley-VCH Verlag GmbH, Weinheim; (d) Matsumoto, K. and Katsuki, T. (2010) in *Catalytic Asymmetric Synthesis* (ed. I. Ojima), 3rd edn., John Wiley & Sons, Inc., Hoboken, NJ, p. 839.

126. (a) Punniyamurthy, T., Velusamy, S., and Iqbal, J. (2005) *Chem. Rev.*, **105**, 2329; (b) Enache, D.I., Edwards, J.K., Landon, P., Solsona-Espriu, B., Carley, A.F., Herzing, A.A., Watanabe, M., Kiely, C.J., Knight, D.W., and Hutchings, G.J. (2006) *Science*, **311**, 362; (c) Piera, J. and Backvall, J.E. (2008) *Angew. Chem. Int. Ed.*, **47**, 3506.

127. (a) Noyori, R., Aoki, M., and Sato, K. (2003) *Chem. Commun.*, 1977; (b) Lane, B.S. and Burgess, K. (2003) *Chem. Rev.*, **103**, 2457.

128. Yamaguchi, T., Matsumoto, K., Saito, B., and Katsuki, T. (2007) *Angew. Chem. Int. Ed.*, **46**, 4729.

129. For selected reviews, see: (a) Kolb, H.C., VanNieuwenhze, M.S., and Sharpless, K.B. (1994) *Chem. Rev.*, **94**, 2483; (b) Johnson, R.A. and Sharpless, K.B. (2000) in *Catalytic Asymmetric Synthesis* (ed. I. Ojima), 2nd edn., Wiley-VCH Inc., New York, p. 357; (c) Kolb, H.C. and Sharpless, K.B. (2004) in *Transition Metals for Organic Synthesis: Building Blocks and Fine Chemicals*, 2nd edn. (eds. M. Beller and C. Bolm), Wiley-VCH Verlag GmbH, Weinheim, p. 275; (d) Kolb, H.C., VanNieuwenhze, M.S., and Sharpless, K.B. (1994) *Chem. Rev.*, **94**, 2483; (e) Ahrgren, L. and Sutin, L. (1997) *Org. Process Res.*, **1**, 425; (f) Salvador, J.A.R., Silvestre, S.M., and Moreira, V.M. (2008) *Curr. Org. Chem.*, **12**, 492.

130. Jensen, K.H., Pathak, T.P., Zhang, Y., and Sigman, M.S. (2009) *J. Am. Chem. Soc.*, **131**, 17074.

131. (a) Chen, M.S. and White, M.C. (2004) *J. Am. Chem. Soc.*, **126**, 1346; (b) Chen, M.S. and White, M.C. (2007) *Science*, **318**, 783; (c) Chen, M.S. and White, M.C. (2010) *Science*, **327**, 566. For recent reviews on C–H bond oxidation, see: (d) Shi, B.F., Engle, K.M., Maugel, N., and Yu, J.Q. (2009) *Chem. Soc. Rev.*, **38**, 3242; (e) Doyle, M.P., Duffy, R., Ratnikov, M., and Zhou, L. (2010) *Chem. Rev.*, **110**, 704; (f) Newhouse, T. and Baran, P.S. (2011) *Angew. Chem. Int. Ed.*, **50**, 3362.

132. (a) Stoltz, B.M. and Ebner, D.C. (2005) in *Handbook of C–H Transformations*, vol. 2 (ed. G. Dyker), Wiley-VCH Verlag GmbH, Weinheim, p. 393; (b) Sigman, M.S. and Jensen, D.R. (2006) *Acc. Chem. Res.*, **39**, 221.

133. (a) Ebner, D.C., Trend, R.M., Genet, C., McGrath, M.J., O'Brien, P., and Stoltz, B.M. (2008) *Angew. Chem. Int. Ed.*, **47**, 6367; (b) Ferreira, E.M. and Stoltz, B.M. (2001) *J. Am. Chem. Soc.*, **123**, 7725.

134. For selected recent reviews, see: (a) Katsuki, T. (2003) in *Comprehensive Coordination Chemistry II*, vol. 9 (ed. J.

McCleverty), Elsevier Science, Oxford; (b) Xia, Q.H., Ge, H.Q., Ye, C.P., Liu, Z.M., and Su, K.X. (2005) *Chem. Rev.*, **105**, 1603; (c) Chatterjee, D. (2008) *Coord. Chem. Rev.*, 252; (d) Díez, D., Núñez, M.G., Antón, A.B., García, P., Moro, R.F., Garrido, N.M., Marcos, I.S., Basabe, P., and Urones, J.G. (2008) *Curr. Org. Synth.*, **5**, 186; (e) Wong, O.A. and Shi, Y. (2008) *Chem. Rev.*, **108**, 3958.

135. For a review, see: (a) Matsumoto, K., Sawada, Y., and Katsuki, T. (2008) *Pure Appl. Chem.*, **80**, 1071. For selected recent examples, see: (b) Matsumoto, K., Kubo, T., and Katsuki, T. (2009) *Chem. Eur. J.*, **15**, 6573; (c) Matsumoto, K., Oguma, T., and Katsuki, T. (2009) *Angew. Chem. Int. Ed.*, **48**, 7432; (d) Egami, H. and Katsuki, T. (2008) *Angew. Chem. Int. Ed.*, **47**, 5171; (e) Egami, H., Oguma, T., and Katsuki, T. (2010) *J. Am. Chem. Soc.*, **132**, 5886.

136. Johnson, R.A. and Sharpless, K.B. (2000) in *Catalytic Asymmetric Synthesis* (ed. I. Ojima), 2nd edn., Wiley-VCH Inc., New York, p. 231.

137. (a) Makita, N., Hoshino, Y., and Yamamoto, H. (2003) *Angew. Chem. Int. Ed.*, **42**, 941; (b) Shi, Y. (2004) *Acc. Chem. Res.*, **37**, 488; (c) Zhang, W., Basak, A., Kosugi, Y., Hoshino, Y., and Yamamoto, H. (2005) *Angew. Chem. Int. Ed.*, **44**, 4389; (d) Zhang, W. and Yamamoto, H. (2007) *J. Am. Chem. Soc.*, **129**, 286; (e) Li, Z., Zhang, W., and Yamamoto, H. (2008) *Angew. Chem. Int. Ed.*, **47**, 7520; (f) Wang, B., Wong, O.A., Zhao, M.X., and Shi, Y. (2009) *J. Org. Chem.*, **73**, 9539; (g) Li, Z. and Yamamoto, H. (2010) *J. Am. Chem. Soc.*, **132**, 7878.

138. For selected reviews, see: (a) Kolb, H.C. and Sharpless, K.B. (2004) in *Transition Metals for Organic Synthesis: Building Blocks and Fine Chemicals* (eds. M. Beller and C. Bolm), 2nd edn., Wiley-VCH Verlag GmbH, Weinheim, p. 309; (b) Bolm, C., Hildebrand, J.P., and Muniz, K. (2000) in *Catalytic Asymmetric Synthesis* (ed. I. Ojima) 2nd edn., Wiley-VCH Inc., New York, p. 412; (c) Schlingloff, G. and Sharpless, K.B. (2001) in *Asymmetric Oxidation Reactions* (ed. T. Katsuki),

Oxford University Press, Oxford, p. 104; (d) Nilov, D. and Reiser, O. (2002) *Adv. Synth. Catal.*, **344**, 1169.

139. For selected recent reviews, see: (a) Chemler, S.R. (2009) *Org. Biomol. Chem.*, **7**, 3009; (b) Muller, T.E., Hultzsch, K.C., Yus, M., Foubelo, F., and Tada, M. (2008) *Chem. Rev.*, **108**, 3795; (c) Aillaud, I., Collin, J., Hannedouche, J., and Schulz, E. (2007) *Dalton Trans.*, 5105; (d) Hii, K.K. (2006) *Pure Appl. Chem.*, **78**, 341.

140. For selected recent examples, see: (a) Du, H., Zhao, B., and Shi, Y. (2007) *J. Am. Chem. Soc.*, **129**, 762; (b) Du, H., Zhao, B., and Shi, Y. (2008) *J. Am. Chem. Soc.*, **130**, 8590; (c) Cardona, F. and Goti, A. (2009) *Nat. Chem.*, **1**, 269; (d) Zhao, B., Du, H., Cui, S., and Shi, Y. (2010) *J. Am. Chem. Soc.*, **132**, 3523.

141. Yamashita, Y., Gopalarathnam, A., and Hartwig, J.F. (2007) *J. Am. Chem. Soc.*, **129**, 7508–7509.

142. For reviews, see: (a) Oestreich, M. (2005) *Angew. Chem. Int. Ed.*, **44**, 2324; (b) Ibrahim, H. and Togni, A. (2004) *Chem. Commun.*, 1147; (c) Castellanos, A. and Fletcher, S.P. (2011) *Chem. Eur. J.*, **17**, 5766. For recent examples, see: (d) Kalyani, D. and Sanford, M.S. (2008) *J. Am. Chem. Soc.*, **130**, 2150; (e) Michael, F.E., Sibbald, P.A., and Cochran, B.M. (2008) *Org. Lett.*, **10**, 793; (f) Wu, T., Yin, G., and Liu, G. (2009) *J. Am. Chem. Soc.*, **131**, 16354; (g) Christie, S.D.R., Warrington, A.D., and Lunniss, C.J. (2009) *Synthesis*, 148; (h) Doroski, T.A., Cox, M.R., and Morgan, J.B. (2009) *Tetrahedron Lett.*, **50**, 5162; (i) Kalyani, D. and Sanford, M.S. (2008) *J. Am. Chem. Soc.*, **130**, 2150.

143. Shibata, N., Kohno, J., Takai, K., Ishimaru, T., Nakamura, S., Toru, T., and Kanemasa, S. (2005) *Angew. Chem. Int. Ed.*, **44**, 4204.

144. (a) Cai, Y., Liu, X., Hui, Y., Jiang, J., Wang, W., Chen, W., Lin, L., and Feng, X. (2010) *Angew. Chem. Int. Ed.*, **49**, 6160; (b) Yeung, Y.Y., Gao, X., and Corey, E.J. (2006) *J. Am. Chem. Soc.*, **128**, 9644.

Commentary Part

Comment 1

Qi-Lin Zhou

Since the beginning of this century, metal-catalyzed asymmetric transformations have continued to be a focal point of extensive research in the field of organic chemistry. The main reasons for this are not only that the 2001 Nobel Prize in Chemistry was awarded to chemists in the fields of metal-catalyzed AH and AO, but also that metal-catalyzed asymmetric reactions have completely revolutionized the approach to organic synthesis. In the last 10 years, tremendous progress has been made in the field of metal-catalyzed asymmetric transformations. Thanks to Sandoval and Noyori, one of the winners of Nobel Prize in Chemistry in 2001, who in this chapter have contributed an excellent overview of the developments in metal-catalyzed asymmetric transformations in last decade, a wealth of information has been made available to anyone who is interested or involved in asymmetric synthesis.

As the carbon–carbon bond forming reaction is the most fundamental transformation for the construction of molecular frameworks in organic chemistry, and the metal-catalyzed asymmetric version of this reaction has undergone rapid developments in recent years, this chapter has a special emphasis on metal-catalyzed asymmetric carbon–carbon bond-forming transformations. For example, the authors illustrated the most prominent breakthroughs in the metal-catalyzed asymmetric cross-coupling reactions including palladium-catalyzed cross-coupling reactions, for which the Nobel Prize in Chemistry in 2010 was awarded. The major achievements in other asymmetric carbon–carbon bond-forming transformations such as hydroformylation, addition, allylic alkylation, cycloaddition, and so on have also been well covered.

Metal-catalyzed AH is the most commonly employed asymmetric catalytic process in the production of chiral compounds. Great progress has been made in the area of metal-catalyzed AH in the last decade. As it is not a focal point of this chapter, the recent achievements in AH reactions are only briefly summarized, so that some of the breakthroughs in AH such as the hydrogenation of nitrogen-containing aromatic heterocycles have not been included.

There have been many achievements in the metal-catalyzed asymmetric formation of carbon–nitrogen, carbon–oxygen, and other carbon–heteroatom bonds that could not be detailed in this short chapter. However, the authors have made a careful selection from a wide range of metal-catalyzed asymmetric transformations and presented us with a very informative overview.

Overall, the authors have described most of the significant advances in the field of metal-catalyzed asymmetric transformations, which should convince us that chiral metal catalysts will continue to play a central role in asymmetric catalysis in the future.

Comment 2

Andreas Pfaltz

Asymmetric catalysis has undergone explosive growth over recent decades, and it has now reached the stage where catalytic asymmetric transformations are routinely used in organic synthesis. Nevertheless, the number of industrial applications is still small. However, it is safe to predict that we will see a substantial increase in industrial catalytic asymmetric processes in the next decade, for several reasons. With the steadily growing number of efficient, practical catalytic methods and the expanding selection of commercially available chiral ligands and catalysts, asymmetric catalysis has become part of the standard repertoire of industrial chemists. In addition, specialized companies have emerged, which offer their help in the development of catalytic asymmetric processes. One can also predict that with the growth of the generic pharmaceutical industry, the need for cost-effective asymmetric transformations will encourage the use of catalytic methods.

A comprehensive overview of asymmetric catalysis was published in 1999 [C1], and since then, the field has grown at rapid pace. Organocatalysis has emerged as an important new subdiscipline, which is treated in a separate chapter in this book. Progress in metal-catalyzed asymmetric transformations is summarized in this chapter, which highlights the many important achievements of the last decade. The well-selected examples of enantioselective reactions clearly show the scope and limitations of the various catalytic methods developed so far. At first sight, the wide range of chiral metal catalysts and the impressive number of enantioselective transformations that are possible today might lead to the conclusion that for essentially all synthetically relevant organic reactions efficient enantioselective catalysts are available. However, a more critical evaluation reveals that the number of truly useful catalyst systems is still limited. Especially catalysts that fulfill the requirements for industrial applications are still rare. In addition to high enantioselectivity, there are several other criteria that are important, such as efficiency (turnover number and frequency), substrate scope, reliability, accessibility and cost of the catalyst, and functional group tolerance. In this respect, most of the current catalyst systems still need to be improved, so the search for new, more efficient catalysts will remain a central concern of future research.

On reading the different sections of this chapter, one can see that various new asymmetric transformations have been developed during the last decade, such as cross-coupling reactions involving sp³ carbon centers, 1,3-DCs of azomethine ylides, reductive coupling of unsaturated reactants to carbonyl compounds and imines, and ring-opening/cross alkene metathesis. Major advances have also been made in the development of new or improved catalysts, which have considerably increased the scope and practicality of established catalytic methods. For allylic substitutions, for example, Ir and Cu catalysts have become available that have strongly enhanced the application range to new substrates and nucleophiles and at the same time have opened up new possibilities for controlling the regioselectivity. For AH, Ir-based catalysts have been developed, which for the first time allowed highly enantioselective reductions of trialkyl-substituted alkenes devoid of coordinating functional groups. The synthetic potential and practicality of conjugate additions to unsaturated carbonyl compounds has been substantially increased with Cu-based catalysts that allow the use of Grignard reagents rather than organozinc reagents. Impressive progress has also been made in enantioselective Rh-catalyzed C–H functionalization by carbenoid or nitrenoid insertion.

Although the aim of the authors was to give an overview of the various metal-catalyzed processes that are currently possible, the material presented also reveals emerging concepts and trends in the development of new catalysts. One can see a strong emphasis on modular, readily accessible ligands, which can be easily modified in order to optimize their structure for a particular application. Good examples are monodentate phosphine ligands derived from BINOL that have found a wide range of applications, most notably in asymmetric hydrogenations on an industrial scale. The high enantioselectivities induced by ligands of this type came as a surprise as it was commonly accepted that chelating diphosphine ligands are generally superior to monophosphines.

Although diphosphine and diamine ligands have dominated for a long time in asymmetric catalysis, mixed donor ligands have attracted increasing attention more recently. Various classes of P,N-ligands based on oxazoline, pyridine, and other heterocycles as N-donors have been developed, which have strongly enhanced the scope of Ir-catalyzed hydrogenation. The list of privileged ligand structures has been augmented by MONOPHOS and by spirocyclic ligands such as SDP (see Figure 9.1). As an alternative to conventional bidentate chelate ligands, the self-assembly of two monodentate ligands through hydrogen bonding or through a bridging metal ion has emerged as a promising approach.

During the last decade, there has been more and more use of combinatorial approaches in asymmetric catalysis, especially in industry, as efficient systems for high-throughput screening became available. Combinatorial strategies have been particularly successful for monodentate phosphine ligands, as they can be readily assembled from simple precursors and, by incorporation of two different monophosphines into a bisphosphine–metal complex, a huge number of different catalysts can be easily generated. At the same time, computational methods have found increasing application, as density functional theory (DFT) calculations today can handle fairly complex catalyst systems even with second- and third-row transition metals and large ligands. Although DFT calculations have been successfully used to study the mechanism of catalytic reactions and to rationalize the observed enantioselectivities, truly rational catalyst design is still out of reach.

How will asymmetric catalysis evolve in the next decade? As the future course of research is unpredictable, no-one really knows (who would have anticipated the sudden dramatic development of organocatalysis?). Nevertheless, it is still possible to identify certain trends and research areas that will likely gain increasing importance. Replacing expensive noble metals by more abundant, inexpensive metals will be one of the challenges of future research. First encouraging results have been obtained with Fe and Cu complexes in AH and AO reactions, although much time and effort will be needed to develop practical catalysts of this type. The search for new, generally applicable ligands will certainly continue. In addition to nitrogen, oxygen, and phosphorus ligands, *N*-heterocyclic carbenes and sulfoxides appear to have considerable potential. Bi- and multimetal complexes or hybrid metal–organocatalysts are also promising catalyst systems that deserve further investigation. Moreover, even well-established methods in asymmetric catalysis still suffer from important gaps in scope that must be addressed by future research. There is no doubt that asymmetric catalysis will continue to be a central area of research in organic chemistry in the next decade and it will be exciting to see the unexpected breakthroughs that will be made during that period.

Comment 3

Xue-Long Hou

Since the first homogeneous metal-catalyzed asymmetric reaction was reported in 1966 by Nozaki and Noyori, the latter a co-author of this chapter, tremendous progress has been made in metal-catalyzed asymmetric reactions, which not only changes the situation of synthetic chemistry totally, but also influences our daily lives deeply because of the importance of chirality in the life sciences, pharmaceuticals, agrochemicals, and so on. This development is still growing at an even more rapid pace, and huge numbers of new protocols and new concepts have emerged in the first decade of this century. The present chapter by Sandoval and Noyori gives an excellent review of the progress in this field during the last decade. The authors focus on the two most important aspects of asymmetric catalysis and also of organic synthesis, carbon–carbon bond forming reactions and reduction/oxidation. By carefully selecting the examples, the breakthroughs and major achievements in metal-catalyzed asymmetric transformations have well been illustrated. In addition to the traditional carbon–carbon bond-formation reactions such as addition and allylation, some emerging areas such as cross-coupling, asymmetric metathesis, hydroformylation, and cycloaddition

are also discussed. Further, hydrogenation, epoxidation, hydrogen transfer reactions, halogenation, and amination are well covered. Some subjects have been recognized by Nobel Prizes in recent years: hydrogenation/epoxidation (2001), alkene metathesis (2005), and palladium-catalyzed cross coupling (2010). Readers from undergraduate students to graduates and also researchers in both academia and industry will gain a very informative overview over this important and challenging field and benefits from this excellent chapter.

As the authors have not aimed to give a comprehensive review, C–X bond formation was only briefly touched upon with some selected examples, some of which, such as carbene insertion of H-X bonds [C2], are examples of recent interest. Some progresses in C–C bond formation, for example, Rh-catalyzed asymmetric carbonyl acylation [C3], are notable examples. Branched product selectivity and application of the α-carbanion of carboxylic acid derivatives as nucleophiles in Pd-catalyzed AAA reactions were not mentioned in the discussion of metal-catalyzed AAA reactions [C4]. The advances in the use of Fe and Au as catalysts in asymmetric catalysis [C5], [C6], the AH of heteroaromatic systems [C7], and asymmetric catalysis via C–H bond activation [C8] are notable achievements in the last decade. Of course, these examples are again not comprehensive, but some of them might encourage further exploration in this fruitful field.

Comment 4

Hisashi Yamamoto

An excellent review is not one which comprehensively covers every detail but rather points out the results of essential milestones and identifies future predominant concepts. It may be difficult to omit less important findings and ideas, but this is essential to producing a valuable article. Unfortunately, many reviews of organic synthesis fall short of this ideal, being either too thorough or making no attempt to summarize properly the actual historical background. In fact, omitting trivial results should be done by taking a bird's-eye look at the entire field of science backed by rational philosophy and sound wisdom.

I believe the present review by Professors Sandoval and Noyori has cleared this difficult bar. The selection of topics for the benefit of human society and concepts reflected by the catalyst design are indeed well balanced with thoughtful comments. The review covers two areas: (i) asymmetric carbon–carbon bond formation and (ii) asymmetric reduction and oxidation, the latter obviously the subject recognized by the 2001 Nobel Prize in Chemistry. Two further Nobel Prizes have been awarded for the asymmetric carbon–carbon bond formation process: alkene metathesis (2005) and cross-coupling reactions (2010). However, rapid progress since that time has expanded knowledge beyond the areas of these two prizes and a great many new programs have appeared. The review has nicely filtered out the classic results and focuses on the important updated advances. I am convinced that this chapter will certainly be very useful for graduate students and postdoctoral fellows who are attempting to enter this challenging area of science.

The only aspect, which seems to be missing is the prediction of coming discoveries in this important field of chemistry. It is challenging to offer such predictions, yet it is just such foresight which will provide encouragement and be useful for younger generations. For example, I believe that selective new AO including C–H functionalization with not only oxygen and nitrogen but also halogens and other heteroatoms can be expected to be significantly developed in the future. I also anticipate that asymmetric and diastereoselective domino and cascade processes for creating multi-stereogenic centers in the molecule to replace the classical long-sequential synthesis of complex molecules may be the next target after the amazing cross-coupling reactions of the last few decades. Replacing toxic metal with nontoxic metal catalysts will be a monumental task. However, even more important will be the unveiling of numerous of unknown reactions hitherto only dreamed of or unexpected for the transformation of organic molecules, which will be done using only metal catalysis.

References

C1. Jacobsen, E.N., Pfaltz, A., and Yamamoto, H. (eds.) (1999) *Comprehensive Asymmetric Catalysis*, Springer, Berlin.

C2. (a) Chen, C., Zhu, S.F., Liu, B., Wang, L.X., and Zhou, Q.L. (2007) *J. Am. Chem. Soc.*, **129**, 12616; (b) Hou, Z., Wang, J., He, P., Wang, J., Qin, B., Liu, X., Lin, L., and Feng, X. (2010) *Angew. Chem. Int. Ed.*, **49**, 4763.

C3. Shen, Z., Khan, H.A., and Dong, V.M. (2008) *J. Am. Chem. Soc.*, **130**, 2916.

C4. (a) You, S.L., Zhu, X.Z., Luo, Y.M., Hou, X.L., and Dai, L.X. (2001) *J. Am. Chem. Soc.*, **123**, 7471; (b) Zhang, K., Peng, Q., Hou, X.L., and Wu, Y.D. (2008) *Angew. Chem. Int. Ed.*, **47**, 1741; (c) Trost, B.M., Lehr, K., Michaelis, D.J., Xu, J., and Buckl, A.K. (2010) *J. Am. Chem. Soc.*, **132**, 8915.

C5. (a) Suzuki, K., Oldenburg, P.D., and Que, L. Jr. (2008) *Angew. Chem. Int. Ed.*, **47**, 1887; (b) Zhu, S.F., Cai, Y., Mao, H.X., Xie, J.H., and Zhou, Q.L. (2010) *Nat. Chem.*, **2**, 546.

C6. For a review of Au-catalyzed asymmetric transformation, see: Widenhoefer, R.A. (2008) *Chem. Eur. J.*, **14**, 5382.

C7. (a) Zhou, Y.G. (2007) *Acc. Chem. Res.*, **40**, 1357; (b) Wang, D.-S., Chen, Q.A., Li, W., Yu, C.B., Zhou, Y.G., and Zhang, X. (2010) *J. Am. Chem. Soc.*, **132**, 8909.

C8. (a) Thalji, R.K., Ellman, J.A., and Bergman, R.G. (2004) *J. Am. Chem. Soc.*, **126**, 7192; (b) Shi, B.F., Maugel, N., Zhang, Y.H., and Yu, J.Q. (2008) *Angew. Chem. Int. Ed.*, **47**, 4882.

10

The Proline-Catalyzed Mannich Reaction and the Advent of Enamine Catalysis

Benjamin List and Sai-Hu Liao

10.1
Introduction

During the last 12 years, *enamine catalysis*, in which carbonyl compounds react via catalytically generated enamines using primary and secondary amine catalysts, has had a significant impact on organic synthesis [1]. Judging from the numerous publications on this topic, one may even get the impression that enamine catalysis is about to become a major approach to asymmetric catalytic transformations of carbonyl compounds. This is surprising if one considers that just a few years ago, when it came to asymmetrically modifying carbonyl compounds in the α-position, the state-of-the-art typically involved chiral auxiliaries and stoichiometric bases to accomplish the generation of a preformed enolate, which would then undergo a reaction with an electrophile. With enamine catalysis, the treatment of an aldehyde or ketone with a catalytic amount of a chiral amine in the presence of another electrophilic reagent directly and generally leads to the desired α-modified carbonyl compound. This indeed constitutes a major advance. Interestingly however, in contrast to most previous advances in the history of chemistry, enamine catalysis has not been an innovation that is based on the *sophistication* of previous methodologies but rather the opposite: an innovation based on their *simplification*. As such, one may argue that enamine catalysis, as a generic activation mode of organocatalysis, should have been invented earlier and not later than the more sophisticated auxiliary/stoichiometric base methods. We find this aspect of the enamine catalysis evolution, in addition to its enormous breadth of use in synthesis, to be a fascinating and thought-provoking topic.

10.2
The Proline-Catalyzed Mannich Reaction

The roots of proline catalysis of the direct asymmetric intramolecular aldol reaction and its generalization as an intermolecular reaction have recently been discussed [2]. The key towards enamine catalysis as a generic activation mode, however, at least in our opinion, had been the discovery of the proline-catalyzed Mannich reaction [3]. The previous extension of the intramolecular aldol reaction to an intermolecular variant

Organic Chemistry – Breakthroughs and Perspectives, First Edition. Edited by Kuiling Ding and Li-Xin Dai.
© 2012 Wiley-VCH Verlag GmbH & Co. KGaA. Published 2012 by Wiley-VCH Verlag GmbH & Co. KGaA.

clearly illustrated that a more general aldolization mechanism must be operative. This was later spectacularly confirmed with the discovery of aldehyde cross-aldolizations and also enol-*exo* and transannular aldolizations [4]. These observations clearly proved that all four types of aldol reactions (intermolecular, enol-*endo*, enol-*exo*, and transannular) can be catalyzed with proline via enamine catalysis. However, that there is even more to proline catalysis than just significantly more aldol chemistry was only demonstrated in the context of the Mannich reaction and the discovery that its three-component variant can be catalyzed by proline with exceptionally high enantioselectivity [3]. For the first time proline catalysis was used in a highly enantioselective non-aldol reaction and with a clear mechanistic hypothesis in mind. Here we focus on this marvelous reaction, which has matured into a truly general and often used methodology, by reviewing its discovery and background and also the astonishing advances in the enamine catalytic asymmetric Mannich reaction during the last dozen years.

The Mannich reaction was originally a three-component coupling of two carbonyl compounds and an amine to furnish β-aminocarbonyl compounds. It proceeds via the *in situ* generation of an imine or iminium ion intermediate from one carbonyl compound, which reacts as an electrophile, with an enolate or its equivalent, which is generated *in situ* from the other carbonyl species. The first asymmetric Mannich reaction was developed by Enders and co-workers in 1985 [5]. Treating the preformed and chiral auxiliary modified imine **1** with preformed chiral enamine **2** gave the corresponding Mannich adduct **3** highly diastereoselectively (Scheme 10.1).

In the following years, a number of other asymmetric Mannich reactions using chiral, enantiopure substrates, either enolate or imine equivalents, or both, have been described. The enantioselective catalysis of Mannich reactions, however, is a fairly recent concept. The first examples were developed by the groups of Tomioka [6], Kobayashi [7], Sodeoka [8], and Lectka [9]. All of these variants require both preformed imine and enol equivalents combined with a metal-based catalyst (for an example, see Scheme 10.2).

It can be considered a breakthrough when Shibasaki and co-workers [10]. designed a direct catalytic enantioselective three-component Mannich reaction of propiophenone (**4**), paraformaldehyde, and pyrrolidine that gave the corresponding product **5** with encouraging enantioselectivity (82:18 *er*) but still in relatively low yield (16%) (Scheme 10.3).

In 2000, when the above situation represented the state-of-the-art in asymmetric Mannich chemistry, we were speculating about the fundamental catalysis principle behind proline catalysis. We realized that the basis of proline catalysis must be the facile

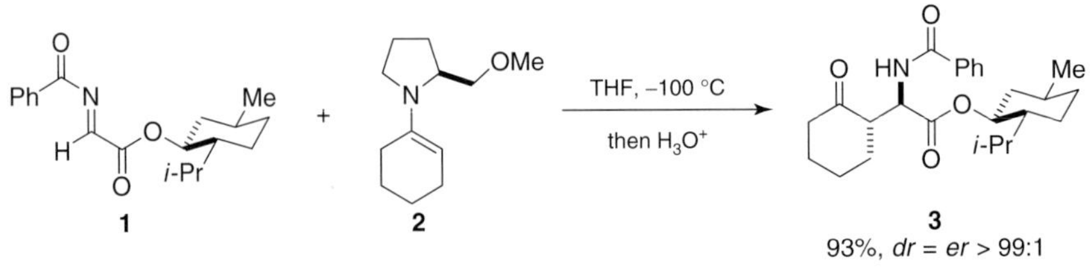

Scheme 10.1 The asymmetric Mannich reaction of a preformed and chiral auxiliary modified imine with a preformed chiral enamine.

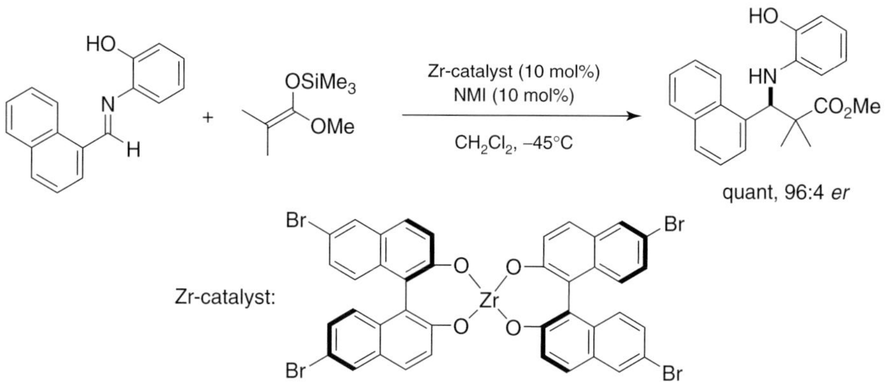

Scheme 10.2 The enantioselective Zr-catalyzed Mannich reaction with a preformed imine.

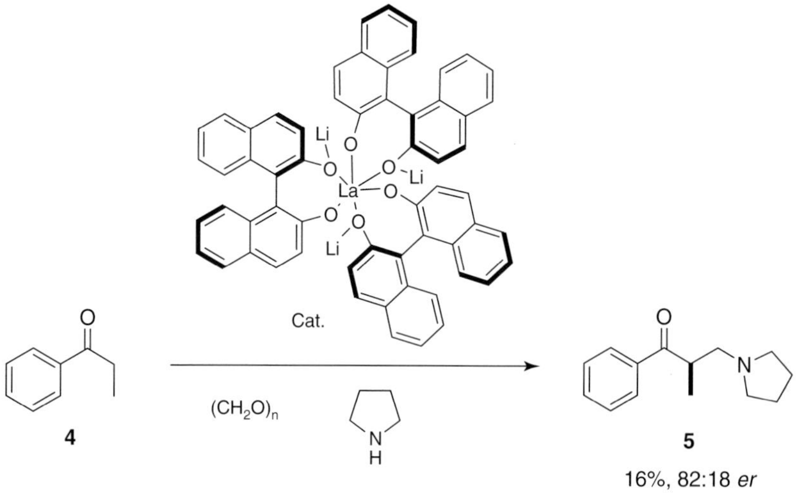

Zr-catalyst:

quant, 96:4 *er*

Scheme 10.3 The direct, catalytic, and enantioselective Mannich reaction with a catalyst.

16%, 82:18 *er*

in situ generation of enamines from ketones and aldehydes, although such intermediates had actually never been detected before. This principle, we reasoned, may well become a generic catalysis activation mode, which we have termed enamine catalysis and which represents a way of merging enolization and enantioselective bond construction in enolate–electrophile-combining reactions. We expected the presumed proline enamines to react not only with carbonyl compounds in aldol reactions but potentially with all types of electrophiles including imines in the corresponding Mannich reactions. We also reasoned that hypothetical proline-catalyzed Mannich reactions may be conducted as direct three-component reactions of ketones, aldehydes, and amines without prior imine formation. Encouragement for this idea came from Manabe and Kobayashi's finding

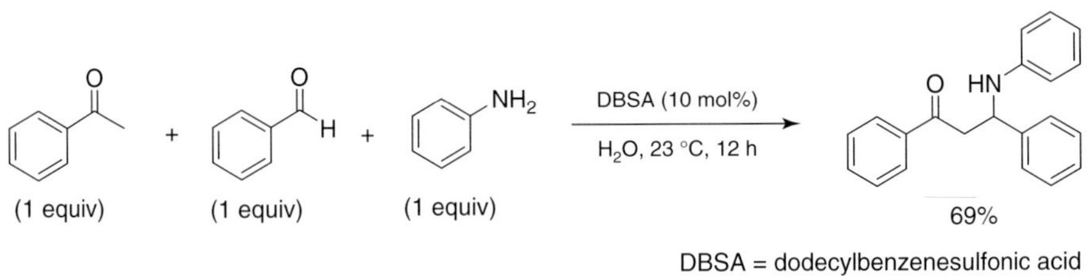
(1 equiv)
(1 equiv)
(1 equiv)
DBSA (10 mol%)
H_2O, 23 °C, 12 h
69%
DBSA = dodecylbenzenesulfonic acid

Scheme 10.4 Direct non-asymmetric Mannich reaction.

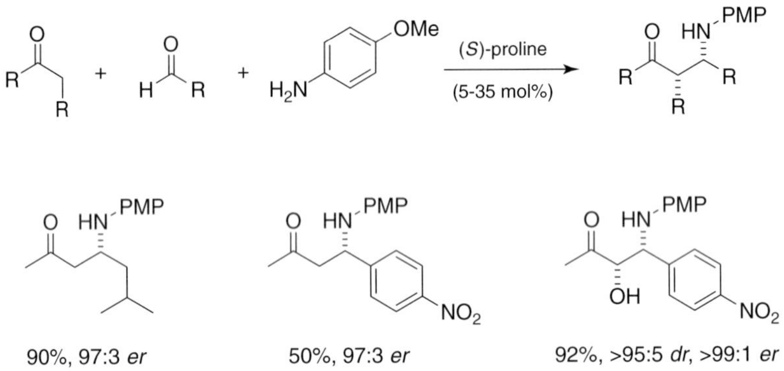
(S)-proline
(5-35 mol%)
90%, 97:3 er
50%, 97:3 er
92%, >95:5 dr, >99:1 er

Scheme 10.5 The first organocatalytic direct enantioselective Mannich reaction.

of very simple and practical reaction conditions for direct non-asymmetric Mannich reactions of ketones with *p*-anisidine and aldehydes (Scheme 10.4) [11].

Indeed, our concept worked well and it was very exciting to find that proline does catalyze several enantioselective Mannich reactions of ketones with aldehydes and *p*-anisidine (Scheme 10.5) [3, 12]. Fortunately we were able to isolate many of the corresponding products with superb enantioselectivities and consequently our results could be published relatively quickly [3]. With α-substituted ketones, the corresponding Mannich products were also formed with high *syn*-diastereoselectivities. For the first time electrophiles other than carbonyls were used in a highly enantioselective proline-catalyzed reaction. However, not only did our process constitute the first catalytic asymmetric three-component Mannich reaction of a free aldehyde (non-formaldehyde) with an unmodified ketone and an amine, it was also the first organocatalytic enantioselective Mannich reaction to be reported.

An interesting aspect concerns the stereochemistry of the proline-catalyzed Mannich reaction. In contrast to the corresponding aldol reaction, which proceeds via nucleophilic addition to the *re*-face of the aldehyde, the proline-catalyzed Mannich reaction proceeds via addition to the imine *si*-face. Keeping Houk and co-workers' aldol transition state in mind [13], we proposed an explanation for this behavior. Accordingly, we hypothesized that proline would form an enamine intermediate, in which the usual *E*-geometry of the C=C-double bond is ensured because of the steric demand of the secondary amine portion of the proline pyrrolidine ring. This enamine would

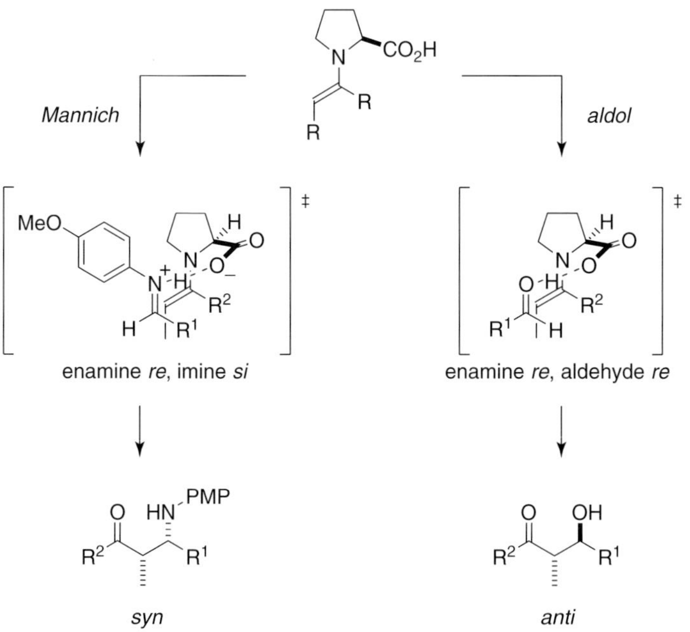

Scheme 10.6 Stereochemical model for *syn*-selective Mannich reaction and *anti*-selective aldol reaction.

then be approached toward its front side (*re*) from either the imine or the aldehyde, directed by the carboxylate of proline. The electrophile faciality in turn would be determined by its geometry: in the Mannich reaction, an (*E*)-imine can reasonably approach the enamine only with its *si*-face, since otherwise stabilizing hydrogen-bonding interactions between proline's carboxylate and the protonated imine would be severely hampered and negative steric interactions would result. In the case of the aldol reaction, the aldehyde will predominantly approach the enamine with its *re*-face such that its substituent would avoid unfavorable interactions with the enamine (Scheme 10.6) [2a, 13].

This model received strong support from computational studies conducted by Bahmanyar and Houk [14]. It predicts that the diastereoselectivity could be controlled by changing the faciality of either the imine or the enamine. Alternatively, the diastereoselectivity may be controllable if the *syn* versus *anti* arrangement of the enamine could somehow be directed (Scheme 10.7).

The latter strategy was later used by Houk and co-workers [15] and also by Maruoka and co-workers [16] in the design of *anti*-selective Mannich catalysts. An elegant although strategically different approach that is based on the requirement of our transition-state model that Z-configured imines would also furnish the corresponding *anti* products, resulting from an *re* attack, was described by Glorius and co-workers [17]. Indeed, the reaction of a cyclic and therefore Z-configured imine with cyclohexanone furnishes essentially only the *anti* product with the expected absolute configuration (Scheme 10.8).

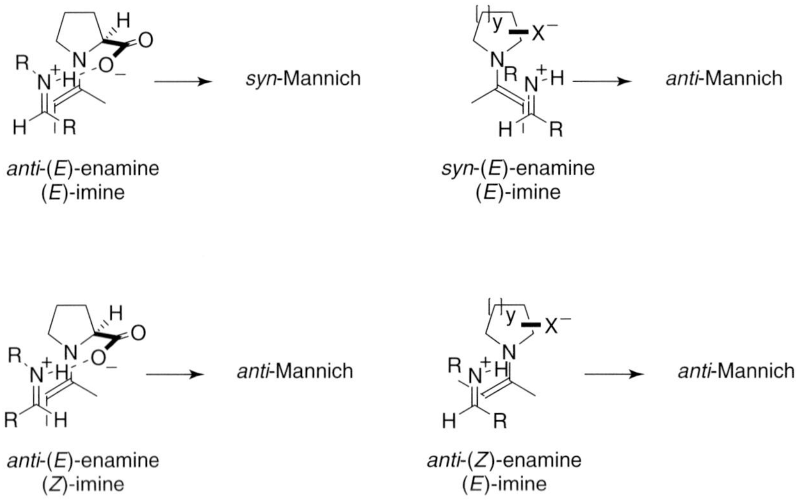

Scheme 10.7 Stereochemical model for *syn*- versus *anti*-selective Mannich reactions.

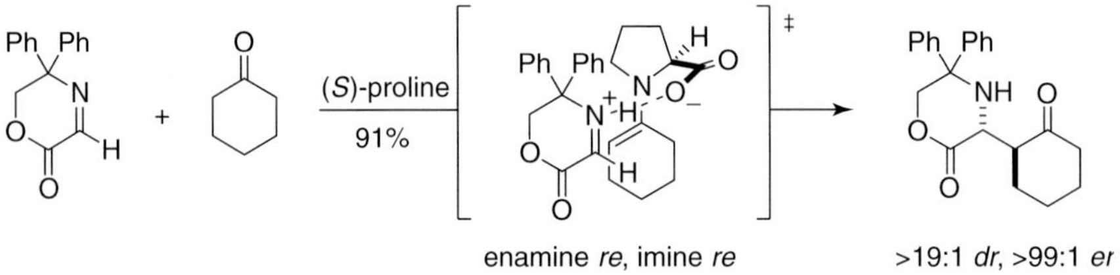

Scheme 10.8 Mannich reaction of a (*Z*)-cyclic imine with cyclohexanone.

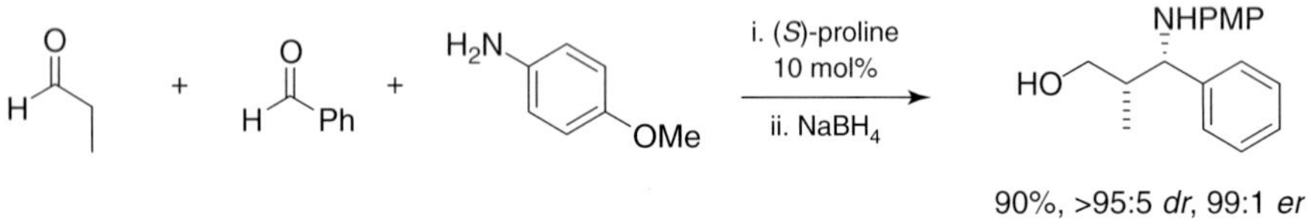

Scheme 10.9 Proline-catalyzed Mannich reaction of two different aldehydes with *p*-anisidine.

Another important advance was the introduction of aldehydes as nucleophiles in proline-catalyzed Mannich reactions [18]. For example, Hayashi *et al.* were able to unite two different aldehydes with *p*-anisidine to give the corresponding *syn*-β-amino aldehyde in a direct asymmetric three-component Mannich reaction (Scheme 10.9) [19].

A potential drawback of the proline-catalyzed Mannich reaction has long been the fact that it required the use of *p*-anisidine as the amine component. This reagent is a very convenient ammonia equivalent that after incorporation into the corresponding Mannich product will readily release the desired primary amine group upon treatment with an oxidant such as ceric ammonium nitrate (CAN). However, *p*-anisidine is also

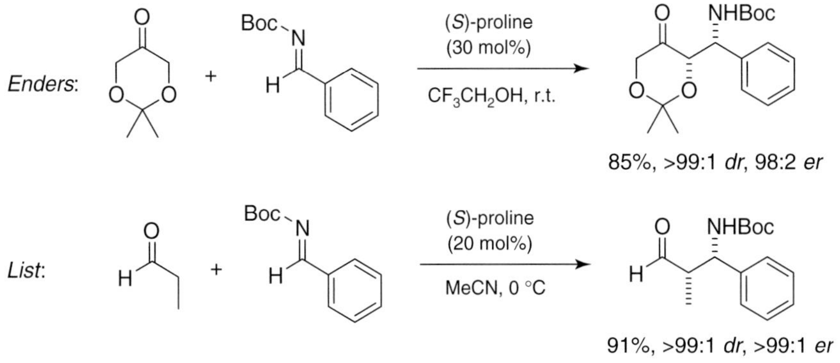

Scheme 10.10 Proline-catalyzed Mannich reaction of N-Boc-protected imines.

relatively toxic and not a particularly atom-economic reagent, and its removal required stoichiometric and relatively expensive reagents [20].

In an alternative approach, Enders *et al.* and our group used *N-tert*-butoxycarbonyl (*N*-Boc)-protected imines with very good results [21]. Whereas Enders *et al.* described the reaction of a ketone with an aromatic *N*-Boc-protected imine, we independently showed that aldehydes can be used as nucleophiles to give the corresponding products with superb enantioselectivities (Scheme 10.10). Remarkably, the desired *syn*-Mannich products typically precipitate directly from the reaction mixtures to give high yields and very high stereoselectivities or alternatively they can be isolated by crystallization induced by trituration with hexanes.

Moreover, using *N*-Boc-protected imines, we were also able for the first time to use acetaldehyde in proline-catalyzed Mannich reactions (Scheme 10.11) [22]. This reaction delivers particularly useful β^3-amino aldehydes, which are precursors of the corresponding amino acid, and of various pharmaceuticals.

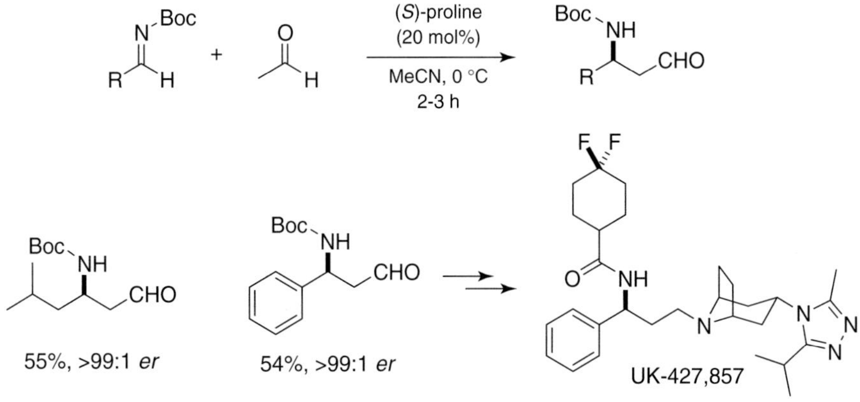

Scheme 10.11 Proline-catalyzed Mannich reaction of N-Boc-protected imines with acetaldehyde.

10.3
Conclusion

In this chapter, we have taken a look at the development of the proline-catalyzed Mannich reaction. Over the years, this particular transformation has advanced significantly into a very general chiral amine-catalyzed Mannich methodology. Both ketones and aldehydes as nucleophiles can be used with either *in situ*-generated imines or preformed imines. Moreover, in either a catalyst- or substrate-controlled way, both possible diastereomers can be accessed highly selectively and typically with very good enantioselectivity. The scope has further been advanced with the use of *N*-Boc-protected imines that introduce a benign and easily removable protecting group into the Mannich products.

Most important for the present discussion is that the proline-catalyzed Mannich reaction really represents the archetype of modern enamine catalysis. This reaction has inspired the development of several dozen new reactions with hundreds of variations. The basis of the discovery was a clear mechanistic hypothesis and the readiness for simplification. Therefore, in this sense, enamine catalysis is indeed the result of a simplification of previous methods and not the other way round. Why is this so unusual in our discipline? What does this tell us about the way in which chemists think about their science? A science that has always been different from other disciplines, in that not only does it demand creativity – we would argue that this is true for all human advances – but also it is specifically based upon it. Could we chemists have a tendency even to be too creative and over-engineer our designs? It appears to us that at least occasionally it is a good idea to take a fresh look at advanced and sophisticated chemical systems with an emphasis on simplification.

References

1. (a) Pihko, P.M., Majander, I., and Erkkila, A. (2010) *Top. Curr. Chem.*, **291**, 29–75; (b) Melchiorre, P., Marigo, M., Carlone, A., and Bartoli, G. (2008) *Angew. Chem. Int. Ed.*, **47**, 6138–6171; (c) Mukherjee, S., Yang, J.-W., Hoffmann, S., and List, B. (2007) *Chem. Rev.*, **107**, 5471–5569; (d) List, B. (2006) *Chem. Commun.*, 819–824; (e) List, B. (2004) *Acc. Chem. Res.*, **37**, 548–557.

2. (a) List, B., Lerner, R.A., and Barbas, C.F. III (2000) *J. Am. Chem. Soc.*, **122**, 2395–2396; (b) List, B. (2010) *Angew. Chem. Int. Ed.*, **49**, 1730–1734.

3. List, B. (2000) *J. Am. Chem. Soc.*, **122**, 9336–9337.

4. (a) Northrup, A.B. and MacMillan, D.W.C. (2002) *J. Am. Chem. Soc.*, **124**, 6798–6799; (b) Pidathala, C., Hoang, L., Vignola, N., and List, B. (2003) *Angew. Chem. Int. Ed.*, **42**, 2785–2788; (c) Chandler, C.L. and List, B. (2008) *J. Am. Chem. Soc.*, **130**, 6737–6739.

5. Kober, R., Papadopoulus, K., Miltz, W., Enders, D., Steglich, W., Reuter, H., and Puff, H. (1985) *Tetrahedron*, **41**, 1693–1701.

6. (a) Fujieda, H., Kanai, M., Kambara, T., Iida, A., and Tomioka, K. (1997) *J. Am. Chem. Soc.*, **119**, 2060–2061; (b) Kambara, T., Hussein, M.A., Fujieda, H., Iida, A., and Tomioka, K. (1998) *Tetrahedron Lett.*, **39**, 9055–9058; (c) Tomioka, K., Fujieda, H., Hayashi, S., Hussein, M.A., Kambara, T., Nomura, Y., Motomu, K., and Koga, K. (1999) *Chem. Commun.*, 715–716; (d) Kambara, T. and Tomioka, K. (1999) *Chem. Pharm. Bull.*, **47**, 720–721; (e) Hussein, M.A., Iida, A., and Tomioka, K. (1999) *Tetrahedron*, **55**, 11219–11228.

7. (a) Ishitani, H., Ueno, M., and Kobayashi, S. (1997) *J. Am. Chem. Soc.*, **119**, 7153–7154; (b) Kobayashi, S., Ishitani,

H., and Ueno, M. (1998) *J. Am. Chem. Soc.,* **120**, 431–432; (c) Ishitani, H., Ueno, M., and Kobayashi, S. (2000) *J. Am. Chem. Soc.,* **122**, 8180–8186.

8. (a) Hagiwara, E., Fujii, A., and Sodeoka, M. (1998) *J. Am. Chem. Soc.,* **120**, 2474–2475; (b) Fujii, A., Hagiwara, E., and Sodeoka, M. (1999) *J. Am. Chem. Soc.,* **121**, 5450–5458.

9. (a) Ferraris, D., Young, B., Dudding, T., and Lectka, T. (1998) *J. Am. Chem. Soc.,* **120**, 4548–4549; (b) Ferraris, D., Young, B., Cox, C., Drury, W.J. III, and Lectka, T. (1998) *J. Org. Chem.,* **63**, 6090–6091; (c) Ferraris, D., Dudding, T., Young, B., Drury, W.J. III, and Lectka, T. (1999) *J. Org. Chem.,* **64**, 2168–2169.

10. (a) Yamasaki, S., Iida, T., and Shibasaki, M. (1999) *Tetrahedron Lett.,* **40**, 307–310; (b) Yamasaki, S., Iida, T., and Shibasaki, M. (1999) *Tetrahedron,* **55**, 8857–8867.

11. Manabe, K. and Kobayashi, S. (1999) *Org. Lett.,* **1**, 1965–1967.

12. List, B., Pojarliev, P., Biller, W.T., and Martin, H.J. (2002) *J. Am. Chem. Soc.,* **124**, 827–833.

13. (a) Bahmanyar, S. and Houk, K.N. (2001) *J. Am. Chem. Soc.,* **123**, 11273–11283; (b) Bahmanyar, S., Houk, K.N., Martin, H.J., and List, B. (2003) *J. Am. Chem. Soc.,* **125**, 2475–2479.

14. Bahmanyar, S. and Houk, K.N. (2003) *Org. Lett.,* **5**, 1249.

15. Mitsumori, S., Zhang, H., Cheong, P.H.-Y., Houk, K.N., Tanaka, F., and Barbas,

C.F. III (2006) *J. Am. Chem. Soc.,* **128**, 1040–1041.

16. (a) Kano, T., Yamaguchi, Y., Tokuda, O., and Maruoka, K. (2005) *J. Am. Chem. Soc.,* **127**, 16408–16409; (b) Kano, T., Yamaguchi, Y., and Maruoka, K. (2009) *Chem. Eur. J.,* **15**, 6678–6687.

17. Hahn, B.T., Fröhlich, R., Harms, K., and Glorius, F. (2008) *Angew. Chem. Int. Ed.,* **47**, 9985–9988.

18. Córdova, A., Watanabe, S., Tanaka, F., Notz, W., and Barbas, C.F. III (2002) *J. Am. Chem. Soc.,* **124**, 1866–1867.

19. Hayashi, Y., Tsuboi, W., Ashimine, I., Urushima, T., Shoji, M., and Sakai, K. (2003) *Angew. Chem. Int. Ed.,* **42**, 3677–3680.

20. (a) Hata, S., Iguchi, I., Iwasawa, T., Yamada, K., and Tomioka, K. (2004) *Org. Lett.,* **6**, 1721–1723; (b) Janey, J.M., Hsiao, Y., and Armstrong, J.D. III (2006) *J. Org. Chem.,* **71**, 390–392; (c) Verkade, J.M.M., van Hemert, L.J.C., Quaedflieg, P.J.L.M., Alsters, P.L., van Delft, F.L., and Rutjes, F.P.J.T. (2006) *Tetrahedron Lett.,* **47**, 8109–8113.

21. (a) Enders, D. and Vrettou, M. (2006) *Synthesis,* 2155–2158; (b) Enders, D., Grondal, C., and Vrettou, M. (2006) *Synthesis,* 3597–3604; (c) Yang, J.W., Stadler, M., and List, B. (2007) *Angew. Chem. Int. Ed.,* **46**, 609–611.

22. Yang, J.-W., Chandler, C., Stadler, M., Kampen, D., and List, B. (2008) *Nature,* **452**, 453–455.

Commentary Part

Comment 1

Seiji Shirakawa and Keiji Maruoka

Although proline-catalyzed asymmetric Robinson annulations were reported in the early 1970s [C1], the real potential of proline as an organocatalyst remained largely unexplored until 2000. The report of List *et al.* in 2000 on proline-catalyzed aldol and Mannich reactions via an enamine catalytic cycle opened a new dimension in proline-catalyzed reactions [C2]. It should be noted that the report of MacMillan and co-workers in 2000 of the discovery of iminium catalysis using

amino acid-derived secondary amine catalysts also accelerated research on asymmetric organocatalysis [C3]. The last 12 years have seen a gold rush of organocatalysis, and proline and its derivatives as organocatalysts have undoubtedly been at the center of the development of organocatalyzed reactions. On the other hand, our group developed axially chiral secondary amine catalysts as artificially designed organocatalysts (Scheme C1) [C4]. These catalysts have shown unique reactivity and selectivity in comparison with proline and its derivatives, and many characteristic features of these catalysts have been reported. The representative features of a designer axially chiral organocatalyst in the Mannich reaction are shown

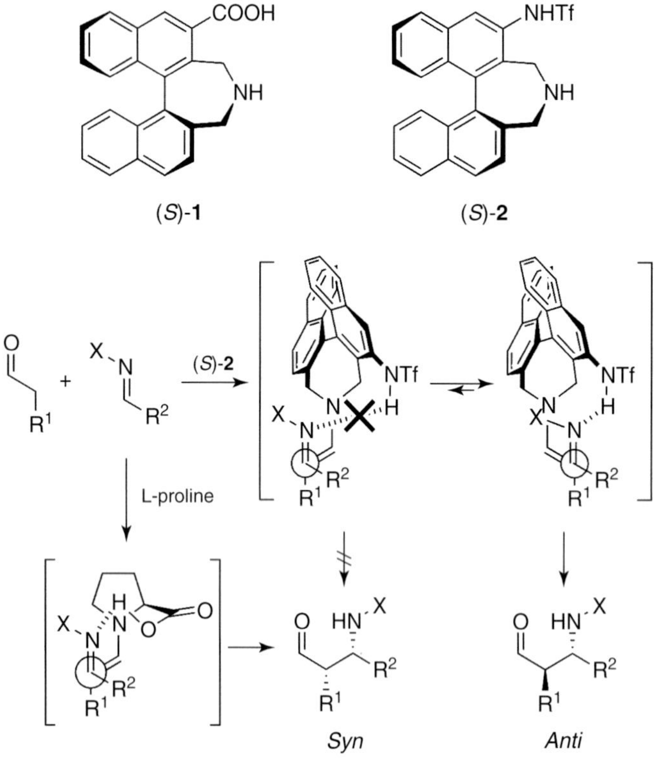

Scheme C1

in Scheme C1 [C5]. In contrast to the *syn*-selective Mannich reaction catalyzed by proline, the use of the axially chiral aminosulfonamide (*S*)-**2** allows an *anti*-selective Mannich reaction with high enantioselectivity. It should be noted that catalyst (*S*)-**2** was also fairly effective for the Mannich reaction between acetaldehyde and *N*-Boc-protected imines [C5b]. These results indicate the high synthetic potential of designer axially chiral organocatalysts. Soon after our report, Barbas and co-workers also reported a similar *anti*-selective asymmetric Mannich reaction catalyzed by 3-pyrrolidinecarboxylic acids [C6].

As other types of nitrogen-containing organocatalysts, cinchona alkaloids [C7] and cinchona alkaloid-derived quaternary ammonium salts [C8] have been utilized as chiral organobase catalysts. Our group has also developed designer axially chiral phase-transfer catalysts (Scheme C2) [C9], and the first-generation catalyst (*S,S*)-**3** (Maruoka Catalyst) was designed in 1999 [C10]. Later, an

even more efficient phase-transfer catalyst (*S*)-**4** (Simplified Maruoka Catalyst), was designed by introducing a combinatorial design approach [C11], which was found to be useful for the further design of various types of mono- and bifunctional phase-transfer catalysts such as (*S*)-**5**, and many efficient asymmetric reactions using these catalysts under phase-transfer conditions have been developed [C9]. Although it was believed for a long period that base additives such as an alkali metal hydroxide or carbonate are essential to promote the phase-transfer reactions, we recently discovered that even without any base additives enantioselective phase-transfer reactions proceeded efficiently in a water-rich biphasic solvent (Scheme C2) [C12]. This simple reaction system has vast potential for the development of environmentally benign practical reactions.

Chiral Brønsted acids such as phosphoric acids [C13] and thioureas [C14] are also important organocatalysts in organic synthesis, and our

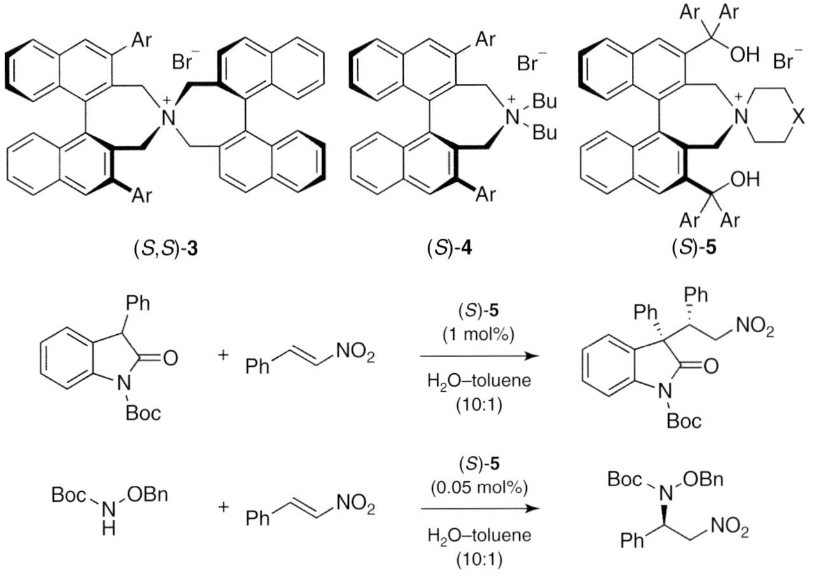

(S,S)-3   (S)-4   (S)-5

Scheme C2

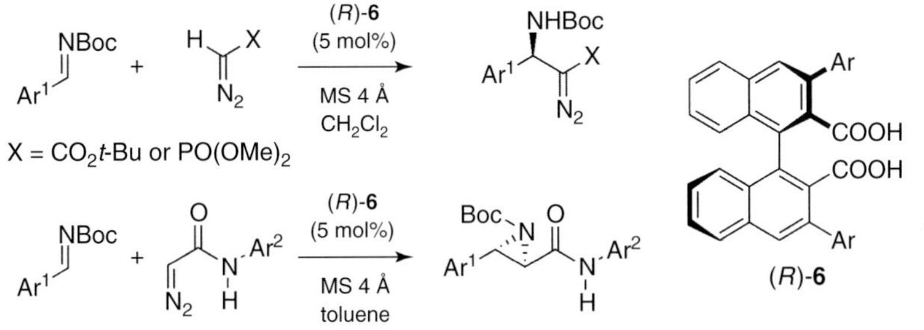

X = CO₂t-Bu or PO(OMe)₂

Scheme C3

group has developed axially chiral dicarboxylic acids of type **6** as promising chiral Brønsted acid catalysts (Scheme C3) [C15].

In the last decade, asymmetric organocatalysis has grown into one of the most useful methods for organic synthesis. However, each well-developed reaction still has some problems from a practical point of view (e.g., high catalyst loading, toxic solvent). Further development of new organocatalysts and transformations is, of course, an important task, and the same importance should be attached to the development of environmentally benign practical reaction systems as the next step in organocatalytic chemistry.

Comment 2

The Early Status of Asymmetric Organocatalysis

Liu-Zhu Gong

Asymmetric reactions under the catalysis of chiral organic molecules, now known as asymmetric organocatalysis, have been reported for more than 100 years, as evidenced by the cinchona alkaloid-catalyzed cyanation reaction [C16]. In the

1970s, proline was found to be able to catalyze effectively the intramolecular aldol reaction, namely Robinson annulation, which is applicable to the synthesis of Wieland–Miescher ketone in high enantiomeric purity [C17]. In 1981, Hiemstra and Wynberg found that the natural alkaloids quinine, quinidine, cinchonine, and cinchonidine were able to promote the addition of aromatic thiols to cycloalkenones, wherein the alkaloid operated as a bifunctional catalyst [C18]. In the same year, Oku and Inoue reported a small peptide-catalyzed enantioselective cyanation of aldehydes [C19]. In 1995, Lu and co-workers reported a [3 + 2]-cycloaddition of allenes to enones, furnishing multiply substituted cyclopentenes under the promotion of phosphines [C20], and its asymmetric version was established by Zhang and co-workers [C21]. These reactions indicated an important generic activation mode of nucleophilic addition catalysis [C22]. In 1996, Yang's and Shi's groups independently established a highly enantioselective epoxidation of alkenes catalyzed by chiral ketones [C23], now used as a robust method to access epoxides with widespread applications in natural product synthesis. Although fundamentally significant, these events were unable to alter the conventional wisdom that asymmetric catalysis was highly dependent on chiral metal and biocatalysts at that time.

Milestone in Asymmetric Organocatalysis

Asymmetric organocatalysis was generalized and accepted as a rising research field thanks to the appearance of the proline-catalyzed intermolecular aldol reaction [C2a] and the iminium-catalyzed Diels–Alder reaction [C3]. The former reaction demonstrates a generic activation mode of enamine catalysis in the functionalization of enolizable ketones and aldehydes and the latter provides a general activation mode for the reactions based on enal and enone substrates.

Enamine Catalysis

In terms of enamine catalysis, the three-component Mannich reaction of aldehydes, ketones, and anilines catalyzed by proline appeared to be the first enamine-catalyzed reaction other than aldolization and could be considered as an important stimulus to generalize the generic activation mode of enamine catalysis [C2b]. Another significantly important event in enamine catalysis is the proline-catalyzed Michael addition reaction [C24]. These two reactions inspired the frequent

development of asymmetric transformations and design of new chiral amine organocatalysts [C25].

Iminium Catalysis

Iminium catalytic reactions have also been reported in the literature for many years. The first iminium-catalyzed conjugate addition reaction of water to enal was reported by Langenbeck and Sauerbier in 1937 [C26]. Yamaguchi *et al.* demonstrated in 1991 that alkali metal salts of proline were able to catalyze the enantioselective Michael addition of malonate to enal [C27]. However, the robustness of this reaction mode had not been recognized in asymmetric catalysis until MacMillan and co-workers uncovered a highly enantioselective Diels–Alder reaction of enals and cyclopentadiene catalyzed by a chiral imidazolidinone [C3]. In addition to cycloaddition reactions, asymmetric iminium catalysis provides a highly general platform for the creation of unprecedented asymmetric conjugate addition reactions to enals and enones [C28].

Domino Reactions by Amine Catalysis

The reversible conversion between enamine and iminium adds more dimensions to chiral amines for the design of new cascade reactions (domino reactions) than other types of organocatalysts. MacMillan and co-workers first combined iminium and enamine catalysis into one sequence, leading to an unprecedented domino reaction for the facile synthesis of structurally diverse and complex chiral molecules with excellent levels of stereoselectivity [C29]. The robustness of the combined enamine and iminium or iminium and enamine catalysis was further illustrated by Enders *et al.* in a three-component domino reaction of aldehydes, α,β-unsaturated aldehydes, and nitroalkenes catalyzed by prolinol derivatives, furnishing cyclohexene derivatives with concomitant creation of four contiguous stereogenic centers with perfect stereocontrol [C30]. This general strategy stimulated the discovery of numerous elegant asymmetric domino reactions that provide unique and easily operative synthetic approaches to access structurally diverse and complex chiral molecules that could not be obtained by using traditional metal-catalyzed asymmetric reactions [C31].

Hydrogen Bonding Catalysis

In addition to enamine and iminium modes of activation, hydrogen bonding activation represents another robust reaction mode for the creation of organocatalytic asymmetric reactions. Jacobsen's and Corey's groups demonstrated the first example hydrogen bonding activation in the Strecker reaction [C32]. Another noteworthy example of hydrogen bonding catalysis in the hetero-Diels–Alder reaction of Danishefsky diene with aldehydes was established by Rawal and co-workers, in which TADDOL functions as an efficient catalyst [C33].

The most significant advance in Brønsted acid catalysis was the discovery of the Binol-based phosphoric acid-catalyzed Mannich reactions reported independently by Akiyama *et al*. [C34] and Uraguchi and Terada [C35] in 2004. The appearance of phosphoric acids prompted the frequent development of a wealth of new reactions [C36]. It is not an overstatement that chiral phosphoric acids are among the most useful organocatalysts discovered so far.

Among hydrogen bonding catalysis, Brønsted acid–Lewis base bifunctional organocatalysts have high ability to enable new reactions involving acidic carbon nucleophiles and heteroatoms. Hiemstra and Wynberg first proposed the reaction mode of Brønsted acid–Lewis base activation [C18]. In 2003, Takemoto and co-workers designed a type of bifunctional organocatalyst with thiourea and tertiary amine functionalities [C37] in a general strategy for the creation of new bifunctional catalysts. Numerous organocatalysts have been prepared on the basis of this generic activation mode [C38].

Conclusion

Undoubtedly, asymmetric organocatalysis appears to be a research field with high importance similar to metal-mediated asymmetric catalysis and has been accepted as the third method to access chiral molecules complementary to metal catalysis and biocatalysis, mainly attributed to the inspiration from the seminal discovery of enamine and iminium catalysis [C2a, C3]. Some workers thought that organocatalysis is functionality chemistry, particularly carbonyl chemistry, but found it difficult to find a solution to the activation of relatively inactive chemical bonds. However, this view should be changed given very recent elegant findings in asymmetric organocatalytic 1,5-hydride shift reactions [C39]. The accommodation of a wide range of functionalities in organocatalysis should be considered an advantage in view of the fact that many organometallic catalytic protocols are sensitive to functional groups and thus require additional protection. The discovery of new organocatalysts to improve catalytic efficacy holds great importance, but finding new activation modes and concepts to address formidable problems is even more appealing. Considering that metal catalysts are efficient in activate inactivating bonds and organocatalysts tolerate functionalities, the combined use of metal and organocatalysts might be a new strategy to develop unprecedented reactions [C40].

Comment 3

Wen-Jing Xiao

Catalysis with a substoichiometric amount of an organic molecule, where a metal atom is not part of the active principle, has been known for more than a century [C41]. The earliest example found in the literature, asymmetric alkaloid-catalyzed cyanohydrin synthesis, was documented by Bredig and Fiske in 1912 [C42]. In 1960, the pioneering work of Pracejus revealed an enantioselective methanolysis of phenylmethyl ketene using a quinuclidine catalyst [C43]. In the early 1970s, much attention was focused on the proline-catalyzed intramolecular aldol reaction, more commonly termed the Hajos–Parrish–Wiechert reaction [C44]. However, organic catalysis or organocatalysis remained largely unexplored in the past century even though several reports demonstrated that some simple organic molecules, such as phosphines, ketones, thioureas, and N-alkylimidazole-containing tripeptides, could be highly effective catalysts in a number of fundamentally important chemical reactions. The early 2000s brought a milestone in the field of organocatalysis, when two seminal publications appeared almost simultaneously: one on enamine catalysis by List *et al*. [C2a] and another one on iminium catalysis by MacMillan and co-workers [C3]. As List and Liao noted in their chapter, the enamine catalysis model is significant because it can not only be applied to intermolecular aldol and Mannich reactions [C2b] but also extended to transformations that have a broader applicability [C41]. Importantly, the work of List *et al*. [C2a] revealed that small organic molecules could catalyze some reactions with almost the same efficiency as large biological molecules through a similar mechanism. Meanwhile, MacMillan and co-workers [C3] clearly introduced the term organocatalysis and established an iminium activation strategy that could be utilized in a broad range of reaction types.

With respect to the rapid and impressive growth in the field of organocatalysis, the establishment of a few general activation modes of organocatalysts should be the most crucial factor. As described in List and Liao's chapter, the enamine activation mode was not exploited for other reactions until 2000, although proline-catalyzed synthesis of Wieland–Miescher ketone was reported in the early 1970s. Since List *et al.* [C2a] first used enamine activation to functionalize ketones, the last 12 years have witnessed an exponential growth in the area of this mode of activation. Perhaps more remarkably, the mode of enamine activation has resulted in the development of bifunctional organocatalysis, which has now been used in many synthetic processes.

The mode of iminium activation is based on the reversible formation of iminium ions from α,β-unsaturated aldehydes and secondary amines in the presence of acids (Scheme C4). The first iminium-catalyzed conjugate addition reaction was reported in 1937 by Langenbeck and Sauerbier [C26]. Moreover, it was found that iminium salts were able to provide significant acceleration in the Diels–Alder reaction [C45]. Despite these advances, it was not until 2000 that MacMillan's group designed a more general catalysis strategy for the Diels–Alder reaction [C3]. More importantly, MacMillan and co-workers were the first to present the iminium catalysis concept by using the term lowest unoccupied molecular orbital (LUMO)-lowering catalysis to describe the activation mode common to both Lewis acids and chiral amine catalysts (Scheme C4). The generality of this concept set in motion the discovery of more than 50 highly enantioselective iminium-catalyzed processes.

In addition to LUMO-lowering activation, MacMillan and co-workers introduced an activation mode of singly occupied molecular orbital (SOMO) catalysis, namely SOMO activation (Scheme C5), in 2007 [C46]. This one-electron mode of activation has permitted the development of several useful transformations [C47]. With the success of organo-SOMO catalysis, MacMillan's group reported the merging of photoredox catalysis with organocatalysis (Scheme C6) in the pursuit of enantioselective α-alkylation, trifluoromethylation, and benzylation of aldehydes [C48]. Although SOMO activation is one of the most recently developed activation modes, the application of

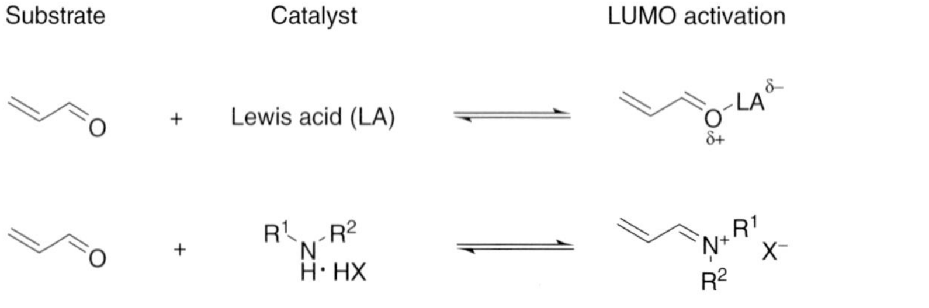

Substrate		Catalyst		LUMO activation

Scheme C4 Iminium activation through LUMO lowering [C3].

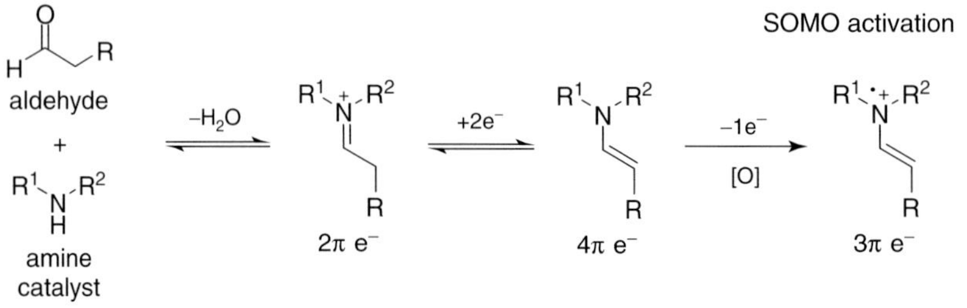

Scheme C5 SOMO activation through a 3p intermediate [C46].

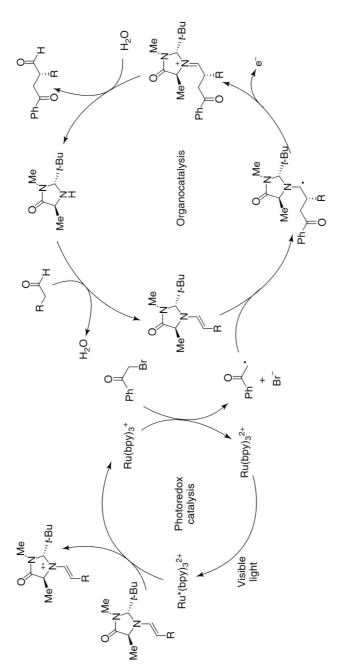

Scheme C6 Merging photoredox catalysis with organocatalysis: enantioselective alkylation of aldehydes [C46a].

this principle will build up new protocols for organocatalytic chemical transformations. Hydrogen bonding catalysis is another important strategy in the area of organocatalysis. Enantioselective synthesis with small-molecule chiral hydrogen bond donors emerged as a frontier of research in the 1980s. In pioneering studies, Hine *et al.* reported that the hydrogen-bonding donors, such as *meta-* and *para-*substituted phenols and biphenylenediols, could be catalysts for the addition reaction of diethylamine to phenyl glycidyl ether [C49]. Subsequent studies by Sigman and Jacobsen [C32a] and Corey and Grogen [C32b] elegantly demonstrated that well-defined hydrogen-bonding interactions could be a general activation mode that permits sufficient activation and direction in enantioselective organocatalysis. This powerful strategy has become the basis of a remarkable number of new enantioselective reactions, providing solutions to challenging transformations of importance to asymmetric synthesis.

Without question, enantioselective organocatalysis has become the third methodology of equal status to organometallic and enzymatic catalysis. A variety of useful asymmetric carbon–carbon and carbon–heteroatom bond-forming reactions have been achieved through enantioselective organocatalysis. The area of organocatalysis will continue to expand steadily with the development of novel catalytic modes and reactions.

References

C1. (a) Eder, U., Sauer, G., and Wiechert, R. (1971) *Angew. Chem. Int. Ed. Engl.*, **10**, 496; (b) Hajos, Z.G. and Parrish, D.R. (1974) *J. Org. Chem.*, **39**, 1615.

C2. (a) List, B., Lerner, R.A., and Barbas, C.F. III (2000) *J. Am. Chem. Soc.*, **122**, 2395; (b) List, B. (2000) *J. Am. Chem. Soc.*, **122**, 9336.

C3. Ahrendt, K.A., Borths, C.J., and MacMillan, D.W.C. (2000) *J. Am. Chem. Soc.*, **122**, 4243.

C4. Reviews: (a) Kano, T. and Maruoka, K. (2008) *Chem. Commun.*, 5465; (b) Kano, T. and Maruoka, K. (2010) *Bull. Chem. Soc. Jpn.*, **83**, 1421.

C5. (a) Kano, T., Yamaguchi, Y., Tokuda, O., and Maruoka, K. (2005) *J. Am. Chem. Soc.*, **127**, 16408; (b) Kano, T., Yamaguchi, Y., and Maruoka, K. (2009)

Angew. Chem. Int. Ed., **48**, 1838; (c) Kano, T., Yamaguchi, Y., and Maruoka, K. (2009) *Chem. Eur. J.*, **15**, 6678.

C6. (a) Mitsumori, S., Zhang, H., Cheong, P.H.-Y., Houk, K.N., Tanaka, F., and Barbas, C.F. III (2006) *J. Am. Chem. Soc.*, **128**, 1040; (b) Zhang, H., Mitsumori, S., Utsumi, N., Imai, M., Garcia-Delgado, N., Mifsud, M., Albertshofer, K., Cheong, P.H.-Y., Houk, K.N., Tanaka, F., and Barbas, C.F. III (2008) *J. Am. Chem. Soc.*, **130**, 875.

C7. Reviews: (a) Kacprzak, K. and Gawroński, J. (2001) *Synthesis*, 961; (b) Tian, S.-K., Chen, Y., Hang, J., Tang, L., McDaid, P., and Deng, L. (2004) *Acc. Chem. Res.*, **37**, 621; (c) Chen, Y., McDaid, P., and Deng, L. (2003) *Chem. Rev.*, **103**, 2965; (d) Gaunt, M.J. and Johansson, C.C.C. (2007) *Chem. Rev.*, **107**, 5596.

C8. Reviews: (a) O'Donnell, M.J. (2004) *Acc. Chem. Res.*, **37**, 506; (b) Lygo, B. and Andrews, B.I. (2004) *Acc. Chem. Res.*, **37**, 518; (c) Maruoka, K. (2008) *Asymmetric Phase Transfer Catalysis*, Wiley-VCH Verlag GmbH, Weinheim; (d) Jew, S.-S. and Park, H.-G. (2009) *Chem. Commun.*, 7090.

C9. Reviews: (a) Hashimoto, T. and Maruoka, K. (2007) *Chem. Rev.*, **107**, 5656; (b) Ooi, T. and Maruoka, K. (2007) *Angew. Chem. Int. Ed.*, **46**, 4222; (c) Maruoka, K., Ooi, T., and Kano, T. (2007) *Chem. Commun.*, 1487; (d) Ooi, T. and Maruoka, K. (2007) *Aldrichim. Acta*, **40**, 77; (e) Maruoka, K. (2008) *Org. Process. Res. Dev.*, **12**, 679; (f) Maruoka, K. (2010) *Chem. Rec.*, **10**, 254.

C10. (a) Ooi, T., Kameda, M., and Maruoka, K. (1999) *J. Am. Chem. Soc.*, **121**, 6519; (b) Ooi, T., Kameda, M., and Maruoka, K. (2003) *J. Am. Chem. Soc.*, **125**, 5139.

C11. (a) Kitamura, M., Shirakawa, S., and Maruoka, K. (2005) *Angew. Chem. Int. Ed.*, **44**, 1549; (b) Wang, X., Kitamura, M., and Maruoka, K. (2007) *J. Am. Chem. Soc.*, **129**, 1038; (c) Kitamura, M., Shirakawa, S., Arimura, Y., Wang, X., and Maruoka, K. (2008) *Chem. Asian J.*, **3**, 1702.

C12. (a) He, R., Shirakawa, S., and Maruoka, K. (2009) *J. Am. Chem. Soc.*, **131**, 16620; (b) Wang, L., Shirakawa, S., and Maruoka, K. (2011) *Angew. Chem. Int. Ed.*, **50**, 5327.

C13. Reviews: (a) Akiyama, T. (2007) *Chem. Rev.*, **107**, 5744; (b) Terada, M. (2008)

Chem. Commun., 4097; (c) Terada, M. (2010) *Synthesis*, 1929.

C14. Reviews: (a) Takemoto, Y. (2005) *Org. Biomol. Chem.*, **3**, 4299; (b) Connon, S.J. (2006) *Chem. Eur. J.*, **12**, 5418; (c) Doyle, A.G. and Jacobsen, E.N. (2007) *Chem. Rev.*, **107**, 5713.

C15. (a) Hashimoto, T. and Maruoka, K. (2007) *J. Am. Chem. Soc.*, **129**, 10054; (b) Hashimoto, T., Hirose, M., and Maruoka, K. (2008) *J. Am. Chem. Soc.*, **130**, 7556; (c) Hashimoto, T., Uchiyama, N., and Maruoka, K. (2008) *J. Am. Chem. Soc.*, **130**, 14380; (d) Hashimoto, T., Kimura, H., and Maruoka, K. (2010) *Angew. Chem. Int. Ed.*, **49**, 6844; (e) Hashimoto, T., Omote, M., and Maruoka, K. (2011) *Angew. Chem. Int. Ed.*, **50**, 3489.

C16. Bredig, G. and Fiske, W.S. (1912) *Biochem. Z.*, **46**, 7.

C17. (a) Eder, U., Sauer, G., and Weichert, R. (1971) *Angew. Chem. Int. Ed. Engl.*, **10**, 496; (b) Hajos, J. and Parrish, D. (1974) *J. Org. Chem.*, **39**, 1615.

C18. Hiemstra, H. and Wynberg, H. (1981) *J. Am. Chem. Soc.*, **103**, 417.

C19. Oku, J.-I. and Inoue, S. (1981) *J. Chem. Soc., Chem. Commun.*, 229.

C20. (a) Zhang, C. and Lu, X. (1995) *J. Org. Chem.*, **60**, 2906–2908; (b) Lu, X., Zhang, C., and Xu, Z. (2001) *Acc. Chem. Res.*, **34**, 535.

C21. Zhu, G., Chen, Z., Jiang, Q., Xiao, D., Cao, P., and Zhang, X. (1997) *J. Am. Chem. Soc.*, **119**, 3836.

C22. (a) Marinetti, A. and Voituriez, A. (2010) *Synlett*, 174; (b) Cowen, B.J. and Miller, S.J. (2009) *Chem. Soc. Rev.*, **38**, 3102; (c) Ye, L.-W., Zhou, J., and Tang, Y. (2008) *Chem. Soc. Rev.*, **37**, 1140.

C23. (a) Yang, D., Yip, Y.C., Tang, M.-W., Wong, M.-K., Zheng, J.-H., and Cheung, K.-K. (1996) *J. Am. Chem. Soc.*, **118**, 491; (b) Tu, Y., Wang, Z.-X., and Shi, Y. (1996) *J. Am. Chem. Soc.*, **118**, 9806.

C24. List, B., Pojarliev, P., and Martin, H. (2001) *J. Org. Lett.*, **3**, 2423.

C25. Mukherjee, S., Yang, J.W., Hoffmann, S., and List, B. (2007) *Chem. Rev.*, **107**, 5471.

C26. Langenbeck, W. and Sauerbier, R. (1937) *Chem. Ber.*, **70**, 1540.

C27. Yamaguchi, M., Yokota, N., and Minami, T. (1991) *J. Chem. Soc., Chem. Commun.*, 1088.

C28. Review: Erkkilä, A., Majander, I., and Pihko, P.M. (2007) *Chem. Rev.*, **107**, 5416.

C29. Huang, Y., Walji, A.M., Larsen, C.H., and MacMillan, D.W.C. (2005) *J. Am. Chem. Soc.*, **127**, 15051.

C30. Enders, D., Hüttl, M.R.M., Grondal, C., and Raabe, G. (2006) *Nature*, **441**, 861.

C31. Enders, D., Grondal, C., and Hüttl, M.R.M. (2007) *Angew. Chem. Int. Ed.*, **46**, 1570.

C32. (a) Sigman, M. and Jacobsen, E.N. (1998) *J. Am. Chem. Soc.*, **120**, 4901; (b) Corey, E.J. and Grogan, M. (1999) *Org. Lett.*, **1**, 157.

C33. Huang, Y., Unni, A.K., Thadani, A.N., and Rawal, V.H. (2003) *Nature*, **424**, 146.

C34. Akiyama, T., Itoh, J., Yokota, K., and Fuchibe, K. (2004) *Angew. Chem. Int. Ed.*, **43**, 1566.

C35. Uraguchi, D. and Terada, M. (2004) *J. Am. Chem. Soc.*, **126**, 5356.

C36. Akiyama, T. (2007) *Chem. Rev.*, **107**, 5744.

C37. Okino, T., Hoashi, Y., and Takemoto, Y. (2003) *J. Am. Chem. Soc.*, **125**, 12672.

C38. Doyle, A.G. and Jacobsen, E.N. (2007) *Chem. Rev.*, **107**, 5713.

C39. (a) Kang, Y.K., Kim, S.M., and Kim, D.Y. (2010) *J. Am. Chem. Soc.*, **132**, 11847–11849; (b) Akiyama, T. *et al.* (2011) *J. Am. Chem. Soc.*, **133**, 6166.

C40. (a) Shao, Z. and Zhang, H. (2009) *Chem. Soc. Rev.*, **38**, 2745–2755; (b) Zhou, J. (2010) *Chem. Asian J.*, **5**, 422.

C41. (a) Berkessel, A. and Groger, H. (2005) *Asymmetric Organocatalysis – from Biomimetic Concepts to Powerful Methods for Asymmetric Synthesis*, Wiley-VCH Verlag GmbH, Weinheim; (b) Dalko, P.I. (2007) *Enantioselective Organocatalysis*, Wiley-VCH Verlag GmbH, Weinheim.

C42. Bredig, G. and Fiske, W.S. (1912) *Biochem. Z.*, **46**, 7.

C43. (a) Pracejus, H. (1960) *Justus Liebegs Ann. Chem.*, **634**, 9; (b) Pracejus, H. (1960) *Justus Liebegs Ann. Chem.*, **634**, 23.

C44. (a) Hajos, Z.G. and Parrish, D.R. (1971) German Patent DE 2102623; (b) Hajos, Z.G. and Parrish, D.R. (1974) *J. Org. Chem.*, **39**, 1615; (c) Eder, U., Sauer, G., and Wiechert, R. (1971) *Angew. Chem. Int. Ed. Engl.*, **10**, 496.

C45. Baum, J.S. and Viehe, H.G. (1976) *J. Org. Chem.*, **41**, 183.

C46. (a) Beeson, T.D., Mastracchio, A., Hong, J.-B., Ashton, K., and MacMillan, D.W.C. (2007) *Science*, **316**, 582; (b) Um, J.M., Gutierrez, O., Schoenebeck, F., Houk, K.N., and MacMillan, D.W.C. (2010) *J. Am. Chem. Soc.*, **132**, 6001.

C47. (a) Jang, H.-Y., Hong, J.-B., and MacMillan, D.W.C. (2007) *J. Am. Chem. Soc.*, **129**, 7004; (b) Kim, H. and MacMillan, D.W.C. (2008) *J. Am. Chem. Soc.*, **130**, 398; (c) Graham, T.H., Jones, C.M., Jui, N.T., and MacMillan, D.W.C. (2008) *J. Am. Chem. Soc.*, **130**, 16494; (d) Conrad, J.C., Kong, J., Laforteza, B.N., and MacMillan, D.W.C. (2009) *J. Am. Chem. Soc.*, **131**, 11640; (e) Wilson, J.E., Casarez, A.D., and MacMillan, D.W.C. (2009) *J. Am. Chem. Soc.*, **131**, 11332; (f) Amatore, M., Beeson, T.D., Brown, S.P., and MacMillan, D.W.C. (2009) *Angew. Chem. Int. Ed.*, **48**, 5121; (g) Jui, N.T., Lee, E.C.Y., and MacMillan, D.W.C. (2010) *J. Am. Chem. Soc.*, **132**, 10015; (h) Rendler, S. and MacMillan, D.W.C. (2010) *J. Am. Chem. Soc.*, **132**, 5027; (i) Mastracchio, A., Warkentin, A.A., Walji, A.M., and MacMillan, D.W.C. (2010) *Proc. Natl. Acad. Sci. U. S. A.*, **107**, 20648.

C48. (a) Nicewicz, D.A. and MacMillan, D.W.C. (2008) *Science*, **322**, 77; (b) Nagib, D.A., Scott, M.E., and MacMillan, D.W.C. (2009) *J. Am. Chem. Soc.*, **131**, 10875; (c) Shih, H.-W., Wal, M.N.V., Grange, R.L., and MacMillan, D.W.C. (2010) *J. Am. Chem. Soc.*, **132**, 13600.

C49. Hine, J., Linden, S.-M., and Kanagasabapathy, V.M. (1985) *J. Am. Chem. Soc.*, **107**, 1082.

11
Recent Topics in Cooperative Catalysis: Asymmetric Catalysis, Polymerization, Hydrogen Activation, and Water Splitting

Motomu Kanai

11.1
Introduction

The field of catalysis requires continuous progress in the twenty-first century because it is the only rational means of producing useful molecules, such as drugs and new materials, in an economical and environmentally benign way on a worldwide scale [1]. Catalysts are also required for the production of clean energy. The development of new catalysts that exhibit groundbreaking efficiency in terms of turnover number, frequency, and selectivity without harming the Earth is the major goal in this field. Exploring conceptually new catalytic substrate activation methods has been mainstream in the history of organic chemistry. Hydrogenation, oxidation, alkene polymerization, cross-coupling reactions, alkene metathesis, and, more recently, C–H functionalization have been developed in this area, and organic chemists must continue to pursue the development of other useful catalytic methodologies. Another very important line of pursuit is the systematic combination of known catalytic activation modes (e.g., activation of electrophiles by a Lewis acid, activation of nucleophiles by a Brønsted base), which leads to qualitative leaps compared with the use of specific individual activation modes. Indeed, this is the approach used in Nature. Natural catalysts, enzymes, comprise as few as 20 natural amino acids without any strong acidity or basicity and yet can generate an extremely diverse structural range of organic molecules due to the systematic cooperation of functional groups.

Two well-known enzymatic reactions, hydrolysis of amide bonds by serine proteases (Scheme 11.1a) [2] and aldol reaction by class II aldolases (Scheme 11.1b) [3], are exemplified here. Amide bonds are generally very stable under neutral conditions at ambient temperature. Serine proteases, such as trypsin and chymotrypsin, smoothly hydrolyze peptides and proteins at a specific amide bond (the C-terminal side of an aromatic-containing amino acid residue) depending on the peptide sequence. The enzymatic hydrolysis begins with nucleophilic attack of the serine-195 hydroxy group of the enzyme to an amide carbonyl carbon of substrate. The carboxylate side chain of asparagine-102 of the enzyme works as a general base, and the anionic charge is relayed to the nucleophilic serine-195 hydroxy group through the imidazole side chain of histidine-57. Those three key functional groups exist in spatial approximation with

Organic Chemistry – Breakthroughs and Perspectives, First Edition. Edited by Kuiling Ding and Li-Xin Dai.
© 2012 Wiley-VCH Verlag GmbH & Co. KGaA. Published 2012 by Wiley-VCH Verlag GmbH & Co. KGaA.

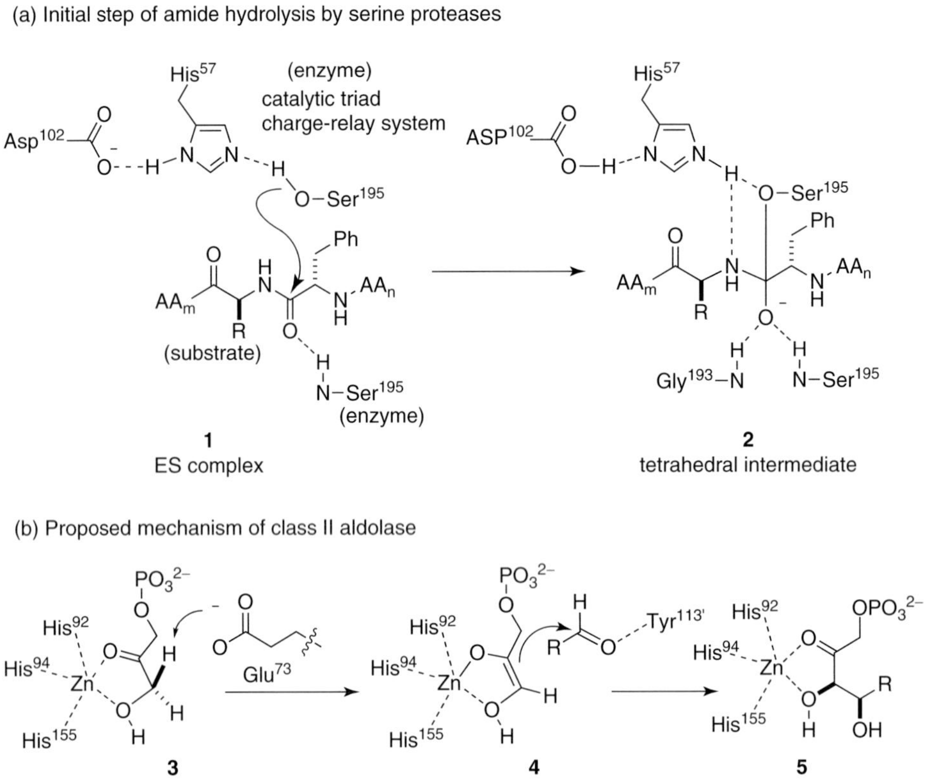

(a) Initial step of amide hydrolysis by serine proteases

1
ES complex

2
tetrahedral intermediate

(b) Proposed mechanism of class II aldolase

3   **4**   **5**

Scheme 11.1   Cooperative catalysis in Nature (enzymes).

each other in the three-dimensionally folded protein structure. This charge relay system of the catalytic triad enhances the nucleophilicity of the hydroxy group of serine-195 via deprotonation (Scheme 11.1a; **1**). Simultaneously, the electrophilicity of the amide carbonyl carbon is enhanced through coordination to the NH proton of the serine-195 backbone acting as a Brønsted acid. Moreover, the anionic charge on the carbonyl oxygen atom that emerges during the hydrolysis (or alcoholysis) is neutralized by the oxyanion hole comprising two NH groups of serine-195 and glycine-193. Tetrahedral intermediate **2** is stabilized by these multiple interactions with the enzyme, leading to the great reaction rate enhancement of amide hydrolysis.

Another seminal example is class II zinc-dependent aldolases that promote the aldol reaction between dihydroxyacetone phosphate (DHAP) and various aldehydes [3]. Based on spectroscopic studies including X-ray crystallography in the presence of inhibitors, the mechanism depicted in Scheme 11.1b was proposed. Chelate coordination of DHAP to a zinc metal supported by three histidine side-chain imidazole groups of the enzyme acidifies the α-proton of the ketone carbonyl group (**3**). The side-chain carboxylate of glutamate-73 deprotonates the *pro-R* α-proton of the activated ketone to generate a zinc enolate species (**4**). The thus-generated zinc enolate reacts with an aldehyde activated by the tyrosine phenolic proton acting as a Brønsted base, affording aldol products **5** with high enantioselectivity. A cooperative mechanism is comprehensively observed in other enzymatic and catalytic antibody-promoted reactions.

This chapter concerns recent topics in artificial cooperative catalysis, focusing on studies achieved in the last decade. Owing to the space limitations, this review is not intended to be comprehensive. Instead, three topics, asymmetric reactions, polymerization, and hydrogen activation/generation, have been selected. They were selected because the concept of cooperative catalysis is an important contributing factor to their progress. Each topic deals with a very limited number of papers. For more details on each topic, readers should refer the original papers cited herein.

11.2
Cooperative Catalysis in Asymmetric Reactions

11.2.1
On the Shoulder of Giants in the Twentieth Century

This section briefly summarizes the pioneering studies that demonstrated the power of cooperativity in designing asymmetric catalysts [4]. In 1986, Ito and co-workers reported that cationic gold(I) complexes prepared from chiral ferrocenylphosphine ligands bearing a tertiary amino group at the terminal position of a pendant chain are effective catalysts for asymmetric aldol reactions of isocyanoacetates and various aldehydes (see Figure 11.1, **6**) [5]. The [substrate]:[catalyst] ratio could be as high as 10 000 : 1 without loss of enantioselectivity (>90% *ee*) in ideal substrate combinations. Based on structure–function (activity and enantioselectivity) relationships using chiral catalysts containing varying linker lengths between the metal and pendant amine, the authors proposed a cooperative transition-state model (**6**), in which the amine functionality works as a Brønsted base to deprotonate the isocyanoacetate and fix the position of the resulting enolate with ion pair formation, while the electrophile (aldehyde) is activated by coordination to the cationic gold metal acting as a Lewis acid.

Noyori and co-workers reported a general enantioselective alkylation of aldehydes using a (−)-3-*exo*-(dimethylamino)isoborneol (DAIB) catalyst and dialkylzinc [6]. Detailed mechanistic studies revealed that this reaction proceeds via a dinuclear zinc species **7** containing the DAIB auxiliary, an aldehyde ligand, and three alkyl groups. The chelated zinc metal works as a Lewis acid to activate aldehydes, while the alkoxide oxygen atom coordinates to dialkylzinc, thus enhancing its nucleophilicity. This model is generally applicable to explain the absolute configuration of the products obtained in numerous related examples [7]. Corey and co-workers reported that chiral oxazaborolidines containing a prolinol skeleton [CBS (Corey–Bakshi–Shibata) catalyst] are extremely general catalysts for the enantioselective reduction of ketones [8]. The reaction proceeds through a dinuclear boron species (**8**), in which the chelated boron atom acts as a Lewis acid to activate ketones while the nitrogen atom of oxazaborolidine acts as a Lewis base to activate borane as a hydride source. Clever electronic communication between two boron atoms exists in dinuclear **8**; coordination of the nitrogen atom to

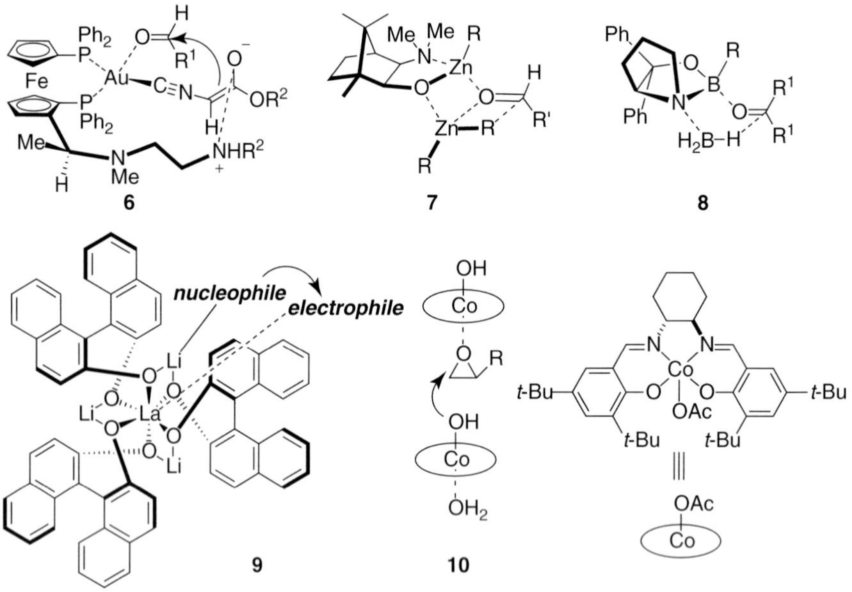

Figure 11.1 Seminal artificial cooperative asymmetric catalysts.

the nucleophilic borane enhances the Lewis acidity of the chelated boron that activates ketones.

Among the most successful bifunctional asymmetric catalysts is Shibasaki and co-workers' lanthanum–lithium–binaphthoxide (LLB) complex, with broad applicability [9]. LLB and its related complexes promote many fundamental C–C, C–S, C–O, and C–N bond-forming reactions, including the nitroaldol reaction, Michael reaction, and direct aldol reaction, to name just a few, with excellent enantioselectivity and broad substrate generality. The lithium naphthoxide moiety acts as a Brønsted base for the generation of nucleophilic species via deprotonation and the central metal lanthanum atom acts as a Lewis acid to activate electrophiles (**9**: Lewis acid–Brønsted base bifunctional asymmetric catalysis). The basic concept has been greatly expanded to various kinds of bifunctional asymmetric catalysts including Lewis acid–Lewis base and Lewis acid–Lewis acid catalysts [9c,f].

Jacobsen and co-workers' cobalt(salen)-catalyzed hydrolytic kinetic resolution of terminal epoxides is another prominent example of cooperative asymmetric catalysis [10]. In ideal cases, terminal epoxides with >99% *ee* can be obtained in yields approaching the theoretical maximum of 50% from racemic starting materials. The reaction exhibits second-order kinetics with respect to the catalyst concentration, indicating simultaneous activation of both the epoxide and water by different monomeric catalysts (**10**). The mechanistic picture led the authors to develop more effective catalyst systems by incorporating a more Lewis acidic component via counterion tuning [10c] or dimerizing the cobalt(salen) component with appropriate linkers [10d,f].

In addition to the above-mentioned cooperative asymmetric catalysts developed in the early stages, there have been many other important contributions [4], including Trost and co-workers' chiral dinuclear zinc complex [11] and asymmetric organocatalysts [12]. The following two subsections briefly describe two recent breakthroughs in the field of asymmetric cooperative catalysis.

11.2.2
Catalyst Higher Order Structure as a Determinant of Function: Catalytic Enantioselective Strecker Reaction of Ketimines by Poly-Rare Earth Metal Complexes

In 2001, Shibasaki's group developed a catalytic enantioselective cyanosilylation of ketones using a gadolinium complex produced from $Gd(O^iPr)_3$ and D-glucose-derived ligand **11** mixed in a 1:2 ratio [Scheme 11.2, Eq. (11.1)] [13]. The related Gd catalyst derived from **12** also promoted a very general enantioselective Strecker reaction of ketimines in the presence of a protic additive [Eq. (11.2)] [14, 15]. Intensive mechanistic studies, including determination of the catalyst composition using electrospray ionization mass spectrometry (ESI-MS), labeling experiments, and kinetic measurements, led to the proposal of a cooperative model (**15**) for the enantio-differentiating cyanation step of the Strecker reaction. Four main issues involved in **15** are (i) Gd:ligand = 2 : 3

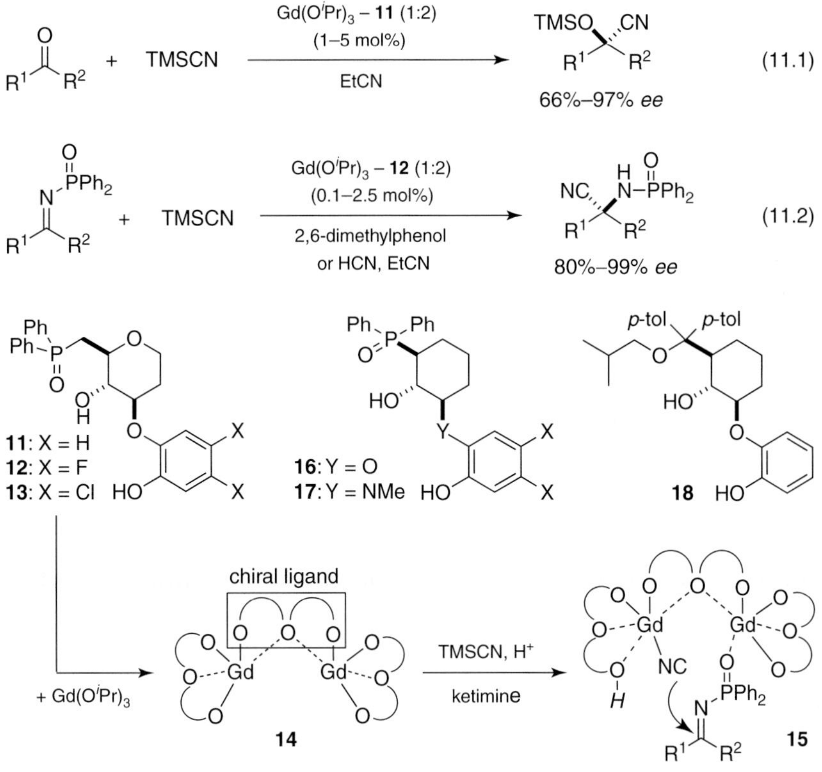

Scheme 11.2 Poly-rare earth metal-catalyzed enantioselective tetrasubstituted carbon-forming reactions.

Table 11.1 Higher order structure of an asymmetric catalyst as a determinant of function: a catalytic enantioselective Strecker reaction of ketimines.

R	2:3 complex 14 (lig. = 13)	4:5+ oxo complex (lig. = 13)
Ph	88% *ee* (S) (0.5 h/Gd = 5 mol%)	96% *ee* (R) (1 h/Gd = 7 mol%)
3-Thiophene	98% *ee* (S) (0.5 h/Gd = 5 mol%)	91% *ee* (R) (2 h/Gd = 7 mol%)
PhCH$_2$CH$_2$	87% *ee* (S) (2 h/Gd = 5 mol%)	87% *ee* (R) (2 h/Gd = 7 mol%)
i-Pr	80% *ee* (S) (2.5 h/Gd = 2.5 mol%)[a]	82% *ee* (R) (12 h/Gd = 7 mol%)
t-Bu	74% *ee* (S) (2 h/Gd = 10 mol%)	98% *ee* (R) (14 h/Gd = 7 mol%)

[a] Ligand 12 was used.

complex to be the active enantioselective catalyst; (ii) a gadolinium isocyanide species generated through transmetallation from trimethylsilyl cyanide (TMSCN) to be the active nucleophile; (iii) the intramolecular transfer of cyanide from a Gd metal to an activated ketimine coordinating to the Lewis acidic Gd; and (iv) the phenolic proton incorporated in the complex (italicized) by a protic additive to act as a catalyst turnover accelerator. In **15**, the positions of both the activated nucleophile and the electrophile are determined by the asymmetric bimetallic complex, thereby affording high enantioselectivity.

To elucidate the three-dimensional enantio-differentiation mechanism of the Gd catalyst, crystallization of the catalytic species was attempted by varying the ligand structure and metal. Colorless, air-stable prisms (**16**: 80% yield) were obtained from a propionitrile–hexane (2:1) solution of the complex prepared from a 2:3 ratio of Gd(O^{i}Pr)$_3$ and ligand **13**. X-ray crystallographic analysis revealed that **16** was a 4:5 complex of Gd and **13** with a μ-oxo atom surrounded by four Gd atoms (Table 11.1) [16]. The tetranuclear structure was maintained in a solution state, and the sole peak observed by ESI-MS on dissolving the crystals in an organic solvent corresponded to the 4:5 complex.

Obviously, the crystals obtained were not the actual enantioselective catalyst (2:3 complex). Even so, the function of the 4:5 complex as an enantioselective catalyst was evaluated by a Strecker reaction of ketimines. Surprisingly, enantioselectivity was completely reversed to excellent levels when the 4 : 5 complex crystals were used as a catalyst, compared with the catalyst prepared *in situ* (Table 11.1) [16]. This dramatic reversal in enantioselectivity was attributed to the change in the assembly state (from 2 : 3 to 4 : 5) of the modules that constitute the chiral polymetallic catalyst. The reaction rate was ~5–50 times slower than that using the catalyst prepared *in situ*. This study clearly demonstrated a new concept in asymmetric catalysis, namely that the higher order assembly structure, not the structure of the chiral ligand, is the determining factor for the function of the asymmetric polymetallic catalyst. Based on this concept, the authors designed new chiral ligands **16–18**, which exhibit excellent functions in several synthetically useful reactions [17–19].

11.2.3
Cooperative Asymmetric Catalysis Involving the Anion Binding Concept

Jacobsen's group reported chiral thiourea-catalyzed enantioselective nucleophilic addition reactions to *N*-acyliminium species, such as the *N*-acyl-Pictet–Spengler reaction [20] and the Mannich-type reaction of *N*-acylisoquinoliniums [21]. Chiral thiourea catalysts were initially intended to function as a hydrogen bond donor to heteroatoms in electrophiles (such as the nitrogen atom of imine substrates) [22]. The interaction mode of simple hydrogen bond donor thiourea catalysts with the cationic enantioselectivity-determining transition states containing weakly hydrogen bond-accepting ability in the above two reactions was not apparent, however. This situation was more recently clarified by Jacobsen and co-workers' proposal in which the chiral thiourea catalyst binds with the anion part of iminium intermediates, thus mediating enantioselective addition to the iminium cation through chiral ion pair formation [23]. The anion binding by the catalyst would also enhance the cationic characteristics of the iminium intermediate by weakening the cation–anion interaction, thereby facilitating the nucleophilic attack.

This anion binding ability of the thiourea functionality was recently incorporated into cooperative asymmetric catalysts as a key function. Catalyst **19** containing thiourea and pyrene moieties produced excellent enantioselectivity in the polycyclization of alkene substrates **20** [Scheme 11.3, Eq. (11.3)] [24, 25]. The mechanism of this interesting reaction was proposed as involving **21**. Substrate hydroxylactam **20** was first converted to the corresponding chlorolactam intermediate in the presence of a catalytic amount of HCl added to the reaction. Hydrogen bond-mediated ionization of the chlorolactam intermediate by the chiral thiourea catalyst generated a catalyst-bound iminium chloride ion pair **21**, which underwent enantioselective cyclization. The pyrene functionality with the polarizable π-electron system of catalyst **19** should stabilize the iminium intermediate through cation–π interaction. Both catalyst activity and enantioselectivity improved with increase in the size of the π system (phenyl < naphthyl < phenanthryl < pyrenyl). The key role of stabilization of cationic intermediates in polyene cyclization by arene π-electrons of protein side chains in enzymatic reactions was proposed previously [26].

Another recent example of cooperative asymmetric catalysis including anion binding of thiourea is the S_N1 α-alkylation of aldehydes with benzhydryl bromides [Scheme 11.3, Eq. (11.4)] [27, 28]. The benzhydryl bromides were ionized and electrophilically activated via abstraction of bromide by the thiourea functionality of the catalyst. The position

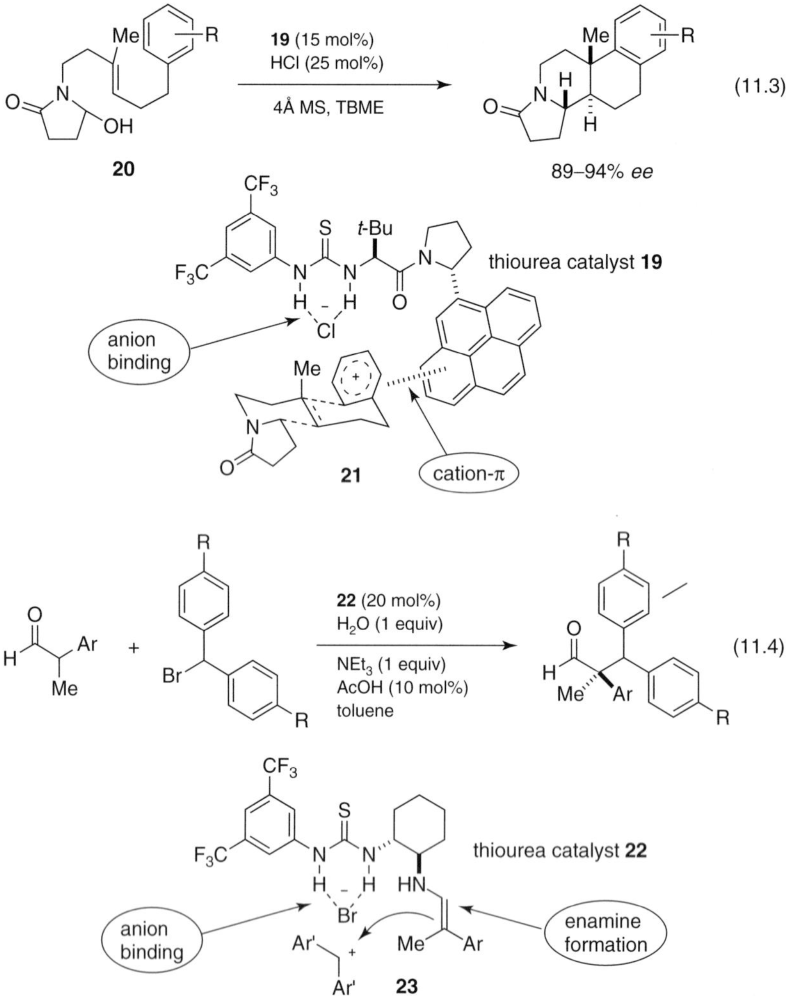

Scheme 11.3 Bifunctional chiral thiourea-catalyzed reactions involving electrophile activation via cation binding.

of the resulting benzhydryl cation was fixed in approximation to the thiourea moiety with ion-pair formation (**23**). Meanwhile, the aldehydes were nucleophilically activated through the formation of enamines with the primary amine functionality of the catalyst. The presence of water and catalytic acetic acid accelerated the reaction, presumably by influencing the imine and enamine formation and imine hydrolysis steps. Stoichiometric triethylamine neutralized HBr generated by the substitution reaction. Interestingly, alkylations using enantio-enriched *p*-chlorobenzhydryl chloride were found to proceed with nearly complete (95%) stereospecificity even though the S_N1 mechanism was operating, which requires that intramolecular addition of the catalyst-associated enamine to the ion-pair intermediate is rapid relative to ion-pair reorganization.

11.3
Cooperative Catalysis in Alkene Polymerization

The introduction of well-defined, single-site organotransition metal alkene polymerization catalysts in the early 1980s highlighted the possibility of controlling and dramatically improving the properties of polymer products based on the catalyst's structural tuning [29]. One of the challenges in the current alkene polymerization research relies on the precise control of regio- and stereoregularities, molecular weights and molecular weight distributions, and comonomer incorporation, which affects polymer properties such as crystallization behavior and mechanical properties. The concept of cooperative catalysis provides a unique platform to achieve this goal.

In 2002, Marks' group reported that the branch content of polyalkene products was dramatically increased in ethylene homopolymerization when bimetallic zirconium catalyst **26** was used compared with monometallic catalyst **24** (Figure 11.2) [30]. Increased effective concentrations of local active sites and bimetallic cooperative effects in **26** were proposed to account for the differences in the microstructures of the product polymers; an eliminated alkene-terminated oligomer fragment might have an enhanced probability of being captured and enchained by a proximate active center of **26** before diffusing away. Furthermore, binuclear zirconium–titanium heterobimetallic catalyst **28** generated by tight ion pairing with a binuclear borate co-catalyst was found to be a markedly superior catalyst to the corresponding monometallic catalysts for the synthesis of linear low-density polyethylene [31]. These studies demonstrate that catalyst/co-catalyst nuclearity significantly affects catalyst activity and linear/branch selectivity in alkene polymerization [32].

Ethylene–styrene copolymers have received great attention due to their impressive viscoelastic behavior, mechanical properties, and compatibilities with other polymeric materials. Copolymerization of ethylene and styrene is challenging [33], however, in part due to the deactivation of the catalyst by arene back-coordination in the styrene 2,1-insertion product. Marks' group reported that bimetallic titanium catalyst **27** affords broad-range controllable styrene incorporation in ethylene–styrene copolymerizations [34]. Moreover, catalyst **29** (Scheme 11.4), containing a C1 linker and thus exhibiting

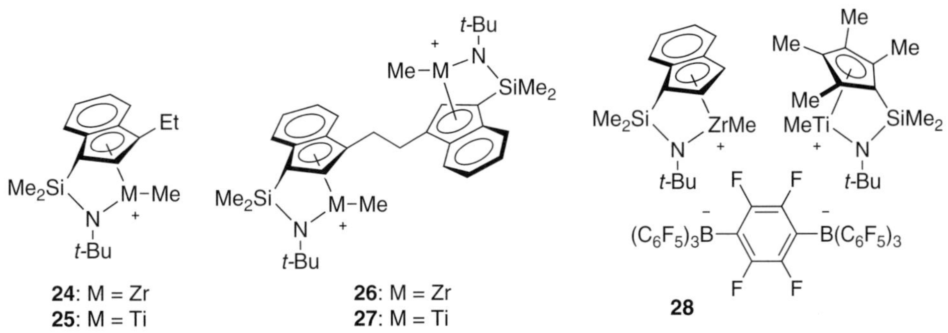

24: M = Zr
25: M = Ti

26: M = Zr
27: M = Ti

28

Figure 11.2 Alkene polymerization catalysts.

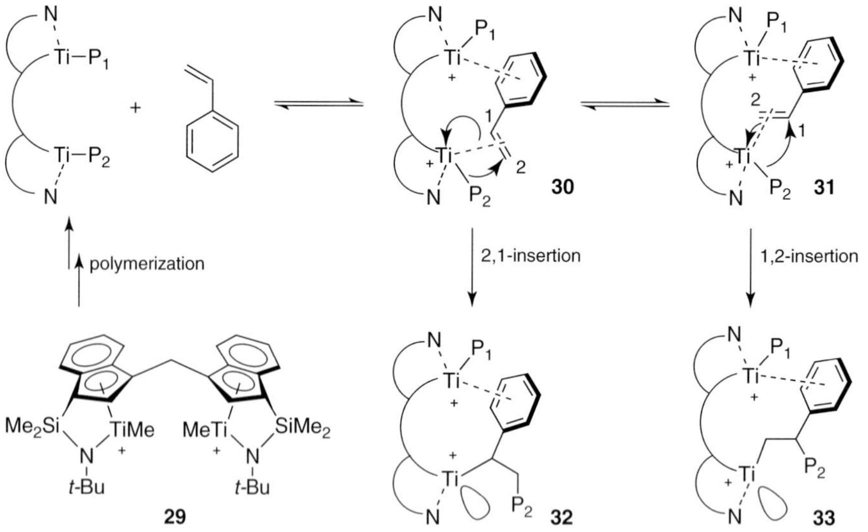

Scheme 11.4 Bimetallic titanium catalyst for efficient copolymerization of ethylene and styrene.

enhanced cooperativity between the two metal centers, produced further improved catalyst activity, and also higher styrene enchainment ability, than **27**. The concentration of coordinating styrene comonomers to the catalyst is increased by interactions between the π-system of the phenyl ring of styrene and the proximate cationic metal center (**30** and **31**). This may enhance the subsequent enchainment probability without poisoning the catalytic center by back-coordination (**32** and **33**). Marks and co-workers' finding indicates that binuclear alkene polymerization catalysts can effect unusual cooperative enchainment processes when comonomers with additional coordination sites are used. Therefore, polynuclear alkene polymerization catalysts offer potential for creating new macromolecular architectures, which conventional monometallic catalysts cannot produce [35].

11.4
Cooperative Catalysis in Hydrogen Activation/Generation

11.4.1
Ligand–Metal Cooperation

In the mid-1990s, Noyori's group discovered a new chemoselective catalytic hydrogenation for the reduction of aldehydes and ketones in the presence of carbon–carbon multiple bonds using combination of a ruthenium–phosphine complex and an ethylenediamine derivative [36]. This finding led to the identification of a catalytic enantioselective transfer hydrogenation of ketones using catalyst **34** [37]. Detailed mechanistic studies revealed that **34** is a precatalyst (Scheme 11.5) [38]; the ruthenium amide works as a Brønsted base to deprotonate i-PrOH (hydride source of transfer hydrogenation), generating ruthenium

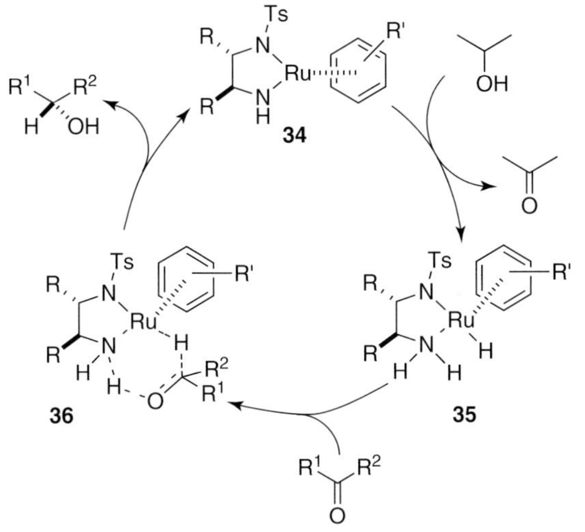

Scheme 11.5 Proposed catalytic cycle of enantioselective transfer hydrogenation of ketones: cooperative catalysis of ligand and metal.

hydride **35**, the active species of the hydrogenation step. The incorporated amine proton of **35** works as an acid to activate substrate ketones, thus facilitating hydride transfer from ruthenium hydride via a six-membered transition state **36** without any direct interaction between the metal and the substrate. The *concerto catalysis* involving simultaneous participation of both metal and ligand to the bond-forming and bond-breaking processes is unique and powerful for other applications [39].

Milstein's group recently reported a ligand–metal cooperated photocatalytic system in which a monomeric ruthenium(II) hydride–dearomatized pincer complex (**37**) promoted consecutive, stoichiometric thermal H_2 and light-induced O_2 evolution from water (Scheme 11.6) [40]. The initial reaction of **37** with water at 25 °C yielded monomeric aromatic Ru(II) hydrido–hydroxo complex **38** through coordination of water at the vacant site on the metal *trans* to the hydride ligand of **37** followed by proton migration to the side arm. This unique water activation process involved cooperation between the metal and the ligand. The metal oxidation state did not change in this process. On further reaction of **38** with water at 100 °C for 3 days, the *cis*-dihydroxo complex **39** was generated with release of H_2. Irradiation of this complex in the 320–420 nm range liberated oxygen and regenerated the starting Ru(II) hydrido–hydroxo complex **37**. The proposed mechanism for this sequence involves a photochemically induced reductive elimination of H_2O_2 from the two hydroxo ligands of **39**, thereby generating unobservable Ru(0) intermediate **40** that quickly undergoes intramolecular proton transfer to regenerate **37**. The eliminated H_2O_2 rapidly disproportionated to H_2O and O_2 in the presence of the metal catalyst. The energy profile of this process based on density functional theory calculation was reported by two groups [41]. Although this process is a stepwise, stoichiometric process and the overall efficiency is not high, it is remarkable that water splitting was achieved basically using sunlight without relying on high-valent metal oxo species. The basic concept of

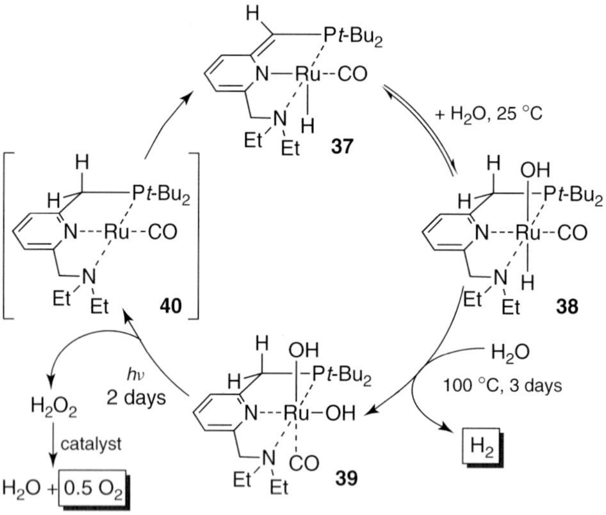

Scheme 11.6 Proposed catalytic cycle of water splitting by a monomeric ruthenium complex: ligand–metal cooperation.

this catalysis has been extended to dehydrogenative ester, amide, and acetal syntheses from alcohols or amines [42], and further expansion can be expected in the future.

11.4.2
Frustrated Lewis Pairs

"Frustrated Lewis pairs" (FLPs) are defined as Lewis acid–base combinations that are sterically prevented from forming strong classical acid–base adducts [43]. An interesting reactivity of FLPs was initially reported by Wittig *et al.* [44] in the 1960s. It was Stephan and co-workers' report in 2006, however, that alerted the chemical community to the true potential of FLPs in various utilities [45]. Stephan's group reported that zwitterionic species **42**, generated from aromatic substitution of $(C_6F_5)_3B$ by dimesitylphosphine followed by treatment of the resulting fluoroborate with a silane, released H_2 upon heating above 100 °C in toluene, forming FLP **41** [Scheme 11.7, Eq. (11.5)]. FLP **41** smoothly absorbed H_2 to regenerate zwitterionic **42** via H–H cleavage at ambient temperature. Therefore, the interconversion between **41** and **42** is reversible. This remarkable finding represents the first non-transition metal system that both releases and takes up dihydrogen.

Identification of the hydrogen activation ability of FLPs led to the development of the transition metal-free catalytic hydrogenation of imines [Scheme 11.7, Eq. (11.6)] and alkenes [46]. Using 5 mol% of **41**, reductions of imines that contained a bulky substituent (R = *t*-Bu or CHPh$_2$) on the nitrogen atom proceeded smoothly (~1 h) in high yields. Imines with electron-withdrawing substituents on the nitrogen (R = SO$_2$Ph) required longer reaction times and/or higher temperatures, suggesting that protonation of the nitrogen atom would be the rate-determining step. The reaction mechanism (**43**) is

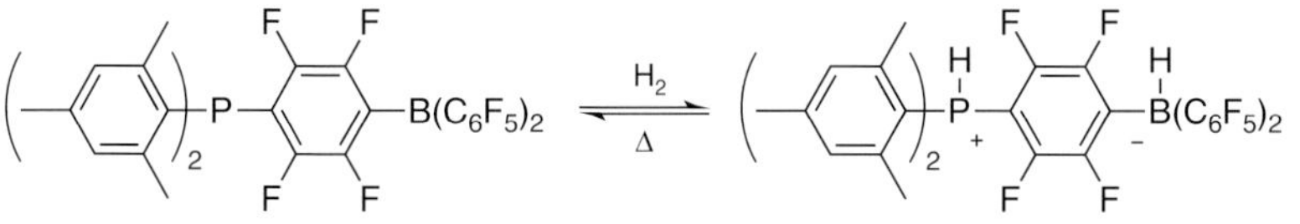

(11.5)

41   **42**

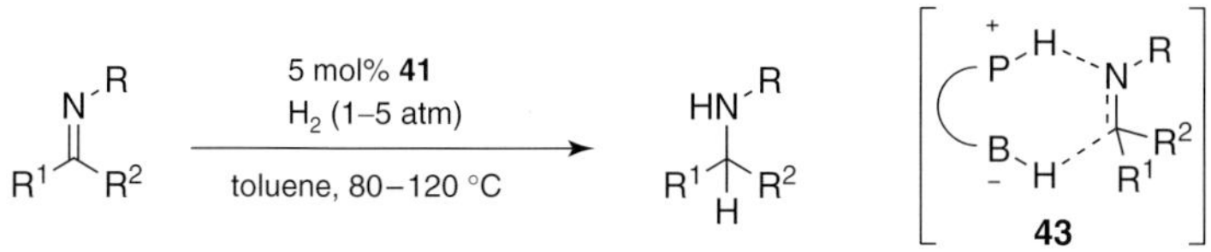

(11.6)

Scheme 11.7 Frustrated ion pairs (FLPs): hydrogen activation and catalytic hydrogenation of imines.

closely related to Noyori and co-workers' proposal for the hydrogenation of ketones (**36** in Scheme 11.5), albeit without involving a transition metal. In addition, extension of the utility of FLP to the activation of ethers, alkenes, alkynes, boranes, and CO_2 has been reported [43, 47].

Elucidation of the mechanism underlying FLP activation of H_2 is one of the most important current objectives in this field. Two alternative scenarios have been proposed based on computational investigations for H_2 activation by FLPs generated from phosphines and electrophilic boron compounds (Scheme 11.8). One view involves the generation of "encounter complex" **44** from the bulky phosphine and $(C_6F_5)_3B$, stabilized by noncovalent $C_6F_5 \cdots t$-Bu interactions and dispersion forces. An electric field is created in the pocket of the FLP, which polarizes the incorporated H_2, leading to heterolytic cleavage of the H–H bond [Scheme 11.8, Eq. (11.7)] [48]. The other view involves Lewis acid activation of H_2 by the highly electrophilic boron center via the formation of van der Waals complex **45**. The initial adduct **46** is formed from **45** by the heterolytic addition of H_2 across the C–B bond. The subsequent sequential proton

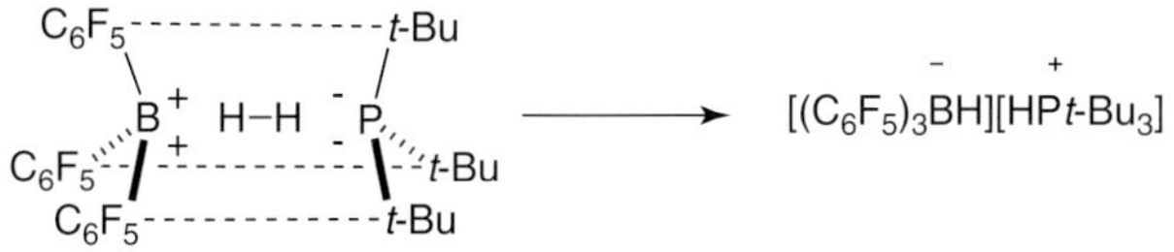

$$[(C_6F_5)_3BH][HPt\text{-}Bu_3]$$

(11.7)

44

(11.8)

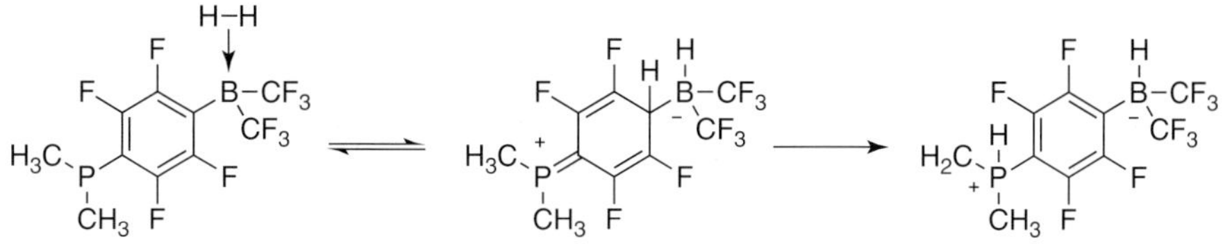

45   **46**   **47**

Scheme 11.8 Proposed mechanisms for hydrogen activation by FLP.

migration reactions produces zwitterion **47** [Scheme 11.8, Eq. (11.8)] [49]. Fundamental understanding of the H₂ activation mechanism based on experiments will be critical to the further development of FLPs' important reactivity.

11.5
Conclusion and Perspectives

Three recent topics in cooperative catalysis were selected and reviewed. The field of asymmetric catalysis has grown rapidly during the past four decades as a very useful synthetic method for pharmaceuticals and structurally complex drug leads. Broadening the reaction scope with truly practical catalyst turnover and enantioselectivity in C–C bond-formation is the current main goal in this field. In this respect, two recent contributions, an enantioselective tetrasubstituted carbon construction (Strecker reaction of ketimines) with functional switching properties of the catalyst and a biomimetic cation cyclization–alkylation with anion binding ability of chiral thiourea catalyst, illustrate the power of asymmetric cooperative catalysis. The alkene polymerization field also has a rich history. Molecular-level control of polymer microstructures based on logical catalyst tuning is the current main goal. Bimetallic catalysts with two catalytic centers cooperating with each other offer a new opportunity to achieve this goal. As a typical example, ethylene–styrene polymerization by a linked bimetallic titanium catalyst is described. Catalytic activation of dihydrogen is a fundamental reaction in organic synthesis. In addition, there is an increasing quest for robust dihydrogen generation methods from water as a clean energy source, due to environmental concerns. Ligand–metal cooperative catalysis might provide a platform concept for the future development of such a method. Transition metal-free dihydrogen activation by the cooperation of a Lewis acid and a Lewis base without adduct formation, that is, FLPs, is a new entry for catalysis and also hydrogen storage. "Cooperative catalysis" continues to be a general basic concept for developing further rich fields.

References

1. Noyori, R. (2009) *Nat. Chem.*, **1**, 5–6.
2. Walsh, C. (1979) *Enzymatic Reaction Mechanisms*, Freeman, New York.
3. Fessner, W.-D., Schneider, A., Held, H., Sinerius, G., Walter, C., Hixon, M., and Schloss, J.V. (1996) *Angew. Chem. Int. Ed. Engl.*, **35**, 2219–2221.
4. For more comprehensive reviews on the topic of cooperative asymmetric catalysis, see: (a) Ma, J.-A. and Cahard, D. (2004) *Angew. Chem. Int. Ed.*, **43**, 4566–4583; (b) Rowlands, G.J. (2001) *Tetrahedron*, **57**, 1865–1882; (c) Walsh, P.J. and Kozlowski, M.C. (2009) *Fundamentals of Asymmetric Catalysis*, University Science Books, Sausalito, CA.
5. (a) Ito, Y., Sawamura, M., and Hayashi, T. (1986) *J. Am. Chem. Soc.*, **108**, 6405–6406; review: (b) Sawamura, M. and Ito, Y. (2000) in *Catalytic Asymmetric Synthesis* (ed. I. Ojima), Wiley-VCH Inc., New York, pp. 493–512.
6. (a) Kitamura, M., Suga, S., Kawai, K., and Noyori, R. (1986) *J. Am. Chem. Soc.*, **108**, 6071–6072; (b) Kitamura, M., Okada, S., Suga, S., and Noyori, R. (1989) *J. Am. Chem. Soc.*, **111**, 4028–4036; (c) Noyori, R.

and Kitamura, M. (1991) *Angew. Chem. Int. Ed. Engl.*, **30**, 49–69 (review).

7. (a) Pu, L. and Yu, H.-B. (2001) *Chem. Rev.*, **101**, 757–824; For the pioneering contribution in this field, see: (b) Oguni, N. and Omi, M. (1984) *Tetrahedron Lett.*, **25**, 2823–2824.

8. (a) Corey, E.J., Bakshi, R.K., and Shibata, S. (1987) *J. Am. Chem. Soc.*, **109**, 5551–5553; (b) Corey, E.J., Bakshi, R.K., Shibata, S., Chen, C.-P., and Singh, V.K. (1987) *J. Am. Chem. Soc.*, **109**, 7925–7926; (c) Corey, E.J. and Helal, C.J. (1998) *Angew. Chem. Int. Ed. Engl.*, **37**, 1986–2012 (review).

9. (a) Sasai, H., Suzuki, T., Arai, S., Arai, T., and Shibasaki, M. (1992) *J. Am. Chem. Soc.*, **114**, 4418–4420; (b) Shibasaki, M., Sasai, H., and Arai, T. (1997) *Angew. Chem. Int. Ed. Engl.*, **36**, 1236–1256 (review); (c) Shibasaki, M., Kanai, M., Matsunaga, S., and Kumagai, N. (2009) *Acc. Chem. Res.*, **42**, 1117–1127 (review); (d) Shibasaki, M., Matsunaga, S., and Kumagai, N. (2008) *Synlett*, 1583–1602 (review); (e) Kanai, M., Kato, N., Ichikawa, E., and Shibasaki, M. (2005) *Synlett*, 1491–1508 (review); (f) Shibasaki, M. and Yamamoto, Y. (eds.) (2004) *Multimetallic Catalysts in Organic Synthesis*, Wiley-VCH Verlag GmbH, Weinheim.

10. (a) Tokunaga, M., Larrow, J.F., Kakiuchi, F., and Jacobsen, E.N. (1997) *Science*, **277**, 936–938; (b) Schaus, S.E., Brandes, B.D., Larrow, J.F., Tokunaga, M., Hansen, K.B., Gould, A.E., Furrow, M.E., and Jacobsen, E.N. (2002) *J. Am. Chem. Soc.*, **124**, 1307–1315; (c) Nielsen, L.P.C., Stevenson, C.P., Blackmond, D.G., and Jacobsen, E.N. (2004) *J. Am. Chem. Soc.*, **126**, 1360–1362; (d) Breinbauer, R. and Jacobsen, E.N. (2000) *Angew. Chem. Int. Ed.*, **39**, 3604–3607; (e) Ready, J.M. and Jacobsen, E.N. (2001) *J. Am. Chem. Soc.*, **123**, 2687–2688; (f) Ready, J.M. and Jacobsen, E.N. (2002) *Angew. Chem. Int. Ed.*, **41**, 1374–1377; (g) Jacobsen, E.N. (2000) *Acc. Chem. Res.*, **33**, 421–431 (review).

11. For recent examples, see: (a) Trost, B.M., Chan, V.S., and Yamamoto, D. (2010) *J. Am. Chem. Soc.*, **132**, 5186–5192; (b) Trost, B.M. and Brindle, C.S. (2010) *Chem. Soc. Rev.*, **39**, 1600–1632 (review).

12. Houk, K.N. and List, B. (eds.) (2004) *Acc. Chem. Res.* (Special Issue on Asymmetric Organocatalysis), **37** (8), 487–631; (b) List, B. (ed.) (2007) *Chem. Rev.* (Special Issue on Organocatalysis), **107** (12), 5413–5883; (c) List, B. and Liao, S.-H. Chapter 10 of this book.

13. Yabu, K., Masumoto, S., Yamasaki, S., Hamashima, Y., Kanai, M., Du, W., Curran, D.P., and Shibasaki, M. (2001) *J. Am. Chem. Soc.*, **123**, 9908–9909.

14. (a) Masumoto, S., Usuda, H., Suzuki, M., Kanai, M., and Shibasaki, M. (2003) *J. Am. Chem. Soc.*, **125**, 5634–5635; (b) Kato, N., Suzuki, M., Kanai, M., and Shibasaki, M. (2004) *Tetrahedron Lett.*, **45**, 3147–3151; (c) Kato, N., Suzuki, M., Kanai, M., and Shibasaki, M. (2004) *Tetrahedron Lett.*, **45**, 3153–3155.

15. For more comprehensive reviews of the catalytic enantioselective Strecker reaction, see: (a) Gröger, H. (2003) *Chem. Rev.*, **103**, 2795–2828; (b) Shibasaki, M., Kanai, M., and Mita, T. (2008) *Org. React.*, **70**, 1–117.

16. Kato, N., Mita, T., Kanai, M., Therrien, B., Kawano, M., Yamaguchi, K., Danjo, H., Sei, Y., Sato, A., Furusho, S., and Shibasaki, M. (2006) *J. Am. Chem. Soc.*, **128**, 6768–6769.

17. Fujimori, I., Mita, T., Maki, K., Shiro, M., Sato, A., Furusho, S., Kanai, M., and Shibasaki, M. (2006) *J. Am. Chem. Soc.*, **128**, 16438–16439.

18. Saga, Y., Motoki, R., Makino, S., Shimizu, Y., Kanai, M., and Shibasaki, M. (2010) *J. Am. Chem. Soc.*, **132**, 7905–7907.

19. Tanaka, Y., Kanai, M., and Shibasaki, M. (2010) *J. Am. Chem. Soc.*, **132**, 8862–8863.

20. Taylor, M.S. and Jacobsen, E.N. (2004) *J. Am. Chem. Soc.*, **126**, 10558–10559.

21. Taylor, M.S., Tokunaga, N., and Jacobsen, E.N. (2005) *Angew. Chem. Int. Ed.*, **44**, 6700–6704.

22. Taylor, M.S. and Jacobsen, E.N. (2006) *Angew. Chem. Int. Ed.*, **45**, 1520–1543.

23. Raheem, I.T., Thiara, P.S., Peterson, E.A., and Jacobsen, E.N. (2007) *J. Am. Chem. Soc.*, **129**, 13404–13405.

24. Knowles, R.R., Lin, S., and Jacobsen, E.N. (2010) *J. Am. Chem. Soc.*, **132**, 5030–5032.

25. For recent examples of catalytic asymmetric polycyclization, see: (a) Ishihara, K., Nakamura, S., and Yamamoto, H. (1999) *J. Am. Chem. Soc.*, **121**, 4906–4907; (b) Sakakura, A., Ukai, A., and Ishihara, K. (2007) *Nature*, **445**, 900–903; (c) Mullen, C.A., Campbell, M.A., and Gagné,

M.R. (2008) *Angew. Chem. Int. Ed.*, **147**, 6011–6014; (d) Rendler, S. and MacMillan, D.W.C. (2010) *J. Am. Chem. Soc.*, **132**, 5027–5029.

26. Review: Wendt, K.U., Schulz, G.E., Corey, E.J., and Liu, D.R. (2000) *Angew. Chem. Int. Ed.*, **39**, 2812–2833.

27. Brown, A.R., Kuo, W.-H., and Jacobsen, E.N. (2010) *J. Am. Chem. Soc.*, **132**, 9286–9288.

28. For other examples of catalytic asymmetric aldehyde alkylation, see: (a) Vignola, N. and List, B. (2004) *J. Am. Chem. Soc.*, **126**, 450–451; (b) Nicewicz, D.A. and MacMillan, D.W.C. (2008) *Science*, **322**, 77–80; (c) Enders, D., Wang, C., and Bats, J.W. (2008) *Angew. Chem. Int. Ed.*, **47**, 7539–7542.

29. Representative reviews of alkene polymerization: (a) Brintzinger, H.H., Fischer, D., Miilhaupt, R., Rieger, B., and Waymouth, R.M. (1995) *Angew. Chem. Int. Ed. Engl.*, **34**, 1143–1170; (b) Britovsek, G.J.P., Gibson, V.C., and Wass, D.F. (1999) *Angew. Chem. Int. Ed.*, **38**, 428–447.

30. Li, L., Metz, M.V., Li, H., Chen, M.-C., Marks, T.J., Sands, L.L., and Rheingold, A.L. (2002) *J. Am. Chem. Soc.*, **124**, 12725–12741.

31. Abramo, G.P., Li, L., and Marks, T.J. (2002) *J. Am. Chem. Soc.*, **124**, 13966–13967.

32. Mononuclear and/or polynuclear rare earth metal complexes are another important class of alkene polymerization catalysts; see: Nishiura, M. and Hou, Z. (2010) *Bull. Chem. Soc. Jpn.*, **83**, 595–608 (review).

33. For other examples of ethylene–styrene copolymerization, see: (a) Luo, Y., Baldamus, J., and Hou, Z. (2004) *J. Am. Chem. Soc.*, **126**, 13910–13911; (b) Son, K.-S. and Waymouth, R.M. (2010) *J. Polym. Sci., Part A: Polym. Chem.*, **48**, 1579–1585.

34. (a) Guo, N., Stern, C.L., and Marks, T.J. (2008) *J. Am. Chem. Soc.*, **130**, 2246–2261; (b) Guo, N., Li, L., and Marks, T.J. (2004) *J. Am. Chem. Soc.*, **126**, 6542–6543.

35. For bimetallic zinc-catalyzed copolymerization of cyclohexene oxide and CO_2, see: Moore, D.R., Cheng, M., Lobkovsky, E.B., and Coates, G.W. (2003) *J. Am. Chem. Soc.*, **125**, 11911–11924.

36. (a) Ohkuma, T., Ooka, H., Hashiguchi, S., Ikariya, T., and Noyori, R. (1995) *J. Am. Chem. Soc.*, **117**, 2675–2676; (b) Ohkuma, T., Ooka, H., Hashiguchi, S., Ikariya, T., and Noyori, R. (1995) *J. Am. Chem. Soc.*, **117**, 10417–10418.

37. Hashiguchi, S., Fujii, A., Takehara, J., Ikariya, T., and Noyori, R. (1995) *J. Am. Chem. Soc.*, **117**, 7562–7563.

38. (a) Haack, K.-J., Hashiguchi, S., Fujii, A., Ikariya, T., and Noyori, R. (1997) *Angew. Chem. Int. Ed. Engl.*, **36**, 285–288; (b) Noyori, R., Yamakawa, M., and Hashiguchi, S. (2001) *J. Org. Chem.*, **66**, 7931–7944.

39. For example: (a) Ikariya, T., Murata, K., and Noyori, R. (2006) *Org. Biomol. Chem.*, **4**, 393–406 (review); (b) Jiang, Y., Jiang, Q., and Zhang, X. (1998) *J. Am. Chem. Soc.*, **120**, 3817–3818; (c) Zweifel, T., Naubron, J.-V., and Grützmacher, H. (2009) *Angew. Chem. Int. Ed.*, **48**, 559–563.

40. Kohl, S.W., Weiner, L., Schwartsburd, L., Konstantinovski, L., Shimon, L.J.W., Ben-David, Y., Iron, M.A., and Milstein, D. (2009) *Science*, **324**, 74–77.

41. (a) Li, J., Shiota, Y., and Yoshizawa, K. (2009) *J. Am. Chem. Soc.*, **131**, 13584–13585; (b) Yang, X. and Hall, M.B. (2010) *J. Am. Chem. Soc.*, **132**, 120–130.

42. (a) Zhang, J., Leitus, G., Ben-David, Y., and Milstein, D. (2005) *J. Am. Chem. Soc.*, **127**, 10840–10841; (b) Gunanathan, C., Ben-David, Y., and Milstein, D. (2007) *Science*, **317**, 790–792; (c) Gunanathan, C., Shimon, L.J.W., and Milstein, D. (2009) *J. Am. Chem. Soc.*, **131**, 3146–3147; (d) Khaskin, E., Iron, M.A., Shimon, L.J.W., Zhang, J., and Milstein, D. (2010) *J. Am. Chem. Soc.*, **132**, 8542–8543.

43. Stephan, D.W. and Erker, G. (2010) *Angew. Chem. Int. Ed.*, **49**, 46–76 (review).

44. (a) Wittig, G., Reppe, H.G., and Eicher, T. (1961) *Liebigs Ann. Chem.*, **643**, 47–67; (b) Tochtermann, W. (1966) *Angew. Chem. Int. Ed. Engl.*, **5**, 351–371 (review).

45. Welch, G.C., Juan, R.R.S., Masuda, J.D., and Stephan, D.W. (2006) *Science*, **314**, 1124–1126.

46. For example: (a) Chase, P.A., Welch, G.C., Jurca, T., and Stephan, D.W. (2007) *Angew. Chem. Int. Ed.*, **46**, 8050–8053; (b) Spies, P., Schwendemann, S., Lange, S., Kehr, G., Fröhlich, R., and Erker, G. (2008) *Angew. Chem. Int. Ed.*, **47**, 7543–7546; (c) Erős, G., Mehdi, H., Pápai, I., Rokob, T.A., Király, P., Tárkányi, G., and Soós, T. (2010) *Angew. Chem. Int. Ed.*, **49**, 6559–6563.

47. Berkefeld, A., Piers, W.E., and Parvez, M. (2010) *J. Am. Chem. Soc.*, **132**, 10660–10661.

48. (a) Rokob, T.A., Hamza, A., Stirling, A., Soós, T., and Pápai, I. (2008) *Angew. Chem. Int. Ed.*, **47**, 2435–2438; (b) Grimme, S., Kruse, H., Goerigk, L., and Erker, G. (2010) *Angew. Chem. Int. Ed.*, **49**, 1402–1405; (c) Rokob, T.A., Hamza, A., Stirling, A., and Pápai, I. (2009) *J. Am. Chem. Soc.*, **131**, 2029–2036.

49. Rajeev, R. and Sunoj, R.B. (2009) *Chem. Eur. J.*, **15**, 12846–12855.

Commentary Part

Comment 1

Takao Ikariya

Many thousands of materials and products, including fuels required by modern society, would not be possible without the existence of catalysts. Catalysts are also crucial for the reduction of water and air pollution and for the reduction of the waste of natural resources and energy. Recent significant advances in green and sustainable science and technology demand more powerful and more sophisticated catalysts, which have a tunable function. Much effort has been devoted to the development of well-designed cooperative molecular catalysts based on the combination of Lewis acidic and basic sites working in concert to activate reacting substrates on the catalyst leading to highly efficient molecular transformations for organic synthesis. Since the cooperative catalysts contain two or more active sites for activation of electrophiles and nucleophiles, they can promote a wide range of molecular transformations by cooperative activation and effective accumulation of reacting substrates on the neighboring active centers in the same molecules. However, they often suffer from acid–base neutralization, mainly for structural reasons, leading to deactivation of the catalysts. Therefore, careful and precise tuning of the structures of the molecular catalysts and also the spatial organization of the functionality is required to achieve the best catalyst performance. Enzymatic systems in Nature can precisely arrange the substrate accumulation around the active center to achieve perfect chemical transformation, as pointed out in this chapter.

In addition, conceptually new metal–ligand cooperating bifunctional catalysts have also recently evolved as cooperative catalysts to steer and control organometallic reactions and catalysis, and now they are also recognized as an alternative and indispensable strategy to realize highly effective molecular transformations. The redox non-innocent ligand therein participates directly in the substrate activation and subsequent bond-forming and -breaking reactions through various secondary interactions, including hydrogen bonding. Therefore, the catalyst deactivation due to acid–base neutralization or destructive aggregation can be minimized.

This unique concept of cooperative and/or bifunctional molecular catalysts leads to high reaction rates and excellent stereoselectivities because the reactions proceed through a tight-fitting assembly of the reactants and chiral catalysts.

In line with recent developments in molecular catalysis, the appearance of a book that contains this chapter dealing with recent achievements in cooperative catalysis is particularly timely. Kanai's chapter is well organized and well written based on his research background and structured to present the three of the currently most interesting topics on cooperative and bifunctional catalysis: (i) asymmetric Strecker reaction with Shibasaki and Kanai's multimetallic cooperative catalysts, (ii) alkene copolymerization with Mark's cooperative bimetallic ionic titanium catalyst, and (iii) hydrogen activation with Noyori and Ikariya's M/NH bifunctional catalyst and hydrogen generation from water with Milstein's pincer-type bifunctional catalysts bearing reversible dearomatization of its pyridyl ligand. In each section, a brief introduction to the cooperative effect relating to the particular topic are given to clarify the cooperative effect in catalysis.

In addition, Stephan's FLPs for activation molecular hydrogen without any metal elements are also very informative for further designing cooperative metal-free and/or metal-based molecular catalysis, although the mechanism of hydrogen activation and heterolytic cleavage with FLPs is still not clear, as mentioned in the text. On the other hand, the activation of molecular hydrogen with metal-based bifunctional catalysts has been well investigated by experimental and computational methods and it has now become clear that it proceeds in a concerted manner, as demonstrated by Ikariya, Andersson, Noyori, and Bergens' groups, whose papers [C1–C4] could have been cited in this chapter. Reversible hydrogen activation and formation with bifunctional molecular catalysts is a crucially important subject in both organic synthesis and future energy technology.

In related to bifunctional catalysis based on the metal–ligand cooperation, Fryzuk *et al.*'s Ru amide chemistry is worth noting in this chapter because their pioneering work inspired the explosive development of bifunctional molecular catalysts. They include Shvo's Ru catalyst, Noyori and Ikariya's Ru catalysts, Milstein's Ru pincer catalysts, and Grützmacher's catalyst systems. These intensive research studies have had an enormous impact on individual elementary reactions in organometallic chemistry and fundamentals in molecular catalysis.

Controlling the stereo- and regiochemistry in monomer incorporation and also copolymerization of simple alkenes are among the most challenging research targets and Kanai's chapter ventures to deal with this exciting research topic based on cooperative bimetallic catalyst systems developed by Marks' group. Unfortunately, the mechanistic details of alkene copolymerization as shown only in Scheme 11.4 do not seem fully understandable. I agree that the concept of cooperative catalysis might be a promising strategy for controlling the comonomer, but more schematic drawings may be needed to clarify the cooperative effect in regioselective alkene copolymerization. Overall, I believe that this chapter by Prof. Kanai will not only be useful to researchers and students already involved in organic synthesis and molecular catalysis but will also attract new people working in various field of chemistry. Some minor comments can be made:

1) believe that the enantioselective alkylation of aldehydes using a dinuclear zinc system was developed independently by Prof. Oguni's and Prof. Noyori's groups. As a pioneering chemist, Prof. Oguni's work should be appreciated in the text.

2) In Table 11.1, the structure of the catalyst **16** should be drawn illustrated. A photograph of the catalysts is not clear.

3) In Section 11.4, although the title of the section is Cooperative Catalysis in Hydrogen

Activation/Generation, no scheme indicating the activation of molecular hydrogen with cooperative catalysts is shown. Instead, Scheme 11.5 shows a catalytic cycle for hydrogen transfer between ketones and alcohols. Of course, the transfer hydrogenation of ketone using alcohols with metal–ligand cooperative catalysts is closely related to the hydrogenation of ketones, but the mechanism of H_2 activation with cooperative catalysts should be discussed.

Comment 2

Takashi Ooi

This chapter well illustrates the importance and great potential of cooperative catalysis in the fields of chemical and material sciences by focusing on three recent topics. The characteristic feature of cooperative catalysis discussed here is that metal centers and/or reactive functionalities are located in a single molecule or a molecular system with appropriate mutual spatial arrangements and that they participate in catalysis in a cooperative manner. In the field of asymmetric catalysis, the design of catalysts possessing both acidic and the basic sites within a single molecular entity is a simple yet fundamental approach towards inducing this type of cooperative catalysis. The thiourea catalyst employed in the anion binding approach by Jacobsen's group is a representative example of organocatalysts. A recent intriguing strategy for imparting additional function to this class of so-called bifunctional catalysts is the exploitation of enthalpy–entropy compensation to control the stereochemistry of carbon–carbon bond-forming reactions using conformationally flexible catalysts. Nagasawa's group reported an enantiodivergent Mannich-type reaction between *N*-Boc-protected imines **2** and malonates **3** catalyzed by chiral guanidinebisthiourea **1** (Scheme C1) [C5]. The absolute configuration of product **4** depended critically on the solvent employed; this dependence is probably attributable to the alteration of the relative conformation with respect to the guanidinium and thiourea moieties on changing the reaction conditions. Kinetic analysis revealed that a single mechanism was operable in the catalytic process occurring in each solvent, and the observed solvent-dependent enantioswitching resided in compensating differences in the enthalpies and

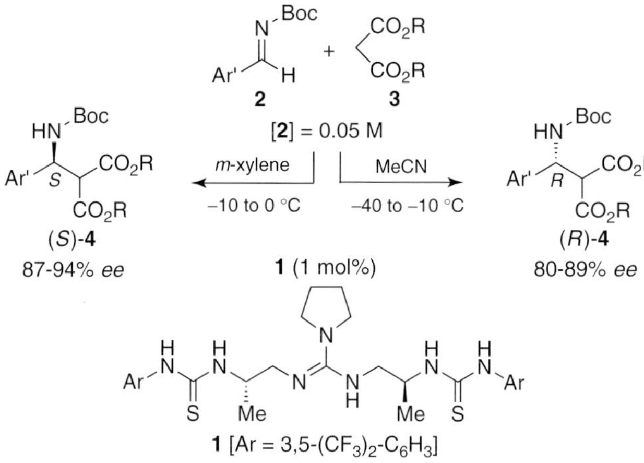

Scheme C1   Enantiodivergent Mannich-type reaction.

entropies of activation. This research can offer new tactics for reaction development based on catalyst design.

Furthermore, *in situ* "switching" of the catalytic function in a predictable manner, particularly by readily available external stimuli, would be one of the key issues to be addressed [C6]. Feringa and Wang introduced a conceptually new approach to switching the ability of an enantiomerically pure catalyst of a single molecular framework for enabling the production of either enantiomer on demand [C7]. They prepared a chiral alkene (5) with similar structural components in each stereoisomer but in a helical orientation about the double bond (Scheme C2). In addition, thiourea and dimethylaminopyridine (DMAP) are attached to the terminus of each component. Starting from the most stable (*P,P*)-*trans*-1, the upper moiety of the molecule underwent a 360° clockwise rotation in one direction, which involved four individual steps; two steps are light-induced *cis–trans* isomerization associated with a 180° rotation around the double bond, and the other two steps are heat-controlled inversion of the helicity. Because the terminal thiourea and DMAP functionalities can only be in close proximity when *cis*-isomers are generated during the rotation, this rotary molecular motor can be used as a dynamic system for inducing and controlling asymmetric cooperative catalysis. In fact, attempts to use each stereoisomer of 5 as a catalyst showed that only (*P,P*)-*cis*-5 and (*M,M*)-*cis*-5 exhibited sufficient catalytic performance, and (*P,P*)-*trans*-5 barely promoted the reaction. More significantly,

a different enantiomer was produced with a nearly identical *ee* of around 50% in those two catalyzed reactions, and the small amount of the product obtained with (*P,P*)-*trans*-5 was a mixture of equal amounts of the enantiomers. These observations indicate the possibility of switching the ability of the enantiopure molecular motor as a catalyst not only to promote the reaction but also to control the preference of the stereochemistry of the product.

On the other hand, two entirely different catalytic processes can be rationally combined in order to realize otherwise difficult stereoselective bond formations. A recent study by Nicewicz and MacMillan showed that a judicious combination of asymmetric enamine catalysis and an $Ru(bpy)_3Cl_2$-catalyzed photoredox process allowed the highly enantioselective direct α-alkylation of aldehydes with α-bromocarbonyl compounds [C8]. In the dual catalysis proposed for this previously elusive transformation (Scheme C3), $Ru(bpy)_3^{2+}$ accepts a photon to populate the metal-to-ligand charge-transfer excited state, $^*Ru(bpy)_3^{2+}$, which would remove a single electron from the sacrificial quantity of enamine to initiate the first catalytic cycle, providing the electron-rich $Ru(bpy)_3^+$. The subsequent rapid single-electron transfer (SET) from $Ru(bpy)_3^+$ to the α-bromocarbonyl substrate would furnish the electron-deficient alkyl radical 7 and regenerate $Ru(bpy)_3^{2+}$. Meanwhile, the second catalytic cycle begins with the condensation of aldehyde with the imidazolidinone catalyst 6 to afford enamine 8, which stereoselectively adds to the alkyl radical 7 (one intersection of the two catalysts) to produce

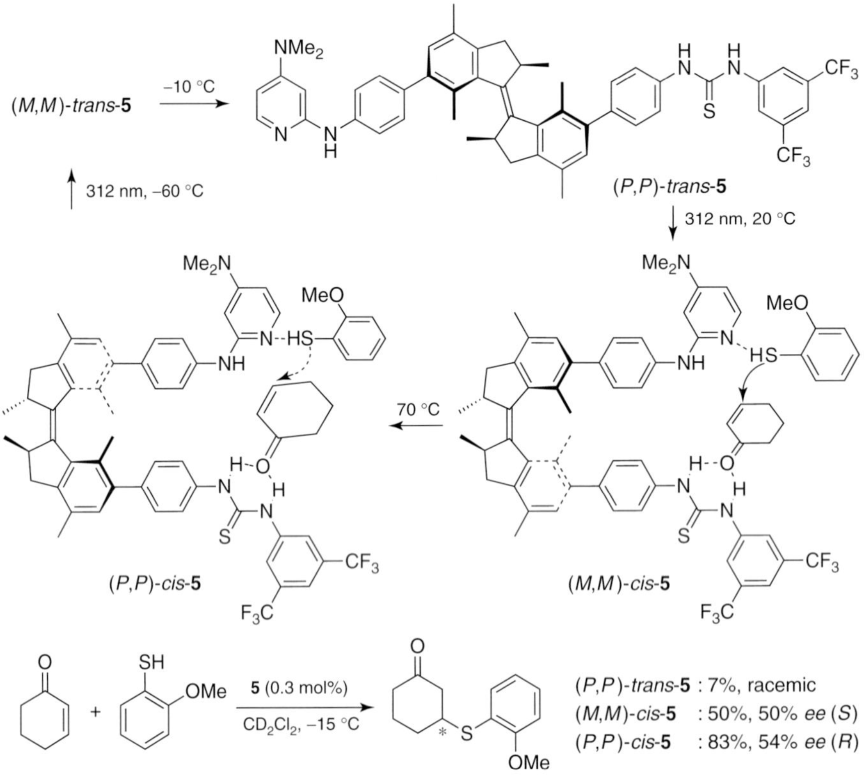

Scheme C2 *In situ* switching of catalytic function by heat and light.

an electron-rich α-amino radical **9**. Then, SET occurs from **9** to *Ru(bpy)₃²⁺ (another intersection) to yield the iminium ion **10** and regenerate Ru(bpy)₃⁺. Finally, the hydrolysis of **10** liberates catalyst **6** and an enantiomerically enriched product. Because the simultaneous operation of the two catalytic processes is a prerequisite for effecting enantioselective alkylation, this system can also be regarded as an asymmetric cooperative catalysis with tremendous synthetic potential.

In addition, Nishibayashi's group achieved a highly enantioselective propargylic alkylation of secondary propargylic alcohols with aldehydes by combined use of a catalytic amount of a thiolate-bridged diruthenium complex (**11**) and a diarylprolinol trimethylsilyl ether (**12**) [C9]. In this system, the ruthenium complex **11** and the chiral secondary amine **12** activated propargylic alcohols and aldehydes, respectively. Thus, the *in situ*-generated chiral enamine **13** added to the ruthenium–allenylidene complex **14** to produce

each diastereomer with a high level of enantioselectivity (Scheme C4).

Notable progress has also been made in the development of relay catalysis, which involves the stepwise, consecutive generation of transient reactive intermediates through several different reactions to give a final product [C10]. Although each step is independently catalyzed by a different catalyst, cooperativity seems to be a crucial element for controlling the entire process in this type of catalysis.

In this regard, on-demand "switching" of the catalytic function appears to be an even more important objective. A new and powerful strategy should be introduced not only for endowing a single catalyst with multiple functions but also for controlling and switching them by using chemical or physical stimuli. Continuous efforts along this line would eventually lead to a point where a chiral catalyst of a single molecular framework conducts multiple tasks in solution in a truly predictable

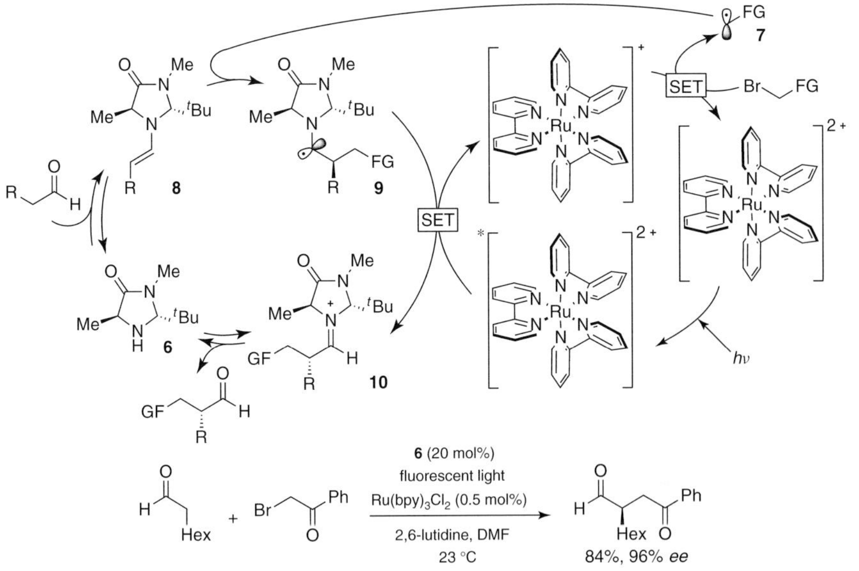

Scheme C3 Dual catalysis: photoredox catalysis and organocatalysis.

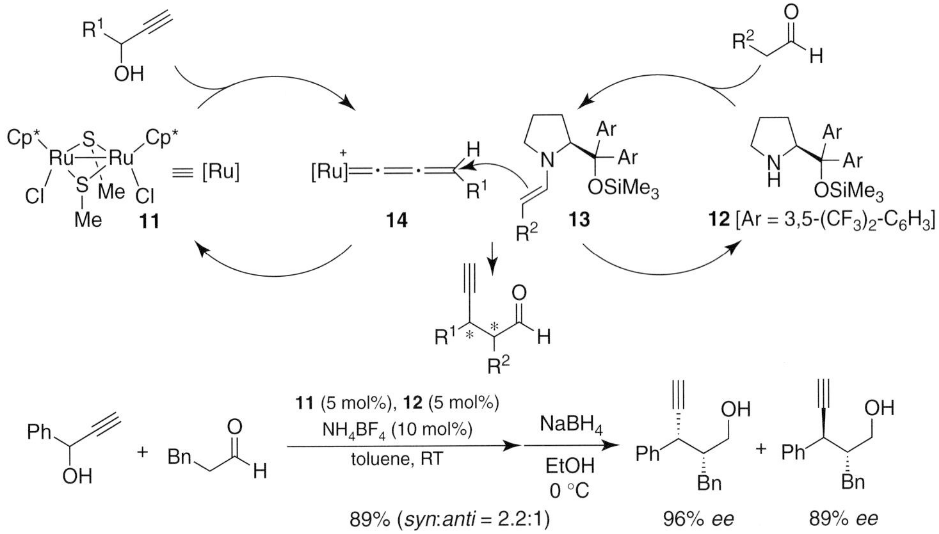

Scheme C4 Combination of transition metal complex and chiral secondary amine.

manner and promotes a requisite set of organic transformations, thereby establishing a rapid and selective assembly of a complex chiral molecule. Taking an actual stride in this direction would contribute greatly to the sustainable chemical synthesis of valuable organic compounds.

Comment 3

Kuiling Ding

Professor Ryoji Noyori emphasized in the first issue of *Nature Chemistry* that "Chemistry has a central role in science, and synthesis has a central role in chemistry" [C1]. In fact, catalysis has been, and will remain, one of the most important research subjects, since this is the only rational means of producing useful compounds in an economical, energy-saving, and environmentally benign way, ideally in 100% yield, with 100% selectivity, and avoiding waste production. In view of the significance of catalysis in synthetic chemistry, the highest level of scientific creativity for catalyst design and insight into the underlying mechanistic issues are critically important. As Nature has served as a dominant source of inspiration in the area of catalysis, it is no surprise that enzymes have served as natural prototypes for the design of catalysts. As a consequence, how to create a catalyst system by learning from Nature's principles and to use this system to mimic or exceed enzymes represent significant challenges for chemists. This chapter by Professor Motomu Kanai highlights the topic of "cooperative catalysis" by focusing on several of the most important and fundamental chemical transformations, including asymmetric catalysis, polymerization, and hydrogen activation/generation, with significant contributions of the concept to their breakthroughs.

The application of cooperative catalysis has also been successful in asymmetric oxidation in addition to asymmetric hydrogenation. In the asymmetric epoxidation of alkenes catalyzed by Ti(salen) complexes using aqueous hydrogen peroxide as the oxidant, the amino proton of salen ligand was found to be critically important for the activation of hydrogen peroxide via hydrogen bonding interaction, indicating the cooperative effect of the functional group of the ligand and the metal center in the catalysis (see Figure C1). This type of catalyst is particularly effective for the asymmetric epoxidation of unactivated alkenes with excellent enantioselectivity [C12].

The most challenging issue remained in catalytic asymmetric carbon–carbon (C–C) bond and carbon–heteroatom (C-X) bond-forming reactions is the efficiency of the homogeneous catalysis, and

accordingly the development of truly practical catalysts with high turnover and enantioselectivity is highly desirable. In addition to the exciting examples presented by Kanai, a further recent effort to address the problem in asymmetric C–C bond formation is the rational design and development of an exceptionally efficient chiral catalyst for enantioselective cyanation of aldehydes using either TMSCN or NaCN as the cyanide source on the basis of the cooperative dual activation concept [C13]. The concept for catalyst development is built on the mechanistic knowledge that bimetallic Ti(salen) complexes are real active species; however, dimeric species of the titanium complexes may dissociate to the catalytically inactive monomeric species and both species exist in a concentration-dependent equilibrium in solution [C14]. Such an equilibrium will obviously reduce the concentration of active dimeric species and accordingly is detrimental to the catalysis. To maximize the amount of bimetallic complex present in solution, appropriate linking of two metallosalen units may overcome the problem of dissociation of the catalytically active dimer, which accordingly facilitates the cooperative activation of both the nucleophile and electrophile in the catalysis. In fact, the optimized catalyst promotes the reaction to afford the corresponding enantioenriched natural or non-natural cyanohydrin derivatives with turnover numbers up to 172 000 and *ee* up to 97% (Scheme C5), which represents a significant advance in the catalytic asymmetric cyanation of aldehydes. The practical features and mechanistic implications of this discovery may extend beyond cyanation itself and provide a useful model for other reaction variants.

Catalytic asymmetric electrophilic vicinal haloamination of alkene derivatives represents one of the most challenging transformations in the difunctionalization of alkenes for C-X bond formation. The breakthrough in this area was realized by Feng and co-workers using an N,N'-dioxide/Sc(III) catalyst (0.5–0.001 mol%) under mild reaction conditions. The catalytic

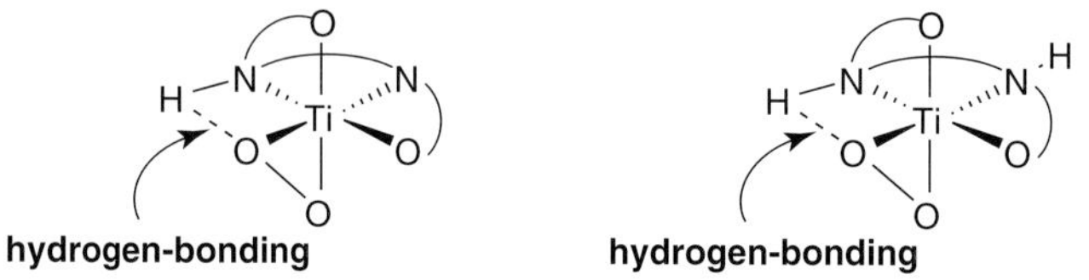

hydrogen-bonding   **hydrogen-bonding**

Figure C1 The proposed active species for Ti-catalyzed oxidations using aqueous hydrogen peroxide as the oxidant.

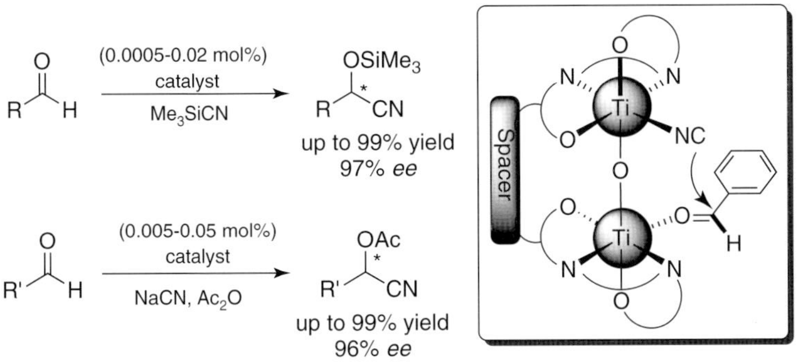

Scheme C5 Cooperative activation of substrates in Ti(salen)-catalyzed asymmetric cyanation of aldehydes.

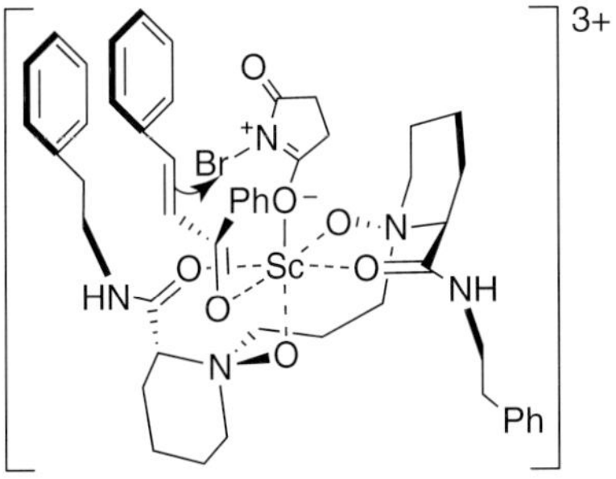

Figure C2 The proposed transition state involving the cooperative action of activated substrates.

asymmetric bromoamination or chloroamination of chalcones or α,β-unsaturated γ-keto esters with sulfonamide and N-bromosuccinimide (NBS) or TsNCl₂ as the reagents afforded the corresponding α-halo-β-amino ketone derivatives in high yields with excellent regio-, diastereo-, and enantiose-lectivities [C15]. The proposed transition state in the catalysis shows that a cooperative action of activated substrates (chalcone and halogenating reagent, see Figure C2) in the catalysis might be critically important for the control of the stereoselectivity, and also its unusually high activity [C16]. Considering the excellent performance of the N,N'-dioxide/Sc(III) catalyst in the present reaction system and many other metal- and organocatalyzed asymmetric reactions [C16], it is obvious that the N,N'-dioxide molecules represent a type of privileged chiral ligands, which will stimulate future efforts to understand the features that account for their broad applicability, and to apply this understanding to seek opportunities for developing new reactions and for solving the problems that remain with existing catalysts.

The strategy of "combined acids" for catalyst design proposed by Yamamoto [C17] can be considered to be closely related to cooperative catalysis. The concept of combined acids, which can be classified into Brønsted acid-assisted Lewis acid (BLA), Lewis acid-assisted Lewis acid (LLA), Lewis acid-assisted Brønsted acid (LBA), and Brønsted acid-assisted Brønsted acid (BBA), has been found to be a particularly useful tool for the design of asymmetric catalysis by taking advantage of the cooperative effect of the component acids, because combining such acids will bring out their inherent reactivity by associative interaction, and also provide better organized structures that allow an effective asymmetric environment. The practice of this stimulating concept was highlighted by Yamamoto and Futatsugi in a review article [C17]. For organocatalysis via hydrogen bonding activation, two examples reported recently by Jacobsen and co-workers should be mentioned here from the viewpoint of cooperative catalysis. One example [C18] is the amidothiourea-catalyzed

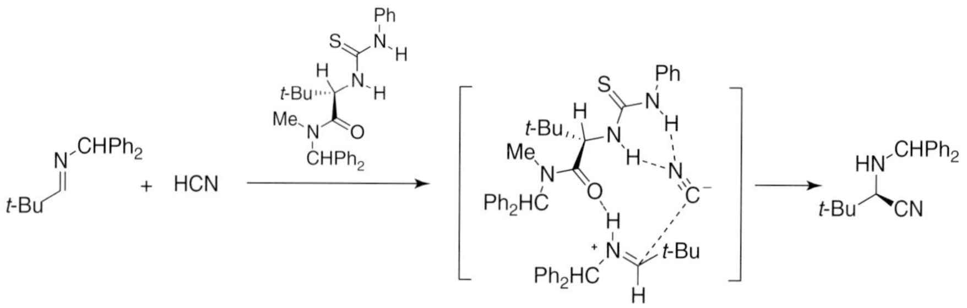

Scheme C6 Cooperative effect of amidothiourea multifunctional moieties in the Strecker reaction.

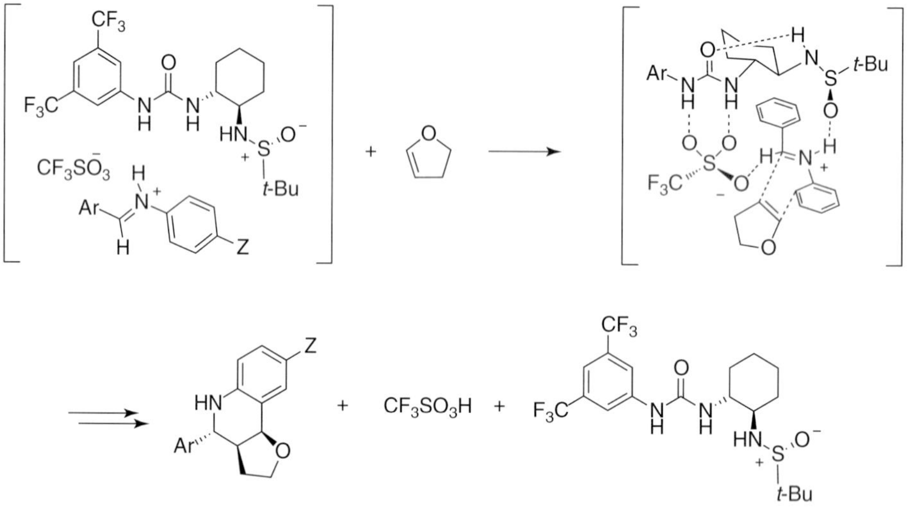

Scheme C7 Asymmetric cooperative catalysis of strong Brønsted acid-promoted Povarov reaction.

Strecker synthesis of highly enantiomerically enriched non-natural amino acids that are not readily prepared by enzymatic methods or by chemical hydrogenation. A mechanism involving initial amidothiourea-induced imine protonation by HCN to generate a catalyst-bound iminium–cyanide ion pair (Scheme C6) was proposed. The collapse of this ion pair and C–C bond formation to form the α-aminonitrile occur in a post-rate-limiting step, indicating the synergistic actions of amidothiourea multifunctional moieties with HCN and imine substrates in the transition-state network.

Another closely related example of asymmetric cooperative catalysis is the strong Brønsted acid-promoted Povarov reaction using chiral ureas (Scheme C7) [C19]. The strategy was based on the precise catalyst–substrate interactions

for inducing enantioselectivity in reactions of protio-iminium ions, wherein a chiral catalyst interacts with the highly reactive intermediate through a network of multiple, specific H-bonding interactions. These noncovalent interactions are maintained in the subsequent stereodetermining cycloaddition events, leading to an attenuation of the reactivity of the iminium ion and allows high enantioselectivity in cycloadditions of electron-rich alkenes. Again, this example illustrates the ability of bifunctional urea catalysts to control precisely the outcome of the reaction through noncovalent interactions alone. Given the known ability of urea and thiourea derivatives to bind a wide range of anions, this strategy will be applicable, in principle, to cationic intermediates with a variety of counter-ion structures.

An alternative strategy conceptually related to co-operative catalysis is "supramolecular catalysis," which has been dominated recently via the assembly of catalyst species through harnessing multiple weak intramolecular interactions with inspiration from enzyme processes [C20]. Such approaches attempt to create an enzyme-like "active site" and have concentrated on reactions similar to those catalyzed by enzymes themselves. The application of supramolecular assembly to the more traditional transition metal catalysis and to small-molecule organocatalysis has been extensively studied. The modularity of self-assembled multicomponent catalysts implies that a relatively small pool of catalyst components can provide rapid access to a large number of catalysts that can be evaluated for industrially relevant reactions. Although still early in its development, already a number of reactions have been covered and several supramolecular catalysts have emerged with properties that surpass those of traditional homogeneous catalysts, showing the strength and potential of such multicomponent catalyst assemblies. The successes of this exciting concept were highlighted recently by Meeuwissen and Reek [C21]. Some other examples, including supramolecular photocatalysis and organocatalysis reported by Bach *et al.* [C22] and Luo and co-workers [C23], respectively, should be given credit from the cooperative catalysis point of view. The field of cooperative catalysis will advance continuously to create breakthrough innovations that have the potential to revolutionize chemical processes, particularly in the pharmaceutical, material, and fine chemical industries. Recent advances in green and sustainable science and technology strongly demand powerful and sophisticated catalysts with a tunable multifunction. The design and development of molecular catalysts on the basis of acid–acid or acid–base cooperation, hetero-multi-metallic cooperation, substrate–media cooperation, and mechanistic aspects of these processes will be of ongoing interest in synthetic organic chemistry. Since these artificial molecular catalysts usually contain two or more active sites for activation of the electrophiles and nucleophiles, they can efficiently assist a wide range of molecular transformations, including C–H, C–C, and C–X bond formation by dual activation and effective accumulation of reacting substrates on the neighboring active centers in the same molecules. Hence the rational design of cooperating ligands which are able to adjust the balance of the electronic factors in bi- and multifunctional catalysts is crucially important to exploit unprecedented catalyst performance. It can be expected that the concept of cooperative catalysis will be a powerful strategy to address the challenging issues, in particular the efficiency, of the catalysis, and accordingly to revolutionize the chemical processes.

Comment 4

David Milstein

As pointed out in this insightful chapter by Kanai, the field of catalysis plays a center-stage, critical role in the development of efficient, selective, and environmentally benign processes of major importance to our society, such as the synthesis of pharmaceuticals, fine chemicals, new materials, and sustainable energy resources. The field of catalysis has experienced remarkable progress in the ability to design catalytic cycles based on fundamental principles, rather than rely on empirical approaches and serendipity (although surprises are, more often than not, part of this trade). In this context, cooperative catalysis is an exciting, rapidly evolving field of high potential for the discovery of novel reactions.

The specific examples of cooperative catalysis chosen by Kanai nicely demonstrate the wide scope and strong impact that cooperative catalysis can have, in various fields, such as asymmetric synthesis, alkene polymerization, hydrogenation, and new approaches to sunlight harvesting. Various cooperation concepts are included.

In the case of enzymatic catalysis, examples of cooperativity of the active catalytic site with the protein backbone are mentioned. As my group's immediate interest is in metal–ligand cooperation, I would like to point out that in metallo-enzymes, cooperative catalysis can also involve direct metal–ligand cooperation, such as with FeFe hydrogenase, in which case cooperation between the iron centers and an amine group, directly attached to the complex, is involved.

The systems covered in this chapter are homogeneous. Heterogeneous systems, such as silica functionalized with different organic groups, can also show cooperative catalysis. For example, silica functionalized with thiols and acids catalyzes the synthesis of bisphenol A [C24]. One advantage of inorganic supports is that they allow the incorporation of incompatible functionalities, such as electrophilic and nucleophilic groups, or acid and base groups, isolated from each other by being tethered to the support, thus permitting dual reactivity in one system.

Within the framework of cooperative catalysis, it is possible to include perhaps also cascade reactions, in which the product of one catalytic cycle serves as an intermediate in another cycle. Of course, this is how biological systems operate. Looking into the future, one can envision the design of several coexisting, integrated catalytic cycles which can perform several parallel and consecutive tasks, a field that can be called "systems catalysis" (in analogy with "systems biology").

Author's Response to the Commentaries

Based on the thoughtful suggestions from the four commentators, I would like to add the following items to Chapter 11.

Section 11.3 **Cooperative Catalysis in Alkene Polymerization.** A mechanism for deactivation of

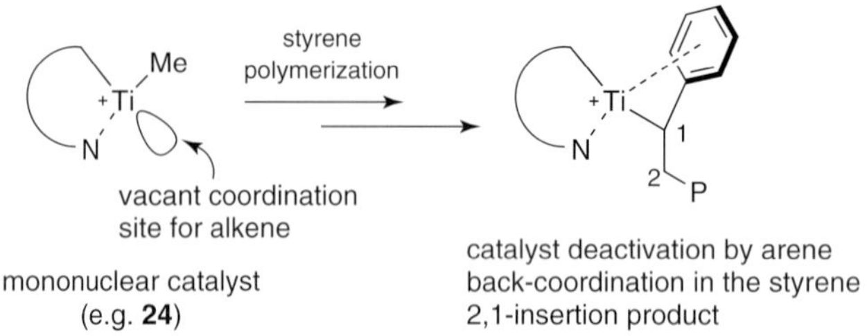

Scheme C8 Plausible origin for deactivation of mononuclear catalysts during styrene polymerization.

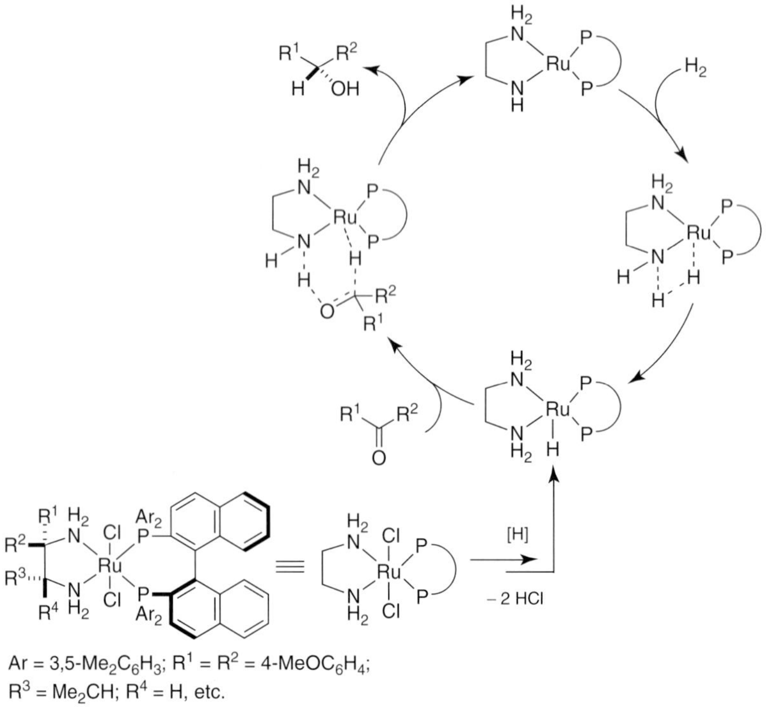

Ar = 3,5-Me$_2$C$_6$H$_3$; R^1 = R^2 = 4-MeOC$_6$H$_4$;
R^3 = Me$_2$CH; R^4 = H, etc.

Scheme C9 Proposed catalytic cycle of asymmetric hydrogenation of ketones.

mononuclear catalysts during styrene polymerization was not clearly described. The catalyst deactivation is thought to be due to saturation of coordination sites on the metal center by arene back-coordination in the styrene 2,1-insertion product (Scheme C8). In the case of binuclear catalysts, however, a vacant coordination site still exists on a metal center even when the arene back-coordination takes place to the other metal in a catalyst (see **32** and **33** in Scheme 11.4). Therefore, binuclear catalysts remain active for styrene polymerization.

Section 11.4 **Cooperative Catalysis in Hydrogen Activation/Generation**, Section 11.4.1 **Ligand–Metal Cooperation**. Although a catalytic cycle of the asymmetric transfer hydrogenation proposed by Noyori and co-workers was shown in Scheme 11.5, a similar catalytic cycle is also operating in the hydrogenation of ketones involving activation of dihydrogen (Scheme C9) [C25]. The ligand–metal (ruthenium) cooperation is key for the chemoselective reduction of polar carbonyl groups in the presence of C=C double bonds. This chemoselectivity is in sharp contrast to conventional hydrogenation reactions in the absence of amine ligands that hydrogenate C=C bonds preferentially to carbonyl groups through the ligation of the C=C bond to the metal. Synthesis of ruthenium(II)–amide complexes and their hydrogen activation ability were originally reported by Fryzuk *et al.* [C26].

References

C1. Ito, M., Hirakawa, M., Murata, K., and Ikariya, T. (2001) *Organometallics*, **20**, 379.

C2. Hedberg, C., Källström, K., Arvidsson, P.I., Brandt, P., and Andersson, P.G. (2005) *J. Am. Chem. Soc.*, **127**, 15083.

C3. Hamilton, R.J., Leong, C.G., Bigam, G., Miskolzie, M., and Bergens, S.H. (2005) *J. Am. Chem. Soc.*, **127**, 4152.

C4. Sandoval, C.A., Yamaguchi,Y., Ohkuma, T., Kato, K., and Noyori, R. (2006) *Magn. Reson. Chem.*, **44**, 66.

C5. Sohtome, Y., Tanaka, S., Takada, K., Yamaguchi, T., and Nagasawa, K. (2010) *Angew. Chem. Int. Ed.*, **49**, 9254–9257.

C6. Stoll, R.S. and Hecht, S. (2010) *Angew. Chem. Int. Ed.*, **49**, 5054–5075.

C7. Wang, J. and Feringa, B.L. (2011) *Science*, **331**, 1429–1432.

C8. Nicewicz, D.A. and MacMillan, D.W.C. (2008) *Science*, **322**, 77–80.

C9. Ikeda, M., Miyake, Y., and Nishibayashi, Y. (2010) *Angew. Chem. Int. Ed.*, **49**, 7289–7293.

C10. Rueping, M., Koenigs, R., and Atodiresei, I. (2010) *Chem. Eur. J.*, **16**, 9350–9365.

C11. Noyori, R. (2009) *Nat. Chem.*, **1**, 5–6.

C12. (a) Matsumoto, K., Sawada, Y., and Katsuki, T. (2008) *Pure Appl. Chem.*, **80**, 1071–1077; (b) Matsumoto, K., Sawada, Y., Saito, B., Sakai, K., and Katsuki, T. (2005) *Angew. Chem. Int. Ed.*, **44**, 4935–4930; (c) Sawada, Y., Matsumoto, K., Kondo, S., Watanabe, H., Ozawa, T., Suzuki, K., Saito, B., and Katsuki, T. (2006) *Angew. Chem. Int. Ed.*, **45**, 3478–3480; (d) Sawada, Y., Matsumoto, K., and Katsuki, T. (2007) *Angew. Chem. Int. Ed.*, **46**, 4559–4561; (e) Matsumoto, K., Oguma, T., and Katsuki, T. (2009) *Angew. Chem. Int. Ed.*, **48**, 7432–7435.

C13. Zhang, Z., Wang, Z., Zhang, R., and Ding, K. (2010) *Angew. Chem. Int. Ed.*, **49**, 6746–6750.

C14. North, M. (2010) *Angew. Chem. Int. Ed.*, **49**, 8079–8081.

C15. (a) Cai, Y.F., Liu, X.H., Hui, Y.H., Jiang, J., Wang, W.T., Chen, W.L., Lin, L.L., and Feng, X.M. (2010) *Angew. Chem. Int. Ed.*, **49**, 6160–6164; (b) Cai, Y.F., Liu, X.H., Jiang, J., Chen, W.L., Lin, L.L., and Feng, X.M. (2011) *J. Am. Chem. Soc.*, **133**, 5636–5639.

C16. Huang, S.X. and Ding, K. (2011) *Angew. Chem. Int. Ed.*, **50**, 7734–7736.

C17. Yamamoto, H. and Futatsugi, K. (2005) *Angew. Chem. Int. Ed.*, **44**, 1924–1942.

C18. Zuend, S.J., Coughlin, M.P., Lalonde, M.P., and Jacobsen, E.N. (2009) *Nature*, **461**, 968–971.

C19. Xu, H., Zuend, S.J., Woll, M.G., Tao, Y., and Jacobsen, E.N. (2010) *Science*, **327**, 986–990.

C20. van Leeuwen, P.W.N.M. (ed.) (2008) *Supramolecular Catalysis*, Wiley-VCH Verlag GmbH, Weinheim.

C21. Meeuwissen, J. and Reek, J.N.H. (2010) *Nat. Chem.*, **2**, 615–621.

C22. For a leading example of supramolecular asymmetric photocatalysis, see: Bauer, A., Westkamper, F., Grimme, S., and Bach, T. (2005) *Nature*, **436**, 1139–1140.

C23. For an excellent new example of asymmetric supramolecular primary amine catalysis in aqueous buffer, see: Hu, S., Li, J., Xiang, J., Pan, J., Luo, S., and Cheng, J.-P. (2010) *J. Am. Chem. Soc.*, **132**, 7216–7228.

C24. Margelefsky, E.L., Zeidan, R.K., and Davis, M.E. (2008) *Chem. Soc. Rev.*, **37**, 1118–1126.

C25. (a) Noyori, R., Koizumi, M., Ishii, D., and Ohkuma, T. (2001) *Pure Appl. Chem.*, **73**, 227–232; (b) Ohkuma, T., Koizumi, M., Doucet, H., Pham, T., Kozawa, M., Murata, K., Katayama, E., Yokozawa, T., Ikariya, T., and Noyori, R. (1998) *J. Am. Chem. Soc.*, **120**, 13529–13530.

C26. Fryzuk, M.D., Montgomery, C.D., and Rettigs, S.J. (1991) *Organometallics*, **10**, 467–473.

12
Flourishing Frontiers in Organofluorine Chemistry

G. K. Surya Prakash and Fang Wang

12.1
Introduction

Organofluorine chemistry concerns molecules with C–F bonds. As an element of extremes, fluorine has been continuously attracting chemists for more than a century [1]. Over the past two decades, fluoroorganics have been receiving increasing attention because of their unique chemical properties and promising applications in biology and materials science. Concomitant with the structural enrichment of fluoroorganics, the applicability of these compounds has been expanding significantly, which in turn triggers the impetus for synthesizing molecules with higher degrees of sophistication. Promoted by this intriguing demand-driven mechanism, organofluorine chemistry has been booming and offering fruitful results to chemists across many disciplines.

Possessing exceptional electronegativity, fluorine can drastically alter the chemical nature of organic molecules via various mechanisms, including electron-withdrawing effects, negative hyperconjugation, field effects, and charge–charge repulsion. On the other hand, owing to the presence of nonbonding electron lone-pairs, fluorine may also behave as an electron-donating substituent through resonance. Although carbon–fluorine bonds are of great thermodynamic stability, the extraordinarily high metal–fluoride lattice energies, compared with those of other metal halides, can lead to facile cleavage of C–F bonds under specific chemical regimes. For these reasons, fluorinating reagents, fluorine-containing building blocks, and the related reaction intermediates always demonstrate unusual inertness and/or unexpected instability, considerably limiting their chemical applicability. The theme of contemporary synthetic organofluorine chemistry is therefore focused on the development of efficient fluorinating reagents, novel fluorinated building blocks, and highly selective synthetic methodologies for the preparation of useful fluoroorganics. In particular, there have been an increasing number of chemical transformations requiring fluorinated organic compounds as privileged catalysts to achieve extraordinary selectivity and productivity. Owing to the substantial weakness of intermolecular interactions with nonfluoroorganics, perfluorinated organic compounds always exhibit low surface tension and unusual miscibility, which have been extensively exploited in

Organic Chemistry – Breakthroughs and Perspectives, First Edition. Edited by Kuiling Ding and Li-Xin Dai.
© 2012 Wiley-VCH Verlag GmbH & Co. KGaA. Published 2012 by Wiley-VCH Verlag GmbH & Co. KGaA.

applications as lubricants and fluorous reaction media in a plethora of synthetic processes [2].

Aside from the above-mentioned synthetic aspects, fluoroorganics have also been extensively utilized as valuable materials owing to their unique biological activities. In brief, the remarkable *"fluorine effects"* are primarily attributed to the combination of fluorine's extreme electronegativity and its steric resemblance to a proton. Biologically, fluorine substitution can lead to fundamental changes in lipophilicity (controlling the absorption and transportation of molecules *in vivo*), acidity, basicity, and protein-binding affinity of organic molecules. Although often believed to be a consequence of the thermodynamic stability of C–F bonds, the outstanding metabolic stability of fluorinated organic molecules (bioavailability) can be *de facto* attributed to the energetic unfavorableness of breaking a C–F bond to form a C–O bond [3]. Owing to the stereoelectronic effect, fluorine shows a profound tendency to adopt a *gauche* position to adjacent heteroatomic substituents. This anomeric/*gauche* effect always causes unexpected variations in bioactivity through stabilization of unusual conformations, and has been utilized as a conformational tool in organic and biological chemistry [4]. The applications of ^{19}F nuclear magnetic resonance (NMR) spectroscopy, ^{19}F NMR–magnetic resonance imaging (MRI) [5], and ^{18}F radiolabeling have become the most promising strategies in *in vivo* and *ex vivo* biological studies.

More importantly, the introduction of fluorinated moieties into catalysts or ligands for chemical transformations can lead to unusual variations in their catalytic activity by manipulating the electronic profiles of these molecules. In fact, a plethora of structurally simple fluorinated small molecules have been developed and thoroughly investigated as strong/superacids since the late nineteenth century [6]. Structurally complicated fluorine-containing molecules, on the other hand, have been extensively employed in metal-based catalytic and organocatalytic processes as ligands or catalysts over the past two decades [7]. Moreover, owing to the considerable increase in the van der Waals radii of fluoroalkyl groups compared with the corresponding non-fluorinated counterparts, fluoroalkylated catalysts can exhibit their stereoselections [8]. Over the past several years, *"antagonistic pairs"* (from the German *antagonistisches paar* [9] or *"frustrated Lewis pairs"* [10], usually composed of fluorinated Lewis acids, have received growing attention. Capable of activating a series of small molecules such as H_2, CO_2, tetrahydrofuran (THF), and alkenes, *"antagonistic pairs"*/*"frustrated Lewis pairs"* have shown immense potential in metal-free catalytic reactions [11].

As mentioned, organofluorine chemistry has become a rather comprehensive subject impacting a broad range of scientific fields. For the sake of brevity, we will refrain from compiling every detail of the development in organofluorine chemistry. Instead, from the viewpoint of synthetic chemists, we will focus the majority of our attention on the innovative achievements in the syntheses of useful fluorinated organic molecules over the past 20 years. We expect this brief review to provide some conceptual insight into the achievements and primary challenges in this field since the end of the last century.

12.2
Synthetic Approaches for the Introduction of Fluorine-Containing Functionalities and Related Chemistry

Fluoroorganics are extremely scarce in Nature. Hence the primary challenge encountered in the early stages of organofluorine chemistry was to construct carbon–fluorine bonds efficiently (Figure 12.1, bond α-β). Historically, attempts to obtain fluorinated organic compounds via synthetic approaches date back to the eighteenth century [12]. The first construction of a C–F bond was achieved by Dumas and Péligot, who successfully prepared methyl fluoride by treating dimethyl sulfate with KF, which marked the beginning of organofluorine chemistry [13]. Half a century later, elemental fluorine was successfully isolated in 1886 by Moissan, who was awarded the Nobel Prize in Chemistry in 1906 for this achievement [14]. As such, in addition to hydrofluoric acid and metallic fluorides, elemental fluorine was added to the toolbox of fluorinating reagents to achieve new fluorination methods. More efficient preparative methods for fluoroorganics of practical interest were then pioneered by Swarts, who first exploited halogen exchange processes under Lewis acid conditions in 1892 [15]. In the early part of the last century, various important fluorine-containing chemicals and materials became accessible on an industrial scale by means of various practical protocols, such as the Balz–Schiemann reaction [16], the cobalt trifluoride process by Fowler *et al.* [17], electrochemical fluorination (ECF) by Simons and co-workers [18], and the Halex process [19]. Since the availability of fluorinated organic molecules has increased dramatically, the investigations of chemical transformations of these compounds eventually evolved as building-block strategies in the mid-twentieth century. A series of synthetic approaches were then adopted as powerful tools for the introduction of fluorinated functionalities (primarily fluoroalkyl moieties) in the 1980s (Figure 12.1, formation of *β-γ* bonds, *γ-δ* bonds, and so on). Since the substrates are "preinstalled" with C–F bonds, the harsh conditions usually involved in C–F bond formation can be avoided, which allows the construction of more complex molecular structures with enhanced selectivity.

Utilizing a variety of fluorination/fluoroalkylation reagents and protocols, organofluorine chemists have been able to prepare fluoroorganics based on nucleophilic, electrophilic, carbene, ylide, electrochemical, and radical pathways [20]. To date, the

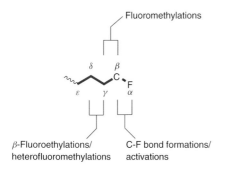

Figure 12.1 Chemical bonds of interest to organofluorine chemists.

majority of the effort has been devoted to the development of versatile reagents and novel synthetic methods applicable under milder conditions with higher efficiency. Aiming at syntheses of stereogenic molecules of potential bioactivity, the asymmetric introduction of fluorinated moieties has become an attractive field. In addition to C–F bond formation reactions, C–F bond activations have also been revisited for their promising utility in the preparation of partially fluorinated molecules [21]. Moreover, transition metal-mediated and -catalyzed fluorinations and fluoroalkylations have burgeoned recently, and have demonstrated prominent synthetic advantages, making them valuable synthetic tools in organofluorine chemistry. Very importantly, the discovery of fluorinase enzyme capable of catalyzing C–F bond formation by O'Hagan and co-workers also paved the way for the biosyntheses of fluorinated organic compounds [22].

12.2.1
Novel Fluorinating Reagents and C–F Bond Formation Reactions

Conceptually, the formation of C–F bonds can be achieved by four mechanistic pathways: radical fluorination (direct fluorination), ECF, nucleophilic fluorination, and electrophilic fluorination (Figure 12.2). Direct fluorination and ECF have been successfully utilized in the preparation of perfluorinated organic compounds in many industrial processes. In the early 1970s, Margrave *et al.* demonstrated the controllable direct fluorination reaction of hydrocarbons (aerosol direct fluorination process) under a radical pathway to afford various useful perfluorochemicals such as perfluorinated alkanes and perfluorinated ethers [23]. Nevertheless, the applicability of these methods is always limited when it comes to fluorine-containing moieties bearing reactive functionalities owing to their unsatisfactory selectivity and harsh operating conditions. In contrast, nucleophilic and electrophilic fluorinations have prevailed in terms of selectivity and functional group compatibility.

12.2.1.1   Nucleophilic Fluorinations
As the first fluorination method to be utilized, nucleophilic fluorination is still among the most important routes to fine fluorochemicals in industry [24]. Apart from large-scale syntheses, its importance has also been reflected in the extensive applications in preparing

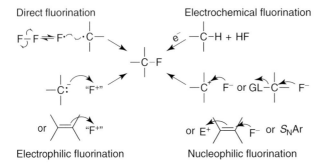

Figure 12.2   Four synthetic pathways for C-F bond formation.

[18]F-radiolabeled compounds for positron emission tomography (PET) [25]. Since nucleophilic fluorinations of halides and pseudohalides undergo S_N2 or S_NAr mechanisms in many cases, the displacement generally allows the introduction of C–F bonds with good regio- and/or stereospecificity. Nevertheless, the low nucleophilicity of the fluoride ion, arising from its poor polarizability, considerably limits the applications of the methodology. In order to tackle this problem, two strategies frequently have been employed, namely (i) the enhancement of the nucleophilicity of the fluoride ion by weakening its solvation, hydrogen bonding, and interactions with counter ions (metallic cations, etc.) and (ii) the activation of the electrophiles with Brønsted or Lewis acids.

For the first strategy, much of the effort has been concentrated on seeking uncoordinating cationic species (such as tetraalkylammonium and tetraphenylphosphonium ions) to render the so-called "naked" fluoride ion as denoted by Liotta and Harris [26]. However, owing to their extremely hygroscopic nature, these reagents are always associated with practical problems, such as storage and handling. Moreover, other than tetramethylammonium fluoride [27], anhydrous tetraalkylammonium fluorides are usually inaccessible in the solid state because of the significantly increased basicity of the fluoride ion, which can lead to severe decomposition of the cations. Surprisingly, the preparation of anhydrous tetra-n-butylammonium fluoride (TBAF$_{anh}$) was achieved by treating tetrabutylammonium cyanide with hexafluorobenzene in dimethyl sulfoxide (DMSO) by Sun and DiMagno [28]. The same group further demonstrated the nucleophilic fluorination of electron-deficient aromatic compounds using this reagent under ambient conditions [29].

Since the late 1970s, instead of modifying the cations, chemists have shifted their attention to stabilizing the fluoride ion by using difluorinated hypervalent silicates and stannates, which have demonstrated superior chemical and physical properties compared with alkali metal and tetraalkylammonium fluorides (Scheme 12.1) [30–32]. Alternatively, probably one of the most impressive findings was achieved in 2006 by Chi and co-workers, who discovered that nonpolar protonic tertiary alcohols can

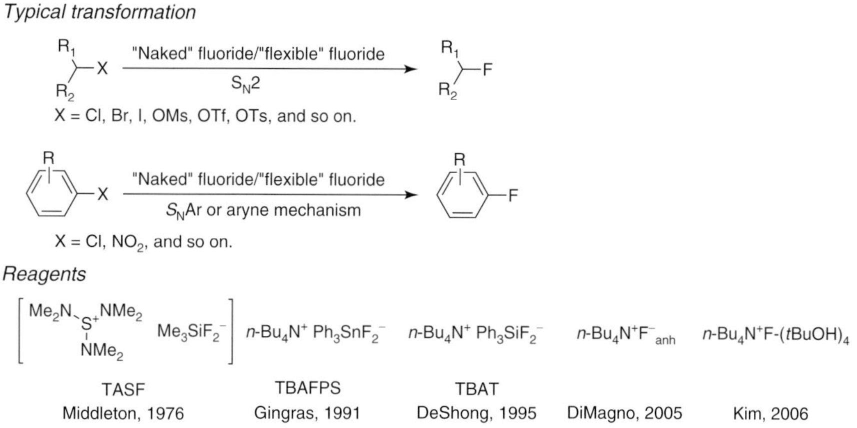

Typical transformation

X = Cl, Br, I, OMs, OTf, OTs, and so on.

X = Cl, NO$_2$, and so on.

Reagents

TASF	TBAFPS	TBAT		
Middleton, 1976	Gingras, 1991	DeShong, 1995	DiMagno, 2005	Kim, 2006

Scheme 12.1 Nucleophilic fluorination reactions and the related reagents.

significantly promote nucleophilic fluorination with alkali metal fluorides [33]. Contrary to conventional theory, the beneficial effects of tertiary alcohols were further elucidated as (i) the solvation of the fluoride ion by weakening ionic metal–fluoride bonding; (ii) formation of more nucleophilic *"flexible"* fluoride by hydrogen bonding with suitable strength; (iii) increasing nucleofugality of leaving groups by their stabilization in the reaction medium; and (iv) inhibiting side reactions such as elimination and intramolecular alkylation by decreasing the basicity of fluoride ion [34]. More recently, Kim *et al.* reported the preparation of tetrabutylammonium tetra(*tert*-butyl alcohol)-coordinated fluoride as a low-hygroscopic fluoride source [35]. As indicated by the X-ray crystal structure, the fluoride ion was confirmed to coordinate with four *tert*-butyl alcohol molecules through hydrogen bonding (Scheme 12.1). Synthetically, the fluorinating power of TBAF–(*t*BuOH)$_4$ was shown to be substantially stronger in comparison with other F$^-$ sources.

As mentioned previously, facile nucleophilic fluorination can be also accomplished through the activation of electrophiles using Brønsted and/or Lewis acids, and such chemistry dates back to the late nineteenth century. Over the past century, a series of important fluorination systems have been developed based on this mechanism, including SbF$_3$ – HF, SbF$_5$ – HF, AlF$_3$ – HF, hypervalent halogen fluorides [36], and amine–hydrogen fluoride (such as 70% HF–pyridine, Olah's reagent [37]). In addition to these systems, gaseous sulfur tetrafluoride (SF$_4$) [38], PhSF$_3$ [39], and its derivatives are among the most versatile nucleophilic fluorinating reagents with significant synthetic usefulness [40]. The treatment of alcohols, carbonyl compounds, and carboxylic acids with these reagents results in the replacement of the C–O bonds with C–F bonds to generate corresponding mono-, di-, and trifluoromethylated products (the deoxofluorination process) (Scheme 12.2). Notably, undergoing an S$_N$2 pathway, deoxofluorination of chiral secondary alcohols may afford products with inverted configurations with high stereoselectivity, which has been shown to be a critical synthetic protocol towards the

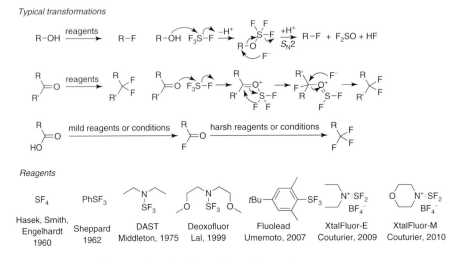

Scheme 12.2 Development of S-F bond-based deoxofluorinating reagents.

formation of stereogenic fluorinated carbon centers [41]. Owing to the toxicity and high volatility of SF$_4$, N,N-diethylaminosulfur trifluoride (DAST) was developed by Middleton as a superior deoxofluorinating agent that has been used for over 30 years [42]. However, the storage and handling of DAST are tedious because of its violently explosive nature and insufficient thermal stability. In order to overcome these deficiencies, intense efforts have been made to develop user-benign alternatives to DAST. Over the past several years, a variety of deoxofluorinating reagents have been introduced with both satisfactory reactivity and enhanced stability, such as bis(methoxyethyl)aminosulfur trifluoride (Deoxo-Fluor) [43], 4-*tert*-butyl-2,6-dimethylphenylsulfur trifluoride (Fluolead) [44], and dialkylaminodifluorosulfinium salts (XtalFluor-E and -M) [45].

It is worth mentioning that, in addition to the above-mentioned fluorinations, a series of fluorine-containing Lewis acids have been used as fluoride sources in ring-opening reactions. BF$_3$ · Et$_2$O was employed by House in the ring opening of epoxides in 1956 to afford the corresponding hydrofluorinated products [46]. The groups of Nakayama, Hu, and Hou expanded this chemistry into the stereoselective ring-opening reactions of aziridines (Scheme 12.3) [47]. Recently, Davies and co-workers reported the efficient ring-opening hydrofluorination of a range of substituted aryl epoxides to *syn*-fluorohydrins [48]. Interestingly, the stereochemical outcomes explicitly indicated substantially different mechanistic pathways during the course of these transformations. As illustrated in Scheme 12.3, in contrast to the ring-opening reactions undergoing an S$_N$2-type pathway to afford configurationally inverted products, carbocation intermediates are

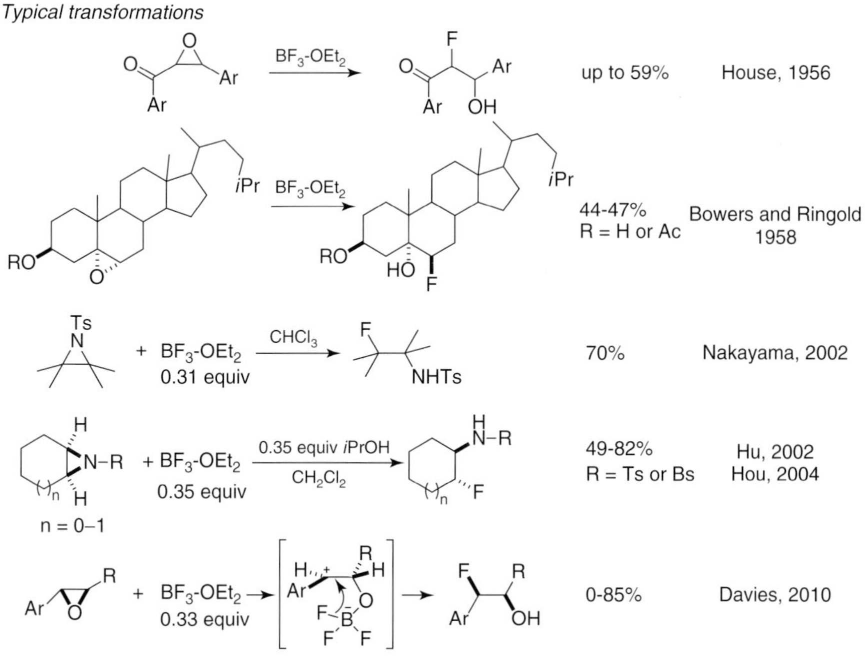

Typical transformations

Scheme 12.3 Ring-opening reactions of aziridines and epoxides using BF$_3$·Et$_2$ as the fluorine source.

involved in the ring opening of epoxides resulting in conservation in stereochemical profiles.

Intriguingly, whereas a large number of reactant-controlled *stereospecific* nucleophilic fluorinations have been demonstrated, only a handful of reagent-controlled stereoselective fluorinating methods based on fluorides are known. Since the S_N2 mechanistic pathway shows a strong tendency to afford configurationally inverted products, kinetic resolution has frequently been involved in asymmetric nucleophilic fluorinations of racemic substrates. Hann and Sampson first described enantioselective deoxofluorination using a stoichiometric amount of an optically enriched DAST analog (**1**) [49] (Scheme 12.4). However, no appreciable stereoselectivity was observed. Instead of using

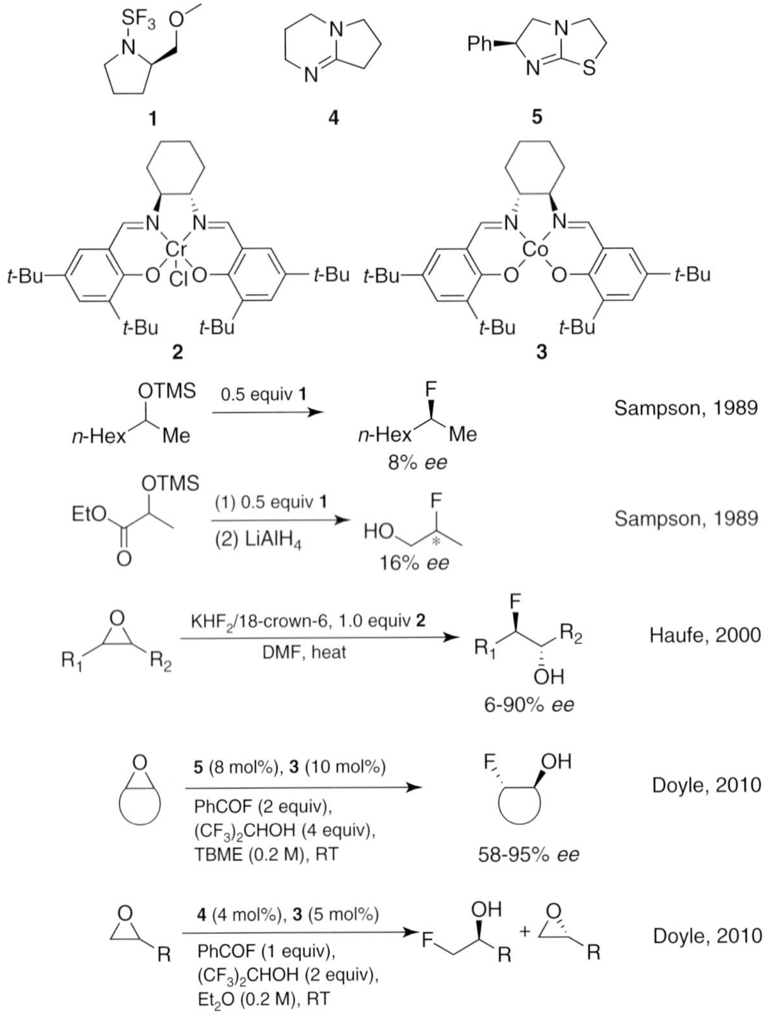

Scheme 12.4 Development of asymmetric nucleophilic fluorinations.

homochiral fluorinating reagents, Bruns and Haufe exploited stoichiometric Jacobsen's (salen) complexes (2) and hydrofluorides to achieve the asymmetric ring opening of epoxides with elevated enantiomeric excess (*ee*) [50] (Scheme 12.4). Recently, a more encouraging approach toward this goal was developed by Doyle *et al.* [51]. Realizing that conventional reaction conditions suffer from competitive background reactions and catalyst inhibition, they adopted specially designed co-catalytic systems to achieve catalytic asymmetric ring opening of epoxides by the fluoride anion.

As mentioned previously, the synthesis of various aryl fluorides was achieved over a century ago, and has been successfully commercialized. In particular, the nucleophilic replacement of electron-deficient aryl halides or pseudohalides (activated) with fluoride can be readily achieved to render regiospecific products. Owing to the difficulty in forming Meisenheimer complexes, nonactivated substrates are always inert under regular non-catalytic nucleophilic fluorinating conditions. To overcome this problem, organometallic fluorine chemistry was introduced and has been vibrant during the last several years [52]. Unlike other aromatic halogenations (Cl, Br, I) mediated by transition metals, the analogous fluorination reactions were a long-standing challenge. Conceptually, the catalytic cycle is composed of three steps: (i) oxidative addition of aryl halides or triflate with Pd(0) complexes; (ii) halogen exchange with fluoride; and (iii) reductive elimination of the fluorinated Pd(II) complexes to release the corresponding aryl fluorides (Figure 12.3).

Although the oxidative addition of Ar–X had been well established over the past several decades, the other two steps were in question (Figure 12.3). The formation of low-valent transition metal–fluoro compounds turned out to be the first obstacle in this catalytic process, primarily due to incompatibility in the hardness of fluoride and transition metal centers [53]. Dixon and McFarland reported the detection of $[(Et_3P)_3Pd^{II}F]^+$ via ^{19}F NMR spectroscopy; however, the attempted isolation of the complex resulted in severe decomposition to yield Et_3PF_2 and Pd(0) complexes (Scheme 12.5) [54]. Grushin's group made pioneering attempts at the catalytic fluorination of arenes using a series of transition metals (including Pd, Rh, Ni, and Pt) in the late 1980s to early 1990s [55]. The unproductive results led to further examination of the mechanistic validity of the proposal by the synthesis of the key reaction intermediates. The synthesis and

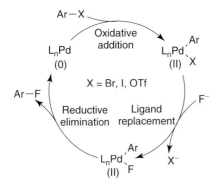

Figure 12.3 Pd-catalyzed nucleophilic fluorination of aryl halides and aryl triflate.

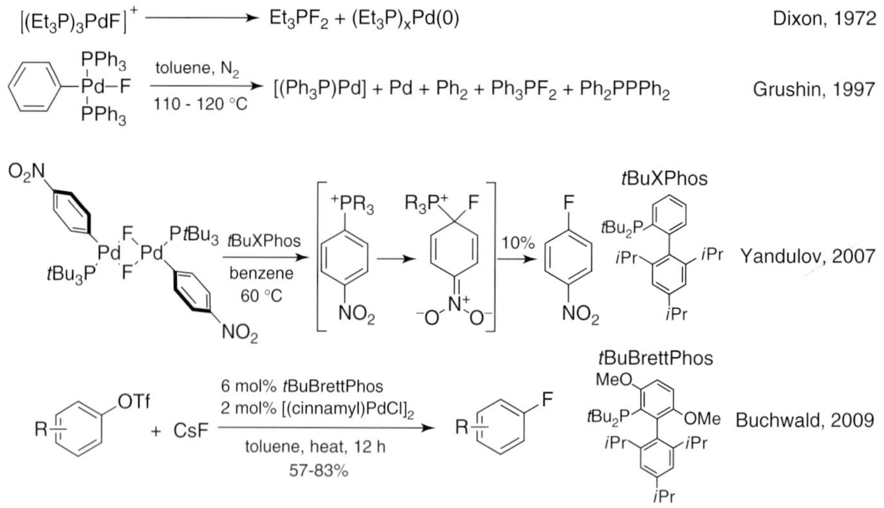

$$[(Et_3P)_3PdF]^+ \longrightarrow Et_3PF_2 + (Et_3P)_xPd(0) \qquad \text{Dixon, 1972}$$

$$\underset{\substack{\text{toluene, N}_2 \\ 110\text{-}120\ °C}}{\longrightarrow} [(Ph_3P)Pd] + Pd + Ph_2 + Ph_3PF_2 + Ph_2PPPh_2 \qquad \text{Grushin, 1997}$$

Yandulov, 2007

Buchwald, 2009

Scheme 12.5 Development of palladium-mediated nucleophilic fluorination of arenes.

isolation of arylpalladium(II) fluoride $[(Ph_3P)_2Pd^{II}(Ph)F]$ were eventually accomplished by ultrasound-promoted I–F exchange or neutralization of M–OH species using $Et_3N \cdot (HF)_3$, which demonstrated the theoretical possibility of halogen exchange [55–57]. On the other hand, the reductive elimination of arylpalladium(II) fluorides to generate aryl fluorides seems to be more difficult because of the facile intramolecular nucleophilic attack of the fluoride on the phosphine ligands (Scheme 12.5) [58]. The decomposition of arylpalladium(II) fluorides was observed to yield a small quantity of aryl fluoride [59]; however, whether this outcome is a consequence of reductive elimination is still questionable (Scheme 12.5) [60]. Although these elegant mechanistic studies have not completely ascertained the catalytic cycle, a palladium-catalyzed nucleophilic fluorination of aryl triflates was achieved by Buchwald and co-workers, who finally put the last piece into this puzzle experimentally (Scheme 12.5) [61]. In addition to good yields and selectivity, the synthetic protocol has also demonstrated satisfactory functional group compatibility and broad substrate scope, which permits the facile syntheses of various fluorinated arenes.

Notably, in 2002, Subramanian and Manzer described the Cu-mediated oxidative fluorination of benzene in gas phase to form fluorobenzene as the exclusive product

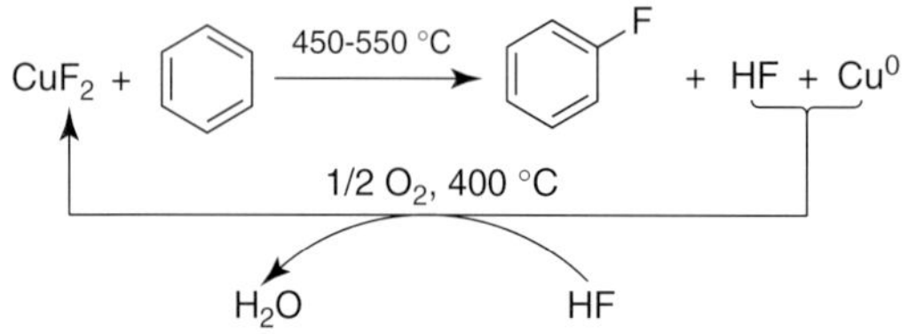

Scheme 12.6 Cu-mediated nucleophilic fluorination of arenes.

(Scheme 12.6) [62]. In spite of a "greener" protocol, the reaction suffered from unsatisfactory yield and poor recyclability of the metal reagent. Janmanchi and Dolbier further improved the process by using nominal $CuAl_2F_8$ as an effective fluorinating agent capable of multiple regeneration [63]. In comparison with the Pd(II)-catalyzed nucleophilic aromatic fluorinations, these synthetic approaches are exclusively restricted to simple aromatics due to severe reaction conditions.

12.2.1.2 Electrophilic Fluorinations

Unlike the fluoride ion, free "F^+" remains unknown in the condensed phase. As a consequence, compared with nucleophilic fluorinations, the electrophilic construction of C–F bonds was established much later owing to the dearth of appropriate "F^+" sources. The first electrophilic fluorination was *de facto* achieved by Inman *et al.* in the reaction of active methylene groups with perchloryl fluoride ($FClO_3$) [64] to afford the corresponding fluorinated compounds [65]. In 1968, Barton and Hesse introduced the concept of electrophilic fluorination for the first time by demonstrating fluoroxytrifluoromethane (CF_3OF) [66] as a versatile reagent for electrophilic olefinic and aromatic fluorinations [67]. Their elegant work further led to the development of a family of electrophilic fluorinating reagents based on the O–F moiety [68], including CF_3CO_2F [69], HOF [70], and $CsSO_4F$ [71]. Furthermore, the application of xenon difluoride (XeF_2) [72] was reported by Yang and co-workers for the fluorination of alkenes and aromatics [73]. In addition to utilizing the labile O–F and Xe–F bonds, the weakness of the F–F bond also allows electrophilic fluorination using elemental fluorine. Cacace and Wolf demonstrated the selective electrophilic aromatic fluorination with highly diluted molecular fluorine [74]. Rozen and Gal later reported the electrophilic fluorination of saturated hydrocarbons using an $F_2-CHCl_3/CFCl_3$ system through a C–H bond insertion pathway, which led to monofluorinated products [75]. Despite their frequent utilization as electrophilic fluorine sources prior to the 1990s, the applicability of these reagents was still restricted by their drawbacks, such as low stability, unsatisfactory selectivity, and limited commercial availability.

Although extensive studies were focused on the O-F-based electrophilic fluorinating reagents from the late 1960s to the early 1980s, the difficulties with their commercialization significantly impeded the applications of these reagents [68]. Instead, N-F-based systems emerged as versatile reagents [76]. Interestingly, the utility of the labile N–F linkage as a fluorinating agent was discovered prior to the O–F reagents [77]. However, because of the poor yields and harsh conditions, for years these reagents did not attract sufficient attention. Since the 1980s, and particularly from the late 1980s to the early 1990s, realizing the potential synthetic usefulness of the N–F class of agents, organofluorine chemists have devoted tremendous efforts to this field. During this N–F reagent era, a range of compounds (neutral R_2NF and R_3N^+F salts) with different fluorinating capabilities [78] were explored, including 1-fluoro-2-pyridone (Tee *et al.* in 1983) [79], N-fluoropyridinium triflate (Umemoto *et al.* in 1986) [80], N-fluoroperfluoroalkylsulfonimides (DesMarteau and co-workers in 1987) [81], and N_2F^+ and NF_4^+ salts [82] (Scheme 12.7). Eventually, two compounds that evolved from this class of reagents were commercialized in the early 1990s, namely N-fluorobenzenesulfonimide (NFSI) (Differding and Ofner in 1991) [83] and 1-chloromethyl-4-fluoro-1,4-diazoniabicyclo[2.2.2]octane bis(tetrafluoroborate)

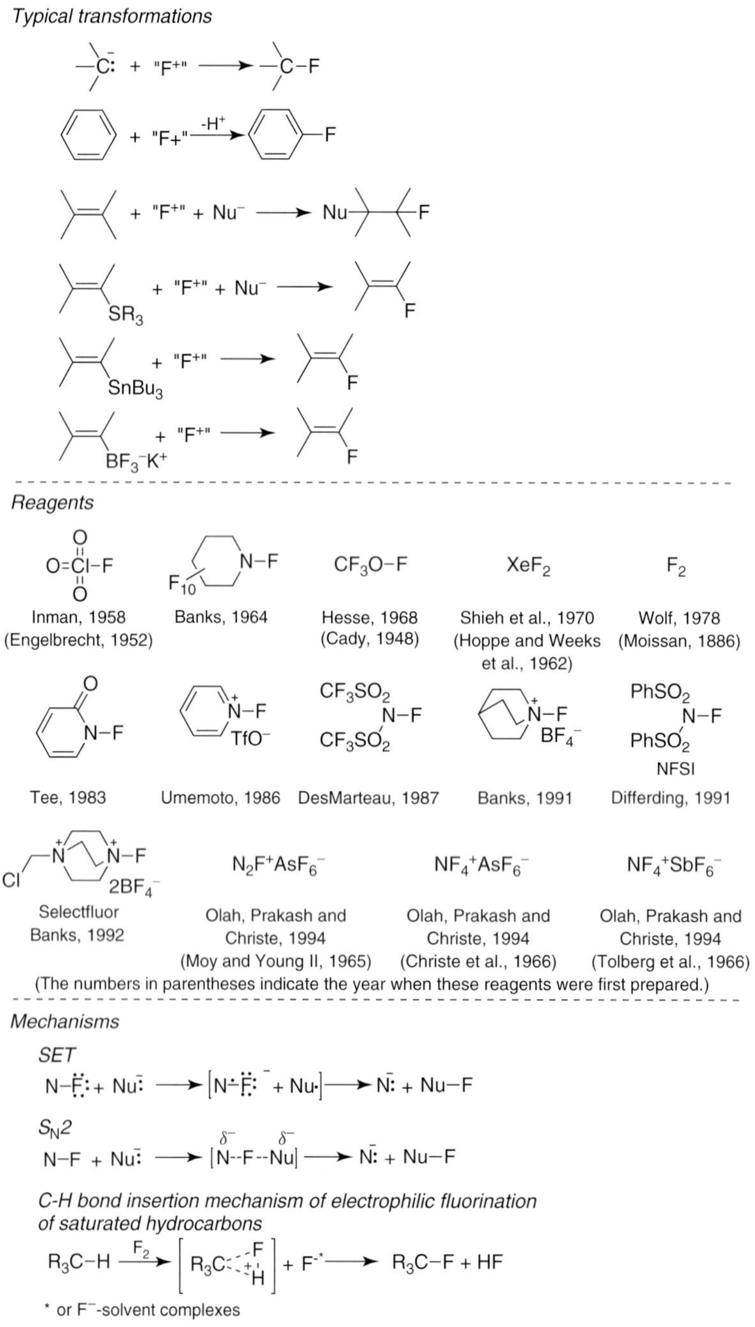

Typical transformations

Reagents

O=CI-F	F₁₀ N–F	CF₃O–F	XeF₂	F₂
Inman, 1958 (Engelbrecht, 1952)	Banks, 1964	Hesse, 1968 (Cady, 1948)	Shieh et al., 1970 (Hoppe and Weeks et al., 1962)	Wolf, 1978 (Moissan, 1886)

Tee, 1983   Umemoto, 1986   DesMarteau, 1987   Banks, 1991   Differding, 1991

Selectfluor
Banks, 1992   Olah, Prakash and   Olah, Prakash and   Olah, Prakash and
                     Christe, 1994   Christe, 1994   Christe, 1994
            (Moy and Young II, 1965)   (Christe et al., 1966)   (Tolberg et al., 1966)

(The numbers in parentheses indicate the year when these reagents were first prepared.)

Mechanisms

SET

$$N-\ddot{F}: + Nu\bar{:} \longrightarrow [N\dot{-}\ddot{F}:^{-} + Nu\cdot] \longrightarrow N\ddot{:} + Nu-F$$

S_N2

$$N-F + Nu\bar{:} \longrightarrow [N--F--Nu]^{\delta^-\delta^-} \longrightarrow N\ddot{:} + Nu-F$$

C-H bond insertion mechanism of electrophilic fluorination of saturated hydrocarbons

$$R_3C-H \xrightarrow{F_2} \left[R_3C\cdots^{+}_{H}^{F} \right] + F^{-*} \longrightarrow R_3C-F + HF$$

* or F⁻-solvent complexes

Scheme 12.7   Electrophilic fluorinations and the related reagents.

(Selectfluor) (Banks *et al.* in 1992) [84]. To date, NFSI and Selectfluor are among the most commonly employed electrophilic fluorinating agents, enabling a variety of C–F bond-forming reactions [76, 85, 86]. Theoretically, there has been much controversy about the mechanistic details of electrophilic fluorinations since the early 1990s [76, 86, 87], (the mechanistic details of electrophilic fluorinations have been thoroughly discussed in [76] and [86]). Based on a series of mechanistic studies, two pathways are usually proposed: single-electron transfer (SET) and nucleophilic substitution (S_N2), depending on the reagents, substrates, and reaction conditions (Scheme 12.7, *Mechanisms*). It is also important to point out that the electrophilic fluorination of saturated hydrocarbons using F_2, N_2F^+, and NF_4^+ salts always proceeds through a C–H bond insertion pathway instead of the two mechanisms mentioned above [75, 88].

As discussed previously, although a few enantioselective protocols have been reported, nucleophilic fluorinations usually provide configurationally inverted products *stereospecifically* via an S_N2 reaction pathway. In comparison, asymmetric electrophilic fluorinations are spectacular. Without the mechanistic restriction (the S_N2 configurational inversion requirement), electrophilic fluorinating reactions are capable of *stereoselective* construction of non-racemic stereogenic fluorinated carbon centers under various chemical regimens [86, 89].

Enantiocontrolled electrophilic fluorination dates back to the late 1980s. α-Fluorination of carbonyl compounds is of particular interest and has been one of the major synthetic goals in asymmetric fluorination. Differding and Lang first documented camphor-based *N*-fluorosultam (**6**) as the homochiral fluorinating reagent in the asymmetric fluorination of metal enolates (Scheme 12.8) [90]. Thereafter, a series of *N*-fluorosultams were synthesized and explored by Davis *et al.* (**7** [91]) and Takeuchi and co-workers (**8** [92], **9** [93]). Although an appreciable increase in *ee* was achieved by appropriate structural modifications, the synthetic usefulness of these approaches was still questionable to a large extent because of their low efficacy. Following the fluorine-transferring reaction of the quinuclidine moieties in cinchona alkaloids [94], Cahard *et al.* (**10** [95]) and Takeuchi and co-workers (**11** [96]) independently discovered the efficient asymmetric fluorinations of enolates by delivering the stereo information from *in situ*-prepared *N*-fluoro-cinchona alkaloid scaffolds. More importantly, a breakthrough was achieved by Hintermann and Togni, who developed the first efficient catalytic asymmetric fluorination method [97]. Instead of using stoichiometric chiral fluorinating agents, TADDOL–titanium complexes (**12**) were exploited as Lewis acid catalysts in combination with Selectfluor in the enantioselective α-fluorination of β-keto esters. Likewise, a large number of chiral Lewis acid complexes, including late-transition metal-based Lewis acids (**13**), have been found to be efficient catalysts in asymmetric fluorination [89d, 98]. In recent years, organocatalysis has emerged as a valuable synthetic tool in asymmetric chemical transformations [99]. In 2002, Kim and Park documented the catalytic enantioselective fluorination of β-keto esters using a cinchonidine-based phase-transfer catalyst (PTC) (**14**) to afford the products in moderate *ee* [100]. Shortly afterwards, in 2005, three research laboratories independently developed asymmetric α-fluorinations of aldehydes via enamine catalysis using different catalysts. Beeson and MacMillan. utilized imidazolidinone (**15**) as the asymmetric catalyst in the transformation to achieve excellent enantioselectivity [101]. Jørgensen and co-workers accomplished the α-fluorination of aldehydes employing the

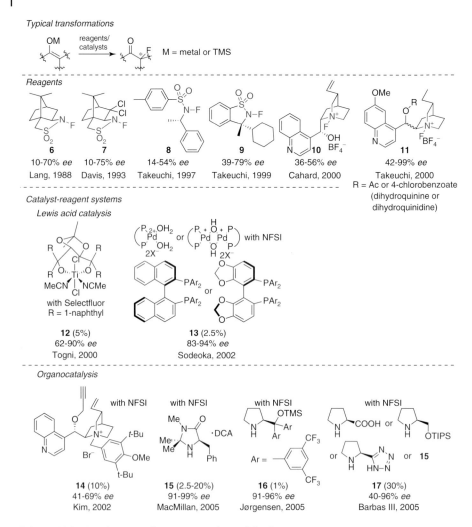

Scheme 12.8 Development of asymmetric electrophilic fluorinations.

remarkably efficient proline derivative **16** as the catalyst to afford the products with good enantiomeric discrimination [102]. Barbas and co-workers, on the other hand, examined a variety of proline analogs and imidazolidinone (**15**), which revealed the latter as the optimal catalyst [103].

Electrophilic aromatic fluorination has been known for several decades since Hesse and co-workers [67b] and Shaw *et al.* [104] reported the reactions between electrophilic fluorinating reagents and arenes in the late 1960s [105]. In general, direct electrophilic fluorination of electron-rich aromatic compounds is feasible, but the fluorination of the electron-deficient arenes which exhibit substantially lower reactivity was achieved under superacid conditions, and Prakash and co-workers reported the electrophilic aromatic fluorination of nonactivated arenes using Selectfluor under superacid conditions

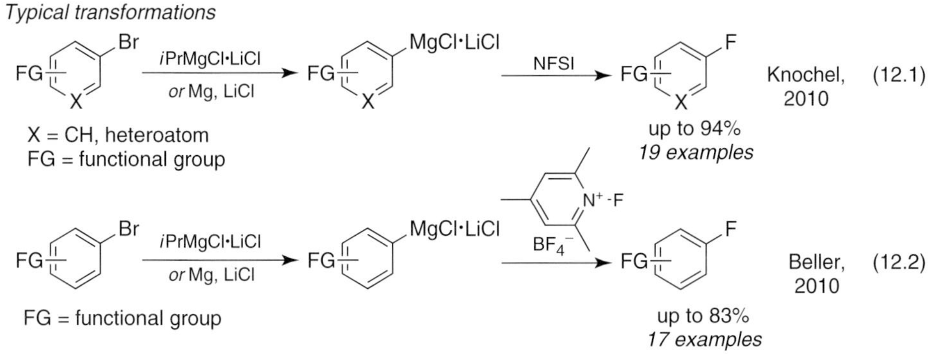

Typical transformations

X = CH, heteroatom
FG = functional group

Knochel, (12.1)
2010

up to 94%
19 examples

FG = functional group

Beller, (12.2)
2010

up to 83%
17 examples

Scheme 12.9 Electrophilic fluorinations of arylmagnesium reagents.

[106]. However, these methods always proceed with unsatisfactory regioselectivity. An alternative synthetic approach treats organometallic reagents with "F$^+$" to afford the corresponding fluoroarenes; however, it is limited to simple substrates [107]. Recently, Knochel and co-workers reported a convenient electrophilic fluorinating protocol for the functionalization of arylmagnesium compounds in the presence of LiCl and NFSI [108]. The methodology has shown remarkable applicability to a broad range of aromatic compounds, including highly functionalized arenes and electron-deficient aromatic and heterocyclic compounds [Scheme 12.9, Eq. (12.1)]. Independently, Beller and co-workers reported a similar procedure using N-fluoro-2,4,6-trimethylpyridinium tetrafluoroborate as the fluorine source to afford various fluoroarenes [109] [Scheme 12.9, Eq. (12.2)].

Apart from those employing a stoichiometric amount of aryl organometallic reagents, transition metal-catalyzed electrophilic aromatic fluorinations have also emerged in recent years [52, 110, 111]. Although the conventional arene fluorination methods will still govern the synthesis of fluorinated aromatics on an industrial scale in the foreseeable future, these novel approaches are believed to be superior in sophisticated syntheses due to their milder reaction conditions. In 2006, Sanford and co-workers demonstrated the first palladium-catalyzed electrophilic aromatic fluorination by means of C–H activation [Scheme 12.10, Eq. (12.3)] [112]. In the presence of directing groups, the method permits the formation of ortho-fluorinated arenes in moderate to good yields. However, the harsh reaction conditions and unsatisfactory regioselectivity are the major obstacles to its practical synthetic utility. Yu and co-workers reported the conceptually similar electrophilic fluorination of N-benzyltriflamides with palladium(II) triflate as the catalyst [Scheme 12.10, Eq. (12.4)] [113]. Although the directing group can be readily converted into other functionalities, the above-mentioned drawbacks remain. In comparison, Ritter and co-workers attempted to achieve facile electrophilic fluorinations undergoing a transmetallation mechanism, which are expected to render enhanced selectivities and synthetic utility. As depicted in Scheme 12.10, aryl–fluorine bonds can be formed by treating aryl stannanes [Eq. (12.5)] [114] or boronic acids [Eq. (12.6)] [115] with "F$^+$" reagents in the presence of stoichiometric amounts of silver triflate [116]. Most recently, a catalytic variant of these methodologies was established by the same group that allows regiospecific transformation of a C–Sn bond into the corresponding C–F bond

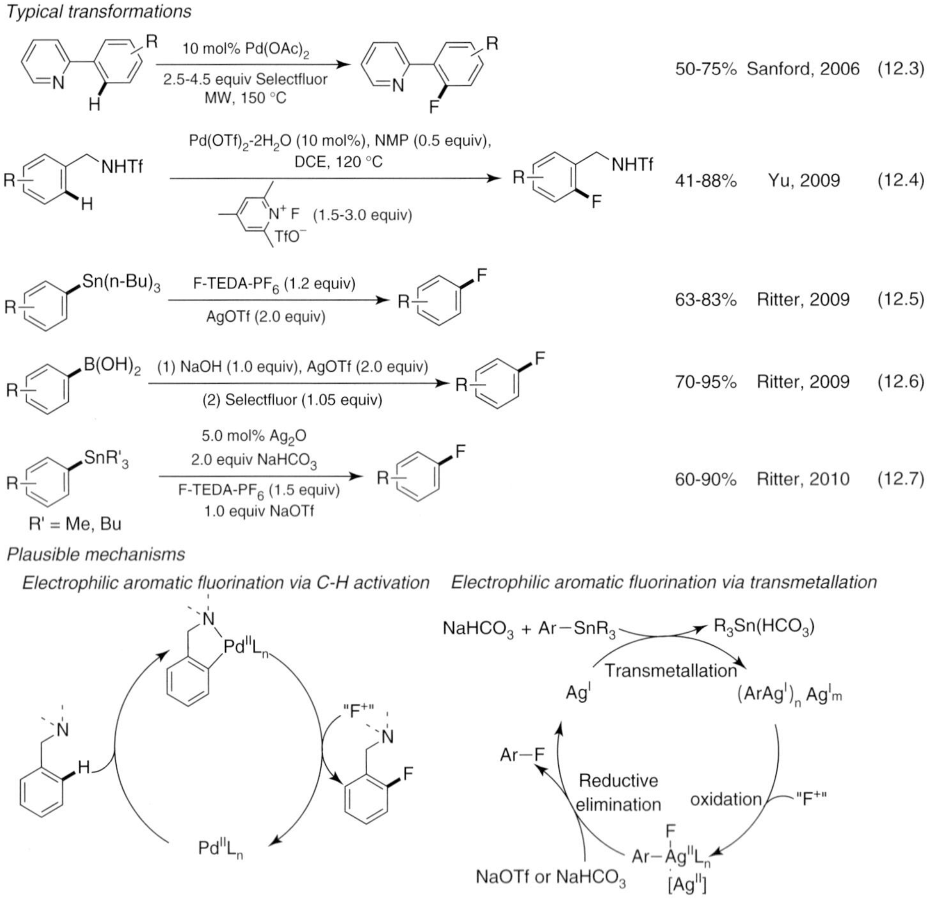

Typical transformations

	50–75%	Sanford, 2006	(12.3)
	41–88%	Yu, 2009	(12.4)
	63–83%	Ritter, 2009	(12.5)
	70–95%	Ritter, 2009	(12.6)
	60–90%	Ritter, 2010	(12.7)

Plausible mechanisms

Electrophilic aromatic fluorination via C-H activation *Electrophilic aromatic fluorination via transmetallation*

Scheme 12.10 Recent developments in transition metal-mediated electrophilic fluorinations of arenes.

[117]. Interestingly, although further experiments remain necessary for ascertaining the mechanism, a silver redox catalytic pathway involving aggregated silver complexes and high-valent silver species was proposed for this cross-coupling reaction.

12.2.2
Efficient Trifluoroalkylation Reactions

Despite the fact that fluoroalkyl moieties can be obtained by direct fluorinations with SF_4 derivatives, fluorides, and/or electrophilic fluorinating agents, utilization of fluoroalkyl synthons prevails unequivocally because of higher efficiency and functional group compatibility. Owing to the lack of versatile fluoroalkylating precursors (particularly fluoromethylating reagents), the introduction of fluorinated motifs by direct construction of C–F bonds monopolized synthetic organofluorine chemistry until the

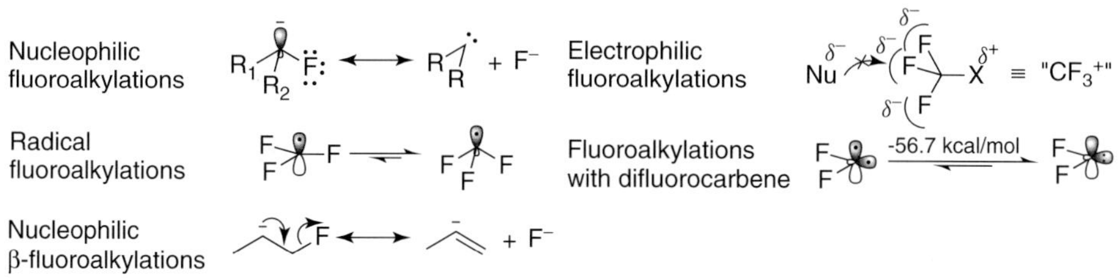

Nucleophilic fluoroalkylations

Electrophilic fluoroalkylations

$\equiv$ "CF_3^+"

Radical fluoroalkylations

Fluoroalkylations with difluorocarbene

−56.7 kcal/mol

Nucleophilic β-fluoroalkylations

Scheme 12.11 Various key reaction intermediates involved in fluoroalkylations.

1950s. Concomitant with the increased number of functionalized fluoroorganics, building block methodologies have become viable synthetic tools in organofluorine chemistry. Mechanistically, a number of key reaction intermediates are involved in fluoroalkylations, including α-fluorocarbanions, α-fluorocarbocations, α-fluorocarbenes, α-fluorinated radicals, β-fluorocarbanions, and others (Scheme 12.11). Because of the steric demand and the extremely strong electronegativity of fluorine, these fluorinated species always demonstrate fundamentally different chemical behavior in comparison with their nonfluorinated counterparts. In this section, the majority of our attention is focused on the most extensively employed nucleophilic and electrophilic α-fluoroalkylation methods for the CF_x-C bond formations. Therefore, nucleophilic β-fluoroalkylations, recently reviewed by Uneyama et al. [118], and also fluoroalkylating protocols based on radicals [119], ylides [120], and carbenes [121], well established in the last century, will not be covered in this section. Moreover, the Julia–Kocienski fluoroolefination, which has also been reviewed thoughtfully [122], also will not be included in this chapter.

12.2.2.1 Nucleophilic Trifluoromethylating Reagents, Trifluoromethyl-Metal Reagents, and Related Chemical Transformations

Defined by the key reaction intermediate, the trifluoromethyl carbanion, nucleophilic trifluoromethylation has been extensively exploited in the syntheses of numerous trifluoromethyl-containing organic molecules of scientific and practical interest. A variety of electrophilic species have been found to be reactive towards the trifluoromethyl anion, including aldehydes, ketones, esters, imines, nitriles, nitrones, and alkyl halides [123–127] (Scheme 12.12). Apart from these transformations, trifluoromethyl organometallic reagents are also applicable in cross-coupling reactions with aromatic halides (the Ullmann reaction), alkenyl halides, and alkynes.

Mechanistically, the repulsive force of the vicinal "anion–lone pair," derived from the adjacent anionic carbon center and non-bonded sp^3 lone pairs on the α-fluorine atom, significantly impairs the stability of the anion, and leads to α-fluoride elimination under many reaction conditions [128] (Scheme 12.12). Even though C–F bonds are among the strongest chemical bonds in organic compounds, stabilization of the CF_3 carbanion proved to be difficult in the presence of metal cations owing to the energy compensation arising from the extraordinary stability of metal–fluoride lattices. The breakthrough was achieved in 1948, when Haszeldine and co-workers first prepared trifluoromethyl iodide (CF_3I), which was employed as the exclusive CF_3 anion precursor through the 1950s to the 1980s [129] (Scheme 12.13). Since the 1980s, several fluoromethyl-containing species

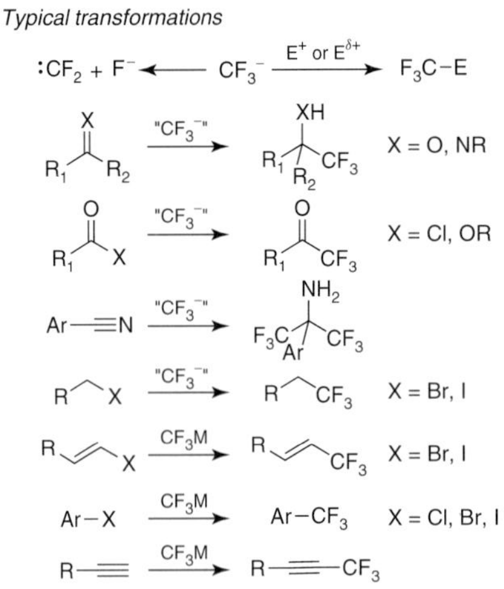

Typical transformations

Scheme 12.12 — positioned at right:

Scheme 12.12 Typical nucleophilic trifluoromethylation reactions.

$:CF_2 + F^- \longleftarrow CF_3^- \xrightarrow{E^+ \text{ or } E^{\delta+}} F_3C-E$

$\underset{R_1 \quad R_2}{\overset{X}{\diagdown C \diagup}} \xrightarrow{\text{"}CF_3^{-}\text{"}} \underset{R_1 \overset{|}{\underset{R_2}{C}} CF_3}{\overset{XH}{}}$ X = O, NR

$\underset{R_1 \quad X}{\overset{O}{\diagdown C \diagup}} \xrightarrow{\text{"}CF_3^{-}\text{"}} \underset{R_1 \quad CF_3}{\overset{O}{\diagdown C \diagup}}$ X = Cl, OR

$Ar-\!\!\equiv\!\!N \xrightarrow{\text{"}CF_3^{-}\text{"}} F_3C\underset{Ar}{\overset{NH_2}{\diagup C \diagdown}}CF_3$

$R \diagdown X \xrightarrow{\text{"}CF_3^{-}\text{"}} R \diagdown CF_3$ X = Br, I

$R \diagup\!\!\diagdown X \xrightarrow{CF_3M} R \diagup\!\!\diagdown CF_3$ X = Br, I

$Ar-X \xrightarrow{CF_3M} Ar-CF_3$ X = Cl, Br, I

$R-\!\!\equiv \xrightarrow{CF_3M} R-\!\!\equiv\!\!-CF_3$

have entered the nucleophilic trifluoromethyl arena, such as CF_2Br_2, trifluoroacetates [130], and CF_3Br [131]. Importantly, Burton and Wieners reported the preparation of trifluoromethyl organometallics using CF_2Br_2 via a difluorocarbene intermediate [132]. Chen and Wu utilized methyl fluorosulfonyldifluoroacetate as a trifluoromethylating reagent for a variety of aryl and alkyl halides in the presence of a catalytic amount of CuI [133]. Notably, although CF_3H is a seemingly obvious trifluoromethyl anion source which was even available prior to the twentieth century [134], its synthetic utility was not achieved until two decades ago [135].

Despite various nucleophilic perfluoroalkylations based on organometallic reagents having been achieved in the early 1950s [136], facile incorporation of the trifluoromethyl motif using similar trifluoromethyl organometallics was a longstanding synthetic challenge [137]. Indeed, extensive attempts to prepare a series of trifluoromethyl-metal species with CF_3I were made in the mid-twentieth century, including bis(trifluoromethyl)mercury (the first trifluoromethyl organometallic

Typical sources of nucleophilic trifluoromethylating species

	Preparation		Application
CF_3I	Haszeldine, 1948	CF_3I/Mg	Haszeldine, 1951
CF_3CO_2M	Swarts, 1939	CF_3CO_2Na/CuI	Matsui et al., 1981
CF_3Br	Simons et al., 1946	CF_3Br/Zn	Ishikawa, 1981
CF_2Br_2	Rathsburg, 1918	CF_2Br_2/Cd or Zn	Burton, 1985
$FSO_2CF_2CO_2Me$	1958	with CuI(cat.)	Chen, 1989
CF_3H	Meslans, 1890	$DMF/\underset{N}{\diagdown}\!\!\diagup\!\!\overset{R_4N^+}{\diagdown O}$ 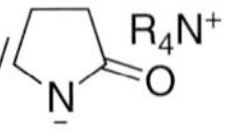	Shono, 1991

Scheme 12.13 Historical development of trifluoromethyl sources for nucleophilic trifluoromethylations.

Hg(CF$_3$)$_2$	CF$_3$Li	CF$_3$MgI	Zn(CF$_3$)$_2$, CF$_3$ZnI, CF$_3$ZnBr	CF$_3$Cu
Haszeldine 1949	Haszeldine Pierce et al. 1949	Haszeldine 1951	Ishikawa 1981	Burton 1986

Scheme 12.14 Landmark applications of trifluoromethyl organometallic reagents and intermediates.

compounds) [138], trifluoromethyllithium [136b, 138], and trifluoromethylmagnesium iodide [139] (Scheme 12.14). Unfortunately, these species are either extremely labile or unexpectedly unreactive, considerably limiting their application in terms of synthetic chemistry. Instead, trifluoromethylcopper [140] and -zinc derivatives [131b, 141] were found to furnish superior reactivity and stability. In particular, CF$_3$Cu, generated *in situ* by treating Cu metal with CF$_3$I or CF$_2$Br$_2$, was developed for a variety of cross-coupling reactions with aryl halides and these are still widely employed for CF$_3$ incorporation (Scheme 12.14).

On the other hand, the emergence of a variety of nonmetal-based trifluoromethylating reagents has ushered a new era since the late 1980s. Hartkopf and De Meijere first reported nucleophilic trifluoromethylation of ketones with *in situ*-generated trialkylsilyl(trifluoromethyl)diazenes to afford the corresponding carbinols in 30–34% yield under harsh conditions [142]. Subsequently, while the preparative approach towards trifluoromethyltrimethylsilane (TMSCF$_3$, the Ruppert–Prakash reagent) was established by Ruppert *et al.* in 1984 [143], the synthetic usefulness of this compound as a nucleophilic trifluoromethylating agent was not realized until Prakash *et al.* made the breakthrough in 1989 [123a]. In comparison with the organometallic reagents and trialkylsilyl(trifluoromethyl)diazenes, the lability of the Si-CF$_3$ bond permits the transformation of the CF$_3$ anion under mild conditions with remarkably high efficacy. Moreover, the reagent has also shown excellent reactivity towards a wide range of functional groups, including aldehydes, ketones, esters, imines, nitriles, nitrones, and alkyl halides and also metallic substrates for *in situ* generation of corresponding organometallic species [144]. Notably, TMSCF$_3$ has demonstrated superior applicability in the asymmetric construction of C-CF$_3$ bonds, thereby significantly facilitating the efficient syntheses of various bioactive chiral fluoroorganics [89e]. In contrast, the synthetic applicability of other trifluoromethytrialkylsilanes has been fairly limited, presumably owing to the relatively low reactivity of these reagents [145]. Similarly to the synthesis of TMSCF$_3$ by Ruppert *et al.*, Pawelke exploited the CF$_3$–tetrakis(dimethylamino)ethylene complex as a trifluoromethylating reagent for the formation of CF$_3$-B and CF$_3$-Si bonds [146]. The chemistry was later extended by Dolbier and co-workers for synthesizing various trifluoromethylated organic molecules [147]. Prakash and co-workers also reported the utility of trifluoromethyl phenyl sulfone (PhSO$_2$CF$_3$), prepared from non-ozone-depleting CF$_3$H, as a versatile trifluoromethylating reagent for carbonyl compounds and imines [148, 149] (Scheme 12.15).

Compared with other precursors (CF$_3$I, CF$_3$Br, CF$_2$Br$_2$, and so on), fluoroform (CF$_3$H) is conceptually the most efficient trifluoromethyl precursor in terms of atom economy; its synthetic utility, however, was unknown for a long time [150]. Pioneered by Shono *et al.*, the incorporation of the CF$_3$ moiety was achieved through deprotonation of CF$_3$H

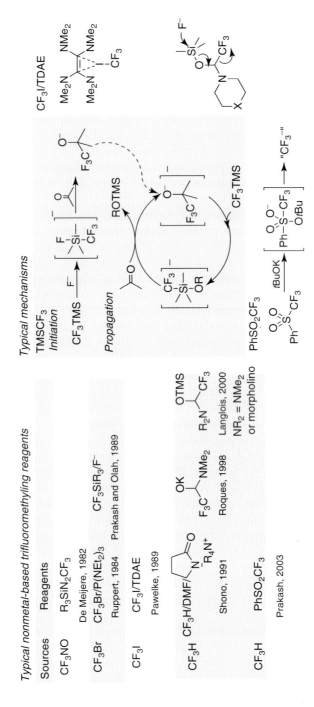

Scheme 12.15 Development of nonmetal-based trifluoromethylating agent.

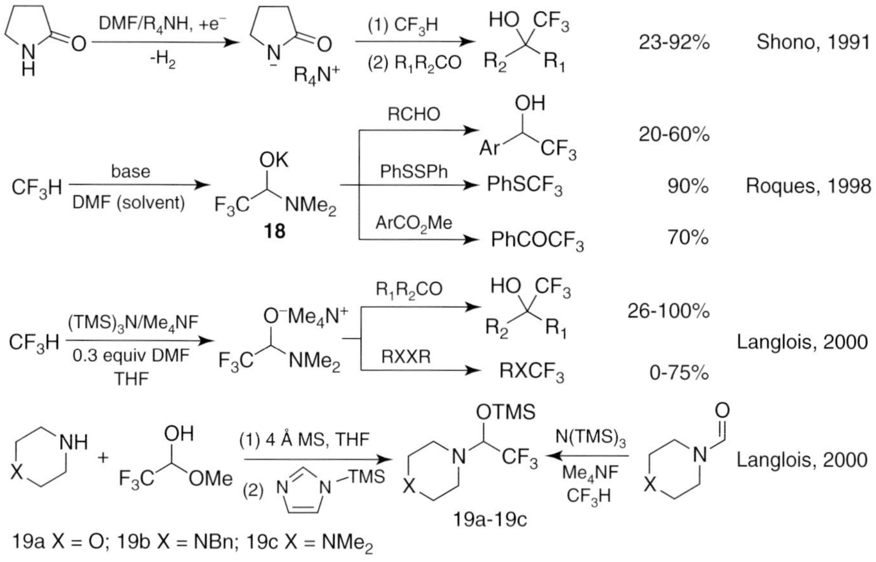

Scheme 12.16 Nucleophilic trifluoromethylation using fluoroform as trifluoromethyl source.

in dimethylformamide (DMF) with electrogenerated bases [135, 151] (Scheme 12.16). Russell and Roques improved the methodologies using readily available bases, and first inferred the formation of the trifluoromethylated hemiaminal intermediate, **18**, responsible for the stabilization of the labile CF_3 anion [152]. Several trifluoromethyl hemiaminals have been prepared thus far by means of the addition of the CF_3 anion to formamides (**19b** and **19c**) or the reaction of fluoral methylhemiacetal with amines (**19a** and **19b**) as trifluoromethylating reagents for a variety of organic substrates [153].

12.2.2.2 Electrophilic Trifluoromethylating Reagents and Reactions

In comparison with trifluoromethylations using an anionic trifluoromethyl equivalent, which date back to the early 1950s, electrophilic trifluoromethylations were achieved only recently [154]. Thanks to the efforts devoted to this field, a variety of efficient reagents have been developed over the past three decades. Electrophilic introduction of the trifluoromethyl moiety has therefore become one of the most promising trifluoromethylating strategies. In addition to the construction of CF_3-C (sp^3, sp^2, or sp) bonds, electrophilic trifluoromethylations have also allowed a large number of heteroatom$-CF_3$ bond-forming reactions, which are of significance in syntheses of compounds possessing unique biological and chemical properties (Scheme 12.17).

Historically, Haszeldine and co-workers first investigated the capability of CF_3I as an electrophilic trifluoromethylating reagent. Treated with potassium hydroxide, CF_3I was converted into fluoroform (CF_3H) as the product, implying the reverse polarization of the C–I bond compared with the nonfluorinated analogs [155]. In 1976, Olah and Ohyama first described the preparation of trifluoromethyl triflate (CF_3OTf) by reaction of triflic acid with fluorosulfuric acid, which evidentially suggested an electrophilic trifluoromethylation mechanism [156]. On the other hand, attempts to trifluoromethylate

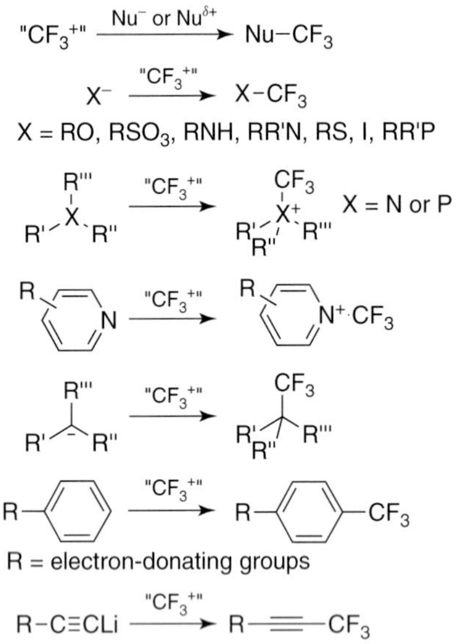

Scheme 12.17 Typical electrophilic trifluoromethylation reactions.

nucleophilic species using trifluoromethyl triflate in which the leaving group ($CF_3SO_3^-$) demonstrates tremendous nucleofugality resulted in nucleophilic attack on the sulfur atom. One may argue that the sluggishness of the electrophilic trifluoromethylations was primarily caused by the extraordinarily strong electronegativity the CF_3 moiety and the difficulties in generation of the cation thereby. However, mass spectrometry confirmed the existence of the trifluoromethyl cation as an abundant species [157]. Surprisingly, a gas-phase ion study suggested that the order of the stability of various methyl cations decreases as $CHF_2^+ > CH_2F^+ > CF_3^+ > CH_3^+$ [158], which can be attributed to a balance between the high electronegativity of fluorine and the back-donation of its lone-pair of electrons to the vacant p-orbital (Scheme 12.18). In fact, the inertness of the CF_3 moiety towards nucleophiles was realized to be a consequence of (i) the reverse polarization of the CF_3–halogen and CF_3-O bonds in nature and (ii) the steric inaccessibility of the CF_3 group for nucleophilic attack (Scheme 12.18).

Although trifluoromethylsulfonium salts were known as early as 1973 [159], the first electrophilic trifluoromethylation was described by Yagupolskii *et al.* using di-aryl(trifluoromethyl)sulfonium salts in 1984 (Scheme 12.19) [160]. Prepared by treating

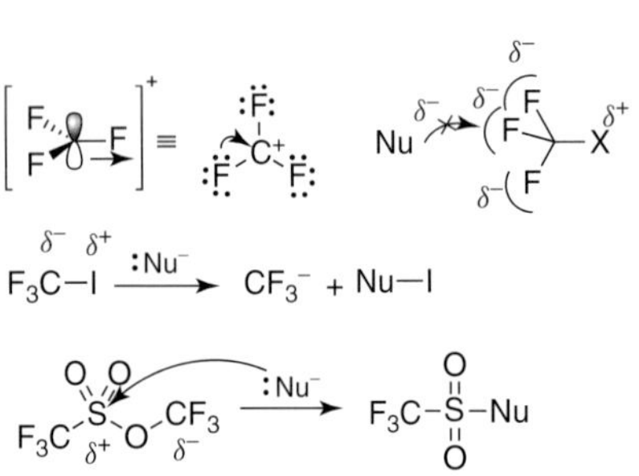

Scheme 12.18 Difficulties in electrophilic trifluoromethylations.

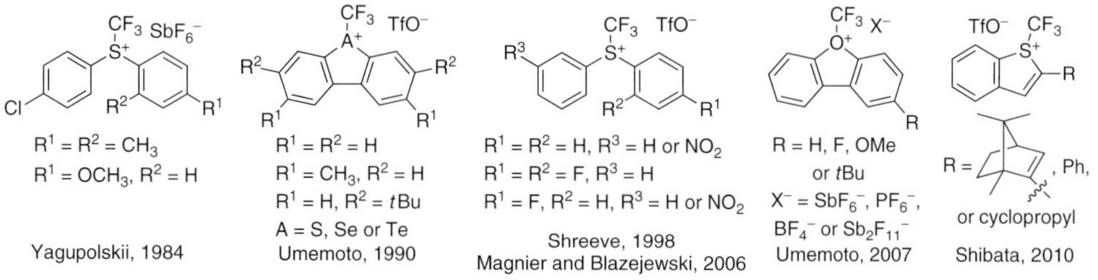

Scheme 12.19 S-(Trifluoromethyl)chalcogen salts as electrophilic trifluoromethylating reagents.

p-chlorophenyl trifluoromethyl sulfoxide with $SF_3^+ SbF_6^-$ followed by anisole or m-xylene, the reagents were capable of transferring CF_3 electrophilically on to p-nitrothiophenolate, rendering the corresponding trifluoromethyl sulfide in moderate yield (Scheme 12.19). However, the inertness of these reagents towards N,N-dimethylaniline at elevated temperature clearly indicates their rather low reactivity. Seeking agents with higher generality and reactivity, Umemoto and Ishihara synthesized a series of (trifluoromethyl)dibenzo chalcogenium salts [161]. With variable trifluoromethylating capability (in the general order S > Se > Te), these reagents demonstrated superior applicability, compared with Yagupolskii et al.'s agents, for trifluoromethylation of various nucleophiles, including carbanions, thiophenolates, phosphines, and iodide. Despite this achievement, the tedious and costly preparative routes significantly limited their synthetic usefulness. Shreeve and co-workers revisited the synthesis and the reactivity of diaryl(trifluoromethyl)sulfonium salts, and exploited them as alternatives to Umemoto's reagents [162]. Further improvements were made by Magnier and co-workers, who achieved straightforward one-pot syntheses of diaryl(trifluoromethyl)sulfonium salts by the treatment of aromatic compounds with potassium trifluoromethanesulfinate and triflic acid [163]. Umemoto et al. reported O-(trifluoromethyl)oxonium salts prepared by photochemical decomposition of corresponding diazonium salts, which permitted direct trifluoromethylation of N- or O-nucleophiles [164]. Importantly, while the chalcogenium salts permit reactions between the CF_3 moiety and nucleophilic species, a standard S_N2 mechanism is adopted only when O-(trifluoromethyl)dibenzofuranium salts are employed [164, 165]. Recently, Shibata and co-workers demonstrated the use of S-(trifluoromethyl)thiophenium salts for electrophilic trifluoromethylations of carbon nucleophiles to give the products in low to high yields [166]. However, attempts to achieve asymmetric incorporation of the CF_3 moiety into a β-keto ester using a camphor-based S-(trifluoromethyl)benzothiophenium salt only afforded a racemate.

In addition to the previously mentioned chalcogenium salts, S-(trifluoromethyl)sulfoximines have also emerged as efficient electrophilic trifluoromethylating reagents over the last decade (Scheme 12.20). Yagupolskii and co-workers first prepared S-trifluoromethylated sulfoximine in 1984 by reaction of trifluoromethyl phenyl sulfoxide and sodium azide in the presence of oleum, but its trifluoromethylating capability was not explored [167]. Adachi and Ishihara patented the cyclic and acyclic trifluoromethyl sulfoximines as viable electrophilic trifluoromethylating reagents for carbon nucleophiles and thiolates to generate the corresponding products in low to moderate yields [168]. Notably, Shibata and co-workers recently reported the utilization

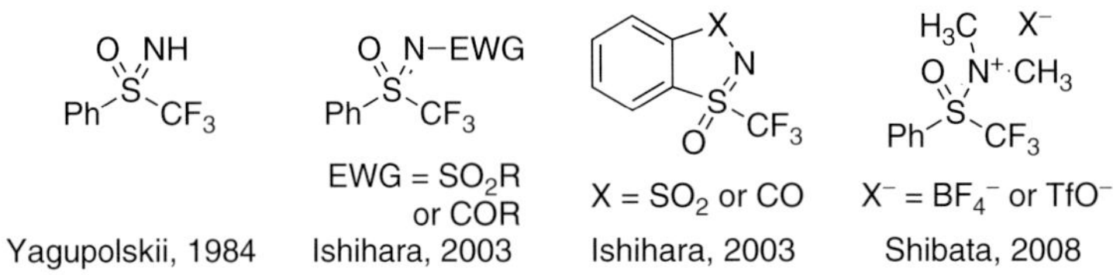

Scheme 12.20 Sulfoximine-based electrophilic trifluoromethylating reagents.

of a trifluoromethylated counterpart of Johnson's reagent [169]. Similarly to the agents mentioned above, this compound permits the trifluoromethylation of various carbon nucleophiles in the presence of 1,8-diazabicyclo[5.4.0]undec-7-ene (DBU) or phosphazene base.

Apart from the electrophilic trifluoromethylating agents based on the CF_3-S bond functionalities, hypervalent iodonium compounds have also been adopted for the introduction of the trifluoromethyl moiety. In comparison with (perfluoroalkyl)aryliodonium triflates, corresponding (trifluoromethyl)iodonium species were not known for many years owing to their thermal instability [154, 170]. In 2006, Togni and co-workers demonstrated, for the first time, a series of stable trifluoromethyl(iodonium) compounds that can be utilized as efficient electrophilic trifluoromethylating reagents under mild conditions (Scheme 12.21) [171]. Intriguingly, the synthetic route was achieved by a formal umpolung of the CF_3 functionality using the Ruppert–Prakash reagent. For the sake of both operative simplicity and expense, an improved synthetic protocol was established by the same group exploiting anhydrous KOAc in a one-pot procedure [172] (Scheme 12.21).

Synthetically, the trifluoromethylating capability of these agents was initially explored by treating **21** with a β-keto ester, which afforded trifluoromethylated product in 67% yield under basic conditions [171] (Scheme 12.22). This chemistry was later extended to other β-keto esters and α-nitro esters [172]. The agents were also found to be reactive

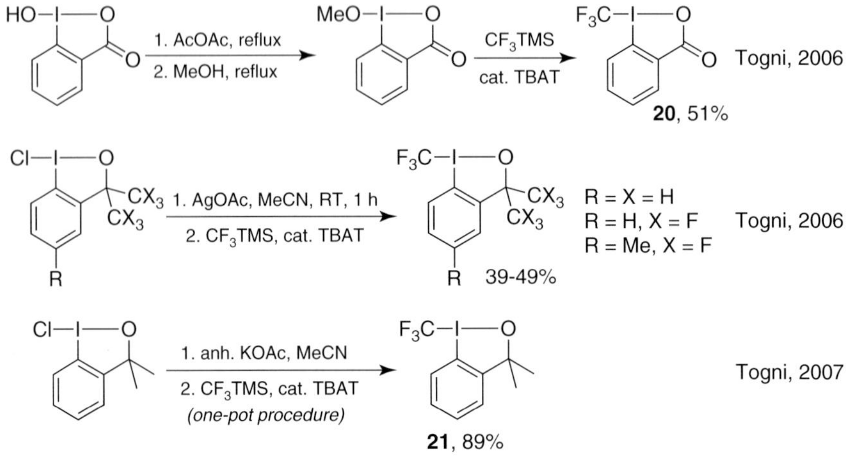

Scheme 12.21 Preparation of (trifluoromethyl)iodonium compounds.

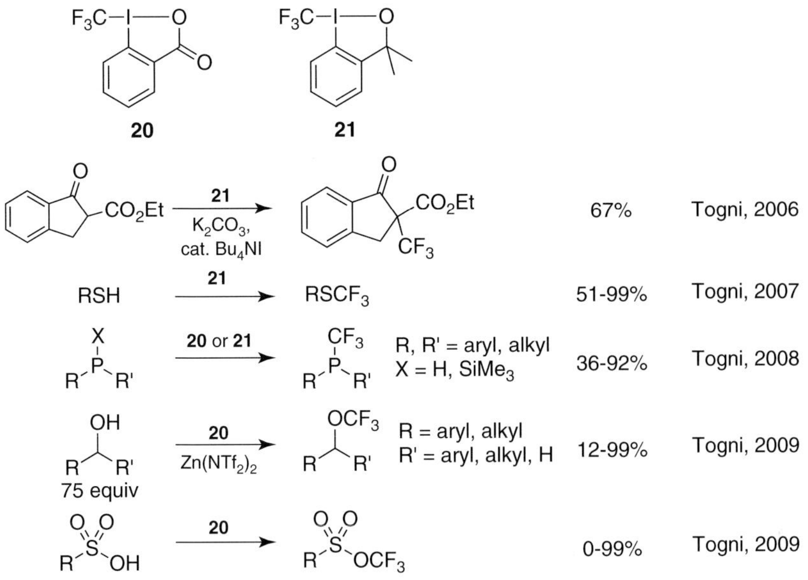

F$_3$C–I–O **20**

F$_3$C–I–O **21**

indanone-CO$_2$Et + **21** → [K$_2$CO$_3$, cat. Bu$_4$NI] → indanone-C(CO$_2$Et)(CF$_3$) 67% Togni, 2006

RSH + **21** → RSCF$_3$ 51–99% Togni, 2007

R–P(X)–R' + **20 or 21** → R–P(CF$_3$)–R' R, R' = aryl, alkyl; X = H, SiMe$_3$ 36–92% Togni, 2008

R–CH(OH)–R' (75 equiv) + **20** [Zn(NTf$_2$)$_2$] → R–CH(OCF$_3$)–R' R = aryl, alkyl; R' = aryl, alkyl, H 12–99% Togni, 2009

R–S(O)$_2$–OH + **20** → R–S(O)$_2$–OCF$_3$ 0–99% Togni, 2009

Scheme 12.22 Electrophilic trifluoromethylation using (trifluoromethyl)iodonium compounds.

towards sulfur [172] and phosphorus [173] nucleophiles. Remarkably, reagent **20** has been successfully applied in the trifluoromethylation of alcohols and sulfonic acids to form the corresponding trifluoromethyl ethers [174] and sulfonates [175], respectively. It is worth mentioning that electrophilic trifluoromethylation of aliphatic alcohols is still challenging. Since a large excess of alcohol (75 equiv.) was necessary for the completion of the reaction, the practicability of the protocol can be argued. Moreover, attempts to treat phenols with reagent **20** led to trifluoromethylation of aromatic rings instead of forming O-CF$_3$ bonds [176] (Scheme 12.22).

12.2.2.3 Recent Developments in the Construction of CF$_3$-C Bonds

Asymmetric CF$_3$-C Bond-Forming Reactions   Trifluoromethylated organic molecules have received increasing attention in recent years owing to their superior biological and physical properties [177]. Although constructions of a stereogenic center bearing the CF$_3$ group from the corresponding prochiral trifluoromethylated carbonyl compounds have been known for decades [178], direct asymmetric introduction of the CF$_3$ moiety into organic compounds, namely stereoselective CF$_3$-C bond formations, remains challenging.

In fact, the first diastereoselective nucleophilic trifluoromethylation of carbonyl compounds was achieved in 1989 when Prakash *et al.* showed the synthetic usefulness of TMSCF$_3$ [123] (Scheme 12.23). Taguchi and co-workers later reported trifluoromethylation of chiral aldehydes using CF$_3$I and zinc under ultrasound irradiation to afford two diastereomers in a ratio of 2.6:1 with moderate yields [179]. Intriguingly, Portella and co-workers reinvestigated the synthesis employing a TMSCF$_3$–fluoride system, giving the two isomers in a 4:1 ratio with excellent yields, which clearly demonstrated

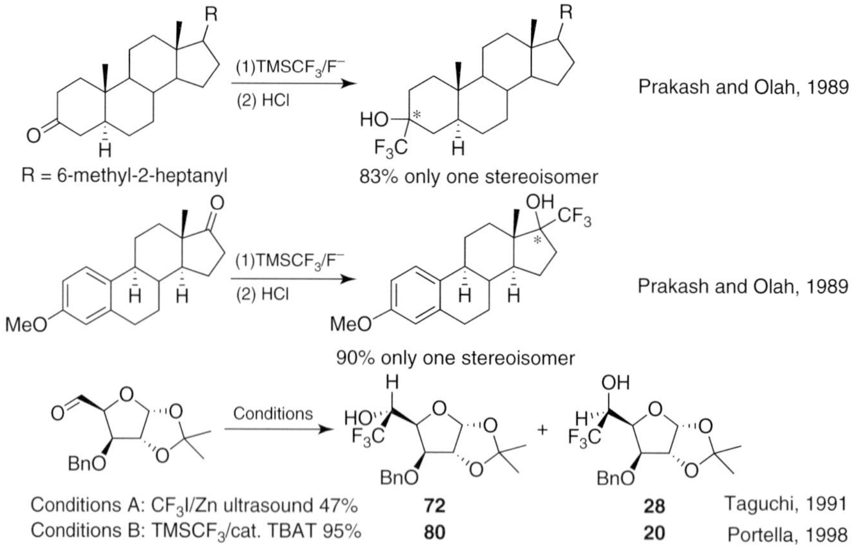

Scheme 12.23 Diastereoselective nucleophilic trifluoromethylation of carbonyl compounds.

the advantages of TMSCF$_3$ [180]. Notably, although nucleophilic trifluoromethylation of chiral carbonyl compounds may render the corresponding products with satisfactory facial discrimination, low stereoselectivities have also been observed under some circumstances [181].

The first catalytic enantioselective trifluoromethylation of carbonyl compounds was reported by Prakash and Yudin, who employed the quinidinium fluoride **22** as the initiator in the trifluoromethylation of 9-anthraldehyde, affording the corresponding product with excellent enantiomeric purity [182] (Scheme 12.24). Shortly afterwards, Iseki *et al.* used cinchonium fluorides (**23** and **24**) as initiators and catalysts, which facilitated the synthesis of α-trifluoromethyl alcohols with low to moderate *ee* [183]. Instead of using fluoride initiators, Fuchikami and co-workers studied the trifluoromethylation of benzaldehyde with trifluoromethyltriethylsilane in the presence of quinine as a Lewis base catalyst, but only low stereoselectivity was achieved [184]. Kuroki and Iseki further attempted to obtain enhanced enantioselection by employing chiral tris(dialkylamino)sulfonium difluorotriphenylstannate (**26**), which, however, did not demonstrate appreciable superiority [185]. Although suffering from lack of generality, a highly efficient asymmetric nucleophilic trifluoromethylation using catalyst **25** was developed by Caron *et al.*, which afforded **31** with 92% *ee* [186]. Promising synthetic approaches utilizing different catalytic systems were established in 2007 by several laboratories. Feng and co-workers reported the use of a new combinatorial catalytic system (**29**) containing disodium (*R*)-binaphtholate and a quinidinium salt for the enantioselective trifluoromethylation of aromatic aldehydes, giving products in moderate *ee* [187]. Meanwhile, Mukaiyama and co-workers also discovered cinchonidium phenoxide (**27**) (an alkoxide-based initiator) as an efficient catalyst furnishing trifluoromethylated alcohols in the trifluoromethylation of ketones [188]. Shortly afterwards, Shibata and co-workers devised a synthetic method based on a

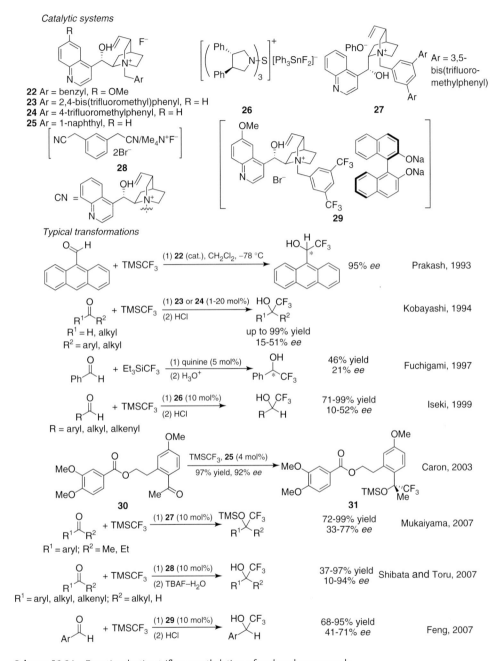

Scheme 12.24 Enantioselective trifluoromethylation of carbonyl compounds.

combination of cinchonium bromide and tetramethylammonium fluoride (**28**) [189]. In comparison with the previous catalytic systems, significantly enhanced stereochemical control and generality were achieved as most of the products were obtained with good *ee* (>70%). Although great efforts have been expended, it is still a significant challenge to develop generally applicable synthetic approaches towards enantiomerically enriched *α*-trifluoromethyl alcohols from the corresponding carbonyl compounds.

Compared with the asymmetric nucleophilic trifluoromethylation of carbonyl compounds, stereoselective syntheses of *α*-trifluoromethylamines from imines have been achieved mainly due to the contributions from Prakash and co-workers over the last decade (Scheme 12.25). In 2001, Prakash *et al.* first reported a novel stereoselective synthetic approach towards chiral trifluoromethylated amines [190]. Utilizing (*S*)-*tert*-butanesulfinyl as a chiral auxiliary and activating functionality for imines, the protocol permitted an efficient synthesis of a variety of chiral trifluoromethylated amines in good yields and with excellent diastereoselectivities. It is worth noting that the removal of the sulfinyl auxiliary can be achieved by treating the products with HCl and MeOH without loss of optical activity (Scheme 12.25). The strategy was further extended by the same group to *α,β*-unsaturated *N*-*tert*-butanesulfinylaldimines, which afforded

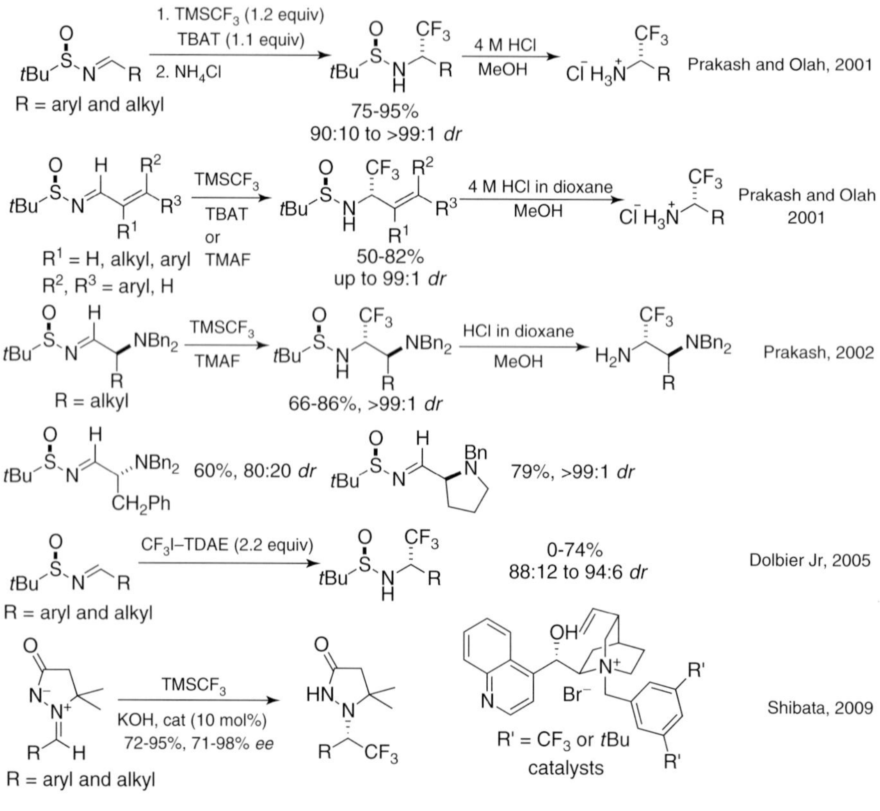

Scheme 12.25 Stereoselective trifluoromethylation of imines.

1,2-adducts with excellent stereoselectivity [191]. Prakash and Mandal also developed the asymmetric synthesis of trifluoromethylated vicinal ethylenediamines using TMSCF$_3$ and α-L-amino-N-*tert*-butanesulfinimines [192]. Intriguingly, when an imine derived from a D-aminoaldehyde was subjected to the reaction, only a slightly decreased diastereoselectivity was observed, which provides an applicable and general synthetic route to both *syn* and *anti* vicinal ethylenediamines. Xu and Dolbier also described a similar methodology employing the CF$_3$I–TDAE [tetrakis(dimethylamino)ethylene] system instead of TMSCF$_3$–fluoride, while the products were obtained in lower yields and diastereoselectivities [147c]. To date, these robust synthetic methods have been extensively utilized in the synthesis of an arsenal of trifluoromethylated amine derivatives [193]. Recently, Shibata and co-workers reported the catalytic enantioselective trifluoromethylation of azomethine imines with TMSCF$_3$ as an alternative approach towards α-trifluoromethylated amines [194]. Employing cinchonium salts in the presence of potassium hydroxide, the trifluoromethyl anion equivalent can be transferred to a variety of azomethine imines, rendering (*S*)-amines with good enantiomeric control. Further treatment of the products with Raney nickel and subsequent decomposition under acidic conditions led to the formation of chiral amines without loss of optical activity.

In addition to the above-mentioned stereoselective nucleophilic trifluoromethylating reactions, asymmetric construction of CF$_3$-C bonds via different mechanistic profiles have also been initiated in recent years. Almost all of these synthetic methodologies have been involved in the trifluoromethylation of enolate derivatives and analogs under radical or electrophilic mechanistic pathways (Scheme 12.26). The first α-trifluoromethylation of carbonyl compounds was described by Kitazume and Ishikawa in 1985 using the optically pure enamine to react with CF$_3$Br or CF$_3$I in the presence of Cp$_2$TiCl$_2$ and ultrasonically dispersed zinc power [141b]. Elliott and co-workers later reported a photochemical intramolecular rearrangement of dienol triflate **36** yielding β-trifluoromethyl enone **37** as a single stereoisomer [195]. Early work by Umemoto and co-workers showed a diastereoselective synthesis of γ-trifluoromethylenones under different conditions [161b, 196]. Accordingly, direct trifluoromethylation of trimethylsilyl enol ether **38** generated products **40** and **41** in a ratio of 1:3.6 (Scheme 12.26). In contrast, the **40:41** ratio was found to be 4:1 when potassium enolate **39** was sequentially treated with borole **44** and electrophilic trifluoromethylating agent **42**, indicating different transition states under these reaction conditions. Moreover, Umemoto and Adachi reported, for the first time, an enantioselective α-trifluoromethylation of ketone mediated by a chiral borepin (**32**), although only a low yield with 45% *ee* was achieved [196]. Notably, Iseki *et al.* established the triethylborane-mediated diastereoselective trifluoromethylation of chiral imide enolates **45** with CF$_3$I via a radical mechanism [197]. Under these reaction conditions, moderate to good diastereomeric discriminations were observed. Strikingly, thereafter chemists paid little attention to this field until recently. In 2006, Itoh and Mikami further extended Iseki *et al.*'s methodology as the first enantioselective radical trifluoromethylation of ketone enolates with a stoichiometric amount of chiral additives [198]. According to Ma and Cahard, an enantioselective trifluoromethylation of β-keto esters was achieved by exploiting Umemoto's reagent **43** with dihydroquinine as chiral base [89e]. Encouragingly, MacMillan and co-workers successfully developed a

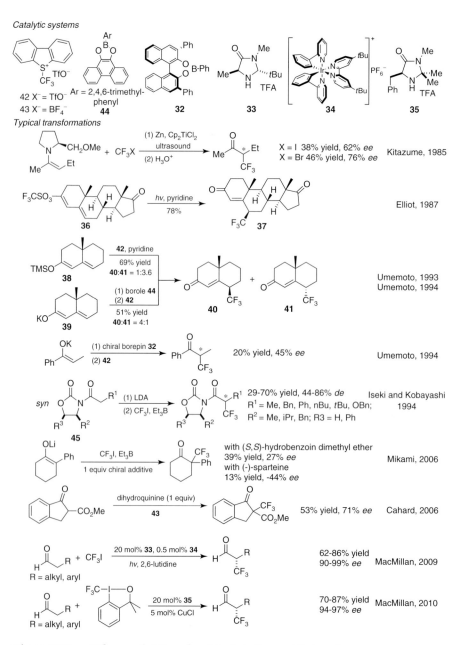

Scheme 12.26 α-Trifluoromethylation of enolate derivatives and their analogs.

highly productive organocatalyzed α-trifluoromethylation of aldehydes via a photoredox pathway [199]. Co-catalyzed by acidified chiral amine catalyst **33** and photocatalyst **34**, aldehydes reacted with CF_3I under radiation from household fluorescent light to generate the corresponding products with high stereoselectivities. The same group further reported a similar transformation by means of electrophilic trifluoromethylation using a (trifluoromethyl)iodonium compound in the presence of imidazolidinone **35** and 5 mol% CuCl [200]. It is worth noting that these two synthetic methods have shown excellent generality in addition to stereochemistry and functional group compatibility, making them viable synthetic approaches.

Transition-Metal-Mediated Aromatic Trifluoromethylations Owing to the immense potential for the preparation of trifluoromethyl-containing aromatic pharmaceuticals, the formation of aryl–CF_3 bonds has received increasing attention. Generally, Cu and Pd metals have frequently been involved in aromatic trifluoromethylations. In fact, the first Cu-mediated aromatic trifluoromethylation was pioneered in 1969 by Kobayashi and Kumadaki, who treated aryl halides with copper powder and a large amount of CF_3I (15 equiv.) at elevated temperatures [140b] (Scheme 12.27). The proposed mechanism of this transformation was established experimentally from the X-ray crystal structures of NHC-stabilized Cu − CF_3 complexes, **46**, which demonstrated effective trifluoromethylating ability [201]. Addressing the demand of Cu-catalyzed aromatic trifluoromethylations, Amii and co-workers developed a remarkable catalytic protocol permitting the recycling of active copper species during the course of the reaction [202]. In previous methods, the slow regeneration of CuI significantly limited the formation rate of Cu−CF_3 complexes, resulting in severe decomposition of the CF_3^- anion. Using a combination of CuI and 1,10-phenanthroline (phen) as the catalytic system, which substantially accelerated the cross-coupling step, the reaction proceeded smoothly to give the products in moderate to good yields. It is worth mentioning that Chu and Qing also utilized a stoichiometric amount of the CuI–phen system in the oxidative cross-coupling reaction between the Ruppert–Prakash reagent and terminal alkynes to yield the corresponding trifluoromethylated products [203].

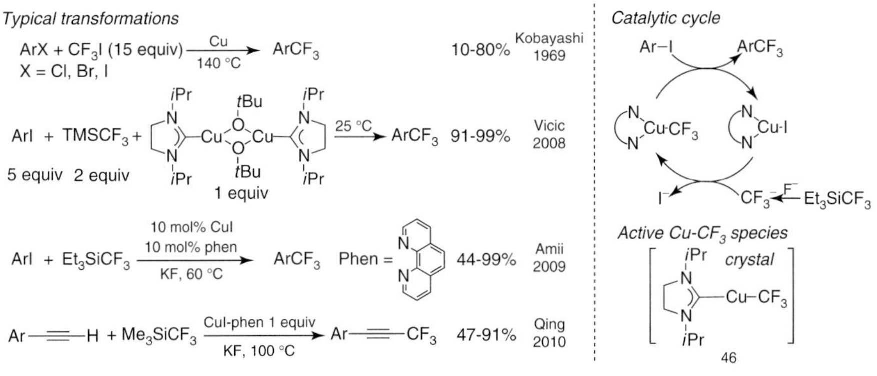

Scheme 12.27 Cu-mediated aromatic trifluoromethylations.

In comparison, palladium-mediated aryl–CF₃ bond-forming reactions are mechanistically more challenging. While a handful of trifluoromethyl palladium compounds had been known prior to the 1980s [137b], the reductive elimination in the catalytic cycle was energetically prohibitive owing to the extraordinary stability of Pd-CF₃ bonds [55]. Kitazume and Ishikawa demonstrated the first palladium-catalyzed trifluoromethylation of aryl iodides employing *in situ*-generated trifluoromethyl zinc compounds to afford the corresponding products in moderate to high yields [204] (Scheme 12.28). Grushin and Marshall first reported a facile trifluoromethylaryl reductive elimination from a xantphos-coordinated Pd(II) center, revealing the mechanistic possibility of Pd-mediated aromatic trifluoromethylations [205]. Sanford and co-workers additionally reported aryl–CF₃ bond formation via reductive elimination from high-valent Pd(IV) complexes [206]. With more practical synthetic utility, Pd(II)-catalyzed *ortho*-trifluoromethylation of

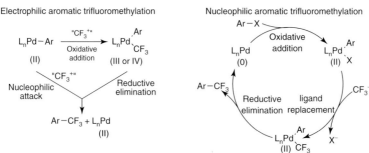

Scheme 12.28 Pd-mediated aromatic trifluoromethylation.

arenes bearing pyridine-directing groups was accomplished by Yu and co-workers in the presence of triflic acid and Umemoto's electrophilic trifluoromethylating reagent [207]. Accordingly, the reaction underwent either direct nucleophilic attack of the "CF_3^+" by the Pd–aryl complexes or an oxidative addition–reductive elimination process. Recently a breakthrough in this field was demonstrated by Buchwald and co-workers, which allows trifluoromethylation of aryl chlorides with a substoichiometric amount of Pd complexes under a regular Pd(0)–Pd(II) catalytic cycle [208]. This process showed satisfactory efficacy and remarkable substrate compatibility, making it a valuable and facile route towards trifluoromethylated arenes.

12.2.3
Novel Methods for the Introduction of Difluoromethyl Motifs

Compared with the continuing enthusiasm for the syntheses of trifluoromethylated organic compounds for over a century, difluoromethylations have received much less attention until recently. To date, difluoromethyl (CF_2H)- and difluoromethylene (CF_2)-containing organic molecules have become of particular interest owing to their diverse applications in materials science, agrochemistry, and the pharmaceutical industry [209]. The CF_2 moiety is known to be isosteric and isopolar with ethereal oxygen in nature, which has led to great potential in the development of bioactive fluorinated analogs of oxygenated molecules such as sugars. The difluoromethyl group, on the other hand, has been found to be a steric and electronic mimic of a hydroxyl group, particularly in terms of its function as a hydrogen bonding donor without developing a negative charge. Owing to these promising applications, organic chemists have devoted their efforts to the investigation of synthetic methods for partially fluorinated molecules bearing CF_2 and CF_2H groups.

Synthetically, a variety of methods for the introduction of *gem*-difluoromethylene moieties into organic compounds were established prior the early 1990s [210]. Broadly, current synthetic protocols are primarily based on three strategies: (i) nucleophilic difluorination of carbonyl compounds and their derivatives; (ii) direct electrophilic *gem*-difluorination of carbanions; and (iii) methodologies using fluorinated synthons by means of nucleophilic, electrophilic, radical, and SET processes [119], and also carbene-based reaction intermediates [121, 211] (Scheme 12.29). In addition to these methods, reductive defluorination from readily available trifluoromethyl-containing precursors has also been described for the synthesis of corresponding difluorinated compounds [21]. In this section, we emphasize building block approaches utilizing difluorinated carbanion and carbocation equivalents, which are the major achievements of the past 20 years. Those interested in difluoromethylations and difluoromethylenations by using radicals, carbene, and organometallic reagents are referred to the excellent review articles cited above.

12.2.3.1 Nucleophilic Difluoromethyl Building Blocks and Approaches
Nucleophilic difluoromethylations are extensively employed in the synthesis of CF_2- and CF_2H-containing organic molecules. As key reaction intermediates, difluoromethyl carbanions are found to react with a series of electrophilic substrates, such as alkyl halides (to form the corresponding substituted alkanes), and carbonyl compounds

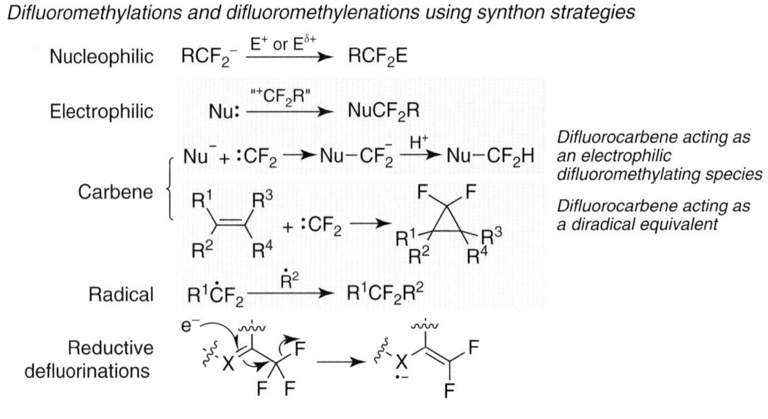

Difluoromethylations and difluoromethylenations using synthon strategies

Scheme 12.29 Difluoromethylations and difluoromethylenations by synthon strategies.

and their analogs (to generate the corresponding difluoromethyl alcohols, amines, ketones, and *gem*-difluorinated alkenes) (Scheme 12.30). Mechanistically, owing to the strong electronegative nature of fluorine, difluoromethyl carbanions usually exhibit a substantial preference for adopting pyramidal geometries over planar structures. Despite the planar conformation usually being assumed by the carbanions in conjugation with nitro or carbonyl groups, an sp^3 hybridization is still energetically favored by the difluorinated counterparts because of increased p-orbital character on the fluorine atoms [212]. Similarly to the CF$_3$ anion, RCF$_2^-$ species also tend to decompose by α-elimination

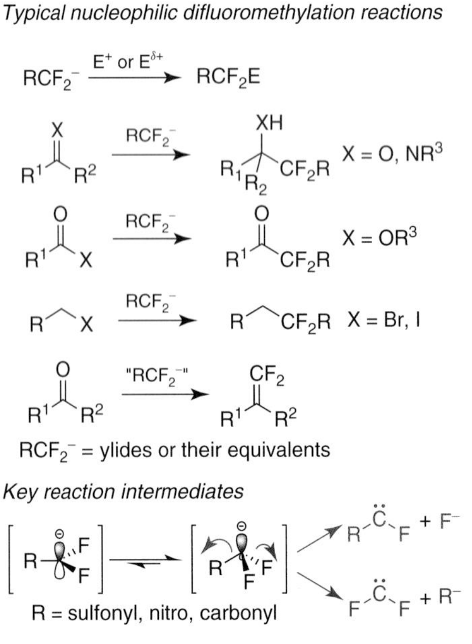

Typical nucleophilic difluoromethylation reactions

Scheme 12.30 Typical nucleophilic difluoromethylations and their reaction intermediates.

of fluoride and auxiliary groups to form more thermodynamically stable carbenes, which can lead to be significant competitive reactions diminishing the synthetic applicability of RCF_2^- anions.

Hence these anionic species are usually functionalized with auxiliary groups possessing electron-withdrawing and charge-delocalizing properties to modulate their stability and reactivity. Many of the auxiliaries also permit further chemical transformations towards useful fluoroorganics. To date, an arsenal of functional groups has been developed as auxiliaries in nucleophilic difluoromethylation reactions for various synthetic purposes. Historically, nucleophilic introduction of the CF_2 moieties was achieved by the use of "P$-CF_2$" derivatives (Scheme 12.31). Pioneered by Burton's group, difluorinated phosphonium salts were first found to be suitable for the synthesis of *gem*-difluorinated alkenes via the Wittig reaction [213]. To increase the nucleophilicity of the ylides, Burton *et al.* further utilized the analogous phosphonate ylide. Mediated by cadmium or zinc metal, diethyl bromodifluoromethylphosphonate was able to react with a large number of electrophiles, such as aldehydes, ketones, acyl chlorides, and allyl bromides [214]. Kondo and co-workers later demonstrated the use of diethyl difluoromethylphosphonate–lithium diisopropylamide and diethyl difluoro(trimethylsilyl)methylphosphonate–cesium fluoride systems as the precursors of the (diethoxyphosphoryl)difluoromethide anion [215]. In fact, owing to the immense importance of bi- and triphosphate functionalities in cellular chemistry, difluoromethylene-containing phosphates and phosphoric acids have been prepared as their analogs, which demonstrated significantly increased biostability,

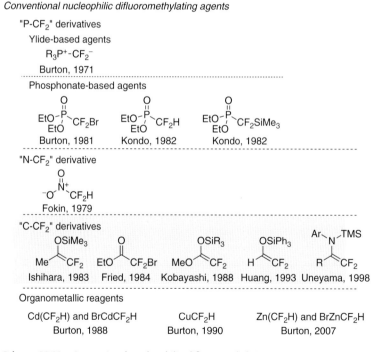

Scheme 12.31 Conventional nucleophilic difluoromethylating reagents.

through the above-mentioned methodologies [216]. Fokin and co-workers reported the Michael addition and the Henry reaction of difluoronitromethane in moderate yields [217]. However, the synthon was not explored further owing to its limited synthetic accessibility [218].

Aside from these "heteroatom–CF_2" based synthons, α,α-difluorinated carbonyl compounds and their derivatives have also been employed as viable building blocks since the early 1980s (Scheme 12.31). Ishihara and co-workers initially demonstrated the synthetic applicability of enol silyl ethers, prepared from chlorodifluoromethyl ketones and chlorotrimethylsilane in the presence of Zn, as efficient difluoromethyl building blocks in the aldol reaction [219]. Under similar conditions, Kobayashi and co-workers prepared 2,2-difluoroketene silyl acetals, which reacted with imines, aldehydes, and ketones, rendering products with remarkable stereoselectivities [220]. Iseki and Kobayashi further developed a catalytic asymmetric aldol reaction of difluoroketene silyl acetal with aldehydes to yield α,α-difluoro-β-hydroxy esters in excellent *ee* [221]. Huang and co-workers reported the preparation of (2,2-difluorovinyloxy)triphenylsilane; however, its synthetic applications remain unexplored [222]. More recently, Uneyama and co-workers documented the electrochemical preparation of N-TMS-β,β-difluoroenamines and the related chemical transformations in an aldol reaction and nucleophilic substitution with alkyl iodide [223]. Intriguingly, the same group also developed a series of preparative methods for α,α-difluoro carbonyl compound derivatives from the readily available trifluoromethylated precursors [224]. Using similar methodology, Prakash *et al.* prepared a series of 2,2-difluorosilyl enol ethers as versatile precursors for the radiochemical synthesis of ^{18}F-labeled trifluoromethyl ketones [225]. In addition to the above-mentioned difluoromethide precursors, commercially available halodifluoroacetates have become one of the most important functionalized difluorinated synthons. First employed in the Reformatskii-type reaction for the synthesis of 2,2-difluoro-3-hydroxy esters [226], the halodifluoroacetate–Zn systems have allowed both diastereoselective [227] and enantioselective [228] addition of α,α-difluorinated esters to aldehydes, ketones, and imines [229]. Similarly to trifluoromethyl organometallic reagents, a handful of difluoromethyl variants have been available since the late 1980s. Primarily contributed by Burton and co-workers, these reagents have been successfully applied in the preparation of difluoromethyl-substituted allenes and in allylation reactions [230].

Over the past decade, a series of novel difluoromethylating reagents derived from "S-CF_2" motifs have been well developed [231] (Scheme 12.32). Remarkably, thanks to the facile reductive S-CF_2 bond cleavage (to form a CF_2H moiety) and their diverse chemical transformations into other difluoromethyl functionalities, utilization of "S-CF_2" building blocks gives superior synthetic advantages. Although difluoromethyl phenyl sulfone ($PhSO_2CF_2H$) was known as early as 1960 [232], a detailed description of $PhSO_2CF_2H$ as a nucleophilic (phenylsulfonyl)difluoromethylation reagent did not appear until much later. Stahly utilized $PhSO_2CF_2H$ in nucleophilic additions to aldehydes to produce the corresponding carbinols in good yield [233]. Shortly afterwards, Sabol and McCarthy reported a multiple-step *gem*-difluoromethylenation of ketones using $PhSO_2CF_2H$ as a difluoromethyl ylide equivalent [234]. However, the related chemistry did not receive particular attention for over a decade until Prakash and co-workers revisited its potential. In 2003, they described an efficient one-pot stereoselective synthesis

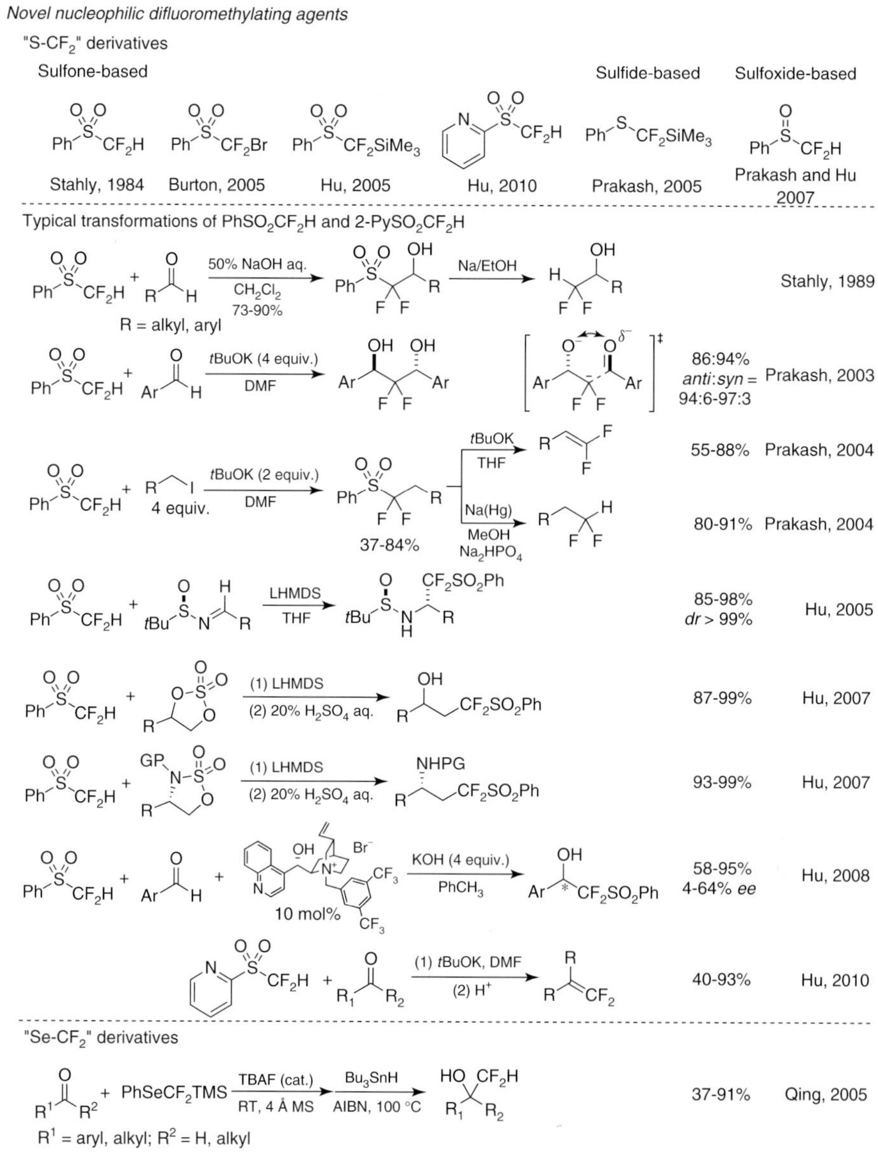

Novel nucleophilic difluoromethylating agents

"S-CF$_2$" derivatives

Sulfone-based

Sulfide-based Sulfoxide-based

Ph–S(O)$_2$–CF$_2$H	Ph–S(O)$_2$–CF$_2$Br	Ph–S(O)$_2$–CF$_2$SiMe$_3$	2-Py–S(O)$_2$–CF$_2$H	Ph–S–CF$_2$SiMe$_3$	Ph–S(O)–CF$_2$H
Stahly, 1984	Burton, 2005	Hu, 2005	Hu, 2010	Prakash, 2005	Prakash and Hu 2007

Typical transformations of PhSO$_2$CF$_2$H and 2-PySO$_2$CF$_2$H

Reaction	Conditions	Yield / Result	Reference
PhSO$_2$CF$_2$H + RCHO (R = alkyl, aryl)	50% NaOH aq. CH$_2$Cl$_2$	73–90% → Na/EtOH	Stahly, 1989
PhSO$_2$CF$_2$H + ArCHO	tBuOK (4 equiv.) DMF	86:94% anti:syn = 94:6–97:3	Prakash, 2003
PhSO$_2$CF$_2$H + R–I (4 equiv.)	tBuOK (2 equiv.) DMF	37–84%	
	tBuOK, THF	55–88%	Prakash, 2004
	Na(Hg), MeOH, Na$_2$HPO$_4$	80–91%	Prakash, 2004
PhSO$_2$CF$_2$H + tBuS-N=CH-R	LHMDS THF	85–98% dr > 99%	Hu, 2005
PhSO$_2$CF$_2$H + cyclic sulfonate	(1) LHMDS (2) 20% H$_2$SO$_4$ aq.	87–99%	Hu, 2007
PhSO$_2$CF$_2$H + GP-N-sultam	(1) LHMDS (2) 20% H$_2$SO$_4$ aq.	93–99%	Hu, 2007
PhSO$_2$CF$_2$H + ArCHO + catalyst (10 mol%)	KOH (4 equiv.) PhCH$_3$	58–95% 4–64% ee	Hu, 2008
2-Py–SO$_2$CF$_2$H + R$_1$C(O)R$_2$	(1) tBuOK, DMF (2) H$^+$	40–93%	Hu, 2010

"Se-CF$_2$" derivatives

Reaction	Conditions	Yield	Reference
R^{1}C(O)R^2 + PhSeCF$_2$TMS R^1 = aryl, alkyl; R^2 = H, alkyl	TBAF (cat.) RT, 4 Å MS Bu$_3$SnH, AIBN, 100 °C	37–91%	Qing, 2005

Scheme 12.32 Development of "S-CF$_2$" bond-based nucleophilic difluoromethylating reagents.

of anti-2,2-difluoropropane-1,3-diols [235]. In the presence of an excess amount of tBuOK, PhSO$_2$CF$_2$H was able to act as a difluoromethylene dianion equivalent. Intriguingly, owing to the charge-charge repulsion during the course of the second addition, the reaction exhibited a substantial preference for generating products with the anti configuration. Additionally, PhSO$_2$CF$_2$H was found to react with primary alkyl iodides through an S_N2 mechanistic pathway [236]. Further treatments of the substituted products under basic and reductive conditions yielded 1,1-difluoro-1-alkenes and 1,1-difluoromethylalkanes, respectively. It is worth noting that the reagent was also able to incorporate the (phenyl-sulfonyl)difluoromethyl moiety into chiral *N*-(*tert*-butylsulfinyl)aldimines with excellent diastereomeric discrimination [237]. Despite these valuable methodologies, $PhSO_2CF_2H$ was found to be inert towards epoxides and aziridines [238]. To achieve the synthesis of β-difluoromethylated or β-difluoromethylenated alcohols and amines, 1,2-cyclic sulfates and sulfamidates exhibiting enhanced electrophilicity in comparison with epoxides and aziridines were utilized by Hu *et al.* to afford the products in good yields [239]. In addition to the reactions mentioned above, $PhSO_2CF_2H$ has been employed in the enantioselective nucleophilic addition of aldehydes catalyzed by cinchona alkaloid-derived ammonium salts, which gave chiral carbinols with 4–64% *ee* [240]. Recently, the same group developed novel difluoromethylating reagents by altering the phenyl group with a series of hete-rocyclic aromatic motifs [241]. Using shelf-stable difluoromethyl 2-pyridyl sulfone, the methodology allowed the facile one-pot *gem*-difluoroolefination of aldehydes and ketones. Notably, although $PhSO_2CF_2H$ can react with enolizable ketones and aldehydes, the syn-thetic route still suffered from low efficiency and harsh reaction conditions [242]. To overcome this problem, Prakash and co-workers further developed fluoride-induced nu-cleophilic (phenylthio)difluoromethylation and (phenylsulfonyl)difluoromethylation of carbonyl compounds using [difluoro(phenylthio)methyl]trimethylsilane ($PhSCF_2TMS$) and [(phenylsulfonyl)difluoromethyl]trimethylsilane ($PhSO_2CF_2TMS$), respectively [243] (Scheme 12.32). Under milder conditions, difluoromethylated carbinols can be obtained in good yields from both enolizable and non-enolizable carbonyl compounds. Inter-estingly, Qing and co-workers adopted [difluoro(phenylseleno)methyl]trimethylsilane ($PhSeCF_2TMS$) as an efficient difluoromethylating agent for enolizable ketones and aldehydes [244].

In addition to the previously described reagents, several other difluoromethyl-containing species have also been utilized in nucleophilic difluoromethylation reactions of carbonyl compounds. In 1995, Hagiwara and Fuchikami first reported a direct fluoride-induced di-fluoromethylation of ketones and aldehydes using (difluoromethyl)dimethylphenylsilane (Scheme 12.33) [245]. Additionally, a series of (1,1-difluoroalkyl)silane derivatives were investigated under similar conditions, affording 1,1-difluoroalkylated products. Prakash and co-workers later employed difluorobis(trimethylsilyl)methane ($TMSCF_2TMS$) as a difluoromethylene dianion equivalent, which readily reacted with aldehydes in the presence of TBAF. It is worth noting that the CF_2H moiety can also be introduced through the partial defluorination of the trifluoromethyl group. By using the Ruppert–Prakash reagent, Portella and co-workers were able to realize the Brook rearrangement to afford 2,2-difluoroenol silyl ethers, which can be converted into various difluoromethylated compounds (Scheme 12.33) [246]. On the other hand, Prakash *et al.* showed that difluoromethylated imines can be achieved through the treatment of aldimines with $TMSCF_3$ in the presence of TMAF [247]. The corresponding amines were obtained by *in situ* reduction using $NaBH_4$ with moderate yields.

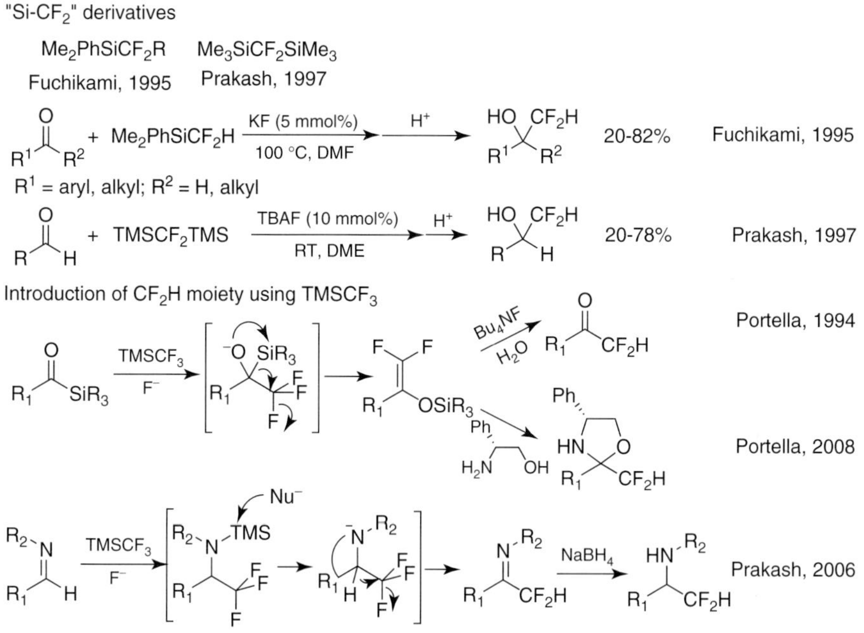

"Si-CF$_2$" derivatives

Me$_2$PhSiCF$_2$R   Me$_3$SiCF$_2$SiMe$_3$

Fuchikami, 1995   Prakash, 1997

R^1 = aryl, alkyl; R^2 = H, alkyl

Introduction of CF$_2$H moiety using TMSCF$_3$

Scheme 12.33 Difluoromethylation using the Ruppert–Prakash reagent and its analogs.

12.2.3.2 Electrophilic Difluoromethyl Reagents and Approaches

Because difluorocarbene can readily react with a variety of nucleophilic species to afford the corresponding difluoromethylated organics, difluorocarbene is *de facto* a "$^+$CF$_2^-$" equivalent. The design and development of other electrophilic difluoromethyl-containing species ("RCF$_2^+$") were therefore not ardently sought and remained unexplored. In 2007, Prakash *et al.* first demonstrated an *S*-(difluoromethyl)diarylsulfonium salt as an electrophilic difluoromethylating agent (Scheme 12.34) [248]. It was found that the reagent permitted the effective difluoromethylation of varied substrates, including oxygen, nitrogen, and phosphorus nucleophiles. The same group further investigated

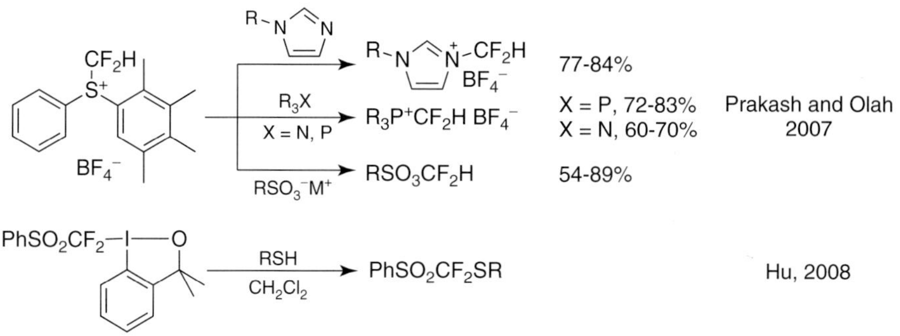

Scheme 12.34 Novel electrophilic difluoromethylating reagents.

difluoromethylation of sulfonic acid salts and imidazoles using a polystyrene-bound *S*-difluoromethylsulfonium reagent which rendered products without the need for purification [249]. More recently, Hu and co-workers reported electrophilic (phenyl-sulfonyl)difluoromethylation with a hypervalent iodine(III)−CF_2SO_2Ph compound [250]. Under mild conditions, the reagent exhibited an excellent difluoromethylating capability towards various thiols to provide phenylsulfonyl difluoromethyl sulfides in good yields (Scheme 12.34).

12.2.4
Catalytic Asymmetric Synthesis of Chiral Monofluoromethylated Organic Molecules via Nucleophilic Fluoromethylating Reactions

As mentioned previously, synthon strategies have substantial synthetic advantages over direct fluorinations for the introduction of the trifluoromethyl and the difluoromethyl motifs. In contrast, both monofluorinated synthons and direct fluorinations are applicable for the selective incorporation of functionalities bearing a single C-F bond depending on the synthetic purpose. In contrast to the extensive utilization of difluorocarbene and difluoromethyl radicals for the introduction of CF_2 moieties, the synthetic utility of monofluorinated radicals and carbenes has received relatively less attention, and the relevant studies can be referred to the excellent review articles cited above. As alternative protocols for monofluoromethyl-containing compounds, monofluoromethylations are achieved primarily in nucleophilic and electrophilic manners [251] (Scheme 12.35). Olah and Pavlath developed the first electrophilic monofluoromethylation of arenes with

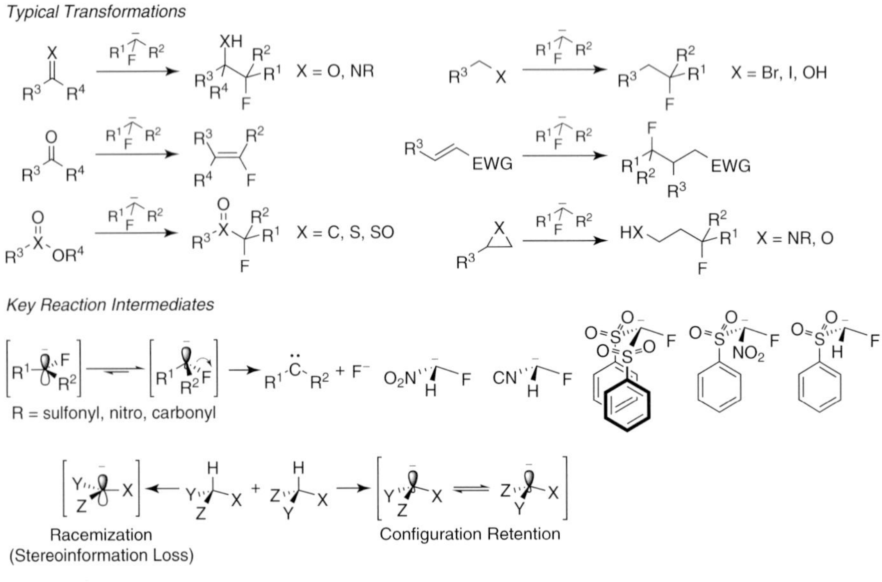

Typical Transformations

Key Reaction Intermediates

Scheme 12.35 Typical transformations and reaction intermediates of nucleophilic monofluoromethylation.

fluoromethanol to yield benzyl fluorides in 1953 [252]. Since then, several reagents have been utilized in the electrophilic monofluoromethylation of a variety of nucleophiles [253]. In addition to these studies, Prakash *et al.* demonstrated the synthetic utility of *S*-(monofluoromethyl)diarylsulfonium salts as a novel electrophilic monofluoromethylating reagent [254]. In contrast to the limited acknowledgment of electrophilic monofluoromethylation, a large number of studies have emphasized the introduction of monofluoromethyl moieties by means of monofluoromethide species. It is found that monofluoromethides are able to react with a broad range of electron-deficient substrates in many types of transformations, including nucleophilic substitutions, aldol reactions, 1,4-addition reactions, Wittig-type reactions, and ring-opening reactions (Scheme 12.35). Recently reported novel agents allowed the further investigation of the monofluoromethylation of other substrates, such as allylic acetates [255], benzynes [256], alcohols [257], and nitrones [258].

Mechanistically, whereas trifluoromethyl and difluoromethyl carbanions exhibit extreme lability, many α-monofluorocarbanions possess reasonable thermostability if their anionic centers are adjacent to electron-withdrawing groups. Hence fluoromethide intermediates can usually be generated *in situ* and readily undergo reactions with electrophilic substrates. However, these carbanions can be rather labile *per se* for isolation owing to the rapid release of fluoride, as depicted in Scheme 12.35. A recent study by Prakash *et al.*, on the other hand, demonstrated or the first time the isolation and structural characterization by X-ray diffraction of an α-fluorinated methide [259]. As predicted by computational studies, α-fluoro(bisphenylsulfonyl)methide was subject to Bent's rule and preferentially assumed a pyramidal geometry [212]. It is important to point out that the pyramidalization of α-fluorocarbanions can play a critical role in the construction of fluorinated stereogenic carbon centers because additional dynamic kinetic resolution of chiral methide intermediates is involved (Scheme 12.35). Moreover, owing to the α-fluorine effect, fluoromethides can exhibit distinct nucleophilicity and reactivity in comparison with their nonfluorinated counterparts [260].

The first nucleophilic fluoromethylation was achieved by Blank and Mager using ethyl fluoroacetate for the preparation of diethyl oxalofluoroacetate [261] (Scheme 12.36). Bergmann and co-workers further systematically studied the reactions between various electrophiles and α-fluoroenolates generated from fluoroacetate, fluoroacetone, and α-fluoro-β-keto esters [262]. They also described the Michael addition of diethyl fluoromalonate to α,β-unsaturated carbonyl compounds and nitroalkenes to yield functionalized α-fluorinated carboxylic acids [263]. In addition to these nucleophilic monofluoromethides obtained by deprotonation, Ishikawa *et al.* reported an ultrasound-promoted Reformatskii-type reaction between trifluoroacetaldehyde and ethyl bromofluoroacetate in the presence of zinc [264]. Since the 1980s, enol silyl ethers [265] and silyl ketene acetals [266] have facilitated a number of chemical transformations for the introduction of monofluoromethyl moieties, such as aldol reactions. In particular, Huang and Chen have also demonstrated the stereoselective synthesis of ethyl α-fluorosilyl enol ether and its synthetic utility in the Mukaiyama aldol reaction [267].

Even though the carbonyl-stabilized fluoromethides have been extensively exploited as valuable fluoromethyl synthons, these reagents are still not applicable in many chemical scenarios owing to the difficult removal of the carbonyl groups by C–C bond cleavage and

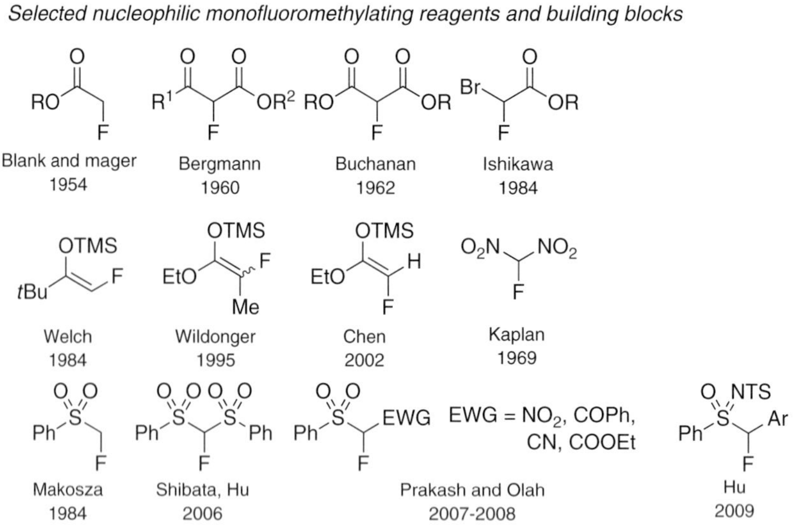

Selected nucleophilic monofluoromethylating reagents and building blocks

Scheme 12.36 Selected nucleophilic monofluoromethylating reagents.

the limited further conversions into other functionalities [251]. To meet this demand, a series of reagents have been designed and employed for incorporating diverse fluoromethyl moieties. Kaplan and Pickard utilized fluorodinitromethane [HCF(NO₂)₂] as a pronucleophile in the Michael addition to methyl acrylate for the demonstration of the α-fluorine effect [260] (Scheme 12.36). The synthetic utility of HCF(NO₂)₂ was described by Gilligan for the synthesis of *N,N*-bis(2-fluoro-2,2-dinitroethyl)-*N*-alkylamines [268]. Nevertheless, the tremendous explosive potential of HCF(NO₂)₂ undermines its synthetic availability and usefulness [269]. In contrast, several sulfur-based reagents have been developed over the past decade. Owing to the facile conversion of the sulfur-containing auxiliaries into other valuable functionalities and also their capability of altering the reactivity of fluoromethides, these reagents have been employed in a large number of fluoromethylating processes. Makosza and Goliński first described the nucleophilic substitution of hydrogen in nitroarenes with fluoromethyl phenyl sulfone (PhSO₂CH₂F) [270]. Shortly thereafter, Peet and co-workers discovered that β-fluoro-alcohols could be obtained by reacting lithiofluoromethyl phenyl sulfone with carbonyl compounds. Further treatment of the alcohols with MeSO₂Cl–NEt₃ or orthophosphoric acid followed by reductive elimination of the sulfone group provided an entry to a series of terminal vinyl fluorides [271]. Hu and co-workers also demonstrated the use of PhSO₂CH₂F in stereoselective syntheses of β-fluoroamines [272]. In recent years, a series of robust reagents have been derived from PhSO₂CH₂F by the incorporation of additional functional groups on the fluorinated carbon (Scheme 12.36). In 2006, Shibata and co-workers reported an asymmetric palladium-catalyzed allylic monofluoromethylation reaction using fluorobis(phenylsulfonyl)methane (FBSM) to yield the corresponding fluoromethylated adducts with high *ee* [255]. Simultaneously, Hu and co-workers independently described the ring-opening reaction of epoxides with FBSM as a pronucleophile [238]. Thanks to its strong electron-withdrawing nature arising from the two sulfonyl groups, the acidity of FBSM and the stability of its anion are significantly enhanced, thereby permitting its application towards various synthetic targets. In addition to FBSM, several functionalized fluoromethyl phenyl sulfone derivatives have been developed by Prakash and co-workers as versatile monofluoromethyl synthons [273]. Recently, Hu and co-workers further reported the remarkable reaction between α-fluorosulfoximines and nitrones to afford (Z)-monofluoroalkenes with high stereoselectivity [258].

Most importantly, catalytic asymmetric syntheses of fluorine-containing organic molecules based on various α-fluorinated pronucleophiles have been recognized as one of the major achievements in synthetic organofluorine chemistry of the last two decades (Scheme 12.37). Generally, these synthetic protocols have emphasized (i) the stereoselective introduction of achiral fluorinated functionalities and (ii) the formation of nonracemic stereogenic fluorocarbon centers [89c, 89f, 274]. Owing to their remarkable reactivity and facile conversions into structurally sophisticated molecules, α-fluorocarbonyl units are among the most extensively utilized synthons in catalytic asymmetric monofluorinating reactions (Scheme 12.37). Lerner and Barbas first described the reaction between fluoroacetone and an aldehyde catalyzed by aldolase catalytic antibody 38C2 via the enamine mechanism [275]. Iseki *et al.* exploited bromofluoroketene ethyl trimethylsilyl acetal as a fluoromethyl synthon to react with various aldehydes in the presence of a catalytic amount of chiral Lewis acid, which afforded aldol adducts with good stereoselectivities [276]. Another example was demonstrated by Kim *et al.*, who utilized diethyl fluoromalonate as a pronucleophile in the Michael addition to chalcone derivatives [277]. Promoted by cinchonium-based PTC, moderate enantiomeric purities could be achieved under mild conditions. In 2004, Barbas and co-workers reported an organocatalyzed asymmetric aldol reaction for the synthesis of *anti-α-fluoro-β-hydroxy* ketones [278]. Using L-prolinol as the catalyst, the protocol permitted a direct C–C bond-forming reaction between fluoroacetone and aldehydes to give the products with satisfactory enantio- and diastereoselectivities. Towards pyrrolidine-based PDE4 inhibitors, Nichols *et al.* reported an enantioselective conjugation addition of diethyl fluoromalonate to *trans*-nitroalkenes in the presence of a Lewis acid and chiral bis(oxazoline) [279]. Moreover, the first asymmetric amination of α-fluoro-β-keto esters with diazodicarboxylates was presented by Togni and co-workers using Cu/Ph-Box catalyst [280]. Likewise, the groups of Maruoka and Kim later achieved the transformation by utilization of chiral quaternary phosphonium salts and chiral nickel complexes as catalysts, respectively [281]. In fact, phase-transfer catalysis was also found to be an effective strategy for the construction of stereogenic fluorinated carbon centers through alkylation of a racemate of α-fluoro-β-keto esters [282]. Thanks to the rapid development of organocatalysis, the groups of Córdova and Rios independently investigated the conjugate addition of fluoromalonates and α-fluoro-β-keto esters to α,β-unsaturated aldehydes undergoing the imine catalytic pathway [283]. Apart from these catalytic systems, chiral hydrogen bonding donors have been used as efficient catalysts for many Michael-type reactions with monofluoromethyl pronucleophiles [284]. In particular, Lu and co-workers explored the asymmetric Mannich reaction of fluorinated keto esters with a tryptophan-derived bifunctional thiourea catalyst, which rendered the products with excellent enantioselectivities and low to good diastereomeric ratios [285]. Meanwhile, Jiang and co-workers

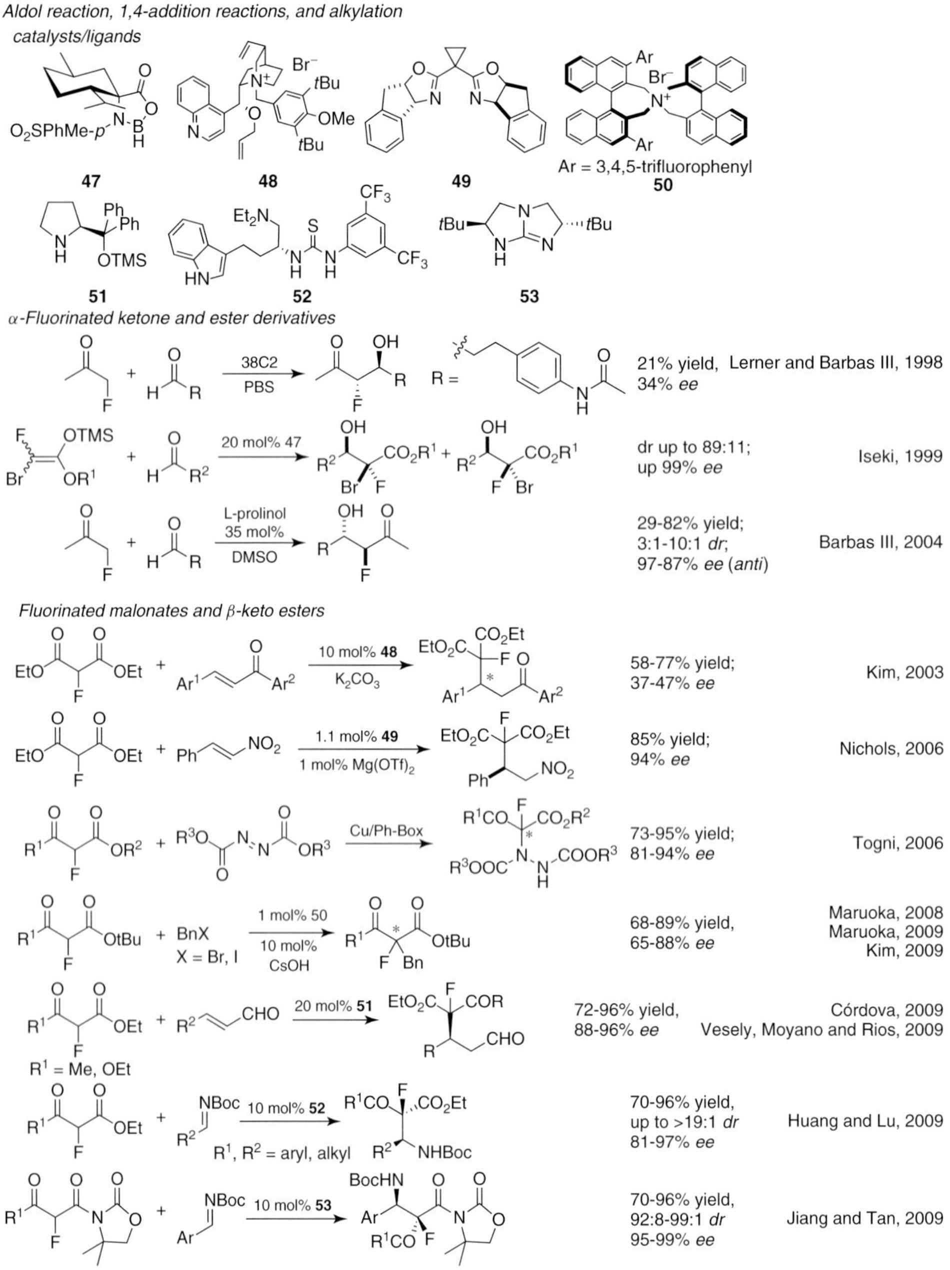

Aldol reaction, 1,4-addition reactions, and alkylation
catalysts/ligands
47
48
49
Ar = 3,4,5-trifluorophenyl
50
51
52
53
α-Fluorinated ketone and ester derivatives
38C2
PBS
21% yield, 34% ee
Lerner and Barbas III, 1998
20 mol% 47
dr up to 89:11; up 99% ee
Iseki, 1999
L-prolinol 35 mol% DMSO
29-82% yield; 3:1-10:1 dr; 97-87% ee (anti)
Barbas III, 2004
Fluorinated malonates and β-keto esters
10 mol% 48
K2CO3
58-77% yield; 37-47% ee
Kim, 2003
1.1 mol% 49
1 mol% Mg(OTf)2
85% yield; 94% ee
Nichols, 2006
Cu/Ph-Box
73-95% yield; 81-94% ee
Togni, 2006
1 mol% 50
10 mol% CsOH
X = Br, I
68-89% yield, 65-88% ee
Maruoka, 2008
Maruoka, 2009
Kim, 2009
20 mol% 51
R1 = Me, OEt
72-96% yield, 88-96% ee
Córdova, 2009
Vesely, Moyano and Rios, 2009
10 mol% 52
R1, R2 = aryl, alkyl
70-96% yield, up to >19:1 dr, 81-97% ee
Huang and Lu, 2009
10 mol% 53
70-96% yield, 92:8-99:1 dr, 95-99% ee
Jiang and Tan, 2009

Scheme 12.37 Asymmetric monofluoromethylations using α-fluorinated carbonyl compounds as pronucleophiles.

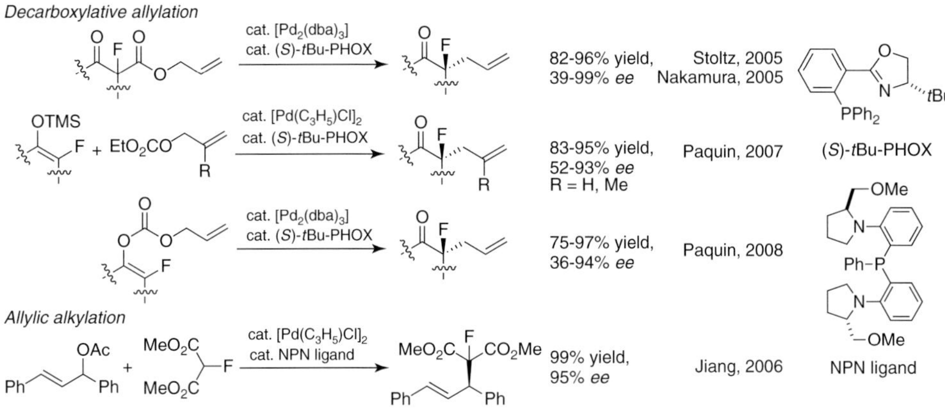

Decarboxylative allylation

82-96% yield,   Stoltz, 2005
39-99% *ee*   Nakamura, 2005

83-95% yield,   Paquin, 2007
52-93% *ee*
R = H, Me

75-97% yield,   Paquin, 2008
36-94% *ee*

(S)-*t*Bu-PHOX

Allylic alkylation

99% yield,   Jiang, 2006
95% *ee*

NPN ligand

Scheme 12.38   Asymmetric monofluoromethylations via decarboxylative allylation and allylic alkylation.

demonstrated a similar synthetic protocol catalyzed by a bicyclic guanidine derivative for the synthesis of α-fluorinated β-amino acids [286].

Intriguingly, several laboratories have been involved in investigations of Pd-mediated asymmetric monofluoromethylation reactions with α-fluorinated carbonyl compounds (Scheme 12.38). In 2005, Stoltz and co-workers demonstrated a Pd-catalyzed enantioconvergent decarboxylative allylation of α-fluoro-β-keto esters [287]. Almost simultaneously, Nakamura *et al.* systematically studied the construction of fluorinated quaternary stereocenters by enantioselective decarboxylation in the presence of an identical catalytic system [288]. It is worth noting that this synthetic route is the first example enabling the asymmetric synthesis of chiral fluorinated carbons via the deracemization of quaternary carbon stereocenters. Paquin and co-workers, on the other hand, reported the enantioselective allylation reaction of fluorinated silyl enol ethers with allyl carbonates in the presence of a catalytic amount of (S)-*t*Bu-PHOX and Pd [289]. Shortly afterwards, a variant intramolecular route was devised by the same group for the synthesis of allylated tertiary α-fluoro ketones [290]. Jiang *et al.* also reported the application of fluoromalonate to a Pd-catalyzed asymmetric allylic alkylation with an N-phenyl-(S)-prolinol-derived P,N-ligand, which gave the product in excellent enantioselectivity and yield [291].

Although a-fluorinated carbonyl compounds have successfully provided access to the syntheses of monofluoromethyl-containing molecules, the difficult removal of the carbonyl moieties to form an unfunctionalized CH$_2$F motif impedes their synthetic applicability. In order to overcome this challenge, FBSM, featuring facile reductive removal of the sulfonyl groups, has been ubiquitously used as a monofluoromethyl equivalent. First developed by the groups of Shibata and Hu in 2006, electrophilic fluorination [238, 255] and/or ECF [292] processes were employed in the conventional preparation of the agent (Scheme 12.39). Hu and co-workers described a superior synthetic route based on the sulfoxidation of fluoromethyl phenyl sulfone followed by oxidation, which avoids the tedious purification necessary in the conventional synthesis [293]. Recently, Prakash *et al.* designed a practical one-step synthesis of FBSM with

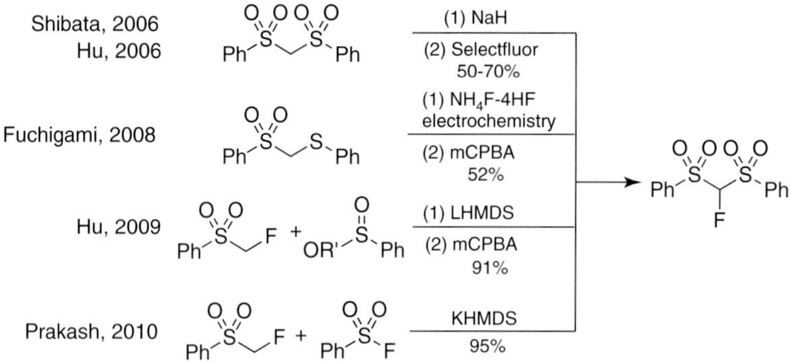

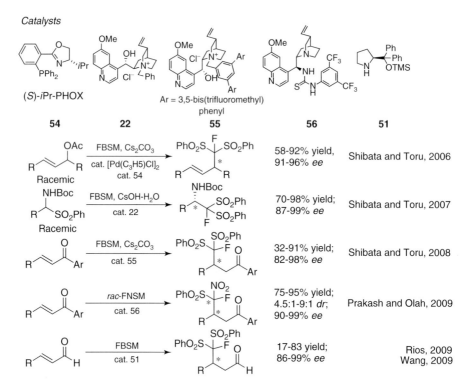

Scheme 12.39 Developments in the synthesis of fluorobis(phenylsulfonyl)methane.

fluoromethyl phenyl sulfone and less costly phenylsulfonyl fluoride, giving FBSM with high efficiency and purity [294].

Mechanistically, FBSM undergoes deprotonation under mild conditions, thereby permitting many chemical transformations that cannot be achieved otherwise. Since its emergence in 2006, FBSM has been explored by Shibata, Hu, Prakash, and many others

Catalysts

Scheme 12.40 Catalytic enantioselective monofluoromethylations using sulfone-based reagents.

as a robust synthon in various chemical reactions, including the ring opening of epoxides [238], Michael addition [256, 273], the Mitsunobu reaction [257], and the fluoromethylation of alkanes [295], arenes [256], and alkynes [296]. In particular, FBSM and its analogs have been deployed as versatile reagents in asymmetric catalysis (Scheme 12.40). Shibata and co-workers first demonstrated FBSM as a pronucleophile in the Tsuji–Trost allylic alkylation [255]. Mediated by [(S)-iPr-PHOX]-ligated Pd complex, FBSM was shown to be capable of reacting with allyl acetates to afford enantiomerically enriched products with moderate to excellent yields. In 2007, the same group extended the application of FBSM to an enantioselective Mannich-type reaction [297]. Using the quinidinium salt **22** as a PTC, the protocol yielded fluorinated *N*-Boc-protected α-fluoro(bisphenylsulfonyl)amines with high *ee*, which could undergo reductive desulfonation to form monofluorinated amines without loss of enantiomeric purity. Under similar conditions, an efficient enantioselective Michael addition of FBSM to α,β-unsaturated carbonyl compounds was achieved by Shibata and co-workers [298]. Almost simultaneously, Prakash and co-workers developed the stereoselective Michael addition of α-fluoro-α-nitro(phenylsulfonyl)methane to chalcone derivatives via a hydrogen-bonding activating mechanism using cinchona alkaloid-derived thiourea catalyst **56** [299]. Intriguingly, owing to the pyramidalization of the fluorinated carbanion intermediate, the observed stereoselectivity was rationalized as a consequence of a dynamic kinetic resolution process. In addition to these two methods, Kim and co-workers later reported the primary amine-catalyzed Michael addition of FBSM to α,β-unsaturated ketones to generate the adducts with excellent enantiomeric enrichment [300]. It is also noteworthy that the groups of Rios and Wang independently explored the conjugate addition of FBSM to enals in the presence of chiral diphenyl-prolinol trimethylsilyl ether **51** [301]. Under almost identical conditions, these protocols allowed the formation of 1,4-adducts with high enantioselectivities.

12.3
Conclusion and Perspectives

Organofluorine chemistry is a flourishing field. In contrast to the abundant naturally occurring hydrocarbons and their derivatives, fluoroorganics are extremely scarce in Nature and are almost exclusively laboratory-made molecules. Thanks to the efforts of numerous dedicated chemists over the past 175 years, a multitude of fluorinated organic compounds, ranging from fairly simple to highly sophisticated, have become a reality, with many practical applications. Possessing unique properties, fluoroorganics offer advantages as privileged materials for chemical, physical, and biological sciences. Over the past 50 years, and particularly over the last two decades, organofluorine chemistry has experienced exponential growth, reflecting the development of various new protocols and robust reagents for the efficient syntheses of useful organofluorine compounds. As a fruitful paradise for chemists, the field has intrigued not only the "specialists" within this community, but also attracted attention from many "outsiders" to bring their concepts, leading to remarkable advances in organofluorine chemistry. As mentioned previously, thanks to the rapid developments in asymmetric synthesis, the stereoselective incorporation of fluorine or fluorinated functionalities has come of age in recent years. Notably, much effort has also been directed towards transition metal-promoted fluorinations and fluoroalkylations that were previously considered arduous transformations.

Even though making predictions is, of course, always risky, we still believe that organofluorine chemistry will remain fascinating and dynamic, just as it has been over the past two decades. Still, the proliferation of efficient synthetic protocols and robust fluorinated synthons will be instrumental to the contribution of fluorinated molecules with immense structural diversity and extraordinary functions. To introduce fluorinated motifs in an asymmetric fashion and/or organometallic pathways should be the central theme in the foreseeable future. It seems clear that longstanding challenges, such as the enhancement of nucleophilicity of fluorides and the stabilization of α- and β-fluorocarbanions, will continue to be the impetus for organofluorine chemists to seek intellectual solutions.

Acknowledgment

Support of our work by the Loker Hydrocarbon Research Institute is greatly acknowledged.

References

1. (a) Banks, R.E. (2000) *Fluorine Chemistry at the Millennium: Fascinated by Fluorine*, Elsevier, New York; (b) Dolbier, W.R. Jr. (2005) *J. Fluorine Chem.*, **126**, 157–163.
2. (a) Zhang, W. (2004) *Chem. Rev.*, **104**, 2531–2556; (b) Zhang, W. (2009) *Chem. Rev.*, **109**, 749–795.
3. Leroux, F., Manteau, B., Vors, J.-P., and Pazenok, S. (2008) *Beilstein J. Org. Chem.*, **4** (13).
4. Hunter, L. (2010) *Beilstein J. Org. Chem.*, **6** (38).
5. (a) Takaoka, Y., Sakamoto, T., Tsukiji, S., Narazaki, M., Matsuda, T., Tochio, H., Shirakawa, M., and Hamachi, I. (2009) *Nat. Chem.*, **1**, 557–561; (b) Terreno, E., Castelli, D.D., Viale, A., and Aime, S. (2010) *Chem. Rev.*, **110**, 3019–3042.
6. Olah, A., Prakash, G.K.S., Molnár, Á., and Sommer, J. (2009) *Superacid Chemistry*, 2nd edn., John Wiley & Sons, Inc., Hoboken, NJ, Chapter 2, pp. 35–82.
7. (a) Uneyama, K. (2006) *Organofluorine Chemistry*, Blackwell, Oxford, Chapter 5, pp. 186–205. (b) the applications of organofluorine compounds have been thoroughly summarized: Kirsch, P. (2004)

Modern Fluoroorganic Chemistry: Synthesis, Reactivity, Applications, Wiley-VCH Verlag GmbH, Weinheim, Chapter 4, pp. 203–277.
8. Schlosser, M. and Michel, D. (1996) *Tetrahedron*, **52**, 99–108.
9. Tochtermann, W. (1966) *Angew. Chem. Int. Ed. Engl.*, **5**, 351–371.
10. Welch, C., Juan, R.R.S., Masuda, J.D., and Stephan, D.W. (2006) *Science*, **314**, 1124–1126.
11. Stephan, W. and Erker, G. (2010) *Angew. Chem. Int. Ed.*, **49**, 46–76, and references therein.
12. Banks, E. and Tatlow, J.C. (1986) *J. Fluorine Chem.*, **33**, 71–108.
13. Dumas, J. and Péligot, E. (1835) *Ann. Pharm.*, **15**, 246.
14. (a) Moissan, H. (1886) *C. R. Acad. Sci.*, **102**, 1543–1544; (b) Lemal, D.M. (2004) *J. Org. Chem.*, **69**, 1–11; (c) Tressaud, A. (2006) *Angew. Chem. Int. Ed. Engl.*, **45**, 6792–6796.
15. Kauffman, G.B. (1955) *J. Chem. Educ.*, **32**, 301.
16. Balz, G. and Schiemann, G. (1927) *Ber. Dtsch. Chem. Ges.*, **60**, 1186–1190.

17. Fowler, R., Buford, W. III, Hamilton, J. Jr., Sweet, R., Weber, C., Kasper, J., and Litant, I. (1947) *Ind. Eng. Chem.*, **39**, 292–298.

18. Chamber, R.D. (2004) *Fluorine in Organic Chemistry*, Blackwell, Oxford, pp. 33–35.

19. Adams, D.J. and Clark, J.H. (1999) *Chem. Soc. Rev.*, **28**, 225–231.

20. For early reviews in mid-1990s, Special Issue on Fluorine Chemistry: Smart, B.E. (ed.) (1996) *Chem. Rev.*, **96**, 1555–1824.

21. Amii, H. and Uneyama, K. (2009) *Chem. Rev.*, **109**, 2119–2183.

22. (a) O'Hagan, D., Schaffrath, C., Cobb, S.L., Hamilton, J.T.G., and Murphy, C.D. (2002) *Nature*, **416**, 279; (b) Dong, C., Huang, F., Deng, H., Schaffrath, C., Spencer, J.B., O'Hagan, D., and Naismith, J.H. (2004) *Nature*, **427**, 561–565.

23. (a) Margrave, J.L., Lagow, R.J., and Conroy, A.P. (1970) *Proc. Natl. Acad. Sci. U. S. A.*, **67**, A8; (b) Adcock, J.L. and Lagow, R.J. (1974) *J. Am. Chem. Soc.*, **96**, 7588–7589; (c) for an excellent review article, see: Adcock, J.L. and Cherry, M.L. (1987) *Ind. Eng. Chem. Res.*, **26**, 208–215.

24. Mascaretti, O.A. (1993) *Aldrichim. Acta*, **26**, 47–58.

25. Kim, W., Jeong, H.-J., Lim, S.T., and Sohn, M.-H. (2010) *Nucl. Med. Mol. Imaging*, **44**, 25–32.

26. Liotta, C.L. and Harris, H.P. (1974) *J. Am. Chem. Soc.*, **96**, 2250–2252.

27. Christe *et al.* first prepared the anhydrous tetramethylammonium fluoride in 1990: Christe, K.O., Wilson, W.W., Wilson, R.D., Bau, R., and Feng, J.-A. (1990) *J. Am. Chem. Soc.*, **112**, 7619–7625.

28. Sun, H. and DiMagno, S.G. (2005) *J. Am. Chem. Soc.*, **127**, 2050–2051.

29. Sun, H. and DiMagno, S.G. (2006) *Angew. Chem. Int. Ed.*, **45**, 2720–2725.

30. Middleton, W.J. (1976) US Patent 3,940,402.

31. Gingras, M. (1991) *Tetrahedron Lett.*, **32**, 7381.

32. Pilcher, S., Ammon, H.L., and DeShong, P. (1995) *J. Am. Chem. Soc.*, **117**, 5166–5167.

33. Kim, D.W., Ahn, D.S., Oh, Y.H., Lee, S., Oh, S.J., Lee, S.J., Kim, J.S., Moon, J.S.D.H., and Chi, D.Y. (2006) *J. Am. Chem. Soc.*, **128**, 16394–16397.

34. Kim, D.W., Jeong, H.J., Lim, S.T., Sohn, M.H., Katzenellenbogen, J.A., and Chi, D.Y. (2008) *J. Org. Chem.*, **73**, 957–962.

35. Kim, D.W., Jeong, H.-J., Lim, S.T., and Sohn, M.-H. (2008) *Angew. Chem. Int. Ed.*, **47**, 8404–8406.

36. Hara, S. (2006) *Adv. Org. Synth.*, **2**, 49–60.

37. Olah, G.A., Welch, J.T., Vankar, Y.D., Nojima, M., Kerekes, I., and Olah, J.A. (1979) *J. Org. Chem.*, **44**, 3872–3881.

38. Hasek, W.R., Smith, W.C., and Engelhardt, V.A. (1960) *J. Am. Chem. Soc.*, **82**, 543–551.

39. PhSF₃ was preliminarily reported: (a) Sheppard, W.A. (1960) *J. Am. Chem. Soc.*, **82**, 4751–4752; and was later employed as a deoxofluorinating reagent: (b) Sheppard, W.A. (1962) *J. Am. Chem. Soc.*, **84**, 3058–3063.

40. Singh, R.P. and Shreeve, J.M. (2002) *Synthesis*, **17**, 2561–2578.

41. For recent examples, see: (a) Hunter, L., Kirsch, P., Slawin, A.M.Z., and O'Hagan, D. (2009) *Angew. Chem. Int. Ed.*, **48**, 5457–5460; (b) Linclau, B., Leung, L., Nonnenmacher, J., and Tizzard, G. (2010) *Beilstein J. Org. Chem.*, **6** (62).

42. Middleton, W.J. (1975) *J. Org. Chem.*, **40**, 574–578.

43. Lal, G.S., Pez, G.P., Pesaresi, R.J., and Prozonic, F.M. (1999) *Chem. Commun.*, 215–216.

44. (a) Umemoto, T. and Xu, Y. (2007) US Patent 7,265,247; (b) Umemoto, T. and Singh, R.P. (2008) US Pat. Appl. US 2008039660.

45. (a) Beaulieu, F., Beauregard, L.-P., Courchesne, G., Couturier, M., LaFlamme, F., and L'Heureux, A. (2009) *Org. Lett.*, **11**, 5050–5053; (b) L'Heureux, A., Beaulieu, F., Bennett, C., Bill, D.R., Clayton, S., LaFlamme, F., Mirmehrabi, M., Tadayon, S., Tovell, D., and Couturier, M. (2010) *J. Org. Chem.*, **75**, 3401–3411.

46. (a) House, H.O. (1956) *J. Am. Chem. Soc.*, **78**, 2298–2302; for a pioneering example of asymmetric ring opening of cholesterol epoxides, see: (b) Bowers, A. and Ringold, H.J. (1958) *Tetrahedron*, **3**, 14–27.

47. (a) Sugihara, Y., Iimura, S., and Nakayama, J. (2002) *Chem. Commun.*, 134–135; (b) Hu, X.E. (2002) *Tetrahedron Lett.*, **43**, 5315–5318; (c) Ding, C.-H.,

Dai, L.-X., and Hou, X.-L. (2004) *Synlett*, 2218–2220.

48. Cresswell, A.J., Davies, S.G., Lee, J.A., Roberts, P.M., Russell, A.J., Thomson, J.E., and Tyte, M.J. (2010) *Org. Lett.*, **12**, 2936–2939.

49. Hann, G.L. and Sampson, P. (1989) *J. Chem. Soc., Chem. Commun.*, 1650–1651.

50. (a) Bruns, S. and Haufe, G. (2000) *J. Fluorine Chem.*, **104**, 247–254; (b) Haufe, G. and Bruns, S. (2002) *Adv. Synth. Catal.*, **344**, 165–171.

51. Kalow, J.A. and Doyle, A.G. (2010) *J. Am. Soc. Chem.*, **132**, 3268–3269.

52. Furuya, T., Kleim, J.E.M.N., and Ritter, T. (2010) *Synthesis*, 1804–1281.

53. Doherty, N.M. and Hoffmann, N.W. (1991) *Chem. Rev.*, **91**, 553–573.

54. Dixon, K.R. and McFarland, J.J. (1972) *J. Chem. Soc., Chem. Commun.*, 1274–1275.

55. Grushin, V.V. (2009) *Acc. Chem. Res.*, **43**, 160–171.

56. Fraser, S.L., Antipin, M.Yu., Khroustalyov, V.N., and Grushin, V.V. (1997) *J. Am. Chem. Soc.*, **119**, 4769–4770.

57. Grushin, V.V. (2002) *Chem. Eur. J.*, **8**, 1006–1014.

58. For selected examples, see: Grushin, V.V. (2000) *Organometallics*, **19**, 1888–1900.

59. Yandulov, D.V. and Tran, N.T. (2007) *J. Am. Chem. Soc.*, **129**, 1342–1358.

60. Grushin, V.V. and Marshall, W.J. (2007) *Organometallics*, **26**, 4997–5002.

61. Watson, D.A., Su, M., Teverovskiy, G., Zhang, Y., García-Fortanet, J., Kinzel, T., and Buchwald, S.L. (2009) *Science*, **325**, 1661–1664.

62. Subramanian, M.A. and Manzer, L.E. (2002) *Science*, **297**, 1665.

63. Janmanchi, K.M. and Dolbier, W.R. Jr. (2008) *Org. Process Res. Dev.*, **12**, 349–354.

64. FClO₃ was first prepared by Engelbrecht's group in 1952: (a) Engelbrecht, A. and Atzwanger, H. (1952) *Monatsh. Chem.*, **83**, 1087–1089; (b) Engelbrecht, A. (1954) *Angew. Chem.*, **66**, 442.

65. Inman, C.E., Oesterling, R.E., and Tyczkowski, E.A. (1958) *J. Am. Chem. Soc.*, **80**, 6533–6535.

66. CF₃OF was prepared by Kellogg and Cady in 1948: Kellogg, K.B. and Cady, G.H. (1948) *J. Am. Chem. Soc.*, **70**, 3986–3990.

67. (a) Barton, D.H.R., Godinho, L.S., Hesse, R.H., and Pechet, M.M. (1968) *Chem.*

Commun., 804–806; (b) Barton, D.H.R., Ganguly, A.K., Hesse, R.H., Loo, S.N., and Pechet, M.M. (1968) *Chem. Commun.*, 806–808.

68. Rozen, S. (1996) *Chem. Rev.*, **96**, 1717–1736.

69. Mulholland, G.K. and Ehrenkaufer, R.E. (1986) *J. Org. Chem.*, **57**, 1482–1489.

70. Migliorese, K.G., Appelman, E.H., and Tsangaris, M.N. (1979) *J. Org. Chem.*, **44**, 1711–1714.

71. Rozen, S. and Gal, C. (1987) *J. Org. Chem.*, **52**, 4928–4933.

72. (a) Hoppe, R., Dähne, W., Mattauch, H., and Rödder, K.M. (1962) *Angew. Chem. Int. Ed. Engl.*, **1**, 599; (b) Weeks, J.L., Chernick, C.L., and Matheson, M.S. (1962) *J. Am. Chem. Soc.*, **84**, 4612–4613.

73. Shieh, T.C., Feit, E.D., Chernick, C.L., and Yang, N.C. (1970) *J. Org. Chem.*, **35**, 4020–4024.

74. Cacace, F. and Wolf, A.P. (1978) *J. Am. Chem. Soc.*, **100**, 3639–3641.

75. (a) Rozen, S. and Gal, C. (1987) *J. Org. Chem.*, **52**, 2769–2779; (b) Rozen, S. and Gal, C. (1988) *J. Org. Chem.*, **53**, 2803–2807; (c) Rozen, S. (1992) Electrophilic fluorination reactions with F₂ and some reagents directly derived from it, in *Synthetic Fluorine Chemistry* (eds. G.A. Olah, R.D. Chamber, and G.K.S. Prakash), John Wiley & Sons, Inc., New York, pp. 143–162.

76. Lal, G.S., Pez, G.P., and Syvret, R.G. (1996) *Chem. Rev.*, **96**, 1737–1756.

77. Banks, R.E. and Williamson, G.E. (1964) *Chem. Ind. (London)*, 1864.

78. Differding, E. and Bersier, P.M. (1992) *Tetrahedron*, **48**, 1595–1604.

79. Tee, O.S., Iyengar, N.R., and Paventi, M. (1983) *J. Org. Chem.*, **48**, 761–762.

80. Umemoto, T., Kawada, K., and Tomita, K. (1986) *Tetrahedron Lett.*, **27**, 4465–4468.

81. Singh, S., DesMarteau, D., Zuberi, S.S., Witz, M., and Huang, H.N. (1987) *J. Am. Chem. Soc.*, **109**, 7194–7196.

82. (a) For the preparation of N₂F⁺AsF₆⁻ salt, see: Moy, D. and Young, A.R. II (1965) *J. Am. Chem. Soc.*, **87**, 1889–1892; (b) for the preparation of NF₄⁺AsF₆⁻ and NF₄⁺SbF₆⁻, see: Christe, K.O., Guertin, J.P., and Pavlath, A.E. (1966) *Inorg. Nucl. Chem. Lett.*, **2**, 83–86; (c) Tolberg, W.E., Rewick, R.T., Stringham, R.S., and Hill,

M.E. (1966) *Inorg. Nucl. Chem. Lett.*, **2**, 79–82.

83. Differding, E. and Ofner, H. (1991) *Synlett*, 187–189.

84. (a) Banks, R.E., Mohialdin-Khaffaf, S.N., Lal, G.S., Sharif, I., and Syvret, R.G. (1992) *Chem. Commun.*, 595–596; (b) Lal, G.S. (1993) *J. Org. Chem.*, **58**, 2791–2796.

85. Singh, R.R. and Shreeve, J.M. (2004) *Acc. Chem. Res.*, **37**, 31–44.

86. Nyffeler, P.T., Durón, S.G., Burkart, M.D., Vincent, S.P., and Wong, C.-H. (2005) *Angew. Chem. Int. Ed.*, **44**, 192–212.

87. For recent investigations on this topic, see: (a) Antelo, J.M., Crugeiras, J., Leis, J.R., and Ríos, A. (2000) *J. Chem. Soc., Perkin Trans. 2*, 2071–2076; (b) Zhang, X., Liao, Y., Qian, R., Wang, H., and Guo, Y. (2005) *Org. Lett.*, **7**, 3877–3880.

88. Olah, G.A., Hartz, N., Rasul, G., Wang, Q., Prakash, G.K.S., Casanova, J., and Christe, K.O. (1994) *J. Am. Chem. Soc.*, **116**, 5671–5673.

89. For selected reviews, see: (a) Ibranim, H. and Togni, A. (2004) *Chem. Commun.*, 1147–1155; (b) Oestreich, M. (2005) *Angew. Chem. Int. Ed.*, **44**, 2324–2327; (c) Prakash, G.K.S. and Beier, P. (2006) *Angew. Chem. Int. Ed.*, **45**, 2172–2174; (d) Brunet, V.A. and O'Hagan, D. (2008) *Angew. Chem. Int. Ed.*, **47**, 1179–1182; (e) Ma, J.-A. and Cahard, D. (2008) *Chem. Rev.*, **108**, PR1–PR43; (f) Cahard, D., Xu, X., Couve-Bonnaire, S., and Pannecoucke, X. (2010) *Chem. Soc. Rev.*, **39**, 558–568.

90. Differding, E. and Lang, R.W. (1988) *Tetrahedron Lett.*, **29**, 6087–6090.

91. Davis, A., Zhou, P., and Murphy, C.K. (1993) *Tetrahedron Lett.*, **34**, 3971–3974.

92. Takeuchi, Y., Satoh, A., Suzuki, T., Kameda, A., Dohrin, M., Satoh, T., Koizumi, T., and Kirk, K.L. (1997) *Chem. Pharm. Bull.*, **45**, 1085–1088.

93. (a) Takeuchi, Y., Suzuki, T., Satoh, A., Shiragami, T., and Shibata, N. (1999) *J. Org. Chem.*, **64**, 5708–5711; (b) Liu, Z. Shibata, N., and Takeuchi, Y. (2000) *J. Org. Chem.*, **65**, 7583–7587.

94. Abdul-Ghani, M., Banks, R.E., Besheesh, M.K., Sharif, I., and Syvret, R.G. (1995) *J. Fluorine Chem.*, **73**, 255–257.

95. Cahard, D., Audouard, C., Plaquevent, J.C., and Roques, N. (2000) *Org. Lett.*, **2**, 3699–3701.

96. Shibata, N., Susuki, E., and Takeuchi, Y. (2000) *J. Am. Chem. Soc.*, **122**, 10728–10729.

97. Hintermann, L. and Togni, A. (2000) *Angew. Chem. Int. Ed.*, **39**, 4359–4362.

98. Hamashima, Y., Yagi, K., Takano, H., Tamas, L., and Sodeoka, M. (2002) *J. Am. Chem. Soc.*, **124**, 14530–14531.

99. Berkessel, A. and Gröger, H. (2005) *Asymmetric Organocatalysis: from Biomimetic Concepts to Applications in Asymmetric Synthesis*, Wiley-VCH Verlag GmbH, Weinheim.

100. Kim, D.Y. and Park, E.J. (2002) *Org. Lett.*, **4**, 545–547.

101. Beeson, T.D. and MacMillan, D.W.C. (2005) *J. Am. Chem. Soc.*, **127**, 8826–8828.

102. Marigo, M., Fielenbach, D., Braunton, A., Kjœrsgaard, A., and Jørgensen, K.A. (2005) *Angew. Chem. Int. Ed.*, **44**, 3703–3706.

103. Steiner, D.D., Mase, N., and Barbas, C.F. III (2005) *Angew. Chem. Int. Ed.*, **44**, 3706–3710.

104. Shaw, M.J., Hyman, H.H., and Filler, R. (1969) *J. Am. Chem. Soc.*, **91**, 1563–1565.

105. Taylor, S.T., Kotoris, C.C., and Hum, G. (1999) *Tetrahedron*, **55**, 12431–12477, and references therein.

106. Shamma, T., Buchholz, H., Prakash, G.K.S., and Olah, G.A. (1999) *Isr. J. Chem.*, **39**, 207–210.

107. (a) Barnette, W.E. (1984) *J. Am. Chem. Soc.*, **106**, 452–454; (b) Snieckus, V., Beaulieu, F., Mohri, K., Han, W., Murphy, C.K., and Davis, F.A. (1994) *Tetrahedron Lett.*, **35**, 3465–3468.

108. Yamada, S., Gavryushin, A., and Knochel, P. (2010) *Angew. Chem. Int. Ed.*, **49**, 2215–2218.

109. Anbarasan, P., Neumann, H., and Beller, M. (2010) *Angew. Chem. Int. Ed.*, **49**, 2219–2222.

110. For electrophilic fluorinations of aryl–palladium complexes and related mechanistic studies, see: (a) Furuya, T., Kaiser, H.M., and Ritter, T. (2008) *Angew. Chem. Int. Ed.*, **47**, 5993–5996; (b) Furuya, T. and Ritter, T. (2008) *J. Am. Chem. Soc.*, **130**, 10060–10061; (c) Ball, N.D. and Sanford, M.S. (2009) *J. Am. Chem. Soc.*, **131**, 3796–3797; (d) Furuya, T., Benitez, D., Tkatchouk, E., Strom, A.E., Tang, P., Goddard, W.A., and Ritter, T. III (2010) *J. Am. Chem. Soc.*, **132**, 3793–3807.

111. Brown, J.M. and Gouverneur, V. (2009) *Angew. Chem. Int. Ed.*, **48**, 8610–8614.

112. Hull, K.L., Anani, W.Q., and Sanford, M.S. (2006) *J. Am. Chem. Soc.*, **128**, 7134–7135.

113. Wang, X., Mei, T.-S., and Yu, J.-Q. (2009) *J. Am. Chem. Soc.*, **131**, 7520–7521.

114. Furuya, T., Strom, A.E., and Ritter, T. (2009) *J. Am. Chem. Soc.*, **131**, 1662–1663.

115. Furuya, T. and Ritter, T. (2009) *Org. Lett.*, **11**, 2860–2863.

116. (a) In fact, the electrophilic fluorination of vinyl stannanes was reported by Mc-Carthy and co-workers in 1993: Matthews, D.P., Miller, S.C., Jarvi, E.T., Sabol, J.S., and McCarthy, J.R. (1993) *Tetrahedron Lett.*, **34**, 3057–3060; (b) the electrophilic fluorination of alkenylboronic acids and trifluoroborates was described by Prakash and co-workers in 1997: Petasis, N.A., Yudin, A.K., Zavialov, I.A., Prakash, G.K.S., and Olah, G.A. (1997) *Synlett*, 606–608.

117. Tang, P., Furuya, T., and Ritter, T. (2010) *J. Am. Chem. Soc.*, **132**, 12150–12154.

118. Uneyama, K., Katagiri, T., and Amii, H. (2008) *Acc. Chem. Res.*, **41**, 817–829.

119. Dolbier, W.R. Jr. (1996) *Chem. Rev.*, **96**, 1557–1584.

120. Burton, D.J., Yang, Z.-Y., and Qiu, W. (1996) *Chem. Rev.*, **96**, 1641–1716.

121. Brahms, D.L.S. and Dailey, W.P. (1996) *Chem. Rev.*, **96**, 1585–1632.

122. (a) For an excellent review on the Julia–Kocienski-type fluoroolefination, see: Zajc, B. and Kumar, R. (2010) *Synthesis*, 1822–1836; (b) Prakash *et al.* recently described the Julia–Kocienski-type monofluoroolefination based on 3,5-bis(trifluoromethyl)phenyl monofluoromethyl sulfones: Prakash, G.K.S., Shakhmin, A., Zibinsky, M., Ledneczki, I., Chacko, S., and Olah, G.A. (2010) *J. Fluorine Chem.*, **131**, 1192–1197.

123. For trifluoromethylation of aldehydes and ketones, see: (a) Prakash, G.K.S., Krishnamurti, R., and Olah, G.A. (1989) *J. Am. Chem. Soc.*, **111**, 393–395; for trifluoromethylation of esters using TMSCF$_3$, see:(b) Krishnamurti, R., Bellew, D.R., and Prakash, G.K.S. (1991) *J. Org. Chem.*, **56**, 984–989.

124. For trifluoromethylation of imines using TMSCF$_3$, see: Blazejewski, J.-C.,

Anselmi, E., and Wilmshurst, M.P. (1999) *Tetrahedron Lett.*, **40**, 5475–5478.

125. For trifluoromethylation of aliphatic halides using CF$_3$Cu, see: (a) Kobayashi, Y., Yamamoto, K., and Kumadaki, I. (1979) *Tetrahedron Lett.*, **42**, 4071–4072; and using TMSCF$_3$, see: (b) Tyrra, W., Naumann, D., Quadt, S., Buslei, S., Yagupolskii, Y.L., and Kremlev, M.M. (2007) *J. Fluorine Chem.*, **128**, 813–817.

126. For trifluoromethylation of nitriles using TMSCF$_3$, see: Huang, A., Li, H.-Q., Massefski, W., and Saiah, E. (2009) *Synlett*, **15**, 2518–2520.

127. For trifluoromethylation of nitrones using TMSCF$_3$, see: Nelson, D.W., Easley, R.A., and Pintea, B.N.V. (1999) *Tetrahedron Lett.*, **40**, 25–28.

128. Adolph, H.G. and Kamlet, M.J. (1966) *J. Am. Chem. Soc.*, **88**, 4761–4763.

129. (a) Banks, A.A., Emeléus, H.J., Haszeldine, R.N., and Kerrigan, V. (1948) *J. Chem. Soc.*, 2188–2190; Simons *et al.* attempted to prepare CF$_3$I by the reaction between pentafluoroiodide and iodoform, but only fluoroform and difluoroiodomethane were detected as the products: (b) Simons, J.H., Bond, R.L., and McArthur, R.E. (1940) *J. Am. Chem. Soc.*, **62**, 3477–3480.

130. Trifluoroacetates were first reported by Swarts: (a) Swarts, F. (1939) *Bull. Soc. Chim. Belg.*, **48**, 176–179; sodium trifluoroacetate was utilized in the *in situ* generation of CuCF$_3$: (b) Matsui, K., Tobita, E., Ando, M., and Kondo, K. (1981) *Chem. Lett.*, 1719–1720.

131. CF$_3$Br was first obtained in 1946: (a) Brice, T.J., Pearlson, W.H., and Simons, J.H. (1946) *J. Am. Chem. Soc.*, **68**, 968–969; and utilized in Zn-mediated trifluoromethylation: (b) Kitazume, T. and Ishikawa, N. (1981) *Chem. Lett.*, 1679–1680.

132. CF$_2$Br$_2$ was first prepared in 1918: (a) Rathsburg, H. (1918) *J. Chem. Soc. Abstr.*, **114**, 333; and was used as a CF$_3$ precursor: (b) Burton, D.J. and Wiemers, D.M. (1985) *J. Am. Chem. Soc.*, **107**, 5014–5015.

133. Chen, Q-Y. and Wu, S.-W. (1989) *Chem. Commun.*, 705–706.

134. Meslans, C. (1890) *C. R. Acad. Sci.*, **110**, 717–719.

135. Shono, T., Ishifune, M., Okada, T., and Kashimura, S. (1991) *J. Org. Chem.*, **56**, 2–4.

136. For perfluoroalkylations using Grignard reagents, see: (a) Henne, A.L. and Francis, W.C. (1951) *J. Am. Chem. Soc.*, **73**, 3518; and using perfluoroalkyllithiums, see: (b) Pierce, O.R., McBee, E.T., and Judd, G.F. (1954) *J. Am. Chem. Soc.*, **76**, 474–478.

137. For an excellent review on perfluoroalkylations using organometallic reagents, see: (a) Burton, D.J. and Yang, Z.-Y. (1992) *Tetrahedron*, **48**, 189–275; for developments in the preparation of trifluoromethyl derivatives of transition metals prior to the 1980s, see: (b) Morrison, J.A. (1983) in *Advances in Inorganic and Radiochemistry*, vol. **27** (eds. H.J. Emeleus and A.G. Sharpe), Academic Press, New York, pp. 293–316.

138. Eméleus, H.J. and Haszeldine, R.N. (1949) *J. Chem. Soc.*, 2948–2952.

139. (a) Haszeldine, R.N. (1951) *Nature*, **167**, 139–140; (b) Haszeldine, R.N. (1954) *J. Chem. Soc.*, 1273–1279.

140. The first cross-coupling reaction using CF$_3$I–Cu was demonstrated by McLoughlin's and Kobayashi's groups simultaneously in 1969: (a) McLoughlin, V.C.R. and Thrower, J. (1969) *Tetrahedron*, **25**, 5921–5940; (b) Kobayashi, Y. and Kumadaki, I. (1969) *Tetrahedron Lett.*, **10**, 4095–4096; the first unequivocal pregenerative route to CF$_3$Cu was reported by Burton and co-workers, who treated Cu with CF$_2$Br$_2$ to yield CF$_3$Cu: (c) Wiemers, D.M. and Burton, D.J. (1986) *J. Am. Chem. Soc.*, **108**, 832–834.

141. For preparation, see: (a) Lange, H. and Naumann, D. (1984) *J. Fluorine Chem.*, **26**, 435–444; for synthetic applications, see: (b) Kitazume, T. and Ishikawa, N. (1985) *J. Am. Chem. Soc.*, **107**, 5186–5191.

142. Hartkopf, U. and De Meijere, A. (1982) *Angew. Chem. Int. Ed. Engl.*, **21**, 443.

143. Ruppert, I., Schlich, K., and Volbach, W. (1984) *Tetrahedron Lett.*, **25**, 2195–2198.

144. For review articles on the application of the Ruppert–Prakash reagent, see: (a) Prakash, G.K.S. and Yudin, A.K. (1997) *Chem. Rev.*, **97**, 757–786; (b) Singh, R.P. and Shreeve, J.M. (2000) *Tetrahedron*, **56**, 7613–7632; (c) Prakash, G.K.S. and Mandal, M. (2001) *J. Fluorine Chem.*, **112**, 123–131; (d) Shibata, N., Mizuta, S. and Kawai, H. (2008) *Tetrahedron: Asymmetry*, **19**, 2633–2644.

145. Prakash, G.K.S. and Wang, F. (2009) Trifluoromethyltriethylsilane, in *e-EROS Encyclopedia of Reagents for Organic Synthesis* (eds. D. Crich, A.B. Charette, P.L. Fuchs, and G. Molander), John Wiley & Sons, Inc., New York, and references therein. doi: 10.1002/047084289X.rn01198.

146. Pawelke, G. (1989) *J. Fluorine Chem.*, **42**, 429–433.

147. For addition to ketones and aldehydes, see: (a) Aït-Mohand, S., Takechi, N., Medebielle, M., and Dolbier, W.R. Jr. (2001) *Org. Lett.*, **3**, 4271–4273; for addition to vicinal diol cyclic sulfates, see: (b) Takechi, N., Aït-Mohand, S., Medebielle, M., and Dolbier, W.R. Jr. (2002) *Org. Lett.*, **4**, 4671–4672; for asymmetric addition to *N*-tosylaldimines and *N*-tolylsulfinimines, see: (c) Xu, W. and Dolbier, W.R. Jr. (2005) *J. Org. Chem.*, **70**, 4741–4745.

148. For preparation, see: (a) Prakash, G.K.S., Hu, J., and Olah, G.A. (2003) *J. Org. Chem.*, **68**, 4457–4463; for synthetic applications as a nucleophilic trifluoromethylating reagent, see: (b) Prakash, G.K.S., Hu, J., and Olah, G.A. (2003) *Org. Lett.*, **5**, 3253–3256.

149. Prakash, G.K.S., Wang, Y., Mogi, R., Hu, J., Mathew, T., and Olah, G.A. (2010) *Org. Lett.*, **12**, 2932–2935.

150. Langlois, B.R. and Billard, T. (2003) *Synthesis*, 185–194.

151. Barhdadi, R., Troupel, M., and Périchon, J. (1998) *Chem. Commun.*, 1251–1252.

152. Russell, J. and Roques, N. (1998) *Tetrahedron*, **54**, 13771–13782.

153. (a) Large, S., Roques, N., and Langlois, B.R. (2000) *J. Org. Chem.*, **65**, 8848–8856; (b) Billard, T., Bruns, S., and Langlois, B.R. (2000) *Org. Lett.*, **2**, 2101–2103; (c) Billard, T., Langlois, B.R., and Blood, G. (2000) *Tetrahedron Lett.*, **41**, 8777–8780.

154. (a) Umemoto, T. (1996) *Chem. Rev.*, **96**, 1757–1777; (b) Shibata, N., Matsnev, A., and Cahard, D. (2010) *Beilstein J. Org. Chem.*, **6**, (65).

155. Banus, J., Emeleus, H.J., and Haszeldine, R.N. (1951) *J. Chem. Soc.*, 60–64.

156. Olah, G.A. and Ohyama, T. (1976) *Synthesis*, 319–320.

157. Olah, G.A., Heiliger, L., and Prakash, G.K.S. (1989) *J. Am. Chem. Soc.*, **111**, 8020–8021.

158. Blint, R.J., McMahon, T.B., and Beauchamp, J.L. (1974) *J. Am. Chem. Soc.*, **96**, 1269–1278.

159. Barrow, M.J., Davidson, J.L., Harrison, W., Sharp, D.W.A., Sim, G.A., and Wilson, F.B. (1973) *Chem. Commun.*, 583–584.

160. Yagupolskii, L.M., Kondratenko, N.V., and Timofeeva, G.N. (1984) *J. Org. Chem. USSR*, **20**, 103–106.

161. (a) Umemoto, T. and Ishihara, S. (1990) *Tetrahedron Lett.*, **31**, 3579–3582; (b) Umemoto, T. and Ishihara, S. (1993) *J. Am. Chem. Soc.*, **115**, 2156–2164.

162. Yang, J.-J., Kirchmeier, R.L., and Shreeve, J.M. (1998) *J. Org. Chem.*, **63**, 2656–2660.

163. (a) Magnier, E., Blazejewski, J.-C., Tordeux, M., and Wakselman, C. (2006) *Angew. Chem. Int. Ed.*, **45**, 1279–1282; (b) Macé, Y., Raymondeau, B., Pradet, C., Blazejewski, J.-C., and Magnier, E. (2009) *Eur. J. Org. Chem.*, 1390–1397.

164. Umemoto, T., Adachi, K., and Ishihara, S. (2007) *J. Org. Chem.*, **72**, 6905–6917.

165. Ono, T. and Umemoto, T. (1996) *J. Fluorine Chem.*, **80**, 163–166.

166. Matsnev, A., Noritake, S., Nomura, Y., Tokunaga, E., Nakamura, S., and Shibata, N. (2010) *Angew. Chem. Int. Ed.*, **49**, 572–576.

167. Kondratenko, N.V., Radchenko, O.A., and Yagupolskii, L.M. (1984) *Zh. Org. Khim.*, **20**, 2250–2251.

168. Adachi, K. and Ishihara, S. (2003) Japanese Patent 20030388769.

169. Noritake, S., Shibata, N., Nakamura, S., and Toru, T. (2008) *Eur. J. Org. Chem.*, 3465–3468.

170. CF$_3$IF$_2$ was prepared as the first (trifluoromethyl)iodonium compound: Schmeisser, M. and Scharf, E. (1959) *Angew. Chem.*, **71**, 524.

171. Kieltsch, I., Eisenberger, P., and Togni, A. (2006) *Chem. Eur. J.*, **12**, 2579–2586.

172. Kieltsch, I., Eisenberger, P., and Togni, A. (2007) *Angew. Chem. Int. Ed.*, **46**, 754–757.

173. Eisenberger, P., Kieltsch, I., Armanino, N., and Togni, A. (2008) *Chem. Commun.*, 1575–1577.

174. Koller, R., Stanek, K., Stolz, D., Aardoom, R., Niedermann, K., and Togni, A. (2009) *Angew. Chem. Int. Ed.*, **48**, 4332–4336.

175. Koller, R., Huchet, Q., Battaglia, P., Welch, J.M., and Togni, A. (2009) *Chem. Commun.*, 5993–5995.

176. Stanek, K., Koller, R., and Togni, A. (2008) *J. Org. Chem.*, **73**, 7678–7685.

177. (a) Kirk, K.L. (2006) *J. Fluorine Chem.*, **127**, 1013–1029; (b) Isanbor, C. and O'Hagan, D. (2006) *J. Fluorine Chem.*, **127**, 303–319.

178. Hub, L. and Mosher, H.S. (1970) *J. Org. Chem.*, **35**, 3691–3694.

179. Hanzawa, Y., Uda, J.-I., Kobayashi, Y., Ishido, Y., Taguchi, T., and Shiro, M. (1991) *Chem. Pharm. Bull.*, **39**, 2459–2461.

180. Lavaire, S., Plantier-Royon, R., and Portella, C. (1998) *Tetrahedron: Asymmetry*, **9**, 213–226.

181. Bansal, R.C., Dean, B., Hakomori, S.-I., and Toyokuni, T. (1991) *Chem. Commun.*, 796–798.

182. (a) Prakash, G.K.S. (1993) Presented at the 29th Western Regional Meeting of American Chemical Society and 32nd Annual Meeting of the Southern California Section of the Society for Applied Spectroscopy, Pasadena, CA, 19–23 October 1993, Paper No. 123; (b) Yudin, A.K. (1996) Thesis, University of Southern California.

183. Iseki, K., Nagai, T., and Kobayashi, Y. (1994) *Tetrahedron Lett.*, **35**, 3137–3138.

184. Hagiwara, T., Kobayashi, T., and Fuchikami, T. (1997) *Main Group Chem.*, **2**, 13–15.

185. Kuroki, Y. and Iseki, K. (1999) *Tetrahedron Lett.*, **40**, 8231–8234.

186. Caron, S., Do, N.M., Arpin, P., and Larivée, A. (2003) *Synthesis*, 1693–1698.

187. Zhao, H., Qin, B., Liu, X., and Feng, X. (2007) *Tetrahedron*, **63**, 6822–6826.

188. (a) Nagao, H., Yamane, Y., and Mukaiyama, T. (2007) *Chem. Lett.*, **36**, 666–667; (b) Nagao, H., Kawano, Y., and Mukaiyama, T. (2007) *Bull. Chem. Soc. Jpn.*, **80**, 2406–2412.

189. Mizuta, S., Shibata, N., Akiti, S., Fujimoto, H., Nakamura, S., and Toru, T. (2007) *Org. Lett.*, **9**, 3707–3710.

190. Prakash, G.K.S., Mandal, M., and Olah, G.A. (2001) *Angew. Chem. Int. Ed.*, **40**, 589–590.

191. Prakash, G.K.S., Mandal, M., and Olah, G.A. (2001) *Org. Lett.*, **3**, 2847–2950.

192. Prakash, G.K.S. and Mandal, M. (2002) *J. Am. Chem. Soc.*, **124**, 6538–6539.

193. Robak, M.T., Herbage, M.A., and Ellman, J.A. (2010) *Chem. Rev.*, **110**, 3600–3740.

194. Kawai, H., Kusuda, A., Nakamura, S., Shiro, M., and Shibata, N. (2009) *Angew. Chem. Int. Ed.*, **48**, 6324–6327.

195. Lan-Hargest, H.-Y., Elliott, J.D., Eggleston, D.S., and Metcalf, B.W. (1987) *Tetrahedron Lett.*, **28**, 6557–6560.

196. Umemoto, T. and Adachi, K. (1994) *J. Org. Chem.*, **59**, 5692–5699.

197. Iseki, K., Nagai, T., and Kobayashi, Y. (1994) *Tetrahedron: Asymmetry*, **5**, 961–974.

198. Itoh, Y. and Mikami, K. (2006) *Tetrahedron*, **62**, 7199–7203.

199. Nagib, D.A., Scott, M.E., and MacMillan, D.W.C. (2009) *J. Am. Chem. Soc.*, **131**, 10875–10877.

200. Allen, A.E. and MacMillan, D.W.C. (2010) *J. Am. Chem. Soc.*, **132**, 4986–4987.

201. Dubinina, G.G., Furutachi, H., and Vicic, D.A. (2008) *J. Am. Chem. Soc.*, **130**, 8600–8601.

202. Oishi, M., Kondo, H., and Amii, H. (2009) *Chem. Commun.*, 1909–1911.

203. Chu, L. and Qing, F.-L. (2010) *J. Am. Chem. Soc.*, **132**, 7262–7263.

204. Kitazume, T. and Ishikawa, N. (1982) *Chem. Lett.*, 137–140.

205. Grushin, V.V. and Marshall, W.J. (2006) *J. Am. Chem. Soc.*, **128**, 12644–12645.

206. Ball, D., Kampf, J.W., and Sanford, M.S. (2010) *J. Am. Chem. Soc.*, **132**, 2878–2879.

207. Wang, X., Truesdale, L., and Yu, J.-Q. (2010) *J. Am. Chem. Soc.*, **132**, 3648–3649.

208. Cho, D.J., Senecal, T.D., Kinzel, T., Zhang, Y., Watson, D.A., and Buchwald, S.L. (2010) *Science*, **328**, 1679–1681.

209. Bégué, J.-P. and Bonnet-Delpon, D. (2008) *Bioorganic and Medicinal Chemistry of Fluorine*, Wiley-VCH Verlag GmbH, Weinheim.

210. Tozer, M.J. and Herpin, T.F. (1996) *Tetrahedron*, **52**, 8619–8683.

211. (a) Dolber, W.R. Jr. and Battiste, M.A. (2003) *Chem. Rev.*, **103**, 1071–1098; (b) Fedoryñski, M. (2003) *Chem. Rev.*, **103**, 1099–1132.

212. Castejon, H.J. and Wiberg, K.B. (1998) *J. Org. Chem.*, **63**, 3937–3942.

213. For the first difluoromethylene olefination reaction using difluorinated phosphonium yildes, see: (a) Naae, D.G. and Burton, D.J. (1971) *J. Fluorine Chem.*, **1**, 123–125; for an excellent review article on the synthetic applications of fluorinated phosphonium salts, see: (b) Burton, D.J. (1993) *J. Fluorine Chem.*, **23**, 339–357.

214. (a) Burton, D.J., Takei, R., and Shin-Ya, S. (1981) *J. Fluorine Chem.*, **18**, 197–202; (b) Burton, D.J., Ishihara, T., and Maruta, M. (1982) *Chem. Lett.*, 755–758.

215. (a) Obayashi, M., Ito, E., Matsui, K., and Kondo, K. (1982) *Tetrahedron Lett.*, **23**, 2323–2326; (b) Obayashi, M. and Kondo, K. (1982) *Tetrahedron Lett.*, **23**, 2327–2328.

216. For recent examples, see: (a) Upton, T.G., Kashemirov, B.A., McKenna, C.E., Goodman, M.F., Prakash, G.K.S., Kultyshev, R., Batra, V.K., Shock, D.D., Pedersen, L.C., Beard, W.A., and Wilson, S.H. (2009) *Org. Lett.*, **11**, 1883–1886; (b) Batra, V.K., Pedersen, L.C., Beard, W.A., Wilson, S.H., Kashemirov, B.A., Upton, T.G., Goodman, M.F., and McKenna, C.E. (2010) *J. Am. Chem. Soc.*, **132**, 7617–7625; (c) Prakash, G.K.S., Zibinsky, M., Upton, T.G., Kashemirov, B.A., McKenna, C.E., Oertell, K., Goodman, M.E., Batra, V.K., Pedersen, L.C., Beard, W.A., Shock, D.D., Wilson, S.H., and Olah, G.A. (2010) *Proc. Natl. Acad. Sci. U. S. A.*, **107**, 15693–15698.

217. (a) Fokin, A.V., Komarov, V.A., Rapkin, A.I., Frosina, K.V., Potarina, T.M., Pasevina, K.I., and Verenikin, O.V. (1978) *Russ. Chem. Bull.*, **27**, 1962–1965; (b) Fokin, A.V. and Voronkov, A.N. (1979) *Russ. Chem. Bull.*, **28**, 1775.

218. (a) Bissell, E.R. (1963) *J. Org. Chem.*, **28**, 1717–1720; (b) Butler, P., Golding, B.T., Laval, G., Loghmani-Khouzani, H., Ranjbar-Karimi, R., and Sadeghi, M.M. (2007) *Tetrahedron*, **63**, 11160–11166.

219. Yamana, M., Ishihara, T., and Ando, T. (1983) *Tetrahedron Lett.*, **24**, 507–510.

220. (a) Kitagawa, O., Taguchi, T., and Kobayashi, Y. (1988) *Tetrahedron Lett.*, **29**, 1803–1806; (b) Taguchi, T., Kitagawa, O., Suda, Y., Ohkawa, S., Hashimoto, A., Iitaka, Y., and Kobayashi, Y. (1988) *Tetrahedron Lett.*, **29**, 5291–5294.

221. Iseki, K., Kuroki, Y., Asada, D., and Kobayashi, Y. (1997) *Tetrahedron Lett.*, **38**, 1447–1448.

222. Jin, F., Xu, Y., and Huang, W. (1993) *J. Chem. Soc., Perkin Trans. 1*, 795–799.

223. Uneyama, K. and Kato, T. (1998) *Tetrahedron Lett.*, **39**, 587–589.

224. (a) Uneyama, K., Maeda, K., Kato, T., and Katagiri, T. (1998) *Tetrahedron Lett.*, **39**, 3741–3744; (b) Uneyama, K., Mizutani, G., Maeda, K., and Kato, T. (1999) *J. Org. Chem.*, **64**, 6717–6723; (c) Amii, H., Kobayashi, T., Hatamoto, Y., and Uneyama, K. (1999) *Chem. Commun.*, 1323–1324.

225. Prakash, G.K.S., Hu, J., and Olah, G.A. (2001) *J. Fluorine Chem.*, **112**, 357–362.

226. Hallinan, E.A. and Fried, J. (1984) *Tetrahedron Lett.*, **25**, 2301–2302.

227. Thaisrivongs, S., Pals, D.T., Kati, W.M., Turner, S.R., and Thomasco, L.M. (1985) *J. Med. Chem.*, **28**, 1555–1558.

228. Braun, M., Vonderhagen, A., and Waldmüller, D. (1995) *Liebigs Ann.*, 1447–1450.

229. (a) Staas, D.D., Savage, K.L., Homnick, C.F., Tsou, N.T., and Ball, R.G. (2002) *J. Org. Chem.*, **67**, 8276–8279; (b) Soloshonok, V.A., Ohkura, H., Sorochinsky, A., Voloshin, N., Markovsky, A., Belik, M., and Yamazaki, T. (2002) *Tetrahedron Lett.*, **43**, 5445–5448.

230. (a) Hartgraves, G.A. and Burton, D.J. (1988) *J. Fluorine Chem.*, **39**, 425–430; (b) Burton, D.J. and Hartgraves, G.A. (1990) *J. Fluorine Chem.*, **49**, 155–158; (c) Burton, D.J., Hartgraves, G.A., and Hsu, J. (1990) *Tetrahedron Lett.*, **31**, 3699–3702; (d) Burton, D.J. and Hartgraves, G.A. (2007) *J. Fluorine Chem.*, **128**, 1198–1215.

231. (a) Prakash, G.K.S. and Hu, J. (2005) New nucleophilic fluoroalkylation chemistry, in *Fluorine-Containing Synthons*, ACS Symposium Series, vol. **911** (ed. V.A. Soloshonok), American Chemical Society, Washington, DC, Chapter 2, pp. 16–56; (b) Prakash, G.K.S. and Hu, J. (2007) *Acc. Chem. Res.*, **40**, 921–930; (c) Hu, J. (2009) *J. Fluorine Chem.*, **130**, 1130–1139.

232. Hine, J. and Porter, J.J. (1960) *J. Am. Chem. Soc.*, **82**, 6178–6181.

233. Stahly, G.P. (1989) *J. Fluorine Chem.*, **43**, 53–66.

234. Sabol, J.S. and MaCarthy, J.R. (1992) *Tetrahedron Lett.*, **33**, 3101–3104.

235. Prakash, G.K.S., Hu, J., Mathew, T., and Olah, G.A. (2003) *Angew. Chem. Int. Ed.*, **42**, 5216–5219.

236. (a) Prakash, G.K.S., Hu, J., Wang, Y., and Olah, G.A. (2004) *Angew. Chem. Int. Ed.*, **43**, 5203–5206; (b) Prakash, G.K.S., Hu, J., Wang, Y., and Olah, G.A. (2004) *Org. Lett.*, **6**, 4315–4317.

237. Li, Y. and Hu, J. (2005) *Angew. Chem. Int. Ed.*, **44**, 5882–5886.

238. Ni, C., Li, Y., and Hu, J. (2006) *J. Org. Chem.*, **71**, 6829–6833.

239. Ni, Y., Liu, J., Zhang, L., and Hu, J. (2007) *Angew. Chem. Int. Ed.*, **46**, 786–789.

240. Ni, C., Wang, F., and Hu, J. (2008) *Beilstein J. Org. Chem.*, **4** (21).

241. Zhao, Y., Huang, W., Zhu, L., and Hu, J. (2010) *Org. Lett.*, **12**, 1444–1447.

242. Prakash, G.K.S., Hu, J., Wang, Y., and Olah, G.A. (2005) *Eur. J. Org. Chem.*, 2218–2223.

243. (a) Prakash, G.K.S., Hu, J., Wang, Y., and Olah, G.A. (2005) *J. Fluorine Chem.*, **126**, 529–534; (b) Ni, C. and Hu, J. (2005) *Tetrahedron Lett.*, **46**, 8273–8277.

244. Qin, Y.-Y., Qiu, X.-L., Yang, Y.-Y., Meng, W.-D., and Qing, F.-L. (2005) *J. Org. Chem.*, **70**, 9040–9043.

245. Hagiwara, T. and Fuchikami, T. (1995) *Synlett*, 717–718.

246. (a) Brigaud, T., Doussot, P., and Portella, C. (1994) *J. Chem. Soc., Chem. Commun.*, 2117–2118; (b) Huguerot, F., Billac, A., Brigaud, T., and Portella, C. (2008) *J. Org. Chem.*, **73**, 2564–2569.

247. Prakash, G.K.S., Mogi, R., and Olah, G.A. (2006) *Org. Lett.*, **8**, 3589–3592.

248. Prakash, G.K.S., Weber, C., Chacko, S., and Olah, G.A. (2007) *Org. Lett.*, **9**, 1863–1866.

249. Prakash, G.K.S., Weber, C., Chacko, S., and Olah, G.A. (2007) *J. Comb. Chem.*, **9**, 920–923.

250. Zhang, W., Zhu, J., and Hu, J. (2008) *Tetrahedron Lett.*, **49**, 5006–5008.

251. (a) Prakash, G.K.S. and Chacko, S. (2008) *Curr. Opin. Drug Discov. Dev.*, **11**, 793–802; (b) Hu, J., Zhang, W., and Wang, F. (2009) *Chem. Commun.*, 7465–7478.

252. (a) Olah, G.A. and Pavlath, A. (1953) *Acta Chim. Acad. Sci. Hung.*, **3**, 203–207; (b)

Olah, G.A. and Pavlath, A. (1953) *Acta Chim. Acad. Sci. Hung.*, **3**, 425–429.

253. Zhang, W., Zhu, L., and Hu, J. (2007) *Tetrahedron*, **63**, 10569–10575, and references therein.

254. Prakash, G.K.S., Ledneczki, I., Chacko, S., Ravi, S., and Olah, G.A. (2008) *J. Fluorine Chem.*, **129**, 1036–1040.

255. Fukuzumi, T., Shibata, N., Sugiura, M., Yasui, H., Nakamura, S., and Toru, T. (2006) *Angew. Chem. Int. Ed.*, **45**, 4973–4977.

256. Ni, C., Zhang, L., and Hu, J. (2008) *J. Org. Chem.*, **73**, 5699–5713.

257. Prakash, G.K.S., Chacko, S., Alconcel, S., Stewart, T., Mathew, T., and Olah, G.A. (2007) *Angew. Chem. Int. Ed.*, **46**, 4933–4936.

258. Zhang, W., Huang, W., and Hu, J. (2009) *Angew. Chem. Int. Ed.*, **48**, 9858–9861.

259. Prakash, G.K.S., Wang, F., Shao, N., Mathew, T., Rasul, G., Haiges, R., Stewart, T., and Olah, G.A. (2009) *Angew. Chem. Int. Ed.*, **48**, 5358–5362.

260. Kaplan, L.A. and Pickard, H.B. (1969) *Chem. Commun.*, 1500–1501.

261. Blank, I. and Mager, J. (1954) *Experientia*, **10**, 77–78.

262. For selected examples, see: (a) Blank, I., Mager, J., and Bergmann, E.D. (1955) *J. Chem. Soc.*, 2190–2193; (b) Bergmann, E.D., Cohen, S., and Shahak, I. (1959) *J. Chem. Soc.*, 3278–3285; (c) Bergmann, E.D. and Shahak, I. (1960) *J. Chem. Soc.*, 5261–5262; (d) Bergmann, E.D. and Shahak, I. (1961) *J. Chem. Soc.*, 4669–4671; (e) Bergmann, E.D. and Cohen, S. (1961) *J. Chem. Soc.*, 3537–3538.

263. (a) Buchanan, R.L., Dean, F.H., and Pattison, L.M. (1962) *Can. J. Chem.*, **40**, 1571–1575; (b) Buchanan, R.L. and Pattison, F.L.M. (1965) *Can. J. Chem.*, **43**, 3466–3468.

264. Ishikawa, N., Koh, M.G., Kitazume, T., and Choi, S.K. (1984) *J. Fluorine Chem.*, **24**, 419–430.

265. Welch, J.T. and Seper, K.W. (1984) *Tetrahedron Lett.*, **25**, 5247–5250.

266. Wildonger, K.J., Leanza, W.J., Ratcliffe, R.W., and Springer, J.P. (1995) *Heterocycles*, **41**, 1891–1990.

267. Huang, X.-T. and Chen, Q.-Y. (2002) *J. Org. Chem.*, **67**, 3231–3234.

268. Gilligan, W.H. (1971) *J. Org. Chem.*, **36**, 2138–2141.

269. Kamlet, M.J. and Adolph, H.G. (1968) *J. Org. Chem.*, **33**, 3073–3080.

270. Makosza, M. and Goliński, J. (1984) *J. Org. Chem.*, **49**, 1488–1494.

271. (a) Inbasekaran, M., Peet, N.P., McCarthy, J.R., and LeTourneau, M.E. (1985) *Chem. Commun.*, 678–679; in fact, McCarthy and co-workers have performed systematic investigations on the utilization of $PhSO_2CH_2F$ as a fluoromethyl Wittig equivalent, for other examples, see: (b) McCarthy, J.R., Matthews, D.P., Stemerick, D.M., Huber, E.W., Bey, P., Lippert, B.J., Snyder, R.D., and Sunkara, P.S. (1991) *J. Am. Chem. Soc.*, **113**, 7439–7440; (c) Chen, C., Wilcoxen, K., Zhu, Y.-F., Kyung, K.-I., and McCarthy, J.R. (1999) *J. Org. Chem.*, **64**, 3476–3482.

272. (a) Li, Y., Ni, C., Liu, J., Zhang, L., Zheng, J., Zhu, L., and Hu, J. (2006) *Org. Lett.*, **8**, 1693–1696; (b) Liu, J., Zhang, L., and Hu, J. (2008) *Org. Lett.*, **10**, 5377–5380.

273. Prakash, G.K.S., Zhao, X., Chacko, S., Wang, F., Vaghoo, H., and Olah, G.A. (2008) *Beilstein J. Org. Chem.*, **4** (17).

274. Shibatomi, K. (2010) *Synthesis*, 2679–2702.

275. Hoffmann, T., Zhong, G., List, B., Shabat, D., Anderson, J., Gramatikova, S., Lerner, R.A., and Barbas, C.F., III, (1998) *J. Am. Chem. Soc.*, **120**, 2768–2779.

276. Iseki, K., Kuroki, Y., and Kobayashi, Y. (1999) *Tetrahedron*, **55**, 2225–2236.

277. Kim, D.Y., Kim, S.M., Koh, K.O., Mang, J.Y., and Lee, K. (2003) *Bull. Korean Chem. Soc.*, **24**, 1425–1426.

278. Zhong, G., Fan, J., and Barbas, C.F. III (2004) *Tetrahedron Lett.*, **45**, 5681–5684.

279. Nichols, P.J., DeMattei, J.A., Barnett, B.R., LeFur, N.A., Chuang, T.-H., Piscopio, A.D., and Koch, K. (2006) *Org. Lett.*, **8**, 1495–1498.

280. Huber, D.P., Stanek, K., and Togni, A. (2006) *Tetrahedron: Asymmetry*, **17**, 658–664.

281. (a) He, R., Wang, X., Hashimoto, T., and Maruoka, K. (2008) *Angew. Chem. Int. Ed.*, **47**, 9466–9468; (b) Mang, J.Y., Kwon, D.G., and Kim, D.Y. (2009) *J. Fluorine Chem.*, **130**, 259–262.

282. Ding, C. and Maruoka, K. (2009) *Synlett*, 664–666.

283. (a) Ullah, F., Zhao, G.-L., Deiana, L., Zhu, M., Dziedzic, P., Ibrahem, I., Hammar, P., Sun, J., and Córdova, A. (2009) *Chem. Eur. J.*, **15**, 10013–10017; (b) Companyo, X., Hejnova, M., Kamlar, M., Vesely, J., Moyano, A., and Rios, R. (2009) *Tetrahedron Lett.*, **50**, 5021–5024.

284. For recent examples, see: (a) Jiang, Z., Pan, Y., Zhao, Y., Ma, T., Lee, R., Yang, Y., Huang, K.-W., Wong, M.W., and Tan, C.-H. (2009) *Angew. Chem. Int. Ed.*, **48**, 3627–3631; (b) Han, X., Luo, J., Liu, C., and Lu, Y. (2009) *Chem. Commun.*, 2044–2046; (c) Cui, H.-F., Yang, Y.-Q., Chai, Z., Li, P., Zheng, C.-W., Zhu, S.-Z., and Zhao, G. (2010) *J. Org. Chem.*, **75**, 117–122.

285. Han, X., Kwiatkowski, J., Xue, F., Huang, K.-W., and Lu, Y. (2009) *Angew. Chem. Int. Ed.*, **48**, 7604–7607.

286. Pan, Y., Zhao, Y., Ma, T., Yang, Y., Liu, H., Jiang, Z., and Tan, C.-H. (2010) *Chem. Eur. J.*, **16**, 779–782.

287. Mohr, T., Behenna, D.C., Harned, A.M., and Stoltz, B.M. (2005) *Angew. Chem. Int. Ed.*, **44**, 6924–6927.

288. Nakamura, M., Hajra, A., Endo, K., and Nakamura, E. (2005) *Angew. Chem. Int. Ed.*, **44**, 7248–7251.

289. Bélanger, É., Cantin, K., Messe, O., Tremblay, M., and Paquin, J.-F. (2007) *J. Am. Chem. Soc.*, **129**, 1034–1035.

290. Bélanger, É., Houzé, C., Guimond, N., Cantin, K., and Paquin, J.-F. (2008) *Chem. Commun.*, 3251–3253.

291. Jiang, B., Huang, Z.-G., and Cheng, K.-J. (2006) *Tetrahedron: Asymmetry*, **17**, 942–951.

292. Nagura H. and Fuchigami T. (2008) *Synlett*, 1714–1718.

293. Ni, C., Zhang, L., and Hu, J. (2009) *J. Org. Chem.*, **74**, 3767–3771.

294. Prakash, G.K.S., Wang, F., Ni, C., Thomas, T.J., and Olah, G.A. (2010) *J. Fluorine Chem.*, **131**, 1007–1012.

295. Prakash, G.K.S., Chacko, S., Vaghoo, H., Shao, N., Gurung, L., Mathew, T., and Olah, G.A. (2009) *Org. Lett.*, **11**, 1127–1130.

296. Ni, C. and Hu, J. (2009) *Tetrahedron Lett.*, **50**, 7252–7255.

297. Mizuta, S., Shibata, N., Goto, Y., Furukawa, T., Nakamura, S., and Toru, T. (2007) *J. Am. Chem. Soc.*, **129**, 6394–6395.

298. Furukawa, T., Shibata, N., Mizuta, S., Nakamura, S., Toru, T., and Shiro, M. (2008) *Angew. Chem. Int. Ed.*, **47**, 8051–8054.

299. Prakash, G.K.S., Wang, F., Stewart, T., Mathew, T., and Olah, G.A. (2009) *Proc. Natl. Acad. Sci. U. S. A.*, **106**, 4090–4094.

300. Moon, H.W., Cho, M.J., and Kim, D.Y. (2009) *Tetrahedron Lett.*, **50**, 4896–4898.

301. (a) Alba, A.-N., Companyó, X., Moyano, A., and Rios, R. (2009) *Chem. Eur. J.*, **15**, 7035–7038; (b) Zhang, S., Zhang, Y., Ji, Y., Li, H., and Wang, W. (2009) *Chem. Commun.*, 4886–4888.

Commentary Part

Comment 1

David O'Hagan

All aspects of the chemical industry demand fluorinated products. Indeed, it is because of the importance of selective fluorination in refining the properties of pharmaceuticals, agrochemicals, and performance materials such as liquid crystals that there will always be a requirement for chemists to turn their attention to the problems of fluorination. What emerges from this overview is just how extensive and vibrant the field is and how impressive the level of innovation and ingenuity

has been. For each category of reagents (e.g., nucleophilic, electrophilic, radical), a chronological approach is taken such that the evolution of the different reagents is tracked. For example, sulfur tetrafluoride (SF_4) was first introduced as a deoxofluorinating reagent in the 1960s by DuPont chemists, who subsequently developed DAST in the 1970s as a liquid alternative to the gaseous reagent. Deoxofluor then emerged as an alternative, owing to its superior thermostability to DAST, in the late 1990s. More recently, non-hygroscopic crystalline solids such as XtalFluor-E and -M and Fluolead have been introduced in an effort to offer advantages over the earlier reagents, although it will require more time to assess their impact on the research community.

Much too in fluorine chemistry has been enabled by the discovery of laboratory-friendly forms of electrophilic fluorine, essentially tamed forms of elemental fluorine (F_2). One of the notable contributions was pioneered by Umemoto and co-workers, who achieved N-fluoropyridinium triflate salts. Electrophilic fluorination reagents until then had been rather unstable for practical utilization and none had really developed as a general reagent. Umemoto and coworkers' innovation was to introduce the first shelf-stable, nonhygroscopic reagent of this class. However, these regents were further enhanced by Differding's (1991) NFSI and also Banks' (1992) Selectfluor. These two reagents remain the most widely used electrophilic fluorination reagents in research laboratories and the pharmaceutical industry throughout the world, and they have truly revolutionized the field, in that the extended chemistry community has access to user-friendly fluorination reagents. For example, their availability has permitted new and important phases of discovery in organometallic fluorinations and asymmetric fluorinations by organocatalysis.

Other important topics are developed in this chapter with a similar historical perspective. Major strides forward have been made in recent years in palladium-mediated C-F bond formation, involving either nucleophilic or electrophilic fluorine sources. For nucleophilic fluorination, the pioneering work of the groups of Dixon and then Grushin is outlined, where stable Pd–F species were made and identified. This was followed more recently by Yandulov's (2007) and then Buchwald's groups (2009), who refined protocols for reductive elimination from these intermediates to generate aryl–F products in an efficient and versatile manner. Importantly, palladium-mediated aryl–F bond-forming protocols have also been developed utilizing electrophilic fluorinating reagents, which were recently contributed by Sanford and Yu and their colleagues. In particular, Ritter and co-workers have developed a complementary methodology to generate aryl–Ag–F intermediates, which undergo reductive elimination to afford aryl–F products.

Prakash has of course a particular association with TMS-CF_3, a reagent known informally as the "Ruppert–Prakash reagent," and it is the most versatile precursor available for the transfer of CF_3^- as a nucleophile. A time-line development of nucleophilic CF_3^- reagents is presented, leading up to the impressive range and versatility of the Ruppert–Prakash reagent in this arena. And then there are the electrophilic CF_3^+ reagents, where a historical perspective is outlined from the seminal $R_2–S^+CCF_3$ reagents of Yagupolskii and then again to Umemoto, through sulfoximine–CF_3 reagents to the more recent and versatile (trifluoromethyl)iodonium reagents [$(CF_3)I^+R_2$] of Togni and their colleagues. The notable contributions of transition metal-catalyzed trifluoromethylation reactions and asymmetric trifluoromethylations are also reviewed. Other topics are covered, outlining developments in both electrophilic and nucleophilic incorporation, including asymmetric achievements for the incorporation of RCF_2 and H_2FC units, largely by building block approaches. The development of innovative fluorination methods is rarely articulated as a high-level objective in modern chemistry, yet this chapter illustrates that practitioners from all of the contributing strands of organic and organometallic chemistry have turned their gaze to the problems of fluorination. This is driven, of course, by the demand for fluorinated entities for industry, and efficient methods are therefore highly valued and of great utility. The continuing demand for more and varied fluorinated molecules will ensure the steady emergence of new and exciting innovations in reagent and methodology development. The value of this chapter by Prakash and Wang is that it provides a concise overview of developments from past to present and is an excellent source for those who want to assess the state-of-the-art. It will perhaps also act as a starting point to encourage chemists to apply their intellectual prowess to this important area of organic chemistry.

Comment 2

Jinbo Hu

Organofluorine chemistry is truly a flourishing research field in modern organic chemistry, as evidenced by the fact that increasingly more scientists have been attracted to study the unique chemistry of fluorine. In fact, fluorine is the only element in the periodic table after which the American Chemical Society named a subdivision – the ACS *Fluorine* Division.

In this chapter, Prakash and Wang give an excellent and fairly comprehensive treatment of the topic, which provides conceptual insight into the achievements and primary challenges in this field since the end of the last century. In the Introduction, they nicely provide an overview of both the unique properties of the fluorine atom and the recent progress in the field of organofluorine chemistry. Thereafter, they focus the majority of their attention on the recent progress in synthetic organofluorine chemistry, which includes C-F bond formation reactions and trifluoromethylation, difluoromethylation, and monofluoromethylation reactions. In each section, the authors highlight the important reagents, reactions, and mechanistic studies of nucleophilic and electrophilic fluorinations and fluoromethylations, with emphasis on the recent breakthroughs in asymmetric reactions and transition metal-promoted reactions. It is particularly impressive that the authors provide a historical background of the related chemistry in each section of the chapter, which also makes the chapter serve as an introductory guide for readers who are the newcomers of the field.

As pointed out by the authors in the Introduction, owing to space restrictions, they refrain from covering every important achievement in the field of organofluorine chemistry. Therefore, several recent important discoveries, such as fluorodesilylation reactions [C1], ^{18}F-labeled fluorination reactions [C2], oxidative trifluoromethylations [C3] (this reaction is included in the chapter as Ref. 203), Pd-catalyzed polyfluoroarylations [C4], CF_3Cu chemistry with electrophilic trifluoromethylating agents [C5], and fluorous phase synthesis [C6], among others, are not included in the chapter.

I fully agree with the authors' perspective that organofluorine chemistry will remain a fascinating and dynamic research field in the coming decades, particularly with the impetus from the highly important applications of fluorinated organic compounds and materials. The development of new fluorinating reagents and fluorinated synthons, and also the discovery of new efficient synthetic protocols for organofluorine compounds (such as in an asymmetric fashion and/or organometallic pathways) will remain the major themes in the field. A more detailed understanding of the unique chemical, physical, and biological properties of organofluorine compounds and materials will also remain a major curiosity of organofluorine chemists. Finally, the development of environmentally friendly organofluorine compounds and materials (possessing the same desired properties as conventional ones) will become an increasingly more urgent task for

organofluorine chemists, since some organofluorine compounds have raised serious environmental concerns.

Comment 3

Kuiling Ding and Li-Xin Dai

The conviction of most people is that a fluorine atom bears a steric resemblance to a proton and a CF_3 group to a CH_3 group. This became the basis of one approach in drug design, namely to replace a proton by F or a CH_3 group by CF_3. Incidentally we found a table of Taft steric substituent constants E_s, for aliphatic substituents in RCOOEt in Uneyama's book *Organofluorine Chemistry* [C7], and selected values are as follows:

R	CH_3	Et	i-Pr	CF_2H	CF_3	s-Bu	t-Bu
E_s	0.00	−0.07	−0.47	−0.67	−1.16	−1.13	−1.54

As E_s is the quantitative measure of steric effects which is widely used in organic chemistry, these values somewhat surprised us. For non-organofluorine chemists, does the above resemblance no longer exist? Or should we consider this mimic more carefully?

Authors' Response to the Commentaries

We are pleased to read the commentaries from Professors David O'Hagan and Jinbo Hu, who have carefully reviewed and provided invaluable suggestions on our chapter. More importantly, they have generously offered their unique perspectives on the state-of-the-art and also for the future prospects for the field. Both commentators have extensively explored and made remarkable contributions to the organofluorine chemistry field in recent years. We believe that such commentaries and perspectives will be extremely helpful for chemists closely associated with organofluorine chemistry and also those who have just started to take an interest in this field.

Organofluorine chemistry is indeed a rapidly developing field involving diverse aspects of many sciences impacting various fields from materials to medicine. Certainly, we were unable to describe all the details of organofluorine chemistry in this brief chapter. Again, we sincerely apologize to all those exceptional chemists whose important contributions to the field had to be omitted for the sake of brevity.

In response to Ding and Dai's comment, even though a fluorine atom is sterically slightly larger than a hydrogen atom, extensive investigations have clearly indicated that the steric resemblance between a trifluoromethyl group and a methyl group is less pronounced based on several different steric scales, such as Taft's steric substituent parameter [C8], the steric parameters introduced by Charton [C9], and others [C10]. Accordingly, the size of the CF_3 group is believed to be similar to that of a *sec*-butyl or an isopropyl group, which has been widely accepted and mentioned by many chemists for general estimation of the steric effect of CF_3. However, using gelatinase B as a probe, it was concluded, very differently, that CF_3 was assessed to be isosteric with an ethyl group, which is much smaller than an isopropyl moiety [C11]. Hence such evaluations can be rather misleading and only marginally applicable to specific steric scenarios. In particular, if the symmetry groups of these functionalities are taken into account, the aforementioned comparison appears to be invalid owing to the lower symmetries of the *sec*-butyl group, the isopropyl group, and also the ethyl group. On the other hand, because CH_3, CF_3, and *t*-Bu possess the same symmetry group (C_{3v}), it can be unequivocally concluded that the order $CH_3 < CF_3 < t$-Bu is presumably applicable for any steric situation. In summary, the statement that the bulkiness of CF_3 is sterically similar to those of Et, *i*-Pr, and *s*-Bu cannot be applied in general unless exhaustive computational and/or experimental studies are performed. Nonetheless, the conclusion that $CH_3 < CF_3 < t$-Bu is unconditionally valid.

References

C1. Greedy, B. and Gouverneur, V. (2001) *Chem. Commun.*, 233–234.

C2. For example: Teare, H., Robins, E.G., Aarstad, E., Sajinder, S.K., and Gouverneur, V. (2007) *Chem. Commun.*, 2330–2332.

C3. Chu, L. and Qing, F.-L. (2010) *J. Am. Chem. Soc.*, **132**, 7262–7263.

C4. Zhang, X., Fan, S., He, C.-Y., Wan, X., Min, Q.-Q., Yang, J., and Jiang,

Z.-X. (2010) *J. Am. Chem. Soc.*, **132**, 4506–4507.

C5. Zhang, C.-P., Wang, Z.-L., Chen, Q.-Y., Zhang, C.-T., Gu, Y.-C., and Xiao, J.-C. (2011) *Angew. Chem. Int. Ed.*, **50**, 1896–1900.

C6. For example: Bejot, R., Fowler, T., Carroll, L., Boldon, S., Moore, J.E., Declerck, J., and Gouverneur, V. (2009) *Angew. Chem. Int. Ed.*, **48**, 586-589.

C7. Uneyama, K. (2006) *Organofluorine Chemistry*, Blackwell, Oxford, p. 83.

C8. Uneyama, K. (2006) *Organofluorine Chemistry*, Blackwell, Oxford, pp. 81–84.

C9. Charton, M. (1975) *J. Am. Chem. Soc.*, **96**, 1552–1556.

C10. Schlosser, M. and Michel, D. (1996) *Tetrahedron*, **52**, 99–108.

C11. Jagodzinska, M., Huguenot, F., Candiani, G., and Zanda, M. (2009) *ChemMedChem*, **4**, 49–51, and reference 2 therein.

Addendum

Owing to the rapid development in the field of organofluorine chemistry over the past year, a list of updated references is included herein.

A. C–F Forming Reactions

1. **Deoxofluorination Reagents**
   (1) For Fluolead, see Singh, R.P. and Umemoto, T. (2011) *J. Org. Chem.* **76**, 3113–3121.
   (2) Deoxofluorination of phenols, see: Tang, P., Wang, W., and Ritter, T. (2011) *J. Am. Chem. Soc.* **133**, 11482–11484.

2. **Electrophilic Fluorination**
   (1) Catalytic asymmetric fluorination of cyclic ketones, see: Kwiatkowski, P., Beeson, T.D., Conrad, J.C., and MacMillan, D.W.C. (2011) *J. Am. Chem. Soc.* 1738–1741.

3. **Transition Metal-Catalyzed C-F Bond Formation**
   (a) Lee, E., Kamlet, A.S., Powers, D.C., Neumann, C.N., Boursalian, G.B., Furuya, T., Choi, D.C., Hooker, J.M., and Ritter, T. (2011) *Science* **334**, 639–642; (b) Chan, K.S.L., Wasa, M., Wang, X., and Yu, J.-Q. (2011) *Angew. Chem. Int. Ed.*

50, 9081–9084; (c) Tang, P. and Ritter, T. (2011) *Tetrahedron* **67**, 4449–4454; (d) Racowski, J.M., Gary, J.B., and Sanford, M.S. (2012) *Angew. Chem. Int. Ed.* 3414–3417.

B. Monofluoromethylation Reactions

1. **Utilization of Fluorobis(phenylsulfonyl) methane (FBSM) and Its Analogs**
Addition of FBSM to aldehydes, see: (a) Shen, X., Zhang, L., Zhao, Y., Zhu, L., Li, G., and Hu, J. (2011) *Angew. Chem. Int. Ed.* **50**, 2588–2592; utilization of FBSM in the Morita-Baylis-Hillman reaction, see: (b) Furukawa, T., Kawazoe, J., Zhang, W., Nishimine, T., Tokunaga, E., Matsumoto, T., Shiro, M., and Shibata, N. (2011) *Angew. Chem. Int. Ed.* **50**, 9684–9688; (c) Yang, W., Wei, X., Pan, Y., Lee, R., Zhu, B., Liu, H., Yan, L., Huang, K.-W., Jiang, Z., and Tan, C.-H. (2011) *Chem. Eur. J.* **17**, 8066–8070; (d) Companyó, X., Valero, G., Ceban, V., Calvet, T., Font-Bardía, M., Moyano, A., and Rios, R. (2011) *Org. Biomol. Chem.* **9**, 7986–7989; employment of FBSM in enantiomeric β-functionalization of aldehydes, see: (e) Zhang, S.-L., Xie, H.-X., Zhu, J., Li, H., Zhang, X.-S., Li, J., and Wang, W. (2011) *Nat. Commun.* **2**, 211–217; utilization of 2-fluoro-1,3-benzodithiole-1,1,3,3-tetraoxide in monofluoromethylation of aldehydes, see: (f) Furukawa, T., Goto, Y., Kawazoe, J., Tokunaga, E., Nakamura, S., Yang, Y., Du, H., Kakehi, A., Shiro, M., and Shibata, N. (2010) *Angew. Chem. Int. Ed.* **49**, 1642–1647; synthesis of fluorobis(phenylsulfonyl)methyl trimethylsilane and its applications in nucleophilic monofluoromethylation of aldehydes, see: (g) Prakash, G.K.S., Shao, N., Zhang, Z., Ni, C., Wang, F., Haiges, R., and Olah, G.A. (2012) *J. Fluorine Chem.* **133**, 27–32.
2. **Electrophilic Monofluoromethylation**
Johnson-type electrophilic monofluoromethylating reagent and its synthetic applications, see: Nomura, Y., Tokunaga, E., and Shibata, N. (2011) *Angew. Chem. Int. Ed.* **50**, 1885–1889.

C. Difluoromethylation Reactions

1. **Difluorocarbene Chemistry**
Addition of difluorocarbene to alkenes or alkynes using TMSCF$_3$ as a difluorocarbene precursor, see: (a) Wang, F., Luo, T., Hu, J., Wang, Y., Krishnan, H.S., Jog, P.V., Ganesh, S.K., Prakash, G.K.S., and Olah, G.A. (2011) *Angew. Chem. Int. Ed.* **50**, 7153–7157; using TMSCF$_2$Cl as a difluorocarbene precursor, see: (b) Wang, F., Zhang, W., Zhu, J., Li, H., Huang, K.-W., and Hu, J. (2011) *Chem. Common.* **47**, 2411–2413; Addition of difluorocarbene to O/S/N/C-nucleophiles using n-Bu$_3$N$^+$CF$_2$HCl$^-$ salt as a difluorocarbene precursor, see: (c) Wang, W. and Huang, J. Hu (2011) *Chin. J. Chem.* **29**, 2717–2721.
2. **Nucleophilic Difluoromethylation**
Facile preparation of TMSCF$_2$H from TMSCF$_3$ via reductive defluorination, see: (a) Tyutyunov, A.A., Boyko, V.E., and Igoumnov, S.M. (2011) *Fluorine Notes* **74**, 1, http://notes.fluorine1.ru/public/2011/1_2011/letters/letter2.html. nucleophilic difluoromethylation of carbonyl compounds using TMSCF$_2$H; see: (b) Zhao, Y., Huang, W., Zheng, J., and Hu, J. (2011) *Org. Lett.* **13**, 5342–5345; preparation of difluorinated sulfonates using difluoromethyl 2-pyridyl sulfone, see: (c) Prakash, G.K.S., Ni, C., Wang, F., Hu, J., and Olah, G.A. (2011) *Angew. Chem. Int. Ed.* **50**, 2559–2563.
3. **Electrophilic Difluoromethylation**
Cu-Catalyzed difluoromethylation of α,β-unsaturated carboxylic acids using a phenylsulfonyl difluoromethyl Togni-type reagent, see (a) He, Z., Luo, T., Hu, M., Cao, Y., and Hu, J. (2012) *Angew. Chem. Int. Ed.*, 10.1002/anie.201200140; Johnson-type electrophilic difluoromethylating reagent and its synthetic applications, see: Prakash, G.K.S., Zhang, Z., Wang, F., Ni, C., and Olah, G.A. (2011) *J. Fluorine Chem.* **132**, 792–798.
4. **Aromatic Difluoromethylation Reactions**
    (1) Radical difluoromethylation of heterocyclics using (CF$_2$HSO$_2$)$_2$ Zn/t-BuOOH/TFA system, see: Fujiwara, Y., Dixon, J.A., Rodriguez, R.A., Baxter, R.D., Dixon, D.D., Collins,

M.R., Blackmond, D.G., and Baran, P.S. (2012) *J. Am. Chem. Soc.* **134**, 1494–1497.

(2) Cu-mediated difluoromethylation of aryl and vinyl iodides using TMSCF$_2$H: Fier, P.S. and Hartwig, J.F. (2012) *J. Am. Chem. Soc.* **134**, 5524–5527.

C. Trifluoromethylation Reactions

1. **Trifluoromethylation Utilizing (Trifluoromethyl)iodonium Compounds (Togni's Reagents)**

(1) Reaction with THF, see: Fantasia, S., Welch, J.M., and Togni, A. (2010) *J. Org. Chem.* **75**, 1779–1782.

(2) Reaction with arenes and N-heteroarenes, see: Wiehn, M.S., Vinogradova, E.V., and Togni, A. (2010) *J. Fluorine Chem.* **131**, 951–957.

(3) Various (trifluoromethyl)iodonium compounds and their reactivity as electrophilic trifluoromethylating reagents, see: Niedermann, K., Welch, J.M., Koller, R., Cvengroš, J., Santschi, N., Battaglia, P., and Togni, A. (2010) *Tetrahedron* **66**, 5753–5761.

(4) For the asymmetric synthesis of α-CF$_3$-substituted carbonyl compounds with chiral auxiliaries, see: Matoušek, V., Togni, A., Bizet, V., and Cahard, D. (2011) *Org. Lett.* **13**, 5762–5765.

(5) Utilization in a Ritter-type reaction, see: Niedermann, K., Früh, N., Vinogradova, E., Wiehn, M.S., Moreno, A., and Togni, A. (2011) *Angew. Chem. Int. Ed.* **50**, 1059–1063.

(6) Cu-catalyzed C$_{(sp3)}$-CF$_3$ bond forming using Togni's reagent, see: Wang, X., Ye, Y., Zhang, S., Feng, J., Xu, Y., Zhang, Y., and Wang, J. (2011) *J. Am. Chem. Soc.* **133**, 16410–16413.

2. **C$_{(sp2/sp3)}$-CF$_3$ Bond Forming Reactions:**

(1) A comprehensive review article on the recent development of aromatic trifluoromethylation reactions, see: Tomashenko, O.A., and Grushin, V.V. (2011) *Chem. Rev.* **111**, 4475–4521.

(2) Preparation of stable Cu-CF$_3$ complexes for aromatic trifluoromethylation, see:

(a) Morimoto, H., Tsubogo, T., Litvinas, N.D., and Hartwig, J.F. (2011) *Angew. Chem. Int. Ed.* **50**, 3793–3798; (b) Tomashenko, O.A., Escudero-Adán, E.C., Belmonte, M.M., and Grushin, V.V. (2011) *Angew. Chem. Int. Ed.* **50**, 7655–7659.

(3) Utilization of CF$_3$H as CF$_3$ source for Cu-catalyzed trifluoromethylation reactions: Zanardi, A., Novikov, M.A., Martin, E., Benet-Buchholz, J., and Grushin, V.V. (2011) *J. Am. Chem. Soc.* **133**, 20901–20913.

(4) Oxidative C$_{(sp2/sp3)}$-CF$_3$ bond forming reactions, see: (a) Senecal, T.D., Parsons, A.T., and Buchwald, S.L. (2011) *J. Org. Chem.* **76**, 1174–1176; (b) Mu, X., Chen, S., Zhen, X., and Liu, G. (2011) *Chem. Eur. J.* **17**, 6039–6042; (c) Jiang, X., Chu, L., and Qing, F.-L. (2012) *J. Org. Chem.* **77**, 1251–1257; (d) Chu, L. and and Qing, F.-L. (2012) *J. Am. Chem. Soc.* **134**, 1298–1304; (e) Zhang, K., Qiu, X.-L., Huang, Y., and Qing, F.-L. (2012) *Eur. J. Org. Chem.* 58–61.

(5) Fe-catalyzed trifluoromethylation of potassium vinyltrifluoroborates, see: Parsons, A.T., Senecal, T.D., and Buchwald, S.L. (2012) *Angew. Chem. Int. Ed.* **51**, 2947–2950.

(6) Radical trifluoromethylation of heterocyclics using CF$_3$SO$_2$Na/t-BuOOH system, see: (a) Ji, Y., Brueckl, T., Baxter, R.D., Fujiwara, Y., Seiple, I.B., Su, S., Blackmond, D.G., and Baran, P.S. (2011) *Proc. Natl. Acad. Sci. U.S.A.* **108**, 14411–14415; (b) Radical aromatic trifluoromethylation using CF$_3$SO$_2$Cl under photoredox catalysis, see: Nagib, D.A. and MacMillan, D.W.C. (2011) *Nature* **480**, 224–228; (c) a similar radical aromatic trifluoromethylation using CF$_3$SO$_2$Na/t-BuOOH/Cu(OTf)$_2$ system, see: Langlois, B.I. and Laurent, E. (1991) *Tetrahedron* 7525–7528.

(7) Comprehensive mechanistic studies on aromatic trifluoromethylation, see: (a) Ye, Y., Ball, N.D., Kampf, J.W., and Sanford, M.S. (2010) *J. Am. Chem. Soc.* **132**, 14682–14687; (b) Ball, N.D., Gary,

J.B., Ye, Y., and Sanford, M.S. (2011) *J. Am. Chem. Soc.* **133**, 7577–7584.

3. **Catalytic Asymmetric Nucleophilic/Electrophilic Trifluoromethylation Reactions, see:** Photoredox catalysis using CF$_3$I as a CF$_3$ radical precursor, see: Pham, P.V., Nagib, D.A., and MacMillan, D.W.C. (2011) *Angew. Chem. Int. Ed.* **50**, 6119–22.

D. CF$_3$X$_n$–C Bond Forming Reactions

1. **Trifluoromethylthiolation**
   (1) Using trifluoromethanesulfanylamides as CF$_3$S$^+$ equivalent, see: Ferry, A., Billard, T., Langlois, B.R., and Bacqué, E. (2009) *Angew. Chem. Int. Ed.* **48**, 8551–8555.
   (2) Pd-catalyzed aromatic trifluoromethylthiolation using AgSCF$_3$ as a SCF$_3$ source, see: Teverovskiy, G., Surry, D.S., and Buchwald, S.L. (2011) *Angew. Chem. Int. Ed.* **50**, 7312–7314.
   (3) Cu-Catalyzed oxidative trifluoromethylthiolation of aryl boronic acids using TMSCF$_3$ and S$_8$, see: Chen, C., Xie, Y., Chu, L., Wang, R.-W., Zhang, X., and Qing, F.-L. (2012) *Angew. Chem. Int. Ed.* **51**, 2492–2495.

2. **Trifluoromethoxylation**
   (1) Using 2,4-Dinitro(trifluoromethoxy) benzene as as a CF$_3$O$^-$ equivalent, see: Marrec, O., Billard, T., Vors, J.-P., Pazenok, S., and Langlois, B.R. (2010) *Adv. Synth. Catal.* **352**, 2831–2837; Marrec, O., Billard, T., Vors, J.-P., Pazenok, S., and Langlois, B.R. (2010) *J. Fluorine Chem.* **131**, 200–207.
   (2) Silver-mediated trifluoromethoxylation of aryl stannanes and arylboronic acids with tris(dimethylamino) sulfonium trifluoromethoxide (TAS-OCF$_3$), see: Huang, C., Liang, T., Harada, S., Lee, E., and Ritter, T. (2011) *J. Am. Chem. Soc.* **133**, 13308–13310.

3. **CF$_3$-C$_n$ Moieties into Organic Compounds**
   (a) Pd-catalyzed β-trifluoromethylation of aryl boronic acids with CF$_3$CH$_2$I, see: (a) Zhao, Y., and Hu, J. (2012) *Angew. Chem. Int. Ed.* **51**, 1033–1036; the Heck coupling reaction between iodoarenes and 1-iodo-3,3,3-trifluoropropane (a trifluoromethylpropene precursor) see: (b) Prakash, G.K.S., Krishnan, H.S., Jog, P.V., Iyer, A.P., Olah, G.A., and George A (2012) *Org. Lett.* **14**, 365–366; an intramolecular decarboxylative allylation of α-trifluoromethyl β-keto esters see: (c) Shibata, N., Suzuki, S., Furukawa, T., Kawai, H., Tokunaga, E., Yuan, Z., and Cahard, D. (2011) *Adv. Synth. Catal.* **353**, 2037–2041.

E. Utilization of Fluorinated Motifs as Mechanistic Tools

1. **Utilization of C–F Bond in Conformational Control**
   (a) Sparr, C., Schweizer, W.B., Senn, H.M., and Gilmour, R. (2009) *Angew. Chem. Int. Ed.* **48**, 3065–3068; (b) Sparr, C., and Gilmour, R. (2010) *Angew. Chem. Int. Ed.* **49**, 6520–6523; (c) Zimmer, L.E., Sparr, C., and Gilmour, R. (2011) *Angew. Chem. Int. Ed.* **50**, 11860–11871; (d) Bucher, C., and Gilmour, R. (2011) *Synlett* 1043–1046; (e) Tanzer, E.-M., Schweizer, W.B., Ebert, M.-O., and Gilmour, R. (2012) *Chem. Eur. J.* **18**, 2006–2013.

2. **Utilization of CF$_3$ Group in Conformational Control**
   (a) Prakash, G.K.S., Wang, F., Ni, C., Shen, J., Haiges, R., Yudin, A.K., Mathew, T., and Olah, G.A. (2011) *J. Am. Chem. Soc.* **133**, 9992–9995; (b) Prakash, G.K.S., Wang, F., Rahm, M., Shen, J., Ni, C., Haiges, R., and Olah, G.A. (2011) *Angew. Chem. Int. Ed.* **50**, 11761–11764.

13
Supramolecular Organic Chemistry: the Foldamer Approach

Zhan-Ting Li

13.1
Introduction

Supramolecular chemistry, or chemistry beyond the molecules, was first conceptualized by Lehn in 1973 [1]. It deals with the chemist's capacity to exploit the structure-directing properties of noncovalent forces for the construction of multicomponent chemical species that, in a similar way to biological systems, exhibit properties not shown by the individual components. Supramolecular organic chemistry is mainly concerned with organic molecules as the interacting components or transition metal ion-free multicomponent systems. After chemists had mastered concepts of covalency and demonstrated their mastery by using these concepts to realize the synthesis of complex natural and non-natural molecules and the correlation of the structures and properties of organic molecules, it became increasingly important and necessary to understand the rules that control the molecular recognition and structures and properties of multi-molecular systems. Pioneering work by Pedersen in 1967 on the complexation of crown ethers with metal ions triggered an upsurge in the context of host–guest chemistry and molecular recognition [2]. With increasing attention to the structures, properties, and formation processes of supramolecular systems as a whole, the increasing sophistication of the supramolecular systems studied, and the blurring roles of individual components as hosts or guests, the concept of self-assembly was also developed rapidly [3]. The last 20 years have witnessed an exponential growth in research in this field, as reflected by the increase in the number of publications with "supramolecular" in their titles (Figure 13.1). A recent search of the SciFinder database revealed that the number of related publications in this area increased from 207 in 1990 to 1625 in 2000 and 5661 in 2009.

The highest level of sophistication of supramolecular chemistry exists in living systems, whose elegant assemblies of organic molecules and macromolecules make up the supramolecular entities with structures and dynamics that enable and support life functions. Thus, the design aspects in supramolecular chemistry, since its establishment, have been, to a great extent, stimulated and inspired by Nature, and all important concepts in supramolecular chemistry can find their origin in Nature. Particularly the concepts of

Organic Chemistry – Breakthroughs and Perspectives, First Edition. Edited by Kuiling Ding and Li-Xin Dai.
© 2012 Wiley-VCH Verlag GmbH & Co. KGaA. Published 2012 by Wiley-VCH Verlag GmbH & Co. KGaA.

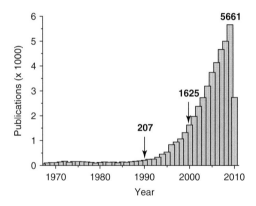

Figure 13.1 Annual publication numbers with "supramolecular" in the titles, obtained from SciFinder database in July 2010.

complementarity and preorganization play key roles in the evolution of the remarkable specificity of enzymes both in the selection of the guest molecules that they bind and in the reactions that they catalyze [4]. The high degree of catalytic selectivity of enzymes results from the complementary binding of a guest to the "active site" of an enzyme, which is realized through the preorganized three-dimensional structure of the amide framework of the enzyme. A large family of cyclic hosts, including crown ethers, cryptands, spherands, and many other kinds of cyclophanes, also utilize their preorganized cyclic structural feature, which substantially reduces the unfavorable negative entropy, to achieve high-stability binding to a guest with complementary size and binding sites. Multivalency is another interacting feature exhibited by life systems, especially proteins and DNA in the formation of their high-grade structures and their binding to other species. For the design of supramolecular recognition or assembly systems, this has also been a key factor to consider, because only cooperative binding by multiple weak noncovalent forces can lead to the formation of stable supramolecular entities.

This chapter is intended to describe the major developments and breakthroughs in the field of supramolecular organic chemistry in the last two decades. As the field has undergone a tremendous explosion in many different respects in that period, to attempt to cover the whole field would either make the chapter impossibly long or the coverage pointlessly superficial. Therefore, we focus here on the supramolecular chemistry of foldamers. Within this confine, we address a selection of the major themes that have preoccupied the emerging discipline in the past two decades and we select important advances of the field. Actually, the rapidly growing research activity in this discipline means that, even with this constraint, much still has to be left out. For example, the biological recognition or interaction of foldamers is not included here. Nonetheless, it is hoped that the chapter will provide a sense of the great potential of preorganized linear organic molecules in the field of supramolecular chemistry.

13.2
Foldamers: the Background

Biomacromolecules adopt distinct conformations to express discrete functions, such as recognition, catalysis, expression and translation, and transport that are essential to life. For example, peptides and proteins form sheet, helix, and turn patterns, and nucleic acids form the double and triple helix and quadruplex motifs. Since the middle of the last century, chemists have learned how to determine the conformational preferences of simple organic molecules and applied the concept to achieve regio- and stereoselective chemical transformations. However, only after the mid-1990s did the control of the three-dimensional conformation of synthetic oligomers become an active topic. It is not unexpected that research along this line was originally inspired by the high-grade structure of proteins and nucleic acids, and has expanded rapidly in recent years, leading to the field of foldamers [5−8].

Foldamers are synthetic oligomers that can spontaneously form a specific compact conformation [6, 7]. This family of unique structures bridges molecular and supramolecular chemistry well because their backbones are constructed covalently, whereas their shapes or conformations are controlled noncovalently. Therefore, foldamers may be considered, to some extent, as "intramolecular" supramolecules. The driving forces for the formation of folded structures include hydrogen bonding, hydrophobic or van der Waals interaction, aromatic stacking or donor−acceptor interaction, and metal ion coordination [1]. The first class of extensively investigated backbones is artificial peptides that are composed of discrete non-natural β-, γ-, and δ-amino acids or their hybrids [6, 7]. By carefully controlling the sequences, the oligomers may produce any of the secondary structural patterns observed for natural peptides. Owing to their similarity to the natural series, the main purpose of creating this class of intramolecular hydrogen bonding-induced foldamers was to develop peptide mimetics or peptidomimetics that may find applications in chemical biology and drug development [9]. In 1995, Lehn and co-workers found that N−N dipolar repulsion in the oligomeric pyridine−pyrimidine strand **1** forced the backbone to adopt a helical conformation [10]. In 1996, Hamilton and co-workers reported that intramolecular hydrogen bonding induced aromatic amide oligomers **2** to adopt a folded or helical conformation [11]. In 1997, Moore and co-workers observed that oligo(*m*-phenyleneethynylene)s **3** undergo hydrophobically driven folding in polar media [12]. Since then, the library of folded structures constructed from aromatic monomers, particularly amide and arylene ethynylene segments, was rapidly expanded.

1

2: R = H or CO$_2$Me

3: n = 12, 14, 16, 18
R = (CH$_2$CH$_2$O)$_3$Me

Owing to the intrinsic rigidity and planarity of aromatic units, foldamers of aromatic backbones usually have well-predictable conformations. By changing the sequence of the aromatic segments and controlling the orientation of their noncovalent sites, the folded or helical backbones may form cavities of distinct width and depth to complex guests of complementary size and shape and/or binding sites. The preorganization of the backbones is expected to favor complexation because, for long backbones, it may considerably reduce the negative entropy caused by the complexing process if the noncovalent force that drives the folding is not very weak. As a result, in many cases, the binding between a foldamer receptor and a guest may reach stability that is comparable to that of cyclic analogs. Therefore, in the past decade, molecular recognition based on foldamers or their modified derivatives have attracted increasing attention [13].

The achievements in foldamer-based molecular recognition also stimulated research on their applications in many other areas of supramolecular chemistry. For example, the stacking behaviors of a variety of folded structures, without or with additional functional groups, have been vigorously investigated. Many new supramolecular species, including organogels, vesicles, and liquid crystals, have thus been assembled [14]. The preorganized conformations of folded or extended backbones also make it possible to control the relative orientation or geometry of appended reaction sites. As a result, a variety of precursors have been designed based on folded segments and utilized to construct complicated macrocyclic architectures in very high yields or even quantitatively [14–16]. A number of examples of catalysis or reaction promotion by foldamer receptors have also been reported [17–19]. Considering the linear feature of foldamers, the advances along this line may lead to the generation of new artificial enzymes.

13.3
Molecular Recognition

In a broad sense, many reported biological and medicinal functions of foldamers that are composed exclusively of non-natural aliphatic amino acid subunits are caused by their binding to a specific biological target [20, 21]. This section covers only the binding phenomena associated with foldamer receptors with quantitative evaluation in the confinement of supramolecular chemistry. The receptors may be roughly divided into two categories. The first category utilizes the cavities formed by the folded backbones themselves to complex distinct guests. The second category may be considered as functionalized or modified foldamers. Their folded moieties mainly serve as preorganized frameworks to converge appended binding sites to maximize their binding efficiency.

13.3.1
m-Phenyleneethynylene Oligomers

The first example of molecular recognition by foldamer receptors was probably that reported by Moore and co-workers in 2000 [22]. They found that in aqueous acetonitrile (60%), dodecamer **4** complexed chiral pinene **5a**, which led to a strong Cotton effect in the wavelength range where the oligomer absorbed. It had been established previously that, in polar solvents, oligomer 4 and its longer analogs were driven hydrophobically to fold to form a tubular cavity [12]. Compound 5a fitted the size of the cavity well and was therefore entrapped by 4, which in turn induced the oligomer to generate a chiral differentiation. The binding adhered strictly to a 1:1 stoichiometry and the binding constant (K_a) was determined as 6830 M^{-1}. Increasing the content of water in the solvent enhanced the binding linearly and thus the K_a was estimated to be 6×10^4 M^{-1} in pure water by means of extrapolation, which further suggested that the binding was a hydrophobically driven process. Foldamer 4 also formed 1:1 complexes with chiral 5b–f. However, their K_as were reduced gradually from 6000 to 1790 M^{-1}, implying that 5a best matched the cavity of the helical oligomer. If methyl groups were introduced to the 4-positions of the benzene rings of 4, the binding capacity of the corresponding foldamer towards 5a–f was decreased dramatically. This observation further supported the entrapping model.

Owing to the intrinsic rigidity of the backbones, the diameter of the cavity formed by this series of foldamers is relatively fixed. They were therefore expected to accommodate only guests of matched sizes, such as 5a–f. Moore's group also prepared oligomers 6, which were incorporated with a 4-dimethylaminopyridine (DMAP) unit at the middle of the backbones to investigate the effect of the folded conformation on their nucleophilicity towards methyl iodide (Figure 13.2) [23]. The methylation of the DMAP unit of the folded oligomer 6 ($n = 6$) by methyl iodide in acetonitrile, producing oligomer 7 ($n = 6$), was accelerated about 400-fold compared with that of the shorter analog 6 ($n = 1$), which could not bind any guest. Preferential solvation of the hydrophobic cavity of the foldamers by methyl iodide was proposed to be responsible for the rate acceleration. When the DMAP unit was moved to the end, the acceleration effect was weakened, which further supported the explanation. On the basis of the above complexation, Smaldone and Moore proposed the concept of a "reactive sieve" (Figure 13.2) [24, 25]. The cavity of the foldamers was initially thought to entrap the ideally sized substrate, causing the consequent reaction to occur at a higher rate than other guests of the same series. Small substrates would be insufficiently bound and react slowly, whereas substrates that are too large would not fit and also react slowly or not at all (Figure 13.2). A systematic kinetic investigation of the reaction of oligomers 6 with different methyl alkanesulfonates to form the N-methylpyridinium derivatives was thus performed. It turned out that the reaction rates did not correspond to the expected optimally sized guests, such as n-BuSO$_3$Me. The system did differentiate between substrates with subtle structural differences, exhibiting

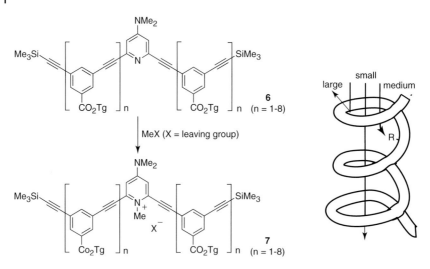

Figure 13.2 Methylation of the DMAP unit in foldamers **6** by methyl iodide and a cartoon illustration of the proposed "reactive sieve" with methyl alkanesulfates as substrates.

a wide range of rate enhancements (45–1600-fold). However, methylation could not be completely inhibited by increasing the substrate size or decreasing the available cavity volume. The result implied that the cavity formed by the foldamers had a large dynamic feature.

The benzene units in oligomers **4** and **6** can be replaced with distinct aromatic units. For example, Inouye and co-workers prepared pyridine-derived oligomers **8** (Figure 13.3) [26]. The oligomers adopted an extended conformation in dichloromethane due to the dipole–dipole repulsion of the adjacent pyridine units. However, in the presence of *n*-octyl β-pyranoside (**9**), the long oligomers ($n = 18, 24$) folded into helicates due to the formation of 1:1 complexes, which were stabilized by intermolecular hydrogen bonds between the nitrogen atoms of the oligomers and the hydroxyl groups of the sugar (Figure 13.3). The K_a of the complex of **8** ($n = 24$) was determined as 1200 M^{-1} in dichloromethane. When the sugar was attached to the end of the oligomers through a

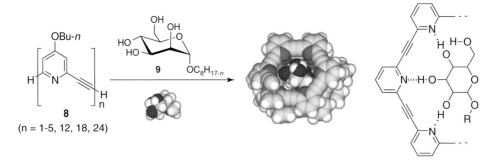

Figure 13.3 Sugar-induced helicate of oligomers **8** in dichloromethane.

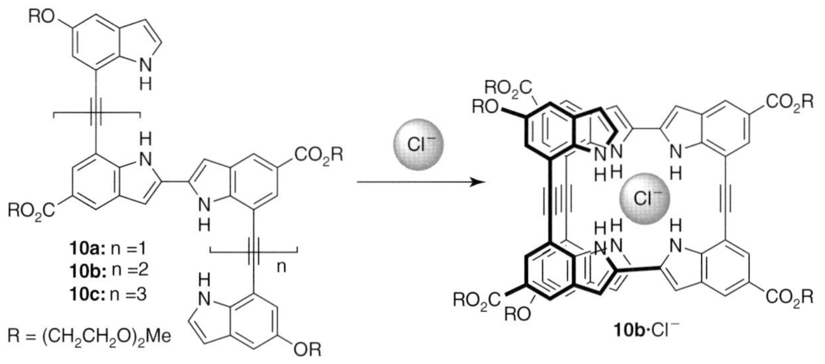

Scheme 13.1 Complexation-induced folding of oligomers **10a–c** by chloride in acetonitrile.

flexible chain of sufficient length, the sugar was entrapped in the cavity of the attached oligomers to give rise to unique intramolecular complexes [27]. The stability of the complexes depended considerably on the length of the oligomers and the structure of the attached sugars, showing that the folded oligomers had a higher-order structure of lowest energy.

Jeong and co-workers found that, when multiple indole segments were connected with ethynylene units, the new oligomers **10a–c** ($n = 1$–3) adopted an extended arrangement and the neighboring indole units maintained an *s-trans* conformation in polar media [28]. In the presence of chloride or other anions, the backbones folded to complex the anion in a 1:1 stoichiometry (Scheme 13.1). Intermolecular N–H$\cdots$ Cl$^-$ hydrogen bonding was evidenced to stabilize the complexes. The K_as of the three complexes in acetonitrile were determined to be as high as 1.3×10^5, 1.1×10^6, and 2.0×10^6 M^{-1}, respectively. The high values reflected high cooperativity of the hydrogen bonds. When the oligomers were completely hydrolyzed to their sodium salts, they became soluble in water [29]. The longest oligomer modestly complexed chloride and fluoride anions in water with $K_a = 65$ and 46 M^{-1}, respectively, whereas the shorter ones did not bind any anions.

13.3.2
Naphthalene-Incorporated Ethylene Glycol Oligomers

Early studies by Vögtle and Weber showed that oligo(ethylene glycols) with two terminal aromatic donors wrapped themselves around a metal ion to acquire a folded conformation in polar solvents [30]. However, ethylene glycol oligomers themselves did not exhibit the capacity due to their high flexibility and hydrophilicity. Li and co-workers reported that oligomers **11a–g** folded into helical conformations in mixtures of chloroform and acetonitrile (Figure 13.4) [31]. When the content of acetonitrile was increased to a certain value, the longer oligomers folded completely owing to the solvophobically driven intramolecular stacking of the naphthalene units. Molecular modeling revealed that the folded backbones afforded a cavity similar to that of 18-crown-6. Long oligomers **11e–g** were thus able to bind ammonium and ethane-1,2-diaminium in 1:1 stoichiometry. The K_as of their complexes with ethane-1,2-diaminium in acetonitrile were determined as

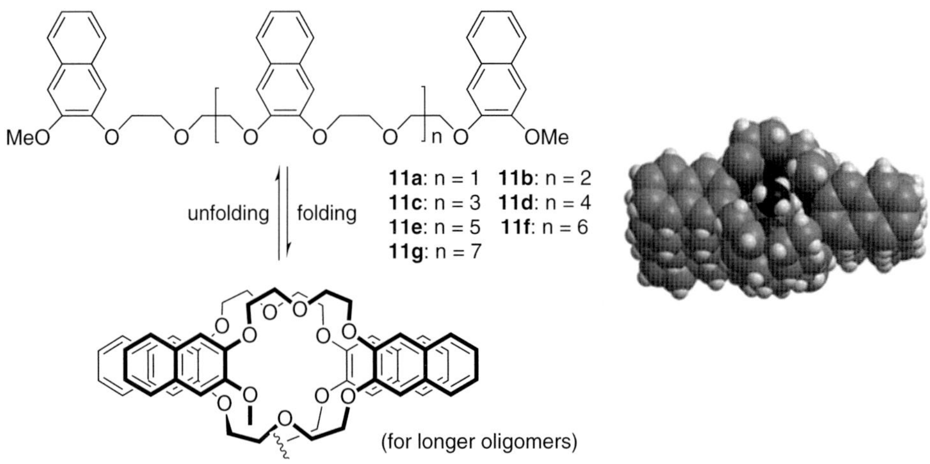

11a: n = 1 11b: n = 2
11c: n = 3 11d: n = 4
11e: n = 5 11f: n = 6
11g: n = 7

unfolding ‖ folding

(for longer oligomers)

Figure 13.4 Solvophobically driven folding of oligomers **11a–g** and the energy-minimized structure of the complex of **11g** with ethane-1,2-diaminium.

$3.5 \times 10^3, 1.0 \times 10^4$, and 2.5×10^4 M^{-1}, respectively. 2D ^{1}H NMR experiments in CD$_3$CN for **11g** revealed strong intramolecular nuclear Overhauser effects (NOEs) between the H1 and H4 signals of the naphthalene units and the OCH$_2$ signals in the presence of the dication. In contrast, similar contacts were not present for the oligomers in the absence of the dication. These observations indicated that the helical conformation of the oligomers became even more compact and rigid after complexation.

13.3.3
Heterocyclic Oligomers

In 2002, Lehn and co-workers reported that oligomer **12** was driven by the N–N dipole repulsion of the neighboring heterocyclic units to fold into a cavity ~3.5 Å in diameter [32]. The convergent arrangement of the strong electric dipoles of the naphthyridine groups made the cavity fairly polar to bind small cationic guests, such as alkali metal ions, hydronium, and guanidinium. The binding further enhanced the intramolecular stacking of the helicate, leading to the formation of cylindrical aggregates which gelated the mixture of chloroform, ethanol, and acetonitrile.

Lehn and co-workers also prepared linear amines **13a–e** [33]. Their fully protonated oligomers were designed to template the controlled oligomerization of the one-turn helicate **12**. Electrospray mass spectrometry (MS) of the mixtures of the identical concentration of **12** with **13a–e** in the presence of trifluoroacetic acid revealed that **13a** and **13c** could easily reach binding saturation (Figure 13.5). That is, they formed a 2:1 complex with **12**. For **13b**, whose three amino units were separated by two methylene groups, even a large excess of **12** (10 equiv.) yielded only minor amount of the fully saturated species $[12_3 \cdot 13bH_2]^{2+}$, whereas 5 equiv. of **12** caused **13d**, whose three amino units were separated by three methylene groups, to form exclusively fully saturated species $[12_3 \cdot 13dH_3]^{3+}$. The mixture of **12** with **13e** (7:1) formed both fully charged

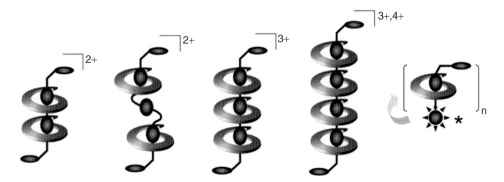

[**12**₄ · **13eH**₄]⁴⁺ and less charged [**12**₄ · **13eH**₃]³⁺. When the ammonium guest was attached with a chiral L-proline group, supramolecular chirality induction was exhibited for the similar multiple-component complexes (Figure 13.5), indicating that the helicates in the oligomeric species adopted an organized orientation around the central strand and amplified the chirality of the L-proline group [34]. The binding was attributed to the intermolecular N–H⋯N hydrogen bonding. The complexes represent new unique pseudorotaxanes with helical subunits as the ring components and, to some extent, mimic the self-assembling pattern of tobacco mosaic virus.

In 2008, Craig and co-workers developed 1,2,3-triazole-based oligomer **14** [35]. The backbone has appreciable conformational freedom and molecular modeling showed no significant preference for any particular conformer. In the presence of tetrabutylammonium chloride, it folded into a helical conformation in acetone. Systematic ¹H NMR

Figure 13.5 Models for the pseudorotaxane-styled supramolecular oligomers between foldamer **12** and amines **13a–e**. The last cartoon represents the supramolecular chiral induction for the same series of complexes by a chiral guest.

studies in acetone-d_6 revealed that a 1:1 complex was formed, which was stabilized by the intermolecular C^5–H$\cdots$Cl$^-$ hydrogen bonding or electrostatic interaction. K_a was estimated to be 1.7×10^4 M^{-1}. The oligomer also complexed Br$^-$ and I$^-$ in acetone, but the complexes were less stable ($K_a = 1.2 \times 10^4$ and 1.3×10^2 M^{-1}, respectively).

Also in 2008, Li and Flood reported the synthesis of 1,2,3-triazole-derived nonamer **15**, which bears three *tert*-butyl groups on the central benzene rings [36]. The binding between **15** and halide anions was investigated in dichloromethane. The binding was confirmed also to be in a 1:1 stoichiometry and associated K_as were determined as 42, 6.9, 268, and 210 M^{-1} for the complexes of F$^-$, Cl$^-$, Br$^-$, and I$^-$, respectively. The values are remarkably lower than that of the complex of **14** with Cl$^-$. It was proposed that the ester group of **14** exerted a strong polarizing effect on the C–H$\cdots$Cl$^-$ hydrogen bond by virtue of its electron-withdrawing character and its ester-linked tetraethylene glycol chains also favored a folded structure, particularly in polar acetone. As a result, **14** displayed a much stronger binding affinity than **15** for halides.

13.3.4
Cholate Oligomers

Owing to its low directionality, it is challenging to utilize the hydrophobic force to construct compact structures from aliphatic oligomers. Early studies by Jiang revealed that long aliphatic amphiphiles fold in aqueous solution [37]. The folded conformations, however, generally lack a precise structural feature. Zhao developed a unique approach by making use of the amphiphilicity of rigid cholate unit [38]. For example, Zhao and Zhong found that cholate hexamers **16** folded into helical structures in mostly nonpolar solvents such as carbon tetrachloride or hexane due to the convergent aggregation of the polar side of the cholate units [39]. The two methionine units were inserted into the linkers to form a bidentate ligand for sensing metal ions. It was revealed that, in t hexane–ethyl acetate–methanol (65:30:5 v/v/v) mixture, **16** complexed Hg^{2+} even at the low Hg^{2+} concentration of 20 nM. The binding stoichiometry was evidenced to be 1:1 and K_a was estimated to be as high as 1.5×10^7 M^{-1}. The binding was highly selective for Hg^{2+} as compared with other divalent transition metal ions. This selectivity was attributed to the binding affinity of Hg^{2+} for the orthionine sulfur. Oligomer **16** could dissolve in the micelles of sodium dodecyl sulfate in water, which allowed for the binding of Hg^{2+} in water. When the concentration of the sulfate salt was 8 mM, the K_a of the complex was estimated to be 6.5×10^5 M^{-1} [40].

16

13.3.5
Aromatic Hydrazide and Amide Oligomers

13.3.5.1 Hydrazide Foldamers
Early work showed that hydrogen bonding-driven aromatic hydrazides have relatively high rigidity and planarity [41]. In 2004, Li and co-workers reported the self-assembly of the first series of hydrazide-based foldamers **17a–c** [42]. The hydrogen bonding-driven preorganization of the frameworks forced half of their C=O groups to be positioned inwards to generate a rigid cavity ~1 nm in diameter. Therefore, alkylated saccharides **18a–d** could be entrapped in the cavity in 1:1 stoichiometry through the formation of multiple intermolecular C=O· · · H–O hydrogen bonds. The binding was evidenced by ^{1}H NMR, fluorescence, and circular dichroism experiments. The largest K_a of 6.9×10^6 M^{-1} was determined for complex **17 c · 18d** in chloroform. This value was about 10-fold higher than that of the complexes of the monosaccharides. The K_as of the complexes of **18c** were generally lower than those of **18a** or **18b** for identical hosts. Clearly, the increase in the hydrogen bonding sites helped to raise the stability of the complexes. As revealed for the complexes between **4** and chiral **5**, all the complexes between **17a–c** and **18a–d** exhibited supramolecular helicity chirality which was induced by the chiral nature of the guests. To investigate the effect of the intramolecular hydrogen bonding on the binding properties, Li and co-workers also prepared two heptamers, which lacked one and two intramolecular hydrogen bonds on the central aromatic units [43]. These oligomers were

found to moderately complex **18a** in chloroform, indicating that fully hydrogen bonded folded structures are much more favorable for complexing the saccharide guests.

17a: n = 1
17b: n = 2
17c: n = 5

R = n-C₈H₁₇

α-D-Glucose (**18a**) α-L-Glucose (**18b**)

β-D-Ribose (**18c**) β-Maltose (**18d**)

The above aromatic hydrazide backbones do not contain any chiral centers. Li *et al.* introduced chiral proline units to the ends of the backbones. Circular dichroism investigations in chloroform showed that the frameworks of the corresponding foldamers **19a–c** generated a helical bias [44]. Molecular dynamic simulations suggested that the favorable and unfavorable conformations had an energetic differentiation of about $10–15\,\text{kcal mol}^{-1}$. These chiral helicates also entrapped saccharides to form both diastereomeric and enantiomeric complexes. The pair of diastereomeric complexes formed between **19a** and **18a** and **18b** displayed the highest 144-fold stability differentiation.

19a: n = 2 (*R, R*)
19b: n = 3 (*R, R*)
19c: n = 3 (*S, S*)
R = (CH₂CH₂O)₂Et

13.3.5.2 Benzamide Foldamers

In 2000, Gong and co-workers reported the first series of benzamide-based foldamers **20a–d** [45], which had a cavity smaller than that of the hydrazide-derived oligomers **17a–c**. Because their C=O groups were all located inwards, in principle, this series of foldamers and related structures could entrap smaller guests through intermolecular hydrogen bonding. In 2009, Gong and co-workers observed that a similar folded hexamer **21** formed a very stable complex with *n*-octylguanidinium tetraphenylborate in chloroform [46]. Direct measurement of its K_a by ¹H NMR spectroscopy could not be performed owing to limited solubility and serious line broadening. By using the extraction method, the K_as of the complexes between the foldamer and guanidinium and octylguanidinium tetraphenylborate in water-saturated chloroform were estimated to be as high as 1.3×10^8 and 1.3×10^7 M^{-1}, respectively.

20a: n = -1
20b: n = 0
20c: n = 1
20d: n = 2

complex of 21 (R = n-C$_{12}$H$_{25}$) and octylguanidinium tetraphenylborate

For benzamide-based oligomers which are connected through two *meta*-substituted amide units, the frameworks should produce cavities similar in size to those of **21**. In 2004, Li and co-workers found that folded heptamer **22** complexed saccharides **18a–d** in chloroform in 1:1 stoichiometry [47]. The K_a values of the complexes were determined as 550–7800 M^{-1} by using fluorescence titration experiments. These values were remarkably lower than those of their hydrazide analog **17b**, which has the same number of benzene units but a larger cavity. Molecular modeling showed that the cavity of the helical **22** was about 0.8 nm in diameter, which is slightly small to entrap a monosaccharide, but the fact that this foldamer still complexed both mono- and disaccharides indicated that the folded structures were also dynamic and thus were able to adapt their size to hold a guest, as observed for oligomer **6** [25].

R = n-C$_8$H$_{17}$
22

23a: n = 0
23b: n = 1
23c: n = 2

When alkoxyl groups, the hydrogen bonding acceptors, are orientated inwards, the corresponding folded frameworks **23a–c** form only very small cavities which are not capable of complexing saccharides in chloroform [48]. Instead, the foldamers exhibited a modest ability to bind primary and secondary alkylammonium ions also in 1:1 stoichiometry, with K_a values being about 17–360 M^{-1}. Intermolecular MeO$\cdots$H–N hydrogen bonds and cation–π interactions were proposed to be the main driving forces for the binding. Clearly, the ether methyl groups combine to impose spatial repulsion for the approach of any guest.

In 2005, Li *et al.* developed a modified scaffold to use fluorine as the hydrogen bonding acceptor [49]. In this way, the steric repulsion caused by the methoxyl groups could be eliminated. The X-ray structures of model compounds and the ^{1}H NMR spectra of oligomers **24a** and **24b** supported the fluorine atoms in this series of molecules being involved in intramolecular five- and six-membered and three-centered F$\cdots$H–N(C=O) hydrogen bonds [50], which forced the oligomers to form a rigid folded conformation. Job plots indicated that the foldamers formed 1:1 complexes with dialkylammoniums **25a** and **25b** in chloroform. The K_as were determined as 4.9×10^6, 8.1×10^6, 2.4×10^5, and 7.3×10^5 M^{-1} for complexes **24a · 25a** and **24b · 25a**, **24a · 25b**, and **24b · 25b**, respectively. NOE contacts were also observed between the ammonium protons and some of the amide hydrogen atoms of the oligomers. The binding of **25b** to the oligomers also caused induced circular dichroism signals at the absorption wavelength of the foldamer. The signals were enhanced with increase in the concentration of the ammonium, but weakened with the addition of polar methanol. These results supported intermolecular N–H$\cdots$F hydrogen bonding or electrostatic interaction being the driving force for the formation of the complexes.

24a

24b

25a

25b

13.3.5.3 Heteroaromatic Amide Foldamers

Helicate **2** did not form a cavity and therefore entrapped no guest. When all the connecting amide units are attached to the 2- and 6-positions of the pyridine segments, the resulting pyridine/dicarboxamide strands preferred to form double, triple, or even quadruple helicates [51–53]. The helical structures may also form a small cavity and thus entrap small guests. For example, in 2002, Huc *et al.* reported that, in the solid state, heptamer

26 encapsulated a water molecule in its cavity [54]. Intermolecular N···H–O(water) and N–H···O(water) hydrogen bonds were the main driving force for the entrapment, although spatial matching should also be a favorable factor.

In 2005, Huc and co-workers prepared oligomer 27a. This heptamer contained two large quinoline segments that were appended to the ends of a pyridine trimer [55]. The X-ray crystal structure revealed that it folded into a closed capsule, the two ends of which were capped by the two shrinking peripheral quinoline units. This closed capsule entrapped a water molecule in the crystal. To detect if this entrapment occurred in solution, ^{1}H NMR experiments were performed in dry and wet chloroform-d, which revealed that drying or wetting the solvent had little effect on the chemical shift of the signals of the peripheral amide protons ($\Delta\delta = 0.2$ ppm), but caused large downfield chemical shifting ($\Delta\delta = 1.31$ ppm) for the signals of most of the central amide protons. These results supported the water molecule being encapsulated in the cavity, which was driven by the formation of strong hydrogen bonding with the inwardly located amide protons. At low temperature ($-50°$C), water gave rise to two different water peaks. One of them showed NOEs with some of the inwardly located amide protons and therefore was assigned to the encapsulated water. An unwinding–rewinding process of the helix was proposed for the binding and release of the guest molecule in the single molecular capsule.

27a: n = 1, R = OBn
27b: n = 2, R = OBu-i
27c: n = 3, R = H

In an attempt to expand the cavity size of the single molecular helix containers, the same group prepared nonameric and undecameric oligomers 27b and 27c, which entrapped one and two water molecules in the crystal [56]. ^{1}H NMR studies in chloroform-d revealed that the capsules could be empty, half-full, or full in solution, as evidenced by distinct signals of water that were in slow exchange on the NMR time scale at low temperature. Although water was the preferred guest, 27a also encapsulated other small polar molecules, such as hydrazine, hydrogen peroxide, and formic acid. Further elongation of the oligomers produced 28, which formed an even longer closed cavity and could entrap longer alkanediols [57]. This closed helicate also encapsulated chiral tartaric acid with exceptional affinity, selectivity, and diastereoselectivity, as revealed by the ¹H NMR spectra in solution and by the X-ray structure in the solid. Impressively, the small chiral guest induced the oligomer capsule to generate chiral bias, which was evidenced in circular dichroism experiments [58].

13.3.5.4 Flexible Arylamide Oligomers

When the aromatic segments of aromatic amide oligomers are not completely intramolecular hydrogen bonded, there will be more than one conformations of low energy due to the rotation of the C–C(Ar) and N–C(Ar) single bonds of the amide units. In 2000, Lehn and co-workers reported one such oligomer, heptamer **29** [59]. The four 2,6-diamidopyridine units formed four-membered hydrogen bonds with the endocyclic pyridine nitrogen atoms as acceptors, but rotation about the (aryl)C–C(=O) bonds might afford as many as 36 quasi-planar/conjugated conformers (Scheme 13.2). ¹H NMR spectra in chloroform-*d* indicated that adding 2 equiv. of cyanurate **30** to its solution caused its framework to form two disk-like entities, which was driven by intermolecular DAD/ADA hydrogen bonds. The 1:2 complex had three possible conformers, depending on the rotation of the central aryl–CO bonds. The two helical segments could further stack to generate column-styled architectures, which aggregated to form supramolecular fibers several microns in length.

In 2009, Zhao and co-workers reported that intramolecular hydrogen bonding-free aromatic amide oligomers **31a–d** that consisted of alternate benzene and naphthalene segments complexed 1,3,5-tricarboxylate anion **32** in 1:1 stoichiometry in highly polar dimethyl sulfoxide (DMSO) [60]. ¹H NMR, fluorescence, and UV–Vis experiments in DMSO indicated that the oligomers did not adopt any defined conformation, even though the large stacking naphthalene units were available. However, adding **32** caused the oligomers to fold completely into helical compact structures because the oligomers exhibited a new set of signals of high resolution in the ¹H NMR spectra (0.77 mM) and the signals of the free oligomers themselves disappeared completely after about 1 equiv. of **32** had been added. Multiple intermolecular NOE contacts were observed, which indicated that intermolecular N–H···O⁻ and C–H···O⁻ hydrogen bonds or electrostatic interactions were the driving forces for the binding (Scheme 13.3). The K_as in DMSO-d_6 were determined to be as high as $2.5 \times 10^4, 8.6 \times 10^5$, and 5.5×10^6 M⁻¹ for complexes **31b · 32, 31c · 32**, and **31d · 32**, respectively, reflecting that these intermolecular hydrogen bonds were highly cooperative and also electrostatic.

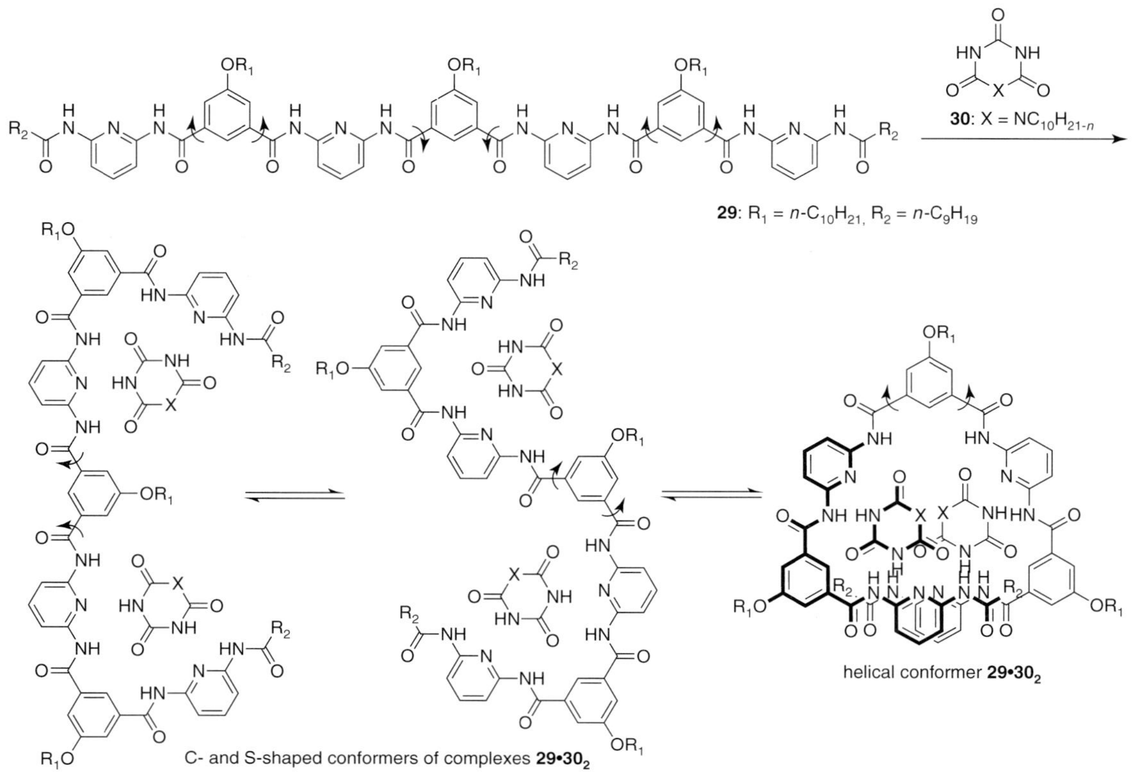

Scheme 13.2 Binding pattern for compounds **29** and **30**, which leads to the formation of a helical conformation (right) for **29** due to intramolecular stacking of its two folded segments.

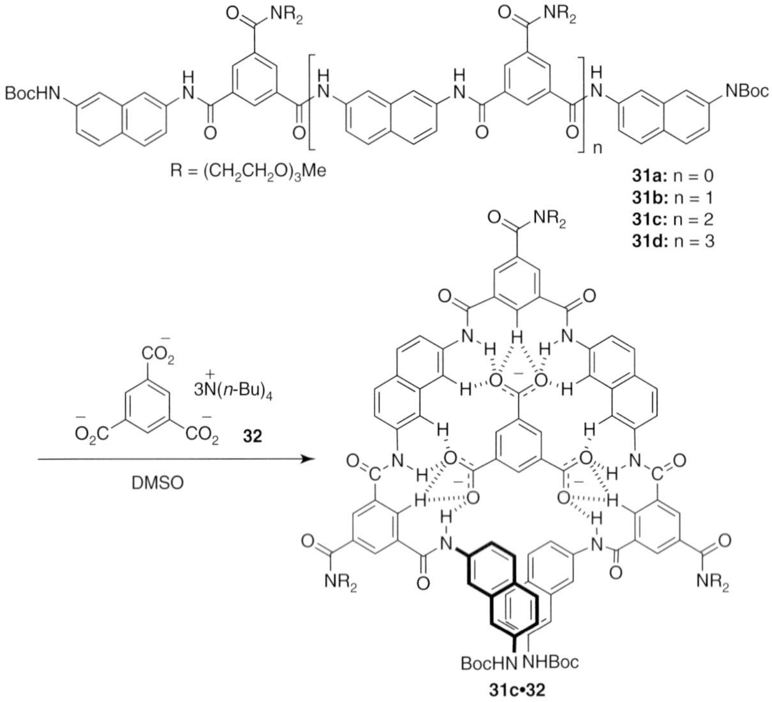

Scheme 13.3 Structures of oligomers **31a–d** and the helical complex between **31c** and **32**, stabilized by intermolecular N–H · · · O and C–H · · · O hydrogen bonds.

13.3.5.5 Modified Arylamide Oligomers: Molecular Tweezers

Short hydrogen-bonded aromatic amide oligomers have well predictable shapes and can be functionalized at distinct positions, especially from the two ends. Li and co-workers developed a number of foldamer tweezers, which complexed guests of matching size and shape [61–68]. For example, compound **33** was induced by four intramolecular hydrogen bonds to adopt a "U"-styled conformation [61]. The two zinc porphyrin units were arranged roughly parallel to each other and thus exhibited high affinity to fullerenes. The K_as of its complexes with C_{60} and C_{70} in toluene were determined as 1.0×10^5 and 1.1×10^6 M^{-1}, respectively, by using the UV–Vis titration method. The binding for chiral C_{60} derivatives caused the bisporphyrin tweezers to generate supramolecular chirality, as evidenced by the formation of induced circular dichroism signals. Compound **34** also has a folded conformation with the appended zinc porphyrin groups being forced to approach and stack with each other [63]. This receptor formed 1:1 complexes with planar electron-deficient molecules, such as **35**, through a unique sandwich-styled binding pattern. The K_a of complex **34·35** was determined as 850 M^{-1} in chloroform. The result showed that, in contrast to the conventional covalently bonded molecular tweezers, which are difficult to adjust their shapes to maximize the binding, hydrogen-bonded tweezers are elastic and are therefore able to change their conformation to improve the binding.

R$_1$ = n-C$_5$H$_{11}$, R$_2$ = n-C$_8$H$_{17}$, Ar = 3,5-di-tert-butylphenyl

36 · 37 · 38

R = n-C$_5$H$_{11}$

Ar = 3,5-di-tert-butylphenyl

33

34 · 35

Extending the rigid framework of the porphyrin tweezers enabled them to complex larger guests. For example, Li and co-workers reported that compound **36** complexed **37** in 1:1 stoichiometry with K_a being 5.7×10^6 M^{-1} in a mixture of chloroform and acetonitrile (4:1 v/v) [67]. Because the ammonium unit in **37** was able to thread through the cavity of 24-crown-8 (**38**) driven by intermolecular N–H···O or C–H···O hydrogen bonding, their 1:1:1 three-component solution (3 mM) gave rise to a dynamic [2]catenane in 55% yield. Adding more **38** or reducing the temperature increased the yield of the dynamic [2]catenane, which could be generated quantitatively at −13 °C when 2 equiv. of **38** was present.

The same group also reported that aromatic amide frameworks could be functionalized from the outside. In this way, they prepared hexamer **39** [68]. The rigid, folded feature of the central aromatic amide framework enabled the appended zinc porphyrin units to separate averagely and thus to interact with structurally matching C$_{60}$ ligand adduct **40** in a "domino" pattern. The apparent K_a of the complex between their zinc porphyrin unit and **40** in chloroform was determined as 3.6×10^4 M^{-1}, which was notably larger than that of shorter analogs. Supramolecular chirality amplification was also exhibited for the complex, because strong induced circular dichroism signals were exhibited in the absorption area of the zinc porphyrin, which were also remarkably stronger than those produced by shorter folded analogs of **39**, indicating that the binding between **39** and the six molecules of **40** was cooperative.

40　　　**39**

PMB = *p*-methoxybenzyl
R = *n*-C$_8$H$_{17}$

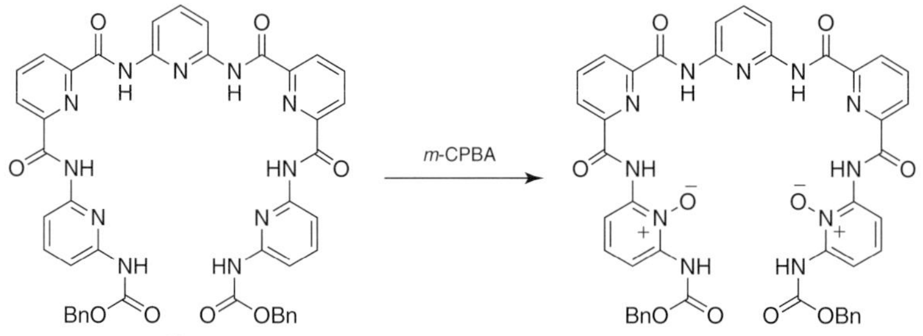

Scheme 13.4 Oxidation of oligomer **41** to **42** by *m*-CPBA.

13.3.5.6 Reaction Acceleration

The folding of backbones of enzymes provides specific active sites for selective catalysis. Moore and co-workers observed that helical *m*-phenyleneethynylene oligomers promoted the methylation of the incorporated pyridine [22, 23]. In 2005, Huc and co-workers reported that the oxidation of the ending pyridine units of **41** and **42** by *m*-chloroperbenzoic acid (*m*-CPBA) was promoted by its folding conformation (Scheme 13.4) [18]. No kinetic studies were carried out. However, compared with the oxidation of short fragment analogs, this oxidation process was remarkably faster. That the ending pyridines instead of the central one were selectively oxidized might be ascribed to the large steric hindrance at the center position. However, the exact mechanism of the oxidation acceleration is still unclear, although the dipolar interaction within the helical conformation and the preassociation of the oxidative reagent in the polar cavity might play roles.

The methoxyl oxygen atoms in the frameworks of long oligomers **23** are arranged to resemble those of crown ethers. In 2007, Li and co-workers reported that the hydrolysis of oliogmers **43a–e**, which bear a nitro group at one end, to phenol derivatives **44a–e** by alkali metal hydroxides was promoted by the long frameworks (Scheme 13.5) [19]. Kinetic investigations in polar dioxane–water (4:1 v/v) at 60–90 °C revealed that, compared with that of the shorter oligomers, the hydrolysis of the longer oligomers **43c–e** was accelerated by up to about fivefold. This acceleration was remarkably weakened when an excessive amount of chloride salt was added. The complexation of the alkoxyl oxygen atoms with the metal ions was proposed to cause the acceleration by activating both the hydrolyzed C–O bond and the hydroxide anion and increasing the effective concentration of the hydroxide anion. This result also indicated that, even in a hot aqueous medium of high polarity, the intramolecular MeO···H–N hydrogen bonding still survives to allow a preorganized conformation for the backbones.

13.4
Homoduplex

In 2000, Lehn and co-workers reported that oligomers **45a–c** folded into helical structures which were driven by continued intramolecular N–H···N(py) hydrogen bonding [51]. In solution, these folded monomers also partially formed helical dimers which were

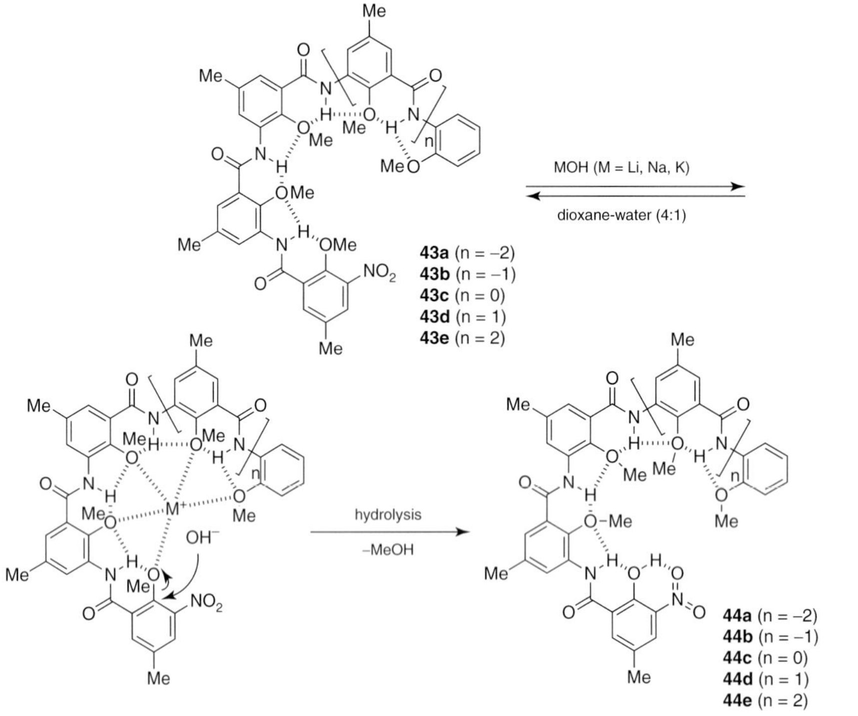

Scheme 13.5 Model for the hydrolysis of foldamers **43a–e** (with **43d** as example) to oligomers **44a–e**.

stabilized by intermolecular hydrogen bonding. For **45c**, single and double molecular helices were generated in the solid state when it was crystallized from a polar or less polar solvent mixture, respectively. Quantitative analysis of the NMR data for **45b** gave a dimerization constant K_d of 6.5×10^4 M^{-1} in CDCl$_3$, which was three orders of magnitude larger than that of **45a** in the same solvent. In toluene-d_8, CD$_2$Cl$_2$, and C$_2$D$_2$Cl$_4$, K_d was found to be 5.5×10^4, 1.0×10^5, and 1.6×10^5 M^{-1}, respectively. Interactions between the side chains and an increase in the interactions between aromatic rings due to the electron-donor character of the decyloxy substituents in **45b** were proposed to rationalize this marked effect. By modifying both the backbones and the side chains, Huc and co-workers also assembled a number of elegant triple and quadruple helices from this series of pyridine/carboxamide strands [52, 53].

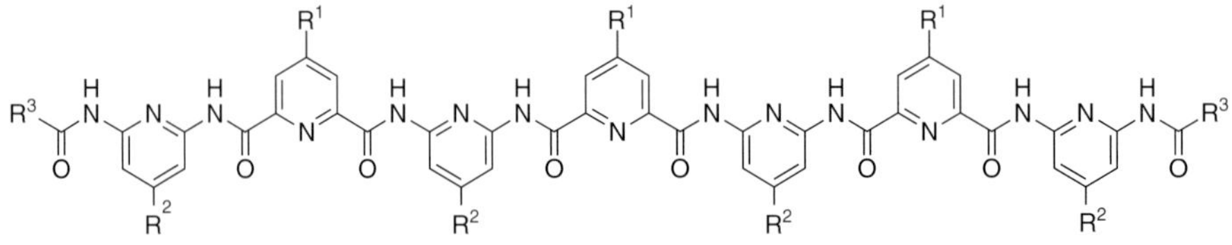

45a: R^1 = OC$_{10}$H$_{21}$; R^2 = H; R^3 = C$_9$H$_{19}$; **45b:** R^1 = R^2 = H; R^3 = C$_9$H$_{19}$; **45c:** R^1 = R^2 = H; R^3 = OBu-t

In 2004, Li and co-workers reported the assembly of extended secondary structures from aromatic amide oligomers **46a–d** [69]. When multiple self-binding sites were introduced to the same side of the backbones, the resulting preorganized monomers might form homoduplex [70, 71]. For example, compounds **47a** and **47b** are attached with two and three amide units. The K_as of their homodimers in chloroform were determined as 3.0×10^3 and $\geq 2.3 \times 10^5$ M^{-1}, respectively, by ^{1}H NMR dilution experiments. In contrast, the K_a of benzamide was about 69 M^{-1} even in less polar benzene. The rapid increase in the binding stability with elongation of the oligomers clearly supports the cooperativity of the intermolecular hydrogen bonding in the formation of the dimeric structures.

46a: n = 0
46b: n = 1
46c: n = 2
46d: n = 3

47a·47a
$K_a = 3000$ M^{-1}

47b·47b
$K_a > 2.3 \times 10^5$ M^{-1}

13.5
Organogels

The stacking of large conjugated systems into three-dimensional cross-linked networks may lead to immobility of the solvent and the formation of an organogel. In this context, discotic aromatic molecules that bear aliphatic chains of suitable length have been vigorously investigated [72]. Aromatic foldamers have crescent or ring-like shapes and relatively large size. Hence their intermolecular stacking may be strong in solution. In 2002, Lehn and co-workers reported that pyridine/pyridazine strand **48** comprising 13 heterocycles gelated dichloromethane [73]. Similarly to oligomer **12**, this compound was also driven by the electrostatic repulsion of the neighboring nitrogen atoms to adopt a helical conformation. The freeze fracture electron micrographs of **48** in a mixture of dichloromethane and pyridine revealed uniform helical fibrils with an diameter of ~55–70 Å, which aggregated into networks of fibrils of micrometer length to form linear and intertwined fibers and macrofibers to gelate the solvent.

48

Aromatic hydrazide-derived foldamers have a relatively large size. In 2008, Li and co-workers found that compounds **49a–f** with seven aromatic rings stacked to gelate many organic solvents of both low and high polarity, including alkanes, arenes, esters, alcohols, and 1,4-dioxane [74]. Large aromatic units were appended to the two ends to enhance their stacking interaction. Because foldamers of the identical backbones complex chiral saccharides in solution [42], it was envisioned that these oligomers should entrap saccharides in the organogel state. This was actually the case. In addition, the gelation for apolar solvents was remarkably enhanced by the addition of **18a** or **18b**. It was proposed that the entrapped saccharides in the hole of the stacking foldamers produced a hydrogen-bonded saccharide chain, which in turn promoted the stacking of the foldamers. The entrapment also caused unique dynamic helicity induction in the organogels. A systematic circular dichroism investigation revealed that the helicity induction was time dependent, taking several minutes to hours to reach equilibrium. In addition, the "sergeants and soldiers" effect was also observed in the organogel phase for all the foldamers [75].

$R = C_{10}H_{21}\text{-}n$

Ar = 49a, 49b, 49c, 49d, 49e, 49f

13.6
Vesicles

When the hydrazide-based foldamers were attached with amphiphilic peptide chains, the corresponding oligomers **50a** and **50b** formed vesicular structures in methanol [76]. The vesicles were confirmed to have a monolayer wall. The stacking of the folded frameworks was proposed to account for the formation of vesicles. The intermolecular hydrogen bonding formed by the amide units of the side chains and the van der Waals force of the peripheral *n*-octyl groups might also make contributions. Interestingly, both compounds also gelated aliphatic hydrocarbons, including *n*-hexane, cyclohexane, *n*-heptane, *n*-octane, and *n*-decane. Scanning electron microscopy images showed that their xerogels formed long fibers. All the above three weak interactions should also exist for driving the formation of the fibers. Further studies showed that the peptide side chains were very robust in inducing other aromatic systems to form vesicles in polar organic solvents [77, 78]. It was proposed that this chain balances well the amphiphilicity of aromatic compounds and promotes their ordered stacking in polar media.

50a   R = CH$_2$CONHCH$_2$CON(*n*-C$_8$H$_{17}$)$_2$

50b

Reverse vesicles are assembling architectures comprising a solvent core of low polarity that is surrounded by a reverse membrane shell. Reverse vesicles have been traditionally assembled from aliphatic amphiphiles in apolar hydrocarbons, in which the hydrophobic interaction of the polar segments is maximized [79]. Li and co-workers showed that the aggregation of triammonium macrocycles **51a** and **51b** led to the formation of reverse vesicles in chloroform [80], which is usually regarded as a "bad" solvent for aggregation of aromatic compounds. These compounds were quantitatively prepared by reducing the related triimine cyclophanes (**72**, see below). When using larger perchloride as the anion, the corresponding salts did not form vesicles. Both **51c**, each ammonium of which bears two larger ethyl groups, and **52** also did not, and X-ray diffraction diagrams of the dried samples of both **51a** and **51b** exhibited a sharp peak at 1.7 nm, which was close to the size of their cyclic framework (1.6 nm). A monolayer stacking mode was therefore proposed for the formation of the vesicles.

51a: R' = *i*-Bu, R" = Me
51b: R' = *n*-C$_8$H$_{17}$, R" = Me
51c: R' = *n*-C$_8$H$_{17}$, R" = Et

52

13.7
Supramolecular Liquid Crystals

When the backbone of tweezer **33** was introduced with long aliphatic chains, the new derivatives **53a–c** became soluble in nonpolar alkanes and gelated the solvents at high concentrations [81]. Adding C$_{60}$ derivatives **54a–c** increased the gelation capacity of the zinc and copper porphyrins due to the formation of 1:1 complexes through enhancement of the stacking of the tweezers. The three tweezers also formed a smectic-C liquid crystal phase with the glassy transition temperature at 45, 64, and 54 °C, respectively. The transition temperature was reduced considerably in the presence of 1 equiv. of the C$_{60}$ derivatives, indicating that the complexation may tune the liquid crystal behavior of the supramolecular architectures.

53a: R^1 = *n*-C$_{12}$H$_{25}$, R^2 = *n*-C$_{16}$H$_{33}$, M = 2H
53b: R^1 = *n*-C$_{12}$H$_{25}$, R^2 = *n*-C$_{16}$H$_{33}$, M = Zn
53c: R^1 = *n*-C$_{12}$H$_{25}$, R^2 = *n*-C$_{16}$H$_{33}$, M = Cu

54a: R^1 = R^2 = *n*-C$_{16}$H$_{33}$
54b: R^1 = *n*-C$_{16}$H$_{33}$, R^2 = Chol
54c: R^1 = R^2 = Chol

chol =

13.8
Macrocycles

13.8.1
Formation of Coordination Bonds

The syntheses of macrocycles are usually of low yields because the formation of linear by-products is usually entropically favorable. Conventional methods for improving the efficiency of macrocyclization reactions include using the high dilution technique and template and the step-by-step approach. Preorganization of precursors can also reduce the negative entropy effect of macrocyclization [82]. The ordered shape of foldamers or folded segments makes them ideal for designing preorganized precursors for selective formation of macrocyclic architectures. One of the early efforts concerned the construction of metallocyclophanes via the formation of the coordination bonds. For example, Li and co-workers prepared compounds **55a** (15%) and **55b** (18%) from the reactions of the related bisalkyne precursors with *trans*-Pt(PEt₃)₂Cl₂ in dichloromethane under the catalysis of CuI [69, 83]. Control experiments showed that similar macrocyclic products were not obtained from the reactions of intramolecular hydrogen bonding-free bisalkynes with *trans*-Pt(PEt₃)₂Cl₂. When hydrogen-bonded straight bisalkyne or bipyridyl precursors were used [84], triangular or square cyclophanes could be prepared. For example, compound **56** was obtained in 15% yield from the CuI-catalyzed reaction of the corresponding straight bisalkyne and (CF₂CO₂)₂Pt(Ph₂PCH₂CH₂CH₂PPh₂) in diethylamine [85]. Hence this approach may be developed as a general method for the construction of very large, rigid macrocyclic systems.

13.8.2
Formation of 1,2,3-Triazoles

N-Benzyl α-peptides prefer to adopt helical conformations with a periodicity of three residues per turn and a helical pitch of about 6.7 Å [86]. In 2007, Kirshenbaum and co-workers reported that this helical secondary structure enabled Cu(I)-catalyzed [3 + 2]-cycloaddition reactions on a solid support to ligate *N*-substituted side chain azide and alkyne functionalities to give rise to macrocycles in high yields [87]. When the reactive groups were introduced at respective sequence positions i and $i + 3$, intramolecular cyclization occurred most favorably (Figure 13.6). For example, the reaction of the corresponding linear precursor afforded monomeric macrocycle **57a** in 80% yield, whereas the dimeric macrocycle **57b** was obtained in only 10% yield. When the reactive groups were attached at respective sequence positions i and $i + 2$ or $i + 4$, the ratio of the two macrocycles was notably decreased. For the i and $i + 5$ oligomeric precursor, the dimeric macrocycle was produced as the main product. The result also supports the periodicity of three residues per turn of the helical backbones.

56

R = *n*-C₆H₁₃

55a: n = 0
55b: n = 1

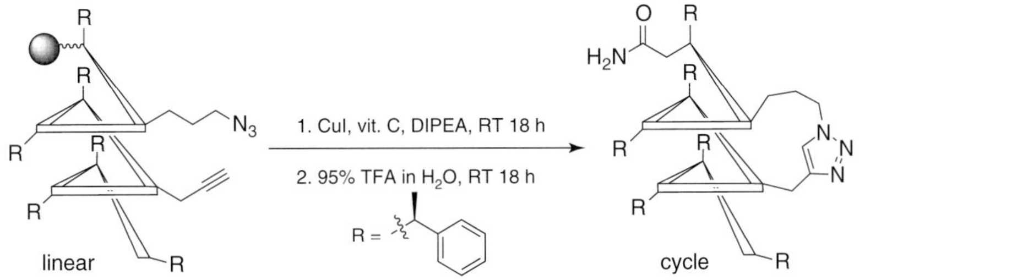

Figure 13.6 Schematic representation of helical *N*-substituted α-peptides, showing the triazole linkage formation between reactive side chains *i* and *i* + 3 which is facilitated by the helical conformation.

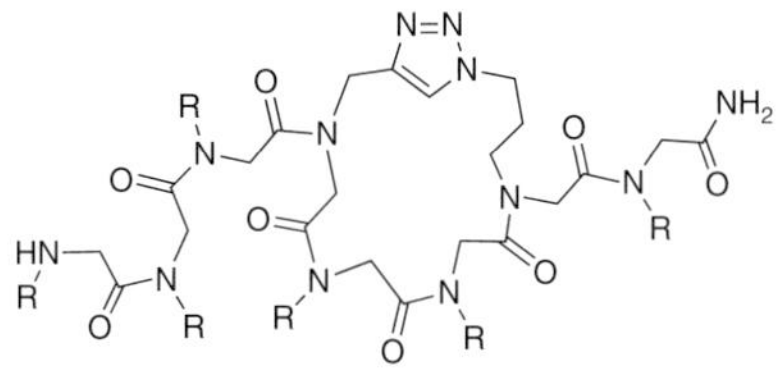

57a

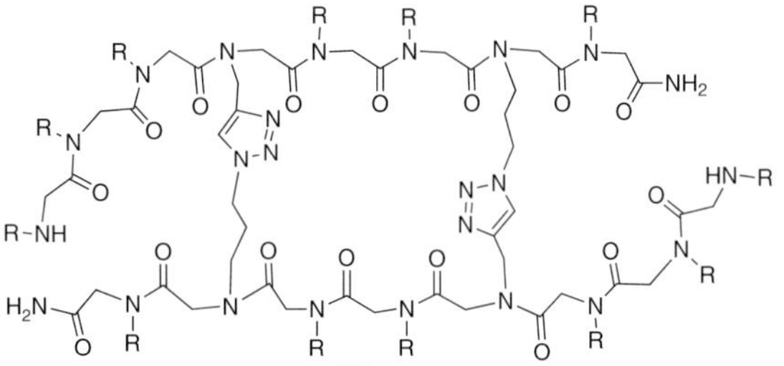

57b

In 2009, Li and co-workers reported that the principle of preorganization also worked well with the formation of macrocycles by the [2 + 3]-cycloaddition of the corresponding hydrogen-bonded bisazides and bisalkynes [88]. For example, compounds **58a** and **58b** were obtained in 82 and 85% yields, respectively, when the reactions were carried out in chloroform in the presence of CuI and diisopropylethylamine (DIPEA). In contrast, macrocycle **58c** of smaller size was prepared in only 20% yield under the same reaction conditions, reflecting that its short but straight biazide precursor was not as efficient as the longer but "U"-shaped precursors in forming the macrocyclic structure.

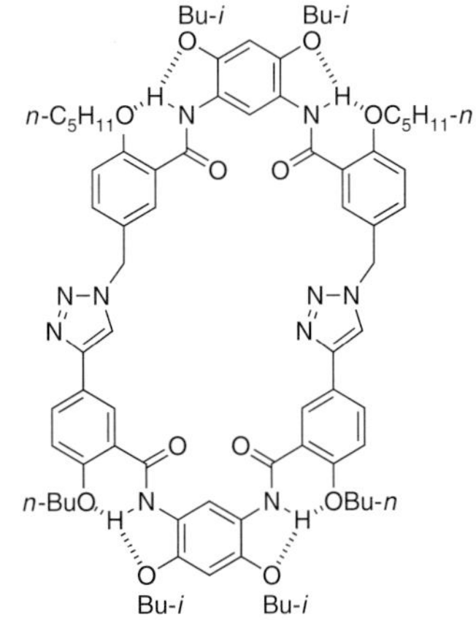

58a
(82%)

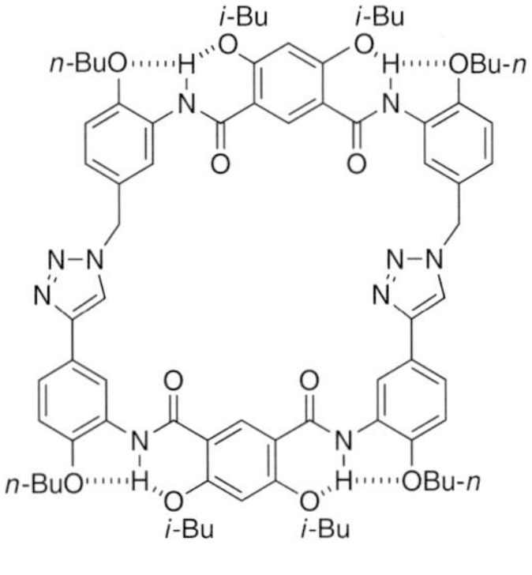

58b
(85%)

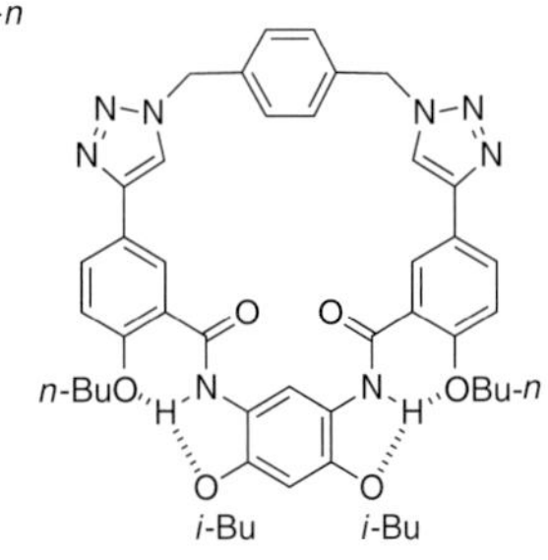

58c
(20%)

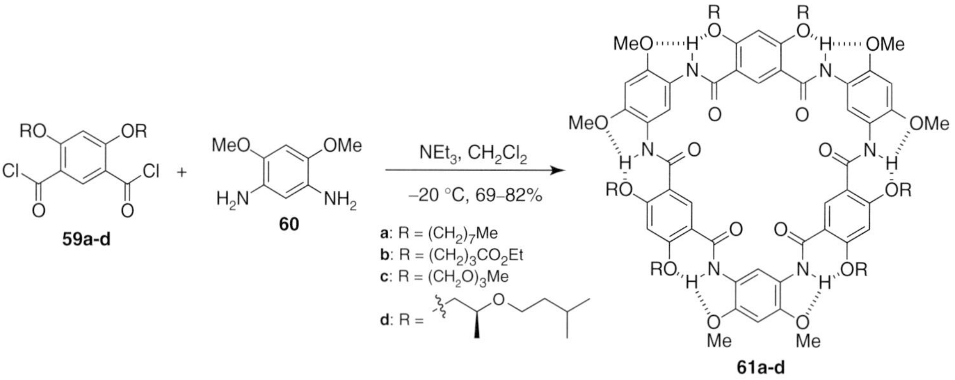

Scheme 13.6 One-step synthesis of macrocycles **61a–d**.

13.8.3
Formation of Amides

Hydrogen-bonded aromatic amide oligomers have readily predictable conformations. Thus, continued formation of amide bonds from simple acyl chloride and aniline precursors would rapidly elongate the backbones in the required direction such that the final cyclization step could occur very efficiently. For example, in 2004, Gong and co-workers reported that macrocycles **61a–d** could be obtained in 69–82% yields from the reactions of diacyl chlorides **59a–d** and diamine **60** in dichloromethane without using the high-dilution technique (Scheme 13.6) [89]. Without the intramolecular hydrogen bonding, even under high-dilution conditions and using calcium chloride as a template, the corresponding 6-mer macrocycle was generated only in 4–11% yield from the reaction of isophthaloyl chloride and *m*-phenylenediamine [90]. Hence the intramolecular hydrogen bonding of the oligomeric intermediates plays a key role in reaching high cyclization yields. Like folded oligomer **21**, these macrocycles are good receptors for guanidinium ion [91].

Replacing *meta*-positioned dichloride **59a–d** with 2,3-dimethoxyterephthaloyl dichloride caused the related coupling reaction to give rise to 16-mer macrocycles **62a–c** in as high as 81% yield [92]. Matrix-assisted laser desorption/ionization time-of-flight (MALDI-TOF) MS revealed that the reactions also afforded 14-, 18-, and 20-mer macrocycles. However, macrocycles **62a–c** were still formed in very high yields. Gong and co-workers also utilized hydrogen-bonded hydrazide segments to build macrocycles. For example, macrocycle **63** could be prepared in 73% yield from a similar one-pot reaction [93]. MALDI-TOF MS of the crude products revealed a dominant signal that corresponded to the [M + Na]$^+$ ion of the 6-mer macrocycle. Based on a similar approach, Li and co-workers [94] and Yuan and co-workers [95] prepared a number of other rigid macrocycles of varying size by using discrete precursors. By using the step-by-step strategy, Zeng and co-workers found that, for the series of foldamers **23**, the formation of 5-mer macrocycles was most favored [96, 97]. For example, pentagon macrocycle **65** was prepared in 60% yield from pentamer **64** after three steps

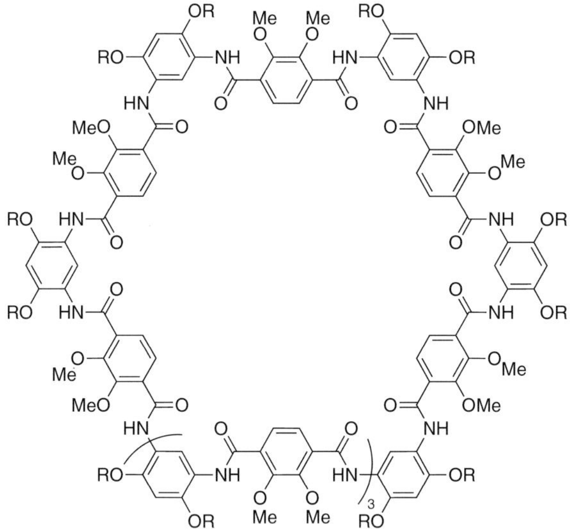

62a: R = *n*-C$_8$H$_{17}$
62b: R = *n*-C$_{12}$H$_{25}$

62c: R =

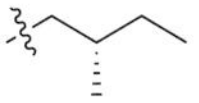

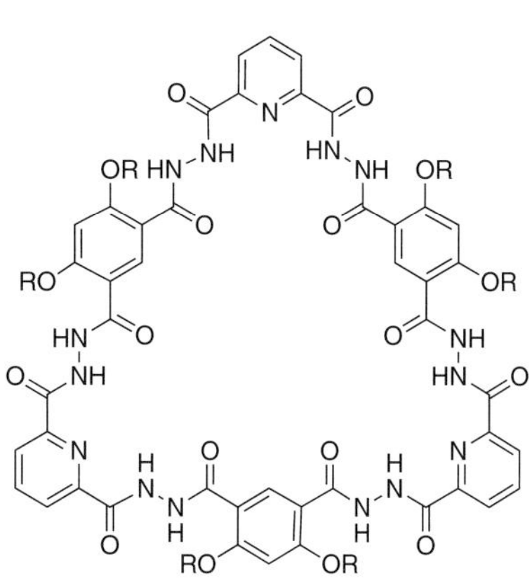

R = CH$_2$CH$_2$CHMe$_2$

63

of reactions (Scheme 13.7). One to three of the methyl groups could be removed with BBr$_3$ to yield phenol derivatives [98]. The new phenol-derived macrocycles were found to be good receptors for alkali metal ions [98], which are potentially useful as ion channels.

Huc and co-workers reported that 8-aminoquinoline-2-carboxylic acid-based oligomers formed stable helical secondary structures, which were also stabilized by continued N···H–N(quin) intramolecular hydrogen bonds [99]. In 2004, the same group reported that this series of folded backbones could also be utilized to promote the formation of macrocycles from the quinoline amino acid [100]. Because the amino group has a low nucleophilicity to couple with acid to form amide, the conventional coupling conditions for amide formation did not work. Therefore, precursor **66** was treated with LiCl and PPh$_3$

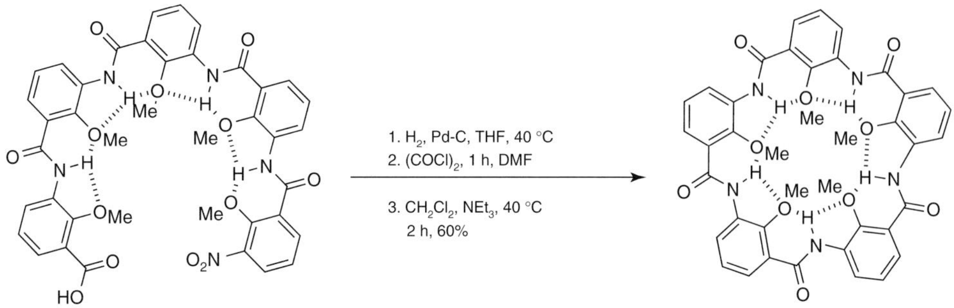

1. H$_2$, Pd-C, THF, 40 °C
2. (COCl)$_2$, 1 h, DMF

3. CH$_2$Cl$_2$, NEt$_3$, 40 °C
2 h, 60%

64

65

Scheme 13.7 Synthesis of pentagon macrocycle **65**.

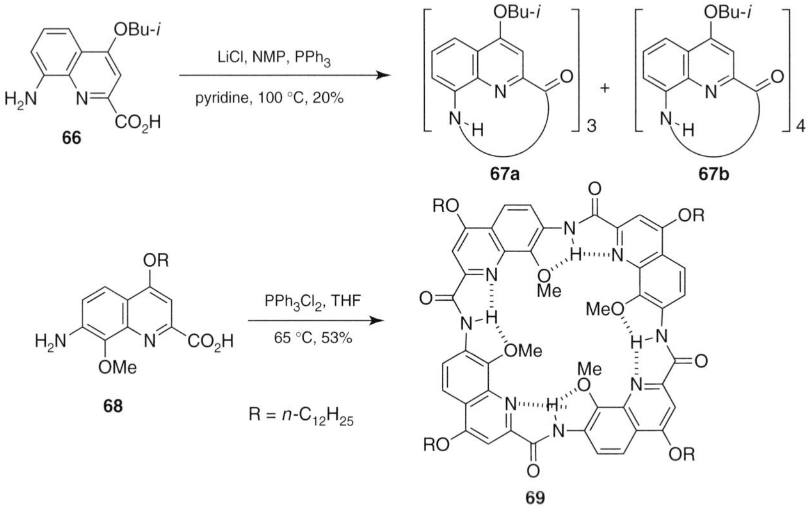

Scheme 13.8 Synthesis of macrocycles **67a**, **67b**, and **69** from quinoline amino acids **66** and **68**, respectively.

in highly polar and hot *N*-methylpyrrolidone–pyridine mixture. Even under these harsh reaction conditions, cyclic 3-mer **67a** and 4-mer **67b** were obtained in 20% total yield (Scheme 13.8). In 2009, Jiang and co-workers designed 7-aminoquinoline-2-carboxylic acid **68**. They found that treatment of this amino acid with PPh_3Cl_2 in refluxing tetrahydrofuran afforded 4-mer macrocycle **69** in 53% yield (Scheme 13.8) [101]. Again, intramolecular hydrogen bonding facilitated the final cyclization.

13.8.4
Formation of Reversible Imine and Hydrazone Bonds

The formation of the amide or 1,2,3-triazole units is a dynamically controlled process. Thus, preorganization of foldamer-derived precursors or intermediates may substantially increase the yields of macrocycles. However, this strategy does not lead to quantitative cyclization. Li and co-workers applied the concept of dynamic covalent chemistry (DCC) to the synthesis of macrocycles from aromatic amide foldamer-derived substrates. The new approach was proven to be very robust for the construction of macrocycles through forming both imine and hydrazone bonds. A number of macrocycles were constructed nearly quantitatively after a sufficient reaction time to allow the reactions to reach equilibrium [78, 102–104]. The reactions were usually performed in chloroform in the presence of an excess of trifluoroacetic acid and the products could be purified by simple crystallization. Compounds **70–74** are representative structures that were prepared by using this approach, which involved the formation of two, six, three, seven, and six imine or hydrazone bonds, respectively. Probably owing to the strong stacking of the rigid backbones and the isomerization and hydrogen bonding exchange of the centrally located triiminobenzenetriol unit, the ^{1}H NMR spectra of **73** in all investigated solvents, including CDCl$_3$ and DMSO-d_6, were of low resolution [104]. However, all of the other four macrocycles exhibited signals of high resolution in their ^{1}H NMR spectra in CDCl$_3$. Hence the reactions could be tracked by using ^{1}H NMR spectroscopy in chloroform-d in the presence of an excess amount of trifluoroacetic acid. It was observed that the reaction mixtures exhibited complicated signals at an early stage, but one set of signals was evolved finally after a sufficiently long time, which depended on the respective reactions [102, 103]. This phenomenon of selective enrichment of required structures is typical for DCC processes, reflecting the conversion of less stable products to the most stable product as a result of the reversibility of the bonds formed. The precursor for two-layered compound **74** consists of two hydrogen-bonded segments which are precursors for **72**. The linker could be modified. The six carbonyl groups of this capsular compound are located inwards and are therefore able to complex diammonium derivatives, which is driven by the intermolecular N–H···O=C hydrogen bonding. New unique pseudo[3]rotaxanes were generated in this manner [103].

13.9
Catalysis

Since natural enzymes are sequence-specific polymers with secondary and higher-grade structures, creating catalysts from bio-inspired folded structures is naturally a topic of interest in the field of foldamer chemistry. The groups of Moore, Huc, and Li reported that several folded *m*-phenyleneethynyne or aromatic amide oligomers exhibited the functions of accelerating or enabling some specific reactions [18, 19, 25]. However, these reactions occurred with the folded structures themselves as substrates and may only be considered as special catalytic processes. In 2009, three groups reported that distinct foldamers could catalyze three kinds of different reactions.

13.9.1
β-Peptides

Gellman and co-workers reported that helical β-peptides presenting arrays of discrete side-chain functional groups could catalyze the retroaldol reaction of sodium 4-phenyl-4-hydroxy-2-oxobutyrate (75) (Scheme 13.9) [105]. β-Peptides 76a–c were designed, which had been found to prefer to form a 14-helix characterized by 14-membered ring hydrogen bonds between the N-H unit of residue i and the C=O unit of residue $i + 2$ [106]. Lactate dehydrogenase (LDH) was used to catalyze the reduction of pyruvate, the retroaldol product, by reduced nicotinamide adenine dinucleotide (NADH). β-Peptide 76a promoted the retroaldol cleavage of 75 with multiple turnovers and significant rate accelerations over the uncatalyzed reaction, whereas 76b and 76c exhibited 20–50-fold lower activity. The cyclically constrained *trans*-2-aminocyclohexanecarboxylic acid (ACHC) has a very high propensity to adopt a helical conformation [107]. Therefore, the authors proposed that 76a and 76b should adopt a more stable 14-helical conformation than 76c. However, 76a has a properly arrayed set of β³-hLys side chains (Scheme 13.10). It was considered that this clustering of β³-hLys residues caused a larger decrease in side-chain ammonium pK_a values than 76b and thus more efficiently facilitated the retroaldol reaction of 75. The retroaldolase reaction catalyzed by 76a also followed Michaelis–Menten

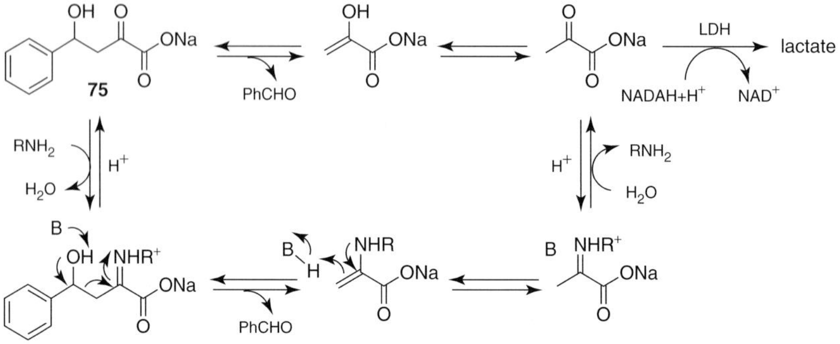

Scheme 13.9 Retroaldol reaction of sodium 4-phenyl-4-hydroxy-2-oxobutyrate (75) and the mechanism of amine catalysis, based on which foldamer catalysts 76a–c were designed.

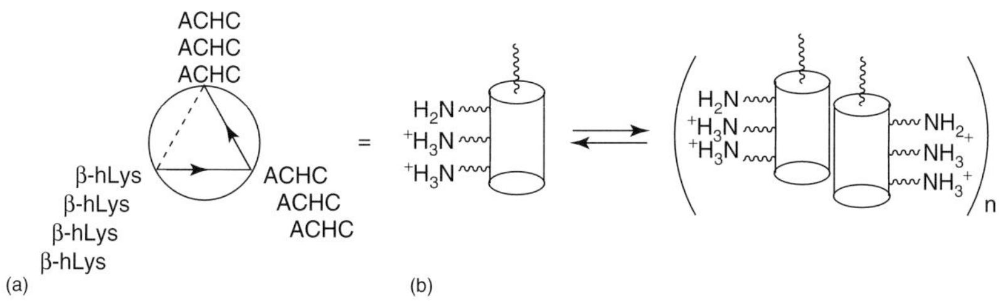

Scheme 13.10 Helical wheel diagram (a) of β-peptide **76a** and a cartoon (b) showing the self-assembly of β-peptide that is partially protonated.

behavior. Comparison of the turnover number of this foldamer with the rate constant for the uncatalyzed retroaldol reaction under identical conditions gave a rate acceleration of $k_{cat}/k_{uncat} = 3000$. Therefore, **76a** was estimated to be roughly twice as efficient as evolutionarily optimized α-peptide aldolases and only eight times less efficient than the best computationally designed 200-residue aldolases. This small difference in rate enhancement between a 10-residue β-peptide and a 200-residue protein underscores the catalytic potential in folded frameworks.

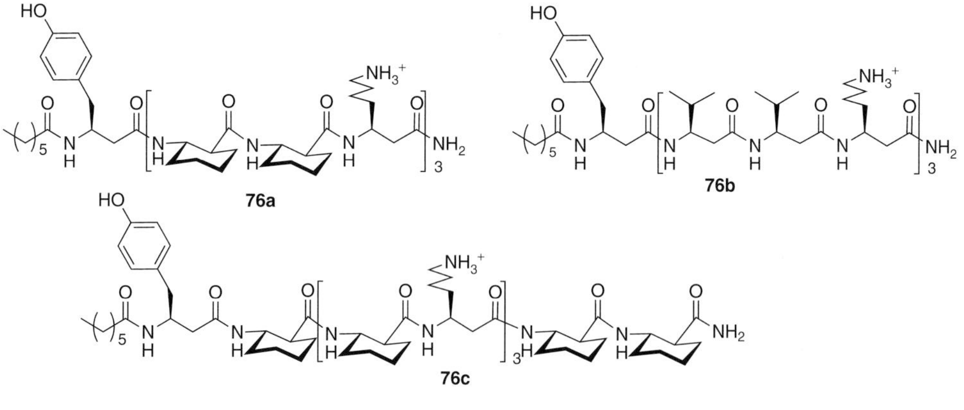

13.9.2
Chiral N-Substituted Glycine Peptoids

Kirshenbaum and co-workers reported that N-(S)- or (R)-phenylethylglycine peptoids that were attached with 2,2,6,6-tetramethylpiperidin-1-oxyl (TEMPO), a well-known catalyst for oxidative transformations, catalyzed the oxidative kinetic resolution (OKR) of 1-phenylethanol **77** (Scheme 13.11) [108]. The chiral catalysts (S)- and (R)-**78** were synthesized on a solid support [109]. The two peptoids exhibited circular dichroism ellipticities (θ) of equal magnitude with opposite sign, indicating the formation of right- and left-handed helices, respectively, with an equivalent degree of helical character. The OKR reactions were performed in a biphasic mixture consisting of the catalytic

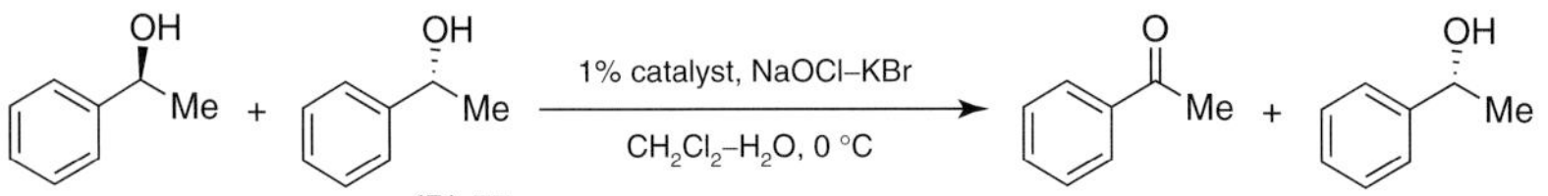

Scheme 13.11 Oxidative kinetic resolution of 1-phenylethanol (**77**) catalyzed by TEMPO-attached *N*-(*S*)- or (*R*)-phenylethylglycine peptoids.

peptoid (1 mol%) dissolved in dichloromethane, an aqueous solution of KBr (0.5 M), and 1-phenylethanol (typically 0.1 mmol). The oxidation reaction commenced upon addition of an aqueous solution containing 0.5 M sodium hypochlorite. A mixture of **79** and **80** was active for oxidation of **77**, but produced no enantioselective transformation. Helical peptoid **81** exhibited greater activity, but still did not produce enantioselective transformation. However, (*S*)-**78** displayed comparable activity but with enantioselectivity. The overall conversion after 2 h was 84% with 60% enantioselectivity for (*S*)-**77**, resulting in 99% enantiomeric excess (*ee*) of the less-reactive (*R*)-**77**. At lower conversions, the apparent selectivity was higher and the *ee* was smaller, as expected for a transformation of less than 100% enantioselectivity. Peptoid (*R*)-**78** exhibited identical catalytic activity and enantioselectivity, but for the opposite enantiomer (*R*)-**77**, producing an *ee* of the less reactive (*S*)-**77**. Although the origins of enantioselectivity in the OKR of secondary benzylic alcohols by the foldamer are not well understood, the fact that chiral information can be transferred from a folded scaffold to an embedded achiral reactive center may be applicable to synthetic foldamers in general.

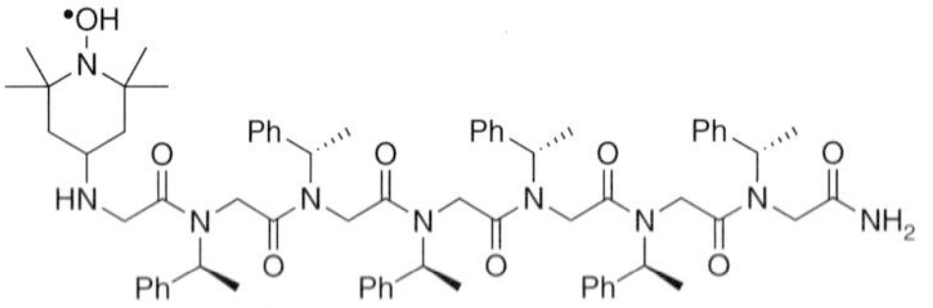

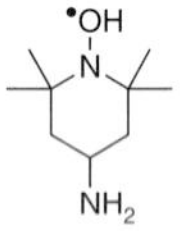

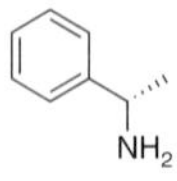

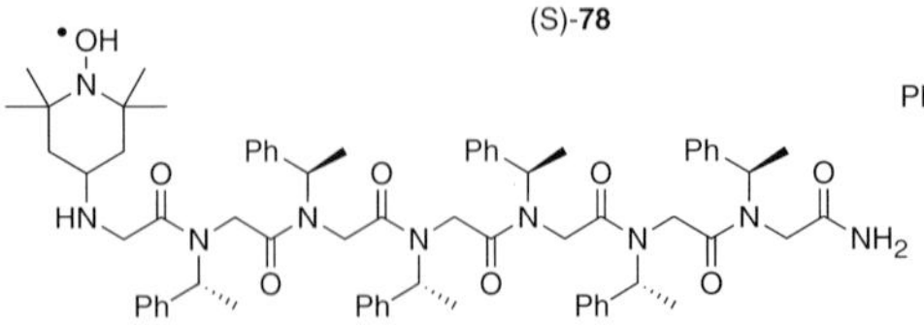

13.9.3
Cholate Oligomers

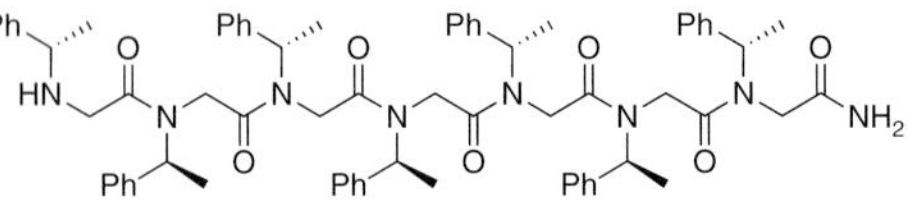

Because of the amphiphilicity of the cholate unit, long cholate foldamers developed by Zhao *et al.* form a hydrophilic cavity 1 nm in diameter in nonpolar solvents and low-percentage polar solvent molecules of are thus enriched in the cavity [110]. In 2009, the same group reported that foldamer **82a** catalyzed the acetylation of alcohols [111]

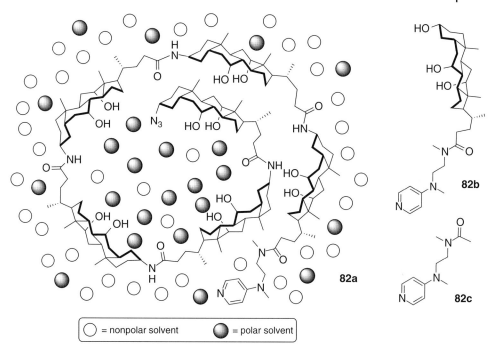

Figure 13.7 Schematic representation of the preferential solvation of folded oligocholate **82** and the structures of control compounds **82b** and **82c**. Polar solvent molecules are enriched in the cavity.

(Figure 13.7). The foldamer bears a 4-dialkylaminopyridine moiety at the chain end, which is an analog of DMAP, a powerful catalyst for acyl transfer. The catalytic activity of **82a** was evaluated using ^{1}H NMR spectroscopy by comparison with **82b** and **82c** for the acetylation of CD_3OD (0.5%, 0.12 M) in a mixture of DMSO and CCl_4 in the presence of DIPEA. It was established that in the presence of a small amount of DMSO, **82a** mainly folded into a helical conformation, whereas a large amount of DMSO caused it to unfold.

A systematic kinetic investigation (Table 13.1) showed that when the amount of DMSO was low, the pseudo-first-order rate constant of **82a** was notably higher than those of both **82b** and **82c**. The values became very close when the amount of DMSO was increased to >15%. These kinetic data are in line with the folding/unfolding of **82a**. It was proposed that, in the folded state, **82a** was able to concentrate methanol into its hydrophilic cavity. Because methanol was revealed to be completely unreactive towards acetic anhydride within the time frame without the DMAP catalyst, acetylation of methanol could only occur after acetic anhydride had reacted with the pyridyl group. Hence the higher activity of the folded oligomer **82a** in comparison with **82b** or **82c** supported a higher effective concentration of methanol in the cavity and near the catalytic pyridine group of **82a**.

Table 13.1 Rate constants for the reaction between methanol and acetic anhydride at 20°C in DMSO–CCl$_4$ mixtures catalyzed by compounds 82a–c.[a]

Entry	DMSO (%)[b]	Catalyst	k (min^{-1})[c]	k_1/k_3[d] (k_2/k_3)[d]
1	1.5	82a	1.0×10^{-2}	2.6 (1.4)
		82b	5.6×10^{-3}	
		82c	3.9×10^{-3}	
2	3.5	82a	8.3×10^{-3}	2.6 (1.4)
		82b	4.6×10^{-3}	
		82c	3.2×10^{-3}	
3	5.5	82a	6.4×10^{-3}	1.7 (1.2)
		82b	4.4×10^{-3}	
		82c	3.8×10^{-3}	
4	7.5	82a	5.4×10^{-3}	1.6 (1.1)
		82b	3.4×10^{-3}	
		82c	3.3×10^{-3}	
5	11.5	82a	2.4×10^{-3}	1.2 (1.1)
		82b	2.2×10^{-3}	
		82c	2.1×10^{-3}	
6	15.5	82a	2.1×10^{-3}	1.1 (1.0)
		82b	2.0×10^{-3}	
		82c	1.9×10^{-3}	
7	19.5	82a	1.7×10^{-3}	1.0 (0.9)
		82b	1.5×10^{-3}	
		82c	1.7×10^{-3}	

[a][Ac$_2$O] = 3.0 mM, [DIPEA] = 20.0 mM, [82a−c] = 20 μM.
[b]Volume percentage of DMSO-d_6 added to control the folding/unfolding of the oligocholate foldamer.
[c]The error for the rate constants determined by ^{1}H NMR spectroscopy was estimated as 10–20%.
[d]k_1, k_2, and k_3 are the rate constants for the reaction catalyzed by 82a, 82b, and 82c, respectively.

13.10
Macromolecular Self-Assembly

It has been established that strong intramolecular stacking exists for hydrogen bonding-driven aromatic amide foldamers, which also makes it difficult to prepare long helical aromatic polymers [112, 113]. As a conceptual extension of the zigzag secondary structures of aromatic amide oligomers [69], Li and co-workers reported the synthesis of polymers P83a−e, whose hydrazide main chains have a rigid extended conformation [114]. Although P83e, which bears amphiphilic triglycol chains, did not form any ordered assembled architectures, P83a−d could form organogels or vesicles, which depended on the side chains and the polarity of the solvents. Control experiments showed that small molecules of the same hydrazide segments with the same side chains did not gelate any solvents or form vesicular structures. Therefore, the enhanced stacking of the long hydrazide main chains of the polymer in both polar and nonpolar solvents was considered to be the main driving force for the formation of both ordered structures, which was further promoted by the side chains through intermolecular hydrogen bonding, van der Waals, and/or solvophobic interactions.

In 2010, Li and co-workers reported that intramolecular hydrogen-bonded aromatic amide foldamers could serve as reversible crosslinks to modulate the thermal and mechanical properties of crosslinked n-butyl methacrylate copolymers [115]. Diamines **84a–c** were thus prepared. Treatment of **84a–c** with copolymer **P85** afforded crosslinked copolymers **P86a–c** through the formation of an imine bond. Nine films of about 0.1 mm in thickness were prepared from these copolymers that contained 0.025, 0.05, and 0.1 mol% of crosslinkers.

The dynamic mechanical analysis together with creep/recovery experiments for the films revealed that, compared with the control polymers that contained similar, but intramolecular hydrogen bonding-free, crosslinkers of the same length, the folded crosslinks in **P86b** and **P86c** substantially improved the mechanical properties of the

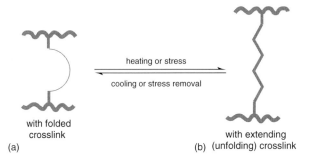

Figure 13.8 Schematic representation of the hydrogen-bonded foldamer crosslinks in *n*-butyl methacrylate copolymers **P86a–c**. Heating or applying a stress causes the folded-state linker (a) to unfold or extend (b), while cooling or removing the stress results in the recovery of the folded state.

copolymers. This improvement was attributed to the ability of the folded cross-links to reveal reversibly their hidden length on extension via dissipative cleavage of the intramolecular hydrogen bonding, as shown in Figure 13.8. Considering the convenient synthesis and modifications of the hydrogen-bonded folded oligomers, the future may witness more applications in modulating the dynamic properties of supramolecular polymers.

13.11
Conclusion and Perspectives

Since the report by Pedersen in 1967 that crown ethers complex alkali metal ions [2], studies on the binding and host–guest behaviors of cyclic structures have been among the central topics of research in supramolecular chemistry. Considering that life systems have been the main sources for chemists to design new molecules for studying molecular recognition and self-assembly, it is reasonable to expect that supramolecular chemistry of foldamers or modified folded structures will attract increasing attention in the coming years, because most biomacromolecules, particularly DNA and proteins, are linear sequences of ordered three-dimensional structures. The achievements described in this chapter are impressive, but more exciting advances should be expected in the future, which, of course, will largely depend on new emerging folding patterns.

We believe that future advances in this area will be multi-directional. In principle, all the recognition and self-assembly properties exhibited by cyclic structures can be observed for folded systems, hence all the functions of linear biomacromolecules can be mimicked by synthetic sequences. Therefore, the potential for applications of ordered linear molecules in the context of supramolecular chemistry should be huge in the future. Here several issues are listed that are considered worthy to receive more attention or to be addressed in the future. Catalytic foldamers will certainly attract more attention. Although a few foldamer catalysts have recently been reported, the types of reactions catalyzed by synthetic sequences are still very limited. Aromatic foldamers of catalytic functions are not yet available. To realize this, a deeper insight into the sequence–structure relationships will be needed. The DCC principle provides an efficient approach for rapidly generating very large macrocycles and capsular structures. Rational design of the preorganized precursors may lead to new functional two- or three-dimensional macrocyclic architectures for molecular encapsulation and catalysis. The development of methods for preparing long folded polymers would also be highly valuable, because such polymers may produce single-stranded tubular structures for encapsulation and transport. To mimic the functions of biomacromolecules more efficiently and extensively, more complicated folding patterns that challenge the complexity of biomacromolecules need to be developed. To do this, a combination of different folding patterns into one sequence may find applications.

Acknowledgments

Financial support was provided by the National Natural Science Foundation and Ministry of Science and Technology of China, Science and Technology Commission of Shanghai Municipality, and Chinese Academy of Sciences. The studies by the author's group were all carried out at the Shanghai Institute of Organic Chemistry.

References

1. Lehn, J.-M. (1973) *Struct. Bond.*, **16**, 1.
2. Pedersen, C.J. (1967) *J. Am. Chem. Soc.*, **89**, 2495.
3. Whitesides, G.M. and Grzybowski, B. (2002) *Science*, **295**, 2418.
4. Grisham, C.M. and Garrett, R.H. (1999) *Biochemistry*, Saunders, Philadelphia, PA, pp. 426–427.
5. Seebach, D. and Matthews, J.L. (1997) *Chem. Commun.*, 2015.
6. Gellman, S.H. (1998) *Acc. Chem. Res.*, **31**, 17.
7. Hill, D.J., Mio, M.J., Prince, R.B., Hughes, T.S., and Moore, J.S. (2001) *Chem. Rev.*, **101**, 3893.
8. Hecht, S. and Huc, I. (eds.) (2007) *Foldamers: Structure, Properties and Applications*, Wiley-VCH Verlag GmbH, Weinheim.
9. Wu, Y.-D. and Gellman, S. (2008) *Acc. Chem. Res.*, **41**, 1231.
10. Hanan, G.S., Lehn, J.-M., Kyritsakas, N., and Fischer, J. (1995) *J. Chem. Soc., Chem. Commun.*, 765.
11. Hamuro, Y., Geib, S.J., and Hamilton, A.D. (1996) *J. Am. Chem. Soc.*, **118**, 7529.
12. Nelson, J.C., Saven, J.G., Moore, J.S., and Wolynes, P.G. (1997) *Science*, **277**, 1793.
13. Juwarker, H., Suk, J.-m., and Jeong, K.-S. (2009) *Chem. Soc. Rev.*, **38**, 3316.
14. Zhao, X. and Li, Z.-T. (2010) *Chem. Commun.*, **46**, 1601.
15. Li, Z.-T., Hou, J.-L., and Li, C. (2008) *Acc. Chem. Res.*, **41**, 1343.
16. Gong, B. (2008) *Acc. Chem. Res.*, **41**, 1376.
17. Stone, M.T., Heemstra, J.M., and Moore, J.S. (2006) *Acc. Chem. Res.*, **39**, 11.
18. Dolain, C., Zhan, C., Léger, J.-M., Daniels, L., and Huc, I. (2005) *J. Am. Chem. Soc.*, **127**, 2400.
19. Yi, H.-P., Wu, J., Ding, K.-L., Jiang, X.-K., and Li, Z.-T. (2007) *J. Org. Chem.*, **72**, 870.
20. Horne, W.S. and Gellman, S.H. (2008) *Acc. Chem. Res.*, **41**, 1399.
21. Claudon, P., Violette, A., Lamour, K., Decossas, M., Fournel, S., Heurtault, B., Godet, J., Mely, Y., Jamart-Gregoire, B., Averlant-Petit, M.-C., Briand, J.-P., Duportail, G., Monteil, H., and Guichard, G. (2010) *Angew. Chem.*, **122**, 343; (2010) *Angew. Chem. Int. Ed.*, **49**, 333.
22. Prince, R.B., Barnes, S.A., and Moore, J.S. (2000) *J. Am. Chem. Soc.*, **122**, 275.
23. Heemstra, J.M. and Moore, J.S. (2004) *J. Am. Chem. Soc.*, **126**, 1648.

24. Smaldone, R.A. and Moore, J.S. (2007) *J. Am. Chem. Soc.*, **129**, 5444.

25. Smaldone, R.A. and Moore, J.S. (2008) *Chem. Eur. J.*, **14**, 2650.

26. Inouye, M., Waki, M., and Abe, H. (2004) *J. Am. Chem. Soc.*, **126**, 2022.

27. Abe, H., Masuda, N., Waki, M., and Inouye, M. (2009) *Chem. Commun.*, 2121.

28. Chang, K.-J., Kang, B.-N., Lee, M.-H., and Jeong, K.-S. (2005) *J. Am. Chem. Soc.*, **127**, 12214.

29. Suk, J.-M. and Jeong, K.-S. (2008) *J. Am. Chem. Soc.*, **130**, 11868.

30. Vögtle, F. and Weber, E. (1979) *Angew. Chem.*, **91**, 813; *Angew. Chem. Int. Ed. Engl.*, **18**, 753.

31. Hou, J.-L., Jia, M.-X., Jiang, X.-K., Li, Z.-T., and Chen, G.-J. (2004) *J. Org. Chem.*, **69**, 6228.

32. Petitjean, A., Cuccia, L.A., Lehn, J.-M., Nierengarten, H., and Schmutz, M. (2002) *Angew. Chem.*, **114**, 1243; *Angew. Chem. Int. Ed.*, **41**, 1195.

33. Petitjean, A., Nierengarten, H., van Dorsselaer, A., and Lehn, J.-M. (2004) *Angew. Chem.*, **116**, 3781; *Angew. Chem. Int. Ed.*, **43**, 3695.

34. Petitjean, A., Cuccia, L.A., Schmutz, M., and Lehn, J.-M. (2008) *J. Org. Chem.*, **73**, 2481.

35. Juwarker, H., Lenhardt, J.M., Pham, D.M., and Craig, S.L. (2008) *Angew. Chem.*, **120**, 3800; *Angew. Chem. Int. Ed.*, **47**, 3740.

36. Li, Y. and Flood, A.H. (2008) *J. Am. Chem. Soc.*, **130**, 12111.

37. Jiang, X.-K. (1988) *Acc. Chem. Res.*, **21**, 362.

38. Zhao, Y. (2007) *Curr. Opin. Colloid Interface Sci.*, **12**, 92.

39. Zhao, Y. and Zhong, Z. (2006) *J. Am. Chem. Soc.*, **128**, 9988.

40. Zhao, Y. and Zhong, Z. (2006) *Org. Lett.*, **8**, 4715.

41. Zhao, X., Wang, X.-Z., Jiang, X.-K., Chen, Y.-Q., Li, Z.-T., and Chen, G.-J. (2003) *J. Am. Chem. Soc.*, **125**, 15128.

42. Hou, J.-L., Shao, X.-B., Chen, G.-J., Zhou, Y.-X., Jiang, X.-K., and Li, Z.-T. (2004) *J. Am. Chem. Soc.*, **126**, 12386.

43. Du, P., Xu, Y.-X., Jiang, X.-K., and Li, Z.-T. (2009) *Sci. China B Chem.*, **52**, 489.

44. Li, C., Wang, G.-T., Yi, H.-P., Jiang, X.-K., Li, Z.-T., and Wang, R.-X. (2007) *Org. Lett.*, **9**, 1797.

45. Zhu, J., Parra, R.D., Zeng, H., Skrzypczak-Jankun, E., Zeng, X.-C., and Gong, B. (2000) *J. Am. Chem. Soc.*, **122**, 4219.

46. Yamato, K., Yuan, L., Feng, W., Helsel, A.J., Sanford, A.R., Zhu, J., Deng, J., Zeng, X.C., and Gong, B. (2009) *Org. Biomol. Chem.*, **7**, 3643.

47. Yi, H.-P., Shao, X., Hou, J.-L., Li, C., Jiang, X.-K., and Li, Z.-T. (2005) *New J. Chem.*, **29**, 1213.

48. Yi, H.-P., Li, C., Hou, J.-L., Jiang, X.-K., and Li, Z.-T. (2005) *Tetrahedron*, **61**, 7974.

49. Li, C., Ren, S.-F., Hou, J.-L., Yi, H.-P., Zhu, S.-Z., Jiang, X.-K., and Li, Z.-T. (2005) *Angew. Chem.*, **117**, 5871; *Angew. Chem. Int. Ed.*, **44**, 5725.

50. Zhu, Y.-Y., Wu, J., Li, C., Zhu, J., Hou, J.-L., Li, C.-Z., Jiang, X.-K., and Li, Z.-T. (2007) *Cryst. Growth Des.*, **7**, 1490.

51. Berl, V., Huc, I., Khoury, R.G., Krische, M.J., and Lehn, J.-M. (2000) *Nature*, **407**, 720.

52. Berni, E., Kauffmann, B., Bao, C., Lefeuvre, J., Bassani, D.M., and Huc, I. (2007) *Chem. Eur. J.*, **13**, 8463.

53. Gan, Q., Bao, C., Kauffmann, B., Grelard, A., Xiang, J., Liu, S., Huc, I., and Jiang, H. (2008) *Angew. Chem.*, **120**, 1739; *Angew. Chem. Int. Ed.*, **47**, 1715.

54. Huc, I., Maurizot, V., Gornitzka, H., and Léger, J.-M. (2002) *Chem. Commun.*, 578.

55. Garric, J., Léger, J.-M., and Huc, I. (2005) *Angew. Chem.*, **117**, 1990; *Angew. Chem. Int. Ed.*, **44**, 1954.

56. Garric, J., Léger, J.-M., and Huc, I. (2007) *Chem. Eur. J.*, **13**, 8454.

57. Bao, C., Kauffmann, B., Gan, Q., Srinivas, K., Jiang, H., and Huc, I. (2008) *Angew. Chem.*, **120**, 4221; *Angew. Chem. Int. Ed.*, **47**, 4153.

58. Ferrand, Y., Kendhale, A.M., Kauffmann, B., Grelard, A., Marie, C., Blot, V., Pipelier, M., Dubreuil, D., and Huc, I. (2010) *J. Am. Chem. Soc.*, **132**, 7858.

59. Berl, V., Krische, M.J., Huc, I., Lehn, J.-M., and Schmutz, M. (2000) *Chem. Eur. J.*, **6**, 1938.

60. Xu, Y.-X., Zhao, X., Jiang, X.-K., and Li, Z.-T. (2009) *J. Org. Chem.*, **74**, 7267.

61. Wu, Z.-Q., Shao, X.-B., Li, C., Hou, J.-L., Wang, K., Jiang, X.-K., and Li, Z.-T. (2005) *J. Am. Chem. Soc.*, **127**, 17460.

62. Liu, H., Wu, J., Jiang, X.-K., and Li, Z.-T. (2007) *Tetrahedron Lett.*, **48**, 7327.

63. Feng, D.-J., Wang, G.-T., Wu, J., Wang, R.-X., and Li, Z.-T. (2007) *Tetrahedron Lett.*, **48**, 6181.

64. Wu, Z.-Q., Li, C.-Z., Feng, D.-J., Jiang, X.-K., and Li, Z.-T. (2006) *Tetrahedron*, **62**, 11054.

65. Li, C.-Z., Zhu, J., Wu, Z.-Q., Hou, J.-L., Li, C., Shao, X.-B., Jiang, X.-K., Li, Z.-T., Gao, X., and Wang, Q.-R. (2006) *Tetrahedron*, **62**, 6973.

66. Li, C.-Z., Li, Z.-T., Gao, X., and Wang, Q.-R. (2007) *Chin. J. Chem.*, **25**, 1417.

67. Wu, J., Hou, J.-L., Li, C., Wu, Z.-Q., Jiang, X.-K., Li, Z.-T., and Yu, Y.-H. (2007) *J. Org. Chem.*, **72**, 2897.

68. Hou, J.-L., Yi, H.-P., Shao, X.-B., Li, C., Wu, Z.-Q., Jiang, X.-K., Wu, L.-Z., Tung, C.-H., and Li, Z.-T. (2006) *Angew. Chem.*, **118**, 810; *Angew. Chem. Int. Ed.*, **45**, 796.

69. Zhu, J., Wang, X.-Z., Shao, X.-B., Hou, J.-L., Chen, X.-Z., Jiang, X.-K., and Li, Z.-T. (2004) *J. Org. Chem.*, **69**, 6221.

70. Zhu, J., Lin, J.-B., Xu, Y.-X., Shao, X.-B., Jiang, X.-K., and Li, Z.-T. (2006) *J. Am. Chem. Soc.*, **128**, 12307.

71. Zhu, J., Lin, J.-B., Xu, Y.-X., Jiang, X.-K., and Li, Z.-T. (2006) *Tetrahedron*, **62**, 11933.

72. van Gorp, J.J., Vekemans, J.A.J.M., and Meijer, E.W. (2002) *J. Am. Chem. Soc.*, **124**, 14759.

73. Cuccia, L.A., Ruiz, E., Lehn, J.-M., Homo, J.-C., and Schmutz, M. (2002) *Chem. Eur. J.*, **8**, 3448.

74. Cai, W., Wang, G.-T., Du, P., Wang, R.-X., Jiang, X.-K., and Li, Z.-T. (2008) *J. Am. Chem. Soc.*, **130**, 13450.

75. Green, M.M., Cheon, K.-S., Yang, S.-Y., Park, J.-W., Swansburg, S., and Liu, W. (2001) *Acc. Chem. Res.*, **34**, 672.

76. Cai, W., Wang, G.-T., Xu, Y.-X., Jiang, X.-K., and Li, Z.-T. (2008) *J. Am. Chem. Soc.*, **130**, 6936.

77. You, L.-Y., Jiang, X.-K., and Li, Z.-T. (2009) *Tetrahedron*, **65**, 9494.

78. Lu, B.-Y., Lin, J.-B., Jiang, X.-K., and Li, Z.-T. (2010) *Tetrahedron Lett.*, **51**, 3830.

79. Tung, S.-H., Lee, H.-Y., and Raghavan, S.R. (2008) *J. Am. Chem. Soc.*, **130**, 8813.

80. Xu, X.-N., Wang, L., and Li, Z.-T. (2009) *Chem. Commun.*, 6634.

81. Xiao, Z.-Y., Hou, J.-L., Jiang, X.-K., Li, Z.-T., and Ma, Z. (2009) *Tetrahedron*, **65**, 10182.

82. Rowan, S.J., Hamilton, D.G., Brady, P.A., and Sanders, J.K.M. (1997) *J. Am. Chem. Soc.*, **119**, 2578.

83. Chen, Y.-Q., Wang, X.-Z., Shao, X.-B., Hou, J.-L., Chen, X.-Z., Jiang, X.-K., and Li, Z.-T. (2004) *Tetrahedron*, **60**, 10253.

84. Wu, Z.-Q., Jiang, X.-K., Zhu, S.-Z., and Li, Z.-T. (2004) *Org. Lett.*, **6**, 229.

85. Wu, Z.-Q., Jiang, X.-K., and Li, Z.-T. (2005) *Tetrahedron Lett.*, **46**, 8067.

86. Kirshenbaum, K., Barron, A.E., Goldsmith, R.A., Armand, P., Bradley, E.R., Truong, K.T.V., Dill, K.A., Cohen, F.E., and Zuckermann, R.N. (1998) *Proc. Natl. Acad. Sci. U. S. A.*, **95**, 4303.

87. Holub, J.M., Jang, H., and Kirshenbaum, K. (2007) *Org. Lett.*, **9**, 3275.

88. Zhu, Y.-Y., Wang, G.-T., and Li, Z.-T. (2009) *Org. Biomol. Chem.*, **7**, 3243.

89. Yuan, L.H., Feng, W., Yamato, K., Sanford, A.R., Xu, D.G., Guo, H., and Gong, B. (2004) *J. Am. Chem. Soc.*, **126**, 11120.

90. Kim, Y.H., Calabrese, J., and McEwen, C. (1996) *J. Am. Chem. Soc.*, **118**, 1545.

91. Sanford, A.R., Yuan, L., Feng, W., Yamato, K., Flowers, R.A., and Gong, B. (2005) *Chem. Commun.*, 4720.

92. Feng, W., Yamato, K., Yang, L.Q., Ferguson, J.S., Zhong, L.J., Zou, S.L., Yuan, S.L.H., Zeng, X.C., and Gong, B. (2009) *J. Am. Chem. Soc.*, **131**, 2629.

93. Ferguson, J.S., Yamato, K., Liu, R., He, L., Zeng, X.C., and Gong, B. (2009) *Angew. Chem.*, **121**, 3196; *Angew. Chem. Int. Ed.*, **48**, 3150.

94. Zhu, Y.-Y., Li, C., Li, G.-Y., Jiang, X.-K., and Li, Z.-T. (2008) *J. Org. Chem.*, **73**, 1745.

95. Yang, L., Zhong, L., Yamato, K., Zhang, X., Feng, W., Deng, P., Yuan, L., Zeng, X.C., and Gong, B. (2009) *New J. Chem.*, **33**, 729.

96. Qin, B., Chen, X., Fang, X., Shu, Y., Yip, Y.K., Yan, Y., Pan, S., Ong, W.Q., Ren, C., Su, H., and Zeng, H. (2008) *Org. Lett.*, **10**, 5127.

97. Yan, Y., Qin, B., Ren, C., Chen, X., Yip, Y.K., Ye, R., Zhang, D., Su, H., and Zeng, H. (2010) *J. Am. Chem. Soc.*, **132**, 5869.

98. Qin, B., Ren, C., Ye, R., Sun, C., Chiad, K., Chen, X., Li, Z., Xue, F., Su, H., Chass, G.A., and Zeng, H. (2010) *J. Am. Chem. Soc.*, **132**, 9564.

99. Jiang, H., Léger, J.-M., and Huc, I. (2003) *J. Am. Chem. Soc.*, **125**, 3448.

100. Jiang, H., Léger, J.-M., Guionneau, P., and Huc, I. (2004) *Org. Lett.*, **6**, 2985.

101. Li, F., Gan, Q., Xue, L., Wang, Z.-M., and Jiang, H. (2009) *Tetrahedron Lett.*, **50**, 2367.

102. Lin, J.-B., Xu, X.-N., Jiang, X.-K., and Li, Z.-T. (2008) *J. Org. Chem.*, **73**, 9403.

103. Xu, X.-N., Wang, L., Lin, J.-B., Wang, G.-T., Jiang, X.-K., and Li, Z.-T. (2009) *Chem. Eur. J.*, **15**, 5763.

104. Lin, J.-B., Wu, J., Jiang, X.-K., and Li, Z.-T. (2009) *Chin. J. Chem.*, **27**, 117.

105. Müller, M.M., Windsor, M.A., Pomerantz, W.C., Gellman, S.H., and Hilvert, D. (2009) *Angew. Chem.*, **121**, 940; *Angew. Chem. Int. Ed.*, **48**, 922.

106. Cheng, R.P., Gellman, S.H., and DeGrado, W.F. (2001) *Chem. Rev.*, **101**, 3219.

107. Raguse, T.L., Lai, J.R., and Gellman, S.H. (2003) *J. Am. Chem. Soc.*, **125**, 5592.

108. Maayan, G., Ward, M.D., and Kirshenbaum, K. (2009) *Proc. Natl. Acad. Sci. U. S. A.*, **106**, 13679.

109. Fafarman, A.T., Borbat, P.P., Freed, H., and Kirshenbaum, K. (2006) *Chem. Commun.*, 377.

110. Zhao, Y., Zhong, Z., and Ryu, E.-H. (2007) *J. Am. Chem. Soc.*, **129**, 218.

111. Cho, H.K., Zhong, Z., and Zhao, Y. (2009) *Tetrahedron*, **65**, 7311.

112. Sinkeldam, R.W., van Houtem, M.H.C.J., Pieterse, K., Vekemans, J.A.J.M., and Meijer, E.W. (2006) *Chem. Eur. J.*, **12**, 6129.

113. Yashima, E., Maeda, K., Iida, H., Furusho, Y., and Nagai, K. (2009) *Chem. Rev.*, **109**, 6102.

114. Zhou, C., Cai, W., Wang, G.-T., Zhao, X., and Li, Z.-T. (2010) *Macromol. Chem. Phys.*, **211**, 2090.

115. Shi, Z.-M., Huang, J., Ma, Z., Guan, Z., and Li, Z.-T. (2010) *Macromolecules*, **43**, 6185.

Commentary Part

Comment 1

Peter J. Stang

A foldamer, a word coined by Gellman in a 1996 *JACS* Communication [C1], refers to any polymer with a strong tendency to adopt a specific compact conformation [C1, C2]. The term *"foldamer"* was further refined by Moore and co-workers in an extensive review, "A field guide to foldamers" [C3]. There are two broad classes of foldamers: (i) biofoldamers such as proteins and nucleic acids and (ii) synthetic foldamers. The field of synthetic foldamers is a relatively new area that has blossomed in the last two decades and deals with the design, synthesis, characterization, and uses of a biological foldamers, having been inspired by the above well-known analogs. Synthetic foldamers, in analogy with their biological analogs, are made up of various oligomeric backbones held together by classical covalent bonds that adopt a particular type of secondary structure in solution (folding) through noncovalent interactions such as hydrogen bondings, $\pi-\pi$ stacking, van der Waals interactions, solvophobic effects, electrostatic interactions, and metal ion coordination. This chapter by Zhan-Ting Li highlights selective developments in synthetic foldamer chemistry in the last two decades, with emphasis on the recent literature. It covers a very eclectic collection of topics from *m*-phenyleneethynylene oligomers through heterocyclic and cholate oligomers to aromatic hydrazide and amide oligomers. Brief discussions of homoduplexes, organogels, vesicles, and supramolecular liquid crystals, and also the formation of macrocycles, catalysis, and macromolecular self-assembly are also provided.

It is evident from the highlights in this chapter that the area of foldamers, a growing part of supramolecular chemistry, is rapidly evolving. However, just as our understanding of proteins and what determines secondary and tertiary structure in biological systems is still incomplete, so is our understanding of the dynamics of foldamers. Likewise, just as in biological systems, where structure and conformation determine function, the challenge in synthetic foldamers is to design rationally structures for desired, predetermined functions such as catalysis, sensing, and encapsulation. There is little doubt that these challenges will be met in the coming decades as chemists unravel the factors that control secondary and tertiary structure and their relationship with function.

Comment 2

Liang Zhao and Mei-Xiang Wang

Introduction

Supramolecular chemistry, referring to the realm of chemistry that goes beyond covalent bonds and individual molecules and focuses on the construction and function of chemical systems made up of assembled molecular components [C4], has flourished immensely since the Nobel Prize in Chemistry was awarded to Cram, Lehn, and Pedersen in 1987. The increased knowledge over noncovalent interactions, varying from weak (electrostatic effects, $\pi-\pi$ interactions, van der Waals forces, and hydrophobic forces) to strong (hydrogen bonding and metal coordination), has culminated in impressive control over molecular self-assembly. Supramolecular synthesis of multicomponent and multidimensional architectures with diverse shapes, compositions, and functionalities is now possible under various conditions in solution and the solid state. Supramolecular studies have been widely extended not only in chemistry, but also in biology and material sciences.

In the history of supramolecular chemistry, macrocyclic compounds have had a significant influence on the discovery and development of molecular recognition, self-assembly, and other supramolecular concepts, and have also facilitated our comprehension of multifarious noncovalent intermolecular interactions, such as $\pi-\pi$ stacking, dipolar–dipolar interactions, and hydrogen bonding. Pedersen's crown ethers [C5], Cram's spherands [C6], and Lehn's cryptands [C7] can be classified as the first generation of macrocyclic molecules in supramolecular chemistry, which boosted the establishment and growth of modern supramolecular chemistry from the initial stage and thereafter served as lasting important building units in molecular self-assembly. The well-known macrocyclic effect of these molecules, which can essentially be ascribed to chelation and entropy effects, contribute to the stabilization of host–guest structures. Along with the

expansion of supramolecular studies, the escalating demand for new host macrocyclic ligands for diverse functional applications impels chemists to design and synthesize new macrocyclic molecules. In addition to the biosynthesized cyclodextrins [C8], the new generation of macrocyclic molecules, such as calixarene [C9], resorcinarene [C10], calixpyrrole [C11], cyclotriveratrylene (CTV) [C12], cucurbituril [C13], and lactam-type cyclophane [C14], have been synthesized in succession during the past 30 years (Figure C1). These macrocyclic molecules push supramolecular chemistry studies forward and expand into new burgeoning research fields. For example, Sessler and co-workers' efforts in investigating pyrrole-containing macrocyclic molecules have not only resurrected chemists' passion in synthesizing related macrocyclic molecules such as calixpyrroles, corroles, and porphycenes, but also initiated their seminal application study in anion recognition, sensing, and transport [C11d,e, C15]. The maturing synthetic methods for easily generating macrocyclic ligands also engender their possible application in generating functional materials via supramolecular synthesis, such as supramolecular polymers derived from a crown ether and paraquat [C16].

The second route to following the progress of modern supramolecular chemistry is the discovery and comprehension of noncovalent interactions. After prolonged endeavors, the list of noncovalent interactions is now on the increase, from van der Waals forces, hydrophobic effect, $\pi-\pi$ stacking, and C-H$\cdots\pi$ and cation$\cdots\pi$ interactions to the recently emerging metallophilic, halogen bonding, and anion$\cdots\pi$ interactions. The research scope in this field is no longer limited to discovering new interactions, but also focuses on exploring the essence of intermolecular noncovalent bonding and how to employ multiple and multifold noncovalent interactions to construct well-shaped and stable supramolecular entities. Nowadays, the understanding of noncovalent interactions plays a vital role in the design and synthesis of functional supramolecules, catalytic reactions, and drug design.

In this commentary, we attempt to sketch the breakthroughs in the field of supramolecular chemistry in the last decade through illustrating three noteworthy macrocyclic molecules and three significant noncovalent interactions. The ingenious design and synthesis of these macrocycles and model molecules used in intermolecular noncovalent interaction studies adequately reflect

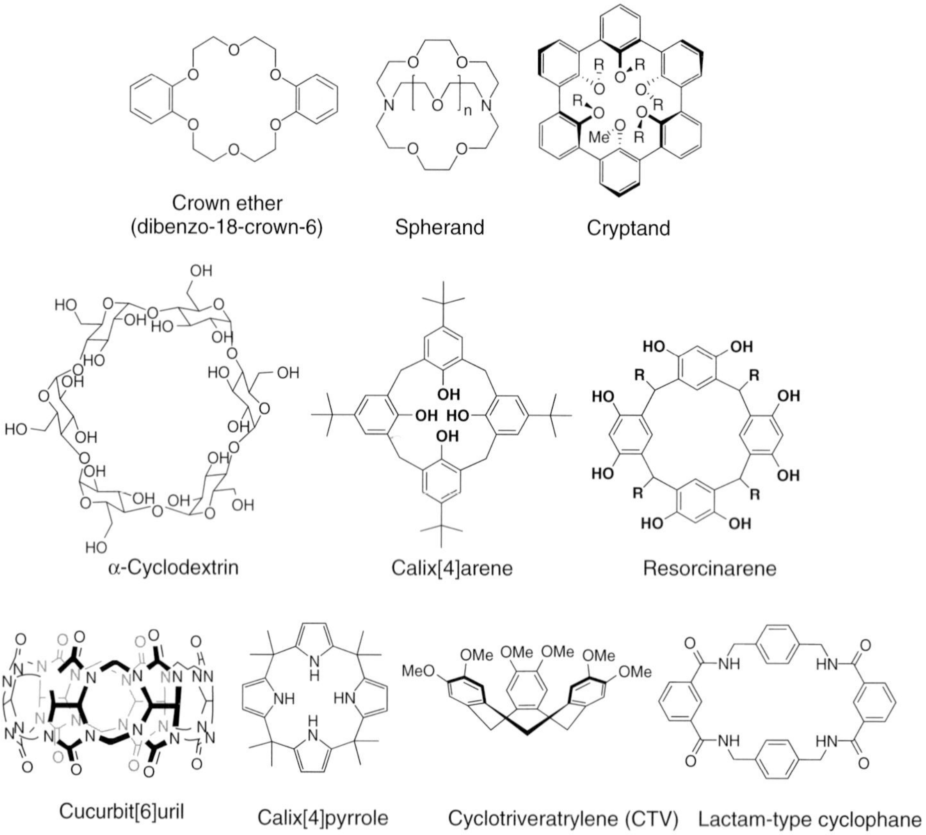

Figure C1 Structures of the common macrocyclic molecules used in modern supramolecular chemistry.

the supportive role of organic synthetic chemistry in supramolecular chemistry. The three new macrocyclic molecules selected will, in our opinion, accelerate progress in molecular recognition, molecular self-assembly, and material sciences, as will the three noncovalent interactions. The exploration of the essence of these interactions and their employment in hierarchical self-assembly will be of intense interest to supramolecular chemists.

Macrocyclic Compounds

Cycloparaphenylenes

The cycloparaphenylene compounds, deemed the shortest possible segment of an armchair carbon nanotube (Figure C2), have been of intense interest to chemists for several decades owing to

their appealing esthetic structures, unique chemical bonding of radially oriented p-orbitals, and potential applications as conducting materials in molecular electronics, organic field-effect transistors (OFETs), and nonlinear optics (NLO) [C17]. Having a well-defined cavity, they could also act as unique macrocyclic host molecules finding wide application in the study of supramolecular chemistry. However, the synthesis of these macrocyclic compounds has always been a formidable challenge owing to the steep increased strain energy upon ring closure. Vögtle's group in 1993 devised several ingenious strategies for the synthesis of [4,5]cycloparaphenylenes, but only a very small amount of a saturated [5]cycloparaphenylene compound could be detected using MS [C18]. In 2008, Bertozzi and co-workers reported the first and elegant synthesis of [9]-,

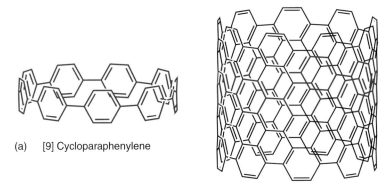

(a)   [9] Cycloparaphenylene

(b)   (9,9) Armchair carbon nanotube

Figure C2   Structure of (a) [9]cycloparaphenylene molecule and (b) its corresponding analog in armchair carbon nanotube.

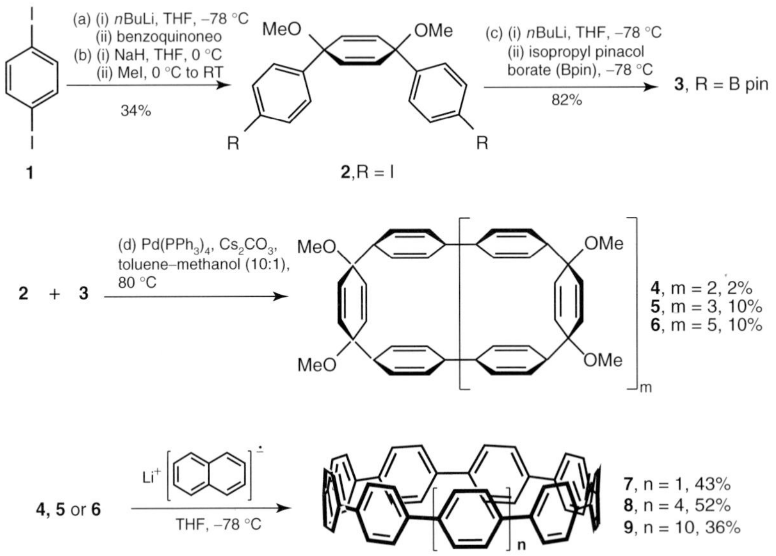

Scheme C1   Bertozzi's synthetic strategy for the synthesis of [9]-, [12]-, and [18]cycloparaphenylenes.

[12]-, and [18]cycloparaphenylenes by using a 3,6-*syn*-dimethoxy-1,4-cyclohexadiene moiety as a building unit (see structure **2** in Scheme C1) [C19]. They initially achieved a mixture of relatively low strained macrocyclic intermediates (**4**–**6** in Scheme C1) arising from sp³-hybridized carbon atom in 2,5-cyclohexadiene-1,4-diol derivatives. Subsequent treatment of each macrocycle with lithium naphthalenide produced the corresponding cycloparaphenylenes in moderate yields. Itami and co-workers subsequently utilized MOM (methoxymethyl ether)-protected *cis*-1,4-dihydroxycyclohexane-1,4-diphenyl as a monomer and judiciously employed a modular and size-selective synthetic approach to synthesize [*n*]cycloparaphenylenes (*n* = 12, 14, 15, and 16) successfully [C20]. Both groups adopt the cyclohexadiene-1,4-diol or cyclohexane-1,4-diol

unit to induce the curvature and rigidity necessary for macrocyclization and the diol units for aromatization in the final step. In contrast, Yamago *et al.* utilized a tetragonal planar bonding platinum center as a curvature unit to construct a C-Pt-C bridged organometallic macrocycle. The reductive elimination of this organometallic macrocycle finally produced the smallest cycloparaphenylene homolog, [8]cycloparaphenylene [C21].

So far, there has been no report of the crystal structures of the above-mentioned cycloparaphenylenes in the solid state and their ^{1}H and ^{13}C NMR spectra exhibit one set of signals in solution. Bertozzi [C19] Itami and co-workers studied the favored conformations of cycloparaphenylenes with computational methods based on density functional theory (DFT) [C22]. Their calculations show a favored staggered configuration in which the dihedral angle between two adjacent phenyl rings alternates within a certain angle. Benzene rings in cycloparaphenylenes can rotate freely at room temperature. The strain energies of [n]cycloparaphenylenes decrease as the diameter of cycloparaphenylene macrocycle increases.

Although the synthesis of aforementioned cycloparaphenylenes was successful in the first stage, further structural and property characterization of cycloparaphenylenes is still in its infancy. It is expected that the functionalization of the cycloparaphenylene ring and the synthesis of heteroatom-incorporated nanohoops will be the next exciting research area in this regard. Moreover, the unique aromatic system of cycloparaphenylenes may endow their organometallic metal–π complexes with special optical and magnetic behaviors. We also look forward to witnessing the extensive application of these amazing macrocyclic [n]cycloparaphenylenes as host molecules in supramolecular chemistry.

Pillar[n]arenes

In contrast to the conventional calixarenes, which are *meta*-bridged phenolic macrocycles and usually form a vase-like structure (Figure C3a), their *para*-bridged analogs have recently attracted much attention because of their unique symmetrical pillar architecture (Figure C3b). These compounds have reactive hydroxy or alkoxy groups at both ends, similarly to cyclodextrins, and their electron-rich hydrophobic cavity is capable of participating in host–guest chemistry. The synthesis of pillar[n]arenes was ignited by a discovery by Ogoshi *et al.* in 2008 [C23]. They found that the condensation of 1,4-dimethoxybenzene with paraformaldehyde catalyzed by an appropriate Lewis acid (such as BF$_3$·Et$_2$O) selectively generated 1,4-dimethoxypillar[5]arene in 22% yield. Cao *et al.* subsequently developed an efficient approach to synthesize pillar[5]arenes in high yield. As shown in Scheme C2, treatment of 2,5-bis(benzyloxymethyl)-1,4-diethoxybenzene in boiling dichloromethane with catalytic amounts of *p*-toluenesulfonic acid produces the cyclic pentamer pillar[5]arene [C24]. Other pillararenes

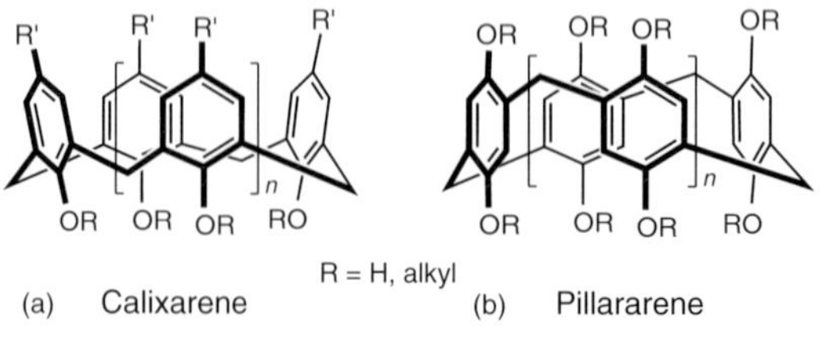

Figure C3 Structures of (a) calixarene and (b) pillararene.

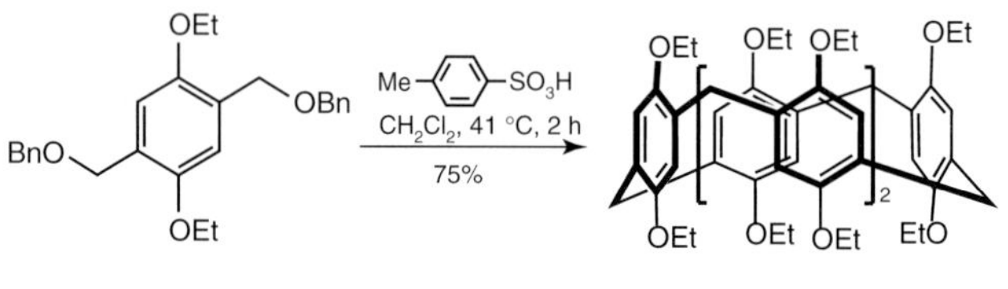

Scheme C2 Synthesis of ethoxy-substituted pillar[5]arene through acid-catalyzed reaction.

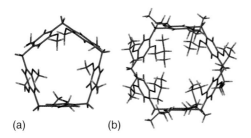

Figure C4 Crystal structures of (a) 1,4-dimethoxypillar[5]arene and (b) 1,4-diisobutoxypillar[6] arene. The core pentagon and hexagon macrocycles are highlighted in blue.

bearing different alkoxy substituents or extending π moieties [C25] are also achieved by virtue of the high reactivity of the hydroxyl groups on pillararene.

The structure of alkoxy- and hydroxy-substituted pillararenes was elucidated by variable-temperature ^{1}H NMR and 2D ROESY NMR spectroscopy. For example, pillar[5]arene exhibits a highly symmetrical structure. The conformation of differently alkoxy-substituted pillararenes is closely related to the alkyl chains. As their length increases, the alkyl substituents pack at the upper and lower rims and thus lower the conformational freedom of the pillararenes. Crystal structures of 1,4-dimethoxypillar[5]arene [C23] and 1,4-diisobutoxypillar[6]arene [C26] are shown in Figure C4.

Pillararene is composed of electron-rich hydroquinones. Therefore, supramolecular chemists embark on its host–guest study by use of cationic species such as pyridinium, viologen, and ammonium salts. Based on such a concept, Ogoshi *et al.* successfully prepared a new type of polyrotaxane composed of pillar[5]arene and viologen polymer [C27].

Although the above synthetic methods are efficient in the synthesis of pillar[5]arenes, other synthetic protocols are necessary for larger pillararenes. The long-tested fragment coupling method should be applied in this field in view of the structural diversity requirement of pillararenes. The synthesis of asymmetric pillararenes with different functional substituents on the upper and lower rims will also afford a new kind of host pillararenes with discernible bonding sites that may be appropriate for hierarchical self-assembly. In addition to electrostatic interactions within the above-mentioned host–guest complexes, we anticipate that the recognition nature of these pillararenes by hydrogen bonding interactions, or both, will attract more attention. Considering the tubular structure of pillararenes, the bridging of these molecules by covalent or noncovalent bonding may induce the formation of nanotube-like structures, which are potentially useful as building units of biosimulated ion channels.

Heteracalixaromatics

Heteracalixaromatics [C28], defined as the next generation of calixarenes with heteroatoms in place of the bridging methylene moieties between aromatics in calixarenes, have been of growing interest owing to their unique electronic features [C29], exceptional structural tenability [C28, C30], and interesting molecule and ion recognition properties [C25] which contribute to their potential applications as nanoelectronic components, metalloenzyme models, nanoscale catalysts, and molecular magnets [C31]. Although the first nitrogen (NH)-bridged calix[4]arene [C32] and oxygen-bridged calix[4]arene [C33] were synthesized in the 1960s, the low yield and arduous synthesis of these macrocyclic compounds impedes further development of this field. In 1997, Miyano and co-workers reported the convenient and practical synthesis of thiacalix[4]arene in 54% yield by heating a mixture of *p-tert*-butylphenol and elemental sulfur in tetraglyme in the presence of sodium hydroxide as a catalyst [C34]. In 2004, Wang's group established a highly efficient stepwise fragment coupling method for the synthesis of various nitrogen- and/or oxygen-bridged calixaromatics, as shown in Scheme C3 [C35]. Spurred by the great development of efficient metal-catalyzed coupling reactions, the synthesis of multifarious heteracalixaromatics has become realizable. It has subsequently opened a new avenue for the study of supramolecular chemistry

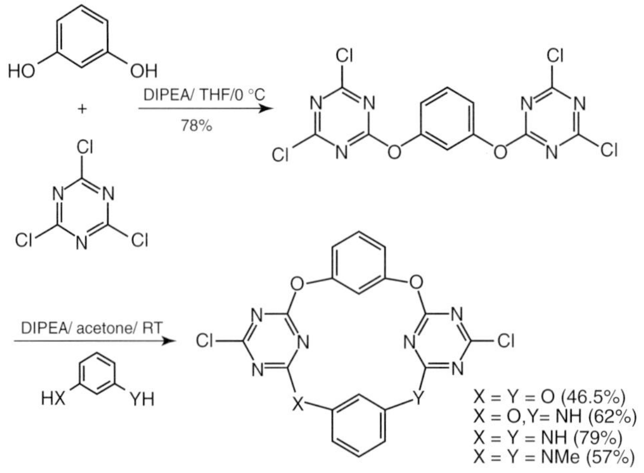

Scheme C3 Wang's fragment coupling synthetic protocol for the synthesis of oxygen- and nitrogen-bridged heteracalixaromatics.

based on heteracalixaromatics. So far, several different heteroatoms, such as oxygen [C28, C36], sulfur [C37], and nitrogen [C28, C38], have been successfully introduced in the macrocyclic skeleton of calixarene.

Structural studies of these heteracalixaromatics have explored unique structural characteristics distinct from conventional methylene-bridged calixarenes, mostly derived from alterable hybrid electron configurations (sp^2 and/or sp^3) of heteroatoms and variable degrees of conjugation between heteroatoms and adjacent aromatics. For example, the crystal structures of methylazacalix[4]pyridine and methylazacalix[8]pyridine (Figure C5) shown that the bridging N-Me groups undergo different degrees of conjugation with adjacent aromatic rings based on the distinct coplanarity situation in the dashed rectangles. Based on NMR studies, the structures of heteracalixaromatics in solution are usually fluxional and interconvert rapidly between several stable conformations, especially for large macrocycles. The substituents on bridging atoms have a significant influence on the conformations of heteracalixaromatics, according to Wang and co-workers' systematic studies [C28, C35].

Function studies of heteracalixaromatics are now focused on their recognition features for complexation with compounds extending from neutral organic molecules to metal cations and anions. Inclusion of organic molecules in the V-shaped cleft of heteracalixaromatics mostly arises from hydrogen bonding, π−π stacking, and hydrophobic

effects, such as the highly selective recognition of various aromatic and aliphatic diols and monools by methylazacalix[4]pyridine [C39]. The high affinity between these heteracalixaromatics and metal cations is rationalized by the electron-donor character of bridging heteroatoms, which enriches the electron density of heterocyclic aromatics and therefore strengthens their coordination ability. For example, the above-mentioned methylazacalix[4]pyridine shows good selectivity for zinc ion and the binding of zinc ion has resulted in a great enhancement of the fluorescence of the intrinsically fluorescent host molecule [C40]. Furthermore, the incorporation of electron-deficient aromatic rings in the heteracalixarene skeleton has facilitated the understanding of an increasing noncovalent intermolecular contact anion−π interaction [C41].

As a new kind of host molecule, the chemistry of heteracalixaromatics has almost boundless potential for development in the future. In addition to the elements O, S, N, and Si that have been introduced as bridging atoms, other nonmetallic elements such as P, As, Se, and Ge may serve as suitable bridging atoms for the synthesis of new heteracalixaromatics. In this regard, efficient synthetic methods of novel and functional heteracalixaromatics should be specifically explored. Physical property studies of these heteracalixaromatics will afford general knowledge about the influence of bridging atoms on their electronic and conformational features. In addition, the construction of more sophisticated

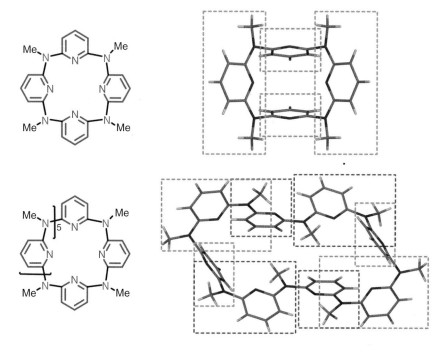

Figure C5 Crystal structures of methylazacalix[4]pyridine and methylazacalix[8]pyridine. The dashed rectangles indicate the conjugation situation of each pyridine ring.

heteracalixaromatic-based supramolecular architectures will be of intense interest, with the prospect of establishing a brand-new molecular recognition system. The incorporation of a catalytic metal center in the inner cavity of functional heteracalixaromatics may generate a kind of supramolecular catalyst.

Noncovalent Interactions

Quadruple Hydrogen Bonding

Hydrogen bonding is one of the most useful and widely used noncovalent interactions in the molecular self-assembly of well-defined supramolecular structures and materials because hydrogen bonds are directional and moderately strong. The linear geometry of a donor–H . . .acceptor motif makes it easy to design appropriate building units for desired structures. The moderate strength of hydrogen bonds makes the self-assembly process reversible and thus guarantees the final achievement of thermodynamically stable products. Also, when stronger hydrogen bonding interactions are required, it is conceivable, for this purpose, to increase the number of hydrogen bonds between components. Inspired by DNA, in which four kinds of nucleic acids associate with their complementary one by two- and threefold hydrogen bonds (Figure C6a) and consequently provide the means to read out and to replicate the information stored therein, Meijer and co-workers devised a novel self-complementary hydrogen bonding motif, 2-ureido-6-methylpyrimidone, by extending the skeletons of the nucleic acid (Figure C6c), thus establishing an ingenious method for constructing quadruple hydrogen bonds in supramolecular architectures [C42].

The strength of quadruply hydrogen-bonded complexes originates not only from the fourfold D-H . . .A contacts, but also from secondary electrostatic interactions between adjacent sites in a complex, which are repulsive between like sites, whereas disparate sites attract each other (Figure C6b). The self-complementary design of quadruple hydrogen bonding offers distinct advantages in self-assembling the DDAA units into discrete supramolecules and even supramolecular polymers. By incorporating such multiple hydrogen bonding units into supramolecular building

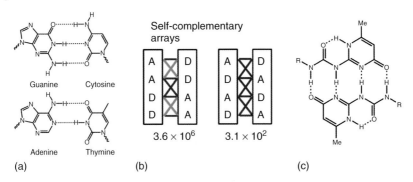

(a) (b) (c)

Self-complementary arrays

3.6×10^6 3.1×10^2

Figure C6 (a) Double and triple hydrogen bonding in DNA. (b) Dimers formed by self-complementary linear arrays of four hydrogen bonding sites, and their predicted stability constants in CDCl$_3$. Hydrogen bonds are shown in blue. Attractive and repulsive secondary interactions are indicated by green and red lines, respectively. (c) Dimeric (DDAA)$_2$ hydrogen bonding in 2-ureido-6-methylpyrimidone.

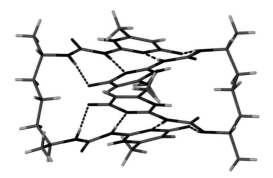

Figure C7 Crystal structure of homochiral "conglomerates" of quadruply hydrogen bonding 2-ureido-4[1*H*]-pyrimidinone derivatives. The red dashed lines indicate the hydrogen bonding.

blocks, a number of functional cyclic oligomeric supramolecular architectures have been achieved, such as enantioselective associated homochiral "conglomerates" of quadruply hydrogen-bonded 2-ureido-4[1*H*]-pyrimidinone derivatives [C43] (Figure C7).

Because of the high association constant, linear polymers built by quadruple hydrogen bonding units have a higher degree of polymerization than the self-assembled systems based on less than four hydrogen bonds. Polymer-like behavior is thus observed in organic solution of such bis-terminated building units. During the past decade, Meijer and co-workers have comprehensively elucidated the self-assembly mechanism of quadruple hydrogen bonding supramolecular polymers and developed multiform synthetic protocols for acquiring functional self-assemblies, for example, optical energy transfer molecular devices [C44]. It is obvious that their intelligent and elaborate design of quadruple hydrogen bonds has revived the study of these traditional noncovalent interactions and shed light on the application of multiple and multifold interactions in self-assembly and particularly in supramolecular polymer synthesis.

Halogen Bonding

Halogen bonding is the noncovalent interaction that occurs between a halogen atom (Lewis acid) and a Lewis base. Halogen bonds can be described in general as D . . .X-Y, where X is the electrophilic halogen atom, D is a donor of electron density

(Lewis base), and Y is a carbon, nitrogen, or halogen atom. In 1863, Guthrie reported the formation of the first halogen bonding adduct compound NH_3I_2 [C45]. In the 1950s, Mulliken developed a detailed theory of electron donor–acceptor complexes, which can be used to describe the mechanism of the halogen bond formation [C46]. The first X-ray crystallographic study of a halogen bond complex, bromine 1,4-dioxanate, was reported by Hassel *et al.* in 1954 [C47]. The O-Br distance in the crystal was measured as 2.71 Å, which is smaller than the sum of the van der Waals radii of oxygen and bromine (3.35 Å). Although Hassel *et al.* demonstrated the importance of halogen atoms in directing molecular self-assembly, the ability of halogen atoms to function as reliable sites via halogen bonding to direct molecular recognition processes long remained underappreciated. In the past decade, numerous papers describing supramolecular architectures formed by iodo and bromo bonding have drawn attention to the ability of these halocarbons to function as versatile building blocks in crystal engineering [C48]. Studies of ion recognition by means of halogen bonding between host molecules and halide anions have also been commenced by several groups recently [C49].

In analogy with the better-known hydrogen bonds, halogen bonds are also strong ($5–180\,kJ\,mol^{-1}$) and show linear directional interactions. Halogens participating in halogen bonding are iodine, bromine, chlorine, and sometimes fluorine. The origin of halogen bonding can be rationalized by the fact that in covalently bonded halogen atoms, the effective atomic radius along the extended C-X bond axis is smaller than in the direction perpendicular to this axis, and therefore a region of positive electrostatic potential is present along the covalent bond. This "σ hole" disposes the lone pair closer towards the halogen atom, and accounts for the orientation of the halogen bonds. Therefore, the halogen bonding strength increases following both the order $F < Cl < Br < I$ and the trend of the electron-withdrawing nature of the carbon atom $C(sp)-X > C(sp^2)-X > C(sp^3)-X$.

One example of a plethora of reported halogen bonding crystal engineering architectures is a linear chain structure constituted by 4,4'-bipyridine or its ethylene analogue 4,4'-ethane-1,2-diyldipyridine with 1,4-diiodotetrafluorobenzene (Figure C8) [C50]. This well-designed structure can be applied in the synthesis of functional materials such as liquid crystals and organic semiconductors. Beer and co-workers recently described their synthesis of a novel bidentate halogen bonding bromoimidazoliophane receptor and employed this receptor for bromide ion recognition [C49]. As shown in Figure C9, the *anti*- and *syn*-conformational receptor can undergo different bonding fashions with bromide ion. The chelating structure of the *syn*-receptor causes its high association constant with halide.

While the study of halogen bonding in molecular self-assembly (especially in crystal engineering) and host–guest chemistry is attracting more attention, another noteworthy subject for supramolecular chemists is the application of halogen bonding in the recognition and transportation of halogenated organic molecules. By some estimates, 50% of compounds in high-throughput drug screens are halogenated because the halogen atoms can help drug candidates increase their lipophilicity and thus favor the intracellular delivery of the drug. We believe that the deep understanding of halogen bonding may afford a viable means for efficient drug delivery.

Anion–π Interaction

The study of the interaction of anions with π-systems, namely anion–π interaction, has received much attention recently because of its importance in biological science and in materials. Anion–π interaction is usually deemed opposite to cation–π interaction, although its underlying principles are still ambiguous. The pioneering theoretical studies of Deyà and Frontera [C51], Mascal and Armstrong [C52], Alkorta [C53]

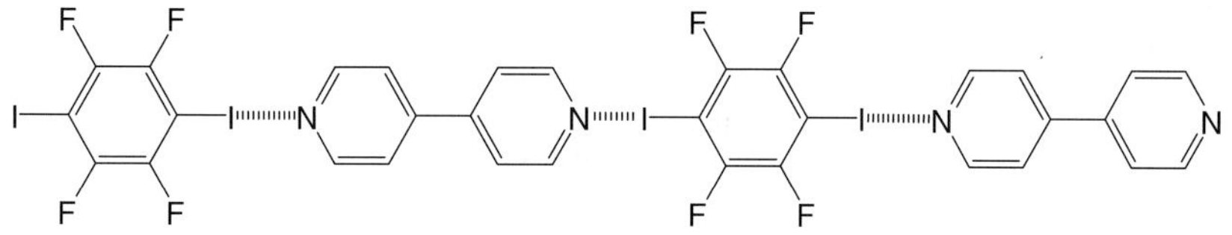

Figure C8 Schematic structure of halogen bonding linear chain built by 4,4'-bipyridine and 1,4-diiodotetrafluorobenzene.

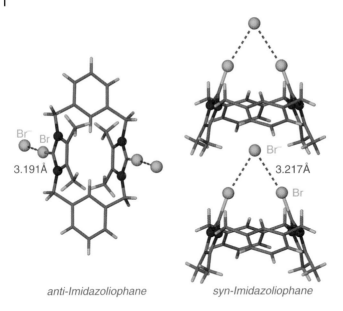

anti-Imidazoliophane   syn-Imidazoliophane

Figure C9 Bromide recognition in bidentate halogen bonding *syn-* and *anti*-bromoimidazoliophane receptors.

and co-workers in 2002 confirm the presence of favorable noncovalent interactions between electron-deficient aromatic rings and anions, with binding energies comparable to those of hydrogen bonds (20–50 kJ mol^{-1}). The electron-deficient aromatic rings include perfluoro-, nitro-, and cyano-substituted benzene, pyridine, pyrazine, and triazine derivatives. Their investigations have also revealed that the anion–π interaction is dominated by electrostatic and anion-induced polarization contributions, the former being correlated with the permanent quadrupole moment of the electron-deficient aromatic ring. Hay and Bryantsev's more recent theoretical work investigated halide–arene interactions and classified them into three types: anion–π, strong σ, and weak σ interactions [C54]. Calculated anion–π complexes with symmetrical arenes all exhibit geometries in which the halide is located exactly above the arene centroid.

Experimental evidence for anion–π interactions has been obtained in several crystalline solids [C55]. However, a formidable challenge in this study is to develop appropriate hosts that exhibit anion recognition in solution but exclude other weak interactions such as hydrogen bonding. Experimental evidence to support the purely non-covalent anion–π interaction with charge-neutral

arenes is very rare. Wang *et al.* synthesized a designed electron-deficient macrocyclic molecule, tetraoxacalix[2]arene[2]triazine, and investigated its complexation with halides through the UV–Vis and fluorescence titration, X-ray crystallography (Figure C10), and calculation methods [C41]. Matile and co-workers designed a series of organic molecules with different electronic features and studied the interaction mechanism of these molecules with other single-atom and polyatomic anions [C56].

Current studies on anion–π interaction are still in the initial phase. Basic principles of anion–π interaction are gradually being established based on theoretical calculations and experimental evidence and should be clarified in more detail. In addition, the recognition system of various anions by the virtue of anion–π interaction needs to be developed by the design and synthesis of highly selective anion receptors. The construction of sophisticated supramolecular architectures assembled through anion–π interaction only or together with other noncovalent interactions necessitates more extensive studies.

Perspectives

The past 40 years and more have witnessed the great promotion effect of macrocyclic compounds

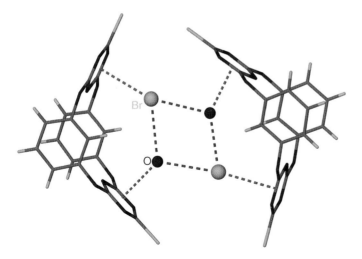

Figure C10 Crystal structure of the anion−π complexes composed of tetraoxacalix[2]arene[2]triazine, water, and bromide.

in supramolecular chemistry. Here we have tried to convince readers that the design and synthesis of new macrocyclic molecules will likewise afford motivation for the expansion of modern supramolecular chemistry. The new macrocycles not only serve as model molecules to discover unraveled noncovalent interactions, but also provide fundamental building units to achieve complicated and functional supramolecules. Moreover, we expect that future supramolecular assemblies will be obliged to use multiple and multifold interactions to generate stable architectures. However, how to combine and integrate multiple noncovalent interactions sites in a single building unit and modulate their multilevel recognition and assembly is still elusive for chemists, and requires our more extensive comprehension of intermolecular noncovalent interactions. The coalescence of the extensive and comprehensive studies between macrocyclic compounds and noncovalent interactions will truly push supramolecular chemistry forwards.

Acknowledgements

We thank the National Natural Science Foundation of China (**21132205**, **21121004**, **21002057**), Ministry of Science and Technology (**2011CB932501**), and Tsinghua University for financial support.

Comment 3

Chen-Ho Tung

Molecular chemistry is concerned with molecules which are constructed on the basis of the formation of covalent bonds between atoms or atomic groups. So far, tens of millions of compounds have been synthesized, which play essential roles in modern civilization and the abundant material world. However, with the increasing accumulation of knowledge on molecular chemistry and substantial intercrossing between chemistry and other areas, it has been recognized that distinctive properties exhibited by molecular materials are controlled not only by the information encoded in single molecules but also by intermolecular interactions and the aggregating behavior. As a result, chemistry has inevitably evolved from the level of the molecule to the level of beyond the molecule, leading to the emergence of supramolecular chemistry, focusing on intermolecular noncovalent interactions and the structures, processes, and properties of aggregates of molecules driven by them. Research in supramolecular chemistry will definitely accelerate the development of chemical science and also promote the disciplinary crossings between chemistry and physics, materials, and life sciences, leading to the creation of new concepts, methods, materials, and devices.

In this context, supramolecular chemistry is regarded as a key source to inspire the innovation of new scientific thoughts and advanced technology in the twenty-first century.

Molecular and site recognition and self-assembly are among the most important scientific issues in supramolecular chemistry. Self-assembly is a process through which molecular building blocks spontaneously form an organized system under certain conditions. Self-assembly is one of the most important tools for creating innovative materials that exhibit novel functions. In 2005, on its 125th anniversary, *Science* magazine explored 25 major questions that face scientific inquiry over the next quarter-century. "How Far Can We Push Chemical Self-Assembly?" is the only one associated with chemistry, reflecting the significance of chemical self-assembly. As is appreciated in the article, the field has made exciting progress, but needs to receive more attention from the scientific community. The author also advised chemists to construct complex and ordered structures via self-assembly, just as Nature does.

One of the key scientific issues in supramolecular chemistry is to understand the additivity, cooperativity, and directionality of weak interactions. Intermolecular interactions include van der Waals forces, hydrophobic–lipophilic interactions, hydrogen bonding, electrostatic interactions, and $\pi-\pi$ stacking. Molecular assembly deals with the additivity and cooperativity of intermolecular interactions, from which directionality and selectivity arise. These factors dictate the result of molecular recognition and site recognition and also push the molecular assembly process to go forward. Understanding how the intermolecular interactions work together to produce additivity and cooperativity and how they lead to the generation of directionality and selectivity is the premise for the development of new assembly methods and for gaining more insight into the functions of assembled systems.

The major goal in studying supramolecular chemistry is to exploit the functions that assembled systems provide. Although the functions of supramolecular systems are manifold, the generation of all the functions can be ascribed to information transfer. A notable feature of supramolecular systems is their responsiveness to external stimuli, which usually result in changes to the structures and functions of the systems. Such changes can be translated into molecular information and then the information can be transported

within the whole system. The essence of information transfer is energy transfer, material transport, and chemical transformation. Therefore, the investigation of electron and energy transfer and chemical transformations inside supramolecular systems is another key scientific problem in this field.

Li's research concerns foldamer-based self-assembly. His present chapter summarizes in detail the progress of foldamer chemistry and the achievements related to supramolecular chemistry. Using molecular folding as an example, the chapter demonstrates how the additivity and cooperativity among various intermolecular interactions (hydrogen bonding, donor–acceptor interactions, hydrophobic interactions, and so on) come into play and where the directionality and selectivity come from. It is the directionality of the intermolecular interactions that leads to the occurrence of molecular recognition and highly ordered assembly (instead of disordered aggregation) and further endows the assembled entities with certain functions. The chapter is well organized with important contents and nicely illustrates the basic scientific problems in supramolecular chemistry, and therefore will be of interest to those working in supramolecular chemistry.

Author's Response to the Commentaries

Reply to Zhao and Wang's Comments

Zhao and Wang briefly introduce the history of macrocyclic chemistry and its leading role in advancing supramolecular chemistry. Clearly, the science of supramolecules originated from studies on the complexation of macrocycles with metal ions. The work of the three winners of the 1987 Nobel Prize in Chemistry in this area was all concerned with macrocycles, that is, crown ethers, cryptands, and spherands. I fully agree that "the design and synthesis of new macrocyclic molecules will likewise afford motivation for the expansion of modern supramolecular chemistry." To some extent, supramolecular chemists are interested in macrocycles because of their stronger binding capacity and higher binding selectivity compared with those of linear analogs. This is crucially important for the application of supramolecular systems in analytical science and chemical biology. The symmetric feature of most of the macrocycles also makes them easier to prepare.

In many cases and in many laboratories, synthesis is a major consideration when starting a supramolecular project.

Compared with macrocycles, linear molecules have attracted relatively less interest. Early studies on podands showed that linear molecules might work as receptors, but they are generally less efficient than the related crown ethers. This is not unexpected considering that the flexibility of the backbones would produce a negative entropic effect upon binding. Foldamers provide an approach to overcoming this problem due to their preorganized conformation, but their synthesis is more time consuming and the structural diversity is limited. As a result, the development of advanced functions such as catalysis and transportation is still a challenge for foldamers. However, considering that natural proteins and nucleic acids evolve their functions from their secondary and higher-grade structures, we may have an expectation for foldamers as synthetic structures to display similar advanced functions if we can render them a comparable complexity.

Reply to Tung's Comments

Supramolecular chemistry will not replace molecular chemistry in the future, but it has greatly extended our knowledge in molecular science. Currently, supramolecular chemists are paying increasing attention to the functions of supramolecules. Molecular recognition is the basis for developing any functions. Although my research interests mainly concern the basic aspects of molecular recognition, investigations "on electron and energy transfer and chemical transformation inside supramolecular systems" are certainly important, because the effort along these lines may lead to practical applications of supramolecules. One area in which I am particularly interested is their catalysis function. Efforts in this direction from many groups have been under way many years, but the examples of challenging the efficiency of enzymes are still limited. I hope to see some new breakthroughs in the near future.

Reply to Stang's Comments

The secondary structures of proteins are the basis for the formation of their higher-grade tertiary and quaternary structures. The structures or structural domains of different grades all contribute for the formation of their functions. Currently, chemists have developed the capacity to construct artificial secondary structures. Many of these structures display functions that can find their origin in natural structures. As commented by Stang, our understanding of protein folding and what determines the secondary and tertiary structures in biological systems is still incomplete. Hence we are still at an early stage in mimicking the structures and functions of large biological molecules. When the understanding of biological systems goes deeper, I believe that chemists will be able to design synthetic structures that challenge the functions of biological systems.

References

C1. Appela, D.H., Christianson, L.A., Karle, I.L., Powell, D.R., and Gellman, S.H. (1996) b-Peptide foldamers: robust helix formation in a new family of -amino acid oligomers. *J. Am. Chem. Soc.*, **118**, 13071–13072.

C2. Manifesto, A. and Gellman, S.H. (1998) Foldamers. *Acc. Chem. Res.*, **31**, 173–180.

C3. Hill, D.J., Mio, M.J., Prince, R.B., Hughes, T.S., and Moore, J.S. (2001) A field guide to foldamers. *Chem. Rev.*, **101**, 3893–4012.

C4. Lehn, J.-M. (1993) *Science*, **260**, 1762.

C5. Pedersen, C.J. (1967) *J. Am. Chem. Soc.*, **89**, 2495; Pedersen, C.J. (1967) *J. Am. Chem. Soc.*, **89**, 7017.

C6. Cram, D.J., Kaneda, T., Helgeson, R.C., and Lein, G.M. (1979) *J. Am. Chem. Soc.*, **101**, 6752.

C7. Lehn, J.M. and Montavon, F. (1978) *Helv. Chim. Acta*, **61**, 67.

C8. (a) Szejtli, J. (1988) *Cyclodextrin Technology*, vol. **1**, Springer, New York' (b) Szejtli, J. (1998) *Chem. Rev.*, **98**, 1743; (c) Szejtli, J. (1998) *Chem. Rev.*, **98**, 1743.

C9. (a) Gutsche, C.D. and Muthukrishnan, R. (1978) *J. Org. Chem.*, **43**, 4905; (b) Gutsche, C.D. (1983) *Acc. Chem. Res.*, **16**, 161.

C10. (a) Högberg, A.G.S. (1980) *J. Org. Chem.*, **45**, 4498; (b) Högberg, A.G.S. (1980) *J. Am. Chem. Soc.*, **102**, 6046; (c) Timmerman, P., Verboom, W., and Reinhoudt, D.N. (1996) *Tetrahedron*, **52**, 2663.

C11. (a) Baeyer, A. (1886) *Ber. Dtsch. Chem. Ges.*, **19**, 2184; (b) Baeyer, A. (1872) *Ber. Dtsch. Chem. Ges.*, **5**, 1094; (c) Gale, P.A., Sessler, J.L., Král, V., and Lynch,

V. (1996) *J. Am. Chem. Soc.*, **118**, 5140; (d) Gale, P.A., Sessler, J.L., and Kral, V. (1998) *Chem. Commun.*, 1; (e) Gale, P.A., Anzenbacher, P.J., and Sessler, J.L. (2001) *Coord. Chem. Rev.*, **222**, 57.

C12. (a) Robinson, G.M. (1915) *J. Chem. Soc.,Trans.*, **107**, 267; (b) Lindsey, A.S. (1965) *J. Chem. Soc.*, 1685; (c) Hardie, M. (2010) *Chem. Soc. Rev.*, **39**, 516.

C13. (a) Behrend, R., Meyer, E., and Rusche, F. (1905) *Justus Liebig's Ann. Chem.*, **339**, 1; (b) Freeman, W.A., Mock, W.L., and Shih, N.-Y. (1981) *J. Am. Chem. Soc.*, **103**, 7367; (c) Kim, K. (2002) *Chem. Soc. Rev.*, **31**, 96.

C14. (a) Joo, K., Ihm, H., and Paek, K. (1995) *Bull. Korean Chem. Soc.*, **16**, 1079; (b) Wack, H., France, S., Hafez, A.M., Drury, W.J., Weatherwax, A., and Lectka, T. (2004) *J. Org. Chem.*, **69**, 4531; (c) Rajakumar, P. and Padmanabhan, R. (2010) *Tetrahedron Lett.*, **51**, 1059.

C15. (a) Sessler, J.L., Camiolo, S., and Gale, P.A. (2003) *Coord. Chem. Rev.*, **240**, 17; (b) Lee, C.-H., Miyaji, H., Yoon, D.-W., and Sessler, J.L. (2008) *Chem. Commun.*, 24.

C16. Niu, Z. and Gibson, H.W. (2009) *Chem. Rev.*, **109**, 6024.

C17. (a) Mayor, M. and Didschies, C. (2003) *Angew. Chem. Int. Ed.*, **42**, 3176; (b) Xu, H., Yu, G., Xu, W., Xu, Y., Cui, G., Zhang, D., Liu, Y., and Zhu, D. (2005) *Langmuir*, **21**, 5391; (c) Marsden, J.A., Miller, J.J., Shirtcliff, L.D., and Haley, M.M. (2005) *J. Am. Chem. Soc.*, **127**, 2464; (d) Mena-Osteritz, E. and Bäuerle, P. (2006) *Adv. Mater.*, **18**, 447; (e) Zhao, T., Liu, Z., Song, Y., Xu, W., Zhang, D., and Zhu, D. (2006) *J. Org. Chem.*, **71**, 7422.

C18. Friedrich, R., Nieger, M., and Vögtle, F. (1993) *Chem. Ber.*, **126**, 1723.

C19. Jasti, R., Bhattacharjee, J., Neaton, J.B., and Bertozzi, C.R. (2008) *J. Am. Chem. Soc.*, **130**, 17646.

C20. (a) Takaba, H., Omachi, H., Yamamoto, Y., Bouffard, J., and Itami, K. (2009) *Angew. Chem. Int. Ed.*, **48**, 6112; (b) Omachi, H., Matsuura, S., Segawa, Y., and Itami, K. (2010) *Angew. Chem. Int. Ed.*, **49**, 10202.

C21. Yamago, S., Watanabe, Y., and Iwamoto, T. (2010) *Angew. Chem. Int. Ed.*, **49**, 757.

C22. Segawa, Y., Omachi, H., and Itami, K. (2010) *Org. Lett.*, **12**, 2262.

C23. Ogoshi, T., Kanai, S., Fujinami, S., Yamagishi, T., and Nakamoto, Y. (2008) *J. Am. Chem. Soc.*, **130**, 5022.

C24. Cao, D., Kou, Y., Liang, J., Chen, Z., Wang, L., and Meier, H. (2009) *Angew. Chem. Int. Ed.*, **48**, 9721.

C25. (a) Ogoshi, T., Umeda, K., Yamagishi, T., and Nakamoto, Y. (2009) *Chem. Commun.*, 4874; (b) Ogoshi, T., Kitajima, K., Aoki, T., Fujinami, S., Yamagishi, T., and Nakamoto, Y. (2010) *J. Org. Chem.*, **75**, 3268; (c) Zhang, Z., Xia, B., Han, C., Yu, Y., and Huang, F. (2010) *Org. Lett.*, **12**, 3285.

C26. Han, C., Ma, F., Zhang, Z., Xia, B., Yu, Y., and Huang, F. (2010) *Org. Lett.*, **12**, 4360.

C27. Ogoshi, T., Nishida, Y., Yamagishi, T., and Nakamoto, Y. (2010) *Macromolecules*, **43**, 7068.

C28. Wang, M.-X. (2008) *Chem. Commun.*, 4541.

C29. (a) Ito, A., Inoue, S., Hirao, Y., Furukawa, K., Katod, T., and Tanaka, K. (2008) *Chem. Commun.*, 3242; (b) Ishibashi, K., Tsue, H., Sakai, N., Tokita, S., Matsui, K., Yamauchi, J., and Tamura, R. (2008) *Chem. Commun.*, 2812.

C30. Fukushima, W., Kanbara, T., and Yamamoto, T. (2005) *Synlett*, 2931.

C31. (a) Morohashi, N., Narumi, F., Iki, N., Hattori, T., and Miyano, S. (2006) *Chem. Rev.*, **106**, 5291; (b) Kajiwara, T., Iki, N., and Yamashita, M. (2007) *Coord. Chem. Rev.*, **251**, 1734.

C32. Smith, G.W. (1963) *Nature*, **198**, 879.

C33. Sommer, N. and Staab, H.A. (1966) *Tetrahedron Lett.*, 2837.

C34. Kumagai, K., Hasegawa, M., Yanari, S.M., Sugawa, Y., Sato, Y., Hori, T., Ueda, S., Kamiyama, H., and Miyano, S. (1997) *Tetrahedron Lett.*, **38**, 3971.

C35. (a) Wang, M.-X. and Yang, H.-B. (2004) *J. Am. Chem. Soc.*, **126**, 15412; (b) Wang, M.-X., Zhang, X.-H., and Zheng, Q.-Y. (2004) *Angew. Chem. Int. Ed.*, **43**, 838.

C36. Maes, W. and Dehaen, W. (2008) *Chem. Soc. Rev.*, **37**, 2393.

C37. Lhotak, P. (2004) *Eur. J. Org. Chem.*, 1675.

C38. Tsue, H., Ishibashi, K., and Tamura, R. (2008) *Top. Heterocycl. Chem.*, **17**, 73.

C39. Gong, H.-Y., Wang, D.-X., Xiang, J.-F., Zheng, Q.-Y., and Wang, M.-X. (2007) *Chem. Eur. J.*, **13**, 7791.

C40. Gong, H.-Y., Zheng, Q.-Y., Zhang, X.-H., Wang, D.-X., and Wang, M.-X. (2006) *Org. Lett.*, **8**, 4895.

C41. Wang, D.-X., Zheng, Q.-Y., Wang, Q.-Q., and Wang, M.-X. (2008) *Angew. Chem. Int. Ed.*, **47**, 7485.

C42. Sijbesma, R.P., Beijer, F.H., Brunsveld, L., Folmer, B.J.B., Hirschberg, J.H.K.K., Lange, R.F.M., Lowe, J.K.L., and Meijer, E.W. (1997) *Science*, **278**, 1601.

C43. ten Cate, A.T., Dankers, P.Y.W., Kooijman, H., Spek, A.L., Sijbesma, R.P., and Meijer, E.W. (2003) *J. Am. Chem. Soc.*, **125**, 6860.

C44. Dudek, S.P., Pouderoijen, M., Abbel, R., Schenning, A.P.H.J., and Meijer, E.W. (2005) *J. Am. Chem. Soc.*, **127**, 11763.

C45. Guthrie, F. (1863) *J. Chem. Soc.*, **16**, 239.

C46. (a) Mulliken, R.S. (1950) *J. Am. Chem. Soc.*, **72**, 600; (b) Mulliken, R.S. (1952) *J. Am. Chem. Soc.*, **74**, 811.

C47. Hassel, O., Hvoslef, J., Vihovde, E.H., and Sörensen, N.A. (1954) *Acta Chem. Scand.*, **8**, 873.

C48. Metrangolo, P., Meyer, F., Pilati, T., Resnati, G., and Terraneo, G. (2008) *Angew. Chem. Int. Ed.*, **47**, 6114.

C49. Caballero, A., White, N.G., and Beer, P.D. (2011) *Angew. Chem. Int. Ed.*, **50**, 1845.

C50. (a) Metrangolo, P., Meyer, F., Pilati, T., Resnati, G., and Terraneo, G. (2007) *Acta Crystallogr., Sect. E*, **63**, o4243; (b) Liantonio, R., Metrangolo, P., Pilati, T., and Resnati, G. (2002) *Acta Crystallogr., Sect. E*, **58**, o575; (c) Blackstock, S.C., Lorand, J.P., and Kochi, J.K. (1987) *J. Org. Chem.*, **52**, 1451.

C51. Quiñonero, D., Garau, C., Rotger, C., Frontera, A., Ballester, P., Costa, A., and Deyà, P.M. (2002) *Angew. Chem. Int. Ed.*, **41**, 3389.

C52. Mascal, M., Armstrong, A., and Bartberger, M.D. (2002) *J. Am. Chem. Soc.*, **124**, 6274.

C53. Alkorta, I., Rozas, I., and Elguero, J. (2002) *J. Am. Chem. Soc.*, **124**, 8593.

C54. Hay, B.P. and Bryantsev, V.S. (2008) *Chem. Commun.*, 2417.

C55. (a) Demeshko, S., Dechert, S., and Meyer, F. (2004) *J. Am. Chem. Soc.*, **126**, 4508; (b) de Hoog, P., Gamez, P., Mutikainen, I., Turpeinen, U., and Reedijk, J. (2004) *Angew. Chem. Int. Ed.*, **43**, 5815; (c) Campos-Fernández, C.S., Schottel, B.L., Chifotides, H.T., Bera, J.K., Bacsa, J., Koomen, J.M., Russell, D.H., and Dunbar, K.R. (2005) *J. Am. Chem. Soc.*, **127**, 12909.

C56. Dawson, R.E., Hennig, A., Weimann, D.P., Emery, D., Ravikumar, V., Montenegro, J., Takeuchi, T., Gabutti, S., Mayor, M., Mareda, J., Schalley, C.A., and Matile, S. (2010) *Nat. Chem.*, **2**, 563.

14
Novel Catalysis for Alkene Polymerization Mediated by Post-Metallocenes: a Gateway to New Polyalkenes

Hiromu Kaneyoshi, Haruyuki Makio, and Terunori Fujita

14.1
Introduction

Polyalkenes, exemplified by polyethylene (PE) and polypropylene (PP), were a notable discovery of the twentieth century. Since PE and PP exhibit many useful properties, such as being cost-effective and lightweight and having excellent mechanical strength, flexibility, processability, chemical inertness, and recyclability, they have today become indispensable materials for everyday life.

The historical discovery of alkene polymerization catalysts was achieved by Ziegler ($TiCl_4$–$AlEt_3$ for PE) and Natta ($TiCl_3$–$AlEt_3$ for PP) in the 1950s [1]. Although the Ziegler catalyst showed relatively low activity ($<1 \, kg \, mmol^{-1}$ Ti) for PE production, in 1968 Kashiwa *et al.* observed that $MgCl_2$-supported $TiCl_4$–$AlEt_3$ catalysts exhibited two orders of magnitude greater activity ($>100 \, kg \, mmol^{-1}$ Ti) [2]. As a consequence, Ziegler–Natta catalysts based on $MgCl_2$ supports currently dominate polyalkene production. However, Ziegler–Natta catalysts produce atactic polypropylene (aPP) together with isotactic polypropylene (iPP), in which the active species are multi-sited.

In 1980, Kaminsky and co-workers reported that bis(cyclopentadienyl)dimethylzirconium (Cp_2ZrMe_2) in conjunction with methylaluminoxane (MAO) displayed a high activity (about $100 \, kg \, mmol^{-1}$ Zr h^{-1}) for the production of PE [3]. A metallocene catalyst is defined as a combination of a group 4 metal complex bearing two cyclopentadienyl (Cp) derivatives with an activator such as MAO and tetrakis[pentafluorophenyl)borate (TPB, $B(C_6F_5)_4^-$] with various counter cations. Since metallocene complexes readily permit the determination of molecular structure and generally afford a single well-defined active species upon activation, metallocenes are called single-site catalysts, unlike Ziegler–Natta catalysts that are ill-defined and generate multiple active sites. The discovery of metallocene catalysts contributed not only to the clarification of alkene polymerization mechanisms, but also to the synthesis of novel polyalkenes. A precise design involving a Cp ligand structure allows control over polymer molecular weight, polymer stereochemistry, and comonomer incorporation, in addition to uniform comonomer distribution derived from the single-site character. As a consequence, metallocene catalysts are used in the commercial production of novel polyalkenes such as linear low-density PEs, syndiotactic polypropylenes (sPPs), and ethylene–α-alkene amorphous copolymers.

Organic Chemistry – Breakthroughs and Perspectives, First Edition. Edited by Kuiling Ding and Li-Xin Dai.
© 2012 Wiley-VCH Verlag GmbH & Co. KGaA. Published 2012 by Wiley-VCH Verlag GmbH & Co. KGaA.

Following the tremendous advances made with metallocene catalysts, researchers have since turned their attention towards the development of a new generation of catalysts that provide ever-higher catalyst productivity and even greater control over polymer microstructures. As a consequence of an enormous amount of academic and industrial research, various post-metallocene catalysts have now been discovered. In this chapter, post-metallocene' is defined as a metal complex without a Cp ligand, "complex" refers to a catalyst precursor that is not activated, and "catalyst" is used to indicate an activated form of the catalyst precursor. Notable post-metallocene single-site catalysts that exhibit unique catalytic performance for alkene polymerization capable of forming novel alkene-based materials are introduced.

14.2
Late Transition Metal Complexes

Numerous reports have revealed that late transition metal complexes can act as ethylene oligomerization catalysts for the production of higher α-alkenes. Since the vacant d-orbital on the metal is at a low energy level, the β-H elimination reaction, which terminates the chain propagation reaction, naturally occurs. Therefore, knowing how to prevent this unfavorable termination reaction is essential when designing alkene polymerization catalysts using late transition metals.

14.2.1
Diimine-Ligated Ni and Pd Complexes

In 1995, Brookhart and co-workers reported that bulky diimine-ligated Ni and Pd complexes **2** and **3** were capable of polymerizing both ethylene and propylene for the production of high molecular weight polymers (Figure 14.1) [4]. This epoch-making discovery was achieved by intelligent ligand design. The introduction of the bulky *N*-aryl group of the diimine ligand protects the vacant axial coordination site of the square-planar complex, leading to the retardation of chain-transfer reactions. Ni complex **1** activated with MAO produces PE with high activity ($\sim$10 kg mmol Ni h^{-1}), and yields PE having a high molecular weight of up to 10^6 g mol^{-1}.

ł A crucial feature of the diimine-ligated Ni and Pd complexes is their versatility for controlling degrees of branching of the resultant PE by changing the ethylene pressure and polymerization temperature. The extent of branching varies from virtually linear (1.2 branches per 1000 carbon atoms) up to 300 branches per 1000 carbon atoms, and the Pd complexes generally form more branched PE under the same polymerization conditions. The proposed mechanism for the production of branched polymers consists of reversible processes involving β-H transfer to the metal, rotation of the coordinated alkene, and metal–hydride insertion in the alkene in the opposite regiochemistry (Scheme 14.1), to which the name "chain walking polymerization" was given since the central metal "walks" on the polymer chain. Chain growth at any given time during the chain walking results in the formation of branching for ethylene polymerization.

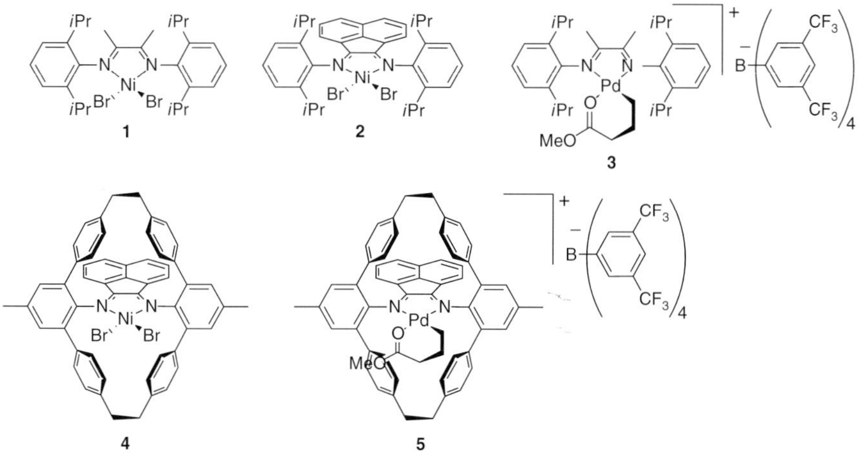

Figure 14.1 Representative diimine-ligated Ni and Pd complexes.

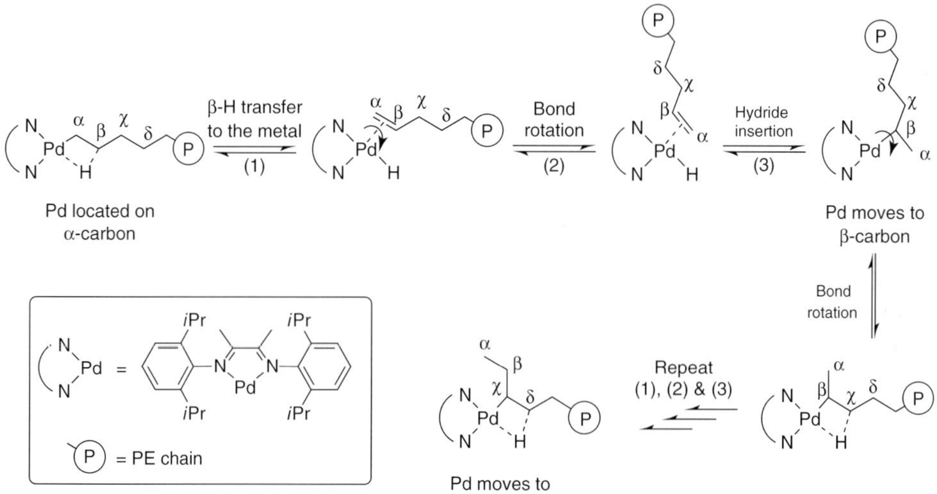

Scheme 14.1 Proposed mechanism for the formation of branched PE.

At lower temperatures below 0 °C, propylene polymerization mediated by **2**–MAO proceeds in a living fashion to yield high molecular weight aPP having fewer methyl branches than expected [300 branches per 1000 C, M_w (weight-average molecular weight) 180 000] with a low polydispersity index [M_w/M_n (number-average molecular weight) < 1.5]. The propylene insertion occurs in a 2,1 fashion in this polymerization, generating a secondary alkyl–metal species. The metal of this species can insert a propylene again in a 2,1 fashion, forming a methyl branch, or alternatively move via the chain walking mechanism along the polymer chain. When the metal goes to the terminal methyl group of the last-inserted propylene monomer, subsequent propylene insertion results in enchainment of the penultimate propylene monomer in a 1,3 fashion. This chain straightening reaction (1,ω-insertion) generally results in fewer branched structures for polymerization of α-alkenes when using the diimine-ligated Ni and Pd catalysts. Additionally, **2**–MAO can prepare di- and triblock copolymers from propylene and α-alkenes [5].

Another important feature is that the diimine-ligated Ni and Pd complexes display excellent functional group tolerance. Importantly, cationic Pd catalyst **3** can copolymerize ethylene with methyl acrylate (MA) to form highly branched random copolymers with ester groups predominantly located at the ends of branches [6]. The copolymer contains MA up to 10 mol%, and has a low polydispersity index ($M_w/M_n < 2$). This is the first example of the copolymerization of ethylene with polar-functionalized vinyl monomers with no spacer groups between the vinyl and the functional group through a coordination-type polymerization using no functional group masking agents.

Guan and co-workers further developed the design of the diimine ligand by the introduction of a cyclophane group (Figure 14.1) [7]. The metal center is located at the core of the cyclophane-based ligand, and therefore vacant axial sites on the metal are efficiently blocked. The robust framework of the ligand enhances catalyst activity ($\sim$40 kg mmol^{-1} Ni h^{-1} by **4**–MMAO (modified methylaluminoxane) and also catalyst thermal stability, resulting in the production of branched PE with high molecular weight ($M_w \approx 400\,000$ g mol^{-1}, $M_w/M_n < 1.8$) at a high polymerization temperature (90 °C). Moreover, the living polymerization of propylene using **4**–MMAO proceeds at a higher temperature (50 °C), yielding less branched aPP (105 branches per 1000 C) relative to the **2**–MAO system. Interestingly, Pd catalyst **5** showed more efficient MA incorporation in copolymerization with ethylene than **3**, with an MA content in the copolymer of up to 22 mol%.

14.2.2
Pyridyldiimine-Ligated Fe and Co Complexes

In 1998, the groups of Brookhart and Gibson independently developed pyridyldiimine-ligated Fe complexes for ethylene polymerization, representing the first report of Fe-based catalysts capable of polymerizing ethylene with high efficiency (Figure 14.2) [8]. Fe complex **6** after activation with MMAO is able to prepare linear PE with very high activity ($\sim$330 g mmol Fe h^{-1}), which is comparable to the activity of the group 4 metallocene catalysts. The catalyst activity of **6**–MMAO increases with increasing ethylene pressure. Interestingly, increasing the steric bulk of the *ortho*-aryl substituent enhances the molecular weight of the PE. For instance, **6**–MMAO (R = *i*Pr) yields PE with an M_w of 71 000 g mol^{-1}, whereas PE formed by **7**–MMAO (R = Me) affords an M_w of 33 000 g mol^{-1}. In addition, the less bulky complex **8** in combination with MAO produces ethylene oligomers ($M_w < 500$ g mol^{-1}). The high polydispersity index of the PE obtained from **6**–MMAO ($M_w/M_n > 5$) is mainly due to the preferential chain transfer reaction to alkylaluminums in MMAO and time-dependent concentration of the available alkylaluminums. The Fe complexes generally exhibit an order of magnitude higher activity and produce higher molecular weight polymer than the Co

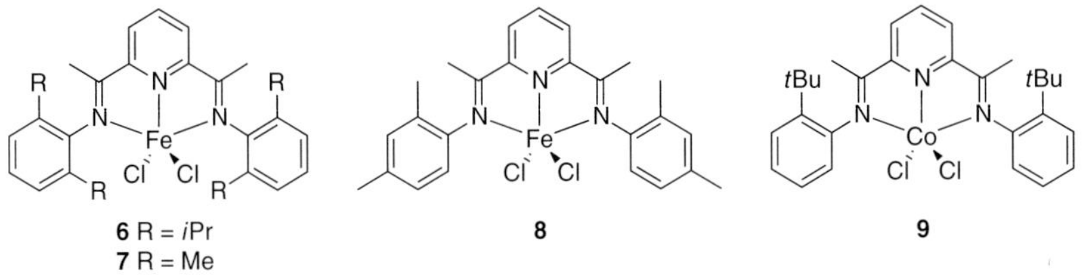

Figure 14.2 Representative pyridyldiimine-ligated Fe and Co complexes.

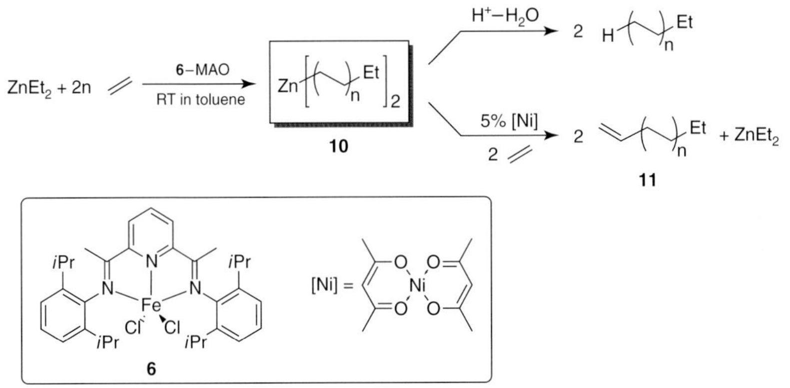

Scheme 14.2 Fe-catalyzed chain growth on Zn.

analogs. Co complex **9** in conjunction with MMAO is able to prepare a high molecular weight PE (M_w 180 000 g mol^{-1}).

Interestingly, linear ethylene oligomers with a low polydispersity index (M_w 800 g mol^{-1}, M_w/M_n 1.1) can be obtained when the ethylene polymerization using **6**−MAO is conducted in the presence of excess ZnEt$_2$ ([Zn]:[Fe] molar ratio = 500) [9]. The catalyst activity of **6**−MAO (1−2 g mmol^{-1} Fe h^{-1}) does not depend on the amount of ZnEt$_2$ with a [Zn]:[Fe] molar ratio ranging from 0 to 500. The molar ratio of the ethylene oligomers obtained to the Zn added is ~2, indicating that long-chain dialkyl-Zn **10** is selectively formed by an Fe-catalyzed chain growth reaction on Zn (Scheme 14.2). The hydrolysis of **10** affords linear ethylene oligomers having a Poisson distribution with fully saturated chain ends. These results demonstrate that Zn facilitates highly efficient chain transfer reversibly to and from the Fe center. Additionally, Zn-terminated ethylene oligomer **10** is available for further modification to obtain vinyl-functionalized ethylene oligomer **11** by an Ni-catalyzed alkene displacement reaction in the presence of ethylene. This methodology is industrially powerful enough to generate a myriad of end-functionalized ethylene oligomers with a low polydispersity index in a cost-effective manner. Further investigation revealed that the group 13 metal alkyl reagents AlR$_3$ (R = Me, Et, iBu) and GaR$_3$ (R = Me, Et, nBu) are also capable of employing highly efficient chain-transfer agents. In the case of AlEt$_3$, the molar mass distribution of the product follows a Schultz−Flory distribution, indicating that the reversible chain transfer is less efficient relative to ZnEt$_2$. Meanwhile, the molar mass distribution of the product obtained with GaMe$_3$ fits neither a Poisson nor a Schultz–Flory distribution, but shows an intermediate character between them, suggesting a better reversible chain-transfer ability than AlEt$_3$.

14.2.3
Phenoxyimine-Ligated Ni Complexes

It is well known that Ni complexes bearing a phosphine-based chelate ligand (Ph$_2$P-CH=C(Ph)O$^-$) act as an alkene oligomerization catalyst [Shell Higher Olefin Process (SHOP)]. In 1998, Grubbs and co-workers reported that the introduction of bulky substituents on the phenoxyimine-ligated Ni complex permitted the production of PE (as did an independent report by DuPont) [10]. Neutral Ni complex **12** is an active catalyst for ethylene polymerization in the presence of 2 equiv. of bis(cyclooctadienyl)Ni [Ni(cod)$_2$] as a phosphine scavenger (Figure 14.3). As a general trend, catalyst productivity and the molecular weight of PE are increased on increasing the steric bulk of the substituents at the 3-position of the phenoxyimine ligand. The catalyst activity of **12** is ~1 g mmol^{-1} Ni h^{-1}, and the M_w of PE is 240 000 g mol^{-1}. Interestingly, Ni complex **13** containing more labile acetonitrile than PPh$_3$ exhibits twice as much activity as **12**, because the equilibrium between the dormant species **14** and **15** shifts toward **15** (Scheme 14.3). The resulting PE has a branched structure (5–55 branches per 1000 C), and the extent of branching can be controlled by the variation of both pressure and temperature.

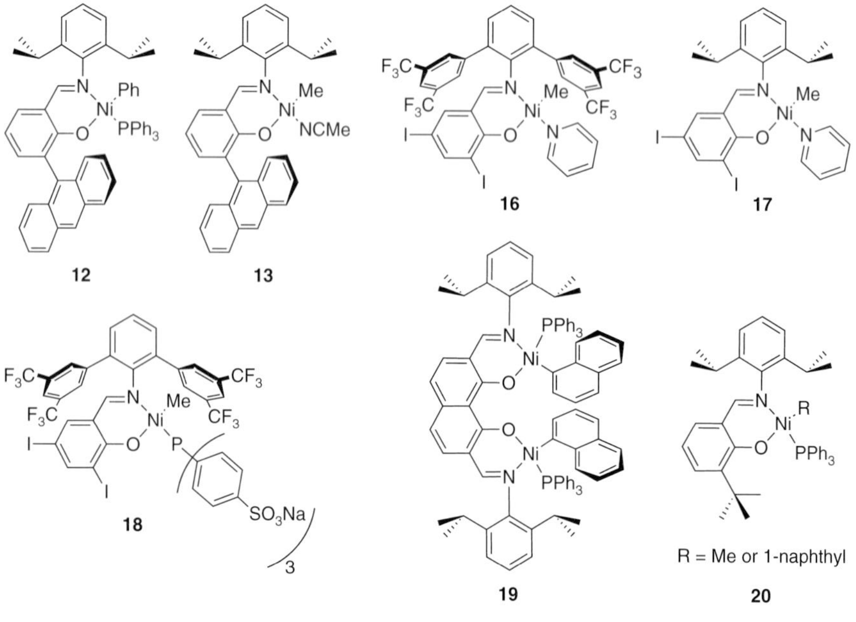

Figure 14.3 Representative salicylaldimine-ligated Ni complexes.

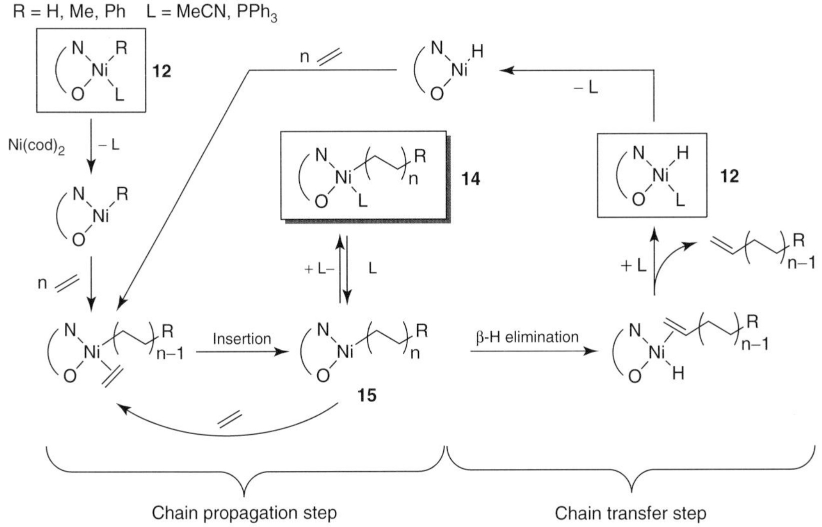

R = H, Me, Ph   L = MeCN, PPh₃

Scheme 14.3 Mechanism of ethylene polymerization mediated by **12**.

Neutral Ni complexes possess excellent tolerance to functional groups. Catalyst **12** maintained high efficiency in ethylene polymerization in the presence of ethers, ketones, and esters. Surprisingly, **12** is capable of polymerizing ethylene even in the presence of 1500 equiv. of water, although significant decreases in both lifetime and activity (<0.1 g mmol^{-1} Ni h^{-1}) are observed. Consequently, catalyst **12** mediates the copolymerization of ethylene with 5-hydroxynorbornene to form a copolymer containing 22 wt% of 5-hydroxynorbornene.

Mecking and co-workers utilized this water tolerance feature for an aqueous emulsion polymerization system [11]. Ethylene polymerization employing novel Ni complex **16** in aqueous emulsion shows an activity of 0.16 g mmol^{-1} Ni h^{-1}, and this is five times higher than the activity shown by *i*Pr-substituted complex **17**. The yield of PE increases linearly with reaction time from 1 to 5 h, indicating the high stability of the catalyst active species in an aqueous emulsion system.

Interestingly, Ni complex **18**, which coordinates a water-soluble phosphine, allows the formation of PE dispersions in the absence of an organic solvent, and a lower polymerization temperature allows higher activity and increases the molecular weight of the PE due to retardation of the irreversible deactivation of **18**. the productivity of **18** at 15 °C in water (0.36 g mmol^{-1} Ni h^{-1}) is higher than that of **16** at 15 °C in toluene (0.02 g mmol^{-1} Ni h^{-1}), although a longer catalyst life is observed for **16**. The complexation equilibrium between hydrophobic active species and hydrophilic phosphine (analogs of **14** in Scheme 14.3) shifts toward the active species (dissociation) in aqueous emulsion because a water-soluble phosphine prefers to be in water after dissociation from **18**, leading to the activation of ethylene insertion around hydrophobic active species. The molecular weight of PE particles is 18 000 g mol^{-1} with a low polydispersity index ($M_w/M_n = 1.9$) and with a low degree of methyl branching (5 branches per 1000 C). The size of the PE particles is extremely small (4 nm measured by dynamic light scattering).

Further successful development of ethylene polymerization in dense carbon dioxide employing a phenoxyimine-ligated Ni–methyl complex was also reported recently [11c].

Marks and co-workers developed a bimetallic neutral Ni complex having a 2,7-diimino-1,8-dioxynaphthalene ligand (Figure 14.3) [12]. Room-temperature ethylene polymerization using bimetallic Ni complex **19** in the presence of 2 equiv. of Ni(cod)$_2$ exhibits a twofold greater productivity along with more frequent methyl branching than the mononuclear analog **20**. This difference may reflect possible enhanced phosphine dissociation due to the encumbered and rigid bimetallic structures. The PE obtained from **19** possesses comparable molecular weight and polydispersity to the corresponding mononuclear complex **20**. Moreover, the **19**-catalyzed copolymerization of ethylene with MA incorporates up to 11 mol% MA, whereas **20** incorporates a negligible amount of MA, probably because the last-inserted MA unit easily forms a stable chelate structure with its carbonyl group, which inactivates the catalyst. The mechanism for a bimetallic catalyst proposes that the carbonyl group of the last-inserted MA unit on one Ni metal atom is trapped by the neighboring Ni metal to prevent catalyst inactivation, resulting in the successful incorporation of MA.

14.3
Early Transition Metal Complexes

Although early transition metals, especially in group 4, played an important role in the historical development of alkene polymerization catalysts, most of the post-metallocenes based on early transition metals came after the late metal post-metallocene complexes in the late 1990s (the post-metallocene era). Some prominent post-metallocene catalysts including group 4 transition metal complexes are discussed below.

14.3.1
Phenoxyimine-Ligated Group 4 Metal Complexes

Bis(phenoxyimine) group 4 transition metal complexes for alkene polymerization (now known as FI catalysts) were discovered by researchers at Mitsui Chemicals in the late 1990s. The phenoxyimine ligand was selected on the basis of a ligand-oriented catalyst design concept, a concept that is founded on the belief that the flexible electronic nature of a ligand is a fundamental requirement for achieving high activity. The notable features of group 4 transition metal complexes bearing a pair of phenoxyimine ligands as alkene polymerization catalysts [13] and a tridentate phenoxyimine-ligated Ti catalyst for the selective ethylene trimerization reaction [14] are outlined below

14.3.1.1 High Activity for Ethylene Polymerization
The original Zr-FI complex **21** (Figure 14.4) activated by MAO displays a very high activity of 520 g mmol^{-1} Zr h^{-1} for ethylene polymerization at 3.3 g mmol^{-1} Ti h^{-1} under an atmospheric ethylene pressure, which is 20 times higher than the activity observed with Cp$_2$ZrCl$_2$–MAO under the same reaction conditions. The Zr-FI–MAO catalyst possesses the highest activity among the three analogs having group 4 metals

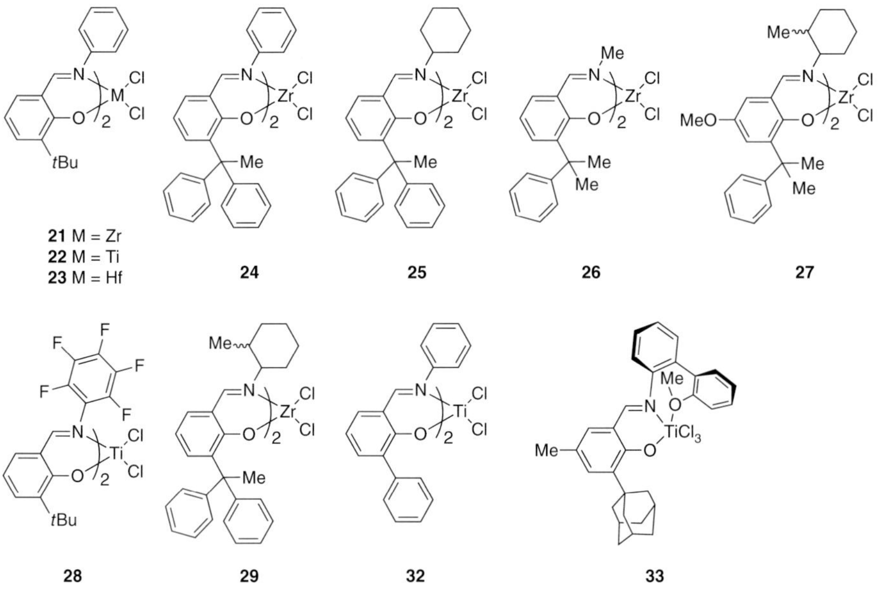

Figure 14.4 Representative phenoxyimine-ligated group 4 metal complexes.

(3.3 g mmol^{-1} Ti h^{-1} for Ti-FI **22**, 6.5 g mmol^{-1} Hf h^{-1} for Hf-FI **23**). The activity of the Zr-FI catalyst increases with increasing steric hindrance of the *ortho*-substituents on the phenoxy moiety. MAO-activated Zr-FI complex **24** having a 1,1-diphenylethyl group at the *ortho*-position to the phenoxy oxygen shows extremely high ethylene polymerization productivity (~2400 g mmol^{-1} Zr h^{-1}), and an additional change of the imino substituents from Ph to cyclohexyl (**25**) achieves an exceptionally high activity of up to ~6600 g mmol^{-1} Zr h^{-1}. This activity is one of the highest catalyst efficiencies for any catalytic reaction ever reported. The bulky substituents on the o-phenoxy fragment protect the metal center or the phenoxy oxygens, thus preventing a side reaction, and may also hinder the formation of a dinuclear complex that is an inactive dormant species.

14.3.1.2 Wide-Ranging Control over the Molecular Weight of the PE

FI catalysts are capable of controlling the molecular weight of PE over a wide range, from about 10^3 to 10^6 g mol^{-1} without adding any chain-transfer reagents. The effect of the substituents at the imine positions is mostly predominant, and introduction of a less bulky group decreases the molecular weight of the resultant PE. For example, ethylene polymerization at 150 °C under high pressure (3 MPa) mediated by **26**–MAO yields low molecular weight PE (M_w 2000 g mol^{-1}) having one vinyl terminus per polymer chain (>95%). The low molecular weight vinyl-terminated PE can be readily converted to epoxy- and diol-terminated PE, and the subsequent synthesis of block copolymers containing PE and poly(ethylene glycol) (PEG) segments (Scheme 14.4). These hybrid materials form a stable water dispersion with no addition of surfactants. For instance, the **AB³** hybrid (average molecular weight of PEG segment 400 g mol^{-1},

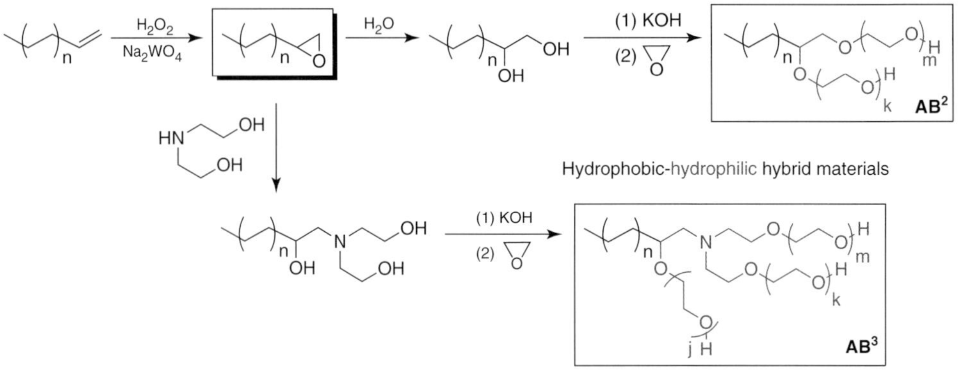

Scheme 14.4 Synthesis of block copolymers using low molecular weight vinyl-terminated PE.

melting temperature $T_m = 121\,^\circ$C) can form a semi-transparent dispersion, consisting of nanoparticles ($\sim$18 nm on average) with a narrow size distribution.

FI catalysts bearing sterically encumbered groups near the polymerization sites enhance the molecular weight of the resultant PE, leading to the production of ultrahigh molecular weight PEs. For example, Zr-FI complex **27** supported on MAO–silica under pressurized ethylene conditions (0.8 MPa at 80 $^\circ$C) affords ultrahigh molecular weight linear PE (viscosity-average molecular weight M_v 4 400 000 g mol^{-1},) with well-controlled spherical morphologies. Density functional theory (DFT) calculations suggest that larger substituents at the imine moiety destabilize the transition state for β-H transfer to the monomer, resulting in an increase in molecular weight.

14.3.1.3 Living Polymerization Mediated by Fluorinated Ti-FI Catalysts

Interestingly, fluorinated Ti-FI catalysts can perform the thermally robust living polymerization of ethylene and the syndiospecific living polymerization of propylene. Ti-FI catalyst **28**–MAO conducts living polymerization with ethylene even at 50 $^\circ$C to afford high molecular weight PE (M_w 420 000 g mol^{-1}) with a low polydispersity index (M_w/M_n 1.1). In addition, **28**–MAO exhibits an exceptionally high turnover frequency (21500 min^{-1} atm^{-1}), which corresponds to the activity observed with a non-living Cp$_2$ZrCl$_2$–MAO catalyst under the same reaction conditions. Propylene polymerization using **28**–MAO at room temperature also proceeds in a living fashion mediated by a repetitive 2,1-insertion mechanism, which yields sPP (*rr* 87%) with a low polydispersity index (M_w/M_n1.1). Based on DFT calculations, the highly controlled living nature of the **28**–MAO system was proposed to originate from weak attractive and electrostatic interactions between the *ortho*-F and a β-H on the growing polymer chain, which suppresses β-H transfer reactions (Figure 14.5). The fluorinated Ti-FI catalysts allow access to di-, tri-, and multi-block copolymers comprised of crystalline and amorphous segments from ethylene, propylene, and higher α-alkenes. The diblock copolymer composed of crystalline PE and amorphous poly(ethylene-*co*-1-hexene) segments possesses a good combination of extensibility and toughness.

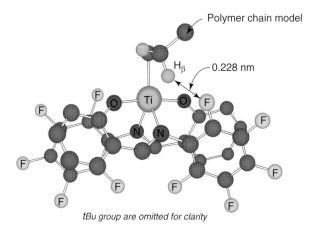

Polymer chain model

H_β — 0.228 nm

tBu group are omitted for clarity

Figure 14.5 Calculated structure of an active species derived from **28**.

14.3.1.4 Effect of Catalyst Activator

The original Zr-FI complex **21** in combination with Al(iBu)$_3$ and [Ph$_3$C]TPB produced remarkably high molecular weight PE (M_v 5 000 000 g mol^{-1}) relative to the case for MAO (M_v 10 000 g mol^{-1}). This unprecedented difference by activators originates from reduction of the imine with Al(iBu)$_3$ to form a bis(phenoxyamine)Zr active species.

Notably, MgCl$_2$–Et$_n$Al(OR′)$_{3-n}$(R′ = 2-ethyl-1-hexyl) can activate and simultaneously immobilize FI catalysts to prepare heterogeneous single-site catalysts, leading to the successful production of noncoherent PE microparticles with excellent morphology and narrow particle size distribution. For instance, Zr-FI complex **29** when reacted with MgCl$_2$–Et$_n$Al(OR′)$_{3-n}$ polymerizes ethylene (80 °C, 0.8 MPa) with high productivity (360 g mmol^{-1}Zr) to form noncoherent spherical PE particles with ultrahigh molecular weight (M_v3 000 000 g mol^{-1}). Importantly, the particle size of the PE is tunable from ultrafine (3 μm) to the submillimeter scale by changing the particle size of MgCl$_2$–Et$_n$Al(OR′)$_{3-n}$.

14.3.1.5 Stereospecific Polymerization of Propylene

Ti-FI catalysts in conjunction with MAO or MgCl$_2$–Et$_n$Al(OR′)$_{3-n}$(R′ = 2-ethyl-1-hexyl) can perform the syndiospecific polymerization of propylene, and the syndioselectivity is proportional to the bulkiness of the *ortho* substituents at the phenoxy fragment. MAO-activated **30** having a trimethylsilyl group generates highly syndiospecific PP (rr = 94%) with a high T_m of 156 °C with a very low polydispersity index (M_w/M_n 1.1), demonstrating living polymerization (Scheme 14.5). Surprisingly, microstructural analyses of the sPP obtained indicated that polymerization proceeded with the consecutive 2,1-insertion of the propylene monomer by a chain-end control mechanism. Interestingly, isoselective polymerization of propylene was also established by the use of Hf-FI complex **31** activated with [Ph$_3$C]TPB and Al(iBu)$_3$. The iPP produced possesses a high molecular weight (M_w360 000 g mol^{-1}) and a high isospecificity (mm = 98%). The T_m of the iPP was 165 °C, which is one of the highest values ever reported, including the use of the heterogeneous Ziegler–Natta catalyst. The polymer structure of the iPP implies

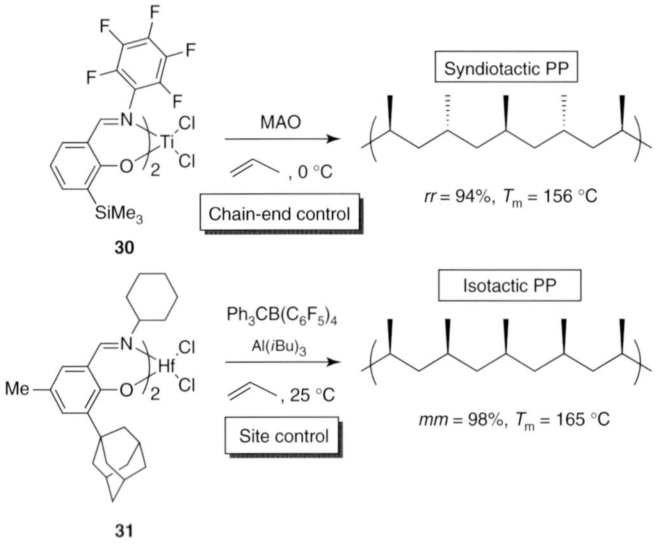

Scheme 14.5 Stereospecific polymerization of propylene catalyzed by FI catalysts.

that propylene monomers were isoselectively enchained by repetitive 1,2-insertion via an enantiomorphic site-control mechanism. As mentioned above, the formation of bis(phenoxyamine)Hf active species probably contributes to the successful isospecific polymerization of propylene.

14.3.1.6 Copolymerization of Ethylene with Cyclic Alkenes

Interestingly, Ti-FI catalysts bearing the phenyl group at the *ortho* position to the phenoxy oxygen can act as an exceptionally high-performance catalyst for the copolymerization of ethylene with cyclic alkenes such as norbornene (NB). For example, Ti-FI complex **32** when activated with MAO displays an activity of (2 g mmol^{-1} Ti h^{-1}), whereas the original Ti-FI complex **22** in conjunction with MAO has no productivity under the same reaction conditions, indicating the beneficial feature of the phenyl group on the *ortho*-phenoxy moiety. Moreover, the ethylene–NB copolymer obtained contains a high NB incorporation ratio (48.4 mol%) with high M_w (200 000) and a low polydispersity index (M_w/M_n 1.1). NMR analyses of the copolymer revealed that the sequence of ethylene and NB is alternating (−NB−E−NB−E−NB− 96%) with a slightly syndiotactic-enriched structure (−NB−E−NB−, *iso:syn* 45:55).

14.3.1.7 Selective Production of 1-Hexene by Ethylene Trimerization

Recently, ligand-oriented catalyst design concept was employed in the discovery of a selective ethylene trimerization catalyst to produce 1-hexene with extremely high activity [14]. Ti complex **33** bearing a tridentate phenoxyimine ligand can selectively convert ethylene to 1-hexene after activation with MAO. The activity of Ti catalyst **33** increased when the ethylene pressure was increased, and 1-hexene productivity reached 6.59 t g^{-1} Ti h^{-1}) with high selectivity (92.3 wt%) when pressurized to 5.0 MPa. The proposed reaction

FI-Ti = tridentate phenoxyimine-ligated Ti

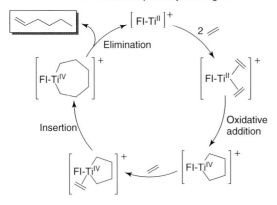

Scheme 14.6 Proposed mechanism for selective ethylene trimerization catalyzed by the **33**–MAO system.

mechanism initiates with the formation of a Ti(II) species. Oxidative addition of two ethylene molecules to the Ti(II) species to form a Ti(IV) metallacyclopentane intermediate, and subsequent insertion of another ethylene molecule in this intermediate yields a Ti(IV) metallacycloheptane species, which is subjected either to stepwise β-H elimination/transfer and reductive elimination or to concerted 3,7-H transfer reaction, which then regenerates the Ti(II) species (Scheme 14.6). Interestingly, catalyst productivity shows a second-order dependence on ethylene pressure, suggesting that the rate-determining step is the formation of the Ti(IV) metallacyclopentane intermediate.

14.3.2
Chelating Bis(phenoxy)-Ligated Group 4 Metal Complexes

A Ti complex bearing a sulfur-bridged chelating bis(phenoxy) ligand s an early example of the linked bis(phenoxy)-ligated metal catalyst for alkene polymerization before the post-metallocene era (Figure 14.6) [15]. Complex **34** in combination with MMAO exhibits good activity (3.1 g mmol^{-1} Ti h^{-1}) for ethylene polymerization, and produces PE with high molecular weight (M_w440 000 g mol^{-1}). Propylene polymerization performed with **34**–MAO yields extremely high molecular weight aPP ($M_w > 8\,000\,000$ g mol^{-1}).

In 2003, Okuda and co-workers reported on another chelating bis(phenoxy)-ligated Ti complex possessing a 1,4-dithiabutanediyl bridge (**35**) (Figure 14.6) [16]. The higher MMAO:**35** molar ratio ([Al] : [Ti] = 5000) resulted in higher ethylene polymerization activity (~24 g mmol^{-1} Ti h^{-1}) and a lower polydispersity index ($M_w/M_n < 3$) for the resulting PE (M_w90 000 g mol^{-1}), indicating that excess MMAO promotes the selective formation of a single-site catalyst. The resulting PE is a mixture of both linear and branched structures (5–10 branches per 1000 C).

Interestingly, the **35**–MAO system is capable of performing isospecific styrene polymerization according to its C_2 symmetric ligand structure [16]. The isotactic polystyrene obtained has ultrahigh molecular weight (M_w2 650 000 g mol^{-1}) with a high T_m (223 °C),

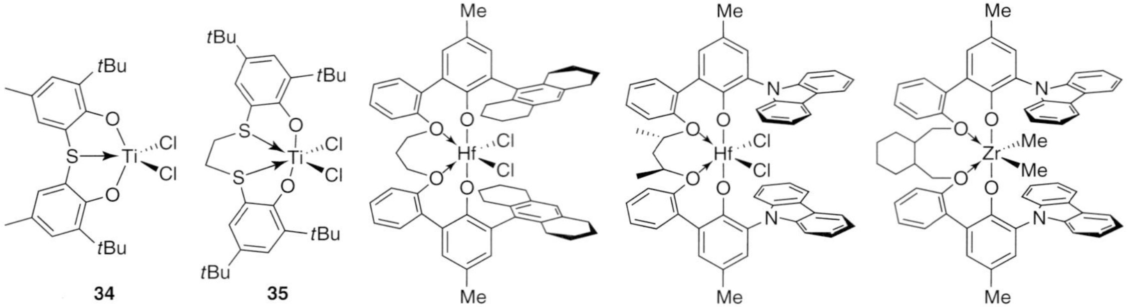

Figure 14.6 Representative chelating bis(phenoxy)-ligated group 4 metal complexes.

and polymer analysis indicated that polymerization proceeds with the secondary insertion of styrene into the Ti-CH$_3$ bond, and there is no detectable regio-inversion in the polymer chain. In the case of copolymerization of ethylene with styrene employing **35**–MAO, poly(ethylene-*co*-styrene)s incorporating styrene up to 68 mol% can form without producing homopolymer byproducts because of the high relative reactivity towards the styrene monomer. Moreover, styrene polymerization mediated by **35**–MAO in the presence of 1,9-decadiene exclusively yields novel vinyl-terminated isotactic polystyrene.

In the meantime, high-throughput-screening (HTS) technology at Symyx discovered novel chelating bis(phenoxy)-ligated Zr and Hf complexes for iPP production, with a C_2-symmetric ligand structure (Figure 14.6) [17]. Propylene polymerization at 50 °C using **36** activated with 1.1 equiv. of TPB and 30 equiv. of AlH(*i*Bu)$_2$ affords iPP with a high T_m of 147 °C. Further modification of substituents on the tetradentate ligand reflects better catalyst performance at a higher polymerization temperature. For instance, propylene polymerization at 110 °C mediated by **37**–1.1TPB–30AlH(*i*Bu)$_2$ shows a high activity ($\sim$460 g mmol^{-1} Hf h^{-1}), and the resulting iPP possesses a high M_w (230 000) with a high T_m of 151 °C. In addition, the copolymerization of ethylene with 1-octene at 130 °C employing **37**–MMAO also displays a high activity ($\sim$60 g mmol^{-1} Hf h^{-1}) under high pressure (6.8 atm), yielding a high molecular weight copolymer (M_w 910 000) with a low polydispersity index (M_w/M_n 2.6). Furthermore, researchers at Dow Chemical observed that the ethylene–1-octene copolymerization catalyzed by **38**–1.2TPB–5MMAO in a continuous solution process (190 °C, 3.45 MPa) shows very high activity ($\sim$690 g mmol^{-1}Zr), and the resulting copolymer has a high molecular weight (M_w71 000 g mol^{-1}) with a low polydispersity index (M_w/M_n 2.0). This is an industrially attractive approach for the cost-effective production of polyalkenes.

14.3.3
Pyridylamine-Ligated Hf Complexes

Researchers at Symyx and Dow Chemical discovered pyridylamine-ligated Hf complexes using HTS technology [18]. Hf complexes possess unique structures that feature the presence of Hf-C(aryl) σ-bonds (**39** and **40** in Scheme 14.7). Surprisingly, **39** activated with 1.1 equiv. of [HNMe$_2$Ph]TPB and 30 equiv. of AlH(*i*Bu)$_2$ is capable of producing iPP having a high molecular weight (M_w 710 000 g mol^{-1}) with a peak high T_m of 141 °C in a high-temperature (90 °C) solution process.

Extensive studies of catalyst activation of the pyridylamine-ligated Hf complex revealed that monomer insertion into the Hf-C(aryl) bond results in the *in situ* modification of the ligand structure, leading to the formation of an active species (Scheme 14.7) [19]. An intermediate **41**, which is stabilized by an Hf–η^2-naphthyl interaction and is inactive to alkenes, is generated by protonation of the σ-aryl–metal bond in **40** with [NHMe$_2$Ph]TPB. The following *ortho*-cyclometallation of the naphthyl group results in the production of monomethyl cationic species **42**, but this species is still a precursor of the real active species. The first monomer insertion to **42** preferentially goes into the Hf–C(aryl) bond instead of the Hf–Me bond to yield real active species **43**, which includes many different diastereomers depending on the regio- and stereochemistry of the first inserted alkene.

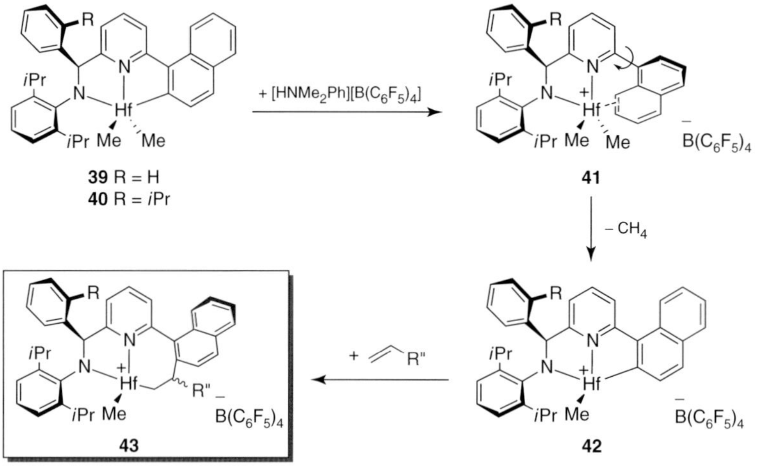

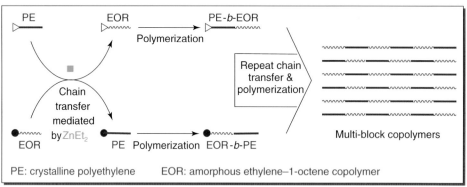

Scheme 14.7 Proposed activation pathway of the pyridylamine-ligated Hf complex.

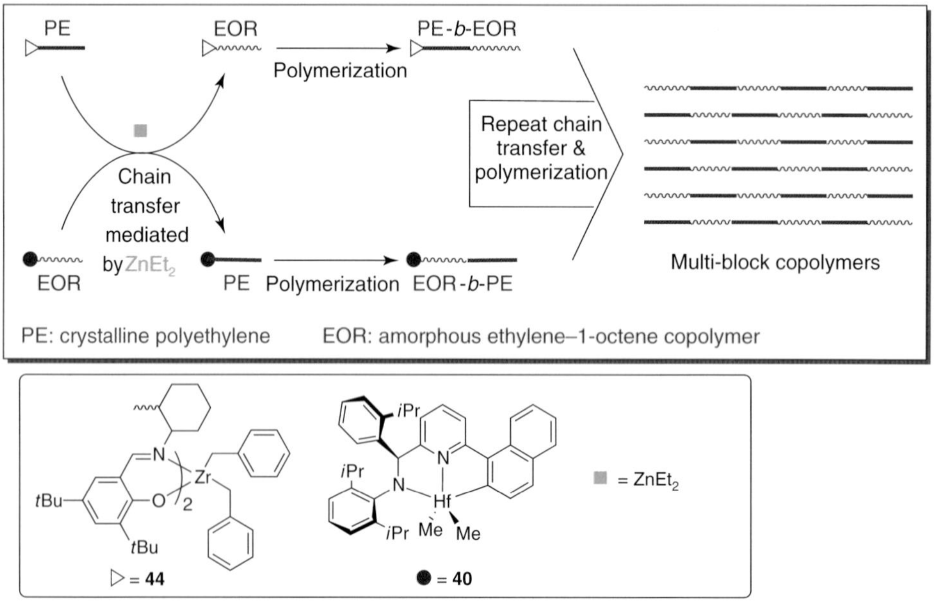

Scheme 14.8 Mechanism for the production of multi-block copolymers.

Notably, researchers at Dow Chemical reported on the production of multi-block copoly-mers comprised of PE and amorphous ethylene–1-octene copolymer (EOR) segments, using a dual catalyst system consisting of Zr-FI complex **44** and pyridylamine-ligated Hf complex **40** in the presence of ZnEt₂ and [HNMe₂Ph]TPB (Scheme 14.8) [20]. Zr-FI catalyst **44** selectively forms crystalline PE even in the co-presence of ethylene and 1-octene, while Hf catalyst **40** produces amorphous EOR. Since both **44** and **40** can efficiently perform the reversible chain-transfer reaction with ZnR_2, unusual active species for both **44** (bound with an EOR chain) and **40** (with a PE chain) simultaneously form after the start of polymerization. Subsequent polymerization of these unusual species produces block linkages (PE-*b*-EOR), and repeating this polymer chain exchange results in the production of multi-block copolymers, (PE-*b*-EOR)$_n$. Both high ethylene selectivity and reversible chain-transfer ability of Zr-FI catalyst **44** permits the formation of unique multi-block copolymers. Unlike the random EOR with similar density, the multi-block copolymer possesses a 40 °C higher melting temperature ($T_m = 120$°C) because of the crystalline PE segments while maintaining excellent elastic properties thanks to the soft EOR segments. Another approach to the efficient synthesis of the block polymer utilizing the **40**–[HNMe$_2$Ph]TPB–ZnEt$_2$ system is the stepwise polymerization methodology in continuous stirred-tank reactors. The primary reactor produces Zn(PE)$_2$ species and, following the copolymerization of ethylene with propylene in the presence of **40**–[HNMe$_2$Ph]TPB and Zn(PE)$_2$ in the secondary reactor, selectively prepares diblock copolymers PE-*b*-(ethylene–propylene copolymer). This is an industrially powerful tool to yield well-defined block copolymers with extraordinarily high efficiency.

14.4
Conclusion and Perspectives

Investigations of the development of new alkene polymerization catalysts, following the development of metallocene catalysts, have resulted in the discovery of a large number of high-performance post-metallocene catalysts based on both early and late transition metals.

The post-metallocene catalysts introduced in this chapter exhibit unique features such as precise control over chain transfers, chain-walking polymerization, high functional group tolerance, polymerization in water, reversible chain-transfer ability, highly syndioselective propylene polymerization despite a C_2 symmetric nature, highly isoselective propylene polymerization at high temperature, and MAO- and borate-free polymerization. These notable features have led to the production of a wide variety of unique alkene-based materials, many of which were previously unavailable with metallocene catalysts. The materials include hyper-branched PE, ultrahigh molecular weight PE, ethylene–MA copolymers, vinyl-terminated PE, ultrafine PE particles, polyolefinic block copolymers from ethylene, propylene, and higher α-alkenes, and PE–PEG hybrid materials.

Additionally, studies of the above post-metallocene catalysts have enhanced our understanding of the causal relationship between catalyst structures, polymerization characteristics, primary and higher order polymer structures, and the chemical and physical properties of the resultant materials.

Therefore, the development and application of post-metallocene catalysts has had an enormous impact on polymerization catalysis and polymer synthesis, and has also given us industrial opportunities to produce value-added alkene-based materials of great benefit to mankind. It is expected that future research on post-metallocene catalysts will expand the range of alkene-based materials and will provide a chance to study catalysis and mechanisms for alkene polymerization.

Acknowledgment

We greatly appreciate the contributions of all those who are active in post-metallocene studies at Mitsui Chemicals, Inc.

References

1. (a) Ziegler, K., Gellert, H.G., Zosel, K., Lehmkuhl, W., and Pfohl, W. (1955) *Angew. Chem. Int. Ed. Engl.*, **67**, 424; (b) Ziegler, K., Holzkamp, E., Breil, H., and Martin, H. (1955) *Angew. Chem. Int. Ed. Engl.*, **67**, 541; (c) Natta, G. (1955) *J. Polym. Sci.*, **16**, 143; (d) Natta, G. (1955) *J. Am. Chem. Soc.*, **77**, 1707.

2. Kashiwa, N., Fujimura, H., and Tokuzumi, Y. (1968) Japanese Patent 1,031,698 (application in Japan on 1 August 1968).

3. (a) Sinn, H., Kaminsky, W., Vollmer, H.J., and Woldt, R. (1980) *Angew. Chem. Int. Ed. Engl.*, **19**, 390; (b) Sinn, H. and Kaminsky, W. (1980) *Adv. Organomet. Chem.*, **18**, 99.

4. (a) Johnson, L.K., Killian, C.M., and Brookhart, M. (1995) *J. Am. Chem. Soc.*, **117**, 6414–6415; (b) Ittel, S.D., Johnson, L.K., and Brookhart, M. (2000) *Chem. Rev.*, **100**, 1169–1203.

5. Killian, C.M., Tempel, D.J., Johnson, L.K., and Brookhart, M. (1996) *J. Am. Chem. Soc.*, **118**, 11664–11665.

6. (a) Johnson, L.K., Mecking, S., and Brookhart, M. (1996) *J. Am. Chem. Soc.*, **118**, 267–268; (b) Mecking, S., Johnson, L.K., Wang, L., and Brookhart, M. (1998) *J. Am. Chem. Soc.*, **120**, 888–899.

7. (a) Camacho, D.H., Salo, E.V., Ziller, J.W., and Guan, Z. (2004) *Angew. Chem. Int. Ed.*, **43**, 1821–1825; (b) Camacho, D.H. and Guan, Z. (2005) *Macromolecules*, **38**, 2544–2546; (c) Popeney, C.S., Camacho, D.H., and Guan, Z. (2007) *J. Am. Chem. Soc.*, **129**, 10062–10063.

8. (a) Small, B.L., Brookhart, M., and Bennett, A.M.A. (1998) *J. Am. Chem. Soc.*, **120**, 4049–4050; (b) Britovsek, G.J.P., Gibson, V.C., Kimberley, B.S., Maddox, P.J., McTavish, S.J., Solan, G.A., White, A.J.P., and Williams, D.J. (1998) *Chem. Commun.*, 849–850; (c) Small, B.L. and Brookhart, M. (1998) *J. Am. Chem. Soc.*, **120**, 7143–7144; (d) Gibson, V.C., Redshaw, C., and Solan, G.A. (2007) *Chem. Rev.*, **107**, 1745–1776.

9. (a) Britovsek, G.J.P., Cohen, S.A., Gibson, V.C., Maddox, P.J., and van Meurs, M. (2002) *Angew. Chem. Int. Ed.*, **41**, 489–491; (b) Britovsek, G.J.P., Cohen, S.A., Gibson, V.C., and van Meurs, M. (2004) *J. Am. Chem. Soc.*, **126**, 10701–10712.

10. (a) Wang, C., Friedrich, S., Younkin, T.R., Li, R.T., Grubbs, R.H., Bansleben, D.A., and Day, M.W. (1998) *Organometallics*, **17**, 3149–3151; (b) Younkin, T.R., Connor, E.F., Henderson, J.I., Friedrich, S.K., Grubbs, R.H., and Bansleben, D.A. (2000) *Science*, **287**, 460–462.

11. (a) Zuideveld, M.A., Wehrmann, P., Röhr, C., and Mecking, S. (2004) *Angew. Chem. Int. Ed.*, **43**, 869–873; (b) Göttker-Schnetmann, I., Korthals, B., and Mecking, S. (2006) *J. Am. Chem. Soc.*, **128**, 7708–7709; (c) Guironnet, D., Göttker-Schnetmann, I., and Mecking, S. (2009) *Macromolecules*, **42**, 8157–8164.

12. (a) Rodriguez, B.A., Delferro, M., and Marks, T.J. (2008) *Organometallics*, **27**, 2166–2168; (b) Rodriguez, B.A., Delferro, M., and Marks, T.J. (2009) *J. Am. Chem. Soc.*, **131**, 5902–5919.

13. (a) Makio, H., Kashiwa, N., and Fujita, T. (2002) *Adv. Synth. Catal.*, **344**, 477–493; (b) Mitani, M., Saito, J., Ishii, S., Nakayama, Y., Makio, H., Matsukawa, N., Matsui, S., Mohri, J., Furuyama, R., Terao, H., Bando, H., Tanaka, H., and Fujita, T. (2004) *Chem. Rec.*, **4**, 137–158; (c) Makio, H. and Fujita, T. (2009) *Acc. Chem. Res.*, **42**, 1532–1544; (d) Makio, H., Terao, H., Iwashita, A., and Fujita, T. (2011) *Chem. Rev.*, **111**, 2363–2449.

14. Suzuki, Y., Kinoshita, S., Shibahara, A., Ishii, S., Kawamura, K., Inoue, Y., and Fujita, T. (2010) *Organometallics*, **29**, 2394–2396.

15. (a) Miyatake, T., Mizunuma, K., Saeki, Y., and Kakugo, M. (1989) *Makromol. Chem. Rapid Commun.*, **10**, 349–352; (b) Miyatake, T., Mizunuma, K., and Kakugo,

M. (1993) *Makromol. Chem. Makromol. Symp.*, **66**, 203–214; (c) Takaoki, K. and Miyatake, T. (2000) *Makromol. Symp.*, **157**, 251–257.

16. (a) Capacchione, C., Proto, A., and Okuda, J. (2004) *J. Polym. Sci. A: Polym. Chem.*, **42**, 2815–2822; (b) Capacchione, C., Proto, A., Ebeling, H., Mülhaupt, R., Spaniol, T.P., Möller, K., and Okuda, J. (2003) *J. Am. Chem. Soc.*, **125**, 4964–4965; (c) Capacchione, C., Proto, A., Ebeling, H., Mülhaupt, R., and Okuda, J. (2006) *J. Polym. Sci. A: Polym. Chem.*, **44**, 1908–1913; (d) Gall, B.T., Pelascini, F., Ebeling, H., Beckerle, K., Okuda, J., and Mülhaupt, R. (2008) *Macromolecules*, **41**, 1627–1633.

17. (a) Boussie, T.R., Brümmer, O., Diamond, G., Goh, C., LaPointe, A.M., Leclerc, M.K., and Shoemaker, J.A. (2003) International Patent Appl. WO2003/091262A1; (b) Boussie, T.R., Brümmer, O., Diamond, G.M., LaPointe, A.M., Leclerc, M.K., Micklatcher, C., Sun, P., and Bei, X. (2007) US Patent 7,241,714, to Symyx Technologies; (c) Konze, W.V. and Vanderlende, D.D. (2007) International Patent Appl. WO2007/136506A2.

18. Boussie, T.R., Diamond, G.M., Goh, C., Hall, K.A., LaPointe, A.M., Leclerc, M.K.,

Murphy, V., Shoemaker, J.A.W., Turner, H., Rosen, R.K., Stevens, J.C., Alfano, F., Busico, V., Cipullo, R., and Talarico, G. (2006) *Angew. Chem. Int. Ed.*, **45**, 3278–3283.

19. (a) Froese, R.D.J., Hustad, P.D., Kuhlman, R.L., and Wenzel, T.T. (2007) *J. Am. Chem. Soc.*, **129**, 7831–7840; (b) Zuccaccia, C., Macchioni, A., Busico, V., Cipullo, R., Talarico, G., Alfano, F., Boone, H.W., Frazier, K.A., Hustad, P.D., Stevens, J.C., Vosejpka, P.C., and Abboud, K.A. (2008) *J. Am. Chem. Soc.*, **130**, 10354–10368; (c) Zuccaccia, C., Busico, V., Cipullo, R., Talarico, G., Froese, R.D.J., Vosejpka, P.C., Hustad, P.D., and Macchioni, A. (2009) *Organometallics*, **28**, 5445–5458; (d) Busico, V., Cipullo, R., Pellecchia, R., Rongo, L., Talarico, G., Macchioni, A., Zuccaccia, C., Froese, R.D.J., and Hustad, P.D. (2009) *Macromolecules*, **42**, 4369–4373.

20. (a) Arriola, D.J., Carnahan, E.M., Hustad, P.D., Kuhlman, R.L., and Wenzel, T.T. (2006) *Science*, **312**, 714–719; (b) Hustad, P.D., Kuhlman, R.L., Arriola, D.J., Carnahan, E.M., and Wenzel, T.T. (2007) *Macromolecules*, **40**, 7061–7064.

Commentary Part

Comment 1

Robert Grubbs

Since the discovery of high-density PE by Ziegler in the 1950s, there have been a number of families of catalysts that have been explored in great detail. The initial Ziegler catalysts produced excellent material and the heterogeneous systems resulting from the reduction of titanium tetrachloride with alkylaluminums continue to be the workhorses of the area. Many advances have been made in these systems to improve and create new polymer properties. The other major heterogeneous catalyst systems are the Phillips systems based on supported chromium.

Breslow's discovery of metallocenes as catalysts and the discovery by Kaminsky of MAO as an activator opened up the area of metallocene systems for polyalkene synthesis. An amazing variety of these systems have been explored and a number

are now being used commercially. As a result of the discovery of Keim that late transition metals such as nickel complexes would produce interesting new PEs under conditions not possible with early systems, a variety on non-metallocene systems have been developed. Through the work of Brookhart and others, the late metal systems have evolved into highly active and controllable systems. It is the evolution of this family of catalysts that is the focus of this chapter. In addition to providing the story of this evolution, the authors provide a concise description of the value and features of this complete family of catalysts. Particularly exciting has been the developments at Mitsui Chemicals, where some of the ligand types that were initially used with early metals have been utilized to produce highly active titanium- and zirconium-based systems. These ligands are easily tailored and can now be designed to provide precise control of the polymer morphology and molecular weight. Although numerous reviews are available on metallocenes, this chapter provides an excellent guide to the non-metallocene family of active catalysts. Of particular importance is the discussion of the discovery of an unusual class of ligand systems for polymerization that were discovered using HTS methods.

In many ways, these new catalysts are an example of the status of modern catalysis. By applying many of the tools of organometallic chemistry, new and useful catalysts can be discovered and understood in detail. These studies have provided an array of polymers with new topologies and properties.

Will there be another generation of polyalkene catalysts that produce polyalkenes with unexpected properties?

Comment 2

Jun Okuda

General

This chapter by Kaneyoshi, Makio, and Fujita compiles the latest developments in the area of post-metallocene or non-metallocene catalysts for alkene polymerization. Despite the paramount importance of polyalkene materials in applications based on the original discovery by Karl Ziegler of the ill-defined "Mischkatalysatoren," it is interesting to note that even after 50 years there still does not exist a clear *molecular* picture about what makes Ziegler catalysts so remarkable and efficient, both in activity and stereoselectivity. Therefore, it is ironical that a class of compounds, discovered at about the same time, coined sandwich complexes or metallocenes, could contribute to alkene polymerization catalysis only after more than 30 years. In particular, the chiral Brintzinger-type *ansa*-zirconocenes not only allowed emulation of the isoselective propene polymerization by molecularly defined catalysts, but also led to an in-depth understanding of the principles governing alkene polymerization catalysis [C1]. Although following the metallocene revolution a certain disappointment set in among some industrial researchers, the true breakthrough brought about by metallocene catalysts pertains to the fundamental insight into the molecular details of the polymerization mechanism. Molecular catalysts can not only reach the performance of heterogeneous catalysts, always vexed by the difficulty of defining and, most importantly, manipulating the active site in a rational way, but also have the potential to surpass

the heterogeneous systems by offering an array of new catalyst structures and thus material properties. PPs synthesized by zirconocene catalysts have begun to occupy niche markets and some of the commercially successful polyalkenes such as linear low-density polyethylene (LLDPE) were first introduced using half-sandwich catalysts. Recently, this class of compounds has expanded their performance to a level of sophistication unreached by heterogeneous or zirconocene catalysts [C2]. Metal complexes without any stabilizing cyclopentadienyl ligand in the ligand sphere have entered the arena of single-site catalysts and now have turned into the mainstream of polymerization catalysis research (Figure C1).

Early Work on Late Metals

As is often the case with major developments, the surprising discovery that late transition metal complexes are active polymerization catalysts had earlier precedents. With hindsight, it is not clear why nickel catalysts, which are exceptionally efficient oligomerization catalyst, (SHOP process) with suitable ligands, cannot polymerize ethylene and 1-alkenes. Therefore, one earlier piece of work deserves mention. Fink [C3] reported in 1985 on a remarkable Ni catalyst that polymerized 1-pentene to give poly(ethylene-*alt*-propylene). The mechanism of this interesting polymerization involves what was later called the chain-walking process by the single-site nickel catalyst center (Figure C2). The historically important fact that nickel contaminants led to Ziegler's epoch-making discovery using $Zr(acac)_3$–$AlEt_3$ also triggered a tremendously rich chemistry by Ziegler's successor, G. Wilke, who has pioneered several homogeneous catalyses involving nickel(0) complexes, for example, butadiene trimerization.

Ligand Design Principles for Post-metallocenes

The body of work by Fujita and co-workers at Mitsui Chemicals Inc. on the so-called FI catalysts reflects the tremendous progress made by post-metallocene catalysts. A fundamental challenge associated with this class of compounds is the mobile properties of the ligands even when they are chelating. As larger group 4 metals can in principle adopt coordination numbers of 6 and above, the possible number of configurational isomers is high. This configurational lability is not too serious in the sterically constrained metallocene wedge, but results in a variety of seemingly

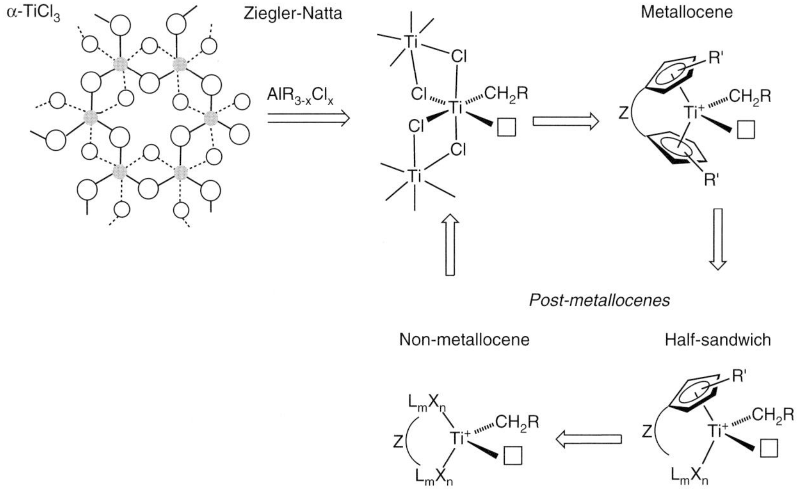

Figure C1   Relation of heterogeneous Ziegler catalysts with homogeneous metallocene, half-sandwich, and non-metallocene catalysts featuring two, one, and no cyclopentadienyl ligands.

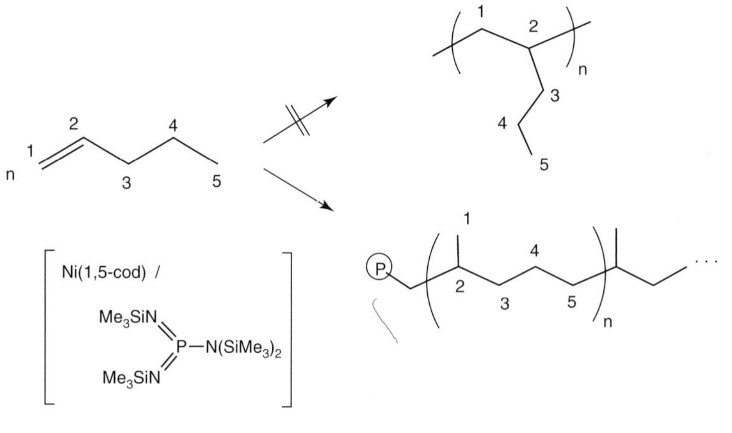

Figure C2   2,ω-Polymerization of 1-pentene to give poly(ethylene-*alt*-propylene).

unrelated and broad types of ligand structures. Supported by high-throughput methodology, research into post-metallocene catalysts gradually will lead to certain design principles that will guide chemists to develop even better catalysts (Figure C3).

In conclusion, post-metallocene-based alkene polymerization catalysts described in this chapter are fine examples of the impact the molecularly defined catalysts can have on ill-defined and difficult to manipulate heterogeneous catalysts, thus complementing multiple-site catalysts used

in industry. In the near future, anchoring molecularly defined single-site catalysts on carefully designed supports will becomes the major challenge in catalysis research.

Comment 3

Eugene Y.-X. Chen

Extensive studies on alkene polymerization catalysis by single-site metallocene and related catalysts, carried out mostly in the 1990s, have

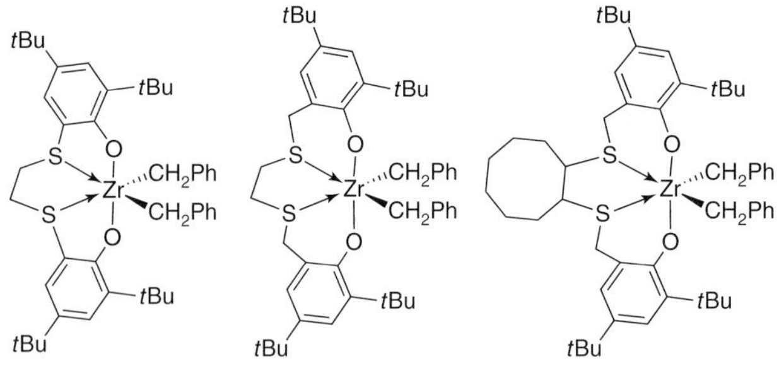

A: oligomer   **B**: atactic   **C**: isotactic

Figure C3   Subtle substituent effect on 1-hexene polymerization found for zirconium benzyl cation catalysts supported by a tetradentate (OSSO)-type ligand. Only precursor **C** affords isotactic poly(1-hexene) [4a], whereas complexes **A** [4b] and **B** [4c] are not suitable.

yielded phenomenal scientific and technological successes in the production of revolutionary polyalkenes. Driven by both the scientific curiosity and the intellectual demand to come up with new catalyst compositions for commercial applications, a large number of academic and industrial researchers have since endeavored to develop catalysts with different metals or metals at different oxidation states and new ligands, especially those devoid of any η^5-cyclopentadienyl, indenyl, or η^3/η^5-fluorenyl moieties (i.e., non-metallocene catalyst systems). Consequently, a great number of non-metallocene alkene polymerization catalyst systems have been discovered, some of which have shown unique catalytic properties compared with metallocene catalyst systems, most notably their high polar functional group tolerance and ability to produce programmable or engineered (co)polymer architectures and topologies through a chain-walking, transfer, or shuttling mechanism.

This chapter admirably captures highlights of recent developments of such non-metallocene catalysts and their derived novel or value-added polyalkene products. The authors are to be congratulated for their careful distillation of a sea of literature on the subject by selecting six catalyst systems based on four types of supporting ligation [i.e., β-diimine, pyridyldiimine (or -amine), chelating bis(phenoxy), and phenoxyimine], three of which are complexes of early transition metals (group 4) whereas the other three are complexes of late transition metals (groups 8–10). These chosen systems represent arguably the most significant, exciting developments in the post-metallocene

era. Adopting the polymerization activity in uniform units of kilograms of polymer per millimole of metal · per hour makes the activity comparison among the systems a straightforward and meaningful task, but the monomer (ethylene and propylene) pressure (often not identified in the review) can drastically alter those catalyst activity values. Also commendably, the authors' description of each catalyst system is concise and insightful, emphasizing only those most unique features of the system.

Not covered in this review chapter is an exciting class of neutral palladium phosphine–sulfonate [P,O^-] catalysts [C5], originally developed by Drent *et al.* [C6], that can copolymerize ethylene with alkyl acrylates into linear copolymers in which acrylate units (up to 17 mol% MA) are incorporated into the PE backbone. Further studies on the [P,O^-]Pd catalyst system carried out chiefly by the research groups of Mecking, Nozaki, Jordan, and Claverie have demonstrated that this catalyst system can copolymerize ethylene with an amazing range of polar vinyl monomers, such as acrylates [C7], acrylonitrile [C8], vinyl ethers [C9], vinyl fluoride [C10], vinyl acetate [C11], vinyl sulfones [C12], and even acrylic acid [C13], and also allyl monomers [C14]. However, the currently greatest issue with this catalyst system is its low activity, and this challenge must be met through the design of next-generation catalysts before commercial utilities of this system can be realized.

Two stereospecific polymerization examples given in the chapter, which show the ability of the specifically designed bulky phenoxyimine–titanium and −hafnium catalysts to produce highly syndiotactic and isotactic poly(propylene)s via chain-end and site control, respectively, address the generally perceived issue of lacking stereochemical control of polymerization by typical non-metallocene catalysts. However, in relative terms, chiral metallocene catalysts, thanks to their large library of various types of readily accessible chiral ligands, are more powerful, precise, and versatile in controlling stereomicrostructures of polyalkenes, especially those of polar functionalized vinyl polymers, as evidenced by the dazzling display of a variety of stereomicrostructures that they can generate [C5a].

A number of potential, and in some cases commercially implemented, applications brought about by the non-metallocene catalysts have stimulated the ever-growing interest in the metal-catalyzed coordination polymerization of alkenes and polar vinyl monomers. Future research in this field will be directed towards addressing the following, largely still unmet, challenges: (i) efficient random copolymerization of polar–nonpolar vinyl monomers and catalytic polymerization of polar alkenes by early metal or lanthanide catalysts; (ii) development of late metal catalysts having high polymerization activity and capable of producing high molecular weight and/or stereoregular polar vinyl polymers and polar–nonpolar vinyl copolymers; and (iii) polymerization or incorporation of naturally renewable polar alkenes for the synthesis of environmentally sustainable polymers.

Authors' Response to the Commentaries

As a result of the four excellent commentaries given by four of the leading investigators in their field, the points that we have tried to make in this review have now become much clearer, for which we are most grateful. The brief but pertinent historical insights into alkene polymerization catalysts given by Professors Okuda and Grubbs are particularly helpful in putting the current post-metallocene developments in perspective. We also thank Professors Chen and Guan for commenting on the palladium phosphine–sulfonate catalysts, which can copolymerize functionalized polar alkenes with nonpolar alkenes into linear random copolymers. Given the significance and recent progress concerning these catalysts, the comments and references provided by these leading authorities are a worthwhile addition for readers.

There is no doubt that one important future direction for transition metal polymerization catalyst development is to expand the scope of polymerizable monomers into conventional vinyl functional alkenes such as those commonly used in radical, anionic, and cationic polymerizations. This expansion must go hand-in-hand with high activity, high stereoselectivity, and compatibility with the polymerization of ordinary nonpolar alkenes such as ethylene and propylene, all of which are possessed by transition metal polymerization catalysts and inherently and uniquely associated with their coordination–insertion mechanisms. Although the late transition metal catalysts have been a mainstream strategy to reach that particular goal, efforts using early metals have also been started recently by several groups, including ourselves [C15].

In addition to all the challenges raised in the commentaries for the next generation of alkene polymerization catalysts, we would also like to point out that the catalysts must establish industrial feasibility for viable commercial applications in an ecologically sustainable manner.

References

C1. (a) Brintzinger, H.H., Fischer, D., Mülhaupt, R., Rieger, B., and Waymouth, R.M. (1995) *Angew. Chem. Int. Ed. Engl.*, **34**, 1143; (b) Resconi, L., Cavallo, L., Fait, A., and Piemontesi, F. (2000) *Chem. Rev.*, **100**, 1253.

C2. Sita, L.R. (2009) *Angew. Chem. Int. Ed.*, **48**, 2464; *Angew. Chem.*, **121**, 2500.

C3. Fink, G. (1985) *Angew. Chem. Int. Ed. Engl.*, **24**, 1001; *Angew. Chem. Int. Ed. Engl.*, **97**, 982.

C4. (a) Ishii, A., Toda, T., Nakata, N., and Matsuo, T. (2009) *J. Am. Chem. Soc.*, **131**, 13566; (b) Capachione, C., Proto, A., Ebeling, H., Mülhaupt, R., Möller, K., Spaniol, T.P., and Okuda, J. (2003) *J. Am. Chem. Soc.*, **125**, 4964; (c) Tshuva, E.Y., Goldberg, I., and Kol, M. (2000) *J. Am. Chem. Soc.*, **122**, 10706.

C5. Reviews: (a) Chen, E.Y.X. (2009) *Chem. Rev.*, **109**, 5157–5214; (b) Nakamura, A., Ito, S., and Nozaki, K. (2009) *Chem. Rev.*, **109**, 5215–5244; (c) Berkefeld, A. and Mecking, S. (2008) *Angew. Chem. Int. Ed.*, **47**, 2538–2542.

C6. Drent, E., van Dijk, R., van Ginkel, R., van Oort, B., and Pugh, R.I. (2002) *Chem. Commun.*, 744–745.

C7. (a) Kryuchkov, V.A., Daigle, J.C., Skupov, K.M., Claverie, J.P., and Winnik, F.M. (2010) *J. Am. Chem. Soc.*, **132**, 15573–15579; (b) Guironnet, D., Rünzi, T., Göttker-Schnetmann, I., and Mecking, S. (2009) *J. Am. Chem. Soc.*, **131**, 422–423; (c) Kochi, T., Noda, S., Yoshimura, K., and Nozaki, K. (2007) *J. Am. Chem. Soc.*, **129**, 8948–8949; (d) Skupov, K.M., Marella, P.R., Simard, M., Yap, G.P.A., Allen, N., Conner, D., Goodall, B.L., and Claverie, J.P. (2007) *Macromol. Rapid Commun.*, **28**, 2033–2038.

C8. Kochi, T., Noda, S., Yoshimura, K., and Nozaki, K. (2007) *J. Am. Chem. Soc.*, **129**, 8948–8949.

C9. Luo, S., Vela, J., Lief, G.R., and Jordan, R.F. (2007) *J. Am. Chem. Soc.*, **129**, 8946–8947.

C10. Weng, W., Shen, Z., and Jordan, R.F. (2007) *J. Am. Chem. Soc.*, **129**, 15450–15451.

C11. Ito, S., Munakata, K., Nakamura, A., and Nozaki, K. (2009) *J. Am. Chem. Soc.*, **131**, 14606–14607.

C12. Bouilhac, C., Rünzi, T., and Mecking, S. (2010) *Macromolecules*, **43**, 3589–3590.

C13. Rünzi, T., Fröhlich, D., and Mecking, S. (2010) *J. Am. Chem. Soc.*, **132**, 17690–17691.

C14. Ito, S., Kanazawa, M., Munakata, K., Kuroda, J.I., Okumura, Y., and Nozaki, K. (2011) *J. Am. Chem. Soc.*, **133**, 1232–1235.

C15. Terao, H., Ishii, S., Mitani, M., Tanaka, H., and Fujita, T. (2008) *J. Am. Chem. Soc.*, **130**, 17636–17637.

15

Chem Is Try Computationally and Experimentally: How Will Computational Organic Chemistry Impact Organic Theories, Mechanisms, and Synthesis in the Twenty-First Century?

Zhi-Xiang Yu and Yong Liang

15.1
Introduction

Computational organic chemistry [1, 2], sometimes also called theoretical organic chemistry, has been revolutionizing organic chemistry since its origin in the 1920s when the Schrödinger equation was first developed. Computational organic chemistry impacted organic chemistry in the last century, mainly on understandings of chemical reactivities and the reaction mechanisms, and the development of theories and concepts, which are still widely used today in both chemistry and biochemistry. In the twenty-first century, computational organic chemistry is continuing this trend, and is expanding to impact synthesis by guiding the design of new reactions, new catalysts, new ligands, and new synthetic strategies. In this chapter, we briefly describe these possible impacts brought by computational organic chemistry. To serve this aim better, we first describe our perception of how computational organic chemistry could impact theory development, mechanism understanding, and synthesis in three separate sections. In order to deliver more convincing statements to chemists about these impacts, in each section we also use some recent computational organic chemistry examples to elucidate these impacts that have already been made by computational organic chemistry.

Before the readers delve further into this chapter, we must emphasize that our choice of examples shown here is partially personal and heavily biased by our research interests and our limited knowledge in this field. In the wider literature, there are many other excellent examples that can also demonstrate these impacts.

15.2
Developing New Theories, Concepts, and Understandings for Organic Chemistry

Chemistry is one of the core sciences, and its technology has been impacting every aspect of our lives. Chemistry has evolved from an experimental science at its birth and in the early days to a science full of both experiments and a plethora of theories and principles, which help us to explain, understand, and even guide experiments. These theories and principles are usually derived from quantum mechanics (QM), which was developed in

Organic Chemistry – Breakthroughs and Perspectives, First Edition. Edited by Kuiling Ding and Li-Xin Dai.
© 2012 Wiley-VCH Verlag GmbH & Co. KGaA. Published 2012 by Wiley-VCH Verlag GmbH & Co. KGaA.

the 1920s. Quantum chemistry (QC), in a simplified expression, is the application of QM to chemical systems. At least in principle, results from QC can be used to understand and predict all kinds of chemistry, as reflected by the comments from the famous physicist and Nobel Laureate P. A. M. Dirac, who said that "The fundamental laws necessary for the mathematical treatment of large parts of physics and the whole of chemistry are thus fully known, and the difficulty lies only in the fact that application of these laws leads to equations that are too complex to be solved."

The 1920s and 1930s were the times before the first computer appeared. The Schrödinger equation is so complex that people without a fast computer at hand in that period felt very pessimistic about solving the equation. As a result, scientists had to apply the Schrödinger equation to solve very simple systems such as H_2, or they solved Schrödinger equations approximately by omitting some integrals, and applied the principles of QC to study chemistry. Examples include Pauling's study of the nature of chemical bonds, Hückel's study of polyene systems, and the development of molecular orbital theory by Hückel and Mulliken. The scientists in the early days can be regarded as the first generation of computational organic chemists, and they provided many theories and concepts for the foundations of chemistry (for example, bonding, antibonding, hybridization orbitals, conjugation, and aromaticity) [2].

QC calculations (in most cases this is called quantum mechanical calculations; here we use both QC and QM to describe this) in the 1960s, 1970s, and 1980s can then be applied to relatively larger systems of up to around 20 heavy atoms, due to the development of many new algorithms and more and more powerful computers, even though the calculation accuracy was not satisfactory in most cases at that time and, for example, semiempirical calculations were usually used. Scientists had applied QC calculations to study some important and very fundamental reactions, such as substitution, elimination, and pericyclic reactions [1], and the important reactive species, such as carbocations, including the nonclassical carbocations, carboanions, radicals, and carbenes. Through studying reaction mechanisms and reactivities of molecules, many theories and concepts, such as the orbital symmetry conservation principle and Frontier molecular orbital (FMO) theories, were developed. These theories and concepts developed in this period, with the aid of computational organic chemistry, enriched our in-depth understanding of organic chemistry and built the cornerstones of this discipline as a science of experiments and theories.

Then in the 1980s and 1990s, the application of density functional theory (DFT) in chemistry greatly expanded the power of computational organic chemistry to more complex systems with high accuracy (within about 5 kcal mol^{-1} calculation error) and many new phenomena were discovered through calculations.

In the 1990s and the beginning of this century, due to the advances in computer science and the development of new theories and methods in computational chemistry and QC, it became possible to solve the Schrödinger equations for a relatively large system with relatively high accuracy. Consequently, our understanding of structure–property–reactivity relationships has been greatly expanded. Many reaction mechanisms have now been well studied using state-of-the-art high-level QC calculations (see Section 15.3). Some of these impacts from computational organic chemistry on organic chemistry can be appreciated from the following examples (details of these researches can be found in the original research papers. Readers are also recommended to consult the book by Bachrach for more examples).

Weak interactions such as C–H· · ·π, C–H· · ·X (X could be oxygen, halogen, or other heteroatom), cation–lone pair, and π−π interactions are important for reactions and for supermolecular and biological chemistry [3]. They were not well understood until recent high-level computational studies could be performed by many chemists. Now it is realized by chemists that such weak interactions are important for enantioselective reactions, for example, in the Noyori hydrogenation reaction, and they have also been applied in reaction design [4].

Aromaticity is a widely used concept in chemistry, but unfortunately it is ill-defined. Through efforts made by many generations of computational organic chemists, numerous new insights into aromaticity have been developed. For example, the nucleus-independent chemical shift (NICS) index developed by Schleyer et al. can be used as a qualitative index to characterize aromatic compounds and transition states (TSs), and to measure their relative aromatic stabilization energies [5].

Hükel aromatic compounds are very common but Möbius aromatic molecules exist only in theory. With the aid of calculations, Herges and co-workers synthesized the first Möbius aromatic compound [6]. Furthermore, through computational study, some reactions with Möbius aromatic TSs have also been found [7].

Every stationary point in a traditional energy surface of a reaction includes only information about its structure and energy, and not its moment. However, computational studies revealed that considering the moment of each stationary point in a reaction energy surface is critical to obtain the correct understanding of the reaction mechanisms [8]. This suggests that nonstatistical dynamics have to be taken into account in interpreting reactions. Such a notion greatly expands our understanding of TS theory and reaction mechanism.

FMO theory, which has been widely applied to explain reactions, has its limitations. Through investigating 1,3-dipolar cycloaddition reactions, Houk and co-workers found that a reaction's activation energy can be decomposed into distortion energy and interaction energy [9a]. They found that the failure of FMO theory in explaining the reactivity and selectivity for a series of similar reactions could be caused by their different distortion energies in the TSs. In contrast, the relative interaction energies in these reactions can be well explained by FMO theory. This method of analyzing activation energies, which has now been widely applied to understanding reactions, gives chemists a fresh angle to comprehend reaction and reactivity [9].

To appreciate the impacts further, we elaborate more in this section to present several examples, which are shown below.

15.2.1
Bifurcations on Potential Energy Surfaces of Organic Reactions

TS theory is the fundamental law guiding our understanding of why a reaction occurs fast or slow. The traditional picture of TS theory is that a TS is connected by its reactant(s) and product(s). However, this picture has changed owing to the computational study of the reaction mechanisms. It was found that a single TS can be shared by two reaction pathways if there is a post-TS bifurcation.

Caramella and co-workers brought the role of bifurcations on potential energy surfaces (PESs) for cycloadditions to the attention of many chemists through the remarkable discovery that the simplest Diels–Alder (DA) reaction of dimerization of cyclopentadiene involve bifurcations on the PESs [10]. For example, the *endo* dimerization of cyclopentadiene was thought to proceed via two competing DA cycloaddition TSs, **TSa** and **TSb**, to give products **A** and **B**, and the Cope rearrangement transition state **TSc** corresponded to the isomerization of cycloadducts **A** and **B** (Scheme 15.1a). However, calculations showed that TSs **TSa** and **TSb** have merged into a single "bis-pericyclic" TS **TSab** in the formation of **A** and **B** (Scheme 15.1b) [10b]. This TS is not connected directly to the products, but to a valley–ridge inflection (VRI) structure, which can give products **A** and **B**. The distribution of products depends on the PES shape and resulting dynamic effects.

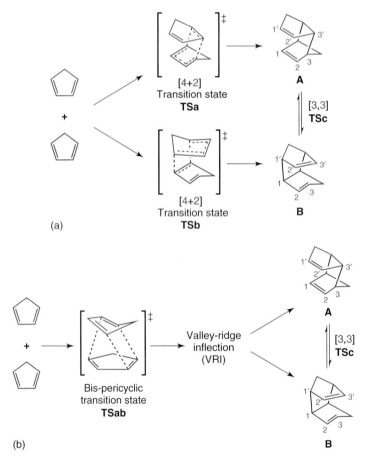

Scheme 15.1   (a) Traditional mechanistic picture of the *endo* dimerization of cyclopentadiene and (b) new mechanistic picture based on the PES bifurcation.

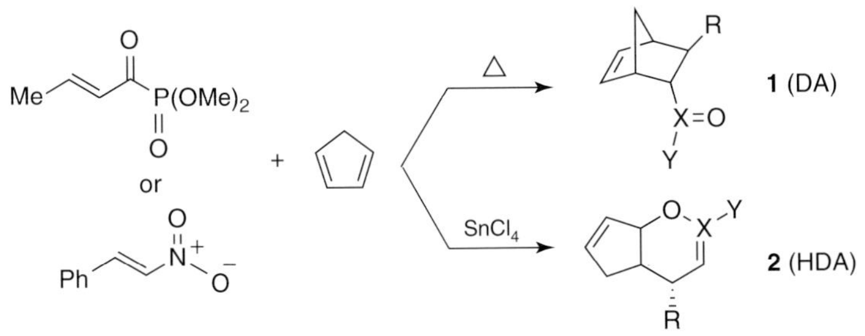

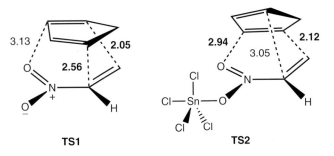

Scheme 15.2 Periselectivity for cycloadditions of cyclopentadiene with α-keto-β,γ-unsaturated phosphonates and nitroalkenes under thermal and Lewis acid-catalyzed conditions.

In 2007, Houk's group found that the periselectivity for cycloadditions of cyclopentadiene with α-keto-β,γ-unsaturated phosphonates or nitroalkenes under thermal and Lewis acid-catalyzed conditions is controlled by a bis-pericyclic TS [11]. Under thermal conditions, cyclopentadiene acts as a 4π-electron component in the [4 + 2]-cycloaddition, and the DA adduct **1** is the major product (Scheme 15.2). In contrast, cyclopentadiene acts as a 2π-electron component in the Lewis acid SnCl$_4$-catalyzed [4 + 2]-cycloaddition, and the hetero-Diels–Alder (HDA) adduct **2** is preferred (Scheme 15.2). Through comparison of the geometries of the bis-pericyclic DA TSs **TS1** and **TS2** (Figure 15.1), a qualitative understanding of the thermal and Lewis acid-catalyzed bifurcating PESs can be obtained. Preference for the DA adduct **1** under thermal conditions is due to the shorter C–C distance (2.56 Å) compared with the C–O interaction (3.13 Å) in the bis-pericyclic TSs **TS1** (Figure 15.1). In contrast, under Lewis acid-catalyzed conditions, SnCl$_4$ coordination decreases the C–O distance (2.94 Å) and increases the C–C distance (3.05 Å) in the bis-pericyclic **TS2** (Figure 15.1), making the formation of a C–O bond in the HDA adduct **2** more favorable.

Singleton and co-workers also showed that a common TS leads to both [4 + 2]- and [2 + 2]-cycloadducts in the reaction of cyclopentadiene and diphenylketene (Scheme 15.3) [12]. A collaborative effort by the Singleton, Foote, and Houk groups was devoted to investigating the mechanism of the singlet-oxygen ene reaction by using ^{13}C kinetic isotopic effects (KIEs) and QM methods [13]. It was also found that this reaction takes

Figure 15.1 Bis-pericyclic DA transition states **TS1** and **TS2** (distances are in ångstroms).

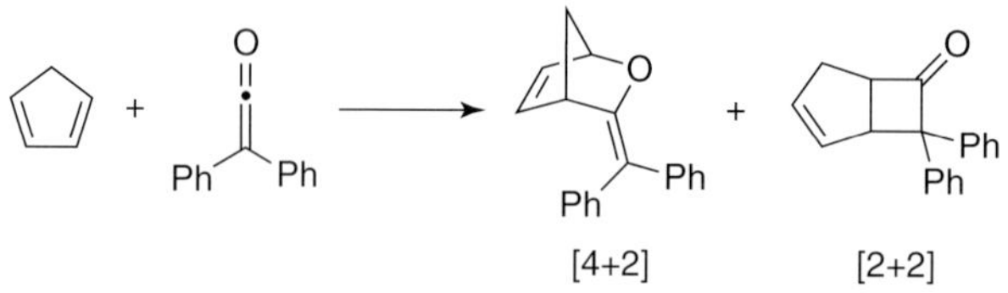

[4+2]   [2+2]

Scheme 15.3 Reaction of cyclopentadiene and diphenylketene.

place through a two-step, no-intermediate mechanism, that is, the PES bifurcates after a common TS.

Such a concept of reaction pathway bifurcations has been found in many organic reactions [14]. This new concept enriches our understanding of reaction mechanism and selectivity and prompts chemists to be prudent when interpreting their experimental results.

15.2.2
Computational Prediction of Carbon Tunneling

QM tunneling is a fundamental physical phenomenon that affects the rates of barrier crossings in chemical reactions. Tunneling effects are usually limited to the hydrogen atom owing to its small mass. Calculations and experiments have demonstrated that tunneling by hydrogen plays an important role in many reaction systems [15]. The very traditional and well-established view of tunneling is that for heavy atoms such as carbon, which have larger masses than hydrogen atom, the tunneling effects are rare [16].

The limited evidence for heavy-atom tunneling was the observation of reactions that occur at cryogenic temperatures, where there is little thermal energy to allow passage over a reaction barrier. In 2003, the Sheridan, Truhlar, and Borden groups reported ring expansion of 1-methylcyclobutylfluorocarbene at 8 K with a measured rate constant of 4.0×10^{-6} s^{-1} in nitrogen, in which carbon tunneling is involved (Scheme 15.4) [17]. Calculations indicated that at this temperature the reaction proceeds via carbon tunneling from a single quantum state of the reactant so that the computed rate constant has achieved a temperature-independent limit (Figure 15.2). When tunneling is not included, the rate constant at 8 K is calculated to be smaller by a factor of 2×10^{152}, but the inclusion of tunneling provides a calculated rate constant of 9.1×10^{-6} s^{-1} that is in good agreement with the experimental value.

In 2008, Borden and co-workers predicted that the ring opening of cyclopropylcarbinyl radical (**3**) to 3-buten-1-yl radical (**4**) below 20 K occurs exclusively by temperature-independent carbon tunneling (i.e., $E_a = 0.0$ kcal mol^{-1}) from the lowest vibrational level of **3**, with the rate constant $k = A = 2.22 \times 10^{-2}$ s^{-1} (Scheme 15.5)

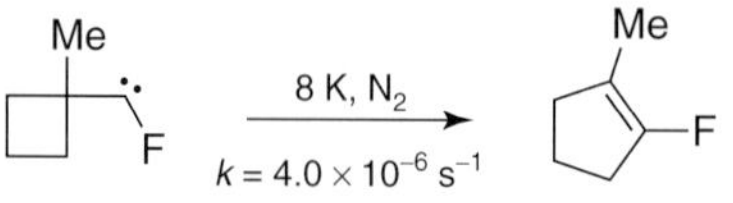

Scheme 15.4 Ring expansion of 1-methylcyclobutylfluorocarbene.

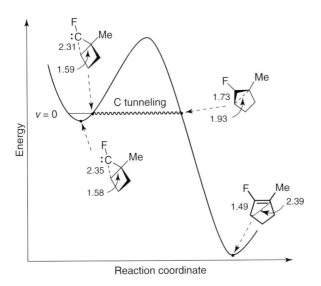

Figure 15.2 Carbon tunneling from a single quantum state.

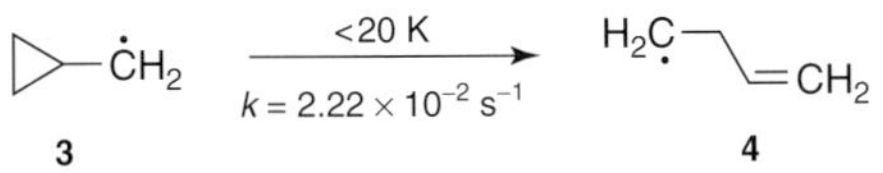

Scheme 15.5 Ring opening of cyclopropylcarbinyl radical.

[18]. Further calculations indicate that the tunneling effect of the carbon atom makes the ring opening of **3** faster at higher temperature. For instance, at 298 K (25 °C), if tunneling is not included, the rate constant for the ring opening of **3** is calculated to be 1.79×10^7 s^{-1}, whereas inclusion of tunneling results in a ~50% higher rate constant of 2.82×10^7 s^{-1}. This suggests that the carbon tunneling in this reaction will be apparent in experimental observations made at routinely accessible temperatures.

Recently, Singleton and co-workers reported experimental evidence for carbon tunneling in the ring opening of cyclopropylcarbinyl radical (**3**) from intramolecular ^{12}C/^{13}C KIEs [19]. To measure the intramolecular KIEs, samples of bromomethylcyclopropane (**5**) at natural abundance were reduced at temperatures ranging from -100 to $80\,°C$ under free-radical conditions (Scheme 15.6). This reaction afforded 1-butene (**6**) as the major product, and the relative ^{13}C content in the C3 versus C4 positions of **6** was analyzed by NMR spectroscopy [20]. In all cases, less ^{13}C was observed at the C4 position of **6** than at C3, reflecting a preference for ^{12}C to undergo the bond-breaking process. Notably, the KIEs are extremely large (Scheme 15.6), and are the largest known at each temperature among conventional ^{13}C KIEs. This indicates that the carbon tunneling effect may play an important role in the ^{13}C KIEs.

To support further the predicted importance of carbon tunneling in cyclopropylcarbinyl ring opening, the KIEs were computed both with and without inclusion of tunneling

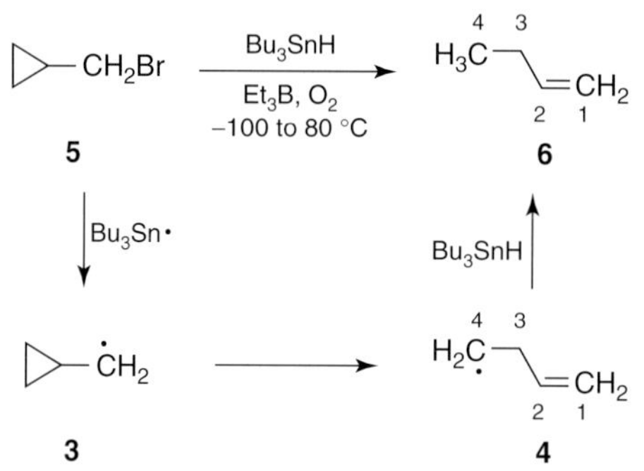

Temperature (°C)	80	0	−100
Experimental KIE	1.062 ± 0.003	1.085 ± 0.003	1.163 ± 0.004
Calculated KIE (without tunneling)	1.048	1.064	1.106
Calculated KIE (with tunneling)	1.057	1.082	1.169

Scheme 15.6 Evidence for carbon tunneling in the ring opening of cyclopropylcarbinyl radical from intramolecular $^{12}C/^{13}C$ KIEs.

(Scheme 15.6). The calculated KIEs without allowance for tunneling drastically underestimate the size of the experimental values. The error in these predictions grows larger as the temperature decreases, rising to as large as 6% at −100 °C. In contrast, the predicted KIEs including tunneling match well with experiments throughout the broad temperature range, with the error never exceeding 0.7%.

The computational prediction and experimental verification of carbon tunneling expand our understanding of tunneling effects of heavy atoms. This finding will stimulate organic chemists to consider more implications about their reactions at low temperature.

15.2.3 Predictions of Contra-Steric Stereochemistry in Cyclobutene Ring-Opening Reactions by the Theory of Torquoselectivity

Today, most computational organic chemistry research involves the explanation of the experimental phenomena that are not well understood. Sometimes the intriguing chemical phenomena cannot be rationalized by existing theories. Computational chemistry may play an important role in developing a new theory that can be used successfully not only for rationalizations but also for predictions. A classic example in this field is Houk and co-workers' theory in electrocyclic reactions, which is referred to as "*torquoselectivity*" or the "*torquoelectronic effect*" [21].

When a 3,4-disubstituted cyclobutene undergoes thermal ring opening to butadiene, the reaction will proceed in a conrotatory manner according to the Woodward–Hoffmann theory. Therefore, *trans*-1,2,3,4-tetramethylcyclobutene (**7**) will give rise to the formation of (*E,E*)-diene (**8**) and/or (*Z,Z*)-diene (**9**) [formation of the (*E,Z*)-diene is forbidden by orbital symmetry rules]. In fact, only the (*E,E*)-diene (**8**) was observed in the experiment (Scheme 15.7). The exclusive formation of **8** was uniformly attributed to repulsive

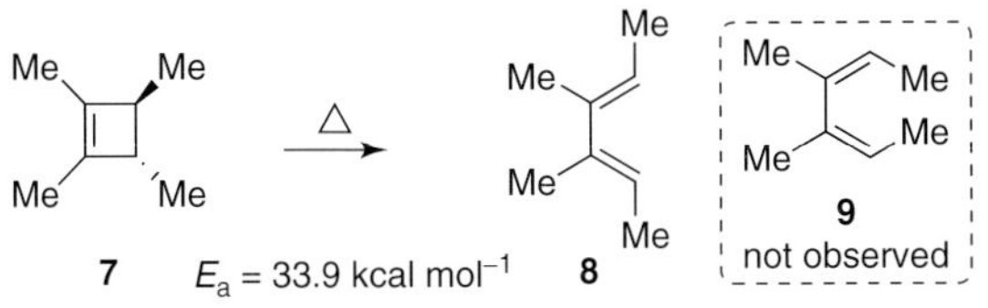

Scheme 15.7 Ring opening of *trans*-1,2,3,4-tetramethylcyclobutene.

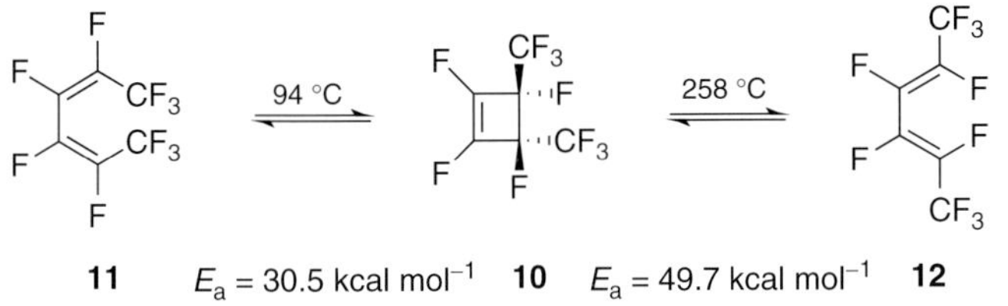

Scheme 15.8 Ring opening of *trans*-perfluoro-3,4-dimethylcyclobutene.

steric interactions that would be present in the TS for the generation of **9** due to two methyl groups rotating simultaneously inwards. This argument had been widely applied to rationalize other similar results in virtually all electrocyclic ring openings of cyclobutenes.

In 1984, Dolbier *et al.* observed a dramatic kinetic difference for the inward versus outward rotation of *trans*-perfluoro-3,4-dimethylcyclobutene (**10**) as shown in Scheme 15.8 [22]. The 19.2 kcal mol^{-1} difference in E_a values for the two competitive processes of **10** leads to a ratio of rates for ring openings to diene **11** versus diene **12** of 1.9×10^9 at 111.5 °C. This clearly shows that the rotation of the much larger CF$_3$ groups inwards and the smaller fluorine substituents outwards is strongly favored, which is obviously contrary to expectations based on the "steric" rationale.

Shortly thereafter, Houk and co-workers reported a related computational study of substituent effects on conrotatory electrocyclic reactions of cyclobutenes, indicating that the energetic preference for substituents at C3 and C4 of cyclobutene to rotate outwards increases as the π-donor nature of the substituent increases [23]. This theoretical study established that the stereochemical trend is subject to electronic control rather than steric control. Following this significant discovery, a comprehensive electronic theory accounting for the stereoselectivity of the cyclobutene ring openings was developed by Houk and co-workers. Also, the term *"torquoselectivity"* was invented to describe the different rotation trends of substituents (inwards versus outwards). It was found that electron-donating substituents at the C3 and C4 positions prefer outward rotation. In contrast, electron-accepting substituents at the C3 and C4 positions prefer inward rotation because delocalization of electron density of the HOMO of the cyclobutene ring opening TS to the electron-accepting substituent stabilizes the inward rotation TS. More importantly, "torquoselectivity" overwhelms the steric effect if these two factors both affect the stereochemistry.

A series of predictions of contra-steric stereochemistry in cyclobutene ring-opening reactions based on Houk and co-workers' theory have been verified by subsequent

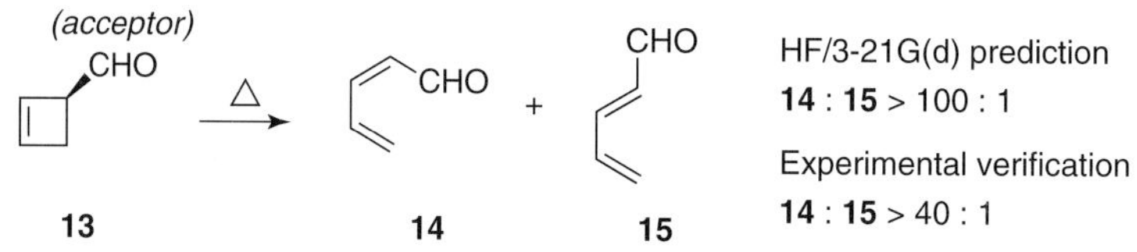

Scheme 15.9 Ring opening of 3-formylcyclobutene with high inward selectivity.

experiments. The first of such predictions was the electrocyclic ring opening of 3-formylcyclobutene (**13**) with high inward selectivity (Scheme 15.9) [24]. Theoretical calculations indicated that a formyl group can accept electron density into its antibonding orbital (π^*) and thus rotates inwards. This prediction was unambiguously demonstrated by the observation of the dominant inward rotation when **13** undergoes ring opening. A vacant boron *p*-orbital can efficiently accept electron density. Therefore, in 1985, Rondan and Houk predicted that a boryl substituent would have a very strong preference for inward rotation [23b]. The inward ring-opening TS of 3-(BMe$_2$)cyclobutene was found to be lower than the outward one by as much as 10.5 kcal mol^{-1}, making inward rotation exclusively. Owing to the difficulty in the preparation of 3-borylcyclobutene, there had been no experimental validation for 20 years. In 2005, Murakami *et al.* successfully synthesized 3-(pinacolatoboryl)cyclobutene (**16**) and verified this prediction experimentally (Scheme 15.10) [25]. Another impressive example that demonstrates the dominance of electronic control over steric control is the ring-opening reaction of 3-methoxy-3-*tert*-butylcyclobutene (**19**) as depicted in Scheme 15.10 [26]. This cyclobutene opens with exclusive inward rotation of the much bulkier *tert*-butyl group because the methoxy group is a strong electron-donating substituent and therefore prefers outward rotation.

This powerful predictive principle for cyclobutene stereoselectivities has been found to be general in various reactions, such as pentadienyl cation cyclization, hexatriene–cyclohexadiene interconversion, electrocyclic ring openings of β-lactone enolates, ketene–imine [2 + 2]-cycloaddition, and so on [27].

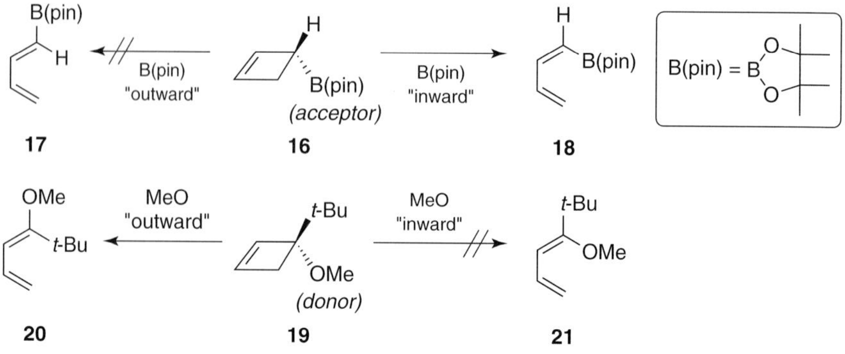

Scheme 15.10 Two classic examples of contra-steric stereochemistry in cyclobutene ring-opening reactions.

From these examples, we can see that our understanding of organic chemistry has gradually increased owing to the contributions from computational organic chemistry, where new ideas and concepts have been developed. In the coming years, computational organic chemistry will continue to help us understand more reactions, and provide new concepts and theories.

15.3
Understanding Reaction Mechanisms

There is no doubt that studying a reaction's mechanism to know why and how this reaction occurs and to understand factors affecting this reaction's performance, such as its yield and regio- and stereoselectivities, is important. The most appealing and complete (maybe ideal) mechanistic scenario of a reaction is that we not only know each elementary step and the sequence of these elemental steps in this reaction, but also have information about the structures and energies of reactants, intermediates, products, and TSs involved in this reaction. An ideal mechanistic picture of a supposed reaction of **A** and **B** with ligand **L** and the catalyst to give product **P** is shown in Figure 15.3.

The importance of studying a reaction's mechanism can be appreciated from the following four points. First, as scientists, we are curious to know how a reaction occurs. Treating a reaction like a black box does not preclude us from using this reaction in synthesis. However, as scientists, understanding how a reaction occurs, especially knowing the PES of this reaction and the structures of the TSs and intermediates, not only can satisfy our curiosity towards nature, but also is a requirement for a researcher to claim that they are a scientist, and not a blacksmith in their workshop.

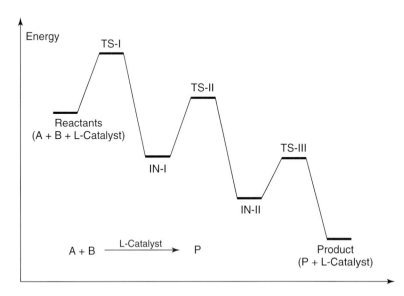

Figure 15.3 Mechanistic picture of a supposed reaction.

Second, understanding a reaction's mechanism is very helpful for optimizing this reaction so that this reaction's yield and selectivity (if exists in this reaction) can be improved. In this case, knowledge of the structures and energies, and information about how substituents and solvents affect this reaction, are very useful for optimizing the reaction conditions and to establish under what circumstance a reaction can work better.

Third, the mechanistic information could provide guidance to designing new reactions and new catalysts. For example, if we know how the reaction of **A** and **B** to give product **P** occurs and know that **TS1** is the stereo-determining TS, then we can compare TSs corresponding to the generation of both *R*- and *S*-products so that we can design chiral ligands by computer with the aid of QM calculations. We can also compute whether other transition metals can reduce the activation energy of this reaction so that a better catalyst could be discovered. If **A** and **B** are not readily available starting materials, maybe we can use other reactants (suppose compounds **C** and **D**) that can give **IN-I** or **IN-II** via a totally different reaction so that we can design a new reaction of **C** + **D** to **P**. In addition, we can add reagent **E** to the reaction system to intercept intermediate **IN-I** or **IN-II** so that a new product can be obtained. Or chemists can introduce some substituent in reactants **A** or **B**, and this substituent in **IN-I** or **IN-II** could then undergo a different reaction pathway to give another product **P'**. There are many possibilities to design new reactions and transformations using a reaction's mechanistic information, and the design is usually determined by a designer's knowledge, ideas, and imagination. The design of new reactions and catalysts based on the studied reaction mechanisms is discussed in Section 15.4.

Fourth, through mechanistic studies, new insights and concepts could be developed and this will stimulate the development of organic chemistry (this has been addressed in Section 15.2).

Figure 15.4 shows how an ideal mechanistic scenario of a complex reaction can be obtained. First, we must propose or conceive the elementary reactions involved and the sequence of these elementary reactions (the proposal step). Actually, this is the most difficult step for a mechanistic study. How to obtain or conceive the reaction mechanism of a complex reaction is beyond the scope of this account, but we believe that this could be obtained by a combination of using experimental results (such as isolation or observation of a critical intermediate), principles of reaction mechanisms, and our imagination.

The second step is to apply QM calculations to compute the structures and energies of all related species to obtain information about structures of all species, kinetics, thermodynamics, rate-, regio-, stereo-determining steps, the solvent effects, and many other factors of interest (the QM calculation step). Sometimes, dynamics calculations have to be carried out to include the effect of nonstatistical effects (see [8] and references therein). The computed results can be then compared with the experimental results to see whether they agree with each other. Here many things from experiments and calculations can be compared, such as activation energies, the regio- and stereochemistry, and so on. If the answer is yes, we can then say that the proposed mechanism is very reasonable. If the answer is no, we have to propose a new reaction mechanism to account for the experimental results. The newly proposed mechanism will also be evaluated by QM calculations to see whether the computed results agree with the experimental results (supposing that high accuracy calculations with real reaction system, not the model

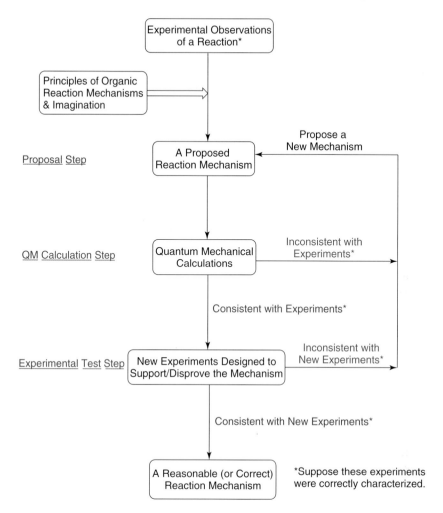

Figure 15.4 Procedure to obtain a reasonable reaction mechanism.

system, are available). This cycle can continue until a mechanism and its computational results agree with the experimental results, then we can temporarily claim that this mechanism is reasonable.

However, if possible, we can design further new experiments based on the mechanism that can explain all previous experiment results to support further or disprove the mechanism (the experimental test step). If the designed experimental results are in exact agreement with those predicted by the mechanism, this suggests that the proposed mechanism is both reasonable and trustworthy. Note that a reaction's mechanism cannot be proved to be correct, but can only be proved to be wrong. Therefore, a reaction mechanism that pass all these procedures can only be labeled as a "reasonable" reaction mechanism. If the designed experimental results are *not* in exact agreement with those predicted by the mechanism, then we have to step back to propose a new mechanism to continue the aforementioned three-step circle until we have a mechanism that agrees with all known and the tested experimental results [28]. After reading Professors Wu and Zhang's comments (see the end of the chapter), we redrew Figure 15.4 to emphasize that all the experiments must be correctly characterized if a reasonable mechanism is to be reached. In some cases, experiments were not correctly or precisely characterized, and calculations helped synthetic chemists to correct their experiments or to give correct mechanistic pathways for the studied reactions.

The traditional mechanistic study in the past usually took actions in the proposal step and experimental test step but not in the QM calculation step. Even though sometimes using only experimental results can give the correct reaction pathway, in-depth information on the reaction's kinetics, rate- and stereo-determining step, structures of all stationary points in the energy surface, the origins of regio- and stereoselectivities, and so on, can only be obtained by QM calculations. In some cases, the QM calculation step is the only way to disprove a proposed mechanism. In Section 15.3.1 below, we show how QM calculations guided the study of the mechanism of the Lu and co-workers' [3 + 2]-cycloaddition reaction.

Over the past several decades, many reactions have been developed and some of them have been widely used in synthesis. However, only a fraction of these reactions have been subjected to mechanistic studies. More and more reactions impacting synthesis will be discovered in the coming years. Therefore, a challenging and essential task for organic chemists in the present century is to obtain quantitative mechanistic pictures of all these important reactions at the molecular level. Both computational and experimental studies of the reaction mechanisms are of equal importance in reaching this goal.

There have been many mechanistic studies of important reactions and new reactions, mainly using QM calculations (in very limited cases, experimental tests were also carried out to support the computational results). These are readily available in the literature and readers are recommended to consult these studies. In addition, we recommend readers to read the book *Strategic Application of Named Reactions in Organic Synthesis* by Kürti and Czakó, where not only the reactions and their applications in synthesis are presented, but also the related computational studies of some of these named reaction mechanisms are referenced [29]. Readers are strongly recommended to read the cited computational studies in this book to gain a full appreciation of the impacts of computational chemistry on organic reaction mechanisms. The book *Computational Organic Chemistry* by Bachrach also provides a good source of many computational studies of reaction mechanisms [1]. Here we present three examples to emphasize further the importance of QM calculations in the mechanistic studies.

15.3.1
Mechanism of Phosphine-Catalyzed [3 + 2]-Reactions of Allenoates and Electron-Deficient Alkenes: Discovery of Water-Catalyzed [1, 2]-Proton Shift

Five-membered carbocycles and heterocycles exist as the backbone of many natural products and other compounds of pharmaceutical significance. Accordingly, much effort has been devoted to the development of new synthetic methods for their construction. Among them, Lu and co-workers' phosphine-catalyzed [3 + 2]-cycloaddition reactions of

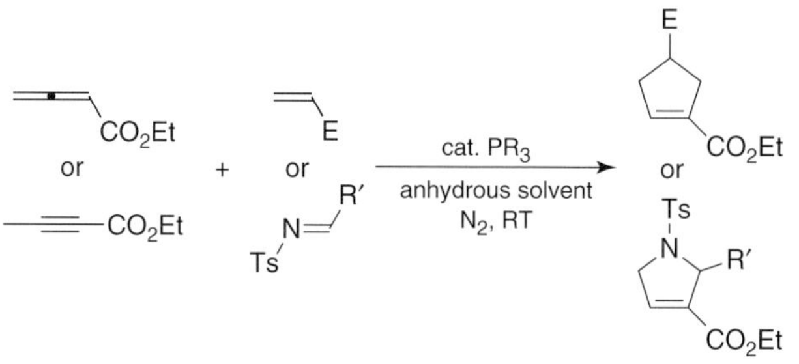

Scheme 15.11 Phosphine-catalyzed [3 + 2]-cycloaddition reactions.

allenoates/2-butynoates with electron-deficient alkenes/imines have been shown to be very efficient in organic synthesis (Scheme 15.11) [30]. This method has been successfully utilized as a key step in the total synthesis of several natural products. In addition, highly asymmetric versions of phosphine-catalyzed [3 + 2]-reactions have also emerged, and further developments and extensions of the chemistry based on nucleophilic phosphine catalysis have been elegantly demonstrated by many groups.

The commonly accepted mechanism for the phosphine-catalyzed [3 + 2]-reaction of allenoates and electron-deficient alkenes is given in Scheme 15.12 [30a,e]. The catalytic cycle starts with formation of a zwitterionic intermediate **A** between allenoate and phosphine, which acts as a 1,3-dipole to react with an electron-deficient alkene to give a five-membered phosphorus ylide **B**. Then a [1,2]-proton shift converts **B** to another zwitterionic intermediate **C**, which gives rise to the final product **D** through the elimination of the phosphine catalyst.

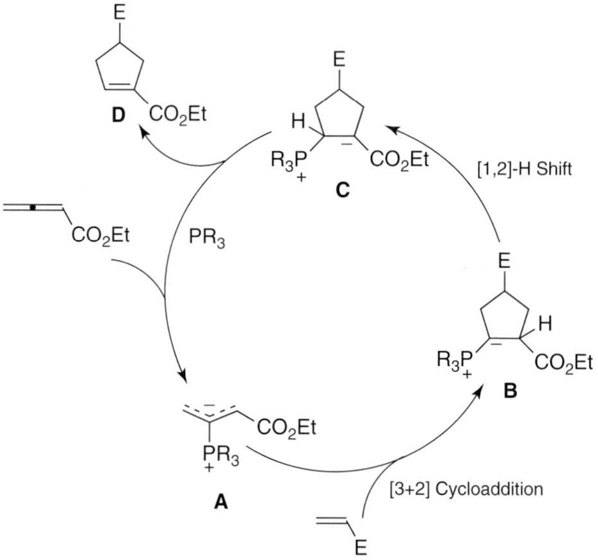

Scheme 15.12 Commonly accepted mechanism for the phosphine-catalyzed [3 + 2]-reaction.

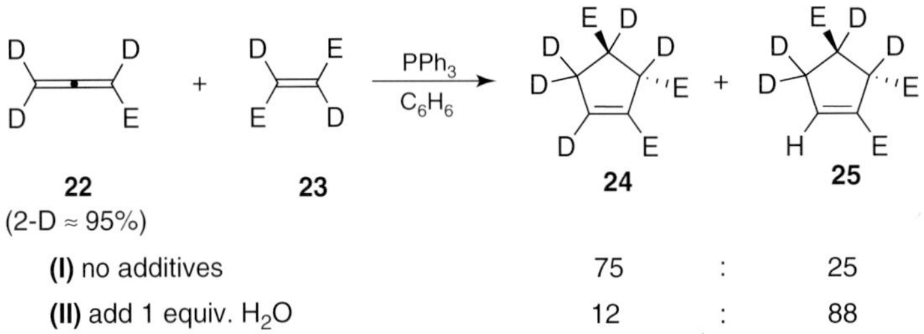

22
(2-D ≈ 95%)

23

24

25

(I) no additives	75	:	25
(II) add 1 equiv. H$_2$O	12	:	88

Scheme 15.13 Isotopic labeling experiments (E $= CO_2$Me).

Based on this proposed reaction mechanism, we computed the energies of intermediates and TSs in the catalytic cycle [31]. DFT calculations indicate that the activation
free energy of a direct [1,2]-proton shift from **B** to **C** is about 40 kcal mol^{-1}. Because
the phosphine-catalyzed [3 + 2]-reaction is usually conducted at room temperature, this
direct intramolecular [1,2]-proton shift pathway requiring such a high activation free
energy is impossible from the viewpoint of kinetics.

To obtain more information about the [1,2]-proton shift process in the
[3 + 2]-cycloaddition, we conducted a PPh$_3$-catalyzed [3 + 2]-cycloaddition of
deuterium-labeled 2,3-butadienoate **22** (the 2-D incorporation level is about 95%) and
fumarate **23** (reaction **I**, Scheme 15.13) in benzene that had been refluxed with Na and
freshly distilled prior to use. It was found that 4-D- and 4-H-substituted products **24** and
25 were obtained in a ratio of 75:25. If the [1,2]-H shift were a simple intramolecular
process, the ratio of 4-D- and 4-H-substituted products **24** and **25** would be about 95:5,
in accord with the 2-D incorporation level of the allenoate **22**. Where does the extra
4-H-substituted product **25** come from? We speculated that the formation of extra **25**
was due to the presence of a trace amount of water in the reaction system. To prove our
hypothesis, we added 1 equiv. of H$_2$O to the [3 + 2]-reaction between **22** and **23** (reaction
II, Scheme 15.13). As expected, the ratio of 4-H-substituted product **25** increased
substantially from 25 to 88%.

Therefore, we proposed a new water-catalyzed [1,2]-proton shift process, which is
shown in Scheme 15.14. Further DFT calculations showed that this water-catalyzed
pathway is very facile, requiring an overall activation free energy of about 8 kcal mol^{-1}.
This well explains the above experimental findings and gives a more reasonable picture
to the [1,2]-H shift process of Lu and co-workers' [3 + 2]-reaction.

In this case, the information obtained from the mechanistic studies of the
phosphine-catalyzed [3 + 2]-cycloaddition reaction, both computationally and experimentally, will greatly deepen our understanding of other related phosphine-catalyzed
reactions and assist the design of novel phosphine-catalyzed reactions. The discovery of
a trace amount of water as the catalyst in [1,n]-proton shifts will prompt chemists to
rethink the role of water and other protic source in organic reactions.

After we published our study of the mechanism of Lu and co-workers' [3 + 2]-reaction,
a computational investigation of the mechanism of the Morita–Baylis–Hillman reaction
was reported [32]. Consistent with our discovery of a water-catalyzed [1,2]-proton shift,

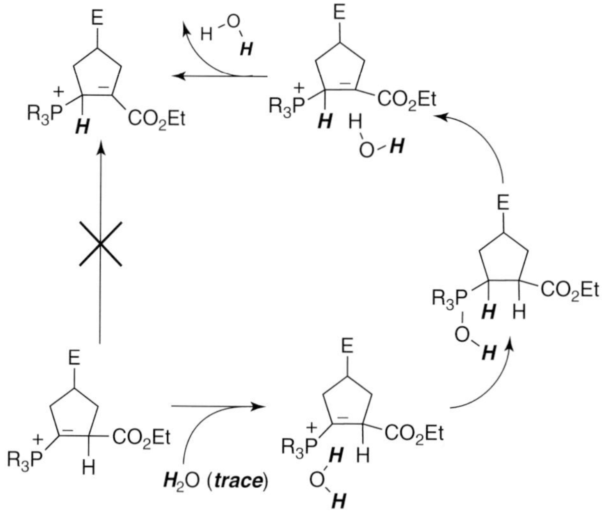

Scheme 15.14 Water-catalyzed [1,2]-proton shift process.

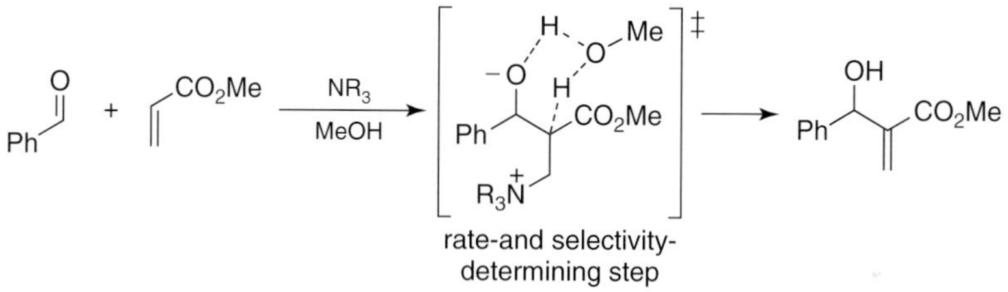

Scheme 15.15 Mechanism of the Morita–Baylis–Hillman reaction in the presence of alcohol.

it was found the methanol can promote [1,3]-proton transfer in the amine-catalyzed Morita–Baylis-Hillman reaction (Scheme 15.15).

Our understanding of Lu and co-workers' [3 + 2]-reaction is also very helpful for optimizing the reaction conditions of asymmetric phosphinothiourea-catalyzed allenoate–imine [3 + 2]-cycloadditions. Fang and Jacobsen found that, in the absence of additive water, the rates of their reactions are slow and a variety of undesired byproducts are formed. In contrast, addition of 0.2 equiv. of water afforded improved the reaction rates and yields (Scheme 15.16) [33].

15.3.2
Mechanism of Metal Carbenoid O–H Insertion into Water: Why Is a Copper(I) Complex More Competent Than a Dirhodium(II) Complex in Catalytic Asymmetric O–H Insertion Reactions?

Transition metal-catalyzed insertions of α-diazocarbonyl compounds into X–H (X = C, O, N, etc.) bonds have attracted considerable interest because these methods provide

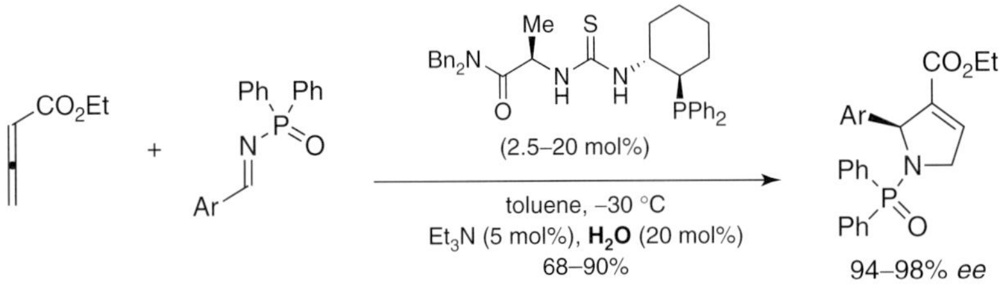

Scheme 15.16 Asymmetric phosphinothiourea-catalyzed allenoate–imine [3 + 2]-cycloadditions.

direct and efficient processes for the creation of carbon–carbon or carbon–heteroatom bonds [34]. In this field, much effort has been devoted to the development of chiral transition metal catalysts for asymmetric versions of these important reactions. Among them, chiral dirhodium(II) catalysts have been proven to be powerful for catalytic asymmetric insertions into C–H bonds, which have been utilized for the syntheses of natural products and pharmaceutical targets [35].

On the other hand, the asymmetric O–H insertion reaction is an ideal synthetic strategy for preparing optically pure α-alkoxy-, α-aryloxy-, and α-hydroxycarboxylic acid derivatives, which are valuable building blocks for the construction of natural products and other biologically active molecules. Surprisingly, although many groups have tried to achieve asymmetric O–H insertions using various chiral dirhodium(II) catalysts, to date there has been no report of a significant level of enantiocontrol (Scheme 15.17). For example, Moody and co-workers tried seven different chiral dirhodium(II) complexes to catalyze the asymmetric insertion of α-diazo esters into alcohols [36]. However, no enantiomeric excess (*ee*) was observed. For the Rh(II)-catalyzed asymmetric insertion of α-diazo esters into water, a maximum *ee* of 8% was reported by Landais and co-workers [37].

Recently, through the use of copper catalysts carrying chiral ligands, Fu's and Zhou's groups independently made remarkable advances in highly enantioselective insertions of α-diazocarbonyl compounds into the O–H bonds of alcohols, phenols, and water (Scheme 15.18) [38]. Why are chiral copper catalysts more competent than chiral dirhodium catalysts in asymmetric O–H insertions? This is an unresolved question. The

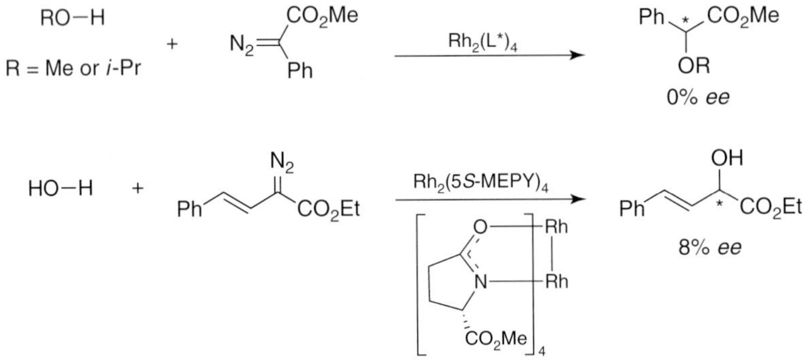

Scheme 15.17 Unsuccessful asymmetric O–H insertion using various chiral dirhodium(II) catalysts.

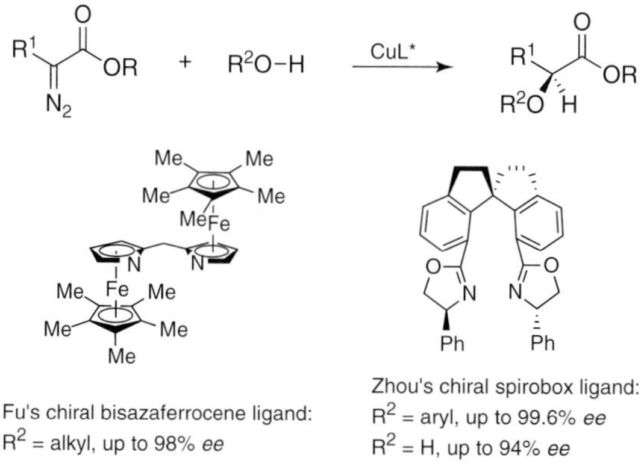

Scheme 15.18 Successful asymmetric O–H insertion using chiral copper catalysts.

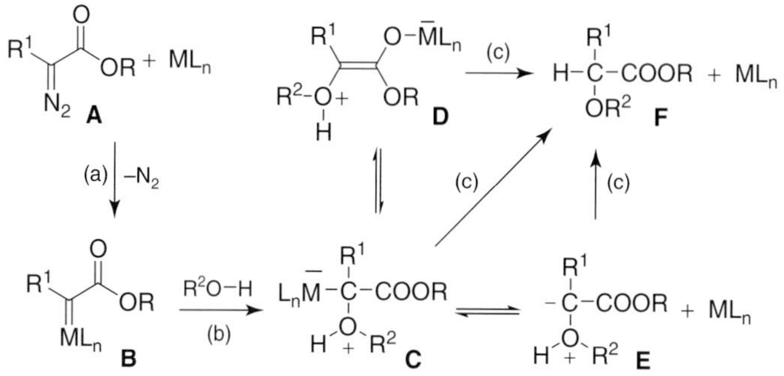

Scheme 15.19 Commonly accepted mechanism for the O–H insertion reaction.

answers to this question will not only provide a comprehensive and deep understanding of the insertions of metal carbenoids into O–H bonds at the molecular level, but also guide further rational design of new catalysts for asymmetric X–H insertions.

The commonly accepted mechanism for the O–H insertion reaction is given in Scheme 15.19 [34, 39]. The whole reaction includes three main processes: (i) metal-catalyzed nitrogen extrusion from α-diazocarbonyl compound **A** to form metal carbenoid **B**, (ii) generation of metal-associated oxonium ylide **C** from metal carbenoid **B** and R^2OH, and (iii) [1,2]-hydrogen shift of metal-associated ylide (**C** or **D**) or free oxonium ylide **E** to give product **F** and regenerate metal catalyst. Although several experimental results have provided evidence of oxonium ylide formation [39], there are still two pivotal problems to be answered: (i) which oxonium ylide is the reactive precursor for the [1,2]-H shift process, a metal-associated one or a metal-free one?; and (ii) how does the [1,2]-H shift occur? In particular, the answer to the first question is critical to asymmetric O–H insertions using chiral metal catalysts. Only when the stereodetermining [1,2]-H shift occurs via the metal-associated ylide pathway can the stereocenter formation be induced by the chirality of the catalyst. This is the prerequisite for achieving enantioselectivity.

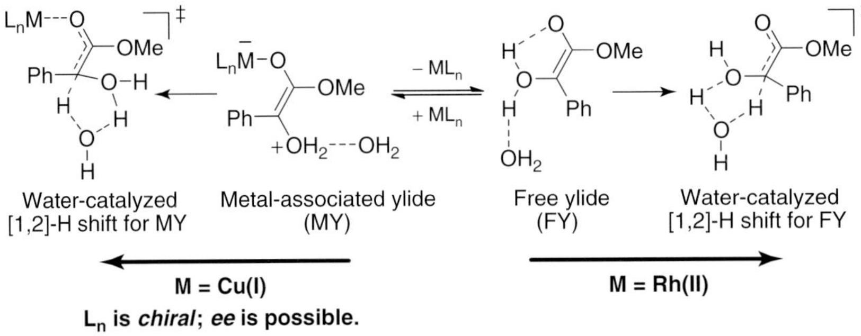

Scheme 15.20 Significant difference in reaction pathways in copper(I)- and dirhodium(II)-catalyzed O–H insertion reactions.

Inspired by computational and experimental studies on the phosphine-catalyzed [3 + 2]-reaction [31], where the water-catalyzed [1,2]-proton shift is much more favored than the direct [1,2]-proton shift, we proved that the reactant water also acts as a proton-transport catalyst to facilitate the [1,2]-H shift in transition metal-catalyzed O–H insertions by using DFT calculations [40].

More importantly, with the aid of DFT calculations, it was found that, in the Cu(I)-catalyzed system, the [1,2]-H shift process (the stereocenter formation step) favors the copper-associated ylide pathway [40]. This ensures that when a chiral copper complex is used as the catalyst, the stereocenter will be formed in a chiral environment, which is a prerequisite for achieving enantioselectivity. In contrast, the free ylide pathway is favored in the Rh(II)-catalyzed system. This implies that when a chiral dirhodium(II) complex is used as the catalyst, the Rh-coordinated ylide will first dissociate to form a free ylide, so the subsequent stereodetermining TS will not be influenced by the chiral ligand. Therefore, it is not surprising that to date there has been almost no enantioselectivity observed when using chiral dirhodium(II) catalysts. The significant difference in reaction pathways renders a copper(I) complex more competent than a dirhodium(II) complex in catalytic asymmetric O–H insertion reactions (Scheme 15.20) [40]. These mechanistic findings will be helpful in understanding other insertion reactions of heteroatom–hydrogen bonds of a wide range of substrates, and provide useful information for the design of new chiral catalysts.

15.3.3
Mechanism of the Nazarov Cyclization of Aryl Dienyl Ketones: Pronounced Steric Effects of Substituents

The Nazarov cyclization is a powerful synthetic transformation for the synthesis of cyclopentane derivatives [41]. Because of its growing popularity in the arena of complex molecular synthesis, the Nazarov reaction has evolved to include several variants. Sarpong and co-workers designed the Nazarov cyclization to construct the A ring of the natural product tetrapetalone A (**26**, Scheme 15.21). Their initial investigations focused on the conversion of readily prepared aryl dienyl ketones **28** into indanone products **27** [42].

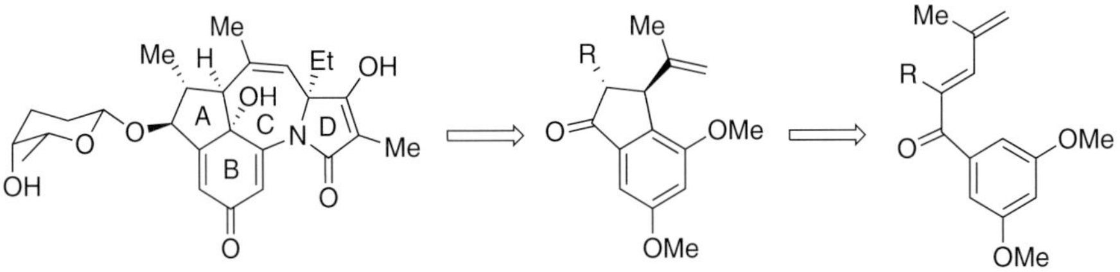

Scheme 15.21 Retrosynthetic approach to tetrapetalone A.

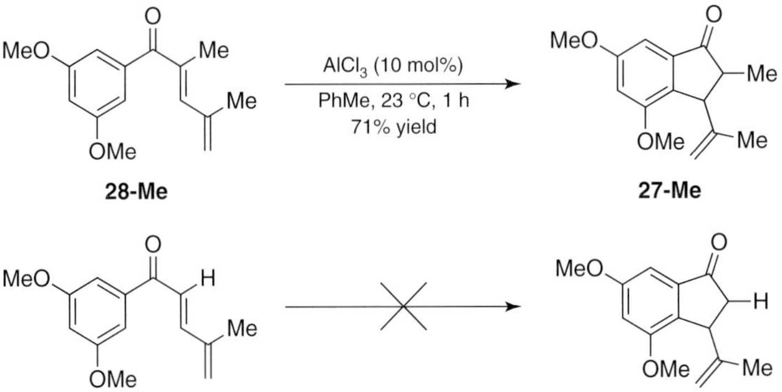

Scheme 15.22 Significantly different experimental results due to the substituent at the α-position of the diene fragment.

Surprisingly, this type of Nazarov reaction is extremely sensitive to the R substituent at the α-position of the diene fragment [42]. As shown in Scheme 15.22, the α,γ-dimethyl-substituted aryldienone **28-Me** reacts smoothly in the presence of catalytic amounts of AlCl₃ to give the desired indanone **27-Me** in 71% yield. However, a screen of various Lewis acids and solvents at different temperatures produced either no reaction or decomposition of γ-methyl-substituted aryl dienone **28-H**.

To understand better the inherent differences that lead to the observed facilitation of the Nazarov cyclization with the introduction of the methyl substituent at the α-position of the diene moiety (**28-Me** versus **28-H**, Scheme 15.22), the Tantillo and Sarpong groups conducted a detailed mechanistic study through the combination of computations and experiments [42]. DFT calculations indicated that the 4π electrocyclization barrier for complex **29** generated from **28-H** and AlCl₃ is 30.6 kcal mol^{-1} in terms of Gibbs free energy (Figure 15.5). However, the 4π electrocyclization of complex **30** generated from **28-Me** and AlCl₃ requires a much lower activation free energy of 24.0 kcal mol^{-1}. This shows that the Nazarov cyclization of **28-Me** is much easier than that of **28-H**, which is consistent with the experimental result. Through the comparison of complexes **29** and **30**, it was found that there is a serious 1,3-allylic interaction between the two methyl groups in **30**, but this interaction is not present in **29** (Figure 15.5). This is also evidenced by a

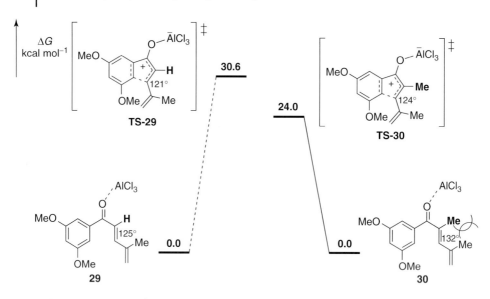

Figure 15.5 Computed structures and relative energies.

distorted C–C–C angle of 132° in the structure of **30** as depicted in Figure 15.5. However, in both cyclization TSs **TS-29** and **TS-30** there are no obvious 1,3-allylic repulsions. As a result, the strain imparted to complex **30** by the 1,3-allylic interaction is significantly reduced as **TS-30** is reached, making the required activation energy from **30** to **TS-30** considerably lower than that from **29** to **TS-29**.

The importance of the α-methyl substituent on the diene to the reactivity of the Nazarov cyclization is further proved by the computational and experimental results shown in Scheme 15.23 [42]. For the reactions of **31-Me** and **31-H**, owing to the replacement of the γ-methyl groups existing in **28-Me** and **28-H** with hydrogen atoms, the activation

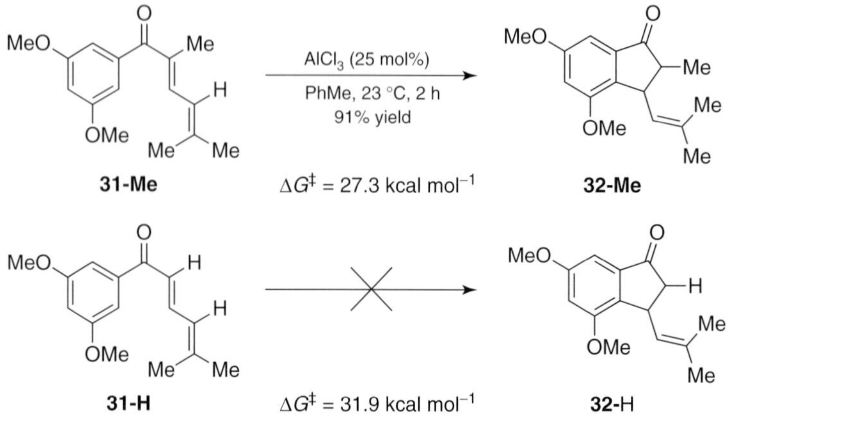

Scheme 15.23 Further computational and experimental results to prove the importance for the reactivity of the α-methyl substituent on the diene.

free energy difference decreases to 4.6 kcal mol^{-1} compared with the 6.6 kcal mol^{-1} difference between the reactions of **28-Me** and **28-H**. This also demonstrates that the 1,3-allylic steric interaction plays an important role in these reactions. Thus Tantillo and co-workers have given us an outstanding example to showcase how a substituent change can remarkably affect the reaction reactivity.

15.4
Computation-Guided Development of New Catalysts, New Reactions, and Synthesis Planning for Ideal Synthesis

As we know, organic chemistry is impacting our lives, in addition to today's science and technology. This is because organic chemistry not only provides knowledge and theories for us to understand how our molecular world works, but also provides synthetic molecules as fine, specialty, and pharmaceutical chemicals and materials used by humans. Today, synthetic organic chemistry is so powerful that, in principle, any complex molecules can be synthesized by using the reactions discovered in the last 200 years. We are very proud of such feats achieved by generations of synthetic organic chemists. However, owing to economic and environmental considerations, the synthesis of any molecules, today and in the future, must be made under the wing of green chemistry and ideal synthesis. Therefore, we must innovate the science of synthesis to have more ideal reactions and catalysts for ideal synthesis (see the definition below). In the discovery of ideal reactions and planning ideal syntheses of molecules, computational organic chemistry can provide guidance and help.

The nine principles of green chemistry [43] set out what we should consider in the synthesis of a target molecule: (i) waste prevention instead of remediation; (ii) atom efficiency (or atom economy according to Trost [44]); (iii) less hazardous/toxic chemicals; (iv) safer products by design; (v) innocuous solvents and auxiliaries; (vi) energy-efficient use by design; (vii) renewable raw materials and solvent; (viii) shorter synthesis (or step economy according to Wender and co-workers [45]); and (ix) a catalytic rather than a thermal process. Ideal synthesis can be regarded as an extreme of green chemistry for the synthesis of a complex target molecule. The ideal synthesis was first proposed by Hendrickson in 1975, who defined "ideal synthesis" as "... to create a complex molecule ... in a sequence of only construction reactions involving no intermediary functionalizations, and leading directly to the target, not only its skeleton but also its correctly placed functionality" [46]. The ideal synthesis proposed by Hendrickson can be described by Scheme 15.24, showing through either a one-step reaction or a one-pot reaction with a sequence of ideal reactions from easily available starting materials to reach the target molecule in 100% yield.

We believe that the ideal synthesis as defined by Hendrickson can be realized only in very limited examples today. However, the concept of ideal synthesis together with the consideration of green chemistry is influencing today and future's synthetic organic chemists, stimulating them to create new reactions and strategies. To meet this aim, researchers will need many ideal or near-ideal reactions. Here we define an ideal reaction as the reaction with the criteria shown in Scheme 15.25.

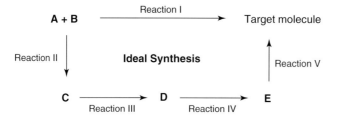

Scheme 15.24 One-pot ideal synthesis using either an ideal reaction or a few ideal reactions in one-pot.

A + B ⟶ P   **An Ideal Reaction**

Criteria of an ideal reaction:
1. **A** and **B** are safe compounds that are easily available or can be obtained by other ideal reactions, which are already available or could be available in the future.
2. The reaction yield is 100% or near 100%.
3. Without catalyst or using a ppb loading of catalyst, or using easily recovered catalyst.
4. Using a benign and reusable solvent that can be easily separated from the product (or without solvent).
5. Easy operation at ambient temperature, not sensitive to moisture or air.
6. Scalable.
7. Broad scope.
8. Short reaction time.

Scheme 15.25 Criteria of an ideal reaction.

We think that a few reactions today, and some reactions in the future, can meet these entire criteria and can be called ideal reactions. However, some reactions today and many reactions in the future meet just a subset of these entire criteria and can be called semi-ideal or near-ideal reactions. For example, a near-ideal reaction could proceed in refluxing solvent for 2 days, or the reaction yield may be 50% but the product can be easily separated from the reaction system. The exact definitions of semi-ideal and near-ideal reactions are still open to practitioners in this field.

Computational organic chemistry impacts the ideal synthesis in two directions: one is to help the design and discovery of new reactions and their catalysts and ligands, and the other is to help the design of the synthesis of target molecules.

The contributions of computational organic chemistry in the design of reactions, catalysts, and ligands include the following:

1) Optimizing known reactions so that these reactions can become ideal or semi-/near-ideal reactions through computational organic chemistry's elucidation of reaction mechanisms and their energy surfaces.

2) Understanding the scope of ideal reactions through computation of the activation barriers and regio- and stereoselectivity when different substituents are used. This is especially important in the synthesis of complex products where many substituents could be present and many side reactions could compete with the major and targeted reaction pathway.

3) Designing new reactions and their catalysts and ligands and optimizing them to become ideal or semi-/near-ideal reactions. This could be accomplished by *de novo* design or with the help of studying the reported reactions' mechanisms and modifying these known (or "lead") reactions (see Section 15.3 concerning mechanism guidance in reaction design).

We must admit that so far today almost all the computational chemistry-designed reactions and catalysts are based on the reported "lead" reactions and "lead" catalysts that were discovered by trial and error and serendipity. There are almost no *de novo*-designed reactions. However, we think that it does not matter too much whether a new reaction is designed *de novo* or is based on "lead" discoveries. The key is to discover new reactions, catalysts, and ligands that can be applied in the ideal synthesis of molecules. The "lead" discoveries from the last two centuries give us countless opportunities and we must take advantage of these discoveries to develop more new reactions, catalysts, and ligands. Certainly our mastery of the knowledge of organic chemistry and our imagination are key for us to design new chemistry with the help of computational chemistry. Even though there are no *de novo*-designed reactions today, with more insights into reaction mechanisms and a deeper understanding of reactivity, together with more and more creative ideas and imagination, more *de novo* reactions and catalysts will be developed in the twenty-first century.

The contributions from computational organic chemistry in designing syntheses can be appreciated in this way as described below. Suppose that there are several possible routes to a target complex molecule and each route includes some ideal, some near-ideal, and perhaps some semi- or non-ideal reactions (Scheme 15.26). How to propose and design these possible synthetic routes is another subject of discussion and is beyond the scope of this chapter. However, we believe that chemical informatics and our knowledge of and experience in synthesis can help. Once we have these designed routes, computational organic chemistry can provide assistance in evaluating them to identify the best route among them, if the selection criteria (such as cost, time, manpower, environmental impact, percentage of ideality, and total yield) are given. How does this proceed?

First, calculations can help to identify whether each reaction in a complex system really works and provide us with the information on this reaction. Usually we have only limited information about a known reaction (no matter whether it is ideal, near-ideal, or non-ideal), since we obtained this information from limited experimental results.

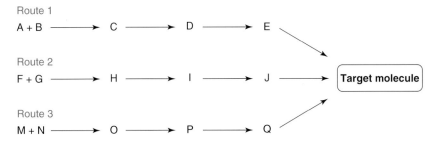

Scheme 15.26 Three routes to a target molecule.

However, in the target molecule synthesis, especially for complex target molecules, many functional groups are present, and the reactions in these routes become complex and could have many side reactions. Therefore, we have to know whether the selected reaction in this specific synthesis really works as expected. Ideally, through calculations, we can establish whether this reaction works and give its correct stereochemistry. Also, calculations can help us to determine the optimal reaction conditions (for example, which solvent is the best) for each step in the selected synthesis route.

Suppose that the reaction of **D** to **E** in route 1 (Scheme 15.26) is a Diels-Alder reaction, but several functional groups are present in the dienes and dienophiles for the reaction of **D** to **E**, and there are several possible side reactions that compete with the DA reaction. In addition, in the synthesis, we expect to obtain an *endo* DA product (stereochemistry issue). We can then compute both the *endo* and *exo* TSs of the DA reaction and the TSs in other possible side reactions. By comparing the energetics of these competing pathways, we can establish whether the *endo* DA reaction can occur and its possible yield. If the results are negative, we can then discard this route or modify it in some way.

Today, in some syntheses, people usually build a three-dimensional or ball-and-stick model to help them understand whether a key reaction in their synthetic route works or not. They are usually concerned about the stereochemistry of its key step, but do not consider the energetics of the key reaction and its possible competing reactions. With the advances in computational chemistry, the evaluation method aided by *ab initio* calculations will be used in the future target molecule synthesis.

Once we have evaluated all possible routes and know the actual possible routes to a complex target molecule, calculations can compute the total yield, cost, time, and other parameters of each route. This information can then be used to choose the best routes to test.

During the experimental tests of these routes, synthetic chemists can also use the experimental feedback to refine their synthetic plan with the help of calculations.

In following, we present several examples to describe how computational organic chemistry helps to design new catalysts and reactions. Readers are also strongly recommended to read a paper by Houk and Cheong [47], where insights into and guidance for the computational design of catalysts are given. Assistance in designing synthetic routes by calculations has not been found, but we believe that this will happen in the coming decades.

15.4.1
Discovery of Catalysts for 6π Electrocyclizations

Pericyclic reactions comprise a major category of reactions of theoretical and synthetic importance. The class of pericyclic reactions for which catalysts have been well developed by far is cycloadditions [48]. Some catalyzed sigmatropic rearrangements have also been described [49]. However, methods for the catalysis of electrocyclizations are very rare except for the Nazarov cyclization [41]. In 2008, the Bergman and Trauner groups discovered for the first time that 6π electrocyclizations can be catalyzed by Lewis acids with the aid of computational chemistry [50].

Previous computational and experimental investigations have shown that the rate of 6π electrocyclizations can be influenced by the electronic effects of the substituents on hexatrienes [51]. It was observed that hexatrienes with electron-withdrawing groups located at the C2 position cyclize with lower energy barriers than those without such substituents. Bergman and co-workers envisioned that if a Lewis acid was introduced as the catalyst, the coordination of a Lewis basic electron-withdrawing group at C2 of the hexatriene to the Lewis acid would further increase the electron-withdrawing effect of the substituent, thereby decreasing the 6π electrocyclization energy barrier. They first assessed the viability of this strategy by using DFT calculations. Computational results predicted that the 6π electrocyclization energy barrier would decrease by 10 kcal mol^{-1} as the 2-carbomethoxy-1,3,5-hexatriene was protonated (Figure 15.6).

Encouraged by the computational result, they subsequently investigated a variety of Lewis acids and found that dimethylaluminum chloride (Me$_2$AlCl) was an efficient catalyst for the reactions of trienes **33** and **34** (Schemes 15.27 and 15.28) [50]. The system of triene **33** has an ester group at the C2 position, whereas the system of triene **35** involves a ketone. It was found that the rates of these 6π electrocyclizations were increased in the presence of Me$_2$AlCl. For example, the thermal electrocyclization of triene **33** has a half-life of 4 h at 50 °C. The addition of 1 equiv. of Me$_2$AlCl makes the half-life decrease to 21 min at the same temperature. For triene **35**, a 55-fold rate acceleration was observed in the presence of 1 equiv. of Me$_2$AlCl at 28 °C. The reaction rates were accelerated when using catalytic amounts of Me$_2$AlCl. Although ~20 mol% or more of the catalyst was necessary to obtain a significant rate acceleration, these results well demonstrated that the substituent-enhancing strategy was successful in the catalysis of 6π electrocyclization reactions. In addition, Eyring plots were employed to determine the activation parameters for these reactions (Schemes 15.27 and 15.28). It was revealed that the activation Gibbs free energies for the catalyzed processes were ~2 kcal mol^{-1} lower than those for the thermal processes, and that these catalyses were primarily enthalpic [50].

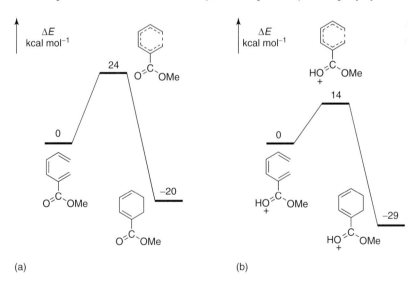

(a)   (b)

Figure 15.6 Relative electronic energies of the (a) thermal and (b) proton-catalyzed 6π electrocyclization pathways.

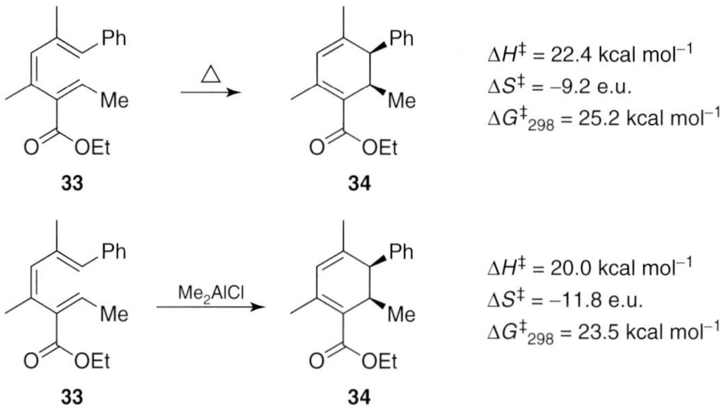
Ph
Me
O
OEt
33
Δ
Ph
Me
O
OEt
34
ΔH‡ = 22.4 kcal mol−1
ΔS‡ = −9.2 e.u.
ΔG‡298 = 25.2 kcal mol−1
Ph
Me
O
OEt
33
Me2AlCl
Ph
Me
O
OEt
34
ΔH‡ = 20.0 kcal mol−1
ΔS‡ = −11.8 e.u.
ΔG‡298 = 23.5 kcal mol−1

Scheme 15.27 Thermal and Me_2AlCl-catalyzed 6π electrocyclizations of triene ester.

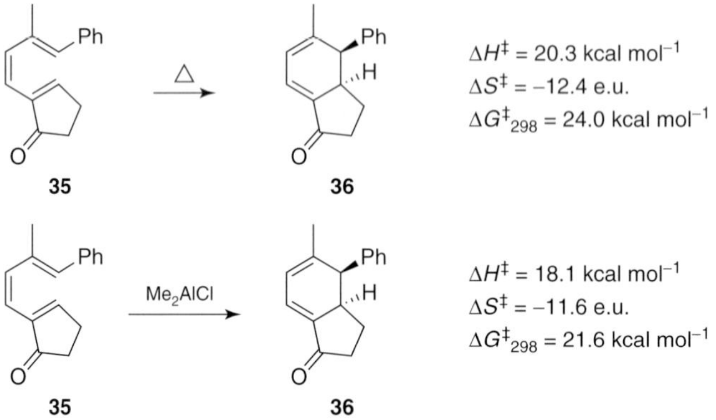
Ph
O
35
Δ
Ph
H
O
36
ΔH‡ = 20.3 kcal mol−1
ΔS‡ = −12.4 e.u.
ΔG‡298 = 24.0 kcal mol−1
Ph
O
35
Me2AlCl
Ph
H
O
36
ΔH‡ = 18.1 kcal mol−1
ΔS‡ = −11.6 e.u.
ΔG‡298 = 21.6 kcal mol−1

Scheme 15.28 Thermal and Me_2AlCl-catalyzed 6π electrocyclizations of triene ketone.

The success of the catalysis of 6π electrocyclizations confirms the predictions of the DFT calculations, which indicate that Lewis acids are potent catalysts for the hexatriene systems substituted at the C2 position with suitable electron-withdrawing groups. The computational and experimental information obtained will provide guidance for developing not only more efficient catalysts, but also chiral Lewis acid catalysts, and organocatalysts, to realize asymmetric 6π electrocyclization reactions [52].

15.4.2
Computational Design of a Chiral Organocatalyst for Asymmetric Anti-Mannich Reactions

Organocatalysis, or the use of small organic molecules to catalyze organic reactions, is a relatively new and popular field within the domain of enantioselective synthesis [53]. For reactions involving *in situ*-generated enamines, proline, and related compounds, have been extensively examined as organocatalysts. Barbas and co-workers reported that, in the reaction of aldehydes with *N*-*p*-methoxyphenyl (PMP)-protected imines catalyzed by the natural amino acid (*S*)-proline, *syn*-Mannich products were obtained with high

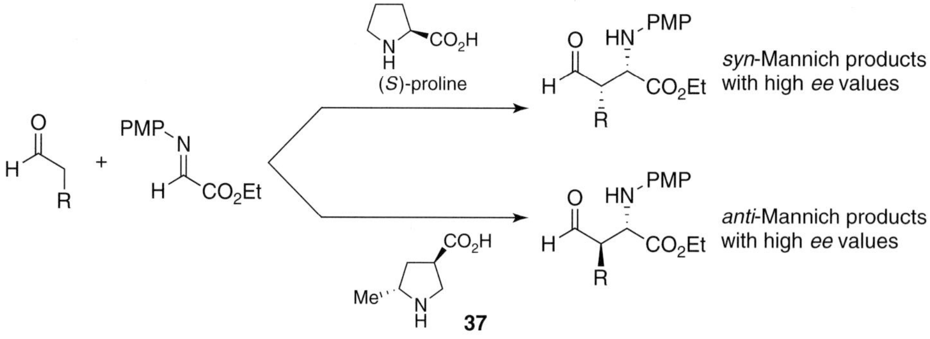

Scheme 15.29 Asymmetric *syn-* and *anti-*Mannich reactions.

enantioselectivities (Scheme 15.29) [54]. Although reactions involving some pyrrolidine derivatives gave *anti-*diastereomers as the major products, the *ee* values were moderate [54]. Therefore, the development of effective highly enantioselective *anti-*Mannich catalysts presented a great challenge at that time. In collaboration with Barbas's group, Houk and co-workers successfully designed amino acid **37** as a highly diastereo- and enantioselective *anti-*Mannich catalyst with the aid of computational chemistry (Scheme 15.29) [55].

The design of catalysts providing *anti-*Mannich products with high *ee* values is based on the key factors that control the diastereo- and enantioselectivities of (*S*)-proline-catalyzed Mannich reactions (Scheme 15.30a) [56]. Four considerations are critical: (i) enamine intermediates exist in an *E* configuration; (ii) owing to the steric repulsions between the enamine and the substituent at the C2 position of the pyrrolidine ring in the s-*cis* conformation of the (*E*)-enamine, the (*E*)-enamine adopts an s-*trans* conformation in the C–C bond formation TS; (iii) the C–C bond formation takes place at the *re* face of the enamine, and this facial selectivity is controlled by the proton transfer from the carboxylic acid to the nitrogen atom of the imine; and (iv) the enamine attacks the *si* face of the (*E*)-imine, which is also controlled by the proton transfer that increases the electrophilicity of the imine.

To realize the highly enantioselective formation of *anti-*Mannich products, it is necessary to reverse the facial selectivity of either the enamine or the imine compared with the proline-catalyzed reactions. It was hypothesized that if the acid functionality was moved to the distal C3 position of the pyrrolidine ring, and a new substituent was introduced at the C5 position of the ring, the facial selectivity of the enamine would change from the *re* face to the *si* face in the C–C bond formation TS, and the enamine would still attack the *si* face of the imine (Scheme 15.30b) [55]. As a result, the *anti-*Mannich products were expected to become dominant. In addition, to avoid steric interactions between the new substituent at the C5 position and the PMP group of the imine in the TS, the substituents at the C3 and C5 positions of the designed catalyst should be in a *trans* configuration.

On this basis, a new chiral organocatalyst, (3*R*,5*R*)-5-methyl-3-pyrrolidinecarboxylic acid (**37**), was designed. Computational studies of the **37**-catalyzed Mannich reaction between propionaldehyde and imine **38** at the HF/6−31G(d) level of theory were used to test this design prior to synthesis [55]. The catalyst **37** was predicted to result in 95:5 *anti:syn* diastereoselectivity and 98% *ee* for the *anti-*Mannich product **39** (Scheme 15.31a).

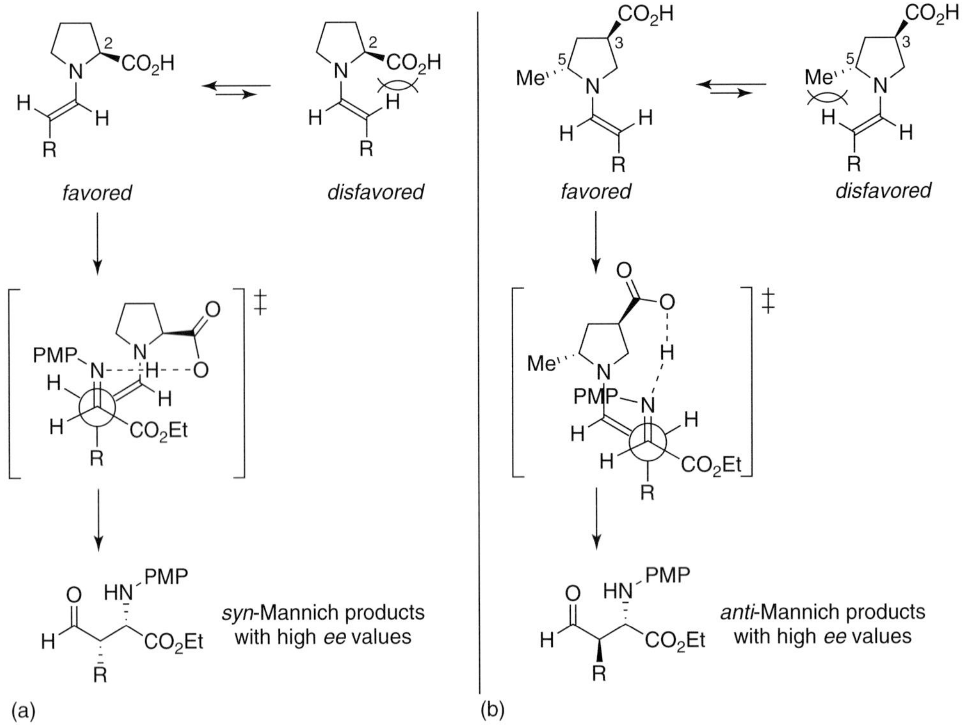

Scheme 15.30 Rationalization of (*S*)-proline-catalyzed *syn*-Mannich reactions and the design of new catalyst for *anti*-Mannich reactions.

Subsequent experiments with this catalyst showed remarkable agreement with computational predictions (Scheme 15.31b).

The production of this excellent organocatalyst for asymmetric *anti*-Mannich reactions well demonstrates that computational design is ready to take its place as an essential component of catalyst design. With the aid of computational methods, a series of potential catalysts with a certain function can be designed and evaluated. Such *in silico* screening will be much faster than preparing each potential catalyst and evaluating it experimentally, so a substantial amount of time can be saved for the generation of an effective catalyst [47].

15.4.3
Computation-Guided Development of Gold-Catalyzed Cycloisomerizations Proceeding via 1,2-Si Migrations

Computational methods not only can help to design catalysts for known reactions but also can predict new reactions. In recent years, homogeneous catalysis by gold complexes attracted much attention [57]. For instance, cycloisomerizations of alkynes and allenes are efficient and versatile methods for the synthesis of cyclic compounds. In this field, Hashmi *et al.* and later Che and co-workers reported Au(III)-catalyzed cycloisomerizations

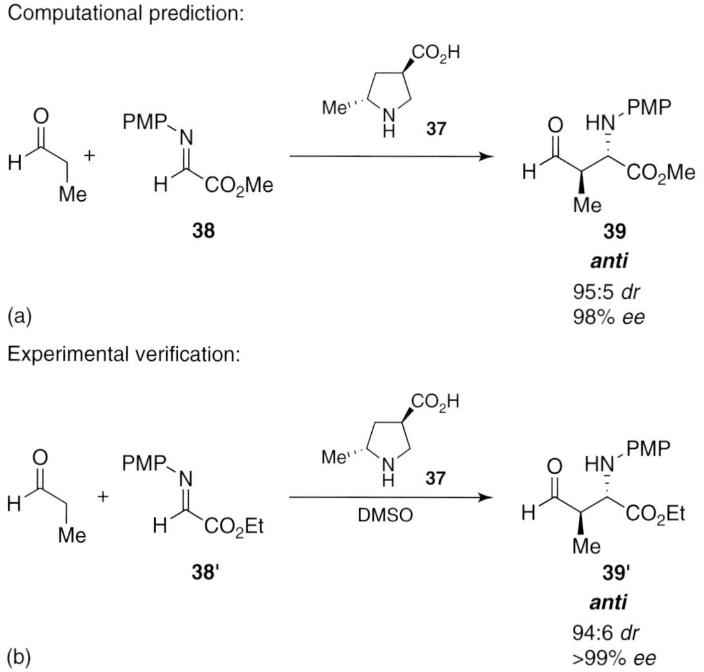

Computational prediction:

(a)

Experimental verification:

(b)

Scheme 15.31 Computational prediction and experimental verification for the performance of the designed new catalyst.

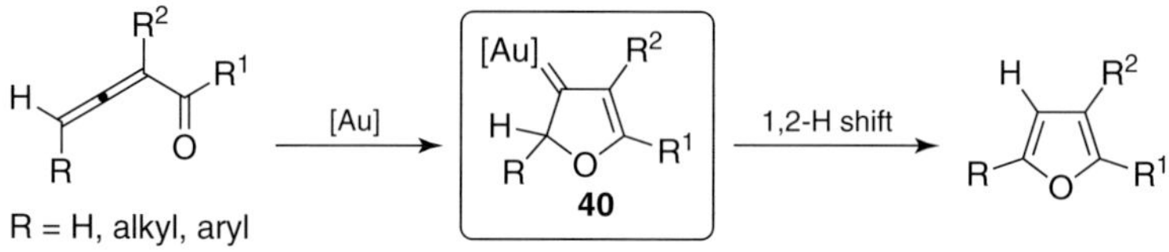

R = H, alkyl, aryl

Scheme 15.32 Gold-catalyzed cycloisomerizations of allenones to furans.

of allenones into furans (Scheme 15.32) [58]. These reactions occur via the formation of the key putative Au–carbene intermediate **40** and a subsequent 1,2-H shift process. Apparently, 1,2-migration of groups rather than the hydrogen atom, taking place via the Au–carbene intermediate, would allow for the synthesis of diversely functionalized furans with substituent patterns that are not easily obtained by using existing methods. Recently, the Li and Gevorgyan groups reported a computation-designed novel gold-catalyzed cycloisomerization proceeding via a 1,2-Si shift for the synthesis of 3-silylfurans [59].

The discovery of this new reaction starts with DFT studies on the migratory aptitudes of hydrogen, alkyl, aryl, and silyl groups in the Au–carbene intermediate **41** (Scheme 15.33) [59]. DFT calculations showed that the 1,2-Si shift was strongly favored over the 1,2-migrations of methyl, phenyl, and even hydrogen in Au–carbene **41**. This suggests that synthetically valuable 3-silylfurans can be generated exclusively in the $AuCl_3$-catalyzed cycloisomerization of silylallenones.

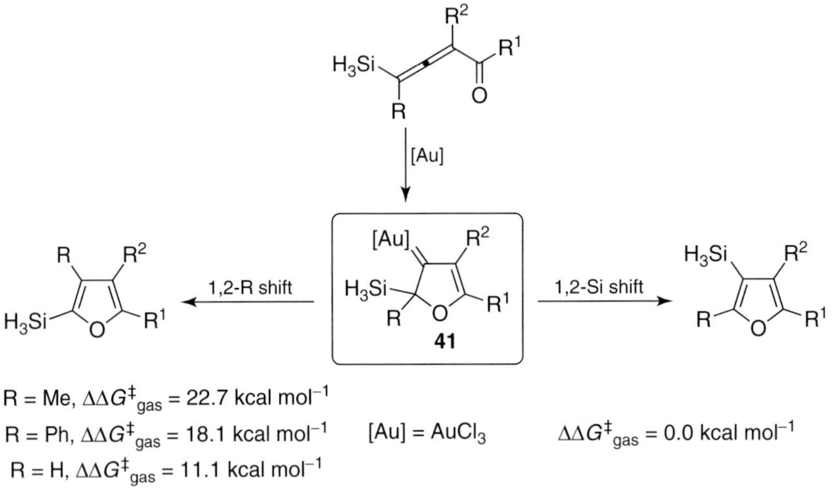
[Au]
1,2-R shift
1,2-Si shift
41
R = Me, ΔΔG‡gas = 22.7 kcal mol−1
R = Ph, ΔΔG‡gas = 18.1 kcal mol−1
R = H, ΔΔG‡gas = 11.1 kcal mol−1
[Au] = AuCl3
ΔΔG‡gas = 0.0 kcal mol−1

Scheme 15.33 Computational prediction for the migratory aptitudes of substituents.

Encouraged by these results, they synthesized several silylallenones and tested the possible cycloisomerizations into 3-silylfurans in the presence of $AuCl_3$ catalyst [59]. To their delight, α-to-Si methyl- and phenyl-substituted allenones underwent 1,2-Si migrations exclusively, affording desired 3-silylfurans in high yields (Scheme 15.34a and b). More importantly, the cycloisomerization of α-to-Si-unsubstituted allenone proceeded very efficiently to give the 3-silylfuran as the sole isomer (Scheme 15.34c). Therefore, the experimentally observed exclusive 1,2-Si shift is in excellent accord with the computation-predicted higher 1,2-shift aptitude of silyl groups over hydrogen, alkyl, and aryl groups.

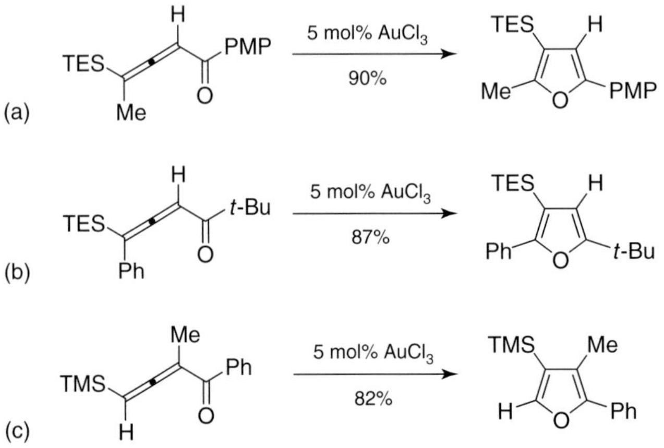
(a)
5 mol% AuCl3
90%
(b)
5 mol% AuCl3
87%
(c)
5 mol% AuCl3
82%

Scheme 15.34 Gold-catalyzed cycloisomerization proceeding via a 1,2-Si shift for the synthesis of 3-silylfurans.

15.4.4
A Computationally Designed Rh (I)-Catalyzed [(5 + 2) + 1] Cycloaddition for the Synthesis of Cyclooctenones

Transition metal-catalyzed cycloaddition reactions constitute an important part of modern organic synthesis. Wender and co-workers developed a series of Rh(I)-catalyzed [5 + 2]-cycloadditions between vinylcyclopropanes (VCPs) and 2π components (such as alkenes, alkynes, and allenes) to afford seven-membered ring products [60]. The rhodium dimer [Rh(CO)$_2$Cl]$_2$ was found to be an especially effective and versatile catalyst for this and many related cycloadditions. However, the intramolecular [5 + 2]-cycloaddition of ene-VCPs was not realized using [Rh(CO)$_2$Cl]$_2$ as catalyst [60b]. Yu *et al.* [61] speculated that the failure of [Rh(CO)$_2$Cl]$_2$ to catalyze the [5 + 2]-reaction of ene-VCPs (relative to yne-VCPs and allenyl-VCPs) could be attributed to the reductive elimination (RE) step in the reaction mechanism (Scheme 15.35). The RE step in the intramolecular [5 + 2]-cycloaddition of ene-VCP would lead to the formation of an C(sp^3)–C(sp^3) bond. However, this kind of RE reaction is not facile compared with the migratory reductive elimination (MRE) of an C(sp^2)–M–C(sp^3) subunit for the formation of an C(sp^2)–C(sp^3) bond [61, 62]. It was hypothesized that the putative intermediate **A** (Scheme 15.35), whose RE to the [5 + 2] product was disfavored, would be converted into intermediate **B** if carbon monoxide was added to the reaction mixture, whose MRE to give a [(5 + 2) + 1] product would be favored by the formation of an C(sp^2)–C(sp^3) bond.

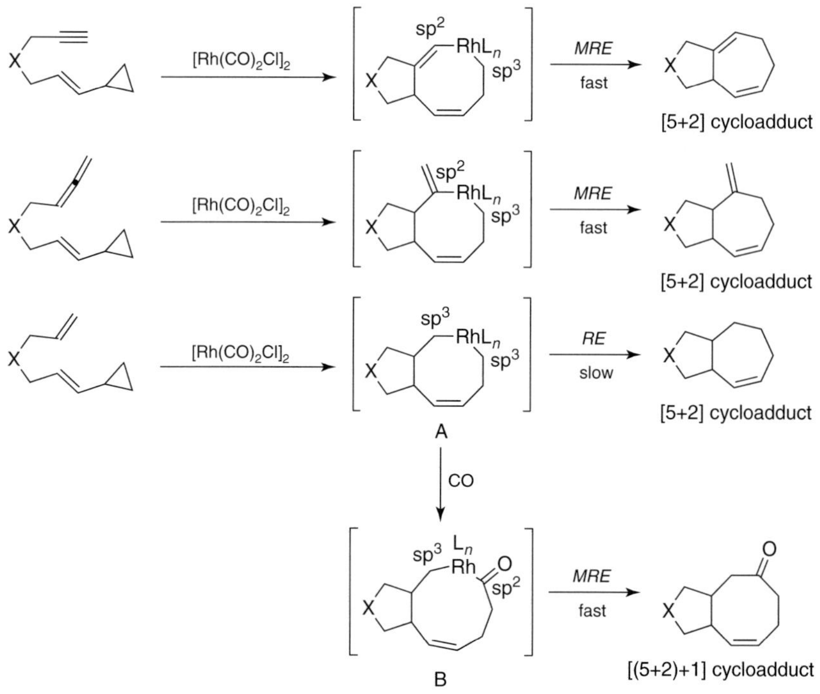

Scheme 15.35 Rationale for the designed two-component [(5 + 2) + 1] reaction.

Before testing the proposed new [(5 + 2) + 1]-reaction experimentally, the energies for the RE, CO insertion, and MRE steps shown in Scheme 15.35 (X = CH$_2$) were computed using the B3LYP method [63]. The activation energies of the RE steps are about 25–30 kcal mol^{-1}, whereas the CO insertion and MRE steps have activation energies of about 13–14 and 23–24 kcal mol^{-1}, respectively. This indicated that the [(5 + 2) + 1] pathway would be favored over the [5 + 2] pathway and the dominant product would be the [(5 + 2) + 1] cycloadduct.

Encouraged by these computational results, we began our experimental test. After the optimization of reaction conditions, a computationally designed Rh(I)-catalyzed two-component [(5 + 2) + 1]-cycloaddition of ene-vinylcyclopropanes (ene-VCPs) and CO was successfully developed (Scheme 15.36) [63]. This reaction provides an efficient way to obtain fused [5–8] and [6–8] ring systems, and this method provides a new opportunity for the total synthesis of natural products (Figure 15.7). We applied a tandem [(5 + 2) + 1]-cycloaddition–aldol reaction strategy to the syntheses of three linear triquinane natural products, (±)-hirsutene, (±)-1-desoxyhypnophilin, and (±)-hirsutic acid C [64]. The [(5 + 2) + 1]-reaction has also proved efficient in the syntheses of (±)-pentalenene and (±)-asterisca-3(15),6-diene [65]. Recently, we also successfully achieved the asymmetric total synthesis of (+)-asteriscanolide based on a chiral substrate induced Rh(I)-catalyzed [(5 + 2) + 1]-cycloaddition reaction to build the [6.3.0] carbocyclic core with high efficiency [66].

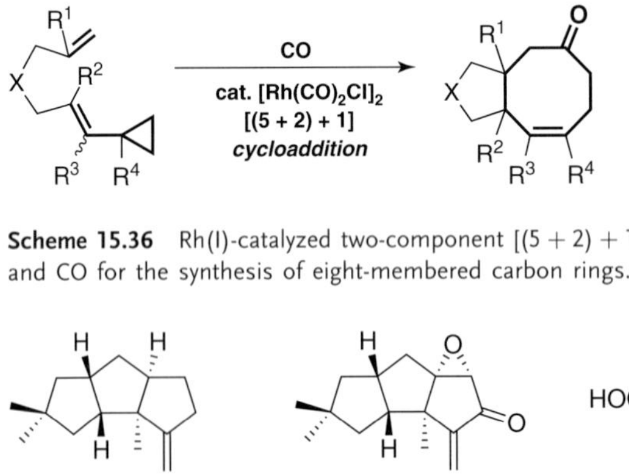

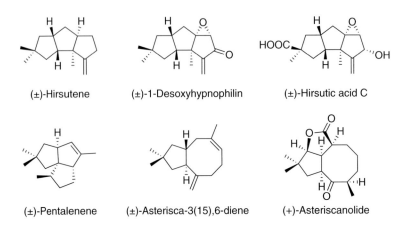

Scheme 15.36 Rh(I)-catalyzed two-component [(5 + 2) + 1]-cycloaddition of ene-vinylcyclopropanes and CO for the synthesis of eight-membered carbon rings.

(±)-Hirsutene   (±)-1-Desoxyhypnophilin   (±)-Hirsutic acid C

(±)-Pentalenene   (±)-Asterisca-3(15),6-diene   (+)-Asteriscanolide

Figure 15.7 Natural products synthesized by using the [(5 + 2) + 1] reaction as the key step.

The above results showcase that computational calculations can complement experiments and provide guidance for the design of new transformations. With the aid of computational methods, the development of new organic reactions, catalysts, and ligands will become more efficient.

15.5
Conclusion

It is clear that computational organic chemistry, which provides new theories and concepts, elucidates reaction mechanisms at the molecular level, and contributes in designing and predicting new catalysts, new reactions, and synthesis strategies, is an essential component of organic chemistry, both in the past and in the twenty-first century. Today it is impossible for organic laboratories to do research without using an NMR instrument. Similarly, it is becoming impossible for organic laboratories to do research without using QM calculations. Computational organic chemistry is just like an enabling eye for chemists to see molecules and their reactions. On the other hand, it is like a compass that guides chemists to carry out research in a more rational way. However, many challenges have to be tackled before this dream can become true.

The accuracy of calculations on a complicated reaction system is critical for computational organic chemistry to exert its impact on organic chemistry as a whole. Today there are computational methods that can model reactions at almost any desired accuracy, provided that sufficient computer resources and time are available. However, it is important that an accuracy of within $1 \, \text{kcal mol}^{-1}$ for a reaction system with several hundred to more than 1000 atoms (which could include some solvent molecules in the calculation system) can be achieved routinely within a reasonably short time. This is very significant in the study of reaction mechanisms, where real reaction systems with explicit solvent molecules included, instead of model systems with simplified reactants and a solvation model, will be used to obtain results that can be compared with those from experiments. This is also important for the future design of the synthesis of a target molecule, where the activation energies of the major reaction pathway and side reaction pathway(s) in each step of a selected synthetic route must be known. This accuracy is also important for understanding asymmetric reactions and for designing chiral ligands, because a $1 \, \text{kcal mol}^{-1}$ energy difference usually results in an *ee* of $\sim$70%.

Almost all organic reactions are carried out in solution. Therefore, modeling of how a solvent affects a reaction is critical to understanding the reaction mechanism, to help in selecting the optimal solvent(s) in a synthesis, and also to computing thermodynamic and kinetic data with high accuracy. It is impossible to include all solvent molecules in QM calculations. In addition to including more solvent molecules in the calculation system (where real reactants, catalysts, ligands, and some solvent molecules are included in the QM calculations), we must also develop more accurate solvation models to describe the solvation effects imposed by bulky solvents that are not included explicitly in the calculations. In this case, calculation results will be more accurate and the direct role of solvent(s) in a studied reaction can be understood in detail at the molecular level.

High accuracy and speed of calculations with many solvent molecules included are still a distant dream but can gradually be achieved. This will mainly be solved by both computer scientists who can develop faster computers and supercomputers, and theoretical chemists who can develop new algorithms and new calculation theories (such as new and accurate solvation models, more precise exchange-correlation functionals, and new quantum physical equations including relativistic effects). Even though computational organic chemists usually do not take part in research in these two fields, they can provide many new insights into and understandings of reaction modes and new theories and principles through their studies so that these theories, principles, and so on can be applied to guide QM calculations. In that case, many unnecessary calculations could be avoided, saving computation resources and time and increasing the efficiency and probability of success.

In addition to the aforementioned challenges involving calculation methods and QC theories, the other hurdle that diminishes the impact of computational organic chemistry on chemistry as a whole comes from the experimental side. No matter how accurate the calculations are, the key factor determining the success of calculations on a chemical problem is whether the computational model used is correct or false. A false model certainly gives the wrong answers and predictions. For instance, today there are many transition metal-catalyzed reactions, and it is very common to use several transition metals and additives to achieve better results. However, the complexity naturally leads to several questions, such as, what is the true catalytic species in this reaction, and what is the role of these additives? Calculations sometimes can help to exclude some proposed catalytic species and hypothesize a reasonable one. However, to increase the credibility of calculations, it is better to use experimental techniques, such as spectroscopic observations or isolation of key catalytic species or intermediates, to identify the catalytic species. Then the calculation results can be considered more reliable. Another example is calculations on reactions using reagents that tend to assemble into clusters (for example, Grignard reagents, alkyllithium reagents). What the actual forms of these reagents are in the reaction systems has to be determined through experiments. Therefore, experimental results, especially for mechanistic studies of reaction mechanisms, are very critical for a successful computational project. For the design of new reactions, catalysts, and ligands, early-stage experimental discoveries are also important for computational organic chemistry in optimizing the reactions, catalysts, and ligands, and for the later design of new ones.

Therefore, the continuing success of organic chemistry in theory, mechanism, and synthesis depends on both experiments and computational organic chemistry. To advance the frontiers of organic chemistry, we should incorporate both experiments and calculations into the fabric of our daily research activities. In other words, chemistry is a science advanced through many "tries," experimentally and computationally. In a concise way to express this notion, this is: Chem is try computationally and experimentally. Here "Chem is try" comes from the word "chemistry." This philosophy is true for all fields of chemistry, certainly including organic chemistry. We are confident that Chem is try computationally and experimentally will greatly accelerate organic chemistry to a higher level of success in the twenty-first century.

We are confident that high-accuracy and high-speed modeling with complex systems will be available sometime during this century and will become indispensable for every organic laboratory. The impacts of computational organic chemistry on chemistry in general will be reinforced by this advancement. Experimental advances and discoveries will certainly continue to enrich the enterprise of organic chemistry. However, we have to keep in mind that science is done by people, not by machines or robots, which are tools for scientists to test their ideas. It is scientists who design experiments and perform calculations. Therefore, a scientist's knowledge, ideas, and imagination are of the highest importance in designing experiments and calculations to discover new chemistry. Therefore, the advance of organic chemistry will depend strongly not only on "tries" from experiments and computation, but also on our knowledge, ideas, and imagination.

Chem is try computationally and experimentally, so is organic chemistry.

Acknowledgments

We thank Peking University and the National Natural Science Foundation of China (**20825205** – National Science Fund for Distinguished Young Scholars) for financial support.

References

1. For an excellent book on computational organic chemistry, see: Bachrach, S.M. (2007) *Computational Organic Chemistry*, John Wiley & Sons, Inc., New York.
2. For excellent and authorative accounts of the history and some excellent examples of computational chemistry and computational organic chemistry, see: (a) Pople, J.A. (1999) *Angew. Chem. Int. Ed.*, **38**, 1894 (b) Hehre, W.J., Radom, L., Schleyer, P.v.R., and Pople, J.A. (1986) *Ab Initio Molecular Orbital Theory*, John Wiley & Sons, Inc., New York (c) Parr, R.G. and Yang, W. (1989) *Density Functional Theory of Atoms and Molecules*, Oxford University Press, New York. . For an excellent introduction to quantum chemistry, see: (d) McQuarrie, D.A. and Simon, J.D. (1997) *Physical Chemistry, a Molecular Approach*, University Science Books, Sausalito, CA.
3. Walsh, P.J. and Kozlowski, M.C. (2008) Nonclassical catalyst–substrate interactions, in *Fundamentals of Asymmetric Catalysis*, University Science Books, Sausalito, CA, Chapter 5.
4. (a) Yamakawa, M., Ito, H., and Noyori, R. (2000) *J. Am. Chem. Soc.*, **122**, 1466 (b) Yamakawa, M., Yamada, I., and Noyori, R. (2001) *Angew. Chem. Int. Ed.*, **40**, 2818 (c) Noyori, R., Yamakawa, M., and Hashiguchi, S. (2001) *J. Org. Chem.*, **66**, 7931 (d) Washington, I. and Houk, K.N. (2001) *Angew. Chem. Int. Ed.*, **40**, 4485 (e) Ribelin, T., Katz, C.E., English, D.G., Smith, S., Manukyan, A.K., Day, V.W., Neuenswander, B., Poutsma, J.L., and Aube, J. (2008) *Angew. Chem. Int. Ed.*, **47**, 6233 (f) Szostak, M., Yao, L., and Aube, J. (2010) *J. Org. Chem.*, **75**, 1235 (g) Bolduc, P., Jacques, A., and Collins, S. (2010) *J. Am. Chem. Soc.*, **132**, 12790.
5. Schleyer, P.v.R., Maerker, C., Dransfeld, A., Jiao, H., and van Eikema Hommes, N.J.R. (1996) *J. Am. Chem. Soc.*, **118**, 6317.
6. Ajami, D., Oeckler, O., Simon, A., and Herges, R. (2003) *Nature*, **426**, 819.
7. Castro, C., Karney, W.L., Valencia, M.A., Vu, C.M.H., and Pemberton, R.P. (2005) *J. Am. Chem. Soc.*, **127**, 9704.
8. For selected reviews and examples, see: (a) Carpenter, B.K. (2005) *Annu. Rev. Phys. Chem.*, **46**, 57 (b) Carpenter, B.K. (1998) *Angew. Chem. Int. Ed.*, **37**, 3340 (c) Doubleday, C., Nendel, M., and Houk, K.N. (1999) *J. Am. Chem. Soc.*, **121**, 4720 (d) Ammal, S.C., Yamataka, H., Aida, M., and Dupuis, M. (2003) *Science*, **299**, 1555 (e) Doubleday, C., Suhrada, C.P., and Houk, K.N. (2006) *J. Am. Chem. Soc.*, **128**, 90 (f) Oyola, Y. and Singleton, D.A. (2009) *J. Am.*

Chem. Soc., **131**, 3130 (g) Xu, L. and Houk, K.N. (2009) *Angew. Chem. Int. Ed.*, **48**, 2746.

9. (a) Ess, D.H. and Houk, K.N. (2007) *J. Am. Chem. Soc.*, **129**, 10646 (b) Schoenebeck, F. and Houk, K.N. (2010) *J. Am. Chem. Soc.*, **132**, 2496.

10. (a) Toma, L., Romano, S., Quadrelli, P., and Caramella, P. (2001) *Tetrahedron Lett.*, **42**, 5077 (b) Caramella, P., Quadrelli, P., and Toma, L. (2002) *J. Am. Chem. Soc.*, **124**, 1130 (c) Quadrelli, P., Romano, S., Toma, L., and Caramella, P. (2002) *Tetrahedron Lett.*, **43**, 8785 (d) Quadrelli, P., Romano, S., Toma, L., and Caramella, P. (2003) *J. Org. Chem.*, **68**, 6035.

11. Çelebi-Ölçüm, N., Ess, D.H., Aviyente, V., and Houk, K.N. (2007) *J. Am. Chem. Soc.*, **129**, 4528.

12. Ussing, B.R., Hang, C., and Singleton, D.A. (2006) *J. Am. Chem. Soc.*, **128**, 7594.

13. Singleton, D.A., Hang, C., Szymanski, M.J., Meyer, M.P., Leach, A.G., Kuwata, K.T., Chen, J.S., Greer, A., Foote, C.S., and Houk, K.N. (2003) *J. Am. Chem. Soc.*, **125**, 1319.

14. For a review, see: Ess, D.H., Wheeler, S.E., Iafe, R.G., Xu, L., Çelebi-Ölçüm, N., and Houk, K.N. (2008) *Angew. Chem. Int. Ed.*, **47**, 7592.

15. For a review, see: Hynes, J.T., Klinman, J.P., Limbach, H.-H., and Schowen, R.L. (2007) *Hydrogen Transfer Reactions*, Wiley-VCH Verlag GmbH, Weinheim.

16. For the discovery of heavy-atom tunneling through experimental and computational work on bond shifting in cyclobutadiene, see: Carpenter, B.K. (1983) *J. Am. Chem. Soc.*, **105**, 1700.

17. Zuev, P.S., Sheridan, R.S., Albu, T.V., Truhlar, D.G., Hrovat, D.A., and Borden, W.T. (2003) *Science*, **299**, 867.

18. Datta, A., Hrovat, D.A., and Borden, W.T. (2008) *J. Am. Chem. Soc.*, **130**, 6684.

19. Gonzalez-James, O.M., Zhang, X., Datta, A., Hrovat, D.A., Borden, W.T., and Singleton, D.A. (2010) *J. Am. Chem. Soc.*, **132**, 12548.

20. Singleton, D.A. and Szymanski, M.J. (1999) *J. Am. Chem. Soc.*, **121**, 9455.

21. Dolbier, W.R., Koroniak, H., Houk, K.N., and Sheu, C. Jr. (1996) *Acc. Chem. Res.*, **29**, 471.

22. Dolbier, W.R., Koroniak, H., Burton, D.J., Bailey, A.R., Shaw, G.S., and Hansen, S.W. Jr. (1984) *J. Am. Chem. Soc.*, **106**, 1871.

23. (a) Kirmse, W., Rondan, N.G., and Houk, K.N. (1984) *J. Am. Chem. Soc.*, **106**, 7989 (b) Rondan, N.G. and Houk, K.N. (1985) *J. Am. Chem. Soc.*, **107**, 2099.

24. Rudolf, K., Spellmeyer, D.C., and Houk, K.N. (1987) *J. Org. Chem.*, **52**, 3708.

25. Murakami, M., Usui, I., Hasegawa, M., and Matsuda, T. (2005) *J. Am. Chem. Soc.*, **127**, 1366.

26. Houk, K.N., Spellmeyer, D.C., Jefford, C.W., Rimbault, C.G., Wang, Y., and Miller, R.D. (1988) *J. Org. Chem.*, **53**, 2125.

27. (a) Liang, Y., Jiao, L., Zhang, S., Yu, Z.-X., and Xu, J. (2009) *J. Am. Chem. Soc.*, **131**, 1542, and references therein (b) Matsuya, Y., Ohsawa, N., and Nemoto, H. (2006) *J. Am. Chem. Soc.*, **128**, 412 (c) Yoshikawa, T., Mori, S., and Shindo, M. (2009) *J. Am. Chem. Soc.*, **131**, 2092.

28. For two selected examples from Wu's group and from our group, see: (a) Chen, B., Hou, X.-L., Li, Y.-X., and Wu, Y.-D. (2011) *J. Am. Chem. Soc.*, **133**, 7668 (b) Chen, Y., Ye, S., Jiao, L., Liang, Y., Sinha-Mahapatra, D.K., Herndon, J.W., and Yu, Z.-X. (2007) *J. Am. Chem. Soc.*, **129**, 10773.

29. Kürti, L. and Czakó, B. (2005) *Strategic Application of Named Reactions in Organic Synthesis*, Elsevier Academic Press, Burlington, MA.

30. (a) Zhang, C. and Lu, X. (1995) *J. Org. Chem.*, **60**, 2906 (b) Xu, Z. and Lu, X. (1997) *Tetrahedron Lett.*, **38**, 3461 (c) Xu, Z. and Lu, X. (1998) *J. Org. Chem.*, **63**, 5031 (d) Xu, Z. and Lu, X. (1999) *Tetrahedron Lett.*, **40**, 549. For reviews, see: (e) Lu, X., Zhang, C., and Xu, Z. (2001) *Acc. Chem. Res.*, **34**, 535 (f) Methot, J.L. and Roush, W.R. (2004) *Adv. Synth. Catal.*, **346**, 1035 (g) Lu, X., Du, Y., and Lu, C. (2005) *Pure Appl. Chem.*, **77**, 1985 (h) Ye, L.-W., Zhou, J., and Tang, Y. (2008) *Chem. Soc. Rev.*, **37**, 1140 (i) Cowen, B.J. and Miller, S.J. (2009) *Chem. Soc. Rev.*, **38**, 3102.

31. (a) Xia, Y., Liang, Y., Chen, Y., Wang, M., Jiao, L., Huang, F., Liu, S., Li, Y., and Yu, Z.-X. (2007) *J. Am. Chem. Soc.*, **129**, 3470 (b) Liang, Y., Liu, S., Xia, Y., Li, Y., and Yu, Z.-X. (2008) *Chem. Eur. J.*, **14**, 4361.

32. Robiette, R., Aggarwal, V.K., and Harvey, J.N. (2007) *J. Am. Chem. Soc.*, **129**, 15513.

33. Fang, Y.-Q. and Jacobsen, E.N. (2008) *J. Am. Chem. Soc.*, **130**, 5660.

34. For reviews, see: (a) Doyle, M.P., McKervey, M.A., and Ye, T. (1998) *Modern Catalytic Methods for Organic Synthesis with Diazo Compounds,* John Wiley & Sons, Inc., New York (b) Zhang, Z. and Wang, J. (2008) *Tetrahedron,* **64**, 6577.

35. For reviews, see: (a) Davies, H.M.L. and Beckwith, R.E.J. (2003) *Chem. Rev.,* **103**, 2861 (b) Davies, H.M.L. and Manning, J.R. (2008) *Nature,* **451**, 417.

36. Ferris, L., Haigh, D., and Moody, C.J. (1996) *Tetrahedron Lett.,* **37**, 107.

37. Bulugahapitiya, P., Landais, Y., Parra-Rapado, L., Planchenault, D., and Weber, V. (1997) *J. Org. Chem.,* **62**, 1630.

38. (a) Maier, T.C. and Fu, G.C. (2006) *J. Am. Chem. Soc.,* **128**, 4594 (b) Chen, C., Zhu, S.-F., Liu, B., Wang, L.-X., and Zhou, Q.-L. (2007) *J. Am. Chem. Soc.,* **129**, 12616 (c) Zhu, S.-F., Chen, C., Cai, Y., and Zhou, Q.-L. (2008) *Angew. Chem. Int. Ed.,* **47**, 932.

39. Lu, C.-D., Liu, H., Chen, Z.-Y., Hu, W., and Mi, A.-Q. (2005) *Org. Lett.,* **7**, 83.

40. Liang, Y., Zhou, H., and Yu, Z.-X. (2009) *J. Am. Chem. Soc.,* **131**, 17783.

41. For reviews, see: (a) Tius, M.A. (2005) *Eur. J. Org. Chem.,* 2193 (b) Pellissier, H. (2005) *Tetrahedron,* **61**, 6479 (c) Frontier, A.J. and Collison, C. (2005) *Tetrahedron,* **61**, 7577.

42. Marcus, A.P., Lee, A.S., Davis, R.L., Tantillo, D.J., and Sarpong, R. (2008) *Angew. Chem. Int. Ed.,* **47**, 6379.

43. (a) Anastas, P.T. and Eghbali, N. (2010) *Chem. Soc. Rev.,* **39**, 301 (b) Horvath, I.T. and Anastas, P.T. (2007) *Chem. Rev.,* **107**, 2167 (c) Horvath, I.T. and Anastas, P.T. (2007) *Chem. Rev.,* **107**, 2169 (d) Hernaiz, M.J., Alcantara, A.R., Garcia, J.I., and Sinisterra, J.V. (2010) *Chem. Eur. J.,* **16**, 9422.

44. (a) Trost, B.M. (1991) *Science,* **254**, 1471 (b) Trost, B.M. (1995) *Angew. Chem. Int. Ed.,* **34**, 259.

45. (a) Wender, P.A. and Miller, B.L. (2009) *Nature,* **460**, 197 (b) Wender, P.A., Verma, V.A., Paxton, T.J., and Pillow, T.H. (2008) *Acc. Chem. Res.,* **41**, 40.

46. (a) Hendrickson, J.B. (1975) *J. Am. Chem. Soc.,* **97**, 5784 (b) Gaich, T. and Baran, P.S. (2010) *J. Org. Chem.,* **75**, 4657.

47. Houk, K.N. and Cheong, P.H.-Y. (2008) *Nature,* **455**, 309.

48. For reviews, see: (a) Kagan, H.B. and Riant, O. (1992) *Chem. Rev.,* **92**, 1007 (b) Gothelf, K.V. and Jøgensen, K.A. (1998) *Chem. Rev.,* **98**, 863 (c) Jøgensen, K.A. (2000) *Angew. Chem. Int. Ed.,* **39**, 3558 (d) Corey, E.J. (2002) *Angew. Chem. Int. Ed.,* **41**, 1650.

49. (a) Hiersemann, M. and Abraham, L. (2002) *Eur. J. Org. Chem.,* 1461 (b) Nubbemeyer, U. (2003) *Synthesis,* 961.

50. Bishop, L.M., Barbarow, J.E., Bergman, R.G., and Trauner, D. (2008) *Angew. Chem. Int. Ed.,* **47**, 8100.

51. Guner, V.A., Houk, K.N., and Davies, I.W. (2004) *J. Org. Chem.,* **69**, 8024.

52. Tantillo, D.J. (2009) *Angew. Chem. Int. Ed.,* **48**, 31.

53. For a review, see: MacMillan, D.W.C. (2008) *Nature,* **455**, 304.

54. Notz, W., Tanaka, F., Watanabe, S., Chowdari, N.S., Turner, J.M., Thayumanuvan, R., and Barbas, C.F. III (2003) *J. Org. Chem.,* **68**, 9624, and references therein.

55. Mitsumori, S., Zhang, H., Cheong, P.H.-Y., Houk, K.N., Tanaka, F., and Barbas, C.F. III (2006) *J. Am. Chem. Soc.,* **128**, 1040.

56. (a) Bahmanyar, S. and Houk, K.N. (2003) *Org. Lett.,* **5**, 1249 (b) Cheong, P.H.-Y., Zhang, H., Thayumanuvan, R., Tanaka, F., Houk, K.N., and Barbas, C.F. III (2006) *Org. Lett.,* **8**, 811.

57. For a review, see: Gorin, D.J. and Toste, F.D. (2007) *Nature,* **446**, 395.

58. (a) Hashmi, A.S.K., Schwarz, L., Choi, J.-H., and Frost, T.M. (2000) *Angew. Chem. Int. Ed.,* **39**, 2285 (b) Zhou, C.-Y., Chan, P.W.H., and Che, C.-M. (2006) *Org. Lett.,* **8**, 325.

59. Dudnik, A.S., Xia, Y., Li, Y., and Gevorgyan, V. (2010) *J. Am. Chem. Soc.,* **132**, 7645.

60. (a) Wender, P.A., Takahashi, H., and Witulski, B. (1995) *J. Am. Chem. Soc.,* **117**, 4720 (b) Wender, P.A. and Sperandio, D. (1998) *J. Org. Chem.,* **63**, 4164 (c) Wender, P.A. and Williams, T.J. (2002) *Angew. Chem. Int. Ed.,* **41**, 4550.

61. Yu, Z.-X., Wender, P.A., and Houk, K.N. (2004) *J. Am. Chem. Soc.,* **126**, 9154.

62. Ozawa, F. and Mori, T. (2003) *Organometallics,* **22**, 3593, and references therein.

63. Wang, Y., Wang, J., Su, J., Huang, F., Jiao, L., Liang, Y., Yang, D., Zhang, S., Wender, P.A., and Yu, Z.-X. (2007) *J. Am. Chem. Soc.,* **129**, 10060.

64. (a) Jiao, L., Yuan, C., and Yu, Z.-X. (2008) *J. Am. Chem. Soc.*, **130**, 4421 (b) Fan, X., Tang, M.-X., Zhuo, L.-G., Tu, Y.Q., and Yu, Z.-X. (2009) *Tetrahedron Lett.*, **50**, 155 (c) Yuan, C., Jiao, L., and Yu, Z.-X. (2010) *Tetrahedron Lett.*, **51**, 5674.

65. Fan, X., Zhuo, L.-G., Tu, Y.Q., and Yu, Z.-X. (2009) *Tetrahedron*, **65**, 4709.

66. Liang, Y., Jiang, X., and Yu, Z.-X. (2011) *Chem. Commun.*, **47**, 6659.

Commentary Part

Comment 1

K. N. Houk

This chapter begins with a clever title that spells out "Chemistry" as "Chem Is Try – Computationally and Experimentally"! The chapter emphasizes how computation and experiment not only advance our understanding of chemistry but also advance the science of synthesis as well. It demonstrates the state-of-the-art in computation applied to mechanisms involved in synthetic reactions. This work constitutes an insightful philosophy of how chemistry research should be done, presented by two of the excellent young scientists who combine computation and synthesis in their research.

Computational methods are assuming an increasing role in synthesis. In the early heyday of synthesis – Woodward, Eschenmoser, Corey, Stork, and others – there were theoretically based insights – stereoelectronics, aromaticity, and the Woodward–Hoffmann rules. Only the onset of computational methods – semiempirical in the 1960s and 1970s, *ab initio* calculations on small model systems in the 1980s and 1990s, and finally DFT in the 1990s and twenty-first century – made possible the type of science described by Zhi-Xiang Yu and Yong Liang in this chapter. Now systems of direct interest to synthetic chemists, not just models, can be calculated.

Unfortunately, synthetic conditions cannot all be computed directly with high accuracy. Solvent models are still only approximate, and many systems have huge numbers of conformations, requiring too many computations to be practical. We will never be able to include adventitious impurities or some of the subtleties common in synthetic processes – heat versus microwave irradiation is an example – in high-accuracy calculations. Finally, to be able to predict truly what will happen under given reaction conditions, all the potential competing reactions need to be computed with

chemical accuracy. This is still only possible for small systems. Morokuma and co-workers have made some progress in doing just this, but still for rather small systems [C1].

In spite of limitations, computation brings new, important elements to synthesis, the ability to test ideas rapidly, at least on model systems, and the ability to understand why something occurs. Finally, when various stereoelectronic factors go in different directions, computations can tell us which factor predominates.

The authors have selected excellent examples of theoretical studies in organic and organometallic chemistry. They highlight the efforts of leading computational organic chemists, including those of Yu's own group. A wide range of phenomena are covered, from the bifurcations on cycloaddition TSs to tunneling in dynamics involving carbon.

Torquoselectivity, one of my favorite discoveries made with Nelson Rondan, is described. This is an early example where some rather daring, but successful, predictions were made. Paul Cheong's work on an *anti*-selective Mannich reaction has a more direct connection to synthesis, and is described very lucidly in this chapter.

The majority of my studies in the last decade have involved both computations and experiment, in collaboration with some of the great synthetic chemists of our time. Zhi-Xiang Yu and Yong Liang describe this whole area of chemistry, and they combine computational excellence with synthetic prowess in their laboratory. It is definitely an approach to synthesis that will gain more practitioners in the future.

Comment 2

Yun-Dong Wu and Xin-Hao Zhang

Computational chemistry developed dramatically in the last few decades on the basis of numerous contributions from pioneers such as John Pople and Walter Kohn. With faster computers and user-friendly software, computational chemistry has become accessible to more and more chemists. Many experimentalists have started to employ computational chemistry as a tool to obtain in-depth information in their laboratories. However, a black-box computational tool does not necessarily guarantee a good choice of method and the correct interpretation of calculation results by a non-expert. On the other hand, an over-simplified model built up *in silico* cannot account for the complexity of a real system. As a result, there is still a gap between computational chemists and experimentalists, especially synthetic chemists. Taking the advantage of performing both computational and experimental research in their own laboratory, Yu and Liang have attempted to bridge this gap between two communities. This chapter is devoted to introducing a synergy between computation and experiment.

The chapter is logically well organized. Between the Introduction and Conclusion, three main sections dealing with concept, mechanism, and synthesis are presented. Each section consists of a general overview with several selected examples. Although, as the authors admit, the choice of examples is focused on their own research interests, the showcase with insightful descriptions will still encourage readers from other fields to appreciate the interplay between theory and experiment. Furthermore, the carefully selected examples achieve some degree of homogeneity that allows readers to follow the thread easily. Instead of mathematical equations, numerous illustrative schemes and figures with chemical structures are presented, which will interest a broader readership.

Section 15.2, entitled "Developing New Theories, Concepts, and Understandings for Organic Chemistry," begins with a historical review of the theories and concepts developed with the aid of computational chemistry from the birth of QC to the end of the last century. Then new impacts were introduced by three seemingly simple but intriguing examples. These fundamental concepts, such as bifurcation and C-tunneling, may change our conventional thinking about reaction processes.

Section 15.3 discusses the roles of both theory and experiment in understanding reaction mechanisms. The authors propose a general protocol to employ QM calculations and designed experiments for a mechanistic study. This is a practical guide for experimentalists to approach computational chemistry. However, in this protocol, we do not fully agree on a minor point. The authors state that when calculations are not consistent with experiment, "we have to propose a new reaction mechanism to account for the experimental results." This statement does not always hold true. Sometimes, such a discrepancy is attributable to experimental results, such as wrong measurements or misinterpretation of experimental data. Instead of endless self-examination, theoreticians should be confident to ask for a re-examination of experimental results at some stage with the accuracy of current computational methodology and carefully designed simulations. The following three cases provide demonstrations of the proposed strategy in specific organic reactions and it is rewarding to read these mechanistic reasonings. Section 15.4 moves on to the impact of theory in the regime of organic synthesis. The authors' passion to pursue an ideal synthesis is fully expressed by their suggestions on *de novo*-designed reactions. Examples in this section cover the hot areas in last decade, for example, organocatalysis and gold-catalyzed reactions. After learning of these successes, one can feel optimistic in predicting that the computation-aided ideal synthesis will be realized in the twenty-first century.

In summary, this chapter severs as a communication between theory and experiment. The authors have done an impressive job in introducing both conceptual insights and practical protocols in combined computational and experimental study. We do hope that some day, chem is no longer a try.

Reference

C1. Maeda, S., Komagawa, S., Uchiyama, M., and Morokuma, K. (2011) *Angew. Chem. Int. Ed.*, **50**, 644–649.

16
Case Study of Mechanisms in Synthetic Reactions

Ai-Wen Lei and Li-Qun Jin

16.1
Introduction

A dramatic rise in the development of organic chemistry, covering the expansion of synthetic methodology and a thorough understanding of the related mechanisms, has been witnessed in recent years [1]. For a transition metal (TM)-catalyzed reaction, the synthetic methodology demonstrates what happens, whereas mechanistic research shows why and how it happens. In the development of organic chemistry, mechanistic studies proceeded in a fundamental manner and allowed a better understanding of reaction processes. These efforts could be beneficial to designing new synthetic reactions, producing new compounds, and improving existing synthetic methods to achieve highly efficient and environmentally benign processes. WE can take the alkene metathesis reaction as an example, for which Chauvin, Grubbs, and Schrock received the 2005 Nobel Prize in Chemistry. Early mechanistic studies by Chauvin elucidated the mechanism as involving a metallacyclobutane intermediate, and clarified the orientation of catalyst development, namely to utilize the metal carbene complex as a catalyst to realize alkene metathesis. The theory carried the alkene metathesis a step forward from an ill-defined catalyst to a well-defined catalyst. Grubbs *et al.* gave insight into the mechanism, revealing that the initial step involves the dissociation of a phosphine ligand from the metal center to produce the active species. To accelerate the dissociation, the so-called second-generation Ru-based Grubbs catalysts, ruthenium–*N*-heterocyclic carbenes, were developed, which have since been commercialized and are now widely employed in cross-metathesis. Understanding the structure and geometry of such intermediates is crucial for the design of more selective and efficient catalysts. Consequently, mechanistic investigations are of great scientific and practical significance.

The investigation of reaction mechanisms by various methods and technologies mainly focuses on elucidating the detailed reaction steps in qualitative and quantitative manners. The following approaches have mostly been applied in recent years: (i) determination of kinetic behavior to obtain the reaction order and propose the rate-limiting step; (ii) isolation or preparation of the reaction intermediates and analyzing them by X-ray diffraction or spectroscopic methods, such as NMR and IR spectroscopy, to investigate the mechanism step by step; (iii) study of stereochemical effects of possible

Organic Chemistry – Breakthroughs and Perspectives, First Edition. Edited by Kuiling Ding and Li-Xin Dai.
© 2012 Wiley-VCH Verlag GmbH & Co. KGaA. Published 2012 by Wiley-VCH Verlag GmbH & Co. KGaA.

intermediates, and also electronic, solvent, and salt effects that may be involved to gain an insight into a reaction.

Of all the methods mentioned above, kinetic studies are the most important approach to achieve a reasonable understanding of reaction mechanisms. The kinetic behavior arises from the original experimental data on the relationship between the concentrations of substrates or products and time. Thermodynamic parameters, such as the rate constant k and activation energy E_a, enable us to understand the rate-influencing factors and the mechanism of the whole reaction process. In this chapter, the importance of kinetic investigations is highlighted.

16.2
Mechanistic Study of Coupling Reactions

Coupling reactions are defined as constructing carbon–carbon or carbon–heteroatom bonds involving homo-coupling and cross-coupling processes [2]. The most studied area is TM (such as Pd and Ni)-catalyzed coupling reactions between organic halides or pseudohalides and nucleophiles, especially organometallic reagents [3]. A series of named reactions, such as Kumada reactions with Grignard reagents [4], Negishi reactions with zinc reagents [5], Hiyama reactions with silicon reagents [6], Stille reactions with organotin reagents [7], and Suzuki reactions with boron reagents [8], have flourished and dominated in this area (Scheme 16.1). It was exciting recent news that Heck, Negishi, and Suzuki were awarded the 2010 Nobel Prize in Chemistry for their remarkable achievements in C–C cross-coupling reactions.

TM catalyzed cross-coupling reactions are expected to proceed by the fundamental steps shown in Scheme 16.2 [3]. Taking palladium-catalyzed coupling reactions as an example, the oxidative addition of an organic halide to palladium(0) produces a palladium(II) species. Nucleophilic attack of the resulting palladium(II) center by organometallic reagent transmetallation is the second step. The resulting pivotal intermediate R^1-Pd-R^2 undergoes reductive elimination to produce the desired product and Pd(0) for the next cycle. For the Heck reaction, the fundamental steps includes oxidative addition, alkene

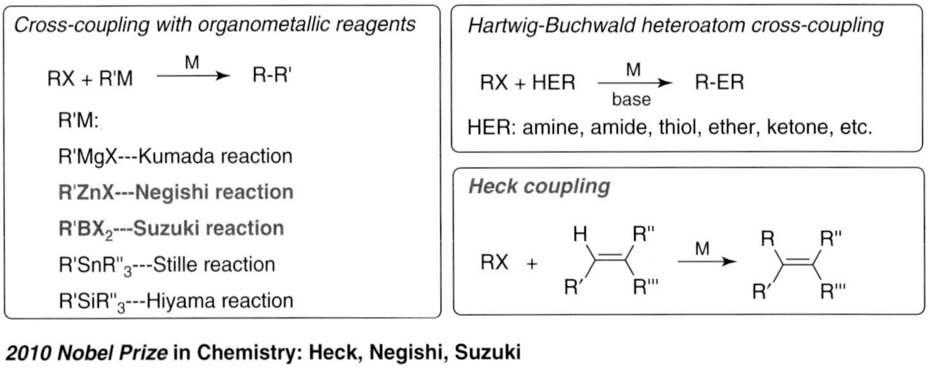

Scheme 16.1 Transition metal-catalyzed cross-coupling reactions.

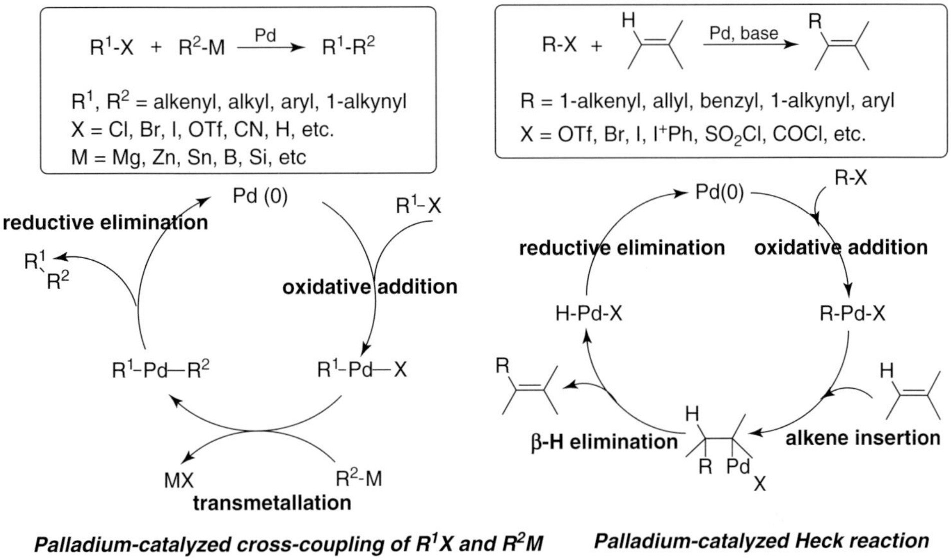

Scheme 16.2 Mechanism of palladium-catalyzed coupling reactions.

insertion, β-H elimination, and reductive elimination. The fundamental steps involved in these reactions seem to be simple; however, they are always accompanied by ligand dissociation, *cis–trans* isomerization, catalyst deposition, and so on.

TM-catalyzed oxidative coupling reactions have been a topic of recent research [9]. The new methodology possesses similar fundamental steps to the traditional coupling model to realize the combination of two nucleophiles. Using the oxidative coupling model to investigate the mechanism, a deeper understanding of the coupling reactions has been obtained.

16.2.1
Oxidative Addition

16.2.1.1 Influence of Ligands on Oxidative Addition

Oxidative addition [Eq. (16.1)] is one of the most important fundamental organometallic reactions in palladium- or nickel-catalyzed systems [2, 3]. It is generally considered to constitute the first step in the catalysis. Owing to the C–M and M–X bond formation, oxidative addition leads to an increase in the covalency and coordination number of the complex [10].

$$L_nM + R\text{-}X \rightleftharpoons ML_m{\overset{R}{\underset{X}{\diagup}}}$$

$$(16.1)$$

In mechanistic studies of cross-coupling reactions, oxidative addition has been investigated relatively more than other steps because the rate-limiting process always involves oxidative addition of an aryl halide or alkyl halide to a metal catalyst. In other words,

	ArI	ArBr	ArCl
ligand free, N, O, and S ligands	RT-80 °C	80–150 °C	usually not possible
triarylphosphines	RT	50–120 °C	>150 °C, if at all
trialkylphosphines and NHcS	RT, very fast	RT	RT-100 °C

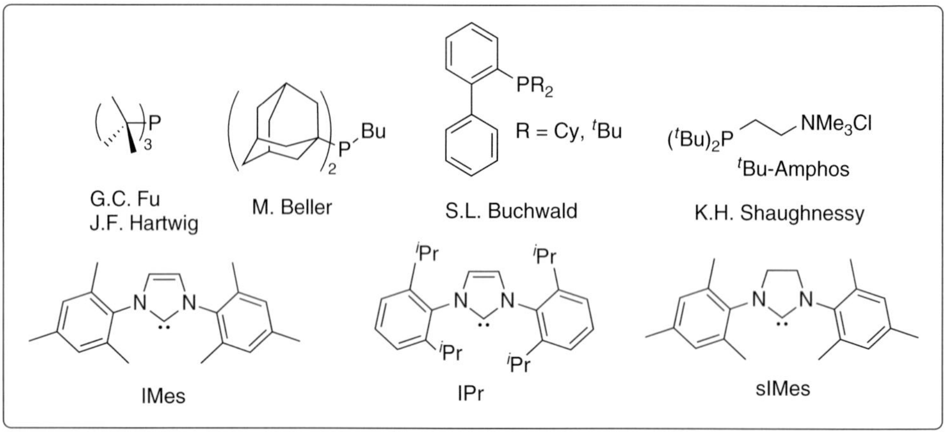

Figure 16.1 Typical ligands used in coupling reactions.

the oxidative addition intermediate is obtained more easily for the investigation of the possible pathway. Moreover, the kinetic behavior of the overall catalytic reaction basically represents the kinetics of the oxidative addition step.

Oxidative addition could be influenced by many factors, and even an imperceptible change could lead to a different mechanism of oxidative addition. In the same catalytic system, different reaction rates could be observed using different substrates in the general sequence RI ≫ RBr ≈ ROTf ≫ RCl [2]. The ligand also plays an important role in palladium-catalyzed coupling reactions (Figure 16.1). P- and N-based ligands occupy an unquestionable primary position. Recently, carbene [11] and pincer ligands were also developed [12]. Of all reported ligands, phosphine ligands are the most popular, including monophosphine ligands (PPh_3, etc.) and bidentate ligands (dppf, etc.). Electron-rich ligands could accelerate or facilitate the oxidative addition as a result of the stabilization of the resulting Pd(II) complex [3]. Thus, Fu and co-workers reported a series of palladium-catalyzed coupling reactions using hindered and electron-rich ligands to solve the problem of the difficult oxidative addition of organochlorides (Scheme 16.3) [13].

Because of the high activity and complicated reaction process, the mechanism of oxidative addition to monophosphine–Pd(0) has been extensively studied [14]. Theoretical and experimental results revealed that Pd(0)L, Pd(0)L$_2$, and Pd(0)L$_3$ are important species and may be involved in the oxidative addition directly or indirectly. In most cases, Pd(0)L$_2$ is considered as the active catalyst in the oxidative addition of aryl and alkyl halides

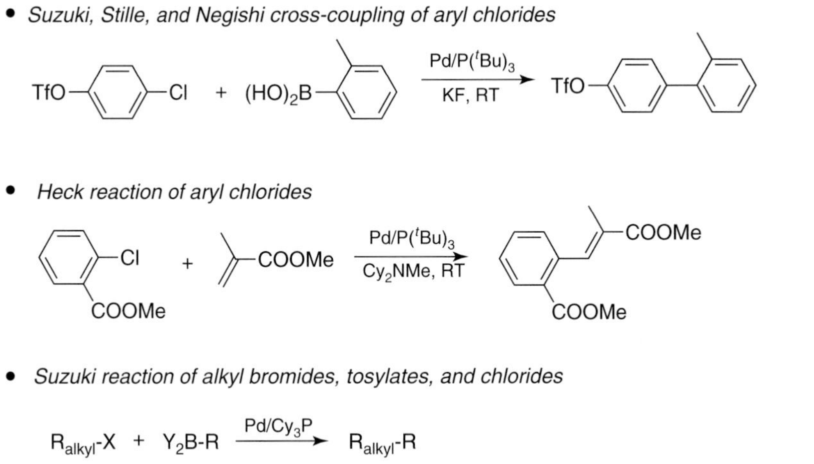
Suzuki, Stille, and Negishi cross-coupling of aryl chlorides
TfO
Cl
(HO)2B
Pd/P(tBu)3
KF, RT
TfO
Heck reaction of aryl chlorides
Cl
COOMe
COOMe
Pd/P(tBu)3
Cy2NMe, RT
COOMe
COOMe
Suzuki reaction of alkyl bromides, tosylates, and chlorides
Ralkyl-X + Y2B-R
Pd/Cy3P
Ralkyl-R

Scheme 16.3 Fu and co-workers' studies on palladium-catalyzed cross-coupling reactions.

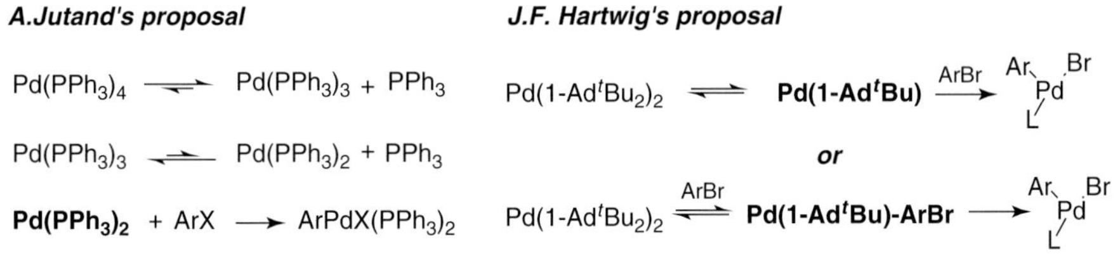
A.Jutand's proposal
Pd(PPh3)4 ⇌ Pd(PPh3)3 + PPh3
Pd(PPh3)3 ⇌ Pd(PPh3)2 + PPh3
Pd(PPh3)2 + ArX → ArPdX(PPh3)2
J.F. Hartwig's proposal
Pd(1-AdtBu2)2 ⇌ Pd(1-AdtBu)
ArBr
Ar Br Pd L
or
Pd(1-AdtBu2)2
ArBr
Pd(1-AdtBu)-ArBr
Ar Br Pd L

Scheme 16.4 Proposals for the mechanism of the oxidative addition step.

(Scheme 16.4). However, the major palladium complex [$Pd(0)L_3$ or $Pd(0)L_2$(solvent)] that exists in the reaction system was less active or inert.

Thus, the ligand dissociation step involved suggests that additional ligand could repress the reaction rate. Hartwig and co-workers found that the type of halide substrate affects the palladium species involved in oxidative addition more significantly than the steric properties of the phosphine ligand [15].

It has been established that when a Pd(II) complex is used as the catalyst precursor, Pd(0) species can be simultaneously formed through an irreversible reduction of the Pd(II) complex by ligands or nucleophiles. In that case, a halide anion or acetate anion should exist in the catalytic reaction system, which has been proved to promote the oxidative addition by forming an ate complex with a palladium center (Scheme 16.5) [16]. However, the oxidative addition in the Heck reaction will be inhibited because the alkene substrate can coordinate to the active species as a π-acceptor ligand [17].

In contrast to the monophosphine-ligated catalyst system, a *cis* palladium(II) intermediate forms in the presence of a bidentate ligand followed by direct transmetallation and reductive elimination leading to the desired product [3]. A 14-electron complex Pd(0)(L–L) is always considered to be the active species no matter whether the catalyst precursor is Pd(0) or Pd(II). Jutand and co-workers made a detailed study of oxidative addition under catalytic conditions using cyclic voltammetry and NMR spectroscopy [18]. It was observed

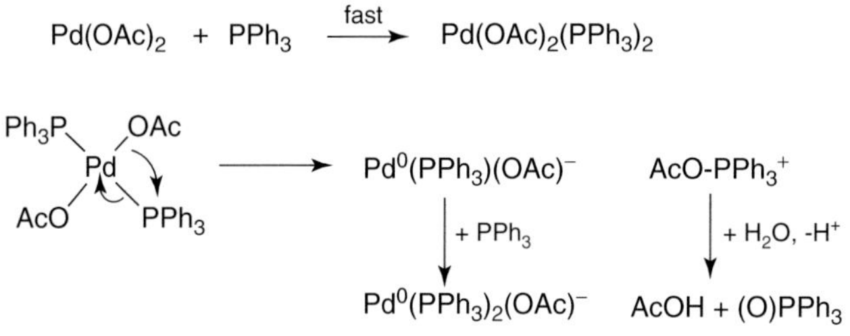

$$Pd(OAc)_2 + PPh_3 \xrightarrow{\text{fast}} Pd(OAc)_2(PPh_3)_2$$

Scheme 16.5 Ate-complex with palladium(0) species.

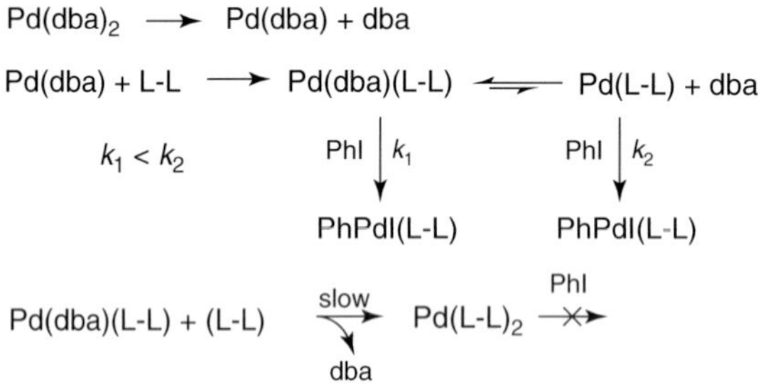

$$Pd(dba)_2 \longrightarrow Pd(dba) + dba$$

$$Pd(dba) + L\text{-}L \longrightarrow Pd(dba)(L\text{-}L) \rightleftharpoons Pd(L\text{-}L) + dba$$

Scheme 16.6 Bidentate ligand in oxidative addition.

that the mixture of Pd(dba)$_2$ + 2(L−L) (L = dppm, dppe, dppp, dppb, dppf, and diop) leads to the formation of Pd(L−L)$_2$, which cannot undergo oxidative addition with aryl iodides. The mixture of Pd(dba)$_2$ and L−L leads to the formation of Pd(dba)(L−L), which is the main complex in solution. However, an endergonic equilibrium is involved with dba and the less ligated complex Pd(L−L). Pd(dba)(L−L) can undergo oxidative addition with aryl iodides but with lower activity than Pd(L−L) (Scheme 16.6). More recently, they found that bidentate thioether-ligated palladium, which is able to react with aryl bromides at low temperature, behaves more actively than the bidentate phosphine ligand system [19].

16.2.1.2 Oxidative Addition of Haloarenes to Trialkylphosphine−Pd(0) Complexes

In the development of cross-coupling reactions, the phosphine ligand played an important role [15]. Especially for the coupling of aryl bromides or aryl chlorides, the hindered alkylphosphine ligands were always involved in resolving the problem of the sluggish oxidative addition process. Hence it is important to elucidate the mechanism of oxidative addition to the L$_2$Pd(0) complex having hindered phosphine ligands.

In the oxidative addition of aryl halides to Pd(0) species with an alkylphosphine as ligand, the 14-electron bisphosphine−palladium complex is considered to be the active species to form a four-coordinate product or the oxidative addition may occur from a 12-electron monophosphine intermediate to form a three-coordinate product (Scheme 16.7). The coordination number of the active palladium species and the product are different with varying ligands. Hartwig and co-workers [15] reported that the mechanism of oxidative addition could be influenced by different aryl halides in

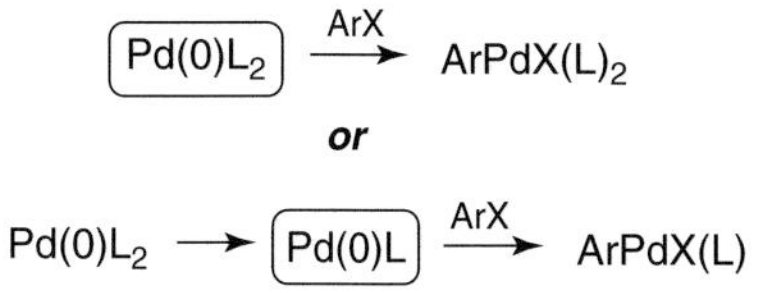

Pd(0)L$_2$ $\xrightarrow{\text{ArX}}$ ArPdX(L)$_2$

Scheme 16.7 Oxidative addition of aryl halide to Pd(0)L$_2$.

or

Pd(0)L$_2$ $\longrightarrow$ Pd(0)L $\xrightarrow{\text{ArX}}$ ArPdX(L)

the same catalytic system. In this case, the oxidative additions of PhX (X = I, Br, Cl) to the complexes Pd(P^{t}Bu$_3$)$_2$ (**1**), Pd(1-AdP^{t}Bu$_2$)$_2$ (**2**), Pd(CyP^{t}Bu$_2$)$_2$ (**3**), and Pd(PCy$_3$)$_2$ (**4**) (1-Ad = 1-adamantyl, Cy = cyclohexyl) were studied to determine the effects of steric properties on the coordination number of the species that undergoes oxidative addition and to determine whether the type of halide affects the identity to this species.

Identification of Oxidative Addition Products with Different Ligands In this section, the oxidative addition of PhX to PdL$_2$ with L (L = P^{t}Bu$_3$, 1-AdP^{t}Bu$_2$, CyP^{t}Bu$_2$, PCy$_3$) to form Pd(II) complexes (**5–19**) is considered. The section is organized by ligand type because the structure of the oxidative addition product depends preferentially on the steric properties of the ligand.

For the oxidative addition of ArI and ArBr to complexes **1** and **2**, the known three-coordinate complexes could be obtained in high yield, whereas it is much more complicated for the reaction of ArCl (Scheme 16.8). The reaction of 1-chloro-2-trifluoromethylbenzene with Pd(P^{t}Bu$_3$)$_2$ (**1**) at 80 °C for 1 h produced a stable complex [P^{t}Bu$_3$Pd(II)ArCl]$_2$ (**10**) in 59% isolated yield, and the reaction of 1-chloro-2-trifluoromethylbenzene with Pd(1-AdP^{t}Bu$_2$)$_2$ (**2**) produced [1-AdP^{t}Bu$_2$Pd(II)ArCl]$_2$ (**12**) in 85% yield. The reactions of PhCl with Pd[0] complexes **1** and **2** formed dimers [LPdPhCl]$_2$ (L = P^{t}Bu$_3$, 1-AdP^{t}Bu$_2$) in low yield with high conversion as a result of decomposition of the product. The yield was up to 80% at 50% conversion with respect to the reacted Pd[0]. The phenylpalladium chloride complexes **9** and **11** were characterized by ^{31}P NMR spectroscopy by comparison with the prepared standard product as depicted in Scheme 16.9. The nuclearity of arylpalladium chloride complexes **10** and **11** is related to their phase. Both of the complexes are dimers in the solid state containing two bridging chlorides, whereas they are predominantly monomeric in solution. The solid-state structures of complexes **10** and **11** were characterized by X-ray diffraction. The actual structures of the oxidative addition products **9–11** in solution were identified by solution molecular weight measurements

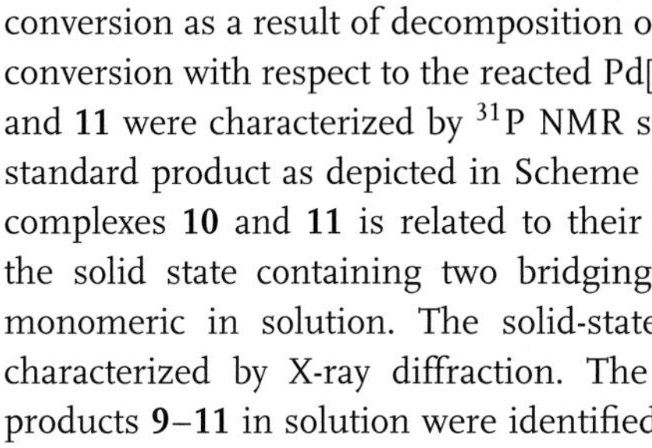

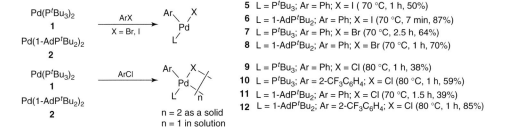

Scheme 16.8 Oxidative addition of ArX to Pd(P^{t}Bu$_3$)$_2$ and Pd(1-AdP^{t}Bu$_2$)$_2$.

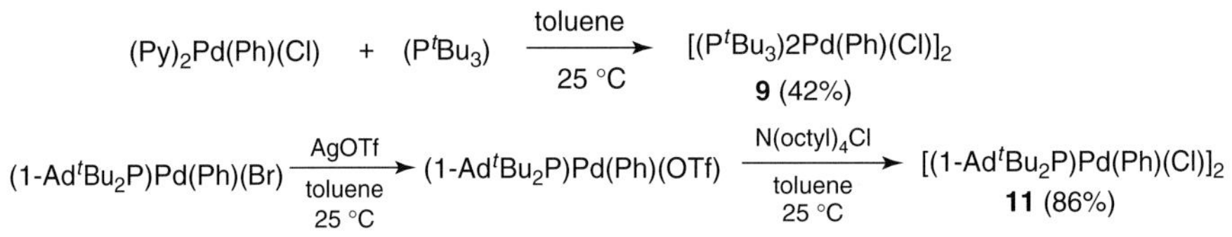

$$(\text{Py})_2\text{Pd}(\text{Ph})(\text{Cl}) \quad + \quad (\text{P}^t\text{Bu}_3) \xrightarrow[\substack{25\,°\text{C}}]{\text{toluene}} [(\text{P}^t\text{Bu}_3)2\text{Pd}(\text{Ph})(\text{Cl})]_2$$

$$\text{9 (42\%)}$$

$$(1\text{-Ad}^t\text{Bu}_2\text{P})\text{Pd}(\text{Ph})(\text{Br}) \xrightarrow[\substack{\text{toluene}\\25\,°\text{C}}]{\text{AgOTf}} (1\text{-Ad}^t\text{Bu}_2\text{P})\text{Pd}(\text{Ph})(\text{OTf}) \xrightarrow[\substack{\text{toluene}\\25\,°\text{C}}]{\text{N(octyl)}_4\text{Cl}} [(1\text{-Ad}^t\text{Bu}_2\text{P})\text{Pd}(\text{Ph})(\text{Cl})]_2$$

$$\text{11 (86\%)}$$

Scheme 16.9 Preparation of complexes **9** and **11**.

and NMR spectroscopy. The molecular weights measured by the Signer method for the more stable complex **10** were 530 and 505 g mol^{-1} in tetrahydrofuran (THF) and benzene, respectively, which were much closer to the calculated molecular weight of the monomer than that of the dimer. In addition, the ^{31}P NMR chemicals shift of complexes **9–11** were similar and were in the range expected to a monomeric species. As a result, Hartwig and co-workers [15] concluded that an equilibrium exists between the dimer and monomer after the oxidative addition of aryl chloride to Pd(0)L$_2$ (L = P^{t}Bu$_3$, 1-AdP^{t}Bu$_2$).

The reaction of CyP^{t}Bu$_2$-ligated palladium complex **3** and aryl halides produced dimeric Pd(II) complexes (**13–16**), as confirmed by ^{31}P NMR spectroscopy, in high yield. However, the known stable *trans* four-coordinate complexes **17–19** were formed in high yield through the oxidative addition of PhI, PhBr, and PhCl to Pd(PCy)$_3$ (**4**) at 25–45 °C, as determined by ^{31}P NMR spectroscopy. The CyP^{t}Bu$_2$-ligated complex **15** was mainly investigated by X-ray diffraction analyses and Signer solution molecular weight measurement. In the solid state, the complex contains two bridging μ$_2$-bromide ligands and *syn*-aryl groups, whereas in solution, the dimer could dissociate predominantly to monomeric species, as confirmed by molecular weight measurements and ^{31}P NMR spectroscopy (Scheme 16.10).

Kinetic Investigation According to the Type of Aryl Halide The kinetics of the oxidative addition of PhI and PhCl to Pd(0) species is simpler than those of PhBr to Pd(0)L$_2$. The kinetic data were plotted as the reciprocal of rate constants and fitted a typical linear equation, corresponding to the reaction occurring through a rapid equilibrium followed by an irreversible step.

Hartwig and co-workers [15] found that the oxidative addition products of aryl bromides to Pd(0) complexes **1** and **2** were unstable and produced a phosphonium salt (L·HX) as a side product, which accelerated the oxidative addition of

$$\text{Pd(CyP}^t\text{Bu}_2)_2 \xrightarrow{\text{ArX}} \left[\begin{array}{c} \text{Ar} \diagdown \diagup \text{X} \\ \text{Pd} \\ \text{L} \diagup \diagdown \end{array} \right]_2$$

3

13 Ar = 3,5-(CF$_3$)$_2$C$_6$H$_3$; X = I (25 °C, 2 h, 98%)
14 Ar = Ph; X = I (60 °C, 40 min, 68%)
15 Ar = Ph; X = Br (70 °C, 1 h, 60-70%)
16 Ar = Ph; X = Cl (80 °C, 1 h, 64-69%)

$$\text{Pd(PCy}_3)_2 \xrightarrow{\text{ArX}} \begin{array}{c} \text{L} \\ | \\ \text{Ar}-\text{Pd}-\text{X} \\ | \\ \text{L} \end{array}$$

4

17 Ar = Ph; X = I (30 °C, 68%)
18 Ar = Ph; X = Br (25 °C, 58%)
19 Ar = Ph; X = Cl (45 °C, 1 h, >95%)

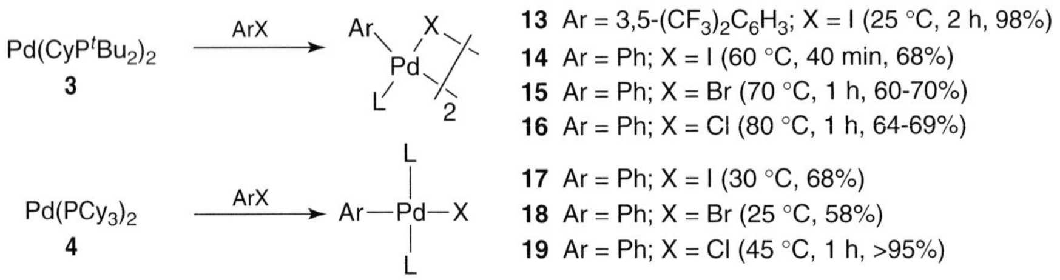

Scheme 16.10 Oxidative addition of ArX to Pd(CyP^{t}Bu$_2$)$_2$ and Pd(PCy$_3$)$_2$.

the aryl bromide (autocatalysis). Consequently, a hindered phosphazene base, *tert*-butyliminotris(pyrrolidino)phosphorane (BTTP), was added to the reaction to suppress the autocatalysis. The reactions of aryl iodides were conducted at lower temperature and the oxidative addition exhibited an exponential decay of the starting Pd(0) species without the autocatalysis irrespective of whether BTTP was added or not. In addition, ^{31}P NMR spectroscopy indicated that Pd(PCy$_3$)$_3$ could be generated from PCy$_3$ ligated complex **4** in the presence of superfluous PCy$_3$.

The rate constants for the reactions of Pd(0) complexes **1–3** were obtained from the decay of the starting Pd(0) complexes, whereas those for the reactions of complex **4** were calculated according to the formation of the product because the spectroscopic signal was broad for complex **4** in the presence of PCy$_3$. For the reaction of PhI with Pd(0) complexes **1–3**, plotting $1/k_{obs}$ versus $1/$[PhI] gave a linear relationship, indicating that the reaction displays a first-order dependence on PhI. The rate constants for the oxidative addition did not change significantly on varying the ligand concentration, suggesting that the reaction is zero order in ligand. Based on these kinetic data, the quantitative rate constants k_{obs} for the reaction of PhI with Pd(0) complexes **1–3** were obtained as $(1.3 \pm 0.2) \times 10^{-3}$, $(3.4 \pm 0.4) \times 10^{-4}$, and $(6.1 \pm 0.6) \times 10^{-4}$ s^{-1}, respectively. The oxidative addition of PhI to PCy$_3$-ligated complexes **4** was conducted at $-80°$C, because the reaction was too fast to be detected at room temperature. The reciprocal of the rate constants for the reactions of the combination of PhI with Pd(0) and PCy$_3$ at various [PhI] and [PCy$_3$] were plotted, indicating that the oxidative addition of PhI to **4** displays first-order kinetics in PhI and inverse first-order in PCy$_3$ at high concentrations of the ligand. As described above, a trisphosphine-ligated palladium complex dominated at low temperature. Hence the reversible dissociation of PCy$_3$ from Pd(PCy$_3$)$_3$ occurs first, followed by the oxidative addition. At low concentrations of the ligand, the major species is Pd(PCy$_3$)$_2$ and the kinetic behavior in PCy$_3$ is similar to that of the other complex, being zero order in the ligand, and the rate constant of the oxidative addition is not dependent on PCy$_3$.

The oxidative addition of PhCl to complexes **1–4** was conducted in toluene at different temperatures and various concentrations of PhCl and ligand. In contrast to the kinetic results with PhI mentioned above, plotting $1/k_{obs}$ versus $1/$[PhCl] revealed a positive dependence of the rate constants on chlorobenzene and plotting $1/k_{obs}$ versus $1/$[ligand] showed an inverse dependence of the rate constants on the ligand.

Hartwig and co-workers [15] focused on the oxidative addition of bromobenzene to 1-Ad^{t}Bu$_2$P-ligated **2** because the oxidative addition product (1-Ad^{t}Bu$_2$P)Pd(Ph)(Br) was much more stable and was formed in high yield in the presence of BTTP. The reactions of bromobenzene with **3** and Cy$_3$P-ligated **4** were conducted in the absence of BTTP. For the oxidative addition of bromobenzene to **2**, the starting **2** decayed exponentially, revealing that the reaction is first order in the Pd(0) complex. The plot of $1/k_{obs}$ versus $1/$[PhBr] was curved, whereas the plot of k_{obs} versus [PhBr] was linear with a positive slope and nonzero y-intercept. Plotting $1/k_{obs}$ versus [1-Ad^{t}Bu$_2$P] suggested that the rate constant for the oxidative addition is independent of the concentration of added 1-Ad^{t}Bu$_2$P. At high concentrations of bromobenzene, the data exhibited some deviations, yet a change in [1-Ad^{t}Bu$_2$P] by a factor of 4 led to a variation in rate of less than a factor of 2. All of these data revealed that two mechanisms could occur simultaneously in the oxidative addition of bromobenzene to complex **2**.

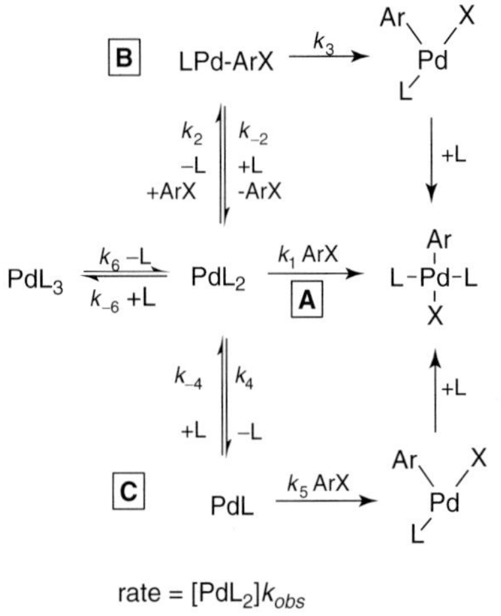

A. Direct reaction to L$_2$Pd

$$k_{obs} = k_1[ArX]$$

B. Associative displacement of the ligand in L$_2$Pd

$$k_{obs} = \frac{k_2 k_3[ArX]}{k_3 + k_{-2}[L]} \qquad \text{If } k_3 \gg k_{-2}[L],$$
$$k_{obs} = k_2[ArX]$$

$$1/k_{obs} = \left(\frac{1}{k_2} + \frac{[L]}{k_2 k_3} \right) \frac{1}{[ArX]}$$

C. Dissociation of ligand from L$_2$Pd

$$k_{obs} = \frac{k_4 k_5[ArX]}{k_5[ArX] + k_{-4}[L]} \qquad \text{If } k_5[ArX] \gg k_{-4}[L],$$
$$k_{obs} = k_4$$

$$1/k_{obs} = \frac{1}{k_4} + \frac{[L]}{k_4 k_5[ArX]}$$

$$\text{rate} = [PdL_2]k_{obs}$$

Scheme 16.11 Three proposed mechanisms for oxidative addition from PdL$_2$.

Discussion According to the kinetic investigation mentioned above, the mechanisms for the oxidative addition of aryl halides to different complexes **1–4** were considered. All of the kinetic data showed that the rate constants for the oxidative addition are positive depending on the concentration of the aryl halides. In that case, three possible mechanisms were proposed by Hartwig and co-workers [15] from the starting palladium complex PdL$_2$ (Scheme 16.11).

In path A, the oxidative addition occurs directly from the aryl halide and Pd(0)L$_2$ to form a four-coordinate arylpalladium halide complex. The reaction will be first order in aryl halide and the rate constant is independent of the concentration of the ligand. Alternatively, the associative displacement of the ligated ligand from Pd(0)L$_2$ by aryl halide could take place to form a monophosphine intermediate coordinated by a haloarene, as shown in path B. Then the cleavage of the C–halide bond will lead to the oxidative addition product of a three-coordinate arylpalladium complex or the final dimerized reaction product. If the initial associative displacement is reversible, the rate constants will depend on the concentrations of both the aryl halide and the ligand, whereas the irreversible associative displacement of the initial step ($k_3 \gg k_{-2}[L]$) will lead to a dependence on the concentration of aryl halide and independence from the concentration of ligand. Consequently, from the kinetic behavior of the aryl halide and the ligand, both paths A and B could be possible. However, the two pathways could be differentiated by the kinetic behavior of their reverse reaction according to the microscopic reversibility model. Finally, the reaction could occur through an initial dissociation of the ligand to produce Pd(0)L, followed by the oxidative addition of the aryl halide resulting in a three-coordinate complex (path C). If the dissociation is reversible, the reaction is dependent on the concentrations of both the aryl halide and the ligand. If the dissociation is effectively irreversible, the rate of the oxidative addition is independent of the aryl halide and ligand, and the rate constant will be equal to k_4. The proposed mechanisms are consistent with the positive dependence on the aryl halide. They could be distinguished by the kinetic behavior of the ligand. When the reaction displays an inverse dependence on the ligand, path A is excluded. Path C could be discarded if the rate constant is dependent on the aryl halide and independent of the added ligand. However, path B would be more complex.

The oxidative addition of ArI to Pd(0)L$_2$ (complexes **1–3**) was first order in [ArI] and zero order in the added ligand, suggesting that the direct oxidative addition (path A) occurs or the irreversible associative displacement of Pd(0)L$_2$ by ArI (path B) takes place. The two mechanisms could not be differentiated from the kinetic behavior of ArI and added ligand, whereas they could be distinguished through the kinetic behavior of their reverse reactions. If the C–X bond cleavage occurs through the direct oxidative addition from PdL$_2$, the microscopic reversibility model indicates that the reverse reaction from ArPd(X)(L) should be first order in the added ligand. If the oxidative addition takes place from an associative displacement product PdL(ArX), the reductive elimination of C–X occurs from a monophosphine–arylpalladium complex and is independent of the ligand. The previous kinetic investigation of the reductive elimination of C–X suggested that the rate of the reductive elimination from (^{t}Bu$_3$P)Pd(Ar)(I) is independent of the added ligand. Therefore, it was concluded that the mechanism of the oxidative addition of aryl iodide to complexes **1** and **2** follows path B, that is, an initial irreversible associative displacement of PdL$_2$ by ArI followed by C–X cleavage to form an arylpalladium complex. The details of the mechanisms of the oxidative addition of complex **3** have not yet been clarified because the reductive elimination of the Cy^{t}Bu$_2$P-ligated complex occurs in a low yield and could not provide solid proof. However, in either case, the kinetic data showed that the rate-limiting step of the reaction between ArI and the Cy^{t}Bu$_2$P-ligated complex involves a bisphosphine complex.

For the oxidative addition of PhI to the Cy$_3$P-ligated complex **4**, the initial species is a triphosphine ligand complex PdL$_3$ at high concentrations of the added ligand. The mechanism may occur through three pathways (Figure 16.2). In either case, the rate of the reaction would display an inverse dependence on the added ligand. When the concentration of the added ligand is low, the major starting palladium complex is PdL$_2$.

A. Direct reaction to L$_2$Pd

$$k_{obs} = \frac{k_6 k_1 [\text{ArX}]}{k_1 [\text{ArX}] + k_{-6}[\text{L}]}$$

$$1/k_{obs} = 1/k_6 + \frac{[\text{L}]}{k_6 k_1 [\text{ArX}]}$$

B. Associative displacement of the ligand in L$_2$Pd

$$k_{obs} = \frac{k_6 k_2 k_3 [\text{ArX}]}{k_2 k_3 [\text{ArX}] + k_3 k_{-6}[\text{L}] + k_{-6}k_{-2}[\text{L}]^2}$$

$$1/k_{obs} = 1/k_6 + \frac{[\text{L}]}{k_6 k_2 [\text{ArX}]} + \frac{[\text{L}]^2}{k_6 k_2 k_3 [\text{ArX}]}$$

C. Dissociation of ligand from L$_2$Pd

$$k_{obs} = \frac{k_6 k_4 k_5 [\text{ArX}]}{k_4 k_5 [\text{ArX}] + k_5 k_{-6}[\text{ArX}][\text{L}] + k_{-6}k_{-4}[\text{L}]^2}$$

$$1/k_{obs} = 1/k_6 + \frac{[\text{L}]}{k_6 k_4} + \frac{[\text{L}]^2}{k_6 k_4 k_5 [\text{ArX}]}$$

Figure 16.2 The rate law of three possible pathways for the oxidative addition of PhI to complex **4**.

The rate constant is independent of the added ligand, indicating that the mechanism of the oxidative addition of ArI to the Cy_3P-ligated complex **4** takes place through the direct oxidative addition of ArI to PdL_2.

All of the kinetic data showed that the rates of the oxidative addition of ArCl to $Pd(0)L_2$ (**2**) and (**3**) are dependent on the concentration of ArCl and inversely dependent on the concentration of added ligand. Hence path A is excluded and the proposed mechanism may contribute to the paths B and C with a reversible dissociation step of the ligand from the starting $Pd(0)L_2$. The two mechanisms could be distinguished by the y-intercept on plotting $1/k_{obs}$ versus $1/[ArCl]$ based on the deduced rate law. If the oxidative addition of ArCl occurs through a reversible associative displacement by ArCl followed by a C–X cleavage, the rate law dictates that the plot of $1/k_{obs}$ versus $1/[ArCl]$ is linear with a zero y-intercept and the plot of $1/k_{obs}$ versus [L] is linear with a nonzero y-intercept. The mechanism of path C with a reversible initial dissociation of ligand predicts that the nonzero y-intercept from the plot of $1/k_{obs}$ versus $1/[ArCl]$ will be identical with the y-intercept from the plot of $1/k_{obs}$ versus $1/[L]$. Hence the kinetic data for the oxidative addition of ArCl to 1-Ad^tBu_2P-ligated complex **2** implied that plotting $1/k_{obs}$ versus $1/[ArCl]$ leads to a distinct nonzero y-intercept line, indicating that a reversible dissociation of ligand could take place. The authors could not distinguish the two mechanisms confidently for the reactions of Cy^tBu_2P-ligated complex **3** and Cy_3P-ligated complex **4** by this analysis because the y-intercepts of the plots of $1/k_{obs}$ versus [ArX] were close to zero. However, the calculated values of k_4k_5 are similar to each other in both cases, suggesting that complexes **3** and **4** react with ArCl via a dissociative mechanism.

The kinetic data for the reaction of PhBr with complexes **2–4** showed that the rate constant is approximately zero-order dependent on the concentration of added ligand. Hence the irreversible step of these oxidative additions involves a bisphosphine–$Pd(0)$ complex. Again, the coordination numbers of the species involved in the irreversible step of the oxidative addition of bromobenzene are identical for all of the $Pd(0)$ species. For the oxidative addition of bromobenzene to 1-Ad^tBu_2P-ligated complex **2** and Cy^tBu_2P-ligated complex **3**, plotting k_{obs} versus [ArBr] led to a line with a nonzero y-intercept, suggesting that two mechanisms exist in the oxidative addition process of bromobenzene, one dependent on the concentration of bromobenzene and the other not. A bisphosphine–palladium complex is involved in the irreversible step in both cases. Of the mechanisms considered, k_{obs} is equal to k_4, corresponding to the initial irreversible dissociation of the ligand from $Pd(0)L_2$, which is represented by the y-intercept. Another mechanism should show a positive dependence on the concentration of bromobenzene and be independent of [L]. The process could be attributed to either direct oxidative addition or irreversible associative displacement by bromobenzene followed by C–X cleavage. The latter is favored because the direct oxidative addition is repressed with the hindered steric ligand. In addition, the microscopic reversibility model, in which the reductive elimination of $(P^tBu_3)Pd(o\text{-}tol)Br$ originates from a three-coordinate intermediate, indicated a small possibility of forming a four-coordinate complex in the presence of 1-Ad^tBu_2P and Cy^tBu_2P. For the oxidative addition of bromobenzene with Cy_3P-ligated complex **4**, the plot of $1/k_{obs}$ versus $1/[PhBr]$ was linear, indicating that the reaction is dependent on the concentration of bromobenzene and weakly dependent on the concentration of added ligand. Only one pathway takes place in the process. The small slope and large y-intercept of the plots of $1/k_{obs}$

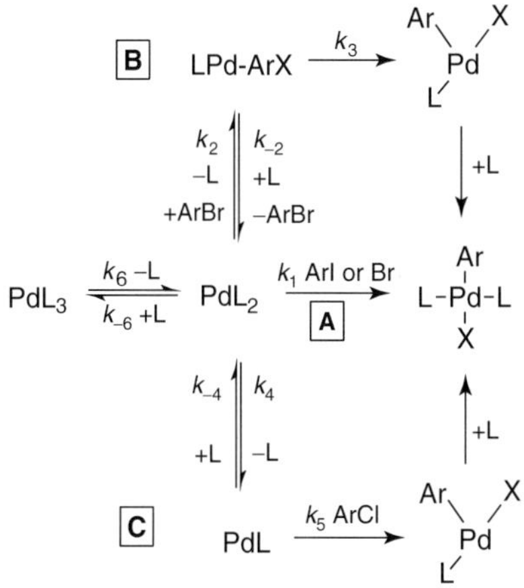

B LPd-ArX $\xrightarrow{k_3}$ $\underset{L'}{\overset{Ar\diagdown X}{Pd}}$

Scheme 16.12 The conclusion of the oxidative addition between $L_2Pd(0)$ and ArX.

k_2 $\quad$ k_{-2}
$-L$ $\quad$ $+L$
$+ArBr$ $\quad$ $-ArBr$

$+L$

PdL_3 $\underset{k_{-6}+L}{\overset{k_6-L}{\rightleftharpoons}}$ PdL_2 $\xrightarrow{k_1 \text{ ArI or Br}}$ $\underset{X}{\overset{Ar}{\underset{|}{L-Pd-L}}}$

A

k_{-4} $\quad$ k_4
$+L$ $\quad$ $-L$

$+L$

C PdL $\xrightarrow{k_5 \text{ ArCl}}$ $\underset{L'}{\overset{Ar\diagdown X}{Pd}}$

versus [L] are consistent with the rate equations corresponding to the direct oxidative addition or nearly irreversible associative displacement of Cy_3P by PhBr. It is difficult to distinguish them merely from the kinetic results.

Conclusion The kinetic data implied that the number of phosphines coordinated to the complex that reacts in the irreversible step of the oxidative addition process for complexes 1–4 depends more on the halide than on the steric properties of the ligands (Scheme 16.12). The rate-limiting step of the oxidative addition of PhI occurs with $L_2Pd(0)$ in all cases, as determined by the lack of dependence of k_{obs} on $[P^tBu_3]$, $[1\text{-}AdP^tBu_2]$, or $[CyP^tBu_2]$ and the inverse dependence of the rate constant on $[PCy_3]$ when the reaction is initiated with $Pd(PCy_3)_3$. The irreversible step of the oxidative addition of PhCl occurs with a monophosphine species in each case, as signaled by an inverse dependence of the rate constant on the concentration of ligand. The irreversible step of the oxidative addition of PhBr occurs with a bisphosphine species, as signaled by the zero-order or small dependence of the rate constant on the concentration of phosphine. Hence the additions of the less reactive chloroarenes occur through lower-coordinate intermediates than additions of the more reactive haloarenes.

16.2.2
Transmetallation

16.2.2.1 General Aspects of the Transmetallation Step
In cross-coupling reactions, transmetallation is another important elementary step after oxidative addition [Eq. (16.2)] [2]. It is always defined as a reaction that transfers an organic group from one metal to another metal. The preparation of organometallic reagents could be regarded as a transmetallation process, for example, organozinc reagent from the corresponding Grignard reagent and $ZnCl_2$. It is established as an excellent way of importing a σ-carbon atom into a coordination sphere of a TM. The thermodynamics are often favorable if M^1 has a lower electronegativity than M^2 and this is often the case if M^1 is a TM and M^2 is a main group metal. Another driving force can be the precipitation

of a main group M^2-X salt from the reaction solution [20].

$$\left[\begin{array}{c} M^1 \overset{R}{\underset{X}{\diamond}} M^2 \end{array} \right]^{\neq}$$

M^1-X + M^2-R $\xrightarrow{\hspace{3cm}}$ M^1-R + M^2-X

M^1 = transition metal, M^2 = main group metal

$$(16.2)$$

The mechanism could be distinctly affected by the ligand, solvent, and nucleophiles. Organometallic reagents as nucleophiles in cross-coupling reactions could be clarified into two kinds: hard nucleophiles, such as Grignard reagents, organolithium reagents, and organozinc reagents, and soft nucleophiles, such as organotin and boron derivatives [3]. Mechanistic studies involving hard nucleophiles are usually difficult because of the fast transmetallation. Recently, much progress has been made through some special theoretical and experimental models [5e, 11d, 20, 21].

In contrast, for a given Pd(0) precursor and a given ligand, cross-coupling of PhI with soft nucleophiles generally requires harsher conditions than that with hard nucleophiles, such as a higher temperature or additional base. Of these reactions, transmetallation is always considered as the rate-limiting step [22]. Taking the Suzuki reaction as an example, additional base is necessary to obtain the desired product. The beneficial effect of added base may arise from coordination between the base and boron to improve the nucleophilicity of the boron reagent or coordination with the metal catalyst center to favor the release of a leaving group in the transmetallation process (Scheme 16.13) [23]. Fluoride derivatives are usually added to the reaction mixture by forming a five-coordinate anion compound with a silicon reagent to promote its transmetallating ability [20, 24]. Transmetallation was always considered to be the rate-limiting step in Stille reactions. In earlier studies, Stille [25] proposed that the transmetallation step involves a transition state (Figure 16.3), which was criticized by Jutand, Hartwig, Espinet, and others [20, 22a,b, 26]. Another possible path was proposed by Espinet's group and will be discussed in the next section.

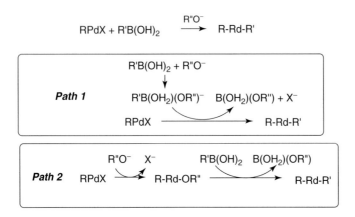

Scheme 16.13 The beneficial effect of added base in the Suzuki reaction.

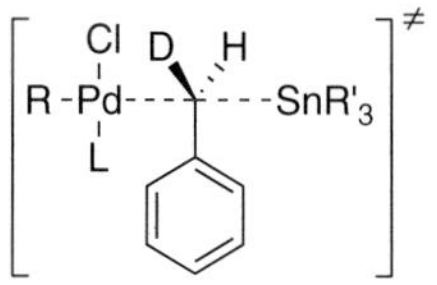

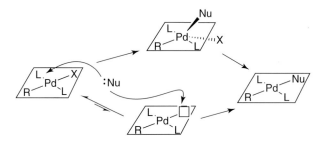

Figure 16.3 Stille proposal for the transition state of transmetallation.

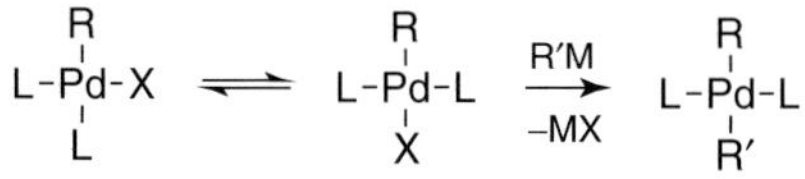

Scheme 16.14 Transmetallation with monophosphine-ligated palladium catalyst.

Scheme 16.15 Association and dissociation paths in transmetallation.

Generally, direct transmetallation occurs after oxidative addition with a bidentate ligand system. The process is relatively simple. For a monophosphine-ligated palladium catalyst, transmetallation takes place after the *cis–trans* isomerization (Scheme 16.14). Maseras and co-workers reported that the *cis*-palladium complex could undergo transmetallation directly in a Suzuki reaction on the basis of theoretical calculations [23f].

After the *cis–trans* isomerization of the oxidative addition adduct to form a tetraco-ordinated square-planar 16-electron species, transmetallation could take place through an association path or a dissociation path (Scheme 16.15). The former involves an 18-electron trigonal bipyramidal complex as an intermediate or transition state; the latter goes through a 14-electron T-shaped intermediate containing a vacant position for the ligand substitution. The reaction becomes complicated when the alkene substrate or solvent coordinates to the metal center to form a 16-electron intermediate [22b].

16.2.2.2 Investigation of the Transmetallation Step in Coupling Reactions

Transmetallation Step in the Stille Reaction The mechanism for the Stille transmetallation reaction has been investigated in detail by Espinet's group [26b]. Previous studies demonstrated that the reaction could be retarded by additional ligand and the kinetic behavior exhibited a first-order dependence on the organotin reagent. Hence the proposed transmetallation mechanism involved a fast reversible dissociation of ligand from a *trans*-Pd(II) complex to form a T-shaped or solvent-coordinated intermediate, and then the nucleophilic attack of the organotin reagent on the palladium complex

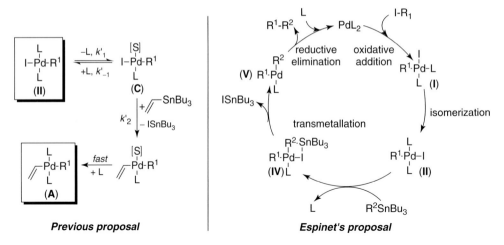

Scheme 16.16 Mechanism of transmetallation in the Stille reaction.

completed the transmetallation, which was always regarded as the rate-limiting step in Stille reactions (Scheme 16.16).

However, Espinet and co-workers considered that the proposal is qualitatively consistent with the observations, while it does not match the quantitative kinetic results. The ligand dissociation constant was calculated to be small, implying that the initial dissociation step is not a fast step. Alternatively, if the dissociation step is the rate-limiting step, the reaction rate is independent of the concentration of organotin reagent, which is contrary to the observations. With regard to the quantitative kinetic data and the qualitative experiment results, a new pathway was proposed involving an associative L for R^2 substitution to form directly a *cis*-R^1/R^2 followed by a fast reductive elimination.

The kinetic studies were supported by the reaction between 1-iodo-3,5-dichlorotrifluoro-benzene and vinyl- or 4-methoxyphenyltributyltin, catalyzed by *trans*-[Pd(C₆Cl₂F₃) I(AsPh₃)₂] [Eq. (16.3) shows the model reaction of Espinet and co-workers' investigation].

$$R^2 = \text{vinyl (21a)} \qquad R^2 = \text{vinyl (23a)}$$
$$R^2 = \text{4-anisyl (21b)} \qquad R^2 = \text{4-anisyl (23b)} \tag{16.3}$$

The overall process of the reaction was monitored by ^{19}F NMR spectroscopy to observe the consumption of the aryl iodide. The kinetic result of the catalytic coupling between **20** and **21a** was analyzed. Plotting $\ln([20]_0 - [20])$ versus time t led to a line with the equation $\ln([20]_0 - [20]) = k_{obs}t$, suggesting that the reaction shows first-order kinetic behavior. Varying the concentration of **20** provided proximate values of the initial rate r_0, while changing the concentration of **21a** led to a different initial rate. Thus, the reaction shows a first-order dependence on **21a**. Otherwise, the slope of plotting $\ln(k_{obs})$ versus $\ln[L]$ was -1.1, indicating a minus first-order kinetic dependence on the ligand. The slope of plotting $\ln(k_{obs})$ versus $\ln(\mathbf{22a})$ was equal to 1, indicating a first-order dependence on the catalyst (**22a**). Numerical analysis of the kinetic data resulted in the rate law for the catalytic Stille reaction as follows:

$$r_{obs} = k_{obs}\,[\mathbf{21a}] = \frac{a\,[\mathbf{22a}]}{[\text{AsPh}_3] + b}\,[\mathbf{21a}] \tag{16.4}$$

with $a = (2.32 \pm 0.09) \times 10^{-5}\ \text{s}^{-1}$ and $b = (6.9 \pm 0.3) \times 10^{-4}\ \text{mol}\,\text{l}^{-1}$.

The temperature dependence of the rate was examined to supply the apparent activation parameters from the plots of $\ln(k_{obs}/T)$ versus T^{-1}. The activation entropy is very negative regardless of the presence or of absence of the added neutral ligand, the type of organotin reagent, and the solvent (Table 16.1). The results suggested that the rate-limiting step involves an associative process. In addition, the reactions of isolated *trans*-$[\text{Pd}(\text{C}_6\text{Cl}_2\text{F}_3)\text{X}(\text{AsPh}_3)]$ (X = Cl, Br, I) with $(\text{CH}_2{=}\text{CH})\text{SnBu}_3)$ were examined to investigate the influence of the halide on the transmetallation rate. The rate dependence of the halide decreases in the order of Cl > Br > I.

In the overall catalytic Stille reaction of $\text{C}_6\text{Cl}_2\text{F}_3\text{I}$ with R_2SnBu_3 ($\text{R}_2 = \text{CH}{=}\text{CH}_2$, **21a**; C_6H_4-4-OMe, **21b**), the only observed palladium intermediate is ArPdX (**22a**). Therefore, the transmetallation is the rate-limiting step and the following reductive elimination is a fast step. Also, the oxidative addition and the subsequent *cis* to *trans* isomerization are faster than the transmetallation step. Consequently, the kinetic outcome of the catalytic Stille reaction could be attributed to that of the transmetallation step.

The true complexity of the rate law reveals the concurrent existence of a set of elemental steps in the transmetallation step. The mechanistic interpretation establishes that the elemental composition in the transition state is $(\mathbf{21a} + \mathbf{22a} - \text{AsPh}_3)$. In other words, the interaction of **21a** and **22a** takes place with release of one molecule of AsPh$_3$. The dissociation of the ligand could occur before or during the interaction of **21a** and **22a**. The general mechanism was described as transmetallation between the organotin reagent and catalyst occurring after the dissociation of the ligand. In that case, if the release of ligand is the rate-limiting step, the rate law of the Stille reaction could be deduced from

Table 16.1 Apparent activation parameters for the coupling of $\text{C}_6\text{Cl}_2\text{F}_3\text{I}$ (**20**, 0.2 mol l^{-1}) with R_2SnBu_3 (**21**, 0.2 mol^{-1}) catalyzed by *trans*-$[\text{Pd}(\text{C}_6\text{Cl}_2\text{F}_3)\text{I}(\text{AsPh}_3)_2]$ (**22a**, 0.01 mol l^{-1}).

R^2	Solvent	$\Delta H^{\neq}_{obs}$ (kJ mol^{-1})	$\Delta S^{\neq}_{obs}$ (kJ mol^{-1})
Vinyl	THF	50 ± 2	-155 ± 7
Vinyl[a]	THF	91 ± 4	56 ± 15
Vinyl	PhCl	70.0 ± 1.7	-104 ± 6
4-Anisyl	THF	72.8 ± 1.1	-100 ± 3

[a] With AsPh$_3$ = 0.02 mol l^{-1}.

a steady-state approximation model:

$$r_{obs,ss} = k_{obs,ss} \, [21a] = \frac{k'_1 k'_2 \, [22a]}{k'_{-1}[AsPh_3] + k'_{-2} \, [21a]} [21a] \qquad (16.5)$$

In the absence of added $AsPh_3$, the concentration of $AsPh_3$ is very low, hence the rate law could be modified to $r_{obs,ss} \approx k'_1 \, [22a]$, suggesting a zero-order kinetic dependence on the concentration of **21a** and the rate constant is independent of the type of organotin reagent employed. Therefore, the hypothesis is excluded because the deduced zero-order kinetic dependence on **[21a]** is contrary to the experimental results showing a first-order dependence.

$$r_{obs,pe} = k_{obs,pe} \, [21a] = \frac{k'_2 K_{dis} \, [22a]_{total}}{K_{dis} + [AsPh_3]} [21a] \, \text{Pr e-equilibrium model} \qquad (16.6)$$

If the dissociation is very fast, the pre-equilibrium model [Eq. (16.6)] with $K_{dis} = k'_1/k'_{-1}$ will be employed. The experimental data analysis gave the value of k'_2 as $0.18 \, s^{-1}$ and K_{dis} as $1.3 \times 10^{-4} \, mol\,l^{-1}$. The value of K_{obs} indicates that 12% of the catalyst **22a** should be dissociated in the absence of added $AsPh_3$. In that case, the dissociation product of **22a** could be detected by ^{19}F NMR spectroscopy. However, only **22a** was observed in the overall process. Therefore, the pre-equilibrium model is not workable.

Consequently, the two dissociative pathways of the ligand from the oxidative addition product for the transmetallation are not feasible. Another associative transmetallation process was proposed (Scheme 16.17) [26b], which leads to the transformation of *trans*-**II** to *cis*-**V** via L-for-R^2 substitution at the coordination plane. A nucleophilic attack of the Pd-coordinated halide on the organotin reagent increases the electrophilicity of the palladium center and the nucleophilicity of the Cα atom of R^2. A five-coordinate intermediate **III** is involved, followed by the release of ligand to produce the palladium species of **IV**. The associative substitution of R^2 for L from a five-coordinate intermediate occurred with preservation of the stereochemistry of the palladium center, assisting

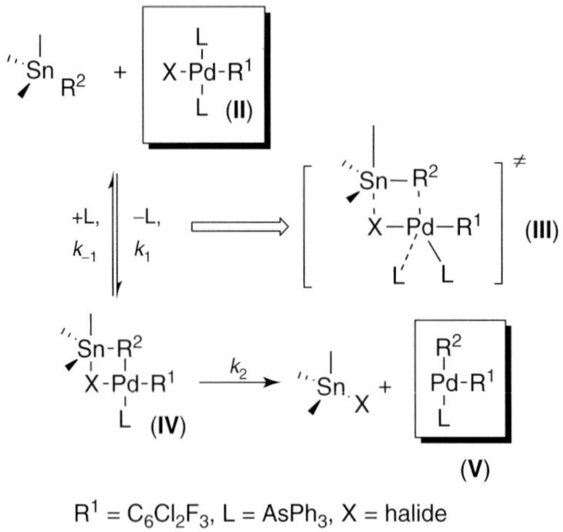

$R^1 = C_6Cl_2F_3$, L = $AsPh_3$, X = halide

Scheme 16.17 Espinet and co-workers' proposal for transmetallation in the Stille reaction.

the formation of a *cis-* R^2 to R^1 transmetallation product and avoiding the *trans* to *cis* isomerization step. Applying the steady-state approximation model [Eq. (16.7)], the rate law is expressed with $k_1 = 0.034 \, mol^{-1} \, 1 \, s^{-1}$ and $k_2/k_{-1} = 6.9 \times 10^{-4} \, mol \, l^{-1}$. As k_1 is dependent on the nature of the organotin reagent, different coupling rates could be observed for different organotin reagents. Furthermore, the large negative value of the entropy could be explained by the associative substitution path and rules out solvent participation, which will produce only a small increase in order in the transition state.

$$r_{obs,ss} = k_{obs,ss} \, [\mathbf{21a}] = \frac{k_1 k_2 \, [\mathbf{22a}]}{k_{-1}[AsPh_3] + k_2} \, [\mathbf{21a}] \quad \text{Steady-state approximation model}$$

$$(16.7)$$

Different halides of *trans*-[Pd(C$_6$Cl$_2$F$_3$)X(AsPh$_3$)$_2$] are expected to have little effect on the entropy of the transmetallation process. Therefore, the variation of the reaction rate mainly relates to the changes in activation enthalpy. Hartwig and co-workers [26a] proposed that a high activation enthalpy is necessary in the formation of the pentacoordinated intermediate **III**. The following nucleophilic attack of R^2 on the palladium center is facilitated as the Pd complex becomes more electrophilic and the Cα more nucleophilic. Moreover, the possible mechanism was also discussed with respect to the electrical properties and steric hindrance.

In conclusion, from these observations, the following mechanism is proposed. Oxidative addition of R_1X to PdL_n gives *cis*-[PdR$_1$XL$_2$], which isomerizes rapidly to *trans*-[PdR$_1$XL$_2$]. This *trans* complex reacts with the organotin compound following an SE_2(cyclic) mechanism, with release of AsPh$_3$ (which explains the retarding effect of addition of L), to give a bridged intermediate **IV**. In other words, an L-for-R$_2$ substitution on the palladium leads R$_2$ and R$_1$ to be situated mutually *cis* to each other. From there, the elimination of XSnBu$_3$ yields a three-coordinate species *cis*-[PdR$_1$R$_2$L], which readily gives the coupling product R_1–R_2 (Scheme 16.17).

Transmetallation Step in the Coupling Reaction of Organozinc Compounds Investigations of transmetallation using hard nucleophilic reagents are rare because the transmetallation step involving them is too fast to be studied in a proper reaction model [21c]. Recently, a series of TM-catalyzed oxidative cross-coupling reactions have been developed by our group [21c]. With the oxidative coupling reaction model, the transmetallation step of the zinc reagent was elucidated as the rate-limiting step, which was successfully clarified qualitatively and quantitatively.

In previous studies, for the mechanism of the oxidative coupling reactions it was proposed that the oxidative addition of an oxidant to the TM produces A–M(II)–B. The resulting intermediate contains two leaving group to undergo transmetallation with two nucleophilic reagents, forming the key intermediate R–M–R, which undergoes reductive elimination directly (Scheme 16.18). For the oxidative cross-coupling reactions, the transmetallation occurs with two different nucleophilic reagents according to their distinct nucleophilicity, such as different metal reagents, alcohols, or amines.

In the overall process, the oxidative addition step could be a fast step with a proper oxidant and the rate-limiting step could be attributed to other elementary processes [21c, 32]. Desyl chloride (2-chloro-1,2-diphenylethanone) was employed as the oxidant.

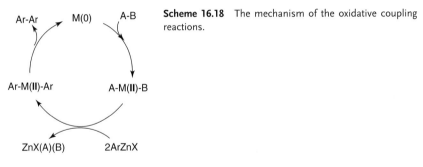

Scheme 16.18 The mechanism of the oxidative coupling reactions.

In the nickel-catalyzed oxidative coupling of the arylzinc reagent, the mechanism was investigated from the kinetic behavior mainly using *in situ* IR spectroscopy [Eq. (16.8) shows the reaction model of Lei and co-workers' investigation of the oxidative coupling reaction of arylzinc compounds].

$$\begin{array}{c} \text{ArZnCl} \\ \text{prepared from} \\ \text{ArLi and ZnCl}_2 \end{array} \xrightarrow[\text{desyl chloride}]{\text{Ni(acac)}_2,\ \text{THF}} \text{Ar-Ar}$$

desyl chloride

(16.8)

First, the kinetics of the catalytic reaction were studied. It was found that the plots of the concentration of desyl chloride versus time were independent on the varying concentrations of desyl chloride in the nickel-catalyzed homocoupling of arylzinc reagent, indicating that the rate constant is independent of the concentration of the oxidant, and the oxidative addition of desyl chloride to nickel is not the rate-limiting step. Subsequently, oxidative coupling reactions with varying concentrations of arylzinc reagents including PhZnCl and *p*-MePhZnCl were conducted. Plotting the initial rates versus the concentration of arylzinc led to a line suggesting that the rate law is described by $r_{obs} = k_{obs}[\text{ArZnX}]$.

The equation revealed that the reaction is first order in arylzinc. In the proposed mechanism mentioned above, arylzinc reagent is only involved in the transmetallation step. Hence the rate-limiting step should be located in the transmetallation step, which could be quantitatively investigated in the catalytic system.

The speculated detailed pathways for the oxidative coupling of phenylzinc include the oxidative addition of Ni^0 with oxidant, RCl, and consequently the double transmetallations of **TM-I** and **TM-II** and the final reductive elimination to obtain biphenyl (Scheme 16.19). The first-order dependence on [ArZnCl] indicated that either **TM-I** or **TM-II** is the rate-determining step. Both intermediates **II-I** and **II-II** resulting from **TM-I** have a phenyl anion as a σ-donor, which should be more electronegative than the intermediate **I**. Hence it was deduced that **TM-I** should be faster than **TM-II**, that is, **TM-II** is the rate-limiting step.

If the reaction operates in path I, **TM(I)-II**, the transmetallation between PhZnCl and Ni-Cl, is the rate-limiting step. This reaction rate would not be affected by varying RCl. If

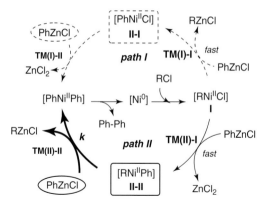

Scheme 16.19 The two proposed pathways of transmetallation in oxidative coupling.

the **TM(II)-II**, the transmetallation between PhZnCl with R−Ni, is the rate-limiting step, the reaction rate should vary when different oxidants R−Cl are employed. Therefore, the kinetic behaviors of the oxidative coupling using different oxidants were investigated (Figure 16.4). The completely different reaction rates clearly reveal that path II operates the reaction, and **TM(II)-II** is the rate-limiting step.

Having confirmed that the transmetallation **TM-II** is the rate limiting step, all of the kinetic studies on the catalytic oxidative coupling should correspond to the transmetallation

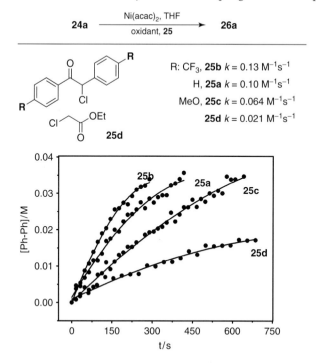

Figure 16.4 Kinetic profiles of oxidative coupling using different oxidants.

of the arylzinc reagent. Therefore, the kinetic investigation of the reactions was conducted at different temperatures and the plots of $\ln(k/T)$ versus $1/T$ gave the activation enthalpy for the transmetallation. The substituent effect of arylzinc reagents on the catalytic system was also considered, and supported the conclusion drawn by kinetic study.

In conclusion, the transmetallation between the arylzinc reagent and ArNi^{II}R was confirmed as the rate-limiting step in the nickel-catalyzed oxidative coupling reactions (Scheme 16.19, path II). It proved to be an excellent model, allowing the first quantitative measurement of the kinetic rate constants of transmetallation from a live catalytic system. The rate constants of different arylzinc reagents and the activation energy were obtained under the operating conditions. The substituent effect of the transmetallation step was also elucidated for the first time from the catalytic reaction.

16.2.3
Reductive Elimination

16.2.3.1 General Aspects of Reductive Elimination
Reductive elimination can be regarded as the reverse reaction of oxidative addition and is crucial for the success of forming C–C or C–heteroatom bonds, because it is an irreversible process to put the catalytic cycle forward, whereas the other steps in coupling reaction are reversible [Eq. (16.9)].

$$ML_m\overset{R}{\underset{R'}{\diagup}} \longrightarrow L_nM + R\text{-}R'$$

$$(16.9)$$

In many cases, the cross-coupling is controlled by the rate and scope of the reductive elimination step [27]. For example, palladium-catalyzed C–O or C–N bond formation is a challenging process because palladium–amido and palladium–alkoxo complexes are reluctant to undergo reductive elimination. It was found that C–O and C–N bond formation could be realized with the development of a suitable catalyst system. Compared with other fundamental reactions, the mechanism of reductive elimination has been relatively less studied as it is always considered a facile process. Related research was focused on alkylpalladium complexes, which undergo the slowest reductive elimination [28]. Theoretical calculations are the primary tool in the exploration of the mechanisms of reductive elimination [27b, 29].

The steric properties and electron effects of ligands and reaction substrates were the main topics considered in mechanistic investigations of reductive elimination. Commonly, reductive elimination occurs through a *cis* square-planar complex or T-shaped species (Scheme 16.20). In the presence of a hindered ligand, the reaction could be promoted through a T-shaped intermediate, which could undergo reductive elimination with a lower activation energy. Moreover, an electron-rich ligand will hamper reductive elimination because the electronegativity of the metal center is increased. At the early stage, Yamamoto and co-workers determined the relationship between reductive elimination and ligands for four-coordinate and T-shaped complexes of different metal catalysts through theoretical calculations [30]. They observed that in the four-coordinate complex, the better the σ-donating capability of the leaving group is, the more facile

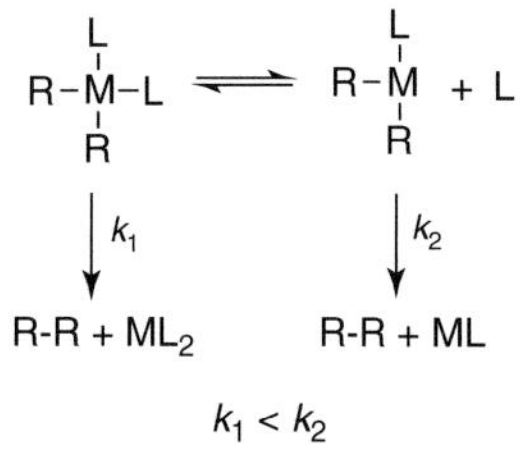

$$R-\underset{\underset{R}{|}}{\overset{\overset{L}{|}}{M}}-L \;\rightleftharpoons\; R-\underset{\underset{R}{|}}{\overset{\overset{L}{|}}{M}} + L$$

$$\downarrow k_1 \qquad\qquad \downarrow k_2$$

$$R\text{-}R + ML_2 \qquad R\text{-}R + ML$$

$$k_1 < k_2$$

Scheme 16.20 Reductive elimination from a *cis* square-planar complex or T-shaped species.

Table 16.2 Calculated activation energy for reductive elimination of $RR'M(PH_3)_2$.

$$RR'Pd(PH_3)_2 \xrightarrow[\text{DFT}]{\text{reductive eliminate}} R\text{-}R' + Pd(PH_3)_2$$

Complex	Calculated activation energy (kcal mol^{-1})
$Pd(Me)_2(PH_3)_2$	25.2
$Pd(CH=CH_2)_2(PH_3)_2$	6.8
$Pd(Ph)_2(PH_3)_2$	11.7
$Pd(C{\equiv}CH)_2(PH_3)_2$	14.6
$Pd(Me)(CH=CH_2)(PH_3)_2$	15.0

the reductive elimination would be. The activation energy for direct elimination of the four-coordinate complex is lower for Ni than for Pd or Pt. T-shaped *trans*-Pd(L)R$_2$ arising from ligand dissociation of PdL$_2$R$_2$ will encounter a substantial barrier to rearrangement to *cis*-Pd(L)R$_2$ followed by reductive elimination. If the leaving groups are poor donors, *cis–trans* isomerization in the three-coordinate complex should be easier than elimination.

Ananikov *et al.* investigated the mechanism and controlling factors for the C–C reductive elimination of $RR'M(PH_3)_2$ with a density functional method (Table 16.2), indicating that the barrier to C–C coupling from the symmetrical complex $R_2M(PH_3)_2$ decreases in the order methyl, ethynyl, phenyl, vinyl [27b, 29b–d]. The activation energies of unsymmetrical coupling reactions are approximately equal to the average values of the corresponding parameters for the symmetrical coupling reactions. Hartwig [27a] also reported the electronic effect of the active and ancillary ligands on the rate of reductive elimination. In contrast, the experiment results suggested that the rate constant for reductive elimination of the asymmetric complex is larger than that for the symmetrical complex. Bo and co-workers [29a] investigated the effects of the bite angle of a chelating bisphosphine ligand and observed that a ligand with a large bite angle could stabilize the reductive elimination intermediate and accelerate the coupling reaction.

With the development of oxidative coupling reactions, studies of reductive elimination with the oxidative coupling model have attracted more attention. Popp and Stahl [31a] and Sanford and co-workers [31b] reported some results for "oxidatively induced" reductive elimination (Scheme 16.21). Sanford and co-workers described the investigation of the mechanism of reductive elimination for $(^tBu_2bpy)Pd^{II}(Me)_2$ and concluded that it

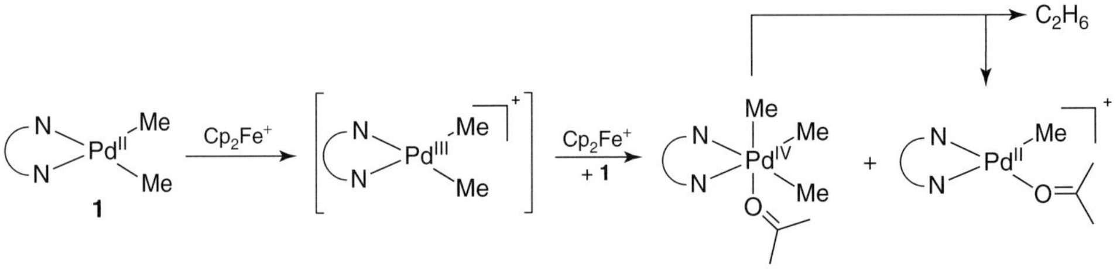

Scheme 16.21 "Oxidatively induced" reductive elimination.

involves Pd(III) and Pd(IV) intermediates with C–C bond formation from the latter in the presence of outer-sphere oxidant ferrocenium. Lei and co-workers investigated the reductive elimination of Csp^2-M-Csp^2 quantitatively with a catalytic oxidative coupling model [32], as described in detail in the next section.

16.2.3.2 Case Study of the Reductive Elimination Step of the Oxidative Coupling Reaction

In the previous introduction to the transmetallation step, desyl chloride could be employed as a suitable oxidant to carry out fast oxidative addition in nickel-catalyzed oxidative coupling reactions of arylzinc reagent [32]. Lei and co-workers found that the preparation of the arylzinc reagent played an important role in the reaction kinetic behavior. Using arylzinc reagent from the corresponding aryllithium reagent, the oxidative coupling behaves with first-order kinetics as mentioned above. However, the reaction employing arylzinc prepared from the corresponding Grignard reagent showed that the rate remains constant throughout the whole catalytic process. This result indicated that the kinetic behavior had changed and the rate-limiting step should not be located in the transmetallation step, but the reductive elimination step is the only elementary step that is independent of the arylzinc reagent and the oxidant (Scheme 16.18). Then, kinetic studies were performed with different concentrations of arylzinc reagent and oxidant. All reactions exhibited zero-order kinetic plots and showed identical rates. The results confirmed that the reductive elimination is indeed a rate-limiting step. It is also noteworthy that the catalytic reaction is first order in $Ni(acac)_2$ [33].

If the Ni species generated from the reductive elimination of Ar-Ni-Ar is different from the Ni species required for oxidative addition, and the rate-determining step is the transformation between these two species, then zero-order kinetics will also be observed. This possibility was excluded by the result that the reaction rate would change when different arylzinc reagents are employed.

Hence all of the kinetic studies of catalytic oxidative coupling should correspond to the reductive elimination of Ar-Ni-Ar. Therefore, kinetic investigations of the reactions were conducted at different temperatures and the plots of $\ln(k/T)$ versus $1/T$ gave the activation enthalpy for the reductive elimination as $9.7\ \mathrm{kcal\ mol^{-1}}$, which reveals the facile nature of the Csp^2-Csp^2 reductive elimination process.

16.3
Mechanistic Study of Aerobic Oxidation

16.3.1
Recent Progress in Aerobic Oxidation

The existence of atmospheric oxygen is vital to maintain the continuing cycle of Nature. Photosynthesis supplies oxygen and respiration provides the essential energy from the complete oxidation process of protein and glucose. The recent *Technology Vision 2020* report published by the Council for Chemical Research highlighted the selective oxidation of organic chemicals as one of the most critical challenges in the chemical industry [34]. Obviously, oxygen is the best choice for the oxidant because it is available at low cost and produces no environmental hazardous byproduct.

How can we achieve efficient oxidation with high selectivity? Undoubtedly, the target will be met by the development of suitable catalysts. In recent years, palladium-catalyzed aerobic oxidation has made great progress, originating from the Wacker process in the 1950s [Eq. (16.10)]. It is an efficient route to produce an aldehyde through the palladium-catalyzed aerobic oxidation of ethylene with Cu(II) as a co-catalyst. This reaction is regarded as a landmark in homogeneous catalysis for industrial application. Despite the early achievements, subsequent advances in palladium catalysis were dominant in cross-coupling reactions. Recently, the development of oxidative coupling reactions and C–H activation has led to a revival of palladium-catalyzed aerobic oxidation.

$$H_2C = CH_2 + O_2 \xrightarrow[\text{H}_2\text{O}]{\text{CuCl}_2,\ \text{PdCl}_2} \text{MeCHO}$$

(16.10)

Palladium-catalyzed aerobic oxidation reactions are considered to be similar to enzymatic oxidases and the catalytic mechanism can be separated into two independent half-reactions: Pd(II)-mediated oxidation of the organic substrates and the oxidation of Pd(0) to Pd(II) with oxygen (Scheme 16.22) [35].

In the overall catalytic process, the co-catalyst Cu(II) was considered to be a medium to combine oxygen with a palladium catalyst. However, its application was hampered by

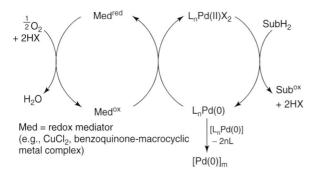

Scheme 16.22 Traditional mechanism for palladium-catalyzed aerobic oxidation reactions. SubH$_2$ is the organic substrate and Sub^{ox} is the oxidized organic product.

employing an equivalent amount of Cu(II) or benzoquinone as the oxidant, which could cause complications and produce side products [36]. Subsequently, a series of catalyst systems have been developed for palladium-catalyzed aerobic oxidation in the absence of a co-catalyst.

Developments in this area originated from the discovery of the Pd(OAc)$_2$–DMSO (dimethyl sulfoxide) system developed by Larock's and Hiemstra's groups independently in the mid-1990s [37] [shown in Eq. (16.11) for the Pd(II)-catalyzed aerobic oxidation of 2-(cyclohex-2-enyl)acetic acid]. This simple catalyst system has been widely applied in alcohol oxidation, intramolecular hetero- and carbocyclization of arenes, and dehydrosilylation of silyl enol ethers [38]. DMSO coordinates with the palladium center to stabilize it.

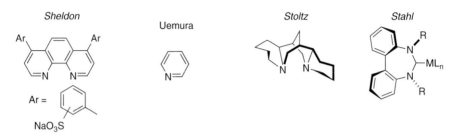

$$\text{(16.11)}$$

90% yield

Stahl and co-workers investigated the mechanism of this system [39]. The crucial factor that depresses the efficiency was found to be catalyst deposition in the Pd(OAc)$_2$–DMSO system. Obviously, a suitable ligand should be applied to stabilize the Pd(II) and prevent the formation of palladium black. However, palladium-catalyzed oxidative reactions were always ligand free because many common ligands are susceptible to oxidative additions.

In the late 1990s, many nitrogen-donor ligands, such as aromatic amines and tertiary alkylamines, were discovered to be compatible and displayed reasonable reactivity in modified reaction conditions (Figure 16.5). For example, Sheldon and co-workers reported an aqueous co-catalyst-free alcohol oxidation and Wacker-type oxidation of terminal alkenes with water-soluble phenanthroline as ligand [40]. The Pd(OAc)$_2$–pyridine catalyst system reported by Uemura and co-workers has been widely employed in oxidative C–C cleavage reactions with tertiary alcohols, and oxidative C–O, C–N, and C–C coupling reactions with alkenes [41]. Stoltz's and Sigman's groups published results on the oxidation of secondary alcohol using (−)-sparteine as the ligand with high enantioselectivity [42]. Stahl's group developed a series of intramolecular and intermolecular aminations of alkenes using Pd(OAc)$_2$–pyridine and Pd(OAc)$_2$–N-heterocyclic carbene systems [35, 43]. All of the results demonstrate exciting prospects for catalytic asymmetric oxidations.

Figure 16.5 Stable nitrogen-donor ligands used in aerobic oxidation.

The detailed investigation of the mechanism of palladium-catalyzed aerobic alcohol oxidation has been reported by Stahl's group [35]. A catalyst system including Pd(OAc)$_2$–pyridine is discussed in the following section.

16.3.2
Mechanistic Characterization of Aerobic Oxidation

16.3.2.1 Kinetic Investigations

The mechanism of the aerobic oxidation of benzyl alcohol catalyzed by Pd(OAc)$_2$–pyridine (5 : 20 mol%) was studied by Stahl and co-workers [Eq. (16.12)] [35]. A computer-interfaced gas-uptake apparatus was constructed to acquire the kinetics of the catalytic reactions *in situ* through monitoring the change in oxygen pressure. They observed a monotonic decrease in the pressure and the lack of an induction period enabled them to obtain the kinetic data via the initial rate method.

$$\text{(16.12)}$$

Plotting the rate of oxygen uptake versus the concentration of different primary reaction components (alcohol, oxygen, and catalyst) led to the initial kinetic results. The kinetics revealed that the reaction rate exhibits a saturation dependence on the concentrations of alcohol and catalyst and is independent of the oxygen pressure above 250 Torr. At reduced pressure, a significantly lower rate could be observed with distinct catalyst deposition of inactive palladium black. Deuterium kinetic isotope effects (KIEs) with PhCD$_2$OH and PhCHDOH were found to be 1.5(3) and 2.6(2), respectively. The reaction rate exhibited a close dependence on [pyridine] with a maximum at a py : Pd ratio of 1 : 1 and decreased strongly at higher py:Pd ratios with a more stable catalyst. At py:Pd ratios below 4 : 1, the relatively high initial rate and low substrate conversion coincided with significant catalyst deposition of palladium black. At a py:Pd ratio of 100 : 1, the reaction rate was low without catalyst deposition even at very low oxygen pressures.

The reaction did not occur in the absence of pyridine, and catalytic turnover could be achieved by adding a catalytic amount of NBu$_4$OAc. A nearly twofold increase in the reaction rate could be achieved by adding 10 mol% of NBu$_4$OAc to the standard catalyst mixture (py:NBu$_4$OAc:Pd $= 4:2:1$). The side product acetic acid could be stabilized by NBu$_4$OAc through the formation of an acid–base complex, which was revealed by X-ray diffraction analysis of a crystal obtained by cooling the catalytic reaction mixture containing NBu$_4$OAc. The catalytic turnover could be retarded at elevated [NBu$_4$OAc]. ^{1}H NMR spectroscopy revealed that an anionic palladium(II) complex, [Pd(py)$_{4-n}$(OAc)$_n$]$^{(n-2)}$ ($n = 3$, 4) resulting from the displacement of ligated pyridine by acetate, which is considered to be maybe a less active catalyst species in the aerobic oxidation of benzyl alcohol, since the addition of acetic acid significantly suppressed the reaction rate.

Kinetic investigations with varying [Pd(OAc)$_2$] at constant [py] (py:Pd $= 12.5$–250) revealed that the reaction rate shows a square root-order dependence on [Pd(OAc)$_2$], which is contrary to the result of a hyperbolic dependence of varying catalyst concentration at a constant py:Pd ratio of 4:1. However, in the presence of a constant ratio of pyridine to acetic acid, a linear dependence of [Pd(OAc)$_2$] was observed. The palladium intermediates of the catalytic reaction were characterized by ^{1}H NMR spectroscopy. The known complex *trans*-(py)$_2$Pd(OAc)$_2$ (**27**) was detected when 2 equiv. of pyridine were added to Pd(OAc)$_2$ in toluene-d_8. If more than 2 equiv. of pyridine were added, complex **27** and free pyridine were observed in the absence of other species even with a further quantity of pyridine.

Addition of benzyl alcohol to the py–Pd(OAc)$_2$(4:1) mixture in toluene-d_8 altered the chemical shifts of both coordinated and free pyridine in the ^{1}H NMR spectra. The changed shift of free pyridine could be attributed to hydrogen bond formation between pyridine and alcohol. However, the coordinated pyridine did not contain a lone electron pair available for hydrogen bonding. The chemical shift changes may originate from the following possibilities: (i) rapid equilibrium formation of a palladium–alkoxide complex, (ii) equilibrium coordination of the alcohol to palladium forming a five-coordinate adduct, or (iii) hydrogen-bonding interactions between the alcohol and the acetate ligands. The integration ratio of the coordinated pyridine and acetate resonances was 2:1, indicating that the possibility (i) can be excluded. The lack of new resonances of the alcohol suggested an equilibrium formation between benzyl alcohol and the complex **27** with the formulation **27·RCH$_2$OH**:

$$(py)_2Pd(OAc)_2 \; + \; RCH_2OH \; \underset{}{\overset{K_{12}}{\rightleftharpoons}} \; [(py)_2Pd(OAc)_2]^\bullet RCH_2OH$$
$$\textbf{27} \qquad\qquad\qquad\qquad\qquad\qquad\qquad \textbf{27}\cdot RCH_2OH$$

$$(16.13)$$

The equilibrium constants at different temperatures could be determined by the integration ratio of complexes **27** and **27·RCH$_2$OH**. Van't Hoff analysis of the equilibrium constants allowed the determination of the thermodynamic parameters of the binding. The equilibrium constants for different alcohols exhibited a linear Hammett relationship with $\rho = +0.37$, revealing that an alcohol bearing that an electron-withdrawing group binds more tightly.

$$(py)_2Pd(OAc)_2 \; + \; PhCOOH \; \underset{}{\overset{K_{12} = 9.3}{\rightleftharpoons}} \; (py)_2Pd(OAc)(O_2CPh) \; + \; AcOH$$
$$\textbf{27} \qquad\qquad\qquad\qquad\qquad\qquad\qquad \textbf{28}$$

$$(16.14)$$

$$(py)_2Pd(OAc)(O_2CPh) \; + \; PhCOOH \; \underset{}{\overset{K_{15} = 3.5}{\rightleftharpoons}} \; (py)_2Pd(O_2CPh)_2 \; + \; AcOH$$
$$\textbf{28} \qquad\qquad\qquad\qquad\qquad\qquad\qquad \textbf{29}$$

$$(16.15)$$

In the aerobic oxidation of benzyl alcohol, a small amount of overoxidation product, benzoic acid, could be produced. The complexes **27** and **27·RCH$_2$OH** were the only species that appeared in the solution during the early period of the reaction. As the reaction proceeded, however, two new sets coordinated-pyridine resonances, attributed to (py)$_2$Pd(OAc)(O$_2$CPh) (**28**) and (py)$_2$Pd(O$_2$CPh)$_2$ (**29**), began to appear. The two species may be produced from the proton-coupled anionic ligand exchange of benzoate for acetate. The equilibrium constants were determined as 9.3 and 3.5, respectively, revealing that benzoate preferentially coordinates to palladium. The reaction profile indicated that the rate of alcohol oxidation decreases more rapidly than expected. The difference between the observed and simulated curves may be due to the lower catalytic

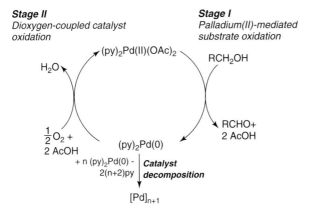

Scheme 16.23 Oxidase-style mechanism of $(py)_2Pd(OAc)_2$-mediated alcohol oxidation.

activity of complexes **28** and **29**. *In situ* formation of **28** and **29** could also release acetic acid, which is known to inhibit the reaction rate. However, the complex **27** was the only palladium species in the oxidation of *sec*-phenethyl alcohol.

A series of *para*-substituted benzyl alcohols and pyridines were employed to investigate the electronic effects on the catalytic turnover rates. Electron-rich substrates react more rapidly and pyridine with an electron-withdrawing substituent could promote the reaction with a higher rate.

The kinetic results and the spectroscopic data provided specific insights into the mechanism of $(py)_2Pd(OAc)_2$-mediated alcohol oxidation. The reaction proceeds by an oxidase-style mechanism in which palladium(II)-mediated substrate oxidation and aerobic oxidation of the catalyst occurs in two independent and sequential stages (Scheme 16.23). The kinetic results indicated that the reaction rate is zero order in oxygen pressure and the rate-limiting step should be located in the palladium(II)-mediated substrate oxidation. The catalytic aerobic oxidation is kinetically invisible in the $(py)_2Pd(OAc)_2$-mediated alcohol oxidation system.

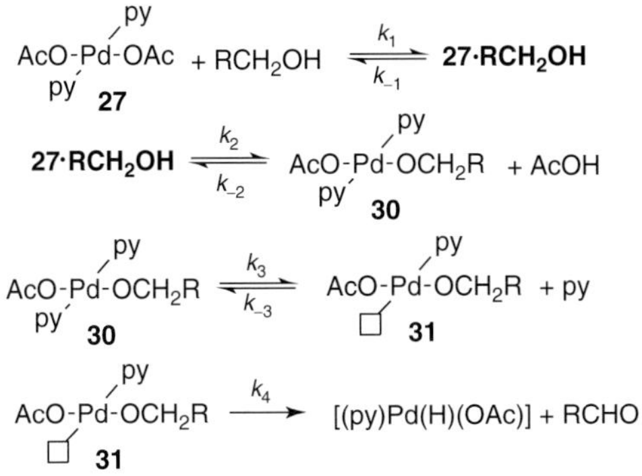

$$\text{AcO-Pd(py)(py)-OAc} \; \mathbf{27} + RCH_2OH \underset{k_{-1}}{\overset{k_1}{\rightleftharpoons}} \mathbf{27 \cdot RCH_2OH}$$

$$\mathbf{27 \cdot RCH_2OH} \underset{k_{-2}}{\overset{k_2}{\rightleftharpoons}} \text{AcO-Pd(py)(py)-OCH}_2R \; \mathbf{30} + AcOH$$

$$\text{AcO-Pd(py)(py)-OCH}_2R \; \mathbf{30} \underset{k_{-3}}{\overset{k_3}{\rightleftharpoons}} \text{AcO-Pd(py)(}\square\text{)-OCH}_2R \; \mathbf{31} + py$$

$$\text{AcO-Pd(py)(}\square\text{)-OCH}_2R \; \mathbf{31} \overset{k_4}{\longrightarrow} [(py)Pd(H)(OAc)] + RCHO$$

Scheme 16.24 Stepwise mechanism of alcohol oxidation by $(py)_2Pd(OAc)_2$.

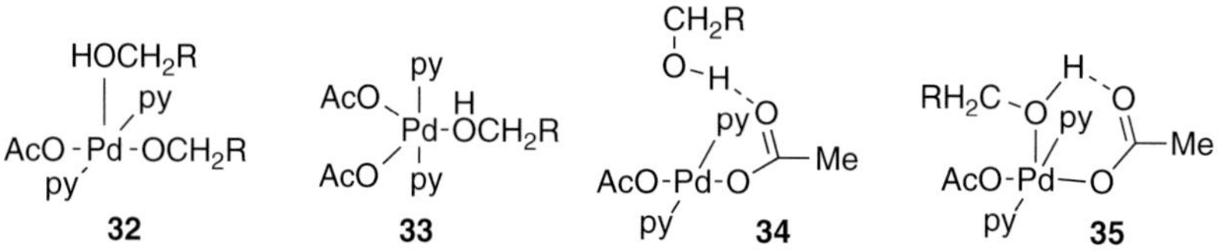

Figure 16.6 Possible structures of the complex **27·RCH₂OH**.

The nonlinear rate dependence on both alcohol and catalyst implicated a multi-step pathway and the whole substrate oxidation process includes at least four steps (Scheme 16.24). The alcohol oxidation begins with the equilibrium interaction between complex **27** and the alcohol to form an unknown complex **27·RCH₂OH**. The resulting adduct could undergo endergonic conversion to a palladium–alkoxide complex **30** with the release of an acetic acid. Dissociation of pyridine from complex **30** generates a three-coordinate palladium(II) complex which permits the irreversible β-H elimination to produce the desired aldehyde and (py)Pd(H)(OAc).

The spectroscopic results implied that the resting species of the catalyst exists exclusively as the complex **27·RCH₂OH** at sufficiently high [RCH₂OH]. With a magnitude defined by the equilibrium constant k_1, higher concentrations of alcohol will confer no incremental benefit on the reaction rate.

Possible structures for the complex **27·RCH₂OH** were proposed (Figure 16.6). Five-coordinate palladium species analogous to the structures **32** and **33** have been speculated as intermediates in palladium(II)–ligand substitution reactions. The coordinated alcohol should possess distinctive spectroscopic properties. However, the NMR spectroscopic data suggested that the ^{1}H NMR resonance of benzylic protons has no obvious shift. Structurally characterized examples of the five-coordinate palladium species generally possess at least one strong π-acceptor, whereas the ligands in complexes **32** and **33** are σ-donors. Therefore, complexes **32** and **33** are discarded. Density functional theory (DFT) calculations indicated that the hydrogen-bonded adduct **34** is the only stable intermediate and the structure **35** could be a transition state along the reaction pathway. In a nonpolar solvent such as toluene, hydrogen bonds could easily be formed and the structure **34** could readily account for the highly fluxional behavior.

The beneficial effect of additional acetate and the inhibitory effect of acetic acid reflect the influence of these reagents on equilibrium formation of alkoxide **30**. The added acetic acid could shift the equilibrium to the left and behaves an inhibitor of the reaction rate. Acetate has been proved to stabilize the acetic acid produced, so exogenous acetate could promote the equilibrium to the right and accelerate the reaction.

The kinetic results suggest that pyridine and acetate can promote the reoxidation of palladium(0) by molecular oxygen. The reaction does not occur without pyridine and palladium black forms rapidly. These observations reflect that direct competition exists between catalyst oxidation and palladium deposition. A beneficial effect of pyridine can be stabilization of palladium(0) to prevent aggregation or enhancement of the palladium(0) oxidation rate with molecular oxygen. Both effects are probably important. The acetate may play a similar role in the reaction as the addition of NBu₄OAc can accelerate the reaction rate in the absence of pyridine.

The catalytic turnover rate can be retarded at higher concentrations of pyridine, probably due to its inhibitory effect on β-hydrogen elimination. Square-planar metal alkyl and alkoxide complexes undergo β-hydrogen elimination through either a three- or a four-coordinate pathway. The former includes a ligand dissociation followed by elimination and a labile ligand would be preferential. The latter pathway always involves a chelating ligand that can hinder ligand dissociation. The inhibition of additional pyridine and the acceleration effect of electron-deficient pyridine on the reaction rate both revealed that the dissociation of pyridine occurs in the process. The steady-state approximation for intermediates **30** and **31** and pre-equilibrium formation of **27·RCH₂OH** provide a rate law. When [PhCH$_2$OH] increases to the point of saturation and K_1[PhCH$_2$OH] $\gg$ 1, the rate law can be simplified, reflecting that acetic acid and pyridine can depress the catalytic turnover rate.

$$\frac{d[RCHO]}{dt} = \frac{K_1 k_2 k_3 k_4 [Pd]_T [PhCH_2OH]}{(1 + K_1[PhCH_2OH])(k_{-2}[AcOH](k_{-3}[py] + k_4) + k_3 k_4)} \tag{16.16}$$

$$\frac{d[RCHO]}{dt} = \frac{k_2 k_3 k_4 [Pd]_T}{k_{-2}[AcOH](k_{-3}[py] + k_4) + k_3 k_4} \quad \text{Simplified rate law} \tag{16.17}$$

The nonlinear rate dependence on the concentration of catalyst can also be explained by the above expression. In the aerobic oxidation of alcohol catalyzed by Pd(OAc)$_2$–bathophenanthroline sulfonate complex and Pd(OAc)$_2$–DMSO, the nonlinear kinetic behavior has been also observed and ascribed to various reasons.

Sheldon and co-workers [40] found that the reaction rate shows a half-order dependence on the catalyst concentration in the aerobic oxidation by palladium–bathophenanthroline. The resting state of the palladium species is a dimer which must break into a monomer to promote the substrate oxidation. The catalyst resting state in the Pd(OAc)$_2$–DMSO system is palladium(II) and the rate-limiting step is located in the palladium reoxidation. Stahl and co-workers reported a competition of catalyst aggregation and decomposition with the reoxidation of palladium(0) by oxygen [39a]. The catalyst aggregation exhibits a bimolecular dependence on [palladium], whereas the catalytic turnover rate is unimolecular in [palladium]. In Pd(OAc)$_2$–pyridine-catalyzed oxidation, the rate dependence on [catalyst] varies under different conditions. The dependence is hyperbolic when [catalyst] is varied at a constant py:Pd ratio of 4 : 1, parabolic when [Pd(OAc)$_2$] is varied in the presence of constant excess [pyridine], and linear when [Pd(OAc)$_2$] is varied in the presence of constant [pyridine] and [AcOH]. These kinetic results could not be explained on the basis of the previously proposed mechanism [39a] because the spectroscopic data preclude the existence of a dimeric palladium species and palladium black was not observed in the reaction process. On the other hand, the newly proposed mechanism (Scheme 16.24) can account for the nonlinear dependence on [catalyst] for the Pd(OAc)$_2$–pyridine system.

In the absence of added acetic acid and pyridine, the steady-state concentrations of acetic acid and pyridine will be equal to the concentrations of intermediates **27** and **28**, respectively. Steps 2 and 3 (Scheme 16.25) reveal a first-order dependence of [catalyst] and the reverse steps exhibit a bimolecular dependence on [catalyst]. Thus, at high concentrations of the catalyst, the reverse steps will gain a competitive advantage over the forward steps. When excess pyridine is added to the reaction mixture, the concentration

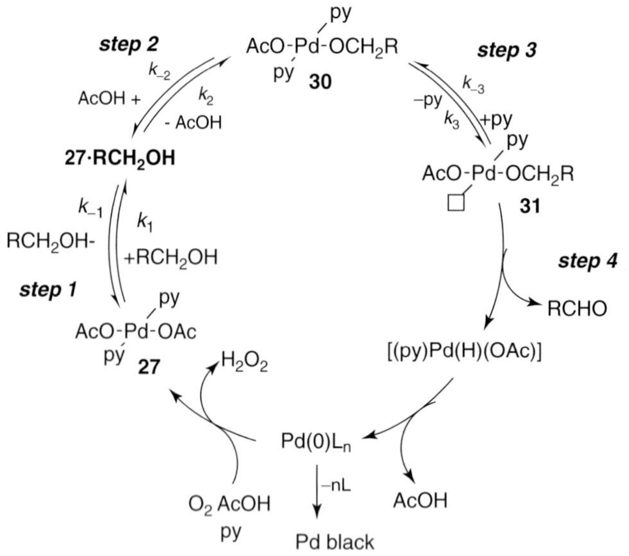

Scheme 16.25 The overall process of (py)$_2$Pd(OAc)$_2$-mediated alcohol oxidation.

of pyridine will not be proportional to the catalyst loading. A half-order dependence on [Pd(OAc)$_2$] is observed under these conditions, originating from the competition of the forward and reverse reactions in step 2. When excess acetic acid and pyridine are added, neither of them is proportional to the catalyst loading and a first-order dependence on catalyst loading is observed. All of these results imply that the nonlinear behavior in [Pd(OAc)$_2$]–pyridine-catalyzed alcohol oxidation may arise from competitive ligand dissociation and recombination.

16.3.2.2 Conclusion

The reaction of Pd(OAc)$_2$–py-mediated alcohol oxidation proceeds through a four-step pathway (Scheme 16.25). The alcohol and the square-planar Pd(II) complex **27** can be combined together in step 1. Then, proton-coupled substitution to produce a palladium alkoxide species **30** occurs in step 2. The following step is the reversible ligand dissociation to generate a three-coordinate Pd(II) intermediate and the aldehyde could be formed by irreversible β-H elimination in step 4. Attempts to increase the catalyst activity by promoting facile ligand dissociation must be balanced by the need to prevent catalyst decomposition.

16.4
Conclusion and Perspective

In the past 10 years, mechanistic studies of synthetic reactions have attracted increasing attention and more and more efforts will be appear in the future. The deep understanding of recently developed alkene metathesis, cross-coupling reactions, and C–H bond activation will become more important, especially C–H bond activation, which has been extensively reported with regard to methodology but relatively less investigated concerning mechanism. The understanding of these reactions, no matter whether right or wrong, will definitely be beneficial in designing new reactions and improving the existing methods for realizing higher efficiency and milder conditions. The basic principle in mechanism studies can be summarized as *"adventurous hypothesis and carefully exploration."* As many different pathways as possible should be proposed, then every hypothesis should be analyzed one by one on the basis of the experimental observations, kinetic behavior, spectroscopic data, and other factors. The most likely path is usually regarded as a reasonable and acceptable mechanism even though there may be many arguments in later studies.

Concerning the methods employed, kinetic investigations are the most important and essential tool, providing significant information about rate-limiting steps and thermodynamic parameters. The rate law can be deduced from kinetic results. Regarding the development of analytical procedures, especially *in situ* instruments such as *in situ* IR, UV, and NMR spectrometers, and so on, have enabled precise kinetic data to be obtained, even under sluggish conditions. Mechanistic studies seem to be much more difficult than the investigation of synthetic reactions. However, we believe that further developments of the analytical instrumentation for mechanistic investigations will be fascinating and more fruitful results of mechanistic studies of synthetic reactions will appear.

Acknowledgments

This work was supported by the National Natural Science Foundation of China (**21025206, 20832003**, and **20972118**) and the "973" Project from MOST, China (**2011CB808600**). In addition, the authors also acknowledge the support from the Fundamental Research Funds for the Central Universities, Program for New Century Excellent Talents in University (NCET), Program for Changjiang Scholars and Innovative Research Team in University (**IRT1030**), and the Academic Award for Excellent PhD Candidates funded by the Ministry of Education of China.

References

1. (a) Hegedus, L.S. (1999) *Transition Metals in the Synthesis of Complex Organic Molecules*, 2nd edn., University Science Books, Sausalito, CA; (b) Larsen, R.D., and Abdel-Magid, A.F. (2004) *Organometallics in Process Chemistry*, Springer, Berlin; (c) Crabtree, R.H. (2009) *The Organometallic Chemistry of the Transition Metals*, 5th edn., John Wiley & Sons, Inc., Hoboken, NJ.

2. de Meijere, A. and Diederich, F. (2004) *Metal-Catalyzed Cross-Coupling Reactions*, 2nd edn., Wiley-VCH Verlag GmbH, Weinheim.

3. Negishi, E. and de Meijere, A. (2002) *Handbook of Organopalladium Chemistry for Organic Synthesis*, Wiley-Interscience, New York.

4. (a) Yang, L., Huang, L., and Luh, T. (2004) *Org. Lett.*, **6**, 1461–1463; (b) Kiso, Y., Tamao, K., and Kumada, M. (1973) *J. Organomet. Chem.*, **50**, C12–C14; (c) Tamao, K., Kiso, Y., Sumitani, K., and Kumada, M. (1972) *J. Am. Chem. Soc.*, **94**, 9268–9269; (d) Tamao, K., Sumitani, K., and Kumada, M. (1972) *J. Am. Chem. Soc.*, **94**, 4374–4376.

5. (a) Zhou, J. and Fu, G.C. (2003) *J. Am. Chem. Soc.*, **125**, 12527–12530; (b) Zhou, J. and Fu, G.C. (2003) *J. Am. Chem. Soc.*, **125**, 14726–14727; (c) Negishi, E., Okukado, N., King, A.O., Van Horn, D.E., and Spiegel, B.I. (1978) *J. Am. Chem. Soc.*, **100**, 2254–2256; (d) King, A.O., Okukado, N., and Negishi, E. (1977) *J. Chem. Soc., Chem. Commun.*, 683–684; (e) Casares, J.A., Espinet, P., Fuentes, B., and Salas, G. (2007) *J. Am. Chem. Soc.*, **129**, 3508–3509.

6. (a) Hatanaka, Y., Fukushima, S., and Hiyama, T. (1989) *Chem. Lett.*, 1711–1714; (b) Lee, J.-Y. and Fu, G.C. (2003) *J. Am. Chem. Soc.*, **125**, 5616–5617; (c) Hiyama, T. (2002) *J. Organomet. Chem.*, **653**, 58–61; (d) Hatanaka, Y. and Hiyama, T. (1988) *J. Org. Chem.*, **53**, 918–920; (e) Hiyama, T., Obayashi, M., Mori, I., and Nozaki, H. (1983) *J. Org. Chem.*, **48**, 912–914.

7. (a) Tang, H., Menzel, K., and Fu, G.C. (2003) *Angew. Chem. Int. Ed.*, **42**, 5079–5082; (b) Menzel, K. and Fu, G.C. (2003) *J. Am. Chem. Soc.*, **125**, 3718–3719; (c) Echavarren, A.M. and Stille, J.K. (1988) *J. Am. Chem. Soc.*, **110**, 4051–4053; (d) Echavarren, A.M. and Stille, J.K. (1987) *J. Am. Chem. Soc.*, **109**, 5478–5486; (e) Milstein, D. and Stille, J.K. (1979) *J. Am. Chem. Soc.*, **101**, 4992–4998.

8. (a) Miyaura, N., Yamada, K., and Suzuki, A. (1979) *Tetrahedron Lett.*, **20**, 3437–3440; (b) Taguchi, T., Itoh, M., and Suzuki, A. (1973) *Chem. Lett.*, 719–722; (c) Brown, H.C., Suzuki, A., Nozawa, S., Harada, M., Itoh, M., and Midland, M.M. (1971) *J. Am. Chem. Soc.*, **93**, 1508–1509; (d) Kirchhoff, J.H., Netherton, M.R., Hills, I.D., and Fu, G.C. (2002) *J. Am. Chem. Soc.*, **124**, 13662–13663; (e) Netherton, M.R., Dai, C., Neuschuetz, K., and Fu, G.C. (2001) *J. Am. Chem. Soc.*, **123**, 10099–10100.

9. (a) Temma, T., Hatano, B., and Habaue, S. (2006) *Tetrahedron*, **62**, 8559–8563; (b) Zhao, Y., Wang, H., Hou, X., Hu, Y., Lei, A., Zhang, H., and Zhu, L. (2006) *J. Am. Chem. Soc.*, **128**, 15048–15049; (c) Jin, L., Zhao, Y., Wang, H., and Lei, A. (2008) *Synthesis*, 649–654; (d) Liu, W., and Lei, A. (2008) *Tetrahedron Lett.*, **49**, 610–613; (e) Cahiez, G., Duplais, C., and Buendia, J. (2009) *Angew. Chem. Int. Ed.*, **48**, 6731–6734; (f) Mo, H. and Bao, W. (2009) *Adv. Synth. Catal.*, **351**, 2845–2849;

(g) Urkalan, K.B. and Sigman, M.S. (2009) *J. Am. Chem. Soc.*, **131**, 18042–18043; (h) Chen, M., Zheng, X., Li, W., He, J., and Lei, A. (2010) *J. Am. Chem. Soc.*, **132**, 4101–4103; (i) Guo, X., Yu, R., Li, H., and Li, Z. (2009) *J. Am. Chem. Soc.*, **131**, 17387–17393; (j) Wang, D., Wasa, M., Giri, R., and Yu, J. (2008) *J. Am. Chem. Soc.*, **130**, 7190–7191; (k) Hamada, T., Ye, X., and Stahl, S.S. (2008) *J. Am. Chem. Soc.*, **130**, 833–835; (l) Guo, Q., Wu, Z., Luo, Z., Liu, Q., Ye, J., Luo, S., Cun, L., and Gong, L. (2007) *J. Am. Chem. Soc.*, **129**, 13927–13938.

10. Rendina, L.M. and Puddephatt, R.J. (1997) *Chem. Rev.*, **97**, 1735–1754.

11. (a) Hadei, N., Kantchev, E.A.B., O'Brien, C.J., and Organ, M.G. (2005) *Org. Lett.*, **7**, 3805–3807; (b) Hatakeyama, T., Hashimoto, S., Ishizuka, K., and Nakamura, M. (2009) *J. Am. Chem. Soc.*, **131**, 11949–11963; (c) Li, Y., Shi, F., He, Q., and You, S. (2009) *Org. Lett.*, **11**, 3182–3185; (d) Zhang, C. and Wang, Z. (2009) *Organometallics*, **28**, 6507–6514; (e) Gao, H., Yan, C., Tao, X., Xia, Y., Sun, H., Shen, Q., and Zhang, Y. (2010) *Organometallics*, **29**, 4189–4192.

12. (a) Wang, L. and Wang, Z. (2007) *Org. Lett.*, **9**, 4335–4338; (b) Li, Z., Gao, Y., Tang, Y., Dai, M., Wang, G., Wang, Z., and Yang, Z. (2008) *Org. Lett.*, **10**, 3017–3020; (c) Liu, J., Wang, H., Zhang, H., Wu, X., Zhang, H., Deng, Y., Yang, Z., and Lei, A. (2009) *Chem. Eur. J.*, **15**, 4437–4445; (d) Moreno, I., SanMartin, R., Ines, B., Herrero, M.T., and Dominguez, E. (2009) *Curr. Org. Chem.*, **13**, 878–895; (e) Vechorkin, O., Proust, V., and Hu, X. (2009) *J. Am. Chem. Soc.*, **131**, 9756–9766; (f) Wang, H., Liu, J., Deng, Y., Min, T., Yu, G., Wu, X., Yang, Z., and Lei, A. (2009) *Chem. Eur. J.*, **15**, 1499–1507.

13. (a) Kirchhoff, J.H., Dai, C., and Fu, G.C. (2002) *Angew. Chem. Int. Ed.*, **41**, 1945–1947; (b) Zhou, J. and Fu, G.C. (2003) *J. Am. Chem. Soc.*, **125**, 12527–12530.

14. (a) Amatore, C., Carre, E., Jutand, A., and M'Barki, M.A. (1995) *Organometallics*, **14**, 1818–1826; (b) Amatore, C. and Jutand, A. (1999) *J. Organomet. Chem.*, **576**, 254–278; (c) Galardon, E., Ramdeehul, S., Brown, J.M., Cowley, A., Hii, K.K.,

and Jutand, A. (2002) *Angew. Chem. Int. Ed.*, **41**, 1760–1763; (d) Stambuli, J.P., Buehl, M., and Hartwig, J.F. (2002) *J. Am. Chem. Soc.*, **124**, 9346–9347; (e) Stambuli, J.P., Incarvito, C.D., Buehl, M., and Hartwig, J.F. (2004) *J. Am. Chem. Soc.*, **126**, 1184–1194; (f) Roy, A.H. and Hartwig, J.F. (2003) *J. Am. Chem. Soc.*, **125**, 13944–13945; (g) Brunel, J.M. (2004) *Mini-Rev. Org. Chem.*, **1**, 249–277; (h) Barrios-Landeros, F. and Hartwig, J.F. (2005) *J. Am. Chem. Soc.*, **127**, 6944–6945; (i) Ahlquist, M., Fristrup, P., Tanner, D., and Norrby, P.-O. (2006) *Organometallics*, **25**, 2066–2073; (j) Ahlquist, M. and Norrby, P.-O. (2007) *Organometallics*, **26**, 550–553; (k) Mitchell, E.A. and Baird, M.C. (2007) *Organometallics*, **26**, 5230–5238.

15. Barrios-Landeros, F., Carrow, B.P., and Hartwig, J.F. (2009) *J. Am. Chem. Soc.*, **131**, 8141–8154.

16. (a) Kalek, M., Jezowska, M., and Stawinski, J. (2009) *Adv. Synth. Catal.*, **351**, 3207–3216; (b) Kozuch, S., Shaik, S., Jutand, A., and Amatore, C. (2004) *Chem. Eur. J.*, **10**, 3072–3080; (c) Amatore, C., Jutand, A., Lemaitre, F., Ricard, J.L., Kozuch, S., and Shaik, S. (2004) *J. Organomet. Chem.*, **689**, 3728–3734; (d) Kalek, M. and Stawinski, J. (2007) *Organometallics*, **26**, 5840–5847.

17. Amatore, C., Carre, E., Jutand, A., and Medjour, Y. (2002) *Organometallics*, **21**, 4540–4545.

18. Amatore, C., Broeker, G., Jutand, A., and Khalil, F. (1997) *J. Am. Chem. Soc.*, **119**, 5176–5185.

19. Paladino, G., Madec, D., Prestat, G., Maitro, G., Poli, G., and Jutand, A. (2007) *Organometallics*, **26**, 455–458.

20. Wendt, O.F. (2007) *Curr. Org. Chem.*, **11**, 1417–1433.

21. (a) Chass, G.A., O'Brien, C.J., Hadei, N., Kantchev, E.A.B., Mu, W., Fang, D., Hopkinson, A.C., Csizmadia, I.G., and Organ, M.G. (2009) *Chem. Eur. J.*, **15**, 4281–4288; (b) Liu, Q., Lan, Y., Liu, J., Li, G., Wu, Y.-D., and Lei, A. (2009) *J. Am. Chem. Soc.*, **131**, 10201–10210; (c) Jin, L.Q., Xin, J., Huang, Z.L., He, J., and Lei, A.W. (2010) *J. Am. Chem. Soc.*, **132**, 9607–9609; (d) Dong, Z., Manolikakes, G., Shi, L., Knochel, P., and Mayr, H. (2010) *Chem. Eur. J.*, **16**, 248–253.

22. (a) Casado, A.L., Espinet, P., and Gallego, A.M. (2000) *J. Am. Chem. Soc.*, **122**, 11771–11782; (b) Espinet, P. and Echavarren, A.M. (2004) *Angew. Chem. Int. Ed.*, **43**, 4704–4734; (c) Albeniz, A.C., Espinet, P., and Lopez-Fernandez, R. (2006) *Organometallics*, **25**, 5449–5455; (d) Nova, A., Ujaque, G., Maseras, F., Lledos, A., and Espinet, P. (2006) *J. Am. Chem. Soc.*, **128**, 14571–14578; (e) Perez-Temprano, M.H., Nova, A., Casares, J.A., and Espinet, P. (2008) *J. Am. Chem. Soc.*, **130**, 10518–10520.

23. (a) Matos, K. and Soderquist, J.A. (1998) *J. Org. Chem.*, **63**, 461–470; (b) Ridgway, B.H. and Woerpel, K.A. (1998) *J. Org. Chem.*, **63**, 458–460; (c) Braga, A.A.C., Morgon, N.H., Ujaque, G., and Maseras, F. (2005) *J. Am. Chem. Soc.*, **127**, 9298–9307; (d) Adamo, C., Amatore, C., Ciofini, I., Jutand, A., and Lakmini, H. (2006) *J. Am. Chem. Soc.*, **128**, 6829–6836; (e) Braga, A.A.C., Morgon, N.H., Ujaque, G., Lledos, A., and Maseras, F. (2006) *J. Organomet. Chem.*, **691**, 4459–4466; (f) Braga, A.A.C., Ujaque, G., and Maseras, F. (2006) *Organometallics*, **25**, 3647–3658.

24. Mateo, C., Fernandez-Rivas, C., Echavarren, A.M., and Cardenas, D.J. (1997) *Organometallics*, **16**, 1997–1999.

25. Stille, J.K. (1986) *Angew. Chem. Int. Ed. Engl.*, **25**, 508.

26. (a) Louie, J. and Hartwig, J.F. (1995) *J. Am. Chem. Soc.*, **117**, 11598–11599; (b) Casado, A.L. and Espinet, P. (1998) *J. Am. Chem. Soc.*, **120**, 8978–8985; (c) Amatore, C., Bahsoun, A.A., Jutand, A., Meyer, G., Ntepe, A.N., and Ricard, L. (2003) *J. Am. Chem. Soc.*, **125**, 4212–4222.

27. (a) Hartwig, J.F. (2007) *Inorg. Chem.*, **46**, 1936–1947; (b) Perez-Rodriguez, M., Braga, A.A.C., Garcia-Melchor, M., Perez-Temprano, M.H., Casares, J.A., Ujaque, G., de Lera, A.R., Alvarez, R., Maseras, F., and Espinet, P. (2009) *J. Am. Chem. Soc.*, **131**, 3650–3657.

28. Culkin, D.A. and Hartwig, J.F. (2004) *Organometallics*, **23**, 3398–3416.

29. (a) Zuidema, E., van Leeuwen, P.W.N.M., and Bo, C. (2005) *Organometallics*, **24**, 3703–3710; (b) Ananikov, V.P., Musaev, D.G., and Morokuma, K. (2002) *J. Am. Chem. Soc.*, **124**, 2839–2852; (c) Ananikov, V.P., Musaev, D.G., and Morokuma, K.

(2005) *Organometallics*, **24**, 715–723; (d) Ananikov, V.P., Musaev, D.G., and Morokuma, K. (2007) *Eur. J. Inorg. Chem.*, 5390–5399.

30. (a) Tatsumi, K., Hoffmann, R., Yamamoto, A., and Stille, J.K. (1981) *Bull. Chem. Soc. Jpn.*, **54**, 1857–1867; (b) Tatsumi, K., Nakamura, A., Komiya, S., Yamamoto, A., and Yamamoto, T. (1984) *J. Am. Chem. Soc.*, **106**, 8181–8188.

31. (a) Popp, B.V. and Stahl, S.S. (2006) *J. Am. Chem. Soc.*, **128**, 2804–2805; (b) Lanci, M.P., Remy, M.S., Kaminsky, W., Mayer, J.M., and Sanford, M.S. (2009) *J. Am. Chem. Soc.*, **131**, 15618–15620.

32. Jin, L., Zhang, H., Li, P., Sowa, J.R., and Lei, A. (2009) *J. Am. Chem. Soc.*, **131**, 9892–9893.

33. Mathew, J.S., Klussmann, M., Iwamura, H., Valera, F., Futran, A., Emanuelsson, E.A.C., and Blackmond, D.G. (2006) *J. Org. Chem.*, **71**, 4711–4722.

34. Stahl, S.S. (2005) *Science*, **309**, 1824–1826.

35. Steinhoff, B.A., Guzei, I.A., and Stahl, S.S. (2004) *J. Am. Chem. Soc.*, **126**, 11268–11278.

36. (a) Neumann, R., Khenkin, A.M., and Vigdergauz, I. (2000) *Chem. Eur. J.*, **6**, 875–882; (b) Sun, H., Harms, K., and Sundermeyer, J. (2004) *J. Am. Chem. Soc.*, **126**, 9550–9551.

37. (a) Larock, R.C. and Hightower, T.R. (1993) *J. Org. Chem.*, **58**, 5298–5300; (b) van Benthem, R.A.T.M., Hiemstra, H., Michels,

J.J., and Speckamp, W.N. (1994) *J. Chem. Soc., Chem. Commun.*, 357–359.

38. (a) Toyota, M. and Ihara, M. (2002) *Synlett*, 1211–1222; (b) Stahl, S.S. (2004) *Angew. Chem. Int. Ed.*, **43**, 3400–3420.

39. (a) Steinhoff, B.A., Fix, S.R., and Stahl, S.S. (2002) *J. Am. Chem. Soc.*, **124**, 766–767; (b) Steinhoff, B.A. and Stahl, S.S. (2006) *J. Am. Chem. Soc.*, **128**, 4348–4355.

40. ten Brink, G., Arends, I.W.C.E., and Sheldon, R.A. (2002) *Adv. Synth. Catal.*, **344**, 355–369.

41. (a) Nishimura, T., Onoue, T., Ohe, K., and Uemura, S. (1998) *Tetrahedron Lett.*, **39**, 6011–6014; (b) Nishimura, T., Ohe, K., and Uemura, S. (1999) *J. Am. Chem. Soc.*, **121**, 2645–2646; (c) Nishimura, T., Onoue, T., Ohe, K., and Uemura, S. (1999) *J. Org. Chem.*, **64**, 6750–6755; (d) Nishimura, T., Ohe, K., and Uemura, S. (2001) *J. Org. Chem.*, **66**, 1455–1465; (e) Nishimura, T., Ohe, K., and Uemura, S. (2001) *J. Org. Chem.*, **66**, 1455–1465; (f) Nishimura, T., Araki, H., Maeda, Y., and Uemura, S. (2003) *Org. Lett.*, **5**, 2997–2999.

42. (a) Jensen, D.R., Pugsley, J.S., and Sigman, M.S. (2001) *J. Am. Chem. Soc.*, **123**, 7475–7476; (b) Mueller, J.A. and Sigman, M.S. (2003) *J. Am. Chem. Soc.*, **125**, 7005–7013; (c) Stoltz, B.M. (2004) *Chem. Lett.*, **33**, 362–367.

43. Konnick, M.M., Guzei, I.A., and Stahl, S.S. (2004) *J. Am. Chem. Soc.*, **126**, 10212–10213.

Commentary Part

Comment 1

Xin Mu, Guo-Sheng Liu, and Qi-Long Shen

Over the past 40 years, TM-catalyzed reactions have developed rapidly, and have been extensively used as powerful tools in organic synthesis. During this period, in-depth mechanistic studies successfully established the general reaction patterns along with the detailed description of each elementary step, which is helpful for understanding the overall reactions. Moreover, the deep understanding of the reactions generally provided some important guidance for further modifications of the catalytic system to achieve

highly efficient conversions. In this chapter, Lei and Jin have presented a powerful method: kinetic analysis in research on the elementary steps of cross-coupling reactions, and also reported mechanistic studies of the aerobic oxidation of alcohols. To emphasize the importance of mechanistic studies in promoting the improvement of catalytic systems, an additional case of palladium-catalyzed cross-coupling of aryl halides and amines is presented in this comment.

Arylamines are important structural motifs in pharmaceutical and materials sciences. Within the last two decades, palladium-catalyzed coupling reactions of amines with aryl halides have emerged that are much more efficient and use milder conditions for the construction of aryl

C–N bonds than traditional methods. Specifically, successive developments of four generations of catalysts have broadened the application of these methods to a variety of substrates with lower temperatures and extremely low catalyst loadings. Detailed mechanistic studies have greatly promoted the understanding of the catalytic cycle and the vital role of Pd–phosphine ligand combinations, which provided accessible approaches to rationalize the design for the next generation of catalysts.

In 1995, Buchwald's and Hartwig's groups simultaneously published the first example of the palladium-catalyzed cross-coupling of aryl bromides with amines without the involvement of toxic and unstable aminostannane reagents. P(o-tolyl)$_3$ was shown to be crucial owing to its sterically demanding properties, which it is believed contribute to the formation of three-coordinate palladium intermediates [C1]. Subsequent kinetic analysis and stoichiometric reactions by Hartwig's group demonstrated that the monophosphine-ligated palladium complexes (three-coordinate) were the active species in oxidative addition, transmetallation, and reductive elimination reactions when P(o-tolyl)$_3$ is utilized [C2]. The Pd-P(o-tolyl)$_3$ catalytic system was mainly effective for coupling of secondary amines with aryl bromides, and is regarded as among the "*first-generation*" catalysts.

A series of kinetic studies were conducted by Hartwig's group on the reductive elimination process of the complex [Pd(PPh$_3$)$_2$(NAr$_2$)(Ph)] [C2d]. They concluded that the elimination from a four-coordinate intermediate occurred in competition with a three-coordinate intermediate. These kinetic results revealed the potential utility of chelating ligands with larger bite angles such as dppe, dppf, and binap, which inhibited the β-hydride elimination and would, therefore, improve the selectivity for amination versus reduction. Based on the mechanistic analysis, (dppf)Pd(Ph)[N(tolyl)$_2$] was synthesized and high-yielding reductive elimination occurred when it was subjected to heating conditions [C3]. Catalytic reactions were even more effective when (dppf)PdCl$_2$ was used as the catalyst for the coupling of aryl bromides or iodides with primary amines. The yields of such reactions were improved significantly over those

involving the first-generation, steric hindered phosphines. Buchwald's group reported a similar Pd$_2$(dba)$_3$–binap system to effect this transformation [C4]. These improved catalytic systems, which also broadened the substrate scope and reduced the side products, are considered as *second-generation* catalysts.

To investigate the chelating catalytic process and develop better catalytic systems, Hartwig's group conducted ^{31}P NMR spectroscopic studies of reactions between o-tolyl or p-tolyl bromide and n-butylamine in the presence of Pd–binap or Pd–dppf under heating conditions [C5]. The formation of Pd(0) L$_2$ (L = binap or dppf) was observed, which indicated that the oxidative addition of aryl bromides is likely involved in the rate-determining step. As a result, the rational design of more active catalysts must promote this process. They proposed that an increased electron density of chelating phosphine ligand might accelerate the reaction rate. To this end, 1,1'-bis(di-*tert*-butylphosphino)ferrocene was prepared to give a fast rate for palladium-catalyzed amination reactions. However, through a series of mechanistic studies, it was later discovered that this bisphosphine underwent P–C bond cleavage and arylation of one cyclopentadienyl group to generate a new kind of sterically hindered pentaphenyl ferrocenyl di-*tert*-butylphosphine ligand (Q-Phos). It is noteworthy that a palladium catalyst containing Q-Phos was remarkably active for C–N, C–C, and C–O bond coupling reactions [C6]. In parallel, hindered alkylphosphine ligands such as P(t-Bu)$_3$ and Buchwald's dialkylphosphino biaryl ligands extended the reaction scope to aryl chlorides under mild conditions. Catalysts containing these hindered alkyl phosphines and also the N-cyclic carbene were termed *third-generation* catalysts [C7].

Third-generation catalysts

The development of *fourth generation* catalysts mainly focused on the limitations of the third-generation catalysts, such as the undesired diarylation byproduct for the coupling of primary alkylamines and the relatively high catalyst loading, which is possibly attributable to the binding of nitrogen nucleophiles to the palladium center, causing deactivation of the palladium catalyst. Hartwig's group reasoned that the selection of ligands should combine steric hindrance, strong electron donation, and tight chelation, and may lead to a catalyst that simultaneously possesses high selectivity, high activity, and a long lifetime. Hence josiphos (CyPF-*t*-Bu), developed for asymmetric hydrogenation, was tested for these purposes [C8]. The excellent substrate scope and extremely low catalyst loading down to 0.05 mol% make this catalytic system a very powerful tool for constructing aryl C–N bonds [C9].

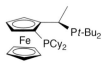

CyPF-*t*-Bu

In conclusion, the evolution of four generations of catalytic systems for the coupling reactions of aryl halides and amines represents an example of the vital role that mechanistic studies play in the development of a synthetic reaction. Similar efforts with mechanistic studies, we believe, have a good chance of appearing in the future to develop new methodology.

Comment 2

Yoshinori Yamamoto

In this chapter, the authors focus on a discussion of the mechanisms of TM-catalyzed coupling reactions, mainly through kinetic studies to elucidate the rate-limiting steps. The deep understanding of mechanistic aspects is very important and helpful for organic synthetic chemists to design and develop new syntheses and to improve the selectivity and efficiency of coupling reactions. In each section, the authors not only describe the fundamental mechanisms of coupling reaction involving oxidative addition, transmetallation, and reductive elimination, but also provide detailed

methods for the identification of the key intermediates and rate-limiting steps through kinetic investigations.

Overall, this is an excellent chapter on the mechanistic aspects of palladium-catalyzed reactions. Readers can obtain a deep insight into the mechanisms of such palladium-catalyzed reactions, and a thorough understanding of these mechanism is needed, especially for students and young researchers, to extend the scope of synthetic applications and to develop future research plans.

Authors' Response to the Commentaries

We thank Prof. Yoshinori Yamamoto for his valuable comments, and also for his suggestions to give proper titles to the four sections and to improve the English, which helped us to polish the text into a more readable and clearer form. All his suggestions have been implemented in the final version of the chapter.

We also thank Xin Mu Guosheng Liu, and Qilong Shen for presenting an interesting description of C–N coupling reactions. This showed us how important mechanistic studies have been in the improvement of catalyst design, generation by generation. We fully agree with their conclusion that: "... efforts with mechanistic studies ... have a good chance of appearing in the future tom develop new methodology."

References

C1. Guram, A.S., Rennels, R.A., and Buchwald, S.L. (1995) *Angew. Chem. Int. Ed. Engl.*, **34**, 1348.

C2. (a) Paul, F., Patt, J., and Hartwig, J.F. (1994) *J. Am. Chem. Soc.*, **116**, 5969; (b) Paul, F., Patt, J., and Hartwig, J.F. (1995) *Organometallics*, **14**, 3030; (c) Paul, F. and Hartwig, J.F. (1995) *J. Am. Chem. Soc.*, **117**, 5373; (d) Driver, M.S. and Hartwig, J.F. (1995) *J. Am. Chem. Soc.*, **117**, 4708.

C3. Driver, M.S. and Hartwig, J.F. (1996) *J. Am. Chem. Soc.*, **118**, 7217.

C4. (a) Wolfe, J.P., Wagaw, S., and Buchwald, S.L. (1996) *J. Am. Chem. Soc.*, **118**, 7215; (b) Wolfe, J.P. and Buchwald, S.L. (2000) *J. Org. Chem.*, **65**, 1144.

C5. Hamann, B.C. and Hartwig, J.F. (1998) *J. Am. Chem. Soc.*, **120**, 7369.

C6. (a) Shelby, Q., Kataoka, N., Mann, G., and Hartwig, J.F. (2000) *J. Am.*

Chem. Soc., **122**, 10718; (b) Kataoka,
N., Shelby, Q., Stambuli, J.P., and
Hartwig, J.F. (2002) *J. Org. Chem.*, **67**,
5553.

C7. (a) Hartwig, J.F., Kawatsura, M., Hauck,
S.I., Shaughnessy, K.H., and Alcazar-Roman,
L.M. (1999) *J. Org. Chem.*, **64**, 5575; (b)

Stambuli, J.P., Kuwano, R., and Hartwig,
J.F. (2002) *Angew. Chem. Int. Ed.*, **41**, 4746.

C8. Shen, Q., Shekhar, S., Stambuli, J.P., and
Hartwig, J.F. (2005) *Angew. Chem. Int. Ed.*,
44, 1371.

C9. Shen, Q., Ogata, T., and Hartwig, J.F.
(2008) *J. Am. Chem. Soc.*, **130**, 6586.

17
Organic Materials and Chemistry for Bulk Heterojunction Solar Cells

Chun-Hui Duan, Fei Huang, and Yong Cao

17.1
Introduction

A solar cell is a photovoltaic device which can convert sunlight into electricity and is considered as one of the most attractive ways to address growing global energy needs using a clean, abundant, and renewable resource. Silicon-based solar cells represent the most mature photovoltaic technology to date with a high power conversion efficiency (PCE) of more than 25% [1]. However, the large-scale commercial development of silicon-based solar cells is limited owing to their relatively high cost. Alternatively, organic photovoltaic (OPV) solar cells have been rapidly developing as one of the most promising candidates for low-cost solar cells, owing to the possibility of fabricating large-area and flexible solar cells by solution processing [2–4]. In 1986, Tang made a great breakthrough in the development of organic solar cells (OSC)s with the invention of a bilayer heterojunction solar cell containing a p-type layer for hole transport and an n-type layer for electron transport, which showed a PCE approaching 1% [5]. In 1992, Sariciftci *et al.* discovered ultrafast photoinduced electron transfer from semiconducting conjugated polymers to fullerenes, which is the basic principle of heterojunction photovoltaic cells based on organic semiconductors [6]. In 1995, Yu *et al.* proposed the concept of the bulk heterojunction (BHJ) solar cell, where a blend of donor and acceptor materials is used as the active layer [7]. It was found that an interpenetrating network with a large donor–acceptor (D–A) interfacial area can be formed by controlling the nanoscale phase separation of the donor and acceptor materials in the blends (Figure 17.1a), resulting in a significantly enhanced charge separation/collection efficiency and hence enhanced PCE. The BHJ structure has been widely used in OSCs and considerable progress has been made in this area, as the PCE has steadily improved from nearly 1% initially to more than 8% recently [8, 9]. Figure 17.1b shows typical current–voltage (J–V) characteristics of a BHJ solar cell, where J_{sc} is the short-circuit current density, V_{oc} is the open-circuit voltage, FF is the fill factor, and P_{in} is the power of the incident light, and then PCE is given by $J_{sc} \times V_{oc} \times FF/P_{in}$. In order to obtain a higher PCE, J_{sc}, V_{oc}, and FF of the solar cells need to be optimized simultaneously. Basic knowledge of the relationship between these properties and the organic D–A materials have been established [10]. For example, it has been shown that J_{sc} is correlated with the absorption of the active

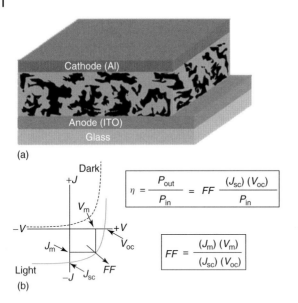

$$\eta = \frac{P_{out}}{P_{in}} = FF \frac{(J_{sc})(V_{oc})}{P_{in}}$$

$$FF = \frac{(J_m)(V_m)}{(J_{sc})(V_{oc})}$$

Figure 17.1 (a) Illustration of device configuration of BHJ solar cell. (b) Current–voltage $(J - V)$ characteristics of BHJ solar cells.

layer and exciton dissociation efficiency, which requires that the materials have a better matched absorption with the solar spectrum and high dielectric constant, respectively. V_{oc} is mainly determined by the energy offset between the highest occupied molecular orbital (HOMO) energy level of the donor materials and the lowest unoccupied molecular orbital (LUMO) acceptor materials. Moreover, the active layers of the solar cells need to have a balanced hole/electron mobility and an ideal morphology with a nanoscale interpenetrating network to obtain a high *FF*. Hence it is critical to develop suitable donor and acceptor materials to meet all these requirements and extensive efforts have been put into the development of new donor and acceptor materials, resulting in a rapid improvement of the performance of BHJ solar cells. Organic chemistry offers a rich variety of tools and opportunities for developing all kinds of BHJ solar cell materials. For instance, the synthesis of conjugated polymers, which are usually used as the donor materials in BHJ solar cells, is mainly based on transition metal-catalyzed cross-coupling reactions, for which Richard F. Heck, Ei-ichi Negishi, and Akira Suzuki were awarded the 2010 Nobel Prize in Chemistry for their contributions in this area [11]. Fullerene chemistry also contributed greatly to the development of acceptor materials and fullerene derivatives, such as [6,6]-phenyl-C_{61}-butyric acid methyl ester (PC$_{61}$BM) and [6,6]-phenyl-C_{71}-butyric acid methyl ester (PC$_{71}$BM) (Figure 17.2), which are still the state-of-the-art acceptor materials in BHJ solar cells [7, 12–14]. Moreover, the newly developed chemistry of the synthesis of block polymers offers possible ways to develop functional conjugated polymers with potentially controllable nanomorphology, which is very important for BHJ solar cells [15]. Thus organic chemistry plays a very important role in the development of BHJ solar cells. In this chapter, typical organic materials for BHJ solar cells and their related chemistry are briefly introduced.

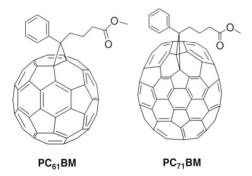

PC$_{61}$BM   PC$_{71}$BM

Figure 17.2 Chemical structures of PC$_{61}$BM and PC$_{71}$BM.

17.2
Molecular Design and Engineering of Donor Materials

17.2.1
Molecular Design and Engineering of Conjugated Polymers

Conjugated polymers not only exhibit unique optical and electronic properties of semiconductors, but also possess the processing advantages and mechanical properties of polymers. As a result, conjugated polymers have been successfully applied in polymer light-emitting diodes (PLEDs) [16], organic field-effect transistors (OFETs) [17], and OSCs [6]. Conjugated polymers have many obvious advantages over inorganic semiconductors, such as ease of fabrication, better mechanical stability, and lower production costs, especially for large-sized devices [18]. In early studies, red- or orange–red-emitting polymers, such as poly {2-methoxy-5-[(2′-ethylhexyl)oxy]-1,4-phenylenevinylene}(MEH-PPV) [7], poly {2-methoxy-5-[(3′,7′-dimethyloctyl)oxy]-1,4-phenylenevinylene}(MDMO-PPV) [19], and poly(3-hexylthiophene) (P3HT) [20] were applied as the donor materials in BHJ solar cells. Later, D–A alternating conjugated polymers were extensively studied, because their electronic properties can be readily tuned by controlling the intramolecular charge transfer (ICT) between the donors and acceptors to meet the requirements for BHJ solar cell application. As discussed previously, nanosized phase-separated morphology of the active layer of the BHJ solar cell is desired for efficient exciton dissociation and charge carrier transport to obtain a current output as large as possible. Block copolymers, which can form self-assembled nanostructures on a molecular scale, were therefore explored for BHJ solar cells. Moreover, block copolymers provide the possibility of integrating multiple electronic properties on one polymer chain via living polymerization of different components [15, 21, 22].

17.2.1.1 Homopolymers
Poly(*p*-phenylenevinylene) (PPV) derivatives were the most widely studied conjugated polymers in the 1990s and were successfully applied in the first PLEDs [16] and polymer-based BHJ solar cells [6]. Many methods have been developed for synthesizing PPV derivatives, including Wessling polycondensation [23], the sulfinyl precursor route

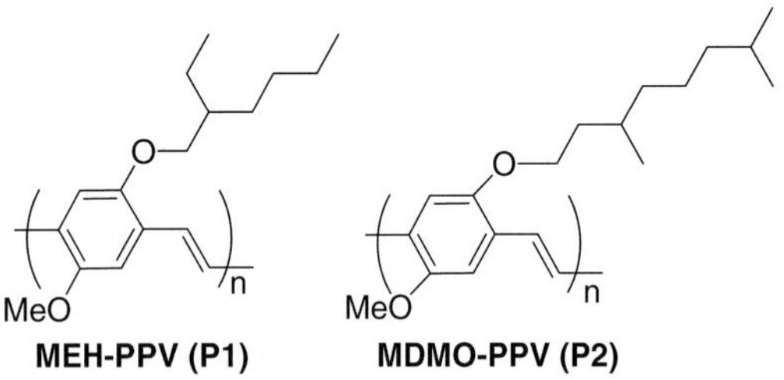

Figure 17.3 Chemical structures of MEH-PPV, MDMO-PPV, and CN-PPV.

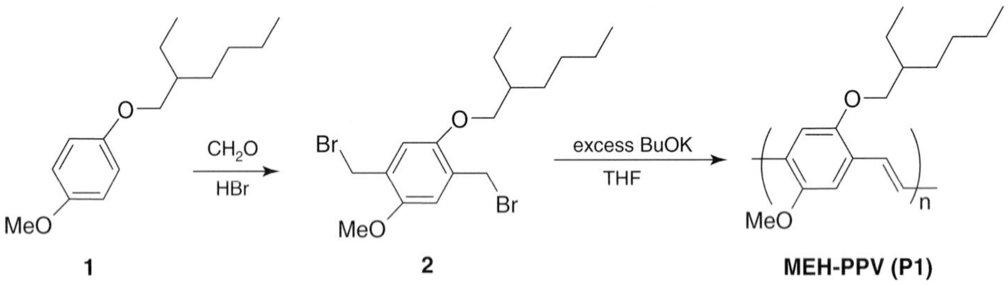

Scheme 17.1 Synthetic route for MEH-PPV.

[24], Gilch polycondensation [25], Wittig polycondensation [26], Heck polycondensation [27], and Knoevenagel polycondensation [28]. Among the various kinds of PPV derivatives, MEH-PPV and MDMO-PPV have been the most studied donor materials in BHJ solar cells owing to their relatively strong absorption in the visible region (Figure 17.3). Although both of them can be prepared by the various methods listed above, Glich polycondensation is the most preferred way in most cases. As shown in Scheme 17.1, Glich polycondensation involves the polymerization of 1,4-bis-halomethylated benzene derivatives in the presence of a large excess of base, through an elimination–polymerization–elimination process, which shortens the overall procedure for preparing the conjugated PPV derivatives compared with the previously widely used Wessling route [25]. Moreover, the reaction conditions of the Glich route are mild and the yield and molecular weight of the resulting polymers are usually high.

MEH-PPV was frequently used as a donor material in the early days of BHJ solar cells, and was even used in bilayer and stacked devices [29]. However, the overall device performance based on MEH-PPV was poor, owing to its narrow absorption spectra and poor compatibility with $PC_{61}BM$. Later, MDMO-PPV was developed and became the most widely used PPV derivative as a donor material for BHJ solar cells. In 2001, an impressive PCE value of 2.5% was achieved for an MDMO-PPV:$PC_{61}BM$ blended composite [19]. In this work, the optimal morphology with suppressed phase separation and optimized microstructure was achieved by using chlorobenzene instead of toluene as the solvent, which led to enhanced hole and electron mobilities in the active layer and ultimately a higher PCE. Subsequently, MDMO-PPV:$PC_{61}BM$ was used as a prototype to investigate various physics and engineering aspects of BHJ solar cells, such as photooxidation [30], stacked cells [31], active layer thickness [32], buffer layer [33], and NMR morphology studies [34]. In addition to $PC_{61}BM$, other acceptors such as $PC_{71}BM$ [13], [6,6]-phenyl-C_{85}-butyric acid methyl ester ($PC_{85}BM$) [35], and even inorganic n-type semiconductors (such as CdSe and ZnO) [36, 37] were also used to blend with MDMO-PPV to fabricate BHJ or hybrid inorganic-organic solar cells. Among these studies, the highest PCE value of 3.0% was achieved for an MDMO-PPV-based BHJ solar cell with $PC_{71}BM$ as acceptor [13].

Polythiophene derivatives are another class of homopolymers that have been extensively studied as donor materials in BHJ solar cells owing to their exceptional optical and electronic properties and good thermal and chemical stability. Especially P3HT was the most promising BHJ solar cell material for a long time and is still the most famous donor material for device physics studies. At an early stage, the synthesis of poly(3-alkylthiophene)s (P3ATs) by oxidation or electrochemical polymerization usually gave rise to random couplings in the P3ATs, due to the presence of three coupling modes, head-to-tail (HT) coupling, head-to-head (HH) coupling, and tail-to-tail (TT) coupling [38, 39]. The low degree of regioregularity can dramatically reduce the electrical conductivity and increase the bandgap of P3ATs. Fortunately, the developments in the synthesis chemistry of regioregular P3ATs paved the way to the use of P3HT as one of the most successful donor materials for BHJ solar cells.

In 1992, McCullough and Lowe first reported the synthesis of HT regioregular P3ATs [40]. As depicted in Scheme 17.2a, the key step of this method is the selective lithiation of 2-bromo-3-alkylthiophene (**3**) with lithium diisopropylamide (LDA) at cryogenic temperatures followed by the addition of $MgBr_2 \cdot Et_2O$ or $ZnCl_2$ to yield the regiospecific organometallic intermediate **4**. The polymerization is subsequently carried out *in situ* using Ni(dppp)Cl$_2$ as catalyst to afford corresponding P3ATs with HT–HT regioregularity of 98–100% [40, 41]. Soon afterwards, Chen and Rieke reported another method for synthesizing regioregular P3ATs. In this approach, by selective oxidative addition of highly reactive zinc to 2,5-dibromo-3-alkylthiophene (**5**), a 2-(bromozincio)-3-alkyl-5-bromothiophene intermediate (**6**) was obtained with a yield

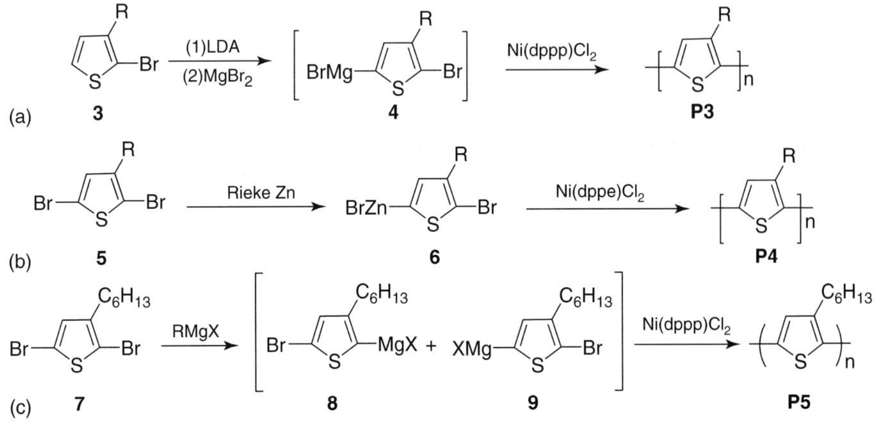

Scheme 17.2 Synthetic routes for regioregular P3ATs: (a) McCullough method; (b) Rieke method; and (c) GRIM method.

of around 90%, which subsequently afforded regioregular P3ATs in the presence of Ni(dppe)Cl$_2$ as catalyst (Scheme 17.2b) [42]. In 1999, an economical method for the synthesis of regioregular P3ATs using Grignard metathesis (GRIM) was reported by Loewe *et al.* [43]. In this method, the use of cryogenic temperatures and a highly reactive metal is not necessary and the preparation of regioregular P3ATs on a kilogram scale is possible. Treating 2,5-dibromo-3-alkylthiophene (**7**) with Grignard reagent gives rise to a mixture of two metallated intermediates in a ratio of 85:15 to 75:25, which is independent of the reaction time, temperature, and Grignard reagent used. Subsequently, the polymerizations takes place in the presence of a catalytic amount of Ni(dppp)Cl$_2$ to yield P3ATs with high regioregularity of over 99% HT coupling (Scheme 17.2c) [43].

The mechanism of nickel-catalyzed polymerization to give regioregular P3ATs was elucidated as a living chain growth process instead of the conventionally accepted step growth polycondensation. The reaction mechanism is presented in Scheme 17.3. First, 2 equiv. of Grignard nucleophile reacted with Ni(dppp)Cl$_2$, generating the organonickel intermediate **12**, which immediately induced reductive elimination involving carbon–carbon bond formation accompanied by Ni migration and insertion into the terminal C–Br bond to yield **13**. Thus, the living chains capable of further reacting with compound **10** were preserved. Growth of the polymer chain was achieved by consecutive coupling between the living polymer chain with a terminal Ni complex and **10** [44–46]. The living nature of the Ni-catalyzed polymerization favors the preparation of regioregular P3ATs with low polydispersity and desired molecular weight. Since then, regioregular P3HT has become the most popular conjugated polymers in OFETs and BHJ solar cells.

P3HT with increased regioregularity exhibits a more coplanar molecular configuration and better packing in the solid state and consequently a red-shifted and intensified absorption, together with enhanced hole transport mobility, which are beneficial for solar cell applications [47]. So far, the best performing BHJ solar cells based on the P3HT:PC$_{61}$BM system have shown PCEs as high as 5.1% [20]. However, the improved regioregularity of P3HT has been reported to be detrimental to the thermal stability of the morphology in the resulteing BHJ solar cell devices. Fréchet and co-workers systematically investigated the influence of the regioregularity of P3HT on the thermal stability of the morphology by comparing three P3HT samples with different regioregularities. It was demonstrated that P3HT with the lowest regioregularity exhibited the best thermal stability, which was attributed to the decreased crystallization-driven phase separation due to the increased disorder in the polymer backbone [48].

In addition to the regioregularity, the molecular weight and distribution of P3HT also play an important role in its electronic properties and ultimately the performance of resulting BHJ solar cells [49–52]. The commonly accepted standpoint of these studies is that a higher molecular weight of P3HT would lead to a larger conjugated length and more effective packing of the polymer chains, which in turn result in broader and more intense absorption and higher carrier transport ability. As a consequence, BHJ solar cells based on high molecular weight P3HT usually exhibited better photovoltaic performance than those based on low molecular weight P3HT. Specifically, it has been reported that it is necessary to use P3HT with a molecular weight of higher than 10 kg mol^{-1} to realize high-performance BHJ solar cells with PCEs of more than 2.5% [51]. Also, P3HT with a narrow distribution is desirable to achieve better device performance [52].

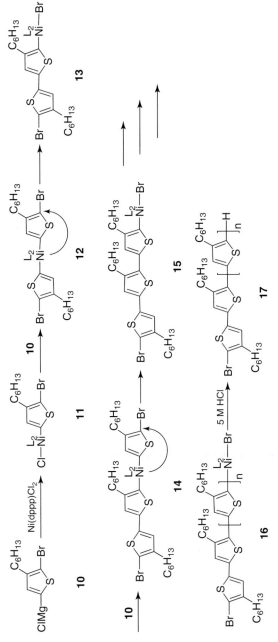

Scheme 17.3 Mechanism of nickel-catalyzed cross-coupling polymerization of P3ATs.

The length of the solubilizing side chains in P3HTs can also significantly influence the behavior of BHJ solar cells based on P3AT donor materials [53, 54]. Nguyen *et al.* reported that side chains with lengths greater than eight carbons would facilitate the diffusion of $PC_{61}BM$ in the polymer matrix during thermal annealing, which results in pronounced phase separation, reduced interfacial area, and thus poorer device performance. On the other hand, poly(3-butylthiophene) lacks sufficient solubility in common organic solvents and thereby suitable processability. As a result, only P3HT can effectively balance the solubility and phase separation behavior in P3ATs and ultimately show the best device performance [53].

Generally, as-cast active layers based on P3HT exhibited poor device performance even though high-quality polymers were used. To improve further the device performance of P3HT-based solar cells, a series of physical optimization methods such as thermal annealing and solvent annealing were also developed, in addition to "molecular" optimization of P3HT. By thermal annealing of the $P3HT:PC_{61}BM$ blend film at 150 °C after the deposition of the cathode, the state-of-the-art solar cells showed PCEs of up to 5.1% [20]. The improved device performance was ascribed to the formation of a nanoscale bi-continuous penetrating network in the active layer, which results in more efficient charge dissociation at the interface and better transport of holes and electrons in individual phases [20]. A similar improvement in device performance was also achieved through solvent annealing or controlling active layer growth rates of $P3HT:PC_{61}BM$ blend film by Yang's group [55]. The formation of a stable nanoscale bi-continuous penetrating network with a maximum interfacial area for exciton dissociation and an average domain size commensurate with the exciton diffusion length ($\sim$10 nm) in the active layer is the ultimate goal of all kinds of morphology optimization methods.

17.2.1.2 Push–Pull Copolymers

One of prerequisites to obtaining a high PCE of BHJ solar cells is that the photoactive materials can harvest as much solar radiation as possible which has a maximum photon flux density in the range 600–700 nm. However, the optical energy bandgaps (E_g) of PPVs and P3ATs are too large to cover this range of solar radiation. Hence the exploitation of new polymeric materials with lower E_g and suitable HOMO and LUMO levels is a critical challenge to improve further the PCEs of polymeric BHJ solar cells. It is now well known that several approaches can reduce the E_g of a conjugated polymer. The first approach is to increase the quinoid population in a polyaromatic conjugated system, which will reduce the bond length alternation and bandgap. The second approach is to enhance the planarization of adjacent aromatic units, which will extend the conjugated length and prompt electron delocalization and consequently reduce the bandgap. The third is introduction of electron-donating units (D) and electron-withdrawing units (A) alternately on the polymer backbone so that the electron delocalization and quinoid population can be enhanced through the push–pull-driven ICT. This D–A molecular design is the most widely studied and most powerful strategy for manipulating the optical and electrical properties of conjugated polymers. Both the bandgap and the energy levels can be tuned by proper selection of the donor unit and acceptor unit. The principle of bandgap reduction by ICT in conjugated polymers can be easily understood by molecular orbital theory, as depicted in Figure 17.4. After covalent bond connection, the HOMO

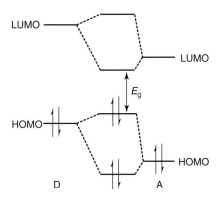

Figure 17.4 Molecular orbital interactions of donor and acceptor units resulting in a lower bandgap in a D–A conjugated polymer.

of the donor moiety will interact with that of the acceptor moiety to generate two new HOMOs, one of them higher and the other one lower than the two initial HOMOs before molecule orbital hybridization. Similarly, two new LUMOs would form after molecule orbital hybridization, one lower and the other higher than the two initial LUMOs of the two moieties. Thus, the overall effect of this redistribution is the formation of a higher positioned HOMO and a lower positioned LUMO in the whole conjugated main chain. As a consequence, the bandgap of the resulting polymer will be effectively narrowed.

The development of transition metal-catalyzed carbon–carbon bond formation reactions greatly facilitated the construction of conjugated polymers for BHJ solar cell application. The originally used carbon–carbon bond formation reactions for the synthesis of conjugated polymers, such as electrochemical and chemical oxidation polymerizations, usually suffer from the ready formation of crosslinked byproducts and defects along the conjugated backbone. The transition metal-catalyzed cross-coupling reaction possesses several advantages, such as more precise control of molecular structures, mild reaction conditions, and tolerance of many functional groups. These benefits of transition metal-catalyzed carbon–carbon coupling reactions provide the convenience of preparing ideal low-bandgap donor materials for BHJ solar cells. Up to now, the most widely utilized transition metal catalysts have been derived from nickel and palladium complexes, although other metals have also been used. Nickel complexes have been extensively used to prepare regioregular polythiophenes, which has been discussed previously. The two most important palladium complex-catalyzed cross-coupling reactions are the Suzuki reaction and Stille reaction. Generally, organic boron and tin reagents are used as nucleophiles to react with electrophilic aryl halides, respectively. The general mechanism of these two transition metal-catalyzed coupling reactions involves oxidative addition, transmetallation, and reductive elimination cycles, as presented in Figure 17.5. Taking the Suzuki reaction as an example, the oxidative addition of aryl halides to a palladium(0) complex yields a palladium(II) complex, which then undergoes transmetallation with an organoboronic nucleophile, followed by reductive elimination of the organic partners resulting in carbon–carbon bond formation and recovery of Pd(0) [56]. It is noteworthy that Suzuki coupling is generally used for polymerizing monomers with boronic groups on the benzene ring, whereas Stille coupling is more widely used for polymerizing monomers with stannyl groups on the thiophene ring. The general synthetic routes of

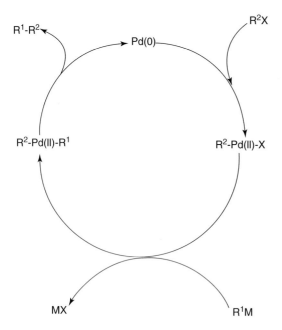

Figure 17.5 Catalytic cycles of transition metal-catalyzed reactions.

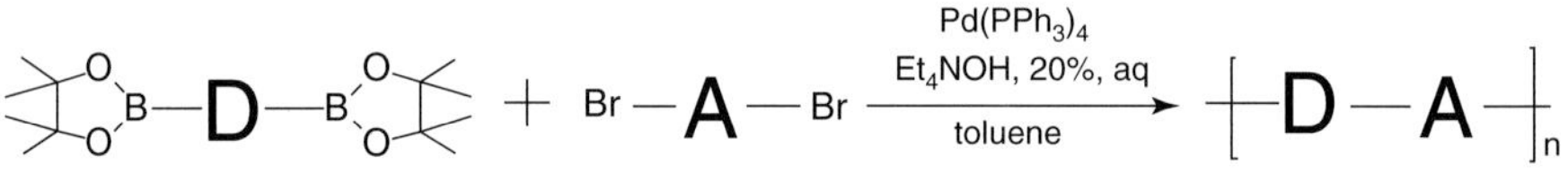

Scheme 17.4 General synthetic route to copolymer via the Suzuki coupling reaction.

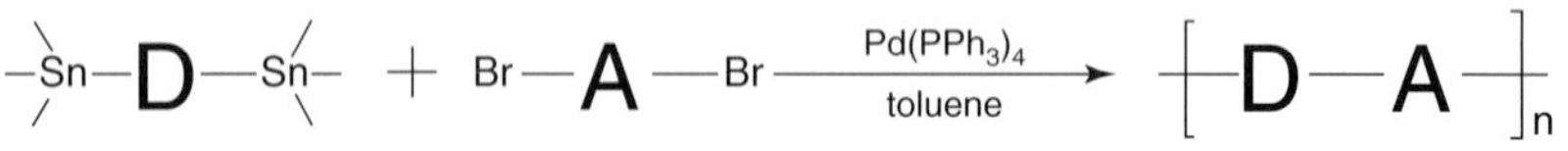

Scheme 17.5 General synthetic route to copolymer via the Stille coupling reaction.

Suzuki coupling and Stille coupling reactions are showed in Schemes 17.4 and 17.5, respectively.

Based on Suzuki coupling reactions, large amounts of low-bandgap conjugated polymers with alternate donor moiety (such as fluorene, silafluorene, carbazole, indenefluorene, indolocarbazole, etc.) and acceptor moiety (such as benzothiadiazole, benzoselenadiazole, quinoxaline, thienopyrazine, diketopyrrolopyrrole, etc.) were developed (Figure 17.6), some of which exhibited promising photovoltaic properties. Fluorene is one of the most widely used segments for constructing organic optoelectronic functional materials for all kinds of device applications, such as PLEDs, OFETs, polymer lasers, and BHJ solar cells. When copolymerized with electron-deficient monomers by Suzuki coupling reactions, the bandgap and energy levels could be readily tuned for fluorene-based D–A conjugated polymers [57]. One advantage of fluorene-based D–A

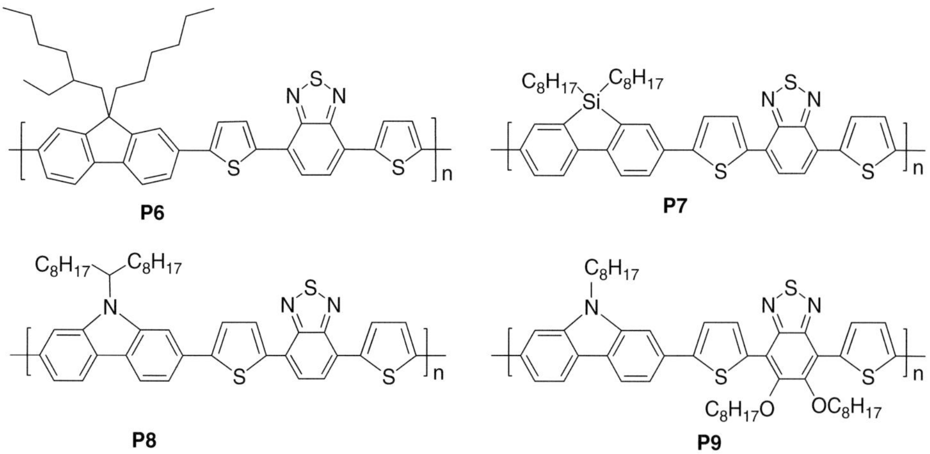

Figure 17.6 Chemical structures of low-bandgap copolymers synthesized by the Suzuki coupling reaction.

low-bandgap polymers is their deep HOMO levels, which is beneficial for obtaining a high V_{oc} for the resulting BHJ solar cells [57]. For example, BHJ solar cells based on **P6**:$PC_{61}BM$ showed a V_{oc} as high as 1.04 V and a PCE of 2.2% [58]. However, the overall device performances of fluorene-based polymers are not very good. When the fluorene at the C9 position is replaced by a silicon atom, a new class of building block, namely, silafluorenes emerges. For example, the alternate D–A copolymer **P7** of silafluorene and dithienylbenzothiadiazole prepared by Suzuki coupling of the corresponding diboronic and dibromide monomers showed enhanced photovoltaic performance. **P8** has a HOMO level of −5.39 eV, an E_g of 1.82 eV, and a field effect transistor (FET) hole mobility of 3×10^{-4} cm^2 V^{-1} s^{-1}. Solar cells fabricated from **P7**:$PC_{61}BM$ (1:2 w/w) blended composite showed a PCE of up to 5.4% with a V_{oc} of 0.9 V, a J_{sc} of 9.5 mA cm^{-2}, and an FF of 0.507 [59].

Carbazole-derived D–A-type low-bandgap conjugated copolymers are another important class of donor materials synthesized by Suzuki coupling reactions. The deep HOMO energy levels, high hole transport ability, and good chemical and environmental stability of carbazole-based polymers make them as promising donor materials for BHJ solar cells. Two typical examples are **P8** and **P9**, which are alternating copolymers based on 2,7-carbazole and dithienylbenzothiadiazole or alkyloxy-substituted dithienylbenzothiadiazole [60–62]. BHJ solar cells made from **P8**:$PC_{61}BM$ blend showed a PCE of 3.6% along with a J_{sc} of 6.92 mA cm^{-2}, a V_{oc} of 0.89 V, and an FF of 0.63. Moreover, a much higher PCE of 6.1% of **P8**-based solar cells was reported by Heeger's group, using $PC_{71}BM$ as the acceptor and titanium sub-oxide (TiOx) as the optical spacer [61]. Soon afterwards, a series of polycarbazole derivatives containing different acceptor units were also developed via the Suzuki coupling reaction by Leclerc's group, and the relationship between polymer structure and photovoltaic properties was systematically studied [63]. In 2009, Qin *et al.* reported another high-performance carbazole-based low-bandgap copolymer **P9** obtained via the Suzuki coupling reaction. **P9** has a planar configuration, which is beneficial for

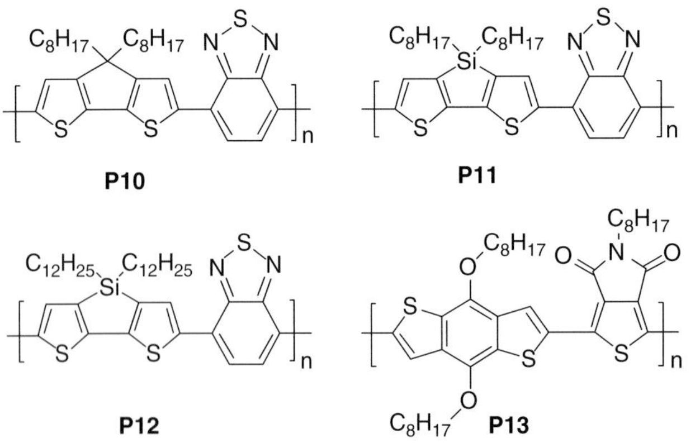

Figure 17.7 Chemical structures of low-bandgap copolymers synthesized by the Stille coupling reaction.

molecular packing in films to facilitate charge carrier transport. BHJ solar cells fabricated from **P9**:PC$_{61}$BM blend showed a high PCE value of 5.4% [62].

Alternatively, the Stille coupling reaction has also been extensively used to build D–A low-bandgap polymers, especially thiophene derived polymers (Figure 17.7). For example, copolymer **P10**, which was synthesized from alkalinized cyclopenta[2,1-*b*:3,4-*b*′]dithiophene (CPDT) and benzothiadiazole via the Stille coupling reaction, has an E_g as low as 1.46 eV. This low bandgap indicates an efficient ICT interaction between CPDT and benzothiadiazole units. A device based on **P10**:PC$_{71}$BM blend showed a high PCE of 3.5% with a J_{sc} of 11.8 mA cm^{-2}, which is mainly attributed to the extended absorption and high hole mobility [64]. Based on the same polymer, Bazan and co-workers further improved the PCE to 5.5% by incorporating a trace amount of additive in the **P10**:PC$_{71}$BM mixture before spin coating, which dramatically optimized the morphology of the photoactive layer [65]. Structurally similar to **P10**, another promising D–A low-bandgap polymer, **P11**, where the bridging atom at the 7-position was changed from carbon atom to silicon atom, was also developed via the Stille coupling reaction. The E_g of **P11** is 1.45 eV, which is close to the optimum bandgap for donor polymers. A BHJ solar cell based on **P11**:PC$_{71}$BM showed a PCE of 5.1% with a J_{sc} of 12.7 mA cm^{-2} and a V_{oc} of 0.68 V [66]. Similarly to P3HT, the device performances of these materials are largely dependent on their molecule weight. Bazan's group developed a series of **P11**s with different molecular weights through microwave-assistant synthesis in combination with screening of the comonomer reactant ratios [67]. Thus, the relationship between the molecular weight of the polymers and their final device performances could be studied. It was clearly shown that the J_{sc} of the resulting devices increased substantially from 4.2 to 17.3 mA cm^{-2}, accompanied by a similarly increase in PCE from 1.2 to 5.9% when the number molecular weight of the polymers increased from 7 to 34 kg mol^{-1}. This work demonstrated the importance of the molecular weight of D–A low-bandgap polymers on their final device behavior and the advantages of new synthetic techniques (such as microwave heating) for obtaining high molecular weight polymers [67].

Benzo[1,2-*b* : 3,4-*b'*]dithiophene (BDT) is another promising building block for the construction of low-bandgap conjugated polymers and has been extensively studied [68, 69]. The symmetry and planarity of the BDT core are perfect, which favors the efficient molecular packing and improved charge carrier transport abilities of the resulting copolymers based on it. Recently, several groups have reported the combination of BDT with thieno[3,4-*c*]pyrrole-4,6-dione (TPD) acceptor to construct the D−A low-bandgap conjugated polymer **P13** via the Stille coupling reaction for BHJ solar cell applications. BHJ solar cells based on **P13** showed PCEs varying from 4.2 to 6.8% with high V_{oc}s of over 0.85 V [70−73].

As discussed above, we can see that the photovoltaic properties of the conjugated polymers were greatly influenced by the building heterocycles. For example, when the C9 fluorene atom was replaced by a silicon or nitrogen atom, or the C7 atom of CPDT was replaced by a silicon atom, the resulting polymers showed pronounced differences in device performance, even though they have similar chemical structures, optical physical properties, and energy levels. This indicates that the rational choice and development of aromatic heterocycles is of practical significance for obtaining outstanding donor materials for BHJ solar cell applications.

17.2.1.3 Conjugated Polymers with Pendant Conjugated Side Chains

With respect to the exploration of D−A-type low-bandgap conjugated polymers, the extension of the absorption spectrum of polythiophene by rational structural modification is also a straightforward approach. Li's group reported a series of polythiophene derivatives with conjugated side chains [such as bi(thienylenevinylene)] (Figure 17.8) to extend the absorption of thiophene-based polymers [74−76]. For example, **P14** shows a broad and intense absorption from 350 to 650 nm, which is broader than that of P3HT. Moreover, the HOMO energy level of polythiophenes with bi(thienylenevinylene) conjugated side chains decreased by about 0.2 eV with respect to that of P3HT, which is beneficial for obtaining a higher V_{oc} of the resulting BHJ solar cells. When **P14** was blended with $PC_{61}BM$ with a weight ratio of 1:1 as the active layer, the resulting solar cell device exhibited a PCE of 3.18%, which was 38% higher than that of a P3HT-based device under the same conditions [74]. Clearly, attaching conjugated side chains on polythiophenes

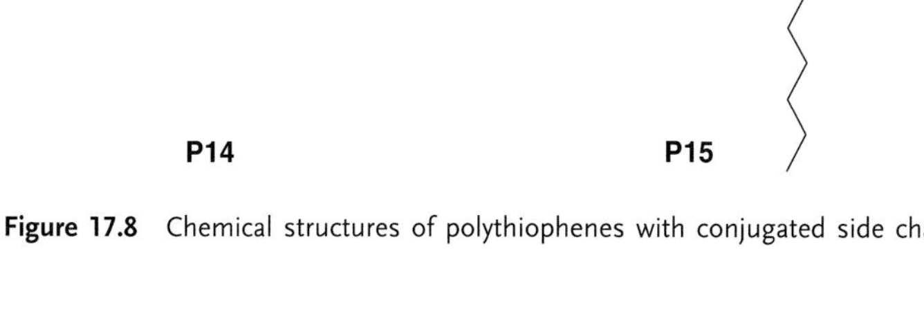

P14   **P15**

Figure 17.8 Chemical structures of polythiophenes with conjugated side chains.

is a simple and efficient approach to enhancing the PCE of polythiophene-based BHJ solar cells. Furthermore, when bi(thienylenevinylene) conjugated side chains were incorporated in poly(thienylenevinylene) derivatives, a much broader absorption band from 350 to 780 nm was achieved for **P15**. However, the performances of the corresponding devices were poor [75].

An alternative strategy for developing low-bandgap conjugated polymers is the design and synthesis of conjugated polymers with pendant D–π–A chromophores as the side chains. In this kind of material, the electron-withdrawing acceptors are attached to the end of the side chains and connected with the electron-donating main chains through the π-bridge, thereby forming a pendant D–π–A side chain. Hence the absorption and bandgap of the polymers could be controlled by manipulating the ICT strength of the pendant side chains, which can be readily tuned by changing the donor/acceptor groups or the π-bridge. In addition, the HOMO levels of the polymers are determined by their main chain, whereas the LUMO levels are determined by the pendant acceptor. Hence the deep HOMO level could be preserved and a high V_{oc} of the resulting BHJ solar cells can be expected if deep HOMO level moieties such as fluorene, silafluorene, and carbazole are incorporated as building blocks on the backbone. Moreover, the two-dimensional-like structural feature of these materials renders good solubility and better isotropic charge-transporting ability compared with traditional linear D–A polymers. Based on this design concept, a variety of novel low-bandgap polymers were developed [77–82]. As shown in Scheme 17.6, an aldehyde-functionalized conjugated polymer **P16** was synthesized from 2,7-fluorene diboronate and a triphenylamine-based dibromide under common Suzuki coupling conditions. The resulting polymers were obtained by a Knoevenagel condensation between the precursor polymer **P16** and different acceptors, such as malononitrile (for **P17**) and diethylthiobarbituric acid (for **P18**), in the presence of pyridine. Both **P17** and **P18** are amorphous and have good solubility in common organic solvents such as chloroform, chlorobenzene, and toluene. As expected, the E_g values were successfully tuned from 1.87 eV for **P17** to 1.76 eV for **P18**, and the HOMO levels were kept at about −5.3 eV by using different acceptors. A PCE of 4.74% was achieved for solar cells based on **P17**:PC71BM blend with a V_{oc} of 0.99 V, a J_{sc} of 9.62 mA cm^{-2}, and an FF of 0.5. The highest PCE value of **P18**:PC71BM solar cells also reaches 4.37% [77]. Later, a series of other D–π–A chromophore-pendent conjugated polymers based on silafluorene, carbazole, phenothiazine, and cyclopentadithiopyrrole (Figure 17.9) were developed by several groups, and some of them also showed promising photovoltaic properties with PCEs of more than 3% [78–82]. Considering its flexibility in fine-tuning the absorption spectra and energy levels of the resultant polymers, this approach provides a new strategy for exploring low-bandgap conjugated polymers for BHJ solar cell applications.

17.2.1.4 Block Conjugated Copolymers

In addition to the intrinsic optical and electronic properties of materials, it is well known that a nanostructured morphology of the active layer with an exciton diffusion length comparable to the domain size and as large an interfacial area as possible is also very important for achieving high-performance BHJ solar cells. Block copolymers made from two or more covalently connected polymer building blocks can spontaneously form

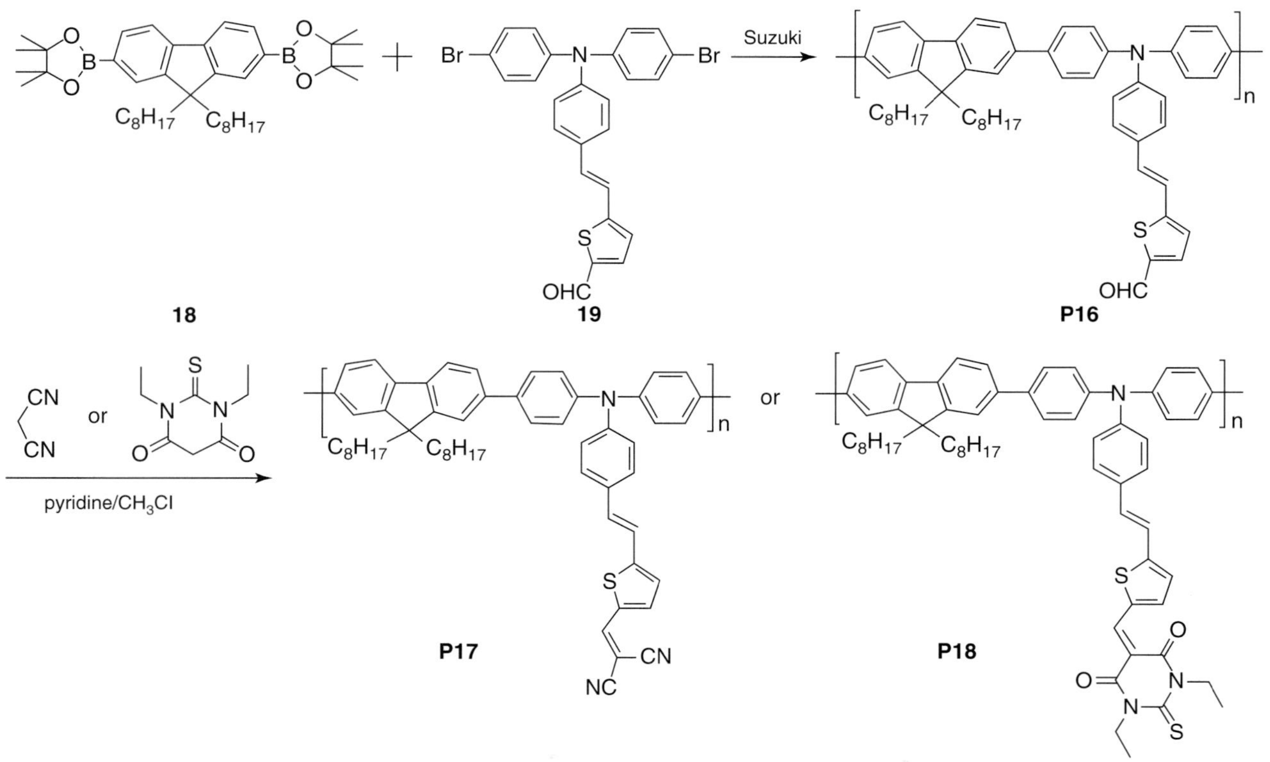

Scheme 17.6 General synthetic route to D–π–A chromophores pendent on conjugated low-bandgap polymers.

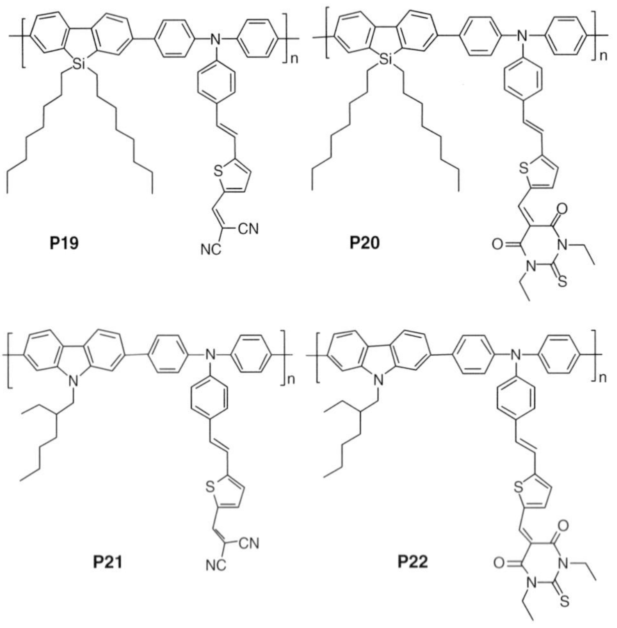

Figure 17.9 Chemical structures of D–π–A chromophores pendent on conjugated low-bandgap polymers.

numerous phase-separated microstructures and nanostructures through tuning the block composition, which can be achieved via chemical methods in molecular dimensions [83], and have shown application potential in many fields, such as surface modifications and compatibilizers of polymer blends. Thus, block copolymers can also be used to control the morphology of the active layer of BHJ solar cells to obtain ideal phase separation for high-performance solar cells [15].

The application of block conjugated copolymers in BHJ solar cells can so far be divided into three main aspects. First, block copolymers have been used as donors with multiple functionalities (e.g., crystallinity, self-assembly, solubility) in BHJ solar cells. For example, Jenekhe and co-workers reported a series of diblock conjugated copolymers, poly(3-butylthiophene)-*b*-poly(3-octylthiophene) (**P23**) (Figure 17.10), which aimed to combine the good crystallinity of polybutylthiophene and the good solubility of polyoctylthiophene [84]. Atomic force microscopy and transmission electron microscopy studies revealed that **P23**:PC$_{71}$BM blend composites had an interpenetrating network with a crystalline polymer domain, which consequently led to a substantially enhanced charge-transporting ability with respect to traditional homopolymers. BHJ solar cells made from **P23** with an equivalent building block showed PCEs of up to 3.0%, which is 1.6–1.9 times higher than those of polybutylthiophene and polyoctylthiophene homopolymer devices fabricated under the same conditions [84].

Further, block copolymers have been used to integrate both electron donors and acceptors on one polymer chain so that the phase aggregation size of donors and acceptors in BHJ solar cells can be controlled at the molecular scale. Thelakkat's group

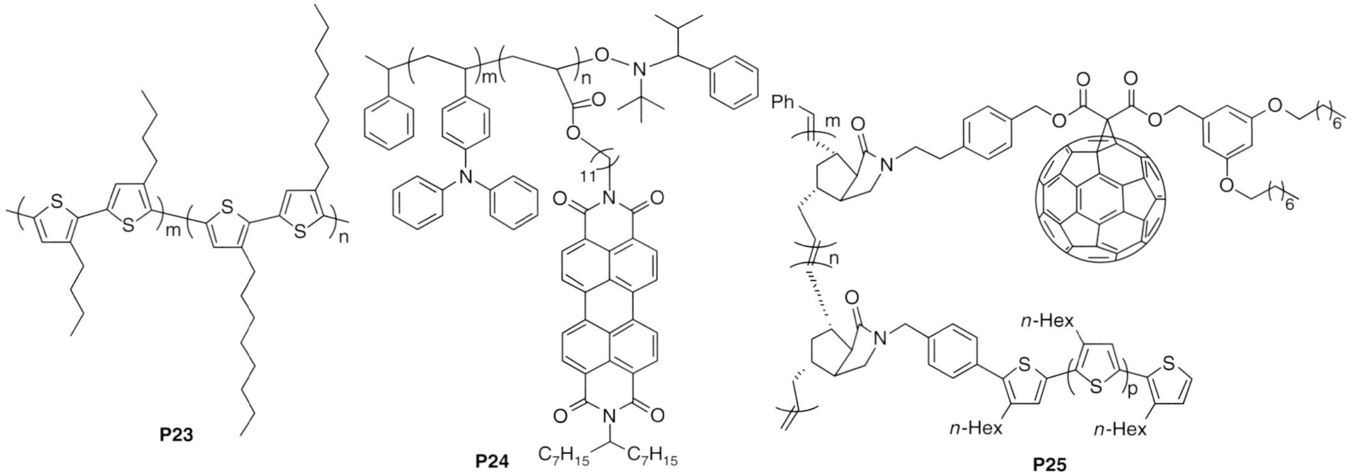

P23 P24 P25

Figure 17.10 Chemical structures of block copolymers of **P23**, **P24**, and **P25**.

integrated both donors and acceptors on the side chain of a block copolymer (**P24**) (Figure 17.10), in which the donor and acceptor moieties were triphenylamine and bisimide acrylate, respectively [85]. The copolymer self-assembled into a nanostructure with a domain size of 10–50 nm to provide as large an interfacial area as possible on the nanoscale for charge separation. A single active layer solar cell of **P24** showed considerably improved performance compared with the corresponding device based on a blend of donor and acceptor polymers [85]. Later, solar cell devices with an improved PCE of 0.32% based on similar block polymers were reported [86]. Russell and co-workers synthesized poly(3-thiophene-*b*-perylenediimide) diblock copolymer and obtained the highest PCE of 0.49% for the corresponding solar cells [87]. Another notable example is a donor–bridge–acceptor–bridge-type block copolymer with both donor and acceptor blocks along the main chain, developed by Sun *et al.*, which presented a dramatically improved performance relative to those of the devices based on the corresponding blend [88].

In addition, block copolymers can be also used as compatibilizers to stabilize the nanoscale morphology of P3HT:PC$_{61}$BM and P3HT:perylenediimide (PDI) blends. In 2006, the amphiphilic diblock copolymer **P25** was developed by Fréchet's group and used as a compatibilizer to reduce the interfacial energy between P3HT and PC$_{61}$BM [89]. On loading 17% of **P25** into P3HT:PC$_{61}$BM blends, no detectable phase separation in the blend film was observed after thermal annealing, whereas pristine P3HT:PC$_{61}$BM blend film showed clearly visible micrometer-sized phase segregation. This reduced phase separation directly resulted in a stable device performance even after thermal treatment [89]. Later, a diblock copolymer, synthesized by "reversible addition–fragmentation chain transfer," was also used as a compatibilizer to control the phase separation of P3HT:PDI blends. With the addition of the compatibilizer, the phase-separated domains of P3HT:PDI blends were smaller than those of P3HT:PDI blends without compatibilizer. Correspondingly, the device performance was also significantly enhanced with the addition of compatibilizer [90]. Similar work was also reported by Wudl and co-workers, who developed rod–coil block copolymers containing P3HT and PC$_{61}$BM using a surfactant to control the nanoscale morphology of P3HT:PC$_{61}$BM blends. On adding 5% of surfactant, a BHJ solar cell showed a PCE of 3.5%, which is about 35% higher than that of the control device without the surfactant [91].

17.2.2
Solution-Processed Small-Molecule Donor Materials

With respect to polymers, small-molecule materials show increased potential in BHJ solar cell applications because of their advantages of ease of synthesis and purification, good reproducibility, and precise molecular structures. It was generally perceived that small molecules cannot form a uniform film and bicontinuous interpenetrating network in combination with fullerene derivative acceptors by a solution process and in turn will produce a poor device performance. However, recent progress of this field showed that solution-processed small-molecule materials can also afford modest photovoltaic performances. A large number of small molecules have been developed, including discotic molecules (such as phthalocyanine, porphyrin, and hexabenzocoronene derivatives)

[92–94], thiophene oligomers and dendrimers [95, 96], soluble fused polycycles [97], push–pull chromophores [98–101], and organic dyes [102, 103]. As several reviews have already systematically discussed the progress in this area [104, 105], we shall just briefly introduce several impressive examples in this section.

To date, the best performing small-molecule donor material is a dihydropyrrolopy-rrolodione (DPP)-based molecule (**20**) (Figure 17.11a) with a terminal benzofuran. DPP is a popular building block for low-bandgap conjugated polymers owing to its easy synthesis, strong light absorption ability and photochemical stability. Via introduction of the terminal benzofuran substituent, the conjugation length and ICT effect of **20** were increased and its absorption was extended. Interestingly, the phase separation behavior of **20**:PC$_{71}$BM could be controlled by thermal annealing. A BHJ solar cell based on the **20**:PC$_{71}$BM system showed a V_{oc} of 0.9 V, a J_{sc} of 10 mA cm^{-2}, an FF of 0.48, and a PCE of 4.4% [99].

In 2009, Nakamura's group reported a trilayer p–i–n OPV device using silyl-methyl[60]fullerene (**21**) and a solution-processable small-molecule tetrabenzoporphyrin precursor (**22**) (Figure 17.11b,c) [93]. The trilayer device was achieved by first spin-coating a solution of tetrabenzoporphyrin precursor, which was thermally converted to an insoluble p-layer. Subsequently, the mixture of donor and acceptor was spin-coated and thermally transformed to form an interdigitated i-layer, which was then covered by an n-layer. Owing to the formation of defined nanoscale columnar BHJ morphology in the

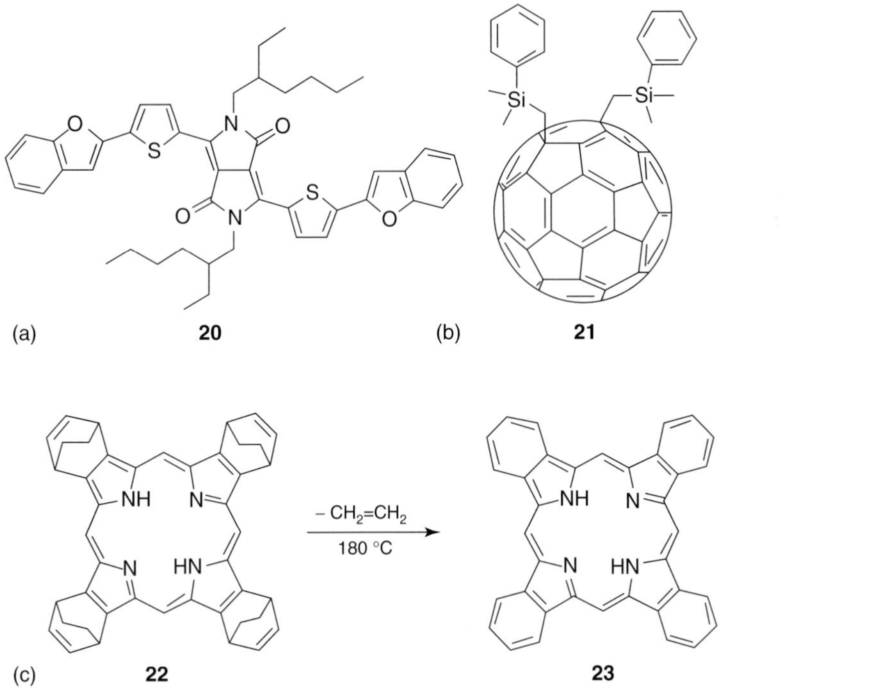

Figure 17.11 (a) Chemical structure of small-molecule donor material **20**. (b) Chemical structure of acceptor material **21**. (c) Thermal retro-Diels–Alder conversion of **22** to **23** at 180 °C.

i-layer, which was revealed by scanning electron microscopy images, the device showed an impressive performance of 5.2% PCE, along with a J_{sc} of 10.5 mA cm^{-2}, a V_{oc} of 0.75 V, and an FF of 0.65 [93]. The key point of this work is the use of thermocleavable small-molecule donor materials to form a stable nanostructure after thermal treatment. So far, this work represents the best result for solution-processed small-molecule solar cells, and provides a prototype device to improve further the efficiency of small-molecule-based solar cells.

17.3
Molecular Design and Engineering of Acceptor Materials

In order to improve substantially the PCE of BHJ solar cells, the importance of the development of electron-acceptor materials is similar to that of light-harvesting electron donor materials. The excellent intrinsic optoelectronic properties (such as excellent electron-accepting capabilities and good electron-transporting abilities) of fullerenes and their derivatives due to their electronegative nature endow them with promising potential as electron-acceptor materials for BHJ solar cell application. Especially the development of the chemistry of the fullerenes allows the precise tailoring of new fullerene derivatives with the desired electronic and chemical properties for BHJ solar cell applications. At present, fullerene derivatives are still the most widely used electron-acceptor materials. In addition, some other n-type materials have also been reported as electron acceptors for BHJ solar cell applications.

17.3.1
Fullerene-Based Acceptors

After the achievement of the synthesis of fullerenes in macroscopic quantities in the 1990s, many chemical modification methods have been developed to explore the special electronic and photophysical properties of fullerenes in various practical applications. It is well known that fullerenes have a large spherical unsaturated carbon network without any hydrogens. This structural characteristic makes the addition reactions and redox reactions on the outside of fullerenes the most effective methods to realize desirable chemical transformations. Among the many kinds of chemical modifications of fullerenes, cycloaddition reactions have been the most intensively investigated and are the most useful methods for obtaining materials with the requisite properties as almost any functional group can be covalently linked to the fullerenes by cycloadditions of suitable addends. Among the most important cycloadditions, [3 + 2]-cycloadditions with 1,3-dipoles and [4 + 2]-cycloadditions such as Diels−Alder reactions are the most widely used methods to synthesize fullerene derivatives as outstanding electron acceptors for BHJ solar cells.

The [3 + 2]-cycloaddition of a diazo compound with a 1,3-dipole to fullerenes involves the formation of intermediate pyrazolines followed by thermal extrusion of N_2 to give two isomeric products, [5,6]-open fulleroid and [6,6]-closed methanofullerene. The former can readily transformed into the latter, which is thermodynamically stable, by thermal or

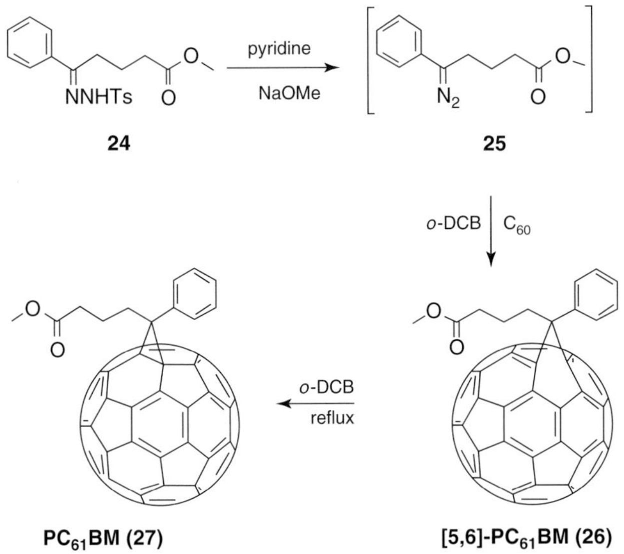

Scheme 17.7 Synthetic route to PC$_{61}$BM.

photochemical means [106, 107]. In 1995, Wudl's group synthesized a fullerene derivative PC$_{61}$BM by a [3 + 2]-cycloaddition reaction, as shown in Scheme 17.7 [108]. They prepared the stable diazo compound **25** *in situ* without purification, and then [5,6]-PCBM (**26**) was prepared via a 1,3-dipolar addition between **25** and C$_{60}$. Subsequently, the conversion of **26** to the [6,6]-closed methanofullerene (**27**) was accomplished by heating. The good solubility in aromatic solvents, strong electron-accepting capability, and excellent electron mobility of PC$_{61}$BM immediately made it the most widely used electron acceptor in BHJ solar cells [7, 20, 109]. PC$_{61}$BM is still considered to be the most successful acceptor.

By side-chain chemistry, a large number of PC$_{61}$BM analogs with different side chains were developed via [3 + 2]-cycloadditions to meet different requirements for photovoltaic applications. In order to improve the miscibility with the donor polymer and optimize the morphology of the active layer of polymer solar cells, Cao's group synthesized a series of PC$_{61}$BM analogs **28** with different ester alkyl chains varying from C$_1$ to C$_{16}$ under similar conditions to PC$_{61}$BM (Figure 17.12). The best device performance was achieved with a butyl-substituted derivative blended with MEH-PPV [110]. In addition, the effects of carbon chain length of PC$_{61}$BM-like molecules on the photovoltaic properties were also investigated by changing the distance of the butyl group of PC$_{61}$BM from three to seven carbon atoms (**29**) (Figure 17.12) [111]. Aimed at improving miscibility with P3HT, a PC$_{61}$BM-like fullerene derivative (**30**) with a thienyl group was synthesized via [3 + 2]-cycloaddition under similar conditions to PC$_{61}$BM (Figure 17.12) [112]. To suppress the crystallization-driven phase separation of the polymer:fullerene BHJ film, much more amorphous fullerene derivatives (**31** and **32**) with a bulkier triphenylamine or 9,9-dimethylfluorene instead of a planar phenyl group-substituted side chain were developed via side-chain chemistry (Figure 17.12). The thermal stabilities of corresponding solar cell devices were greatly enhanced [113]. A

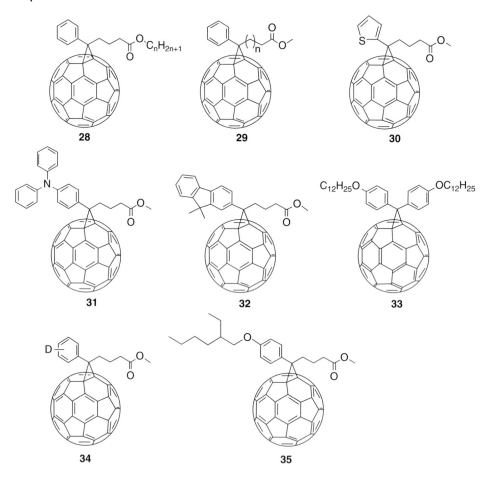

Figure 17.12 Chemical structures of C_{60} derivatives synthesized from [3 + 2]-cycloaddition reactions.

symmetrical $PC_{61}BM$-like fullerene derivative **33** (Figure 17.12) containing two benzene rings on the methanofullerene synthesized by [3 + 2]-cycloaddition exhibited interesting photovoltaic behavior. The V_{oc} of **33**-based devices is 0.1 V above that of the $PC_{61}BM$-based devices under the same conditions despite the identical HOMO energy level values of **33** and $PC_{61}BM$ [114]. The influence of electron-donating and -withdrawing groups on the phenyl ring of $PC_{61}BM$ on their electronic properties was successfully investigated, profiting from the facile synthesis of different side chains (**34** and **35**) (Figure 17.12) [115].

Like diazo compounds, azides can also be used as 1,3-dipoles for [3 + 2]-cycloaddition to prepare soluble fullerene derivatives for use as n-type semiconductors for BHJ solar cells. More interestingly, both the iminofullerene isomers [5,6]-open azafulleriod (**38**) and [6,6]-closed aziridinofullerene (**39**) could be obtained owing to the much more electronegative nitrogen atom that can stabilize the former (Scheme 17.8). Hence the

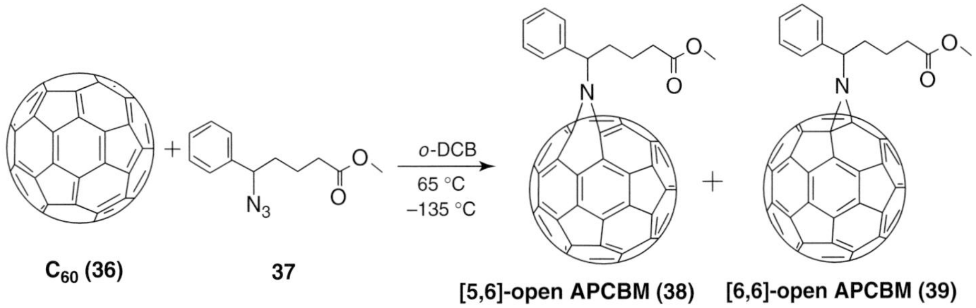

Scheme 17.8 Synthetic route to aziridinofullerene via [3 + 2]-cycloaddition reaction.

study of the influence of the nitrogen atom on the electronic properties of the fullerene cage and, more importantly, the comparative study between the 60 π-electron system and the 58 π-electron system of fullerene derivatives can readily be carried out. Interesting, solar cell devices based on **38** showed enhanced photocurrent and PCE compared with **39**, which could be attributed to the lower degree of symmetry of open **38** and consequently enhanced absorption [116, 117].

Higher fullerenes such as C_{70} and C_{84} can be also modified via 1,3-dipolar [3 + 2]-cycloadditions using a similar procedure to that for their C_{60} analogs such as $PC_{61}BM$ (Figure 17.13) [13, 35]. One of the advantages of higher fullerene derivatives compared with their C_{60} counterparts is that their low degree of symmetry makes the lowest-energy transitions become allowed and remarkably enhanced light absorptions are realized. Therefore, an enhanced photocurrent can be expected when $PC_{61}BM$ is replaced by $PC_{71}BM$ as acceptor [118]. Actually, almost all state-of-the-art devices based on new low-bandgap polymers make use of $PC_{71}BM$ instead of $PC_{61}BM$ as acceptor material as the absorption valley of low-bandgap polymers in the visible region could be effectively complemented by the latter while it has similar energy levels as the former [118]. To date, the reported best performing BHJ solar cell device which showed the highest PCE of 7.73% is a $PC_{71}BM$-based device [119]. Nevertheless, $PC_{85}BM$-based devices showed very poor performance, which was attributed to the lower LUMO energy levels, diminished solubility, and high unbalanced charge carrier mobility in $PC_{85}BM$:MOMO-PPV composite of these higher fullerene derivatives [35]. In addition, endohedral fullerene derivatives such as **42** have also been reported as acceptors for photovoltaic applications, and a much higher V_{oc} of the resulting BHJ solar cells than that of $PC_{61}BM$ was obtained [120]. It should be noted that the chemical reactivity of endohedral fullerene $Lu_3N@C_{80}$ is different from that of empty-cage fullerenes, and much more rigorous modification conditions are needed to obtain soluble derivatives in high yield. Anyway, the main obstacle to the large-scale photovoltaic application of higher fullerene derivatives and endohedral fullerene derivatives is their high cost and low abundance.

It should be pointed out that the chemical transformations of fullerenes always give rise to multi-adduct byproducts, such as bis-adducts, tris-adducts, and so on. However, multi-adducts of fullerene derivatives have not attracted much attention from device researchers, possibly because multi-adducts generally are a mixture of a number of regioisomers and there are difficulties with purification [121]. In 2008, Blom and

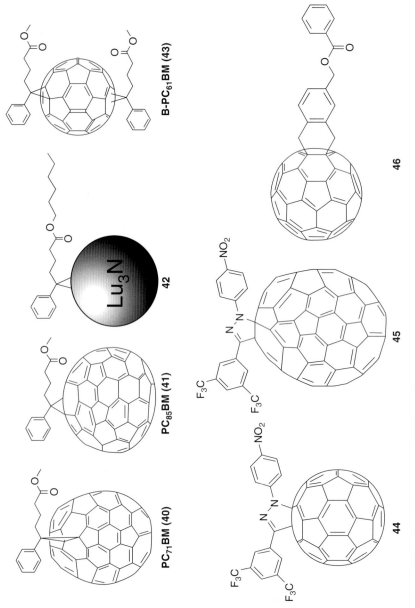

Figure 17.13 Chemical stuctures of fullerene derivatives.

co-workers first reported the use of the bis-adduct analog B-PC$_{61}$BM (**43**) of PC$_{61}$BM as an efficient acceptor material for BHJ solar cells [122]. The LUMO level of **43** is about 0.1 eV higher than that of PC$_{61}$BM, which results in an enhanced V_{oc} for the corresponding devices with P3HT as donor material. More importantly, it was reported that the bis-adduct isomer mixture can be used directly without further separation of the individual isomers, which is convenient for device applications. The highest PCE of 4.5% was achieved with a V_{oc} of 0.73 V for the **43**:P3HT blend, whereas the values for PC$_{61}$BM:P3HT were only 3.8% and 0.58 V, respectively.

Pyrazolino[60]fullerenes are generally the intermediates of 1,3-dipolar cycloadditions of 1,3-nitrile imines to fullerenes, and they have hardly been exploited owing to their poor thermal stability [123]. Inganäs's group first utilized these fullerene derivatives (**44** and **45**) (Figure 17.13) as electron acceptors in combination with an infrared (IR)-absorbing low-bandgap polymer as donor for BHJ solar cell applications [124]. They found that **44** showed better device performance than PC$_{61}$BM, which was ascribed to the much lower LUMO level of **44** than PC$_{61}$BM owing to the electronegative nature of the nitrogen atom linked to the C$_{60}$ cage [125] and, as a consequence, a better match of LUMO energy levels of donor and acceptor materials for efficient charge separation [124]. To compensate for the absorption valley of the IR-absorbing low-bandgap polymer in the visible region, C$_{70}$ derivative **45** was further developed for use as an electron acceptor material. Correspondingly, the J_{sc} value was improved from 1.76 mA cm^{-2} for **44** to 3.4 mA cm^{-2} for **45** [126].

The [6,6]-double bonds of fullerenes can act as dienophiles and therefore enable fullerenes to be functionalized through [4 + 2]-cycloadditions such as Diels–Alder reactions. However, the nature of the reversibility of Diels–Alder reactions restricted the practical application of fullerene-based materials derived from this kind of reaction, owing to their generally poor thermal stabilities of the resulting materials. One effective approach to obtain thermally stable Diels–Alder adducts of fullerenes is the use of *o*-quinodimethane derivatives as dienes, which can prepared *in situ* through several steps [106]. Fréchet's group first synthesized a series of C$_{60}$ derivatives as acceptors in photovoltaic applications based on Diels–Alder reactions [127]. The key compound was synthesized by the direct reaction of C$_{60}$ with an *o*-quinodimethane formed *in situ* by iodide-induced bromine elimination from methyl (3,4-dibromomethyl)benzoate. Through esterification of the dihydronaphthyl fullerene benzyl alcohol, a series of functional fullerene derivatives with electron-donating or electron-withdrawing side chains could be obtained and used to evaluate their photovoltaic application potential. The phenyl-substituted derivative **46** (Figure 17.13) presented the best device performance with a PCE of 4.5% in a blend with P3HT, which represented the best result for a non-PCBM-based BHJ solar cell at that time [127].

It was reported that the Diels-Alder [4 + 2]-cycloaddition reaction of C$_{60}$ with isoindene, generated thermally from indene *in situ*, can also result in thermally stable fullerene derivatives [128]. With a similar method, Li and co-workers synthesized an indene–C$_{60}$ bis-adduct IC$_{60}$BA (**49**) with excess of indene in 1,2,4-trichlorobenzene under reflux, as depicted in Scheme 17.9a. IC$_{60}$BA showed a better solubility in common organic solvents (>90 mg ml^{-1} in chloroform) and a 0.17 eV higher LUMO energy level than PC$_{61}$BM. More importantly, an exciting photovoltaic performance of the P3HT-based

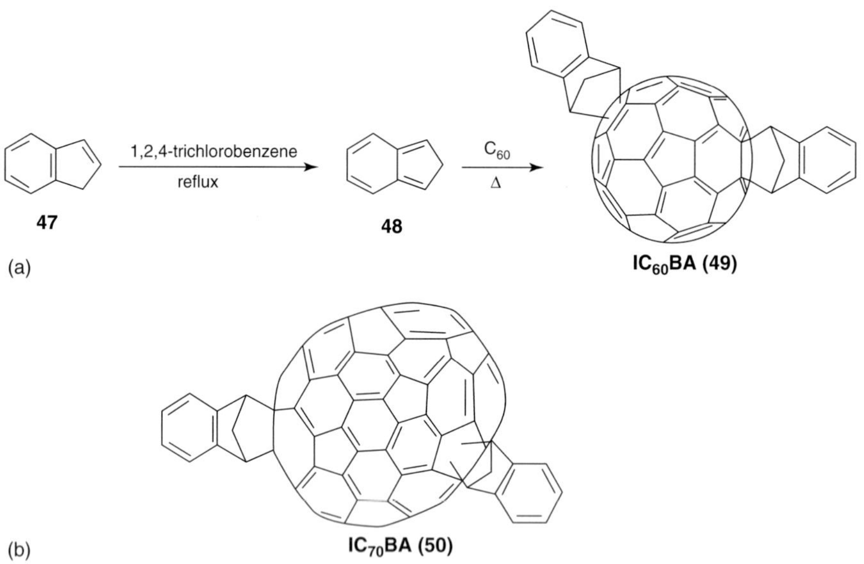

(a)

IC$_{60}$BA (49)

(b)

IC$_{70}$BA (50)

Scheme 17.9 (a) Synthetic route to IC$_{60}$BA via Diels–Alder [4 + 2]-cycloaddition reaction. (b) Chemical structure of IC$_{70}$BM.

solar cell with IC$_{60}$BA as acceptor was achieved. The PCE and V_{oc} of the IC$_{60}$BA-based device were 5.44% and 0.84 V, respectively, whereas those of a PC$_{61}$BM-based device were only 3.88% and 0.58 V. These results were consistent with those reported in a patent [129]. Obviously, this significantly improved device performance should be ascribed to the much higher LUMO level of ICBA in comparison with PC$_{61}$BM [12]. After device optimization with solvent annealing and prethermal annealing at 150 °C for 10 min, the P3HT:IC$_{60}$BA-based devices demonstrated a PCE as high as 6.48% with a V_{oc} of 0.84 V, a J_{sc} of 10.61 mA cm^{-2}, and an FF of 0.727 [130]. Soon afterwards, an indene-C$_{70}$ bis-adduct IC$_{70}$BA (**50**) (Scheme 17.9b) was also developed with a high yield of 58% via a similar method to that for ICBA in order to increase the visible absorption. The device based on P3HT:IC$_{70}$BA presented PCE and V_{oc} values of 5.64% and 0.84 V, respectively, whereas those of P3HT:PC$_{71}$BM were only 3.96% and 0.58 V, respectively [14]. Up to now, ICBA and IC$_{70}$BA represent the best performing acceptor system for P3HT-based solar cells, but there are no reports of more efficient solar cells with highly efficient low-bandgap polymers as donor based on this acceptor system.

In addition to cycloadditions, nucleophilic additions such as the Bingel–Hirsch reaction are another kind of important reaction used in the synthesis of methanofullerenes as acceptors for BHJ solar cells [131, 132]. The advantages of this method for modifying fullerenes are the easy preparation and wide selectivity of side chains and the high yield, as shown in Scheme 17.10. By changing the substituent groups of malonate, the PCE of the BHJ solar cells with MEH-PPV as donor was enhanced from 0.49 to 1.33%, which was ascribed to the improved compatibility between MEH-PPV and fullerene derivatives (**52**) [131, 132]. An interesting study was of a solar cell fabricated by layer-by-layer (LBL) alternating deposition of water-soluble anionic conjugated polyelectrolytes and cationic

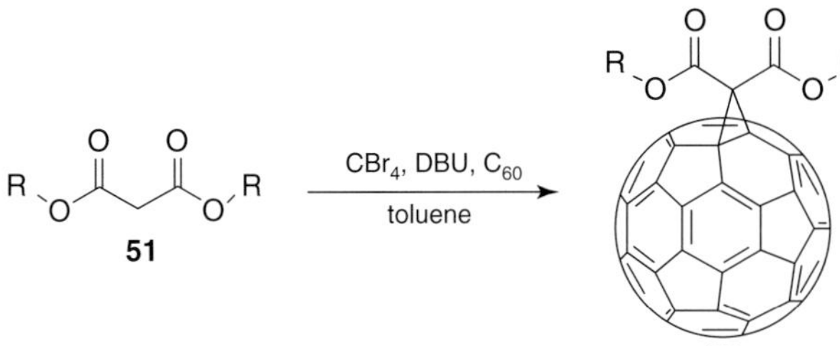

Scheme 17.10 General synthetic route for fullerene derivatives via the Bingel–Hirsch reaction.

C_{60} derivatives [133]. The introduction of two polar headgroups for every malonate rendered the required water solubility of C_{60} fullerenes and ultimately the possibility of fabrication of solar cells by LBL deposition. However, the device performance was poor, possibly due to the inefficient charge transport in these devices [133].

17.3.2
Non-Fullerene-Based Acceptors

Although fullerene derivatives have many excellent optoelectronic properties and even dominate the field of acceptors for BHJ solar cells, they still suffer from limited solubility, lengthy synthesis routes, low yields, and difficult purification processes. Therefore, the exploration of non-fullerene-based acceptors is necessary.

PDI is a typical building moiety for n-type semiconductor materials, which have good electron affinities and electron mobilities [134, 135], strong absorption in the visible and near-IR regions, and good optical and thermal stability. Hence PDI was readily used to construct acceptor materials, including PDI-derived small-molecule materials and polymers for OSC application. For example, **53** was used as an acceptor material in the first bilayer heterojunction solar cell reported by Tang in 1986 (Figure 17.14) [5]. In 2000, Friend's group fabricated a P3HT-based BHJ solar cell using **54** (Figure 17.14) as acceptor material. The external quantum efficiency (EQE) increased from 0.2% for a pristine P3HT device to 7% for the P3HT:**54** device at 495 nm, indicating an effective charge dissociation between P3HT and **54** [136]. Subsequently, they reported a solar cell with an EQE of 34% at 490 nm when **54** was blended with a discotic liquid crystalline hexa-*peri*-hexabenzocoronene derivative [94]. In 2007, Zhan *et al.* reported a PDI-derived n-type polymer (**P26**) with a high electron mobility of 1.3×10^{-2} cm^{-2} V^{-1} s^{-1} and broad absorption extending to the near-IR region. When blended with **P14**, an average PCE as high as 1% was achieved for a BHJ solar cell with **P26** as acceptor [137]. Other PDI-derived materials, such as two-cable polymers and block copolymers, have also been developed for solar cell applications [138, 139]. Notably, the performances of solar cells with PDI-derived acceptors are not comparable to those of fullerene-based devices at the present stage.

CN-PPVs (**P27**) (Figure 17.14) contain strong electron-withdrawing cyano groups on their conjugated vinylene bridges, which endow them with a low LUMO level and greatly increase the electron affinities and electron-transporting abilities. As a result, CN-PPVs

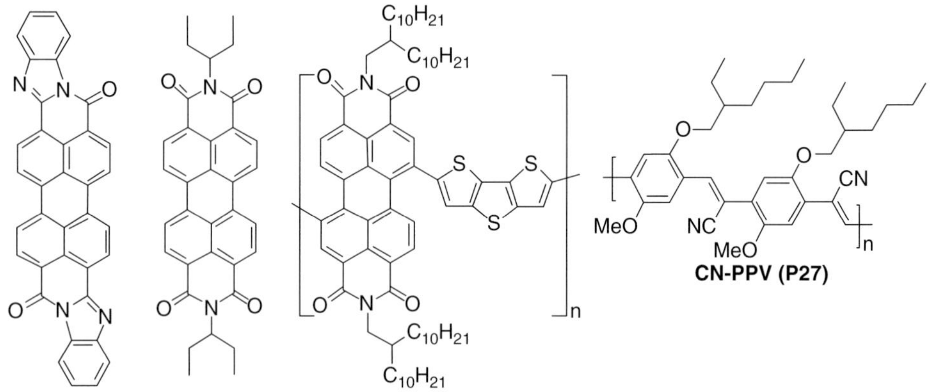

Figure 17.14 Chemical structures of PDI derivatives and CN-PPV.

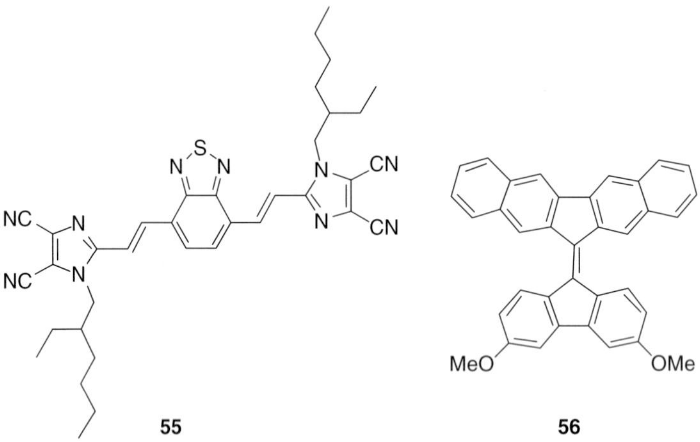

Figure 17.15 Chemical structures of non-fullerene-based acceptors.

are widely used as polymer-based electron acceptors in solar cells. For example, a bilayer device with CN-PPV as electron acceptor layer and poly[3-(4-octylphenyl)thiophene] as electron donor layer was successfully fabricated by a lamination technique followed by thermal treatment, and a PCE of up to 1.9% was achieved [140].

Sellinger and co-workers developed a series of non-fullerene-based acceptors through Heck coupling of 2-vinyl-4,5-dicyanoimidazoles with different dibromoaromatics (Figure 17.15). The optical properties and energy levels of this class of material could be tuned by choosing the proper aromatic core [141]. BHJ solar cells made from **55** and poly[N-(2′-decyltetradecyl)carbazole]-2,7-diyl showed a PCE of 0.75% and a V_{oc} of 1 V [142].

Recently, Wudl's group also reported a series of non-fullerene small-molecule acceptors based on a 9,9′-bifluorenylidene core, as shown in Figure 17.15 [143]. For the

9,9′-bifluorenylidene core, there is a strong repulsive interaction between the H1−H1′ and H8−H8′ protons in the ground state because owing the 9−9′ double bond strength, the dimer takes a coplanar conformation. When an electron is added to this dimer, the steric strain will be released and a 14 π-electron Hückel aromatic system will be obtained. Therefore, the 9,9′-bifluorenylidene core can be used to construct electron acceptors for BHJ solar cell applications. When **56** was used as electron acceptor with the combination of P3HT as donor material to fabricate BHJ solar cells, the highest PCE of around 2% with a V_{oc} of 1.1 V was achieved [143].

17.4
Conclusion and Outlook

Since the invention of BHJ solar cells, significant progress has been achieved, including the development of photoactive materials, device construction, and elucidation of device mechanisms, which has resulted in PCEs surpassing 8%. In this great achievement, the development of photoactive materials, including both donor and acceptor materials, has played an important role. The advances in organic chemistry are driving this field in many respects, such as the design and construction of promising donor polymers via various transition metal-catalyzed carbon−carbon coupling reactions, exploration of fullerene-based acceptor materials using various advanced fullerene modification methods, and design and preparation of block copolymers by living polymerizations to control the nanostructured morphology of the active layer of BHJ solar cells in molecular dimensions. In spite of the great achievements in this area, there are still many challenges to improve further the performance of BHJ solar cells with the aim of 10% PCE for commercial applications. The development of new materials, including low-bandgap polymers to harvest more solar radiation, the design of new soluble acceptors with ideal energy levels to match donor materials, and the synthesis of new materials for device interface modification would be effective strategies to drive the progresses in this field further.

Acknowledgments

The work was financially supported by the Natural Science Foundation of China (No. **50990065**, **51010003**, **51073058**, and **20904011**) and the Ministry of Science and Technology, China (MOST) National Research Project (No. **2009CB623601**).

References

1. Zhao, J.H., Wang, A.H., Green, M.A., and Ferrazza, F. (1998) *Appl. Phys. Lett.*, **73**, 1991–1993.
2. Günes, S., Neugebauer, H., and Sariciftci, N.S. (2007) *Chem. Rev.*, **107**, 1324–1338.
3. Thompson, B.C. and Fréchet, J.M.J. (2008) *Angew. Chem. Int. Ed.*, **47**, 58–77.
4. Cheng, Y.J., Yang, S.H., and Hsu, C.S. (2009) *Chem. Rev.*, **109**, 5868–5923.

5. Tang, C.W. (1986) *Appl. Phys. Lett.*, **48**, 183–185.
6. Sariciftci, N.S., Smilowitz, L., Heeger, A.J., and Wudl, F. (1992) *Science*, **258**, 1474–1476.
7. Yu, G., Gao, J., Hummelen, J.C., Wudl, F., and Heeger, A.J. (1995) *Science*, **270**, 1789–1791.
8. Konarka Technologies (2011) Konarka®, *http://www.konarka.com* (last accessed 31 January 2012).
9. Solarmer Energy (2011) Solarmer Energy, Inc.,*http://www.solarmer.com* (last accessed 31 January 2011).
10. Scharber, M.C., Wuhlbacher, D., Koppe, M., Denk, P., Waldauf, C., Heeger, A.J., and Brabec, C.J. (2006) *Adv. Mater.*, **18**, 789–794.
11. Nobelprize.org (2011) Nobelprize.org, the Official Web Site of the Nobel Prize, *http://nobelprize.org* (last accessed 31 January 2011).
12. He, Y., Chen, H.Y., Hou, J., and Li, Y. (2010) *J. Am. Chem. Soc.*, **132**, 1377–1382.
13. Wienk, M.M., Kroon, J.M., Verhees, W.J.H., Knol, J., Hummelen, J.C., van Hal, P.A., and Janssen, R.A.J. (2003) *Angew. Chem. Int. Ed.*, **42**, 3371–3375.
14. He, Y.J., Zhao, G.J., Peng, B., and Li, Y.F. (2010) *Adv. Funct. Mater.*, **20**, 3383–3389.
15. Segalman, R.A., McCulloch, B., Kirmayer, S., and Urban, J.J. (2009) *Macromolecules*, **42**, 9205–9216.
16. Burroughes, J.H., Bradley, D.D.C., Brown, A.R., Marks, R.N., Mackay, K., Friend, R.H., Burn, P.L., and Holmes, A.B. (1990) *Nature*, **347**, 539–541.
17. Tsumura, A., Koezuka, H., and Ando, T. (1986) *Appl. Phys. Lett.*, **49**, 1210–1212.
18. Zhong, C., Duan, C., Huang, F., Wu, H., and Cao, Y. (2010) *Chem. Mater.*, **23**, 326–340.
19. Shaheen, S.E., Brabec, C.J., Sariciftci, N.S., Padinger, F., Fromherz, T., and Hummelen, J.C. (2001) *Appl. Phys. Lett.*, **78**, 841–843.
20. Ma, W.L., Yang, C.Y., Gong, X., Lee, K., and Heeger, A.J. (2005) *Adv. Funct. Mater.*, **15**, 1617–1622.
21. Sommer, M., Lang, A.S., and Thelakkat, M. (2008) *Angew. Chem. Int. Ed.*, **47**, 7901–7904.
22. Barrau, S., Heiser, T., Richard, F., Brochon, C., Ngov, C., van der Wetering, K., Hadziioannou, G., Anokhin, D.V., and Ivanov, D.A. (2008) *Macromolecules*, **41**, 2701–2710.
23. Wessling, R.A. (1985) *J. Polym. Sci. Polym. Symp.*, **72**, 55.
24. Burn, P.L., Kraft, A., Baigent, D., Bradley, D.D.C., Brown, A.R., Friend, R.H., Gymer, R.W., Holmes, A.B., and Jackson, R.W. (1993) *J. Am. Chem. Soc.*, **115**, 10117–10124.
25. Gilch, H.G. and Wheelwright, W.L. (1966) *J. Polym. Sci., Part A: Polym. Chem.*, **4**, 1337–1349.
26. Liao, L., Pang, Y., Ding, L., and Karasz, F.E. (2001) *Macromolecules*, **34**, 6756–6760.
27. Bao, Z., Chen, Y., Cai, R., and Yu, L. (1993) *Macromolecules*, **26**, 5281–5286.
28. Moratti, S.C., Cervini, R., Holmes, A.B., Baigent, D.R., Friend, R.H., Greenham, N.C., Gruner, J., and Hamer, P.J. (1995) *Synth. Met.*, **71**, 2117–2120.
29. Alam, M.M. and Jenekhe, S.A. (2004) *Chem. Mater.*, **16**, 4647–4656.
30. Pacios, R., Chatten, A.J., Kawano, K., Durrant, J.R., Bradley, D.D.C., and Nelson, J. (2006) *Adv. Funct. Mater.*, **16**, 2117–2126.
31. Kawano, K., Ito, N., Nishimori, T., and Sakai, J. (2006) *Appl. Phys. Lett.*, **88**, 073514.
32. Lenes, M., Koster, L.J.A., Mihailetchi, V.D., and Blom, P.W.M. (2006) *Appl. Phys. Lett.*, **88**, 243502.
33. Park, J., Han, S.H., Senthilarasu, S., and Lee, S.H. (2007) *Sol. Energy Mater. Sol. Cells*, **91**, 751–753.
34. Mens, R., Adriaensens, P., Lutsen, L., Swinnen, A., Bertho, S., Ruttens, B., D'Haen, J., Manca, J., Cleij, T., Vanderzande, D., and Gelan, J. (2008) *J. Polym. Sci., Part A: Polym. Chem.*, **46**, 138–145.
35. Kooistra, F.B., Mihailetchi, V.D., Popescu, L.M., Kronholm, D., Blom, P.W.M., and Hummelen, J.C. (2006) *Chem. Mater.*, **18**, 3068–3073.
36. Sun, B.Q., Marx, E., and Greenham, N.C. (2003) *Nano Lett.*, **3**, 961–963.
37. Beek, W.J.E., Wienk, M.M., Kemerink, M., Yang, X.N., and Janssen, R.A.J. (2005) *J. Phys. Chem. B*, **109**, 9505–9516.

38. Wei, Y., Chan, C.C., Tian, J., Jang, G.W., and Hsueh, K.F. (1991) *Chem. Mater.*, **3**, 888–897.

39. Maior, R.M.S., Hinkelmann, K., Eckert, H., and Wudl, F. (1990) *Macromolecules*, **23**, 1268–1279.

40. McCullough, R.D. and Lowe, R.D. (1992) *J. Chem. Soc., Chem. Commun.*, 70–72.

41. McCullough, R.D., Lowe, R.D., Jayaraman, M., and Anderson, D.L. (1993) *J. Org. Chem.*, **58**, 904–912.

42. Chen, T.A. and Rieke, R.D. (1992) *J. Am. Chem. Soc.*, **114**, 10087–10088.

43. Loewe, R.S., Khersonsky, S.M., and McCullough, R.D. (1999) *Adv. Mater.*, **11**, 250–253.

44. Yokoyama, A., Miyakoshi, R., and Yokozawa, T. (2004) *Macromolecules*, **37**, 1169–1171.

45. Sheina, E.E., Liu, J., Iovu, M.C., Laird, D.W., and McCullough, R.D. (2004) *Macromolecules*, **37**, 3526–3528.

46. Iovu, M.C., Sheina, E.E., Gil, R.R., and McCullough, R.D. (2005) *Macromolecules*, **38**, 8649–8656.

47. Kim, Y., Cook, S., Tuladhar, S.M., Choulis, S.A., Nelson, J., Durrant, J.R., Bradley, D.D.C., Giles, M., Mcculloch, I., Ha, C.S., and Ree, M. (2006) *Nat. Mater.*, **5**, 197–203.

48. Woo, C.H., Thompson, B.C., Kim, B.J., Toney, M.F., and Fréchet, J.M.J. (2008) *J. Am. Chem. Soc.*, **130**, 16324–16329.

49. Zen, A., Pflaum, J., Hirschmann, S., Zhuang, W., Jaiser, F., Asawapirom, U., Rabe, J.P., Scherf, U., and Neher, D. (2004) *Adv. Funct. Mater.*, **14**, 757–764.

50. Kline, R.J., McGehee, M.D., Kadnikova, E.N., Liu, J.S., Fréchet, J.M.J., and Toney, M.F. (2005) *Macromolecules*, **38**, 3312–3319.

51. Schilinsky, P., Asawapirom, U., Scherf, U., Biele, M., and Brabec, C.J. (2005) *Chem. Mater.*, **17**, 2175–2180.

52. Koppe, M., Brabec, C.J., Heiml, S., Schausberger, A., Duffy, W., Heeney, M., and McCulloch, I. (2009) *Macromolecules*, **42**, 4661–4666.

53. Nguyen, L.H., Hoppe, H., Erb, T., Günes, S., Gobsch, G., and Sariciftci, N.S. (2007) *Adv. Funct. Mater.*, **17**, 1071–1078.

54. Wu, P.T., Xin, H., Kim, F.S., Ren, G., and Jenekhe, S.A. (2009) *Macromolecules*, **42**, 8817–8826.

55. Li, G., Shrotriya, V., Huang, J.S., Yao, Y., Moriarty, T., Emery, K., and Yang, Y. (2005) *Nat. Mater.*, **4**, 864–868.

56. Miyaura, N. and Suzuki, A. (1995) *Chem. Rev.*, **95**, 2457–2483.

57. Beaupré, S., Boudreault, P.L.T., and Leclerc, M. (2010) *Adv. Mater.*, **22**, E6–E27.

58. Svensson, M., Zhang, F.L., Veenstra, S.C., Verhees, W.J.H., Hummelen, J.C., Kroon, J.M., Inganas, O., and Andersson, M.R. (2003) *Adv. Mater.*, **15**, 988–991.

59. Wang, E.G., Wang, L., Lan, L.F., Luo, C., Zhuang, W.L., Peng, J.B., and Cao, Y. (2008) *Appl. Phys. Lett.*, **92**, 033307.

60. Blouin, N., Michaud, A., and Leclerc, M. (2007) *Adv. Mater.*, **19**, 2295–2300.

61. Park, S.H., Roy, A., Beaupre, S., Cho, S., Coates, N., Moon, J.S., Moses, D., Leclerc, M., Lee, K., and Heeger, A.J. (2009) *Nat. Photonics*, **3**, 297–303.

62. Qin, R., Li, W., Li, C., Du, C., Veit, C., Schleiermacher, H.F., Andersson, M., Bo, Z., Liu, Z., Inganas, O., Wuerfel, U., and Zhang, F. (2009) *J. Am. Chem. Soc.*, **131**, 14612–14613.

63. Blouin, N., Michaud, A., Gendron, D., Wakim, S., Blair, E., Neagu-Plesu, R., Belletete, M., Durocher, G., Tao, Y., and Leclerc, M. (2008) *J. Am. Chem. Soc.*, **130**, 732–742.

64. Zhu, Z., Waller, D., Gaudiana, R., Morana, M., Muhlbacher, D., Scharber, M., and Brabec, C. (2007) *Macromolecules*, **40**, 1981–1986.

65. Peet, J., Kim, J.Y., Coates, N.E., Ma, W.L., Moses, D., Heeger, A.J., and Bazan, G.C. (2007) *Nat. Mater.*, **6**, 497–500.

66. Hou, J., Chen, H.Y., Zhang, S., Li, G., and Yang, Y. (2008) *J. Am. Chem. Soc.*, **130**, 16144–16145.

67. Coffin, R.C., Peet, J., Rogers, J., and Bazan, G.C. (2009) *Nat. Chem.*, **1**, 657–661.

68. Hou, J., Park, M.H., Zhang, S., Yao, Y., Chen, L.M., Li, J.H., and Yang, Y. (2008) *Macromolecules*, **41**, 6012–6018.

69. Liang, Y. and Yu, L. (2010) *Acc. Chem. Res.*, **43**, 1227–1236.

70. Zou, Y.P., Najari, A., Berrouard, P., Beaupre, S., Aich, B.R., Tao, Y., and Leclerc, M. (2010) *J. Am. Chem. Soc.*, **132**, 5330–5331.

71. Zhang, Y., Hau, S.K., Yip, H.L., Sun, Y., Acton, O., and Jen, A.K.Y. (2010) *Chem. Mater.*, **22**, 2696–2698.

72. Piliego, C., Holcombe, T.W., Douglas, J.D., Woo, C.H., Beaujuge, P.M., and Fréchet, J.M.J. (2010) *J. Am. Chem. Soc.*, **132**, 7595–7597.

73. Zhang, G., Fu, Y., Zhang, Q., and Xie, Z. (2010) *Chem. Commun.*, **46**, 4997–4999.

74. Hou, J.H., Tan, Z.A., Yan, Y., He, Y.J., Yang, C.H., and Li, Y.F. (2006) *J. Am. Chem. Soc.*, **128**, 4911–4916.

75. Hou, J.H., Tan, Z., He, Y.J., Yang, C.H., and Li, Y.F. (2006) *Macromolecules*, **39**, 4657–4662.

76. Li, Y.F. and Zou, Y.P. (2008) *Adv. Mater.*, **20**, 2952–2958.

77. Huang, F., Chen, K.S., Yip, H.L., Hau, S.K., Acton, O., Zhang, Y., Luo, J., and Jen, A.K.Y. (2009) *J. Am. Chem. Soc.*, **131**, 13886–13887.

78. Duan, C.H., Cai, W.Z., Huang, F., Zhang, J., Wang, M., Yang, T.B., Zhong, C.M., Gong, X., and Cao, Y. (2010) *Macromolecules*, **43**, 5262–5268.

79. Duan, C.H., Chen, K.S., Huang, F., Yip, H.L., Liu, S.J., Zhang, J., Jen, A.K.Y., and Cao, Y. (2010) *Chem. Mater.*, **22**, 6444–6452.

80. Zhang, Z.G., Liu, Y.L., Yang, Y., Hou, K., Peng, B., Zhao, G., Zhang, M., Guo, X., Kang, E.T., and Li, Y. (2010) *Macromolecules*, **43**, 9376–9383.

81. Hsu, S.L., Chen, C.M., and Wei, K.H. (2010) *J. Polym. Sci., Part A: Polym. Chem.*, **48**, 5126–5134.

82. Sahu, D., Padhy, H., Patra, D., Huang, J.H., Chu, C.W., and Lin, H.C. (2010) *J. Polym. Sci., Part A: Polym. Chem.*, **48**, 5812–5823.

83. Jenekhe, S.A. and Chen, X.L. (1998) *Science*, **279**, 1903–1907.

84. Ren, G., Wu, P.T., and Jenekhe, S.A. (2010) *Chem. Mater.*, **22**, 2020–2026.

85. Lindner, S.M., Hüttner, S., Chiche, A., Thelakkat, M., and Krausch, G. (2006) *Angew. Chem. Int. Ed.*, **45**, 3364–3368.

86. Sommer, M., Lindner, S.M., and Thelakkat, M. (2007) *Adv. Funct. Mater.*, **17**, 1493–1500.

87. Zhang, Q., Cirpan, A., Russell, T.P., and Emrick, T. (2009) *Macromolecules*, **42**, 1079–1082.

88. Sun, S.S., Zhang, C., Ledbetter, A., Choi, S., Seo, K., Bonner, C.E., Drees, M., and Sariciftci, N.S. (2007) *Appl. Phys. Lett.*, **90**, 043117.

89. Sivula, K., Ball, Z.T., Watanabe, N., and Fréchet, J.M.J. (2006) *Adv. Mater.*, **18**, 206–210.

90. Rajaram, S., Armstrong, P.B., Kim, B.J., and Fréchet, J.M.J. (2009) *Chem. Mater.*, **21**, 1775–1777.

91. Yang, C., Lee, J.K., Heeger, A.J., and Wudl, F. (2009) *J. Mater. Chem.*, **19**, 5416–5423.

92. Petritsch, K., Dittmer, J.J., Marseglia, E.A., Friend, R.H., Lux, A., Rozenberg, G.G., Moratti, S.C., and Holmes, A.B. (2000) *Sol. Energy Mater. Sol. Cells*, **61**, 63–72.

93. Matsuo, Y., Sato, Y., Niinomi, T., Soga, I., Tanaka, H., and Nakamura, E. (2009) *J. Am. Chem. Soc.*, **131**, 16048–16050.

94. Schmidt-Mende, L., Fechtenkotter, A., Müllen, K., Moons, E., Friend, R.H., and MacKenzie, J.D. (2001) *Science*, **293**, 1119–1122.

95. Sun, X., Zhou, Y., Wu, W., Liu, Y., Tian, W., Yu, G., Qiu, W., Chen, S., and Zhu, D. (2006) *J. Phys. Chem. B*, **110**, 7702–7707.

96. Ma, C.Q., Fonrodona, M., Schikora, M.C., Wienk, M.M., Janssen, R.A.J., and Bäuerle, P. (2008) *Adv. Funct. Mater.*, **18**, 3323–3331.

97. Winzenberg, K.N., Kemppinen, P., Fanchini, G., Bown, M., Collis, G.E., Forsyth, C.M., Hegedus, K., Singh, T.B., and Watkins, S.E. (2009) *Chem. Mater.*, **21**, 5701–5703.

98. Roquet, S., Cravino, A., Leriche, P., Aleveque, O., Frere, P., and Roncali, J. (2006) *J. Am. Chem. Soc.*, **128**, 3459–3466.

99. Walker, B., Tamayo, A.B., Dang, X.D., Zalar, P., Seo, J.H., Garcia, A., Tantiwiwat, M., and Nguyen, T.Q. (2009) *Adv. Funct. Mater.*, **19**, 3063–3069.

100. Sun, M., Wang, L., Zhu, X., Du, B., Liu, R., Yang, W., and Cao, Y. (2007) *Sol. Energy Mater. Sol. Cells*, **91**, 1681–1687.

101. Zhang, J., Deng, D., He, C., He, Y., Zhang, M., Zhang, Z.G., Zhang, Z., and Li, Y. (2010) *Chem. Mater.*, **23**, 817–822.

102. Silvestri, F., Irwin, M.D., Beverina, L., Facchetti, A., Pagani, G.A., and Marks, T.J. (2008) *J. Am. Chem. Soc.*, **130**, 17640–17641.

103. Rousseau, T., Cravino, A., Bura, T., Ulrich, G., Ziessel, R., and Roncali, J. (2009) *Chem. Commun.*, 1673–1675.

104. Roncali, J. (2009) *Acc. Chem. Res.*, **42**, 1719–1730.

105. Walker, B., Kim, C., and Nguyen, T.Q. (2011) *Chem. Mater.*, **23**, 470–482.

106. Hirsch, A. (1994) *The Chemistry of the Fullerenes*, Georg Thieme, Stuttgart.

107. Thilgen, C. and Diederich, F. (2006) *Chem. Rev.*, **106**, 5049–5135.

108. Hummelen, J.C., Knight, B.W., LePeq, F., Wudl, F., Yao, J., and Wilkins, C.L. (1995) *J. Org. Chem.*, **60**, 532–538.

109. Liang, Y., Feng, D., Wu, Y., Tsai, S.T., Li, G., Ray, C., and Yu, L. (2009) *J. Am. Chem. Soc.*, **131**, 7792–7799.

110. Zheng, L.P., Zhou, Q.M., Deng, X.Y., Yuan, M., Yu, G., and Cao, Y. (2004) *J. Phys. Chem. B*, **108**, 11921–11926.

111. Zhao, G., He, Y., Xu, Z., Hou, J., Zhang, M., Min, J., Chen, H.Y., Ye, M., Hong, Z., Yang, Y., and Li, Y. (2010) *Adv. Funct. Mater.*, **20**, 1480–1487.

112. Popescu, L.M., van't Hof, P., Sieval, A.B., Jonkman, H.T., and Hummelen, J.C. (2006) *Appl. Phys. Lett.*, **89**, 213507.

113. Zhang, Y., Yip, H.L., Acton, O., Hau, S.K., Huang, F., and Jen, A.K.Y. (2009) *Chem. Mater.*, **21**, 2598–2600.

114. Riedel, I., von Hauff, E., Parisi, H., Martin, N., Giacalone, F., and Dyakonov, V. (2005) *Adv. Funct. Mater.*, **15**, 1979–1987.

115. Kooistra, F.B., Knol, J., Kastenberg, F., Popescu, L.M., Verhees, W.J.H., Kroon, J.M., and Hummelen, J.C. (2007) *Org. Lett.*, **9**, 551–554.

116. Yang, C., Cho, S., Heeger, A.J., and Wudl, F. (2009) *Angew. Chem. Int. Ed.*, **48**, 1592–1595.

117. Park, S.H., Yang, C., Cowan, S., Lee, J.K., Wudl, F., Lee, K., and Heeger, A.J. (2009) *J. Mater. Chem.*, **19**, 5624–5628.

118. Yao, Y., Shi, C., Li, G., Shrotriya, V., Pei, Q., and Yang, Y. (2006) *Appl. Phys. Lett.*, **89**, 153507.

119. Chen, H.Y., Hou, J.H., Zhang, S.Q., Liang, Y.Y., Yang, G.W., Yang, Y., Yu, L.P., Wu, Y., and Li, G. (2009) *Nat. Photonics*, **3**, 649–653.

120. Ross, R.B., Cardona, C.M., Guldi, D.M., Sankaranarayanan, S.G., Reese, M.O., Kopidakis, N., Peet, J., Walker, B., Bazan, G.C., Van Keuren, E., Holloway, B.C., and Drees, M. (2009) *Nat. Mater.*, **8**, 208–212.

121. Fromherz, T., Padinger, F., Gebeyehu, D., Brabec, C., Hummelen, J.C., and Sariciftci, N.S. (2000) *Sol. Energy Mater. Sol. Cells*, **63**, 61–68.

122. Lenes, M., Wetzelaer, G.J.A.H., Kooistra, F.B., Veenstra, S.C., Hummelen, J.C., and Blom, P.W.M. (2008) *Adv. Mater.*, **20**, 2116–2119.

123. Wang, G.W., Li, Y.J., Peng, R.F., Liang, Z.H., and Liu, Y.C. (2004) *Tetrahedron*, **60**, 3921–3925.

124. Wang, X.J., Perzon, E., Delgado, J.L., de la Cruz, P., Zhang, F.L., Langa, F., Andersson, M., and Inganäs, O. (2004) *Appl. Phys. Lett.*, **85**, 5081–5083.

125. Espildora, E., Delgado, J.L., de la Cruz, P., de la Hoz, A., Lopez-Arza, V., and Langa, F. (2002) *Tetrahedron*, **58**, 5821–5826.

126. Wang, X.J., Perzon, E., Oswald, F., Langa, F., Admassie, S., Andersson, M.R., and Inganäs, O. (2005) *Adv. Funct. Mater.*, **15**, 1665–1670.

127. Backer, S.A., Sivula, K., Kavulak, D.F., and Fréchet, J.M.J. (2007) *Chem. Mater.*, **19**, 2927–2929.

128. Puplovskis, A., Kacens, J., and Neilands, O. (1997) *Tetrahedron Lett.*, **38**, 285–288.

129. Laird, D.W., Stegamat, R., Richter, H., Vejins, V., Scott, L., and Lada, T.A. (2008) Patent WO 2008/018931 A2.

130. Zhao, G., He, Y., and Li, Y. (2010) *Adv. Mater.*, **22**, 4355–4358.

131. Li, J.X., Sun, N., Guo, Z.X., Li, C.J., Li, Y.F., Dai, L.M., Zhu, D.B., Sun, D.K., Cao, Y., and Fan, L.Z. (2002) *J. Phys. Chem. B*, **106**, 11509–11514.

132. Zheng, L., Zhou, Q., Deng, X., Fei, W., Bin, N., Guo, Z.X., Yu, G., and Cao, Y. (2005) *Thin Solid Films*, **489**, 251–256.

133. Mwaura, J.K., Pinto, M.R., Witker, D., Ananthakrishnan, N., Schanze, K.S., and Reynolds, J.R. (2005) *Langmuir*, **21**, 10119–10126.

134. Chen, Z., Debije, M.G., Debaerdemaeker, T., Osswald, P., and Würthner, F. (2004) *ChemPhysChem*, **5**, 137–140.

135. Struijk, C.W., Sieval, A.B., Dakhorst, J.E.J., van Dijk, M., Kimkes, P., Koehorst, R.B.M., Donker, H., Schaafsma, T.J., Picken, S.J., van de Craats, A.M., Warman, J.M., Zuilhof, H., and

Sudholter, E.J.R. (2000) *J. Am. Chem. Soc.*, **122**, 11057–11066.

136. Dittmer, J.J., Marseglia, E.A., and Friend, R.H. (2000) *Adv. Mater.*, **12**, 1270–1274.

137. Zhan, X.W., Tan, Z.A., Domercq, B., An, Z.S., Zhang, X., Barlow, S., Li, Y.F., Zhu, D.B., Kippelen, B., and Marder, S.R. (2007) *J. Am. Chem. Soc.*, **129**, 7246–7247.

138. Neuteboom, E.E., Meskers, S.C.J., van Hal, P.A., van Duren, J.K.J., Meijer, E.W., Janssen, R.A.J., Dupin, H., Pourtois, G., Cornil, J., Lazzaroni, R., Bredas, J.L., and Beljonne, D. (2003) *J. Am. Chem. Soc.*, **125**, 8625–8638.

139. Neuteboom, E.E., van Hal, P.A., and Janssen, R.A.J. (2004) *Chem. Eur. J.*, **10**, 3907–3918.

140. Granstrom, M., Petritsch, K., Arias, A.C., Lux, A., Andersson, M.R., and Friend, R.H. (1998) *Nature*, **395**, 257–260.

141. Shin, R.Y.C., Kietzke, T., Sudhakar, S., Dodabalapur, A., Chen, Z.K., and Sellinger, A. (2007) *Chem. Mater.*, **19**, 1892–1894.

142. Shin, R.Y.C., Sonar, P., Siew, P.S., Chen, Z.K., and Sellinger, A. (2009) *J. Org. Chem.*, **74**, 3293–3298.

143. Brunetti, F.G., Gong, X., Tong, M., Heeger, A.J., and Wudl, F. (2010) *Angew. Chem. Int. Ed.*, **49**, 532–536.

Commentary Part

Comment 1

Niyazi Serdar Sariciftci

In this chapter Chunhui Duan *et al.* compile a large amount of data and review recent progress in the synthesis and application of materials for solar energy conversion in OPV devices. Such a chapter could never be totally comprehensive since a very large body of work has been published in the literature and the progress in recent years has been immense. However, the general construction of the chapter and the different materials discussed therein fit well with the state-of-the-art in this field.

Generally, in an interdisciplinary research area such as "organic photovoltaic research," it is extremely difficult to bring material synthesis, material characterization, device fabrication, device characterization, and spectroscopy scientists together. These fields all belong to different scientific disciplines such as chemistry, physics, and electronic engineering. Even communication is difficult owing to the different nomenclatures that these disciplines use that may not be completely understood by other disciplines.

However, in the case of the development of OPV materials, this has been successful, as illustrated in this chapter. If we look to the most important physical insights of the "photon harvesting" problem, we can clearly illustrate this interdisciplinary progress.

So, what is the photon harvesting problem in OPV diodes? As described in Figure C1, the absorption of OPV materials such as poly(*p*-phenylenevinylene)s covers only a limited part of the spectrum, mostly $\lambda < 600$ nm. All the solar photons that have lower energy photons in the IR region are excluded from the photoconversion process.

Here in this problem of "photon harvesting," interdisciplinary science worked in an exemplarily successful way. It is obvious from Figure C1 that we need photoactive polymers for OPV devices, which absorb energies at higher wavelengths, that is, "low-bandgap materials." This interdisciplinary insight was articulated by many workers in this field and, after several years of discussion and modeling, the synthetic organic chemists came up with strategies for designing molecular structures to achieve this low-bandgap (high-wavelength) absorption.

This chapter gives a beautiful insight into these strategies. Especially the push–pull strategy using D–A-type units in the conjugated polymer backbone worked well. According to this strategy, the combination of electron-donating and electron-withdrawing units in the same conjugation would result in a low-bandgap polymer (see Figure 17.14).

The experimental results are more than encouraging, since the new materials described in this chapter have higher and higher wavelengths for their absorption edge. Given the fact that some of these materials are also very good charge carrier conductors and perform a photoinduced electron

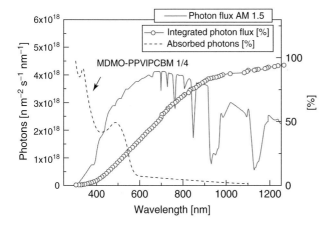

Figure C1 The solar spectrum and the optical density of PPV-based organic solar cell materials.

transfer with the usual acceptors such as PCBM, the expected increase in photovoltaic efficiency was clearly observable.

Today, polymers for OPV with the highest efficiency are based on this molecular strategy and have absorption edges around 800–900 nm with an overall PCE above 8%. According to Figure C1, if we can get the absorption edge down to 1000 nm we can expect a doubling of the efficiency due to a simple photon harvesting increase. This would lead to materials that perform with around 12–15% PCEs. Using different strategies, synthetic organic chemists will reach these materials in due time and thus such efficiencies are within reach.

My great satisfaction with this progress is the materials chemistry strategies behind it. This shows that the understanding is increasing and giving a profound roadmap for success.

Of course, the acceptor side of the materials also needs more attention in the future. We do not see any fundamental problem why conjugated organic polymers should not substitute the acceptor molecule PCBM used in such devices. In our original patent in 1992, which described the BHJs in claim No. 3, there is also a special mention of polymer–polymer solar cells that can perform in the future as well as polymer–fullerene mixtures. Here there is a great need for electron-withdrawing and electron-conducting conjugated polymers as n-type materials. Organic material chemists should attack this issue also with great energy as they did with the low-bandgap materials.

My overall view of this progress in the synthetic organic chemistry of materials is very positive and the future looks bright for OSC research.

Comment 2

Yongfang Li

BHJ OSCs or OPVs, which are based on solution-processable organic semiconductor donor and acceptor materials, have attracted great attention in recent years because of their advantages of easy fabrication, simple device structure, low cost, light weight, and flexibility in comparison with traditional inorganic semiconductor-based solar cells. Current studies of OSCs are mainly focused on increasing the PCE of the devices, and the key point in increasing the PCE is the design and synthesis of high-efficiency organic donor and acceptor materials in the OSCs [C3–C6]. Obviously, organic chemistry plays a key role in the development of new organic semiconductor molecules, including conjugated polymers and organic small-molecule semiconductors.

The PCE of the solar cells is proportional to short-circuit current density (J_{sc}), open-circuit voltage (V_{oc}), and fill factor (FF) of the devices. To increase the photovoltaic efficiency, the following five requirements should be kept in mind in the molecular design of OPV materials:

1) Broad and strong absorption in the visible and near-IR region to match the solar spectrum to increase the J_{sc} of the devices.

2) Suitable LUMO and HOMO energy levels to facilitate the exciton dissociation at the donor–acceptor interface (electron transfer from donor to acceptor and hole transfer from acceptor to donor) and to obtain a higher V_{oc} of the OSC devices.

3) High charge carrier mobility (high hole mobility for the donors and high electron mobility for the acceptors) to enhance the charge-transfer efficiency (so as to increase J_{sc}) and increase the *FF* of the devices.

4) High solubility for solution processing in fabrication of the OSCs.

5) Preferable morphology and self-assembly of the interpenetrating network of the donor–acceptor blend active layer, which influence the J_{sc}, V_{oc}, and *FF* of the OSCs significantly.

These five requirements are not independent of each other. For example, tuning the LUMO and HOMO energy levels in item (2) will change the energy bandgap, influencing the absorption in item (1), and improving the solubility of the molecules in item (4) by attaching large alkyl side chains will influence the morphology in item (5) and the charge carrier mobility in item (3) of the OSCs. Therefore, we need to strike a balance between the above five issues and optimize the molecular structure for high photovoltaic performance. In addition, for conjugated polymers, high purity and a high molecular weight with a narrow molecular weight distribution are also crucial in photovoltaic applications. Organic chemistry also plays an important role in the synthesis and purification of high-quality conjugated polymer photovoltaic materials.

This chapter written by Chunhui Duan, Fei Huang, and Yong Cao describes the history of the development of BHJ OSCs and the basic requirements of the organic donor and acceptor materials in the Introduction section. Then the authors focus their attention on the molecular design, molecular engineering, and synthesis of the organic donor and acceptor molecules. On the basis of the authors' strong background in organic semiconductors, this chapter elucidates the principles of the molecular design of and the synthetic routes to the OPV materials very clearly, and addresses the five requirements mentioned above deeply in the molecular design. The contents of this chapter include the most important aspects of OPV molecules, including the molecular design, synthesis, and photovoltaic properties of various types of conjugated polymer donors, solution-processable organic small-molecule donors, fullerene derivative acceptors, conjugated polymer acceptors, and D–A double cable and D–A block copolymers. This chapter should attract the attention of organic chemists for the design and synthesis of new OPV molecules, and will greatly promote progress in research on OSCs.

In the following, I would like to emphasize the importance of the electronic energy level engineering of organic semiconductor photovoltaic materials. Let us start by analyzing the electronic energy levels of the active blend layer of P3HT and [6,6]-phenyl-C_{61}-butyric acid methyl ester (PC$_{60}$BM). P3HT and PC$_{60}$BM are the most representative donor and acceptor materials, respectively, but the energy level matching is not ideal. Figure C2 shows the electronic energy levels of P3HT and PC$_{60}$BM [C6]. Higher lying LUMO and HOMO levels of the donor than those of the acceptor are necessary for the exciton dissociation at the donor–acceptor interface

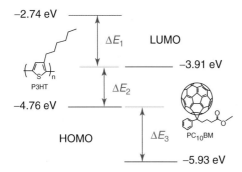

Figure C2 Electronic energy levels of P3HT and PC$_{60}$BM.

to overcome the binding energy of the excitons, which is commonly about 0.3–0.5 eV. The V_{oc} of the OSCs is proportional to the difference (ΔE_2) between the LUMO of the acceptor and the HOMO of the donor. Obviously, the LUMO energy offset ΔE_1 and the HOMO energy offset ΔE_3 in the P3HT–PCBM system, which are greater than 1 eV, are too large. The high ΔE_1 and ΔE_3 result in a lower value of ΔE_2, which leads to a lower V_{oc} (about 0.6 V) of the OSCs based on P3HT–PCBM. In addition, the absorption properties depend on the bandgap (E_g, the difference between its LUMO and HOMO) of the organic semiconductors, and the stability of the donor materials against oxidation is related the HOMO energy level position. Therefore, tuning the electronic energy levels of the donor and acceptor materials is crucial for improving their photovoltaic properties. Based on the electronic energy levels in Figure C2, we need to decrease (move downwards) the HOMO level of the donor or to increase (move upwards) the LUMO level of the acceptor to increase the V_{oc} of the devices, and we need to reduce the E_g of the donor to enhancing sunlight harvesting.

In electronic energy level engineering, the D–A copolymerization described extensively in this chapter is a successful strategy for reducing the E_g value and decreasing the HOMO level of the organic donor. Another approach to tune the electronic energy levels is to attach electron-donating (electron-rich) or electron-withdrawing (electron-deficient) functional substituents. Usually, electron-donating substituents (such as alkoxy, aryl, and alkyl groups) shift the HOMO and LUMO energy levels upwards, and electron-withdrawing substituents (such as fluorine, cyano, nitro, carboxy, and carboxylate groups) shift the LUMO and HOMO energy levels downwards. For example, the LUMO and HOMO energy levels of MEH-PPV were decreased by about 0.6 eV after substitution of a cyano group on the double-bond carbon of the polymer [C7].

For conjugated polymer photovoltaic donor materials, their HOMO levels need to be decreased to increase the V_{oc} of the devices, as mentioned above. Hou and co-workers successfully tuned the LUMO and HOMO energy levels of the narrow-bandgap copolymers of BDT and thienothiophene (TT) by using a stronger electron-withdrawing substituent on the TT unit (Figure C3) [C8]. On changing the alkoxylate group in PBDTTT-E to a stronger electron-withdrawing

carboxy group in PBDTTT-C, the LUMO and HOMO energy levels of the polymer decreased by 0.11 eV. The electronic energy levels of the polymer were shifted further downwards by 0.1 eV after attaching a fluorine on the TT unit in PBDTTT-CF. The decrease of the HOMO energy level while maintaining the same E_g of the polymers led to a higher V_{oc} and high PCE of OSC devices based on the polymers as donor. An OSC based on PBDTTT-CF showed a PCE of over 7% with a high V_{oc} of 0.76 V [C8].

Another interesting example is the electronic energy level tuning of poly(thienylenevinylene) (PTV) derivatives. PTVs possess broad absorption spectra in the visible region with an absorption edge at about 750 nm, which is attractive for application as a photovoltaic material. However, the photovoltaic performance of alkyl-substituted PTVs is very poor (the PCE is about 0.2%) [C9], which may be related to the nonfluorescent property of the polymer. By attaching a carboxylate group on the PTV polymer, the HOMO level was decreased from −5.05 to −5.26 eV (the molecular structures and electronic energy levels of the PTV derivatives are shown in Figure C4.) More interestingly, the carboxylate-substituted P3CTV showed photoluminescence, although weak. A polymer solar cell based on P3CTV as donor and PC$_{60}$BM as acceptor demonstrated a PCE of 2.01% with a high V_{oc} of 0.86 V [C10]. The PCE of P3CTV is an order of magnitude higher than that of the common PTV derivative P3HTV.

For organic semiconductor acceptors blended with P3HT, an upwards shifted LUMO level is preferred for a higher V_{oc} of the OSCs, as discussed previously. Electron-donating groups could shift the LUMO energy level upwards. A successful example is the electron-rich indene addition to fullerenes, such as indene-C$_{60}$ bis-adduct (IC$_{60}$BA) [C11] and indene-C$_{70}$ bis-adduct (IC$_{70}$BA) [C12] (see Scheme 17.9). The LUMO energy level of IC$_{60}$BA is shifted 0.17 eV higher than that of PC$_{60}$BM [C11]. After device optimization, the PCE of the polymer solar cell based on P3HT–ICBM reached 6.48% with a high V_{oc} of 0.84 V [C13].

In summary, the design and synthesis of organic semiconductor photovoltaic materials represent an exciting new research field. There are great opportunities and challenges for organic chemists and materials scientists in designing new molecular structures, developing new synthetic methods, and obtaining high-efficiency organic donor and acceptor materials. I believe that the research on

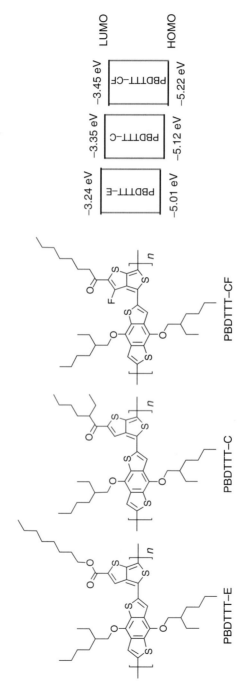

Figure C3 Molecular structures and electronic energy levels of PBDTTT copolymers [C8].

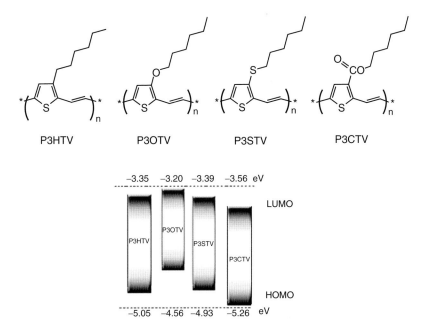

Figure C4 Molecular structures and electronic energy levels of poly(thienylenevinylene) derivatives [C10].

OPV materials will make significant progress and the PCEs of BHJ OSCs will reach 10% in a few years.

Comment 3

Guillermo C. Bazan

Conjugated polymers and small molecules for OSCs constitute a centerpiece of contemporary organic chemistry. The work accumulated thus far is vast in terms of molecular design, desired catalytic transformations, how best to configure molecules into solid-state active layer films, device integration, and the evaluation of how molecular structure control can be taken advantage of to convert sunlight into electricity. It would be impossible to summarize succinctly this abundant literature, but this chapter by Duan, Huang, and Cao provides an excellent entry into the general classes of polymers and molecular components. One finds here a proper historical context, together with the major ongoing challenges and opportunities for organic chemists.

Molecular engineering principles are provided so that the optical properties are optimized to match the characteristics of the terrestrial solar spectrum. Such requirements have been successfully met by the design of narrow-bandgap D–A polymers and structures with quinoidal characteristics. At the same time, one needs to fine-tune carefully the orbital energy levels of the donor and acceptor components in the active layers so that the open-circuit voltage of the device is maximized and thereby attain larger PCEs. A thorough survey of the most significant molecular structures for structural modulation is given. These are, however, very basic requirements. Proper solar cell management requires control over the mobilities of charge carriers, which are typically transported across the bulk via hopping-type transport. There is a good discussion on how best to control polymer structure and processing of thin films so that these electronic properties are attained. For example, thermal annealing, solvent exposure, and the presence of additives at the point of film formation can lead to morphological characteristics that can benefit or deteriorate the overall performance. These simple methods for processing are extremely important for verifying that the molecular design afforded by physical organic principles is properly taken advantage of in the fabrication of devices.

Some of the most subtle effects of molecular structure are also revealed, for example, the regioregularity of the repeat units in polythiophenes and the impact on the average degree of polymerization. Transition metal-mediated polymerization procedures are the best synthetic methodologies to attain specific structures, and this point is well discussed. It is also worth pointing out that there are excellent opportunities for organometallic chemists to develop more efficient polymerization catalysts and initiators to achieve better control over molecular weight and polydispersities.

Small molecules based on fullerene-type structures are of paramount importance in the photoinduced electron transfer event, whereby charge carriers are created. More recently, there has been a fair amount of effort to see if donor molecules can also be considered. It is unclear, however, at this stage of the technological push, which particular broad classes of materials will emerge as being superior.

Reading beyond the contents of the chapter, one can see that the science of organic materials for BHJ devices extends beyond what is required for technological innovation. Unlike the inorganic counterparts, the final arrangement of the organic solid state is governed by weak forces and, for the typical materials used, van der Waals forces, for which very few guidelines exist. Moreover, the desired BHJ arrangement that leads to optimal performance may not be at a thermodynamic minimum. Understanding the kinetic control of film formation is only just emerging and how organic chemists can influence these processes via molecular control remains unclear at present. Successful creation of OSCs will require molecular design that influences behavior beyond the isolated molecule and can manage the collective response by integrating self-assembly principles. These are remarkable challenges in pursuit of important societal problems.

Comment 4

Xiong Gong

This chapter overviews and summarizes the roadmap of developed materials, both electron donors and electron acceptors for OPV cells, in particular, BHJ polymer solar cells. It is a great review and can be published as it is. However, it would be much better if the authors could address following issues:

1) The fundamental physics is relatively less described. Could the authors describe more physics of OPV and BHJ polymer solar cells in the Introduction?
2) What are the general rules for the design of molecules that probably give high device performance?
3) What are the advantages and disadvantages of each of the molecules that the authors present in this chapter?
4) In the Conclusion and Outlook section, could the authors list what the criteria are for molecules that relate to the device performance?

Authors' Response to the Commentaries

It was very helpful to send our chapter to the various reviewers for comments and suggestions. We highly value their comments and fully agree with their suggestions to emphasize several important aspects in the design and synthesis of materials for polymer BHJ solar cells. Based on their suggestions, some revisions were made and references were updated. Although limitation on space meant that we could always list all the important discoveries in this exciting research field and also that we were unable to touch on device physics in depth as suggested by Prof. Gong, we encourage readers of this chapter to read the comments of the reviewers in order to gain a fuller understanding of the design principles in the molecular design of BHJ materials and their recent progress.

References

C1. Sariciftci, N.S. and Heeger, A.J. (1992) US Patent 5,331,183.

C2. Heeger, A.J., Sariciftci, N.S., and Namdas, E.B. (2010) *Semiconducting and Metallic Polymers*, Oxford University Press, New York.

C3. Li, Y.F. and Zou, Y.P. (2008) *Adv. Mater.*, **20**, 2952–2958.

C4. Chen, J.W. and Cao, Y. (2009) *Acc. Chem. Res.*, **42**, 1709–1718.

C5. Cheng, Y.J., Yang, S.H., and Hsu, C.S. (2009) *Chem. Rev.*, **109**, 5868–5923.

C6. He, Y.J. and Li, Y.F. (2011) *Phys. Chem. Chem. Phys.*, **13**, 1970–1983.

C7. Li, Y.F., Cao, Y., Gao, J., Wang, D.L., Yu, G., and Heeger, A.J. (1999) *Synth. Met.*, **99**, 243–248.

C8. Chen, H.Y., Hou, J.H., Zhang, S.Q., Liang, Y.Y., Yang, G.W., Yang, Y., Yu, L.P., Wu, Y., and Li, G. (2009) *Nat. Photonics*, **3**, 649–653.

C9. Hou, J.H., Tan, Z.A., He, Y.J., Yang, C.H., and Li, Y.F. (2006) *Macromolecules*, **39**, 4657–4662.

C10. Huo, L.J., Chen, T.L., Zhou, Y., Hou, J.H., Yang, Y., and Li, Y.F. (2009) *Macromolecules*, **42**, 4377–4380.

C11. He, Y.J., Chen, H.Y., Hou, J.H., and Li, Y.F. (2010) *J. Am. Chem. Soc.*, **132**, 1377–1382.

C12. He, Y.J., Zhao, G.J., Peng, B., and Li, Y.F. (2010) *Adv. Funct. Mater.*, **20**, 3383–3389.

C13. Zhao, G.J., He, Y.J., and Li, Y.F. (2010) *Adv. Mater.*, **22**, 4355–4358.

18
Catalytic Utilization of Carbon Dioxide: Actual Status and Perspectives

Albert Boddien, Felix Gärtner, Christopher Federsel, Irene Piras, Henrik Junge, Ralf Jackstell, and Matthias Beller

18.1
Introduction

Sensible resource management including the recycling of waste products is a prerequisite for the sustainable development of mankind. In 2008, the total anthropogenic CO_2 emissions accounted for $\sim$29 Gt, which is double the amount of emissions in 1973, and this increased CO_2 emission is mainly caused by the transportation and industrial sectors [1]. Considering the current industrial demand for CO_2 to produce chemicals such as urea, salicylic acid, ethylene, propylene carbonate (PC), and polycarbonates, it is clear that these processes cannot be considered to consume the tremendous amounts of CO_2 produced by humans.

In general, the production of CO_2 is closely connected to the generation of energy by the combustion of fossil resources. Clearly, the world's energy demand is approximately 100 times larger than the consumption of chemicals. Thus, apart from reducing the CO_2 output, carbon recycling from CO_2 towards the formation of gaseous or liquid fuels represents an attractive alternative to reducing CO_2 emissions. Large-scale CO_2 utilization deals with various problems, which are mainly caused by:

- the presence of clean CO_2 as feedstock
- CO_2 separation from air
- the availability of a pure and renewable reducing agent, for example, hydrogen source
- energy costs
- the availability of highly active and stable catalysts for CO_2 conversion.

The utilization of CO_2 as a *C1*-feedstock for the production of fuels is chemically an issue of reduction of the very stable and inert CO_2 gas molecule. The conversion of CO_2 in order to produce chemicals is outperformed using high-energy reactants such as epoxides (ring tension), hydrogen, and unsaturated compounds to overcome thermodynamic barriers. In other words, this means that for coupling of CO_2 with unreactive compounds, high-energy input (heat, light, electricity) is necessary to overcome energy barriers. Apart from this, the conversion of CO_2 can be enhanced if the product formed is removed

Organic Chemistry – Breakthroughs and Perspectives, First Edition. Edited by Kuiling Ding and Li-Xin Dai.
© 2012 Wiley-VCH Verlag GmbH & Co. KGaA. Published 2012 by Wiley-VCH Verlag GmbH & Co. KGaA.

Table 18.1 Enthalpies for various chemical transformations using CO_2 as starting material.

Reaction	$\Delta H_{298\ K}$ (kJ mol^{-1})	Ref.
$CO_2 + H_2 \rightarrow HCOOH$	-31.2	[3, 4]
$CO_2 + H_2 + NH_3 \rightarrow NH_4{}^+ + HCOO^-$	-84.3	[3, 4]
$HCOOH \rightarrow CO + H_2O$	$+28.7$	[3]
$CO_2 + 3H_2 \rightarrow CH_3OH + H_2O$	-49.4	[5–7]
$CO_2 + H_2 \rightarrow CO + H_2O$	$+41.1$	[7, 8]
$CH_4 + CO_2 \rightarrow 2CO + 2H_2$	$+247$	[6, 8]
$CH_4 + H_2O \rightarrow CO + 3H_2$	$+206$	[6, 8]
$CO + 2H_2 \rightarrow CH_3OH$	-90.8	[5–7]

from the process (shift of equilibrium). Nonetheless, it is clear that a variety of valuable products, which are of the outmost importance for both the chemical feedstock and energy sectors, can be assumed using CO_2 as a *C1*-source (Table 18.1) [2].

18.2
Catalytic Reductions of CO₂ to Formic Acid and Methanol

Catalysis has proven to be a key technology for the utilization of CO_2. In this section, we deal with the catalytic conversion of CO_2 to reduced products such as methanol and formic acid (FA). Made via CO_2 reduction, both chemicals can be assumed to be sustainable materials, which can contribute to the paradigm change away from fossil-based raw materials towards energy vectors based on renewables. In this respect, carbon capture and storage (CCS) is vital for reducing anthropogenic CO_2 emissions, to attenuate the impact of global warming. CCS is viable to reduce CO_2 emission, caused by the energy sector, by 20% and currently this technology is in the up-scale stage [9]. Several methods, such as membranes, physical and chemical adsorption, and solvent absorption to strip CO_2 from the processes, have been developed for post-combustion (predominantly $CO_2–N_2$ separation), precombustion ($CO_2–H_2$), oxyfuel ($CO_2–H_2O$), and natural gas sweetening ($CO_2–CH_4$) [10]. Overall, the key factor is the development of suitable materials to separate CO_2 from exhaust gas. In addition to the endeavor to store separated CO_2 permanently in deep-level saline aquifers or vast oil and gas fields, chemistry provides significant opportunities to utilize it as a *C1*-carbon source [11, 12] (Figure 18.1).

18.2.1
Electrochemical CO₂ Reduction

The electrochemical conversion of CO_2 has been studied for decades and several reviews have been published in this area [13]. Proton-coupled multi-electron steps are in general more favored than single-electron reductions because more thermodynamically stable products are produced. According to the reaction thermodynamics listed in Table 18.2,

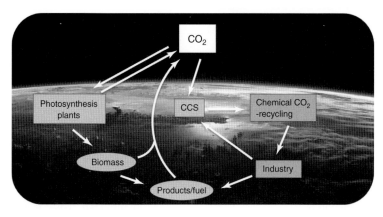

Figure 18.1 CO_2 – the base of a green carbon cycle.

Table 18.2 Possible half-cell reactions for the electrocatalytic conversion of carbon dioxide [14–16].

Reaction equations	$E°$ (V)[a]
$2CO_2 + 2H^+ + 2e^- \rightleftharpoons \quad H_2C_2O_4$	−0.475
$CO_2 + 2H^+ + 2e^- \rightleftharpoons \quad HCO_2H$	−0.199
$CO_2 + 2H^+ + 2e^- \rightleftharpoons \quad CO + H_2O$	−0.109
$CO_2 + 4H^+ + 4e^- \rightleftharpoons HCHO + H_2O$	−0.071
$CO_2 + 6H^+ + 6e^- \rightleftharpoons CH_3OH + H_2O$	+0.030
$2CO_2 + 12H^+ + 12e^- \rightleftharpoons C_2H_4 + 4H_2O$	+0.079
$CO_2 + 8H^+ + 8e^- \rightleftharpoons CH_4 + 2H_2O$	+0.169

[a] $E°$ versus NHE at 298 K.

full conversion of carbon dioxide can theoretically be achieved in all the listed reactions. Hence the final products of the electrochemical reduction depend on the selected reaction conditions such as applied potential, electrode material, temperature, pressure, and electrolyte/solvent. Owing to the mass transfer of CO_2, the solubility of the latter is of great importance. For instance, up to 30 mM of CO_2 can be diluted in water at 101 kPa, and in aqueous electrolytes the concentration can exceed 10 M. It was found that selectivity and insufficient derived product concentrations were an obstacle for the reasonable utilization of CO_2 through electro-reduction. In addition, low catalytic activity, stability, and product selectivity also remain as major challenges.

In Scheme 18.1, a general approach for an electrocatalytic system is presented [17]. Catalysts are applied on the electrodes in order to improve the overall activity and selectivity of the derived system. An optimal electrocatalyst must display a good thermodynamic match between the redox potential ($E°$) for the electron-transfer reaction and the catalytic performance of the chemical reaction.

Research on suitable transition metal catalysts and their redox potentials, current efficiencies, and kinetics (chemical, electron transfer) has attracted increasing attention

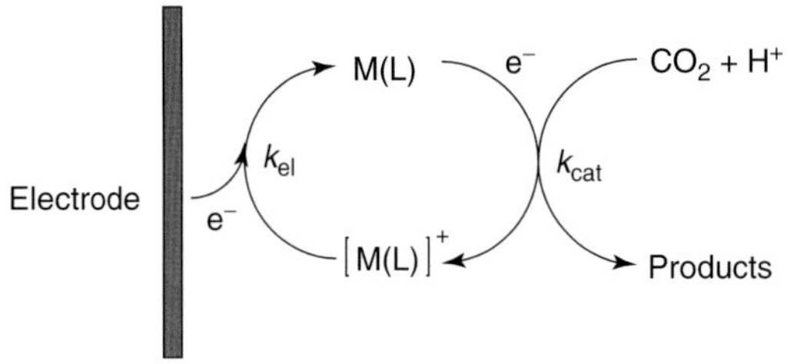

Scheme 18.1 General setup for catalytic electroreduction of carbon dioxide.

in the last 40 years. Mo, Cu, Pb, In, Hg, Bi, Sb, and Sn electrodes were applied successfully in heterogeneous catalysis for the reduction of CO_2 to FA/formates with high selectivity. The competing reduction to CO and water could be kinetically inhibited, which is due to the high hydrogen overpotential of these materials. Homogeneous catalysts can be divided into three subcategories: (i) catalysts with macrocyclic ligands, (ii) catalysts with bipyridyl ligands, and (iii) catalysts with coordinated phosphine ligands. In 1980, Fisher and Eisenberg reported tetraazomacrocyclic complexes of cobalt and nickel which showed high current efficiencies (up to 98%), but a low turnover number (TON) of only nine [18]. Sauvage and co-workers studied Ni^{II}(cyclam) complexes, which were found to be stable, highly efficient, and selective [19]. Non-precious metal catalysts were also studied and Savéant's group reported on iron(0) porphyrins, which are capable of reducing CO_2 to CO at -1.8 V versus the normal hydrogen electrode (NHE) in dimethylformamide (DMF) [20]. Since the 1980s, several metal complexes bearing bipyridine ligands have been investigated. Here, mainly rhenium [21] and ruthenium [22] complexes were found to reduce CO_2 at -1.49 and -1.40 V vs. NHE, respectively. Rhodium and iridium complexes were studied by Meyer's group, but relatively poor efficiencies (64% for HCO_2^-) and turnover numbers (6.8–12.3) were found. Interestingly, no CO was detected during the experiments, hence high selectivity can be concluded [23]. Slater and Wagenknecht reported in 1984 a Rh(dppe)$_2$Cl [dppe = 1,2-bis(diphenylphosphino)ethane] complex which showed current efficiencies of ∼42% for CO_2 to formate at -1.55 V vs. NHE [24]. Palladium complexes were studied most extensively for CO_2 reduction. For instance, [Pd(triphos)(solvent)][BF$_4$]$_2$ was found to form a Pd(I) intermediate under the reaction conditions that subsequently reacts with CO_2 to form a five-coordinate CO_2 adduct. After reduction and double protonation of the Pd–CO_2 complex, CO dissociation was observed. These classes of Pd complexes have shown catalytic rates in the range $10–300$ M^{-1} s^{-1} with current efficiencies exceeding 90% for CO production [25]. Recently, organocatalysts have also gained considerable attention. In this respect, Bocarsly and co-workers used simple pyridinium derivatives to synthesize FA, formaldehyde, and methanol from CO_2 [26]. At hydrogenated Pd electrodes, current efficiencies of up to 11% for FA and

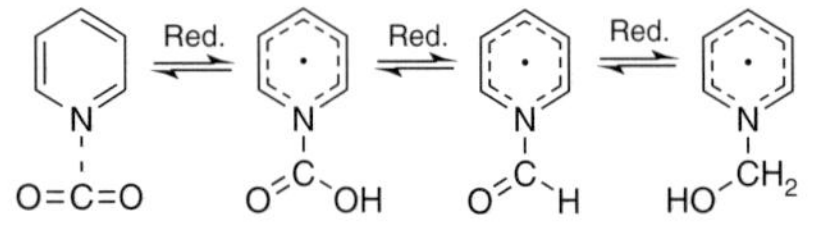

Scheme 18.2 Proposed species associated with the pyridinium-catalyzed CO_2 electro-reduction.

22% for methanol production were observed. By the use of cyclic voltammetry (CV) and also molecular orbital calculations and NMR studies, they obtained details of the mechanism. The data indicated that a complete reduction of carbon dioxide to methanol occurred sequentially through $2e^-$- and $4e^-$-reduced intermediates FA and formaldehyde, respectively, and that pyridinium radicals play a key role in that cycle (Scheme 18.2) [27].

18.2.2
Photochemical CO₂ Reduction

Significant energy input is needed to promote the conversion of thermodynamically stable CO_2. Except for nuclear power and renewable energy, almost all energy is derived from fossil-based raw materials with concomitant production of CO_2. Obviously, an almost unlimited source of energy is sunlight, which provides more than 2.9×10^{24} J per year (calculated from the ASTM G137 air mass 1.5 spectra; data are available from the Renewable Energy Resource Center of the National Renewable Energy Laboratory of the US Department of Energy at *http://rredc.nrel.gov/solar/spectra/am1.5/*) [28]. Hence it is highly desirable to use photocatalysis for the reduction of CO_2 [29, 30]. The formal reduction of CO_2 to $CO_2{}^-$ is highly unfavorable, having a formal reduction potential of -2.14 V [29]. Proton-assisted electron-transfer processes are an attractive alternative to reduce CO_2 at much lower potentials. The general task for a catalytic photosystem is the generation of highly reduced species with the help of light irradiation. Transfer of the reduction equivalents on to CO_2 permits reduction to, for example, CO or FA (Scheme 18.3).

The reaction principle is related to natural photosystems (PSI, PSII), which have been developed by evolutionary processes [31, 32]. Here reduction equivalents are generated by light excitation in PSI and PSII. The electrons for this process are taken from water, which is oxidized to oxygen in the presence of $CaMn_4$ clusters in the oxygen-evolving complex. Inspired by the beauty of nature, researchers try to mimic these systems (Scheme 18.4). In homogeneous multicomponent systems (type I, Scheme 18.4) photosensitizers (PSs) are used to harvest light. Upon light excitation, the excited state is reduced by a sacrificial reagent (SR), mostly amines. The reduced state PS^- is formed. From the latter, electrons are transferred onto metal catalysts (cat), which interact with CO_2 and reduce the latter to CO or FA. In addition to the classical $Ru(bpy)_3{}^{2+}$ [30, 33] and derivatives thereof, phenazine [34] and *p*-terphenyl [35] have also been used as PSs. As catalysts, tetraaza macrocyclic complexes of nickel and cobalt or Re–bpy complexes have been explored [29, 35, 36]. Re-based catalysts are most active, with turnover numbers of 48 [37].

Scheme 18.3 Principle of photocatalytic reduction of CO_2.

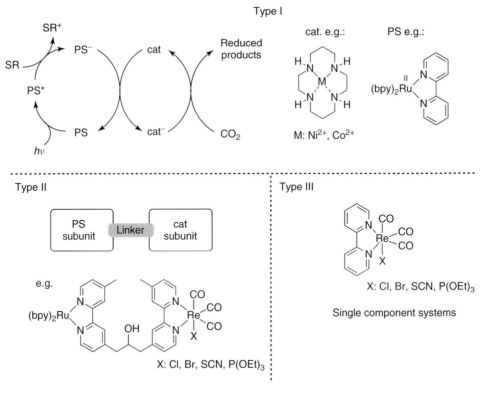

Scheme 18.4 Multicomponent (type I), dyad (type II), and single-component systems (type III) for photocatalytic CO_2 reduction.

Multicomponent systems for CO_2 reduction are related to systems for the reduction of aqueous protons in water reduction systems [38–40].

In supramolecular devices (type II, Scheme 18.4), the PS and catalyst (cat) are coupled via a linker unit in one molecule [41, 42]. This strategy leads to faster electron transfer and improved stabilities and activities. The highest productivities for such dyad systems are TON 232 and a quantum efficiency of 0.21 for an Ru–Re dyad [29]. Instead of multicomponent systems, single-component systems (type III, Scheme 18.4) were also developed. Here, light interaction and CO_2 reduction are carried out at the same metal center. The most prominent examples are complexes of the type $[(N^\wedge N)Re(CO)_3X]$ $[N^\wedge = 4,4'$-R-2,2'-bipyridine, 1,10-phenanthroline; X $=$ Cl, Br, SCN, P(OEt)$_3$] [40]. Other work is based on metallomacrocyclic catalysts such as metalloporphyrins (TONs range from 40 to 300) [29, 43, 44]. Notably, all these systems are able to reduce CO_2 in the presence of light and an SR, although the activities are still not very high and need to be further improved. However, also if much higher activity could be established, a major challenge remains: the coupling of CO_2 reduction with a reasonable oxidation reaction (e.g., water oxidation to oxygen) to avoid the application of SRs. This is the barrier that has to be surmounted for "real-world" applications. Keeping this in mind, photocatalysis can contribute to a great extent to developing useful methods for recycling of CO_2 and the production of "solar fuels." Recently, the direct photocatalytic hydrogenation of CO$_2$ was achieved via a rhenium(I) phenanthroline–polyoxometallate hybrid [45]. Notably, this dyad uses no SR. Key to success is the Pt-catalyzed activation of hydrogen by oxidative addition on a polyoxometallate. Thus, the origin of electrons in this system is hydrogen. However, the reported activities (TON 23, quantum yield 0.011) are very low.

Heterogeneous photocatalysts for the reduction of CO$_2$ are also under intense investigation. Most of them are based on TiO$_2$ and due to the bandgap of 3.2 eV they are only active under UV irradiation. These systems have been reviewed elsewhere [46, 47] and will not be discussed here.

18.2.3
Catalytic Reduction of CO$_2$ to Methanol

With an annual production of more than 40 Mt in 2007, methanol is an important bulk scale product of the chemical industry [6]. The recent industrial production of methanol is based on fossil resources. In a catalytic reaction, syngas (a mixture of CO and H$_2$) is converted to CH$_3$OH using heterogeneous catalysts at high temperature (>200 °C) (Scheme 18.5). Interestingly, additional CO$_2$ is often introduced in order to improve the catalyst activity. Applied catalysts are based on copper supported on ZnO, ZrO$_2$, Al$_2$O$_3$, TiO$_2$, or SiO$_2$. The typical reaction conditions are 5.0–10.0 MPa and 473–523 K [48].

As shown in Scheme 18.6, methanol is used for the production of various other large-scale chemicals such as acetic acid (Monsanto process), formaldehyde, and methyl *tert*-butyl ether (MTBE). Additionally, methanol is considered as a sustainable platform for the post-fossil fuel age [49]. It has excellent combustion characteristics and can be used in direct methanol fuel cells to produce energy. Dimethyl ether (DME), which is easily produced from methanol by dehydration over acidic catalysts, has been proposed as a diesel substitute.

In order to meet the goal of sustainable methanol production, reorganization of the starting materials from fossil-based CO towards atmospheric CO$_2$ is essential. A benign perspective is the hydrogenation of CO$_2$ in the exothermic reaction shown in Scheme 18.7.

$$CO + 2 H_2 \longrightarrow CH_3OH \quad \Delta H_{298\,K} = -90.8 \text{ kJ mol}^{-1}$$

Scheme 18.5 Thermodynamics of CO reduction with H$_2$.

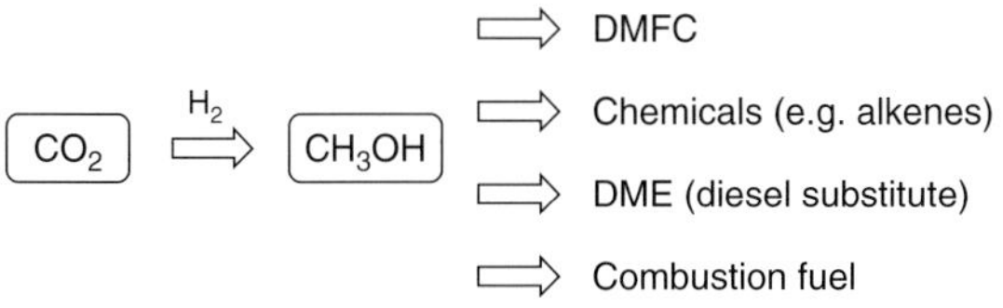

Scheme 18.6 Methanol as chemical and energy platform.

$$CO_2 + 3\,H_2 \longrightarrow CH_3OH + H_2O \qquad \Delta H_{298\,K} = -49.4\ \text{kJ mol}^{-1}$$

Scheme 18.7 Thermodynamics of methanol production via carbon dioxide hydrogenation.

Despite the capture of dilute CO_2 from the atmosphere (380 ppm), the issues of availability and green hydrogen and also large amounts of energy and highly active catalysts have to be addressed. In view of these challenges, it is interesting that the implementation of a commercial CO_2 to methanol plant was successfully demonstrated in Iceland. Key to the success was geothermal energy for water electrolysis and as a CO_2 source, that was used to establish a sustainable CO_2 to liquid fuel process [50]. However, so far such initiatives (examples from Japan, Australia, and Europe are also known) are pilot plants, because they cannot compete economically with classical methanol production from syngas. However, this may change as a result of CO_2 taxes or strong environmental policies in general. Therefore, efforts are being made to explore possible catalysts that may lead to more active systems and a deeper understanding of the reactions.

18.2.3.1 Heterogeneous Catalysis

The development of heterogeneous catalysts for the production of methanol from CO_2 is highly desirable, since heterogeneous catalyst systems are, in terms of applicability in industrial bulk processes, often superior to homogeneous systems. Recent developments in catalyst design have focused on modification of copper catalysts, which are used for industrial methanol synthesis from CO and H_2. The majority of the systems studied contain copper and zinc. Different additives (VO_x, MnO_x, and MgO) and the influence of different metal oxides (Al_2O_3, TiO_2, Ga_2O_3) have been investigated [51]. Notably, a number of large-scale processes and pilot plants were demonstrated [8, 48]. Especially $Zr-ZrO_2$ doping of $Cu-ZnO_2$ led to improved stability of the catalysts [48, 52, 53]. Watanabe and co-workers reported a multicomponent catalyst [54] ($Cu-ZnO-ZrO_2-Al_2O_3-SiO_2$) with remarkably high selectivity (99.8%) and a lifetime of less than 1 year [54–56]. In a test plant, up to 50 kg per day of MeOH were produced at 250 °C and 5 MPa [55]. In the field of heterogeneous catalyst, zeolite catalysts were investigated via theoretical methods by Chan and Radom [57, 58].

The main challenge for the design of new heterogeneous catalysts is selectivity. Often the competitive reverse water gas shift (RWGS) reaction determines the methanol selectivity. In general, catalysts capable of catalyzing methanol production from CO_2 also promote the RWGS reaction and permit a second reaction pathway for methanol production via CO hydrogenation (Scheme 18.8). A serious debate has started on the mechanism and so far it is not clear whether methanol production starts from CO or CO_2 [59].

$$CO_2 + 3\,H_2 \longrightarrow CH_3OH + H_2O \qquad -49.4\ \text{kJ mol}^{-1}$$

$$\text{RWGS} \quad CO_2 + H_2 \longrightarrow CO + H_2O \qquad +41.1\ \text{kJ mol}^{-1}$$

$$CO + 2\,H_2 \longrightarrow CH_3OH \qquad -90.8\ \text{kJ mol}^{-1}$$

Scheme 18.8 Possible RWGS reaction pathways.

18.2.3.2 Homogeneous Catalysis

The main products of homogeneous CO_2 reduction are CO and FA/formates (see Section 18.2.4). However, owing to its high potential as a fuel substitute and as a platform to produce other chemicals, it is highly desirable to find suitable catalysts that will allow the production of methanol from CO_2, and some interesting reactions have been reported. In 1993, Saito and co-workers reported a catalyst system based on $Ru_3(CO)_{12}$ capable of the hydrogenation of CO_2 to a mixture of CO, methanol, and methane. A halide additive was used in order to prevent Ru metal formation. A TON of 95 (for the production of MeOH) was obtained using an $Ru_3(CO)_{12}-KI$ system in N-methylpyrrolidone (NMP) at 240 °C [60]. A reaction mechanism was proposed, as shown in Scheme 18.9.

A major challenge for the reduction of CO_2 is to overcome the C–O bond enthalpy (532 kJ mol^{-1}). A number of approaches have resolved this issue by the formation of strong B–O, Al–O, and Si–O bonds and conversion of CO_2 to reduced products. With the catalyst Ir(CN)(CO)dppe, hydrosilylation of CO_2 to methoxy species was observed as early as 1989 by Eisenschmid and Eisenberg [62]. Similar reactions also occurred with ruthenium, although the final product was a silyl formate [63].

In the last few years, new concepts for CO_2 reduction to methanol have been reported (Scheme 18.10). Stephan and co-workers reported the frustrated Lewis pair (FLP) [64] $P^tBu_3 - B(C_6F_5)_3$ for the reversible activation of CO_2 under mild conditions (1 atm, room temperature) [65]. Shortly thereafter, a related concept was used for the selective hydrogenation of CO_2 to methanol using an FLP based on 2,2,6,6-tetramethylpiperidine and $B(C_6F_5)_3$ by Ashley *et al.* [66]. A combination of AlX_3 (X = Cl, Br) and $PMes_3$ (Mes = 2,4,6-$C_6H_2Me_3$) was used for the same transformation using ammonia borane (H_3NBH_3) as the reducing agent [67, 68]. A second concept without the use of metals as catalysts was developed on the basis of N-heterocyclic carbenes (NHCs). TONs of up to 1840 [turnover frequency (TOF) 25.5 h^{-1}] were reported by Zhang and co-workers for the hydrosilylation of CO_2 using NHCs as organocatalysts [69]. Additionally, the role of these organocatalysts was investigated theoretically [70].

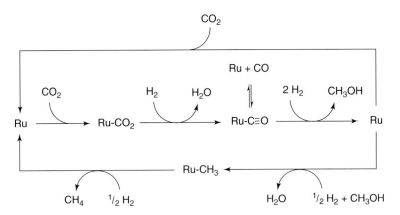

Scheme 18.9 Catalytic reduction of CO_2 to CO, CH_3OH, and methane using an $Ru_3(CO)_{12}$ catalyst [61].

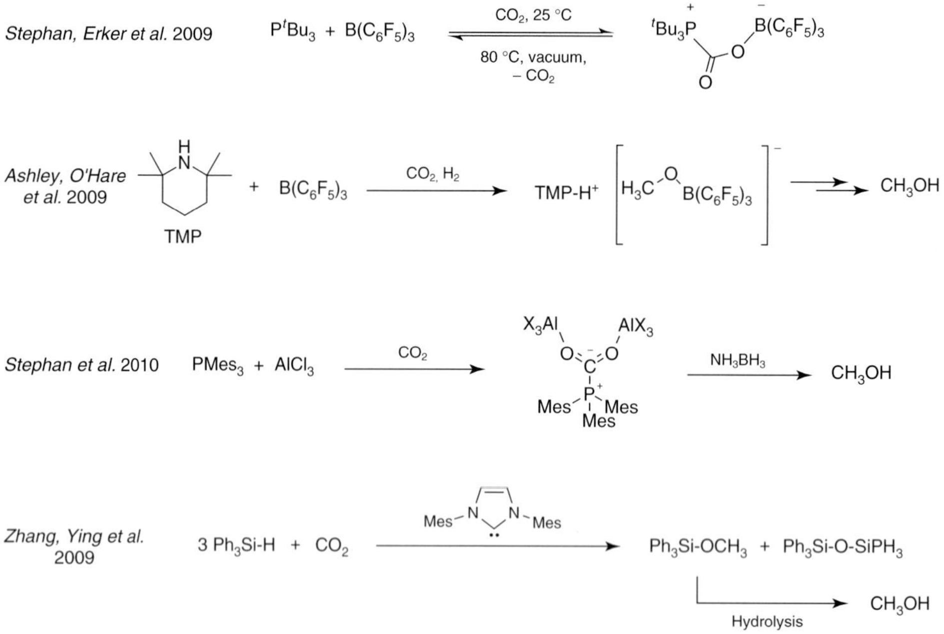

Stephan, Erker et al. 2009 P^tBu_3 + $B(C_6F_5)_3$ ⇌ (CO_2, 25 °C / 80 °C, vacuum, $-CO_2$) $^tBu_3P^+$ ...O... $B(C_6F_5)_3^-$

Ashley, O'Hare et al. 2009   TMP + $B(C_6F_5)_3$ → (CO_2, H_2) TMP-H$^+$ [H_3C–O...$B(C_6F_5)_3$]$^-$ → CH_3OH

Stephan et al. 2010 $PMes_3$ + $AlCl_3$ → (CO_2) [X_3Al...O=C...O...AlX_3, Mes_3P^+] → (NH_3BH_3) CH_3OH

Zhang, Ying et al. 2009   3 Ph_3Si-H + CO_2 → (Mes-N...N-Mes carbene) Ph_3Si-OCH_3 + Ph_3Si-O-$SiPH_3$ → (Hydrolysis) CH_3OH

Scheme 18.10 Reduction of carbon dioxide with frustrated Lewis pairs and NHC catalysts.

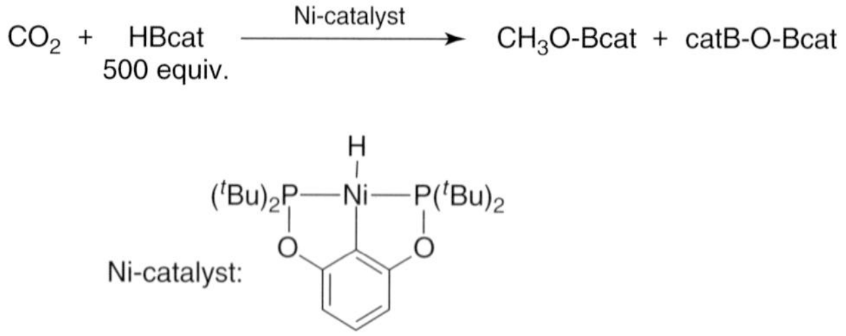

CO_2 + HBcat (500 equiv.) → (Ni-catalyst) CH_3O-Bcat + catB-O-Bcat

Ni-catalyst: ($^tBu)_2P$—Ni(H)—$P(^tBu)_2$ pincer complex

Scheme 18.11 CO_2 reduction by a nickel–hydrido pincer complex [71].

Nickel catalysts for the reduction of CO_2 to methoxide were reported by Guan and co-workers [71]. Using catecholborane (HBcat) as hydroboration reagent, a TOF of 495 h^{-1} (based on HB) was obtained in 1 atm CO_2 at room temperature (Scheme 18.11).

The presented examples are nice demonstrations of new concepts in the field of CO_2 reductions. Nevertheless, one has to clearly state that all of them make use of expensive reduction reagents (e.g., boranes, silanes) and produce lots of waste. Hence, practical applications which makes use of these reactions are not being expected.

18.2.3.3 Enzymatic Approaches

Since the matter of life is based on CO_2 as raw material, it is not surprising that biocatalytic solutions for the reduction CO_2 to methanol are considered. For example, Obert and Dave reported an enzymatically coupled sequential reduction of CO_2

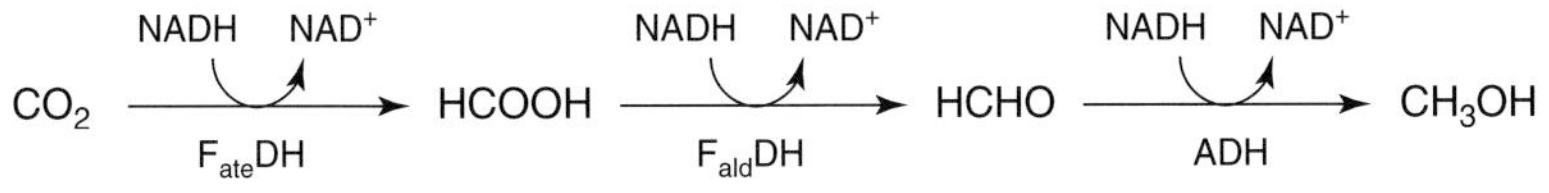

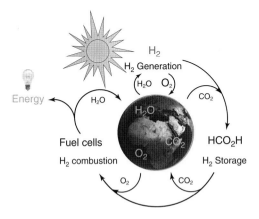

Scheme 18.12 Sequential enzymatic reduction of CO_2 by the three different dehydrogenases: formate dehydrogenase ($F_{ate}DH$), formaldehyde dehydrogenase ($F_{ald}DH$), and alcohol dehydrogenase (ADH).

using three different dehydrogenases (Scheme 18.12) [72]. Initial reduction of CO_2 by formate dehydrogenase is followed by conversion of formate to formaldehyde by formaldehyde dehydrogenase. The production of methanol from formaldehyde is catalyzed by alcohol dehydrogenase. This elegant system is based on NADH as electron donor, the final electron source for the three reductions. Encapsulation of the system results in enhanced activities [72].

18.2.4
Catalytic Reduction of CO₂ to Formic Acid

Formic acid is used in the food (preserving agent), leather (etching), and agricultural industries and also for decalcination. In addition, formates (e.g., $NaHCO_2$) are applied to defrost airstrips [73]. Organic chemistry uses FA as a reducing reagent, for example, in transfer hydrogenation reactions [74]. Recently, we and others have proposed FA as a potential hydrogen storage material.

A potential hydrogen-based energy scenario using abundant water, oxygen, and carbon dioxide is depicted in Scheme 18.13. (i) Hydrogen is generated by photocatalytic water splitting producing H_2 and O_2 [75]. (ii) A so far unsolved problem is the storage of hydrogen. We and others [76] have suggested using FA or formates as hydrogen storage material. Carbon dioxide can be converted into formates by catalytic hydrogenation. Formic acid (4.4 wt% H_2) or formates (1.4 wt% H_2 for sodium formate) are both nontoxic

Scheme 18.13 Sustainable energy cycle based on photocatalytic water splitting, FA as hydrogen storage material, and fuel cells as energy supply.

	$\Delta G°$ kJ mol^{-1}	$\Delta H°$ kJ mol^{-1}	$\Delta S°$ J mol^{-1}
CO_2 (g) + H_2 (g) $\longrightarrow$ HCO_2H (l)	+32.9	−31.2	−215
CO_2 (aq) + H_2 (aq) $\longrightarrow$ HCO_2H (l)	−4	n.a	n.a
CO_2 (g) + H_2 (g) + NH_3 (aq) $\longrightarrow$ NH_4^+ (aq) + HCO_2^- (aq)	−9.5	−84.3	−250
CO_2 (g) + H_2 (g) + $HNMe_2$ (l) $\longrightarrow$ $HC(O)NMe_2$ (l) + H_2O (l)	n.a	−239	n.a
HCO_2H $\longrightarrow$ CO + H_2O	−12.4	28.7	+138

Scheme 18.14 Thermodynamic data for the hydrogenation of carbon dioxide to formates and FA derivatives.

and stable compounds that can easily be stored and transported as liquids (FA) or solids (formates). (iii) Notably, FA can be selectively decomposed, giving a hydrogen carbon dioxide mixture (1:1). (iv) Hydrogen can effectively be used in fuels cells, producing energy and water as the only by-product. Capture of carbon dioxide and back-transformation into FA/formates closes the energy cycle of the proposed "formic acid economy." A convenient production of FA can be achieved via catalytic hydrogenation of carbon dioxide. However, this reaction possesses unfavorable thermodynamics (negative entropy) – two gases react to give a liquid – and high pressures have to be used during the reaction. However, with either the use of amines or the formation of formates in basic media and also suitable solvents, FA can be stabilized and the whole process can be exergonic (Scheme 18.14).

Therefore, reactions are performed in water as solvent or co-solvent as hydrated starting materials and products reduce the standard free energy from +33 to −4 kJ mol^{-1} [77]. If water is applied, the CO_2–bicarbonate–carbonate equilibrium has to be considered and the pH of the reaction solution and the solubility of CO_2 have a great influence on the process. However, crucial in all cases is the solubility of H_2 because it is comparably low. To enhance the solubility of H_2 and mass transport processes, often supercritical CO_2 (scCO_2) was applied and higher activities and stabilities were observed. In the following section we consider heterogeneous and homogeneous catalysis for CO_2 hydrogenation to FA and formates. Additionally, different mechanistic proposals are presented and discussed.

18.2.4.1 Heterogeneous Catalysis

While significant efforts have been made to study the homogeneous hydrogenation of CO_2 to FA derivatives, research on the heterogeneous catalysis of CO_2 hydrogenation has been scarce. This is due to the inconvenient reaction conditions for typical heterogeneous catalysis of CO_2 reduction to FA. Gas-phase reactions and high temperatures (CO_2 hydrogenation is exothermic) lead to low chemoselectivity towards FA/formates. Hence most of the known heterogeneous catalysts hydrogenate CO_2 to CO, methanol, and methane more easily. Notably, one of the first reports on the synthesis of FA

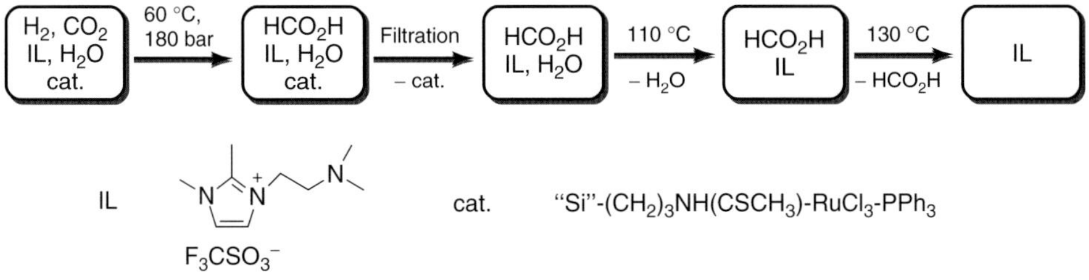

Scheme 18.15 Reaction and isolation protocol for heterogenized ruthenium catalyst.

from CO_2 was appeared as early as 1935 during various studies of different water gas shift catalysts! Farlow and Adkins reported that CO_2 was hydrogenated in the presence of Raney nickel at 80 °C and 200–400 atm or less if an amine was introduced into the reaction mixture [78]. At higher temperatures they found that the produced formate dehydrated, resulting in substituted formamidines. Subsequent studies focused on cobalt, iron, nickel, and copper–alumina and their alloys as catalysts for the reduction of CO_2 to various products (mainly methanol and methane) [79]. In 1988, Wiener *et al.* applied Pd on charcoal to reduce carbonates to FA under mild conditions [80]. They observed a strong influence of carbonate concentration on the initial rates. However, no complete conversion was observed. Heterogeneous catalysts that are prepared by immobilizing ruthenium on silica and polystyrene resin have also been used for this reaction. Applying an "Si"–(CH_2)–$NH(CH_2)_3CH_3$–Ru catalyst together with NEt_3, activities (TOF) of more than 1384 h^{-1} with 100% selectivity at 80 °C and 160 bar were obtained [81]. In 2008, Han and co-workers presented an immobilized ruthenium catalyst of the type "Si"–$(CH_2)_3$ $NH(CSCH_3)$–$RuCl_3$ – PPh_3, which was applied in combination with an ionic liquid (IL), 1-(*N,N*-dimethylaminoethyl)-2,3-dimethylimidazolium trifluoromethanesulfonate ([mammim][TfO]) [82]. In aqueous solution, both components catalyze the hydrogenation of CO_2 with fairly good activity (TOF = 103 h^{-1} at 60 °C and 180 bar) and high selectivity. The highest conversion, a molar ratio of 0.96 (FA:IL), was obtained after 10 h under these conditions. Notably, this system possesses some unique features such as reusability of catalyst and IL and allows the relatively easy isolation of FA via a simple distillation procedure (Scheme 18.15).

18.2.4.2 Homogeneous Catalysis

The homogeneous hydrogenation of CO_2 to FA derivatives has been of interest for more than 40 years and the first reports are from the mid-1970s [83]. Since then, many improvements have been achieved and highly active catalysts are known today. Several reviews have summarized the results over the years [77, 84]. In Table 18.3, a summary of the most active catalyst systems known to date is given.

As shown in Table 18.3, highly active catalyst systems have been achieved for the CO_2 and also bicarbonate hydrogenation. Especially the recent work of Nozaki's group has marked a significant breakthrough. The use of a defined iridium–pincer trihydride complex [Ir(III)PNP] (Scheme 18.16) containing two diisopropylphosphino substituents and KOH as base led to an outstandingly high TON of 3 500 000 with a TOF of 73 000 h^{-1}

Table 18.3 Catalytic hydrogenation of CO_2 and $NaHCO_3$ with precious and non-precious metal catalysts.

Catalyst precursor	Solvent	Base/ additive	p_{H_2/CO_2} (bar)	T (°C)	t(h)	TON	TOF (h^{-1})	Ref.
$[Ir^{III}PNP]$	H_2O	KOH	30/30	120	48	3.5×10^6	73 000	[85]
$[Cp^*Ir(phen)Cl]Cl$	H_2O	KOH	30/30	120	48	222 000	33 000	[86]
$[RuCl_2(OAc)(PMe_3)_4]$	$scCO_2$	NEt_3/C_6F_5OH	70/120	50	0.3	31 667	95 000	[87]
$[RuH_2\{PMe_3\}_4]$	$scCO_2$	MeOH	80/120	50	0.5	2000	4000	[88]
$[RuH_2\{PMe_3\}_4]$	$scCO_2$	$NEt_3/dppb$	85/120	50	47	7200	150	[89]
$[RuClCp^*(Dhphen)]Cl$	H_2O	KOH	30/30	120	24	15 400	642	[90]
$RuCl_2(p$-cymene$)(PTA)^a$	H_2O	–	100/0	70	n.a.^{b}	n.a.	287	[91]
$[RuCl_2(benzene)]_2/dppm^a$	H_2O/THF	–	80/0	70	2	1374	687	[93]
$[RuCl_2(benzene)]_2/dppm^a$	H_2O/THF	–	50/35	70	2	2518	1259	[92]
$[RuCl_2(tppms)_3]^a$	H_2O	–	60/35	80	n.a.	n.a.	9600	[93]
$[Rh(acac)(dcpb)]$	DMSO	NEt_3	20/20	RT^c	0.2	267	1335	[94]
$[RhCl(tppts)_3]$	H_2O	$NHMe_2$	20/20	RT	12	3439	287	[95]
$[RhCl(cod)]_2$	DMSO	$Net_3/dppb$	20/20	RT	22	1150	52	[96]
$[RhCl(PPh_3)_3]$	MeOH	Net_3/PPh_3	20/40	25	20	2700	125	[97]
$PdCl_2$	H_2O	KOH	110/n.a	160	3	1580	530	[98]
$Fe(BF_4)_2 \cdot 6H_2O/PP_3{}^a$	MeOH	–	60/0	80	20	610	30.5	[99]
$FeCl_3$, dcpe (1:1.5)	DMSO	DBU	40/60	50	n.a	113	15.1	[100]
$NiCl_2$, dcpe (1:1.5)	DMSO	DBU	40/160	50	216	4400	n.a.	[100]

^{a} $NaHCO_3$ is used as C1 source.
^{b} n.a., Not applicable.
^{c} RT, room temperature.

[85]. The rates obtained open up the possibility of applying CO_2 hydrogenation on an industrial scale, although scale-up has not been demonstrated so far.

The strong coordination between the metal and the tridentate ligand prevents the latter from being easily removed from the metal center. In addition, the pyridine-based ligand possesses an interesting redox chemistry and opens up a unique type of cooperative catalysis [101]. However, only in a highly diluted system (100 ppb) were the remarkable activity and a maximum FA:base ratio of 0.70 obtained. Further development towards the gram-scale production of formates is needed. Apart from the success of precious metal catalysts capable of hydrogenating carbon dioxide or bicarbonates, non-noble metal catalysts have recently attracted considerable attention. With the use of 0.14 mol% of an *in situ* catalyst composed of $Fe(BF_4)_2 \cdot 6H_2O$ and the tetraphos ligand (PP_3), sodium formate was formed from bicarbonates in excellent yield (88%) and with significant activity (TON = 610). No additional CO_2 pressure or base was needed [99].

If aqueous carbonate solutions are used, the yield of FA in general does not exceed 1 mol of FA per mole of base. In organic solvents [83b] or $scCO_2$, yields of up to 1.9 (FA:base) can be obtained [88, 89]. Recently, Fachinetti and co-workers obtained an FA–TEA (triethylamine) azeotrope of 1.78 resulting after distillation in an FA:TEA molar ratio of 2.35 [102]. Key to the success was incorporation of CO_2 and H_2 in premixed

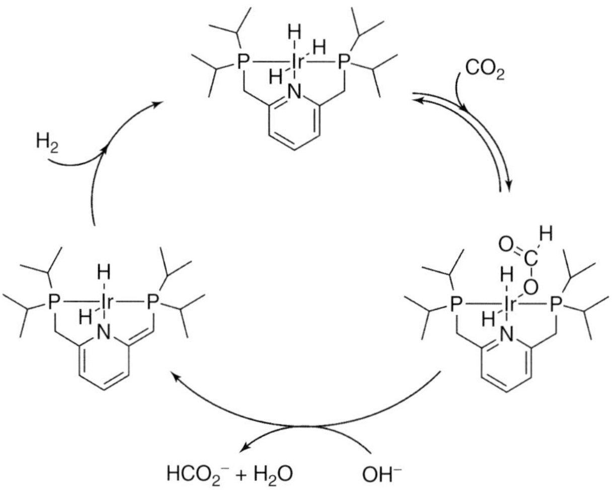

Scheme 18.16 CO₂ hydrogenation by Nozaki and co-workers.

FA–TEA solution with an [RuCl$_2$(PMe$_3$)$_4$] precursor at 40 °C and 120 bar. In most reported hydrogenations of carbon dioxide to FA, an excess of hydrogen and carbon dioxide is necessary for high conversions. Recently, our group has shown that nearly full conversion, with respect to CO$_2$, can be achieved, if bicarbonates are hydrogenated (without additional CO$_2$ pressure) [92, 99]. Moreover, the solvent plays a crucial role in all catalytic systems. Owing to the better solvation properties of formates in polar solvents such as dimethyl sulfoxide (DMSO), they are superior to nonpolar solvents for entropic reasons. Even in aprotic solvents, or scCO$_2$, at least a small quantity of water or an alcohol is needed for this purpose. In addition to their ability to solvate the products, solvents can also play a key role in the catalytic cycle by stabilizing reactive intermediates, and especially for water several interactions have been proposed [103]. In addition to the use of polar additives, more often bases (with water or organic solvents) are used. Apart from TEA, bases with a higher pK_A [e.g., 1,8-diazabicyclo[5.4.0]undec-7-ene (DBU)] often provide higher activity and higher FA:base ratios [87]. The reason for the differences between the bases is yet not clear, but many options can be discussed. In addition to the formation of FA–base educts, which have drastic effects on the thermodynamics, an active role of the base can also be assumed. Probably ammonium species (e.g., HNEt$_3^+$) influence the hydrolysis of the metal formates, or promote the dissociation of FA from the complex. Moreover, they could assist the CO$_2$ insertion into a metal–hydride bond (discussed later). Another option for the enhanced activity, especially in the case of amidine or guanidine bases, is the possibility of binding CO$_2$ effectively and therefore enhancing the mass transport. Effective bases in organic solvents normally have a pK_A of 8–12. Weaker bases such as DMF and pyridine showed lower activity and therefore smaller FA:base ratios [77, 88, 89, 104]. In the presence of water, bases will also promote the formation of bicarbonates. Apart from water, alcohols have also been used as additives. In this regard, Munshi *et al.* investigated a large number of alcohols as co-catalysts in the presence of [RuCl(OAc)(PMe$_3$)$_4$] as catalyst. They observed that the most active alcohols were those with a pK_A below that of the protonated amine (10.7 for $HNEt_3^+$) and a potentially coordinating conjugated base [e.g., triflic acid and $(CF_3)_2C_6H_3OH)$] [87]. The effect of alcohols was further investigated and a concerted ionic mechanism, in which a hydride from the metal and a proton from the alcohol or protonated base are transferred to carbon dioxide in one step, was developed [87, 105, 106].

In addition to the given explanation for the effects of different additives, several other mechanisms [84], especially for rhodium and ruthenium complexes, have also been proposed. Apart from the ionic hydrogenation [107] and concerted ionic hydrogenation mechanisms [105], insertion and bicarbonate hydrogenation mechanisms are also known. CO_2 readily undergoes an insertion reaction with metal hydrides to form a metal formate complex [108, 109], and this is in general considered to be the first step in the catalytic cycle and is often regarded as the rate-determining step (Scheme 18.17) [77]. The mechanism of this insertion reaction is a recent research topic and mainly two possibilities are likely (Scheme 18.18): in path (a) CO_2 insertion takes place without prior CO_2 coordination and in path (b) a ligand dissociates from the metal complex and CO_2 coordinates to the metal prior to insertion.

Matsubara and Hirao suggested that path (a) is predominant for arene–Ru complexes in the presence of bidentate phosphine ligands, for example, dppm [bis(diphenylphosphino)methane] [106, 110]. On the other hand, *ab initio* calculations on $RuH_2(PH_3)_4$ performed by Musashi and Sakaki [111] support path (b), where prior to coordination of CO_2 the dissociation of a ligand is needed for the insertion of CO_2 in the metal–hydride bond.

The liberation of FA can further proceed via three different paths: (i) hydrogenolysis, (ii) hydrolysis or alcoholysis, and (iii) reductive elimination and H_2 activation (Scheme 18.19). Further hydrogenolysis can proceed either through oxidative addition of H_2 (homolytic H_2 splitting), or coordination of H_2 followed by elimination of FA (heterolytic H_2

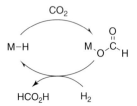

Scheme 18.17 Insertion of CO_2 into metal hydride species.

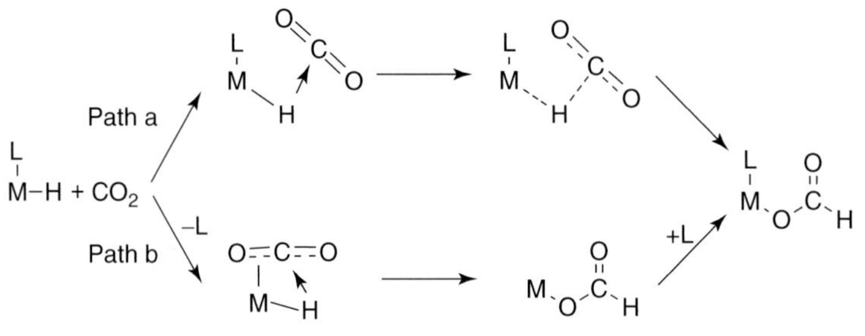

Scheme 18.18 Different possibilities for CO_2 insertion into a metal hydride bond.

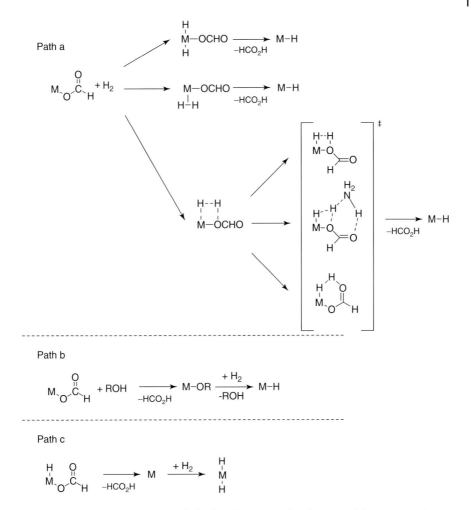

Scheme 18.19 Mechanistic proposals for the elimination of FA from metal formate complexes.

splitting) or via a "sigma-bond metathesis" mechanism. The last mechanism can include a four-centered and base-assisted or six-centered transition state without prior coordination of H_2 (Scheme 18.19) [84]. However, so far it is not predictable which mechanism plays the predominant role for a selected catalyst system. Studies on the decomposition of FA derivatives concluded that the insertion mechanism should operate for Ru(II) catalysts [112, 113]. In addition, the present base (e.g., amines) also played a crucial role [114].

The hydrogenation of bicarbonates or $CO_2 +$ base in water is shown in Scheme 18.20. Several groups [84, 93, 115] have also studied the mechanism of this hydrogenation in amine-free aqueous solutions. In general, the pH of the reaction solution must be significantly basic (pH > 8), otherwise a decrease in activity is observed. A simplified reaction pathway for the bicarbonate hydrogenation is given in Scheme 18.21.

$$CO_2 + H_2O \xrightleftharpoons[]{\text{Base}} HCO_3^- + H^+ \xrightleftharpoons[]{H_2} HCO_2^- + H^+ + H_2O$$

Scheme 18.20 Hydrogenation of bicarbonates to formates.

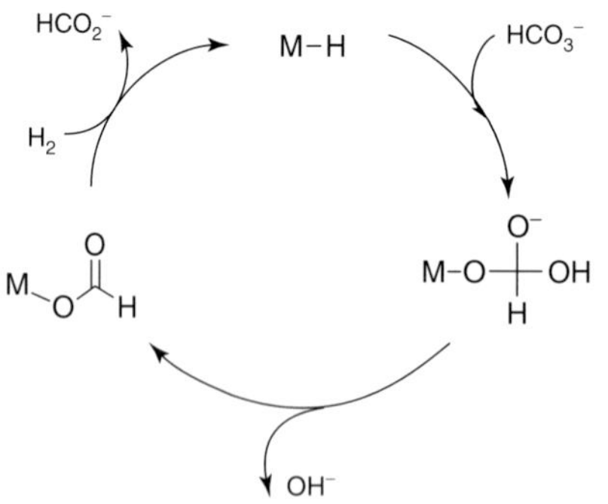

Scheme 18.21 Catalytic cycle for hydrogenation of bicarbonates to formates.

18.3
CO$_2$ as a C1-Building Block in C–C Coupling Reactions

The final product of any combustion of carbon-containing material is carbon dioxide and water. CO_2 represents an energy sink and considerable energy input is needed in order to produce reduced products. However, the bonding of nucleophiles (e.g., OH$^-$ from water to produce bicarbonate) is energetically favored. The reaction with other nucleophiles [116], although kinetically a challenge, opens up a variety of reactions involving C–N (e.g., carbamates), C–O (e.g., carbonates), and C–C (e.g., carboxylations) bond formation. Established industrial processes such as the Kolbe–Schmitt reaction (C–C formation, salicylic acid) or the production of urea (C–N formation) rely on this principle and use CO_2 as a *C1*-building block.

Classical organic synthesis uses CO_2 as reactant for Grignard or organolithium reagents for the production for carboxylic acids (Scheme 18.22). Inspired by the numerous cross-coupling reactions [117], a number of studies have dealt with the use of CO_2 as an electrophile. In general, transition metal-catalyzed reactions show higher functional group tolerance than additions to Grignard or organolithium compounds. Protocols for the generation of carboxylic acids [118] were developed for various substrates, including boronic acids [119], stannates [120, 121], and zinc reagents [122]. Typically, catalysts are based on nickel, rhodium, palladium, and copper. Protocols for the coupling of aromatic and also alkenyl and alkyl groups are available [119–122]. The coupling of alkynes can be achieved via copper-catalyzed C–H activation [123, 124]. Recently, the coupling of aromatic moieties and CO_2 via C–H activation was also discovered by different groups using copper [125] and gold [126] catalysts. Additionally, a protocol without the use of any metal is available, although the substrate scope is limited to CH-acidic heteroarenes [127]. Valuable products such as acrylic acids, propiolic acids, and benzoic acids are formed during these reactions. The most economically important product is acrylic acid.

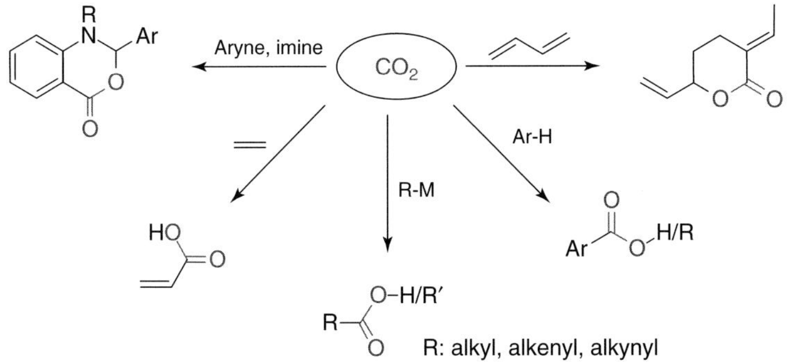

Scheme 18.22 Selected examples of recent developments using CO_2 as a *C1*-building block in C-C bond formation reactions.

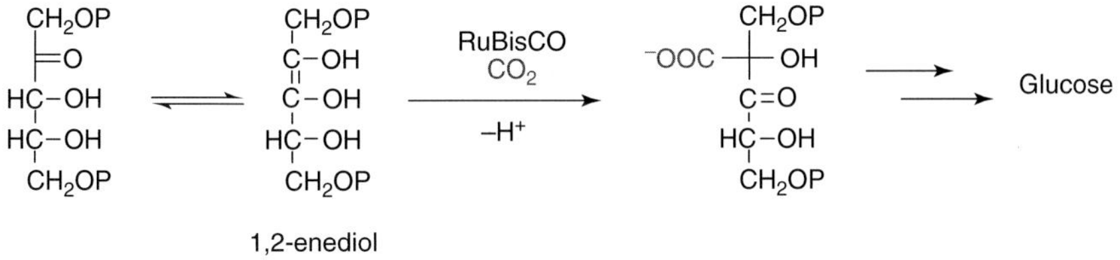

Scheme 18.23 Carboxylation of ribulose-1,5-diphosphate by RuBisCO.

Although interesting theoretical [128] and first experimental results by Aresta *et al.* [129] have been reported for the generation of this compound, an economically viable synthesis from ethylene and carbon dioxide is still far away.

The reaction of carbon dioxide and 1,3-butadiene was studied in the presence of mainly palladium catalysts [116]. This reaction leads to interesting C_9 products such as lactones, acids, and esters. Continuous mini-plant scale-up was also carried out successfully [130], although the product selectivity needs to be improved.

Obviously, all green plants use CO_2 to build up the matter of life. During the evolution process, a biochemical system for CO_2 fixation was developed. Here, ribulose-1,5-bis(phosphate)carboxylase oxidase (RuBisCO) catalyzes the reaction of CO_2 with the 1,2-enediol ribulose-1,5-diphosphate (Scheme 18.23) [131].

18.4
Catalytic C–O Bond Formation Utilizing Carbon Dioxide

From a thermodynamic point of view, it makes most sense to maintain the oxidation state of CO_2. Hence promising targets for refinement of carbon dioxide are especially carbonates, which represent an important class of compounds in industry for the synthesis of polycarbonates. In general, carbonates are divided into linear and cyclic carbonates [132]. Industrially important examples include dimethyl carbonate (DMC), diphenyl carbonate (DPC), ethylene carbonate (EC), and PC (Scheme 18.24).

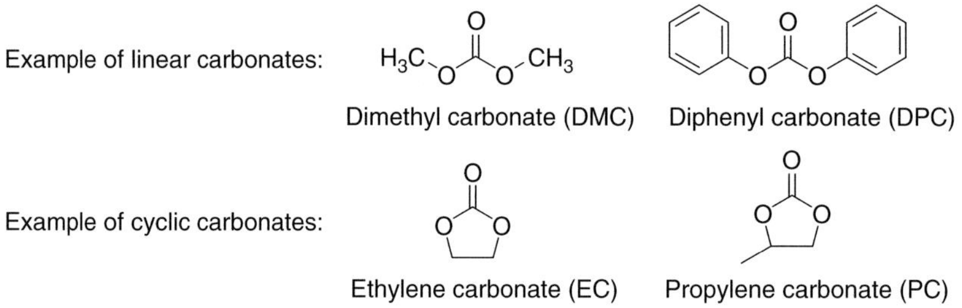

Example of linear carbonates:

Dimethyl carbonate (DMC)   Diphenyl carbonate (DPC)

Example of cyclic carbonates:

Ethylene carbonate (EC)   Propylene carbonate (PC)

Scheme 18.24 Important examples of organic carbonates.

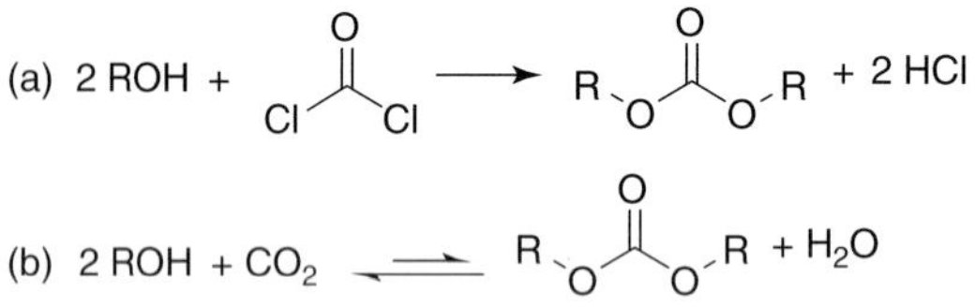

(a) 2 ROH + (phosgene) → carbonate + 2 HCl

(b) 2 ROH + CO_2 ⇌ carbonate + H_2O

Scheme 18.25 Production of polycarbonates using phosgene (a) or CO_2 (b).

The total use of organic carbonates in the world market is relatively small with respect to the amount of available carbon dioxide – about 100 000 tons per year. However, this demand is expected to increase in the coming years [133]. DMC, EC, and DPC are useful intermediates when preparing polycarbonates, which are used for engineering plastics, for example, in DVDs and airplane windows. The annual production of polycarbonates is ~3 million tons and this demand is also expected to increase further. However, today still 80–90% of polycarbonates are produced by using the toxic and corrosive phosgene (Scheme 18.25a).

In producing polycarbonates via CO_2, water is the only byproduct, making this reaction route more environmentally friendly (Scheme 18.25b) [134]. EC, PC, and DMC can also used be as electrolytes in lithium ion batteries and as aprotic polar solvents. Organic carbonates, mainly DMC, have been proposed as fuel additives and addition of carbonates to gasoline results in improved octane values [135].

18.4.1
Synthesis of Linear Carbonates

There are several possibilities for the synthesis of linear carbonates using CO_2 (Scheme 18.26), but only one method proceeds directly in one step (Scheme 18.26d) making this route the most desirable. However, so far only route (a) has been commercialized for the production of DMC. The main problem with this approach is that the reaction proceeds via cyclic carbonates and that the starting material (ethylene oxide) is also toxic. Yields of DMC up to 97% using this method have been reported using DBU as catalyst [136]. In route (b), urea is first formed by the reaction of CO_2 and ammonia, which then undergoes transesterification with alcohol to form the carbonate. The ammonia can be recycled and only CO_2 and alcohol are consumed. Unfortunately, the equilibrium is

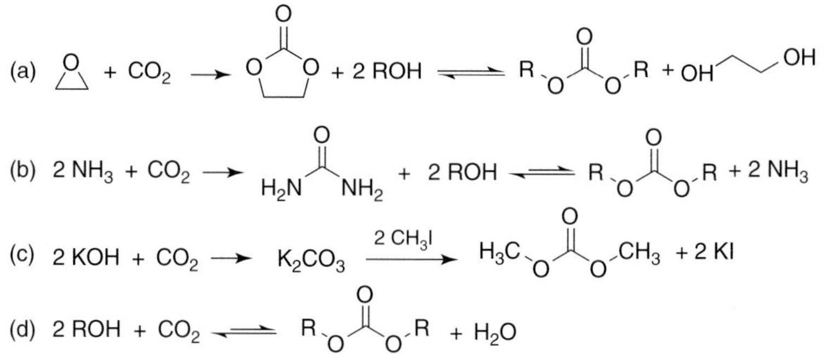

Scheme 18.26 Utilizing CO_2 to produce linear carbonates.

unfavorable. Nevertheless, yields of up to 90% have been reported for this method by combining weak Lewis acids and bases [137].

DMC can also be prepared by reacting CO_2 and methanol in the presence of methyl iodide and base (Scheme 18.26c). The major disadvantage is the salt waste that is produced as byproduct. The TONs based on CH_3I are ≤ 1 and it seems from deuterium experiments that CH_3I acts as a reactant rather than as a catalyst in the reaction [138]. To improve the reaction, microwave irradiation has been used whereby DMC can be synthesized in a shorter time and at a lower temperature compared with normal conditions [139].

The major drawback in producing carbonates directly from alcohols and CO_2 (Scheme 18.26d) is the equilibrium, which is shifted to the left of the reaction. Since this reaction has the highest potential for syntheses of linear carbonates based on CO_2 (due to the availability and price of the starting materials, the byproducts obtained, and the number of reaction steps), much effort has been made to shift the unfavorable equilibrium. So far, two methods have been developed for this purpose: (i) increasing the CO_2 concentration by using $scCO_2$ and (ii) removing the water that is formed as a byproduct by using dehydrating agents. To remove the water produced, the use of nonrecyclable dehydrating agents such as orthoesters or Mitsunobu reagent and also the utilization of recyclable dehydrating agents such as molecular sieves, zeolites, and acetals, and the use of separation membranes, have been investigated. Notably, without removing the water the observed yield for DMC is very low, only 1–2% based on MeOH (Table 18.4) [140].

Table 18.4 DMC synthesis using different types of dehydrating agents and catalysts.

Entry	Catalyst	Dehydrating agent	Additive	Conditions	DMC yield (%)	Ref.
1	$Bu_2Sn(OMe)_2$	None	None	$100\,°C$, 30 bar CO_2	TON 0.1	[139a]
2	$Bu_2Sn(OMe)_2$	None	None	$100\,°C$, 50 bar CO_2	TON 0.98	[139b]
3	$Bu_2Sn(OMe)_2$	Molecular sieves	None	$150\,°C$, 25 bar H_2	TON 3.2	[140]
4	$Bu_2Sn(OMe)_2$	MS 3A	None	$180\,°C$, 300 bar CO_2	45^a	[141]

^{a}Yield based on MeOH.

Advantages using inorganic dehydrating agents are their easier separation and recycling compared with stoichiometric organic dehydration reagents such as orthoesters. However, at the typical high reaction temperatures (150 °C), the addition of zeolites or molecular sieves does not improve the carbonate yield [141]. In order to reach high dialkyl carbonate yields in the presence of inorganic dehydrating agents (>50% DMC yield), a process has been developed in which the reaction mixture is warmed to the necessary temperature and then circulated to a room temperature dehydrating agent solution containing molecular sieves [142].

Furthermore, DMC has been produced by removing the water formed with membranes such as mesoporous silica, polyimide–silica, and polyimide–titania hybrid membranes [143]. The yields obtained, however, are far from being as good as when organic dehydrating agents such as acetals and orthoesters are employed.

The most intensively studied catalysts for the synthesis of linear organic carbonates are tin alkoxides because they show the best activity. As shown in Scheme 18.27, tin alkoxide **1** (L-Sn-OMe) should react with CO_2 to form L-Sn-OCO$_2$Me (**2**). Insertion of CO_2 in the tin-oxygen bond proceeds rapidly at room temperature [144]. Then, DMC is obtained after thermolysis of **2** at 180 °C in the presence of methanol and CO_2. Attempts to regenerate **1** from the formed L-Sn-OH (**3**) or L-Sn=O (**4**) yielded **5**, [Bu$_2$(MeO)SnOSn(MeO)Bu$_2$]$_2$, instead of **1**, Bu$_2$Sn(OMe)$_2$, indicating that a catalytic cycle based on **5** is more likely than a cycle based on **1** [145].

An alternative process for DMC proceeds via cyclic carbonates and subsequent transesterification. Here, CO_2 reacts initially with epoxides to form cyclic carbonates. Ester

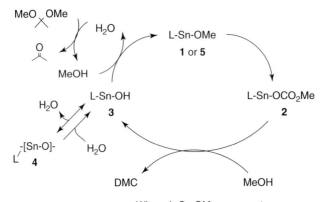

Scheme 18.27 Proposed mechanistic cycle for DMC synthesis with 2,2-dimethoxypropane as dehydrating agent.

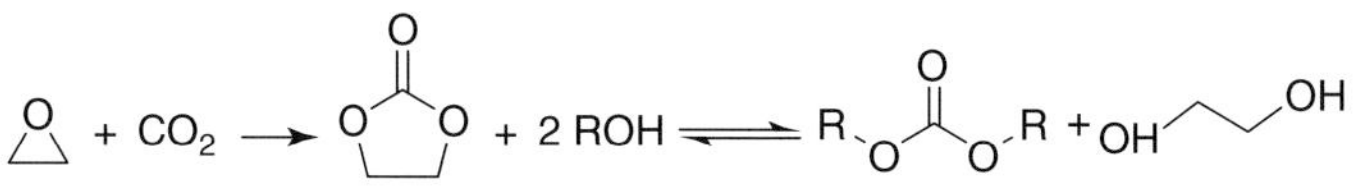

Scheme 18.28 Synthesis of DMC via cyclic carbonates.

Table 18.5 Examples of the synthesis of DMC via cyclic carbonates.

Entry	Catalyst	Epoxide	MeOH (mmol)	Conditions	DMC yield (%)	Ref.
1	MgO	PO^a	200	150 °C, 80 bar CO$_2$	14	[145]
2	Mg[smectite]	PO^a	200	150 °C, 80 bar CO$_2$	34	[145]
3	KI/K$_2$CO$_3$	EO^b	562	140 °C, 150 bar CO$_2$	73	[146]
4	DBU	PEPO^c	40	150 °C, 150 bar CO$_2$	97	[135]

^{a}Propylene oxide.
^{b}Ethylene oxide.
^{c}1,2-Epoxy-3-phenoxypropane.

exchange with alcohols leads to the linear carbonate (Scheme 18.28). The main advantage of using this method is that no dehydrating agents are necessary since water is not produced as byproduct. The disadvantage, however, is the large excess of alcohols required for ester exchange. In addition, side reactions such as ring opening of the starting material are possible. In Table 18.5, known catalytic systems are summarized. The best catalysts such as DBU and K$_2$CO$_3$ are used for both the formation of the cyclic carbonates and the ester exchange [146, 147].

In conclusion, there are three key issues for producing linear organic carbonates efficiently. (i) When using dehydrating agents, they should be recyclable and nontoxic. In this regard, acetals are superior to orthoesters, which are more difficult to recycle, whereas molecular sieves do not adsorb water as well at higher reaction temperatures (>150 °C). (ii) It is important to use active catalysts, So far, the best systems are based on tin [for example, Bu$_2$Sn(OMe)$_2$]. Addition of acidic co-catalysts such as Sc(OTf)$_3$ or RNH$_3$(OTf) improves the yield when using acetals as dehydrating agent. (iii) The influence of the CO$_2$ pressure is also important.

18.4.2
Synthesis of Cyclic Carbonates

Cyclic five- and six-membered carbonates are an important class of compounds, useful, for example, as electrolytes in lithium batteries and as intermediates for polycarbonate formation (see above). In general, they are easily obtained by reaction of CO$_2$ and oxiranes or oxetanes (Scheme 18.29) [148].

Typical catalysts for cyclic carbonate synthesis using oxiranes are onium or halide salts such as tetraethylammonium bromide (Et$_4$NBr) and potassium iodide (KI). Especially KI is preferred as a catalyst in industry because it is highly soluble in the carbonate

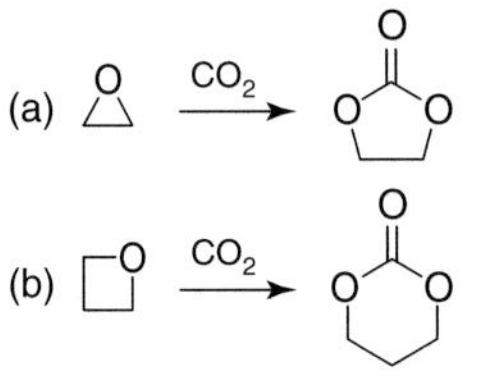

Scheme 18.29 Formation of five- and six-membered cyclic carbonates starting from CO_2 and oxiranes (a) and oxetanes (b).

Table 18.6 Selection of catalysts for the production of cyclic carbonates.

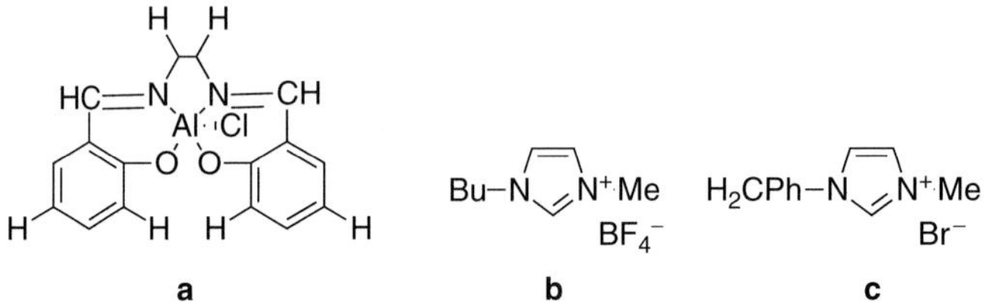

Entry	Catalyst	Epoxide	Conditions	TON	Ref.
1	NaI/PPh$_3$/PhOH	PO^a	120 °C, 40 bar CO_2	48	[157]
2	ZnBr$_2$(PMe$_2$Ph)$_2$	PO^a	100 °C, 34 bar CO_2	1640	[158]
3	Pd$_2$(dba)$_3$ + dppe	Propargylic oxirane	50 °C, 1 bar CO_2	14	[159]
4	6a/[Bu$_4$NBr]	PO^a	25 °C, 6 bar CO_2	504	[148a]
5	6a/[18-crown-6-KI]	PO^a	25 °C, 6 bar CO_2	463	[148b]
6	6b	PO^a	110 °C, 25 bar CO_2	40	[149a]
7	6c + ZnBr$_2$	PO^a	100 °C, 15 bar CO_2	5580	[149b]
8	MgO	PO^a	135 °C, 20 bar CO_2	41% yield	[151a]
9	SmOCl	PO^a	200 °C, 140 bar CO_2	99% yield	[151b]
10	[SiO$_2$−C$_3$H$_6$−PPh$_3$Br]	PO^a	90 °C, 10 bar CO_2	99% yield	[152]

^{a}Propylene oxide.

produced, which acts as a solvent in the reaction, and it is a cheap and stable catalyst compared with other salts. However, in addition to KI, various salts of Sn, Ni, Na, Zn, Cu, Ru, Pd, Co, Re, and so on, combined with amines, crown ethers, and phosphines, have also been reported as catalysts (Table 18.6).

Another group of highly active catalysts are complexes of phthalocyanines and salens [149]. In addition, the use of ILs has been reported for the production of cyclic carbonates as catalysts or together with other metal halides [150]. As some ILs are not soluble in scCO$_2$, separation of the product and catalyst mixture might become easier [151]. A different approach for catalyst separation is the use of heterogeneous metal oxide-based catalysts: Examples of active metal oxides are MgO, SmOCl, ZnO−SiO$_2$, and Cs−P−SiO$_2$ [152]. Another way to achieve high activity and easy recycling is to use immobilized molecular catalysts such as polymer- and silica-supported catalysts. It has been shown that silica-supported onium salts increase the activity [153]. Also, the use of synergistic immobilized catalysts such as quaternary phosphonium halides on silica has been reported to increase the catalytic activity further [154].

Increasing the pressure of CO_2 and the use of $scCO_2$ both improved cycloaddition and simplified the separation process, because the starting material is soluble in $scCO_2$ whereas the cyclic carbonates are not soluble. In this regard, the development of catalysts that are also soluble in $scCO_2$ is an interesting task [155].

Regarding the mechanism, it is known that the active anion X attacks preferentially the less hindered side of the oxirane to open the ring. The resulting alkoxide reacts with CO_2 to produce the alkoxycarbonate anion, which then undergoes an S_N2 reaction and forms the five- or six-membered ring. Release of the anion X forms the active catalyst again (Scheme 18.30) [156].

Interestingly, cyclic organic carbonates can be obtained from the reaction of CO_2 with alkenes via oxidative carboxylation using Nb as catalyst (Scheme 18.31) [160]. Another way is to use Bu_2SnO for the dehydrative reaction of 1,2-diols (Scheme 18.32) [161]. Furthermore, it is possible to use iron or copper catalysts to react $scCO_2$ with cyclic ketals (Scheme 18.33) [162]. Finally, a more special synthesis has been described involving reaction of CO_2 with propargyl carbonates using a palladium catalyst (Scheme 18.34) [163].

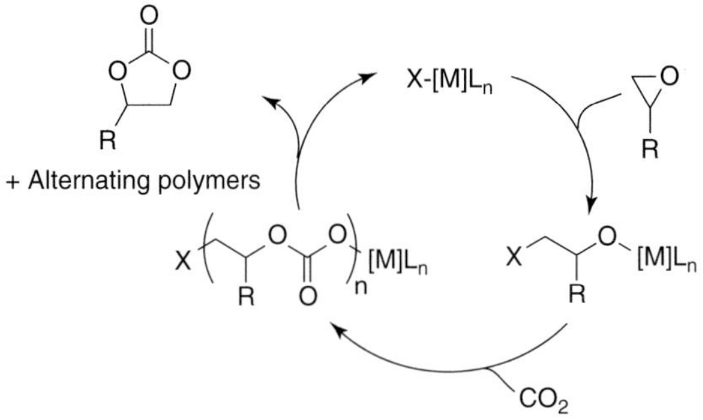

Scheme 18.30 Postulated catalytic cycle for reaction of CO_2 and oxiranes.

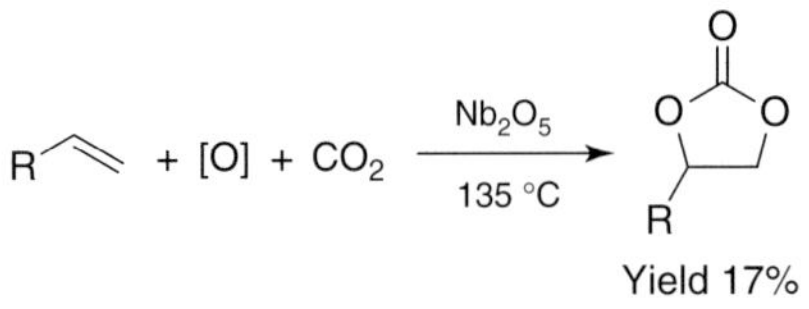

Yield 17%

Scheme 18.31 Cyclic carbonates starting from CO_2 and alkenes.

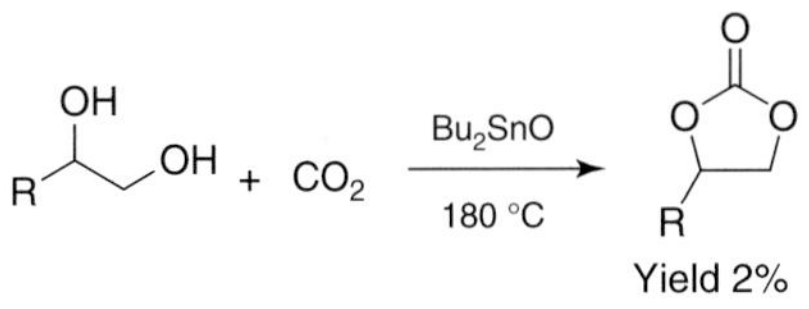

Yield 2%

Scheme 18.32 Cyclic carbonates from CO_2 and 1,2-diols.

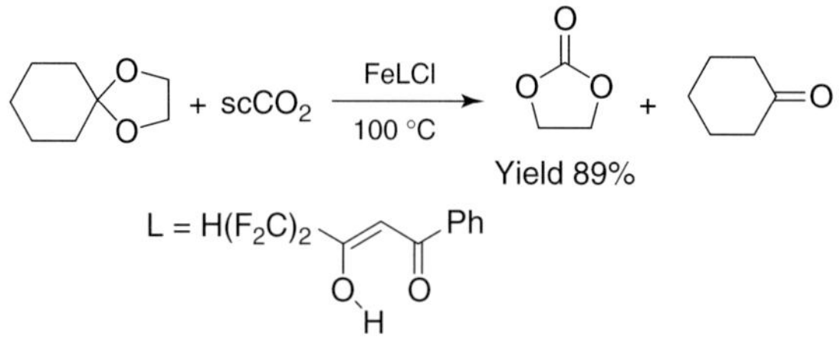

Scheme 18.33 Formation of cyclic carbonates starting from CO_2 and a cyclic ketal.

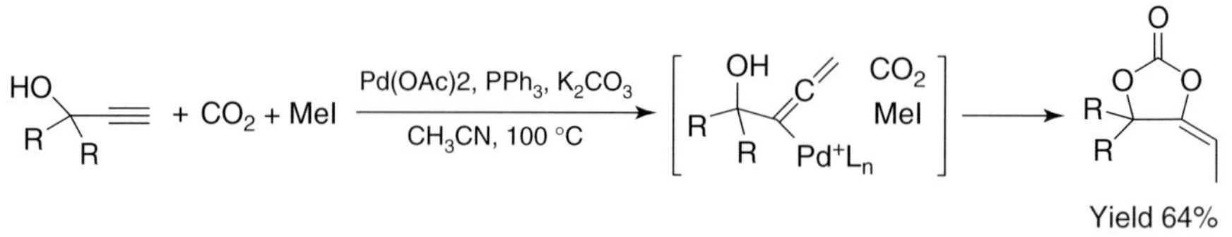

Scheme 18.34 Cyclic carbonates starting from CO_2 and propargyl alcohols.

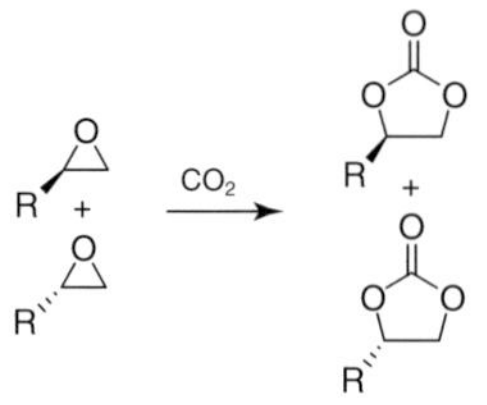

Scheme 18.35 Synthesis of chiral cyclic carbonates.

From an academic point of view, the synthesis of optically active cyclic carbonates by applying chiral cobalt–salen complexes as catalysts is also worth noting. High enantiomeric excesses (*ees*) up to >90% have been achieved by using multiple-bonded substrates such as aminoepoxides (Scheme 18.35, Table 18.7).

18.5
Current Industrial Processes Using CO_2

Despite its advantages, so fare few large-scale industrial processes exist that employ this valuable feedstock. The limited number of applications is partially explained by thermodynamic reasons. It is generally accepted that a significant energy supply is necessary to convert CO_2 into useful products since it presents the thermodynamically most favorable oxidation state of the carbon atom. Nevertheless, by using appropriate conditions, reactions with negative Gibbs free energy can be achieved [167].

More specifically, the current industrial use of CO_2 is ~115 Mt per year, which does not include the large amount of CO_2 used for enhanced oil recovery. However, this amount represents merely 0.5% of its current anthropogenic emission of 24 Gt per year [168]. Apart from chemical applications, CO_2 is applied in food production for carbonation of beverages, controlled-atmosphere packaging, and as a cryogenic fluid in

Table 18.7 Stereoselective synthesis of cyclic carbonates.

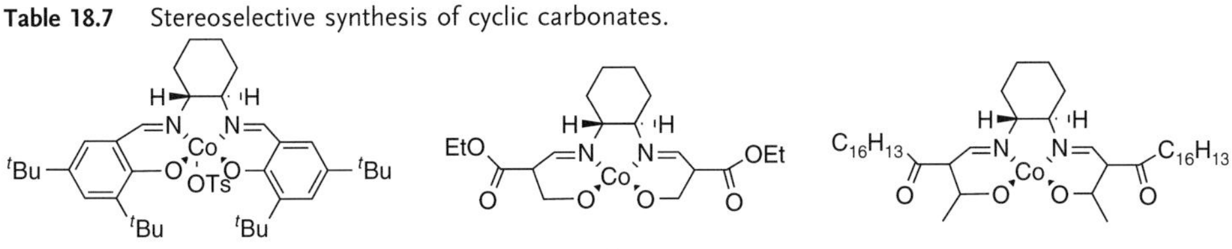

Entry	Catalyst	Epoxide	Reaction conditions	ee (%)	TON	Ref.
1	**7a**	PO[a]	−40 °C, 1 bar CO₂	83	400	[164]
2	**7b**	EO[b]	30 °C, 1 bar CO₂	86	25	[165]
3	**7c**	N-Carbozolylmethyloxirane	30 °C, 20 bar CO₂	92	22.5	[166]

[a] Propylene oxide.
[b] Ethylene oxide.

freezing procedures, or as dry-ice for temperature control during storage and transport. As an easily accessible supercritical fluid, CO₂ is an important extraction solvent in the decaffeination of coffee [169] and tea and also in the separation and purification of volatile flavors and fragrances [170]. Other applications include welding systems, fire extinguishers, water treatment processes, horticulture, and environmental protection in the metal industry.

In the chemical industry, CO₂ is mainly used for the production of urea. Other smaller scale applications include the synthesis of organic monomers, polycarbonates, and salicylic acid (Scheme 18.36). Each of these processes is discussed below in more detail.

Notably, inorganic carbonates such as Na₂CO₃, K₂CO₃, and BaCO₃ are produced starting from CO₂. Furthermore, CO₂ is employed as an additive in the manufacture of methanol. The distribution of the main applications is shown in Figure 18.2.

18.5.1
Urea

The synthesis of urea constitutes the main industrial conversion of carbon dioxide, with an annual production of 100 Mt [171]. Urea is primarily used as a nitrogen-rich fertilizer

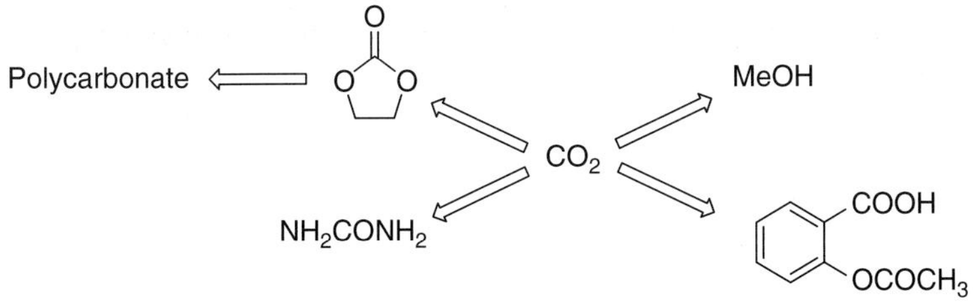

Scheme 18.36 Industrial transformations of CO₂.

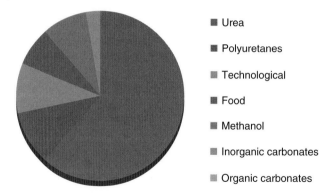

- ■ Urea
- ■ Polyuretanes
- ■ Technological
- ■ Food
- ■ Methanol
- ■ Inorganic carbonates
- ■ Organic carbonates

Figure 18.2 Distribution of carbon dioxide applications.

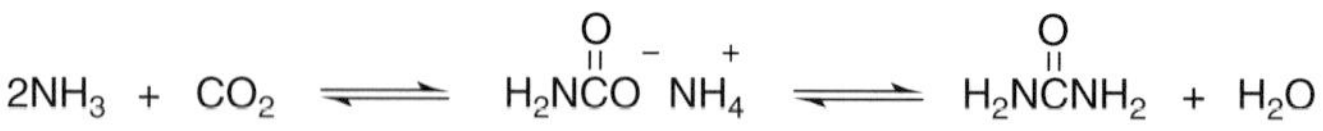

$$2NH_3 + CO_2 \rightleftharpoons H_2N\overset{\overset{\displaystyle O}{\|}}{C}O^- \; NH_4^+ \rightleftharpoons H_2N\overset{\overset{\displaystyle O}{\|}}{C}NH_2 + H_2O$$

Scheme 18.37 Synthesis of urea from ammonia and CO_2.

for agricultural applications and it is an important intermediate for the manufacture of resins for plastics and adhesives. Moreover, it is also utilized as a component of animal feed. Urea is produced starting from ammonia and carbon dioxide in a high-pressure reactor, typically a stainless-steel vessel containing trays to enhance the mixing capacity. The process is divided into two steps (Scheme 18.37). In the first reaction (exothermic) ammonium carbamate is formed, which is then dehydrated to the desired product in a second endothermic step. The resulting aqueous solution consists of urea, ammonia, and ammonium carbamate. Elevated pressure, from 150 to 250 bar, and high temperatures in the range 150–200 °C are employed as process parameters.

The original synthetic procedure was discovered by Bassarov in 1870 [172]. The basic process was first developed by BASF in 1922 and was called the Bosch–Meiser urea process after its inventors [173]. Generally, urea plants comprise a high-pressure synthesis section and a lower pressure recovery section that recycles intermediates and unreacted ammonia. A further sector for urea purification can be provided downstream of the recovery section. The various processes usually differ in the recycling and recovery parts. Most of the current plants use stripping processes, aimed at recovering the carbamate from the high-pressure reactor to the so-called high-pressure loop. In a stripping plant, the synthesis reactor, the stripper, and the condenser usually form the high-pressure synthesis system. The procedures can be substantially divided into the carbon dioxide stripping process and ammonia self-stripping process. In the former case, the carbamate dissociates into ammonia and carbon dioxide by addition of fresh carbon dioxide in a steam-heated tubular heat exchanger, and it is later recovered in a condenser. In the latter case, the stripping agent is the gaseous ammonia generated from the thermal dissociation of carbamate in the reactor. The manufacturing process is designed to minimize the formation of biuret byproduct (Scheme 18.38), which is often toxic towards the plants that have to be fertilized.

$$2 \, H_2NCONH_2 \; \rightleftharpoons \; H_2NCONHCONH_2 \; + \; NH_3$$

Scheme 18.38 Biuret formation.

$$3 \, CO \; + \; 9 \, H_2 \; + \; CO_2 \; \longrightarrow \; 4 \, CH_3OH \; + \; H_2O$$

Scheme 18.39 Synthesis of methanol from CO, H$_2$, and addition of CO$_2$.

18.5.2
Methanol

Methanol is produced in ~40 Mt per year as a primary raw material for the chemical industry [6]. Methanol is synthesized starting from H$_2$ and CO via coal or natural gases which contain small quantities of carbon dioxide (Scheme 18.39). The total volume of carbon dioxide used in this process is estimated as ~14 Mt. Today, natural gas is the preferred feedstock because it contains a higher amount of hydrogen and lower amounts of impurities, and it offers the opportunity to consume less energy [174]. Nevertheless, coal-based syngas is still mostly used in countries where this resource is abundant, predominantly in China and South Africa. In industrial processes, copper/zinc dioxide-based catalysts are employed at 250–300 °C and 5–10 MPa (ICI process) with very high performance.

CO$_2$ addition considerably improves the efficacy of the process and it is considered to be directly converted without the *in situ* formation of CO [175]. Recently, the direct synthesis of methanol from H$_2$ and CO$_2$ has attracted increasing interest. A first pilot plant for the synthesis of methanol from H$_2$ and CO$_2$ has been built in Japan on a scale of 50 kg per day [176]. A production rate of up to 600 g l^{-1} h^{-1} is achieved with excellent selectivity at 250 °C and 3–7 MPa by using an SiO$_2$-modified copper/zinc oxide catalyst. However, this process is still not economically viable, although it is technically suitable. Nevertheless, Mitsui Chemicals recently built a pilot plant with the capacity to synthesize ~100 t of methanol per year starting form CO$_2$. A catalyst based on Cu/ZnO/Al$_2$O$_3$ is used in this process [177].

18.5.3
Carboxylic Acids

Carbon dioxide can be used for the conversion of potassium and sodium phenolates to the corresponding carboxylic acids. This type of reaction is called the Kolbe–Schmitt process and has been used since 1874 for the industrial production of salicylic acid which is an intermediate in the manufacture of acetylsalicylic acid (aspirin) [178]. It proceeds simply by heating the phenolate salt in an autoclave at 125 °C under 5–7 bar CO$_2$ pressure (Scheme 18.40). By work-up with H$_2$SO$_4$, the corresponding acid is obtained. This reaction has also been applied industrially for the preparation of different pharmaceuticals and colorants such as *p*-aminosalicylic acid, 2-hydroxynaphthalene-3-carboxylic acid, and 2-hydroxycarbazole-3-carboxylic acid.

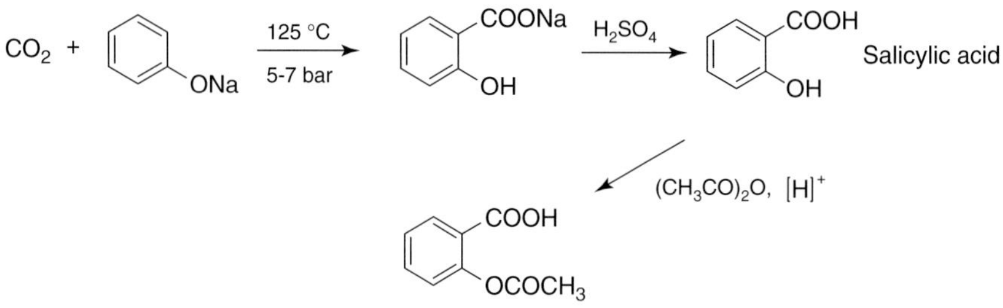

Scheme 18.40 Synthesis of aspirin.

18.5.4
Carbonates

As discussed in Section 18,4 the most important industrial carbonates are EC, PC, DMC, and DPC. The industrial use of CO_2 as a building block for these chemicals represents a significant environmental improvement in comparison with the conventional procedure, which involves the utilization of phosgene. The first applications of these reactions date back to the second half of the twentieth century [179]. The key discovery was the rapid cycloaddition of carbon dioxide to epoxides, which form cyclic carbonates in the presence of quaternary ammonium halides (Scheme 18.41).

Union Carbide successfully used this reaction for the synthesis of ethylene glycol as the first step followed by hydrolysis of EC [180]. PC extraction solvents are produced by Arco with the same procedure. Texaco filed various patents on acid–base catalysts for the final step in the synthesis of DMC, which involve the transesterification of the cyclic carbonate with methanol [181]. The production of organic carbonates amounts to about 100 000 t per year, but it is expected to increase considerably owing to the increase in polycarbonate manufacture. Polycarbonates have been synthesized from carbon dioxide since 1969 [182]. The same starting materials as for the carbonates are used [183]. These products are typically employed as degradable surfactants or as an alternative to polyurethanes as engineering plastics which find application in mobile phones, cameras, computers, automobiles, CDs, DVDs, and other commodities [184]. Polycarbonates were produced in an annual amount of $\sim$3.2 Mt in 2008, increasing by 8–10% each year.

In 2002, Asahi Kasei developed and industrialized the world's first polycarbonate process using CO_2 as starting material for the synthesis of polycarbonates based on bisphenol-A without phosgene and methylene chloride [134]. Until then, all polycarbonates had been manufactured by using carbon monoxide in the so-called "phosgene process" (Idemitzu Petrochemicals, Teijin, Bayer, GE, SABIC Innovative Plastics, Mitsubishi, and Dow). Here phosgene is formed from CO and chlorine. The new technology

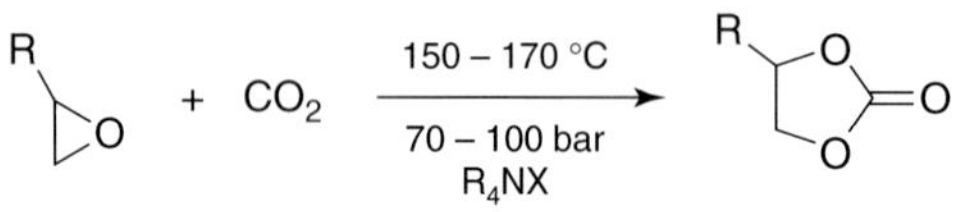

Scheme 18.41 Synthesis of carbonates starting from CO_2 and epoxide.

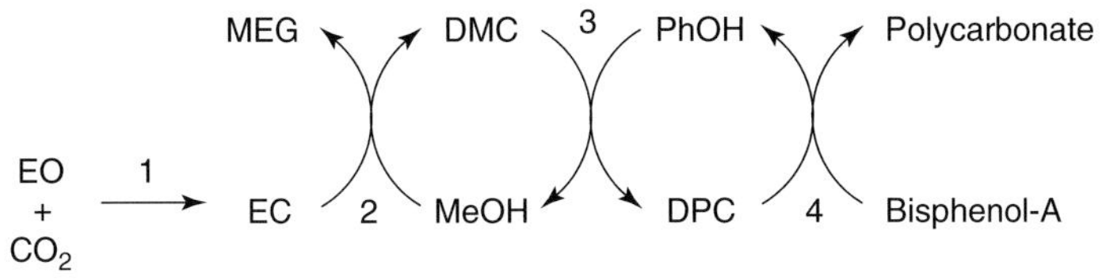

MEG ⤡ ⤢ DMC   3   PhOH ⤡ ⤢ Polycarbonate

EO
+   ──1──→   EC   2   MeOH   DPC   4   Bisphenol-A
CO₂

Scheme 18.42 Polycarbonate process of Asahi Kasei.

not only shows economic and environmental advantages, but also allows energy to be conserved and CO_2 emission to be reduced (173 t per 1000 t of product). The first commercial plant was launched in Taiwan by Chimei-Asahi, a joint project between Chi Mei and Asahi Kasei. Initially production was 50 000 t per year, which rapidly increased to 150 000 t in 2006 [185]. Today, three additional commercial plants, in Korea (two plants of 65 000 t per year) and in Russia (65 000 t per year), are productively working with the Asahi Kasei process and a plant of 260 000 t per year was expected to be launched in 2010 [186]. The manufacture of polycarbonates consists of four steps, as illustrated in Scheme 18.42.

In the first step, EC is synthesized from ethylene oxide (EO) and in liquid or supercritical CO_2, which is recovered from the EO production [187]. In a continuous multistage distillation apparatus, EC and methanol, present in excess, react at 65–100 °C in the presence of both heterogeneous and homogeneous catalysts to form DMC and monoethylene glycol (MEG) [188]. Different catalysts are efficient, such as resins containing quaternary ammonium chloride, carbonates, hydroxides, alkali/alkaline earth metal hydroxides and alkoxides, tin and titanium precursors, tertiary amines, cyclic amidines, and so on. In the following step, the monomer DPC is synthesized in two reactions from DMC and PhOH: the transesterification of DMC with PhOH to form methylphenyl carbonate (MPC) and the disproportionation (or self-transeserification) of MPC to DPC. The use of several catalysts such as $Ti(OPh)_4$, Bu_2SnO, and $Pb(OPh)_2$ has been reported [189]. The two reactions with product separations occur contemporaneously in two continuous multistage distillation columns connected to each other, which operate at ~200 °C. The DMC and methanol coproduced are easily recycled. In the last step, DPC reacts with bisphenol-A to produce the desired polycarbonates and PhOH. A new type of polymerization reactor has been developed, which avoids the complications due to the very high viscosity in the melt polymerization of DPC and bisphenol-A. This reactor is based on the Earth's gravity, does not utilize any mechanical mixing system, and works at temperatures below 270 °C. The post-polymerization reaction of PC proceeds easily while the prepolymer flows down along the vertical guides set up in the reactor.

18.6
Conclusion and Outlook

Apart from biomass, carbon dioxide is the only sustainable carbon source for the chemical industry. Therefore, increased use of CO_2 as a carbon feedstock in the chemical industry is highly desirable. However, in general, utilization of CO_2 especially for carbon–carbon bond formation needs significant energy or energy-rich coupling partners. Often, the overall assessment of such reactions does not make them environmentally benign and economically viable. Therefore, for the short-term future for industry, more promising reactions utilizing CO_2 are the syntheses of both linear and cyclic organic carbonates and polycarbonates. Apparently, organic carbonates might find increased applications, for example, as electrolytes in lithium batteries in electric vehicles and as intermediates when producing polycarbonates without using toxic phosgene. On the other hand, for the mid- to long-term future, the development of so-called artificial photosynthesis processes in which carbon dioxide is transformed into higher carbon building blocks with the help of renewable energy offers interesting possibilities but definitely needs more scientific progress. Direct photochemical reductions of carbon dioxide are also very interesting in this regard. Clearly, more research on suitable catalysts is needed to drive such reactions efficiently.

Finally, it should be made clear that chemical utilization of CO_2 will not decrease the so-called greenhouse effect. Even though the use of CO_2 as a carbon source is important and can be considered to be a "green carbon source," the potential amount of CO_2 that can used by the chemical industry is at best only 1–5% of the amount of CO_2 released. Further, each chemical that is produced from CO_2 will emit costly and undesirable CO_2 at the disposal stage. Hence research in this area continues to be important.

References

1. International Energy Agency (2010) *Key World Energy Statistics 2010*, International Energy Agency, Paris.

2. (a) Darensbourg, D.J. (2010) *Inorg. Chem.*, **49**, 10765–10780; (b) Aresta, Arakawa, H., Aresta, M., Armor, J.N., Barteau, M.A., Beckman, E.J., Bell, A.T., Bercaw, J.E., Creutz, C., Dinjus, E., Dixon, D.A., Domen, K., DuBois, D.L., Eckert, J., Fujita, E., Gibson, D.H., Goddard, W.A., Goodman, D.W., Keller, J., Kubas, G.J., Kung, H.H., Lyons, J.E., Manzer, L.E., Marks, T.J., Morokuma, K., Nicholas, K.M., Periana, R., Que, L., Rostrup-Nielson, J., Sachtler, W.M.H., Schmidt, L.D., Sen, A., Somorjai, G.A., Stair, P.C., Stults, B.R., and Tumas, W. (2001) *Chem. Rev.*, **101**, 953–996; (c) Aresta, M. (2010) *Carbon Dioxide as Chemical Feedstock*, Wiley-VCH Verlag GmbH, Weinheim; (d) Aresta, M. and Dibenedetto, A. (2007) *Dalton Trans.*, 2975–2992.

3. Loges, B., Boddien, A., Gärtner, F., Junge, H., and Beller, M. (2010) *Top. Catal.*, **53**, 902–914.

4. Enthaler, S., von Langermann, J., and Schmidt, T. (2010) *Energy Environ. Sci.*, **3**, 1207–1217.

5. Rahimpour, M.R. (2008) *Fuel Process. Technol.*, **89**, 556–566.

6. Olah, G.A., Goeppert, A., and Prakash, G.K.S. (2009) *J. Org. Chem.*, **74**, 487–498.

7. Zahedi, G., Elkamel, A., and Lohi, A. (2007) *Energy Fuels*, **21**, 2977–2983.

8. Centi, G. and Perathoner, S. (2009) *Catal. Today*, **148**, 191–205.

9. Metz, B., Davidson, O., de Coninck, H., Loos, M., and Meyer, L. (2005) *Special Report on Carbon Dioxide Capture and Storage*, Cambridge University Press, Cambridge.

10. D'Alessandro, D.M., Smit, B., and Long, J.R. (2010) *Angew. Chem. Int. Ed.*, **49**, 6058–6082.

11. Orr, F.M. (2009) *Energy Environ. Sci.*, **2**, 449–458.

12. Haszeldine, R.S. (2009) *Science*, **325**, 1647–1652.

13. (a) Dubois, D.L. (2006) in *Encyclopedia of Electrochemistry* (eds A.J. Bard and M. Stratmann), Wiley-VCH Verlag GmbH, Weinheim, pp. 202–225; (b) Frese, J.K.W.

(1993) in *Electrochemical and Electrocatalytic Reactions of Carbon Dioxide* (eds. B.P. Krist and H.E. Guard), Elsevier, Amsterdam, pp. 145–216.

14. Oloman, C. and Li, H. (2008) *ChemSusChem*, **1**, 385–391.

15. Lee, J., Kwon, Y., Machunda, R.L., and Lee, H.J. (2009) *Chem. Asian J.*, **4**, 1516–1523.

16. Gattrell, M., Gupta, N., and Co, A. (2007) *Energy Convers. Manage.*, **48**, 1255–1265.

17. Benson, E.E., Kubiak, C.P., Sathrum, A.J., and Smieja, J.M. (2009) *Chem. Soc. Rev.*, **38**, 89–99.

18. Fisher, B. and Eisenberg, R. (1980) *J. Am. Chem. Soc.*, **102**, 7361–7363.

19. Beley, M., Collin, J.P., Rupert, R., and Sauvage, J.P. (1984) *J. Chem. Soc., Chem. Commun.*, 1315–1316.

20. Hammouche, M., Lexa, D., Momenteau, M., and Savéant, J.M. (1991) *J. Am. Chem. Soc.*, **113**, 8455–8466.

21. Hawecker, J., Lehn, J.M., and Ziessel, R. (1984) *J. Chem. Soc., Chem. Commun.*, 328–330.

22. Ishida, H., Tanaka, K., and Tanaka, T. (1987) *Organometallics*, **6**, 181–186.

23. Bolinger, C.M., Story, N., Sullivan, B.P., and Meyer, T.J. (1988) *Inorg. Chem.*, **27**, 4582–4587.

24. Slater, S. and Wagenknecht, J.H. (1984) *J. Am. Chem. Soc.*, **106**, 5367–5368.

25. (a) DuBois, D.L., Miedaner, A., and Haltiwanger, R.C. (1991) *J. Am. Chem. Soc.*, **113**, 8753–8764; (b) Raebiger, J.W., Turner, J.W., Noll, B.C., Curtis, C.J., Miedaner, A., Cox, B., and DuBois, D.L. (2006) *Organometallics*, **25**, 3345–3351.

26. Seshadri, G., Lin, C., and Bocarsly, A.B. (1994) *J. Electroanal. Chem.*, **372**, 145–150.

27. Cole, E.B., Lakkaraju, P.S., Rampulla, D.M., Morris, A.J., Abelev, E., and Bocarsly, A.B. (2010) *J. Am. Chem. Soc.*, **132**, 11539–11551.

28. See also: Wöhrle, D. (1998) in *Photochemie* (eds. D. Wöhrle, M.W. Tausch, and W.-D. Stohrer), Wiley-VCH Verlag GmbH, Weinheim, pp. 113–177.

29. Morris, A.J., Meyer, G.J., and Fujita, E. (2009) *Acc. Chem. Res.*, **42**, 1983–1994.

30. Inakaki, A. and Akita, M. (2010) *Coord. Chem. Rev.*, **254**, 1220–1239.

31. Lubitz, W., Reijerse, E.J., and Messinger, J. (2008) *Energy Environ. Sci.*, **1**, 15–31.

32. Barber, J. (2008) *Inorg. Chem.*, **47**, 1700–1710.

33. Hawecker, J., Lehn, J.-M., and Ziessel, R. (1985) *J. Chem. Soc., Chem. Commun.*, 56–58.

34. Ogata, T., Yamamoto, Y., Wada, Y., Murakoshi, K., Kusaba, M., Nakashima, N., Ishida, A., Takamuku, S., and Yanagida, S. (1995) *J. Phys. Chem.*, **99**, 11916–11922.

35. Matsuoka, S., Yamamoto, K., Ogata, T., Kusaba, M., Nakashima, N., Fujita, E., and Yanagida, S. (1993) *J. Am. Chem. Soc.*, **115**, 601–609.

36. Tinnermans, A.T.A., Koster, T.P.M., Thewessen, D.H.M.W., and Mackor, A. (1984) *Recl. Trav. Chim. Pays-Bas*, **130**, 288–295.

37. Hawecker, J., Lehn, J.M., and Ziessel, R. (1983) *J. Chem. Soc., Chem. Commun.*, 536–538.

38. For a review on biomimetic systems, see: Wang, M. and Sun, L. (2010) *ChemSusChem*, **3**, 551–554.

39. For an example from our own group, see: Gärtner, F., Sundararadju, B., Surkus, A.-E., Boddien, A., Loges, B., Junge, H., Dixneuf, P.H., and Beller, M. (2009) *Angew. Chem. Int. Ed.*, **48**, 9962–9965; *Angew. Chem.*, **121**, 10147–10150.

40. Takeda, H. and Ishitani, O. (2010) *Coord. Chem. Rev.*, **254**, 346–354.

41. Sato, S., Koike, K., Inoue, H., and Ishitani, O. (2007) *Photochem. Photobiol. Sci.*, **6**, 454–461.

42. Kimura, E., Wada, S., Shionoya, M., and Okazaki, Y. (1994) *Inorg. Chem.*, **33**, 770–778.

43. Grodkowski, J., Behar, D., and Neta, P. (1997) *J. Phys. Chem. A*, **101**, 248–254.

44. Behar, D., Dhanasekaran, T., Neta, P., Hosten, C.M., Ejeh, D., Hambright, P., and Fujita, E. (1998) *J. Phys. Chem. A*, **102**, 2870–2877.

45. Ettedgui, J., Diskin-Posner, Y., Weiner, L., and Neumann, R. (2011) *J. Am. Chem. Soc.*, **133**, 188–190.

46. Kočí, K., Obalová, L., Matějová, L., Plachá, D., Lacný, Z., Jirkovský, J., and Šolcová, O. (2009) *Appl. Catal. B: Environ.*, **89**, 494–502.

47. Kitano, M., Matsuoka, M., Ueshima, M., and Anpo, M. (2007) *Appl. Catal. A: Gen.*, **325**, 1–14.

48. Yang, C., Ma, Z., Zhao, N., Wie, W., Hu, T., and Sun, Y. (2006) *Catal. Today*, **115**, 222–227.

49. Olah, G.A., Goeppert, A., and Prakash, G.K.S. (2005) *Beyond Oil and Gas: the Methanol Economy*, Wiley-VCH Verlag GmbH, Weinheim.

50. Shulenberger, A.M., Jonsson, F.R., Ingolfsson, O., and Tran, K.-C. (2007) US Patent Appl. 2007/0244208A1.

51. Yu, K.M.K., Curcic, I., Gabriel, J., and Tsang, S.C.E. (2008) *ChemSusChem*, **1**, 893–899, and references therein.

52. Wambach, J., Baiker, A., and Wokaun, A. (1999) *Phys. Chem. Chem. Phys.*, **1**, 5071–5080.

53. Lachowskaa, M. and Skrzypeka, J. (2004) *Stud. Surf. Sci. Catal.*, **153**, 173–176.

54. Omae, I. (2006) *Catal. Today*, **115**, 33–52.

55. Kubota, T., Hayakawa, I., Mabuse, H., Mori, K., Ushikoshi, K., Watanabe, T., and Saito, M. (2001) *Appl. Organomet. Chem.*, **15**, 121–126.

56. Wu, J., Luo, S., Toyir, J., Saito, M., Takeuchi, M., and Watanabe, T. (1998) *Catal. Today*, **45**, 215–220.

57. Chan, B. and Radom, L. (2010) *J. Am. Chem. Soc.*, **130**, 9790–9799.

58. Chan, B. and Radom, L. (2006) *J. Am. Chem. Soc.*, **128**, 5322–5323.

59. Baiker, A. (2000) *Appl. Organomet. Chem.*, **14**, 751–762.

60. Tominaga, K., Sasaki, Y., Kawai, M., Watanabe, T., and Saito, M. (1993) *J. Chem. Soc., Chem. Commun.*, 629–631.

61. Jessop, P.G., Ikariya, T., and Noyori, R. (1995) *Chem. Rev.*, **95**, 259–272.

62. Eisenschmid, T. and Eisenberg, R. (1989) *Organometallics*, **8**, 1822–1824.

63. Süss-Fink, G. and Reiner, J. (1981) *J. Organomet. Chem.*, **221**, C36–C38.

64. Stephan, D.W. and Erker, G. (2010) *Angew. Chem. Int. Ed.*, **49**, 46–76; *Angew. Chem.*, **122**, 50–81.

65. Mömming, C.M., Otten, E., Kehr, G., Fröhlich, R., Grimme, S., Stephan, D.W., and Erker, G. (2009) *Angew. Chem. Int. Ed.*, **48**, 6643–6646; *Angew. Chem.*, **121**, 6770–6773.

66. Ashley, A.E., Thompson, A.L., and O'Hare, D. (2009) *Angew. Chem. Int. Ed.*, **48**, 9839–9843; *Angew. Chem.*, **121**, 10023–10027.

67. Ménard, G. and Stephan, D.W. (2010) *J. Am. Chem. Soc.*, **132**, 1796–1797.

68. Roy, L., Zimmermann, P.M., and Paul, A. (2011) *Chem. Eur. J.*, **17**, 435–439.

69. Rudiuan, S.N., Zhang, Y., and Ying, J.Y. (2009) *Angew. Chem.*, **121**, 3372–3375; *Angew. Chem. Int. Ed.*, **48**, 3322–3325.

70. Hunang, F., Lu, G., Zhao, L., Li, H., and Wang, Z.-X. (2010) *J. Am. Chem. Soc.*, **132**, 12388–12396.

71. Chakraborty, S., Zhang, J., Krause, J.A., and Guan, H. (2010) *J. Am. Chem. Soc.*, **132**, 8872–8873.

72. Obert, R. and Dave, B.C. (1999) *J. Am. Chem. Soc.*, **121**, 12192–12193.

73. Reutemann, W. and Kieczka, H. (2009) Formic acid, in *Ullmann's Encyclopaedia of Industrial Chemistry*, 7th edn., Wiley-VCH Verlag GmbH, Weinheim.

74. Gladiali, S. and Alberico, E. (2006) *Chem. Soc. Rev.*, **35**, 226–236.

75. (a) Kudo, A. and Miseki, Y. (2009) *Chem. Soc. Rev.*, **38**, 253–278; (b) Osterloh, F.E. (2008) *Chem. Mater.*, **20**, 35–54; (c) Maeda, K. and Domen, K. (2007) *J. Phys. Chem. C*, **111**, 7851–7861.

76. (a) Loges, B., Boddien, A., Gärtner, F., Junge, H., and Beller, M. (2010) *Top. Catal.*, **53**, 902–914; (b) Johnson, T.C., Morris, D.J., and Wills, M. (2010) *Chem. Soc. Rev.*, **39**, 81–88; (c) Jiang, H.-L., Singh, K., Yan, J.-M., Zhang, X.-B., and Xu, Q. (2010) *ChemSusChem*, **3**, 541–549; (d) Makowski, P., Thomas, A., Kuhn, P., and Goettman, F. (2009) *Energy Environ. Sci.*, **2**, 480–490; (e) Enthaler, S., von Langermann, J., and Schmidt, T. (2010) *Energy Environ. Sci.*, **3**, 1207–1217; (f) Makowski, P., Thomas, A., Kuhn, P., and Goettman, F. (2009) *Energy Environ. Sci.*, **2**, 480–490; (g) Hamilton, C.W., Baker, R.T., Staubitz, A., and Manners, I. (2009) *Chem. Soc. Rev.*, **38**, 279–293.

77. Jessop, P.G., Joó, F., and Tai, C.-C. (2004) *Coord. Chem. Rev.*, 2425–2442.

78. Farlow, M.W. and Adkins, H. (1935) *J. Am. Chem. Soc.*, **57**, 2222–2223.

79. (a) Ipatieff, V.N. and Monroe, G.S. (1945) *J. Am. Chem. Soc.*, **67**, 2168–2171; (b) Stowe, R.A. and Russel, W.W. (1954) *J. Am. Chem. Soc.*, **76**, 319–323; (c) Cratty, L.R. and Russel, W.W. Jr. (1957) *J. Am. Chem. Soc.*, **80**, 767–773.

80. Wiener, H., Blum, J., Feilchenfeld, H., Sasson, Y., and Zalmanov, N. (1988) *J. Catal.*, **110**, 184–190.

81. Zhang, Y., Fei, J., Yu, Y., and Zheng, X. (2004) *Catal. Commun.*, **5**, 643–464.

82. Zhang, Z., Xie, Y., Li, W., Hu, S., Song, J., Jiang, T., and Han, B. (2008) *Angew. Chem. Int. Ed.*, **47**, 1127–1129.

83. (a) Jezoswska-Trebiatowska, B. and Sobota, P. (1974) *J. Organomet. Chem.*, **80**, C27–C28; (b) Inoue, Y., Izumida, H., Sasaki, Y., and Hashimoto, H. (1976) *Chem. Lett.*, 863–864.

84. (a) Leitner, W. (1995) *Angew. Chem. Int. Ed.*, **34**, 2207–2221; *Angew. Chem.* **107**, 2391–2405; (b) Jessop, P.G., Ikariya, T., and Noyori, R. (1995) *Chem. Rev.*, **95**, 259–272; (c) Himeda, Y. (2007) *Eur. J. Inorg. Chem.*, 3927–3941; (d) Jessop, P.G. (2007) in *Handbook of Homogeneous Hydrogenation* (eds. J.G. de Vries and C.J. Elsevier), Wiley-VCH Verlag GmbH, Weinheim, pp. 489–511.

85. Tanaka, R., Yamashita, M., and Nozaki, K. (2009) *J. Am. Chem. Soc.*, **131**, 14168–14169.

86. Himeda, Y., Onozawa-Komatsuzaki, N., Sugihara, H., and Kasuga, K. (2007) *Organometallics*, **26**, 702–712.

87. Munshi, P., Main, A.D., Linehan, J.C., Tai, C.C., and Jessop, P.G. (2002) *J. Am. Chem. Soc.*, **124**, 7963–7971.

88. Jessop, P.G., Hsiau, Y., Ikariya, T., and Noyori, R. (1996) *J. Am. Chem. Soc.*, **118**, 344–355.

89. Jessop, P.G., Ikariya, T., and Noyori, R. (1994) *Nature*, **368**, 231–233.

90. Himeda, Y., Onozawa-Komatsuzaki, N., Sugihara, H., Arakawa, H., and Kasuga, K. (2003) in Proceedings of the 50th Symposium on Organometallic Chemistry, Osaka, Japan.

91. Horváth, H., Laurenczy, G., and Kathó, Á. (2004) *J. Organomet. Chem.*, **689**, 1036–1045.

92. Federsel, C., Jackstell, R., Boddien, A., Láurenczy, G., and Beller, M. (2010) *ChemSusChem*, **3**, 1048–1050.

93. Elek, J., Nádasdi, L., Papp, G., Laurenczy, G., and Joó, F. (2003) *Appl. Catal. A*, **255**, 59–67.

94. Angermund, K., baumann, W., Dinjus, E., Fornika, R., Görls, H., Kessler, M., Krüger, C., Leitner, W., and Lutz, F. (1997) *Chem. Eur. J.*, **3**, 755–764.

95. Gassner, F. and Leitner, W. (1993) *J. Chem. Soc., Chem. Commun.*, 1465–1466.

96. Graf, E. and Leitner, W. (1992) *J. Chem. Soc., Chem. Commun.*, 623–624.

97. Ezhova, N.N., Kolesnichenko, N.V., Bulygin, A.V., Slivinskii, E.V., and Han, S. (2002) *Russ. Chem. Bull. Int. Ed.*, **51**, 2165–2169.

98. Kudo, K., Sugita, N., and Takezaki, Y. (1977) *Nippon Kagaku Kaishi*, 302–309.

99. Federsel, C., Boddien, A., Jackstell, R., Jennerjahn, R., Dyson, P.J., Scopelliti, R., Laurenczy, G., and Beller, M. (2010) *Angew. Chem Int. Ed.*, **49**, 9777–9780.

100. Tai, C.C., Chang, T., Roller, B., and Jessop, P.G. (2003) *Inorg. Chem.*, **42**, 7340–7341.

101. van der Vlugt, J.I., and Reek, J.N.H. (2009) *Angew. Chem. Int. Ed.*, **48**, 8832–8846; *Angew.Chem.*, **121**, 8990–9004.

102. Preti, D., Squarcialupi, S., and Fachinetti, G. (2010) *Angew. Chem. Int. Ed.*, **49**, 2581–2584; *Angew. Chem.*, **122**, 2635–2638.

103. Ohnishi, Y.-Y., Nakao, Y., Sato, H., and Sakaki, S. (2006) *Organometallics*, **25**, 3352–3363.

104. (a) Ikariya, T., Jessop, P.G., and Noyori, R. Japan Tokkai 5-274721; (b) Thomas, C.A., Bonila, R.J., Huang, Y., and Jessop, P.G. (2001) *Can. J. Chem.*, **79**, 719–724; (c) Gao, Y., Kuncheria, J.K., Jenkins, H.A., Puddephatt, R.J., and Yap, G.P.A. (2000) *J. Chem. Soc., Dalton Trans.*, 3212–3217; (d) Zhang, J.Z., Li, Z., and Wang, H. (1996) *J. Mol. Catal. A: Chem.*, **112**, 9–14.

105. Ng, S.M., Yin, C., Yeung, C.H., Chan, T.C., and Lau, C.P. (2004) *Eur. J. Inorg. Chem.*, 1788–1793.

106. Matsubara, T. (2001) *Organometallics*, **20**, 19–24.

107. Burgemeister, T., Kastner, F., and Leitner, W. (1993) *Angew. Chem. Int. Ed. Engl.*, **32**, 739–741; *Angew. Chem.*, **105**, 781–783.

108. Ito, T. and Yamamoto, A. (1982) in *Organic and Bioorganic Chemistry of Carbon Dioxide* (eds. S. Inoue and N. Yamazaki), Kodansha, Tokyo.

109. Behr, A. (1983) *Catalysis in C1 Chemistry*, Reidel, Dordrecht, p. 169.

110. Matsubara, T. and Hirao, K. (2001) *Organometallics*, **20**, 5759–5768.

111. Musashi, Y. and Sakaki, S. (2000) *J. Am. Chem. Soc.*, **122**, 3867–3877.

112. Fellay, C., Yan, N., Dyson, P.J., and Laurenczy, G. (2009) *Chem. Eur. J.*, **15**, 3752–3760.

113. Loges, B., Boddien, A., Junge, H., Noyes, J.R., Baumann, W., and Beller, M. (2009) *Chem. Commun.*, 4185–4187.

114. Junge, H., Boddien, A., Capitta, F., Loges, B., Noyes, J.R., Gladiali, S., and Beller, M. (2009) *Tetrahedron Lett.*, **50**, 1603–1606.

115. See also the work of the groups of Joó and Laurenczy, Fukuzumi, and Leitner, e.g.: (a) Laurenczy, G., Joó, F., and Nádasdi, L. (2000) *Inorg. Chem.*, **39**, 5083–5088; (b) Ogo, S., Kabe, R., Hayashi, H., Haradaa, R., and Fukuzumi, S. (2006) *Dalton Trans.*, 4657–4663; (c) Dedieu, A., Eichberger, M., Fornika, R., and Leitner, W. (1997) *J. Am. Chem. Soc.*, **119**, 4432–4443.

116. (a) Aresta, M. and Dibenedetto, A. (2007) *Dalton Trans.*, 2975–2992; (b) Sakakura, T., Choi, J.C., and Yasuda, H. (2007) *Chem. Rev.*, **107**, 2365–2387.

117. Wu, X.-F., Anbarasan, P., Neumann, H., and Beller, M. (2010) *Angew. Chem. Int. Ed.*, **49**, 9047–9050; *Angew. Chem.*, **122**, 9231–9234.

118. Correa, A. and Martín, R. (2009) *Angew. Chem. Int. Ed.*, **48**, 6201–6204; *Angew. Chem.*, **121**, 6317–6320.

119. (a) Ukai, K., Aoki, M., Takaya, J., and Iwasawa, N. (2006) *J. Am. Chem. Soc.*, **128**, 8706–8707; (b) Ohishi, T., Nishiura, M., and Hou, Z. (2008) *Angew. Chem.*, **120**, 5876–5879; *Angew. Chem. Int. Ed.*, **47**, 5792–5795; (c) Takaya, J., Tadami, S., Ukai, K., and Iwasawa, N. (2008) *Org. Lett.*, **10**, 2697–2700.

120. Shi, M. and Nicholas, K.M. (1997) *J. Am. Chem. Soc.*, **119**, 5057–5058.

121. Aresta, M., Nobile, C.F., Albano, V.G., Forni, E., and Manassero, M. (1975) *J. Chem. Soc., Chem. Commun.*, **15**, 636–637.

122. (a) Yeung, C.S. and Dong, V.M. (2008) *J. Am. Chem. Soc.*, **130**, 7826–7829; (b) Ochiai, H., Jang, M., Hirano, K., Yorimitsu, H., and Oshima, K. (2008) *Org. Lett.*, **10**, 2681–2683.

123. Zhang, W.-Z., Li, W.-J., Zhang, X., Zhou, H., and Lu, X.-B. (2010) *Org. Lett.*, **12**, 4748–4751.

124. Yu, D. and Zhang, Y. (2010) *Proc. Natl. Acad. Sci. U. S. A.*, **107**, 20184–20189.

125. (a) Zhank, L., Cheng, J., Ohishi, T., and Hou, Z. (2010) *Angew. Chem. Int. Ed.*, **49**, 8670–8673; *Angew. Chem.*, **122**, 8852–8855; (b) Boogaerts, I.I.F., Fortman, G.C., Furst, M.R.L., Cazin, C.S.J., and Nolan, S.P. (2010) *Angew. Chem. Int. Ed.*, **49**, 8674–8677; *Angew. Chem.*, **122**, 8856–8859.

126. Boogaerts, I.I.F. and Nolan, S. (2010) *J. Am. Chem. Soc.*, **132**, 8858–8859.

127. Vechoorkin, O., Hirt, N., and Hu, X. (2010) *Org. Lett.*, **12**, 3567–3569.

128. Schubert, G. and Pápai, I. (2003) *J. Am. Chem. Soc.*, **125**, 14847–14858.

129. Aresta, M., Pastore, C., Giannoccaro, P., Kovacs, G., Dibenedetto, A., and Papai, I. (2007) *Chem. Eur. J.*, **13**, 9028–9034.

130. Behr, A. and Becker, M. (2006) *Dalton Trans.*, 4607–4613.

131. Gutteridge, S. and Gatenby, A.A. (1995) *Plant Cell*, **7**, 809–819.

132. Shaikh, A.G. and Sivaram, S. (1996) *Chem. Rev.*, **96**, 951–976.

133. Sakakura, T. and Kohno, K. (2009) *Chem. Commun.*, 1312–1330.

134. Fukuoka, S., Kawamura, M., Komiya, K., Tojo, M., Hachiya, H., Hasegawa, K., Aminaka, M., Okamoto, H., Fukawa, I., and Konno, S. (2003) *Green Chem.*, **5**, 497–507.

135. Pacheco, M.A. and Marshall, C.L. (1997) *Energy Fuels*, **11**, 2–29.

136. Kishimoto, Y. and Ogawa, I. (2004) *Ind. Eng. Chem. Res.*, **43**, 8155–8162.

137. Ball, P., Füllmann, H., and Heitz, W. (1980) *Angew. Chem. Int. Ed. Engl.*, **19** (9), 718–720.

138. (a) Fujita, S., Bhanage, B.M., Ikushima, Y., and Arai, M. (2001) *Green Chem.*, **3**, 87–91; (b) Fang, S.N. and Fujimoto, K. (1996) *Appl. Catal. A*, **142**, L1–L3.

139. Chun, Y., He, Y.G., and Zhu, J.H. (2001) *React. Kinet. Catal. Lett.*, **74**, 23–27.

140. (a) Sakai, S., Fujinami, T., Yamada, T., and Furusawa, S. (1975) *Nippon Kagaku Kaishi*, 1789–1794; (b) Yamazaki, N., Nakahama, S., and Higashi, F. (1978) *Rep. Asahi Glass Found. Ind. Technol.*, **33**, 31–45.

141. Kizlink, J. and Pastucha, I. (1995) *Collect. Czech. Chem. Commun.*, **60**, 687–692.

142. Choi, J.C., He, L.N., Yasuda, H., and Sakakura, T. (2002) *Green Chem.*, **4**, 230–234.

143. Li, C.F. and Zhong, S.H. (2003) *Catal. Today*, **82**, 83–90.

144. Choi, J.C., Sakakura, T., and Sako, T. (1999) *J. Am. Chem. Soc.*, **121**, 3793–3794.

145. (a) Ballivet-Tkatchenko, D., Jerphagnon, T., Ligabue, R., Plasseraud, L., and Poinsot, D. (2003) *Appl. Catal., A*, **255**, 93–99; (b) Ballivet-Tkatchenko, D., Burgat, R., Chambrey, S., Plasseraud, L., and Richard, P. (2006) *J. Organomet. Chem.*, **691**, 1498–1504.

146. Bhanage, B.M., Fujita, S., Ikushima, Y., Torii, K., and Aria, M. (2003) *Green Chem.*, **5**, 71–75.

147. Cui, Y.H., Wang, T., Wang, F.J., Gu, C.R., Wang, P.L., and Dai, Y.Y. (2003) *Ind. Eng. Chem. Res.*, **42**, 3865–3870.

148. (a) Coates, G.W. and Moore, D.R. (2004) *Angew. Chem. Int. Ed.*, **43**, 6618–6639; (b) Darensbourg, D.J. (2007) *Chem. Rev.*, **107**, 2388–2410.

149. (a) Lu, X.B., Zhang, Y.J., Liang, B., Li, X., and Wang, H. (2004) *J. Mol. Catal. A: Chem.*, **210**, 31–34; (b) Chang, T., Jing, H.W., Jin, L., and Qiu, W.Y. (2007) *J. Mol. Catal. A: Chem.*, **264**, 241–247.

150. (a) Peng, J.J. and Deng, Y.Q. (2001) *New J. Chem.*, **25**, 639–641; (b) Li, F.W., Xiao, L.F., Xia, C.G., and Hu, B. (2004) *Tetrahedron Lett.*, **45**, 8307–8310.

151. Ji, D.F., Lu, X.B., and He, R. (2000) *Appl. Catal. A*, **203**, 329–333.

152. (a) Yano, T., Matsui, H., Koike, T., Ishiguro, H., Fujihara, H., Yoshihara, M., and Maeshima, T. (1997) *Chem. Commun.*, 1129–1130; (b) Yasuda, H., He, L.N., and Sakakura, T. (2002) *J. Catal.*, **209**, 547–550.

153. Sakai, T., Tsutsumi, Y., and Ema, T. (2008) *Green Chem.*, **10**, 337–341.

154. Takahashi, T., Watahiki, T., Kitazume, S., Yasuda, H., and Sakakura, T. (2006) *Chem. Commun.*, 1664–1666.

155. (a) He, L.N., Yasuda, H., and Sakakura, T. (2003) *Green Chem.*, **5**, 92–94; (b) Sako, T., Fukao, T., Sahashi, R., Sone, M., and Matsuno, M. (2002) *Ind. Eng. Chem. Res.*, **41**, 5353–5358.

156. Lu, B.X., Liang, B., Zhang, Y.J., Tian, Y.Z., Wang, Y.M., Bai, C.X., Wang, H., and Zhang, R. (2004) *J. Am. Chem. Soc.*, **126**, 3732–3733.

157. Huang, J.W. and Shi, M. (2003) *J. Org. Chem.*, **68**, 6705–6709.

158. Kim, H.S., Bae, J.Y., Lee, J.S., Kwon, O.S., Jelliarko, P., Lee, S.D., and Lee, S.H. (2005) *J. Catal.*, **232**, 80–84.

159. Yoshida, M., Murao, T., Sugimoto, K., and Ihara, M. (2007) *Synlett*, 575–578.

160. Aresta, M., Dibenedetto, A., and Tommasi, I. (2000) *Appl. Organomet. Chem.*, **14**, 799–802.

161. Du, Y., Kong, D.L., Wang, H.Y., Cai, F., Tian, H.S., Wang, J.Q., and He, L.N. (2005) *J. Mol. Catal. A: Chem.*, **241**, 233–237.

162. Aresta, M., Dibenedetto, A., Dileo, C., Tommasi, I., and Amodio, E. (2003) *J. Supercrit. Fluids*, **25**, 177–182.

163. Uemura, K., Kawaguchi, T., Takayama, H., Nakamura, A., and Inoue, Y. (1999) *J. Mol. Catal. A: Chem.*, **139**, 1–9.

164. Berkessel, A. and Brandenburg, M. (2006) *Org. Lett.*, **8**, 4401–4404.

165. Tanaka, H., Kitaichi, Y., Sato, M., Ikeno, T., and Yamada, T. (2004) *Chem. Lett.*, **33**, 676–677.

166. Yamada, W., Kitaichi, Y., Tanka, H., Kojima, T., Sato, M., Ikeno, T., and Yamada, T. (2007) *Bull. Chem. Soc. Jpn.*, **80**, 1391–1401.

167. Yang, H., Xu, Z., Fan, M., Gupta, R., Slimane, R.B., Bland, A.E., and Wright, I. (2008) *J. Environ. Sci.*, **20**, 14–27.

168. (a) Metz, B., Davidson, O., De Coninck, H.C., Loos, M., and Meyer, L.A. (eds.) (2005) *Intergovernmental Panel on Climate Change Special Report on Carbon Dioxide Capture and Storage*, Cambridge University Press, Cambridge; http://www.ipcc.ch/pdf/special-reports/srccs/srccs_wholereport.pdf (last accessed 3 February 2012); (b) Aresta, M. and Tommasi, I. (1997) *Energy Convers. Manage.*, **38**, S373–S378.

169. (a) Hunt, A.J., Sin, E.H.K., Marriott, R., and Clark, J.H. (2010) *ChemSusChem*, **3**, 306–322; (b) Chang, C.J., Chiu, K.-L., Chen, Y.-L., and Chang, C.-Y. (2000) *Food Chem.*, **68**, 109–113; (b) Ramalakshmi, K. and Raghavan, B. (1999) *Crit. Rev. Food Sci. Nutr.*, **39**, 441–456.

170. (a) Zekovic, Z., Pfaf-Sovljanski, I., and Grujic, O. (2007) *J. Serb. Chem. Soc.*, **72**,

81–87; (b) Leeke, G., Gaspar, F., and Santos, R. (2002) *Ind. Eng. Chem. Res.*, **41**, 2033–2039; (c) Vollbrecht, R. (1982) *Chem. Ind. (London)*, 397–405.

171. Messen, J.H. and Petersen, H. (2003) in *Ullmann's Encyclopedia of Industrial Chemistry* (eds. M. Bohnet and F. Ullmann), 6th edn., vol. **37**, Wiley-VCH Verlag GmbH, Weinheim, pp. 683–718.

172. Fromm, D. and Lützov, D. (1979) *Chem. Unserer Zeit*, **13**, 78–81.

173. Bosch, C. and Meiser, W. (1922) (to BASF), US Patent 1,429,483.

174. (a) Weissermel, K. and Arpe, H.-J. (eds.) (2003) *Industrial Organic Chemistry*, Wiley-VCH Verlag GmbH, Weinheim; (b) Cheng, W.H. and Kung, H.H. (eds.) (1994) *Methanol Production and Use*, Marcel Dekker, New York.

175. (a) Rozovskii, A.Y. and Lin, G.I. (2003) *Top. Catal.*, **22**, 137–150; (b) Saito, M., Fujitani, T., Takeuchi, M., and Watanabe, T. (1996) *Appl. Catal.*, **138**, 311–318; (c) Rozovskii, A. (1989) *Russ. Chem. Rev.*, **58**, 41–56.

176. Ushikoshi, K., Mori, K., Watanabe, T., Takeuchi, M., and Saito, M. (1998) in *Advances in Chemical Conversions for Mitigating Carbon Dioxide* (eds. T. Inui, M. Anpo, K. Izui, S. Yanagida, and T. Yamaguchi), Elsevier, Amsterdam, pp. 357–362.

177. (a) An, X., Li, J., Zuo, Y., Zhang, Q., Wang, D., and Wang, J. (2007) *Catal. Lett.*, **118**, 264–269; (b) Saito, M. and Murata, K. (2004) *Catal. Surv. Asia*, **8**, 285–294; (c) Takeuchi, M., Fujitani, T., Toyir, J., Luo, S., Wu, J., Mabuse, H., Ushikoshi, K., Mori, K., and Watanabe, T. (2000) *Appl. Organomet. Chem.*, **14**, 763–772; (d) Saito, M. (1998) *Catal. Surv. Jpn.*, **2**, 175–184.

178. (a) Kolbe, H. (1860) *Liebig's Ann.*, **113**, 125–127; (b) Schmitt, R. (1885) *J. Prakt.*

179. (a) Peppel, W.J. (1958) *Ind. Eng. Chem.*, **50**, 767–770; (b) Vierling, K. (1943) (to I. G. Farbenindustrie), German Patent DE 740366.

180. (a) Pacheco, M.A. and Marshall, C.L. (1997) *Energy Fuels*, **11**, 2–29, and references therein.

181. (a) Knifton, J.F. (1988) (to Texaco), US Patent 4,734,518; (b) Knifton, J.F. (1987) (to Texaco), US Patent 4,661,609; (c) Duranleau, R.G., Nieh, E.C.Y., and Knifton, J.F. (1986) (to Texaco), US Patent 4,691,041; (d) Knifton, J.F. and Duranleau, R.G. (1991) *J. Mol. Catal.*, **67**, 389–399.

182. (a) Inoue, S., Koinuma, H., and Tsuruta, T. (1969) *Makromol. Chem.*, **130**, 210–220; (b) Inoue, S., Koinuma, H., and Tsuruta, T. (1969) *J. Polym. Sci., Part B: Polym. Phys.*, **7**, 287–292.

183. Sakatura, T., Choi, J.-C., and Yasuda, H. (2007) *Chem. Rev.*, **107**, 2365–2387.

184. Beckman, E.J. (2004) *J. Supercrit. Fluids*, **28**, 121–191.

185. Fukuoka, S., Tojo, M., Hachiya, H., Aminaka, M., and Hasegawa, K. (2007) *Polym. J.*, **39**, 91–114.

186. Fukuoka, S., Fukawa, I., Tojo, M., Oonishi, K., Hachiya, H., Aminaka, M., Hasegawa, K., and Komiya, K. (2010) *Catal. Surv. Asia*, **14**, 146–163.

187. Okamoto, H. and Someya, K. (2007) US Patent 7,199,253.

188. Fukuoka, S., Miyaji, H., Hachiya, H., and Matsuzaki, K. (2008) European Patent EP 1,953,131.

189. Fukuoka, S., Tojo, M., and Kawamura, M. (1993) US Patent 5,210,268.

Chem., **31**, 397–411; (c) Ota, K. (1974) *Bull. Chem. Soc. Jpn.*, **47**, 2343–2344; (d) Lindsey, A.S. and Jeskey, H. (1957) *Chem. Rev.*, **57**, 583–620.

Commentary Part

Comment 1

Gábor Laurenczy

The authors of this chapter on the catalytic utilization of CO_2 definitely focus our attention on one of the central questions of mankind, which is linked with several aspects of energy sources, supply, and utilization in the future. The increasing global energy demand, coupled with depletion of fossil fuel resources, and the environmental impact of these current fuels, and the atmospheric concentration of carbon dioxide (which has risen from 278 ppm during the preindustrial era to the current level of 388 ppm) these are just some of the different, but strongly linked, phenomena of this global problem of the twenty-first century. On the other hand, carbon dioxide is the primary carbon source in the atmosphere and constitutes the basis for all organic matter on Earth and, hence, life.

The authors analyze deeply all the facets of this greenhouse gas, carbon dioxide, linked with the necessity to decrease its concentration in the atmosphere. As they summarize, the actual industrial possibilities and demands are very far from the real needs: it is impossible today to convert the annually emitted $\sim$29 Gt of carbon dioxide into useful, technically demanded chemicals. Nevertheless, transformation of the omnipresent carbon dioxide into important industrial commodities, such as urea, methanol, acetic acid, polycarbonates, FA, calcium carbonate, silicon building materials, and so on, cannot be neglected or underestimated; these utilizations also contribute to the reduction of global CO_2 emission. Hence the increased exploitation of CO_2 as a starting material by the chemical industry is highly desirable. In addition to novel applications in organic carbonates, the combination of CO_2 and hydrogen constitutes a versatile replacement for toxic CO, which is widely used as a raw material in bulk-scale chemical processes. A clear drawback is that a considerably higher energy input is needed for such a combination, compared with the more reactive CO.

As the authors develop in detail, the general solution for CO_2 utilization and recycling could be the transformation of the CO_2 into liquid fuels. Nevertheless, owing to the relative chemical inertness and the poor reactivity of CO_2, being in a deep "valley" in thermodynamic terms, these reactions require a lot of energy, highly active reactants, assistance of metal-based catalysts, and so on. The solution could be to use renewable energy sources, for example, solar energy, and hydrogenate CO_2 into FA, this being the first step on the way to complete reduction of CO_2 to methane. Hence a promising approach to overcoming the low reactivity of CO_2 is its activation by catalytic hydrogenation, to form FA or its derivatives.

Matthias Beller and his co-authors develop this possibility in detail concerning the CO_2–FA cycle in energy/hydrogen storage and delivery, as his group is among the most important contributors to this field. FA has several advantages over other liquid fuels derived from carbon dioxide: pure FA

is a liquid with a flash point–ignition temperature of 69 °C, much higher than that of the gasoline (-40°C) and ethanol (13 °C); it contains 53 g l^{-1} of hydrogen at room temperature and atmospheric pressure, which is twice as much as compressed hydrogen gas can attain at 350 bar pressure, and 85% FA solution is not flammable. Concerning the "energy recuperation" in the form of hydrogen, the catalytic decomposition of FA takes place rapidly, over a wide range of pressures (up to 1250 bar); FA splitting leads only to gaseous products (H_2/CO_2), and the straightforward reduction of CO_2 leads to the formation of FA. Hence there is significant potential for the development of a viable "joined-up" system of hydrogen generation, storage, and transport based on the use of FA, which is safer to use than high-pressure hydrogen gas. The reaction can be carried out in aqueous solution, and the high-pressure gases formed were found not to inhibit the decomposition of FA.

To summarize, this chapter highlights very clearly the increasing necessity to use CO_2 as a primary carbon resource, using renewable energy sources, to produce (new) energy carriers. Carbon dioxide reduction/hydrogenation is discussed in detail, specifically with focus on research.

The conclusion of this topic shows clearly that artificial photosynthesis and direct photochemical reductions for transformation of carbon dioxide are the most attractive possibilities, but definitely need more scientific progress.

Comment 2

Min Shi

Carbon dioxide, the Earth's most abundant carbon resource, is remarkably little used as a chemical feedstock. As the output of CO_2 from combustion into the environment continues to rise, threatening a global environmental crisis, the development of practical and useful methods for regenerating organic compounds from CO_2 has become of increasing importance. Indeed, the utilization of CO_2 as a *C1*-feedstock for the production of organic compounds is a difficult subject because it is chemically an issue of reduction of the very stable and inert gas molecule CO_2. As mentioned in this chapter, the conversion of CO_2 in order to produce chemicals is outperformed by using high-energy reactants such as epoxides (ring tension), hydrogen, and unsaturated compounds to overcome thermodynamic barriers. This chapter sets out recent developments in catalytic reductions of CO_2 to formic acid and methanol, CO_2 as a *C1*-building block in C-C-coupling reactions, catalytic C–O bond formation utilizing CO_2, and current industrial processes using CO_2. Although a comprehensive list of challenges cannot be provided here, I outline a few of them. First, the catalytic reduction of CO_2 to methanol under heterogeneous and homogeneous catalysis and also enzymatic approaches has been fully described. This is a practically useful process to regenerate organic chemicals or energy from CO_2 because methanol can be used for the production of various other large-scale chemicals such as acetic acid (Monsanto process), formaldehyde, and MTBE. Additionally, methanol is considered to be a sustainable platform for the post-fossil fuel age. The authors clearly point out the efficiency, drawbacks, and perspectives in this field. Another interesting aspect is that the authors outline CO_2 as a *C1*-building block in C–C coupling reactions. Using CO_2 as an electrophile, transition metal-catalyzed reactions show higher functional group tolerance compared with additions to Grignard or organolithium compounds, producing a variety of carbonyl group-containing products in good yields along with good regio- and chemoselectivities.

Overall, after reading this chapter, I believe that transition metal-catalyzed CO_2 fixation and direct photochemical reductions of CO_2 are very interesting in this regard. As pointed out by the authors, clearly more research for suitable catalysts is needed to drive such reactions efficiently. Concerning the transition metal-catalyzed fixation of CO_2, several recent reports are relevant (e.g., [C1, C2]). Moreover, several recent reports on the NHC-promoted fixation of CO_2 could be cited (e.g., [C3–C5]).

References

C1. Yoshida, S., Fukui, K., Kikuchi, S., and Yamada, T. (2010) *J. Am. Chem. Soc.*, **132**, 4072.

C2. Williams, C.M., Johnson, J.B., and Rovis, T. (2008) *J. Am. Chem. Soc.*, **130**, 14936.

C3. Nair, V., Varghese, V., Paul, R.R., Jose, A., Sinu, C.R., and Menon, R.S. (2010) *Org. Lett.*, **12**, 2653.

C4. Gu, L.-Q., and Zhang, Y.-G. (2010) *J. Am. Chem. Soc.*, **132**, 914.

C5. Riduan, S.N., Zhang, Y., and Ying, J.-Y. (2009) *Angew. Chem. Int. Ed.*, **48**, 3322.

19
Synthetic Chemistry with an Eye on Future Sustainability

Guo-Jun Deng and Chao-Jun Li

19.1
Introduction

Sustainability refers to the ability to provide a healthy, satisfying, and high-quality life for all people on Earth, now and for generations to come. Increasingly, stringent environmental legislation has generated a pressing need for cleaner methods of chemical products; for instance, technologies that can reduce or, preferably, eliminate the generation of waste and avoid the use of toxic and/or hazardous solvents [1]. This is also known as the "green chemistry" [2]. In 1987, a UN report defined sustainable development as "development that meets the needs of the present without compromising the ability of future generations to meet their own needs" [3]. Chemistry is a central science in energy conversions and substance transformations. Hence chemists are uniquely qualified to provide a molecular-level approach to sustainable development of our society. Sustainable development instituted a paradigm shift of chemists from the traditional concept of process efficiency that mainly focuses on chemical yield, to the one that checks the balance of all materials used in chemical transformations. In 1991, Trost presented a new guiding principle for evaluating the efficiency of specific chemical processes, termed the "atom-economy," which has subsequently been incorporated into the "Twelve Principles of Green Chemistry" and has altered the way in which many chemists design and plan their syntheses [4]. The "Twelve Principles of Green Chemistry" can be paraphrased as follows:

1) waste prevention instead of remediation
2) atom efficiency
3) less hazardous/toxic chemicals
4) safer products by design
5) innocuous solvents and auxiliaries
6) energy efficient by design
7) preferably renewable raw materials
8) shorter syntheses (avoiding derivatization)
9) catalytic rather than stoichiometric reagents
10) design of products for degradation

Organic Chemistry – Breakthroughs and Perspectives, First Edition. Edited by Kuiling Ding and Li-Xin Dai.
© 2012 Wiley-VCH Verlag GmbH & Co. KGaA. Published 2012 by Wiley-VCH Verlag GmbH & Co. KGaA.

11) analytical means for pollution prevention
12) inherently safer processes and chemical products.

If sustainable development is the goal, then green chemistry is the key means to achieve it. Atom economy (atom efficiency or atom utilization) is an extremely useful tool for the rapid evaluation of the amounts of inherent waste that will be generated by alternative processes. Hence to judge the relative greenness of a reaction, it can be calculated by dividing the molecular weight of the product by the sum total of the molecular weights of all substances formed in the stoichiometric equation for the reaction involved. For example, the atom economy of oxidizing a secondary alcohol to the corresponding ketone by O_2 is illustrated in Scheme 19.1.

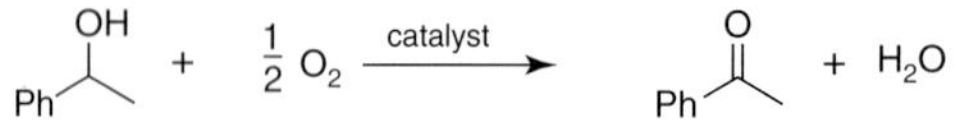

$$\text{Atom economy (\%)} = \frac{\text{molecular weight of products}}{\text{molecular weight of all products}} \times 100\% = \frac{120}{138} \times 100\% = 87\%$$

Scheme 19.1 Example of atom economy calculation.

Since its birth about two decades ago, the field of green chemistry has placed specific emphasis on innovations in chemical synthesis. At the center, there has been a renewed focus on the age-old pursuit of organic chemists to design synthetic reactions in terms of efficiency, with atom and step economy being a major goal [5]. Great progress has been made to achieve such a goal since the emergence of green chemistry [6]. Another key area of research of green chemistry is the study of cleaner and more environmentally benign solvents used in various chemical processes, with great advances achieved in using water, ionic liquids, and supercritical fluids in addition to other emerging alternatives. Progress also has been made in the use of renewable feedstocks instead of non-renewable petroleum for the production of both fine and bulk chemical products and for energy applications.

19.1.1
Chemical Feedstocks

At present, the main feedstock for the chemical industry and energy uses stems from nonrenewable crude oil (petroleum) and natural gas. Continuing questions about the endurance and stability of petroleum supplies, and also the environmental impacts of its production and use, have driven the development of alternative chemical feedstocks production and fuel production based on renewable resources. On the other hand, Nature produces the vast amount of 170 billion tonnes of biomass per year by photosynthesis, 75% of which can be assigned to the class of carbohydrates [7]. Biomass carbohydrates are the most abundant renewable resources and are currently viewed as a feedstock for the green chemistry of the future. However, a major obstacle to using renewable biomass as feedstock is the need for novel chemistry to transform the large amounts of biomass selectively and efficiently, in its natural state, without extensive functionalization, defunctionalization, or protection. It will be a long journey to achieve this goal although great progress has been made during the past decade [8].

19.1.2
Green Solvents

Solvents are auxiliary materials that play an important role in chemical production and synthesis. By employing a solvent, reactions may proceed faster and more smoothly and selectively. Sometimes, the solvent acts as a heat transfer medium which makes the reaction safer, especially for large-scale production. Organic reactions are generally performed in organic solvents, which generate the largest amount of "auxiliary waste" in most chemical productions. Removal of residual solvent from products is usually achieved by evaporation; therefore, leakage and evaporation of highly volatile solvents inevitably lead to atmospheric pollution. Polar aprotic solvents such as N,N-dimethylformamide (DMF) and dimethyl sulfoxide are the choice of solvents for many nucleophilic substitutions. They have high boiling point and are not easily removed by distillation. They are also water miscible, which allows their separation by washing with water. Unfortunately, this inevitably leads to contaminated aqueous effluent. As part of green chemistry efforts, various relatively nontoxic/nonhazardous and cleaner solvents have been evaluated as replacements.

The best solvent is no solvent [9], but if one has to use a solvent then water should be the best choice (given that no contaminated water will be released!). Water is the most abundant and the only natural solvent on Earth, and hence readily available and environmentally benign. Since Rideout and Breslow reported that Diels–Alder reactions could be accelerated by using water as solvent [10], there has been increasing recognition that organic reactions can proceed well in aqueous media and offer advantages over those occurring in organic solvents. Much effort has been dedicated to the development of organic reactions in water and great progress has been made in the past three decades. In addition to Diels–Alder reactions, many other reactions, and even "water-sensitive" reactions, have been successfully carried out in water [11]. Another recent significant development in the field of aqueous organic reactions is the development of organic reactions "on water" by Sharpless and co-workers [12]. Water not only is used as a greener reaction medium but also, in most "on water" reactions, plays a crucial role in achieving large rate accelerations and selectivity enhancement [13].

Various features of supercritical carbon dioxide ($scCO_2$) make it an interesting green solvent for chemical synthesis [14]. CO_2 is renewable, nonflammable, and can be readily converted to its supercritical fluid state. The $scCO_2$ solvent can be easily separated by depressurization and recapture after the reaction is completed. $scCO_2$ is emerging as an important medium for chemical synthesis [15]. Another feature of liquid CO_2 and $scCO_2$ is their high miscibility with gases, which offers high efficiency in reactions such as hydrogenations with hydrogen gas [16].

In addition to the above two "natural green solvents," various non-natural ones such as ionic liquids [17] and fluorous solvents [18] have also been intensively studied as greener alternatives. These solvents have been extensively studied in the last decade as media for organic synthesis and catalysis.

19.1.3
Reactions

Reactions play the most fundamental role in synthesis. The key to waste minimization in fine chemical manufacture is the widespread substitution of classical organic synthesis employing preactivated functional groups and stoichiometric amounts of organic or inorganic reagents with more direct, cleaner, and catalytic alternatives [19]. An example is C–C bond formation based on direct C–H activation. This methodology can directly transform the C–H bonds of organic molecules into desired structures without extra chemical transformations and functionalizations and represents an important class of desirable greener reactions for the future. Moreover, the direct conversion of C–H bonds into C–C bonds leads to more efficient syntheses of complex products with reduced synthetic operations (Scheme 19.2) [20]. Let us consider biaryl synthesis as an example. Classical synthetic routes such as Grignard-type reactions require four steps and use stoichiometric and hazardous reagents, whereas biaryl synthesis based on double C–H activation requires only one step in the presence of a stoichiometric oxidant with a catalytic amount of catalyst (when O_2 is used as the terminal oxidant, water is the only byproduct). Hence the new route makes the synthesis simpler and reduces waste generation significantly. Recently, great progress has been made in transition metal-catalyzed activation and further reaction of C–H bonds.

Protection–deprotection of functional groups is widely used in organic synthesis. This strategy increases the number of steps in synthesizing the desired products. Novel chemistry is needed to perform organic synthesis without protection and deprotection. Recently, progress has been made in this area, especially for total synthesis [21]. Also of fundamental importance to greener syntheses is to develop tandem and cascade reaction processes that incorporate as many reactions as possible in order to give the final product in one operation. Performing sequential reactions in a flow reactor is also an alternative way to simplify operations.

During the past two decades, many book chapters and review papers on green chemistry and sustainable development have been published. Great progress has been made in

C–C bond formation based on classic Grignard-type cross-coupling

C–C bond formation based on double C–H activation

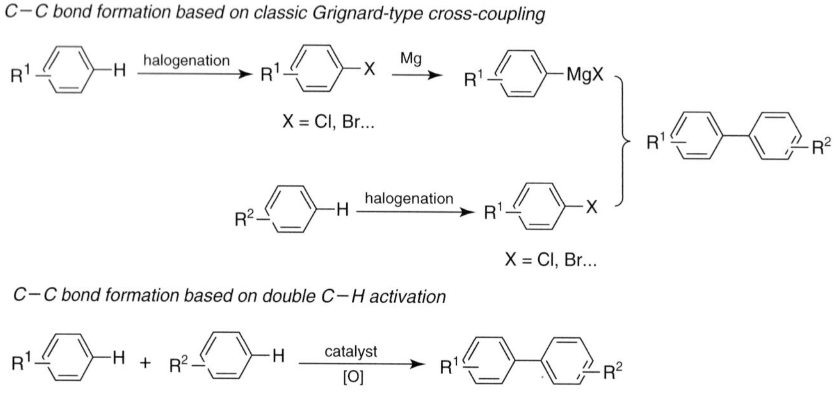

Scheme 19.2 Synthetic route comparison of C–C bond formation.

recent years, hence it is very difficult to review all of the progress that has been made from every aspect. In this chapter, we mainly describe our progress in research on cross-dehydrogenative coupling (CDC) and nucleophilic addition of terminal alkynes in water during the past decade as illustrative examples on this topic.

19.2
Cross-Dehydrogenative Coupling

Carbon–carbon bond formation is central to organic chemistry and is a very important method to convert simple molecules to more complex compounds. It has attracted much attention to develop efficient synthetic routes. For decades, well-known reactions such as the Mizoroki–Heck and the cross-coupling reaction (Hiyama, Kumada, Negishi, Still, Suzuki–Miyaura) have been the methods of choice to form new C–C bonds [22]. These methods greatly extended the scope and increased the efficiency of C–C bond formation in modern organic chemistry. However, these methods must use prefunctionalized starting materials and/or stoichiometric amounts of metals (Scheme 19.3a). It was therefore highly desirable to develop greener synthetic methods to construct C–C bonds. One of them is transition metal-catalyzed direct C–H bond activation and subsequent C–C bond formation. Since the pioneering work by both Murai *et al.* [23] and Fujiwara and co-workers [24] on C–C bond-forming reactions through the catalytic cleavage of C–H bonds, this research area has undergone rapid development and is becoming an increasingly viable alternative [25]. Nevertheless, these reactions still require a functionalized partner to generate the desired C–C bond formation product (Scheme 19.3b). In 2003, our laboratory first specifically proposed and has subsequently been exploring the possibility of developing a methodology to construct functional molecules by only using C–H bonds, which we termed cross-dehydrogenative coupling (CDC) (Scheme 19.3c), irrespective of the specific reaction mechanism and substrate types [19, 26]. Compared with traditional methods, CDC reactions afford C–C bonds in fewer steps by avoiding substrate prefunctionalization, which obviously will result in less waste generation. Hence less waste and good atom economy are features of this new methodology. Great progress has also been made recently in arene–arene coupling via the reaction of sp^2 C–H/sp^2 C–H bonds by several other laboratories [27].

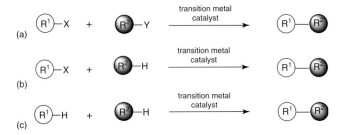

Scheme 19.3 The evolution of direct C–C bond formation.

19.2.1
CDC Reaction of the α-C–H Bond of Nitrogen in Amines

19.2.1.1 Alkynylation (sp³–sp Coupling)

Nitrogen-containing compounds are of great importance as building blocks for pharmaceuticals, agrochemicals, dyes, and ligands [28]. Much effort has been made to develop efficient synthetic methods to synthesize nitrogen-containing compounds. Direct alkynylation of the methyl group adjacent to nitrogen is an attractive method for the following reasons: (i) propargylamines are of great pharmaceutical interest and are useful intermediates for synthesizing nitrogen-containing compounds [29]; (ii) the sp³ C–H bond α to nitrogen in amines can be activated readily to generate iminium ions intermediate via a single-electron-transfer (SET) process (Scheme 19.4) [30]. We speculated that the CDC reaction of the sp³ C–H bond α to nitrogen with a terminal alkyne should therefore occur similarly under oxidative conditions.

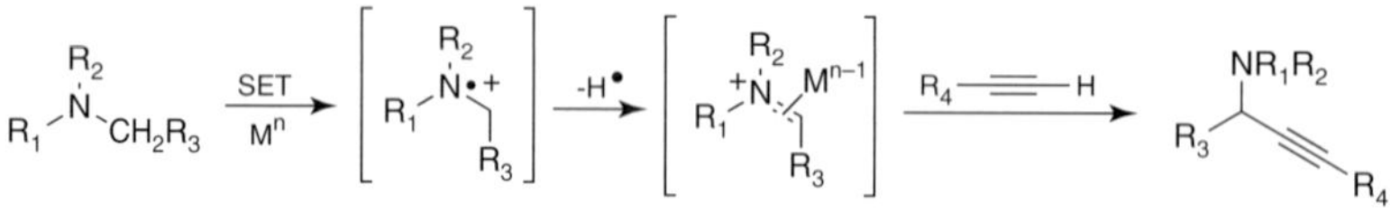

Scheme 19.4 Proposed alkynylation of the C-H bond α to nitrogen in an amine.

Commercially available phenylacetylene and *N,N*-dimethylaniline were used as the model substrates for our initial investigation. No reaction was observed in the absence of copper catalyst or oxidant. Various copper salts were examined as catalysts by using *tert*-butyl hydroperoxide (TBHP) as oxidant. CuBr, CuBr₂, CuCl, and CuCl₂ were all found to be highly effective for direct alkynylation of *N,N*-dimethylaniline. The best yield was obtained when the *N,N*-dimethylaniline:alkyne:TBHP ratio was 2:1:1. Decreasing the amount of *N,N*-dimethylaniline and TBHP both decreased the reaction yields. In the presence of CuBr and TBHP, various alkynes reacted smoothly with *N,N*-dimethylaniline derivatives to give the desired alkynylation products in 12–82% yields (Scheme 19.5) [31]. The direct alkynylation showed good regioselectivity when *N,N*-dimethylbenzylamine was reacted with phenylacetylene. The alkynylation takes place preferentially at the alkyl rather than the benzyl position, and no minor product was isolated. Similar results were also obtained when tetrahydroisoquinoline was reacted with phenylacetylene.

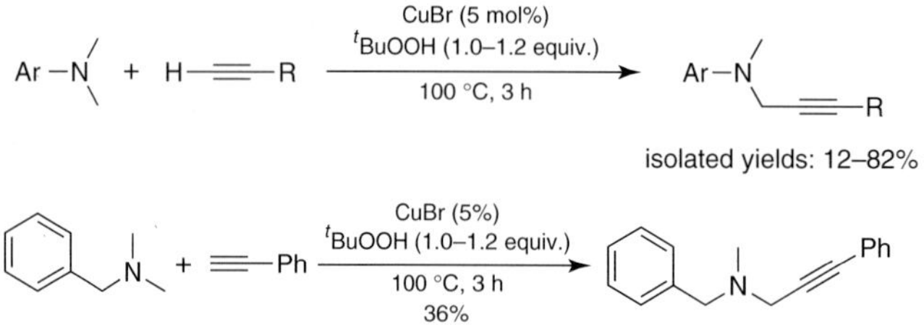

Scheme 19.5 Copper-catalyzed direct alkynylation of anilines and *N,N*-dimethylbenzylamine.

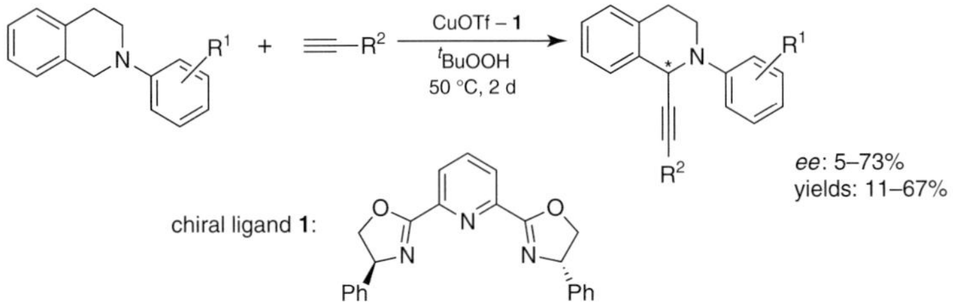

Scheme 19.6 Asymmetric alkynylation of tetrahydroisoquinolines with terminal alkynes.

Tetrahydroisoquinoline alkaloids with a stereocenter at C1 carbons exist widely in nature and are compounds of extensive interest owing to their biological and pharmacological properties [32]. It would be interesting to see if it is possible to achieve enantioselective C–C bond formations based on the direct alkynylation of prochiral CH_2 groups by introducing chiral catalysts. Various copper salts together with chiral bisoxazolines as ligands were investigated as catalysts at low temperatures. CuOTf–pybox (**1**) showed the best enantioselectivity together with good regi-selectivity (Scheme 19.6) [33]. Various solvents can be used and the highest enantioselectivity was obtained using tetrahydrofuran (THF) as solvent. Other chiral phosphine ligands showed lower enantioselectivity for this reaction.

The synthesis of non-natural (synthetic) amino acids has attracted much attention and general methods for the direct modification of natural peptides (a class of renewable feedstocks) are highly desirable [34]. The direct α-alkynylation of natural peptides takes advantage of existing structure and can potentially provide rapid access to diverse new peptides. We found that various *p*-methoxyphenyl (PMP) glycine amides can be directly alkynylated with phenylacetylene at room temperature via the CDC reaction without protection of the free N–H protons (Scheme 19.7) [35]. CuBr was found to be the most efficient catalyst for this reaction, and good to excellent yields were obtained. The reaction is highly selective, and no reaction was observed with the corresponding glycine ester. This methodology provides a versatile method to synthesize homophenylalanine

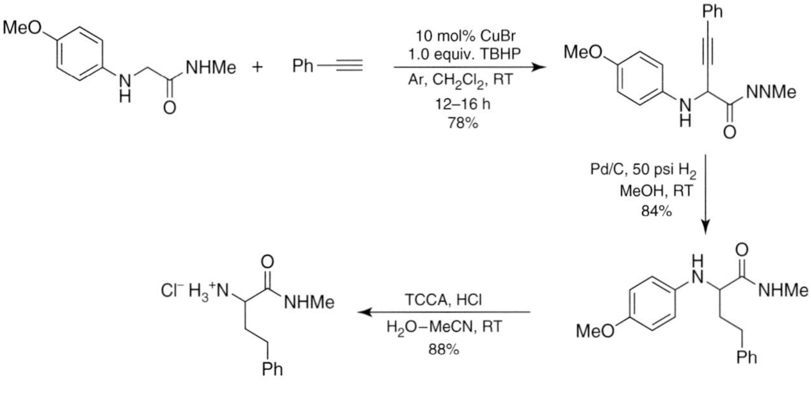

Scheme 19.7 Synthesis of a homophenylalanine derivative via CDC. TCCA, trichloroisocyanuric acid.

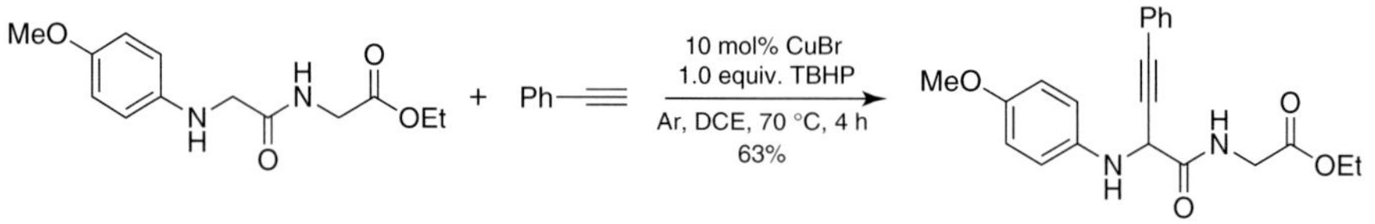

Scheme 19.8 Site-specific alkynylation of a dipeptide.

derivatives, an important synthon in many important angiotensin-converting enzyme inhibitors, via the hydrogenation of alkyne and removal of PMP (Scheme 19.7) [36].

This method was also successfully used for direct and site-selective peptide functionalizations by using dipeptides as model substrates (Scheme 19.8). The coupling reaction proceeded smoothly at 70 °C in dichloroethane, affording the alkynylation product selectively in 63% isolated yield, at the glycine amide site only [37]. Aryl, vinyl, and indolyl can also be introduced selectively to the α-position of the terminal glycine moieties under modified conditions. In the meantime, the configurations of other stereocenters in the peptide are maintained [38]. Since the functionalized peptides could easily be deprotected and carried on to the next coupling process, this approach provides a useful tool for peptide-based biological research.

19.2.1.2 Arylation (sp³–sp² Coupling)

It was thought that an iminium intermediate would be formed during the direct alkynylation of the α-sp³ C–H bond of nitrogen in amines via CDC. This implies that other C–H-based nucleophiles besides terminal alkynes can also couple with the α-sp³ C–H bond of nitrogen in amines via CDC in a similar way. To improve our hypothesis, electron-rich indoles were tested for this kind of coupling reaction. Indeed, the reaction of *N*-phenyltetrahydroisoquinoline with free indoles using the CuBr–TBHP system produced the desired product in good to excellent yields (Scheme 19.9) [39]. The reaction was not sensitive to moisture or air. Even when the reaction was carried out in water under an atmosphere of air, the desired product was obtained in moderate yield. The yield was improved from 45 to 75% when the temperature was increased from room temperature to 50 °C. Under the optimized conditions, various indoles were coupled with *N*-phenyltetrahydroisoquinoline. Other electron-rich aromatic compounds such as 2-naphthol would also couple with tetrahydroisoquinolines under similar conditions.

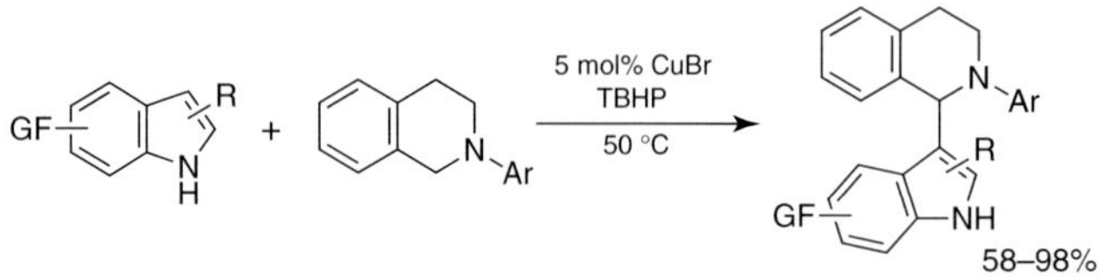

Scheme 19.9 CDC reactions of indoles with tetrahydroisoquinolines.

19.2.1.3 Alkylation (sp³–sp³)

The direct cross-coupling reaction between sp³ C–H and sp³ C–H was our next target to be investigated after the success of the CDC reactions between sp³ C–H and sp C–H, and

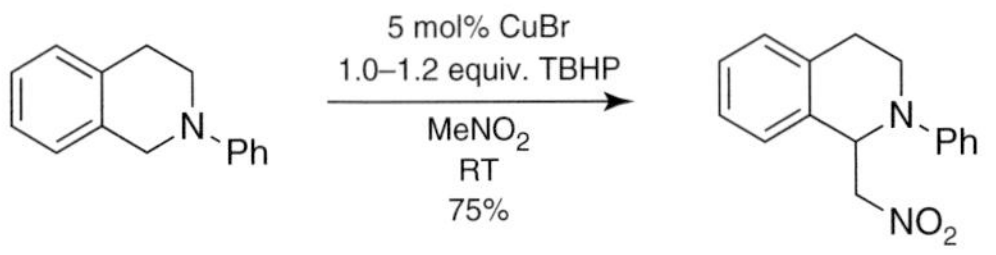

Scheme 19.10 CDC reaction of tetrahydroisoquinoline with nitromethane.

also between sp^3 C–H and sp^2 C–H. The reaction of 1,2,3,4-tetrahydroisoquinoline with nitromethane was chosen as the model reaction, in which nitromethane was also used as solvent. Various copper catalysts were examined at ambient temperature. The desired product was obtained in all cases, and the best yield was obtained when CuBr was used as the catalyst. Under the optimized conditions, various β-nitroamine derivatives were generated by this new methodology (Scheme 19.10) [40]. The use of nitroethane instead of nitromethane gave the desired compounds with good isolated yields. This method provides an alternative approach to preparing vicinal diamines which were traditionally prepared by the nitro-Mannich (aza-Henry) reaction.

Dialkyl malonates have relatively reactive CH$_2$ groups and were therefore examined as coupling reagents under similar reaction conditions. The reaction of tetrahydroisoquinolines with various dialkyl malonates in the presence of 5 mol% CuBr at room temperature gave the desired CDC products in high yields (Scheme 19.11). The desired product was obtained in 72% isolated yield even with only 0.5 mol% CuBr catalyst. Sodeoka and co-workers reported the reaction of malonate with N-Boc-protected tetrahydroisoquinolines asymmetrically by using a chiral palladium catalyst [41]. 2,3-Dichloro-5,6-dicyanobenzoquinone (DDQ) was used as oxidant and was added to the reaction solution slowly. The reaction generated the desired product in 86% enantiomeric excess (ee) together with 82% isolated yield (Scheme 19.12). When dehydroisoquinoline was used as the starting material, the enantioselectivity could be improved to 97% ee and no oxidant is necessary.

Molecular oxygen is a green and renewable oxidant widely used in many reactions. The only byproduct generated in the reactions is water. Hence the replacement of peroxides by molecular oxygen (and in water) would offer a more atom-economical and safer

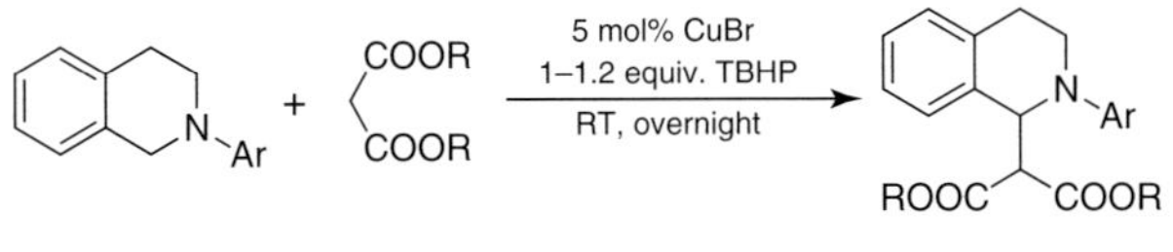

Scheme 19.11 CDC reaction of tertiary amines with malonates.

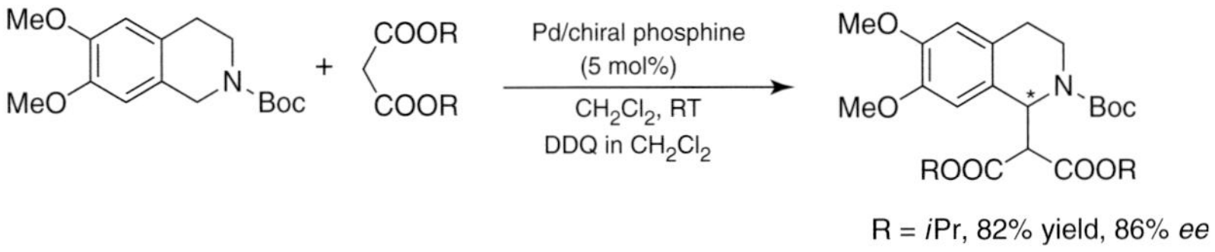

R = iPr, 82% yield, 86% ee

Scheme 19.12 Catalytic asymmetric addition of malonate to isoquinoline.

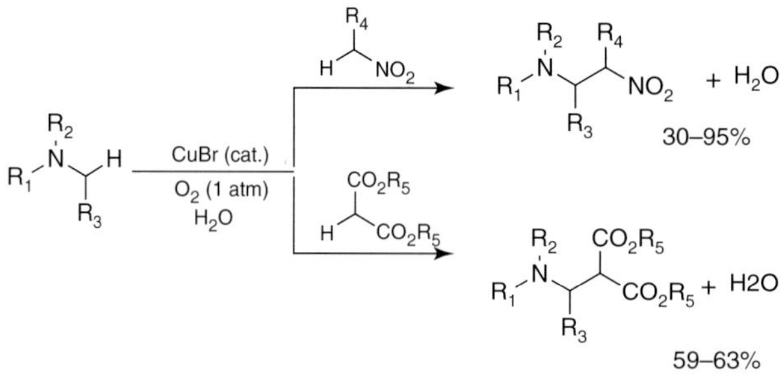

Scheme 19.13 CDC reaction of tertiary amines with oxygen in water.

process. Interestingly, molecular dioxygen can serve as the hydrogen acceptor efficiently when water is used as solvent. Both the nitroalkane reaction and the malonate reaction gave the corresponding CDC products via the reaction of two sp^3 C–H bonds catalyzed by copper bromide under an oxygen atmosphere in water (Scheme 19.13) [42].

19.2.2
CDC Reaction of α-C–H Bond of Oxygen in Ethers (sp^3–sp^3)

The synthesis of oxygen-containing compounds is of great interest in industrial and academic research, because these compounds are important structural units in many natural products and biologically active compounds [43]. The key to achieving successful CDC reactions of the α-C–H bond of nitrogen in amines is the formation of iminium intermediates. Compared with the α-C–H bond of nitrogen in amines, however, the functionalization of the α-C–H bond of oxygen in ethers is a more challenging task owing to the higher oxidation potential. Therefore, a stronger oxidant than peroxide or oxygen, such as DDQ, would be required. By using the combination of indium and copper as catalyst in the presence of DDQ, CDC reaction of the sp^3 C–H bond adjacent to an oxygen atom with an sp^3 C–H bond in pronucleophiles proceeds efficiently at room temperature (Scheme 19.14) [44]. The CDC reaction of the α-C–H bond of oxygen in ethers provides a simple and efficient catalytic method to construct β-diester ethers.

With the success of the CDC reaction between the α-C–H bond of oxygen in ethers and malonates, we next investigated the CDC reaction between the α-C–H bond of oxygen in ethers and less active simple ketones under similar conditions. Unfortunately, the In–Cu–DDQ reaction system is not effective in CDC reactions between benzyl ethers and simple ketones. However, we found that CDC reactions between benzyl ethers and simple ketones proceeded smoothly when mediated by DDQ without using any metal

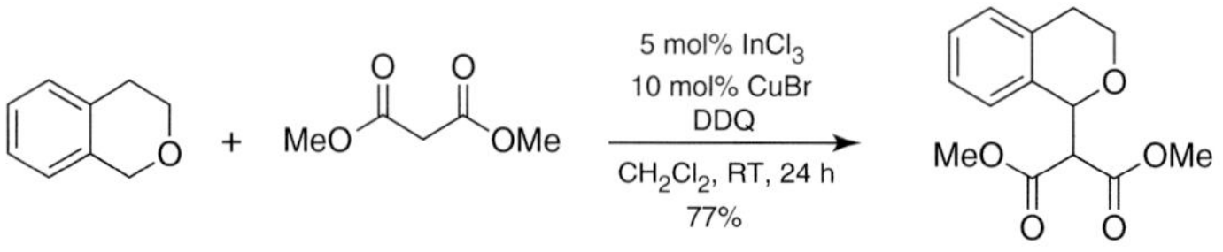

Scheme 19.14 CDC reaction of isochroman with dimethyl malonate.

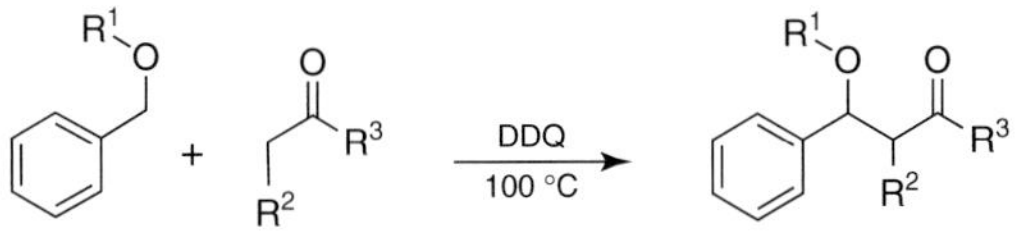

Scheme 19.15 CDC reaction between benzyl ethers and simple ketones.

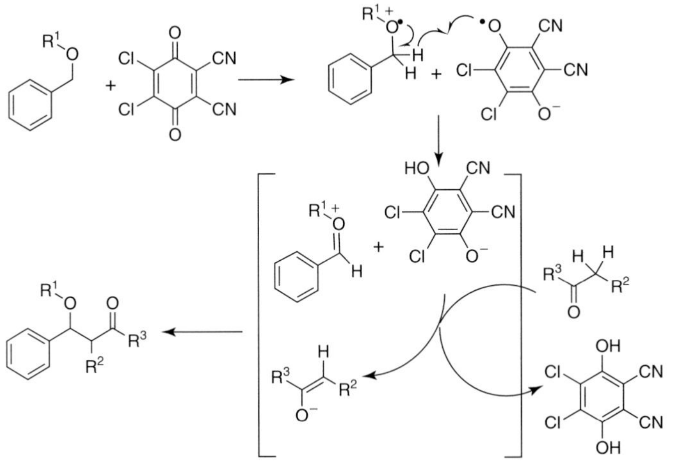

Scheme 19.16 Proposed mechanism for the CDC reaction of benzyl ethers with ketones.

catalyst at higher temperature (Scheme 19.15) [45]. The desired products were obtained in moderate to good yields when the reactions were carried out at 100 °C.

A tentative mechanism for the coupling is proposed in Scheme 19.16. A SET from benzyl ether to DDQ generates a radical cation and a DDQ radical anion. The radical oxygen of the DDQ radical anion then abstracts an H-atom from the radical cation and generates a benzoxy cation, and the anionic oxygen of DDQ radical anion then abstracts an α-hydrogen from the ketone to generate an enolate. Finally, the attack of the enolate on the benzoxy cation generates the CDC product and the quinone derivative.

19.2.3
CDC Reaction of Allylic and Benzylic C–H Bonds

19.2.3.1 Allylic Alkylation (sp³–sp³)
Palladium-catalyzed allylic alkylation (the Tsuji–Trost reaction) (Scheme 19.17a) is an important reaction for constructing C–C bonds in modern organic synthesis [46]. As a general protocol, a carboxylate (or another leaving group) is required at the allylic position, which is activated by a palladium catalyst during the reaction with a pronucleophile. The direct utilization of an allylic C–H bond rather than an allylic functional group would avoid the need to synthesize the allylic functional group, thus leading to reduced synthetic steps (Scheme 19.17b).

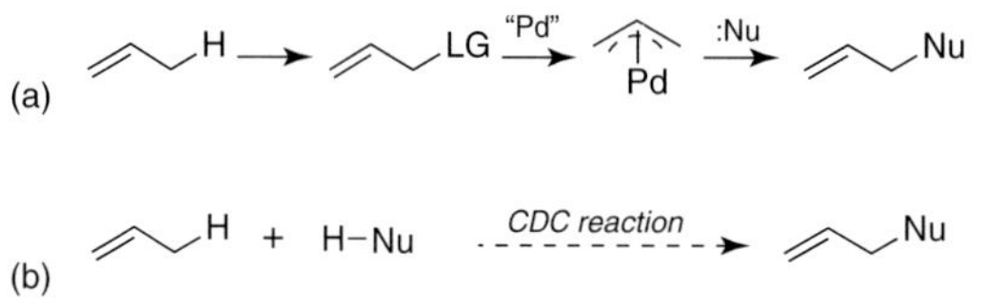

(a)

(b)

Scheme 19.17 (a) Tsuji–Trost reaction and (b) allylic CDC reaction.

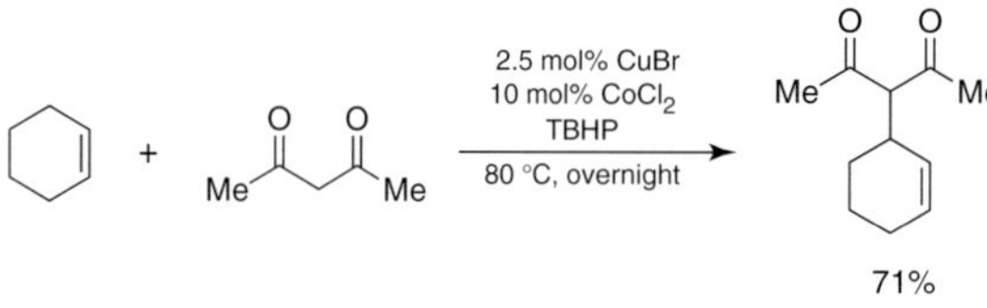

71%

Scheme 19.18 Allylic alkylation via CDC reaction.

To begin our study, cyclohexene and 2,4-pentanedione were chosen as the standard substrates to search for potential catalysts and suitable reaction conditions. We found that by using the combination of CuBr (2.5 mol%) and $CoCl_2$ (10 mol%) as catalyst, the CDC reaction proceeded smoothly and gave the desired product in 71% yield (Scheme 19.18) [47].

19.2.3.2 Benzylic Alkynylation (sp^3–sp)

Previously we described the efficient use of DDQ for C–C bond formation between benzyl ethers and simple ketones through activation of benzylic C–H bonds. The benzylic C–H bond lacking an adjacent heteroatom is more difficult to activate [48]; however, Shi *et al.* recently reported the use of DDQ for the arylation of diphenylmethane derivatives [49]. We found that the direct alkynylation of diphenylmethane can also be achieved in the presence of a copper catalyst using DDQ as the oxidant (Scheme 19.19). Various alkynes can react with diphenylmethane derivatives effectively and give the desired products in good to excellent yields [50].

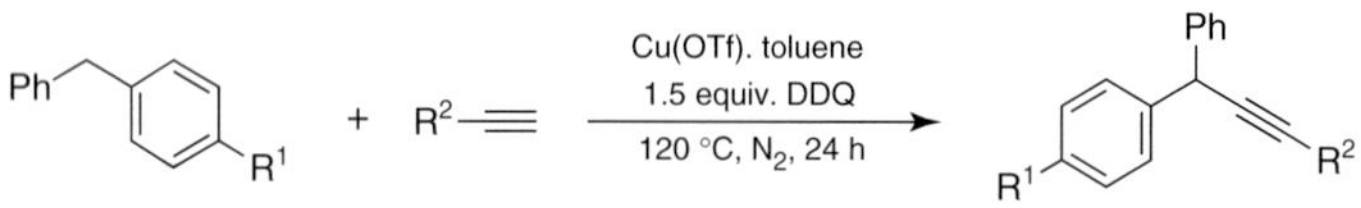

Scheme 19.19 CDC reaction of alkynes and diphenylmethane derivatives.

19.2.3.3 Benzylic Alkylation (sp^3–sp^3)

Although transition metal catalysts are highly effective for direct CDC reactions, the use of more readily available and nontoxic catalysts instead of expensive and sensitive catalysts is highly desirable. The numerous advantages of iron make it highly attractive as a catalyst or reagent for chemical synthesis [51]. We developed an $FeCl_2$-catalyzed oxidative activation of a benzylic C–H bond which is followed by a cross-coupling reaction with diketones to form C–C bonds (Scheme 19.20) [52]. Good to excellent yields were obtained when the reactions were performed at 80 °C by using *tert*-butyl peroxide as oxidant. The reaction was also found to proceed efficiently at room temperature.

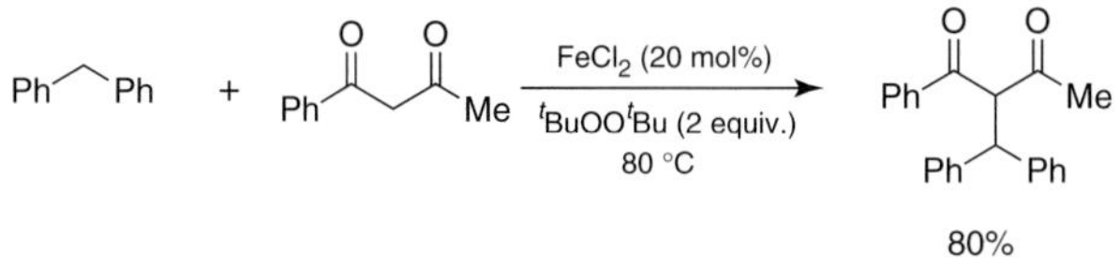

Scheme 19.20 FeCl$_2$-catalyzed alkylation of a benzylic C–H bond.

This kind of catalyst was also efficient for CDC reactions between the α-C–H bond of oxygen in ethers and diketones [53].

19.2.4
CDC Reaction of Alkane C–H Bonds

19.2.4.1 Alkane Alkylation (sp^3–sp^3)

Various direct cross-coupling reactions have been successfully developed in our group and others recently. In these cases, the C–H bond adjacent to a heteroatom or C=C bond was activated and subsequently reacted with activated methylene compounds to form new C–C bonds. However, it is a great challenge to use simple alkanes (without any functional groups) to form C–C bonds by such an oxidative approach. Historically, the Fenton chemistry [54] and the Gif processes [55] established the conversion of aliphatic C–H bonds into C–O bonds under mild conditions by using peroxides as oxidant and catalyzed by various iron catalysts [56]. We rationalized that the process may be used for C–C bond formation with the interception of intermediates in such reactions. Indeed, it was found that by using 10 mol% FeCl$_2$·4 H$_2$O as the catalyst and *tert*-butyl peroxide as the oxidant at 100 °C for 12 h under an atmosphere of nitrogen, various activated methylene substrates reacted with cycloalkanes in good yields in most cases (Scheme 19.21) [57].

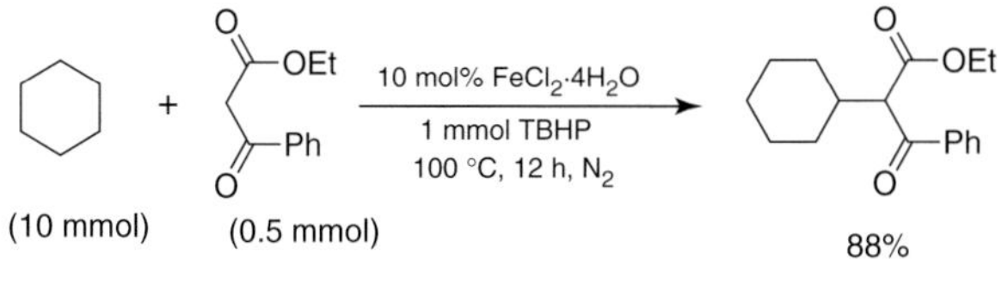

Scheme 19.21 Simple alkane alkylation via CDC.

19.2.4.2 Alkane Arylation (sp^3–sp^2)

After successful C–C bond formation using simple alkanes and activated methylenes via CDC, our next target was direct cross-coupling of simple alkanes and inactivated aryl C–H bonds. Recently, the Pd-catalyzed direct functionalization of aryl C–H to form C–C bonds has been investigated intensively [25c]. To begin our study, we chose cyclohexane and 2-phenylpyridine as the model substrates. Various palladium salts were investigated under different conditions. To our surprise, only a trace amount of the desired product was obtained, with methylation as the major reaction. Then reaction of 2-phenylpyridine with dicumyl peroxide in the absence of cyclohexane was further investigated. With 10 mol% Pd(OAc)$_2$ as the catalyst at 130 °C under an

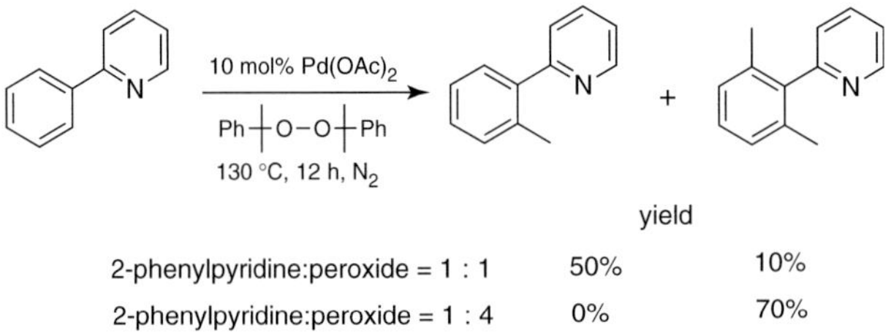

	yield	
2-phenylpyridine:peroxide = 1 : 1	50%	10%
2-phenylpyridine:peroxide = 1 : 4	0%	70%

Scheme 19.22 Methylation of 2-phenylpyridine with dicumyl peroxide.

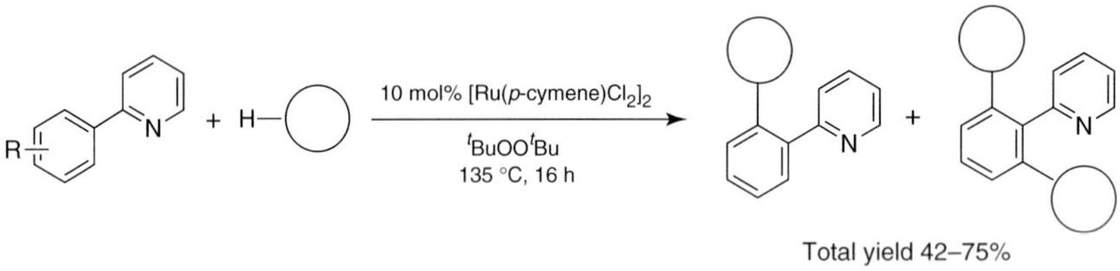

Total yield 42–75%

Scheme 19.23 CDC reaction between phenylpyridines and cycloalkanes.

atmosphere of nitrogen, mono- and bismethylation products were obtained in good yields (Scheme 19.22) [58]. When 1 equiv. of the peroxide was used, the major product was the monomethylation compound. With 4 equiv. of the peroxide, the bismethylation compound was the only product, obtained in 70% yield.

At the beginning of our methylation reaction of 2-phenylpyridine, we found a trace amount of alkylation product. We therefore thought that it might be possible to improve the reaction yield by using other transition metals. Various transition metal catalysts were examined and we found that ruthenium salts were effective for the direct alkylation of 2-phenylpyridine with cyclooctane. The best yield was obtained when 10% [Ru(*p*-cymene)Cl$_2$]$_2$ was used as the catalyst in the presence of 2 equiv. of *tert*-butyl peroxide at 135 °C under an atmosphere of air (Scheme 19.23) [59].

The mechanism of the reaction was proposed to involve a ruthenium-catalyzed aryl C–H activation followed by an H–alkyl exchange [60]. Reductive elimination of this intermediate generated arene–cycloalkane coupling product and regenerated the active ruthenium catalyst (Scheme 19.24). Deuterium isotope experiments showed a large negative kinetic isotope effect, which suggested that the ruthenium-catalyzed aryl C–H activation is a fast equilibrium and the H–alkyl exchange is the rate-limiting step.

Under similar conditions, however, unactivated pyridines or quinolines did not react with simple alkanes even on increasing the amount of *tert*-butyl peroxide to 3 equiv. Alternatively, we rationalized that it might be possible to increase the acidity of the C2–H bond on these heteroaromatic rings by using a Lewis acid catalyst and, thus, to increase the reactivity of pyridine derivatives. We found that quinoline reacted with cycloalkanes in the presence of *tert*-butyl peroxide catalyzed by Sc(OTf)$_3$ and gave the bisalkylation product in 93% NMR yield (Scheme 19.25) [61]. Two equivalents of peroxide are necessary to achieve high reaction yields.

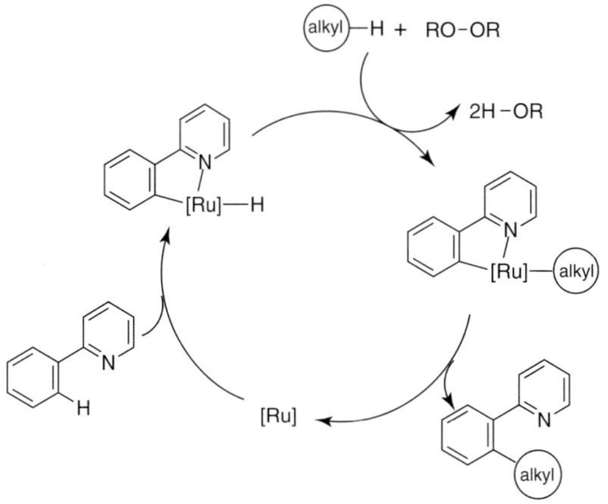

Scheme 19.24 Proposed mechanism for the Ru-catalyzed cycloalkylation of arenes.

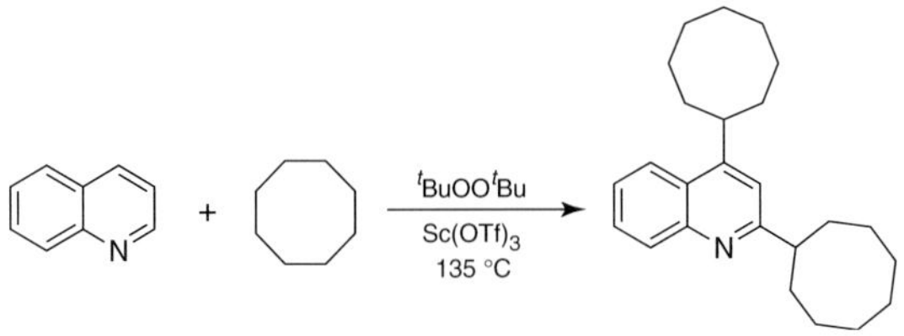

Scheme 19.25 CDC reaction between quinoline and cyclooctane.

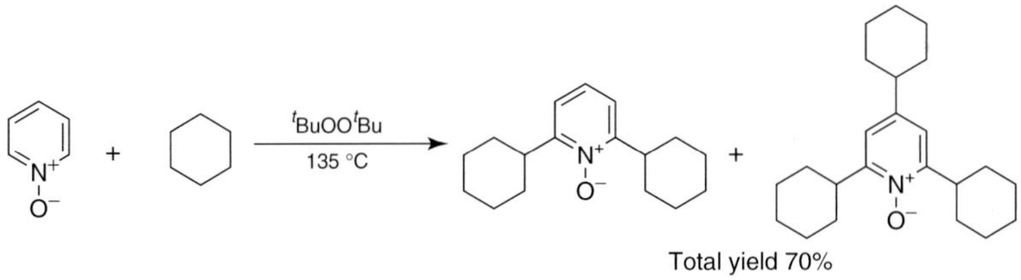

Total yield 70%

Scheme 19.26 CDC reaction between pyridine *N*-oxide and cyclohexane.

To our delight, no catalyst was required when the more active pyridine *N*-oxide was used as the substrate. The additional *N*-oxide moiety enhanced the reactivity significantly. Pyridine *N*-oxide reacted with cyclohexane in the presence of ^{t}BuOO^{t}Bu and gave di- and trialkylation products in 70% total yield (Scheme 19.26) [62]. The reactions also took place with norbornane and 1,4-dioxane to give structurally interesting molecules in good yields.

19.2.5
CDC Reaction of Aryl C–H Bonds

The CDC reaction between two aryl C–H bonds can be a potential method to prepare biaryl structures. Sanford and co-workers reported that 2-arylpyridines underwent a highly regioselective Pd-catalyzed oxidative C–C coupling at room temperature with oxone as a terminal oxidant [63]. On the other hand, Yu and co-workers discovered that such a reaction could be carried out with an *in situ* iodination followed by a $Cu(OAc)_2$-mediated Ullmann coupling to give homodimerized product [64]. During our recent studies on the ruthenium-catalyzed CDC of 2-phenylpyridine with cycloalkanes mediated by peroxides, we observed a trace amount of the oxidative homo-coupling product of 2-phenylpyridine. Subsequently, various conditions regarding the ruthenium catalyst and the oxidant were examined to optimize the formation of this homo-coupling product. We found that using $FeCl_3$ instead of peroxide as the stoichiometric oxidant increased the reaction yield. The best yield was obtained by using $[Ru(p\text{-cymene})Cl_2]_2$ as catalyst and 0.8 equiv. $FeCl_3$ as oxidant at 110 °C in chlorobenzene (Scheme 19.27) [65].

Alkynylindoles are important precursors for bioactive natural molecules, pharmaceuticals, and molecular organic materials. The palladium-catalyzed direct alkynylation of indole halides is one of the most versatile methods to achieve this kind of compound [66]. We developed a palladium-catalyzed direct oxidative alkynylation of indoles with terminal alkynes under oxygen (Scheme 19.28). The reaction of 1,3-dimethylindole with phenylacetylene was chosen as the model reaction. Various palladium catalysts were examined under an atmosphere of oxygen. The desired product was obtained in all cases, and the best yield was obtained when K_2PdCl_4 was used as catalyst at 80 °C. Other 1,3-dialkylindoles with different substituents were also applicable to this direct oxidative alkynylation and the desired products were obtained in moderate to good yields [67]. At the same time, Haro and Nevado developed a similar reaction using gold as catalyst and $PhI(OAc)_2$ as oxidant. Indoles and substituted arenes were coupled with electron-deficient alkynes and gave the desired products in moderate yields [68].

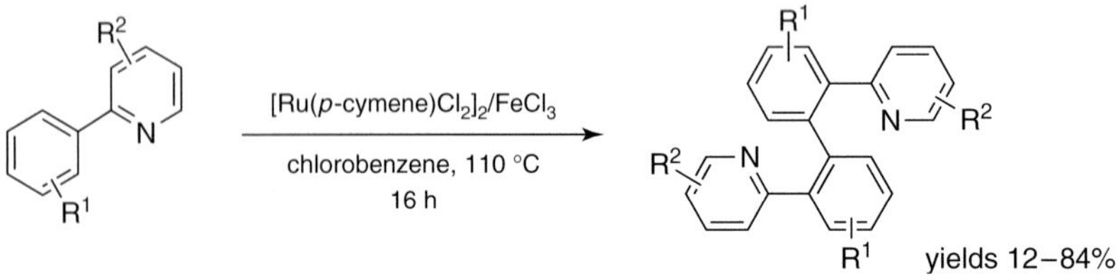

Scheme 19.27 Ruthenium-catalyzed homo-coupling of 2-arylpyridines.

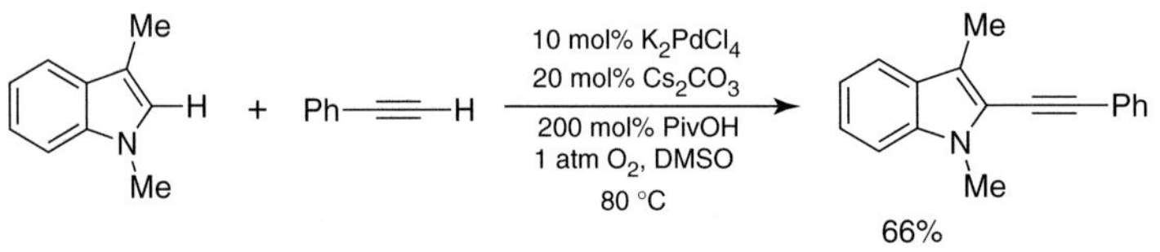

Scheme 19.28 CDC reaction of 1,3-dimethylindole with phenylacetylene.

19.3
Nucleophilic Addition of Terminal Alkynes in Water

Functionalized alkynes are useful intermediates and can transform into a wide range of structures. The nucleophilic addition of metal acetylides to unsaturated electrophiles is one of the classical reactions to generate functionalized alkyne products [69]. However, this classical methodology requires the presynthesis of metal acetylides from terminal alkynes and stoichiometric organometallic reagents. In addition, acids need to be added to quench the reaction at the end. Hence the overall process requires multiple steps, generates stoichiometric waste, and suffers from low atom economy [70].

The catalytic and direct alkynylation of unsaturated electrophiles can be a green alternative to achieve functionalized alkynes by avoiding presynthesis of metal acetylides. Trost *et al.* pioneered a palladium-catalyzed conjugated direct addition of terminal alkynes to electron-deficient alkynes [71]. Highly efficient, direct, and asymmetric conjugated additions of terminal alkynes to electron-deficient alkenes have been developed more recently [72]. Further, direct addition of terminal alkynes to aldehydes was reported by Yamaguchi *et al.* [73], Han and Huang [74], and Carreira *et al.* [75]. For the addition of terminal alkynes to C=N bonds, Miura *et al.* reported the addition of acetylene to nitrones through the initial formation of a dipolar cycloaddition [76], while Carreira and co-workers used a Zn(II)-catalyzed process in CH_2Cl_2 for the addition of terminal alkynes to nitrones [77].

In most cases, these reactions have mainly been carried out in organic solvents with limited substrate scope. For the past two decades, we and others have been studying Barbier–Grignard-type nucleophilic addition reactions in water [78, 79]. In addition, we have developed various reaction conditions for the direct alkynylation of aldehydes, imines, and alkenes and also alkynes by using terminal alkynes as starting materials in water [80].

19.3.1
Direct Nucleophilic Addition of Terminal Alkynes to Aldehydes

Propargylic alcohols are versatile building blocks in organic synthesis. Their traditional synthesis involves the alkynylation of ketones and aldehydes using a stoichiometric amount of organometallic reagents in an anhydrous organic solvent. To achieve the direct alkynylation of aldehydes in water, various metal salts were examined. We found that the direct addition of terminal alkynes to aldehydes in water can be realized by using a ruthenium–indium bi-catalyst system (Scheme 19.29) [81]. In this combination, the $In(OAc)_3$ presumably plays the role of a Lewis acid tolerated in water and activated the carbonyl, whereas the ruthenium chloride converts the alkynyl C–H bond to an alkynylruthenium intermediate. Since the metal ions can be regenerated, only a catalytic amount of the catalysts is required. The addition of a base and morpholine improved the yield of the reaction significantly.

In comparison with other transition metals, silver has been less studied as a catalyst for coupling purposes. Alkynylsilver compounds are known to form readily from terminal alkynes and silver salts in water. However, simple alkynylsilver reagents are too stable

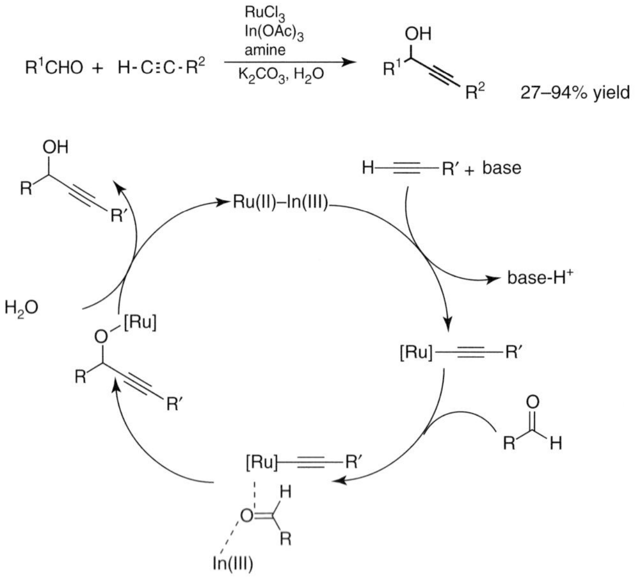

R^{1}CHO + H-C≡C-R^2 $\xrightarrow[\text{K}_2\text{CO}_3,\ \text{H}_2\text{O}]{\substack{\text{RuCl}_3 \\ \text{In(OAc)}_3 \\ \text{amine}}}$ R^1 ⟍⟋ R^2 27–94% yield

Scheme 19.29 Proposed mechanism of the alkynylation catalyzed by Ru–In in water.

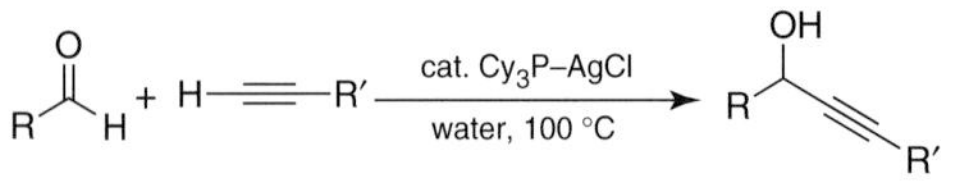

Scheme 19.30 Silver-catalyzed alkynylation of aldehydes in water.

to participate in nucleophilic additions to carbonyls. We postulated that by using a strong electron-donating phosphine ligand, it would serve two purposes: (i) weaken the C–Ag bond and (ii) the increase the Lewis acidity of the [Ag] center. Based on this, various silver–phosphine catalysts were investigated for the direct alkynylation of aldehydes in water. A highly efficient alkynylation of aldehydes was developed by using an Cy$_3$P–AgCl complex as catalyst (Scheme 19.30) [82]. The reaction was dually promoted by the electron-donating phosphine ligand and water (yield 63–98%). Almost no reaction was observed in toluene. For electron-withdrawing aldehydes such as 4-trifluoromethylbenzaldehyde, the reaction proceeded smoothly at room temperature.

Later, we found that silver complexes were also useful as catalysts for the synthesis of (*Z*)-2,3-dihydrobenzofuran-3-ol derivatives, which are key intermediates for the synthesis of biologically active aurones [83]. We developed a highly efficient alkynylation–cyclization of terminal alkynes with salicylaldehydes leading to substituted 2,3-dihydrobenzofuran-3-ol derivatives by using a Cy$_3$P–Ag complex as catalyst in water. The *Z*/*E* selectivity of the reaction was controlled by the counter anions in the silver complex (Scheme 19.31). Furthermore, after simple extraction, the organic phase could then be oxidized directly without any further purification to give aurones [84].

Isochromenes are common structural units in natural products and exhibit interesting biological activities such as antibiotic properties [85]. Among the methods

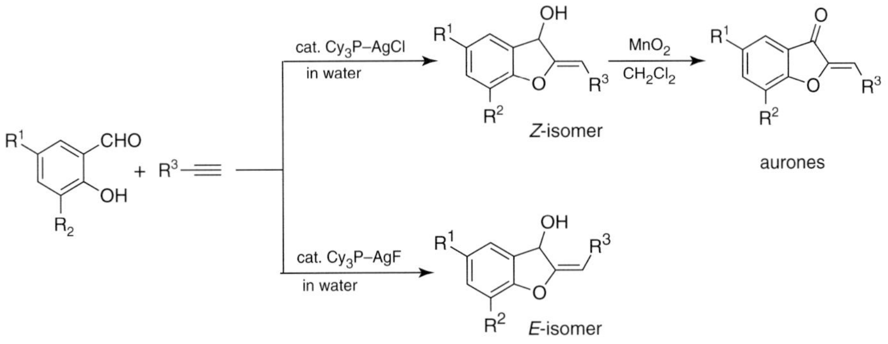

Scheme 19.31 Synthesis of 2,3-dihydrobenzofuran-3-ol and aurone derivatives.

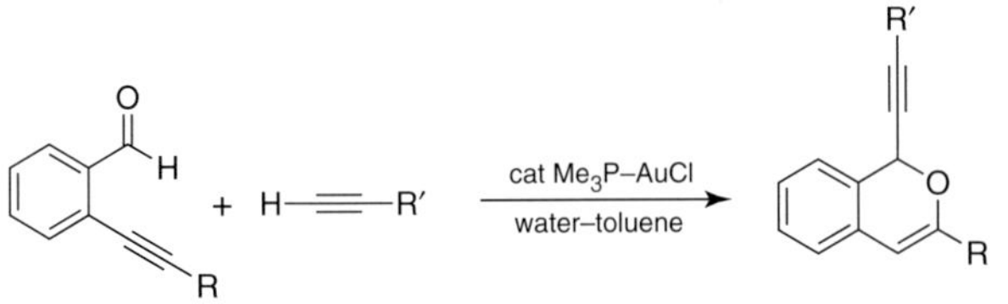

Scheme 19.32 Gold(I)-catalyzed cascade alkynylation–cyclization.

to construct such structures, nucleophilic addition to alkynylarylaldehydes followed by alcohol–alkyne cyclization is the most useful synthetic route. By using a strategy similar to the silver-catalyzed cascade reaction, we found that highly efficient alkynylation–cyclization of terminal alkynes with *o*-alkynylarylaldehydes could be achieved by using a gold–phosphine complex as the catalyst in water–toluene mixture. The desired products, 1-alkynyl-1*H*-isochromenes, were obtained in good yields (Scheme 19.32). Water played a key role in this reaction and no reaction was observed in the absence of water [86].

19.3.2
Direct Addition of Terminal Alkynes to Ketones in Water

Usually, ketones are much less reactive for both steric and electronic reasons. However, one exception is trifluoromethyl ketones. The electron-withdrawing trifluoromethyl group adjacent to a carbonyl group can activate the latter. We found that highly effective direct alkynylation of a trifluoromethyl ketone in water or in organic solvents could be achieved with silver as the catalyst (Scheme 19.33). Trifluoropyruvate reacted with terminal alkynes efficiently in water at room temperature [87]. The best yield was obtained by using the Cy_3P–AgF complex as catalyst. It is important to note that Shibasaki and co-workers reported the asymmetric alkynylation of activated trifluoromethyl ketones using a chiral bidentate phosphine ligand or pybox and a catalytic copper salt in the presence of base [88].

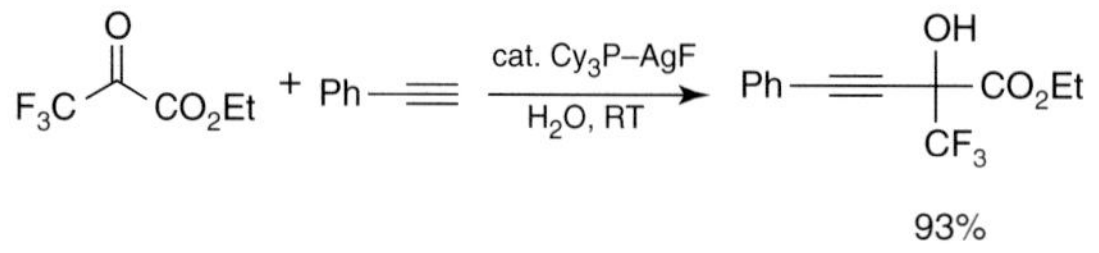

Scheme 19.33 Silver-catalyzed direct alkynylation of trifluoropyruvate in water.

19.3.3
Addition of Terminal Alkynes to Imines, Tosylimines, Iminium Ions, and Acyliminium Ions

19.3.3.1 Imines

Propargylamines are important synthetic intermediates for the synthesis of various nitrogen compounds and a structural feature of many biologically active compounds and natural products [89]. The direct addition of terminal alkynes to imines can be the most efficient and green route to generate propargylamines, and the only byproduct is water. However, the reaction conditions need to be modified since simple imines are less reactive than carbonyls. After screening different metal catalysts, we found a highly efficient addition of acetylenes to various imines (aldehyde–alkyne–amine coupling, which we termed A^3-coupling) by using an Ru–Cu catalyst in water or under neat conditions (Scheme 19.34) [90]. A wide range of alkynes, aldehydes, and aromatic amines can be converted readily into synthetically useful propargylamines in one pot in high yields. At the same time, Ishii and co-workers [91] and Carreira and Fischer [92] reported a similar coupling by using an iridium catalyst in anhydrous toluene under an inert atmosphere. However, the reactions were limited to trimethylsilylacetylene.

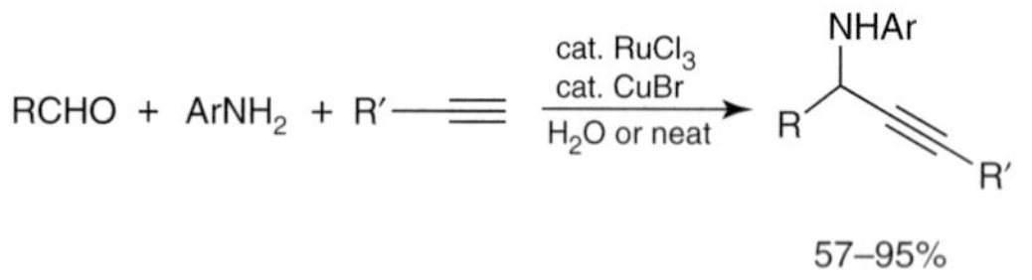

Scheme 19.34 Ru–Cu-catalyzed addition of imines with alkynes.

During our investigation, we noticed that the conversion was much lower in the absence of ruthenium catalyst. We therefore hypothesized that the low conversion by using CuBr as the catalyst might in fact provide us with an opportunity to develop highly enantioselective alkyne–imine additions. To achieve this goal, various chiral ligands were investigated. Finally, optically active propargylamines were obtained by direct addition of terminal alkynes to imines catalyzed by tridentate bis(oxazolinyl)pyridines–Cu(OTf) in water and organic solvents at room temperature (Scheme 19.35) [93]. The enantioselectivity was up to 99.6% *ee* in organic solvents and 84% *ee* in water [94].

19.3.3.2 Iminium Ions

Iminium ions can be readily generated from secondary amines and aldehydes *in situ* and are more reactive than imines. Prior to our own investigations, Youngman and Dax reported a copper-catalyzed solid-phase Mannich condensation of secondary amines, aldehydes, and alkynes via the formation of iminium ions [95]. Pagni and co-workers

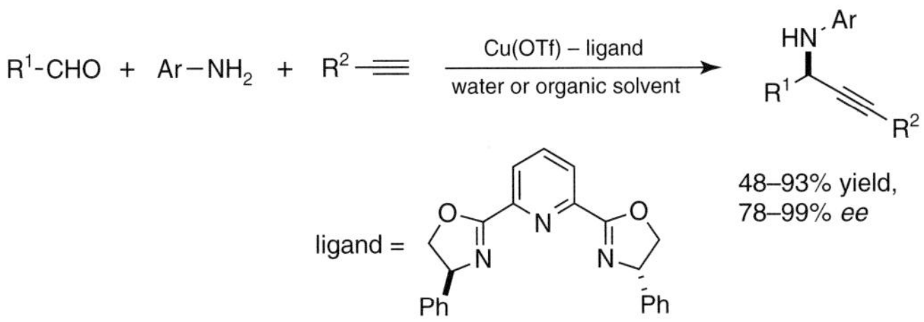

Scheme 19.35 Asymmetric addition of imines with alkynes in water and organic solvent.

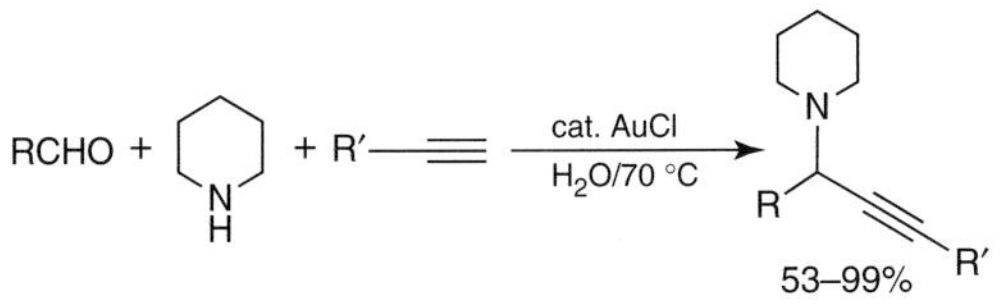

Scheme 19.36 Gold-catalyzed A^3 coupling reactions in water.

reported a microwave-assisted reaction of terminal alkynes, amines, and formaldehyde in the presence of Cu(I) on an Al_2O_3 support [96].

Our previous methods in water were largely limited to imines generated from arylamines and aromatic aldehydes. When secondary amines such as piperidine were used instead of aromatic amines, the desired product was obtained in only 10% yield under similar conditions. Various transition metals were examined to improve the reaction yield. We found that gold salts can efficiently catalyze the direct coupling of aldehydes, alkynes, and secondary amines in water (Scheme 19.36) [97]. In the presence of 1 mol% of AuBr$_3$, the A^3 coupling product was achieved in quantitative yield. AuCl, AuI, and AuCl$_3$ were all effective for this kind of reaction. The nature of the reaction medium significantly affected the reaction. Water was shown to be the best solvent for the process. Organic solvents, such as THF, toluene, and DMF, gave low conversions and more byproducts. Subsequently, related gold-catalyzed A^3 reactions in water were developed by Che and co-workers [98]. The reaction gave very high diastereoselectivity with proline as the secondary amine.

The high efficiency of gold as a catalyst for the three-component coupling of aldehydes, secondary amines, and alkynes in water led us to examine Ag catalysts for related couplings. Various water-soluble or partially soluble silver salts such as AgNO$_3$, Ag$_2$O, AgOAc, Ag$_2$SO$_4$, AgOTf, and AgBF$_4$ were examined first and all were found to catalyze the reaction of benzaldehyde, piperidine, and phenylacetylene with low conversions ($\sim$25–45%). To our surprise, water-insoluble silver salts such as AgCl, AgBr, and AgI all showed good catalytic activity for the three-component coupling, and AgI was found to be the most effective (Scheme 19.37) [99]. Both aromatic and aliphatic aldehydes were able to undergo addition to afford the corresponding three-component propargylamines effectively. Subsequently, Maggi *et al.* found that silver-nanoparticles and supported silver can also catalyze the A^3 reaction efficiently under various conditions [100]. Rueping *et al.* reported the asymmetric silver-catalyzed alkyne iminium addition reaction [101].

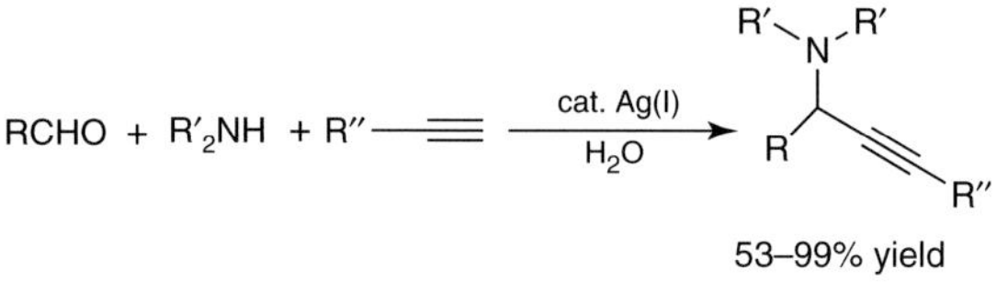
RCHO + R'$_2$NH + R''
cat. Ag(I)
H_2O
R'
R'
N
R
R''
53–99% yield

Scheme 19.37 Silver-catalyzed A^3 coupling in water.

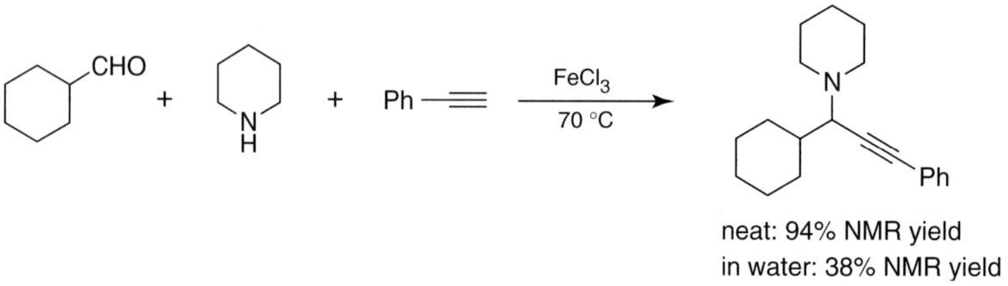
CHO
+
N
H
+
Ph
$FeCl_3$
70 °C
N
Ph
neat: 94% NMR yield
in water: 38% NMR yield

Scheme 19.38 Fe-catalyzed A^3 coupling reactions.

Iron catalysts are readily available, inexpensive, and environmentally friendly, with special reactivities. These features make them valuable and advantageous in synthetic processes. As a continuing effort to seek green catalysts, various iron salts were tested for the A^3 coupling reaction. We found that $FeCl_3$ is an effective catalyst for the three-component coupling of aldehydes, alkynes, and amines with good yields under neat conditions and in air. Other iron salts were also efficient for this reaction. However, the reaction was sensitive to water, and a much lower yield was obtained when the reaction was carried out in water (Scheme 19.38) [102]. The same reaction can be achieved by using Fe_3O_4 nanoparticles in water and organic solvents in air. The desired products were obtained in moderate to high yields under mild conditions. The magnetic Fe_3O_4 nanoparticles can be separated and recovered by simple precipitation [103].

Great progress was made in asymmetric A^3 coupling reactions shortly after our report on the Cu-catalyzed asymmetric addition of terminal alkynes to imines. Knochel and co-workers reported a highly efficient asymmetric addition of terminal alkynes to enamines catalyzed by a Cu–QUINAP complex (Scheme 19.39) [104]. Carreira and co-workers found that the use of Cu–PINAP as catalyst could improve the enantioselectivity [105]. Tu and co-workers observed that secondary amine–aldehyde–alkyne additions could be

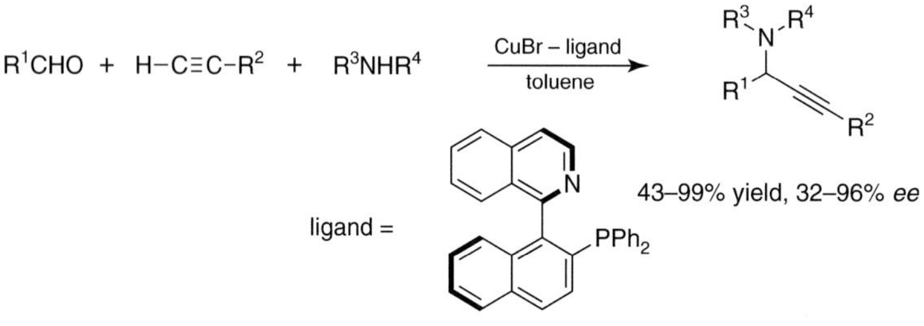
R^1CHO + H–C≡C–R^2 + R^3NHR^4
CuBr – ligand
toluene
R^3
R^4
N
R^1
R^2
N
43–99% yield, 32–96% *ee*
ligand =
PPh_2

Scheme 19.39 Cu–QUINAP-catalyzed asymmetric A^3 coupling.

catalyzed by CuI in water under microwave irradiation [106]. Formaldehyde can also serve as the aldehyde component for CuI-catalyzed A^3 coupling in aqueous solution [107].

19.3.3.3 Acylimine and Acyliminium Ions

Acylimine and acyliminium ion compounds can be generated conveniently *in situ* from a variety of methods and the products can be modified easily for various synthetic purposes. We found that under ultrasonic irradiation, CuBr (10–30 mol%) can catalyze the addition of terminal alkynes to both acylimine and acyliminium ions in water in moderate yields [108]. The reaction yield can be improved by increasing the amount of CuBr (Scheme 19.40). Under similar reaction conditions, the coupling between α-methoxy-*N*-(alkoxycarbonyl)pyrrolidine and phenylacetylene in the presence of a stoichiometric amount of CuBr afforded the desired product in good yields. Subsequently, we and others found that acyliminium ions can be generated *in situ* from imines and acyl chloride can be coupled with alkynes under anhydrous conditions in good yields [109].

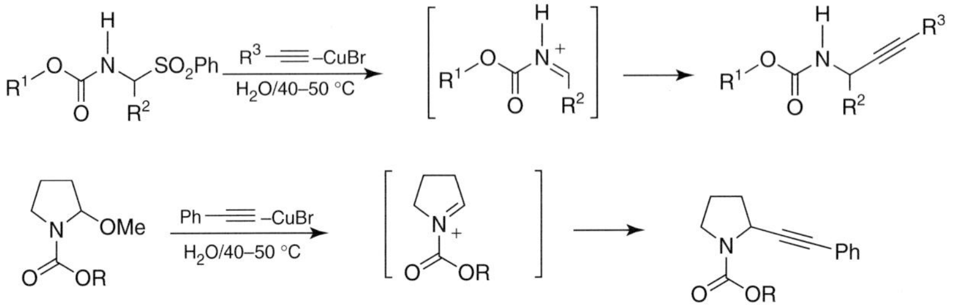

Scheme 19.40 Addition of alkynes to acylimines and acyliminium ions.

19.3.3.4 Multiple and Tandem Addition of Terminal Alkynes to C=N Bonds

Tandem/cascade/domino reaction processes have received a good deal of attention over the last decade owing to the efficiency and step economy inherent in the methodology [110]. Dipropargylamines are important components of biologically active compounds and widely used as synthetic intermediates [111]. However, despite their widespread use, currently reported methods for their synthesis have serious limitations. Ishii and co-workers reported an iridium-catalyzed five-component double A^3 coupling in organic solvents leading to dipropargylamines. However, with this system, the terminal alkyne is limited to trimethylsilylacetylene [91b]. Using our ruthenium–copper co-catalyst system, a five-component double aldehyde–alkyne–amine coupling was developed to synthesize various dipropargylamines from a range of simple amines, aldehydes, and alkynes in one pot under mild conditions in water under an atmosphere of air (Scheme 19.41) [112]. The reaction yield was better than in toluene at room temperature.

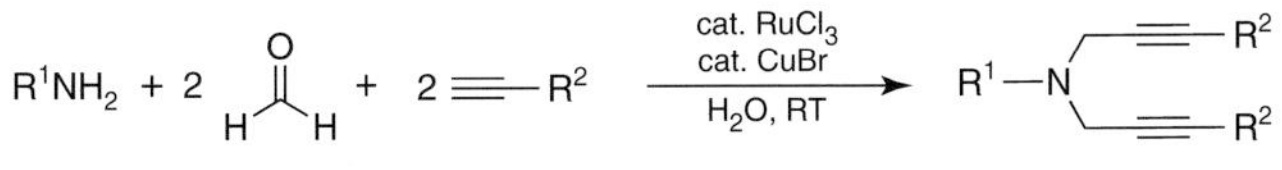

Scheme 19.41 Synthesis of dipropargylamines via double A^3 coupling.

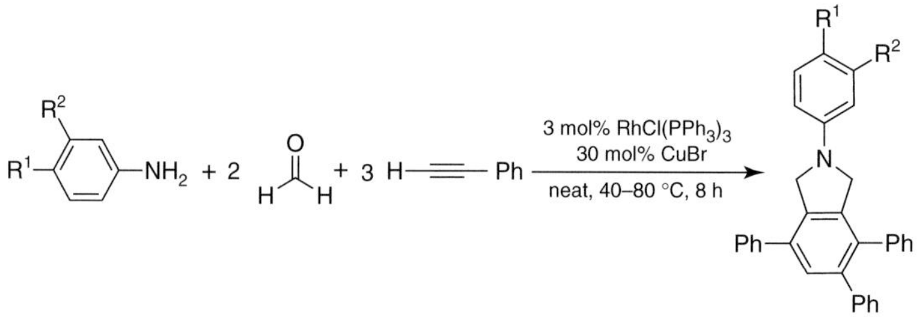

Scheme 19.42 Rh–Cu-catalyzed substituted isoindolines from an alkyne, an aldehyde, and an amine.

Subsequently, this strategy was successfully applied to the efficient preparation of biologically active tetrasubstituted isoindolines. Tetrasubstituted isoindolines can be synthesized readily from three alkyne units, two aqueous formaldehyde units, and a primary amine via three consecutive reactions, two A³ couplings and a final [2 + 2 + 2]-cycloaddition, in a single synthetic operation by using a combined copper–rhodium catalyst (Scheme 19.42) [113]. The A³ couplings were catalyzed by copper bromide and the cycloaddition was catalyzed by Wilkinson's catalyst.

19.3.4
Direct Conjugate Addition of Terminal Alkynes in Water

Various efficient methods for the direct addition of terminal alkynes to C=N or C=O bonds have been developed by us and other groups, whereas conjugate addition of terminal alkynes to C=C double bonds in water has only been explored recently owing to the lower electrophilicity of C=C bonds. In 2003, Carreira and Knopfel reported the first conjugate addition reaction of terminal alkynes to unsaturated carbonyl acceptors catalyzed by copper in water at room temperature. Good to excellent yields were obtained under standard conditions (Scheme 19.43) [114]. Asymmetric conjugate addition of terminal alkynes to unsaturated carbonyl acceptors was also developed by the same

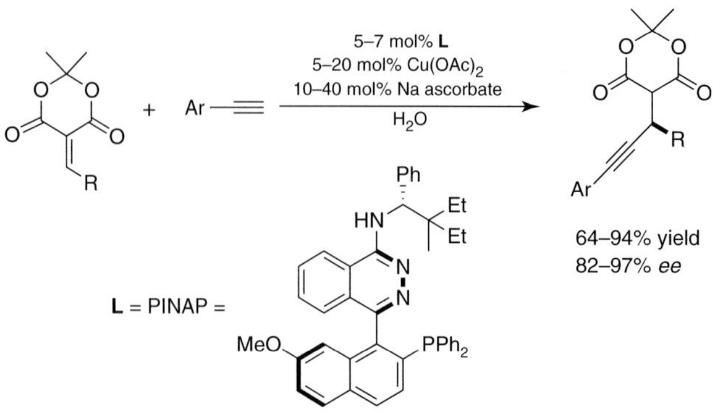

Scheme 19.43 Cu-catalyzed addition of terminal alkynes to C=C bonds.

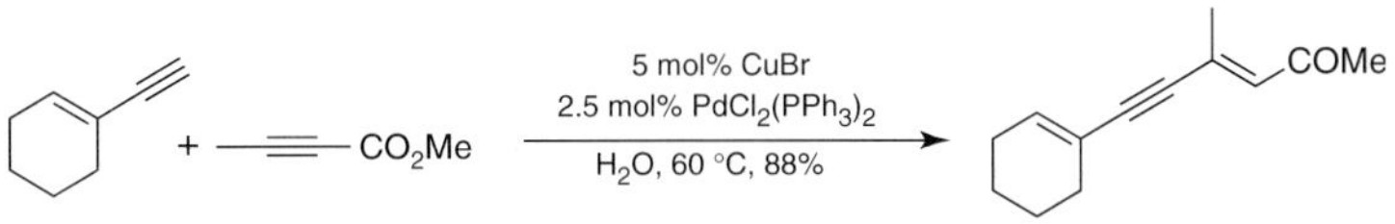

Scheme 19.44 Alkyne addition to alkyne.

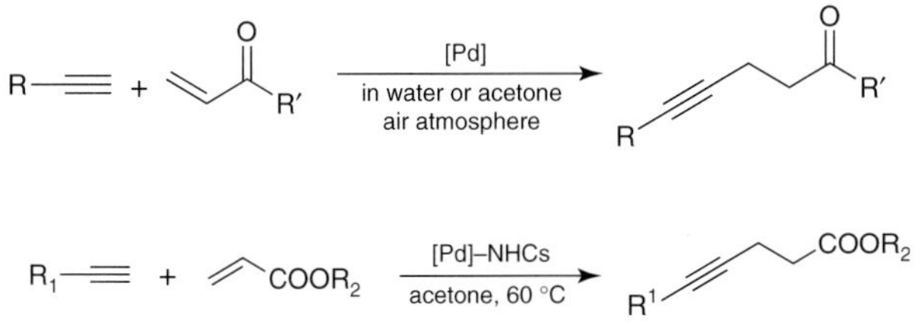

Scheme 19.45 Alkyne addition to vinyl ketones and acrylate esters.

group. When Cu(OAc)$_2$–PINAP complex was used as the chiral catalyst, the desired addition products were obtained in 82–97% *ee* with good yields [115].

In our own studies, we developed a facile and selective copper–palladium-catalyzed addition of terminal alkynes to electron-deficient alkynes in water and under an air atmosphere (Scheme 19.44) [116]. The reaction performed better in water than in toluene. For example, an addition of phenylacetylene to 4-phenyl-3-butyn-2-one at room temperature in water gave a 63% yield of the desired product whereas less than 10% of the desired product was obtained when the reaction was carried out in toluene.

We also developed a palladium-catalyzed 1,4-addition of terminal alkynes to vinyl ketones in water (Scheme 19.45) [117]. The nature of the phosphine ligands affected the reaction significantly. A yield of less than 10% was obtained when PPh$_3$ was used as ligand. Electron-rich ligands improved the reaction yields significantly. The best yield was obtained with trimethylphosphine, and the desired product was obtained in 77% yield. In most cases, better yields were obtained in water than in organic solvents such as acetone. Using highly electron-donating *N*-heterocyclic carbenes (NHCs) as ligands further increased the catalytic activity of palladium and allowed the direct addition of terminal alkynes to acrylate esters in water with low yield (Scheme 19.45) [118]. Much higher yields were obtained when the reaction was carried out in acetone. Lerum and Chisholm also reported Rh-catalyzed direct addition of terminal alkynes to vinyl ketones in aqueous dioxane solutions [119].

19.4
Conclusion and Perspectives

Taking into consideration both chemical efficiency and atom economy, the direct and selective transformation of an unactivated C–H bond into a C–C bond is one of the most powerful tools for introducing molecular complexity into organic molecules. The development of organic transformations based on the reaction of C–H bonds in water is a highly challenging subject because it is necessary to overcome some of the most fundamental concepts and theories such as classical reactivity and bond strength. Nevertheless, as shown in this chapter, reactions of C–H bonds in water can indeed occur, albeit still in a very limited way at present. Various C–C bonds were formed directly from C–H and C–H bonds in water and in organic solvents under oxidative conditions. Dioxygen can serve as a potential green and mild oxidant for this kind of reaction. Thus the only byproduct generated from the CDC reaction could be water. Moreover, the direct alkynylation of unsaturated electrophiles with terminal alkynes showed comparable or even better reactivities in water than in organic solvents. From an economic or ecological point of view, it can be expected that C–H activations and reactions, especially multi-C–H bond reactions, have a bright future for chemical syntheses in water under mild conditions. These reactions provide a green alternative to classical organic reactions.

Acknowledgments

Present and past co-workers in our laboratory, whose names are included in the list of references, are gratefully acknowledged for their hard work. We also thank the E. B. Eddy Endowment Chair Fund (McGill University), Canada Research Chair (Tier I) Foundation (to C.J. Li), the CFI, NSERC, FQRNT, the US NSF CAREER Award, the CIC (Merck Frosst/Bohringer Ingelheim/AstraZeneca), the GCI-Pharmaceutical Roundtable, and the US NSF–EPA Joint Program for a Sustainable Environment for partial support of our research. The National Natural Science Foundation of China (**20902076**) is also acknowledged.

References

1. (a) Anastas, P.T. and Warner, J.C. (1998) *Green Chemistry: Theory and Practice*, Oxford University Press, Oxford; (b) Sheldon, R.A., Arends, I., and Hanefeld, U. (2007) *Green Chemistry and Catalysis*, Wiley-VCH Verlag GmbH, Weinheim.

2. (a) Anastas, P. and Farris, C. (eds.) (1994) *Benign by Design: Alternative Synthetic Design for Pollution Prevention*, American Chemical Society, Washington, DC; (b) Dunn, P.T., Wells, A.S., and Williams, M.T. (2010) *Green Chemistry in the Pharmaceutical Industry*, Wiley-VCH Verlag GmbH, Weinheim; (c) Clark, J. and Macquarrie, D. (2002) *Handbook of Green Chemistry and Technology*, Blackwell Science, Oxford.

3. World Commission on Environment and Development (1987) *Our Common Future*, Oxford University Press, Oxford.

4. Trost, B.M. (1991) *Science*, **254**, 1471–1477.

5. Trost, B.M. (2002) *Acc. Chem. Res.*, **35**, 695–705.

6. (a) Horváth, I.T. and Anastas, P.T. (2007) *Chem. Rev.*, **107**, 2169–2173; (b) Li, C.J. and Trost, B. M. (2008) *Proc. Natl. Acad. Sci. U. S. A.*, **105**, 13197–13202.

7. Corma, A., Iborra, S., and Velty, A. (2007) *Chem. Rev.*, **107**, 2411–2502, and references therein.

8. For some recent reviews on biomass, see: (a) Argyropoulos, D.S. (ed.) (2007) *Materials, Chemicals, and Energy from Forest Biomass*, American Chemical Society, Washington, DC; (b) Navarro, R.M., Peòa, M.A., and Fierro, J.L.G. (2007) *Chem. Rev.*, **107**, 3952–3991; (c) George, W.H., Iborra, S., and Corrma, A. (2006) *Chem. Rev.*, **106**, 4044–4098.

9. Tanaka, K. (ed.) (2003) *Solvent-Free Organic Synthesis*, Wiley-VCH Verlag GmbH, Weinheim.

10. Rideout, D.C. and Breslow, R. (1980) *J. Am. Chem. Soc.*, **102**, 7816–7817.

11. (a) Li, C.J. and Chan, T.H. (1997) *Organic Reactions in Aqueous Media*, John Wiley & Sons, Inc., New York; (b) Herrerias, C.I., Yao, X.Q., Li, Z.P., and Li, C.J. (2007) *Chem. Rev.*, **107**, 2546–2562; (c) Li, C.J. (2005) *Chem. Rev.*, **105**, 3095–3166; (d) Li, C.J. (1993) *Chem. Rev.*, **93**, 2023–2035; (e) Li, C.J. and Chan, T.H. (2007) *Comprehensive Organic Reactions in Aqueous Media*, John Wiley & Sons, Inc., New York.

12. Narayan, S., Muldoon, J., Finn, M.G., Fokin, V.V., Kolb, H.C., and Sharpless, K.B. (2005) *Angew. Chem. Int. Ed.*, **44**, 3275–3279.

13. Chanda, A. and Fokin, V.V. (2009) *Chem. Rev.*, **109**, 725–748.

14. Jessop, P.G. and Leitner, W. (eds.) (1999) *Chemical Synthesis Using Supercritical Fluids*, Wiley-VCH Verlag GmbH, Weinheim.

15. (a) Jessop, P.G., Ikariya, T., and Noyori, R. (1994) *Nature*, **368**, 231–233; (b) DeSimore, J.M., Guan, Z., and Elsbernd, C.S. (1992) *Science*, **257**, 945–947; (c) Leitner, W. (2002) *Acc. Chem. Res.*, **35**, 746–756.

16. Liu, H.Z., Jiang, T., Hang, B.X., Liang, S.G., and Zhou, Y.X. (2009) *Science*, **326**, 1250–1252.

17. Wasserscheid, P. and Welton, T. (eds.) (2002) *Ionic Liquids in Synthesis*, Wiley-VCH Verlag GmbH, Weinheim.

18. Gladysz, J., Curran, D., and Horváth, I. (eds.) (2004) *Handbook of Fluorous Chemistry*, Wiley-VCH Verlag GmbH, Weinheim.

19. Sheldon, R.A. (2000) *Pure Appl. Chem.*, **72**, 1233–1246.

20. (a) Yu, J.Q. and Shi, Z.J. (eds) (2010) *C–H Activation*, Springer, Berlin; (b) Dyker, G. (ed.) (2005) *Handbook of C–H Transformations*, Wiley-VCH Verlag GmbH, Weinheim.

21. (a) Baran, P.S., Maimore, T.J., and Richter, J.M. (2007) *Nature*, **446**, 404–408; (b) Chan, T.H. and Li, C.J. (1992) *J. Chem. Soc., Chem. Commun.*, 747–748.

22. (a) Diederich, F., Stang, P.J. (1998) *Metal-Catalyzed Cross-Coupling Reactions*, Wiley-VCH Verlag GmbH, Weinheim; (b) Geissler, H. (1998) in *Transition Metals for Organic Synthesis* (eds. M. Beller and C. Bolm), Wiley-VCH Verlag GmbH, Weinheim; Chapter 2.10, pp. 158–183;

(c) Miyaura, N. (ed.) (2002) *Cross-Coupling Reactions*, Springer, Berlin.

23. Murai, S., Kakiuchi, F., Sekine, S., Tanaka, Y., Kamatani, A., Sonoda, M., and Chatani, N. (1993) *Nature*, **366**, 529–531.

24. Jia, C., Piao, D., Oyamada, J., Lu, W., Kitamura, T., and Fujiwara, Y. (2000) *Science*, **287**, 1992–1995.

25. (a) Goldberg, K.I. and Goldman, A.S. (eds.) (2004) *Activation and Functionalization of C–H bonds*, ACS Symposium Series, vol. **885**, American Chemical Society, Washington, DC.; (b) Lyons, T.W. and Sanford, M.S. (2010) *Chem. Rev.*, **110**, 1147–1169; (c) Chen, X., Engle, K.M., Wang, D.H., and Yu, J.Q. (2009) *Angew. Chem. Int. Ed.*, **48**, 5094–5115; (d) Ashenhurst, J.A. (2010) *Chem. Soc. Rev.*, **39**, 540–548; (e) Johansson, C. and Colacot, T.J. (2010) *Angew. Chem. Int. Ed.*, **49**, 676–707; (f) Campeau, L.C., Stuart, D.R., and Fagnou, K. (2007) *Aldrichim. Acta*, **40**, 35–41.

26. Li, C.J. (2009) *Acc. Chem. Res.*, **42**, 335–344.

27. (a) Cai, G., Fu, Y., Li, Y., Wan, X., and Shi, Z. (2007) *J. Am. Chem. Soc.*, **129**, 7666–7673; (b) Stuart, D.R. and Fagnou, K. (2007) *Science*, **316**, 1172–1175; (c) Dwight, T.A., Rue, N.R., Charyk, D., Josselyn, R., and DeBoef, B. (2007) *Org. Lett.*, **9**, 3137–3139; (d) Hull, K.L. and Sanford, M.S. (2007) *J. Am. Chem. Soc.*, **129**, 11904–11905; (e) Xia, J.B. and You, S.L. (2007) *Organometallic*, **26**, 4869–4871; (f) Jia, C., Kitamura, T., and Fujiwara, Y. (2001) *Acc. Chem. Res.*, **34**, 633–639; (g) Stuart, D.R., Villmure, E., and Fagnou, K. (2007) *J. Am. Chem. Soc.*, **129**, 12072–12073; (h) Dohi, T., Ito, M., Morimoto, K., Iwata, M., and Kita, Y. (2008) *Angew. Chem. Int. Ed.*, **47**, 1301–1304; (i) Do, Q. and Daugulis, O. (2009) *J. Am. Chem. Soc.*, **131**, 17052–17053.

28. (a) Lawrence, S.A. (2004) *Amines: Synthesis Properties and Applications*, Cambridge University Press, Cambridge; (b) Landquist, J.K. (1984) in *Comprehensive Heterocyclic Chemistry* (eds. A.R. Katritzky and C.W. Rees) Pergamon Press, Oxford; (c) Patai, S. (1996) *The Chemistry of Amines, Nitroso, Nitro and Related Groups*, John Wiley & Sons, Ltd., Chichester.

29. Nakamura, H., Kamakura, T., Ishikura, M., and Biellmann, J.F. (2004) *J. Am. Chem. Soc.*, **126**, 5958–5959.

30. (a) Leonard, N.J. and Leubner, G.W. (1949) *J. Am. Chem. Soc.*, **71**, 3408–3411; (b) Murahashi, S.I., Komiya, N., and Terai, H. (2005) *Angew. Chem. Int. Ed.*, **44**, 6931–6933.

31. (a) Li, Z. and Li, C.J. (2004) *J. Am. Chem. Soc.*, **126**, 11810–11811; (b) Li, Z.P. and Li, C.J. (2006) *Pure Appl. Chem.*, **78**, 935–945.

32. Chrzanowska, M. and Rozwadowska, M.D. (2004) *Chem. Rev.*, **104**, 3341–3370.

33. (a) Li, Z.P. and Li, C.J. (2004) *Org. Lett.*, **6**, 4997–4999; (b) Li, Z.P., Macleod, P.D., and Li, C.J. (2006) *Tetrahedron: Asymmetry*, **17**, 590–597.

34. Twyman, R.M. (2004) *Principles of Proteomics*, BIOS Scientific Publishers, New York.

35. Zhao, L. and Li, C.J. (2008) *Angew. Chem. Int. Ed.*, **47**, 7075–7078.

36. Wyvratt, M.J. (1988) *Clin. Physiol. Biochem.*, **6**, 217–229.

37. For reviews, see: (a) Meyers, A.I. (1985) *Aldrichim. Acta*, **18**, 59–68; (b) Beak, P., Zajdel, W.J., and Reitz, D.B. (1984) *Chem. Rev.*, **84**, 471–523.

38. Zhao, L., Basle, O., and Li, C.J. (2009) *Proc. Natl. Acad. Sci. U. S. A.*, **105**, 4106–4111.

39. (a) Li, Z.P. and Li, C.J. (2005) *J. Am. Chem. Soc.*, **127**, 6968–6969; (b) Li, Z.P., Bohle, D.S., and Li, C.J. (2006) *Proc. Natl. Acad. Sci. U. S. A.*, **103**, 8928–8933.

40. Li, Z.P. and Li, C.J. (2005) *J. Am. Chem. Soc.*, **127**, 3672–3673.

41. Sasamoto, N., Dubs, C., Hamashima, Y., and Sodeoka, M. (2006) *J. Am. Chem. Soc.*, **128**, 14010–14011.

42. Basle, O. and Li, C.J. (2007) *Green Chem.*, **9**, 1047–1050.

43. (a) Blunt, J.W., Copp, B.R., Munro, M.H.G., Northcote, P.T., and Prinsep, M.R. (2005) *Nat. Prod. Rep.*, **22**, 15–61; (b) Dembitsky, V.M., Gloriozova, T.A., and Poroikov, V.V. (2005) *Mini Rev. Med. Chem.*, **5**, 319–336; (c) Simmons, T.L., Andrianasolo, E., McPhail, K., Flatt, P., and Gerwick, W.H. (2005) *Mol. Cancer Ther.*, **4**, 333–342; (d) Newman, D.J. and Cragg, G.M. (2004) *J. Nat. Prod.*, **67**, 1216–1238; (e) Bongiorni, L. and Pietra, F. (1996) *Chem. Ind. (London)*, 54–58.

44. Zhang, Y.H. and Li, C.J. (2006) *Angew. Chem. Int. Ed.*, **45**, 1949–1952.

45. Zhang, Y.H. and Li, C.J. (2006) *J. Am. Chem. Soc.*, **128**, 4242–4243.

46. (a) Tsuji, J. (2000) *Transition Metal Reagents and Catalysts: Innovations in Organic Synthesis*, John Wiley & Sons, Inc., New York, Chapter 4, pp. 109–168; (b) Trost, B.M. and Grawley, M.L. (2003) *Chem. Rev.*, **103**, 2921–2944.

47. Li, Z.P. and Li, C.J. (2006) *J. Am. Chem. Soc.*, **128**, 56–57.

48. (a) Borduas, N. and Powell, D.A. (2008) *J. Org. Chem.*, **73**, 7822–7825; (b) Song, C.X., Cai, G.X., Farrell, T.R., Jiang, Z.P., Li, H., Gan, L.B., and Shi, Z.J. (2009) *Chem. Commun.*, 6002–6004; (c) Correia, C.A. and Li, C.J. (2010) *Tetrahedron Lett.*, **51**, 1172–1175.

49. Shi, Z.J., Li, S., Lu, X.Y., Lu, B.J., and Li, Y.Z. (2009) *Angew. Chem. Int. Ed.*, **48**, 3817–3820.

50. Correia, C.A. and Li, C.J. (2010) *Adv. Synth. Catal.*, **352**, 1446–1450.

51. (a) Bolm, C., Legros, J., Le Paih, J., and Zani, L. (2004) *Chem. Rev.*, **104**, 6217–6254; (b) Furstner, A. and Martin, R. (2005) *Chem. Lett.*, **34**, 624–629.

52. Li, Z.P., Cao, L., and Li, C.J. (2007) *Angew. Chem. Int. Ed.*, **46**, 6505–6507.

53. Li, Z.P., Yu, R., and Li, H. (2008) *Angew. Chem. Int. Ed.*, **47**, 7497–7500.

54. (a) Sawyer, D.T., Sobkowiak, A., and Matsushita, T. (1996) *Acc. Chem. Res.*, **29**, 409–416; (b) Walling, C. (1998) *Acc. Chem. Res.*, **31**, 155–157.

55. Barton, D.H.R. and Doller, D. (1991) *Pure Appl. Chem.*, **63**, 1567–1576.

56. Bolm, C., Legros, J., Paih, J.L., and Zani, L. (2004) *Chem. Rev.*, **104**, 6217–6254.

57. Zhang, Y.H. and Li, C.J. (2007) *Eur. J. Org. Chem.*, 4654–4657.

58. Zhang, Y.H., Feng, J.Q., and Li, C.J. (2008) *J. Am. Chem. Soc.*, **130**, 2900–2901.

59. Deng, G.J., Zhao, L., and Li, C.J. (2008) *Angew. Chem. Int. Ed.*, **47**, 6278–6282.

60. Chatani, N., Ie, Y., Kakiuchi, F., and Murai, S. (1997) *J. Org. Chem.*, **62**, 2604.

61. Deng, G.J. and Li, C.J. (2009) *Org. Lett.*, **11**, 1171–1174.

62. Deng, G.J., Ueda, K., Yanagisawa, S., Itami, K., and Li, C.J. (2009) *Chem. Eur. J.*, **15**, 333–337.

63. (a) Hull, K.L., Lanni, E.L., and Sanford, M.S. (2006) *J. Am. Chem. Soc.*, **128**, 14047–14048; (b) Whitfield, S.R. and Sanford, M.S. (2007) *J. Am. Chem. Soc.*, **129**, 15142–15143.

64. Chen, X., Dobereiner, G., Hao, X.S., Giri, R., Maugel, N., and Yu, J.Q. (2009) *Tetrahedron*, **65**, 3085–3089.

65. Guo, X.Y., Deng, G.J., and Li, C.J. (2009) *Adv. Synth. Catal.*, **351**, 2071–2074.

66. For examples, see: (a) Facoetti, D., Abbiati, G., D'Avolio, L., Ackerman, L., and Rossi, E. (2009) *Synlett*, **14**, 2273–2276; (b) Yue, D.W. and Larock, R.C. (2004) *Org. Lett.*, **6**, 1037–1040; (c) Zhang, H. and Larock, R.C. (2002) *J. Org. Chem.*, **67**, 7048–7056.

67. Yang, L., Zhao, L., and Li, C.J. (2010) *Chem. Commun.*, **46**, 4184–4186.

68. Haro, T. and Nevado, C. (2010) *J. Am. Chem. Soc.*, **132**, 1512–1513.

69. (a) Stang, P.J. and Diederich, F. (eds.) (1995) *Modern Acetylene Chemistry*, Wiley-VCH Verlag GmbH, Weinheim; (b) Diederich, F., Stang., P.T., and Tykwinski, R.R. (eds.) (2005) *Acetylene Chemistry – Chemistry, Biology and Materials Science*, Wiley-VCH Verlag GmbH, Weinheim; for an excellent review on the subject, see: (c) Trost, B.M. and Weiss, A.H. (2009) *Adv. Synth. Catal.*, **351**, 963–983.

70. Trost, B.M. (1995) *Angew. Chem. Int. Ed. Engl.*, **34**, 259–281.

71. Trost, B.M., Sorum, M.T., Chan, C., Harms, A.E., and Ruhter, G. (1997) *J. Am. Chem. Soc.*, **119**, 698–708.

72. Nishimura, T., Takatsu, T., Shintani, R., and Hayashi, T. (2007) *J. Am. Chem. Soc.*, **129**, 14158–14159.

73. Yamaguchi, M., Hayashi, A., and Minami, T. (1991) *J. Org. Chem.*, **56**, 4901–4902.

74. Han, Y. and Huang, Y.Z. (1995) *Tetrahedron Lett.*, **36**, 7277–7280.

75. Carreira, E.M., Frantz, D.E., Fässler, R., and Tomooka, C.S. (2000) *Acc. Chem. Res.*, **33**, 373–381.

76. Miura, M., Enna, M., Okuro, K., and Nomura, M. (1995) *J. Org. Chem.*, **60**, 4999–5004.

77. Frantz, D.E., Fässler, R., and Carreira, E.M. (1999) *J. Am. Chem. Soc.*, **121**, 11245–11246.

78. Li, C.J. (1996) *Tetrahedron*, **52**, 5643–5668.

79. Li, C.J. and Zhang, W.C. (1998) *J. Am. Chem. Soc.*, **120**, 9102–9103.

80. (a) Li, C.J. (2010) *Acc. Chem. Res.*, **43**, 581–590; (b) Chen, L. and Li, C.J. (2006) *Adv. Synth. Catal.*, **348**, 1459–1484.

81. Wei, C.M. and Li, C.J. (2002) *Green Chem.*, **4**, 39–41.

82. Yao, X. and Li, C.J. (2005) *Org. Lett.*, **7**, 4395–4398.

83. Boumendjel, A. (2003) *Curr. Med. Chem.*, **10**, 2621–2630.

84. Yu, M., Skouta, R., Zhou, L., Jiang, H.F., Yao, X., and Li, C.J. (2009) *J. Org. Chem.*, **74**, 3378–3383.

85. Dyker, G. (1999) *Angew. Chem. Int. Ed.*, **38**, 1698.

86. Yao, X. and Li, C.J. (2006) *Org. Lett.*, **8**, 1953–1955.

87. Deng, G.J. and Li, C.J. (2008) *Synlett*, 1571–1573.

88. Motoki, R., Kanai, M., and Shibasaki, M. (2007) *Org. Lett.*, **9**, 2997–3000.

89. (a) Kauffman, G.S., Harris, G.D., Dorow, R.L., Stone, B., Parsons, R., Pesti, J., Magnus, N., Fortunak, J., Confalone, P., and Nugent, W. (2000) *Org. Lett.*, **2**, 3119–3121; (b) Huffman, M.A., Yasuda, N., DeCamp, A.E., and Grabowski, E.J. (1995) *J. Org. Chem.*, **60**, 1590–1594.

90. Wei, C. and Li, C.J. (2002) *Chem. Commun.*, 268–269.

91. (a) Satoshi, S., Takashi, K., and Ishii, Y. (2001) *Angew. Chem. Int. Ed.*, **40**, 2534–2536; (b) Sakaguchi, S., Mizuta, T., Furuwan, M., Kubo, T., and Ishii, Y. (2004) *Chem. Commun.*, 1638–1639.

92. (a) Carreira, E.M. and Fischer, C. (2001) *Org. Lett.*, **3**, 4319–4321; (b) Fischer, C. and Carreira, E. (2004) *Synthesis*, 1497–1503.

93. Wei, C. and Li, C.J. (2002) *J. Am. Chem. Soc.*, **124**, 5638–5639.

94. Wei, C., Mague, J.T., and Li, C.J. (2004) *Proc. Natl. Acad. Sci. U. S. A.*, **101**, 5749–5754.

95. Youngman, M.A. and Dax, S.L. (1997) *Tetrahedron Lett.*, **38**, 6347–6350.

96. Kabalka, G.W., Wang, L., and Pagni, R.M. (2001) *Synlett*, 676–678.

97. (a) Wei, C. and Li, C.J. (2003) *J. Am. Chem. Soc.*, **125**, 9584–9585; (b) Skouta, R. and Li, C.J. (2008) *Tetrahedron*, **64**, 4917–4938.

98. Lo, V., Kung, K., Wong, M., and Che, C.M. (2009) *J. Organomet. Chem.*, **694**, 583–591, and references therein.

99. Wei, C.M., Li, Z.G., and Li, C.J. (2003) *Org. Lett.*, **5**, 4473–4475.

100. Maggi, R., Bello, A., Oro, C., Sartori, G., and Soldi, L. (2008) *Tetrahedron*, **64**, 1435–1439.

101. Rueping, M., Antonchick, A.P., and Brinkmann, C. (2007) *Angew. Chem. Int. Ed.*, **46**, 6903–6906.

102. Chen, W.W., Nguyen, R., and Li, C.J. (2009) *Tetrahedron Lett.*, **50**, 2895–2898.

103. Zeng, T.Q., Chen, W.W., Cirtiu, C.M., Moores, A., Song, G.H., and Li, C.J. (2010) *Green Chem.*, **12**, 570–573.

104. Gommermann, N., Koradin, C., Polborn, K., and Knochel, P. (2003) *Angew. Chem. Int. Ed.*, **42**, 5763–5766.

105. Knopfel, T.F., Aschwanden, P., Ichikawa, T., Watanabe, T., and Carreira, E.M. (2004) *Angew. Chem. Int. Ed.*, **43**, 5971–5973.

106. Shi, L., Tu, Y.Q., Wang, M., Zhang, F.M., and Fan, C.A. (2004) *Org. Lett.*, **6**, 1001–1003.

107. Bieber, L.W. and da Silva, M.F. (2004) *Tetrahedron Lett.*, **45**, 8281–8283.

108. Zhang, J., Wei, C., and Li, C.J. (2002) *Tetrahedron Lett.*, **43**, 5731–5734.

109. (a) Black, A.D. and Arndtsen, B.A. (2004) *Org. Lett.*, **6**, 1107–1110; (b) Fischer, C. and Carreira, E.M. (2004) *Org. Lett.*, **6**,

1497–1499; (c) Wei, C. and Li, C.J. (2005) *Lett. Org. Chem.*, **2**, 410–414.

110. For reviews, see: (a) Nicolaou, K.C., Edmonds, D.J., and Bulger, P.G. (2006) *Angew. Chem. Int. Ed.*, **45**, 7134–7186; (b) Padwa, A. (2004) *Pure Appl. Chem.*, **76**, 1933–1952; (c) Parsons, P.J., Penkett, C.S., and Shell, A.J. (1996) *Chem. Rev.*, **96**, 195–206; (d) Bunce, R.A. (1995) *Tetrahedron*, **51**, 13103–13159.

111. Devries, V.G., Bloom, J.D., Dutia, M.D., Katocs, A.S. Jr., and Largis, E.E. (1989) *J. Med. Chem.*, **32**, 2318–2325.

112. Bonfield, E.R. and Li, C.J. (2007) *Org. Biomol. Chem.*, **5**, 435–437.

113. Bonfield, E.R. and Li, C.J. (2008) *Adv. Synth. Catal.*, **350**, 370–374.

114. Carreira, E.M. and Knopfel, T.F. (2003) *J. Am. Chem. Soc.*, **125**, 6054–6055.

115. (a) Knopfel, T.F., Zarotti, P., Ichikawa, T., and Carreira, E.M. (2005) *J. Am. Chem. Soc.*, **127**, 9682–9683; (b) Fujimori, S.M. and Carreira, E.M. (2007) *Angew. Chem. Int. Ed.*, **46**, 4964–4967.

116. Chen, L. and Li, C.J. (2004) *Tetrahedron Lett.*, **45**, 2771–2774.

117. Chen, L. and Li, C.J. (2004) *Chem. Commun.*, 2362–2364.

118. Zhou, L., Chen, L., Skouta, R., and Li, C.J. (2008) *Org. Biomol. Chem.*, **6**, 2969–2977.

119. Lerum, R.V. and Chisholm, J.D. (2004) *Tetrahedron Lett.*, **45**, 6591–6594.

Commentary Part

Comment 1

Roger A. Sheldon

The Brundtland Report, "Our Common Future," published in 1987, defined sustainable development as

> ... development that meets the needs of the present generation without compromising the ability of future generations to meet their needs.

It consists of two key elements: the concept of "*needs*," particularly the essential needs of the world's poor, and the concept of "*limitations*" imposed by the state of technology and social organization on the environment's ability to meet present and future needs.

Simultaneously with the adoption of the principles of sustainability, the late 1980s witnessed the emergence of the principles of *clean chemistry* and *waste minimization* [C1] and the underlying concepts of *atom utilization* [C1] or *atom economy* [C2] and the *Environmental or E factor* [C3]. Atom economy and the E factor were subsequently widely accepted as useful and complementary metrics for assessing the (potential) environmental impact of chemical processes. The former, defined as the molecular weight of the product divided by the total molecular weight of all products formed in the stoichiometric equation, is a theoretical number that is very useful for comparing various possible synthetic routes before any experiment is performed. The latter, defined as the kilograms of waste divided by the kilograms of product, is the actual amount of waste formed in the process. It takes the chemical yield into account and includes all reagents, solvent losses, process aids and, in principle, even fuel. This waste consists primarily of inorganic salts that are formed in the reaction or in downstream processing. The E factor increases dramatically on going from bulk to fine chemicals and pharmaceuticals. This is partly explained by the fact that the products are generally more complicated than bulk chemicals and require multi-step syntheses, but is also a consequence of the widespread use of classical stoichiometric reagents such as Brønsted and Lewis acids and bases and inorganic or organic oxidants and reductants.

In the late 1990s, Anastas and Warner [C4] coined the terms *benign by design* and *green chemistry* to describe the deliberate design of chemical processes and products for minimal environmental impact and the latter term has since been widely adopted by industry and academia. Green chemistry can be succinctly defined as follows:

> Green chemistry efficiently utilizes (preferably renewable) raw materials, eliminates waste, and avoids the use of toxic and/or hazardous solvents and reagents in the manufacture and applications of chemical products.

It embraces three key objectives: (i) efficient utilization of raw materials (including energy), (ii) avoidance of toxic/hazardous substances, and (iii) replacement of fossil feedstocks by renewable resources. The key to efficient utilization of raw materials is substitution of classical organic syntheses with atom- and step-economic [C5] catalytic alternatives [C6]. Another major source of waste in the pharmaceutical industry is solvent losses. Moreover, many of the solvents commonly used in organic syntheses are toxic and/or hazardous materials and should be avoided. Hence a second trend in green chemistry involves the introduction of alternative reaction media. As we noted elsewhere [C7], the best solvent is no solvent, but if a solvent must be used then water and scCO$_2$ have excellent credentials, as discussed in this chapter by Deng and Li. Similarly, ionic liquids and fluorous media have definite advantages in this context, as also mentioned. The authors also note that much can be gained by the employment of step-economic, catalytic cascade processes and

performing sequential reactions in flow reactors to simplify the overall operation.

The third trend in green chemistry and sustainable technology involves the transition from a world economy that is largely based on nonrenewable fossil feedstocks to a sustainable one that is based on renewable resources. Among various sustainable energy options, only biomass is a source of carbon-based fuels and chemicals. Here again, clean catalytic technologies – both chemo- and biocatalytic – have a key enabling role to play [C8]. As Deng and Li note, a major obstacle to the utilization of renewable biomass is the need for novel chemistry to transform large amounts of biomass, selectively and efficiently, without the requirement for extensive functionalization, defunctionalization, or protection, that is, by developing atom- and step-economic chemo- and biocatalytic methodologies.

Many key transformations in organic synthesis involve the formation of new carbon–carbon bonds and, in their further discussion of cleaner catalytic alternatives for classical organic syntheses, Deng and Li first focus on this very important class of reactions. For example, classical formation of C–C bonds, such as in biaryls by cross-coupling, involves initial introduction of a halogen atom followed by conversion to an organometallic species such as a Grignard reagent. The latter is then coupled to another halogenated aromatic molecule. Overall, the reaction comprises four steps and uses hazardous stoichiometric reagents and solvents and generates stoichiometric amounts of halide salts. Many improvements have been developed in the last few decades in these so-called CDC reactions, as exemplified by the Mizoroki–Heck, Suzuki–Miyaura, and Negishi couplings, and Heck, Negishi, and Suzuki were awarded the 2010 Nobel Prize in Chemistry for their pioneering contributions to this important area. Nonetheless, these methods all involve the introduction of a halogen atom that is subsequently removed in the coupling step, involving C–X moieties, and ends up as waste. Hence an improvement avoids the prefunctionalization of one of the components and involves the metal-catalyzed coupling of a C–X bond to a C–H bond with loss of HX (as a salt). Further evolution in greenness is achieved in a so-called catalytic CDC whereby two C–H moieties are coupled, ideally using a green oxidant such as molecular oxygen. Much progress has been made in this area in recent years and Deng and Li present a comprehensive overview of the different combinations that are possible, for example, coupling of C–H to C–H adjacent to N in amines or C–H adjacent to O in ethers and also simple C–C couplings involving allylic or benzylic C–H bonds or simple alkanes as one of the partners. An added benefit is that many of these direct oxidative coupling reactions can be performed in water as the solvent. The reactions generally utilized copper, ruthenium, or palladium as catalysts in conjunction with oxygen or TBHP as the stoichiometric oxidant. Here I would have expected (more) references to the pioneering work of Kharasch and Murahashi on copper- and ruthenium-catalyzed dehydrogenative couplings. The second class of reactions that is reviewed by Deng and Li in the context of green and sustainable chemistry is the nucleophilic addition of terminal alkynes to unsaturated electrophiles, such as aldehydes, ketones, imines, and electron-deficient carbon–carbon double or triple bonds. The classical methodology for performing these transformations involves the presynthesis of a metal acetylide from the terminal alkyne and an organometallic reagent, leading to stoichiometric amounts of inorganic waste after quenching with acid. In these reactions, there is no overall change in redox state of the reactants and products and, consequently, no oxidant is required. An important improvement is achieved by designing green catalytic systems that operate in water as the solvent. Examples include ruthenium–indium and ruthenium–copper bimetallic systems and palladium, silver, and gold complexes. From both environmental and economic viewpoints, the use of copper or, better still, iron catalysts stands out. For example, $FeCl_3$ was found to be an effective catalyst for the three-component coupling of an aldehyde, an alkyne, and an amine. Interestingly, the reaction could be performed in water as solvent, in air, using magnetic Fe_3O_4 nanoparticles as the catalyst, which could in principle, be separated and recovered by magnetic decantation.

Summarizing, as the authors note, this chapter presents a very useful overview of recent developments in the area of direct and selective transformations of unactivated C–H bonds into C–C bonds, one of the most powerful tools for developing molecular complexity in organic molecules. To conduct such reactions in water is particularly challenging as it conflicts with traditional concepts of reactivity and bond strength. Many pioneering reactions are described that can form the basis for green and sustainable alternatives to classical synthetic methodologies. Presumably, some of these reactions can, in the future, also be applied to transformations of biomass-derived carbohydrates, thereby incorporating the third objective of green and sustainable chemical technology: the transition to renewable resources.

Comment 2

Tak Hang Chan

This chapter, "Synthetic Chemistry with an Eye on Future Sustainability," provides an important perspective on the future of synthetic organic chemistry. By the latter part of the twentieth century, with the successful syntheses of vitamin B_{12} [C9], palytoxin [C10], and many other complex natural products [C11], some may have been wondering about the future of organic synthesis [C12]. It has been argued that "the competitive nature of the pharmaceutical and biotechnology industries and their drive to discover and produce new cures for disease will demand new and sharper tools for organic synthesis" [C12]. This chapter offers an additional, perhaps more compelling, argument for the future development of synthetic chemistry: new chemical reactions and syntheses must be discovered to meet the challenge of sustainability as illustrated by the precepts of green chemistry [C13]. Deng and Li discuss in detail two general

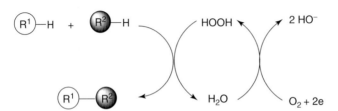

Scheme C1 *In situ* generation of oxidant for CDC reaction.

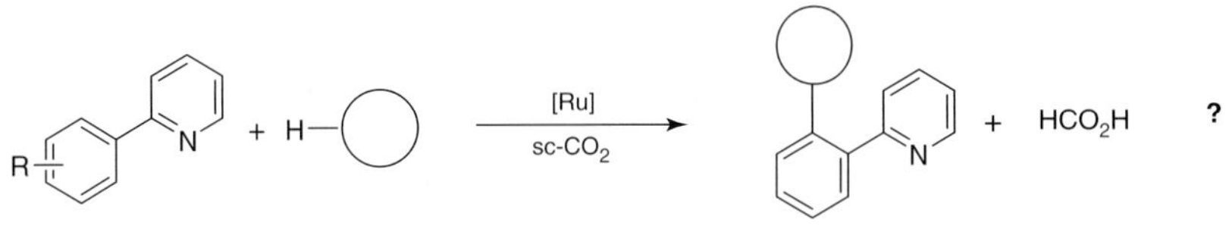

Scheme C2 CDC reaction between phenylpyridines and cycloalkanes in scCO$_2$.

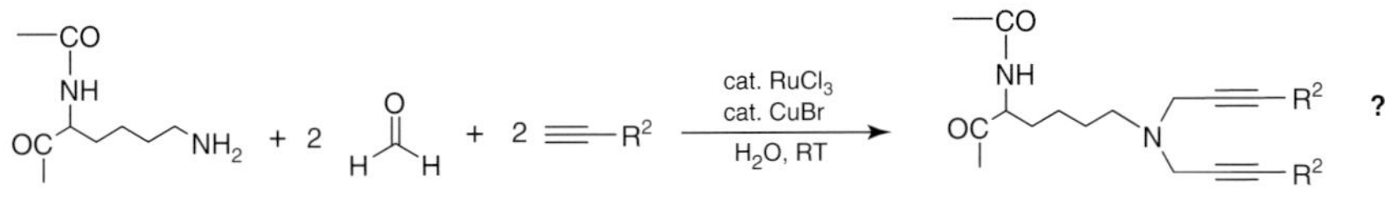

Scheme C3 Modification of a lysine-containing peptide with double A^3 coupling.

types of reactions, (i) CDC reactions and (ii) catalytic nucleophilic additions of terminal alkynes in water, to showcase the kind of new reactions that may provide a "greener" alternative to the classical organic reactions. This chapter complements other chapters in this book, Chapter 8 on click chemistry, Chapter 7 on C–H activation, and Chapters 9–11 and 14 on catalysis, in providing a full picture of the future direction of synthetic chemistry within the theme of sustainability.

The task of developing new synthetic reactions which meet all the requirements of green chemistry is not an easy one. The progress may be incremental in that it may provide improvements in only one or two aspects of the "twelve principles of green chemistry" [C13]. In the C–C bond formation reactions as outlined in Scheme 19.3, the CDC reactions (c) offer improvements over many of the traditional reactions in (a) and (b) because prefunctionalization of H to X or Y is not necessary, thus reducing the number of steps required and, as a consequence, reducing waste. On the other hand, it is evident that the CDC reaction requires a stoichiometric amount of oxidant. If the oxidant turns out to be oxygen with water as the side product, the "greenness" of the reaction is still fairly high because both oxygen and water are innocuous to the environment and considered sustainable. On the other hand, some of the CDC reactions require oxidants such as ^{t}BuOOH, ^{t}BuOO^{t}Bu, or DDQ, and further improvements to eliminate the need for these stoichiometric reagents would seem desirable. A possible approach would be to couple the CDC reaction to

the *in situ* generation of hydrogen peroxide electrochemically from oxygen [C14], as illustrated in Scheme C1.

Another possible approach is to couple the CDC reaction to a catalytic hydrogenation reaction. A particularly fascinating possibility is to use the ruthenium-catalyzed hydrogenation of carbon dioxide to formic acid [C15]. Since ruthenium catalyst is known to catalyze a number of CDC reactions (e.g., Scheme 19.23), one may wonder whether, if the reaction were to be carried out in scCO$_2$, a concomitant formation of formic could in principle result, obviating the need for extraneous oxidant, as illustrated in Scheme C2.

The second class of reactions discussed in this chapter is the direct conjugate addition of terminal alkynes in water. Like other addition reactions, the atom efficiency of such reactions is 100%. That these reactions can be carried out in water provides another "green" advantage. An interesting mechanistic aspect is that some reactions (e.g., Scheme 19.36) proceeds better in water than in organic solvents. This is reminiscent of the classical observation by Rideout and Breslow of the acceleration of Diels–Alder reactions in water [C16].

An important possible application of this class of reactions is to modify water soluble biomolecules directly without the need for derivatization. One can imagine a sugar molecule replacing the aldehyde component in the A^3 coupling reaction as outlined in Scheme 19.36. Equally intriguing is the possibility of modification of peptides containing amino groups in the side chains, for example, lysine, as illustrated in Scheme C3 [C17]. These reactions may open up a host of opportunities in chemical biology.

References

C1. Sheldon, R.A. (1992) in *Industrial Environmental Chemistry* (eds. D.T. Sawyer and A.E. Martell), Plenum Press, New York, pp. 99–119.

C2. (a) Trost, B.M. (1991) *Science*, **254**, 1471; (b) Trost, B.M. (1995) *Angew. Chem. Int. Ed. Engl.*, **34**, 259.

C3. (a) Sheldon, R.A. (1992) *Chem. Ind. (London)*, 903; (b) Sheldon, R.A. (1997) *Chem. Ind. (London)*, 12; (c) Sheldon, R.A. (2007) *Green Chem.*, **9**, 1273–1283; (d) Sheldon, R.A. (2008) *Chem. Commun.*, 3352.

C4. Anastas, P.T. and Warner, J.C. (eds.) (1998) *Green Chemistry: Theory and Practice*, Oxford University Press, Oxford.

C5. Wender, P.A., Handy, S.T., and Wright, D.L. (1997) *Chem. Ind. (London)*, 765–769.

C6. Sheldon, R.A., Arends, I.W.C.E., and Hanefeld, U. (2007) *Green Chemistry and Catalysis*, Wiley-VCH Verlag GmbH, Weinheim.

C7. Sheldon, R.A. (2000) *Pure Appl. Chem.*, **72**, 1233.

C8. Sheldon, R.A. (2011) *Catal. Today*, **167**, 3–13.

C9. (a) Woodward, R.B. (1973) *Pure Appl. Chem.*, **33**, 145; (b) Eschenmoser, A. and Wintner, C.E. (1977) *Science*, **196**, 1410.

C10. Armstrong, R.W., Beau, J.M., Cheon, S.H., Christ, W.J., Fujioka, H., Ham, W.H., Hawkins, L.D., Jin, H., Kang, S.H., Kishi, Y., Martinelli, M.J., McWhorter, W.W. Jr., Mizuno, M., Nakata, M., Stutz, A.E., Talamas, F.X., Taniguchi, M., Tino, J.A., Ueda, K., Uenishi, J., White, J.B., and Yonaga, M. (1989) *J. Am. Chem. Soc.*, **111**, 7530.

C11. Nicolaou, K.C. and Sorenson, E.J. (1996) *Classics in Total Synthesis*, VCH Publishers, New York.

C12. Nicolaou, K.C., Vourloumis, D., Winssinger, N., and Baran, P.S. (2000) *Angew. Chem. Int. Ed.*, **39**, 44.

C13. Anastas, P.T. and Warner, J.C. (1998) *Green Chemistry: Theory and Practice*, Oxford University Press, Oxford.

C14. Tang, M.C.Y., Wong, K.Y., and Chan, T.H. (2005) *Chem. Commun.*, 1345.

C15. Jessop, P.G., Ikariya, T., and Noyori, R. (1994) *Nature*, **368**, 231.

C16. Rideout, D.C. and Breslow, R. (1980) *J. Am. Chem. Soc.*, **102**, 7816.

C17. Bonfield, E.R. and Li, C.J. (2007) *Org. Biomol. Chem.*, **5**, 435–437.

20

Organic π-Conjugated Molecules for Organic Semiconductors and Photovoltaic Materials

De-Qing Zhang, Xiao-Wei Zhan, Zhao-Hui Wang, Jian Pei,
Guan-Xin Zhang, and Dao-Ben Zhu

20.1
Introduction

Organic π-conjugated molecules are one category of organic compounds. A key feature of π-conjugated molecules that renders them useful for many applications is the electronic structure of π-systems, which can lead to extensive delocalization of electrons throughout the molecule. Organic π-conjugated molecules can be classified as electron-rich and electron-deficient types, also referred as to electron donors and acceptors, respectively. Such a structural feature of organic π-conjugated molecules endows them not only with special reactivity but also with interesting physical properties, for which they have been intensively investigated with a view to obtaining functional materials. For instance, they usually show strong absorption throughout the UV–visible and near-infrared parts of the electromagnetic spectrum. Moreover, both the strength and position of the absorption bands can be tuned in a rational manner by appropriate substitution of the molecules with electron donors and acceptors. This is very important for their application in organic photovoltaic (OPV) systems to harvest photons with wavelengths as high as ∼900 nm.

Substitution of the molecules with electronic donors tends to destabilize the highest occupied molecular orbital (HOMO), thereby lowering the ionization potential of the molecules and thus facilitating injection of holes from electrodes (specifically the anode). Substitution of the molecules with electron acceptors tends to stabilize the lowest unoccupied molecular orbital (LUMO), thereby increasing the electron affinity of the molecules and thus facilitating injection of electrons from an electrode (specifically the cathode). As a result, molecules with donors are often used as hole-transport materials and molecules with acceptors are often used as electron-transport materials. Another very important characteristic of π-conjugated molecules is how molecular entities are organized in the solid state with respect to one another (or with other molecules) to ensure that the correct electronic coupling between molecules can facilitate electron (or hole) transfer processes. As a result, much work has been focused on developing organic semiconductors that have the correct electronic structure for injection (or collection) of charges, relatively low reorganization energies, and the appropriate intermolecular packing arrangements.

Organic Chemistry – Breakthroughs and Perspectives, First Edition. Edited by Kuiling Ding and Li-Xin Dai.
© 2012 Wiley-VCH Verlag GmbH & Co. KGaA. Published 2012 by Wiley-VCH Verlag GmbH & Co. KGaA.

Some of these π-conjugated molecules as electron donors and acceptors exhibit reversible redox reactions, that is, they can be reversibly oxidized or reduced to the respective cations or anions. Tetrathiafulvalene (TTF) and its derivatives belong to this type of organic π-conjugated molecule, and they can be reversibly transformed into the corresponding radical cations (TTF$^+$) and dications (TTF^{2+}) by either electrochemical or chemical means [1–5]. By taking advantage of this feature of TTF, Stoddart and co-workers have built a beautiful chemistry of TTF-based rotaxanes and catenanes [6–11]. They have successively demonstrated the application of these rotaxanes and catenanes to construct molecular devices [8–11]. Other elegant switchable systems with electroactive units have also been reported in recent years [12].

A number of organic π-conjugated molecules show responsiveness to light irradiation. One of the most investigated phenomena is light-induced emission. π-Conjugated molecules are usually emissive in solution, but they become almost nonemissive in the solid state. However, Tang, Zhu, and co-workers have discovered aggregation-induced or enhanced emission for a few π-conjugated molecules [13–16]. The use of π-conjugated systems as chemical sensors within the concept of the so-called "molecular wire approach" has been extensively investigated [17, 18].

Some π-conjugated molecules undergo structural transformation upon light irradiation. Irie and other groups have demonstrated various elegant light-driven switchable systems [19–25] featuring photoresponsive units such as diarylethenes that are highly durable. Feringa and co-workers developed a series of molecular rotary motors based on overcrowded alkenes for which the photoinduced *cis–trans* isomerization of C=C is crucial for realizing the unidirectional rotation [24, 25].

Some π-conjugated molecules possess significant nonlinear optical (NLO) properties which enable them to have potential applications in optical signal processing, all-optical switching, and optical computing [26]. Multiphoton absorption has been reported for π-conjugated molecules and such multiphoton absorption materials can be applied in 3D data storage, 3D microfabrication, optical limiting, optical pulse suppression, frequency up-converted lasing, bio-imaging, and photodynamic therapy [27–31].

In this chapter, owing to limitations on space, we cannot discuss the developments in all aspects of organic π-conjugated molecules, but rather focus on the design and synthesis of organic π-conjugated molecules for organic semiconductors with high mobility. Organic semiconductors can be classified as p-type and n-type, which transport preferably holes and electrons, respectively. Organic ambipolar semiconductors are also available. However, we will concentrate just on the discussion of organic p- and n-type semiconductors. For application in OPV systems, both electron donor and acceptor conjugated molecules are needed for the generation of charge carriers upon light irradiation. For this reason, we also discuss organic π-conjugated molecules for OPV materials.

20.2
Conjugated Molecules for p-Type Organic Semiconductors

Acenes are an important class of organic π-conjugated molecules. In recent years, larger acenes such as hexacene, heptacene, octacene, and nonacene have attracted increasing

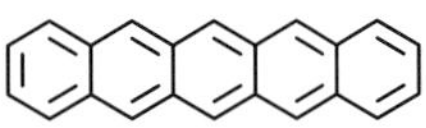

Figure 20.1 Pentacene.

1

attention since they are interesting for applications in organic electronics. They exhibit small bandgaps and potentially high charge-carrier mobilities, but they are also far more prone to oxidative degradation, a characteristic that limits their opportunities for use in organic field effect transistor (OFET) applications. Recent decades witnessed a surge in the synthesis of relatively soluble and stable higher order acene derivatives, and several new synthetic methods have been developed.

Pentacene [32, 33] (**1**, Figure 20.1) is among the best known of the acenes as a p-channel semiconductor and is still used as the benchmark material because of its excellent semiconducting behavior, which is comparable to that of hydrogenated amorphous silicon. However, the crystal structures of acenes such as pentacene adopt the so-called "herringbone" motif with only minimal π-stacking, since a cofacial π-stacking structure is expected to provide more efficient orbital overlap and thereby facilitate carrier transport. Numerous attempts have been made to disrupt this herringbone structure and promote π-stacking in pentacene by introducing one or more relatively bulky groups into the *peri*-positions, and also accompanied with improved solubility and stability.

Anthony *et al.* reported a series of 6,13-disubstituted pentacenes, in which a rigid alkyne spacer is used to hold the bulky groups away from the aromatic core to allow exquisite control over the solid-state arrangement of the acene molecules (Scheme 20.1), along with dramatic increases in stability and solubility [34]. These silylethynyl-substituted pentacenes typically adopt either a 1D "slipped-stack" or 2D "bricklayer" arrangement especially for TIPS-pentacene (**2**, Figure 20.2); TIPS = triisopropylsilyl in the solid state. The thin films of TIPS-pentacene exhibit hole mobilities as high as 0.4 cm^2 V^{-1} s^{-1} [35], whereas OFETs based on crystalline microribbons formed by solution deposition show hole mobilities as high as 1.42 cm^2 V^{-1} s^{-1} and an on:off ratio of 10^5 [36].

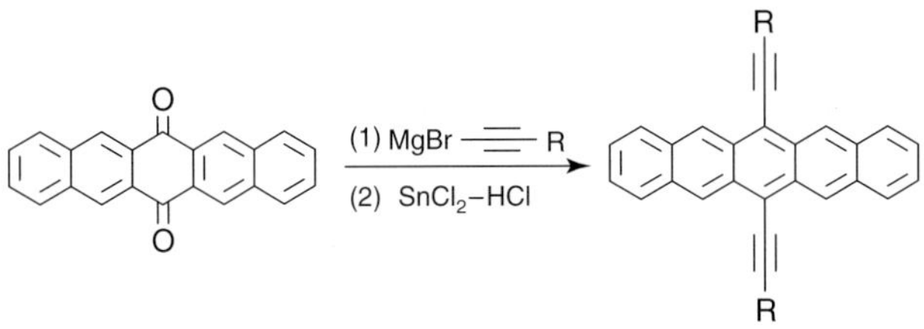

Scheme 20.1 Synthesis of 6,13-bis[(trialkylsilyl)ethynyl]pentacene.

6,13-Bis(alkylthio)pentacenes (**3–5**) were synthesized by Kobayashi *et al.* [37]. It was observed that the introduction of methylthio groups into pentacene at the 6, 13-positions in **4** changed the packing structure from a herringbone to a cofacial π-stacking motif. Hexathiapentacene (**6**) [38], prepared by reaction with pentacene and sulfur, has been the subject of recent device studies. X-ray crystallographic analysis of **6** showed that it adopts strongly π-stacked arrays in two directions with very close S–S contacts between neighboring sulfur atoms. Top-contact OFETs fabricated from evaporated thin films of **6**

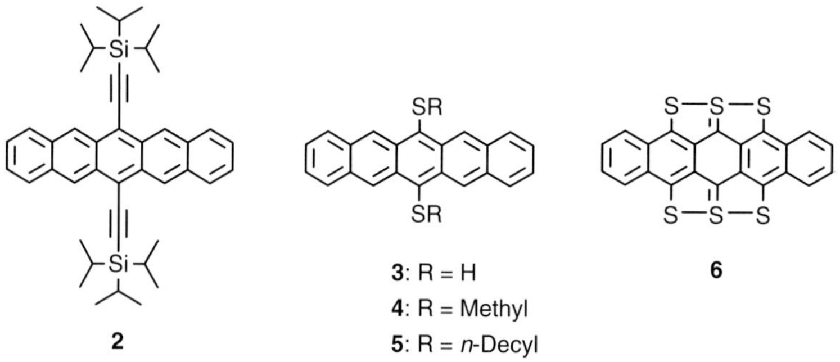

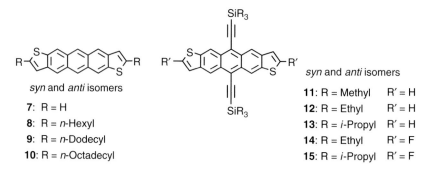

Figure 20.2 TIPS-pentacene and 6,13-bis(alkylthio)pentacenes and Hexathiapentacene.

yield mobilities on the order of 10^{-2} cm^2 V^{-1} s^{-1} and on:off ratio of 10^7 [39]. For devices based on its single nanowires by solution processing, the carrier mobilities can reach 0.27 cm^2 V^{-1} s^{-1} [40].

One of the most thoroughly studied heteropentacenes is anthradithiophene (ADT), which was prepared as inseparable mixtures of *syn*- and *anti*-isomers by Katz and co-workers (Figure 20.3) [41]. ADT (**7**) exhibits only moderate mobility in an OFET device of 0.09 cm^2 V^{-1} s^{-1}, whereas the thiophene units of ADT substituted with alkyl chains can dramatically improve film quality.

Starting with the same *syn*- and *anti*-anthradithiophenequinone mixture, a variety of trialkylsilylacetylenes were appended to the ADT core to afford several solid-state arrangements, with small changes in the size of the alkyne substituent leading to significant changes in the crystalline order (Figure 20.3) [42]. The triethylsilyl derivative **12** adopts a 2D π-stacking motif very similar to that of TIPS-pentacene. The device performance is improved for OFETs based on the solution-deposited films of **11–13** by increasing the crystalline order. Drop-cast films of **12** yielded devices with mobilities as high as 1.0 cm^2 V^{-1} s^{-1} and on:off ratios of 10^7, both values being well within the range necessary for useful commercial devices.

The addition of a few fluorine substituents does not alter p-type behavior but dramatically improves thermal and photostability and induces solid-state interactions that accelerate crystallization. Fluorine-substituted **14** in particular is a robust semiconductor

Figure 20.3 Anthradithiophene (ADT) and its derivatives.

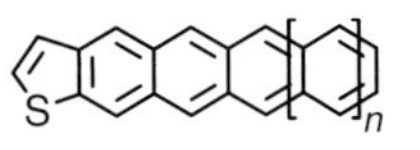

Si i-Pr$_3$ Si i-Pr$_3$

16: n = 0
17: n = 1
18: n = 2

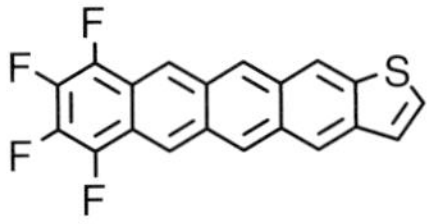

19

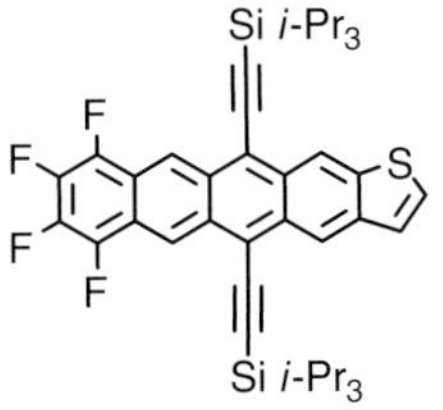

Si i-Pr$_3$

20

Si i-Pr$_3$

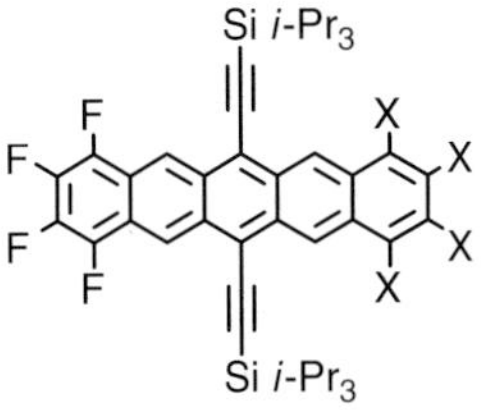

21: X = H
22: X = F

Figure 20.4 The dissymmetric linear acenes.

that easily and reproducibly forms stable, high-quality thin films, allowing detailed studies, which is not possible with nonfluorinated **11** with a herringbone packing arrangement [43].

Dissymmetric linear acenes containing fused thiophene units were prepared by Bao and co-workers [44, 45]. Tetraceno[2,3-*b*]thiophene (**17**, Figure 20.4) has a slightly higher oxidation potential than pentacene, and the mobility of its thin film reaches 0.47 cm^2 V^{-1} s^{-1}. Introduction of fluorine atoms in the terminal ring of **19–22** reduces the π-π stacking distance between the conjugated acene backbones. All the molecules **20–22** pack in the highly desirable 2D bricklayer structure, which is characteristic of TIPS-pentacene-type molecules [46, 47].

Benzodithiophene derivatives **23–25** (Figure 20.5) were synthesized by Takimiya *et al.* [48]. Crystallographic analysis of diphenylbenzodiselenophene (**24**) revealed that the phenyl substituents are coplanar with the heteroacene and that the material packs with extensive edge-to-face interactions similar to unsubstituted pentacene. Detailed studies of the physicochemical properties of the heterocycles **26** and **27** reveal the perturbation effect of fused benzene rings on the electronic structure, suggesting that the control of the HOMO level can be regarded as a key factor for the stabilization of p-channel organic semiconductors in OFET applications [49].

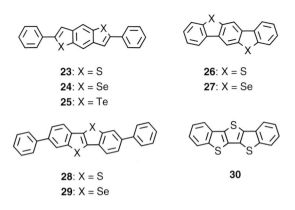

23: X = S
24: X = Se
25: X = Te

26: X = S
27: X = Se

28: X = S
29: X = Se

30

Figure 20.5 Fused chalcogenophenes.

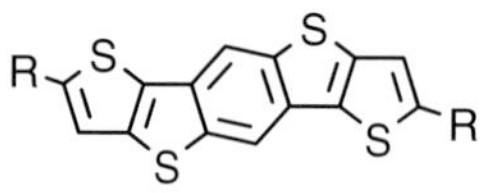

31: R = CH$_3$
32: R = C$_6$H$_{13}$
33: R = C$_9$H$_{19}$
34: R = C$_8$H$_{17}$

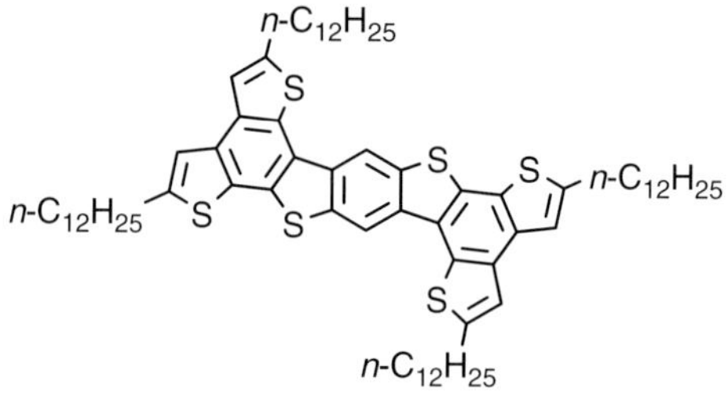

Figure 20.6 Thiophene-containing analog of pentacene (DTBDT) and condensed benzothiophene.

Compounds **28** and **29** have proven the utility of fused chalcogenophene compounds as high-performance OFET materials [50, 51]. It should be noted that OFETs based on the selenophene analog **29** showed high stability both in the shelf lifetime test and in the operation lifetime test in air, indicating that **29** is a valuable OFET material possessing both high performance and stability [50].

Dibenzo[*d,d′*]thieno[3,2-*b*;4,5-*b′*]dithiophene (**30**, Figure 20.5) is an analog of pentacene that possesses a similar rigid, linear, and coplanar conjugated structure [52]. Its crystal structure shows a highly ordered herringbone packing structure. The field-effect mobility was measured to be around 0.51 cm^2 V^{-1} s^{-1} and the on:off ratio of 4.5 × 10^6 is one of the highest mobility values reported for organic semiconductors with ionization potentials higher than 5.5 eV [52, 53].

Dithienos[2,3-*d*;2′,3′-*d′*]benzo[1,2-*b*;4,5-*b′*]dithiophene(DTBDTs) (**31–34**, Figure 20.6), with four symmetrically fused thiophene-ring units substituted by different alkyl groups, were studied by Müllen's group [54]. Solution-processed OFETs based on the hexyl-substituted derivative (**32**) revealed a high charge-carrier mobility of up to 1.7 cm^2 V^{-1} s^{-1} and an on:off ratio of 10^7 for long-range ordered thin films. This result indicates that small heteroacene molecules possessing an extended aromatic core and solubilizing alkyl chains are extremely promising candidates for solution-processed organic electronic devices.

A novel planar condensed benzothiophene was synthesized by Pei and co-workers [55]. This molecule (**35**) with alternate benzene and thiophene units combines the elements of oligoacenes and peripherally rich sulfur atoms, which favors π−π stacking to form 1D nano- or microstructures. The individual microwire transistors with carrier mobilities as high as 2.1 cm^2 V^{-1} s^{-1} were easily fabricated utilizing this molecule by a direct solution process [56].

Wang and co-workers reported novel high-performance organic semiconductors based on an annelated β-trithiophene (Figure 20.7, **36**), which possesses an extraordinary compressed packing mode combining edge-to-face π−π interactions and S−S interactions in the solid state and showing good device performance [57]. Incorporation of two hexyl, phenyl, and thiophenyl substituents, one at each end of molecule **36** (**37–39**), is crucial to investigate the effect of different end groups on the packing motif and thus the device

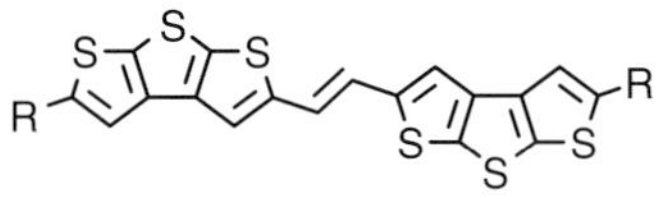

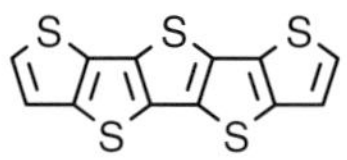

36: R = H
37: R = *n*-Hexyl
38: R = Phenyl
39: R = Thiophenyl

40

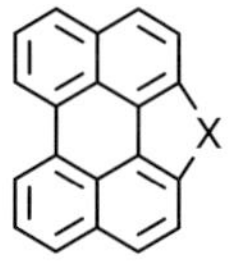

41: X = S
42: X = Se

43

Figure 20.7 Derivatives of β-trithiophenes (**36**), pentathienoacene (**40**), sulfur- and selenium-containing perylenes (**41** and **42**), and dithioperylene (**43**).

performance. The phenyl-substituted derivative, **38**, demonstrates a remarkably high thin-film OFET performance, with mobility up to 2.0 cm^2 V^{-1} s^{-1} and an on:off ratio of up to 10^8. In addition, the devices show good environmental stability, even after storage in air for several months [58].

The linearly condensed fused-ring pentathienoacene **40** was synthesized by Liu and co-workers [59]. In contrast to pentacene, **40** has a larger bandgap and therefore is expected to exhibit good stability in device applications. A moderate mobility of 0.045 cm^2 V^{-1} s^{-1} was achieved based on thin-film transistors owing to the significant π-face separation between molecules in the solid state.

Wang and co-workers presented a crystal engineering strategy to control the organization of heteroatom-annulated perylenes in the solid state that enforces cofacial π-stacking with channels for hole transport directed by chalcogen–chalcogen interactions [60]. Double-channel 1D charge transport superstructures have been achieved involving sulfur- or selenium-containing perylenes (Figure 20.7, **41** and **42**), which facilitates charge transport [61, 62]. The integration of two sulfur atoms instead of one into the perylene skeleton to afford dithioperylene (**43**) induces a compressed, highly ordered packing mode directed by marked S–S contacts of 3.45 Å. The mobilities were up to 2.13 cm^2 V^{-1} s^{-1} for a dithioperylene 1D single-crystalline nanoribbon by direct solution processing make it particularly attractive for electronic applications [63].

As extended conjugated molecules, well-defined nano-graphene polycyclic aromatic hydrocarbons (PAHs) have been intensively studied. Hexa-*peri*-hexabenzocoronenes (HBCs) and their derivatives are typical examples. Self-assembly of certain HBCs leads to functionalized nanotubes which exhibit potential applications in a number of areas [64, 65].

Discotic liquid crystals (LCs) are a promising class of materials for molecular electronics thanks to their self-organization and charge-transporting properties.

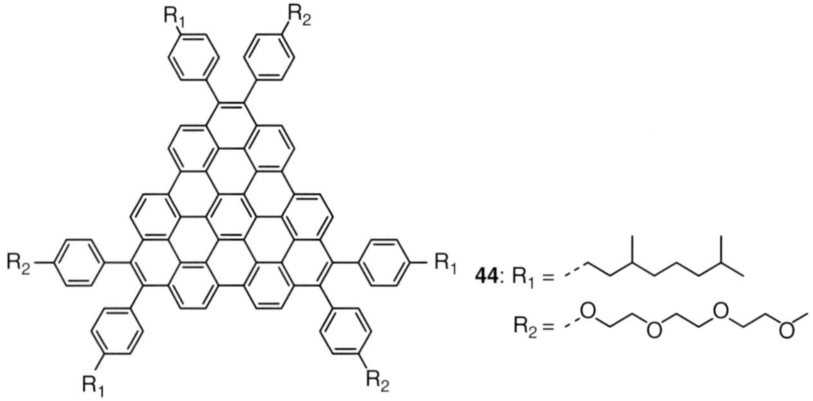

Figure 20.8 A carbon-rich compound with C_3 symmetry and alternating hydrophilic–hydrophobic side chains (**44**).

Müllen and co-workers reported molecule **44** with C_3 symmetry and alternating hydrophilic–hydrophobic side chains, possessing the desired helical microstructure, which was found to exhibit a high mobility of $0.2 \text{ cm}^2 \text{ V}^{-1} \text{ s}^{-1}$ by time-resolved microwave conductivity measurements [66] (Figure 20.8). They also outlined the criteria for achieving high charge-carrier mobilities by rational design of PAHs, paving the way for the broad application of discotic semiconductor materials in the future. It was also reported that alkyl-substituted HBCs were very promising semiconducting materials for OFETs and photovoltaic devices [67].

20.3
Conjugated Molecules for n-Type Organic Semiconductors

Above we discussed recent progress in the design and synthesis of conjugated molecules for p-type organic semiconductors with high hole mobility. In fact, n-type organic semiconductors are highly desirable for applications such as p–n junction diodes, complementary logic bipolar transistors, and OPV devices. Although numerous studies have indicated that typical p-type semiconductors such as pentacene and rubrene can also exhibit respectable electron mobilities, the corresponding device fabrication and measurement need to be performed very carefully; active electrodes such as Ca are usually used, and proper steps to prevent exposure to moisture and oxygen have to be taken. Consequently, such n-type semiconductor devices are not stable under ambient conditions.

Electron-deficient conjugated molecules have relatively low LUMO levels and relatively high electron affinities; as a result, electron injection from metal electrodes will become more efficient, and the corresponding anions will become more stable, thus improving the environmental stability of n-type semiconductors. One attractive way to obtain

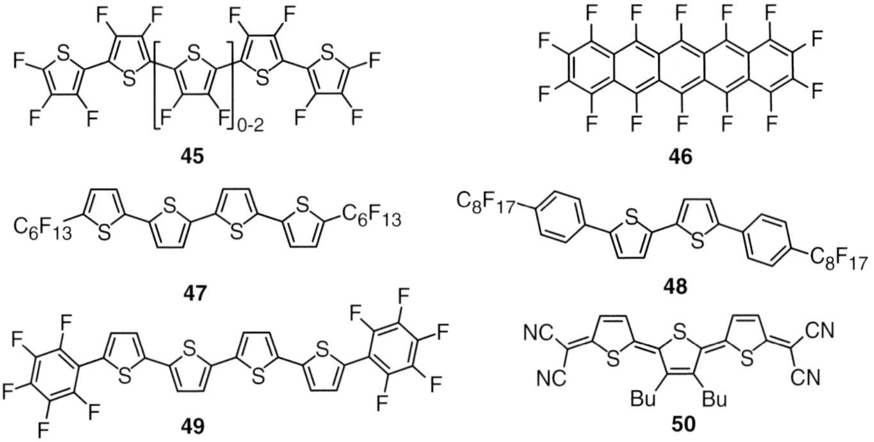

Figure 20.9 Electron-deficient conjugated molecules with strong electron-withdrawing groups such as fluorine or cyano.

electron-deficient conjugated molecules is to attach strong electron-withdrawing groups such as fluorine and cyano to π-conjugated molecules. Sakamoto and co-workers directly fluorinated the conjugated core of oligothiophenes, pentacene, and tetracene (**45** and **46**, Figure 20.9) [68–72]. As expected, the redox potentials of perfluorinated derivatives are shifted positively relative to the nonfluorinated conjugated molecules. Perfluoropentacene (**46**) is an excellent example of this strategy; the crystal packing of the material is not significantly changed compared with that of pentacene, while the electron-deficient nature of the chromophore allows untreated SiO_2 to be used as a dielectric for evaporated films of this material. OFETs with perfluoropentacene (**46**) gave high electron mobilities of $0.11–0.22$ cm^2 V^{-1} s^{-1} with on:off ratios of 10^5 [70–72]. However, it is noteworthy that while the materials have been stabilized to the point that fabrication on untreated SiO_2 is possible, perfluoropentacene is not sufficiently stable for the OFET device to be operated in air.

The addition of perfluoroalkyl chains can also switch the semiconductors from p-type to n-type. Facchetti *et al.* were the first to show that n-type semiconductors can be generated in oligothiophenes through functionalization with perfluoroalkyl chains [73]. The quarter-thiophene **47** with two perfluoroalkyl groups was found to show high electron mobility (up to 0.2 cm^2 V^{-1} s^{-1}) with an on:off ratio of 10^6 [74]. Facchetti *et al.* also investigated a series of perfluoroalkylated thiophene–phenylene oligomers (**48** and **49**, Figure 20.9). Compound **48** was the best-performing semiconductor in the series. The corresponding OFET with **48** as the active layer shows an electron mobility of 0.074 cm^2 V^{-1} s^{-1} with an on:off ratio of 10^6 [75]. A terthiophene-based compound **50** capped by two dicyanomethylene groups at each end was prepared as a potential n-type semiconductor. Optimized device measurements gave electron mobilities as high as 0.2 cm^2 V^{-1} s^{-1} with an on:off ratio of 10^5 under vacuum conditions [76].

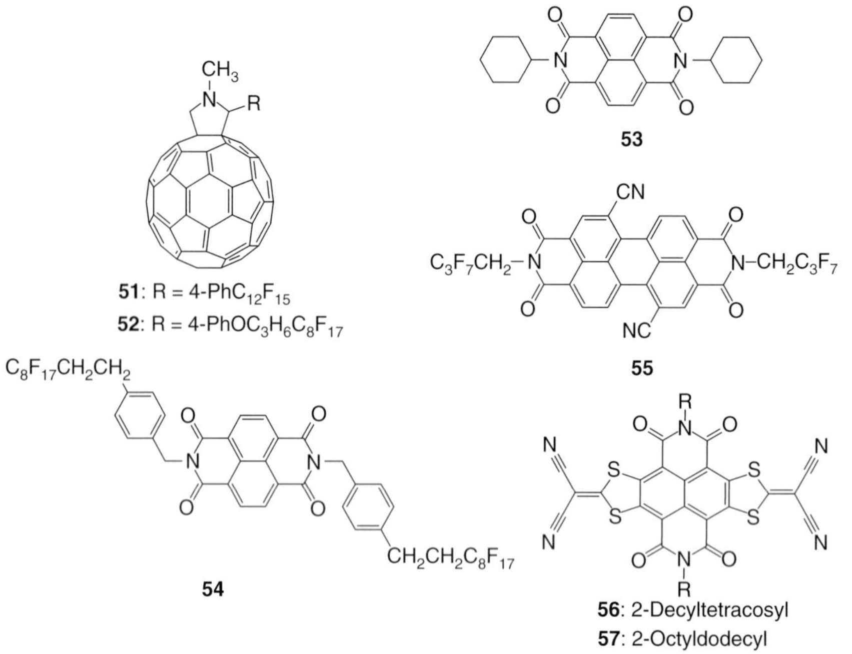

51: R = 4-PhC$_{12}$F$_{15}$
52: R = 4-PhOC$_3$H$_6$C$_8$F$_{17}$

53

55

54

56: 2-Decyltetracosyl
57: 2-Octyldodecyl

Figure 20.10 Fullerenes, NDI, and PDI molecules for n-type semiconductors.

By considering the fact that C$_{60}$ and its analogs and derivatives can be reduced at accessible potentials, they are potential n-type semiconductors. In fact, fullerenes, in particular C$_{60}$ derivatives, are widely employed in organic solar cell (OSC) devices as electron acceptors. However, the OFET devices with C$_{60}$ and its derivatives are typically less stable in the environment. Later, two groups reported that the attachment of perfluoroalkyl chains (**51** and **52**, Figure 20.10) further enhanced both electron mobility and stability under ambient operating conditions. The unencapsulated transistors made from **51** and **52** could be operated in air for several days following an initial decrease in performance. The high packing density of the perfluoroalkyl chains was thought to inhibit the diffusion of oxygen and water within the film and improve stability [77, 78].

Rylene diimides such as naphthalene diimide (NDI) and perylene diimide (PDI) exhibit relatively high electron affinities, high electron mobilities, and excellent chemical, thermal, and photochemical stabilities. These materials are electron deficient due to the substitution of an aromatic core with two sets of electron-accepting imides groups that are mutually conjugated. Among the first n-type semiconductors reported were NDI and PDI. Among NDI derivatives, the highest mobility in inert atmosphere was found to be 6.2 cm^2 V^{-1} s^{-1} for **53** (Figure 20.10) [79], and the highest mobility under ambient conditions was found to be 0.57 cm^2 V^{-1} s^{-1} for **54** [80]. OFETs based on **55** as one PDI derivative exhibited electron mobilities of 6.0 and 3.0 cm^2 V^{-1} s^{-1} in vacuum and in air, respectively, which are the highest reported for PDI derivatives [81].

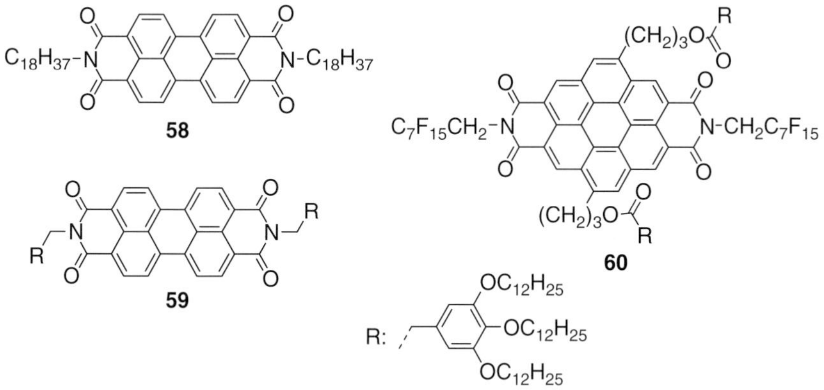

Figure 20.11 Substituted PDI molecules for n-type semiconductors.

Ambient stability in rylene diimide devices has been reported through the introduction of fluorocarbon substituents at the N,N-positions and/or introduction of electron-withdrawing core substituents such as CN and F which lower the LUMO energy level with respect to vacuum. For instance, Jones et al. found that the LUMO energy was effectively lowered and the solubility was increased by core substitution of PDI with cyano groups [82]. Gao et al. recently described core-expanded NDI derivatives **56** and **57** fused with 2-(1,3-dithiol-2-ylidene)malonitrile groups. Compounds **56** and **57** show good solubility in common solvents and the corresponding solution-processed OFETs, operating under ambient conditions, exhibit high electron mobilities of up to 0.51 cm^2 V^{-1} s^{-1} [83].

Modifications of the groups linked to the N-positions of NDI and PDI derivatives have also been carried out for the development of n-type semiconductors with high electron mobilities. For instance, groups that can lead to the formation of discotic LCs are linked to PDIs such as compounds **58** and **59** (Figure 20.11). Discotic LCs can self-assemble into highly ordered one-dimensional columnar stacks; intermolecular π-π orbital overlap within the stacks can lead to increased mobility compared with that in amorphous materials. Mobilities of up to 0.1 cm^2 V^{-1} s^{-1} have been measured by pulse radiolysis time-resolved microwave conductivity (PR-TRMC) techniques for **58** [84]. An et al. found that **59** displays a space charge-limited current (SCLC) mobility as high as 1.3 cm^2 V^{-1} s^{-1} under ambient conditions [85], and even higher values (up to 6.7 cm^2 V^{-1} s^{-1} for **60**) were found for closely related coronene-2,3:8,9-tetracarboxylic diimides [86].

20.4
Conjugated Molecules for Photovoltaic Materials

Recently, OSCs have attracted much attention owing to their unique advantages such as low cost, flexibility, light weight, and large-area device fabrication [87]. Encouraging power conversion efficiencies power conversion efficiency (PCEs) as high as 7–8% have been achieved from blends of polymer donors and fullerene acceptors [88]. Small

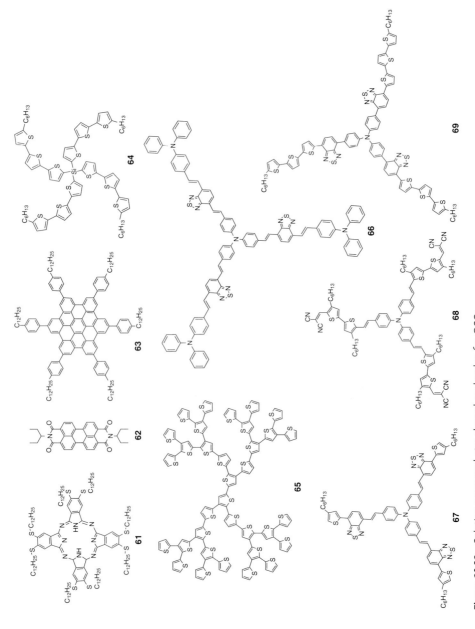

Figure 20.12 Solution-processed star-shaped molecules for OSCs.

conjugated molecules offer potential advantages over conjugated polymer counterparts in terms of defined molecular structure, definite molecular weight, high purity, easy purification, easy mass-scale production, and good batch-to-batch reproducibility [89]. Now the highest PCE of solution-processed bulk heterojunction (BHJ) solar cells based on small conjugated molecules is up to 4.4% [90].

The pioneering work on conjugated molecule OSCs is based on solution-processed phthalocyanine or HBC–PDI blends [91, 92]. Devices based on a blend of phthalocyanine (**61**, Figure 20.12) and PDI (**62**) showed very poor performance [short-circuit current density (J_{SC}) = 0.75 μA cm^{-2}, open-circuit voltage (V_{OC}) = 0.11 V] relative to known polymer BHJ systems at that time [91]. Devices based on the donor **63** and acceptor **62** also did not demonstrate very impressive performance (J_{SC} = 33.5 μA cm^{-2}, V_{OC} = 0.69 V) [92]. Hence the original small-molecule solution-processed BHJs did not show much promise as solar cell materials.

Roncali *et al.* reported 2D and 3D conjugated networks of star-shaped oligothiophenes used in solution-processed OSCs [93]. The first solar cell device with a blend of **64** and phenyl-C$_{61}$-butyric acid methyl ester (PC$_{61}$BM) could not compare with polymer solar cells at that time [93]. The most effective results with star-shaped oligothiophenes were reported by Bauerle and co-workers in 2008 using an oligothiophene dendrimer (**65**, Figure 20.12) blended with PC$_{61}$BM; optimized devices showed a J_{SC} of 4.19 mA cm^{-2}, a V_{OC} of 0.97 V, a fill factor (FF) of 0.42, and a PCE of 1.7% [94].

In order to extend the absorptions to the visible region and hence improve the efficiency of OSCs, new star-shaped conjugated molecules with electron-donating and -accepting units have been designed and investigated. Compound **66** is one of these star-shaped molecules containing triphenylamine (TPA) and benzothiadiazole as the electron-donating and -accepting units, respectively [95]. Replacement of the end group of TPA in **66** with 4-hexylthiophene led to **67** and the PCE of the solar cell fabricated with **67** and [6,6]-phenyl-C$_{71}$-butyric acid methyl ester (PC$_{71}$BM) reached 2.39% [96]. Li and co-workers further modified the chemical structure by choosing dicyanovinyl as the end group and 4,4′-dihexyl-2,2′-bithiophenevinylene as π-bridge leading to **68**; the corresponding solar cell based on **68** and PC$_{71}$BM showed an efficiency of 3.0% [97]. Zhan and co-workers reported a new 3D, star-shaped, D–A–D organic small molecule with TPA as core and donor (D) unit, benzothiadiazole as bridge and acceptor (A) unit, and oligothiophene as arm and donor unit (**69**, Figure 20.12). Compound **69** exhibits excellent thermal stability, broad and strong absorption, relatively high mobility, and a relatively low HOMO level. Solution-processed OSCs based on a blend of **69** and PC$_{71}$BM afforded a PCE as high as 4.3% without any post-treatments such as thermal annealing, solvent annealing, or additive addition [98].

Linear conjugated electron donor–acceptor molecules are also promising for solar cells. Zhan and co-workers reported a conjugated molecule (**70**) containing TPA, thiophene, and benzothiadiazole units (Figure 20.13); a blend of **70** and PC$_{71}$BM yielded a solar cell with a PCE of 2.86% [99]. Chen and co-workers prepared a heptathiophene with dicyanovinyl end groups (**71**), which exhibited broad absorption up to 750 nm and a relatively high hole mobility of 1.5 × 10^{-4} cm^2 V^{-1} s^{-1}) [100]. An optimized solar cell based on the blend of **71** and PC$_{61}$BM exhibited a PCE of 3.7% [101]. Nguyen and co-workers reported a soluble 3,6-diaryl-2,5-dihydropyrrolo[3,4-*c*]pyrrole-1,4-dione

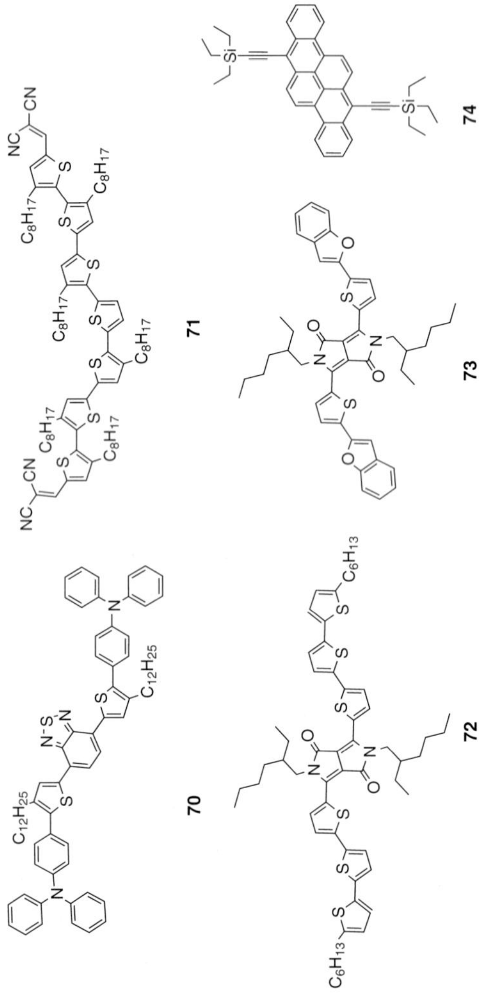

Figure 20.13 Linear conjugated electron donor–acceptor molecules.

(DPP)-based oligothiophene (**72**); a solar cell based on a blend of **72** and PC$_{71}$BM showed a relatively high PCE of 3.0% [102]. The solar cell performance of DPP-based materials was further improved by replacing terminal bithiophene with benzofuran (**73**) [90].

Polycyclic arenes and acenes such as pentacene have attracted considerable attention as electron donors in OSCs due to broad absorptions and high hole carrier mobilities [103]. Winzenberg *et al.* reported a spin-cast thin-film solar cell with dibenzochrysene (**74**, Figure 20.13) and PC$_{61}$BM, showing a PCE of 2.3% [104].

In the past several years, many small molecular materials have been reported. Most of the design principles that apply to polymer solar cells in terms of energy levels and device properties also are suitable for small-molecule BHJ OSCs. The energy levels, absorption characteristics, and charge-carrier mobilities can be controlled well through the design of small molecules. For most of the reported small-molecule BHJ OSCs, the V_{OC} tends to be higher than that observed in analogous polymer devices, whereas the FF and J_{SC} are generally lower. Hence the major challenge in small-molecule BHJs may be improving the J_{SC} and FF. Also, the ability to fine-tune the phase separation and self-assembly of small molecules via synthetic design may allow them to compete with polymers as the preferred type of organic semiconductors for high-efficiency solar cells.

20.5
Conclusion and Outlook

Organic electronics has attracted increasing attention as a newly emerging field of science and technology that covers chemistry, physics, and materials science. The design and synthesis of new materials, characterization of their structures and properties, and fabrication of devices have been the major subjects of recent extensive studies. The development of novel conjugated molecules is continuing at a rapid pace, driven by the rapid growth of interest in the fields of OPV, OFETs, organic light-emitting diodes (OLEDs), and so on. Significant progress has been made, but further improvements are required in terms of design and synthesis, purification, toxicity, and scale-up of the conjugated molecules, in addition to device performance, operational stability, and ease of fabrication. Organic chemistry offers an opportunity to control these aspects at the molecular level, but the use of molecular design guidelines is necessary to tailor function further owing to the vast parameter space available. Based on the studies surveyed in this chapter, we suggest that areas that may deserve further attention include the following:

1) further development of air-stable, high-mobility, n-type semiconductors, in particular solution-processable molecules
2) in addition to 1D (linear) and 2D (macrocyclic or star-shaped) molecular structures, possible strategies for increasing the dimensions towards 3D small molecular architectures
3) conjugated molecules containing heteroatoms other than sulfur, nitrogen, boron, and selenium, such as silicon and even metal atoms.

References

1. Yamada, J. and Sugimoto, T. (eds.) (2004) *TTF Chemistry: Fundamentals and Applications of Tetrathiafulvalene*, Springer-Verlag, Heidelberg.
2. Shibaeva, R.P. and Yagubskii, E.B. (2004) *Chem. Rev.*, **104**, 5347.
3. Bryce, M.R. (1999) *Adv. Mater.*, **11**, 11.
4. Canevet, D., Sallé, M., Zhang, G., Zhang, D., and Zhu, D. (2009) *Chem. Commun.*, 2245.
5. Wu, H., Zhang, D., Su, L., Ohkubo, K., Zhang, C., Yin, S., Mao, L., Shuai, Z., Fukuzumi, S., and Zhu, D. (2007) *J. Am. Chem. Soc.*, **129**, 6839.
6. Olson, M.A., Botros, Y.Y., and Stoddart, J.F. (2010) *Pure Appl. Chem.*, **82**, 1569.
7. Fang, L., Olson, M.A., Benítez, D., Tkatchouk, E., Goddard, W.A. III, and Stoddart, J.F. (2010) *Chem. Soc. Rev.*, **39**, 17.
8. Badjić, J.D., Balzani, V., Credi, A., Silvi, S., and Stoddart, J.F. (2004) *Science*, **303**, 1845.
9. Liu, Y., Flood, A.H., Bonvallet, P.A., Vignon, S.A., Northrop, B.H., Tseng, H.-R., Jeppesen, J.O., Huang, T.J., Brough, B., Baller, M., Magonov, S., Solares, S.D., Goddard, W.A., Ho, C.-M., and Stoddart, J.F. (2005) *J. Am. Chem. Soc.*, **127**, 9745.
10. Green, J.E., Choi, J.W., Boukai, A., Bunimovich, Y., Johnston-Halperin, E., Delonno, E., Luo, Y., Sheriff, B.A., Xu, K., Shin, Y.S., Tseng, H.R., Stoddart, J.F., and Heath, J.R. (2007) *Nature*, **445**, 414.
11. Nguyen, T.D., Liu, Y., Saha, S., Leung, K.C.F., Stoddart, J.F., and Zink, J.I. (2007) *J. Am. Chem. Soc.*, **129**, 626.
12. Ceroni, P., Credi, A., and Venturi, M. (eds.) (2010) *Electrochemistry of Functional Supramolecular Systems*, John Wiley © Sons, Inc., Hoboken, NJ.
13. Hong, Y., Lam, J.W.Y., and Tang, B.Z. (2009) *Chem. Commun.*, 4332.
14. Wang, M., Zhang, G.X., Zhang, D.Q., Zhu, D.B., and Tang, B.Z. (2010) *J. Mater. Chem.*, **20**, 1858.
15. Yu, G., Yin, S.W., Liu, Y.Q., Chen, J.S., Xu, X.J., Sun, X.B., Ma, D.G., Zhan, X.W., Peng, Q., Shuai, Z.G., Tang, B.Z., Zhu, D.B., Fang, W.H., and Luo, Y. (2005) *J. Am. Chem. Soc.*, **127**, 6335.
16. Luo, J.D., Xie, Z.L., Lam, J.W.Y., Cheng, L., Chen, H.Y., Qiu, C.F., Kwok, H.S., Zhan, X.W., Liu, Y.Q., Zhu, D.B., and Tang, B.Z. (2001) *Chem. Commun.*, 1740.
17. Swager, T.M. (1998) *Acc. Chem. Res.*, **31**, 201.
18. McQuade, D.T., Pullen, A.E., and Swager, T.M. (2000) *Chem. Rev.*, **100**, 2537.
19. Irie, M. (2000) *Chem. Rev.*, **100**, 1685.
20. Myles, A.J. and Branda, N.R. (2002) *Adv. Funct. Mater.*, **12**, 167.
21. Balzani, V., Credi, A., and Venturi, M. (2009) *Chem. Soc. Rev.*, **38**, 1542.
22. Ma, X. and Tian, H. (2010) *Chem. Soc. Rev.*, **39**, 70.
23. Muraoka, T., Kinbara, K., and Aida, T. (2006) *Nature*, **440**, 512.
24. van Delden, R.A., Ter Wiel, M.K.J., Pollard, M.M., Vicario, J., Koumura, N., and Feringa, B.L. (2005) *Nature*, **437**, 1337.
25. Vicario, J., Katsonis, N., Ramon, B.S., Bastiaansen, C.W.M., Broer, D.J., and Feringa, B.L. (2006) *Nature*, **440**, 163.
26. Marder, S.R., Kippelen, B., Jen, A.K.-Y., and Peyghambarian, N. (1997) *Nature*, **388**, 845.
27. He, G.S., Tan, L.S., Zheng, Q., and Prasad, P.N. (2008) *Chem. Rev.*, **108**, 1245.
28. Rumi, M., Barlow, S., Wang, J., Perry, J.W., and Marder, S.R. (2008) *Adv. Polym. Sci.*, **213**, 1.
29. Kawata, S., Sun, H.B., Tanaka, T., and Takada, K. (2001) *Nature*, **412**, 697.
30. Zhou, W.H., Kuebler, S.M., Braun, K.L., Yu, T.Y., Cammack, J.K., Ober, C.K., Perry, J.W., and Marder, S.R. (2002) *Science*, **296**, 1106.
31. He, G.S., Markowicz, P.P., Lin, T.C., and Prasad, P.N. (2002) *Nature*, **415**, 767.
32. Kelley, T.W., Muyres, D.V., Baude, P.F., Smith, T.P., and Jones, T.D. (2003) *Mater. Res. Soc. Symp. Proc.*, **771**, 169.
33. Heringdorf, F.J.M.Z., Reuter, M.C., and Tromp, R.M. (2001) *Nature*, **412**, 517.
34. Anthony, J.E., Eaton, D.L., and Parkin, S.R. (2002) *Org. Lett.*, **4**, 15.
35. Sheraw, C.D., Jackson, T.N., Eaton, D.L., and Anthony, J.E. (2003) *Adv. Mater.*, **15**, 2009.
36. Kim, D.H., Lee, D.Y., Lee, H.S., Lee, W.H., Kim, Y.H., Han, J.I., and Cho, K. (2007) *Adv. Mater.*, **19**, 678.
37. Kobayashi, K., Shimaoka, R., Kawahata, M., Yamanaka, M., and Yamaguchi, K. (2006) *Org. Lett.*, **8**, 2385.

38. Goodings, E.P., Mitchard, D.A., and Owen, G. (1972) *J. Chem. Soc., Perkin Trans. 1*, 1310.

39. Briseno, A.L., Miao, Q., Ling, M.-M., Reese, C., Meng, H., Bao, Z., and Wudl, F. (2006) *J. Am. Chem. Soc.*, **128**, 15576.

40. Briseno, A.L., Mannsfeld, S.C.B., Lu, X., Xiong, Y., Jenekhe, S.A., Bao, Z., and Xia, Y. (2007) *Nano Lett.*, **7**, 668.

41. Laquindanum, J.G., Katz, H.E., and Lovinger, A.J. (1998) *J. Am. Chem. Soc.*, **120**, 664.

42. Payne, M.M., Parkin, S.R., Anthony, J.E., Kuo, C.C., and Jackson, T.N. (2005) *J. Am. Chem. Soc.*, **127**, 4986.

43. Subramanian, S., Park, S.K., Parkin, S.R., Podzorov, V., Jackson, T.N., and Anthony, J.E. (2008) *J. Am. Chem. Soc.*, **130**, 2706.

44. Tang, M.L., Okamoto, T., and Bao, Z. (2006) *J. Am. Chem. Soc.*, **128**, 16002.

45. Tang, M.L., Mannsfeld, S.C.B., Sun, Y., Becerril, H.A., and Bao, Z.N. (2009) *J. Am. Chem. Soc.*, **131**, 882.

46. Tang, M.L., Reichardt, A.D., Miyaki, N., Stoltenberg, R.M., and Bao, Z. (2008) *J. Am. Chem. Soc.*, **130**, 6064.

47. Swartz, C.R., Parkin, S.R., Bullock, J.E., Anthony, J.E., Mayer, A.C., and Malliaras, G.G. (2005) *Org. Lett.*, **7**, 3163.

48. Takimiya, K., Kunugi, Y., Konda, Y., Niihara, N., and Otsubo, T. (2004) *J. Am. Chem. Soc.*, **126**, 5084.

49. Ebata, H., Miyazaki, E., Yamamoto, T., and Takimiya, K. (2007) *Org. Lett.*, **9**, 4499.

50. Takimiya, K., Kunugi, Y., Konda, Y., Ebata, H., Toyoshima, Y., and Otsubo, T. (2006) *J. Am. Chem. Soc.*, **128**, 3044.

51. Takimiya, K., Ebata, H., Sakamoto, K., Izawa, T., Otsubo, T., and Kunugi, K. (2006) *J. Am. Chem. Soc.*, **128**, 12604.

52. Gao, J.H., Li, R.J., Li, L.Q., Meng, Q., Jiang, H., Li, H.X., and Hu, W.P. (2007) *Adv. Mater.*, **19**, 3008.

53. Locklin, J., Ling, M.M., Sung, A., Roberts, M.E., and Bao, Z. (2006) *Adv. Mater.*, **18**, 2989.

54. Gao, P., Beckmann, D., Tsao, H.N., Feng, X., Enkelmann, V., Baumgarten, M., Pisula, W., and Müllen, K. (2009) *Adv. Mater.*, **21**, 213.

55. Zhou, Y., Liu, W., Ma, Y., Wang, H., Qi, L., Cao, Y., Wang, J., and Pei, J. (2007) *J. Am. Chem. Soc.*, **129**, 12386.

56. Zhou, Y., Lei, T., Wang, L., Pei, J., Cao, Y., and Wang, J. (2010) *Adv. Mater.*, **22**, 1484.

57. Tan, L., Zhang, L., Jiang, X., Yang, X.D., Wang, L.J., Wang, Z., Li, L., Hu, W., Shuai, Z., Li, L., and Zhu, D. (2009) *Adv. Funct. Mater.*, **19**, 272.

58. Zhang, L., Tan, L., Wang, Z., Hu, W., and Zhu, D. (2009) *Chem. Mater.*, **21**, 1993.

59. Xiao, K., Liu, Y., Qi, T., Zhang, W., Wang, F., Gao, J., Qiu, W., Ma, Y., Cui, G., Chen, S., Zhan, X., Yu, G., Qin, J., Hu, W., and Zhu, D. (2005) *J. Am. Chem. Soc.*, **127**, 13281.

60. Jiang, W., Qian, H., Li, Y., and Wang, Z. (2008) *J. Org. Chem.*, **73**, 7369.

61. Sun, Y., Tan, L., Jiang, S., Qian, H., Wang, Z., Yan, D., Di, C., Wang, Y., Wu, W., Yu, G., Yan, S., Wang, C., Hu, W., Liu, Y., and Zhu, D. (2007) *J. Am. Chem. Soc.*, **129**, 1882.

62. Tan, L., Jiang, W., Jiang, L., Jiang, S., Wang, Z., Yan, S., and Hu, W. (2009) *Appl. Phys. Lett.*, **94**, 153306.

63. Jiang, W., Zhou, Y., Geng, H., Jiang, S., Yan, S., Hu, W., Wang, Z., Shuai, Z., and Pei, J. (2011) *J. Am. Chem. Soc.*, **133**, 1.

64. Hill, J.P., Jin, W., Kosaka, A., Fukushima, T., Ichihara, H., Shimomura, T., Ito, K., Hashizume, T., Ishii, N., and Aida, T. (2004) *Science*, **304**, 1481.

65. Jin, J., Yamamoto, Y., Fukushima, T., Ishii, N., Kim, J., Kato, K., Takata, M., and Aida, T. (2008) *J. Am. Chem. Soc.*, **130**, 9434.

66. Feng, X., Marcon, V., Pisula, W., Hansen, M.R., Kirkpatrick, J., Grozema, F., Andrienco, D., Kremer, K., and Müllen, K. (2009) *Nat. Mater.*, **8**, 420.

67. Wu, J., Pisula, W., and Müllen, K. (2007) *Chem. Rev.*, **107**, 718.

68. Sakamoto, Y., Komatsu, S., and Suzuki, T. (2001) *J. Am. Chem. Soc.*, **123**, 4643.

69. Osuna, R.M., Ortiz, R.P., Delgado, M.C.R., Sakamoto, Y., Suzuki, T., Hernandez, V., and Navarrete, J.T.L. (2005) *J. Phys. Chem. B*, **109**, 20737.

70. Inoue, Y., Sakamoto, Y., Suzuki, T., Kobayashi, M., Gao, Y., and Tokito, S. (2005) *Jpn. J. Appl. Phys., Part 1*, **44**, 3663.

71. Sakamoto, Y., Suzuki, T., Kobayashi, M., Gao, Y., Fukai, Y., Inoue, Y., Sato, F., and Tokito, S. (2004) *J. Am. Chem. Soc.*, **126**, 8138.

72. Sakamoto, Y., Suzuki, T., Kobayashi, M., Gao, Y., Inoue, Y., and Tokito, S. (2006) *Mol. Cryst. Liq. Cryst.*, **444**, 225.

73. Facchetti, A., Deng, Y., Wang, A., Koide, Y., Sirringhaus, H., Marks, T.J., and Friend, R.H. (2000) *Angew. Chem.*, **39**, 4547.

74. Facchetti, A., Mushrush, M., Yoon, M.H., Hutchison, G.R., Ratner, M.A., and Marks, T.J. (2004) *J. Am. Chem. Soc.*, **126**, 13859.

75. Facchetti, A., Letizia, J., Yoon, M.H., Mushrush, M., Katz, H.E., and Marks, T.J. (2004) *Chem. Mater.*, **16**, 4715.

76. Chesterfield, R.J., Newman, C.R., Pappenfus, T.M., Ewbank, P.C., Haukaas, M.H., Mann, K.R., Miller, L.L., and Frisbie, C.D. (2003) *Adv. Mater.*, **15**, 1278.

77. Chikamatsu, M., Itakura, A., Yoshida, Y., Azumi, R., and Yase, K. (2008) *Chem. Mater.*, **20**, 7365.

78. Wobkenberg, P.H., Ball, J., Bradley, D.D.C., Anthopoulos, T.D., Kooistra, F., Hummelen, J.C., and de Leeuw, D.M. (2008) *Appl. Phys. Lett.*, **92**, 143310.

79. Shukla, D., Nelson, S.F., Freeman, D.C., Rajeswaran, M., Ahearn, W.G., Meyer, D.M., and Carey, J.T. (2008) *Chem. Mater.*, **20**, 7486.

80. See, K.C., Landis, C., Sarjeant, A., and Katz, H.E. (2008) *Chem. Mater.*, **20**, 3609.

81. Molinari, A.S., Alves, H., Chen, Z., Facchetti, A., and Morpurgo, A.F. (2009) *J. Am. Chem. Soc.*, **131**, 2462.

82. Jones, B.A., Ahrens, M.J., Yoon, M.H., Facchetti, A., Marks, T.J., and Wasielewski, M.R. (2004) *Angew. Chem. Int. Ed.*, **43**, 6363.

83. Gao, X., Di, C., Liu, Y., Yang, X., Fan, H., Zhang, F., Liu, Y., Liu, H., and Zhu, D. (2010) *J. Am. Chem. Soc.*, **132**, 3697.

84. Struijk, C.W., Sieval, A.B., Dakhorst, J.E.J., van Dijk, M., Kimkes, P., Koehorst, R.B.M., Donker, H., Schaafsma, T.J., Picken, S.J., van de Craats, A.M., Warman, J.M., Zuilhof, H., and Sudhölter, E.J.R. (2000) *J. Am. Chem. Soc.*, **122**, 11057.

85. An, Z., Yu, J., Jones, S.C., Barlow, S., Yoo, S., Domercq, B., Prins, P., Siebbeles, L.D.A., Kippelen, B., and Marder, S.R. (2005) *Adv. Mater.*, **17**, 2580.

86. An, Z., Yu, J., Domercq, B., Jones, S.C., Barlow, S., Kippelen, B., and Marder, S.R. (2009) *J. Mater. Chem.*, **19**, 6688.

87. Zhan, X.W. and Zhu, D.B. (2010) *Polym. Chem.*, **1**, 409.

88. Cheng, Y.J., Yang, S.H., and Hsu, C.S. (2009) *Chem. Rev.*, **109**, 5868.

89. Walker, B., Kim, C., and Nguyen, T.-Q. (2008) *Chem. Mater.*, **20**, 32.

90. Walker, B., Tamayo, A.B., Dang, X.-D., Zalar, P., Seo, J.H., Garcia, A., Tantiwiwat, M., and Nguyen, T.-Q. (2009) *Adv. Funct. Mater.*, **19**, 3063.

91. Petritsch, K., Dittmer, J.J., Marseglia, E.A., Friend, R.H., Lux, A., Rozenberg, G.G., Moratti, S.C., and Holmes, A.B. (2000) *Sol. Energy Mater. Sol. Cells*, **61**, 63.

92. Schmidt-Mende, L., Fechtenkotter, A., Mullen, K., Moons, E., Friend, R.H., and MacKenzie, J.D. (2001) *Science*, **293**, 1119.

93. Roncali, J., Frere, P., Blanchard, P., de Bettignies, R., Turbiez, M., Roquet, S., Leriche, P., and Nicolas, Y. (2006) *Thin Solid Films*, **511**, 567.

94. Ma, C.Q., Fonrodona, M., Schikora, M.C., Wienk, M.M., Janssen, R.A.J., and Bauerle, P. (2008) *Adv. Funct. Mater.*, **18**, 3323.

95. He, C., He, Q., Yi, Y., Wu, G., Bai, F., Shuai, Z., and Li, Y. (2008) *J. Mater. Chem.*, **18**, 4085.

96. Zhang, J., Yang, Y., He, C., He, Y., Zhao, G., and Li, Y. (2009) *Macromolecules*, **42**, 7619.

97. Zhang, J., Deng, D., He, C., He, Y., Zhang, M., Zhang, Z., Zhang, Z., and Li, Y. (2011) *Chem. Mater.*, **23**, 817.

98. Shang, H., Fan, H., Liu, Y., Hu, W., Li, Y., and Zhan, X.W. (2011) *Adv. Mater.*, **23**, 1554.

99. Fan, H.J., Shang, H.X., Li, Y.F., and Zhan, X.W. (2010) *Appl. Phys. Lett.*, **97**, 133302.

100. Liu, Y., Wan, X., Yin, B., Zhou, J., Long, G., Yin, S., and Chen, Y. (2010) *J. Mater. Chem.*, **20**, 2464.

101. Yin, B., Yang, L., Liu, Y., Chen, Y., Qi, Q., Zhang, F., and Yin, S. (2010) *Appl. Phys. Lett.*, **97**, 023303.

102. Peet, J., Tamayo, A.B., Dang, X.D., Seo, J.H., and Nguyen, T.Q. (2008) *Appl. Phys. Lett.*, **93**, 163306.

103. Sundar, V.C., Zaumseil, J., Podzorov, V., Menard, E., Willett, R.L., Someya, T., Gershenson, M.E., and Rogers, J.A. (2004) *Science*, **303**, 1644.

104. Winzenberg, K.N., Kemppinen, P., Fanchini, G., Bown, M., Collis, G.E., Forsyth, C.M., Hegedus, K., Singh, T.B., and Watkins, S.E. (2009) *Chem. Mater.*, **21**, 5701.

Commentary Part

Comment 1

Seth R. Marder

Organic semiconductors have emerged as potentially important materials for a variety of applications ranging from OLEDs through transistors to OPV materials. This chapter provides readers with a overview of recent advances in the field of hole and electron transport materials that are central to all the applications referred to above. In addition, π-conjugated molecules have important roles as dyes and pigments, which can be used in many applications as diverse as biological imaging, paints for cars, and NLO applications. A key feature of π-conjugated molecules that renders them useful for all of these application is the electronic structure of π-systems, which can lead to extensive delocalization of electrons throughout the molecule. This in turn can impart various properties, including strong absorption throughout the UV–visible and near-infrared regions of the electromagnetic spectrum. Both the strength and position of the absorption bands can be tuned in a rational manner by appropriate substitution of the molecules with electron donors and electron acceptors. Substitution of the molecules with electron donors tends to destabilize the HOMO, thereby lowering the ionization potential of the molecules and thus facilitating injection of holes from electrodes (specifically the anode). Substitution of the molecules with electron acceptors tends to stabilize the LUMO, thereby increasing the electron affinity of the molecules and thus facilitating injection of electrons from an electrode (specifically the cathode). As a result, molecules with donors are often used as hole-transport materials and molecules with acceptors are often used as electron-transport materials. In addition, molecules with both electron donors and electron acceptors can have low optical gaps and strong absorption that can extend past the visible into the near-infrared region of the electromagnetic spectrum, which is important for applications in OPV systems to harvest photons with corresponding wavelengths as high as about 900 nm. For all the applications discussed, the presence of a π-system and substitution with either or both donors and/or acceptors can impart some of the important properties for the molecules to be useful as semiconductors and light absorbers for

OPV systems. Other very important characteristics involve how molecules are organized in the solid state with respect to one another (or with other molecules) to ensure that the correct electronic coupling between molecules can facilitate electron (or hole) transfer processes. As a result, much work has focused on developing semiconductors that have the correct electronic structure for injection (or collection) of charges, relatively low reorganization energies, and the appropriate intermolecular packing arrangements.

The chapter highlights some of the key systems that have been examined by various groups, in which the molecular engineering approaches have resulted in materials with state-of-the-art properties, and it summarizes some of the strategies employed to achieve this level of performance. As the field of organic semiconductors rapidly advances, it is important for readers of this chapter to try to understand the concepts that have helped to impart the properties and not simply use it as a review article since, by the time it is published, the materials will no longer represent the state-of-the-art. However, it is likely that people will, for an extended period of time, be able to benefit from the strategies described. I hope that readers will find this chapter a helpful introduction to organic semiconductors and some of their applications.

Comment 2

Tien Yau Luh

A chapter on organic π-conjugated systems for functional materials is without any doubt timely since the first volume on this subject was published more than a decade ago [C1]. With limited space, comprehensive coverage of the recent advances in this area is definitely not an easy task. This chapter focuses on both p- and n-type organic semiconducting molecules. In particular, large acenes and heteroaromatic fused acene analogs are discussed in detail. The importance of HBCs and related nanographene PAHs is briefly summarized. Self-assembly of certain functionalized HBCs leading to control of the molecular ordering and their potential applications is worth noting [C2, C3]. The π-π stacking of conjugated systems plays a pivotal role in cooperative aggregation in supramolecular

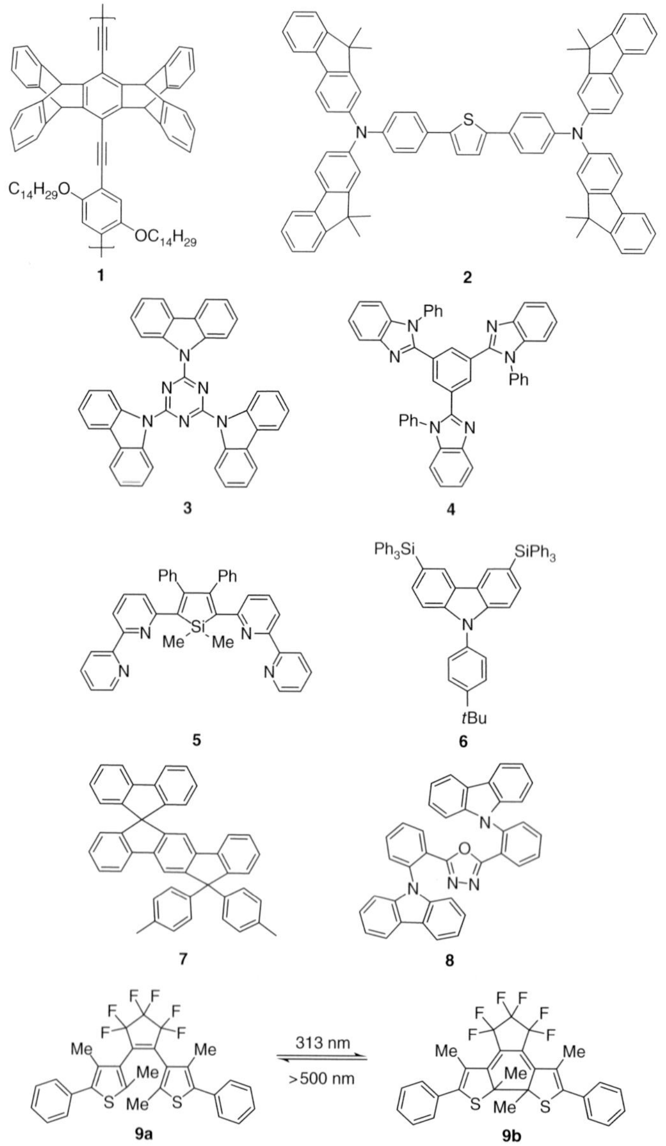

Figure C1 Examples of molecules with (π)-conjugated systems used as chemical sensors created using the so-called "molecular wire approach".

assembly [C4]. Although cyclophane chemistry has been extensively studied, incorporation of these moieties into polymers also appears to be interesting [C5]. DNA molecules can be considered as multiple layers of cyclophanes. Synthetic analogs known as polymeric ladderphanes have

been demonstrated to exhibit significant interactions between adjacent planar aromatic linkers [C6].

Owing to π-π interactions and a regular stacking order, conjugated aromatic compounds play important roles in LCs. For example, self-organized soft materials can be fabricated from functional LC assemblies [C7]. Further, tailor-made synthesis of discotic LCs allows the materials to be applied to plastic electronics [C8]. Conducting discotic LCs have the capacity to act as active electronic components in organic devices [C9]. It is noteworthy that discotic columnar LCs are known have high exciton diffusion lengths and charge carrier mobilities along the columns compared with conventional disordered materials [C10]. In particular, alkyl-substituted HBCs were reported to be very promising semiconducting materials in OFETs and OPV devices [C11]. Recently, formation of a face-on oriented bilayer of two discotic columnar LCs for organic donor–acceptor heterojunctions has been demonstrated [C12].

The use of π-conjugated systems as chemical sensors within the concept of the so-called "molecular wire approach" has practical applications [C13, C14]. This concept takes advantage of collective properties such as the conductivity and fluorescence of the systems that are sensitive to very minor perturbations. One particular example is the TNT sensor (1) [C15]. Upon exposure to TNT vapor at the ppb level, the fluorescence is quenched by more than 50% in 1 min. The Langmuir–Blodgett technique is used to control the conformations or aggregation of conjugated polymer based on the electronic properties of large π-conjugated systems [C16]–[C20] (Figure C1).

The multilayer OLED initiated by Tang and van Slyke [C17] has successfully paved the way towards practical commercialization of this technology. The device efficiency has been dramatically improved because of balanced charge transportation and efficient charge recombination by adopting a complicated device configuration, which is usually composed of different functional materials. Typical hole-transporting materials include triarylamine-based molecules and electron-rich heteroarenes, such as thiophene derivative 2 [C22]. Molecules with electron-deficient heteroarenes, such as 3 and 4 [C23], are representative electron-transporting materials. Owing to significant σ^*-π^* interaction, the silole-containing oligomer 5 is a new type of electron-transporting material [C24]. For the emitting materials, the manipulation of donor–acceptor electronic interactions is adopted for tailoring the emission colors, especially for long-wavelength emissions such as green, orange, and red. However, the stability and efficiency of emitters with long-wavelength fluorescence are still unsatisfactory. First, the

achievable 100% internal quantum efficiency is based on the use of a host–guest strategy with triplet emitters (mainly transition metal complexes as guest) dispersed into organic host materials with high triplet energy, such as carbazole derivatives 6 [C25]. More recently, the pure hydrocarbon π-conjugated molecule 7 [C26] and bipolar molecules 8 [C27], which combine both hole- and electron-transport features, have emerged as new host materials.

Diarylethenes such as 9a and 9b and related photochromic materials are worth mentioning [C28]. These molecules are highly durable; coloration–decoloration cycles can be repeated more than 10^4 times while the photochromic properties remain essentially unchanged.

Authors' Response to the Commentaries

Following the valuable comments of Prof. Marder, we modified the Introduction by emphasizing the characteristics of π-conjugated molecules following his suggestions. It is hoped that the modifications are appropriate.

Prof. Luh is also thanked for his detailed comments. As he mentioned, organic π-conjugated systems aimed at functional materials have attracted significant attention in recent years. However, owing to the limitations on space, we could not discuss the developments in all aspects of organic π-conjugated molecules, but rather we have focused on organic π-conjugated molecules for organic semiconductors and OPV materials. For this reason, we did not discuss the conjugated molecules relevant to OLEDs. However, we emphasized fluorescent sensors (from Swager and others) and photochromic compounds (from Irie and others) according to Prof. Luh's suggestions. In addition, we modified the discussion of HBC studies (from Aida and Müllen). Again, it is hoped that these modifications are appropriate.

References

C1. Müllen, K. and Wegner, G. (eds.) (1998) *Electronic Materials: the OligomerApproach*, Wiley-VCH Verlag GmbH, Weinheim

C2. Hill, J.P., Jin, W., Kosaka, A., Fukushima, T., Ichihara, H., Shimomura, T., Ito, K., Hashizume, T., Ishii, N., and Aida, T. (2004) *Science*, **304**, 1481.

C3. Jin, J., Yamamoto, Y., Fukushima, T., Ishii, N., Kim, J., Kato, K., Takata, M., and Aida, T. (2008) *J. Am. Chem. Soc.*, **130**, 9434.

C4. Palmans, A.R.A. and Meijer, E.W. (2007) *Angew. Chem. Int Ed.*, **46**, 8948.

C5. Morisaki, Y. and Chujo, Y. (2008) *Prog. Polym. Sci.*, **33**, 346.

C6. Chou, C.-M., Lee, S.-L., Chen, C.-H., Biju, A.T., Wang, H.-W., Wu, Y.-L., Zhang, G.-F., Yang, K.-W., Lim, T.-S., Huang, M.-J., Tsai, P.-Y., Lin, K.-C., Huang, S.-L., Chen, C.-C., and Luh, T.-Y. (2009) *J. Am. Chem. Soc.*, **131**, 12579.

C7. Kato, T., Mizoshita, N., and Kishimoto, K. (2005) *Angew. Chem. Int. Ed.*, **45**, 38–68.

C8. Laschat, S., Baro, A., Steinke, N., Gies-selmann, F., Hagele, C., Scalia, G., Judele, R., Kapatsina, E., Sauer, S., Schreivogel, A., and Tosoni, M. (2007) *Angew. Chem. Int. Ed.*, **46**, 4832.

C9. Novoselov, K.S., Geim, A.K., Moro-zov, S.V., Jiang, D., Katsnelson, M.I., Grigorieva, I., Dubonos, S.V., and Firsov, A. (2005) *Nature*, **438**, 197.

C10. Sergeyev, S., Pisula, W., and Geerts, Y.H. (2007) *Chem. Soc. Rev.*, **36**, 1902.

C11. Wu, J., Pisula, W., and Müllen, K. (2007) *Chem. Rev.*, **107**, 718.

C12. Thiebaut, O., Bock, H., and Grelet, E. (2010) *J. Am. Chem. Soc.*, **132**, 6886.

C13. Swager, T.M. (1998) *Acc. Chem. Res.*, **31**, 201.

C14. McQuade, D.T., Pullen, A.E., and Swager, T.M. (2000) *Chem. Rev.*, **100**, 2537.

C15. Yang, J.-S. and Swager, T.M. (1998) *J. Am. Chem. Soc.*, **120**, 11864.

C16. Kim, J. and Swager, T.M. (2001) *Nature*, **411**, 1030.

C17. Nesterov, E.E., Zu, Z., and Swager, T.M. (2005) *J. Am. Chem. Soc.*, **127**, 10083.

C18. Andrew, T.L. and Swager, T.M. (2011) *J. Polym. Sci., Part B: Polym. Phys.*, **49**, 476.

C19. Zahn, S. and Swager, T.M. (2002) *Angew. Chem. Int. Ed.*, **41**, 4226.

C20. Zhu, Z. and Swager, T.M. (2002) *J. Am. Chem. Soc.*, **124**, 9670.

C21. Tang, C.W. and van Slyke, S.A. (1987) *Appl. Phys. Lett.*, **51**, 913.

C22. Okumoto, K., Ohara, T., Noda, T., and Shirota, Y. (2001) *Synth. Met.*, **121**, 1655.

C23. Inomata, H., Goushi, K., Masuko, T., Konno, T., Imai, T., Sasabe, H., Brown, J.J., and Adachi, C. (2004) *Chem. Mater.*, **16**, 1285.

C24. Tamao, K., Uchida, M., Izumizawa, T., Furakawa, K., and Yamaguchi, S. (1996) *J. Am. Chem. Soc.*, **118**, 11974.

C25. Tasi, M.-H., Lin, H.-W., Su, H.-C., Wu, C.-C., Fang, F.-C., Liao, Y.-L., Wong, K.-T., and Wu, C.-I. (2006) *Adv. Mater.*, **18**, 1216.

C26. Chi, L.-C., Hung, W.-Y., Chiu, H.-C., and Wong, K.-T. (2009) *Chem. Commun.*, 3892.

C27. Tao, Y., Wang, Q., Yang, C., Wang, Q., Zhang, Z., Zou, T., Qin, J., and Ma, D. (2008) *Angew. Chem. Int. Ed.*, **47**, 8104.

C28. (a) Feringa., B. L. (ed.) (2001) *Molecular Swiches*, Wiley-VCH Verlag GmbH, Weinheim; (b) Irie, M. (2000) *Chem. Rev.*, **100**, 1685.

21
The Future of Organic Chemistry – an Essay
Ronald Breslow

21.1
Introduction

In this concluding chapter, I will predict where organic chemistry will be going in the future, both by extrapolating from the interesting and pioneering work that is already in the field and also from the unmet needs that will have to be satisfied by future discoveries in organic chemistry. This is an essay and not a review. I will from time to time mention people whose work illustrates some of the points, but I will not include references. Interested readers can find the work of those people in the usual way.

21.2
The Field of Organic Chemistry Will Broaden

21.2.1
Synthesis

At its earliest foundations, organic chemistry was distinguished from inorganic chemistry in the sense that organic chemistry was related to life processes. That distinction is no longer relevant. One can describe the difference between the fields by saying that organic chemistry is largely concerned with the elements at the top of the periodic table, with some exceptions for non-transition metals, halogens, sulfur, and phosphorus. So organic chemists use some of the other elements in the periodic table, including transition metals, as elements in catalysis, as components of reagents, and as components of final target molecules. Inorganic chemistry "owns" the rest of the periodic table, focusing on transition metals. The boundaries between these two fields are breaking down, as organic chemists increasingly become involved with matters that were previously the province only of inorganic chemists, while inorganic chemists increasingly know enough organic chemistry to be able to make components of molecules that contain both an inorganic metal and an organic ligand, for instance.

In the future, we will increasingly teach synthesis and do research across the boundary. One possibility is to collaborate, as happens now. More desirably, we will educate our

Organic Chemistry – Breakthroughs and Perspectives, First Edition. Edited by Kuiling Ding and Li-Xin Dai.
© 2012 Wiley-VCH Verlag GmbH & Co. KGaA. Published 2012 by Wiley-VCH Verlag GmbH & Co. KGaA.

students more broadly so that they can work on both sides of the disappearing boundary between these fields. The whole periodic table will be our playground. The field of *synthetic chemistry* will be called just that. It will still require the knowledge needed for success in organic chemistry, but will also require the knowledge central to inorganic chemistry.

21.2.2
Reactions Mechanisms and Theory

The tools to determine reaction mechanisms are the same for organic and inorganic chemistry, so mechanism courses will increasingly use examples from both areas. For example, in elementary organic chemistry courses we will discuss the hydrolysis of carboxylic esters, but we will also contrast the mechanisms with those for the hydrolysis of phosphate esters, and perhaps arsenate esters. In addition, both areas will in the future use more and more of *chemical theory and computation.*

The goal of mechanistic studies is to describe detailed pathways and the energies along all those pathways for chemical and biochemical reactions. In a sense, we would like to make a movie showing exactly how the reactions occur, and at what speed at all the points along the way. Experimental mechanism studies cannot do this. We can get the energy of a transition state, but what detailed path gets us there from the starting materials? Theory can give a precise description, calculating every point, but is it correct? All practical computational theories involve approximations, such as ignoring the polarizabilities of atoms and groups. The function of experimental reaction mechanism work is to provide tests of salient points to see if the theories get them right, in particular with evidence on the detailed structures of transition states and their energies. Thus organic chemists of the future will know a lot of theory. Pioneers in the area today – organic chemists who have become expert theorists as part of their research programs – include Ken Houk and Paul Schleyer.

A knowledge of reaction mechanisms is critical for chemists inventing new organic or inorganic reactions. They will increasingly use *computation* as a tool as well, perhaps in collaboration. I collaborated with my colleague Richard Friesner recently to show that modern theory can reliably predict which catalysts for Shi oxidation can afford enantiomeric excesses, and are therefore of practical interest. Increasingly, organic chemists will test their new ideas against modern theoretical methods, and select the reactions and reagents that theory supports as being likely to be effective. This will replace a lot of trial and error in trying to devise effective catalysts for selective reactions, for instance.

21.2.3
Physical Chemistry

The study of the modern field of physical chemistry has significant interaction with organic chemistry, indeed there is a field called *physical organic chemistry* that includes photochemistry, nanochemistry, theoretical chemistry, and important instrumental chemistry including nuclear magnetic resonance (NMR), electron spin resonance (ESR), single-molecule spectroscopy, and laser methods. Increasingly, physical chemists are investigating systems that are relevant to organic, inorganic, and biological chemistry. The quantitative results from such studies will help us understand the properties of new molecules and new reactions and also those complex reactions in biology that still require further understanding.

21.2.4
Biology

Organic and inorganic chemists can interact with modern biology in two ways. In one, we chemists develop tools to help solve biological problems. A good example is the creation by chemists of fluorescent probes to examine processes in living cells. Many people with education as organic or inorganic chemists are pioneering this field, including Carolyn Bertozzi and Laura Kiessling. Organic chemist George Trainor pioneered the invention of fluorescent methods for DNA sequencing, an invention that made the process feasible. Jacqueline Barton has used her inorganic expertise to explore electrical conduction through DNA. This major area of organic and inorganic chemistry interacting with biology will continue to grow as we furnish tools to learn how memory works in the brain, for instance, and address other problems. The next discusses medicinal chemistry, obviously a place where chemistry and biology meet.

The second way in which chemists interact with biology is to derive inspiration from Nature's chemistry, a field that we have named *biomimetic chemistry* (some prefer the term "bio-inspired" with the same meaning). As chemists and biologists working in the field of biochemistry have learned more about how enzymes work, for instance, organic and inorganic chemists have invented new chemistry using the principles that enzymes have taught us. Other chemists study artificial membranes to mimic those of the cell. Polymer chemists imitate the special properties that the large molecules of biology have. As we learn more and more about biology, the field of biomimetic chemistry will grow with the invention of new catalysts and new multicomponent structures imitating some aspects of the living cell.

The education of organic and inorganic chemists will include enough modern biology to make it possible to work in these two aspects of the chemistry–biology interface.

21.2.5
Medicinal Chemistry

From the beginning, organic chemists have pursued the branch called medicinal chemistry. In one area, the chemistry of *natural products* involved examining leaves, roots, and so on, that were known to have useful biological properties. Very often these came from folk medicine, and in China there is a very strong history of such materials that were known to be useful as medicinal compounds for various diseases. Organic chemists isolated and purified these substances to obtain the active component from the natural materials, and then these substances were considered to be interesting and important synthetic targets. This activity still goes on, with one recent aspect being the exploration of natural materials from the oceans, where chemical warfare between bacteria and their targets are carried out by the synthesis of antibacterial compounds in soft corals and other ocean components. The toxic compounds produced by some fish can also be good leads for medicinal compounds. The exploration of natural products for medicinal purposes still continues, and is likely to be useful in the future.

Another common way of inventing a medicine is to find the natural substances in living systems that bind to biological receptors, and then to imitate those substances so as to modify the responses of the organisms. A recent approach developed to do this is to find materials that modulate the transcription of genes, so that drugs can be found that will not be poisons but will instead be able to change biological behavior in a useful direction. One example of this class of compounds is the materials called *histone deacetylase inhibitors*, of which the earliest one now in human medicinal use was invented by myself and my collaborator Paul Marks. There is great interest in developing compounds that would modify biological behavior in this way, and one can expect this to continue into the future.

Although the advances made in medicinal chemistry are important and exciting, and have increased life expectancy over the past century by more than 50%, there are still many diseases for which appropriate medicinal compounds to cure them are not yet fully available. One is *cancer* in its various forms, one of the major killers of humans. A lot of work is under way to build on the successes that have been achieved in this field, to produce medicinal treatments so effective that cancer would no longer be a threat to people. This is likely to be a continuing target for a long time to come. Another surprising target is *bacterial infections*. Some of the earliest successes in medicinal chemistry were antibiotics that could cure many bacterial diseases, but new resistant organisms have now developed that are no longer subject to destruction by the existing antibacterials; there is now an emergency effort to find new antibacterials that can deal with these resistant organisms. This war is likely to continue into the future as new drugs are developed to deal with the resistant organisms only to find that new types of resistance appear, and once again it will be necessary to find new drugs to deal with those resistant organisms.

Among the most challenging problems in the medicinal chemistry area is to find good treatments for *viral diseases*. Viruses are not living organisms, and the standard antibacterials are not effective against them. Unless much better ways of dealing with viruses are found by the invention of new antiviral medicines, the danger to humanity is substantial. As the recent Hollywood movie *Contagion* emphasizes, a lethal virus that can be easily transmitted among people could cause a major threat to human life, and some scenarios are even more bothersome. A particularly alarming possibility would be at hand if some of the most lethal viruses could be passed on by mosquito bites. West Nile virus can be transmitted in that way, but luckily it is not as dangerous as some other viruses that so far are not transmissible by mosquito bites. The dangerous ones would include the AIDS virus. Imagine a world in which a simple mosquito bite could spread AIDS to innocent victims. Even more worrisome is the virus that causes Ebola hemorrhagic fever. The Ebola virus is so lethal that there is really no good treatment for it except to isolate the victims so they do not infect other people, but that is hardly a treatment. Many pharmaceutical companies are working vigorously on antivirus programs, but the challenge is great. We can expect this to be a continuing challenge for many years into the future.

There are some diseases for which we have treatments but not cures. AIDS is one of them. Even with the effective cocktail of drugs that can deal with the virus when it is exposed, about 20% of the AIDS viruses are latent and inactive, thus unaffected by these drugs. A recent promising discovery is that certain compounds may cause the AIDS viruses to become active, when they can be destroyed by the normal cocktail of drugs, and if this holds up it will perhaps promise a cure and not just a treatment for AIDS. There are many other diseases for which we have no such cure. A good example is *schizophrenia*, where the deranged behavior of those with the disease can be controlled if they take their drugs, but not cured. Thus, when they stop taking their drugs, they become a danger to themselves and others. We need a cure for schizophrenia. We also need a cure for *diabetes*, since again this is a disease that is treated but not cured. *Alzheimer's disease* is another disease for which there may be some palliative ways to treat it, but no good way to cure or prevent it. Hence medicinal organic chemists will be at work for into the future inventing new drugs to handle these and other remaining challenges.

Sometimes people wonder whether the field of synthesis of new molecules has been so heavily worked on that we may be coming to the limit of the possibilities. However, the best estimates of the number of compounds that could exist with the typical elements that are in medicines and with the typical size of medicines is approximately a 1 with 40 zeros after it. Hence a clear prediction is that organic chemists will be making new compounds for years, as long as one can imagine, and that within those new classes of compounds being made we can expect to find more and more useful medicinal compounds with the efforts of synthetic organic chemists.

I have emphasized organic chemistry here because almost all the drugs that are currently in use are organic, but there have been several examples of inorganic discoveries that are important. One of them is the drug cisplatin, discovered by an inorganic chemist, which has use in anticancer treatments. Also, some medicinal compounds have been made by substituting other metals in heme for the iron that is naturally present. There are many enzymes which have metals in them as part of their reactive sites and also as part of their structural components, so we can expect that there will be continuing progress in developing new medicinal compounds that also have an inorganic component in them.

21.2.6
Multimolecular Systems

Until fairly recently, chemists have been mostly concerned with the properties of pure homogeneous systems. We synthesize and purify single molecules to determine their properties, and take complex mixtures from Nature and resolve them in such a way as to obtain the properties of individual components. However, there is a new approach to chemistry, involving both synthesis and mechanism studies, that focuses instead on the properties of *organized systems* consisting of more than one molecular type. The most obvious stimulus for this is biomimetic chemistry. The properties of a living cell are the result of the complex mixture of molecules within it, not the properties of any single molecular type. One of the goals of chemistry in the future, and in particular organic chemistry, will be to create molecular systems that imitate some of the properties of life. This is both a challenge to our abilities and a goal, since molecular systems are likely to have useful properties not available from single molecular components. Simple mixtures could be of interest, but even more interesting are *organized* molecular systems, either by directed synthesis or by spontaneous self-assembly.

There are major efforts to develop ways to harness the Sun, in order to generate useful *solar energy*. In one version, systems are being created that can take sunlight and imitate the photosynthetic mechanisms of life to generate molecules whose energy can then be captured later, perhaps by combustion. With this scheme, carbon dioxide would be captured and fixed. Here we are in direct competition with Nature, in which sunlight grows living plants, from trees to agricultural crops all the way down to algae. It remains to be seen whether we can make systems that compete with this process, and whether they will be economically viable. Another version is to use sunlight directly to convert water to oxygen and hydrogen, separate the two and then use the hydrogen either to manufacture something or to be incorporated in a fuel cell in which energy is harvested by running the photolytic cleavage in reverse and regenerating water. There is a lot of interest in systems that can do this, and they have both organic and inorganic components in them.

Perhaps the most interesting challenge is to produce chemical systems that are effective in a process directly generating electricity from sunlight, using *photovoltaic materials*. The best of these photovoltaic materials so far are largely inorganic, but consist of multiple layers in order to capture the full spectrum of solar radiation. Organic materials play a role in separating the layers, and perhaps in producing photovoltaic materials based entirely on organic compounds. One advantage of this plan is that it does not compete directly with food production, since it can and should be done in deserts. There is some work going on this field, but much more can be expected. It addresses not the medical needs of humanity but the energy needs, specifically the need to generate electricity from sunlight, the ultimate source, rather than using sunlight of the past that generated hydrocarbons or carbohydrates and then burning them to get the energy.

One way to be inspired in the creation of such new molecular systems is to examine in detail those which Nature has already produced. For instance, the simple use of sunlight in photosynthesis to generate electrons involves several different components, and we would be well advised to try to generate the same kinds of multicomponent systems to take advantage of different properties – one being the ability to generate an electron of reasonably high energy and the other to find mechanisms by which that electron can be passed away from the original source to other parts of a molecular system where the electron cannot then simply be returned to the excited molecule and quenched by recombination with it. The materials that one needs to create for this are both the original compounds that mimic chlorophyll and also the molecular wires that can carry current away from the excited chlorophyll to other components of an overall system. Hence there is significant interest in creating molecular wires, an area in which I have an active program.

Some impressive photovoltaic devices are already being invented, including some that can exceed 40% utilization of solar energy to generate electricity. Most estimates are that this degree of efficiency is sufficient to make the systems truly practical. However, they would work best in a desert where there are no clouds and little rain, so the problem is what to do with the energy that is being generated. This is a place where a chemical system could play a role. If we want simply to transfer the electricity to places where it will be used, it would be very helpful if chemists could invent practical superconducting wires that would be able to carry large currents at room temperature while still remaining resistance free. This is a goal for the combination of organic and inorganic chemistries.

Another plan is to take the electricity generated in the desert at a photovoltaic plant and directly perform a chemical reaction with that electricity to produce an energy-rich material that can be transported to places where its energy could then be captured. This is one description of a storage battery, and we still need much better batteries, consisting of multiple chemical components, to serve as practical ways to store the energy generated in a desert. A challenge for chemists is to come up with materials – either organic or inorganic, or both – that can effectively store energy and be transported in a normal way, and then perhaps utilized in a fuel cell to generate electricity where it is needed or to use the materials themselves in some way. For instance, an efficient process for generating aluminum and then using it in a battery to regenerate the electricity has a lot of theoretical attraction, but the problem is that the oxide layer on the surface of the aluminum makes efficient energy generation and capture and recharging difficult. A challenge for chemists, organic or inorganic, is to come up with some way to make an aluminum air battery truly reversible with very little energy loss. Organic chemists also have the challenge to come up with an appropriate way in which available materials could be converted into high-energy feedstocks right at the desert site, so they could then be transported in the normal way to places where they would be used. This points to the desirability of significant progress in organic electrochemistry, one of the other branches of our field.

Another need is batteries that will make electric vehicles truly practical. This means that they should not be excessive in weight or cost and that they should be able to carry a large amount of electricity and be quickly and easily recharged. The capacity of the batteries is not yet good enough but it is close, but recharging is a major problem since there are really no good practical batteries that can be recharged in the 10 min that one that would be hooked up to wires in a filling station. This is a practical challenge that will determine whether we can truly go to an electric vehicle fleet for all of our needs rather than continue to use petroleum. A non-chemical approach to solving this charging problem is the Israeli plan to have batteries interchangeable in our vehicles, so the discharged battery would be exchanged for a charged one in the "filling station." Then the discharged batteries could be recharged slowly in the filling station itself, from electric power.

21.2.7
Pollution and Toxicity

The chemical manufacturing industry is extremely large, all over the world. However, one of the problems with chemical manufacturing is the production of unwanted waste and the resulting pollution of land, water, and air. One of the challenges for organic chemists in the future is to develop processes to manufacture useful chemicals – materials and medicines – without such pollution. The importance of this field is already recognized in the existence of a movement called "green chemistry," but there is still plenty that needs to be done. The byproducts and also the products themselves must not be toxic in any way. It is a challenge for us to invent new pesticides able to kill the insects that are a problem without killing bees, for instance, or causing other damage. The DDT saga reminds us. Another problem is to generate agrochemicals such as weed killers that will perform their job but then disappear harmlessly from the environment. At one point it was thought desirable to have such chemicals last as long as possible so they did not have to be reapplied, but now we recognize that at least as desirable is to have such useful chemicals with limited persistence, so they disappear harmlessly in the environment. It is a challenge for organic chemists to come up with substances to replace the materials we use now that will be just as effective, but will be harmless to the environment including all the creatures in it, which includes, of course, humans. Insects such as mosquitoes are disease vectors, and we have to control them.

There is another problem which has come to our attention, and that we need to think about in considering challenges for the future – the importance that materials we put into the air will be harmlessly destroyed after their use. Our most recent example of this is the problem with chlorofluorocarbons – destroying the ozone layer that protects us all from high-energy, short-wavelength ultraviolet solar radiation. However, we also need to be aware of the potential harm from other materials liberated into the air or into the water by current manufacturing processes. Organic chemists will need to invent new ways to manufacture useful chemicals, including medicines. This is a challenge that chemists can welcome as yet another requirement for their synthetic inventions and procedures.

One aspect of this is to invent catalysts that are so selective that the reactions we perform in manufacturing are complete, with the production of no side products. The invention of new catalysts, especially those showing high chemical selectivity in terms of the products produced, will continue to be a challenge in the future. With the right catalysts we can also use much less energy in manufacturing, so this will contribute to solving the problem of having enough energy to run our modern civilizations.

One of the challenges for chemists is to learn how to detect and detoxify materials that are purposely spread by terrorists. We need good ways to detect explosives, and relying on the sensitivity of a dog's nose to do this certainly indicates how far we have to go in the area of the analytical chemistry of organic materials. We also have to be able to develop good ways to detoxify any poisons or biological agents that are purposely used in terrorist attacks.

As another challenge, if we are going to continue to use nuclear reactors to generate energy, we have to learn how to deal with the products of the processes. Currently we bury large amounts of material containing small amounts of highly dangerous radioactive materials, so one of the challenges is to learn how to separate the dangerous material from the inert matrix that does not need to be isolated. This is a challenge to develop effective selective binding compounds that can be used to capture and isolate some of the radioactive components of nuclear waste, and it is a challenge that will make a huge difference in whether nuclear energy really becomes practical. It would also be helpful in the case of a nuclear accident, where again isolating the dangerous material from the water that is in the plant would make it much easier to protect people from exposure to nuclear contaminants that resulted from a nuclear accident.

21.3
Conclusion

These are only some of the challenges facing us that organic chemists and their cousins inorganic chemists should be able to address and solve. The results of solutions to these problems could be enormously positive in improving the human condition, and so these are problems that certainly deserve the attention and creative inventiveness of organic chemists. As one more prediction, creative inventive chemists will think of other areas where they can make a difference, and other solutions to problems that are important, that are not included in this brief essay. Organic chemistry is a creative and useful field, whose future is as exciting as its past.

Index

Organic Chemistry – Breakthroughs and Perspectives, First Edition. Edited by Kuiling Ding and Li-Xin Dai.
© 2012 Wiley-VCH Verlag GmbH & Co. KGaA. Published 2012 by Wiley-VCH Verlag GmbH & Co. KGaA.